Computing and Systems Technology Division

Core Programming Topic at the 2011 AIChE Annual Meeting

Minneapolis, Minnesota, USA
16-21 October 2011

Volume 1 of 2

ISBN: 978-1-61839-731-7

Printed from e-media with permission by:

Curran Associates, Inc.
57 Morehouse Lane
Red Hook, NY 12571

CURRAN ASSOCIATES INC. proceedings.com

Some format issues inherent in the e-media version may also appear in this print version.

Printed by Curran Associates, Inc. (2012)

For permission requests, please contact AIChE
at the address below.

AIChE
3 Park Avenue
New York, NY 10016-5991

Phone: (203) 702-7660
Fax: (203) 775-5177

www.aiche.org

Additional copies of this publication are available from:

Curran Associates, Inc.
57 Morehouse Lane
Red Hook, NY 12571 USA
Phone: 845-758-0400
Fax: 845-758-2634
Email: curran@proceedings.com
Web: www.proceedings.com

TABLE OF CONTENTS

(12a) An Overview of CAST Division Activities

Mahmoud El-Halwagi, The Artie McFerrin Department of Chemical Engineering, Texas A&M University, College Station, TX and **B. Wayne Bequette**, Chemical and Biological Engineering, Rensselaer Polytechnic Institute, Troy, NY

An overview of CAST-related activities will be presented.

See more of this Session: CAST Division Plenary
See more of this Group/Topical: Computing and Systems Technology Division

Paper Unavailable

(12b) Computer Aided Solvent Selection and Design Framework

Igor Mitrofanov, Elisa Conte, Jens Abildskov, Grkan Sin and Raifqul Gani, CAPEC, Department of Chemical and Biochemical Engineering, Technical University of Denmark, Kgs. Lyngby, DK-2800, Denmark

Solvents are widely used as a reaction medium, as a reactant or as carrier at some stages of the manufacturing chain in products from the chemical, fine chemical, pharmaceutical, food, and agrochemical industries. Solvent are either required for processing after which they are removed or they are part of the final product formulation. Therefore, solvents are playing an important role in product synthesis and formulation, product delivery, separation processes etc [1]. On the other hand, solvent selection and design is a complex problem, which requires decision making in several levels for indentifying the best candidates depending on different multi-objective criteria namely environment, health, safety, process feasibility and economics.

One of the criteria of solvent selection is the environmental impact because of the excessive consumption and utilization in a wide range of industries, millions of tons solvents have to be wasted every year [2]. Therefore, it becomes important to minimize and optimize the use of organic solvents as much as possible, to satisfy the "Green Chemistry Principles" [3]. Another challenge is that currently solvent selection relies very much on previous experiences, trial and error with different solvent candidates. Such heuristic approach while valuable on their own, however arguably are not fit to deal with a complex multi-criteria optimization and search problem, which is the case for solvent selection.

Therefore the purpose of this contribution is to develop a systematic framework and implement it as software for selection and design of solvents for many applications including organic synthesis, complex reaction systems and solvent-based separations. The solvent selection framework is based on a combination of knowledge from industrial practice and computer-aided tools and methods for property prediction and computer-aided molecular design (CAMD) principles. This framework is applicable for solvent selection and design in product design as well as process design.

The first module of the framework is dedicated to the solvent selection and design for organic synthesis. This part uses the solvent selection methodology of Gani et al. [4,5], which has been extended to handle multi-step chemical syntheses as well as solvent substitution for specific reaction steps in existing processes. The methodology for organic synthesis accordingly involves five steps for each reaction:

1. Problem identification. Finding an objective for given system, identifying actual functions of the solvent.
2. Search criteria definition. The solvent functions that satisfy the operational needs of the process are defined in terms of a set of search criteria (R-indices), defined in terms of: physical and

chemical properties (solvent-pure properties); Environment, Health and Safety (EHS) characteristic (solvent-EHS properties); operational properties (solvent–solute properties).

3. Performing the search. The search step consists of two stages. The first is a generation and property identification of solvent candidates using special software ProCAMD and ProPred, which are the implementations of computer-aided molecular techniques. The second consists of assigning the RS-indices following the reaction–solvent and then consulting the known solvent database and identifying the set of solvents that satisfy search criteria.
4. Score table assignment. A list of feasible solvents needs to be created. The scores are assigned from the calculated values of RS indices. The scores give a weight to each of the calculated RS indices.
5. Matrix of solvents. After the scores table has been generated, a short list of feasible solvents is obtained for each reaction step.

This methodology has been evaluated with several practical application examples including a single reaction solvent screening problem, a solvent replacement problem and a solvent selection for a multi-stage system.

The second module of the framework is dedicated to solvent selection for separation processes. One of the important tasks in separation processes is an identification of a pure solvent or anti-solvent for a specific Active Pharmaceutical Ingredient, which is a problem of major concern for the pharmaceutical R&D departments. Solvents, lipids and other compounds are commonly employed in product formulation as well as in APIs processing. In addition, the design of solvent mixtures that sometimes show improved characteristics of solubility toward a particular API could bring several advantages.

The framework includes a model-based method for solvent screening and solvent mixtures design for pharmaceutical applications. The problems that can be solved with this method are the following: given the API, select a good solvent or a good antisolvent; design a solvent mixture that improves the solubility if the solubility target cannot be reached with a single solvent; identify the optimal antisolvent that added in mall amount to the mixture API-solvent gives the highest precipitation of API (for crystallization processes).

The method is based on the theory of the conceptual segments (hydrophobic, polar and hydrophilic) introduced by Chau-Chyun and Song (2004) with the NRTL-SAC method [6]. In order to make the method completely independent from the availability of experimental data, a Marrero and Gani (2001) Group Contribution model (M&G GC^+) [7] for the prediction of the conceptual segments is developed in this work. With the above model, the API molecule can be described in terms of conceptual segments (first step of the method) and, then, the solvent power of the pure solvents can be computed and ranked from the best to the worst (anti-solvent). Hence, the design of a solvent mixture that enhances the API solubility can be performed by identifying the conceptual segments combination that maximizes the excess properties. An anti-solvent that mixed with the best solvent gives a large decrease in the API solubility can be identified in the same way as before, but this time excess properties of mixing have to be minimized.

References:

1. R. Gani, C. J. Gonzalez, A. Kate, P. A. Crafts, A Modern Approach to Solvent Selection. Chemical Engineering Journal (2006), 30-43.
2. P. K. W. Lau, A. Koenig, Management, disposal and recycling of waste industrial organic solvents in Hong Kong. Chemosphere (2001), 44(1), 9-15
3. P.T. Anastas, J.C. Warner, Green Chemistry: theory and practice, Oxford: Oxford University Press, U.K., 1998
4. R. Gani, C .Jimenez-Gonzalez, D.J.C. Constable, Method for Selection of solvents for promotion of organic reactions. Computers and Chemical Engineering 29 (2005) 1661-1676
5. M. Folic, R. Gani, C. Jimnez-González, D.J.C. Constable, Systematic Selection of Green Solvents for Organic Reaction Systems. Chinese Journal of Chemical Engineering. 16(3) (2008) 376-383
6. Chen C.-C. and Song Y. Solubility Modeling with a Nonrandom Two-Liquid Segment activity coefficient model. Ind. Eng. Chem. Res. 2004, 43, 8354-8362
7. Marrero J. and Gani R. Group contribution based estimation of pure component properties. Fluid Phase Equilibr. 2001, 183-184, 183-208

See more of this Session: CAST Division Plenary
See more of this Group/Topical: Computing and Systems Technology Division

(12c) Classification of Chemical Production Scheduling Problems and Approaches, and a General Solution Framework

Sara Zenner and **Christos Maravelias**, University of Wisconsin, Madison, WI

Classification of ChemicalProduction Scheduling Problems and Approaches, and a General Solution Framework

Sara Zenner, Christos T. Maravelias

Thepresent talk consists of three parts:

(1)A classification of chemical production scheduling problems

(2)A critical review of the types of modeling approaches proposed in theliterature

(3)A general representation and solution method for chemical production scheduling

ProblemClassification

Scheduling problems appear in a number of sectorsand applications, ranging from the oil industry to the pharmaceutical andspecialty chemical sectors, and from consumer goodsand the food industry to metal manufacturing. In addition, there is a largenumber of different processing characteristics and restrictions (e.g., batchvs. continuous processing, transportation vs. conversion processes, mixingrules, storage constraints, etc) which lead to a wide range of schedulingproblems. In this paper we present a general classification of schedulingproblems based on a series of attributes, including: (i)the industrial/market environment within which a company operates, (ii) theinteraction of the scheduling problem with the other planning functions of theenterprise, (iii) the characteristics of the production environment, (iv) thestructure of the facility, (v) the production recipe, and (vi) the detailedprocessing restrictions. We then discuss how three specific aspects(bill-of-materials requirements, mixing/splitting rules, batch integrityrestrictions) can lead to problem classes that existing modeling approachescannot address and we critically review the shortcomings of existingrepresentations and modeling approaches.

Review ModelingApproaches

Based on the insights provided by our problemclassification we review existing scheduling approaches in terms of a broadspectrum of attributes. Most existing method classifications are based on asubset of these attributes (e.g., facility structure and time representation).In this talk we attempt to synthesize all previous approaches and propose aroadmap that includes all known modeling approaches. Specifically, we propose amulti-dimensional view where the"universe" of scheduling approaches is treated as a multi-dimensional spacewhere each dimension corresponds to a key modeling feature. The primarydimensions are:

a) *Major entity(ies) modeled*:these typically include orders (or batches) and materials/resources.

b) *Optimization decisions*: the major decisions are batching (or lot-sizing),assignment, and sequencing and timing; all approaches involve a subset of thesethree.

c) *Time representation*: this includes the selection of the type of time grid (discrete vs.continuous), as well as more subtle choices (e.g., unit-specific vs. commontime grid, local vs. global precedence variables, etc.).

GeneralModeling and Solution Framework

Based on our representation of modeling approaches,we revisit the classes of problems that cannot be addressed using existingmethods. We explain why this is the case and highlight the "holes" in terms ofproblem representation. We then present a general framework, which is ageneralization of the approach of Sundaramoorthy and Maravelias (2011), to address some of these shortcomings.Finally, we show how our approach can be used to represent problems that havebeen addressed using different modeling approaches, and how other modelingapproaches can be "reduced" to ours. We close with some computationalresults.

Sundaramoorthy, A.; Maravelias,C. T. A GeneralFramework for Process Scheduling. *AIChE J.*, 57(3), 695-710, **2011**. (DOI: 10.1002/aic.12300).

See more of this Session: CAST Division Plenary
See more of this Group/Topical: Computing and Systems Technology Division

(12d) Model Development and Control Strategies for Multicomponent Alloy Solidification Processes and Its Application to Solar Cell Manufacturing

German A. Oliveros[1], Ruochen Liu[1], B. Erik Ydstie[1] and Sridhar Seetharaman[2], (1)Chemical Engineering, Carnegie Mellon University, Pittsburgh, PA, (2)Materials Science and Engineering, Carnegie Mellon University, Pittsburgh, PA

In this work we present the mathematical description and a control formulation of solidification processes and their application to the manufacturing of high purity silicon wafers suitable for solar cells. We apply the results to prove a novel continuous process to produce the wafers from the melt. The theoretical work is part of a larger research program that also involves experimental and computational work. The conceptual design consists of producing a thin film of silicon on top of the melt, making use of the fact that solid silicon floats on its melt just like ice floats on water. These ideas serve as our main motivation to put our investigation in the context of renewable energies. Model development provides a significant aid in understanding transport phenomena taking place in the system and developing techniques to control it.

The process of solidification involves the segregation of impurities into the melt, which, from a process systems perspective, is a purification process. This is beneficial because we are producing the desired material while purifying it. Purity requirements of solar grade silicon are very stringent in order to achieve optimal solar cell efficiency, and the current processes to produce it need to be carefully managed and controlled [1]. Mathematical models can be used to calculate the extent of purification that can be achieved through solidification of a thin silicon film and its dependence on the system dynamics, specifically, the velocity of the crystallization front and the temperature gradients at the solid-liquid interface, which are intrinsically coupled. At high interfacial velocities, the formation of a solute enriched boundary layer next to the crystallization front leads to potential interfacial instabilities that affect the quality of the final product [2].

Previous studies [3] provided the initial insights into many aspects of the proposed process through the decoupling of convection in the melt and solidification dynamics for pure silicon. In this work we extend the formulation to metallurgical grade silicon (multicomponent alloy). Using a finite-difference model, it is shown that for a simple silicon-aluminum alloy, there is an optimal interface velocity where the segregation of impurities achieves complete mixing of the melt and the disappearance of the boundary layer. The model is validated via sensitivity analysis. Process control strategies are also developed to achieve an optimal solute distribution in the melt, by adjusting the heat released and supplied to the system.

[1] Brown,R.A, Theory of Transport Processes in Single Crystal Growth from the Melt, AIChE Journal, 34,881-911,1988.

[2] Mullins, W.W., Sekerka, R.F. Stability of a Planar Interface during solidification of a dilute binary alloy, Journal of Applied Physics, 35,444-451, 1964.

[3] Ruochen Liu, German Oliveros, Sridhar Seetharman and B. Erik Ydstie, Multiscale Modeling of a Silicon Solar Wafer Manufacturing Process, 21st European Symposium on Computer-Aided Process Engineering, 2011, Accepted.

See more of this Session: CAST Division Plenary
See more of this Group/Topical: Computing and Systems Technology Division

(12e) Modeling and Control of Aggregate Thin Film Surface Morphology Using Stochastic PDEs and a Patterned Deposition Rate Profile

Xinyu Zhang, **Jianqiao Huang**, Gangshi Hu, G. Orkoulas and Panagiotis D. Christofides, Department of Chemical and Biomolecular Engineering, University of California, Los Angeles, Los Angeles, CA

This work focuses on modeling and control of aggregate thin film surface morphology for improved light trapping using a patterned deposition rate profile. The dynamics of the evolution of the thin film surface height profile are modeled by an Edwards-Wilkinson-type equation (a second-order stochastic partial differential equation) in two spatial dimensions. The thin film surface morphology is characterized in terms of aggregate surface roughness and surface slope. These variables are computed with respect to appropriate visible light-relevant characteristic length scales and defined as the root-mean-squares of height deviation and slope of aggregate surface height profiles, respectively. Analytical solutions of the expected aggregate surface roughness and surface slope are obtained by solving the Edwards-Wilkinson equation and are used in the controller design. The model parameters of the Edwards-Wilkinson equation can be estimated from kinetic Monte-Carlo simulations using a novel parameter estimation procedure. This parameter dependence on the deposition rate is used in the formulation of the predictive controller to predict the influence of the control action on the surface roughness and slope at the end of the growth process. The cost function of the controller involves penalties on both aggregate surface roughness and mean slope from set-point values as well as constraints on the magnitude and rate of change of the control action. The controller is applied to the two-dimensional Edwards-Wilkinson equation. Simulation results show that the proposed controller successfully regulates aggregate surface roughness and slope to set-point values at the end of the deposition that yield desired levels of thin film reflectance and transmittance levels.

See more of this Session: CAST Division Plenary
See more of this Group/Topical: Computing and Systems Technology Division

(82a) Himmelblau Award Presentation

David A. Kofke, Chemical and Biological Engineering, University at Buffalo, The State University of New York, Buffalo, NY

The recipient of the David Himmelblau Award for Innovations in Computer-Based Chemical Engineering Education will deliver a lecture based on the work recognized by the award.

See more of this Session: Computing and Technology In Chemical Engineering Education I
See more of this Group/Topical: Computing and Systems Technology Division

Paper Unavailable

(82b) Expertise In Chemical Process Modeling

Paul M. Mathias, Fluor Corporation, Aliso Viejo, CA

Chemical process modeling is based upon a small set of very general equations, and then particular problems are treated as a special case of these fundamental equations. This quantitative approach has resulted in engineers with strong mathematical skills, and the development of powerful software that provides rigorous solutions to highly complex problems. But an unintended consequence is that chemical engineers have lost a tacit understanding of physical properties and process behavior, and, as a result, process designs have suffered and the trouble-shooting skills of chemical engineers have diminished. In other words, chemical engineers are not attaining expertise.

So, how do we improve the expertise of chemical engineers? This presentation will build upon previous work of the author [1, 2, 3], and demonstrate that internalizing the patterns of physical properties and process behavior is a vital component to attain process-modeling expertise. Interestingly, key solutions to the problem are enabled by software advances, namely the easy availability of comprehensive property databases and advances in graphics and visualization.

1 Paul M. Mathias, "Interactive Visualization of the McCabe-Thiele Diagram," Chemical Engineering Progress, December, 105(12), 36-44 (2009).

2 Paul M. Mathias and Suphat Watanasiri, "Some Patterns of Fluid Phase Behavior," J. Chem. Eng. Data, Article ASAP, 15 March 2011.

3 Paul M. Mathias and Suphat Watanasiri, "Patterns of Chemical Process Modeling," presented at AspenTech Global Conference, Washington, D.C., 23-25 May 2011.

See more of this Session: Computing and Technology In Chemical Engineering Education I
See more of this Group/Topical: Computing and Systems Technology Division

(82c) Students Learn Fundamentals of Process Operations and Control Using Dynamic Simulator of An Integrated Gasification Combined Cycle (IGCC) Plant with CO2 Capture

Debangsu Bhattacharyya[1], Richard Turton[1] and Stephen E. Zitney[2], (1)Department of Chemical Engg., West Virginia University, Morgantown, WV, (2)Collaboratory for Process and Dynamic Systems Research, National Energy Technology Laboratory, U.S. Department of Energy, Morgantown, WV

While today's undergraduate chemical engineering students often learn about the importance of dynamic simulation and process control, they typically do not spend considerable time gaining practical experience in plant-wide operations and control. At West Virginia University's (WVU) National Research Center for Coal and Energy, students have access to a real-time dynamic simulator for an integrated gasification combined cycle (IGCC) plant with carbon capture and compression. Fueled with coal, petroleum coke, and/or biomass, the gasification island of the IGCC plant consists of two oxygen-blown, downward-fired, entrained-flow, slagging gasifiers with radiant syngas cooler, two-stage sour shift reactors followed by a dual-stage acid gas removal process. The combined cycle island consists of two F-class gas turbines, steam turbine, and a heat recovery steam generator with three-pressure levels. The IGCC dynamic simulator is operated under the auspices of the National Energy Technology Laboratory's (NETL) Advanced Virtual Energy Simulation Training And Research (AVESTAR™) Center.

Using the IGCC operator training system (OTS), students gain hands-on, interactive, simulator-based experience with normal base load operations, cold, warm, and hot startups, plant shutdown, load shedding and following, and fuel switching. The students also investigate abnormal situations in real-time, including equipment malfunctions and failures, together with changes initiated through actions from the instructor station or from field operators. In addition, students learn how the plant responds dynamically to changes in the manipulated inputs, as well as how the control system impacts plant performance, stability, robustness and disturbance rejection characteristics. The students study the step and ramp response of the simple linear, lower order processes along with more complicated higher-order and nonlinear processes. The students observe the "real-life" transient response of the processes with non-minimum phase behavior and face the challenge of controlling these systems with PID only control. Various open-loop unstable processes are studied and the role of feedback mechanism in making the system closed-loop stable is explained by demonstrating a number of familiar systems in this simulator environment. Other than the conventional control structures, the simulator is utilized for studying more complex control structures such as cascade control, feedforward control, ratio control, override control, split-range control, etc. The system is also capable to monitor, evaluate, and report the performance of the students through a trainee performance monitoring (TPM) module. The students are challenged by asking them to control the plant within a set of operating limits in the face of the disturbances and simulated process faults.

See more of this Session: Computing and Technology In Chemical Engineering Education I
See more of this Group/Topical: Computing and Systems Technology Division

(82d) Teaching Design Using the CACHE Learning Resource Center

Warren D. Seider, Chemical and Biomolecular Engineering, The University of Pennsylvania, Philadelphia, PA

The initial version of a "Guide to Teaching Design with Internet Links" will be introduced. It describes a framework of materials for teaching product and process design with links to recommended materials from many sources including textbooks, monographs, completed design case studies, design problem statements, design software, POWER POINT lectures, and related materials prepared by design instructors. Several area editors have assembled materials in 10 design areas including product design, process integration, process simulation, the environment, safety, design optimization, ... This presentation is intended to introduce some of the features in the guide and show how to use them effectively.

See more of this Session: Computing and Technology In Chemical Engineering Education I
See more of this Group/Topical: Computing and Systems Technology Division

(82e) Peer Evaluation In Chemical Engineering Design Through Wikis

Caryn L. Heldt, Chemical Engineering, Michigan Technological University, Houghton, MI

Practicing engineers are put into situations where their work will be critically evaluated by others and they are also expected to critically evaluate the work of others. To help practice these skills, we used peer evaluation through wikis as a method for the chemical engineering design teams to both evaluate other's work and to be evaluated. Students, in groups of four, were asked to design a process that will manufacture products ranging from acetone and CO_2 scrubbing solvents to ibuprofen and glucose. After four weeks, a progress report was due. This progress report was submitted to the instructor for grading, as well as posted on a wiki site. The students were assigned two groups to assess and they were given the suggestion to look at both the technical and communication aspects of the report. The groups were assigned to allow the students to read a variety of projects from different sections of the class and also for the strong groups to be paired with weaker groups, and vice versa. The learning objectives of the exercise were to (1) enhance written communication skills due to (i) peer review of their work and (ii) requirement to express opinions in a written format, (2) develop critical thinking skills when evaluating other projects, and (3) obtain an in-depth understanding of their own projects from peer-review comments. Surveys were also done to assess the students' attitudes toward the wiki assignment. At the conclusion of the project, the individual section instructors were all impressed with the evaluations done by the students. The students were interested and engaging in their comments and often came up with ideas that were not thought of by the instructor. This paper will present some of the positive and negative aspects of using wiki's in engineering education and explore student reaction to this educational model.

See more of this Session: Computing and Technology In Chemical Engineering Education I
See more of this Group/Topical: Computing and Systems Technology Division

(94a) The Minnesota Years: An Advisor's Perspective

George Stephanopoulos, Massachusetts Institute of Technology, Cambridge, MA

Some perspectives will be provided on Manfred's activities as a Ph.D. student at the University of Minnesota.

See more of this Session: In Honor of Manfred Morari's 60th Birthday I- Papers by Invitation Only
See more of this Group/Topical: Computing and Systems Technology Division

Paper Unavailable

(94b) The Wisconsin Years: IMC and CAD

W. Harmon Ray, Chemical and Biological Engineering, University of Wisconsin - Madison, Madison, WI

Some perspectives are provided on Manfred Morari's activities at the University of Wisconsin.

See more of this Session: In Honor of Manfred Morari's 60th Birthday I- Papers by Invitation Only
See more of this Group/Topical: Computing and Systems Technology Division

Paper Unavailable

(94c) The Caltech Years: Resiliency and Controllability

Ignacio E. Grossmann, Department of Chemical Engineering, Carnegie Mellon University, Pittsburgh, PA

This presentation will provide an overview of the resiliency and controllability literature.

See more of this Session: In Honor of Manfred Morari's 60th Birthday I- Papers by Invitation Only
See more of this Group/Topical: Computing and Systems Technology Division

Paper Unavailable

(94d) Nonlinear IMC to NMPC

Lorenz T. Biegler, Chemical Engineering, Carnegie Mellon University, Pittsburgh, PA

An overview and perspectives are provided on the development of Internal Model Control and Model Predictive Control for nonlinear systems.

See more of this Session: In Honor of Manfred Morari's 60th Birthday I- Papers by Invitation Only
See more of this Group/Topical: Computing and Systems Technology Division

Paper Unavailable

(94e) The ETH Years: Hybrid Systems

Tariq Samad, Honeywell Technology Center, Honeywell International, Minneapolis, MN

Perspectives are given on the impact of hybrid systems in industrial applications.

See more of this Session: In Honor of Manfred Morari's 60th Birthday I- Papers by Invitation Only
See more of this Group/Topical: Computing and Systems Technology Division

Paper Unavailable

(94f) Perspectives On Manfred Morari's Contributions

Thomas F. Edgar, Chemical Engineering, The University of Texas at Austin, Austin, TX

This presentation will provide some perspectives on Manfred Morari's contributions to process systems engineering.

See more of this Session: In Honor of Manfred Morari's 60th Birthday I- Papers by Invitation Only
See more of this Group/Topical: Computing and Systems Technology Division

(104a) Particle Based Multi-Scale Modeling of the Dynamic Response of Hexahydro-1,3,5-Trinitro-s-Triazine (RDX): Constant Energy Dissipative Particle Dynamics with Coarse-Grained Density Dependent Potentials

Joshua D. Moore[1], Sergei Izvekov[1], Martin Lisal[2] and John K. Brennan[1], (1)Weapons and Materials Research Directorate, U.S. Army Research Laboratory, Aberdeen Proving Ground, MD, (2)E. Hala Laboratory of Thermodynamics, Academy of Sciences of the Czech Republic, Prague, Czech Republic

Nanocomposites are inherently heterogeneous, containing several species types of disparate shapes and sizes. For example, energetic materials are often comprised of micro-sized crystallites (or grains) held together by a small amount of polymer binder, and in some instances contain metal nanoparticles. Accordingly, mechanical stimulation (e.g., shock or shear) of these materials can incite responses over a wide range of spatial and temporal scales. Localized regions of elevated thermal energy, or hot spots, often occur near microscale defects in the material such as voids and grain boundaries. However, some energy transfer processes are atomistically governed, for which field-based microscale, mesoscale, and continuum modeling techniques are currently unable to provide the required level of fidelity. Modeling the thermal and mechanical response of these materials atomistically, despite growing computational power and resources, remains a challenge due to the length and time scales required, as often billions of molecules are necessary to be modeled on micron-size scales.

To overcome some of these challenges, we have implemented multiscale techniques to bridge the atomistic and mesoscale levels of description. The multiscale coarse-graining method (MS-CG) [1,2], a *bottom-up* approach based on force-matching, has been utilized to derive particle-based coarse-grained potentials for hexahydro-1,3,5-trinitro-s-triazine (RDX), where one RDX molecule is mapped onto a single interaction site. To improve transferability of the MS-CG model (required to accurately capture thermodynamic responses during mechanical and thermal stimulation), explicit MS-CG mapping from the atomistic to the microscale is made from ambient to high pressures [2]. The resulting density dependent MS-CG model reproduces several atomistic properties within reasonable agreement from ambient to high pressures for the molecular crystal including the lattice structure, melting point, elastic and vibrational properties, and the phase diagram [3]. For the molten state, the 2-body and often multi-body structure are also well described with the model [3].

Despite the successes of the MS-CG model of RDX to describe static properties, it inevitably cannot account for accurate energy and momentum exchange in response to mechanical stimulation using traditional molecular dynamics. This is due to intramolecular degrees of freedom (i.e., intramolecular bonding) being coarse-grained out, and thus, the heat capacity is underestimated by nearly an order of magnitude [3]. To correct this deficiency, we account for momentum and energy transfer under mechanical shock by utilizing the constant-energy Dissipative Particle Dynamics method (DPD-E) [4-6]. Unlike traditional Dissipative Particle Dynamics [7-9], which conserves only momentum, the DPD-E method conserves both energy and momentum by assigning an internal energy to each particle, which for the MS-CG RDX case, helps to minimize the effect of coarse-graining the intramolecular degrees of freedom.

In this talk, we will present results for mechanical and thermal shock loading of our MS-CG model of RDX using DPD-E. Various modeling parameters have been investigated for sensitivity including spatial interaction cutoffs used to determine local densities for the density dependent potentials, the mesoparticle equation of state that accounts for the particle internal energy and the overall dynamics through energy dissipation. Results will be compared directly to atomistic simulation for both mechanical shock and the response due to thermal heating of the material.

This effort is supported by the recently established Institute for Multi-Scale Reactive Modeling of Insensitive Munitions (MSRM), which is a multi-team effort led by the U.S. Army Research Laboratory and the U.S. Army Armament Research, Development, and Engineering Center, involving various other national laboratories and academic groups totaling over 20 scientists. J.D.M. gratefully acknowledges support, in part, by an appointment to the Internship/Research Participation Program for the U.S. Army Research Laboratory administered by the Oak Ridge Institute of Science and Education through an agreement between the U.S. Department of Energy and U.S. ARL (Project ID: 201003211).

[1] S. Izvekov and G.A. Voth, *J. Chem. Phys.*, **123**, 134105 (2005).

[2] S. Izvekov, P.W. Chung, and B.M. Rice, *J. Chem. Phys.*, **133**, 064109 (2010).

[3] S. Izvekov, P.W. Chung, and B.M. Rice, *Submitted* (2011).

[4] J. Bonet Avalos and A.D. Mackie, *Europhys. Lett.*, **40**, 141 (1997).

[5] P. Español, *Europhys. Lett.*, **40**, 631 (1997).

[6] M. Lísal, J.K. Brennan, and J. Bonet Avalos, *Submitted* (2011).

[7] P.J. Hoogerbrugge and J.M.V.A. Koelman, *Europhys. Lett.*, **19**, 155 (1992).

[8] J.M.V.A. Koelman and P.J. Hoogerbrugge, *Europhys. Lett.*, **21**, 363 (1993).

[9] R.D. Groot and P.B. Warren, *J. Chem. Phys.*, **107**, 4423 (1997).

See more of this Session: Multiscale Modeling II

(104b) Temperature-Accelerated Molecular Dynamics Reveals That Insulin Can Undergo Large-Scale Conformational Reorganization On Binding to Its Receptor

Harish Vashisth, Chemistry and Biophysics, University of Michigan, Ann Arbor, MI and Cameron F. Abrams, Drexel University, Philadelphia, PA

Insulin regulates blood glucose levels in higher organisms by specifically binding to and activating a transmembrane glycoprotein known as the insulin receptor (IR). Detailed photo-crosslinking studies have suggested that insulin molecule reorganizes on binding to IR due to the displacement of its flexible B-chain C-terminus in the presence of a tandem hormone binding structural motif called CT-peptide. Although a detailed understanding of how insulin binding leads to receptor activation remains elusive, our recently proposed all-atom structural models of insulin/IR complexes provide information valuable for exploring structure-function repertoire of IR. However, the large-scale conformational change in the insulin was not studied in these structural models due to the absence of then-unresolved CT-peptide, and also the lack of adequate simulation techniques. We have proposed a new conformational sampling algorithm, temperature-accelerated molecular dynamics (TAMD) (2010, *PNAS*, **107**, 4961-4966), that provides an immediate opportunity to use all-atom simulation to directly address these important gaps in our understanding of insulin binding to IR. Using a combination of Metropolis Monte-Carlo (MC) and molecular dynamics (MD) simulations, we have constructed all-atom structural models of insulin/IR complexes in the presence of (now-resolved) CT-peptide. We further apply TAMD to the C-terminus of IR-bound (T and R) insulin molecules at a fictitious thermal energy of 6 kcal/mol, and observe that while insulin remains stably-bound to IR, it can undergo large-scale conformational change in the B-chain C-terminus at ~50-ns time-scale resulting in the exposure of hidden hydrophobic core of insulin. These results are significant because we demonstrate at the atomistic-scale for the first time that insulin molecules can undergo large-scale reorganization on binding to IR in presence of CT-peptide. Moreover, we suggest TAMD as a viable simulation technique to study such large-scale conformational changes in proteins.

(104c) Multiscale Modeling for Phospholipid Bilayer Simulations

Emily Curtis, Chemical & Biomolecular Engineering, North Carolina State University, Cary, NC

Multiscale Modeling for PhospholipidBilayer Simulations

A multiscale modelingapproach was used to develop an implicit-solvent intermediate-resolution modelto simulate the behavior of the lipid, 1,2-dipalmitoyl-*sn*-glycero-3-phosphocholine (DPPC) inwater. By reducing the DPPCrepresentation to 14 coarse-grained sites with unique properties, treatingsolvent implicitly, and employing discontinuous molecular dynamics, a very fastalternative to traditional molecular dynamics, we are able to simulate thespontaneous formation of a bilayer from a random solution of 512 DPPC moleculesin several hours on a fast workstation. Data used to calculate the coarse-grained model parameterswas obtained by running united-atom simulations with GROMACS. Radial distribution functions wereplotted for all pairs of non-bonded coarse-grained sites to estimate thehardsphere and square-well diameters. The interaction energy between each pair of non-bonded coarse-grainedsites was calculated using the Boltzmann inversion scheme. The relative stiffness of each lipid ismaintained by imposing pseudobonds, which limit the bond lengths to fluctuateby the same amount observed in the united atom simulations. In the model each coarse-grained sitehas its own realistic mass. Simulationresults show that the model accurately reproduces structural properties of theDPPC bilayer including the area per lipid, bilayer thickness, bond order andmass density profiles. Currentwork is focused on extending the model to lipid mixtures and simulating thephase separation that occurs as a bilayer composed of DPPC and 1,2-dipalmitoyl-*sn*-glycero-3-phosphate (PA) changes froman initially-homogeneous state (at neutral pH) to a phase-separated state atlow pH.

See more of this Session: Multiscale Modeling II
See more of this Group/Topical: Computational Molecular Science and Engineering Forum

(104d) A Coarse Grained Model of Inhibitors to Scavenger Receptor Uptake to LDL

Michael Tomasini and M. Silvina Tomassone, Chemical and Biochemical Engineering, Rutgers University, Piscataway, NJ

The buildup and oxidation of low-density lipoproteins (LDL) in the walls of arterial blood vessels results in atherosclerotic lesions and arterial damage. The immune response of this arterial damage is to recruit macrophages expressing scavenger receptors specific for the uptake of oxidized LDL. The result is the formation of foam cells and fatty streaks, the first signs of atherosclerotic plaques. One possible method to combat the formation of atherosclerotic plaques is through the use of inhibitors which prevent the uptake of oxidized LDL by the scavenger receptors. Recently, Tian et al. 2004 [1] developed a class of molecules termed amphiphilic scorpion-like macromolecules (AScM) comprised of a mucic acid head group with attached aliphatic chains and a long polyethylene glycol (PEG) tail used specifically to block uptake of oxidized LDL by scavenger receptors. In this work, we use computational simulations to explore the structure and interactions of AscM micelles with the specific scavenger receptor – scavenger receptor A (SR-A) in order to shed light on an optimal inhibitor of oxidized LDL uptake.

In our simulations, we use a coarse-grained model based on the MARTINI force field [2] to describe the AScM molecules as well as the collagenous portion of SR-A where binding is assumed to take place [3]. As the interaction of oxidized LDL is thought to be mediated by electrostatic interactions, we vary the number (0, 1, or 2) of carboxylate groups at different positions on the AScM (near the mucic acid head or at the PEG tail) to determine the effect on both AScM micellation (which it though necessary for SR-A binding) and AScM association with SR-A to determine the best inhibitor.

References:

[1] Tian L, Yam L, Zhou N, Tat H, and Uhrich KE. Amphiphilic Scorpion-like Macromolecules: Design, Synthesis, and Characterization. *Macromolecules*. 2004; **37**: 538-543.

[2] Monticelli L, Kandasamy SK, Periole X, Larson RG, Tieleman DP, and Marrink SJ. The MARTINI Coarse-Grained Force Field: Extension to Proteins. *J. Chem. Theory and Comput.* 2008; **4**(5): 819-834.

[3] Platt N and Gordon SJ. Is the Class A Macrophage Scavenger Receptor (SR-A) Multifunctional? – The Mouse's Tale. *J. Clin. Invest.* 2001; **108**(5): 649-654.

See more of this Session: Multiscale Modeling II
See more of this Group/Topical: Computational Molecular Science and Engineering Forum

(104e) Multiscale Molecular Modeling of Fullerol-Dendrimer Complexes

Seung Ha Kim and Monica H. Lamm, Department of Chemical and Biological Engineering, Iowa State University, Ames, IA

Dendrimers are large, synthetic macromolecules that have regular and highly branched architectures. The multivalency of dendrimers permits them to bind with other species and makes them attractive platforms for a wide range of applications, including medicine, surface science, catalysis, and water treatment. To better understand the physical mechanisms of dendrimer-guest assembly, we have developed a multiscale molecular model for simulating flexible macromolecules in solution, based on the force matching method for coarse graining. We illustrate this approach by simulating the complexation between poly(amidoamine) (PAMAM) dendrimers and $C_{60}(OH)_{20}$ (fullerol) in aqueous solution. Specifically, we systematically examine the effects of concentration, and of dendrimer molecular weight, on the resulting PAMAM-fullerol complexes and compare our model predictions to recent experimental measurements.

See more of this Session: Multiscale Modeling II
See more of this Group/Topical: Computational Molecular Science and Engineering Forum

(104f) Multiscale Modeling of Perfluoropolyether Lubricants with Functional Endgroups

Robert Smith, Pil Seung Chung, Lorenz T. Biegler and Myung S. Jhon, Chemical Engineering, Carnegie Mellon University, Pittsburgh, PA

Physical properties of nanoscale perfluoropolyether (PFPE) films, which are generally used as a perfluoro linear co-polymer lubricant in hard disk drives (HDDs) has been previously studied in disparate molecular/mesoscale simulations. The PFPE molecules must exhibit strong adhesion to the carbon overcoat surface as well as rapid self-healing properties; therefore, a fundamental understanding of nanoscale characteristics of lubricant structure which contribute to desired macroscale performance must be established for successful nanotechnology design.

Due to the hydroxyl functional groups at each chain end of well-known PFPE (e.g., Zdol and Ztetraol), PFPE on an amorphous carbon surface shows peculiar characteristics (i.e., layering structure and entanglement on the surface). Via coarse-grained (CG) molecular dynamics (MD), we qualitatively validated molecular scale phenomena related to nanostructured conformation and dynamic behavior of molecularly thin lubricant film [1, 2]. To validate this coarse-grained model, a procedure in multi-scale modeling [3] which links atomic scale phenomena to CG MD is established thereby providing a form of synergy between the atomistic and molecular scales. Accurate intra / intermolecular force field parameters for PFPE and carbon surfaces are implemented in CG MD to enable improvements in our molecular scale model.

In this study, we use *ab initio* methods to determine intramolecular force field parameters for PFPE molecules and their intermolecular interaction with other PFPEs as well as with various carbon overcoat surfaces. Calculations based on *ab initio* methods are used to determine the intramolecular (stretching, bending, and torsional) force field parameters via solving the *ab initio* Hessian matrix eigenvalue problem. The intermolecular interaction between endgroups in pure and blended PFPEs is quantified using the supermolecular approach to molecular interaction. In addition, we explore novel carbon overcoat technology in HDD by calculating PFPE interaction with graphene. The implementation of atomically thin graphene as a carbon overcoat has the potential for a factor of two improvement in areal recording density. We compare the PFPE interaction with the traditional carbon overcoat versus graphene. We also evaluate PFPE-PFPE and PFPE-carbon surface geometry as a function of the PFPE structures particularly for new-generation lubricants including DDPA. The force field is evaluated by comparing interaction energies and molecular geometries of equilibrium configurations obtained using *ab initio* calculations to those obtained via the classical force field analysis.

[1] Q. Guo, L. Li, Y.-T. Hsia, and M.S. Jhon, "A Spreading Study of Lubricant Film via Optical Surface Analyzer and Molecular Dynamics", *IEEE Trans. Magn.* 42(10), 2528-2530 (2006).

[2] P.S. Chung, H. Chen, and M.S. Jhon, "Molecular dynamics simulation of binary mixture lubricant films", *J. Appl. Phys.*, 103, 07F526 (2008).

[3] D. Kim, P.S. Chung, P. Jain, S.H. Vemuri, and M.S. Jhon, "Multiscale modeling of head disk interface", IEEE Trans. Magn., 46, 2401-2404 (2010).

See more of this Session: Multiscale Modeling II
See more of this Group/Topical: Computational Molecular Science and Engineering Forum

(104g) Nanoparticles At the Water-Decane Interface: Evidence of Emergent Behavior From Equilibrium Multi-Scale Simulations

Heng Fab, School of Chemical, Biological and Materials Engineering, The University of Oklahoma, Norman, OK and Alberto Striolo, School of Chemical Biological and Materials Engineering, The University of Oklahoma, Norman, OK

Pickering emulsions (emulsions stabilized by solid nanoparticles rather than by surfactants) are finding practical applications in several areas, including cosmetics, drug-delivery, and even in-situ upgrade of bio-oil to fungible bio-fuels [Science 327 (2010) 68]. It is known that solid particles adsorb at water-oil interfaces to reduce the contact area between the two immiscible phases. Stable emulsions are obtained when the particles strongly adsorb at the interfaces. We will report here equilibrium molecular dynamics simulation results employed to differentiate the behavior of silica-based nanoparticles used to stabilize either water-in-decane or decane-water emulsions. We will compare nanoparticles with similar surface properties, but different distribution of the surface groups. Particular attention will be given to Janus as opposed to homogeneous nanoparticles [Langmuir 27 (2011) 5264-5274]. The all-atom simulation results have been used to parameterize dissipative particle dynamics simulations. Using the latter tool, we have obtained interesting morphologies depending on nanoparticle concentration, and water-decane molar ratio in the various simulated systems. The simulation results will be discussed in relation to available experimental data.

See more of this Session: Multiscale Modeling II
See more of this Group/Topical: Computational Molecular Science and Engineering Forum

(116a) Heuristic Design of Reaction/Separation Processes

William L. Luyben, Chemical Engineering, Lehigh University, Bethlehem, PA

The dominant trade-off in chemical process design is between reactor size and recycle flowrate. Big reactors require larger capital investments in vessels and catalyst, but they result in smaller recycle flowrates for a given yield, which means lower capital investments and energy costs in the separation section of the process. Small reactors have reverse effects. Therefore an economic optimum exists that balances the costs of the reaction section and the separation section of the process.

This paper presents a heuristic approach to quickly determine this optimum trade-off during preliminary conceptual design. The basic idea is to start with a very large reactor and find the recycle flowrate required to meet some specified conversion/yield/selectivity criterion. This is the minimum recycle flowrate. Then a heuristic similar to that used in distillation column design is employed. The actual recycle flowrate is set equal to 1.2 times the minimum, and the reactor and separation sections are designed with this recycle. The heuristic recycle ratio has some dependence on the phase equilibrium (decreases as relative volatility decreases), catalyst cost (increases as catalyst cost increases) and the number of recycle streams and distillation columns.

Two different cases are explored using hypothetical components. In the first, the process has a CSTR, two distillation columns and one recycle stream. Two consecutive reactions ($A + B \rightarrow C$ and $A + C \rightarrow D$) produce a desired product C and an undesired product D. Achieving high selectivity requires low concentrations of A and C, so there is a large recycle of mostly B. Relative volatilities in this first case are assumed to be $\alpha_A > \alpha_B > \alpha_C > \alpha_D$, so there is one recycle from the overhead of the first distillation column containing unreacted A and B. The second column separates C and D.

In the second case, relative volatilities are assumed to be $\alpha_A > \alpha_C > \alpha_B > \alpha_D$, so there are three distillation columns and two recycle streams. One recycle containing unreacted A goes overhead in the first distillation column. Product C goes overhead in the second column. A second recycle stream of unreacted B goes overhead in the third column with product D going out the bottom.

An actual process (production of methoxy-methyl-heptane) using real components is also studied to verify the results of the proposed heuristic procedure.

See more of this Session: Process Design I
See more of this Group/Topical: Computing and Systems Technology Division

THERMAL INTEGRATION OF REBOILERS POSITIONED CLOSE TO THE PINCH POINT OF A PROCESS

ID# 226627

[1]Dionicio Jantes Jaramillo, [1]Graham Thomas Polley, [2]Truls Gundersen and [1]Martín Picón Núñez

[1]Universidad de Guanajuato, Guanajuato, México

[2]Norwegian University of Science and Technology, Trondheim, Norway

Abstract

Reboiler integration has become a common alternative for energy savings in industrial processes. Operational and capital costs of distillation columns are directly affected by large changes in the operating pressure (i.e. if the design pressure increases the heat load in reboilers and the number of equilibrium stages required for the separation increase as well). Moderate changes in pressure allow modifying the internal temperature profile in the column and keep these costing parameters about constant. Moreover, reboiler design strongly depends on the outlet vapor fraction achieved by this unit. Then, operating pressure of distillation columns and the level of vaporization can be selected in order to "thermally accommodate" a reboiler stream into the composite curves of a process. In this work, the best trade-off energy/capital of the heat exchanger network is determined when a reboiler stream is integrated close to the Pinch Temperature. This paper examines the integration of both vertical and horizontal thermo-siphon reboilers.

Introduction

Distillation represents one of the largest energy consuming processes in the process industries. This situation has encouraged a large number of researchers to develop clever strategies which optimize the use of energy. For many years, Pinch Technology has been successfully applied in several situations to save energy by maximizing the process-to-process heat exchange (Linnhoff and Flower, 1978). In recent years, Mathematical Programming has also proven to be a powerful alternative to tackle the problem of high energy consumption even for large problems (Anantharaman and Gundersen, 2006). Other strategies also exist that focus on the design of heat integrated distillation sequences by considering the operating pressure as an optimization variable (Rathore et al., 1974). In 1995, Rajah and Polley presented a general methodology for the design of an integrated separation system by synthetizing the global problem into the Heat Load Table.

This paper looks at the integration of reboilers (i.e. vertical and horizontal thermo-siphons reboilers) when the temperature of the column's bottoms is close to the Pinch Temperature of a process. This situation is particularly common when a condensing stream is also involved into the energy analysis. A simple case study of a four streams process (2 hot streams and 2 cold streams) is used to demonstrate the beneficial consequences when the operation of a distillation column is subject to changes in pressure and/or manipulation of exit reboiler quality from an energy/capital savings viewpoint.

Heat Load-Temperature Characteristics

Consider the vaporisation of a binary mixture. If one ignores (for the moment) the temperature difference required to initiate vapour generation, the vaporisation of a liquid mixture commences once the "bubble point" of the mixture is exceeded. As vapour is generated the concentration of the less volatile component increases and this leads to an increase in saturation temperature (see Figure 1).

If the design of the reboiler is such that liquid and vapour are in contact throughout the vaporization process (achieved in vertical tubes) then the vaporization process could continue until all of the liquid is evaporated at which point the saturation temperature has reached the "dew point" of the mixture.

If the design of the reboiler is such that liquid and vapour flows are separated (as occurs in horizontal tube bundles (see Vazquez–Ramirez et al., 2011), then vaporisation is completed at the boiling point of the less volatile component.

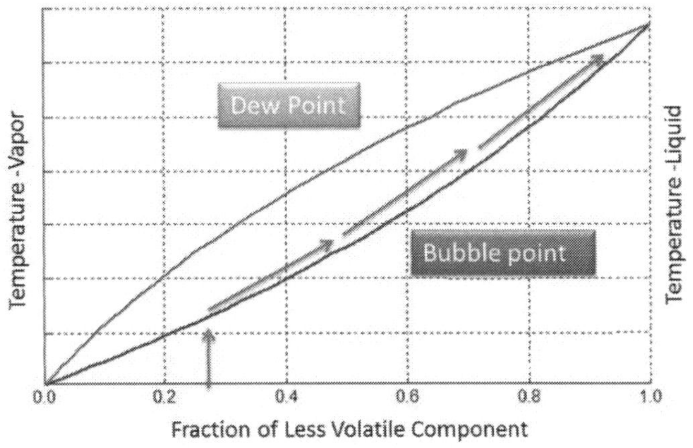

Figure 1. Vapor-Liquid Equilibrium

These are limiting values. In practice, even in once through designs, the feed to a reboiler does not equate with the quantity of material that is to be vaporised. With a vertical thermo-siphon reboiler the maximum amount will rarely exceed 20% of the feed and in most applications will be less than 10% of the feed. For horizontal thermo-siphons, the fraction of feed vaporised should be lower.

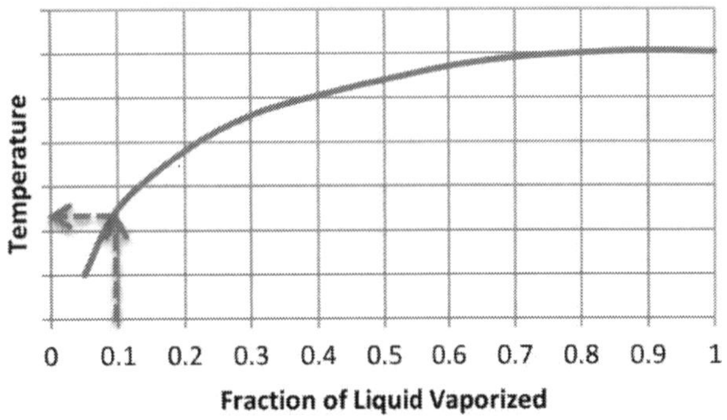

Fiqure 2. Temperature vs. Vapour

A reasonable temperature-heat load relationship for evaluating how a reboiler can be integrated into a process would be a heat load equal to the required boil-up absorbed at a temperature calculated for 10% vaporisation of the mixture being evaporated. This temperature is determined from the heat demand curve for the mixture (see Figure 2).

Integration of Reboiler Streams Close to the Pinch

The condition of a reboiler stream close to the Pinch Temperature is commonly found due to the presence of a condenser stream in the system (see Figure 3). Notice that reboiler and condenser streams do not correspond to the same distillation column.

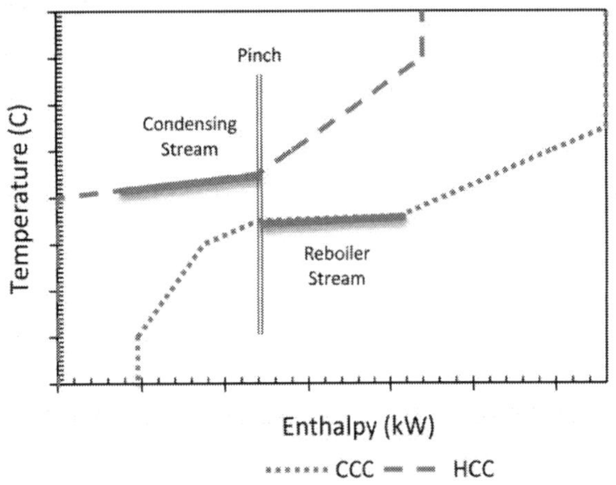

Figure 3. Reboiler Close to Pinch Temperature

During the column design, once the top pressure has been specified, the bottoms pressure is calculated by accounting for specific pressure drop through the column. Then, an increment in the top pressure raises the pressure in the bottoms and vice versa. Therefore, pressure can be selected in order to accommodate a reboiler stream around the Pinch Temperature defined by process.

Three options can be examined: a) if the reboiler stream is positioned below Pinch, the available heat of the process satisfies the heat demand recently incorporated to the system (i. e. according to Pinch rules; heating is not allowed below Pinch). Thus, the process-to process heat exchange is maximized, b) if the reboiler stream is placed above Pinch, the horizontal temperature profile of this unit enlarges the driving forces of this region (in other words; capital cost is minimized), c) a double benefit can be expected when the reboiler stream is placed across Pinch. This third option involves reboilers installed in parallel arrangement.

Case Study

Consider the system shown in Table 1. The Composite Curves Diagram for this system is shown in Figure 4. In this diagram, reboiler stream has not yet been included.

Table 1. Process Streams

Stream	Tin (°C)	Tout (°C)	Q (kW)
H1	140	115	1000
H2	115	110	1200
C1	80	100	400
C2	100	125	1625

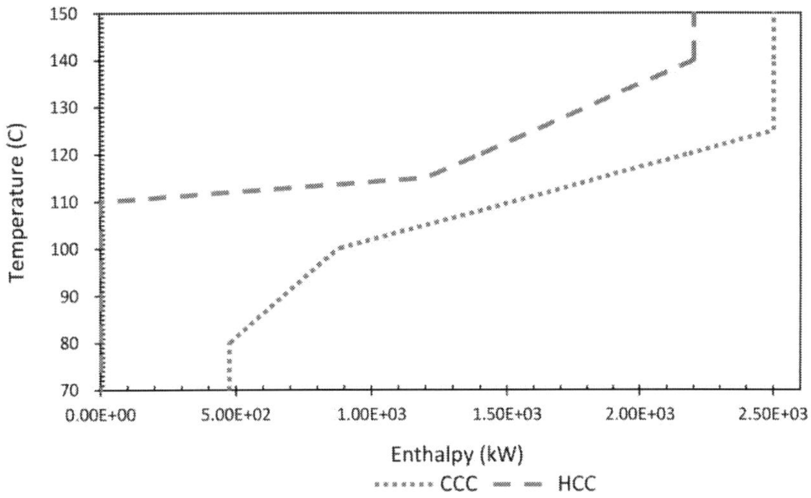

Figure 4. Composite Curves

The reboiler stream data considered for integration is shown in Table 2. In all locations, the heat load required in the column is assumed to be constant while temperature change in the reboiler stream is only 1°C. In this manner the energy/cost profit is only reflected over the heat exchanger network.

Table 2. Reboiler Stream

Location	Tin (°C)	Tout (°C)	Q (kW)
Below	100	101	800
Across	103	104	800
Above	105	106	800

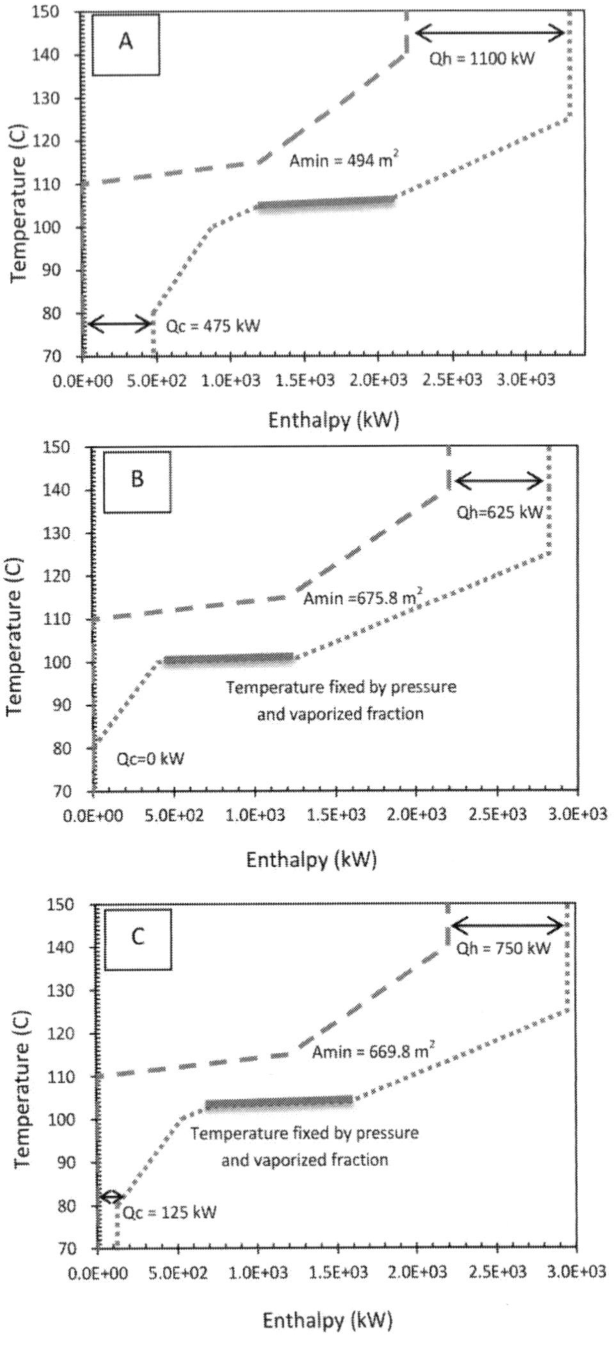

Figure 5. Reboiler integration options

Results and Conclusions

The three different locations of the reboiler stream are shown in the Figure 5. In option A (i.e. a unit placed above Pinch), large temperature differences ensures the lowest heat transfer area of the network (including a small area in the reboiler). However, the temperature at which the hot liquid enters the reboiler can affect the two-phase conditions within this unit. This can lead to operability problems such as throttling in vertical thermo-siphons, excessive bypassing in horizontal thermo-siphon with horizontal baffle cuts and low bundle submergence in horizontal thermo-siphon with vertical baffle cuts. In addition, the critical heat flux condition can be easily reached in units operating at high temperature differences.

For option B (i.e. unit placed below Pinch), the cold composite curve is horizontally approaching to the hot composite curve. This change decreases the hot utility about 43% compared with option A; however the minimum heat transfer area of the network increases 36.8%. Finally, for option C (i.e. unit placed across Pinch) the energy savings are about 31.2% by increasing the minimum heat transfer area about 35.6%. So, once the energy requirements are determined, the project manager can explore different alternatives and decide the best option for energy/capital savings. These alternatives are defined as a function of the current Pinch Point of the process.

Acknowledgements

We would like to thank the Research Council of Norway (RCN) and the Consejo Nacional de Ciencia y Tecnología (CONACYT) México, for financial support.

References

1. B. Linnhoff and J.R. Flower, "Synthesis of heat exchanger networks: I. Systematic generation of energy optimal networks". AIChE Jl 24 (1978), pp. 633–642.
2. Anantharaman R. and Gundersen T. "Developments in the Sequential Framework for Heat Exchanger Network Synthesis for solving Industrial size Problems", Proceedings from the joint PSE'2006 and ESCAPE'16, July 10-13, Garmisch-Partenkirchen, Elsevier, vol. 21A, pp. 725-730, 2006.
3. Rathore, R. N. S., Van Wormer, K. A. and Powers, G. J. "Synthesis of distillation systems with energy integration". AIChE Journal 20 (1974) b, pp. 940-950.
4. W. Rajah and G.T. Polley, "Synthesis of practical distillation schemes". Trans. IChemE 73/A (1995), pp. 953–966.
5. Vazquez–Ramirez E.E., Riesco Avila M. and Polley G.T. "Two Phase Flow through Horizontal Tube Bundles: Implications for Heat Exchanger Design", AIChE Spring Meeting, 2011.

(116c) Exergetic Temperatures for Exergy Targeting

Danahe Marmolejo-Correa and Truls Gundersen, Energy and Process Engineering, Norwegian University of Science and Technology, Trondheim, Norway

This work presents a methodology for targeting heat and power consumption in energy intensive processes (such as refrigeration cycles, power plants and heat recovery systems). The methodology uses process integration and exergy analysis as tools to obtain minimum exergy consumption and minimum losses (including exergy destruction). By decomposing the thermo-mechanical (physical) exergy of material streams into their pressure and temperature based exergies, it is possible to independently calculate the targets for power and heating systems. Besides, the identification of exergy sources and sinks becomes evident and easy to calculate. Exergy is neither linear with temperature nor with pressure. Actually, pressure is involved in rather complicated relations with temperature and exergy; however, it can be simplified by assuming ideal gas conditions. Under these circumstances, the decomposition of thermo-mechanical (physical) exergy is sharp and there is no dependency of temperature in the pressure based exergy. The targeting method utilizes this sharp split to independently manipulate the exergy components in order to get a linear relation with the specific heat capacity and an Exergetic Temperature. Therefore, changes from one exergetic temperature to another (i.e. for heat exchange without pressure drop) will be represented by a straight line in an Exergetic Temperature – Exergy diagram (T^e - E). By definition, the exergy of a system is the maximum ability to produce work, thus it measures the thermodynamic quality of energy. Parameters, such as the Carnot factor, give indication of the level of the exergy; large figures indicate high exergy levels and consequently low figures indicate low exergy levels. The Exergetic Temperatures are also indicators of exergy level. Thus, T^e – E diagrams illustrate the level and availability of exergy in material streams, allowing better understanding of the exergy transfer, and this insight can be used to improve process systems.

For processes where heat and power are intimately related, such as Low Temperature Processes (LTPs), exergy seems to be a perfect evaluator for the conceptual design procedure. A new Exergy Transfer Effectiveness (ETE) is proposed for the calculation of a Coefficient of Exergy Performance (CEP) for processes above, below and across ambient temperature. Previous CEPs will be briefly commented. The ETE utilizes the simple concept of sources and sinks; however, due to the special behavior of temperature based exergy in some operating conditions (below and across ambient temperature), the recognition of these sources and sinks can be difficult. A general rule is presented for the identification of exergy sources and sinks in processes, where only thermo-mechanical (physical) exergy is under transformation. For any process stream subject to change in exergy, from state a (T_a, p_a) to state b (T_b, p_b) without crossing ambient temperature, it is established that all negative changes in exergy ($\Delta E_{a\text{-}b} = E_b - E_a$), or in its components, are categorized as *exergy sources* (except for the change in pressure based exergy due to pressure drop), while positive changes in exergy represent *exergy sinks*. If the operating conditions of the process or unit are across T_0 (i.e. compressors in refrigeration cycles), an intermediate state i (T_0, p_i) is introduced, resulting in the existence of *one source* (from state a to i) and *one sink* (from state i to b) for the change in temperature based exergy.

Two simple case studies are used to illustrate the methodology and its capabilities. The first case is a simple heat recovery system with two hot streams and two cold streams where pressure drop is neglected. The heat exchange takes place above ambient conditions (T_0, p_0). The second case study is a reverse Brayton cycle utilized for the liquefaction of natural gas.

See more of this Session: Process Design I
See more of this Group/Topical: Computing and Systems Technology Division

Economic objective function in process flow sheet optimization – do we know everything about it?

Zorka Novak Pintarič, Mihael Kasaš, Zdravko Kravanja
University of Maribor, Faculty of Chemistry and Chemical Engineering
Smetanova 17, SI-2000 Maribor, Slovenia

Introduction

Design and synthesis of process flow sheets by using mathematical programming approach are considered difficult, especially if large-scale industrial processes are concerned or additional challenges are included, e.g. uncertainty for flexible design. Over the last three decades, the process engineering community put tremendous effort into developing a) modeling approaches to deriving appropriate mathematical formulations of such processes, and b) solution strategies for solving mathematical models efficiently to optimum solutions. Less attention has been paid to the type and correctness of the economic criteria used in the objective functions of flow sheet optimization models.

The annual profit, the total annual cost, the payback time, the net present value (NPV), and the internal rate of return (IRR) are the economic criteria that are more often applied to flow sheet design and synthesis. Many authors have observed that these measures generate different optimal flow sheet solutions when used as decision criteria, e.g. Faria and Bagajewicz (2009), Bagajewicz (2008). In our previous work (Novak Pintaric and Kravanja, 2006) it was established that economic criteria can be roughly classified into three groups regarding the optimal results generated: quantitative criteria which are expressed in monetary units (e.g. profit, cost), qualitative criteria which are expressed in non-monetary units (e.g. IRR, payback time), and compromise criteria, e.g. the NPV, which are the compromise between qualitative and quantitative measures.

In recent year, a further research has been carried out which revealed some fundamentally new understandings about the correctness of various economic criteria, and the consequences of their use in optimization-based flow sheet synthesis and design (Kasas et al. 2011a, 2011b). To the best of our knowledge, some of the following questions have been cleared up for the first time: a) which economic criteria are the correct ones, b) what is the reason for the differences between those optimal solutions obtained by different criteria, and c) what are the consequences of using inappropriate criteria in the objective functions of flow sheet optimization models.

The correctness of economic criteria in process flow sheet optimization

In the financial theory, projects are classified into independent and mutually exclusive ones. Both groups of projects have specific approaches of how to select the best alternative(s). It is important to realize that flow sheet design and synthesis by means of mathematical programming belong to the selection between mutually exclusive alternatives. The capital investment is very often considered unconstrained, as it is explicitly assumed that funds are available for realization of selected optimal design. According to the finance and accounting theory, the net present value is the correct criterion for such type of investment projects, as it uses all the cash flows of the project and discounts the cash flows properly. Other criteria are not totally correct, e.g. mutually exclusive alternatives cannot be compared by their individual IRRs, as

they do not take into account the level of investment, and mistakenly favor projects with small investment. A similar effect was observed in the case of the payback time criterion. Incremental analysis of IRR is needed, where the IRR of each increment of investment must satisfy the minimum acceptable rate of return (MARR) requirement. Criteria like total cost and profit are the accounting measures, which are not based on the cash flows, and are thus inappropriate for investment decision making.
The above mentioned criteria would, in general, generate different optimal results when used in flow sheet optimization problems as the objective functions. Moreover, the consequences of using inappropriate criteria could be far more serious than it might seem at first glance.

The source of the differences between optimal flow sheet designs

The most obvious reason for differences between optimal flow sheet designs obtained by using various economic criteria is that the stationary conditions of these criteria are different. For example, the maximum NPV at unconstrained investment is obtained at an investment level where the incremental NPV equals zero, and incremental IRR equals the MARR. The maximum profit before taxes and the maximum IRR are obtained at those investment levels where the incremental profit and IRR, respectively, become equal to 0. It could be shown, by analyzing stationary conditions regarding the relation between the cash flow and investment, that qualitative criteria, e.g. IRR, generate optimal solutions at the lowest investment levels, while the quantitative criteria, e.g. profit and cost, at the highest investment levels. The investment levels of those solutions obtained by compromise criteria like the NPV are in-between, as the NPV establishes a balance between the profitability of investment and long-term steady generation of cash flow.
In spite of well-defined differences in stationary conditions, some process flow sheet models produce negligible differences when optimized with different objectives, while other models produce substantially-different optimal designs. A source of this apparent contradiction is in the accuracy and preciseness of flow sheet model regarding the trade-offs between the investment and cash flow generated. These trade-offs are often insufficient if simple and aggregated models are used, e.g. stoichiometric model for the reactor or simple component splitter for the separation. More precise trade-offs are present in more detailed modeling where increased investment level results in increased benefit, e.g. higher conversion in the kinetic reactor, higher purity of main product in more precise separator model etc. The more precise correlations between the investment and resulting gains are included in the model, the larger differences in optimal results could be expected.
It was shown (Kasas et al. 2011a) that the steepness of the cash flow derivative curve vs. investment is the main responsible for the magnitude of the differences between optimal results obtained by different economic criteria. The more gentle derivative function of cash flow vs. investment, the larger differences would be obtained. Less precise models generate unimodal cash flow functions with maximum, whose derivative curves are very steep. Optimal solutions obtained by different criteria are thus similar or equal. More precise models generate monotonically increasing cash flow functions and flat derivative curves, optimal solutions are thus further apart.

The consequences of using different economic criteria

The consequences of using different economic criteria in process flow sheet optimization are numerous, and can be classified into economic, operational and environmental. Economic consequences follow straightforwardly from the facts given in previous subsection. The investment levels and cash flows of optimal solutions increase from the qualitative criteria (IRR, payback time) over the compromise (NPV) to the quantitative criteria (profit, cost).

The operational consequences can be evaluated through the overall operational efficiency of optimal solutions, which could be expressed as a financial ratio between the net income and sales. It was proved that operational efficiencies grow from the IRR, over the NPV, to the profit solution (Kasas et al. 2011b). IRR criterion leads to inefficient utilization of reactant and utilities, which reflects in lower conversion in the reactor, lower purity of the product, lower level of heat recovery etc. Maximization of profit produces the most efficient solutions with the lowest operating costs, while the operational efficiencies of optimal NPV solutions are in-between.

Optimal solutions are different also regarding the environmental performance. The latter can be evaluated using various indicators, e.g. global warming, acid rain potential etc. Mixed integer nonlinear programming (MINLP) syntheses of heat integrated processes (methanol, dimethyl ether, and HDA) showed that profit criterion generates environmentally most conscious solutions, while the environmental impacts of the IRR solutions are the highest. Optimal NPV designs are the environmental compromises between the other two solutions.

Conclusions

This contribution summarizes the outcomes of the research on the economic criteria used in process flow sheet optimization, and points out some important findings that have been often overlooked by now. The NPV is the correct criterion for optimization-based flow sheet design and synthesis, while other measures like profit and internal rate of return are not applicable decision criteria. The insufficient trade-offs in too simplified and aggregated models could prevent from generating proper optimal solutions even with the correct optimization criterion. By using the NPV criterion in precisely defined process flow sheet models, a well-balanced compromise between the economic performance, operational efficiency, and environmental impacts can be achieved during a single-criterion optimization. Such solutions represent correct compromise designs in the financial and environmental sense, as well as in the sense of efficient revenue generation and operating cost control.

References

Bagajewicz M. On the use of net present value in investment capacity planning models. Ind. Eng. Chem. Res. 2008, 47, 9413-9416.

Faria D.C., Bagajewicz M.J. Profit-based grassroots design, retrofit of water networks in process plants. Comput. Chem. Eng. 2009, 33, 436-453.

Kasaš M., Kravanja Z., Novak Pintarič Z. Suitable modeling for process flow sheet optimization using the correct economic criterion. Ind. Eng. Chem. Res. 2011a, 50, 3356-3370.

Kasaš M., Kravanja Z., Novak Pintarič Z. Achieving profitably, operationally and environmentally compromise flow-sheet designs by a single-criterion optimization. Submitted to AIChE J. 2011b.

Novak Pintarič Z., Kravanja Z. Selection of the economic objective function for the optimization of process flow sheets. Ind. Eng. Chem. Res. 2006, 45, 4222-4232.

(116f) Heat Exchanger Network Synthesis with Non-Isothermal Mixing Using a Stagewise Superstructure

Kefeng Huang[1], Eid Almutairi[2] and Iftekar A. Karimi[1], (1)Department of Chemical & Biomolecular Engineering, National University of Singapore, Singapore, Singapore, (2)Chemical Engineering Department, King Fahd University of Petroleum & Minerals, Dhahran, Saudi Arabia

Energy is a global concern. Depleting fossil energy resources, soaring energy demands, stringent environmental regulations, alarming climate changes, growing international competition and market globalization, etc. highlight the importance of energy conservation. As a holistic approach to utilize energy efficiently and economically in a chemical process, heat exchanger network synthesis (HENS) tries to exchange heat optimally between the hot and cold process streams and utilities within a process to reduce energy consumption. Because of its extensive energy usage, the chemical process industry has given serious and significant attention to HENS in both research and practice for several decades now.

Many optimization models for the simultaneous HENS have employed superstructures (Yee & Grossmann, 1990; Ciric & Floudas, 1991) of possible 2-stream matches of HEs. Yee & Grossmann (1990) proposed a multistage superstructure of HEs for each stream. They assumed that all split substreams of a stream exiting a stage have the same temperature. The literature has called this assumption isothermal mixing, which makes the energy balances for substream mixing at each stage linear. However, as Yee & Grossmann (1990) explained, the stagewise superstructure misses several HEN configurations, and the assumption of isothermal mixing obviously omits configurations involving non-isothermal mixing. Yee & Grossmann (1990) remarked that isothermal mixing may restrict the area trade-offs between the exchangers and overestimate the area cost, when stream splits with non-isothermal mixing may give superior HENs. In spite of these known limitations of isothermal mixing, non-isothermal mixing in stagewise superstructure has attracted limited attention in the HENS literature. Yee & Grossmann (1990) proposed an effective heuristic strategy to improve the suboptimal HENS arising from isothermal mixing. While this strategy overcomes partially the impact of isothermal mixing, the overall approach still cannot guarantee a global solution due to the inherent nonconvex nature of the simultaneous HENS. Bjork & Westerlund (2002) compared HENS with and without isothermal mixing in the stagewise superstructure of Yee and Grossmann (1990) using a global optimization strategy. Recently, Hasan et al. (2009) slightly modified the stagewise superstructure and generalized simultaneous HENS to address non-isothermal phase changes, nonlinear T-H profiles, and non-isothermal mixing.

In this work, we modify the Synheat model of Yee and Grossmann (1990) to address HENS with non-isothermal mixing. Specifically, we focus on the limitations of literature models for handling non-isothermal mixing, and propose new and improved bounds and logical constraints to obtain superior

HENs. We demonstrate the effectiveness of our modifications and solution algorithms by presenting examples in which our approach yields better HENs compared to the existing literature. We show that even the global optimization algorithm of Bjork & Westerlund (2002) for non-isothermal mixing fails to obtain globally best HENs in case of some large problems. We also study the impact of using LMTD approximations in HENS formulations.

References

1. Ciric AR, Floudas CA. Heat exchanger network synthesis without decomposition. *Computers & Chemical Engineering.* 1991;15(6):385-396.
2. Yee TF, Grossmann IE. Simultaneous optimization models for heat integration--II. Heat exchanger network synthesis. *Computers & Chemical Engineering.* 1990;14(10):1165-1184.
3. Björk K-M, Westerlund T. Global optimization of heat exchanger network synthesis problems with and without the isothermal mixing assumption. *Computers & Chemical Engineering.* 2002;26(11):1581-1593.
4. Hasan MMF, Jayaraman G, Karimi IA, Alfadala HE. Synthesis of heat exchanger networks with nonisothermal phase changes. *AIChE Journal.* 2009;56(4):930-945.

See more of this Session: Process Design I
See more of this Group/Topical: Computing and Systems Technology Division

(135a) Global Optimization of Mixed-Integer Quadratically-Constrained Quadratic Programs (QCQP) Through Piecewise-Linear and Edge-Concave Relaxations

Ruth Misener and Christodoulos A. Floudas, Department of Chemical and Biological Engineering, Princeton University, Princeton, NJ

Nonconvex quadratically-constrained quadratic programs (nonconvex QCQP) and those admitting integer variables (MIQCQP), are ubiquitous in process systems applications including heat integration networks, separation systems, reactor networks, reactor-separator-recycle systems, and batch processes. Among these applications are the pooling problems, which represent the optimization challenge of maximizing profit subject to feedstock availability, intermediate storage capacity, demand, and product specification constraints [9, 17, 18, 19]. The pooling problem, which is a MIQCQP, has important practical applications to many process systems engineering domains, including petroleum refining, water systems, supply-chain operations, and communications [4, 11, 16, 26].

We previously addressed large-scale pooling problems to ε-global optimality using a disjunctive relaxation formulation that activates appropriate under- and overestimators in specific domain segments and integrated this relaxation scheme into a branch-and-bound global optimization algorithm. Similar underestimators have been exploited in an array of process network applications [7, 10, 12, 13, 15, 20, 21, 27]. Here we address generic MIQCQP by extending these results beyond the special structure of MIQCQP representing process network problems. Because there is an inherent trade-off between relaxation tightness and solving time, we seek to elucidate the conditions under which these relaxations offer an advantage over the commonly-used termwise hull relaxation.

Complementing the piecewise-linear relaxations, we investigate polyhedral facets generated through edge-concave aggregations [2, 5, 8, 14, 23, 24, 25]. Edge-concave functions admit a vertex polyhedral envelope and therefore have a convex hull consisting entirely of linear facets [14, 23, 24, 25]. We detect possible low-dimensional ($n \leq 3$) edge-concave aggregations of two or three variables in MIQCQP, determine the convex hull of the aggregated terms, and integrate facets that strictly dominate the termwise relaxation into our underestimation scheme at every node of a branch-and-bound tree. The facets that do not strictly dominate the termwise relaxation are still useful for bounds reduction within the context of branch-and-bound global optimization.

The proposed global optimization approach, whose novel contributions are rooted in the underestimators, determines an appropriate relaxation scheme employing piecewise linear underestimators and edge-concave relaxations and integrates the resulting mixed-integer linear program

(MILP) into a branch-and-bound global optimization algorithm. We also incorporate standard techniques such as Reformulation-Linearization Technique feasibility-based range reduction [3, 22], limited optimality-based bounds tightening, and reliability branching [1, 6]. Extensive computational studies are presented for point packing problems, quadratic programming problems, and standard and generalized pooling problems.

[1] T. Achterberg, T. Koch, and A. Martin. Branching rules revisited. Oper. Res. Lett., 33(1):42–54, 2005.

[2] K. M. Anstreicher and S. Burer. Computable representations for convex hulls of low-dimensional quadratic forms. Math. Program., 124 (12):33–43, 2010.

[3] C. Audet, P. Hansen, B. Jaumard, and G. Savard. A branch and cut algorithm for nonconvex quadratically constrained quadratic programming. Math. Program., 87(1):131 – 152, 2000.

[4] M. Bagajewicz. A review of recent design procedures for water networks in refineries and process plants. Comput. Chem. Eng., 24:2093-2113, 2000.

[5] X. Bao, N. V. Sahinidis, and M. Tawarmalani. Multiterm polyhedral relaxations for nonconvex, quadratically-constrained quadratic programs. Optimization Methods and Software, 24(4-5):485 – 504, 2009.

[6] P. Belotti, J. Lee, L. Liberti, F. Margot, and A. Wachter. Branching and bounds tightening techniques for non-convex MINLP. Optimization Methods and Software, 24(4-5):597–634, 2009.

[7] M. L. Bergamini, I. Grossmann, N. Scenna, and P. Aguirre. An improved piecewise outer-approximation algorithm for the global optimization of MINLP models involving concave and bilinear terms. Comput. Chem. Eng., 32(3):477 – 493, 2008.

[8] S. Burer and A. N. Letchford. On nonconvex quadratic programming with box constraints. SIAM J. Optim., 20(2):1073–1089, 2009.

[9] C. E. Gounaris, R. Misener, and C. A. Floudas. Computational comparison of piecewise-linear relaxations for pooling problems. Ind. Eng. Chem. Res., 48(12):5742 – 5766, 2009.

[10] M. M. F. Hasan and I. A. Karimi. Piecewise linear relaxation of bilinear programs using bivariate partitioning. AIChE J., 56(7):1880 – 1893, 2010.

[11] J. Jezowski. Review of water network design methods with literature annotations. Ind. Eng. Chem. Res., 49(10):4475 – 4516, 2010.

[12] R. Karuppiah and I. E. Grossmann. Global optimization for the synthesis of integrated water

systems in chemical processes. Comput. Chem. Eng., 30:650 – 673, 2006.

[13] J. Li, R. Misener, and C. A. Floudas. Continuous-time modeling and global optimization approach for scheduling of crude oil operations. AIChE Journal, 2011. In Press.

[14] C. A. Meyer and C. A. Floudas. Convex envelopes for edge-concave functions. Math. Program., 103 (2):207–224, 2005.

[15] C. A. Meyer and C. A. Floudas. Global optimization of a combinatorially complex generalized pooling problem. AIChE J., 52(3):1027 – 1037, 2006.

[16] R. Misener and C. A. Floudas. Advances for the pooling problem: Modeling, global optimization, and computational studies. Applied and Computational Mathematics, 8(1):3 – 22, 2009.

[17] R. Misener and C. A. Floudas. Global optimization of large-scale pooling problems: Quadratically constrained MINLP models. Ind. Eng. Chem. Res., 49(11):5424 – 5438, 2010.

[18] R. Misener, C. E. Gounaris, and C. A. Floudas. Mathematical modeling and global optimization of large-scale extended pooling problems with the (EPA) complex emissions constraints. Comput. Chem. Eng., 34(9):1432 – 1456, 2010.

[19] R. Misener, J. P. Thompson, and C. A. Floudas. APOGEE: Global optimization of standard, generalized, and extended pooling problems via linear and logarithmic partitioning schemes. Comput. Chem. Eng., 35(5):876 – 892, 2011.

[20] V. Pham, C. Laird, and M. El-Halwagi. Convex hull discretization approach to the global optimization of pooling problems. Ind. Eng. Chem. Res., 48:1973 – 1979, 2009.

[21] Y. Saif, A. Elkamel, and M. Pritzker. Global optimization of reverse osmosis network for wastewater treatment and minimization. Ind. Eng. Chem. Res., 47(9):3060 – 3070, 2008.

[22] H. D. Sherali and C. H. Tuncbilek. New reformulation linearization/convexification relaxations for univariate and multivariate polynomial programming problems. Oper. Res. Letters, 21(1):1 – 9, 1997.

[23] F. Tardella. On a class of functions attaining their maximum at the vertices of a polyhedron. Discret. Appl. Math., 22:191–195, 1988/89.

[24] F. Tardella. On the existence of polyhedral convex envelopes. In C. A. Floudas and P. M. Pardalos, editors, Frontiers in Global Optimization, pages 563–573. Kluwer Academic Publishers, 2003.

[25] F. Tardella. Existence and sum decomposition of vertex polyhedral convex envelopes. Optim. Lett.,

2:363–375, 2008.

[26] V. Visweswaran. MINLP: Applications in blending and pooling. In C. A. Floudas and P. M. Pardalos, editors, Encyclopedia of Optimization, pages 2114 – 2121. Springer Science, 2 edition, 2009.

[27] D. S. Wicaksono and I. A. Karimi. Piecewise MILP under-and overestimators for global optimization of bilinear programs. AIChE J., 54(4):991 – 1008, 2008.

See more of this Session: Advances In Optimization I
See more of this Group/Topical: Computing and Systems Technology Division

(135b) Multi-Variate, Multi-Term, and Multi-Constraint Relaxations for Global Optimization of Nonconvex NLPs and MINLPs with BARON

Nick Sahinidis, Chemical Engineering, Carnegie Mellon University, Pittsburgh, PA and Aida Khajavirad, Carnegie Mellon University, Pittsburgh, PA

The development of the BARON algorithms and software for global optimization began nearly twenty years ago. Initially, the software integrated factorable nonlinear relaxations within a domain-reduction-based branch-and-bound algorithm. BARON became more widely used and recognized only after nonlinear relaxations were replaced by polyhedral ones that increased its robustness and performance. This talk focuses on the third generation of branch-and-reduce algorithms that recently started to emerge. Based on multi-variate, multi-term, and multi-constraint relaxation techniques, these algorithms sharpen lower bounds and enhance performance. The talk will describe a large number of novel convex/concave envelopes for such functions, and present extensive computational results with the implementation of these envelopes in BARON.

See more of this Session: Advances In Optimization I
See more of this Group/Topical: Computing and Systems Technology Division

(135c) Black-Box Optimization Via Global Optimization of Surrogate Models

Satyajith Amaran, Carnegie Mellon University, Pittsburgh, PA and Nick Sahinidis, Chemical Engineering, Carnegie Mellon University, Pittsburgh, PA

In many practical scientific applications, there is a need to find optimal parameters without knowledge of the underlying functional form of what we are optimizing. Such problems, known as black-box optimization problems, commonly occur when there is an attempt to optimize over a simulation or to perform optimal experimental design. Though traditional derivative-free approaches such as genetic algorithms and Nelder-Mead simplex algorithms have been used for a long time, only in the last decade have there been attempts to systematically address such problems through mathematical programming ideas with an intent to ensure convergence and to solve larger problems. Comprehensive reviews of current methods to solve such problems are found in [1,2].

The specific problem we address is the minimization of a smooth, continuous, unconstrained objective function which is expensive to evaluate and has no inherent noise. Our aim then becomes to improve the quality of our solution with the least number of function evaluations. Toward this end, we first sample the function in a certain region at a relatively few number of points. Then, we build a surrogate model to uniquely interpolate the function at these points. Derivative-based approaches can now be used to minimize this surrogate model. To find progressively better points that approach a local minimum, this entire step is embedded in an iterative trust-region framework in which the algorithm terminates on the satisfaction of specific criteria.

Most trust-region algorithms solve the subproblem by finding a point that gives a sufficient decrease in the value of the objective. The novelty of our approach is that it takes advantage of the fact that our surrogate models are capable of capturing non-convexities of the the underlying objective function. As function evaluations are expensive, globally minimizing this surrogate model using deterministic global optimization techniques is considerably less expensive. We demonstrate that finding global minima of surrogates reduces the number of iterations of the overall algorithm to converge to a stationary point.

The algorithm is first applied to standard test problems from literature and compared against existing algorithms as proof-of-concept. We then consider problems of practical interest from the chemical engineering literature. We follow this with a discussion of the successes and the limitations of our approach.

[1] A. R. Conn, K. Scheinberg, and L. N. Vicente, Introduction to derivative-free optimization, MPS-SIAM Series on Optimization, SIAM, Philadelphia, 2009.

[2] L. M. Rios and N. V. Sahinidis, Derivative-free optimization: A review of algorithms and

comparison of software implementations, http://thales.cheme.cmu.edu/dfo/, 2011.

See more of this Session: Advances In Optimization I
See more of this Group/Topical: Computing and Systems Technology Division

(135d) Convergence Rate of Convex Relaxations

Agustn Bompadre, Graduate Aerospace Laboratories, California Institute of Technology, Pasadena, CA and **Alexander Mitsos**, Mechanical Engineering, Massachusetts Institute of Technology, Cambridge, MA

Optimization is a widely used tool in process systems engineering, but often the optimization problems have multiple suboptimal local minima. Deterministic global optimization algorithms can solve such problems, typically employing convex/concave relaxations of the objective and constraints. Several methods have been proposed for the construction of convergent relaxations, including the McCormick [1] and the alphaBB relaxations [2] and the use of auxiliary variables [3]. To ensure finite termination, these relaxations must converge to the nonconvex functions in the limit, i.e., as the diameter of the host sets vanishes (reduction to singleton), e.g., through branching or subdivision. In interval extensions [4] this convergence is characterized using the convergence order which compares the rate of convergence of the estimation error to the rate of the decrease of the range of the function.

In this talk, first the notion of convergence order is extended from interval extensions to convex relaxations in the pointwise metric and Hausdorff metric. The main part of the talk develops theory for the well-known McCormick relaxations of factorable functions [1]. Convergence rules are established for the addition, multiplication and composition operations. The convergence order of the composite function depends on the convergence order of the relaxations of the factors. No improvement in the order of convergence compared to that of the underlying bound calculation, e.g., via interval extensions, can be guaranteed unless the relaxations of the factors have pointwise convergence of high order. Additionally, the McCormick relaxations are compared with the alphaBB relaxations by Floudas and coworkers [2], which guarantee quadratic pointwise convergence. Illustrative and numerical examples are given and hybrid methods are discussed. The implication of the results are discussed for practical bound calculations as well as for convex/concave relaxations of factors commonly found in process systems engineering models.

[1] G. P. McCormick. Computability of global solutions to factorable nonconvex programs: Part I. Convex underestimating problems. Mathematical Programming, 10(2):147--175, 1976.

[2] I. P. Androulakis, C. D. Maranas and C. A. Floudas. alphaBB: A Global Optimization Method for General Constrained Nonconvex Problems, Journal of Global Optimization, 7(4): 337-363, 1995.

[3] M. Tawarmalani and N. V. Sahinidis. Convexification and Global Optimization in Continuous and Mixed-Integer Nonlinear Programming. Nonconvex Optimization and its Applications. Kluwer Academic Publishers, Boston, 2002

[4] R. Moore. Methods and Applications of Interval Analysis. SIAM, Philadelphia, PA, 1979

(135e) Tight LP Relaxations for Optimization Problems with Nonlinear Parametric ODEs

Benoit Chachuat, Department of Chemical Engineering, Imperial College London, London, United Kingdom

The ability to compute tight enclosures for the solutions of nonlinear, parametric ordinary differential equations (ODEs) such as

$$\dot{\mathbf{x}}(t) = \mathbf{f}(t, \mathbf{x}(t), \mathbf{p}), \quad \mathbf{x}(t_0) = \mathbf{h}(\mathbf{p}), \quad \mathbf{p} \in P \subset \mathbb{R}^{n_p},$$

(1)is central to deterministic global optimization methods for nonconvex dynamic optimization. These enclosures are needed to compute lower or upper bounds for general functionals such as

$$\mathcal{J}(\mathbf{p}) = \phi(\mathbf{x}(t_f), \mathbf{p}) + \int_{t_0}^{t_f} \psi(\mathbf{x}(t), \mathbf{p}) dt,$$

which in turn can be exploited by branch-and-bound algorithms and their variants. Other related applications are in the field of mixed-integer dynamic optimization (MIDO) and bilevel dynamic optimization.

Two main classes of methods have been reported in the literature to construct the desired convex/concave relaxations. The first class proceeds by constructing an auxiliary system of ODEs, the solutions of which describe convex/concave functions [2,1] with respect to the parameters **p**, pointwise in the integration variable t. The second class builds upon verified solution methods for ODEs [3,4], whereby the integration domain is discretized into a finite number of steps, and it was shown in [5,6] how convex/concave bounds can be propagated at each step via a two-phase mechanism that uses high-order Taylor series expansion: compute an a priori enclosure of the solutions over the current integration step in the first phase; then, obtain convex/concave bounds on the solutions at the end of the current step in the second phase.

The aforementioned methods rely on the ability to propagate convex/concave bounds for factorable functions, e.g., using the McCormick technique [7]. This technique is appealing in that it does not require the introduction of auxiliary variables, yet it comes with two important limitations: (i) the resulting relaxations are typically nonsmooth; and (ii) the relaxation procedure has to be applied multiple times in order to obtain convex/concave bounds at multiple parameter values **p** in P. While both deficiencies can be addressed by deriving affine relaxations from subgradient information [8], this leads to weaker relaxations.

In deterministic global optimization of nonconvex NLP, much emphasis has been on the construction of LP relaxations, which provide lower/upper bounds on the global solution value both reliably and efficiently using state-of-the-art LP technology [9]. Following similar ideas, this paper presents a general approach for constructing LP relaxations for the solutions of the parametric ODEs (1). The proposed algorithm builds upon two-phase methods for verified ODE solution and is rigorous in its accounting of truncation errors. The first phase remains unchanged and determines both a finite integration step and an a priori enclosure of the ODE solutions over that step. In the second phase, in addition to propagating interval bounds for the ODE solutions at the end of the current step, affine cuts that enclose these solutions are appended to an LP relaxation model. A variant in which Taylor models are propagated in lieu of interval bounds is also considered. Both approaches have been implemented in C++ using an object-oriented approach and operator overloading. The resulting LP relaxations are illustrated by way of numerical examples, and comparisons with other relaxations methods for parametric ODEs are made to demonstrate their efficiency.

Bibliography

1. A. B. Singer and P. I. Barton. Bounding the solutions of parameter dependent nonlinear ordinary differential equations. *SIAM J Sci Comput*, 27(6):2167-2182, 2006.
2. J. K. Scott, B. Chachuat, and P. I. Barton. Nonlinear convex and concave relaxations for the solutions of parametric ordinary differential equations. In revision: *SIAM J Control Optim*, 2010.
3. N. S. Nedialkov, K. R. Jackson, and G.F. Corliss. Validated solutions of initial value problems for ordinary differential equations. *Appl Math Comput*, 105:21-68, 1998.
4. Y. Lin and M. A. Stadtherr. Validated solutions of initial value problems for parametric ODEs. *Appl Numer Math*, 57(10):1145-1162, 2007.
5. A. M. Sahlodin and B. Chachuat. Discretize-then-relax approach for convex/concave relaxations of the solutions of parametric ODEs. *Appl Numer Math*, 61(7):803-820, 2011.
6. A. M. Sahlodin and B. Chachuat. Convex/Concave Relaxations of Parametric ODEs using Taylor Models. *Comput Chem Eng*, 35(5):844-857, 2011.
7. G. P. McCormick. Computability of global solutions to factorable nonconvex programs: Part I - Convex underestimating problems. *Math Program*, 10:147-175, 1976.
8. A. Mitsos, B. Chachuat and P. I. Barton. McCormick-based relaxations of algorithms. *SIAM J Optim*, 20:573-601, 2009.
9. M. Tawarmalani and N. V. Sahinidis. *Convexification and Global Optimization in Continuous and Mixed-Integer Nonlinear Programming: Theory, Algorithms, Software, and Applications*. Nonconvex Optimization and Its Applications. Boston, MA: Kluwer Academic Publishers, 2002.

See more of this Session: Advances In Optimization I
See more of this Group/Topical: Computing and Systems Technology Division

(135f) Deterministic Global Optimization with Differential-Algebraic Equations Embedded

Joseph K. Scott, Chemical Engineering, Massachusetts Institute of Technolgy, Cambridge, MA and Paul I. Barton, Process Systems Engineering Laboratory, Dept. of Chemical Engineering, Massachusetts Institute of Technology, Cambridge, MA

A fully deterministic method is proposed for the globalsolution of optimization problems with nonlinear, semi-explicit, index-onedifferential-algebraic equations (DAEs) embedded. Detailed models of a widevariety dynamic chemical processes, which may include material and energybalances, chemical reaction kinetics, thermodynamic relationships, andsemi-empirical process models, are naturally described by systems of nonlinearDAEs. Accordingly, many important process engineering problems take the form ofoptimization problems with DAEs embedded, including determination of optimaloperating conditions and/or control profiles, as well as safety verificationproblems. Another important application is parameter estimation fordifferential-algebraic models, which takes the form of a least-squaresminimization with DAEs embedded.

Research on dynamic optimization in the chemical engineeringliterature alone has produced a number of representative examples which are nonconvex and display multiple suboptimal local minima.Moreover, the absence of an analytical solution to the embedded dynamic systemin all but trivial cases makes it extremely difficult to determine whether agiven problem is convex or not. Theseconsiderations motivate the development of global optimization algorithms forproblems with dynamic systems embedded. This need is further compounded by theobservation that local optima do not provide meaningful information for manyimportant applications, including safety verification problems and parameterestimation problems in the context of model falsification.

Previous literature has primarily treated optimizationproblems with DAEs embedded using local methods or stochastic global methodswhich cannot guarantee global optimality. Another simple approach is to discretize the embedded DAEs in order to reduce the original dynamic optimizationproblem to a standard nonlinear program, which can be solved to globaloptimality using existing methods. However, an accurate representation of the embeddeddynamics requires a fine discretization, which results in the addition of toomany variables for practical global optimization, even for very simpleproblems. A more customized approach, whichdoes not require discretization, uses a branch-and-bound optimization strategyin conjunction with a theory for constructing convex underestimating programsusing second order sensitivities of the embedded DAEs. Though this method appears to produce theglobal solution for simple test cases, it employs a sampling procedure toobtain global information about the second order sensitivities, and hencecannot guarantee global optimality in the general case.

The method presented in this work uses recent extensions ofMcCormick's relaxation theory to construct convex and concave relaxations ofthe solutions of nonlinear systems of DAEs. Using these relaxations, existingtechniques can be used to construct convex underestimating programs, which inturn enables the use of branch-and-bound global optimization techniques.Because no discretization is required, the proposed algorithm operates in theoriginal decision space, without the introduction of additional variables.Moreover, global information concerning solutions of the dynamic system isobtained via the proposed relaxation theory, rather than through sampling.Accordingly, the proposed algorithm is fully deterministic.

This presentation describes the proposed algorithm, withparticular focus on the construction of convex and concave relaxations of thesolutions of the embedded DAEs, and performance is analyzed through chemicalengineering examples.

See more of this Session: Advances In Optimization I
See more of this Group/Topical: Computing and Systems Technology Division

A debottlenecking study for a dehydration column

Ahmed A. Youssef and Mohan Khadilkar

SABIC Innovative Plastics, Mt. Vernon, USA*

1. Abstract

This study presents a case study of the debottlenecking of a distillation column used in the dehydration of a process stream. Due to recent developments on the process, limitations on the water stripping performance arose. Therefore, a debottlenecking study was conducted to alleviate the load on the distillation column, allow improved capacity and prevent the operation under the flooding regime. A thorough investigation of the system's hysteresis and operating conditions (flow rates, temperatures, pressures, compositions, etc.) was carried out. This showed, for instance, an increase in the pressure drop across the column height. An ASPEN Plus®* model was developed, validated and utilized to identify the associated bottlenecks, their interactions and their impact on the performance of the water removal column. A set of recommendations with respect to the design and operating parameters including adjustments to the feed temperature are presented to showcase the potential solutions of the problem.

* ASPEN Plus is a trademark of AspenTech.

2. Introduction

The utilization of a distillation step in order to accomplish a dehydration process is a common practice in industrial operations. Depending on the nature of the process and the operating conditions of the distillation columns, a number of issues can impact their performance. Such malfunction can be manifested in, but not limited to, a decrease in the separation efficiency and irregular heat effects around the condenser and/or the reboiler for example. Some of the typical root causes can be sought by investigating the degree of flooding in the column, the pressure drop across its sections, the feed stream composition and the column's internals (packing, trays, distributors, etc.). In the current study, a detailed description of the system in use, the "symptoms" of the observed malfunction and the route for tackling the problem, along with the conclusions and recommendations will be outlined.

3. Case description

3.1 Background

The dehydration section in discussion consists of two distillation columns. The feed to the first column (C-41) is a "wet" adduct solution containing a given amount of water governed by the operating conditions. Water is removed in C-41 (also will be referred to as the water stripper) as overheads while the "dry" adduct solution separates out in the bottoms stream. The water stripper's distillate proceeds to the second column (C-48 and also called the azeotrope column) for further processing. The azeotrope column's feed is, basically, a vapor phase and both of its overheads and bottoms streams exiting the dehydration section in discussion. The second column is required to yield further concentration of C-41's overheads.

A Process Flow Diagram (PFD) highlighting the dehydration section is given in Figure 1.
It is noteworthy that C-41 operates with no condenser while C-48 has a condenser and a reboiler. The water stripper is 6' 6" in diameter and contains three packing sections. The packing material is Mellapak plus of varying characteristics. Internally, C-41 also has a liquid and a gas distribution plates. As for the azeotrope column, it is 4' in diameter and houses two Mellapak packing sections.
There are several controllers around the dehydration section that maintain its desired operation. These include the temperature of the section's feed, the temperature at the bottom of C-41 that correlates to the steam consumption directly, and the azeotrope column's reflux ratio.

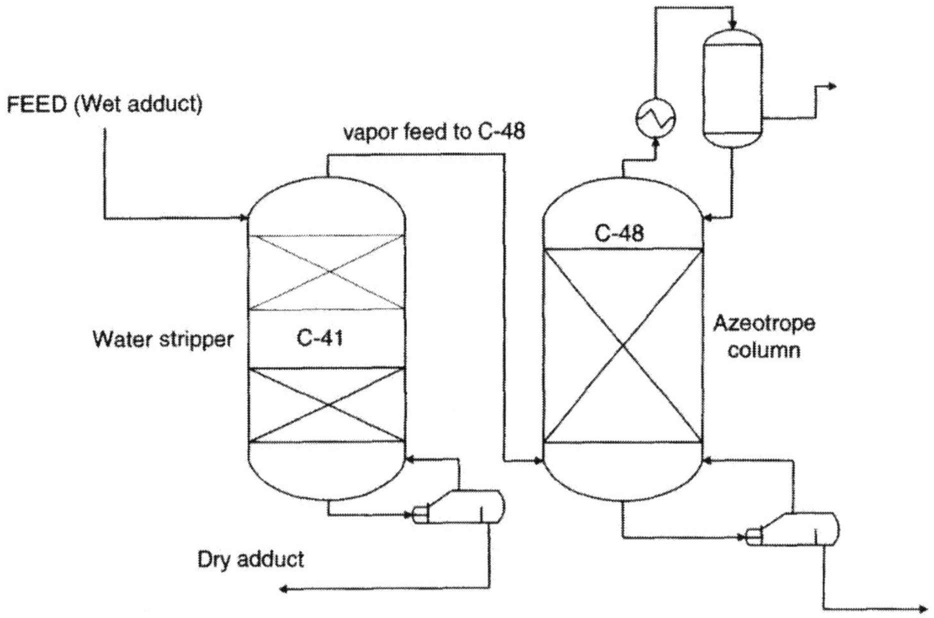

Fig. 1 – Dehydration section PFD

3.2 Problem description

A new technological addition developed within Sabic Innovative Plastics was introduced to the process. This resulted in a significant increase in the percent water fed to the dehydration section. At low operational rates, the dehydration section was able to maintain similar performance compared to prior to the implementation of the new modification. That state of operation continued for ~9 months. However, as the plant operation was ramped up towards its maximum, a sharp increase in the pressure drop across the water stripper column was observed (see Figure 2) corroborated with an increase in the normalized reboiler steam usage (see Figure 3).

Amid fears of a flooding condition inside the C-41 column, adjustments were made to cut the amount of water in the feed. Despite that effort, the continuous increase of the operation capacity showed no improvement in the above trends.

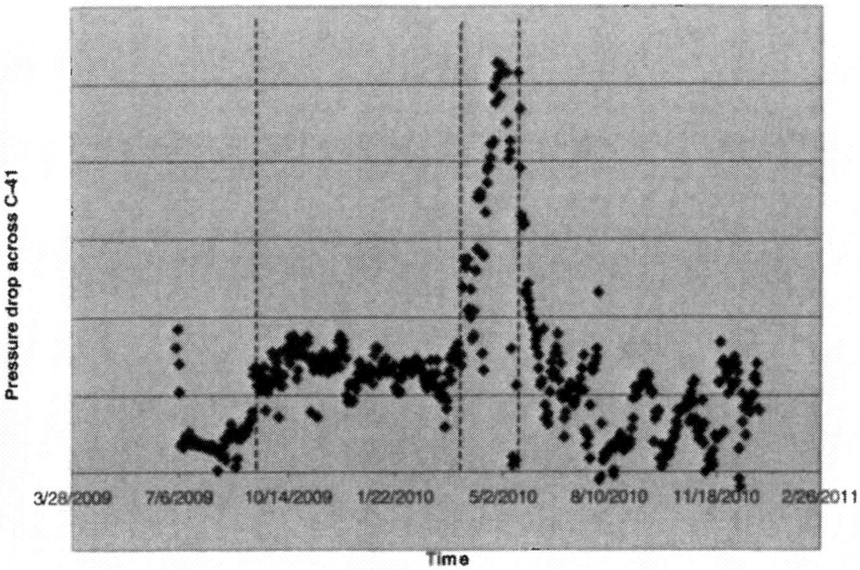

Fig. 2 – Pressure drop across C-41

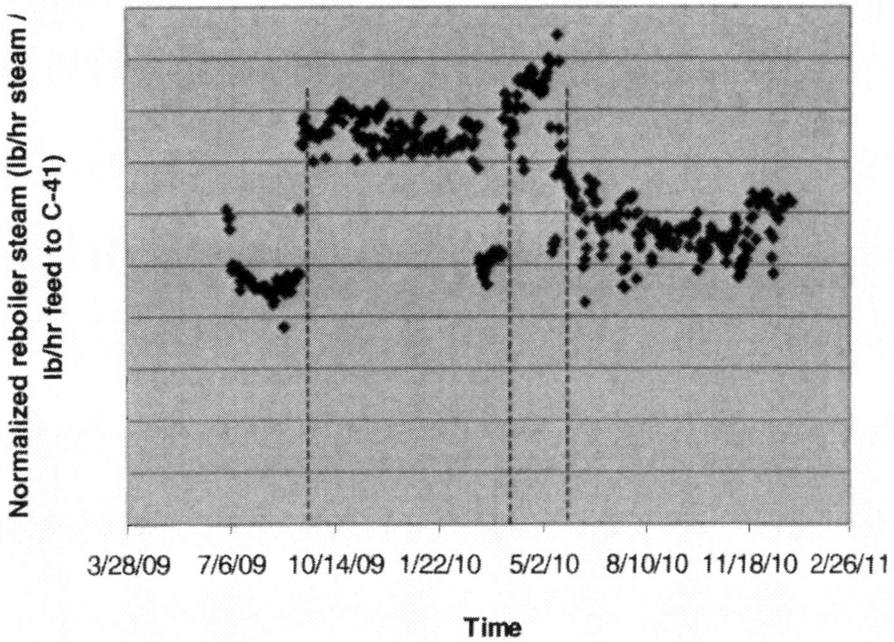

Fig. 3 – Normalized C-41 reboiler steam to feed flow rate

Hence, the above indicators drove a halt on the utilization of the new technology and necessitated the restoring of the original conditions until a better understanding of the situation is reached and a debottlenecking study is completed.

It is noteworthy that the modification to the process limits the plant to 70% of its capacity. The debottlenecking would result in a saving in the raw material usage that translates to ~ $2 MM/y. In addition, the modified process yields a product of higher quality.

In order for this to be accomplished, a mass balance was performed around the dehydration section (see Figure 4). The latter was limited by the availability of the composition data available from plant sampling.

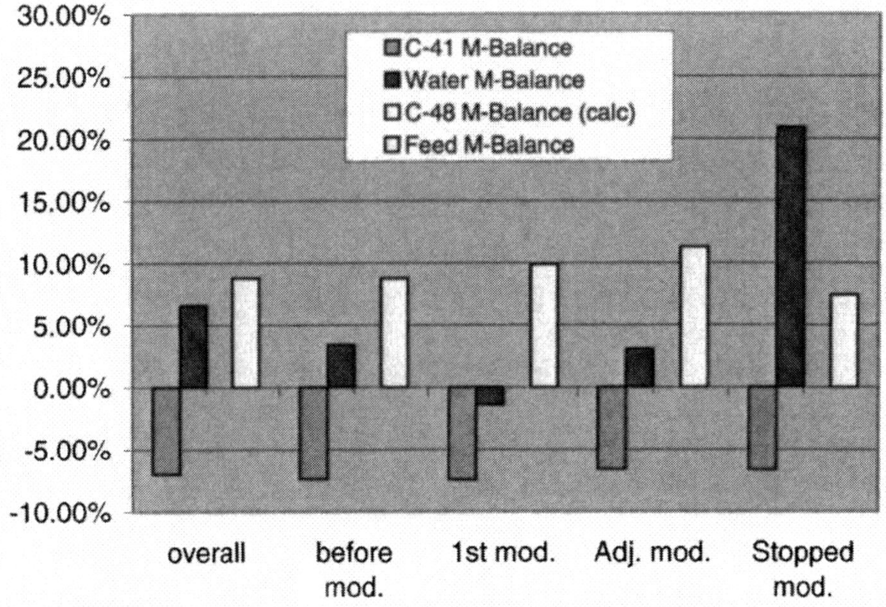

Fig. 4 - Mass balance closure around the dehydration section

4. Root Cause Investigation

The water content in the dehydration column bottoms is a critical indicator of the efficiency of separation. Historical data for the water content during the period of 2009 through 2010 when the modification was employed is shown in Figure 5.

Throughout the time frame of concern, and in spite of the change in the production of the plant, minor effect on the water level in the bottoms of C-41 is observed. Potential modifications in operating conditions can cause such trend as will be highlighted later.

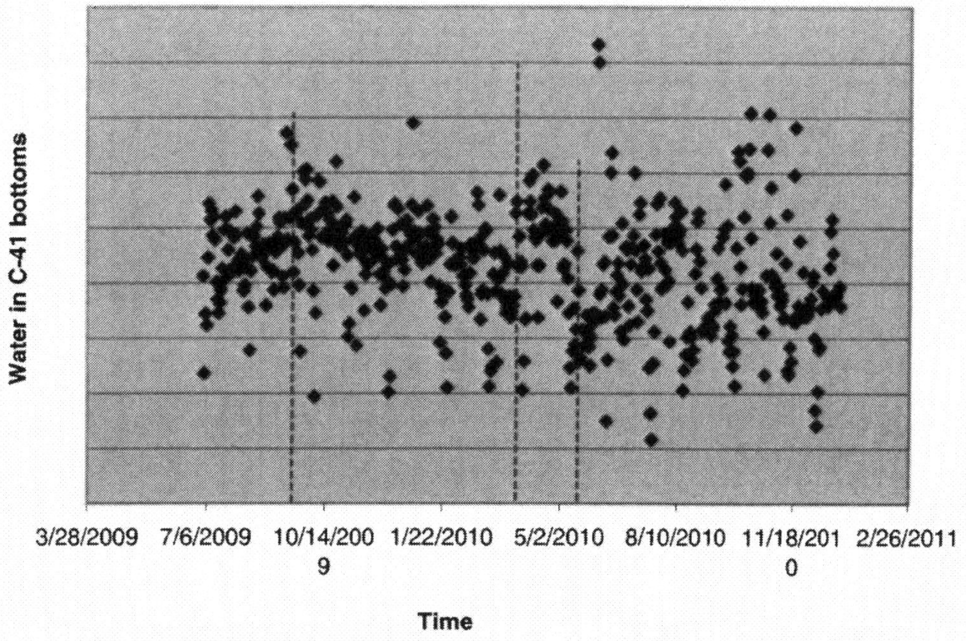

Fig. 5 - Water content in C-41 bottoms

One of the key relationships around the dehydration section is the reboiler steam requirement for C-41 when plotted against the pressure drop in the column. This is particularly important since it correlates one of the main variables (bottoms temperature) that govern the water content in the bottom stream of the water stripper (hence, its efficiency) to the pressure drop, which is an indicator of any malfunction in the operation of the column and its capacity. This relationship is shown in Figure 6. It is clear from Figure 6 that as the reboiler steam usage in the water dehydration column increases (potentially as a result of higher feed flow rate and/or higher bottom's temperature), the pressure drop increases accordingly.

The second key component of the factors that would impact the performance in the dehydration section is the inlet temperature. Higher feed temperature increases the amount of flash yielding to lower work done by the column yielding more separation efficiency and taking the system away from flooding regime [1].

However, plotting the feed temperature vs. water content in bottom showed no obvious trend in the range of feed temperature evaluated for the current study on the water content in the bottom of the water stripper (Figure 5).

The latter can be attributed to the small range of variability in the feed temperature employed within the dehydration section, which does not allow for a wide range of data to estimate the impact of changing the feed temperature.

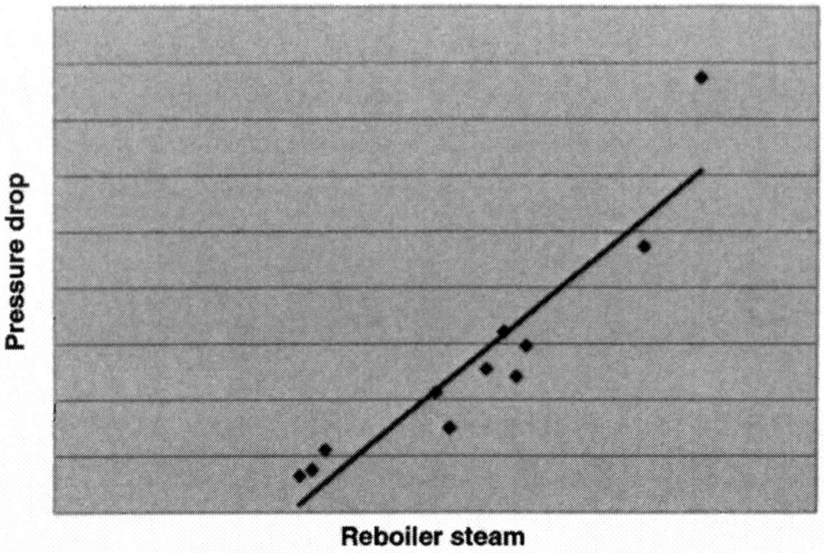

Fig. 6 - Reboiler steam vs. pressure drop in C-41

Moreover, a comparison of such operating conditions against a sister plant site of SABIC Innovative Plastics shows that a much broader range of feed temperature was employed.
The question of whether additional heating capacity can be utilized for the feed stream needs to be answered. The dehydration section's current capability allows flexibility in supplying hotter feed to the C-41 column. However, the benefits from such practice need to be evaluated which will be accomplished in the next section via the simulations using the developed ASPEN model.

5. ASPEN model development

The above representation implies that the list of factors that influence/indicate the dehydration section performance include:

- Feed stream properties
- Feed temperature
- Pressure drop in the columns
- Packing details
- Steam usage in the reboiler

5.1. ASPEN Model setup

The flow diagram of the dehydration model is shown in Figure 7. The model input and specifications for the two columns can be listed as follows:
Water stripper (C-41):

- Radfrac with no condenser
- Input: Feed flow rate and composition, pressure profile.
- Design spec: Bottoms' water content (vary distillate to feed ratio).
- Packing – 3 sections

- No specific efficiency (Murphree) specified.

Azeotrope column (C-48):

- Operating with a fixed Reflux Ratio as input.
- Input: Pressures and distillate to feed ratio.
- Design spec: Bottoms' temperature.

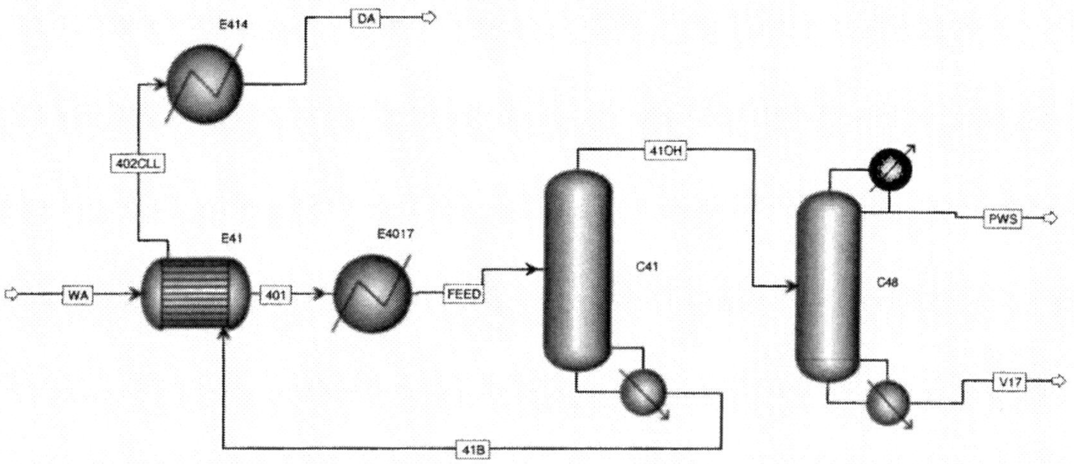

Fig. 7 – ASPEN model diagram

5.2. Model validation and sensitivity

As discussed above, the data period under investigation where the system modifications were utilized in the plant includes a variety of plant productivities as well as a wide range of operating conditions. This is crucial for the selection of the validation periods.

A total of 10 validation periods were utilized.

The current dehydration section model does:

- Predict the C-41 top temperature well as well as the C-48 top temperature
- Match reasonably the C-41 bottom temperature
- Meet the flow rate at the C-48 overheads
- Predict the composition in C-48 overheads fine
- Agree with the C-41 reboiler duty (more deviation at higher loads/water content) well

On the other hand, the ASPEN model showed a mismatch of the split across C-41, hence, underestimate the C-48 feed stream and its bottoms. The latter may be attributed to the fact that the mass balance demonstrated in Figure 5 was not fully closed but rather at ~6.5%. This small deviation can attribute to the mismatch noted above. Another explanation of the mismatch can be due to an entrainment effect in C-41 that can also lead to a certain difference in the reboiler duty at higher capacities.

Higher feed temperature is known to yield flashing. This relieves the load of the C-41 column allowing better separation. This is, predictably by the model, desirable. A sensitivity analysis on that variable is shown in Figure 8.

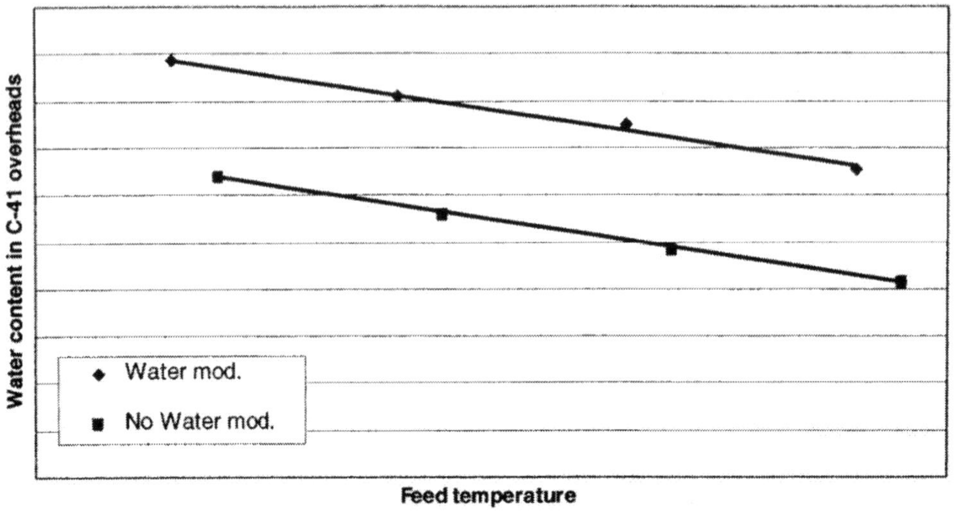

Fig. 8 –ASPEN predicted sensitivity analysis of the feed temperature

6. A discussion on flooding and pressure drop

The investigation of the water stripper's internals revealed that in addition to the packing, liquid and gas distributors also are present. A big relief and a decrease in the pressure drop can be gained from the removal of the gas distributor and hence it is recommended to check the condition of the gas distributor and evaluate the removal to ensure the benefits. This practice has proved harmless in similar columns operating under similar conditions in our sister plants.

A column operated under flooding conditions before gas distributor removal showed a drastic decrease in the top temperature afterwards (~70 °F step).

Corresponding top temperatures in our current case <u>does not</u> exhibit more than ~10 °F difference between extreme cases with the most extreme modifications in the process.

7. Conclusions and recommendations

In conclusion, the study for debottlenecking the dehydration section in our plant was accomplished by analyzing the plant data and evaluating optimized operating conditions with an ASPEN model.

The main recommendations from this work can be listed as follows:

- Design an EVOP for a gradual increase of the water content in the feed of the system to better assess the impact on the dehydration section performance.
- Consider the removal of the gas distributor in the water stripper; a practice that proved harmless on the operation of the column yet efficient in tackling pressure drop issues.
- Including additional compositional sampling in extra positions around the system.

8. Acknowledgments

The authors are grateful for the support provided by the Manufacturing team. In addition, special thanks go to the managers of the Center of Excellence for the initiation of the project.

*Sabic Innovative Plastics is a trademark of SABIC Holding Europe B.V..

Reference

[1] Erik Delsman, Internal Sabic Innovative Plastics technical memo report, 2009-SABICIP-0102.

(140b) Fast NMPC Applied to Industrial High Purity Propylene Distillation

Gilvan A. G. Fischer, Henrique Lage Refinery, Petrleo Brasileiro S.A. Petrobras, So Jos dos Campos, Brazil and Lorenz T. Biegler, Chemical Engineering, Carnegie Mellon University, Pittsburgh, PA

Distillation is one of the most used separation processes. Distillation columns are largely studied on process control because they are directly related with product quality, energy consuming and control difficulties due to nonlinearities. With increasing competitiveness, there is a demand to push the process to most profitable operating point, generally closer to the boundary of the admissible operating region, without violating the operating limits and satisfying product specifications. To guarantee the achievement of these requisites the controller must explicitly consider process nonlinearities and constraints. These problems motivate the use of nonlinear model predictive control (NMPC) with rigorous dynamic models. The NMPC is able to improve operation stability and increase profit.

In this work a detailed propylene distillation model of an industrial unit is derived. The model is represented by tray-by-tray equations consisting of mass balances, energy balances, phase equilibrium, hydraulic and summation equations. This DAE system contains the rigorous description of the plant, bound constraints for the differential, algebraic states and for the manipulated variables. Here, we use a simultaneous collocation-based approach. The infinite dimensional problem was transcribed to a finite element discretization at Radau collocation points.

This industrial-size distillation first principle model incorporates the discretized dynamic as algebraic constraints and has thousands (40,000) of variables which is challenging to solve. One advantage of this dynamic model is that it has a sparse structure. In order to reduce computational complexity, the NLP solver must have the ability to exploit this sparse structure. Another advantage is that the dynamic model is completely algebraic and first and second order derivative information can be obtained directly from modeling platforms. To solve the resulting NLP problem we use the interior-point solver IPOPT (Wächter & Biegler, 2006). To handle the bound constraints IPOPT add logarithmic barrier terms to the objective function. IPOPT applies Newton's method to the KKT conditions of system to solve each barrier problem. The factorization of the KKT matrix is the most expensive step at each iteration in the algorithm. For large systems, this factorization step can take a significant amount of computation time.

A barrier for these optimization-based strategies of NMPC is long time feedback delay due to on-line solution. In order to solve this problem have been developed advanced-step NMPC strategies (Zavala & Biegler, 2007) that use the sensitivity equation to get the fast updated solution. The control action is extracted from the approximate solution vector and applied on plant. The asNMPC has the same nominal stability properties of the ideal NMPC (Zavala et al, 2008). Success on the application of this strategy for an air separation unit (Huang et al, 2009) leads to the study of other industrial problems.

Industrial data from a propylene unit localized at REVAP (Henrique Lage Refinery, Petrobras-Brasil) were used to determination of system parameters for the propylene distillation model. The main uncertainty sources are the tray efficiencies and heat transfer coefficients. In this work, we are verifying the performance of the MHE/asNMPC maintaining the composition on desired level and maximizing the valuable product on a modeled industrial case, dealing with measured disturbances as plant feed, unmeasured disturbances as well as the influence of the model-plant mismatch.

References

A. Wächter and L. T. Biegler, On the Implementation of a Primal-Dual Interior Point Filter Line Search Algorithm for Large-Scale Nonlinear Programming, Mathematical Programming 106(1), pp. 25-57, 2006

Zavala, V. M.; and Biegler, L.T. The Advanced Step NMPC Controller. Proceedings of the IEEE Workshop on Advanced Process Control Applications for Industry - Vancouver, Canada. 2007

V. Zavala, C. Laird, L. Biegler, A fast moving horizon estimation algorithm based on nonlinear programming sensitivity, Journal of Process Control 18 (2008) 876–884.

R. Huang, V.M. Zavala and L.T. Biegler. Advanced step nonlinear model predictive control for air separation units. J. of Process Control,19(4) 678-685, 2009.

See more of this Session: Applications of Optimization In Industrial Practice
See more of this Group/Topical: Process Development Division

MULTIOBJECTIVE OPTIMIZATION OF A SPOUTED BED REACTOR

Ghanim M. Alwan[1], **Stoyan Nedeltchev**[2], **Shreekanta Aradhya**[3] and Muthanna Al-Dahhan[3], (1)Chemical Engineering, University of Technology, Baghdad, Iraq, (2)Department of Chemical and Biological Engineering, Missouri University of Science and Technology, Rolla, MO, (3)Chemical and Biological Engineering, Missouri University of Science and Technology, Rolla, MO

Abstract:

Gas-solid spouted beds are either cylindrical bed with cone base or the whole bed is in a cone shape where the gas enters as a jet. The gas forms a spout region that carries the solids upward in a diluted phase which forms a fountain at the top of the bed where the solids fall down and move downward in the annular region. These beds have found wide and efficient applications as reactors, dryers, coaters, etc. Various design and operating variables affect the performance of these beds. It seems that the performance of the gas-solid spouted bed benefit from solids uniformity structure along with lower pressure drop (PD). Therefore, the focus of this work is to perform multiobjective optimization on gas-solid spouted bed, to maximize the uniformity index (UI) of the solid particles across the spouted bed and to minimize the PD along the bed. Hence, UI and PD are considered as the objective functions. Four decision variables were selected which affect the objective functions of the process. These decision variables are: gas velocity, pressure, particle density and particle diameter. The dynamics of the spouted bed reactor are highly interactive nonlinear processes. Thus, the optimization technique is a powerful tool to obtain the desired design parameters and set of operating conditions. This would guide the experimental work and reduce the risk and cost of the design and operation. It also yields the best range of the operating conditions which could improve the efficiency of the spouted bed reactor. Various optimization algorithms are available. However, the reliability of the optimized solutions depends on the formulation of the objective functions and the selected optimization technique.

In this work, genetic algorithm has been used which is considered the best stochastic search technique for highly nonlinear objective functions. The accuracy and the reliability of the optimization search could be enhanced by using adaptation operators. It has been found that maximum UI could be obtained with high density particles, while minimum PD could be obtained with lower particle density. The results of the optimization have been compared with the experimental data using sophisticated optical probe and high accuracy pressure transducers.

Keywords: Genetic algorithm, Optimization, Particle Technology, Spouted bed

Nomenclatures:

C: Concentration of solid particles,[-]

dp: Diameter of solid particle, [mm]

P: Pressure into spouted bed, [Kpa]

UI: Uniformity index of solid particles,[-]

Vg: Superficial velocity of gas, [m/s]

Greek letters:

*ƒ*s: Density of solid particles,[Kg/m^3]

ÂP: Pressure drop a cross the spouted bed,[Kpa]

€: Porosity of bed, [-]

Ø: Sphersity of the solid particle, [-]

Introduction:

A spouted bed is a special case of fluidization.It is an effective means of contacting gas with coarse solid particles. There is increasing a pplication of spouteds as reactors such as; coating, desulfurization, CO2 capture, combustion and gasification of coal and biomass (Limtrakuletal.2004).The spouted bed is a kind of high performance reactor for fluid-solid particles reaction, also it is a hybrid fluid-solid contacting system (Wang etal.2001).

The conical spouted bed reactor may be an alternative to the fluidized bed reactor for obtaining olefins by polyolefin pyrolysis, and moreover, it is also suitable for obtaining high yields of wax, which is a raw material that is handled easily in commercial cracking units, either by steam cracking or by catalytic cracking in FCC units (Aguado etal.2002).

Conversion and coating of particle nuclear fuel is performed in spouted bed reactor, in order to achieve uniform particle coating and prevent agglomeration of the particle bed. Spouted bed reactors are specialized fluidized bed reactors with unique fluidization characteristics and particle handling advantages (Wendt etal.2007).

Figure (1) illustrates the spouted bed schematically.The bed is made up of three distinct regions: the dilute core called the spout, the surrounding annular dense region called the annulus, and the the dilute fountain region above the bed surface. Solid particles are carried up rapidly with the fluid (usually gas) in the spout to the fountain and fall down onto the surface of the annulus by gravity.In the annulus, the particles move slowly downward and, to some extent, inward as aloose packed bed(Epstein and Grace, 1984.

Plasma when coupled with spouted bed reactor of gas-solid reaction brings in several advantages such as high rate of heat & mass transfer (Sathiyamoorthy, 2010).However, it is better to develop the design of the spouted bed reactor to overcome the large pressure drop and instability of the operation and also improve the uniformity of the products resulting from the chemical or physical treatment due to the elimination of the back mixing (Prachayawarakorn etal.2005).

Spouted bed coating reactor helps in uniform exposure of surface to the precursor by systematic circulation of particles within the reactor. The spouted bed reactor is characterized of its stable solid circulation which is responsible to enhance heat and mass transfer (Sathiyamoorthy etal. 2010).

Most studies, concering solid behavior and hydrodynamic correlations of bed reactors, have been done at ambient temperature. The lack of studies at higher temperatures is due difficulties associated with measuring techniques under these conditions. In extrapolating hydrodynamic parameters derived at ambient temperature to higher temperatures, only the physical property changes of gas and solid phases, such as density and viscosity are taken into consideration (Radmanesh, etal.2005).

The mimicking with the spouted bed nuclear particles coating reactor was investigated in the present work using the steel and glass beads as the solid particles and the air as the spouted gas at ambient temperature. Steady state study was done to select the effective process variables. The uniformity of the solids particle was considered as the new key parameter in the scale-up study of the present spouted reactor. It seems that the performance of the spouted bed reactor benefit from solids uniformity structure along with lower pressure drop.

Therefore the objective was to maximize the uniformity index of the solid particles and minimize the pressure drop across the bed. Optimization technique is a powerful tool to obtain the desired design parameters and set of operating conditions. This would guide the experimental work and reduce the risk and cost of the design and operation. It also yields the best range of the operating conditions which could improve the efficiency of the spouted bed reactor.

Multiobjective optimization technique was needed to solve the conflicted optimization problem of the system. Two advanced optimization techniques were implemented with the present spouted bed reactor which were Genetic and Pattern Search Algorithms.

Experimental set-up:

The experimental set-up was designed and constructed in the best way to collect the data as explained in Figure(2).The cylindrical spouted bed made of Plexiglas was employed in this study which was the major part of the experimental set-up. The bed is (3 inches in diameter and 36 inches in height).At the bottom of the bed, there is a 60° cone-shaped distributor (3 inches in height), on which 20 holes (0.25 inch in diameter) are perforated horizontally. The distributor was made of stainless steel (316L) with thickness of(0.25 inch).The spouting nozzle (0.25 inch in diameter) locates in the center of the distributor.

The spouted gas was air from the air compressor and the gas flow rate was controlled by the pressure regulator and measured by the flow meter (Figure 2). The solid particles used were steel and glass beads with different average diameters and properties (Table 1).

Holes are drilled at vertical intervals of (1.86 inch) along the column wall with diameter of (0.5 inch), in which the optical probe was placed at different radial positions: 1.5, 1.25, 1.0,0.75,0.5,and 0.25 inch also at axis positions:7.5, 5.5 and 3.5 inches above the distributor.

The newly optical probe (manufactured by the Institute of Chemical Metallury, Chinese, and Academy of Sciences) can be used to measure both solids concentration and solids velocity and their fluctuation. The Particle Velocity analyzer (PV6) which manufactured by the 'Institute of Chemical Metallurgy, Chinese, Academy of Science')consists of; photoelectric converter and amplifying circuits, signal pre-processing circuits, high-speed A/D interface card and its software PV6, is adapted to the optical probes.

The pressure drop across the bed is measured by the pressure transceducer (Type: PX309-002G5V) with the range of (0-2.0 psig) manufactured by omega company . The pressure into the spouted bed adjusting within the desired values by using the inverted circular stabilizer, 60 mm in diameter was installed at the top of the bed column. This was preventing the spout fountain from swaying.

The structure of the fountains depended upon the value of the air velocity which entered to the spouted bed. Figure (3-a) illustrates the fountain in the spouted bed for different spouted gas superficial velocities. The Computed Tomography (CT) as a radio active technique based on gamma ray was used in the present work to monitor the solid holdup distribution into the radial axis of the spouted bed reactor (Figure 3-b).Also Computational Fluid Dynamic (CFD) simulation on the spouted bed was used to appear the solid hold up into the axial direction(Figure 4). These techniques were used to confirm the measurements of the optical probes. The red color indicates to high concentration of sol ids (solid bulk) while the blue color indicates to very low concentration of particles (gas bulk).

The effective process variables are affecting the solid concentration and pressure drop into the reactor are; gas velocity, pressure, particle density and particle diameter ((Wendt etal.2007). The operating conditions of the process are explained in Table (2).

The calibration of the optical probe needed to about (30-40 miutes).The interval time of each measurement was within 3.0 minutes.The warming time of the experimental rig was 30.0 minutes.However each experiment needed to about two hours and for validity the experiment was repeated twice.

Optimization Technique:

Deterministic algorithms for function optimization are generally limited to convex regular functions. However, many functions are either not differentiable or needed a lot of difficult mathematical treatment: discretiation, decomposition, sensitivity computing…etc.Therefore, stochastic sampling methods could used instead to optimize such functions.

In the present work two optimization techniques were used:

-Genetic Search Algorithm

-Pattern Search Algorithm

The genetic algorithm (GA) has been applied to investigate the optimum conditions of the process.GA is search algorithm based on mechanics of natural selection and natural genetics.GA is based on Darwins′

theory of ‘survival of the fittest’.Thereare several genetic operators.Such as; population, selection, crossover, mutation…etc.Each chromosome represents a possible solution to the problem being optimized, and each bit (or group of bits) represents a value of variable of the problem (gene).A population of chromosomes represents a set of possible solution.these solutions are classified by an evaluation function, giving better values, or fitness, to better solution(Gupta and Srivastova, 2006).

Pattern Search algorithm was used to confirm the optimal results obtained by the Genetic algorithm also when implemented with the genetic algorithm process to switch to Hook-Jeeves algorithm.The Pattern Search algorithm was used to refine the decision variables (Palonen etal.2009).In the present work, the Genetic & Pattern search were applied with the aid by the Tool Box of MATLAB language (version 7.10.0.499-R2010a).

Result and Discussion:

The stability conditions and the solid particles concentration distribution were studied in the spouted bed reactor. Also the uniformity index of the solid particles and the pressure drop across the bed were focused in the present work.

Flow Regimes:

The flow regimes in the present spouted bed reactor were studied to limit the stable gas velocity. (Xu etal. 2009) were previously mentioned to that as follows:

-$Vg < 0.785$ (m/s)	packed bed flow regime
- $0.785 < Vg < 0.887$ (m/s)	unstable spouting flow regime
- $0.887 < Vg$ (m/s)	stable spouting flow regime

All the parameters of the reactor process are difficulty to be measured; also the efficiency of the spouted reactor is dropped at the unstable conditions. A stable spouted bed is observed when the particles in the bed fluidize homogenously and they are not discharged from the apparatus.

Uniformity Index:

In the present work, the uniformity index relation was developed and implemented regarding to the Nedeltchev and Zhang relations (Nedeltchev etal.2000 and Zhange tal.2011). For steady state condition, the equation can be written as follows:

UI= (Cav-Cmin)/ (Cmax-Cmin) (2)

Where;
Cav: Average solid concentration, [-]

Cmin: Minimum solid concentration [-

Cmax: Maximum solid concentration [-]

UI: Uniformity index of the solid particles, [-]

The results of CT and CFD techniques confirmed the validity of equation (2).Figure (5-a) illustrates the effect of the velocities of air on the uniformity index (UI)of the steel and glass beads.The UI decreased with increasing the air velocity (Vg) as a result of increasing the disperation of the solid particles due the kinetic energy of the solid particles was increased.Also two regions appeared in these curves which represented the transition region from the packed bed flow regime at Vg of 0.74 m/s to the stable spouting flow regime at Vg of 0.95 and 1.0 m/s.The axial mixing is more vigorous than the radial mixing with the increase of spouting air velocity (Zhang etal.2007).The static bed height has important effect on the uniformity of the final mixtures. with increasing the static bed height, the degree of mixing axially decreased then increased, but in the lateral direction decreased (Hao etal.2008).Also with the increase in the spouting gas velocity the coating thickness linearly decreased (Sathiyamoorthy etal.2010).TheUI of the steel beads is higher than that of glass beads (Figure 5-a) due to high scattering occurred with the particles of low density. So the low concentration of solid particles was obtained for all positions with the glass beads system this affected directly on the values of UI. Figure (5-b) explains the effect of the spouted bed pressuer on the UI.As the pressure increased the stochastic motion and the scattering of particles will be increased, then the uniformity of solid particles decreased.

The density of the solid particles has the positive effect on the UI of the spouted bed reactor (Figuer6-a).Since the increasing of the particles density tend to get the bed more strength and resistance to the motion and dispersion against the vortex of the fountain. So it gives the solid particles more stability and uniformity.The denser material is continued to spout in the centeral region while the less dense particle formed atorriodal vortex around the central spout (Halow etal.2010).The solid beads diameter has negative effect on the UI of the solid particles as shown in Figure (6-b).As the diameter of the particles decreased, the free area (mean free path) into the bed increased. Then the particles of small size are helpful to raise the mixing speed (Hao etal.2008).This gets the smaller particles more uniformity in the radial distribution. However, the uniformity index is good indication for improve the mass, heat and the conversion of reactants into the nuclear spouted bed reactor.

Pressure Drop:

Figures (7&8) explain the effect of the process variables (air velocity, bed pressure, density and diameter of solid particles) on the pressure drop across the reactor bed.The pressure drop increased with increasing the velocity of air for both steel&glass beads due to increasing the kinetic energy and the interaction of the solid particles (Correa etal.1999).

In the case of steel beads the pressure drop is higher than that with glass beads because of high strength and friction with steel beads (Figure 7-a).The effect of the spouted bed pressure on the pressure drop across the bed was very low. Then the pressure drop increased slightly with increasing the inlet pressure of the air to the spouted bed (Figure 7-b).The dense particles(steel) create more resistance and friction against the air flow and then tend to raise the pressure drop across the bed (Figure 8-a).The porosity of the bed increased with the large particles diameter, so that the strength of the solids to the air flow then reduced.This tend to reduced the pressure drop across the spouted bed (Figure 8-b).These conclusions also confirmed by (Zhong etal.2006).

Optimization Problem:

The spouted bed is a hybrid fluid solid contacting system (Wang etal.2001).So that the dynamic modeling for optimization requires model that describe the essential dynamic characteristics of the process under study. For the present complex nonlinear spouted bed process, a simple and effective model may be difficult to derive. Process measurements are the most important source of information of a specific process.While first principles provide general information about the behavior of the process, process measurements are required to identify the process model (Roffel and Betlem, 2006).

In the present work, it is recommended to formulate the objectives i.e. the uniformity index of the solid particles (UI) and the pressure drop (ΔP) across the spouted bed reactor with the effective (decision) process variables. The decision variables are; air velocity, bed pressure, density and diameter of particles. The objective of optimization search is to maximize UI and minimize ΔP.

Two advanced nonlinear regression optimization algorithms were used as: Newton-Quasi and Hook-Jeevs pattern moves to formulate the developed nonlinear objective functions with the aid of the computer program (Statistica version6).

The empirical correlations are:

Maximize $UI=0.184*Vg^{-0.214}*P^{-0.301}*\int s^{0.316}*dp^{-0.267}$ (3)

With correlation coefficient of(R=0.99).

Minimize $\Delta P=0.037*Vg^{0.381}*P^{0.05}*\int s^{0.407}*dp^{-0.221}$ (4)

With correlation coefficient of (R=0.94)

Subject to Inequality constraints of;

$$0.74\leq Vg\leq 1.0$$
$$101.0\leq P\leq 312.0 \quad (5)$$
$$2400.0\leq \int s\leq 7400.0$$
$$1.09\leq dp\leq 2.18$$

From equations (3 and 4), one can conclude that the spouted bed is highly nonlinear interacted process. Equation (3) confirms that the air velocity, reactor pressure and particle diameter have the negative effect on the uniformity index (UI) while the density of solids has the positive effect as described in the previous section. Also the density of the solid particles seem the more effective variable which have higher power of (0.316).

From equation (4), one can conclude that the air velocity, reactor pressure and density of the solid particles have positive effect on the preesure drop across the bed, while the diameter of particles has negative effect. The density of solid particles also is the more effective variable.

In extrapolating hydrodynamic parameters derived at ambient temperature to higher temperatures, only the physical property changes of gas and solid phases, such as density and viscosity are taken into consideration (Radmanesh, etal.2005).

However, the spouted bed reactor is a hybrid fluid-solid contacting system.So that the optimization problem of the present work represented by the two conflicted equations(equations 3&4) with inequality constraints (equation 5).

Optimization Search:

Optimization technique is a powerful tool to develop the design of the spouted bed reactor to overcome the large pressure drop, the instability of the operation and also improve the uniformity index of the solid particles.

In the present work, the stochastic techniques (Genetic and Pattern search) were used to obtain the optimal solutions of the process depended upon the process models of the system (equations 3, 4&5).The objective of the present work is to maximize the uniformity index (UI) of the solid particles and minimize the pressure drop (Δp) across the bed.

Two optimization searches were used:

-Single objective search:

In this search, the optimization technique was applied separately for each objective functions (equations 3&4) with inequality constraints (equation 5).Several experiments were applied to investigate the optimal solution of the system.Also the operators of Genetic algorithm and pattern search were tunned to get the best solution. Tables (3&4) explain the best operators of the two algorithms for the two objectives. Figures (9&10) illustrate the outputs of the algorithms excutions.GA was implemented with the pattern search by using the hybrid function to refine the decision variables (Table3).The pattern search technique (Figure 10) confirms the Genetic algorithm and get approximately the same optimum values as shown in Table (6).The results of the optimization searches are agree since the values of the objectives and the continuous &discrete variables within the the limits of the operating conditions. Also the experimental results confirm the optimal results that the maximum UI obtained with high density material (steel beads) while the minimum (ΔP) obtained with low density (glass beads).This confirms the mathematical model of the process (equations 3&4) that the density of solids was the effective variable on UI and ΔP.Also, the accuracy of the optimization techniques are high due to it can capture the discrete variables (density and and diametr of solid particles) with the integer values and the velocity of air (Vg) at the stable region (Table 6).The objective function of UI needed to (55) numbers of generation which is more than those for the objective function of ΔP which operated with (51) generations.This is due to that the function of UI (equation 3) is more complex than ΔP function (equation 4).However, the maximum UI was obtained with the steel beads system while the minimum ΔP obtained with the glass beads system for the same operating conditions. The accuracy of the optimization searches could be increased by the adaptation of the Genetic and Pattern search algorithms operators because of highly interacted and nonlinears objective functions.

-MultiObjective Search:

In this technique the two objective functions were used together.Also several experiments were done to select the best operators of GA as shown in Table (5).Since the objective functions are conflicted, so that 26 chromosomes were created (Figure 11-a).Each chromosome represents the possible solution to the optimization problem, and each bit represents the value of the variables of the problem.This means that the optimization search created 26 sets for the process design and operating conditions. This means that we can operate the spouted bed reactor with others new materials and operating conditions. optimization search can reduce the number of experimental runs, also the consumed cost for design and operation. The designer or operator can select the best solution regarding to the limits of the operating conditions. Figure (11-b) illustrates the output of the algorithm excution.The optimum values of UI are between (0.447 to 0.755) while for ΔP within (0.895-1.672).The best optimum results are approximately closed that obtained by the single search (Table 6).This indicates to the good selection of the algorithm operators (Table 5).Due to the two objectives functions are conflicted then the number of generations is equal to 109.This high when compared with the single objective techniques which have numbers of generation within (51 to 55) as shown in Table (3).The optimum results by the multiobjective search were stay in the limits of the operating conditions of the process (equation 5).Also this technique confirms the resuls obtained by the single objective techniques as explained in Table (6).This indicates to high accuracy of the optimization techniques used in the present work.However, the multiobjective search is more accurate than the single objective technique since it treated with the overall process variables of the system and could give the globle optimum results. The improvement was exactly obtained by using the optimization techniques.The maximum UI was obtained by the experimental work conditions were 0.485 and 0.55for steel&glass beads respectively as shown in Figure (5).The value of UI could be increased to 0.75 by the optimization techniques within the limits of the operatingconditions.Also the improvement for Δp function was occurred when compared the experimental results (Figure 7) with the optimum result of 0.89Kpa obtained within the same limits of the operating conditions. Table (7) explains the improvement of the optimization technique on the objectives of the spouted bed reactor when compared the best experimental results with that obtained by the optimum search within the limits of operating conditions. The increasing of UI will be obtained by the steel beads and the decreasing of ΔP obtained by the glass beads. This means that the density of solid particles have the major effect on the UI and ΔP.

The success of the optimization search depended on the best construction of the objective correlations with the decision variables and on the good selection of the optimization algorithm (Palonen etal.2009).In the present work, the stochastic GA technique was successful with the interacted non linear system.The accuracy of the optimization search can be increased by adaptation the algorithm operators.

Conclusions:

1-Spouted-gas bed is a hybrid fluid-solid contacting reaction system. The spouted reactor was highly interacted nonlinear process system.

2- The uniformity index of the solid particles get good indication for enhances the conversion of reactants and coating of particles. The reducing of the pressure drop across the bed within the stable flow dropped the risk of the dissipated pumping energy.

3-The effective variables are; gas velocity, pressure, particle density and particle diameter.

4-The density of solid particles was the effective variable on the uniformity index of solid and the pressure drop across the spouted bed.

5-It was better to correlate the objectives directly to the decision variables of the process. The empirical model based optimization could be implemented to find the optimum parameters.

6- It is better to develop the design of the spouted bed reactor to overcome the large pressure drop, the instability of the operation and also improve the uniformity of the products. The optimization technique was the powerful tool for generate developed designs and sets of operating conditions for the spouted bed.

7-The optimal results can reduced the number of experimental runs and the cost consumed for design and operation. Also it limits the best range of operating conditions which improve the efficiency of the spouted reactor.

8-Maximum uniformity index obtained with high density beads (steel) while minimum pressure drop obtained with lower particles density (glass).

9-Good validity obtained when compared the optimal solutions with the experimental results.

10-The success of optimization search related to the formulation of the objective function also the selection of the suitable searching technique.

11-Genetic algorithm was the suitable stochastic search for the nonlinear functions. The accuracy of the search could be increased with the adaptation of the algorithm operators.

References:

1-Limtrakul, S., Boonsrirat, A.and Vatanathm, T.'DEM Modeling andSimulation of a Catalytic Gas-Solid Fluidized Bed Reactor: a Spouted Bed as aCase Study', Dept.ofChem.Engng.kasetsar University, Bangkok, Thailand, Elsevier Ltd, (2004).

2-Wang, Z., Chen, P., Li, H., Wu, C.andChen, Y.'Study on the Hydrodynamics of a Spouting-Moving Bed', Ind.Engng.Chem.Res, 40, pp.4983-4989, (2001).

3-Aguado,R.,Olazar,M.,Prieto,R. and Bilbao,J.,'Kinetic Study of Polyolefin Pyrolysis in a Conical Spouted Bed Reactor',Ind.Eng.Chem.Res.41(18),pp.4559-4566,(2002).

4-Wendt, D.S, Bewley, R.L. and Wides, W.E.,'A Spouted Bed Reactor Monitoring System For Particulate Nuclear Fuel', proceeding of space nuclear conference, Boston, Massachusetts, June 24-28, paper 2023, (2007).

5-Epstein, N., and Grace, J.R.,'Spouting of Particulate Solids', Hand Book of Powder Science and Technology, Chapter 11,2nd, Van Nostrand Reinhold Co., N.Y.pp.509-536,(1984).

6-Sathiyamoorthy, D.,'Plasma Spouted-Fluidized Bed for Materials processing, Journal of Physics, Conference Series 208, (2010).

7-Prachayawrakorn, S., Reuengnarong, S.and Soponronnarit, S.'Characteristics of Heat Transfer in Two-Dimensional Spouted Bed', School of Energy and Materials, University of Technology, Bangkok, Thailand,(2005).

8-Radmanesh,R.,Mabrouk,R.,Chaouki,J.,and Guy,C.,'Effect of Temperature on Solids Mixing in a Bubbling Fluidized Bed Reactor', International Journal of Chemical Reactor Engineering,Vol.3,A16,(2005).

9-Gupta, A.K. and Srivastava, R.K.,'Integral Water Treatment Plant Design Optimization: a Genetic Algorithm Based Approach',IE Journal,vol.87,September,(2006).

10-Palonen, M., Hasan, A. and Siren, K.,'A Genetic Algorithm for Optimization of Build-ing Envelope and HVAC system Parameters', Eleventh International IBPSA Conference, July 27-30, Glasgow, Scotland, (2009).

11-Xu,J.,Bao,X.Wei,W.,Bi,H.T.,Grace,J.R.,and Lim,C.J.'Chaotic Characteristics of Pressure Fluctuation in a Gas Spouted Bed', the Canadian Journal of Chemical Engineering, vol.87, April, (2009).

12-Nedeltchev,S.,Ookawara,S.,and Ogawa,K.,'Time-dependent Mixing Behaviors in Lower andUpper Zones of Bubble Column', J.chem.Engng.of Japan, Vol.33, No.5, pp.761-767, 2000).

13-Zhang, Y., Zhong, W.and Jin, B.,'New Method for the Investigation of Particle Mixing Dynamic in a Spout-Fluid Bed', Powder Technology, 24, January, (2011).

14-Zhong, W., Zhang, M., Jin, B.and Xiao, R.,'Experimental and Model Investigations on Gas Mixing Behaviors of Spout-Fluid Beds,' International Journal of Chemical Reactor Engineering, vol.5, Article A32, (2007).

15-Hao, H., Guoxin, H.and Fengchao, W.,'Experimental Study on Particles Mixing in an Annular Spouted Bed, Energy and Conversion Management, 49,257-266, Elsevier, (2008).

16-Halow, J.S, Holsopple, K.andFinng, E.A,'Segregation in a Spouted Fluidized Bed, AIChE, Annual Meeting, Salt Lake City, U.S.A, (2010).

17-Correa, N.A., Freire, J.T and Correa, R.J,'Improving Operability of Spouted Beds Using a Simple Optimization Control Structure', Brazilian.J.Chem.Engng.vol.16, no.4, Dec.Saopaulo, (1999).

18-Zhong, W., Chen, X.and Zhang, and M.'Hydrodynamic Characteristics of Spout-Fluid Bed: Pressure Drop and Minimum Spouting/Spout-Fluidizing Velocity', Chemical Engineering Journal, 118, 37-46, (2006).

19-Roffel, B. and Betlem, B.,'Process Dynamics and Control: Modeling for Control and Prediction', Jhon-Wiley, Sons Ltd, 3rd edition, (2006).

Table (1)-Properties of the particulate materials.

Material	dp(mm)	∫s(Kg/m^3)	€	Ø
Steel beads	1.09	7400.0	0.42	1.0
Glass beads	1.09	2450.0	0.42	1.0
Glass beads	2.18	2400.0	0.41	1.0

Table (2)-Operating conditions of the system.

Air velocity(m/s)	Pressure(Kpa)	Temperature(K)
0.74	101	298
0.95	101	298
1.0	312	298

Table (3)-The best operators of the single objective GA.

Operator	Objective(UI)	Objective(ΔP)
Population type	Double vector	Double vector
Population size	80	100
Creation function	Feasible population	Feasible population
Scaling function	Rank	Rank
Selection function	Roulette	Stochastic uniform
Crossover function	Scattered	Scattered
Crossover fraction	0.8	0.7
Mutation function	Adaptive feasible	Adaptive feasible
Migration direction	Forward	Both
Migration fraction	0.1	0.2
Hybrid function	Pattern search	Pattern search
Number of generation	55	51
Function tolerance	1.0E-6	1.0E-6

Table (4)-The best operators of the single objective Pattern Search.

Operator	Objective(UI)	Objective(ΔP)
Poll method	Gps positive basis 2N	Gps positive basis 2N
Polling order	Consecutive	Consecutive
Search	Complete search	Complete search
Initial mesh size	1.0	1.0
Scale	On	On
Expansion factor	2.0	2.0
Contraction factor	0.5	0.4
Penalty factor	100	100
Cache tolerance	1.0E-4	1.0E-4
Mesh tolerance	1.0E-6	1.0E-6

Table (5)-The best operators of the multiobjective GA.

Operator	Type and values
Population type	Double vector
Population size	80
Selection function	Tournament
Crossover function	Scattered
Crossover fraction	0.8
Mutation function	Adaptive feasible
Migration direction	Forward
Migration fraction	0.2
Hybrid function	Fgoalattain
Number of chromosome	26
Number of generation	109
Function tolerance	1.0E-6

Table (6)-Optimal solutions by the optimization Techniques.

	Objective(UI)		Objective(ΔP)		MOGA	
Decision variables	SOGA	PS	SOGA	PS	Max.	Min.
Gas velocity(m/s)	0.89	0.889	0.89	0.88	0.89	0.8
Bed pressure(Kpa)	101	101	101	101	101	101
Density of solid(Kg/m^3)	7400	7400	2400	2400	7400	2400
Diameter of particle(mm)	1.09	1.09	2.18	2.18	1.09	2.18
Value of objective	0.76	0.77	0.88	0.89	UI=0.45-0.75 ΔP=0.89-1.76	

PS: Pattern Search

SOGA: Single Objective Genetic Algorithm

MOGA: Multi Objective Genetic Algortim

Table (7): Comparison between the best experimental and the optimal results.

Materials	Max.UI(exp)	Optimal UI	Min.ΔP(exp)	Optimal ΔP
Steel beads	0.53	0.75	1.82	1.67
Glass beads	0.42	0.45	0.94	0.89

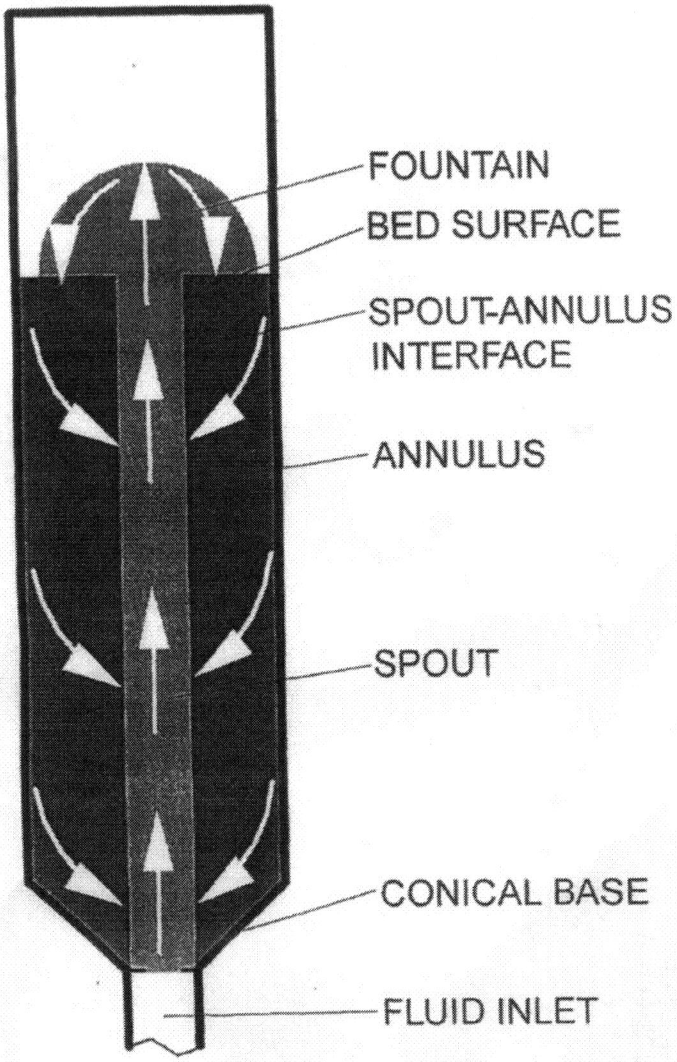

Figure(1):Spouted bed.

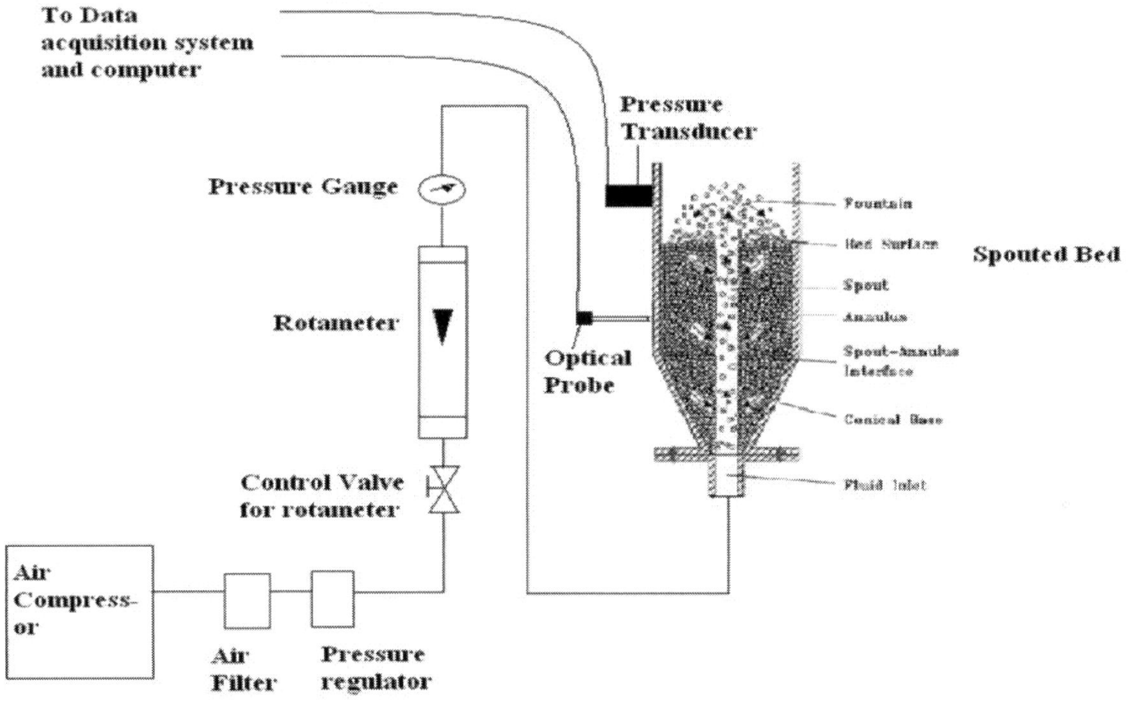

Figure (2): Experimental set-up

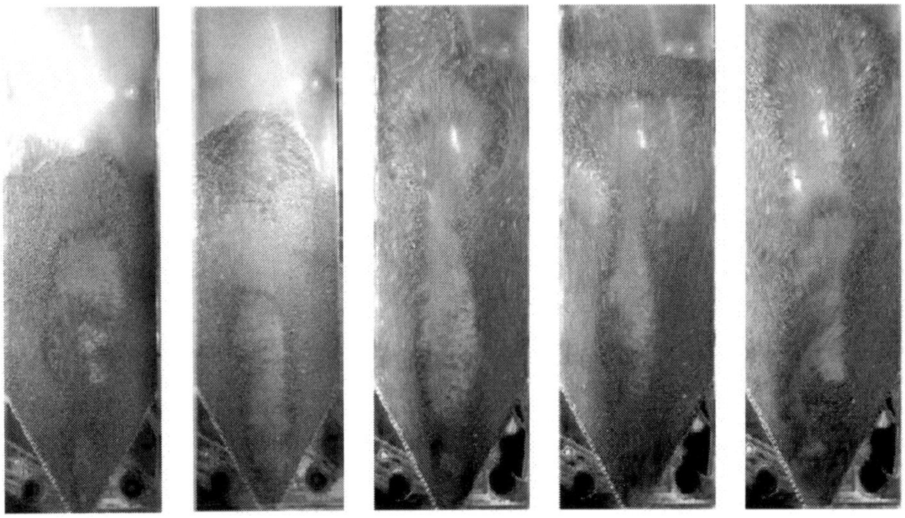

(a)

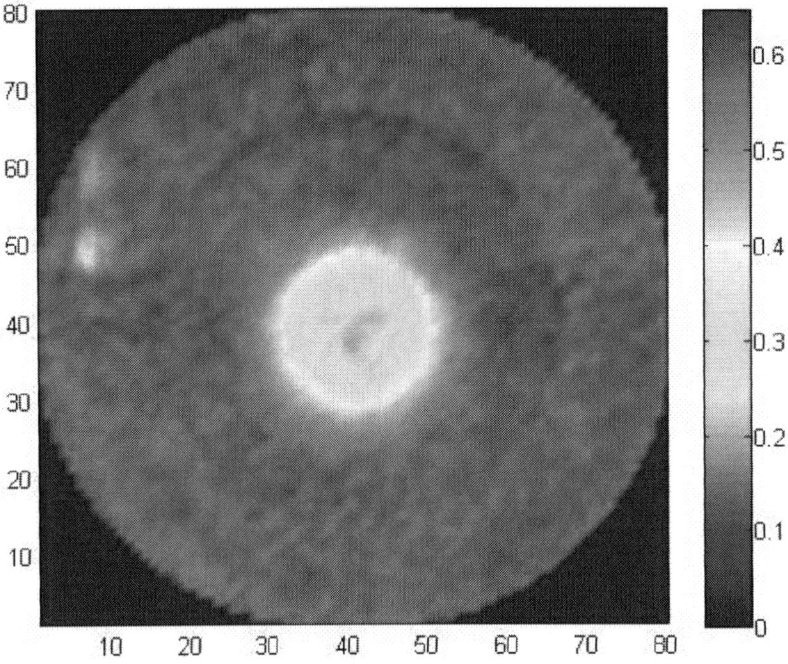

(b)

Figure (3): (a)-Flow regimes into the spouted bed for different spouted gas velocities.
(b)-Computed Tomography of solid holdup distribution into the spouted reactor.

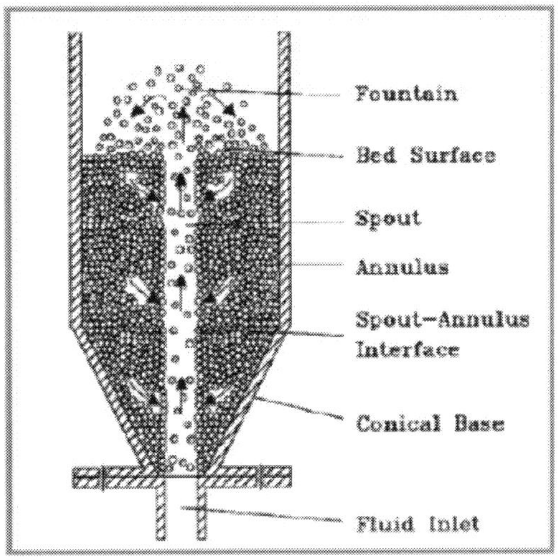

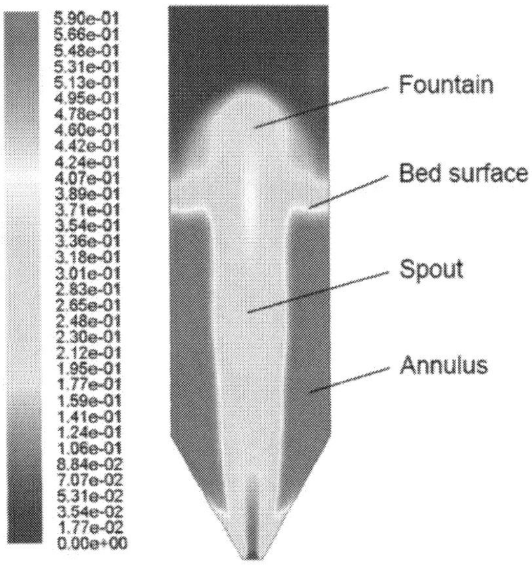

Figure(4):CFD simulation on the spouted bed appears the solid particles hold up.

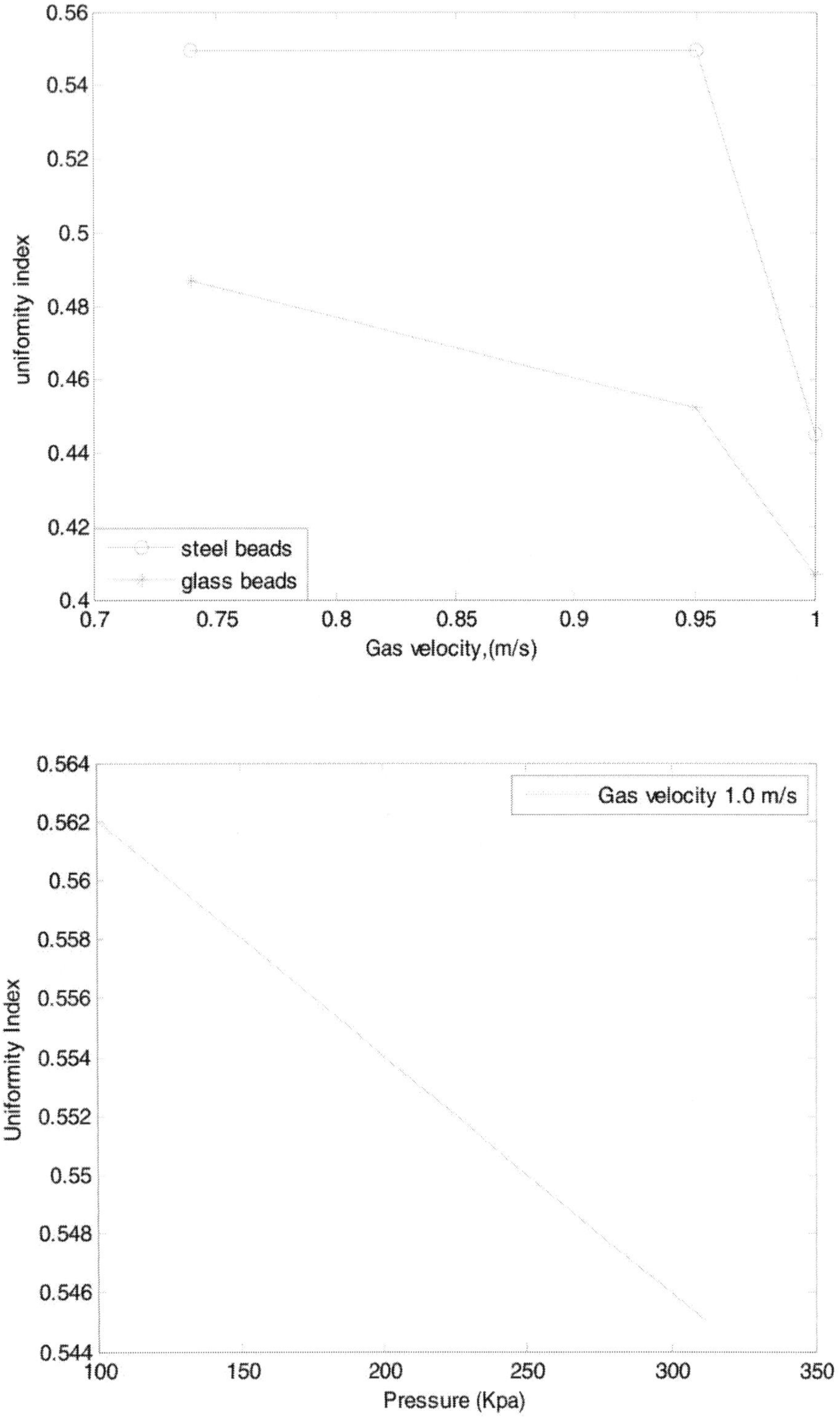

Figure (5): Uniformity index against :(a)gas velocity,(b)bed pressure.

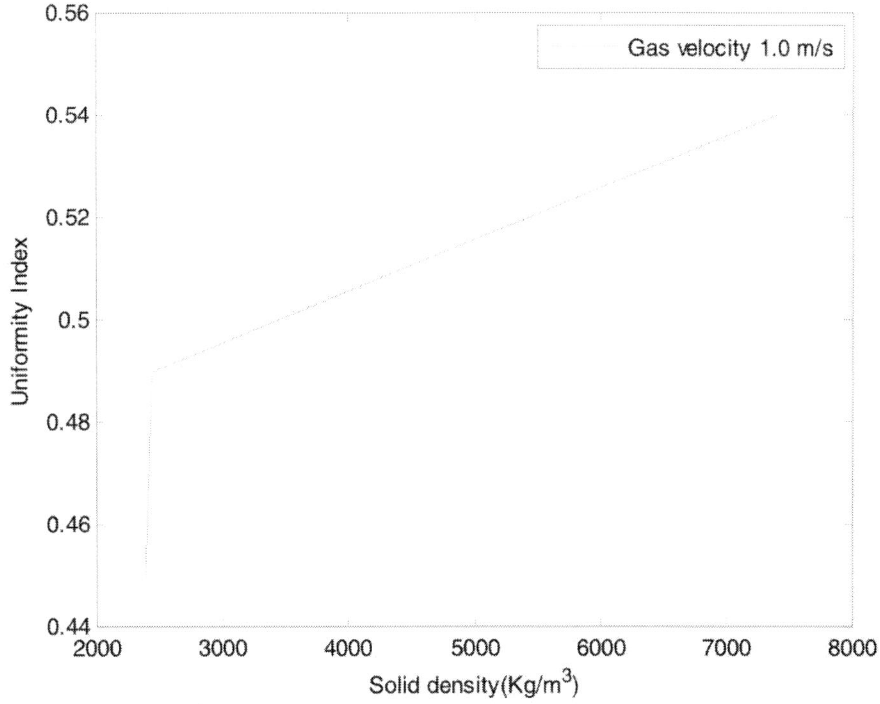

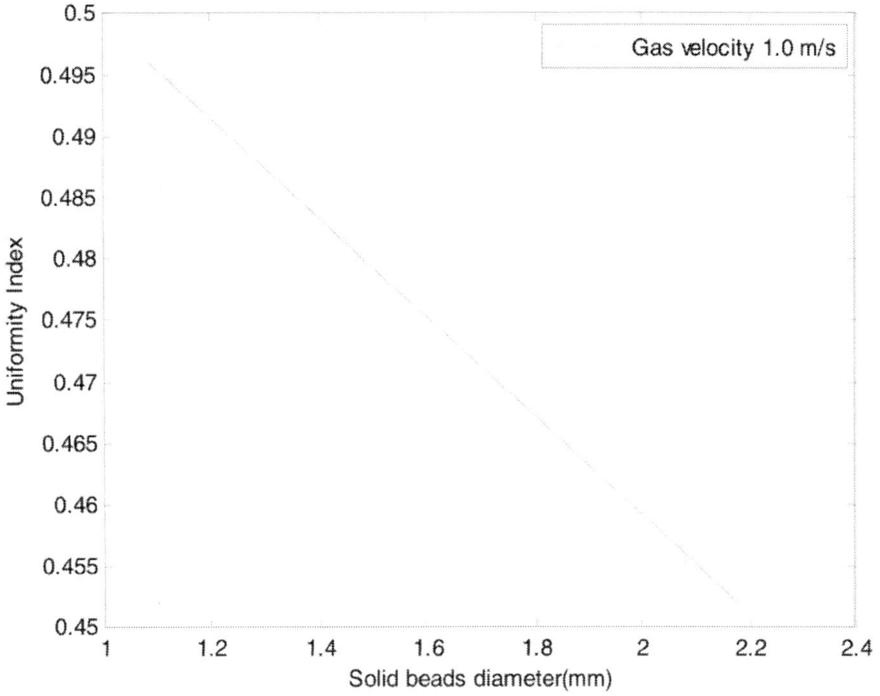

Figure (6): Uniformity index against ;(a)solid density,(b)solid diameter.

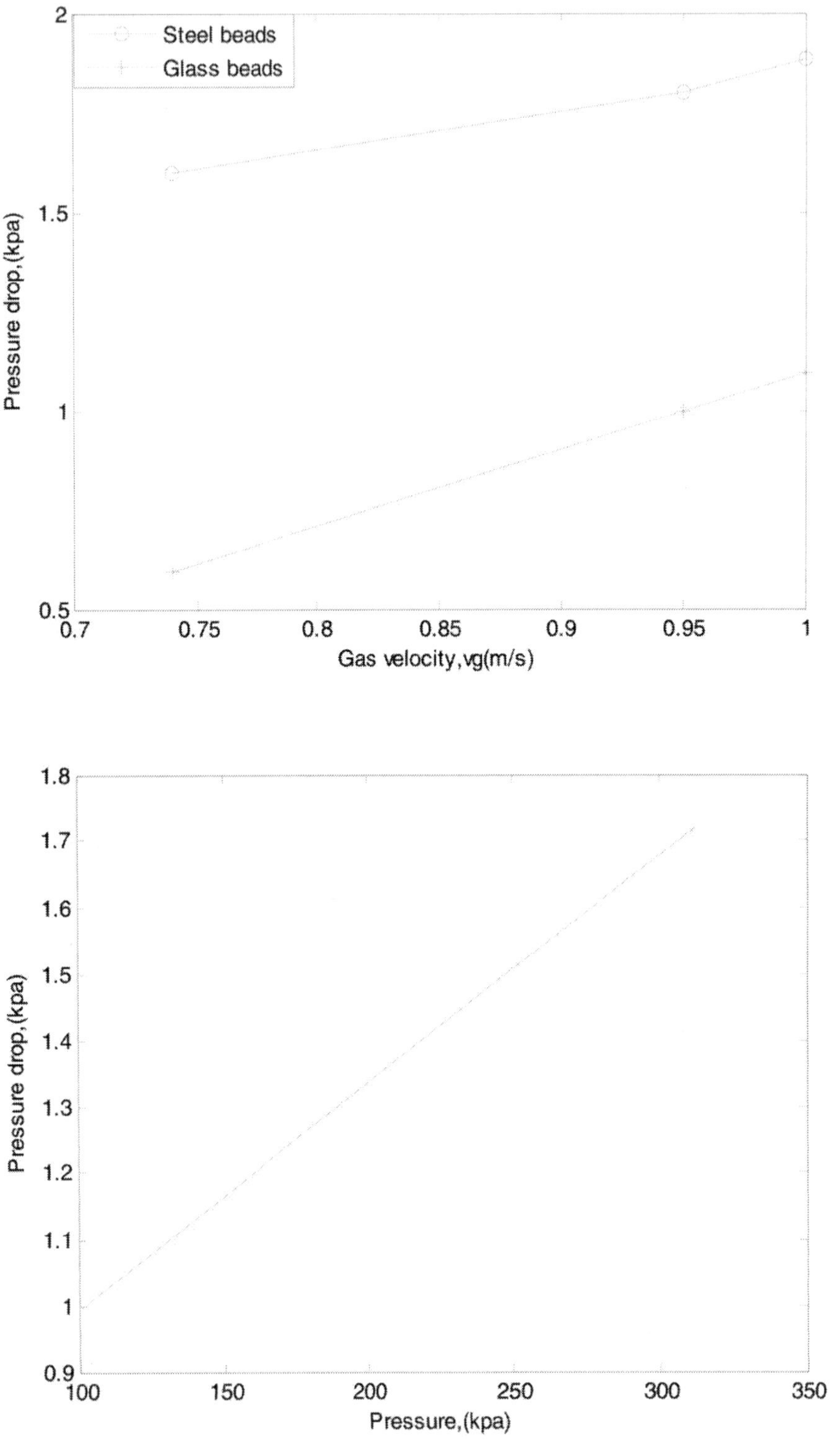

Figure (7): Pressure drop against ;(a)gas velocity,(b) bed pressure.

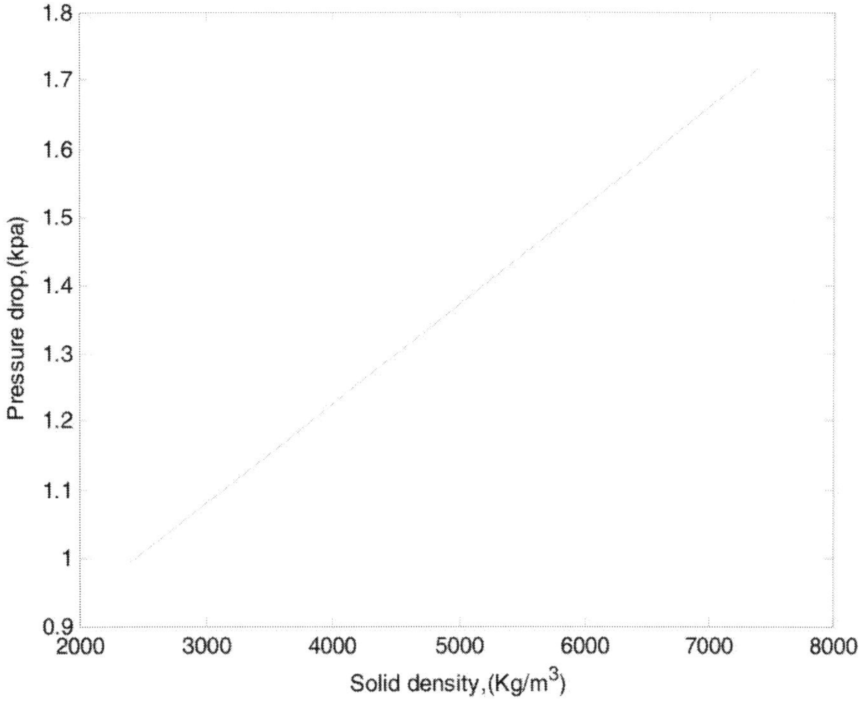

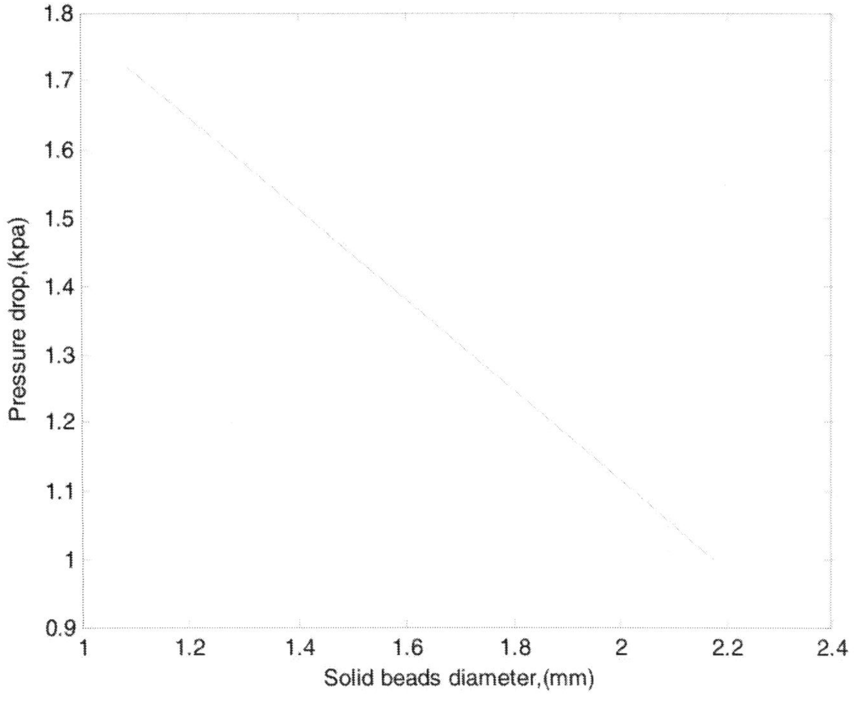

Figure (8): Pressure drop against ;(a)solid density,(b)solid diameter at vg=1.0.

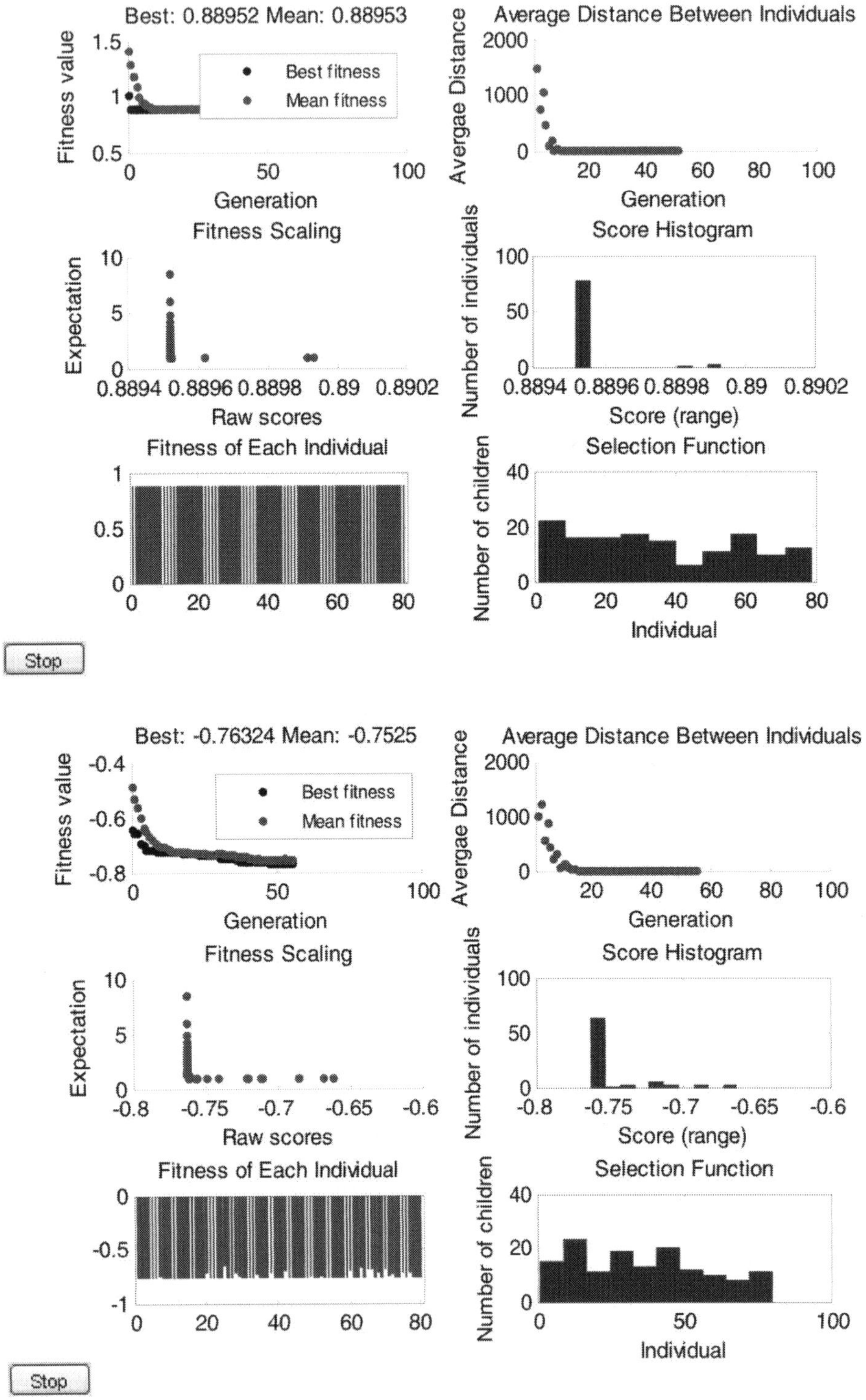

Figure (9): Output of single objective genetic algorithm for ;(a) pressure drop, (b) uniformity index.

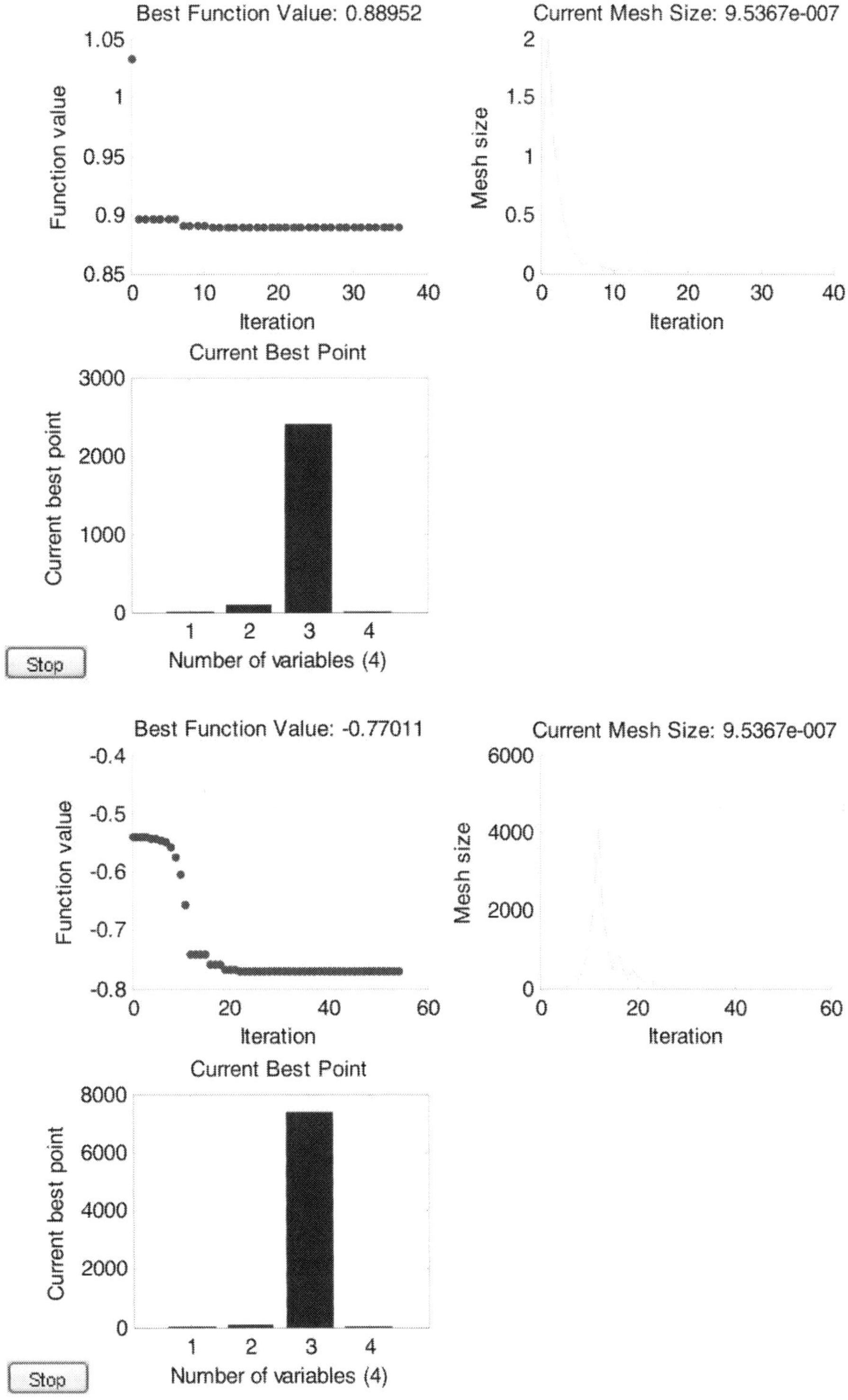

Figure (10): Output of single objective search pattern algorithm for ;(a) pressure drop,(b)uniformity index.

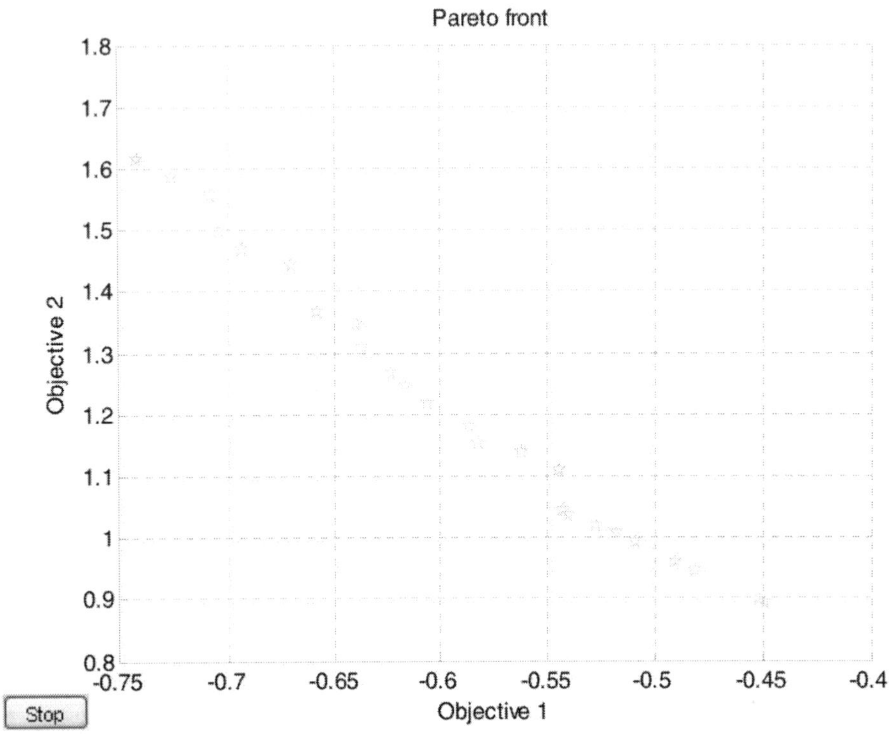

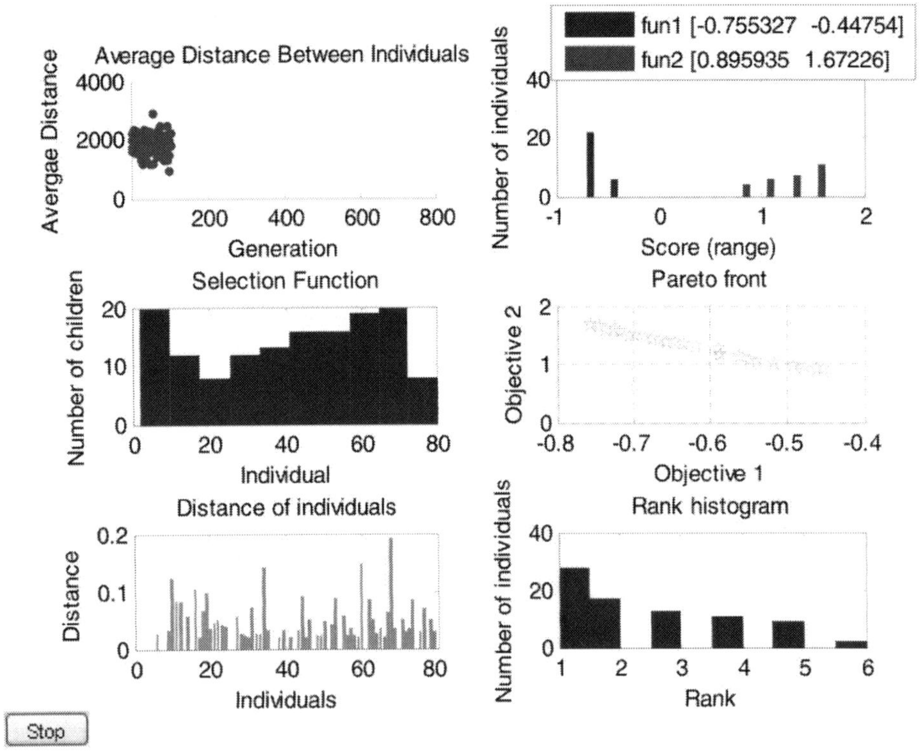

Figure (11): Output of multiobjective genetic algorithm ;(a) Pareto front, (b) genetic parameters.

(140d) A Multifaceted Experimental Design Investigation of Nodule Formation In Electroformed Parts In Compact Disc Plating Operations

Z. Otero Gephardt, Chemical Engineering, Rowan University, Glassboro, NJ and John Heinzel, NSWCCD Code 985, Naval Sea Systems Command - Philadelphia, Philadelphia, PA

A multifaceted experimental design combining chemical mixture and operational parameters was developed to identify the factors affecting nodule formation in electroformed parts in compact disc plating operations. The mixture variables investigated were the bath wetting agent (surface tension reducer), bromide salt (anode corrosion) and grain refiner (stress reducer) concentrations. The anode nickel level and plating head rotational speed were the operating variables studied. A disc quality score to account for and quantify nodule size and location was developed, and optimal operating parameters and bath mixture concentrations were identified. The analysis indicated that high level interactions among mixture variables and operating parameters have a significant effect on the disc quality score. All plating bath components and, to a lesser extent, rotational speed had a significant effect on disc quality for the variable ranges investigated. The concentrations of wetting agent and stress reducer in the plating bath had the most significant impact on nodule formation. The location of the nodules on the disc was not influenced by plating bath or operating variables.

See more of this Session: Applications of Optimization In Industrial Practice
See more of this Group/Topical: Process Development Division

MOLECULAR MODELS USING VBA TO OPTIMIZE THE PROCESS OF PETROLEUM REFINING

Tatiane G. Vilas, University of São Paulo, São Paulo, Brazil

Jadson P.A. Silva, Petrobras Distribuidora, Rio de Janeiro, Brazil

Cláudio O. Nascimento, University of São Paulo, São Paulo, Brazil

Abstract

This work makes the combination between information from mass spectroscopy analysis and algorithm search in VBA (Visual Basic for Applications) for identify a specific molecule in groups of known molecules. The method proposed in this work is based on classifying compounds by their molecular weights, i.e., the compounds present in a mixture will be separated into classes by a mass difference as small as you want. You can thus see which group of molecules previously generated an oil compound belongs by its molecular weight. As these groups were generated by compounds which are of a known composition, it is known that composition of a given compound will have between the possible compositions. To generate molecules based on knowledge of components of oil, we used the formulas of general organic compounds, which are formed by homologous series. Homologous series are series of organic compounds with similar chemical properties whose components differ by a constant reactive molecular weight (CH2). These series are generated and their molecular weights are calculated and arranged in increasing order of mass to be generated by the groups developed a computational procedure in VBA.

Introduction

Petroleum is a complex mixture of hydrocarbons with thousands of different molecular components. To determine the molecular composition of the oil components and products of the refinery is not a feasible task with the current analytical techniques because of the large number of components. Modeling the composition and behavior of these complex mixtures in refinery processes has over the years, evolved into a concept called lumping (Quann, 1992). The lumping models are quick and easy to formulate and solve, making them more dominant in the process control, design and optimization. In lumping models, thousands of individual constituents in a complex raw material are grouped into a broad but measurable category of classes of compounds or boiling point range (Quann, 1996).

The lumping methods have been applied to model almost all the reaction processes of refining, such as catalytic cracking, hydrotreating, etc. These methods are helpful when the main requirement is more comprehensive information, such as yield or octane number of gasoline. However lumping models cannot provide detailed molecular information, such as paraffins, olefins, benzene, sulfur, which are important for refining operations (Hu, 2001).

The molecular basis for refining model is to characterize the composition of the oil. Modern analytical techniques have been used, such as gas chromatography (GC), high-performance liquid chromatography (HPLC) and mass spectrometry (MS), nuclear magnetic resonance spectroscopy (NMR) to characterize the oil fractions. Techniques allowing obtaining the molecular composition of petroleum streams (Aye, 2005).

In this work we developed a method to describe the molecular composition of samples of asphalt. Using data obtained from analytical techniques to obtain molecular properties, along with a database generated by homologous series is possible to systematize the identification of components of oil.

Metodology

The method proposed in this work is based on classifying compounds by their molecular weights, i.e., the compounds present in a mixture will be separated into classes by a mass difference as small as you like. It is possible thus know which group of molecules before generating a given compound of oil by its molecular weight. As these groups were generated by compounds, which are of a known composition, that composition of a given compound will be known among the possible compositions.

To generate molecules based on knowledge of components of oil, we used the general formulas of organic compounds, which are formed by homologous series. The series is homologous series of organic chemical compounds with similar chemical properties whose components differ by a constant reactive molecular mass (CH_2). These series were generated and their molecular weight were calculated and arranged in order of increasing mass groups to be generated by a computational routine developed in VBA language. Below, in Table 1, are some examples of functions from organic compounds present in oil, such as simple hydrocarbons, nitrogen compounds, sulfur compounds, oxygenates, aromatics and saturated compounds closed chains among others, determined to make up the database of molecules used in this work.

Table 1. Organic compounds homologous series.

Organic Groups	Organic Compounds General Formula	Carbon number
Alkanes or Paraffins	C_nH_{2n+2}	
Alkadienes	C_nH_{2n-2}	n>=2
Benzene rings	C_nH_{2n-6}	n>=6
Olefins	C_nH_{2n}	n>=2
Alkines	C_nH_{2n-2}	n>=2
Alcohol single bonds	$C_nH_{2n+2}O$	

Alcohol one double bond	$C_nH_{2n}O$	
Carboxylic acids with single bonds, acyclic	$C_nH_{2n}O_2$	
Carboxylic acids with one Double Bond, acyclic	$C_nH_{2n-2}O_2$	n>=3
Ketones with single bonds, acyclic	$C_nH_{2n}O$	n>=3
Ketones with one double bonds, acyclic	$C_nH_{2n-2}O$	n>=4
Ether with single bonds, acyclic	$C_nH_{2n+2}O$	n>=2
Ether with one double bonds, acyclic	$C_nH_{2n}O$	n>=3
Ester with single bonds	$C_nH_{2n}O_2$	n>=2
Ester with one double bond	$C_nH_{2n-2}O_2$	n>=4
Amines single bonds	$C_nH_{2n+3}N$	
Amines with one double bond	$C_nH_{2n+1}N$	n>=2
Amides single bonds	$C_nH_{2n+1}ON$	n>=3
Mercaptans	$C_nH_{2n+2}S$	
Carbon disulfides	$C_nH_{2n+2}S_2$	n>=2
Benzothiofenes	$C_nH_{2n-10}S$	
Di-Benzothiofenes	$C_nH_{2n-16}S$	
Polysulfides	$C_nH_{2n+2}S_3$	

With the help of mass spectrometry samples of asphalt were analyzed, which were determined the molecular weight of compounds. These compounds had their masses determined by comparison with the similar series through a computer system developed in VBA. And thus classified into groups according to a criterion of difference in mass, as illustrated in the diagram below.

Figure 1. Illustrative scheme to search for groups of molecules

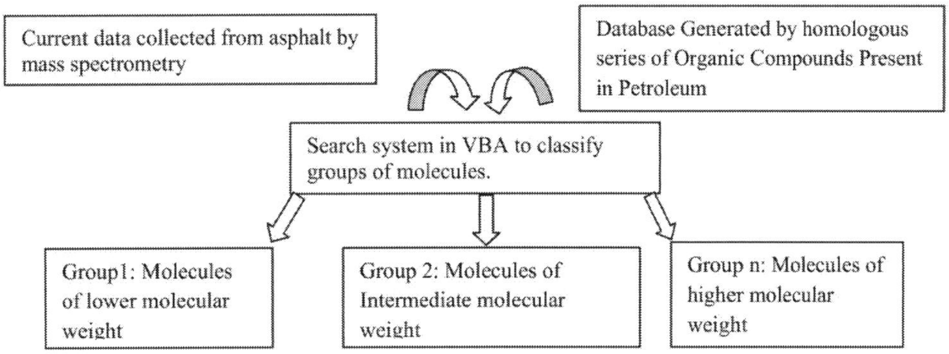

The search system was generated with algorithms in VBA language, the operation is as follows: since the homologous series was generated and terminated their molecular weight compared with the molecular weight obtained from mass spectrometry. Each theoretical molecular weight (generated by homologous series) will be compared with those obtained from experimental analysis, and groups are formed when the difference between the masses is less than a predetermined threshold.

Figure 2: Schematic illustration of the operation of the search system

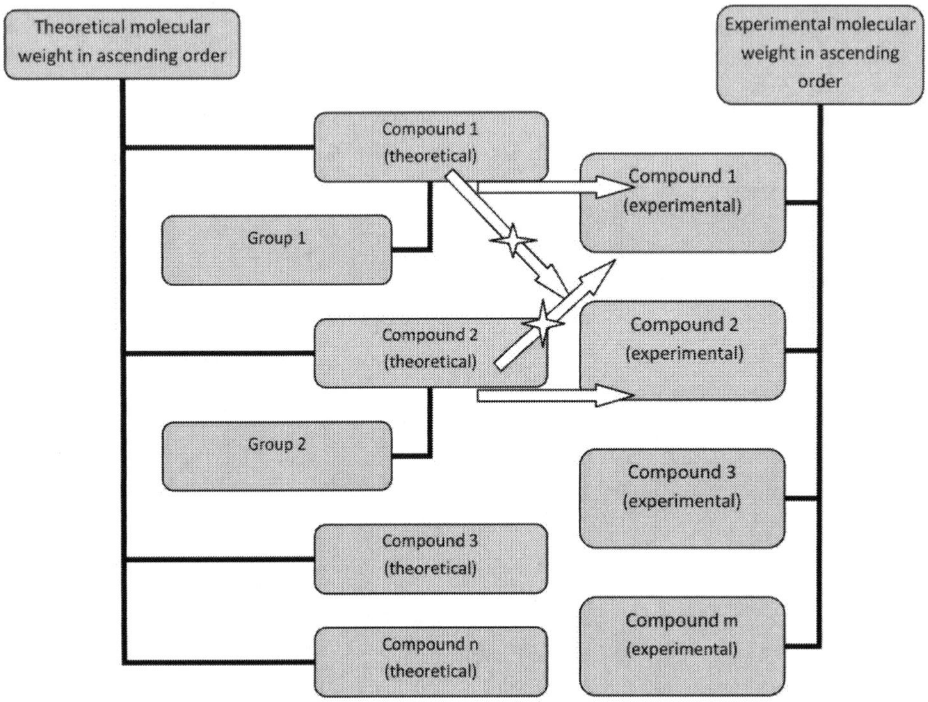

Results and Discussion

For the classification of the components of asphalt sample was generated a database of molecular weight of the homologous series, whose examples are found in the methodology above. Sample analysis were made of asphalt in the mass spectrometer and obtained their molecular weight with high accuracy. To systematize the classification and separation of the groups, we used a search algorithm developed using VBA (Visual Basic for Applications).

Table 2: Asphalt compounds

Groups generated by homologous series	Molecular weight of the sample (g/mol)
276, 1582 (C20 H20 O)	276, 1145
	276, 1383
	276, 1508
	276, 1747
	276, 1872
276, 2089 (C18 H28 O2)	276, 2321
	276, 2447

	276, 2685
	276, 2811

Table 3: Asphalt compounds with high molecular weight

Groups generated by homologous series	Molecular weight of the sample (g/mol)
332, 2191 (C24 H28 O)	332, 2008
	332, 2134
	332, 2372
	332, 2498
	332, 2517
	332, 2948
332, 3079 (C23 H40 O)	332, 3074
	332, 3436

Tables 2 and 3 above show the groups generated by the proposed method, such groups are of structural isomers. Grouping structural isomers reduces the complexity of the system for thousands perhaps millions of molecules. Structural isomers with the same number of carbon are compounds that have similar physical and chemical properties which allow correlating them (Hu, 2002). These correlations, together with kinetic properties and thermodynamics are possible to obtain molecular models, to be attached to the models of existing processes in the refinery.

Conclusions

Current oil has a large number of different molecular components, but to represent the molecules or to build only a small number of molecular groups is required. Having a set of atomic elements such as C, H, O, S, N, which are the molecules of oil, this set can be thought of as a structure group, with a molecular formula to promote the representation of the molecule. The molecular formula is limited and can only provide molecular weight of molecules and cannot distinguish between a large set of stoichiometric isomers.

From the standpoint of modeling the refining process requires a set of molecular information such as presence of: double bond, branching, sulfur, oxygen, nitrogen, or molecular arrangement simplified to provide a convenient representation in the fundamental chemical processes.

References

Aye,M.; Zhang,N.(2005). A Novel Methodology in Transforming Properties of Refining Streams into Molecular Information. Chemical engineering Science, 60:6702-6717.

Hu,S.; Towler,G.; Zhu,F.(2002). Combine Molecular Modeling with Optimization to Stretch Refinery Operation. Ind. Eng. Chem.,Res.41:825-841.

Hu,S.; Zhu,X.X.(2001). A General Framework for Incorporating Molecular Modelling into overall Refinery Optimisation. Applied Thermal Engineering 21:1331-1348.

Quann,R.; Jaffe,S. (1992). Structure-Oriented Lumping: Describing the Chemistry of Complex Hydrocarbon Mixtures, Ind. Eng. Chem.,Res.31:2483-2497.

Quann,R.J.; Jaffe,S.B. (1996). Building Useful Models of Complex Reaction Systems in Petroleum Refining. Chemical Engineering Science,51:1615-1635.

(140f) Optimization and Thermodynamic Analysis of Mixed Refrigerant System In Ethylene Plants

Karthik Krishnadevarajan[1], Jian Zhang[1] and Qiang Xu[2], (1)Dan F Smith Department of Chemical Engineering, Lamar University, Beaumont, TX, (2)Dan F. Smith Department of Chemical Engineering, Lamar University, Beaumont, TX

Optimization and Thermodynamic Analysis of Mixed Refrigerant System in Ethylene Plants

Karthik Krishnadevarajan, Jian Zhang and Qiang Xu

Dan F Smith Department of Chemical Engineering

Lamar University, Beaumont, TX 77710

Abstract

Refrigeration is one of the most important and crucial system in chemical plants. A refrigeration system generally works by removing heat from low temperature streams through vapor compression cycles at the expense of mechanical work. Since a refrigeration system can cool down a process stream below the ambient temperature, it is indispensable to cryogenic cooling and used for separation process in many chemical industries, such as large scale productions of ethylene, oxygen, nitrogen and liquefied natural gas.

Usually refrigeration systems employ single component as refrigerant as long as it is thermodynamically desirable and operationally feasible. But a multi component mixture can also be used as refrigerant, because a mixed refrigerant system provides a continuum of temperatures with smaller temperature differences even at the lower temperatures. This leads to a low energy loss in the system. Mixed refrigerant system has many inherent advantages than the traditional single component refrigerant system.

In an ethylene plant, usually refrigeration is provided by cascading two or three single-component refrigeration systems. This process is not only expensive but also inefficient because it needs more facilities in comparison to single component refrigeration systems. The process streams in ethylene plant vary from 40 to -140 °C. For a single component refrigerant it is impossible to provide constant refrigeration at a particular temperature. Mixed Refrigerant is used in order to improve refrigeration process by using four components as mixed refrigerant.

The present work deals with developing a mixed refrigerant system for an ethylene plant. The process is developed using a rigorous simulation model. Also the process is optimized for the optimal design and operating parameters of the system. The design and operating parameters include compressor work and total flow rate of the mixed refrigerants. This model has few challenging factors. First, the minimum temperature difference in the heat exchangers is taken as 2 ^{o}C. Second, the heat ------duty of all the heat exchangers in the process is fixed based on the plant data. Finally, the process is non linear so optimization is more complex. To demonstrate this work, process streams thermodynamics are also analyzed using thermodynamic analysis. The base case results are compared with optimized results based on the rigorous simulation and optimization methods.

See more of this Session: Applications of Optimization In Industrial Practice
See more of this Group/Topical: Process Development Division

(140g) Real Time Optimization of Industrial Gas Plants and Networks

Tong Li, Air Liquide, Newark, DE

Industrial gases are widely used in many areas including the chemical, refining, metallurgy, environmental, and semiconductor industries. The industrial gases used in large quantities typically include oxygen, nitrogen, argon, hydrogen, and carbon monoxide (CO).

The air gases – oxygen, nitrogen and argon – are usually produced through cryogenic distillation, when a large amount of them are needed in high purity. In a cryogenic air separation unit (ASU), air is first liquefied at a very low temperature and then separated into different components through distillation. This is a very energy intensive process. For example, the Air Liquide Group runs about 400 ASUs worldwide, and its electricity consumption in 2010 was more than one thousandth of the total world electricity consumption. Steam methane reforming (SMR) is one important process to produce hydrogen and CO. In this process, natural gas is reacted with steam in a heated furnace at the presence of catalysts to produce syngas, which is a mixture of hydrogen, CO, methane, and other gases. This mixture is then further processed to produce pure hydrogen, CO, or certain mixtures. These products are often transported to customers through a pipeline, which connects multiple SMR plants and customers. SMR processes also consume a lot of energy, mainly in term of natural gas. Around half of Air Liquide's annual energy bill is for the purchase of natural gas.

It can be seen that improving the efficiency of these industrial gas plants and pipelines is important in improving an industrial gas company's profit and competitiveness while reducing its energy consumption and carbon footprint. Real time optimization (RTO) is an important technique to achieve such a goal. A real time optimizer takes market dynamics such as changes in energy price and client demands into account in real time, and calculates the most profitable mode of operating the plants by using the plant models with energy consumption equations and process constraint equations. Then the calculated optimal set points are downloaded to the plant control system to ramp the plant or sent to operators as references.

In this work, two RTO applications in industrial gas production are studied: one ASU plant and one SMR pipeline. The important aspects of the applications are addressed including the general IT structure and functions of its different software components, important steps in completing such projects, and major challenges and corresponding solutions in mathematical optimization.

See more of this Session: Applications of Optimization In Industrial Practice
See more of this Group/Topical: Process Development Division

ID# 221166

THE ROLE OF PHYSICAL PROPERTY DATABASES IN CH. E. EDUCATION

Mordechai Shacham, Ben-Gurion University of the Negev, Beer-Sheva, Israel

Michael B. Cutlip, University of Connecticut, Storrs, CT

Michael Elly, Intel Corp., Qiryat Gat, Israel

Introduction

Numerical problem solving in Chemical Engineering typically requires a mathematical model of the problem along with physical and thermodynamic data and correlations (equations) of the chemical substances involved. The required pure compound properties are usually divided into constant properties and temperature and/or pressure dependent properties. In most ChE textbooks, the preparation of the mathematical model associated with a particular problem is typically emphasized. The property data required for solution of the problems are usually provided in tabular or graphical form in the appendix of the textbook and/or in a CD associated with the book (see for example, Felder and Rousseau, 2000 and Himmelblau and Riggs, 2004).

Recently, various databases that contain extensive physical property data for a large number of compounds have become available. Typical examples are the DIPPR® database (Rowley et al., 2010) and the NIST database (http://webbook.nist.gov/chemistry/) databases. The use of the data available in these databases for problem solving has significant benefits over the use of the data provided in the textbooks. Some of the more important advantages are:

1. *In current engineering practice, the databases are used as a principal source of property data and correlations.* Thus it is important that students become experienced with the application of these sources during their educational programs.
2. *The databases provide consistent sets of correlations for temperature-dependent properties enabling the use of solution techniques independent of the format in which the property data is provided.* If temperature-dependent data are provided in tabular or graphical forms, for example, this prevents the use of standard numerical methods for problem solution, and data format dependent ad hoc solution techniques have to be used.
3. *The data provided in databases are continuously updated with the new data as they become available.* Property data in textbooks are often taken "as is" from references that may be as much as half a century old. The textbook data may be incorrect or even contradictory. It is important that students become accustomed to using reliable data.
4. *The data available in the databases are evaluated and in cases when multiple values are available for the same property, one recommended value is selected by the database professionals.* There may be very substantial differences between property values reported by different investigators. Brauner et al., 2005, for example mention the case of the melting point of 4-methyloctane for which the recommended value is 159.95 K while one reported experimental value is 219.62 K. The students must be made aware of the fact that several property values may be available and learn how to find and utilize the value with the highest confidence in its correctness.
5. *The databases usually provide uncertainty values (upper limit on experimental error) which enable estimation of the uncertainty of the problem solution using error*

propagation analysis. An important goal of chemical engineering education is to impart to students that the numerical "solution" of a problem almost always has some associated uncertainty.

A convenient option for chemical engineering educators to incorporate the use of databases into their teaching has been integrated into the POLYMATH package. This widely-used computational tool for numerical problem solving now contains a "sample database" subset of the DIPPR© Database[1]. The "sample database" contains 34 constant properties and 16 temperature dependent property correlations for 112 compounds that are most frequently used in chemical engineering textbooks. A special interface POLYMATH Database Interface (PDI) now enables a convenient search of the sample database for selected compounds. The desired properties can be selected, and the data and correlations outputted in a format that can be copied and pasted directly into a computer code with all the significant figures given in the database. The formats that are currently supported are for POLYMATH[2], MATLAB[3] and Excel[4].

In the following example, the use of the POLYMATH interface to the DIPPR Database will be demonstrated for an adiabatic reactor design problem that will illustrate and highlight the database usage.

Production of Acetic Anhydride – An Example

Process models based on material and energy balances usually include some constant and temperature dependent data. A typical example of this type is presented by Fogler (1992). This example involves the vapor phase cracking of acetone to ketene and methane

$$CH_3COCH_3 \rightarrow CH_2CO \quad + \quad CH_4$$
$$\text{Acetone (A)} \rightarrow \text{Ketene (B)} + \text{Methane (C)}$$

as a key step in manufacturing acetic anhydride. The reaction is first order with respect to acetone (A) and the specific reaction rate can be expresses by

$$\ln(k) = 34.34 - \frac{34222}{T} \tag{1}$$

where k is the specific reaction rate (1/s) and T is the temperature (K). It is desired to design a tubular reactor which yields 20% conversion. The case considered here involves adiabatic operation of the reactor, feed flow rate of F_{A0} = 8000 kg/hr acetone, inlet temperature of T_0 = 1035 K and pressure P = 162 kPa.

The mole balance equations for this system can be written

$$\frac{dF_A}{dV} = r_A; \frac{dF_B}{dV} = -r_A \text{ and } \frac{dF_C}{dV} = -r_A \tag{2}$$

---

[1] The Design Institute for Physical Properties and its acronym DIPPR® are registered trademarks of the American Institute of Chemical Engineers (AIChE®)
[2] POLYMATH is a product of POLYMATH Software, http://www.polymath-software.com
[3] MATLAB is a trademark of The Math Works, Inc. http://www.mathworks.com
[4] Excel is a trademark of the Microsoft Corporation, http://www.microsoft.com

where F_A, F_B and F_C are the molar flow rates (in mol/s) of compounds A, B and C, respectively, V is the volume (m^3), and r_A is the rate of the reaction mol/(m^3*s) with respect to A:

$$r_A = -kC_A \tag{3}$$

Here C_A is the concentration of A (mol/m^3) and it can be calculated using the ideal gas law.

$$C_A = \frac{F_A}{(F_A + F_B + F_C)} \frac{P}{RT} \tag{4}$$

The enthalpy balance equation yields

$$\frac{dT}{dV} = -\frac{-r_A\{-[\Delta H^0_{Rx}(T_R) + \Delta H_Q(T)]\}}{F_A C_{PA} + F_B C_{PB} + F_C C_{PC}} \tag{5}$$

where $\Delta H^0_{Rx}(T_R)$ is the heat of the reaction at a reference temperature (T_R, usually 25 °C) and $\Delta H_Q(T)$ is the enthalpy change that results when the temperature is raised from T_R to some temperature T. C_{PA}, C_{PB} and C_{PC} are the heat capacities (J/mol*K) of components A, B and C, respectively.

Fogler (1992) rewrote these equations in terms of conversion (X = 1-F_A/F_{A0}) as the use of conversion enables obtaining graphical solution for the adiabatic case. Rewriting the mole balance equations in terms of the conversion yields:

$$\frac{dX}{dV} = \frac{-r_A}{F_{A0}} \tag{2A}$$

Since the flow-rates of the various compounds not calculated explicitly, Eq. 4 cannot be used for calculation of C_A as it needs to be expressed in terms of the original concentration of A in the reactor feed C_{A0}, the conversion in the reactor X, and the temperature change in the reactor relative to the feed temperature T_0.

$$C_A = -\frac{C_{A0}(1 - X)T_0}{(1 + X)T} \tag{4A}$$

The use of the conversion requires revision of the enthalpy balance equation.

$$\frac{dT}{dV} = -\frac{-r_A\{-[\Delta H^0_{Rx}(T_R) + \Delta C_P(T - T_R)]\}}{F_{A0}(C_{PA} + X\Delta C_P)} \tag{5A}$$

where ΔC_P is the overall change in the heat capacity per mol of A reacted.

Solution of the Example Problem Using Traditional Techniques

Fogler (1992) used the Simpson graphical technique to find the volume of the reactor and the exit temperature for conversion of 20% of acetone data. Equations (2A), (3), (4A) and (5A) were used in the solution. Fogler (1992) provided standard heats of formation and 2nd order polynomial equations for calculating C_P for the three compounds. As the example is based on Jeffrey's (1964) publication, it is assumed that this publication is the source of the data used by Fogler. The set of equations and data (in POLYMATH format) as presented in the 3rd edition of the Fogler book is shown in Table 1. The row numbers shown in Table 1 are not part of the program; they were added as references for the explanations that follow.

The model is in a compact form as many of the calculations were carried out separately. In line 3 the value of F_{A0} is given in (mol/s). The value of C_{A0} in mol/m^3, calculated using the ideal gas law is shown in line 5. The equation for C_{PA} is shown explicitly for compound A (in line 7) however for the rest of the compounds the equations are not shown and only the calculated expression for ΔC_P is presented in line 8. The heats of formations of the individual compounds are not shown either, only the resultant numerical value of $\Delta H^0_{Rx}(T_R)$ is presented.

Figures 1 and 2 show the change of the conversion and temperature, respectively, in the tubular-flow adiabatic reactor as function of the reactor volume. The curves that were obtained by the model shown in Table 1 are marked as "Jeffrey's data". The conversion of 20% is obtained with the reactor volume $V = 1.191$ m^3 and the exit temperature in this case is $T = 937$ K. Fogler (1992) obtained reactor volume of 1.27 m^3 with the very same outlet temperature using Simpson graphical integration technique.

Extracting Physical Properties from the DIPPR Database Using the POLYMATH Database Interface (PDI)

The physical property data and correlations that needed for the enthalpy balance include heat of formation data and equations for calculating heat capacity and enthalpy values at temperature T for the reactant and the two products. Current, up to date data and equations for these properties can be found in the DIPPR database (Rowley et al., 2010). This database can be accessed through the POLYMATH Database Interface (PDI) software.

The interface program enables searching the database for a particular compound, marking the desired properties and obtaining as output the necessary data and correlations in a format that can be copied and pasted directly into a computer code.

The information that is provided by the interface program in the POLYMATH format is demonstrated in Table 2 for the properties of acetone that are required for the example problem. This table shows most of the information as provided by the interface when the POLYMATH format output is requested. The code that is generated by the interface program includes correlation equations, definition of constant values and comments. Lines 1-11 can be generated at once by selecting the desired "Compounds and Properties" and the "Basic Report." The "Report Level" determines the amount of information that is to be included as comments.

Lines 1 through 4 contain the information related to the heat of formation of acetone. In line 1, the full name of the property and the full name of the compound are shown (as a comment: text that starts with the "#" sign and ends with the end of the line). In line 2, the family of the compound and in line 3 its chemical formula are presented. In line 4, the property value related information is displayed. The units of the property and its uncertainty (< 1% in this case) are included as comments. The variable into which the value of the property is entered made up from the symbol of the property (*HFOR* in this case) and the chemical formula of the compound involved.

The information provided for ideal gas heat capacity includes the full name of the property and the compound (line 5), the compound's family (line 6) and its chemical formula (line 7). For temperature dependent properties DIPPR provides the information regarding range of validity for the temperature (line 8) and range of validity for the property (line 9).

The ideal gas enthalpy and heat capacity (lines 10 – 11) deserve special discussion. In most cases DIPPR provides the coefficients for the Aly and Lee (1981) equation for this property. The integrated form of this equation, for calculating ideal gas enthalpy is considerably different than the heat capacity equation. For the benefit of the users, whenever the ideal gas heat capacity is requested by the user, the equation for ideal gas enthalpy is also included, as a comment (see line 10). Note that the enthalpy equation includes an integration constant *Hcon*. This constant depends on the standard state selected for the enthalpy calculation. For example, selecting as standard state pure gaseous component at 25 °C (298.15 K) yields *Hcon* as ideal gas enthalpy at 298.15 K. This value has to be subtracted from the enthalpy of ideal gas at temperature *T*.

Solution of the Example Problem Using Basic Model Variables and DIPPR Property Data

The traditional solution technique that includes the use of "conversion" in the differential equations, requires extensive manipulation of the equations. This extra effort is not required when numerical solution techniques are used. Furthermore, in order to provide clear and complete documentation of the problem, it is preferable to include all the original data and variables in the model and carry out the calculations, even the simplest ones, with the computer model. The POLYMATH model for solution of the example problem, using these principles and data from the DIPPR database is shown in Table 3. For most of the variables the units are provided in forms of comments.

The principal model equations are shown in lines 2-13. The mole balance equations (Eq. 2) are in lines 2-4 while the enthalpy balance equation (Eq 5) is in line 5. The notation ICP_C3H6O, ICP_C2H2_O and ICP_CH4 is used for representing C_{PA}, C_{PB} and C_{PC}, respectively. The division by 1000 is used for converting the C_P units from J/kmol*K (as provided by the DIPPR correlation) to J/mol *K for consistency with the mole balance equations. The variables r_A (Eq. 3), C_A (Eq. 4) and k are calculated in lines 6, 8, and 12, respectively and the term: $\Delta H^0_{Rx}(T_R)+\Delta H_Q(T)$ is calculated in line 13. The conversion X is not required for the solution, using this formulation; however, it is calculated (in line 7) as the reactor volume is specified in terms of the desired conversion.

In lines 15–33, the heat of formation data and the equations for calculating the heat capacity and enthalpy of the reactant and the products are presented. Only the information which is essential for the problem solution is shown. This includes the property values, units, upper limit on the uncertainty of the property values and temperature range of applicability of the correlation equations. The range of applicability is important in order to ensure that the solution of the problem does not require the use of one or more of the equations outside of their range of validity. The uncertainty level of the data is important in order to determine what size of "safety factor" has to be added to the calculated reactor volume, so that the desired conversion can be achieved even in the "worst case" situation.

The problem specific constants, initial and final values are grouped together in lines 35-44. The inclusion of all the data and calculations needed for the problem solution and grouping together the model equations, physical property data and problem specific data provide clear and complete documentation of the model. This in turn enables easy and efficient debugging, modification and extension of the model of the problem. The POLYMATH problem solution is obtained after the problem equations are automatically ordered just prior to the solution.

The graphical solution obtained using the model presented in Table 3 is shown in Figures 1 and 2 (the curves marked as "DIPPR data"). Observe that the temperature and the conversion in the reactor are significantly lower in this case than when using the Jeffreys data. The volume required in order to achieve 20% conversion is 2.052 m^3, thus the reactor size is larger by 72% than the size obtained using the Jeffreys data. The model shown in Table 1 cannot be used for locating the causes of difference because most of the original data was omitted, and it is necessary to consult the detailed description provided by Fogler (1992). It turns out that they main source of the difference is the heat of formation value of ketene which is provided by Jeffreys (-61.09 kJ/mol) and the one that provided by DIPPR (-47.5 kJ/mol). The source of the DIPPR heat of formation is Padley et al. (1986) where Jeffrey's publication is from 1964. This demonstrates the importance of the use of the most up to date physical property data from reliable sources in process design calculations.

Conclusions

The example presented here demonstrates some of the advantages of the use of property databases and mathematical models that contain accurate and documented data typically needed in chemical engineering problem solving. This work demonstrates, in particular, that using physical property information from dated sources may lead to considerable errors in design. The inclusion of the original data in the model is shown to make the debugging, modification and extension of detailed modeling problems more easily accomplished.

It is important for educators to introduce students to available and convenient problem solving tools. Additionally students should be exposed to:

- Use of consistent physical property data with documented uncertainty and range of applicability values, extracted from reliable property databases.
- Documentation of the problem, that includes a complete set of the original equations, property data and correlations and problem specific data, in the computer model, clearly marking the units and uncertainties of the variables and temperature range of applicability of the correlations.

References

1. Aly, F. A., and L. L. Lee, "Self-Consistent Equations for Calculating the Ideal Gas Heat Capacity Enthalpy, and Entropy". Fluid Phase Equilib. 6, 169 (1981)
2. Brauner, N., Shacham, M., Cholakov, G. St. and Stateva, R. P., "Property Prediction by Similarity of Molecular Structures – Practical Application and Consistency Analysis", Chem. Eng. Sci. 60, 5458 – 5471 (2005)
3. Felder, R. M. and Rousseau, R. W., *Elementary Principles of Chemical Processes*,3rd Ed, John Wiley & Sons, Inc, Hoboken, New-Jersey, 2000.
4. Fogler, H. S., *Elements of Chemical Reaction Engineering*, 2nd Ed, Prentice Hall, Englewood Cliffs, New Jersey, 1992.
5. Himmelblau D. M., and Riggs, J. B., Basic principles and Calculations in Chemical Engineering, 7th Ed., Prentice-Hall, Upper Saddle River, New-Jersey, 2004.
6. Jeffreys, G.V., A Problem in Chemical Engineering Design: The Manufacture of Acetic Anhydride, 2nd ed. (London: Institute of Chemical Engineers, 1964)
7. Pedley, J.B.; Naylor, R.D.; Kirby, S.P.; Thermochemical Data of Organic Compounds; Chapman and Hall, London, 1986.
8. Rowley, R. L.; Wilding, W. V.; Oscarson, J. L.; Yang, Y.; Zundel, N. A., DIPPR Data Compilation of Pure Chemical Properties, Design Institute for Physical Properties, (http//dippr.byu.edu), Brigham Young University, Provo, Utah, 2010.

Table 1. POLYMATH Solution of the Example Problem Using the Equations Presented by Fogler (1992)

No.	Equation/ # Comment
1	d(X) / d(V) = -ra/Fao
2	d(T) / d(V) = -ra*(-deltaH)/(Fao*(Cpa+X*delCp))
3	Fao = 38.3
4	To = 1035
5	Cao = 18.8
6	Tr = 298 # K
7	Cpa = 26.63+0.183*T-45.86*10^(-6)*T^2
8	delCp = 6.8-11.5*10^(-3)*T-3.81*10^(-6)*T^2
9	k = 8.2*10^14*exp(-34222/T)
10	ra = -k*Cao*(1-X)/(1+X)*To/T
11	deltaH = 80770+6.8*(T-Tr)-5.75*10^(-3)*(T^2-Tr^2)-1.27*10^(-6)*(T^3-Tr^3)
12	V(0) = 0
13	V(f) =1.191
14	X(0) = 0
15	T(0) = 1035

Table 2. Physical properties of acetone as obtained from the DIPPR database by the POLYMATH interface

No.	Equation/ # Comment
1	# Enthalpy of Formation of Ideal gas at 298.15 K and 100000 Pa of ACETONE
2	# Compounds Family: KETONES
3	# Compounds Structure: CH3COCH3
4	HFOR_C3H6O = -215700000 # J/kmol (Uncertainty < 1%)
5	# Ideal Gas Heat Capacity of ACETONE
6	# Compounds Family: KETONES
7	# Compounds Structure: CH3COCH3
8	# Valid Temperature Range: 200K to 1500K
9	# Valid Value Range: 60487 to 188200 (J/kmol*K)
10	# HIG_C3H6O = 57040*T+163200*1607*coth(1607/T)-96800*731.5*tanh(731.5/T) +HCON_C3H6O # J/kmol
11	ICP_C3H6O = 57040 + 163200 * (1607 / T / sinh(1607 / T)) ^ 2 + 96800 * (731.5 / T / cosh(731.5 / T)) ^ 2 # J/kmol*K (Uncertainty < 1%)

Table 3. POLYMATH Solution of the Example Problem Using DIPPR Physical Properties

No.	Equation/ # Comment
1	# Model equations
2	d(FA)/d(V) = rA
3	d(FB)/d(V) = -rA
4	d(FC)/d(V) = -rA
5	d(T)/d(V) = (-(deltaH)) * (-rA) / ((FA * ICP_C3H6O + FB * ICP_C2H2O + FC * ICP_CH4)/1000)
6	rA = -k * CA
7	XA = (FA0 - FA) / FA0
8	CA = yA * P / (8.31 * T)
9	yA = FA / (FA + FB + FC)
10	yB = FB / (FA + FB + FC)
11	yC = FC / (FA + FB + FC)
12	k = 8.2E14 * exp(-34222 / T)
13	deltaH = (HFOR_C2H2O +HFOR_CH4 -HFOR_C3H6O+HIG_C2H2O +HIG_CH4 -HIG_C3H6O)/1000 #J/mol
14	#
15	# Thermodynamic data and correlations from the DIPPR database
16	# ACETONE
17	HFOR_C3H6O = -215700000 # J/kmol (Uncertainty < 1%)
18	HCON_C3H6O = 57040*Tref+163200*1607*coth(1607/Tref)-96800*731.5*tanh(731.5/Tref) # J/kmol
19	HIG_C3H6O = 57040*T+163200*1607*coth(1607/T)-96800*731.5*tanh(731.5/T)-HCON_C3H6O # J/kmol
20	# Valid Temperature Range:200K to 1500K
21	ICP_C3H6O = 57040 + 163200 * (1607 / T / sinh(1607 / T)) ^ 2 + 96800 * (731.5 / T / cosh(731.5 / T)) ^ 2 # J/kmol*K (Uncertainty < 1%)
22	# KETENE
23	HFOR_C2H2O = -47500000 # J/kmol (Uncertainty < 5%)
24	HCON_C2H2O = 36940*Tref+68650*1490*coth(1490/Tref)-45240*640*tanh(640/Tref) # J/kmol
25	HIG_C2H2O = 36940*T+68650*1490*coth(1490/T)-45240*640*tanh(640/T)-HCON_C2H2O # J/kmol
26	# Valid Temperature Range:150K to 1500K
27	ICP_C2H2O = 36940 + 68650 * (1490 / T / sinh(1490 / T)) ^ 2 + 45240 * (640 / T / cosh(640 / T)) ^ 2 # J/kmol*K (Uncertainty < 1%)
28	# METHANE
29	HFOR_CH4 = -74520000 # J/kmol (Uncertainty < 1%)
30	HCON_CH4 = 33298*Tref+79933*2086.9*coth(2086.9/Tref)-41602*991.96*tanh(991.96/Tref) # J/kmol
31	HIG_CH4 = 33298*T+79933*2086.9*coth(2086.9/T)-41602*991.96*tanh(991.96/T)-HCON_CH4 # J/kmol
32	# Valid Temperature Range: 50K to 1500K
33	ICP_CH4 = 33298 + 79933 * (2086.9 / T / sinh(2086.9 / T)) ^ 2 + 41602 * (991.96 / T / cosh(991.96 / T)) ^ 2 # J/kmol*K (Uncertainty < 1%)
34	#
35	# Constants, initial and final values
36	Tref=298.15 #K
37	P = 162 * 1000 # Pa
38	FA0 = 38.3 # mol/m^3
39	FB(0) = 0
40	FA(0) = 38.3
41	FC(0) = 0
42	T(0) = 1035 #K
43	V(0) = 0
44	V(f) = 2.052 # m^3

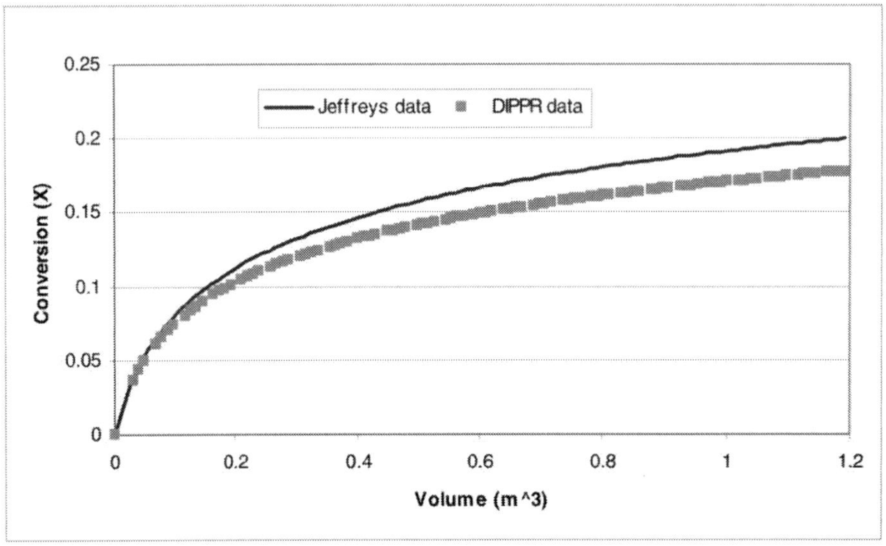

Figure 1 – Conversion of acetone as function of reactor volume

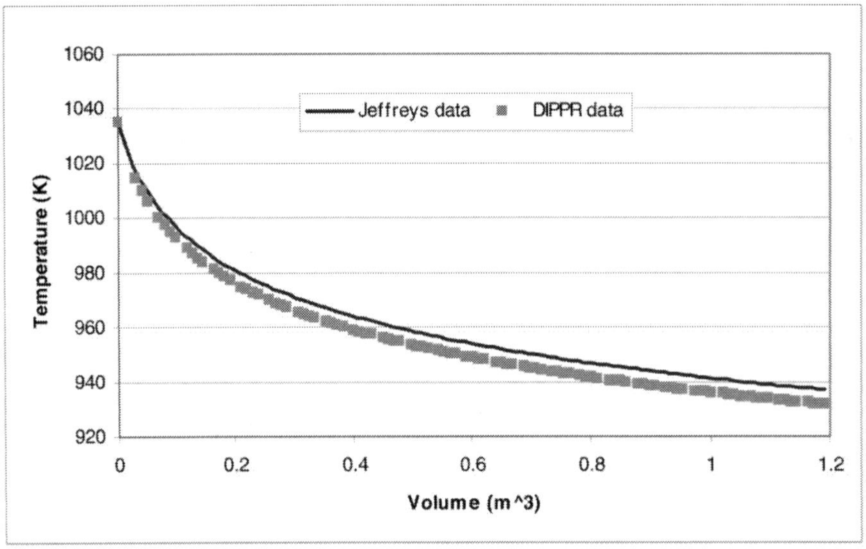

Figure 2 – Temperature change in the reactor as function of volume

(148b) Molecular Simulation Modules for Instruction In Thermodynamics, Transport, Kinetics, and Materials

David A. Kofke and Andrew J. Schultz, Chemical and Biological Engineering, University at Buffalo, The State University of New York, Buffalo, NY

We describe the results of our project to develop and use molecular simulation modules for undergraduate and graduate instruction. These modules have been developed to support instruction of molecular concepts appropriate to a range of courses in the chemical engineering curriculum. The aim is to provide students with tools that permit them to see and manipulate models for the molecular behavior that gives rise to the macroscopic properties they learn about in thermodynamics, transport, and other courses.

Each module consists of an interactive molecular simulation and written supporting material related to it. The molecular simulation is written in the Java programming language, and in most cases it can be run as an applet in a web browser. In all cases the simulation can instead be downloaded and run as an application on the user's machine. The supporting material provides an introduction to the concepts demonstrated by the applet, with some detailed background about what is being modeled. There are also examples describing the use of the simulation, and exercises suitable to use as homework assignments. We have developed a consistent web-based interface that organizes all of the material in each module.

The simulations have been used in classrooms at several institutions, and independent assessments have been performed. Results show that students enjoy the simulations and consider that they help them to learn the material.

See more of this Session: Computing and Technology In Chemical Engineering Education II
See more of this Group/Topical: Computing and Systems Technology Division

(148c) Teaching Molecular Dynamics and Monte Carlo Simulations: Lessons Learned From the Statistical Thermodynamics Workshops At the School of Advanced Studies In Applied Thermodynamics, Rio De Janeiro, Brazil

Edward J. Maginn[1], Frederico W. Tavares[2], Charlles R.A. Abreu[3], Jindal K Shah[4] and Craig Tenney[4], (1)Chemical & Biomolecular Engineering, University of Notre Dame, Notre Dame, IN, (2)Chemical Engineering Program COPPE, Federal University of Rio de Janeiro, Rio de Janeiro, Brazil, (3)Chemical Engineering, Universidade Estadual de Campinas, Campinas, Brazil, (4)Chemical and Biomolecular Engineering, University of Notre Dame, Notre Dame, IN

The authors presented two weeklong workshops on theory and applications of molecular dynamics and Monte Carlo simulations at the Federal University of Rio de Janeiro, Brazil in the summers of 2010 and 2011. The workshops were sponsored by the Chemical Engineering Program of the Alberto Luiz Coimbra Institute (COPPE/UFRJ/Brazil). Participants ranged from new graduate students in a diverse range of fields of study to experienced faculty working in related areas who were interested in learning about molecular simulations. In this talk, we will discuss how the course was designed, how we overcame some of the challenges involved in teaching such a course, and suggestions for how the course could be adopted to other audiences.

See more of this Session: Computing and Technology In Chemical Engineering Education II
See more of this Group/Topical: Computing and Systems Technology Division

Paper #: 238570

LIQUID/VAPOR EQUILIBRIUM VIA EQUATIONS OF STATE FOR FIRST SEMESTER SOPHOMORE STUDENTS

By
Daniel Forciniti
Chemical Engineering Department
Missouri University of Science and Technology
Rolla, MO 65401

INTRODUCTION

The availability of powerful computers and of "ready to use" software for the estimation of thermodynamic properties of pure components or mixtures have transformed some basic thermodynamic courses. Many textbooks come with software already developed or with enough information for the students to write simple programs. Are we forgetting concepts by emphasizing calculations? An affirmative answer to this question will definitely weaken the quality of our programs. It is obvious that some compromise is needed between the teaching of concepts and the temptation of overusing computers.

I started to include more and more complex (computer based) calculations in the two thermodynamic classes that I teach as the accessibility of computers increased (I have been doing this for 20 years). Soon I realized that many of the homework problems were not didactic all because the students were becoming just "calculating machines". The fast pace of homework assignments left little room to explore conceptual issues. A few years ago I started with the idea of "mini project" in which the students must use an equation of state to generate a table of thermodynamic properties for a particular compound.

They use an equation of state that I select from the open literature (we use from 6 to 10 different E.O.S. each semester but no two groups have exactly the same problem because they use different compounds). They have to derive expressions for enthalpy and entropies for that particular equation of state and use those values to find the saturation pressure for a given temperature. In addition they have to generate H-S, and P-V plots and compare their calculations against experimental data that they have to find too (an excellent tool to increase the national membership). The parameters of the equation of state are not explicitly given but some references are made available.

Some specific tasks are: 1) Derivation of the expressions for enthalpies and entropies for the equation of state. This is done by hand and they have to check their solution using Maple. 2) Solving the cubic equation of state for a particular temperature (sub-critical) at various pressures. 3) Determining the position of the spinodal and binodal curves by finding points of divergence of the isothermal compressibility and by finding the pressure at which the condition of chemical equilibrium is satisfied. 4) Preparing the required plots and including experimental values.

The "project" has become a staple for the Thermo 1 course and to a lesser extent of Thermo 2. We have incorporated this into our ABET strategy. The work done by the students is used as a performance indicator for the program outcome: "The S&T Chemical Engineering program graduate will have an ability to apply knowledge of mathematics, science, and engineering fundamentals".

1

EXAMPLE

A class of approximately 60 students was divided in groups of 2 according to their cumulative GPA. They were asked to prepare two plots: P vs. V and H vs. S using an equation of state (E.O.S. assignments are below). The P-V plot must have at least three sub-critical isotherms and one supercritical whereas the H-S plot must have at least two isobars. They needed to present their results (calculated) as well as experimental values in those plots. Specific tasks were:

1) Derivation the expressions for enthalpies and entropies for your equation of state.
2) Solution of the cubic equation of state for a particular temperature (sub-critical) at various pressures. They needed to have at least three solutions in the liquid branch, three in the vapor branch and enough points in the two phase region to determine the position of the spinodal and bimodal.
3) Calculation of entropies and enthalpies at two different pressures.
4) Preparation of plots and inclusion of, properly referenced, experimental values.

Table 1. Example of E.O.S. used in the project. No single group has exactly the same project. Cheating is nearly impossible.

Group	Compound	E.O.S.
A	Methane	Kedge and Trebble (Ref. 5), non cubic with critical correction
B	Decane	Esmaeilzadeh and M. Roshanfekr (Ref. 6)
C	Propene	Ishikawa, W.K. Chung (Ref. 7)
D	Methane	Kedge and Trebble (Ref. 5), cubic with critical correction
E	Dodecane	Esmaeilzadeh and M. Roshanfekr (Ref. 6)
F	Propane	Ishikawa, W.K. Chung (Ref. 7)
G	Methane	Kedge and Trebble (Ref. 5), non cubic without critical correction
H	Carbon dioxide	Ishikawa, W.K. Chung (Ref. 7)
I	Benzene	Esmaeilzadeh and M. Roshanfekr (Ref .6)
J	Methane	Kedge and Trebble (Ref. 5), cubic without critical correction
K		K.H. Kumar and K.E. Starling. (Ref. 8)
L	Carbon Dioxide	Boston-Mathias (Ref 2)
M	Carbon Monoxide	Twu with SRK (generic) (Ref 1 and 2)
N	Decane	K.H. Kumar and K.E. Starling. (Ref. 8)
P	Propylene	Twu with Peng Robinson (generic) (Ref 1 and 2)
Q	Ethanol	Twu with TST (individual component paramaters) (Ref 1 and 2)
R	Ethanol	K.H. Kumar and K.E. Starling. (Ref. 8)
S	Ethylene	Twu with Peng Robinson (generic) (Ref 1 and 2)
T	Ethylene	Kozoi (Ref 4)
U	n-Octane	Twu with TST (generic) (Ref 1 and 2)
V	n-Octane	Harmens (Ref 3)
W	Toluene	Harmens [Ref 3)
Y	Benzene	K.H. Kumar and K.E. Starling. (Ref. 8)
Z	Benzene	Esmaeilzadeh and M. Roshanfekr (ref. 6)

2

1) Twu, C.H., Sim, W.D., and Tassone, V. Getting a handle on advanced cubic equations of state. Chem. Eng. Progress, November 2002.
2) Neau et al. The Soave, Twu and boston-Mathias alpha functions in cubic equations of state. Fluid Phase Equilibria, 276:156-164 (2009)
3) Harmens and Knapp. Three-Parameter cubic equation of state for normal substances. Ind. Eng. Chem. Fundam. 19:291-294 (1980)
4) Koziol. Quintic equation of state for pure substances in sub- and supercritical range. Fluid Phase Equilibria 263:18-25 (2008)
5) C.J. Kedge and M. A. Trebble. An empirical near –critical correction for cubic and non-cubic equations of state. Fluid Phase Equilibria, 194-197: 401-409 (2002)
6) F. Esmaeilzadeh and M. Roshanfekr. A new cubic equation of state for reservoir fluids. Fluid Phase Equilibria, 239: 83-90 (2006)
7. T. Ishikawa, W.K. Chung and B.C. –Y. Lu. AIChE J., 26:372 (1980).
8. K.H. Kumar and K.E. Starling. Ind. Eng. Chem. Fundam., 21:255 (1982).

The project was graded according to the rubric shown in Table 2. Approximately 10 % of the students returned superb reports. 20% of the student returned acceptable report. Approximately 30 % of this class failed Thermo 1. The students that passed this class joined me the following semester for a heat transfer class. Their thermodynamics background was excellent.

3

Table 2. Example of a rubric used to evaluate this class.

Rubric: Project

Topic	Poor (0 Points)	Fair (40 Points)	Good (80 Points)	Excellent (100 points)	Points
Experimental Data	Experimental data are missing	Few experimental data are included. Reference is missing	Sufficient experimental data are included. Reference is included.	Experimental data covering of regions of interest are included. References are completed and properly cited.	0.01x
PV calculations	PV calculations are incomplete. The parameters of the equation of state are not calculated properly (if applicable).	PV calculations are incomplete. The parameters in the equation of state are properly calculated. No documentation is included. A sample calculation is not included	PV calculations are complete. The parameters in the EOS are properly calculated. Documentation is incomplete. A sample calculation is included but it is incomplete	PV calculations are complete. The parameters in the EOS are properly calculated. Documentation is complete: Spread Sheets, MathCad pages or equivalent are included and properly explained. A sample calculation is included.	0.06x
Derivation of residual enthalpies and entropies	The derivations were not done or they were incorrectly. Maple solutions may or may not be present	The derivations were done but they are not correct. Maple solutions may or may not be present.	The derivations are correct but details are missing. Maple solutions are present and used to compare own.	The derivations are correct and enough details are included. Maple solutions are present and used to compare own.	0.02x
Calculation of Entropies and Enthalpies	No calculations were done.	Calculations were done but they are incorrect. The ideal gas contribution was not considered. Reference state is not	Calculations were done and they are correct. Calculations include a limited number of points.	Calculations were done and they are correct. Calculations include a large number of points. The ideal gas contribution was included.	0.06x

		well defined.	The ideal gas contribution was included. Reference state is not well defined.	Reference state is well defined.	
PV and HS plots	Both or one the two plots are missing.	Both plots are present but they are not complete. Experimental data is not included. Plots are of poor quality.	Both plots are present and they are complete. Comparison with experimental data is not satisfactory. Plots are of fair quality. The binodal curve is present but not the spinodal	Both plots are present and they are complete. Comparison with experimental data is satisfactory. Plots are of excellent quality. Both binodal and spinodal curves are present.	0.06x
Presentation of Work	The work done is not explained.	The work done is explained but not in detail. Overall presentation is poor.	The work done is explained in detail. Overall presentation is fair.	The work done is very well explained. The grammar and style are of good quality. Overall presentation is very professional	0.04x
Total percent points (max 25 points)					-----

(148f) Integrating Computational Transport Phenomena Into the Undergraduate Engineering Curriculum

Charles A. Petty[1], Satish Muthu[1] and Andre Benard[2], (1)Chemical Engineering and Materials Science, Michigan State University, East Lansing, MI, (2)Mechanical Engineering, Michigan State University, East Lansing, MI

The field of Computational Transport Phenomena (CTP) or, equivalently, computational fluid dynamics (CFD), has advanced to the point where commercial software can readily be integrated into the undergraduate STEM curriculum. CTP methods provide a powerful and intuitive means to understand the consequences of common assumptions used to analyze the transport of momentum, energy, and mass in complex and simple geometries. Although current CTP technologies can simulate a wide variety of physical problems encountered in engineering and in other disciplines, the use of advanced computational methods in the undergraduate engineering curriculum is not as widespread as it should be.

The proposed presentation will discuss how CTP methods can be used to support student learning and intuition of transport phenomena by simulating the consequences of balance principles, constitutive models, boundary conditions, and initial conditions. Selected examples will be presented where CTP has been employed in the undergraduate curriculum to support specific learning objectives. The presentation will also discuss technical and practical barrier problems that limit the introduction and use of advanced computational methods into the undergraduate curriculum.

See more of this Session: Computing and Technology In Chemical Engineering Education II
See more of this Group/Topical: Computing and Systems Technology Division

(150a) Engineering the Lymph Node Microenvironment with Controlled Release Polymer Particles for Enhanced Vaccination

Christopher M. Jewell, Departments of Bioengineering, Materials Science and Engineering, and The Ragon Institute of MGH, MIT, and Harvard, Koch Institute for Integrative Cancer Research, Massachusetts Institute of Technology, Cambridge, MA, Sandra C. Bustamante Lopez, Departments of Bioengineering, Materials Science and Engineering, Koch Institute for Integrative Cancer Research, Massachusetts Institute of Technology, Cambridge, MA and Darrell J. Irvine, Departments of Bioengineering, Materials Science and Engineering, Koch Institute for Integrative Cancer Research, Ragon Institute of MGH, MIT, and Harvard, and the Howard Hughes Medical Institute, Massachusetts Institute of Technology, Cambridge, MA

Recent studies have demonstrated a simple, potentially universal strategy to enhance vaccines via intra-lymph node (*i.LN.*) injection. Lymph nodes (LNs) are a key site of immune response generation, where dendritic cells (DCs) process and present antigens to resident T cells and B cells, which then migrate to the periphery to carry out mechanisms of adaptive immune response. Since direct administration of antigens or adjuvants in LNs concentrates vaccines at this key site, *i.LN.* immunization with soluble vaccines has demonstrated great promise in recent human clinical trials. However, these strategies require complex, multiple injection regimens owing to rapid vaccine clearance from LNs. We hypothesized that combining *i.LN.* injection with controlled release biomaterials permitting sustained, concentrated dosing to local LN environments might augment the function of adjuvants or other drugs, providing a route for improved vaccination or immunotherapy. In this talk I will discuss two components of a new materials-based vaccination strategy aimed at 1) increasing vaccine potency and 2) tailoring specific aspects of immune response.

We first encapsulated poly(inosinic:cytidylic acid) (polyIC) – a potent nucleic acid adjuvant undergoing clinical trials – in biodegradable poly(lactide-co-glycolide) (PLGA) microparticles (MPs) or nanoparticles (NPs) designed to release polyIC in LNs over several days. We hypothesized particle size would allow control over intra-LN particle fate, with NPs being internalized by LN-resident cells, and MPs remaining extracellular and releasing encapsulated cargo within the LN. MPs and NPs were synthesized by double emulsion and lipids were used to stabilize the aqueous/organic interface, and to incorporate functionalized lipids (e.g., poly(ethylene glycol) to control aggregation). Properties such as particle size (0.2-15 μm), polyIC encapsulation (0.5-4% wt/wt), and release kinetics (2-25 days) were tunable by controlling emulsion conditions. To identify the target LN injection sites we administered a non-toxic tracer at the tail base of mice. Trafficking of the dye to the draining LN allowed subsequent identification of vaccine injection sites. Following injection of fluorescently-labeled particles, live animal imaging and histological analysis confirmed particles were localized within LNs. We used this approach to inject polyIC *i.LN.* in soluble form or encapsulated in MPs (MP-polyIC) or NPs (NP-polyIC) and discovered that MP-polyIC conferred a 4.5-fold increase in polyIC persistence four days after injection compared to soluble polyIC, with polyIC-NPs mediating an intermediate enhancement less than MP-polyIC, but greater than soluble polyIC. The enhancement observed with MP-polyIC mediated a dramatic increase in polyIC uptake by LN-resident DCs over 4 days, in contrast to soluble polyIC which was flushed away one day after injection. Mice vaccinated with antigen mixed with MP-polyIC exhibited antigen-specific T cell (CD8+) levels as high as 18% (7.6-fold greater than MP-polyIC given intramuscularly, *i.m.*; 4.0-fold greater than soluble polyIC given *i.LN.*). This increase indicates potent "arming" of the immune system, unmatched by *i.LN.* injection of antigen mixed with the same dose of NP-polyIC, or of a 10-fold greater doses of soluble polyIC. When T-cells were restimulated with antigen, MP-polyIC enhanced secretion of key cytokines (i.e., IFN-γ, TNF-α) involved in generating immunity, suggesting a functional or protective capacity that we are now exploring.

We have extended this initial approach to tailor immune response by sustained delivery of small molecule immunomodulators – molecules able to bias the specific nature or duration of immune response. A number of such agents have been described, but the potential of these compounds is limited by poor water solubility and rapid clearance *in vivo*. Thus treatment with these drugs typically involves high doses administered at frequent intervals. Since many immunomodulators act in LNs, polymeric particles designed to slowly release encapsulated small molecules in the LN could allow sustained stimulation, enabling this class of drugs to be effectively employed without continuous re-dosing. Daily, low-dose peripheral injections of Rapamycin –a potent small molecule drug – have recently been found to greatly enhance T cell response. Using modified solvent evaporation, we prepared rapamycin PLGA MPs (MP-rapa) and conducted biodistribution studies using *i.m.* or *i.LN.* injections of soluble rapamycin or MP-rapa. HPLC analysis of drug extracts prepared from blood and excised LNs demonstrates MP-rapa permits continuous dosing of rapamycin in LNs for 1 week, whereas injections of soluble rapamycin are undetectable at times longer than 24 hours after injection. In vaccination studies with model antigens, MP-rapa significantly increased the duration of CD8 T cell response, and induced a 6-fold increase in central memory T cells – cells critical for long-term vaccine protection. Strikingly, this enhancement was mediated by just three *i.LN.* injections of MP-rapa, and these benefits were unmatched by 150 daily peripheral injections of rapamycin in soluble form – the "gold" standard injection regimen for rapamycin. In preliminary studies with clinically-relevant HIV gag antigens, MP-Rapa induced a 2.2-fold increase in CD8 memory populations compared with soluble drug formulations, and these studies are currently being extended to small animal challenge models of HIV.

See more of this Session: Disease Models and Therapies
See more of this Group/Topical: Food, Pharmaceutical & Bioengineering Division

(150b) Design and Characterization of Self-Assembling, Peptide-Based Biomaterials As Immunotherapeutics

Amanda Trent, Biomolecular Science and Engineering, University of California, Santa Barbara, Santa Barbara, CA, Matthew J. Black, Chemical Engineering, University of California, Santa Barbara, Santa Barbara, CA, Colleen Olive, Immunity and Vaccinology Laboratory, The Queensland Institute of Medical Research, Herston, Queensland, Australia and Matthew Tirrell, Institute for Molecular Engineering, University of Chicago, Chicago, IL

With the goal of exploring strategies for the rational design of immunotherapeutics, we propose using peptide amphiphile (PA) molecules as building blocks for novel peptide antigen delivery systems. PAs can form versatile macromolecular constructs that combine the biofunctionality of peptides with a controllable self-assembly character obtained by synthetic chemical conjugation of peptide head groups with hydrophobic hydrocarbon tails (i.e. a fatty acid). In aqueous conditions, the hydrophobic effect drives PAs to self assemble into hierarchical structures, including spherical, cylindrical and worm-like micelles. PAs confer a number of attractive properties for peptide-based immunotherapy, including the multivalent presentation of peptide antigens, the enhanced peptide secondary structure often seen upon aggregation, and the potential of the hydrophobic tails to act as adjuvants. Peptides derived from the model tumor antigen ovalbumin and from an antigenic sequence found in Group A Streptococcus have been incorporated into two different PA vaccines. Their physical characterization, as well as their immunogenicity both *in vitro* and in mouse studies will be presented.

See more of this Session: Disease Models and Therapies
See more of this Group/Topical: Food, Pharmaceutical & Bioengineering Division

(150c) siRNA Based Gene Therapy In Alzheimer's Disease

Aparna Chaudhary and **Sanjeev Garg**, Chemical Engineering, IIT Kanpur, Kanpur, Uttar Pradesh, India

Alzheimer's disease (AD) is one of the most common forms of dementia. Existing therapeutics in AD are devoted to symptomatic relief. However, after the emergence of amyloid hypothesis, disease modifying therapeutics are being explored. According to amyloid hypothesis, a fundamental pathological event in AD is the sequential cleavage of the trans-membrane amyloid precursor protein (*APP*) by β-Site APP-Cleaving Enzyme1 (***BACE1***) and γ-*secretase*, generating toxic C-terminal fragments (*CTF*) and *Amyloid*β (Aβ). Aβ deposition occurs in the extracellular space which leads to plaque formation.

The present work aims to inhibit the expression of gene responsible for production of BACE1 enzyme through RNA interference. RNA interference is an intracellular mechanism that can be used as a powerful tool for gene silencing mediated by small interfering RNAs (siRNA). Suitable siRNA is the critical factor for effective silencing which can be achieved by a design tool. A siRNA design tool was developed and its viability was established with the reported siRNA constructs for genes implicated in cancer. In AD, it is proposed to inhibit the cascade of reaction which is responsible for Aβ aggregation. The siRNA constructs generated were compared with validated datasets for BACE1. The results obtained for BACE1 were better in thermodynamic score, GC content, positional preferences as compared to validated datasets. The off-target silencing effects were also comparable with the published results. It is to be noted that the results generated are *in silico* and need to be verified on mammalian cell lines where *BACE1* is expressed.

See more of this Session: Disease Models and Therapies
See more of this Group/Topical: Food, Pharmaceutical & Bioengineering Division

(150d) Identifying Novel Bacterial Proteases Involved In Enhanced Virulence During Chronic Infection

Logan MacDonald[1], Justin Nice[1], Sean O'Keefe[1] and **Bryan W. Berger**[2], (1)Bioengineering, Lehigh University, Bethlehem, PA, (2)Chemical Engineering and Program in Bioengineering, Lehigh University, Bethlehem, PA

Host protease-antiprotease balance has emerged as a major factor in regulating the extent of inflammatory response by neutrophils, macrophages and endothelial cells to bacterial infections. In response to pro-inflammatory cytokines, activated neutrophils release a series of extracellular proteases, including cathepsins, MMPs and neutrophil elastase (NE), all of which exhibit strong elastolytic activity, the largest percentage of which is attributed to cathepsins. Activated cathepsins are inhibited by cystatins, and dysregulation of the cathepsin-cystatin balance has been strongly linked to chronic bronchial inflammation during *P. aeruginosa* infection, especially in cystic fibrosis (CF) patients. We recently identified a family of conserved bacterial peptidases present in several obligate pathogens, including *P. aeruginosa*, that we believe play a critical role in disturbing the cystatin-cathepsin balance during chronic infection. We plan to investigate peptidase expression levels from clinical isolates to correlate expression with severity of chronic infections, and utilize a novel chemical biology approach to design peptide-based inhibitors of protease function as a potential therapeutic and diagnostic for *P. aeruginosa* infection. Our work will advance the understanding of how chronic inflammation in the endobronchial space, exacerbated by the dominant neutrophil inflammatory response, can lead to host-induced tissue damage caused by pathogens such as *P. aeruginosa*.

See more of this Session: Disease Models and Therapies
See more of this Group/Topical: Food, Pharmaceutical & Bioengineering Division

(150e) Interplay of Flow and Oncotically Active Solute Transport Across the Arterial Endothelium: Hydraulic Conductivity Masking and Relevance to Atherogenesis

Shripad D. Joshi, Chemical Engineering, City College and Graduate Center of City University of New York, New York, NY, Kung-Ming Jan, College of Physician and Surgeons, Columbia University, Bronx, NY and David S. Rumschitzki, Department of Chemical Engineering, The City College of The City University of New York, New York, NY

Macromolecular transport across the vessel wall, apparently the earliest pre-atherosclerotic event, is known to occur due to advection by transmural (i.e., across the vessel wall) pressure-driven plasma (water and solutes) transport through leaky junctions around rare endothelial cells that are temporarily widened and leaky, some because they are either dying or dividing. A wall parameter, the hydraulic conductivity (L_p), defined as the ratio of the transmural water flux to (hydrostatic-oncotic, the latter universally taken as zero in large artery walls) pressure difference, is central to the understanding of this transmural water transport in detail. On one hand, it advects macromolecules like low-density lipoprotein (LDL) into the arterial subendothelial intima (SI) through these leaks and spreads it there, thereby giving it the chance to bind to extracellular matrix and possibly to trigger the start of atherosclerotic lesion formation. On the other hand, the overall water flow across the normal (non-leaky) endothelium dilutes the lo-cal SI LDL concentration, thereby likely slow-ing binding kine-tics, and washing not-yet-bound lipid further into the wall.

Our group's discovery of the presence of Aquaporin-1 (AQP) in rat aortic endothelial cells suggests a new possibility of water transport across the endothelial cell, alongside the generally accepted paracellular route. Interestingly, we found that chemically blocking AQPs changes rat aortic wall Lp in a strongly pressure dependent manner. We then proposed a new theory that agrees well with this otherwise perplexing experimentally-observed pressure-dependence. It suggests that AQPs contribute about 30% to the phenomenological endothelial Lp at low transmural pressures. However, since the lumen is isotonic, an AQP-mediated, transendothelial cell, pure water flow into the arterial SI should set up an oncotic pressure gradient $\Delta\Pi$ that opposes and quickly halts the ΔP-driven flow through the cell. How then could trans-AQP flow persist for many hours, as our chemical blocking of AQP experiments show? We have developed a filtration and advection-diffusion model for water and the transport of the small, oncotically-active solute albumin across the glycocalyx layer, across endothelial interface and through the entire vessel wall. The filtration flow and the mass transfer problem are intimately coupled: The flow advects tracer while albumin differences across the endothelium set up a $\Delta\Pi$ that sucks fluid across the endothelium. And it is nonlinear: the advective term is the product of two dependent variables, velocity and (the gradient of) the concentration. We simultaneously solve the combined steady state problem as the long-time solution of an unsteady problem that we attack with finite difference methods. We use parameters taken from the literature, mostly from experiment.

Our results show that, since the media retardation coefficient ~ 0.3, the media acts to strongly filter the transport of the solute albumin relative to the water solvent, causing a build-up of albumin in the media that backs up into the SI by the time steady state is achieved. In contrast to our naïve suggestion above, this increased albumin concentration results in *higher* osmotic pressures in the SI than in the glycocalyx, which drives the water flow into the SI from the luminal side of the EC, and not the other way around. Thus, even though oncotic gradients are normally assumed negligible across the walls of large arteries, we find that they indeed play a role *within* those vessel walls. We show to what extent these oncotic gradients mask the true wall and layer hydraulic conductivities. We look at the effects of vesicular albumin transport and of varying the junctional boundary conditions on these results. Given AQP's presence in the high-pressure cardiac endocardium, heart, lung and renal epithelia, resolving and understanding how mass-transfer-induced osmotic gradients interact with the flows that advect those solutes may have far broader and illuminating implications.

See more of this Session: Disease Models and Therapies
See more of this Group/Topical: Food, Pharmaceutical & Bioengineering Division

(150f) Molecularly Imprinted Polymeric Nanoparticles for Treating Viral Infections - A Synthetic Antibody Approach

Yen Wah Tong, Department of Chemical and Biomolecular Engineering, National University of Singapore, Singapore, Singapore and Niranjani Sankarakumar, Chemical and Biomolecular Engineering, National University of Singapore, Singapore, Singapore

Molecular imprinting is an effective method that has been used commercially for more than a decade in chemical purification, analysis and sensor technologies. The working principle behind this technique is the creation of recognition sites in polymers through the specific 3-dimensional arrangements of monomers around the template molecules to be imprinted. This makes the imprint highly specific in recognizing the template molecule, and this behaviour is a chemical analog of biological antibodies.

In this research, our goal is to create molecular imprints of viruses on the surfaces of polymeric nanoparticles to "capture" and remove any active, disease-causing viruses. Virus transmitted diseases such as chikungunya, dengue and hepatitis pose significant social and economic problems throughout the world every year. Hence, the removal of viruses could be the ultimate approach for a vast range of virally transmitted diseases. Typical virus removal methods include ultrafine filtration or high speed centrifugation which is usually slow and expensive. Therefore, virus specific molecularly imprinted polymers may be an attractive and cheaper alternative to existing techniques involving the design of the "**viral sponge**" which would specifically and selectively remove or kill pathogenic virus from blood or sera.

We prepared surface imprinted polymers for viruses and tested its capability of specifically adsorbing viruses and preventing viral infections under *in vitro* conditions. For this proof of concept study, a simple bacteriophage was used as the model virus template. Virus imprinted (vMIPs) and Non-imprinted polymeric nanoparticles sized about 40nm were synthesized using miniemulsion polymerization. Methyl methacrylate (MMA) and ethylene glycol dimethacrylate (EGDMA) were employed as the functional and cross-linker monomers, respectively. The imprinted polymers exhibited higher preferential binding to template virus as compared to controls. *In vitro* experiments in a model host-virus system proved that our vMIPs was able to partially inhibit phage infection in *E.coli*. This research is the first of its kind to demonstrate the usage of molecularly imprinted polymers to catch viruses in the presence of its host cells.

See more of this Session: Disease Models and Therapies
See more of this Group/Topical: Food, Pharmaceutical & Bioengineering Division

(150g) Breast Cancer CSC Display Novel In Vivo Imaging Features, Reminiscent of Development, During Early Tumor Progression

Natesh Parashurama[1], Neethan Lobo[2], Michael Clarke[2] and Sanjiv Sam Gambhir[3], (1)Department of Radiology- Molecular Imaging Program at Stanford, Stanford University, Stanford, CA, (2)Institute for Stem Cell Biology and Regenerative Medicine, Stanford University, Stanford, CA, (3)Radiology and Molecular Imaging Program, Stanford University Schools of Engineering and Medicine, Stanford, CA

Virtually nothing is known about how nascent tumors form from cancer stem cells (CSC) and mammary stroma, yet this is essential for advancing breast cancer stem cell biology and developing molecular imaging and therapeutic strategies directed towards CSC. With this in mind, we developed a novel in vivo intravital imaging system designed to both image across the entire mammary gland and stroma. Importantly, this allowed us to identify new, rapidly developing, in vivo, quantitative imaging features at high resolution/sensitivity, not appreciable with conventional techniques. Using the developing mouse as a model for early tumor growth, we imaged and mapped the same fluorescent mammary gland at 3 and 4.5 weeks, and demonstrated rapid, spatially restricted growth; certain branches displayed a 400% increase in area over 1.5 weeks and maintained branching architecture, while neighboring ducts within 500μm displayed a 200% decrease. While examining stromal elements responsible for duct growth, we identified that normal mammary buds in virgin females displayed frequent microvascular interactions, and surprisingly, 77.7 % ± 6.4 (mean ± SD) of buds strongly interacted with the microvasculature (n=3, total terminal end buds =120). Notably, this stroma imaging feature was "conserved" in the 3 week old developing mouse, where we found 33/40 (83%) of mammary buds displayed analogous microvascular interactions. We then applied these imaging features to interpret a CSC tumor model designed to simulate early tumor formation from CSC. Two weeks after direct injection of 2500 CSC (or Tumorigenic) or non-CSC (non-Tumorigenic) cells from FvB MMTV-Wnt1 mice derived tumors into the developing GFP mammary gland, we observed new imaging features from the host mammary glands and host vasculature of the CSC populations only, including a 170% increase ($p < 0.28$) in side branching of endogenous, eGFP+ ducts, with maintained host microvascular-bud interactions. Lastly, two weeks after transplantation of dye-labeled CSC and non-CSC , we observed a 240% increase in overall angiogenesis ($p < 0.027$), and quantifiable imaging features of normal growth ($p < 0.005$), including selective bud and duct outgrowths supplied with vasculature, and invasion by loosely organized neovasculature. Overall, our data suggests that CSC populations display several rapidly developing imaging features involving CSC and the underlying stroma, some of which are seen in developing mammary gland, which can be potentially useful for early radiologic identification of CSC in tissues, identification of new targets for molecular probes, monitoring therapy, or understanding the biology of CSC-stromal interactions.

See more of this Session: Disease Models and Therapies

(155a) The Kinetics of Redox Reactions of Hexavalent Chromium and Trichloroethylene In Wastes Containing Both Contaminants

Jacob Mlusu and Ramesh Chawla, Chemical Engineering, Howard University, Washington, DC

This paper presents the feasibility of using oxidizing and reducing agents for the simultaneous remediation of heavy metals and chlorinated hydrocarbons in aqueous and soil systems. Chelsea soil was selected as a model soil while trichloroethylene (TCE) and hexavalent chromium (Cr(VI)) were selected as model contaminants. Ferrous ion, Fe(II) was used as a reducing agent while permanganate, (MnO_4^-), was used as an oxidizing agent. Bench-scale experiments were conducted to determine the reaction kinetics and extent of redox reactions of Cr(VI) and TCE in separate and combined systems in both aqueous and soil systems. In the redox reaction, Cr(VI) is reduced to the nonhazardous Cr(III), and Fe (II) is oxidized to ferric ion, Fe(III); while TCE is oxidized to chloride, hydrogen, and carbon dioxide and permanganate is reduced to manganese dioxide (MnO_2.). In the experimental analysis, the Cr(VI)-Fe (II) reaction yielded 100% and 99% Cr(VI) reduction when Fe(II) was used in water and Chelsea soil, respectively. The overall reaction was determined to be second order, first order with respect to both Cr (VI) and Fe(II). The TCE-$KMnO_4$ reaction yielded 100%, and 82% TCE oxidation by excess MnO_4^- in water and Chelsea soil, respectively. The overall reaction was determined to be second order reaction, first order with respect to both TCE and MnO_4^- in both the aqueous and soil systems. The Cr(VI)-TCE reaction yielded 92% Cr(VI) reduction in water when excess TCE was used in the absence of Fe(II) and $KMnO_4$. The overall reaction was determined to be second-order overall, first order with respect to both Cr(VI) and TCE. TCE was found to have a significant effect on the reduction of Cr(VI) and vice versa. When TCE was present in the aqueous system, the Cr(VI)-Fe(II) reaction with the same experimental conditions as before yielded 94% Cr(VI) reduction. Presence of TCE was advantageous to the reduction of Cr(VI), as it acted as an additional reductant. When Cr(VI) was present in the aqueous system, TCE-KMnO4 reaction with the same conditions as before yielded 97% TCE oxidation.

See more of this Session: Fundamentals of Environmental Process and Reaction Engineering
See more of this Group/Topical: Environmental Division

(155b) Kinetic Modeling of Gas-Phase Mercury Oxidation by Chlorine and Bromine In Combustion Effluents During Oxy-Combustion

Itsaso Auzmendi-Murua, Chemical, Biological, and Pharmaceutical Engineering, New Jersey Institute of Technology, Newark, NJ and Joseph W. Bozzelli, New Jersey Institute of Technology, Newark, NJ

Around 29 Gt of CO_2 are emitted annually worldwide. The main sources of these emissions are transportation and electricity-generating plants, which use coal and fuel combustion. Since CO_2 capturing techniques are still expensive, in order to reduce CO_2 emissions from fuel combustion, oxy-combustion has been proposed as a solution, with the aim of obtaining combustion flue gases with higher concentrations of CO_2 that are easier to remove. The implementation of oxy-combustion affects the speciation of the rest of the combustion products, including the speciation of mercury. Different techniques have been proposed for the control of mercury emissions, due to its high toxicity. In this work, the addition of halogens (chlorine and bromine) to the flue gases have been studied in order to oxidize mercury from its elemental form to oxidized form – HgX_2. As these mercury halides are water soluble, it is feasible to capture the HgX using conventional air pollution control devices. An elementary reaction mechanism has been developed for modeling the oxidation of mercury by the addition of halides, in both regular combustion and oxy-combustion conditions. Thermochemical properties and kinetic parameters have been calculated for Hg and Halogen – NO_x species using B3LYP and CBS-QB3 methods. The elementary reaction mechanism has been constructed with use of chemical activation analysis for association and addition reactions with quantum RRK analysis for k(E) and Master Equation analysis for fall-off. Results indicate that the presence of NO_x inhibits Hg oxidation by chlorine addition, through several catalytic cycles for loss of Cl. As for bromine addition, similar catalytic cycles are observed, but in this case, the overall thermochemistry favours more mercury oxidation. Therefore, results show that during oxy-combustion conditions, Hg oxidation is enhanced, since flue gases contain lower concentrations of NO_x.

See more of this Session: Fundamentals of Environmental Process and Reaction Engineering
See more of this Group/Topical: Environmental Division

A Computational Model to Evaluate Solar Offset and Water Generation of Various Cooling Systems Using Location-Specific Annual Climate Data

Dia Milani[a], Ali Abbas[a], Matteo Chiesa[b]

[a]Chemical and Biomolecular Engineering School, The University of Sydney, Sydney, Australia

[b]Masdar, Abu Dhabi Future Energy Company, Abu Dhabi, UAE

Abstract

An algorithm has been developed to address conventional vs. renewable dilemma. The debate of using renewable energy to replace conventional sources in various cooling systems to partially compensate this energy staggering operation and improve cooling efficiency is still everlasting. Because of numerous uncertainties associated with climatic data and the effect of climate change, the accuracy of predicting the trend of solar cooling's performance is not sufficient; and therefore, the confidence over capital vs. operational expenditure and payback time calculation was diluted. In this project, annual statistical weather data of any location can be analyzed to predict the possible offset that renewable solar can offer to slash energy requirement of a particular cooling or dehumidification system. Matching that analysis against total cost of installation and operation will provide solid ground for investment decision making criteria. The project takes into account location specific boundary conditions such as the cooling demand time series, solar resource availability, climatic conditions, component cost and performance parameters. This methodology evaluates the techno-economic performance of the solar collector systems include solar thermal and solar electric cooling options and account for water generation from ambient moisture as another significant product to be offset from total cost. Testing this algorithm for hot spot locations such as Abu Dhabi revealed an interesting potential for solar cooling investment, although more validation of these parameters are needed.

Keywords: Solar Cooling, Cooling Generation Cost, Photovoltaic, Adsorption Chillers, Thermoelectric Cooler, Absorption Chillers

(155d) Toxicity of Flue Gas In Microalgae CO2 Mitigation Systems

Adriana Pacheco, Mario M. Alvarez and Javier Lara-Gil, Biotechnology, Tecnologico de Monterrey, Monterrey, Mexico

Microalgal-based CO_2 capture from flue gas is an attractive mitigation strategy for industry. However, other gases are also present that may be harmful to microalgae growth, such as NO_x and SO_x. In this study, we simulated flue gas by adding to the culture medium salts of nitrite, sulfite, or bisulfite. Two high CO_2 tolerant species were evaluated: an environmental isolate (Rio) and *Scenedesmus* sp. (UTEX 1589). Toxicity tests were performed in 96-well microplates under an atmosphere of 2% (v/v) CO_2 to determine the concentration range of trace gases for microalgae growth. Both algae were able to tolerate concentrations of nitrite (>1600 ppm) and sulfite (400 ppm) higher than those reported in effluent gas. However, bisulfite inhibited growth at 50 ppm. This effect was associated with a drastic decrease in medium pH. As a result, we explored the use of cement dust and $CaCO_3$ to lessen the decline in pH. We used concentrations of 300 ppm and 240 ppm of cement dust and $CaCO_3$, respectively, to increased pH from 2.6 to 5.0. This increment in pH resulted beneficial to the growth of both species, tolerating concentrations of 400 ppm bisulfite (2X upper limit concentration found in flue gas). In all tests, environmental isolate (Rio) generated more biomass than the collection strain. Therefore, our microalgae isolate shows a better tolerance to the flue gas constituents, where cement dust might be a beneficial component of the mitigation system.

See more of this Session: Fundamentals of Environmental Process and Reaction Engineering
See more of this Group/Topical: Environmental Division

(155e) Reducing Air Quality Impact From Plant Start-up Emissions by Integrating Air Quality Modeling and Plant Start-up Simulation

Jian Zhang, Qiang Xu and Thomas C. Ho, Dan F. Smith Department of Chemical Engineering, Lamar University, Beaumont, TX

Chemical industry concentrated regions experience highly localized and transient air pollution events. For instance, chemical plant flaring activities, especially intensive flaring during plant startup process, emit huge amounts of volatile organic compounds (VOC) and nitrogen oxides (NO_X), causing elevated ground-level ozone concentrations that violate the National Ambient Air Quality Standards. Emissions are inevitable during plant startups, their air quality impacts, however, are different under different meteorological conditions. This provides opportunities to identify cost-effective control strategies for reducing air quality impacts from plant start-up emissions.

In this paper, detailed process system knowledge of plant startup emission and speciation is integrated with air quality modeling and simulation for the first time to quantitatively study regional air quality impact from ethylene plant startup emissions. The general process has three major steps: emission source characterization, flaring emission speciation and quantification, and air quality simulation and decision support. In the first step, plant-wide dynamic simulation and optimization are employed to minimize start-up emissions and identify benchmark emission dynamic profiles during a plant start-up. In the second step, simplified flaring emission speciation and quantification are conducted to identify dynamic profiles of ozone precursors, such as various VOC and NOx. Then, the flaring precursor information is input to the air quality model, CAMx. In the third step, multiple air quality simulations are performed for start-up emission impact analysis and decision support. This is the most complex step. Firstly, a background air quality simulation is performed. The simulation needs atmospheric reaction mechanism, geological domain information, emission inventory data, meteorology information, initial conditions and boundary conditions. The simulation gives temporal and spatial background ozone and other air pollutants distributions. It also gives different ozone formation mechanisms for different regions. Secondly, the startup procedures are arranged into different time windows, and start-up emissions are integrated into the background air quality model to generate different cases. Through multiple air quality simulations, regional air pollutant increments can be obtained. Through comparison and analysis, the best start-up time window will be identified.

A typical ethylene plant startup procedure has been studied. Results show that local maximum ozone increment can be significant due to an ethylene start-up emission. However, if appropriate start-up time window is selected, the maximum ozone increment can also be substantially reduced comparing with the worst case.

(155f) Modeling and Optimization of a Modified Claus Process As Part of An Integrated Gasification Combined Cycle (IGCC) Plant with CO2 Capture

Debangsu Bhattacharyya[1], **Dustin D. Jones**[2], Richard Turton[1] and Stephen E. Zitney[3], (1)Department of Chemical Engg., West Virginia University, Morgantown, WV, (2)Chemical Engineering, West Virginia University, Morgantown, WV, (3)Collaboratory for Process and Dynamic Systems Research, National Energy Technology Laboratory, U.S. Department of Energy, Morgantown, WV

The modified Claus process is one of the most common technologies for sulfur recovery from acid gas streams. In the case of an Integrated Gasification Combined Cycle (IGCC), the feeds to the Claus process are the off-gas from the sour water stripper (SWS) and the acid gas from the stripper of the acid gas removal (AGR) unit. Because of the stringent environmental regulations considered in this work, the tailgas from the Claus process is further treated in a hydrogenation unit, compressed from about 1 atm to 50 atm, and recycled back to the AGR unit for recapture. Important design criteria for the Claus unit, when part of an IGCC power plant, are the ability to destroy ammonia completely and deep sulfur recovery from a relatively low purity acid gas stream without sacrificing flame stability. Due to these criteria, modifications are often required to the process, resulting in a modified Claus process. For studies discussed here, these modifications include the use of a 95% pure oxygen stream as the oxidant, a split flow configuration, and preheating of the feeds with the intermediate pressure steam generated in the waste heat boiler (WHB).

In the future, for IGCC plants with CO_2 capture, the Claus unit must satisfy the emission standards without sacrificing the plant efficiency in the face of typical disturbances of an IGCC plant such as rapid change in the feed flowrates due to load-following and wide changes in the feed composition because of changes in the coal feed to the gasifier. The Claus unit should be adequately designed and its control design should be appropriately implemented to satisfy these objectives. A representative dynamic model is needed for these studies. Even though the Claus process has been commercialized for decades, most papers treat the key reactions in a Claus process as equilibrium reactions. Such models may be validated by manipulating the temperature approach to equilibrium for a set of steady-state operating data, but are of limited use for dynamic studies.

In a Claus process, especially in the furnace and the WHB, a number of reactions take place along with the widely-considered Claus reaction. In this work, a set of linearly independent reactions has been identified and kinetic models of the flame zone, anoxic zone, WHB, and catalytic reactors have been developed. Various approaches are taken to derive the kinetic rate expressions which are either missing in the open literature or found to be inconsistent. A set of plant data is used for optimal estimation of the kinetic parameters while a set of plant data is used for the model validation. The final model agrees well with the published plant data.

Using the developed kinetics models of the Claus reaction furnace, WHB, and catalytic stages, sensitivity studies are completed to examine whether important process variables could be controlled in the face of the typical process disturbances from the IGCC plant. Additionally, a study is carried out to examine the optimal operation of a Claus plant considering three, highly interactive process variables in the context of an IGCC power plant with CO_2 capture. The first variable is the single pass conversion of H_2S in the Claus unit. Because of a very high operating pressure of the AGR process in this IGCC plant, considerable compression cost is associated with recycling the unconverted tail gas. The second variable is the flow of hydrogen to the AGR unit from the Claus unit. Hydrogen sent to the AGR unit can be converted to electric power more efficiently by the gas turbine than burning it in the Claus furnace and raising steam. The third variable is the consumption of oxygen in the Claus unit. Considerable operating cost is incurred in the air separation unit for producing oxygen. An optimization study examines the tradeoffs of these process variables along with the net power requirement of the Claus unit.

See more of this Session: Fundamentals of Environmental Process and Reaction Engineering
See more of this Group/Topical: Environmental Division

(161a) Internal Model Control, Model Predictive Control, and Shell

Carlos E. Garcia, Port Arthur Refinery, Motiva Enterprises LLC, Port Arthur, TX and Daniel E. Rivera, School for Engineering of Matter, Transport, and Energy, Arizona State University, Tempe, AZ

Perspectives are presented on the early development of internal model control, model predictive control, and advanced control implementations at Shell.

See more of this Session: In Honor of Manfred Morari's 60th Birthday II- Papers by Invitation Only
See more of this Group/Topical: Computing and Systems Technology Division

Paper Unavailable

(161b) MIMO Controllability and Decentralized and Plantwide Control

Sigurd Skogestad, Department of Chemical Engineering, Norwegian University of Science and Technology, Trondheim, Norway

This presentation describes Manfred's early work in controllability analysis and the robust design of decentralized and plantwide control systems.

See more of this Session: In Honor of Manfred Morari's 60th Birthday II- Papers by Invitation Only
See more of this Group/Topical: Computing and Systems Technology Division

Paper Unavailable

(161c) State-Space and Robust Model Predictive Control

Mayuresh V. Kothare, Chemical Engineering, Lehigh University, Bethlehem, PA and Jay Lee, Georgia Institute of Technology, Atlanta, GA

This presentation will describe Manfred's contributions to state-space and robust model predictive control.

See more of this Session: In Honor of Manfred Morari's 60th Birthday II- Papers by Invitation Only
See more of this Group/Topical: Computing and Systems Technology Division

Paper Unavailable

(161d) The Influence of MPC On Biomedical Control

Francis J. Doyle III, Chemical Engineering, University of California, Santa Barbara, Santa Barbara, CA

This talk will describe the influence of model predictive control on biomedical control.

See more of this Session: In Honor of Manfred Morari's 60th Birthday II- Papers by Invitation Only
See more of this Group/Topical: Computing and Systems Technology Division

Paper Unavailable

(161e) Control of Hybrid Systems

Francesco Borrelli, Mechanical Engineering, University of California at Berkeley, Berkeley, CA

This presentation will describe advances in the control of hybrid systems.

See more of this Session: In Honor of Manfred Morari's 60th Birthday II- Papers by Invitation Only
See more of this Group/Topical: Computing and Systems Technology Division

Paper Unavailable

(161f) Rebuttal

Manfred Morari, Control Laboratory, Swiss Federal Institute of Technology Zurich, ETHZ, Zurich, Switzerland

Closing comments for the session.

See more of this Session: In Honor of Manfred Morari's 60th Birthday II- Papers by Invitation Only
See more of this Group/Topical: Computing and Systems Technology Division

Paper Unavailable

(200a) Using Convex Nonlinear Relaxations In the Global Optimization of Nonconvex Generalized Disjunctive Programs

Juan P. Ruiz, Dept of Chemical Engineering, Carnegie Mellon University, Pittsburgh, PA and Ignacio Grossmann, Chemical Engineering, Carnegie Mellon University, Pittsburgh, PA

Many problems in engineering can be represented by a set of algebraic expressions using continuous and discrete variables, leading to a Mixed-integer Nonlinear Program (MINLP). In order to represent accurately the behavior of physical, chemical, biological, financial or social systems, nonlinear expressions are often used. In general, this leads to an MINLP where the solution space is nonconvex and hence difficult to solve since this may give rise to local solutions. In the last decade, many global optimization algorithms for nonconvex problems have been proposed [1,2]. The efficiency of these methods heavily relies on having tight relaxations and that is why many of the contributions in this area have been related to this subject. However, in general, finding the global optimum of large-scale nonconvex MINLP models in reasonable computational time remains a largely unsolved problem.

Raman and Grossmann [4] presented an alternative way to represent discrete/continuous optimization problems by using not only algebraic expressions but also disjunctions and logic propositions giving rise to Generalized Disjunctive Programming (GDP), which can be regarded as an extension of Disjunctive Programming [5]. This higher abstraction level representation not only facilitates the modeling, but also keeps the underlying logical structure of the problem, which can be exploited to find stronger relaxations and hence, develop more efficient solution methods. With this in mind, we have recently presented a method to solve nonconvex Generalized Disjunctive Programs that involve bilinear and concave functions [6]. The main idea behind this methodology to find strong relaxations relies on a two step procedure. In the first step the nonconvex functions are replaced with polyhedral relaxations leading to a Linear GDP. In the second step the results of Sawaya and Grossmann [7] are used. These authors showed that a stronger relaxation of Linear GDPs can be obtained by the application of basic steps, a process that consists of intersecting disjunctions to obtain equivalent disjunctive sets whose hull relaxation is tighter. Even though the method we presented showed significant improvements when compared to traditional approaches, the efficiency of the method depends on the strength of the polyhedral relaxations of the nonconvex functions.

In this paper we address the global optimization of GDP problems that in addition to bilinear and concave terms, involve other terms such as linear fractional for which nonlinear convex relaxations have shown to provide rigorous convex envelopes that are much tighter than linear relaxations. The use of nonlinear convex relaxations leads to a nonlinear convex GDP which relaxation can be strengthen by using recently results from the work of Ruiz and Grossmann [8].

We first define the general nonconvex GDP problem that we aim at solving and review the traditional method to find relaxations. Second, we show how we can strengthen the relaxation of the traditional approach by presenting a systematic procedure to generate a hierarchy of relaxations based on the application of basic steps to nonlinear disjunctive convex sets. We outline a set of rules that avoids the exponential transformation to the Disjunctive Normal Form leading to a more efficient implementation of the method. Finally we assess the performance of the method by solving to global optimality several instances of process network, reactor network and heat exchanger network problems. It is shown that the proposed strong nonlinear relaxations lead to stronger lower bound predictions and a decrease in the required number of nodes when used within an NLP-based spatial branch and bound framework, as a result requiring reduced computational times.

REFERENCES

[1] Floudas, C. A., Deterministic global optimization: Theory methods and applications., Dordrecht, The Netherlands: Kluwer Academic Publishers, 2000.

[2] Tawarmalani, M. and Sahinidis, N., *Convexification and Global Optimization in Continuous and Mixed-Integer Nonlinear Programming.*, Kluwer Academic Publishers , 2002.

[3] Horst, R. and Tuy, H., *Global Optimization deterministic approaches (3rd Ed)*, Berlin: Springer-Verlag, 1996.

[4] Raman R. and Grossmann I.E., *Modelling and Computational Techniques for Logic-Based Integer Programming*, Computers and Chemical Engineering, **18**, 563, 1994.

[5] Balas E., *Disjunctive Programming*, Annals of Discrete Mathematics, **5**, 3-51, 1979.

[6] Ruiz, J.P. and Grossmann, I.E., *Strengthening lower bounds for the global optimization of non-convex Generalized Disjunctive Programs*, Computers and Chemical Engineering, **34**:6, 914-930, 2010.

[7] Sawaya N., *Thesis: Reformulations, relaxations and cutting planes for generalized disjunctive programming*, Carnegie Mellon University, 2006.

[8] Ruiz, J.P. and Grossmann, I.E., *A hierarchy of relaxations for convex Generalized Disjunctive Programs*, Submitted for publication.

See more of this Session: Advances In Optimization II
See more of this Group/Topical: Computing and Systems Technology Division

(200b) Parallel Solution of Large-Scale, Block-Structured, Nonlinear Programming Problems with Significant Coupling

Jia Kang, Artie McFerrin Department of Chemical Engineering, Texas A&M University, College Station, TX, Johan Akesson, Department of Automatic Control, Lund University, Lund, Sweden and Carl D. Laird, Chemical Engineering, Texas A&M University, College Station, TX

Efficient solution of parameter estimation problems is important to all fields of science and engineering where decisions are made based on results from simulation. When describing the physics of extensive engineering and biological problems, we observe that these large-scale nonlinear parameter estimation problems frequently possess block angular structure or can be transformed to such a structure. The dominant computational expense in an interior-point based NLP solver is the solution of a linear set of equations arising from the application of a modified Newton's method to the first-order optimality conditions. The structure of the parameter estimation problem induces structure in this linear system, and one can exploit this structure to obtain more efficient solutions of the nonlinear parameter estimation problem in parallel.

In previous work, we have developed a nonlinear interior-point algorithm with a parallel Schur-complement decomposition approach that is designed to allow parallel block-based linear algebra operations to exploit the problem structure. However, as the number of coupling variables increases, this approach becomes less competitive as the Schur-complement increases in size, and the number of backsolves required to form the Schur-complement increases. In our current work, we show that this bottleneck can be overcome and solve the Schur-complement equations using a preconditioned conjugate gradient method. This new algorithm avoids forming and factorizing the Schur-complement explicitly. The parallel scalability of this approach is demonstrated on randomly generated parameter estimation problems. Furthermore, significant parallel speedup is possible on two parameter estimation problems over large-scale water distribution networks. The performance of this approach is significantly affected by the choice of preconditioner. We test this technique with different preconditioning approaches and demonstrate that significant speed improvement is possible compared with previous approaches.

Finally, we have interfaced this parallel solver to the python-based PYOMO modeling language. This approach allows for straightforward formulation of the structured problem, and provides parallel evaluation of the problem objective, constraints, gradients, and hessians.

See more of this Session: Advances In Optimization II
See more of this Group/Topical: Computing and Systems Technology Division

(200c) COLIN: Optimization Infrastructure for Hybrid Algorithms

John D. Siirola and William E. Hart, Sandia National Laboratories, Albuquerque, NM

While the optimization practitioner has a large library of optimization algorithms and approaches at their disposal, there are many challenging optimization problems (notably large-scale non-convex optimization and non-convex multi-objective optimization) that can not be reliably addressed by a single deterministic approach. In these cases, an alternative approach is to turn to stochastic search routines like simulated annealing and genetic algorithms. Past work has shown that it is possible to significantly improve the performance of these stochastic search routines by hybridizing these routines with local search capabilities in both structured [1] and unstructured [2] arrangements. However, the construction of these hybrid systems is a tedious and *ad hoc* process.

In this presentation, we present COLIN, a Common Optimization Library INterface for general optimization problems. COLIN differs from other integrated optimization platforms (e.g., DAKOTA, GAMS, or AMPL) in that it does not require client problems and optimization algorithms to adhere to a strict (nominally MINLP) interface. Instead, COLIN allows each problem and algorithm to implement only the interface that they directly support, using native data types of their choosing. COLIN then provides the necessary functionality to dynamically resolve and transform both data types and entire problem representations as necessary to enable seamless communication between the problem and solver. Similar transformations leverage concepts from Polymorphic Optimization [3] to support a global multi-solver, multi-problem annotated solution repository. This flexible interface allows COLIN to support the rapid development and evaluation of structured and unstructured hybrid optimization algorithms.

[1] W.E. Hart and R.K. Belew. Optimization with genetic algorithm hybrids that use local search. In R. K. Belew and M. Mitchell, editors, *Adaptive Individuals in Evolving Populations: Models and Algorithms, volume 26 of Santa Fe Institute studies in the science of complexity*, pages 483-496. Addison-Wesley, 1996.

[2] J.D. Siirola, S. Hauan, and A.W. Westerberg. Computing Pareto fronts using distributed agents. *Computers & Chemical Engineering*, 29(1):113-126, 2004.

[3] J.D. Siirola and S. Hauan. Polymorphic Optimization. *Computers & Chemical Engineering*, 31 (10):1312-1325, 2007.

See more of this Session: Advances In Optimization II
See more of this Group/Topical: Computing and Systems Technology Division

(200d) Optimal Sensitivity Based On IPOPT

Rodrigo Lopez-Negrete[1], Hans Pirnay[2] and Lorenz T. Biegler[1], (1)Chemical Engineering, Carnegie Mellon University, Pittsburgh, PA, (2)RWTH Aachen University, Aachen, Germany

Optimal Sensitivity Based on IPOPT

The sIPOPT library is an open source implementation that provides optimal sensitivity of solutions of nonlinear programming (NLP) problems. It has been adapted to a fast solver based on a barrier NLP method. The program evaluates the sensitivity of the KKT system with respect to model parameters. The code is integrated to the NLP solver IPOPT, and thus we can use the factorized KKT matrix at the solution provided by the solver so that sensitivities to parameters are determined with minimal computational cost. Aside from estimating sensitivities for parametric NLPs, the program provides approximate NLP solutions that can be used for nonlinear model predictive control and state estimation. Moreover, we also provide the elements of a QP-based sensitivity approach through the use of a "fix-relax strategy" to account for active set changes. These are enabled by prefactored KKT matrices and a fix-relax strategy based on Schur complements. In addition, reduced Hessians are obtained at minimal cost and these are particularly effective to approximate covariance matrices in parameter and state estimation problems. The sIPOPT program is demonstrated using the AMPL interface on a simple example. In addition sIPOPT applications to parameter estimation, state estimation and nonlinear model predictive control are highlighted.

Literature cited

Pirnay, H., Lopez-Negrete, R., and Biegler, L.T., Optimal Sensitivity Based on IPOPT, submitted to Math. Prog. Comp., 2011.

Wächter, A. and Biegler, L.T., On the Implementation of a Primal-Dual Interior Point Filter Line Search Algorithm for Large-Scale Nonlinear Programming. Mathematical Programming 106(1), 2006, pp. 25–57.

Fiacco, A.V., Introduction to Sensitivity and Stability Analysis in Nonlinear Programming, volume 165 of Mathematics in Science and Engineering. Academic Press, 1983.

Zavala, V.M., Computational Strategies for the Operation of Large-Scale Chemical Processes. Ph.D. thesis, Carnegie Mellon University, 2008.

See more of this Session: Advances In Optimization II
See more of this Group/Topical: Computing and Systems Technology Division

(200e) Advances In Multi-Parametric Mixed Integer Linear Programming

Martina Wittmann-Hohlbein and Efstratios N. Pistikopoulos, Department of Chemical Engineering, Centre for Process Systems Engineering, Imperial College, London, United Kingdom

The presence of uncertainty in mixed integer linear programming (MILP) models employed in widespread application fields, including planning/scheduling, process synthesis and hybrid control, significantly increases the complexity and computational effort in retrieving optimal solutions. A particular difficulty arises when uncertainty is simultaneously present in the coefficients of the objective function (OFC), the constraint matrix (LHS) and the constraint vector (RHS), which transforms the original linear model into a nonlinear and non-convex one with respect to both the optimization variables and the parameters.

Here, we discuss two novel methods for the solution of the general multi-parametric (mp) MILP problem with OFC-, RHS- and LHS-uncertainty. The first approach is a two-stage method, [1], which efficiently obtains approximate solutions by combining suitable robust optimization and multi-parametric programming techniques yielding a tight upper bound on the optimal objective function value. In the first step a partial immunization of the model against LHS-uncertainty is performed, whereas in the second step explicit solutions of the partially robust counterpart are derived using a recently proposed mp-MILP algorithm [2].

The second approach deals with the global solution of the general mp-MILP problem. Extending the ideas of [3] to our framework, we outline the steps of a parametric branch and bound algorithm for the global solution of mp-LP problems with LHS-uncertainty, which play a key role in the overall global solution of mp-MILP problems.We exploit the special structure of the mp-MILP problem to construct competitive multi-parametric under- and overestimating problems employing McCormick-type relaxations of bilinear terms. Furthermore, we discuss the alternative of embedding novel piecewise affine relaxations of bilinear terms [4,5] in the proposed procedure. The optimal solution of the general mp-MILP problem is approximated by piecewise affine functions.

References

[1].Wittmann-Hohlbein, M., Pistikopoulos, E.N. A robust optimization based approach to the general solution of mp-MILP problems. *Accepted for publication in Proceedings of 21st European Symposium on Computer Aided Process Engineering*.

[2].Faísca, N.P., Kosmidis, V.D., Rustem, B., Pistikopoulos, E.N. Global optimization of multi-parametric MILP problems. *Journal of Global Optimization* 45, 131–151 (2009).

[3].Dua, V., Papalexandri, K.P., Pistikopoulos, E.N. Global optimization issues in multiparametric continuous and mixed-integer optimization problems. *Journal of Global Optimization* 30, 59-89 (2004).

[4].Wicaksono, D.S., Karimi, I.A. Piecewise MILP under- and overestimators for global optimization of bilinear programs. AIChE Journal 54, 991-1008 (2008).

[5].Gounaris, C.E., Misener, R., Floudas, C.A. Computational comparison of piecewise–linear relaxations for pooling problems. Industrial & Engineering Chemical Research 48, 5742-5766 (2009).

(200f) From Time Representations In Scheduling to Hybrid Spatial Representations In 2-D Allocation Problems

Pedro M. Castro, Modeling and Optimization of Energy Systems, LNEG, Lisbon, Portugal and Ignacio E. Grossmann, Department of Chemical Engineering, Carnegie Mellon University, Pittsburgh, PA

There have been remarkable advances in scheduling approaches by the Process Systems Engineering community in the last 20 years. Major breakthroughs were: (i) unified frameworks for process representation that allow the systematic modeling of a process flowsheet or production recipe in terms of the virtual entities that are the state, tasks and resources; (ii) alternative time representations ranging from discrete-time to continuous-time, where a few alternative concepts can be employed (e.g. single time grid, multiple time grids, a.k.a. unit-specific, and immediate or general precedence sequencing variables).

In the context on N-dimensional allocation problems [1], scheduling problems can be viewed as being 1-D, in which the relevant dimension is time. Cutting and packing start at 2-D and one common aspect is that small items have to be packed into, or cut from, one or more objects, often called strips or bins. In the paper industry for example, the objective is cutting jumbo reels of paper into smaller reels so as to minimize trim losses.

Existing MILP models for 2-D allocation problems (e.g. strip packing) rely on a full continuous-space representation featuring continuous variables to determine the (x,y) coordinates of a given point of the rectangle in the strip [2-3]. The binary variables identify the relative position of a rectangle with respect to another and appear in the big-M, no overlap constraints. It is also straightforward to come up with a full discrete-space model using a mesh consisting of a 1x1 squares [4].

The novelty of this paper consists of two new MILP models that combine alternative spatial representation concepts into the x and y axis of the 2-D problem. It can be viewed as the follow-up of last year's presentation that involved a discrete-space representation in the x-axis and a continuous-space representation with multiple time grids in the y-axis [4]. Now, we propose using sequencing variables in the y-axis whereas for the x-axis we can either use: (i) a discrete-space representation or (ii) a continuous-space representation with a single grid [5]. Note that while the former can be classified as a hybrid discrete/continuous-space model, the latter is a full continuous-space model.

The computational performance has been tested in a set of 29 test instances taken from the literature including some of the well-known *cgcut*, *ngcut* and *ht* instances. Interestingly all 5 MILPs tested (one discrete-space, two hybrid discrete/continuous and two continuous-space models) are valid approaches in the sense that they are the best approach for at least one problem. Some require specialized solution strategies in the search for the global optimal solution (e.g. when using spatial grids consisting of multiple event points, a problem inherited from their scheduling counterparts) while for others solving the model just once is enough.

[1] Westerlund J, Papageorgiou LG, Westerlund T. A MILP model for N-dimensional allocation. Comput. Chem. Eng. 2007; 31: 1702-1714.

[2] Castillo I, Westerlund J, Emet S, Westerlund T. Optimization of block layout design problems with unequal areas: A comparison of MILP and MINLP optimization models. Comput. Chem. Eng. 2005; 30: 54-69.

[3] Sawaya NW, Grossmann IE. A Cutting Plane Method for Solving Linear Generalized Disjunctive Programming Problems. Comput. Chem. Eng. 2005; 29: 1891-1913.

[4] Castro PM, Oliveira JF. Scheduling Inspired Models for Two-Dimensional Packing Problems. European Journal of Operational Research. Under Review.

[5] Castro PM, Barbosa-Póvoa AP, Matos HA,; Novais AQ. Simple Continuous-Time Formulation for Short-Term Scheduling of Batch and Continuous Processes. Ind. Eng. Chem. Res. 2004; 43: 105-118.

See more of this Session: Advances In Optimization II
See more of this Group/Topical: Computing and Systems Technology Division

(214a) Passivity-Based Plantwide Control Design by Flowsheet Decomposition

Timothy P. McFarland and B.Erik Ydstie, Chemical Engineering, Carnegie Mellon University, Pittsburgh, PA

Design of control structures for chemical plants typically occurs at the unit operations level. Controllers can be designed for desired closed loop stability and performance properties for each unit. The units, when placed in series retain these properties, but the presence of recycle streams changes the dynamics of the entire system. Control of recycle streams has been studied in detail throughout the control literature. However, these studies usually do not include global stability proofs for integrated systems which may have many nested recycles. Furthermore, current methods are useful for analysis but can't be easily used to construct plantwide control systems without analyzing the dynamics of the entire system.

In this paper, we utilize passive systems theory to develop a constructive procedure to design control laws for complex networks of chemical process systems. We use a natural storage function for passivity design that is derived from the second law of thermodynamics. The method has many similarities with the tangent plane method developed by Gibbs to show stability of multiphase and reactive equilibrium systems. We show that the storage function can be generalized to non-equilibrium systems and that chemical process systems are made passive by controlling one extensive variable for each independent phase and extent of reaction. The control laws are implemented locally and individual units become passive. If all units in a chemical processing plant are designed to be locally passive, the entire network will be passive and retain desirable properties like bounded input-output stability and Lyapunov stability, with the proper choice of inter-connection variables.

We present a method to construct process control systems for large scale and complex networks of chemical processes with a minimum amount of process information. We utilize extensive variable inventory control and identify the minimum number of controlled inventories to assure passivity of the entire network. The remaining degrees of freedom are available for flow control and process optimization. The paper has a theoretical and a practical component. New results on passivity of complex networks and graph-decomposition of process flowsheets are presented. The chemical looping combustion process is used as an illustrative example for the procedure.

See more of this Session: Control of Large Scale and Networked Systems
See more of this Group/Topical: Computing and Systems Technology Division

(214b) A Graph Theoretic Approach for Model Reduction and Control of Complex Energy Integrated Process Networks

Sujit S. Jogwar, Seongmin Heo, **Srinivas Rangarajan** and Prodromos Daoutidis, Chemical Engineering and Materials Science, University of Minnesota, Minneapolis, MN

Complex process networks, consisting of interconnections of numerous reaction-separation units via material recycle and energy integration, represent a key feature of modern chemical plants. The design of such complex networks, motivated by process limitations (such as reaction kinetics/thermodynamics) and economic benefits (reduced material and/or energy consumption), has received significant attention over the past few years. These complex process networks, however, are quite difficult to operate, especially in the context of transitions between different operating points which are fairly common in the current environment of frequent changes in market conditions and economic objectives. Specifically, a disturbance in one process unit is often propagated over a large portion of the network owing to the coupling between various process units. Though a system specific analysis can prove beneficial for a particular case, a generic network level analysis allows for the identification of characteristic features exhibited by a class of complex networks, thus broadening the range of applicability of the established results.

In our previous research, we have identified several such fundamental structures, such as networks with large material recycle, networks with large energy recycle, networks with large energy throughput, which capture essential features of some of the relatively simpler integrated process networks (for example, reactor-distillation column systems with low per pass conversion, reactor-feed effluent heat exchanger networks, energy integrated distillation columns, etc). Each of these fundamental structures exhibits unique dynamic features and allows for the reduction (in terms of size and complexity) of the underlying process model.

In the case of complex process networks, the need for model reduction becomes more imperative, and at the same time, more challenging to execute. A preliminary analysis of some of the complex networks [1] has revealed that their underlying structure is composed of combinations of one or more fundamental structures (or building blocks) identified previously. However, the interconnection of (a large number of) these building blocks can result in interactions and dynamic phenomena being exhibited by these networks that are more intricate than can be anticipated by considering the mere association of these building blocks.

Graph theory provides a powerful framework to analyze the structural properties of such complex networks, responsible for their ensemble behavior. The modular structure of these complex networks lends themselves naturally to a graph theoretic analysis, whereby in addition to identifying the fundamental structures, one can achieve the reduction of the overall network through successive application of such identification steps.

In this work, we focus on complex energy integrated process networks characterized by a segregation of the energy flows in the network. This is typical in many industrial systems such as networks of integrated distillation columns, cryogenic systems, etc. We have developed a suite of algorithms within a graph-theoretic framework for identifying the time scales exhibited by the network, generating scaled dynamic equations in each time scale and classifying different control objectives and manipulated inputs to be used in the controllers in each time scale. Specifically, a complex network is represented as an *energy flow graph* with nodes corresponding to the process units and edges corresponding to the energy flows connecting different units. Equivalent representations of the fundamental structures are identified (for example, recycle being represented by a cycle or throughput being represented by a path) and several cycle finding and graph re-writing algorithms are implemented on the energy flow graph to arrive at equivalent (induced subgraph) representations of the original complex network in each time scale. The nodes in each of these subgraphs represent the units evolving in that time scale. Application of graph traversing algorithms (e.g. depth-first search, breadth-first search) provides scaled equations for the underlying dynamic equations as well as the corresponding reduced order models (if any) in each time scale. Furthermore, it also allows for the classification of the various control objectives to be pursued in each time scale and provides a pool of manipulated inputs to be considered for the same. This framework is generic and parallels the mathematically rigorous model reduction using singular perturbations.

The proposed framework presents an efficient tool for the analysis of complex networks, without performing a rigorous dynamic analysis. The analysis results can aid controller design (input-output pairing, controller hierarchies) for plant-wide control problems. The advantages of the proposed framework are illustrated via application to an integrated naphtha reforming example which utilizes a network of process-to-process heat exchangers to recover energy from the hot reformer exit stream and recycle it back to preheat the cold feed, resulting in multiple interconnected energy recycle loops.

[1] S.S. Jogwar and P. Daoutidis. Energy flow patterns and control implications for integrated distillation networks. *Ind. Eng. Chem. Res.*, 49:8048 – 8061, 2010.

See more of this Session: Control of Large Scale and Networked Systems
See more of this Group/Topical: Computing and Systems Technology Division

(214c) Robust Quasi-Decentralized Control of Networked Process Systems Under Communication Scheduling

Yulei Sun and Nael H. El-Farra, Department of Chemical Engineering and Materials Science, University of California, Davis, Davis, CA

Control of large-scale plants with distributed interconnected processing units is a fundamental problem in process control that has been the subject of significant research work in both the academic and industrial circles. Methodological frameworks that have been pursued to address this problem range from traditional centralized and decentralized control methods to distributed cooperative model predictive control formalisms to agent-based systems and supervisory control formulations. Recently, a quasi-decentralized networked control framework that offers a compromise between the complexity of centralized control designs and the performance limitations of fully decentralized controllers, has been developed in [1,2] for multi-unit plants whose subsystems communicate over a resource-constrained communication medium. In this framework, the transfer of information between the distributed control systems is kept to a minimum by allowing the local controllers to rely on the predictions of locally-embedded models of the other subsystems, and forcing communication to take place only at discrete time instances to provide the necessary correction to model prediction errors. This approach aims to provide a balance between the need to enforce the desired stability and performance properties on the one hand, and the need to conserve network resources, on the other. This is an important consideration, especially in view of the growing emphasis on augmenting process control systems with wireless sensor networks in order to expand their capabilities.

Beyond suspending and re-establishing communication, an important means of conserving network resources is to allow only a subset of the plant units to communicate with the rest of the plant at any given time according to a certain communication schedule. Forcing the different subsystems to transmit their data at different times (rather than simultaneously) creates an opportunity for providing a more targeted correction to the models' estimation errors, whereby the models with the largest uncertainties could receive more timely updates than would be feasible in the absence of scheduling. The successful implementation of this approach, however, requires that certain practical implementation issues, such as the presence of time-varying external disturbances, which arise frequently in process operations, be explicitly accounted for. For quasi-decentralized networked control systems, external disturbances, if not accounted for, may not only harm the stability and performance properties of the local control systems, but can also alter the optimality properties of the communication and scheduling policies, and ultimately interfere with the objective of balancing the inherent tradeoff between stability and network utilization.

The objective of this work is to develop a robust scheduled quasi-decentralized networked control framework for multi-unit plants that exchange information over a shared communication network and are subject to time-varying external disturbances. The framework integrates local model-based feedback control with sensor scheduling to robustly stabilize the plant while minimizing the rate at which each subsystem must collect and disseminate its data to the rest of the plant. The plant-wide control structure consists of a collection of robust feedback controllers that enforce an arbitrary degree of disturbance attenuation. Communication between the controllers is suspended periodically for time intervals during which each controller relies on models of the plant units to generate the necessary control action. Communication is then re-established at discrete time instances according to a certain schedule that determines the order and times at which the subsystems transmit the data needed to update the states of the models embedded in the target units. An explicit characterization of the interdependence between the minimum allowable communication rate, the scheduling strategy for sensor transmissions, the size of the disturbances, the accuracy of the models, the choice of control laws, and the achievable degree of disturbance attenuation, is obtained. Finally,, the results are illustrated using a reactor-separator plant example.

References:

[1] Sun, Y. and N. H. El-Farra, "Quasi-decentralized Model-Based Networked Control of Process Systems," *Computers and Chemical Engineering*, **32**, 2016-2029, 2008.

[2] Sun, Y. and N. H. El-Farra, "A Quasi-decentralized Approach for Networked State Estimation and Control of Process Systems," *Industrial and Engineering Chemistry Research*, **49**, 7957–7971, 2010.

See more of this Session: Control of Large Scale and Networked Systems
See more of this Group/Topical: Computing and Systems Technology Division

(214d) Model Predictive Control of Nonlinear Singularly Perturbed Systems: Application to a Large-Scale Process Network

Xianzhong Chen[1], Mohsen Heidarinejad[2], Jinfeng Liu[3], Muoz de la Pea[3] and Panagiotis D. Christofides[3], (1)Chemical and Biomolecular Engineering, Univ. of California, Los Angeles, Los Angeles, CA, (2)Electric Engineering, University of California, Los Angeles, Los Angeles, CA, (3) Department of Chemical and Biomolecular Engineering, University of California, Los Angeles, Los Angeles, CA

Chemical processes and plants are characterized by nonlinear behavior and strong coupling of physico-chemical phenomena occurring at disparate time-scales. Examples
include fluidized catalytic crackers, distillation columns, biochemical reactors as well as chemical process networks in which the individual processes evolve in a fast timescale
and the network dynamics evolve in a slow time-scale. Two-time-scale processes can be conveniently described using nonlinear singularly perturbed systems. Model predictive control (MPC) is a practically-important control framework which can be used to design and coordinate control systems and can explicitly handle input and state constraints. MPC utilizes a model to predict the future evolution of the plant at each sampling time according to the current state over a given prediction horizon. In the context of MPC of
singularly perturbed systems, most of the efforts have been dedicated to linear systems or to MPC of specific classes of two-time-scale processes, and general stability results are not available.

This work focuses on model predictive control of nonlinear singularly perturbed systems. A composite control system using multirate sampling (i.e., fast sampling of the fast
state variables and slow sampling of the slow state variables) and consisting of a "fast" feedback controller that stabilizes the fast dynamics and a model predictive controller that stabilizes the slow dynamics and enforces desired performance objectives in the slow subsystem is designed. Using stability results for nonlinear singularly perturbed systems, the closed-loop system is analyzed and sufficient conditions for stability are derived. A large-scale nonlinear reactor-separator process network which exhibits two-time-scale behavior is used to demonstrate the controller design including a distributed implementation of the predictive controller.

See more of this Session: Control of Large Scale and Networked Systems
See more of this Group/Topical: Computing and Systems Technology Division

(214e) Globally Optimal Control of Nonlinear Process Networks: Easier Than Globally Optimal Control of Nonlinear Processes

Zayna Alhusseini, chemcial and biomolecular Engineering, UCLA, Los Angeles, CA and Vasilios Manousiouthakis, Chemical Engineering Department, University of California at Los Angeles, Los Angeles, CA

Optimal control methods for chemical processes abound. The nonlinear behavior of chemical process however limits the success of these methods. Calculus of variations formulations give rise to boundary value problems with possibly multiple solutions. Control parametrization formulations give rise to nonlinear programs which can only be locally solved. In this work, we demonstrate that for large classes of chemical processes the solution of the optimal control problem over both control strategies and process networks is easier (in that it can be solved globally) than the control problem over the control strategy of a single process. We illustrate this surprising discovery by considering an optimal control problem for a particular process network.

See more of this Session: Control of Large Scale and Networked Systems
See more of this Group/Topical: Computing and Systems Technology Division

(214f) A Safe-Parking Approach to Handle Partial Plant Shutdown

Siam Aumi, Miao Du and Prashant Mhaskar, Chemical Engineering, McMaster University, Hamilton, ON, Canada

A typical chemical plant consists of transporting material through numerous connected processing units (and buffer tanks) in order to produce the final product with a desired quality. The primary control objective in the plant is to maintain the production rate and quality at their given set-points in the presence of disturbances; however, this control objective is jeopardized when an intermediate process unit is shutdown (i.e., no flows into or out of the unit) for a finite duration of time due to equipment failure or to perform necessary maintenance operations. This scenario is classified as a partial shutdown of the plant and typically does not allow continuation of operation at the nominal operating point. As a result, the plant is either completely shutdown, which has significant economic ramifications, or ad-hoc modifications to the pre-shutdown control policy are made. Adopting these ad-hoc approaches, however, can also potentially result in temporarily shutting down the entire plant due to the onset of hazardous conditions. In this work, we propose a systematic approach to reconfigure the plant control strategy during shutdown in order to achieve optimal operation during the shutdown and a smooth transition to the nominal operation once the shutdown unit is recovered.

The key to successful plant operation during a partial shutdown is to effectively utilize the capacities of the processing units and carefully placed buffer tanks. In general, buffer tanks in a chemical plant play an important role by attenuating or dampening the effects of disturbances along the production line. During a partial shutdown, buffer tanks become particularly important since their available capacities can be exploited to maintain plant operation to some extent. To understand this, suppose there is a buffer tank preceding and following the shutdown unit. During the shutdown period, the preceding buffer tank can serve as a holdup tank, allowing for continued operation of the upstream units while the buffer tank following the shutdown unit can serve as a feed tank for the downstream units. In operating the upstream and downstream plants, however, the finite capacities of the buffer tanks and the processing units together with the knowledge of the shutdown period length must be accounted for in any control strategy implemented during the shutdown. In addition, since the primary control objective remains to operate the plant at its nominal operating point, the shutdown controller must also steer the upstream and downstream plants in such a way that they can be efficiently recovered back to their nominal operating points once the shutdown unit is reinserted into the network. In [1], this problem was addressed using a model predictive control (MPC) design where the control objective during the partial shutdown was to maximize the plant profit while satisfying a restoration constraint that (implicitly) required all the process states at the end of the shutdown period to be in a region from where the nominal operation can be resumed. However, the feasibility of this optimization problem was assumed but not guaranteed.

In this work, by recognizing that the plant is essentially decomposed into two decoupled batch subsystems (one comprised of the upstream units and the other comprised of the downstream units) during the shutdown, we recast the partial shutdown problem as two separate batch control problems wherein the control objective in each batch subsystem is to reach a desired end-point neighborhood from where the nominal operation of the plant can be resumed at the end of shutdown. With this interpretation of the partial shutdown problem, we unite the safe-parking and safe-steering frameworks proposed in [2] and [3] to come up with a predictive controller for each subsystem (independently) with guarantees on recovering the process to its nominal operating point. The controller design for each subsystem begins with explicitly characterizing the stability region of the nominal operating point in each unit of the subsystem (using the tools as in [2]) and then using a subset of these regions to define the desired end-point neighborhood. Next, we use the concept of reverse-time reachability regions (RTRRs) (defined as the set of states from where a batch process can be driven to the desired end-point neighborhood by batch termination subject to input constraints) to formulate a computationally efficient predictive controller that maintains the subsystem states within its corresponding RTRRs for the duration of the shutdown period. By doing so, the desired end-point neighborhood based on which the RTRRs are generated is guaranteed to be reachable. The implication of this is that we can transition back to the nominal operating point at the end of the shutdown period since the desired end-point neighborhood was defined using the stability regions of the individual units.

To demonstrate the efficacy of the proposed approach, we consider a multi-unit nonlinear process with carefully placed buffer tanks subject to input constraints and a partial shutdown in one of the intermediate units. The partial shutdown scenario considered precludes the possibility of maintaining operation at the nominal operating point during the entire shutdown period, and calls for a reconfiguration of the plant controllers. By implementing RTRR-based predictive controllers for the two resulting batch subsystems, the downstream and upstream batch plants were operated such that the nominal plant operation was resumed efficiently once the shutdown unit was recovered.

[1] Chong Z, Swartz CLE. Model-based control of multi-unit systems under partial shutdown conditions. In: Proceedings of the 2009 American Control Conference. 2009; pp. 160–165.

[2] Gandhi R, Mhaskar P. A Safe-Parking Framework for Plant-Wide Fault-Tolerant Control. Chem Eng Sci. 2009;64:3060–3071.

[3] Aumi S, Mhaskar P. Safe-steering of Batch Processes. AIChE J. 2009;55:2861–2872.

See more of this Session: Control of Large Scale and Networked Systems
See more of this Group/Topical: Computing and Systems Technology Division

(214g) Distributed Model Predictive Control for UK CO2 Capture, Transport and Storage Network

Alicia Arce, MATGAS Research Center, Barcelona, Spain, Niall Mac Dowell, Chemical Engineering, Imperial College London, London, United Kingdom, Alejandro J. del Real, Research and Development, IDENER, Seville, Spain, Nilay Shah, Centre for Process Systems Engineering, Department of Chemical Engineering, Imperial College London, London, United Kingdom and Lourdes F. Vega, MATGAS Research Center and Carburos Metlicos, Air Products Group, Barcelona, Spain

The large scale deployment and operation of CO_2 capture, transport and storage (CCS) infrastructure will be extremely capital intensive. Although it is possible, using spatially and temporally explicit methods, to design a CO_2 capture and transport network which will minimise the lifetime costs associated with the network , it is important to realise that the CO_2 sources will themselves operate in a dynamic fashion. Indeed, the largest fixed point source emitters of CO_2, i.e., power stations, operate in a very dynamic fashion. Thus, efficient operation of the CCS network is key to minimising the costs associated with the operation of the network. For this, a state-of-the-art control technology is required.

The CCS network on which our current work is based on was optimally designed for deployment UK which and is comprised of 33 sources and 6 offshore sinks with three onshore terminals. All sources, sinks and terminals are connected via pipeline. In this contribution, we develop a dynamic distributed model predictive control which concurrently maximises the CO_2 storage capacity and minimizes the operation cost whilst meeting infrastructure physical constraints such as availability, storage and pipeline capacity.

Owing to the high computational burden associated to such a dynamic optimization problem, the control framework utilized herein is based on the distribution of the overall network control effort among local agents presented in previous work. Such an approach allows very large network simulation and management to be performed through parallel computation with little hardware requirements. As it is based on Lagrange-based distributed predictive control, the proposed algorithm uses a dynamic model of the CCS network taking explicitly into account system constraints while optimizing a multiobjective control function of the overall network, dividing this smart grid into subnetworks and assigning them to local control agents. Consequently, the computed set of local control moves is equivalent to those corresponding to the optimal solution of the overall network optimization problem. The dynamic model utilized is based on the so-called network hubs, which are defined as interfaces between input and output flows and the transportation infrastructure.This concept offers a number of very convenient features that makes it a very flexible and standard formulation in order to model and analyze general complex networks including multiple flows and storage.

Mac Dowell, N., Murthy, N. V. S. N., Alhajaj, A. and Shah, N. Multiscale whole-systems design and analysis of CO2 capture and transport networks, ESCAPE21, 2011 (Accepted)

"*Development of an integrated CO_2 Capture, Transportation and Storage Infrastructure for the UK and North Sea using an Optimisation Framework*". Paolo Prada. M.Sc. Thesis. Imperial College London. September, 2010.

"*An Integrated Framework for Distributed Model Predictive Control of Large Scale Networks: Applications to Power Networks*". Alejandro J. del Real. Ph.D. Thesis. University of Seville. May, 2011.

Optimal Power Dispatch of Energy Networks Including External Power Exchanges". Alejandro del Real Torres, M.d Galus, Carlos Bordons Alba, G. Andersson. Proceedings of the European Control Conference 2009. Pag. 3616-3621. Budapest, Hungría. 2009.

See more of this Session: Control of Large Scale and Networked Systems
See more of this Group/Topical: Computing and Systems Technology Division

(240a) Reducing Specific Energy Consumption In Reverse Osmosis (RO) Water Desalination: An Analysis From First Principles

Mingheng Li, Department of Chemical and Materials Engineering, California State Polytechnic University, Pomona, CA

The previously derived characteristic equation of RO in Li (Ind. Eng. Chem. Res., 49, 1822-1831, 2010) is used to describe single- or multi-stage ROs with/without an energy recovery device (ERD). Analysis is made at both the theoretical limit (with analytical solutions provided if possible) and practical conditions (using constrained nonlinear optimization). It is shown that reducing specific energy consumption (SEC) normalized by feed osmotic pressure, or NSEC in ROs can be pursued using one or more of the following three independent methods: (1) increasing a dimensionless group $\gamma = A_{total}L_p\Delta\pi_0/Q_f$, (2) increasing number of stages, and (3) using an ERD. When γ increases, the feed rate is adversely affected and the NSEC reduces but flattens out eventually. Using more stages not only reduces NSEC but also improves water recovery. However, The NSEC flattens out when the number of stages increases and ROs with more than five stages are not recommended. Close to the thermodynamic limit where γ is sufficiently large, the NSEC of ROs up to five stages approaches 4, 3.60, 3.45, 3.38 and 3.33 respectively. The ERD can significantly reduce the NSEC, theoretically to 1, while the corresponding recovery approaches zero. The NSEC becomes larger when the required water recovery increases. It is found that a combination of all three methods can significantly reduce the NSEC while maintaining a high recovery and a reasonable feed or permeate rate. An NSEC around 2.5–2.8 with an 80% water recovery may be possible using 3–5 RO stages and an ERD of 90% efficiency operated at a γ about 3–5 (or $Q_f = 0.2–0.3\ A_{total}L_p\Delta\pi_0$).

See more of this Session: Modeling and Computation In Energy and the Environment
See more of this Group/Topical: Computing and Systems Technology Division

(240b) Study On Energy Consumption and Emission Generation for A Chemical Plant Under Different Start-up Strategies

Jie Fu, Chuanyu Zhao, Xiongtao Yang, Chaowei Liu, Qiang Xu and Kuyen Li, Dan F. Smith
Department of Chemical Engineering, Lamar University, Beaumont, TX

Chemical plant start-ups are very critical dynamic operations, which consume huge amount of energy for heating, cooling, compression and other usages; meanwhile, they also generates large quantities of off-spec products for flaring, causing significant and intensive air emissions. Many studies have been conducted on the energy consumption minimization or energy efficiency improvement for normal chemical process production. However, quantitative studies of the energy consumption and emission generation analysis on chemical plant start-ups are still lacking. In this study, plant-wide dynamic simulations are employed to study energy consumption and emission generation for a chemical plant under different start-up strategies, so that the quantitative technological support can be provided to identify the most desirable start-up operation.

In order to examine impact from different start-up strategies for the sake of flare minimization, two start-up designs with different recycling stream patterns and two start-up operating strategies with different time arrangement are performed. Besides the plant-wide model, rigorous dynamic simulations are also conducted for critical units to examine their operating safety. Based on the multiple dynamic simulations, the dynamic profiles of energy and emission during the plant start-up can be obtained and summarized. Details of energy consumption on cooling and heating duties of distillation towers, auxiliary heat exchanger duties, and power consumption for compressor system are provided for each start-up case. Simultaneously, the identifications of emission sources, including the speciation and the percentage information, as well as the dynamic emission profiles during an ethylene plant start-up are obtained. Through comprehensive analysis and comparison, the most desirable start-up solution is identified. This study not only characterizes the emission generation during the plant start-up, supporting flare minimization activities that benefits environmental sustainability, but also enhances critical research on energy and raw material consumptions during plant-start-up that finally benefits the industrial sustainability.

See more of this Session: Modeling and Computation In Energy and the Environment
See more of this Group/Topical: Computing and Systems Technology Division

(240c) Dynamic Optimal Well Placement In Oil Reservoirs

Mohammad Sadegh Tavallali, Department of Chemical & Biomolecular Engineering, National University of Singapore, Singapore, Singapore, Kwong Meng Teo, Department of Industrial & Systems Engineering, National University of Singapore, Singapore, Singapore and Iftekar A. Karimi, National University of Singapore, Singapore, Singapore

DynamicOptimal Well Placement in Oil Reservoirs

M. S. Tavallali[1], K. M. Teo[2], I. A.Karimi[1*]

[1]Department of Chemical &Biomolecular Engineering, 4 Engineering Drive 4

[2] Department of Industrial &Systems Engineering, 1 Engineering Drive 2
National University of Singapore, Singapore 117576

Abstract

OPEC predicts the oil industry to increase itsproduction by almost 23% of the current level to be able to fulfill the worldoil demand in the next 20 years. However, the hydrocarbon resources are limited,and hence it is highly crucial to exploit the new, marginal and existing maturefields in an optimal way. In this situation, optimal well placement can play animportant role with long-term financial impact. However, determining optimalwell locations is inherently a complex nonlinear and combinatorial problem dueto the spatial-temporal continuity equations and their non-differentiability withrespect to well location, nonlinearities of the multiphase flow, existence ofdifferent uncertainties and a myriad of related decisions such as the number ofpotential well positions, their types, functionalities (producer / injector),trajectories, inclinations, drilling schedules, etc.

In this work, we address thischallenging problem using a combination of mathematical programming and local searchfor locating optimum drilling sites. While mathematical programming is a potentialtool for analyzing such problems, very few contributions have used this approachfor well placement (Rosenwald and Green 1974; Ierapetritou,Floudas et al. 1999; Cullick,Vasantharajan et al. 2004). Even these efforts have usedonly static features of the system. Therefore, to use mathematical programmingand bridge the gap between the static and dynamic information, one must addresstwo key issues. First, one must capture the dynamics of the multiphase flow andtemporal behavior over the production horizon. Second, one needs a specializedalgorithm that exploits the model structure to solve such a large and complexnonlinear, nonconvex, discrete optimization problem. These goals motivated the presentstudy.

For model development, we assume thatall deterministic geological and PVT information are supplied and try tomaximize the profit of oil production over a planning horizon. The modelinvolves three sets of constraints (physical, logical and operational) relatedto the structure and dynamics of a reservoir over time. These constraintsresult in a multiperiod, dynamic, mixed integer nonlinear (MINLP) model withdiscretized partial differential equations. For model solution, we havedeveloped an outer approximation algorithm based on the work of (Viswanathan and Grossmann 1990), which exploits themodel structure. While the master problem identifies new promising well locations,the primal problem assesses their quality. Since the primal problem is alsolarge, PDE-constrained, and highly nonlinear, we further decompose it into aseries of smaller subproblems over the planning horizon to improve solutionspeed.

This algorithm while effective inidentifying regions of optimal well locations, terminated prematurely withoutspotting actual locations rigorously. Therefore, we made two furthermodifications. For the model, we considered convective flows amongstneighboring cells and the concept of upstream flow to relax some of thestructural constraints. For the algorithm, we added a local neighborhood searchafter the termination. We probe the neighborhood for possible improvement andthen continue the search using the KKT points of a neighbor with higherobjective value (if any). We also simplified our linearization scheme toimprove solution time and avoid eliminating potential solutions.

We use two examples to successfully demonstratethe impact of the above modifications. The examples involve up to threeproducer wells to be placed in an oil reservoir with two existing producer andinjector wells. We consider a planning horizon of 1470 days. The resultsconfirm that the decomposition of the primal problem and simplification inlinearization are essential to solve this large scale PDE constraint MINLPmodel. However, since it is a nonlinear nonconvex problem, guaranteeing aglobal optimal solution is difficult. However, our hybrid algorithm certainlyimproves the solution and likelihood of reaching a global optimum. For example,for the case of mono well placement, we compared our results with completeenumeration, using a commercial reservoir simulator (ECLISPE). Our algorithm reachedthe global optimum successfully.

Acknowledgement: We would like to thank NationalUniversity of Singapore and SINGA program for the financial support of thisresearch, and also Schlumberger Company for granting the academic license of ECLIPSE.We are also thankful to Mr. David Baxendale from RPS Energy Limited for hisvaluable industrial insights.

References

Cullick, A. S., S. Vasantharajan, et al. (2004). DeterminingOptimal Well Locations From A 3D Reservoir Model. EUROPEAN PATENTAPPLICATION.

Ierapetritou, M. G., C. A. Floudas, etal. (1999). "Optimal location of vertical wells: Decompositionapproach." AIChE Journal **45**(4): 844-859.

Rosenwald, G. W. and D. W. Green (1974)."Method For Determining The Optimal Location Of Wells In A Reservoir UsingMixed-Integer Programming." Soc Pet Eng AIME J **14**(1): 44-54.

Viswanathan, J. and I. E.Grossmann (1990). "A combined penalty function and outer-approximationmethod for MINLP optimization." Computers and Chemical Engineering **14**(7):769-782.

*Corresponding author: Tel.: +65 6516-6359, Fax: +65 6779-1936, Email cheiak@nus.edu.sg

See more of this Session: Modeling and Computation In Energy and the Environment
See more of this Group/Topical: Computing and Systems Technology Division

(240d) Optimal Synthesis for the Feed-Water-Heater Network of a Pulverized Coal (PC) Power to Minimize Water Consumption

Juan M. Salazar and Urmila Diwekar, Vishwamitra Research Institute, Center for Uncertain Systems: Tools for Optimization and Management, Clarendon Hills, IL

Coal-fired power plants contribute to almost 50% of the United States' total electric power production. At the same time, pulverized coal (PC) power plants are large water consumers since residual heat from the steam cycle is removed by evaporating water at the cooling tower. Therefore, reduction of such residual heat implies reduction of water consumption. A potential approach for the reduction of the residual heat in the steam cycle of the PC process is formulating an optimal design of the feed water heat exchange network (HEN). Conventional methodology for such HEN synthesis is based on equal enthalpy change on the feed water stream for each of the feed water heaters. Turbine discharge pressure and temperature as well as the mass flow rates of the steam bleeding streams are calculated based on such heat load distribution. In this work, we employ a rigorous-simulation pinch-technology refined approach for HEN synthesis. The methodology uses Aspen Energy Analyzer (AEA) to determine the mass flowrates of the bleeding streams while generating alternative designs that can potentially reduce the water consumption by reducing the total cooling requirement. The resulting alternative designs are simplified models of HENs and does not reflect realistic water consumption. Therefore, we generated a superstructure that merges all designs in Aspen Plus to obtain the HENs rigorous simulation. Finally, the optimization approach to process synthesis is applied resulting in optimal configuration and designs.

See more of this Session: Modeling and Computation In Energy and the Environment
See more of this Group/Topical: Computing and Systems Technology Division

(240e) Dynamic Modeling for Ionic Liquid-Based Absorption Refrigeration Systems

D. Andrei Ma˘ ces• and Mark A. Stadtherr, Department of Chemical and Biomolecular Engineering, University of Notre Dame, Notre Dame, IN

Residual heat is a common waste product of many industrial processes, and often low quality enthalpyis available from natural sources (e.g., geothermal springs, solar collectors, etc). Absorptionrefrigeration cycles are able to utilize such energy to produce useful cooling, with littleelectricity required (Srikhirin et al., 2001). In many developing areas, compression-based refrigerationmay not be feasible due to weaknesses in the electrical grid, and sources of cheap residual heatmay be present, both arguments for absorption refrigeration.

Certain ionic liquids (ILs) have inherent advantages as absorbents, such as negligible vapor pressureand ease of refrigerant separation, chemical inertness, and good transport properties (Shiflett andYokozeki, 2006). In this presentation, we develop a rigorous dynamic model to study selected IL-waterabsorption refrigeration systems and the global sensitivity of cycle parameters to aid in design andoptimization.

Traditionally, absorption refrigeration systems are rated through their coefficient of performance (COP), the ratio between evaporator cooling duty and the energy spent (generator heat duty and pump work), usually at steady state. However, when designing or optimizing absorption refrigeration systems, thetransient behavior of the cycle is also important. Nevertheless, most literature is focused on steady-statemodeling, and rigorous dynamic models for IL-based systems are lacking.

The challenges of creating rigorous dynamic models include the need for good qualitythermodynamic and transport data (Kelkar and Maginn, 2007; Rodríguez and Brennecke, 2006), themultitude of states to be tracked and numerical stiffness. Using the basic framework of Cai et al.(2007), we set out to develop a comprehensive lumped-compartment model as a system of ordinary differentialequations (ODEs) solved as an initial value problem (IVP). This structure readily lends itself to studyingglobal sensitivity, using stochastic or deterministic tools. The states tracked are compartment massesand compositions, flow rates, and temperatures, while other transient parameters (e.g., enthalpy flows,performance, etc) are derived from these ODE states.

Systems of interest are based on 1-ethyl-3-methylimidazolium ([emim]) salts, such as trifluoroacetate ([TFA]), trifluoromethanesulfonate (or triflate) ([OTf]), and ethylsulfate ([EtSO_4]). Weassume a geothermal heat source, generating 1 ton of refrigeration (12,000 BTU/h), with a space to becooled at 10 °C. Example results for these systems include dynamic behavior of the equipment,performance, and cooling duty, as well as their sensitivity with respect to thermodynamic and transportproperties. Advantages of selected approach include the use of established numerical modeling tools,such as

MATLAB, and easy coupling with stochastic or deterministic sensitivity tools. Control algorithmscan be also readily studied.

References

Cai, W., Sen, M. & Paolucci, S. (2007). Dynamic modeling of an absorption refrigeration system usingionic liquids. *2007 ASME International Mechanical Engineering Congress and Exposition*, Proceedings of IMECE2007.

Srikhirin, P., Aphornratana, S. & Chungpaibulpatana, S. (2001). A review of absorption refrigerationtechnologies. *Renewable & Sustainable Energy Reviews*, *5*, 343–372.

Kelkar, M. S. & Maginn, E. J. (2007). Effect of temperature and water content on the shear viscosityof the ionic liquid 1-ethyl-3-methylimidazolium bis(trifluoromethanesulfonyl)imide as studied byatomistic simulations. *Journal of Physical Chemistry B*, *111*, 4867–4876.

Rodrı́ guez, H. & Brennecke, J. F. (2006). Temperature and composition dependence of the density andviscosity of binary mixtures of water + ionic liquid. *Journal of Chemical Engineering Data*,*51*, 2145–2155.

Shiflett, M. B. & Yokozeki, A. (2006). Absorption cycle utilizing ionic liquid as working fluid. *U.S. Pat.2006/0197053 A1*.

See more of this Session: Modeling and Computation In Energy and the Environment
See more of this Group/Topical: Computing and Systems Technology Division

NUMERICAL SIMULATION OF FLOW AND COKING CHARACTERISTIC IN AN FCC DISENGAGER

Juan Wang*, Yu Mao, Jiangyun Wang
State Key Laboratory of Heavy Oil Processing, China University of Petroleum
Beijing, China

Introduction

In recent years, it becomes more serious and common that the coking problem of reactor-regenerator system in the fluid catalytic cracking process (FCC), because the deterioration and higher density of the crude oil processed becomes more and more seriously in the FCC.[1] From the process viewpoint, pressure drop after coking from the reactor to the fractionator will be higher. Thus the pressure has to raise in the regenerator to keep pressure balance and normal operation, which would increase the reaction depth, and increase the difficulty of control the product quality and component distribution. From the fluid mechanics viewpoint, the space of oil gas flow will be reduced because of the presence of coke block after the coking grows to a certain degree. The coke can even block the flowing passage, such as the narrow area of the dipleg of cyclone separator and other passages. If this happens, steps have to be taken for coke cleaning. From the equipment viewpoint, the coke block will be crushed or dropped out under the role of scour of oil gas. The smaller coke block may flow in the disengager following the oil gas and may abrase the equipment. The big coke block may be jammed to the dipleg of cyclone separator or enter the baffle plate of stripper section to disturb its normal operation of relevant equipment. In a word, the coking has threatened the operation of process equipment under the normal and long periodic situation, even lead to shut down the equipment unplannedly and also makes more difficulties of equipment-repairing. All of this may result in considerable economic loss for refineries. Also it is an urgent problem for refineries to solve. Currently, each refinery is studying on new anti-coking technology aimed at process viewpoint mostly, such as increasing the stripping steam, injecting anti-coking steam, enlarging catalyst-oil ratio, keeping stable feeding property, optimizing feeding atomization effect, and controlling reaction depth etc.[2-4] Also some refineries introduce some new patent technique into the process, such as UOP's VSS, CUP's VQS and so on to solve coking problem.[5,6] But, from the fluid flowing characteristics viewpoint, there is less report about coking reasons to analyze the detail of oil gas and catalyst flowing condition in the full disengagement space completely.

Generally, FCC disengager contains stripping section, reacting section, disengagement space, separator system. Its structure is very complicated and its size is also large. In the actual process operation course, only some parameters in certain part can be measured and regulated to ensure the normal process. The residence time, the distribution of oil gas and catalyst inside the disengager cannot be measured with simple method. In this paper, full-scale model in all flowing area in the disengager was established, and RNG k-ε and Eulerian Model was adopted in the computational process. All details of fluid flow in all disengagement spaces were obtained. The reason of coking from the fluid flowing characteristics viewpoint was analyzed. Theoretical guidance to optimize the operation and design of disengager was made clearly.

The main regions and basic reasons for coking in FCC disengager

There are coking problems in many FCC disengagers in China more or less. This has become one of the main factors which affect catalytic cracking unit to operate for long period. The main regions of coking in the disengager are as follows.

(1) Inner wall in disengager shell.
(2) External wall of dipleg in primary and secondary cyclone separator.
(3) External wall of uprising pipe and supporting beam in cyclone separator.
(4) The top of gas collection chamber.

Coking in those regions of FCC disengager is closely relative to the flow characteristics. Coke may form gradually in these regions with low velocity, reversed flow or vortex flow. Also the long average residence time of oil gas is another main reason for coking. Therefore, it is necessary to understand flowing condition, distribution of oil gas and catalyst, residence time of oil gas to optimize disengager design and to reduce the possibility of coking.

Geometric model of disengager

Figure1 shows the structure of disengager in catalytic cracking unit using in a refinery with an annual production capacity of 1.4 Mt/a. The diameter of shell is 7 m, and the total height is 21 m. The separation system of the disengager consists of two primary cyclone separators and four secondary cyclone separators. The type of linkage between primary and secondary cyclone separators is open type. There are two rectangular passages on the left and right of discharge pipe of the primary cyclone separators, which direct to two inlets of secondary cyclone separators respectively. In this paper, the full size three dimensional mathematical models without any simplification was build up and calculating grid was generated according to actual structure of disengager in service.

The main process of fluid flow in the disengager can be described as follow. The mixture of oil gas and catalyst after reacting in the riser goes through horizontal pipe(1) into primary cyclone(2), the oil gas separated primarily out of discharge pipe of primary cyclone separators gets into secondary cyclone separators(3) to be separated again, and then the oil gas is discharged from secondary cyclone separators(4) in the end; the catalyst separated two times gets into dense phase layer of the disengager along with the dipleg of cyclone separators, at the end, the catalyst(6) gets out of the lower export after the catalyst stripper(5) removes most of these entrained hydrocarbons by displacing them with steam.

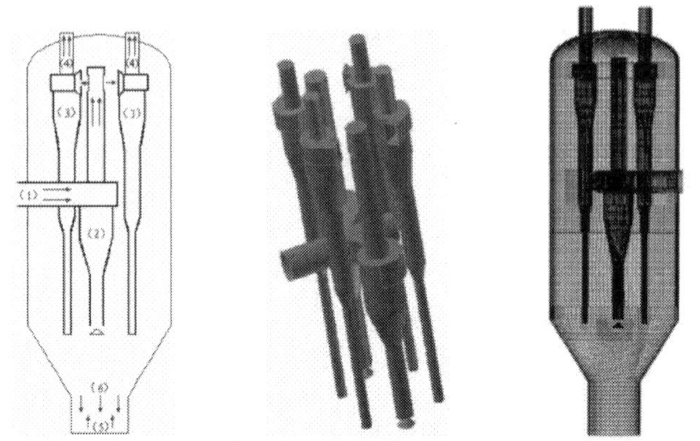

(a) Flow chart (b) Structure sketch (c) grid generation

Figure1 FCC disengager

Computational models and boundary conditions

The gas solid flow in the FCC disengager is very complicated. The concentration of catalyst is lower in the primary and secondary cyclone separator, top of disengager, while that of catalyst is very higher in the stropping section. The oil gas with catalyst strong rotational flow in the separator system and around linkage of every separator stage, but its flow is very slow in the middle part of disengagement space.

It was on the assumption that the flow in the disengager was incompressible flow with constant temperature since there was heat preservation layer out of the disengager. RNG k-ε Model was chosen to simulate the turbulent flow because of its priority to the rotational condition of the average flow process.[7] Eulerian Model was chosen to simulate the gas solid two phase flow. Discrete phase model was chosen to simulate the flow track of oil gas and catalyst.[8]

In the process of calculation, boundary conditions were exerted according to practical operation condition. The horizontal inlet of disengager and the inlet at the bottom of stripping section were defined as the boundary of velocity inlet which the quantity of oil gas and stripping steam were the same as the practical production. The outlet of discharge pipe in secondary cyclone was exerted pressure outlet boundary condition of full development. The other outlets and inlets in primary cyclone and secondary cyclone were considered as the interior flow boundaries under the pressure balance function inside the disengager. Wall boundary conditions were exerted on disengager shell and cyclone separator wall and so on.

Result and analysis of simulation

The analysis of path lines of oil gas

Figure 2 shows the path lines of oil gas, which enters into primary cyclone separator from horizontal pipe inlet of disengager, then comes to secondary cyclone from discharge pipe of primary cyclone separators.

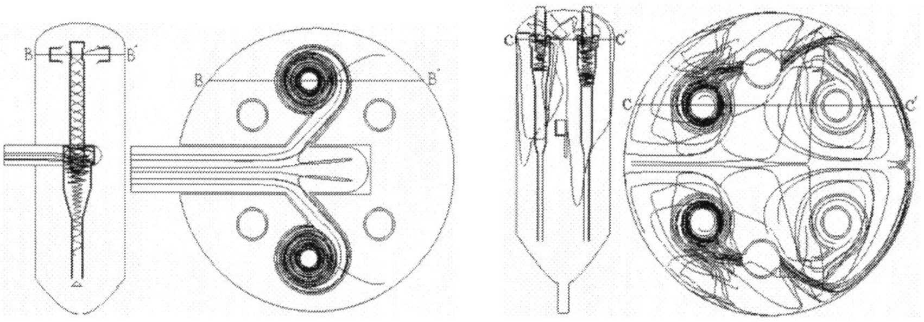

(a)Primary cyclone (b)Secondary cyclone

Figure 2 Path lines in primary and secondary separator

After the mixture of oil gas and catalyst entering the two primary cyclone separators distributed symmetrically, the catalyst and a little of oil gas flow downward along with inner wall of primary cyclone separator in the way of spiral under the centrifugal force and outer vortex, the mixture enters into the dense phase layer of disengager from discharge outlet of primary cyclone separator. A great deal of oil gas enters central negative pressure area under the effect of inner vortex and flows upward along with inner wall of exhaust pipe spirally. Though the geometric center of rectangular outlet of primary cyclone separator's discharge pipe is opposite to the inlet of secondary cyclone separator's straightly, the center of liquid flowing at the outlet can't point to the geometric center of inlet on the secondary cyclone separator, and there are axial and tangential velocity when the oil gas exhausts from discharge pipe. These factors induce that oil gas discharged from rectangular outlet at the outlet of primary cyclone separator can't directly enter secondary cyclone separator. Thus quite a lot of oil gas enters into the disengagement space. The flow velocity is very small, which is beneficial to catalyst particles carried by oil gas to disengage down the dense phase layer, but it leads to long average residence time of part oil gas under high temperature.Oil gas discharged from the outlet of primary cyclone separator enter into straightly into disengagement space, which can cause molecule of oil gas to reaction unduly, especially, heavy oil with great density and high viscosity. All this may be the main reason for coking in the inner wall of disengager, the shell of primary and secondary cyclone separator. After the second separation, catalyst particles and a little of oil gas enter the dense phase layer along with the dipleg of primary cyclone separators, oil gas purified leaves the disengager for the next process.

Velocity distribution of the fluid flowing

Figure 3 shows the tangential and axial velocity distribution separator along the central axial straight line through the central inlet line of horizontal section in the primary and secondary cyclone. Both the tangential velocity and axial velocity are very low near the surface of volute wall in the central straight line B-B' of primary cyclone separator, and the value of velocity is less than 1 m/s. The tangential and axial velocity of oil gas are very large in the inner of primary cyclone separator where the maximum tangential velocity may be 12 m/s and the maximum axial velocity may be 30 m/s. The velocity distribution is not symmetric, mainly because the shape of volute inlet is not symmetric completely. In one side of primary cyclone separator, the flowing momentum is larger when the liquid entering the passage of primary cyclone separator. The tangential velocity appears a wave trough and the axial velocity appearing a wave crest in the place X=0.3~0.8 m at the outer part of volute inlet in the structure of axis symmetry. In the central straight line C-C' of secondary cyclone separator, the distribution of the tangential and axial velocity is almost the same inner of the two secondary cyclone separator. The quantity of liquid entering into secondary cyclone separator close to the discharge pipe of the primary cyclone separator is bigger and the liquid momentum is larger, so the peak value of tangential velocity is slightly higher. Actually, the distance between primary cyclone separator's discharge outlet and every secondary cyclone separator's inlet is not equal. The oil gas velocity is very small (<1 m/s) in the disengagement space. The flowing velocity is bigger relatively in the oil gas inlets of horizontal pipe, primary cyclone and secondary cyclone separators.

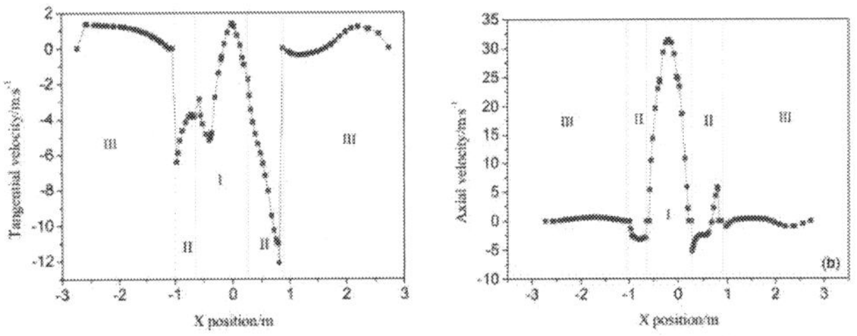

Figure 3 Velocity distribution on the line B-B'

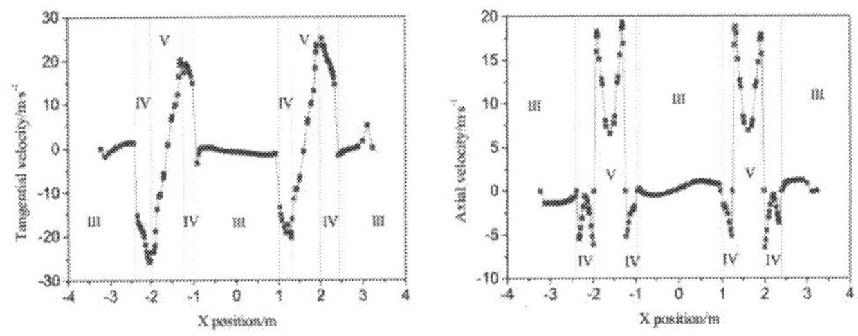

Figure 4 Velocity distribution on the line C-C'

I-exit tube of primary cyclone separator; II-primary cyclone separator; III-disengagement space; IV-secondary cyclone separator; V-exit tube of primary cyclone separator.

The distribution of oil gas and catalyst

Figure 5 and Figure 6 show the distribution of two phases in primary and secondary cyclone separator. The material level of oil gas and catalyst in the dipleg of the primary cyclone separator is lower than dense phase layer level in the disengager, because the operation pressure is positive in the primary cyclone separator; it is opposite in the secondary cyclone separator, the negative pressure operation makes material level higher than the dense phase layer level. a little of oil gas discharged from rectangular outlet of primary cyclone separator enters into the disengagement space but secondary cyclone separator straightly, and distributes everywhere equably. Also some catalyst carried by this part of oil gas entering disengagement space and adheres to shell inner wall in disengager, the outer wall of the primary and secondary cyclone separator's discharge pipe. All the above are the main coking reasons for those parts.

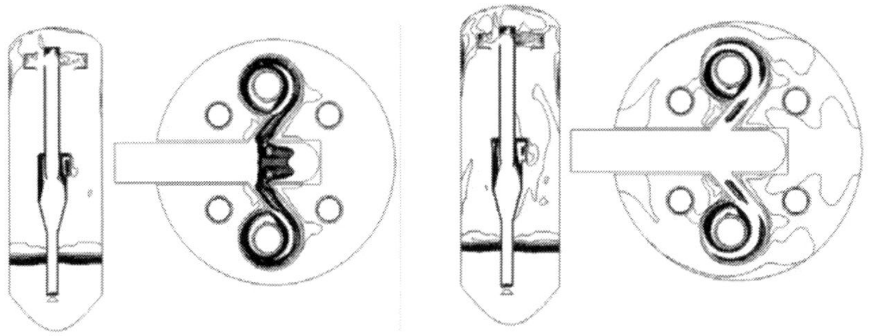

(a) Distribution of oil gas (b) Distribution of catalyst

Figure5 Distribution of two phases in primary cyclone

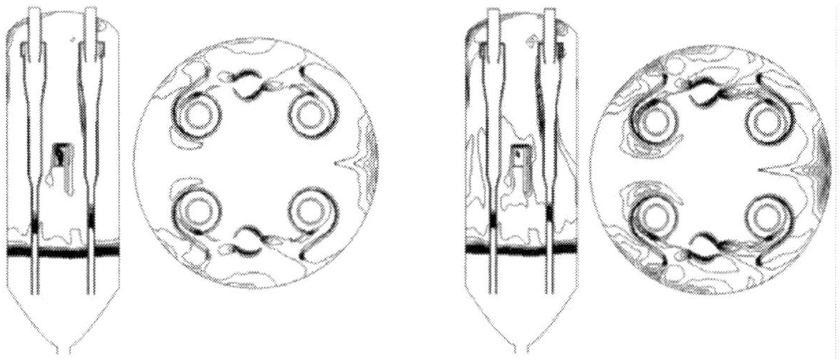

(a) Distribution of oil gas (b)Distribution of catalyst

Figure 6 Distribution of two phases in secondary cyclone

Conclusions

The analysis of flowing and coking factor in the disengager can come to the conclusions as follows.

(1) There is tangential and axial velocity upward when the oil gas discharges from the outlet pipe of the primary cyclone separator. The reason that oil gas enters the disengagement space directly but secondary cyclone separators is that flowing center of fluid can't correspond to the geometric center of the secondary cyclone separator's inlet entrance.

(2) The oil gas velocity is very small in the disengagement space, which is beneficial to catalyst particles in oil gas to disengage down the dense phase layer. But it leads to long residence time of oil gas under high temperature which is discharged from the outlet of the secondary cyclone separator going straightly into disengagement space. Therefore, it causes molecule of oil gas to reation unduly, especially, heavy oil with great density and high viscosity. This is the main reason for coking in the inner wall of disengager, the shell of cyclone separators.

(3) Part of catalyst enters into disengagement space straightly followed by oil gas discharged from the outlet of the primary cyclone separator and may adhere to the shell inner wall in disengager, the outer wall of the primary and secondary cyclone separator's outlet pipe. Especially when processing

the heavy oil with high viscosity, the catalyst adheres and cumulates more easily, which make the coking become worse and worse.

(4) It can relieve coking condition by the way of arranging the relative position properly between primary and secondary cyclone separators, improving the linkage between primary and secondary cyclone separators, decreasing the possibility of oil gas discharged from the outlet of the primary cyclone separator entering into the disengagement space.

(5) RNG k-ε Model and Eulerian Model can be applied to simulate the two phase flow condition of oil gas and catalyst in the disengager successfully.

References

1. Investigation Group of Long Period Running of FCCU, SINOPEC (1998), "Investigation on Long Period Running of FCCU," *Petroleum Refinery Engineering* , 28(2), pp. 6-13.
2. Luo Qiang, Zhao Yupeng, Gao Sheng (2003), "Causes of Coking in the Settler of FCC Unit and the Countermeasures", *Petroleum Refinery Engineering*, 33(6), pp.1-5.
3. Ji Genzhong, Zhu Hongqi (2002), "Exploration on Causes of Coking in FCCU Reaction System," *Petroleum Refinery Engineering*, 32(6), pp.6-9.
4. Ye Xiaodong, Xu Wuqing, Liu Jingxiang (2003), "Analysis of Coking Reason for Different Structure Heavy Oil FCCU and Its Prevention Measures," *Petroleum processing and Petrochemicals*, 34(2), pp.21-25.
5. Xiao Zuohua, Xia Yinhou (2001), "Using UOP's Techniques to Revamp FCCU," *Petroleum Refinery Engineering*, 31(6), pp.19-22.
6. Lu Chunxi, Cai Zhi, Shi Mingxian (2004), "Experimental Study and Industry Application of a New Votex Quick Separation System at FCCU Riser Outlet," *Acta Petrolei Sinica (Petroleum Processing Section)*, 20(3), pp.24-29.
7. Zhou Lixing (1994), "Theory and Simulation of Gas-particle Two-phase Flow and Combustion," Science Press, Beijing.
8. Wang Fujun (2004), "Computational Fluid Dynamics Analysis- Theory and Application of CFD software," Tsinghua University Press, Beijing.

(240g) CFD Modeling of a Molten Slag Jet Free Surface Flow During Mineral Wool Fiberization

Dimitrios I. Gerogiorgis, Dimitrios Panias and Ioannis Paspaliaris, National Technical University of Athens (N.T.U.A.), Athens, Greece

Red mud fiberization is a process with remarkable industrial importance, alleviating environmental pressure by ensuring the transformation of this major by-product of aluminium production into useful mineral wool products. The most widespread mineral wool production process is molten rock (in this case, molten red mud) fiberization, either by means of fast rotating spinning wheels (Sirok et al., 2008), or via an impinging air jet; the latter is a promising method which avoids mechanical wear as well as rotating parts, but has not been adequately elucidated. The molten slag which remains after pig iron casting enters via a siphon neck into a heated homogenization reservoir, flows out of a heated ladle orifice at high temperature (1400 °C) and adjustable flowrate, and forms a free-falling vertical jet which visibly radiates its excessive heat: at a given distance, a high-velocity impinging air jet meets the vertical melt jet perpendicularly (or at an angle), inducing intensive droplet generation and breakup. Melt droplets (briefly connected by surface tension and viscous forces to each other) emerge at a high velocity, and nascent fibers are rapidly ejected away from the vertical melt jet, experience rapid cooling along their trajectory and then accumulate in a fiber collection chamber, to undergo further processing towards mineral wool.

Molten red mud fiberization is a free-boundary problem which can be tackled via advanced CFD methods. Typically, a free-boundary problem consists of a set of elliptic partial differential equations (PDE) which must be satisfied within a bounded domain, together with the necessary momentum and heat flux boundary conditions. The actual axisymmetric flow field is not known, but is assumed to be well within a cylindrical bounded domain. A rigorous mathematical formulation has been derived and published by Epikhin et. al. (1981), while other studies focus on multiphase phenomena during jet breakup (Silaev, 1967; Kuan, 2009) and fiber formation (Kulago, 1985). Thermophysical and transport properties of the molten melt have been studied extensively; however, it is unclear how both geometric characteristics (reservoir/ladle/orifice dimensions, melt jet height, impingement angle) as well as operational parameters (reservoir/ladle/ambient temperature, ambient relative humidity, melt flowrate and composition, air jet velocity) affect both the performance and robustness of this novel process, as well as the quality characteristics of the product (fiber size distribution, mass and size distribution of unfiberized droplets).

This paper focuses on high-fidelity CFD modeling of the molten jet free surface flow under external cooling. The CFD model encompasses all interrelated physicochemical phenomena (melt laminar flow, radiative cooling) and considers temperature-dependent transport properties (density, viscosity, surface tension) for the molten slag, in order to understand how the foregoing geometric degrees of freedom (manipulated process variables) affect the shape and temperature of the resulting flow field until the fiberization zone, where it meets the impinging air jet. Sensitivity analyses with respect to the key geometric variables provide further insight and operational guidelines. Experimental validation is also anticipated via data from a novel NTUA pilot plant which is under construction. The validated CFD model will be used for optimization and control studies towards achieving optimal operation.

REFERENCES

Balomenos, E., Giannopoulou, I., Panias, D., Paspaliaris, I., ENEXAL: Novel technologies for enhanced energy and exergy efficiencies in primary aluminium production industry, *Metalurgija - J. Metall.* **15**(4): 203-217 (2009).

Epikhin, V.E., Kulago, A.F., Shkadov, V.Y., Influence of convection and thermal radiation on the cooling of a vertically incident jet of melt, *J. Eng. Phys.* **41**(4): 1091-1099 (1981).

Kuan, B., CFD modelling of liquid jet and cascade breakup in crossflows, Proceedings of the 7th International Conference on CFD in the Minerals and Process Industries (CD), CSIRO, Melbourne, Australia (2009).

Kulago, A.E., Dependence of the length of mineral fibers on the physical properties of the melt, *Glass Ceram.* **42**(7): 317-320 (1985).

Silaev, A.F., Mechanism of disintegration of molten metal streams by a gas jet, *Powder Metall. Met. C.* **6** (5): 350-353 (1967).

Sirok, B., Blagojevic, B., Pullen, P., *Mineral Wool – Production and Properties*, Woodhead Publishing, Cambridge (2008).

See more of this Session: Modeling and Computation In Energy and the Environment
See more of this Group/Topical: Computing and Systems Technology Division

(240h) Scalable Stabilized FE Formulations for Simulating Turbulent Reacting Flows In Light Water Reactors

Roger P. Pawlowski[1], John N. Shadid[2], Tom M. Smith[2] and Eric C. Cyr[2], (1)Computational Sciences, Sandia National Laboratories, Albuquerque, NM, (2)Sandia National Laboratories, Albuquerque, NM

This presentation will discuss progress towards developing a large-scale parallel CFD capability using stabilized finite element formulations to simulate turbulent reacting flow and heat transfer in light water nuclear reactors (LWRs). Numerical simultation plays a critical role in the design, certification, and operation of LWRs. The Consortium for Advanced Simulation of Light Water Reactors is a U. S. Department of Energy Innovation Hub that is developing a virtual reactor toolkit that will incorporate science-based models, state-of-the-art numerical methods, modern computational science and engineering practices, and uncertainty quantification (UQ) and validation against operating pressurized water reactors. It will couple state-of-the-art fuel performance, neutronics, thermal-hydraulics (T-H), and structural models with existing tools for systems and safety analysis and will be designed for implementation on both today's leadership-class computers and next-generation advanced architecture platforms.

We will first describe the finite element discretization utilizing PSPG, SUPG, and discontinuity capturing stabilization. We will then discuss our initial turbulence modeling formulations (LES and URANS) and the scalable fully implicit, fully coupled solution methods that are used to solve the challenging systems. These include globalized Newton-Krylov methods for solving the nonlinear systems of equaitons and preconditioned Krylov techniques. The preconditioners are based on fully-coupled algebraic multigrid and approximate block factorization preconditioners. We will discuss how these methods provide a powerful integration path for multiscale coupling to the neutronics and structures applications. Initial results on scalabiltiy will be presented. Finally we will comment on our use of embedded technology and how this capbaility impacts the application of implicit methods, sensitivity analysis and UQ.

See more of this Session: Modeling and Computation In Energy and the Environment
See more of this Group/Topical: Computing and Systems Technology Division

(306a) Determination of An Accurate Reduction Method for Stochastic Chemical Kinetic Models with Applications In Synthetic Biological Design

Patrick Smadbeck, Chemical Engineering and Materials Science, University of Minnesota - Twin Cities, Minneapolis, MN and Yiannis N. Kaznessis, Chemical Engineering and Materials Science, University of Minnesota, Minneapolis, MN

Production of new biological devices within synthetic biology has been primarily driven by experimental research in recent years. Forward engineering of these biological devices, however, will require the prediction of function and thus may be driven by computational modeling.

The main benefit of simulation to predict biological function is in screening potential gene networks. As genetic devices are combined to produce complex outputs, the number of potential gene configurations will increase rapidly. Experimentally constructing each configuration would be time consuming and expensive. Biological simulations may aide in screening these potential gene networks to determine promising configurations.

Unfortunately, the simulation of complex gene networks is challenging and often inaccurate. A fundamental source of inaccuracies is the inability to determine many of the kinetic parameters in the chemical reaction network. As models become more complex the number of unknown kinetic parameters will increase dramatically. One solution to such a problem is to reduce the number of unknown kinetic parameters by combining reactions using steady-state assumptions and the use of complex rate laws.

The present work focuses on a general reduction method for biological systems and the issues surrounding model reduction in stochastic simulations. Numerous reduction methods exist, but only for limiting cases (e.g. a stochastic quasi-steady state approximation requires a near zero molecule count for components to be removed from the simulation). The present work differs from previously developed reduction methods by exploring the consequences of reducing models outside of these limiting cases. The method takes in a complex chemical kinetic model which functions far from the thermodynamic limit (i.e. stochasticity is important). Sets of reactions within the model are then reduced to single kinetic rate laws (e.g. transcription with repression can be represented by a Hill-type kinetic rate law). By utilizing complex rate laws a conceptually simpler model is produced, and by accounting for the consequences of such approximations accuracy can be maintained.

See more of this Session: Mathematical Approaches In Systems Biology
See more of this Group/Topical: Food, Pharmaceutical & Bioengineering Division

(306b) Dynamical Pathway Sensitivity Analysis for Biological Systems

Thanneer Malai Perumal, Department of Chemical and Biomolecular Engineering, National University of Singapore, Singapore, Singapore and **Rudiyanto Gunawan**, Institute for Chemical and Bioengineering, ETH Zurich, Zurich, Switzerland

The normal functioning of biological systems relies on the coordinated actions of a multitude of components (i.e., from genes to proteins to metabolites and small molecules). The complexity that is often associated with cellular response to a particular stimulus, limits human intuition in understanding their functional dynamics and emergent behaviors. This has motivated the use of quantitative dynamical models to represent the underlying behavior of biological systems[1]. To this end, various modeling paradigms and their associated systems analysis tools, such as the one presented here and elsewhere[2,3], have been developed to elucidate the underlying mechanistic details that are responsible for giving the observed behavior. Existing model analysis tools, based on perturbations on system dynamics, include structured singular values[4], bifurcation analysis[5], and parametric sensitivity analysis[6] and its variants, such as dynamic Green's Function Matrix[3] and impulse parametric sensitivity Analysis[2]. Most of these analyses introduce static or dynamic perturbations to system states or parameters one at a time, and quantify the resulting change in system behavior, based on which mechanistic details of system dynamics are inferred.

On the other hand, there is increasing evidence that the structure of biological networks is closely related to their functions[7]. Hence, qualitative structural analysis of biological networks is an alternative tool used to understand system functionality and dynamics[8]. Existing structure analysis methods, mainly borrowed from graph theory concepts, range from simple degree centrality[9] to spectral[10] and communicability measures to SigFlux method[11]. Given the network topology, these methods identify how important a particular molecule or a reaction is to the networks connectivity and system functionality.

But as said earlier, cells often rely not on a single molecule or reaction, rather on a group of molecules and reactions that gives rise to the observed behavior[12]. Hence it is equally important to analyse pathways, comprising multiple molecules and reactions. To this end, we have created a new pathway sensitivity analysis tool that combines both the above said, structural and dynamic information of the network. The analysis presented here consists of two steps: (1) identification of elementary signaling modes[8] (structure) and (2) introduction of impulse perturbations on the parameters associated with each individual modes (dynamics).

The efficacy of the present method is demonstrated using two biological applications for: (i) understanding the competing mechanisms of type-I/II apoptosis in Jurkat cell lines, which gives raise to the caspase-3 cleavage[13], and (ii) identifying the robustness causing mechanism of central carbon metabolism in E. coli[14]. The pathway sensitivity analysis of Fas-induced apoptosis in Jurkat cell lines showed that the initial activation of caspase-3 is due to mitochondrial independent pathway, which is later taken over by mitochondrial dependent pathway. On the other hand, the pathway sensitivity to E. coli central carbon metabolism revealed the 6pg-ribu5p pathway, including the Tkb conversion of xyl5p to f6p, as the responsible mechanism that promotes the robustness in pyruvate production against external perturbations.

References:

1. Szallasi Z, Stelling J, Periwal V. *System Modeling in Cell Biology From Concepts to Nuts and Bolts*. Cambridge, Massachusetts: The MIT Press; 2006.

2. Perumal TM, Gunawan R. Understanding dynamics using sensitivity analysis: caveat and solution. *BMC Syst Biol.* Mar 15 2011;5(1):41.

3. Perumal TM, Wu Y, Gunawan R. Dynamical analysis of cellular networks based on the Green's function matrix. *J Theor Biol.* Nov 21 2009;261(2):248-259.

4. Shoemaker JE, Doyle III FJ. Identifying Fragilities in Biochemical Networks: Robust Performance Analysis of Fas Signaling-Induced Apoptosis. *Biophys J.* 2008.

5. Battogtokh D, Tyson JJ. Bifurcation analysis of a model of the budding yeast cell cycle. *Chaos.* Sep 2004;14(3):653-661.

6. Varma A, Morbidelli M, Wu H. *Parametirc Sensitivity in Chemical Systems*: Cambridge University Press, Cambridge, UK; 1999.

7. Stelling J, Klamt S, Bettenbrock K, Schuster S, Gilles ED. Metabolic network structure determines key aspects of functionality and regulation. *Nature.* Nov 14 2002;420(6912):190-193.

8. Wang RS, Albert R. Elementary signaling modes predict the essentiality of signal transduction network components. *BMC Syst Biol.* 2011;5:44.

9. Wasserman S, Faust K. *Social network analysis : methods and applications*. Cambridge ; New York: : Cambridge University Press; 1994.

10. Perra N, Fortunato S. Spectral centrality measures in complex networks. *Phys Rev E Stat Nonlin Soft Matter Phys.* Sep 2008;78(3 Pt 2):036107.

11. Liu W, Li D, Zhang J, Zhu Y, He F. SigFlux: a novel network feature to evaluate the importance of proteins in signal transduction networks. *BMC Bioinformatics.* 2006;7:515.

12. Bhalla US, Iyengar R. Emergent properties of networks of biological signaling pathways. *Science.* 1999;283(5400):381--387.

13. Hua F, Hautaniemi S, Yokoo R, Lauffenburger DA. Integrated mechanistic and data-driven modelling for multivariate analysis of signalling pathways. *J R Soc Interface.* 2006;3(9):515--526.

14. Chassagnole C, Noisommit-Rizzi N, Schmid JW, Mauch K, Reuss M. Dynamic modeling of the central carbon metabolism of Escherichia coli. *Biotechnol Bioeng.* Jul 5 2002;79(1):53-73.

See more of this Session: Mathematical Approaches In Systems Biology
See more of this Group/Topical: Food, Pharmaceutical & Bioengineering Division

(306c) Uncovering Hidden Variability and Dynamic Patterns: Strategies for Analyzing High-Dimensional Data Sets

Mary M. Staehle[1], Babatunde A. Ogunnaike[2], James Schwaber[3] and Rajanikanth Vadigepalli[3], (1) Rowan University, Glassboro, NJ, (2)Chemical Engineering, University of Delaware, Newark, DE, (3) Department of Pathology, Anatomy and Cell Biology, Thomas Jefferson University, Philadelphia, PA

In the field of systems biology, new technologies have enabled the acquisition of large amounts of data with high-dimensional experimental designs. The resulting data sets provide an opportunity to examine multiple experimental variables simultaneously and to uncover "hidden," physiologically-relevant sources of variability.

To introduce the analytical strategies we have developed for analyzing high-dimensional data sets, we will utilize a specific set of high-dimensional gene expression data that was developed for systematic comparisons of cellular responses to chronic alcoholism and alcohol withdrawal. This data set includes 195 rat samples from two brain regions, three alcohol states, and five experimental time points, with at least five replicates per condition. For each of the 195 samples, the gene expression of 145 transcripts was measured with Fluidigm's BioMark™ high-throughput quantitative reverse transcription PCR platform.

Our analysis of this data set revealed "hidden" changes in gene expression across the three times of the day that the samples were collected, indicating that gene expression varies in dynamic, diurnal patterns. We developed methods for characterizing these as both qualitative and quantitative patterns, and we found that, in some cases, the patterns are brain region specific. Furthermore, principal component analysis (PCA) of the transcriptomic data revealed that the two brain regions have distinct molecular phenotypes and therefore must be analyzed separately in order to capture perturbation-induced changes in expression. The "hidden" sources of variability in gene expression data provide additional insight into the overall functionality of these brain regions, but they also obscure the desired dynamics. When the variability due to diurnal patterns and intra-brain region differences is removed, PCA of the data illuminates dynamic patterns of gene expression that evolve during alcohol withdrawal. The translation of these analytical strategies to characterize "hidden" variability and analyze gene expression patterns in other high-dimensional data sets would lead to systems-level insight in biological systems.

See more of this Session: Mathematical Approaches In Systems Biology
See more of this Group/Topical: Food, Pharmaceutical & Bioengineering Division

(306d) An Improved De Novo Peptide Sequencing Framework Using Decomposition and Integer Linear Optimization

Zukui Li, Richard C. Baliban and Christodoulos A. Floudas, Chemical and Biological Engineering, Princeton University, Princeton, NJ

Peptide and protein identification is of fundamental importance in the study of proteomics. Specifically, peptide sequencing and identification forms a basis for peptide identification. Of the two most frequent computational approaches for peptide sequencing, de novo methods have distinct advantages over database methods in that they can analyze peptides not present in a protein database and are more amenable to identifying post-translational modifications.

Tandem mass spectrometry (MS/MS) coupled with high performance liquid chromatography (HPLC) has emerged as a powerful protocol for high-throughput and high sensitivity peptide and protein identification experiments. They have served as an impetus for the recent development of numerous computational approaches formulated to sequence peptides robustly and efficiently with particular emphasis on the integration of sequencing algorithms into a high throughput computational framework for proteomics.

PILOT is an integer linear optimization based de novo peptide sequencing framework developed in our group [1,2]. The performance of PILOT has been shown to be good on high quality spectrums. However, its accuracy is still restricted for low quality spectrums characterized by the following factors: large amount of noise peak, doubly charged ions in low precision spectrum (e.g., ion trap), because of the selection of false fragment ions peaks.

An improved de novo peptide sequencing framework has been developed in this work. The algorithm is based on a spectrum decomposition strategy and the integer linear optimization technique, and it is composed by four stages. In the first stage of the algorithm, the whole spectrum is considered and different sets of high confidence fragment peak solutions are identified by sequencing b-ion and y-ion series simultaneously using integer linear optimization. In the second stage, a MS/MS spectrum is decomposed into left hand side (LHS) and right-hand-side (RHS) by the m/z value (mass charge ratio) of the doubly charged precursor ions (y_N^{++}). The LHS and RHS spectra are sequenced separately by considering the high confidence solution obtained from the first stage (through fixing the usage of or setting high intensity of the tag peaks). Up to twenty sets of LHS solutions and up to twenty sets of RHS solutions are obtained for the LHS and RHS of the spectrum, respectively. Using integer cuts, all the LHS solutions and the RHS solutions are distinct. In the third stage, the full sequence is generated by combining the LHS and RHS sequence with a set of predefined combination rules, which include the complementarity restriction, LHS and RHS connection through single or double amino acid mass, and limits on number of mass gaps. In the final stage of the algorithm, the candidate sequences are ranked by cross-correlating the theoretical spectra of the candidate sequences with the experimental tandem mass spectrum and finally a predefined number of top ranked sequences are reported.

The proposed de novo sequencing framework is tested on various data sets generated from both low precision Ion Trap and high precision Orbi Trap mass spectrometers. Furthermore, the method has also been tested by combining the de novo and database techniques in a hybrid framework [3] which utilizes the de novo sequences to generate large "sequence tags" which are further used to query a protein database. Computational results on the de novo peptide identification and hybrid method based identification, as well as comparisons with state of the art methods will be presented.

[1] P.A. DiMaggio and C.A. Floudas. A mixed-integer optimization framework for de novo peptide identification. AIChE Journal, 53(1), 160-173 (2007).

[2] P.A. DiMaggio and C.A. Floudas. De novo peptide identification via tandem mass spectrometry and integer linear optimization. Anal. Chem., 79, 1433-1446 (2007).

[3] P.A. DiMaggio, C.A. Floudas, B. Lu and J.R. Yates. A Hybrid Method for Peptide Identification using Integer Linear Optimization, Local Database Search, and QTOF or OrbiTrap Tandem Mass Spectrometry. J. Proteome Res., 7, 1584-1593 (2008).

See more of this Session: Mathematical Approaches In Systems Biology
See more of this Group/Topical: Food, Pharmaceutical & Bioengineering Division

(306e) A Fitness Index for Transplantation of Perfused DCD Rat Livers

Sinem Perk[1], Maria-Louisa Izamis[1], Herman Tolboom[2], Basak Uygun[1], Francois Berthiaume[1], Martin L. Yarmush[1] and Korkut Uygun[1], (1)Center for Engineering in Medicine, Massachusetts General Hospital, Harvard Medical School, Shriners Hospital for Children, Boston, MA, (2)Division of Cardiac and Vascular Surgery, University Hospital Zurich, Zurich, Switzerland

An estimated 4,000 patients perish every year due to lack of transplantable livers. The estimated pool of donor after cardiac death (DCD) livers that are marginally damaged is approximately 6,000 grafts per year, which can be used to resolve the crisis in organ availability. In DCD organs, the presence of cardiac death means that the organs recovered go through a phase of absence of blood supply and anoxia, also called ischemia. Ischemia is a very severe mode of injury and for the liver beyond 30 min graft survival in the recipient falls dramatically precluding the use of most DCD organs in transplantation. Preclinical studies with machine perfusion systems indicate that DCD livers can be successfully recovered from ischemia.

The aim of this work is the development of an index of fitness for transplantation for perfusion-recovery of marginally damaged DCD livers based on metabolic activity during machine perfusion.

A particular mode of machine perfusion, Normothermic extracorporeal Machine Perfusion (NMP) aims to mimic the in vivo conditions of the organ in order to maximize treatment efficacy during perfusion-preservation. It is exceptionally well-suited to monitor the dynamics of recovery from ischemic injury to evaluate the viability of the graft for transplantation; this is because the perfused organ is in a fully metabolically active state, and this activity can be evaluated and compared to an ideal organ to determine the likelihood of success. Such a quality-control step is a critical need in utilization of DCD organs to ensure that the perfused organ can be transplanted with minimal concerns.

Currently the graft quality is assessed based on gross morphology and histology, which are subjective and ineffective to distinguish between viable and nonviable DCD livers. A quantitative and objective index of fitness for transplantation relative to ischemic injury that is based on organ function is highly desirable. A suitable framework for developing such an index relative to ischemic condition is multivariate statistical process monitoring (SPM). The liver metabolism is a complex network which features many correlated metabolites. SPM look at the whole system, identify commonalities between different perfusions, correlations among variables and trends in time and hence are ideal.

A multiway principal component analysis liver perfusion model captures the correlation structure between the metabolites during perfusion of fresh livers that were later successfully transplanted and defines the confidence limits based on the metabolite concentrations in the perfusion medium. The differences in functioning of DCD livers are then evaluated by projecting the metabolite concentrations for the damaged livers onto the fresh liver model; this can be done on an online basis so that the recovery from ischemic injury is evaluated quantitatively during perfusion allowing interventions to the organ if necessary. The squared prediction error statistic quantifies the similarities of ischemic livers to fresh livers and hence constructs an index of ischemia.

Further, an online multiway partial least squares discriminant analysis model predicts the end-of-perfusion quality of a liver during perfusion. This complements the ischemia index by classifying any given liver as ischemic of fresh and forms the basis for index of transplantation of DCD livers. Once the quality of a given liver is determined as ischemic; its fitness for transplantation is assessed based on the confidence limits for successfully transplanted ischemic livers. The effectiveness of multivariate techniques on modeling dynamic metabolic data and the decision algorithm for the evaluation of the ischemic condition and the fitness for transplantation are demonstrated and tested on independent livers.

See more of this Session: Mathematical Approaches In Systems Biology
See more of this Group/Topical: Food, Pharmaceutical & Bioengineering Division

(306f) Comparative Analysis of Transcriptome and Lipidome of RAW 264.7 and Primary Macrophages

Mano R. Maurya[1], Shakti Gupta[2], Eoin Fahy[1], Ashok Reddy Dinasarapu[1], Sean Li[1], Manish Sud[1] and Shankar Subramaniam[3], (1)Department of Bioengineering, University of California San Diego, La Jolla, CA, (2)Department of Bioengineering, University of California, San Diego, CA, (3)Department of Bioengineering, San Diego Supercomputer Center, Department of Chemistry & Biochemistry, University of California San Diego, La Jolla, CA

Macrophages play essential roles in immunity and lipid homeostasis. Studies of macrophage biology have been significantly advanced by the availability of cell lines such as RAW264.7 cells. However, it is unclear how these cell lines differ from primary macrophages, such as thioglycolate-elicited peritoneal macrophages (TGEM) and bone marrow-derived macrophages (BMDM). We used the inflammatory stimulus Kdo_2-Lipid A (KLA), a selective Toll-like receptor 4 agonist, to stimulate RAW264.7 and TGEM cells. Temporal changes of lipid and gene expression levels were concomitantly measured. We have performed a systems-level analysis of the lipidomic and transcriptomic data. In this work, we present a comprehensive comparison between the two cell types.

Upon KLA treatment, Both RAW264.7 and TGEM cells show strong inflammatory response on the gene expression level. However, relative to RAW264.7 cells, TGEM cells show more rapid and intense inflammatory response. The increase in prostaglandin levels in both the primary cells is substantially less as compared to that in RAW264.7 cells. Sphingolipid-1-phosphate increases considerably in RAW264.7 cells. Cholesterol *de novo* synthesis increases considerably in RAW264.7 cells, but 25-hyrdoxycholesterol, an endogenous agonist of liver X receptor, increases considerably in the TGEM cells. Specific commonalities and differences are also observed in energy metabolism related processes such as TCA cycle, beta oxidation and oxidative phosphorylation.

Our findings demonstrate that, although there are basic similarities in the responses of the three cell types, significant temporal and quantitative differences exist. Hence, caution is advised in using RAW264.7 cells as a model to study the kinetics of inflammatory response, lipid metabolism, and related disorders such as atherosclerosis.

Acknowledgement: The experimental data used in this work was provided by the Lipid Metabolite and Pathway Strategy (LIPID MAPS) Consortium experimental laboratories and is available on the LIPID MAPS web site (www.lipidmaps.org). LIPID MAPS is funded by the National Institute of General Medical Sciences.

(306g) Modeling the Synergistic Effects of Combination Therapy for Glioblastoma Multiforme

Matthew Browe, Chemical Engineering, Pennsylvania State University, State College, PA, Francisco G. Vital-Lopez, Department of Chemical Engineering, The Pennsylvania State University, University Park, PA and **Antonios Armaou**, Chemical Engineering, The Pennsylvania State University, University Park, PA

Glioblastoma multiforme (GBM) is among the most aggressive and challenging human cancers, accounting for 50-60% of primary brain tumors in adults (Mohin, 2006) and associated with a median survival of 10-12 months after diagnosis (Alford, 1991). Aside from its high resistance to conventional cancer treatments such as radiotherapy (Kleinsmith, 2006) and vulnerable region of genesis, the invasiveness of GBM tumor cells is primarily what has rendered it a very deadly disorder and why even seemingly complete surgical resection of the tumor fails uniformly. *In vivo* experimental results have verified this characteristic, as glioma cells have been identified throughout the central nervous system within 7 days of tumor xenograft implantation in a rat brain (Silbergeld, 1997).

The current presentation will focus on our efforts to capture these effects more effectively in a relevant mathematical description and implement current treatment methods. Specifically, a hybrid discrite/ continuous model of a glioblastoma multiforme (GBM) tumor that was developed in (Vital-Lopez, 2010) was modified to include haptotactic migration strategies and radiotherapeutic treatment methods. Haptotactic migration strategies in the model showed little improvement in overall tumor growth and this component likely will involve further modification with the parameters describing the dynamics of the coupled effects of extracellular matrix degradation and repair. Radiotherapy methods were analyzed against typical fractionation schemes with and without quiescent cell killing assumed and the model was tested against fractionation schemes *in vivo* as well. Assumption of quiescent cell killing provided the greatest fidelity with other models, but neither method accurately depicted the relative effect of high-dose fractionation schemes against other schemes of equal cumulative dosage. Lastly, analysis of the model against combination therapies *in vivo* confirmed a synergistic effect with concomitant temozolomide and radiotherapy and provides a template for future work investigating specific synergism parameters.

References

Alford, E. C., et al. (1991). *The Pathology of the Aging Human Nervous System.* Philadelphia: Lea and Febiger

Kleinsmith, L. J., et al. (2006). Principles of Cancer Biology. San Francisco: Pearson Education.

Mohin, G., et al. (2006). Glioblastoma Multiforme: advances in postsurgical management. *Community*

Oncology , *3*, 678-684.

Silbergeld, D., et al. (1997). Isolation and characterization of human malignant glioma cells from histologically normal brain. *Journal of Neurosurgery* , *86*, 525-531.

Vital-Lopez, F., et al. (2010). Modeling the effect of chemotaxis on glioblastoma tumor progression. *American Institute of Chemical Engineers Journal* , *57*, 778-792.

See more of this Session: Mathematical Approaches In Systems Biology
See more of this Group/Topical: Food, Pharmaceutical & Bioengineering Division

(316a) Nonsmooth Dynamic Optimization for Bioreactors with Flux Balance Models Embedded

Kai Hoeffner, Process System Engineering Laboratory, Department of Chemical Engineering, Massachusetts Institute of Technology, Cambridge, MA, Stuart Harwood, Process Systems Engineering Laboratory, Deptartment of Chemical Engineering, Massachusetts Institute of Technology, Cambridge, MA and Paul I. Barton, Process Systems Engineering Laboratory, Dept. of Chemical Engineering, Massachusetts Institute of Technology, Cambridge, MA

A nonsmooth dynamic optimization algorithm for fed-batch biochemical reactor models based on dynamic flux balance analysis (DFBA) is presented.

Large-scale production of renewable liquid fuels through biochemical conversion is a growing market. The development of optimal process control strategies for this and similar processes is an important research problem. Optimal model-based fed-batch control strategies rely on the predictive capabilities of the process model. Dynamic flux balance analysis is a promising modeling framework that has these desired capabilities. DFBA assumes that the intracellular species are at equilibrium with the extracellular environment. The resulting underdetermined stoichiometric model is solved under the assumption of a cellular objective such as growth rate maximization. The cell model is coupled with the dynamic mass balance equations of the extracellular environment via expressions for the rates of substrate uptake and product excretion, which imposes additional constraints on the linear program (LP) defined by growth rate maximization of the cell. The linear program is embedded in the dynamic model of the bioreactor, giving an accurate model of the substrate consumption and biomass production during operation. The overall DFBA model is a system of ordinary differential equations for which the evaluation of the right-hand side requires not only function evaluations, but also the solution of one or more linear programs. The resulting dynamic optimization problem is a bilevel optimization problem, where the upper level aims to maximize the final desired product yield of the batch, while the lower level, representing the cell model, is optimizing the biomass generation.

Optimal fed-batch control for DFBA models faces two challenges: First, numerical integration of DFBA models requires a fast, scalable and reliable algorithm that takes into account the special structure of the system, since the naive approach of solving the LP at each time step is inaccurate, due to numerical noise from the LP solver, and inefficient, due to the computational cost of solving the LP, especially for large problems. Second, due to the nonsmoothness property the sensitivities with respect to the control inputs are possibly set-valued. This has to be accounted for during integration and optimization.

The proposed optimization algorithm overcomes both challenges. A novel algorithm, which efficiently integrates DFBA models, called DSL48LPR, is considered here. Using KKT theory, the DFBA model is rewritten as a hybrid differential-algebraic equation (DAE) system such that the embedded LP only has to be solved once at events where the hybrid mode switches. The hybrid modes can be characterized as regimes in which the intracellular kinetics are limited by different combinations of uptake constraints, e. g., aerobic and anaerobic growth phase.

Using the concept of the generalized Jacobian, a set-valued map, a nonsmooth bundle solver in combination with the DSL48LPR yields an optimal solution. Due to the semismoothness property of the generalized Jacobian, we can compute an element of the generalized Jacobian to be passed to the nonsmooth bundle solver.

See more of this Session: Modeling, Optimization and Control of Sustainable Processes
See more of this Group/Topical: Computing and Systems Technology Division

(316b) Optimal Operation of a CO2 Capturing Plant for a Wide Range of Disturbances

Mehdi Panahi, Chemical Engineering, Norwegian University of Science and Technology (NTNU), Trondheim, Norway and Sigurd Skogestad, Chemical Engineering, Norwegian Univeristy of Science and Technology, Trondheim, Norway

Combustion of the fossil fuels in power plants is one of the major sources of producing CO_2. Amine-based absorption/stripping CO_2 capturing processes are the dominant method in removing CO_2 content from the combustion flue gas streams. The stripping column uses a large fraction, about 15-30% of the net power generated in the power plant. Optimal operation of these processes is therefore very important. In particular, we study how to minimize the energy requirement during operation in the presence of unexpected disturbances while satisfying operational constraints.

In this study, the optimal operation of a post-combustion CO_2 capturing process is considered using the systematic plantwide procedure of Skogestad (2004) where the economical objective is to minimize sum of the energy costs and the penalty on the CO_2 released to the air.

The self-optimizing procedure is applied to select the best controlled variables (CVs) in different operational regions (in presence of large disturbances). These are the variables, which when they are kept constant in their optimal nominal setpoints, indirectly give operation close to the optimum. Operational regions are defined as the sets of optimally active constraints that are encountered as the disturbances (e.g., feed flue gas flowrate coming from the power plant) change over the expected operating window.

Dynamic simulation of the process is used to design the control layer and implement the final control structures for different regions. In the dynamic mode, first "stabilizing" CVs are identified and their control loops are closed to avoid instability or drift in the process when disturbances occur. In the supervisory control layer, one may use decentralized (PID controllers) or multivariable control (MPC). To analyze the feasibility of decentralized control, we use the relative gain array (RGA) as a tool to select pairings of the self-optimizing CVs with the manipulated variables. The RGA analysis gives some surprising insights into the dynamic behaviour of this process.

The performance of different decentralized control configurations and MPC (model predictive control) are compared. Finally, we propose a simple control structure which handles the disturbances close to optimal in all operational regions.

See more of this Session: Modeling, Optimization and Control of Sustainable Processes

(316c) Minimizing the Energy and Demand Cost of Building Energy Systems Based On Economic MPC

Jingran Ma[1], **S. Joe Qin**[1] and Tim Salsbury[2], (1)Mork Family Department of Chemical Engineering and Materials Science, University of Southern California, Los Angeles, CA, (2)Johnson Controls, Milwaukee, WI

A simulation-based control system was established in our previous work to show the effectiveness of model predictive control in reducing the energy and demand costs for HVAC systems of commercial buildings. An economic objective function was introduced in the MPC framework to account for the electricity cost under time-of-use rate and demand charges.

In this work, we improve the configuration and iterative procedure of optimization in MPC to achieve larger economic profits. Soft constraints are applied to handle the infeasibility problem and the relation between cost savings and constraint violations is investigated. In the perspective of modeling, we utilize variables from air handling units (AHU) such as sensible cooling rate and air mass flow rate in addition to zone temperature setpoints to enhance the accuracy of model prediction. Besides, the robustness of the control system is studied by carrying out sensitivity analysis on ambient temperatures.

See more of this Session: Modeling, Optimization and Control of Sustainable Processes
See more of this Group/Topical: Computing and Systems Technology Division

(316d) Hierarchical Control of Networks Featuring Large Solvent Recycle

Sujit S. Jogwar, **Ana I. Torres** and Prodromos Daoutidis, Chemical Engineering and Materials Science, University of Minnesota, Minneapolis, MN

Motivated by the cost benefits resulting from the reduced consumption of make-up solvent and the environmental regulations demanding minimal discharge of toxic chemicals, solvent recovery and recycle is one of the essential features of integrated chemical processes. Examples of such processes include those in which a non-aqueous solvent is used as the reaction media, or as an extractant to purify a product or remove a contaminant. A characteristic common to most of these cases is the presence of a large solvent recycle when compared to the fresh solvent feed.

In our previous work (Baldea and Daoutidis, 2007; Kumar and Daoutidis, 2002) it was demonstrated that networks featuring flows of different orders of magnitude give rise to multi-time scale dynamics, and thus present significant control challenges. In this work, the defining characteristics of processes with significant solvent recycle are identified and a systematic framework for their analysis and control is developed. As a first step, a generic prototype network that captures the essential features of such processes is proposed. It is shown that the material balance model of this network can be recast in a singularly perturbed form, resulting in two-time scale dynamics. A model reduction framework is then developed to obtain reduced order models that capture the dynamics of this network in the fast and slow time scales. Finally, a hierarchical control strategy resulting from this time scale multiplicity is proposed. Specifically, the control objectives related to the individual units are addressed in the fast time scale, whereas those associated with the solvent recycle loop are addressed in the slow time scale.

A process for the production of 5-hydroxymethylfurfural (HMF), a key intermediate in the production of biomass derived fuels and chemicals, is considered as a case study. HMF synthesis is based on the acid catalyzed dehydration of sugars, and is the intermediate product of consecutive reactions. In order to improve its yield, HMF needs to be separated from the reaction medium with solvent extraction being a commonly used approach. In previous work (Torres et al., 2010) a continuous process for the production of HMF using solvent extraction was designed and optimized. The optimal steady state was characterized by a large solvent recycle flow when compared to the fresh solvent feed, a feature representative of the prototype network identified earlier. The model reduction and hierarchical control strategy developed for the prototype network is applied to this system, and model based controllers are derived to address control objectives (product purity and holdups) in each time scale. The effectiveness of the control scheme is demonstrated through closed loop simulations of the system under various operating scenarios.

See more of this Session: Modeling, Optimization and Control of Sustainable Processes
See more of this Group/Topical: Computing and Systems Technology Division

(316e) Mathematical Modeling and Steady-State Analysis of a Proton-Conducting Solid Oxide Fuel Cell

Mona Bavarian and Masoud Soroush, Chemical and Biological Engineering, Drexel university, Philadelphia, PA

Generally, there are two types of solid oxide fuel cells based on the electrolyte used in the cell [1]. The electrolyte can be an oxygen ion-conducting or a proton-conducting material. The typical oxygen ion-conducting electrolyte is yttria-stabilized zirconia (YSZ), which requires the cell to be operated at high temperatures from 800 to 1200°C to exhibit high ionic conductivity. This is due to the high activation enthalpies of their conductivity [2]. Because of their high operation temperature, SOFCs with oxygen ion-conducting electrolytes have two major drawbacks: long start-up and shut-down, and the high cost of materials that stand the high temperatures, in the manufacture of the fuel cells [3]. By lowering the operation temperature to 600-800°C, these challenges can be addressed to some extent. One way to develop a low temperature SOFC is to utilize a proton conducting electrolyte. Perovskite-type proton conductors, such as $SrCeO_3$ and $BaCeO_3$, are doped with low valence cations such as Y^{3+}or Yb^{3+} to create oxide ion vacancies, which are required for the formation of protonic defects [2, 4].

This study deals with a solid oxide fuel cell with $SrCe_{0.95}Yb_{0.05}O_{3-\alpha}$ (SCY) electrolyte and two platinum electrodes. A mathematical model of the proton-conducting SOFC is first developed. The model captures electrochemical processes as well as the transport phenomena. The model is validated with the experimental results of Iwahara [5] obtained under isothermal conditions. The fuel cell model is then used to study the existence of multiplicity in the cell under non-isothermal conditions. Simulation results show that a multiple steady states region exists at low inlet fuel and air temperatures under non-isothermal conditions. The occurrence of ignition and extinction in the cell solid (electrolyte, anode and cathode) temperature is observed. This result is in agreement with the studies on oxygen ion-conducting SOFCs in which the existence of multiplicity is attributed to the dependence of electrolyte oxygen-ion conductivity on temperature. The results show that the region of -steady-state multiplicity disappears as the inlet fuel and air temperatures increase.

References:

[1] A. Arpornwichanop, Y. Patcharavorachot, S. Assabumrungrat, Chemical Engineering Science, 65 (2010) 581-589.

[2] K. Kreuer, Ann. Rev. Mater. Res., 33 (2003) 333-359.

[3] D. Brett, A. Atkinson, N. Brandon, S. Skinner, Chem. Soc. Rev., 37 (2008) 1568-1578.

[4] H. Matsumoto, Y. Furuya, S. Okada, T. Tanji, T. Ishihara, Science and Technology of Advanced Materials, 8 (2007) 531-535.

[5] H. Iwahara, Solid State Ion., 28 (1988) 573-578.

See more of this Session: Modeling, Optimization and Control of Sustainable Processes
See more of this Group/Topical: Computing and Systems Technology Division

(316f) Adaptive Subspace System Identification for CO2 Capture Processes

Ricardo Dunia, Chemical Engineering Department, The University of Texas at Austin, Austin, TX, Gary Rochelle, Chemical Engineering, University of Texas at Austin, Austin, TX and S. Joe Qin, Mork Family Department of Chemical Engineering and Materials Science, University of Southern California, Los Angeles, CA

The CO_2 capture process with amine solvent based absorption and stripping is the most significant industrial method for the removal of carbon dioxide from coal and natural gas-fired power plants [1]. Several dynamic models have been developed for the CO_2 capture process [2][3]. However, most of these models require reaction kinetic and thermodynamic properties of the species considered in the process. Therefore, to obtain the simulation results, rigorous property calculations are programmed in specialized software that runs in off-line hardware. Such a modeling implementation approach is useful for model based control design and optimization strategies. Nevertheless, the application of such type of models during process operations is limited because simulation convergence, time and computational resources are limited in industrial setups.

A practical approach to obtain a dynamic model from a process plant is to use data-driven empirical models, where the model is made to match the process measurements. Among empirical models, the subspace system identification algorithms have been well accepted in industry not only because of their simplicity and robustness, but also because they provide state space form models that are very convenient for prediction, process monitoring and model based control [4] [5] [6]. However, CO_2 capture process operating changes are subject to fluctuations in electric power consumption as supplementary fuel is burned during peak hours of electric energy usage. Therefore, empirical models should adapt accordingly to different process loads, including startup and shutdown procedures that are particularly common in CO_2 capture operations.

This work applies adaptive subspace system identification [7] to a CO_2 recovery pilot plant. The empirical model is updated by collecting data at every sampling time from more than seventy sensors and making typical operational changes to perturb the process. Such changes are expected to excite the different modes of operation to determine the number of states required to describe the dynamic response of the process, which can change depending of the process operating mode. The development of such an adaptive empirical model represents a valuable tool for advanced process control applications, where minimal energy consumption and optimal CO_2 recovery should be achieved at different operating circumstances.

References

[1] J. M. Plaza, D. Van Wagener, G. T. Rochelle, Modeling CO2 Capture with Aqueous Monoethanolamine, in: Gale, J and Herzog, H and Braitsch, J (Ed.), GREENHOUSE GAS CONTROL TECHNOLOGIES 9, Vol. 1 of Energy Procedia, 2009, pp. 1171–1178, 9th International Conference on Greenhouse Gas Control Technologies, Washington, DC, NOV 16-20, 2008. doi:10.1016/j.egypro.2009.01.154.

[2] S. Ziaii, G. T. Rochelle, T. F. Edgar, Dynamic Modeling to Minimize Energy Use for CO2 Capture in Power Plants by Aqueous Monoethanolamine, INDUSTRIAL & ENGINEERING CHEMISTRY RESEARCH 48 (13) (2009) 6105–6111. doi:10.1021/ie801385q.

[3] H. M. Kvamsdal, J. P. Jakobsen, K. A. Hoff, Dynamic modeling and simulation of a CO2 absorber column for post-combustion CO_2 capture, CHEMICAL ENGINEERING AND PROCESSING 48 (1) (2009) 135–144. doi:10.1016/j.cep.2008.03.002.

[4] B. Wahlberg, M. Jansson, T. Matsko, M. A. Molander, Experiences from subspace system identification - Comments from process industry users and researchers, Vol. 364 of LECTURE NOTES IN CONTROL AND INFORMATION SCIENCES, 2007, pp. 315–327.

[5] W. Favoreel, B. De Moor, P. Van Overschee, Subspace state space system identification for industrial processes, JOURNAL OF PROCESS CONTROL 10 (2-3) (2000) 149–155, 5th IFAC Symposium on Dynamics and Control of Process Systems (DYCPOPS 5), CORFU, GREECE, JUN 08-10, 1998.

[6] M. Basseville, M. Abdelghani, A. Benveniste, Subspace-based fault detection algorithms for vibration monitoring, AUTOMATICA 36 (1) (2000) 101–109.

[7] K. Kameyama, Akira Ohsumi, Subspace-based prediction of linear time-varying stochastic systems, AUTOMATICA 43 (2007) 2009-2021.

See more of this Session: Modeling, Optimization and Control of Sustainable Processes
See more of this Group/Topical: Computing and Systems Technology Division

(316g) Systematic Methods for the Elimination of Redundant Life Cycle Assessment Metrics In the Multi-Objective Optimization of Industrial Processes

Nagore Sabio[1], Carlos Pozo[1], **Gonzalo Guilln-Goslbez**[1], Laureano Jimnez[1], Ramkumar Karuppiah[2], Venkatesh Vasudevan[2], Nicolas Sawaya[2] and John T. Farrell[2], (1)Departament d'Enginyeria Quimica, Universitat Rovira i Virgili, Tarragona, Spain, (2)Corporate Strategic Research, ExxonMobil Research and Engineering, Annandale, NJ

Multi-objective optimization (MOO) is increasingly being used for environmental applications. One of the main advantages of this technique is that it treats environmental performance indicators as individual objectives rather than as constraints imposed on the system. This allows identification of process alternatives that can lead to the minimization of environmental impacts.

The selection of a suitable environmental performance indicator is a key issue in the application of MOO to environmental problems. Amongst environmental indicators that are in use today, those based on life cycle assessment (LCA) principles (Guine et al., 2002) are becoming the preferred choice. However, the computational burden of existing solution methods for MOO grows rapidly with the number of objectives. This limitation is critical in environmental applications, particularly in those based on the combined use of MOO and LCA where the simultaneous analysis of a wide range of environmental metrics is sought. Furthermore, increasing the number of objectives in multi-objective optimization models leads to solutions that are not only difficult to generate, but also to visualize and interpret.

Previous attempts to ameliorate this difficulty focused on aggregation methods that relied on the use of single indicators calculated by attaching weights to the individual environmental metrics considered in the analysis. These weights, which are typically defined by a panel of experts, may not necessarily reflect the decision-maker's preference. Furthermore, aggregation methods modify the Pareto structure of the problem in a manner that some parts of the search space might be left out of the analysis.

We investigate the use of two systematic methods, one based on a rigorous mixed-integer linear programming (MILP) formulation, and the other on principal component analysis, to reduce the dimensionality of MOO problems for environmental applications. We introduce a novel heuristic procedure that can be coupled with these techniques to improve their computational performance. Our final goal is to keep the size of the MOO model at a manageable level while still preserving its Pareto structure to the extent possible. An additional objective of our study is to enhance our knowledge of the Pareto structure of a given problem. This would facilitate analysis of the Pareto frontier by the decision-maker, by extracting the relationships between the different environmental impacts associated with a process. The capabilities of these computational strategies are highlighted by applying them to a generic industrial network. For this case study, the most relevant environmental metrics are chosen and redundant metrics are eliminated using the proposed approaches.

References

Guine, J. B.; Gorre, M.; Heijungs, R.; Huppes, G.; Kleijn, R.; de Koning, A.; van Oers, L.; Sleeswijk, A. W.; S. Suh, S.; de Haes, H. A. U.; de Bruijn, H.; van Duin, R.; Huijbregts, M. A. J. Handbook on Life Cycle Assessment. Operational Guide to the ISO Standards; Kluwer Academic Publishers: Dordrecht, The Netherlands, 2002.

See more of this Session: Modeling, Optimization and Control of Sustainable Processes
See more of this Group/Topical: Computing and Systems Technology Division

(317a) Application of the Path Integration Formalism to Analysis of Activated Processes In Complex Molecular Systems

Dmitry I. Kopelevich, Dept. of Chemical Engineering, University of Florida, Gainesville, FL

One of the challenges in modeling of dynamics of complex molecular systems is prediction of rates of rare evens, i.e. events involving activated transitions between metastable states. The timescale of such events is often inaccessible to direct molecular dynamics simulations. Therefore, these events are analyzed using a reduced stochastic model, such as the Langevin equation, which explicitly accounts for dynamics of a small number of key degrees of freedom (reaction coordinates) while modeling all other degrees of freedom as a thermal bath. The rates of rare events can then be obtained by solving the obtained stochastic equation(s). A common method of analysis of rare events in systems with multiple reaction coordinates is based on the assumption that the system follows a minimum energy path (MEP), i.e. a path such that the free energy is minimized in all directions transversal to the path direction. This assumption effectively reduces the multi-dimensional stochastic process to a quasi-one-dimensional process confined to the MEP.

However, the MEP assumption is not always valid. In particular, our group has demonstrated that substantial deviations from the MEP may occur even in processes involving motion on relatively simple free energy landscapes, such as molecular transport across a liquid-liquid interface or a lipid membrane. These deviations from the MEPs occur due to non-adiabatic coupling between the translational motion of a solute molecule and thermal fluctuations of the interface.

In this talk, we present a theoretical analysis of such non-adiabatic processes. This analysis is based on applying the path integration formalism to solution of the Langevin equation. This formalism yields a distribution of paths connecting metastable states. In particular, this approach enables us to identify the most likely path (MLP) to be taken by the system. We demonstrate that the MLP may significantly deviate from the MEP. Once the MLP is identified, we obtain the rate of the activated events by computing the path integral. The path integral is computed using a harmonic (WKB-like) approximation in the neighborhood of the MLP.

See more of this Session: Molecular and Mesoscopic Systems: Methods and Applications
See more of this Group/Topical: Computing and Systems Technology Division

(317b) Coupling Ballistic Transport and Surface Reaction Models for Simulation of Atomic Layer Deposition Processes

Curtisha D. Travis and Raymond A. Adomaitis, Department of Chemical & Biomolecular Engineering; Institute for Systems Research, University of Maryland, College Park, MD

This paper examines the connections that exist between precursor transport and surface reaction modeling elements in atomic layer deposition process simulators. In particular, we consider nanoporous materials modified by ALD, taking advantage of ALD's potential for conformal deposition in high-aspect-ratio structures. Because of the very large Knudsen number associated with these processes, the diffusion of gas-phase species inside the nanostructures takes place in a purely ballistic manner. Closed-form expressions for the pore-feature precursor transmission probabilities describing the fluxes between the pore internal surfaces are developed and then compared to previous deterministic modeling studies (e.g., those described and cited in [1]).

The transport model elements then are coupled to ALD surface reaction models that describe precursor chemisorption, desorption, and surface reactions that form the ALD film and gas-phase products [2]. Mean-field and simple Monte-Carlo surface reaction models will be described. We will show that a distinct mathematical structure describes this coupling; understanding this structure is crucial to formulating the simulator in such a way that modifications to the surface reaction elements or transport geometry are simple to make. The combined system is spatially discretized and integrated over each precursor exposure and purge period to determine the spatio-temporal features of the reaction surface states during a complete ALD cycle, a computational task presenting its own set of numerical challenges.

Predictions from our dynamic model will be compared to previously published studies of conformal ALD [3]-[7] to validate our simulator and to gain further insight into the physical mechanisms at work in ALD processes. The consistency of our predictions relative to previous modeling work will be quantitatively assessed, and new perspectives on the cited studies will be provided when possible. Overall, our goal is to establish a baseline from which more refined models can be developed, to explore more intricate deposition surface topographies and more realistic surface reaction mechanisms and kinetic expressions. Ultimately, the physically based models of the type we develop can be exploited to determine optimal precursor exposure levels for ALD-based nanomanufacturing operations.

References

[1] Cale, T. S., M. O. Bloomfield, and M. K. Gobbert, Two deterministic approaches to topography evolution, Surface & Coatings Tech. 201 8873-8877 (2007).
[2] Elliott, S. D., Models for ALD and MOCVD growth of rare earth oxides, Rare Earth Oxide Thin

Films, Topics Appl. Physics 106 7386 (2007).
[3] Dendooven, J., D. Deduytsche, J. Musschoot, R. L. Vanmeirhaeghe, and C. Detavernier, Modeling the conformality of atomic layer deposition: The effect of sticking probability, J. Electrochem. Soc., 156 (4) 63-67 (2009).
[4] Elam, J. W., D. Routkevitch, P. P. Mardilovich, and S. M. George, Conformal coating on untrahigh-aspect-ratio nanopores of anodic alumina by atomic layer deposition, Chem. Mater. 15 3507-3517 (2003).
[5] Gordon, R. G., D. Hausmann, E. Kim, and J. Shepard, A kinetic model for step coverage by ALD in narrow holes or trenches, Chem. Vap. Deposition 9 (2) 73-78 (2003).
[6] Granneman, E., P. Fischer, D. Pierreux, H. Terhorst, and P. Zagwijn, Batch ALD: Characteristics, comparison with single wafer ALD, and examples, Surface & Coatings Tech. 201 8899-8907 (2007).
[7] Perez, I. E. Robertson, P. Banerjee, L. Henn-Lecordier, S. J. Son, S. B. Lee, and G. W. Rubloff, TEM-based metrology for HfO2 layers and nanotubes formed in anodic aluminum oxide nanopore structures, Small 4 1223-1232 (2008).

See more of this Session: Molecular and Mesoscopic Systems: Methods and Applications
See more of this Group/Topical: Computing and Systems Technology Division

(317c) Substrate-Induced Oscillatory Activity of An Enzyme Regulated by Multisite Phosphorylation

Ping Liu, Ioannis G. Kevrekidis and Stanislav Y. Shvartsman, Department of Chemical and Biological Engineering, Princeton University, Princeton, NJ

Despite the fact that the MAPK signaling cascade has been well studied before due to its fundamental importance in life science, studies of oscillatory behavior associated with it are relatively few.

One of these studies [1] shows that oscillatory activity can arise *proactively* from the backward coupling between a bistable double phosphorylation stage and a previous upstream stage in the cascade.

Recent studies [2] demonstrated that substrates can control the level of MAPK phosphorylation *retroactively* by competing with phosphatases.

In this work, we demonstrate that oscillatory activity can also be induced *retroactively* in a double phosphorylation stage by a *downstream* substrate, whose degradation is catalyzed by the end product of the stage (the double phosphorylated enzyme form); this constitutes a forward coupling between the double phosphorylation stage and a downstream stage in the cascade.

We propose a simple form of such a system, and present parameter values for which oscillatory activity arises.

We then reduce the full model (using time-scale separation) to a simplified one, which qualitatively agrees with the full model at the appropriate limit.

We quantify the behavior in the form of a two-parameter diagram: one parameter is related to the level of *input* to the cascade, while the other is related to the *load* (the amount of downstream substrate).

The resulting monostable, bistable and oscillatory regimes are marked on this diagram. Based on our observations we hypothesize the existence of two necessary (even if not sufficient) conditions for the oscillatory activity to arise in this system: (1) an associated subsystem – keeping the substrate level fixed – is bistable; and (2) a separation exists between the time scale of substrate degradation (slow) and those associated with the double phosphorylation stage (fast).

References

1. Qiao, L., R. B. Nachbar, I.G. Kevrekidis, S. Y. Shvartsman. 2007. Bistability and oscillations in the Huang-Ferrell model of MAPK signaling. *PLOS Comput. Biol.* 3:1819-1826.
2. Kim, Y., Z. Paroush, K. Nairz, E. Hafen, G. Jimenez, and S. Y. Shvartsman. 2011. Substrate-dependent control of MAPK phosphorylation in vivo. *Mol. Syst. Biol.* 7:467.

See more of this Session: Molecular and Mesoscopic Systems: Methods and Applications
See more of this Group/Topical: Computing and Systems Technology Division

(317d) Stochastic Modeling and Monte Carlo Simulation of Bacterial Disinfection: Generalized Approach

Andres Argoti[1], L. T. Fan[1], Ronaldo G. Maghirang[2] and S. T. Chou[3], (1)Department of Chemical Engineering, Kansas State University, Manhattan, KS, (2)Department of Biological and Agricultural Engineering, Kansas State University, Manhattan, KS, (3)Department of Finance and Banking, Kun Shan University, Yung-Kang City, Taiwan

A generalized stochastic model for bacterial disinfection is introduced herein; its formulation is based on a highly non-linear intensity of transition given in terms of the product of the number concentration of bacteria and a power function of time. The exhaustive disinfection of bacteria is of paramount importance in various industries, especially those involving food and pharmaceuticals or those involving cattle-feeding operations; it is essential in eliminating pathogenic bacteria to guard against their deleterious effects on humans and animals. The disinfection of bacterial populations in fluid media entails the elimination or attenuation of vast numbers of microorganisms, which are discrete and mesoscopic in nature. These microorganisms exhibit incessant and irregular motion as well as complex non-linear behavior: Their motion is self-propelled due to their motility and is strongly affected by the flow of the surrounding fluid media and also by collision among themselves as well as with the surrounding vessel surfaces, and mixing devices. Consequently, various attributes of the bacteria during disinfection, such as their number concentration and size, will exhibit random, or stochastic, fluctuations as time progresses. These fluctuations will tend to be especially intense at the tail end, or termination period, of disinfection when the number of bacteria becomes exceedingly small. It is, therefore, highly desirable that the resultant random fluctuations be explored via stochastic paradigms; however, relatively little has been accomplished so far in this regard. The generalized, non-linear stochastic model proposed in the current contribution for bacterial disinfection aims at filling such a void. The model leads the master equation, which can be analytically solved for particular instances by conventional mathematical approaches or by rational approximation methods, e.g., system-size expansion. Nevertheless, the master equation has been simulated most frequently via the Monte Carlo method to circumvent the inherent complexity of solving it analytically or numerically by conventional techniques. To illustrate, the mean, variance (standard deviation), and coefficient of variation of the number concentration of bacteria during disinfection have been evaluated from their analytical expressions whenever possible and/or through Monte Carlo simulation otherwise. The results are in line with the available experimental data as well as with those computed from the corresponding deterministic models.

See more of this Session: Molecular and Mesoscopic Systems: Methods and Applications
See more of this Group/Topical: Computing and Systems Technology Division

(317e) Molecular Thermodynamics Modeling of Water Equilibrium and Heat of Sorption In Human Stratum Corneum

Victor R. Vasquez, O. Hanbury and Charles Coronella, Chemical Engineering, University of Nevada, Reno, Reno, NV

Understanding the thermodynamic water interactions in biological materials plays an important role for many scientific and engineering applications. In particular, preservation in artificial environments or skin care using moisturizers and other drugs. The stratum corneum (SC) is part of the human skin, which serves as an important barrier to maintain the physical and chemical equilibrium with the environment for many compounds, in particular water. Due to SC's rate-limiting diffusion characteristics, it is a good model to study water-skin interactions, in general. The description of water in SC is commonly performed using extensions of the Langmuir and the Brunauer-Emmett-Teller (BET) adsorption models developed mainly to describe the adsorption of gases in hard surfaces such as metals. A common extension of this approach is the Guggenheim-Anderson-deBoer (GAB) isotherm model. These models do not take into account other important water chemical activity effects in soft-biological materials such as mixing and elastic contributions. In this work, we use a molecular thermodynamics framework to describe the equilibrium moisture content (EMC) in SC. The parameters of the model are obtained using experimental data reported in the literature. The proposed modeling approach, in addition to describe EMC, is easily expanded to obtain the water heat of sorption. This molecular-based model provides a a more fundamental view on the water interactions in biological materials from a classical thermodynamics perspective. This is achieved by building water free energy expressions with physical parameters that capture better the contributions of adsorption, energies of mixing, and elastic or stretching of the material. The estimation of water heat of sorption is done naturally by calculating the difference in enthalpy between the SC and vapor phases -- extracted from the free energy expressions. The results of the model show very good agreement for EMC and heat of sorption of previously reported data on this system, with model parameters describing water molecular thermodynamic interactions more naturally.

See more of this Session: Molecular and Mesoscopic Systems: Methods and Applications
See more of this Group/Topical: Computing and Systems Technology Division

(317f) Predator-Prey Targeting In Porous Medium by Chemical Swarm Robots

Peter Grancic, Chemical Robotics Laboratory, Department of Chemical Engineering, Institute of Chemical Technology, Prague, Prague, Czech Republic and Frantisek Stepanek, Department of Chemical Engineering, Institute of Chemical Technology, Prague, Prague, Czech Republic

The multi-body dynamics of a system of composite Brownian particles (here referred to as the chemical swarm robots) in a porous environment is investigated. In general, the structure of the particles allows to encapsulate specific chemical agents (e.g., valuable chemical cargo or chemical signals) into their internal compartments. The encapsulated chemicals can be delivered towards a given target location and released in response to external stimuli (e.g., local change in a given solute concentration or temperature). Two types of particles are considered (1) prey particles characterised by increased motility, pre-programmed to release chemical signals (chemo-attractant) in the proximity of a given target and (2) predator particles undergoing chemotaxis with respect to the chemical signal concentration gradients, responsible for the delivery of valuable chemical cargo. The ability of the particles to coordinate their movement is determined by a multitude of individual attributes of the particles, such as diffusivity, rates of the chemical signal release, gradient sensing strength, initial distance and signalling threshold. For a number of applications, such as distributed chemical processing and targeted drug delivery, the understanding of factors that govern the collective behaviour of the particles, especially their ability to localise a given target is of immense importance.

A predator-prey model based on the combination of Brownian dynamics method for discrete particle movement and diffusion problem for continuous mass transport coupled through the chemotaxis phenomena is analysed in terms of the ability of the particles to localise a given target. The results demonstrate that the selected performance criteria (i.e., the mean localisation time, the success rate and the mean target residence time) can be improved when appropriate signalling process is chosen. Furthermore, for an optimum target localisation strategy, the topological complexity of the porous environment needs to be reflected. The performed series of parametric studies represent a computer-aided design of real chemical swarm robots, fully based on chemical principles.

See more of this Session: Molecular and Mesoscopic Systems: Methods and Applications
See more of this Group/Topical: Computing and Systems Technology Division

(330a) Global Optimization of Thermochemical-Based Coal, Biomass, and Natural Gas to Liquids Processes Via Lograthmic Partioning Schemes

Richard Baliban, Josephine A. Elia, Ruth Misener and Christodoulos A. Floudas, Chemical and Biological Engineering, Princeton University, Princeton, NJ

Thermochemical conversion of biomass, coal, and natural gas to produce liquid fuels often includes the production of Fischer-Tropsch (FT) liquids utilizing a synthetic gas (syngas) intermediate [1]. Specifically, coal gasification, biomass gasification or natural gas reforming is used to create the syngas which is then converted to liquid hydrocarbons in the FT reactors [2,3]. The raw liquid product is subsequently upgraded to produce the necessary hydrocarbon fuels. Current studies generally focus on a small set of process designs with specific operating conditions to evaluate the cost of crude oil that would make the design competitive with petroleum derived fuels. However, it remains unclear whether any one of the designs represents the topology of a refinery that can produce transportation fuels at the lowest cost. Additionally, even if the process topology is correct, the selection of process operating conditions in the given process design may be suboptimal, leading to a higher cost of liquid fuels. To address both of these issues, a process synthesis approach can be used which can assess a large feasible set of process flowsheets to find the one(s) that yield a minimal liquid fuels cost [4]. A mixed-integer nonlinear non-convex optimization based mathematical model (MINLP) that seeks to minimize the cost of liquid fuels production would allow for analysis of the distinct topologies and operating conditions.

To ensure a high-quality solution to the MINLP, a global optimization approach is proposed to find the CBGTL refinery system topology that provides the lowest levelized cost of transportation fuel production. Continuous variables are used to model stream flow rates, process operating conditions, and thermodynamic properties while binary variables are used to model the logical use of a process unit or a process stream. The MINLP model includes simultaneous heat and power integration utilizing heat engines to recover electricity from the process waste heat. A branch-and-bound scheme is introduced that utilizes piecewise linear underestimators to determine tight lower bounds on the optimal solution. The capability of the method is demonstrated using two large case studies that feature 2,266 nonconvex terms, 21 binary variables, 15,232 continuous variables, and 15,533 constraints.

[1] National Academy of Sciences, National Academy of Engineering, and National Research Council. Liquid Transportation Fuels from Coal and Biomass: Technological Status, Costs, and Environmental Issues. Prepublication. Washington, D. C., EPA, 2009.

[2] R. C. Baliban, J. A. Elia, and C. A. Floudas. Toward Novel Biomass, Coal, and Natural Gas Processes for Satisfying Current Transportation Fuel Demands, 1: Process Alternatives, Gasification Modeling, Process Simulation, and Economic Analysis. Ind. Eng. Chem. Res., 49:7343-7370, 2010.

[3] J. A. Elia, R. C. Baliban, and C. A. Floudas. Toward Novel Biomass, Coal, and Natural Gas Processes for Satisfying Current Transportation Fuel Demands, 2: Simultaneous Heat and Power Integration. Ind. Eng. Chem. Res., 49:7371-7388, 2010.

[4] R. C. Baliban, J. A. Elia, and C. A. Floudas. Optimization Framework for the Simultaneous Process Synthesis, Heat and Power Integration of a Thermochemical Hybrid Biomass, Coal, and Natural Gas Facility. Comp. Chem. Eng., doi:10.1016/j.compchemeng.2011.01.041,2011.

See more of this Session: Process Design II
See more of this Group/Topical: Computing and Systems Technology Division

(330b) A New Algorithm for the Global Optimization of Property-Based Water Integration In Eco-Industrial Parks

Eusiel Rubio-Castro[1], Jos Mara Ponce-Ortega[2], M.M. El-Halwagi[3] and Medardo Serna-Gonzlez[2], (1) Chemical Engineering, Universidad Michoacana de San Nicols de Hidalgo, Morelia, Michoacan, Mexico, (2)Chemical Engineering Department, Universidad Michoacana de San Nicols de Hidalgo, Morelia, Michoacan, Mexico, (3)The Artie McFerrin Department of Chemical Engineering, Texas A&M University, College Station, TX

Most of the reported methodologies for the wastewater inter-plant integration (i.e., eco-industrial parks) using mathematical programming approaches have three major disadvantages: simplified superstructure, based on composition and do not guarantee the global optimal solution. Therefore, this paper presents an optimization MINLP model based on properties for the wastewater integration in eco-industrial parks considering a new superstructure that involves all configurations of interest like water integration in a single plant and interplant, the existence of sheared treatment units and treatment units exclusive for each plant. The interplant water network considers the regeneration of the streams to improve their properties through a treatment system, whose selection is optimized to meet the property constraints in the process sinks in each plant and the constraints for the environmental discharge. The problem consists in finding the configuration with the minimum total annual cost (i.e., treatment units cost, fresh water cost, piping cost) given a set of process effluents with flowrates and properties known, a set of process sinks with flowrates known and permissible limits of properties. This paper proposes a new strategy for the global optimization of the bilinear terms using a spatial search that solves iteratively two models for the lower bound (using a convex relaxed model) and for the upper bound (using a convex integer model) until the gap between these two models is smaller than a given tolerance. To obtain a valid lower bound, a piecewise under estimator based on the properties has been proposed; however, this increases the problem size but the advantage in the proposed model is that every bilinear term depends on the conditions (flowrates and property operator) in the interceptors, then if the property operators are transformed to parameters we can see the effect in the overall process with only vary the value of the property operator in the regeneration zone, and it is easier than to analyze the effect of all possible flowrates in the superstructure. In addition, this work proposes a new strategy for the branching variables, which is based on the property interceptors too. The model was implemented in the software GAMS and it was applied to several examples without numerical problems. The results show that the optimal solution is reached in a few iterations.

See more of this Session: Process Design II
See more of this Group/Topical: Computing and Systems Technology Division

(330c) Path Flow Indicators As Optimization Drivers In a Systematic Retrofit Methodology for Chemical Batch Processes

Stavros Papadokonstantakis[1], Andrea Anja Bumann[1] and Konrad Hungerbhler[2], (1)Chemistry and Applied Biosciences, Institute of Chemical- and Bioengineering, Swiss Federal Institute of Technology (ETHZ), Zurich, Switzerland, (2)Chemistry and Applied Biosciences, Institute for Chemical- and Bioengineering, Swiss Federal Institute of Technology, Zurich (ETHZ), Zurich, Switzerland

In a competitive globalized economy retrofit design of chemical plants is typically motivated by economic incentives to meet dynamic market requirements in cost-effective ways subjected to environmental and societal constraints. Especially for batch processes producing low volume-high value chemicals retrofitting is of high importance due to inherent process layout flexibility, uncertainty in product demand and shorter time-to-market periods leading to limited time for process development and operation under non-optimal conditions. It is therefore highly important to develop systematic methodologies, conceptually robust yet easily applicable in industrial practice, in order to enhance the often empirically based retrofitting. In this perspective, a good starting point can be systematic methods originally developed for continuous process design (i.e., heat and mass integration, superstructure optimization, flowsheet decomposition etc.) with proper adjustments for the intrinsic special features of chemical batch processes.

The present study applies a source-sink oriented path flow decomposition method, originally developed for continuous processes, to the field of batch process retrofitting. In particular, it is intended to enhance the concept of path flow indicators targeting at process improvement potential by revealing the extent to and the conditions under which they can serve as multi-objective optimization drivers. The study also focuses on the boundary conditions of the batch process retrofitting problem formulation, and in particular on the batch-to-batch variability effect for base case definition and interpretation of the retrofitting results.

The systematic retrofit methodology consists of three stages summarized below. In the first stage, the base case scenario is defined. First, process boundaries, objectives and path flow indicators are selected. Then, the real process performance is investigated based on historical process data. After this evaluation, it has to be decided whether process performance is sufficiently described by an average batch or whether batch-to-batch variability has to be included. For instance, batch-to-batch variability can be represented as a set of reference batches that describe the base case scenario. In a next step, process modeling is conducted and mass and energy balances of the base case(s) have to be derived. At the end of this stage, the objective functions of the base case(s) are calculated, which serve as reference in the next stages.

The second stage comprises a procedure that systematically generates promising retrofit alternatives for the respective base case(s). The procedure decomposes the whole flowsheet of the base case into component path flows, for which various path flow indicators are calculated that quantitatively or qualitatively describe the process performance. In this work, a new path flow indicator, called "renting cost indicator", is introduced that enlarges the scope of batch process retrofit alternatives by considering the occupancy times of the component path flows in the process units. All path flow indicators are linked to heuristics which in combination with sensitivity analysis result in parametric and/or structural retrofit alternatives of the base case(s) targeting at improved process performance with respect to these indicators.

In the last stage of the methodology, the retrofit alternatives are classified and prioritized. The objective is not restricted to economic process improvement as ecological aspects related to process energy consumption are also investigated and Pareto-front analysis is conducted. More importantly, the generated retrofit alternatives are also classified according to path flow indicator differences compared to the base case(s). This enables a more detailed analysis of the diverse effects a retrofit alternative can have. In particular, it is investigated in what extent the indicator values are shifted after applying the respective retrofit alternative. Since the indicator values of the base case(s) are typically used as drivers to define retrofit alternatives, the quantification of the effect of retrofit alternatives on the indicators is crucial for the robustness of the methodology. In this context, the objective function differences of the retrofit alternatives are compared to the path flow indicator differences in order to analyze the trends. For the analysis of the trends a new approach is proposed that considers batch-to-batch variability to cluster similarly performing retrofit alternatives, providing a more robust interpretation of the retrofitting results.

The presented retrofit methodology was applied to an industrial mono-product batch process for a specialty chemical. Based on the process data treatment results the methodology was first evaluated for an average batch, and then a set of reference batches was selected to describe batch-to-batch variability. For all scenarios, the path flow indicator matrix of the base case(s) was calculated and with the help of applicable heuristics and sensitivity analysis on process parameters promising retrofit alternatives were defined. The newly introduced renting cost indicator identified the potential for unit occupancy time reduction and was demonstrated to be a valuable complement to the rest of the path flow indicators. Moreover, trends have been revealed between the impact of retrofit alternatives on well-defined sections of the path flow indicator matrix and their effect on single objectives and multi-criteria Pareto-fronts. This is an important step towards a better understanding of the conditions under which a retrofit action originally inspired by path flow indicator values will be indeed successful from an overall system perspective. Finally, batch-to-batch variability innovatively coupled with a fuzzy clustering approach to provide a differentiation measure for retrofit alternatives was shown to smoothen the trendlines between changes in path flow indicator matrix sections due to retrofit alternatives and their multi-objective Pareto front relative positions. Overall, the application of the method has been able to reveal retrofit alternatives leading to 5-10% reduction in the environmental impact of an industrial batch process already running for a significant time period.

See more of this Session: Process Design II

(330d) Computer Aided Flowsheet Design Using a Group Contribution Based Approach

Susilpa Bommareddy and Mario Richard Eden, Department of Chemical Engineering, Auburn University, Auburn, AL

There are many different approaches to process synthesis including expert systems, optimization or algorithmic methods, and conceptual methods based on physical insights. This paper highlights a novel hybrid method for Computer Aided Flowsheet Design (CAFD) that combines physical insights with algorithmic reverse design approaches to enable systematic identification of feasible flowsheets at significantly reduced computational expense.

The framework presented in this paper is based on the process group (PG) contribution approach developed by d'Anterroches and Gani (2005). The CAFD approach was inspired by the group contribution based methods for Computer Aided Molecular Design (CAMD), which includes building blocks (atoms and functional groups) to generate and represent molecules; group contribution (GC) based property models to predict target properties; a standard molecular structure notation system (such as SMILES) to store and visualize the molecular structure information; and a synthesis method to generate and screen molecules that match the target (design) properties. Analogous to CAMD, in the CAFD approach, flowsheets are generated and represented by functional process groups; process group contribution based property models are employed to predict flowsheet properties; a notation system (called SFILES) is used for storing the flowsheet structural information; and a synthesis method is used to generate and identify the feasible flowsheets.

Like functional groups in molecules that are characterized by atoms and their molecular weight, each process group is characterized by the type of unit operation/process and their corresponding driving force. Each PG contributes to the flowsheet (performance) properties, which can be calculated once a feasible configuration has been identified. The candidate flowsheets are ranked based on performance criteria like energy consumption, amount (mass) of external agents used and/or cost/profit. Once a set of near optimal flowsheet alternatives have been identified, rigorous simulation is used to verify the predicted performance and select the best option.

See more of this Session: Process Design II
See more of this Group/Topical: Computing and Systems Technology Division

(330e) A Modeling Approach to for Chemical Process Synthesis

Carlos A. Henao, Chemical and Biological Engineering, University of Wisconsin - Madison, Madison, WI and Christos T. Maravelias, Department of Chemical and Biological Engineering; DOE Great Lakes Bioenergy Research Center, University of Wisconsin-Madison, Madison, WI

Introduction

For more than thirty years, advances in discrete-continuous optimization formulations and solution techniques have contributed to the development of systematic superstructure-based methodologies for the design of chemical process technology. In theory, such approach has the potential to yield a true optimal solution to the process synthesis problem since it pursues a simultaneous determination of the best process structure and operational conditions, thus accounting for all the complex interactions between design decisions. However, the practical use of superstructure optimization presents its own challenges in terms of the time required to set up the realistic optimization models in a way that is also consistent with the capabilities and limitations of current modeling tools and numerical solvers.

In this work, a refined methodology for the formulation of superstructure optimization models is presented, including a new way to generate superstructures, as well as a modular implementation of models supported by the use of set handling capabilities present in common mathematical programming languages (e.g. GAMS), and the incorporation of surrogate models to reduce the inherent mathematical complexity, allowing the application of current optimization solvers (e.g. DICOPT, BARON). We show how such approach leads to rich and yet solvable models whose final implementation can be created, modified and set up for solution in a very short time.

Superstructure generation

The generation of a proper superstructure is the first key step in the successful application of the mentioned optimization-based process synthesis methods. In fact, since the proposed superstructure supports the formulation of the optimization model, it ultimately determines its search space as well as its mathematical complexity. The richer the superstructure (in terms of the number of unit operations and interconnections), the wider the search space and the higher the probability of finding the best process configuration; however, richer superstructures also lead to more complex optimization models.

Within our methodology, superstructures are build from three basic element types: Processing units, unit ports and streams. Processing units, include reaction and separation operations only. Here, ports are used as interfaces to connect a unit to the rest of the superstructure. Inlet and outlet unit ports are considered stream mixers and splitters, respectively, allowing multiple units connected to every port. Streams are regarded as connections between an outlet unit port and an inlet unit port, being capable of performing a general conditioning operation (i.e. temperature and pressure conditioning). Given an initial set of operations, a high connectivity between ports results in a very rich superstructure. In this work, we discuss guidelines for the selection of a proper set of operations, as well as rules to avoid counterproductive connections. Ultimately, the goal is to create a rich but reasonable superstructure based on engineering judgment.

Effective superstructure modeling

Independent of the graphic representation and the mathematical programming strategy used (e.g. Mixed Integer Non-linear Programming.), a large superstructure model includes the models of its unit operations and streams, along with complementary constraints involving discrete variables that allow the simultaneous evaluation of multiple operating conditions and process configurations. In this way, the practical implementation of a large superstructure optimization model requires for the use of a modular approach, where building block models are developed independently for every superstructure element type (e.g. units, unit ports, and streams) with constrains and variables indexed over proper control sets.

For our work, we have implemented these standard element models using GAMS. A distinct model was build per unit operation type in terms of connection port variables and the unit operational variables. Raw material and unit outlet ports are modeled as multi-stream splitters. Product and unit inlet ports are modeled as multi-stream mixers enforcing pressure balance on their inlet streams. Streams are modeled as general conditioning units combining compression/expansion and heating/cooling capabilities. The stream model can be as simple or complicated as required (e.g. including multiple cooling/heating and isentropic/adiabatic compression/expansion stages) providing flexibility and the possibility of additional modeling developments to account for process heat and power integration. All the constraints (i.e. equations and inequalities) in these models are indexed by one or more of the following fundamental sets: components (user supplied), units (calculated from a user supplied parameter indicating the number of units per type included in the superstructure), ports (calculated automatically from the set of units and a predefined list of in-out ports per unit type) and streams (declared as outlet-inlet port pairs and calculated automatically from a user supplied parameter).

The mentioned modular approach, combined with set indexing, allows the set up and modification of complex superstructure optimization models in a very short time by specifying a short list of key parameters: the component list, number of units per type, and list of connections.

Mathematical complexity and the use of surrogate models

Most of the mathematical complexity in a superstructure optimization model comes from its discrete nature and the highly non linear equations used to realistically model component mixtures and chemical processing units. In principle, the inherently large number of equations, variables and types of nonlinearities (including non-convexities) lead to models that are difficult to solve, even when using state of the art optimization solvers. To reduce such complexity, we propose the replacement of complicated and highly non-linear equations with surrogate models. This replacement can be done at different levels, going from individual equations (e.g. thermodynamic properties calculations and reaction kinetics) and parts of unit operation models, to entire unit models, and even entire plant subsystem models.

The mentioned approach leads to the creation of unit surrogate models composed by many of the linear equations in the original unit models (e.g. material and energy balances, etc.) and a set of dimensionally reduced mappings which help establishing the correct non-linear relationship between the surrogate model variables. The final surrogate implementation involves the construction of non-linear mappings as multivariable surfaces built by fitting data generated using commercial process simulators. We have also established that the use of surrogate thermodynamics (e.g. a surrogate model that allow the calculation of key thermodynamic properties from mixture state variables) is key in our approach. For this project, a MATLAB code including a MATLAB-(APEN PLUS) interface has been developed to automatically generate samples of the independent variables, execute APEN PLUS simulation cases, retrieve simulation results and create the mentioned mappings via data fitting. Several examples are included to present the key modeling ideas as well as the final implementation and computational efficiency of the methodology.

See more of this Session: Process Design II
See more of this Group/Topical: Computing and Systems Technology Division

(330f) Biomass to Bio-Chemicals: Extractive-Reaction Processes for the Production of 5-Hydroxymethylfurfural

Ana I. Torres[1], Prodromos Daoutidis[1] and Michael Tsapatsis[2], (1)Chemical Engineering and Materials Science, University of Minnesota., Minneapolis, MN, (2)Chemical Engineering & Materials Science, University of Minnesota., Minneapolis, MN

Furanic compounds such as 5-hydroxymethylfurfural (HMF) have the potential to serve as biomass based substitutes of petroleum derived building blocks and are thus key intermediates in the production of biomass derived fuels and chemicals. HMF synthesis is based on the acid catalyzed dehydration of sugars, mainly hexoses, a reaction that is highly non-selective when taking place in aqueous media. In order to improve HMF yield, several laboratory scale reaction- separation processes have been reported, with their common characteristic being the presence of multiple processing steps and interconnected unit operations. Establishing their feasibility from an economic and energy efficiency perspective, and choosing the best processing path requires the evaluation of these processes from a systems perspective. Yet, studies in this direction are still scarce.

This work focuses on the design and optimization of processes that consider the solvent extraction approach as a way to improve HMF yield. These processes consist of a biphasic reactor in which HMF decomposition is prevented by the addition of an organic solvent that selectively removes the HMF produced in the aqueous phase. In order to improve HMF recovery, a liquid-liquid extractor usually needs to be coupled to the reactor (Roman-Leshkov et al., Science, 2006).

Here, we develop a novel process in which extraction and reaction are combined in a single unit, the extractive-reactor. These type of extractive-reaction processes have been broadly applied to the production of metals and their application to the production of chemicals has been mentioned in the literature (Minottti et al., IECR, 1998). The proposed process consists of a tubular biphasic reactor-extractor in which the aqueous (reactive) phase containing the sugar (fructose) and catalyst and the organic (extractive) phase are fed counter-currently. The HMF extracted by the organic phase is later on separated from the solvent by evaporation. Both the evaporated solvent and the aqueous solution containing unreacted fructose are recycled back to the extractive-reactor. For this process, we formulate an optimization problem in order to find the minimum cost of HMF that balances raw material, energy and capital costs. The effect of different solvents, fructose prices and kinetic scenarios is also addressed.

The results obtained showed an improvement over those published in our previous work (Torres et al., Energy & Environmental Science, 2010) in which a continuous process consisting of a separate reactor and extractor was studied. A comparison of performance, capital and manufacturing costs of both optimal processes is presented at the end of the study.

(361a) Molecular Simulation Studies On the Rheological Properties of Silica Nanoparticles Embedded In a Polyethylene Melt

Yangyang Shen and M. Silvina Tomassone, Chemical and Biochemical Engineering, Rutgers University, Piscataway, NJ

The addition of nanoparticles to polymer composites has been shown to significantly influence their mechanical, optical, and electrical properties. We use a combination of thermodynamics theory, Molecular Dynamics (a coarse grained approach) and atomistics simulations to develop a molecular model for Nanoparticle Composites. In our coarse-grained model, eight methylene groups of polyethylene are represented by one soft bead. The nanoparticles are modeled as spherical clusters of beads kept together by rigid harmonic bonds. The particles move in a simulation box of dimensions 30nmx30nmx16nm. Using this coarse-grained model, we explore the diffusion of particles in the melt and rheological properties of the polymeric fluids and nanoparticle composites under shear flow. Specifically we investigate the effects of polymer chain length, nanoparticle dimension and filling fraction on viscosities. Over the range of shear rates studied, the shear viscosity is strongly dependent on the shear rate. These trends are quantified by fitting the power-law behavior. For all the polymer matrices, the shear viscosity consistently increases with filler concentration, and increases up to 2 orders of magnitude with respect to the viscosity of pure polymer. The monotonic increase of viscosity with filler concentration is commonly observed in experimental systems. We also demonstrate that the addition of nanoparticle fillers leads to a more pronounced shear thinning behavior, as a result of increasing shear rates in the gaps between the filler particles. The results of our simulations are compared with an experimental analysis of a polyethylene filled with spherical silica nanoparticles of various sizes. The viscosity of polymer nanocomposite is sensitive to the size of the filler particles. We found that decreasing the size of spherical filler particles while keeping the filler volume fraction constant leads to an increase of the viscosity. This is mainly due to the aggregation of small nanoparticles reduces the chain movements and slows down the relaxation of polymer, as a result, less material that flows, leading the melt viscosity to increase. For these polymer composites where fillers are in the nanometer scale, Einstein and Batchelor equations greatly under predict the data. These findings qualitatively agree with the experimental studies reported in the literature.

See more of this Session: Composites: Modeling of Composites
See more of this Group/Topical: Materials Engineering and Sciences Division

(361b) Long Glass Fiber Orientation In Extensional Flow

Kevin J. Meyer, Kevin C. Ortman and D. G. Baird, Chemical Engineering, Virginia Tech, Blacksburg, VA

Long Glass Fiber Orientation in Extensional Flow

Kevin J. Meyer, Kevin C. Ortman andDonald G. Baird

Department of ChemicalEngineering, Virginia Tech, Blacksburg VA 24061

With the need for alternative lightweightmaterials growing daily, significantly more research is being conducted in thearea of reinforced composites. A cheap and effective way of increasing theproperties of a polymer is to embed both short ($L < 1mm$) and long glassfibers ($L > 1mm$) in the matrix. The bulk properties of the composite materialin question are highly governed by the orientation of the embedded fibers. Whilesignificant effort and emphasis has been placed on studying the orientation offibers in injection molding, which are shear-dominated flow fields, the flowsare usually very complicated with final fiber orientation being governedprimarily by the processing technique. In extensional flow (shear-free flow) wepresent a way to govern the fiber orientation kinetics completely independentof processing technique. Under biaxial deformations, due to isotropic stressesoccurring during squeezing, the fibers retain their initial orientation. Withthis knowledge, initial fiber orientation can be chosen a priori thus governingfinal material properties. Currently, modeling of fiber orientation inthermoplastic composites is done by way of the widely used Folgar-Tucker modelmodified for slow orientation kinetics. Due to the over prediction of fiberorientation in shear and extensional flow fields, a different modeling approachfirst provided by Strautins and Latz,is presented as an alternative to modeling thermoplastic composites. More specifically,we present a way to obtain model parameters used in the orientation equationsfrom rheological data only. The orientation parameters are obtained by usingthe Dinh-Armstrong variation of the Lipscomb stressmodel for concentrated suspension. This technique provides a way to model fiberorientation without any knowledge of transient orientation data to predictfinal orientation. It is the purpose of this paper to apply current theoreticalmodels and compare these models to experimentally obtained orientation data togain insight into the orientation kinetics of long glass fibers in shear-freeflow fields for predictive ability of composite properties of compressionmolded samples.

See more of this Session: Composites: Modeling of Composites
See more of this Group/Topical: Materials Engineering and Sciences Division

(361c) Hybrid Model and Application In the Study of Interfacial Properties of Nanocomposites

Jie Feng and Sadhan C. Jana, Polymer Engineering, The University of Akron, 44325, OH

All-atom model provides accurate results for simulation of large systems, but is time-consuming. Coarse-grained model, on the other hand, accelerates the simulation but loses accuracy in many situations. To alleviate this dilemma, we propose a "concern oriented" hybrid approach (COHA), which achieves a balance between accuracy and simulation time. In this approach, all atom resolution is only kept for the parts that are of utmost importance, while coarse-grained resolution is applied for the rest playing minor roles. The proposed model will be elaborated using a few examples of the binding phenomenon of polymers on the inorganic surfaces in solutions. In these cases, the atomistic details for polymers and surfaces are kept, but the structures of the solvent molecules are coarse-grained. The coarse-grained force field for solvents has been identified via verification with experimental values of physical quantities, such as density and evaporation energy.

See more of this Session: Composites: Modeling of Composites
See more of this Group/Topical: Materials Engineering and Sciences Division

(361d) A More Realistic Model for the Study of Thermal Conductivity of Nanocomposites

Khoa N. D. Bui and Dimitrios V. Papavassiliou, School of Chemical, Biological and Materials Engineering, The University of Oklahoma, Norman, OK

Calculating the thermal conductivity (K_{eff}) for carbon nanotube-based nanocomposites is a challenging task due to the presence of thermal boundary resistance (TBR) as well as the randomness of dispersion state and of the geometry of carbon nanotubes (CNTs). It has been found that the TBR of carbon nanotube to carbon nanotube ($TBR_{CNT-CNT}$) can be higher than the TBR between a nanotube and the surrounding matrix [1-3]. The effective medium theory is capable of calculating K_{eff} assuming that the CNTs are in a perfect dispersion state and excluding the $TBR_{CNT-CNT}$ effect [4]. However, our previous work with CNTs modeled as stiff cylinders showed that the tendency of CNTs to create bundles and the $TBR_{CNT-CNT}$ can dramatically suppress K_{eff} at high volume fraction of CNTs [5]. The CNTs are observed to have "worm-like" shape in reality [4] and no prior work has been reported in studying the effect of the curvature (expressed in terms of the persistence length of the CNTs) on the K_{eff} of nanocomposites.

In this work, we will report on the development of a new algorithm allowing the generation of CNTs with worm-like geometry in 3D, and with different persistence length. The use of these geometries in conjunction with off-lattice Monte Carlo simulations [3,5] in order to study the effective thermal properties of nanocomposites will be discussed.

References

[1] J.E. Peters, D.V. Papavassiliou, B.P. Grady, Macromolecules, 41 (2008) 7274-7277.

[2] S. Maruyama, Y. Igarashi, Y. Taniguchi, J. Shiomi, J. Therm. Sci. Tech., 1 (2006) 138-148.

[3] K. Bui, H. M. Duong, A. Striolo, and D. V. Papavassiliou, J. Phys. Chem. C, 115 (2011)10.

[4] C. W. Nan, G. Liu, Y. Lin, and M. Li, Applied Physics Letters, 85 (2004) 16.

[5] K. Bui, B. P. Grady and D. V. Papavassiliou, Chemical Physics Letters, doi:10.1016/j.cplett.2011.04.005 (2011).

(361e) New Insight Into the Reaction Mechanism of SiC Oxidation and Nitridation: A Density Functional Theory Study of β-SiC (001) Surface

Satyender Goel and Linda J. Broadbelt, Chemical and Biological Engineering, Northwestern University, Evanston, IL

Wide interest in detailed investigations of SiC can be divided into two major areas: exploring their potential as high-frequency materials in electronic applications and assessing their continuous stability when exposed to high temperature environments. One common factor in both types of applications is the importance of SiC oxidation. The mechanism of this process likely proceeds via a complex set of serial and parallel reaction steps, and while predicting all of them is challenging mapping a comprehensive set computationally would offer significant insight into the dominant adsorption and reaction chemistry of gaseous species interacting with SiC surfaces. In this work, reaction pathways leading to active/passive oxidation of SiC surfaces were investigated systematically using density functional theory. First, adsorption of gaseous species (small molecules containing O, C, H and N) interacting in different orientations and positions on the β-SiC (001) surface was studied. Based on the most favorable adsorption configurations, various reaction channels were explored by identifying transition state structures. Interactions with both the carbon and silicon layers of SiC were analyzed. Interactions with the carbon layer were found to be important for NO dissociation leading to Si-C bond breaking by insertion. The observations made for atomic and multiatomic species adsorption on the SiC surface were extended to form reaction pathways resulting in SiC oxidation products, including CO, SiO_2 and SiO.

Rate constants were calculated with transition state theory, and microkinetic modeling was carried out to identify the dominant pathways that were mapped. The thermodynamic data was interpreted to understand the role of temperature in the formation of SiO instead of SiO_2.

See more of this Session: Composites: Modeling of Composites
See more of this Group/Topical: Materials Engineering and Sciences Division

(367a) Optimization and Control of Heteroepitaxial Surface Morphologies

Jacob McGill, Nasser Mohieddin Abukhdeir, Babatunde A. Ogunnaike and Dionisios G. Vlachos, Chemical Engineering, University of Delaware, Newark, DE

Recent discoveries have shown that surface pattern formation in the heteroeptitaxial growth of bimetallic thin films can greatly influence the properties of novel bimetallic catalytic materials and other thin-films technologies. In particular, for catalytic applications, nano-scale surface morphology has been shown to influence the reactivity and specificity of bimetallic catalysts and, thus, a method to control surface morphology effectively at nanoscopic length scales would provide a means to "tune" the reactive properties of a material for different applications. Currently, the only available technologies for directly influencing and controlling nano-scale pattern features involve "top-down" manufacturing approaches, such as lithography techniques, which are limited to feature sizes in the range of 30 nm. However, heteroepitaxial surface patterns form with characteristic lengths as small as approximately 1 nm. Self-assembly, the autonomous arrangement of molecules into nano-scale patterns due to intermolecular forces, has the potential provide a "bottom-up" approach to control surface pattern morphologies at various length scales ranging from about 1 nm to several μ m.

While the benefits of scalable "bottom-up" manufacturing over current "top-down" manufacturing are clear, practical implementation of controlled self-assembly has been hindered by the inherent nonlinearity and stochasticity of self-assembly dynamics. Model-based control techniques provide an efficient theoretical framework for overcoming these obstacles and, with the recent development of models that accurately capture the physics of pattern formation during self-assembly, they have become a viable option for designing and implementing strategies for controlling surface morphologies through self-assembly. Our ultimate goal is to create an efficient model-based control scheme harnessing self-assembly dynamics for controlling nano-scale surface morphologies. However, before such a control scheme can be implemented, a careful analysis of the controllability and sensitivity of a pattern with respect to a very limited set of manipulated variables (MVs) must be performed.

In this work, we present an analysis of a continuum model describing heteroepitaxial self-assembly of Pb on Cu(111) (Chatterjee and Vlachos, 2007; Abukhdeir, et al, 2011) where various metrics for quantifying patterns are studied in response to changes in a limited set of MV. The focus of our study is the controllability and sensitivity of the surface morphology of Pb patterns based on manipulation of a limited number of MVs that can be adjusted at the macroscopic scale, such as surface temperature, (sub-monolayer) Pb coverage, and Pb/Pb interaction strength, since, in a manufacturing setting, these will be the only MVs available to control the surface. In particular, this presentation will discuss how the surface morphology in several stable pattern regimes is affected by changes in the MVs, what the most effective metrics for quantifying surface patterns are, and to what degree the densities of several classes of defects can be minimized based on the imperfect information provided by these metrics. The implications of these findings on model-based control strategies for controlling surface morphologies will also be discussed.

Chatterjee, A. & Vlachos, D. G. Systems tasks in nanotechnology via hierarchical multiscale modeling: Nanopattern formation in heteroepitaxy. Chem. Eng. Sci., 2007

Abukhdeir, N. M., Vlachos, D. G., Katsoulakis, M. & Plexousakis, M. Long-time integration methods for mesoscopic models of pattern-forming systems. J. Comp. Phys., 2011

See more of this Session: Emerging Control Applications
See more of this Group/Topical: Computing and Systems Technology Division

Valuation of Chemical Processes Subject to Stochastic Price Uncertainty

Patrick Mousaw, Fanghui Fan, Fernando Garcia, and Jeffrey Kantor*

August 14, 2011

Abstract

The valuation of commodity chemical processes is a complex function of the potential to generate economic return in a complex stochastic environment. Unlike the financial options traded in the commodity markets, chemical processes provide the owner with considerable flexibility to adjust feedstocks subject to standard stoichiometric and thermodynamic balances. We have previously reported results on a new class of thermodynamically valid steady-state models for energy utilities and solutions to the resulting valuation and hedging problems. Here we extend our approach to two new classes of processes that include steady-state stoichiometric models and commodity storage.

Our approach is to use stochastic price models for chemical commodities coupled with standard stoichometric and energy balances. Process flexibilty is modeled as degrees of freedom in a linearly constrained mass and energy balances and, additionally, the operation of a commodity storage facility. Stochastic price models include multivariate Ito-type models such as mean-reverting Ornstein-Uhlenbeck model. To maximize expected value, we employ standard results to construct the Hamilton-Jacobi-Bellman equation to assign valuation and to resolve and process degrees of freedom.

In this paper we will outline the following results –

- Contrary to the case of simple financial options on commodities, the valuation of flexible chemical processes cannot be projected onto an underlying 'replicating portfolio' of market prices. Thus the standard assumptions of market completeness do not apply to flexible chemical processes. We therefore use arbitrage arguments to construct and upper (seller's value) and lower bounds (buyer's value) for a flexible chemical process. To the best of knowledge, this is the first report discussing the problem of incomplete markets for chemical process valuation.
- The valuation of commodity storage results in feedback control policy for optimal economic operation. We outline a generic economic model for commodity storage and demonstrate key analytical results.

References

*Author to whom correspondence should be addressed: Kantor.1@nd.edu

(367c) MPC and the Smart Grid: Optimal Operation and Transmission Expansion

Donald J. Chmielewski, Chemical and Biological Engineering, Illinois Institute of Technology, Chicago, IL

In the US, an overwhelming portion of electric energy is generated in environmentally unfriendly fossil fuel based power plants. While this has been accepted for decades, society is beginning to demand a more sustainable approach. Unfortunately, it is not just a matter of replacing sources MW by MW. While fossil based plants can be located near load centers, the efficiency of renewable sources is location dependent. Unfortunately, transmission infrastructure from locations with high renewable potential is either of low capacity or nonexistent. Furthermore, the uncertain power output of renewable sources suggests that dispacthable fossil based plants will need to be utilized to a much greater extent. Again, the existing infrastructure is ill-equipped to deal with such an increase in power flow variability.

To address these issues a number of approaches may be employed: increased usage of power flow control devices, construction of energy storage systems, and the addition of new transmission lines. The first provides additional flexibility in designating how power will flow though the grid. (For example power from an overloaded line may be routed to a less loaded line.) However, not all lines will need control infrastructure, and judicious allocation of these capital investments will be prudent. The second approach allows for available power to be time-shifted to periods of high demand. Again, the capital cost of these systems is significant, and location of a storage facility is typically limited by geographic considerations. The question of adding new transmission lines is dominated by identifying the appropriate location of the line (to maximally exploit the previous two actions), as well as design capacity and system type (AC or DC). Additional questions include the sizing and placement of new generation sources as well as determining if these sources should be fossil or renewable.

This paper will provide a review of the current methods used for optimal energy dispatch and how MPC can be used to enhance operation under the conditions of more renewable sources as well as the availability of new power flow control devices and energy storage systems. In addition, the procedure of transmission infrastructure expansion will be discussed in the context of smart grid operation.

See more of this Session: Emerging Control Applications
See more of this Group/Topical: Computing and Systems Technology Division

(367d) Microalgae Bioreactor Control by Model Predictive Techniques

Javad Abdollahi, Chemical and Materials Engineering, University of Alberta, Edmonton, AB, Canada and Stevan Dubljevic, Chemical & Materials Engineering, University of Alberta, Edmonton, AB, Canada

Bio-fuel is a sustainable replacement for fossil fuels because of less pollution and its renewablity, however production costs makes it not competeble with traditional fuel sources [1]. Today, industrial biodiesel production is based on oleaginous crops. Cultivation of these crops needs vast land and high amount of water. By the ndings from [2], the total farmland area of the earth is needed to produce biodiesel to support current power plant and transportation fuel consumption. Also, it is expected that the world will need almost 60% more energy in 2030 if current policies are held [3]. Bio-fuel production from phototrophic and heterotrophic microalgae is a novel method of oil production that is going to replace bio-fuel production from crops, see [1]. In this method there is no need for vast land, high amounts of water and in the case of heterotrophic microalgae there is no need for sunlight, see [1]. Furthermore, increasing oil production rate can ensure the economic feasibility of biodiesel production. In this work we explore model predictive techniques to maximize oil production rate.

We built this work on an existing heterotrophic microalgae fed-batch model [4] which describes the algae growth and oil production. The fed-batch microalgae reactor is fed by two nutrients, glucose as carbon source and glycine as nitrogen source. The experimental reactor setup is realized such that the temperature and pH are regulated at constant values by a conventional controller and the reactor is monitored by Raman spectroscopy laser which measures certain species and substrate concentrations. In addition from the model [4], growth, nitrogen uptake and oil production rates are highly nonlinear and the maximum value of rate functions are time varying [4].

In this work moving horizon estimator (MHE) and model predictive control (MPC) are used to estimate the states and parameters of microalgae bioreactor and consequently optimize and control the oil production rate. The highly nonlinear rate functions, time varying unknown parameters, unavailable full state measurements and uncertainties in modeling make the observer and controller design challenging. Unknown state/parameters estimation is realized by the moving horizon estimator which is necessary to be applied to microalgae bioreactor system since it is suitable for constrained state estimation [5, 6]. In particular concentrations in the bioreactor admit positive values and also the estimated parameters should be maintained in the reported ranges, see [7]. The control objective is to maximize the oil production rate, since feasibility studies showed a potential of about 20 times more production compared to the traditional cultivation of phototrophic microalgae [4]. Finally, the moving horizon estimator and model predictive control is implemented in order to solve the state/parameter estimation and control problem.

References

[1] Yusuf Chisti, Biodiesel from microalgae, Biotechnology Advances 25 (2007) 294-306

[2] Martin I. Hoert, et al, Advanced Technology Paths to Global Climate Stability: Energy for a Greenhouse Planet, Science, vol. 298 1 November (2002), 981-987

[3] Vishwanath Patil, et al, Towards Sustainable Production of Biofuels from Microalgae, International journal of molecular Science (2008), 9, 1188-1195

[4] H. De la Hoz Siegler, A. Ben-Zvi, R.E. Burrell, W.C. McCarey, The dynamics of heterotrophic algal cultures, Bioresource Technology 102 (2011) 5764-5774

[5] C.V.Rao, J.B. Rawlings, D. Q. Mayne, Constrained State Estimation for Nonlinear Discrete-Time Systems: Stability and Moving Horizon Approximations, IEEE Transaction on Automatic control, Vol. 48, No. 2, February 2003

[6] A. Alessandri, M. Baglietto, G. Battistelli, Moving-horizon state estimation for nonlinear discrete-time systems:new stability results and approximation schemes, Automatica 44 (2008) 1753-1765

[7] K. Surisetty, H. De la Hoz Siegler, W. C. McCarey, A. Ben-Zvi, Model reparametrization and output prediction for a bioreactor system, Chemical Engineering Science 65 (2010) 4535-4547

See more of this Session: Emerging Control Applications
See more of this Group/Topical: Computing and Systems Technology Division

(367e) Nanoscale Process Control Engineering

Raj Chakrabarti, Chemical Engineering, Purdue University, WestLafayette, IN

Most of our current chemical technologies are based on control systems geared to the laws of classical physics and chemical kinetics. Richard Feynman, in his famous 1959 essay "There's Plenty of Room at the Bottom," predicted the unfolding of a whole new era of industrial revolution if we knew how to control the dynamics of individual atoms, molecules and nanostructures – where timescales range from femtoseconds to microseconds. Such systems occupy a unique place in the hierarchy of dynamical systems to which control engineering may be applied, because of the transition that occurs between classical and quantum physical laws. The burgeoning field of nanoscale process control engineering entails the development of methods to steer atomic and molecular dynamics using lasers and other types of coherent inputs, which can exploit the wave-particle duality of matter to access reaction pathways – ranging from selective bond cleavage and rearrangement reactions to coherent energy transfer - and product yields that are impossible via conventional catalytic routes. Whereas nature has evolved sophisticated strategies for manipulating process dynamics on molecular length and timescales – e.g., in the coherent control of photosynthetic solar energy conversion, which is nearly 100% efficient -- our ability to control these process dynamics is only beginning to develop. Realizing the potential of this emerging field, the National Research Council (NRC) has put Controlling the Quantum World at the forefront of its 10-year agenda in atomic and molecular physics and the US Department of Energy (DOE) has based four of its Five Grand Challenges in energy science on nanoscale process control. This session will aim to introduce the methods of nanoscale process control engineering to the CHE systems community, review important recent developments and identify fruitful synergies for future research.

See more of this Session: Emerging Control Applications
See more of this Group/Topical: Computing and Systems Technology Division

(367f) Control of a Motor Intended Neural Prosthetic Finger Using a Network of Cortical Motor Neurons

Gautam Kumar[1], Nitish V. Thakor[2] and Mayuresh V. Kothare[1], (1)Chemical Engineering, Lehigh University, Bethlehem, PA, (2)Biomedical Engineering, Johns Hopkins University, Baltimore, MD

Neurophysiological systems, such as natural limbs, are partially driven by neural activities of the primary motor cortex (M1) projected to the spinal motor units [1]. Experimental evidence indicates that these M1 neural activities contain sufficient information which can be extracted to drive motor intended neural prostheses, such as artificial limbs. An interface that connects a motor intended neural prosthesis, wired or wireless, to the M1 is called brain-machine interface (BMI).

Current formulations of BMI are insufficient in providing feedback information, such as kinematic and dynamic senses of the prosthetic device movement to the brain. This information is necessary for the online error correction during the voluntary movement of the device. Micro-stimulation techniques, relatively new in BMIs application, have emerged as a promising approach in providing artificial proprioceptive feedback by stimulating the appropriate sensory areas of the brain. The aim here is to incorporate missing natural feedbacks, such as proprioception and tactile, available from the motor intended prosthetic device back to the brain. This may provide us an enhanced control over prosthetic limbs similar to our natural limbs. Although these techniques are promising for developing next generation BMIs, the experimental trial and error approach in designing appropriate sensory input currents may damage a part of the brain or modify the natural functionality of the brain. Therefore, it is appropriate to take a control-theoretic approach. To do this, we can use well established optimal feedback control theory as a parallel alternative towards developing next generation BMIs. This approach promises to provide flexibility in the design of optimal stimulating sensory input currents in addition to providing detailed analysis of the closed-loop neural prosthetic system under various conditions. Motivated by these facts as well as our previous work on the control of a prosthetic finger using a single M1 neuron [2], we extend our earlier work to provide a more generalized control-theoretic framework. Within this framework, neural activities of a network of M1 neurons are used to control a prosthetic device. We focus here on closed-loop control of a prosthetic finger about a single joint which is controlled by a small network of M1 neurons. Our framework consists of: a) a network of M1 neurons connected through seed neurons b) a decoder model which takes the instantaneous firing rate of M1 neurons and an ensemble based inter-spike interval (ISI, the time difference between two consecutive instantaneous ensemble firing of M1 neurons) and generates the encoded torque information, c) a prosthetic finger model which uses the torque to drive the finger about a single joint, and finally d) a receding horizon based optimal control strategy which uses system models as well as available feedback information, visual and proprioceptive, from the finger to generate optimal input currents needed to stimulate seed neurons.

In this presentation, we limit ourselves to consider a small network of neurons consisting of 4 seed neurons and 10 M1 neurons. Experimental evidence in neuroscience suggests that the M1 neurons interact with neurons from various other cortical areas of the brain and receive motor intended information by making synaptic connections with them. To keep our mathematical and computational formulation simple, we represent an ensemble of neurons within each cortical area of the brain by a seed neuron. Each seed neuron represents an individual cortical area namely pre-motor, somatosensory, post parietal and cerebellum. These cortical areas projects neural activities to M1 neurons by establishing synaptic connections. Each seed neuron captures firing rate behaviors of an ensemble of neurons. These seed neurons are activated by external stimuli, designed by the receding horizon based optimal control strategy using the feedback information available from the prosthetic system. Dynamic behavior of each M1 neuron is described by the Izhikevich single neuron model. Dynamic behavior of each seed neuron is described by a population based model. We define synaptic connectivity from seed neurons to M1 neurons using a well-defined feedforward model. Synaptic connectivity among M1 neurons are defined by a well-defined recurrent model. We assume that there is no synaptic connectivity among seed neurons. With time invariant synaptic connectivity and corresponding weights, we model the synaptic current to a M1 neuron in the form of the sum of alpha functions. These functions capture the effect of variability in spiking behaviors of other M1 neurons and seed neurons connected to the M1 neuron in the network. We further assume that all neurons in the network are excitatory. This means that each action potential in a presynaptic neuron, which occurs during the firing of a presynaptic neuron, makes contribution towards an increment in its postsynaptic neuron membrane potential. Thus, in terms of an input-output model, we define a small network of neurons where the inputs to the network are four current stimuli which stimulate each of the seed neurons. Outputs of the network are the number of M1 neurons emitting actions potentials at a given instant of time and the ensemble based ISI. These outputs are then used by a dynamical decoder model, developed by us, which decodes the torque information and thus actuates the prosthetic finger. The rest of the control-theoretic framework follows directly from our previous work on the design and control of an index finger using cortical activities of a single M1 neuron [2].

We consider a minimum time control problem in the receding horizon framework. The cost function for the controller is the total time required to reach the desired angular target space defined by a terminal set constraint. This cost is minimized over stimulating input currents of four seed neurons subject to two system constraints. The first constraint ensures that the resultant movement of the prosthetic finger is continuous and smooth. In terms of the firing activities of neurons, this means that the next firing of ensemble of motor neurons occurs before the initial movement generated by previous firings dies out. The second constraint defines a terminal set constraint which ensures that the goal directed voluntary movement terminates once the finger is in the region of target space defined by the terminal set constraint. At each time varying sampling time, defined as the ensemble based ISI, the controller computes four stimulating input currents. These computations utilize the information about the current position of the finger, which is available to the controller through the visual feedback. The overall control problem, at each sampling time, leads to a nonconvex nonlinear constrained optimization problem. This problem is solved numerically by implementing the infeasible interior point algorithm along with the Markov filter method.

Proprioceptive feedback carries information such as the sense of movement, position, force and effort required to move the finger which together provides the perception of the limb position to our brain. In the absence of natural proprioceptive pathways, we capture this feature of perception in a form of an artificial proprioceptive feedback current model. This model is a linear function of the torque applied at the joint and the velocity of the prosthetic finger. We project the feedback current induced by this model directly to all M1 neurons in the network such that each M1 neuron receives the same proprioceptive information. By varying time invariant gains of the proprioceptive feedback current model, we study the effect of proprioception on the activities of M1 neurons under the closed-loop framework. This study leads to an important conclusion that the firing rates of M1 neurons increase with the increase in external load opposing the direction of movement. This conclusion is consistent with the experimental observations made by Evarts and Fetz in early 70's [1]. The adaptation of M1 neuron activities subject to an external load would be impossible in the absence of proprioceptive feedback because the visual feedback alone cannot make distinction between external loads. This observation emphasizes the necessity of including artificial proprioceptive feedback in designing next generation BMIs. The second conclusion of this talk is the importance of visual feedback in rejecting internal noises. It is now widely accepted in the neuroscience community that neurons, in general, possess stochastic characteristics due to internal fluctuations while transmitting synaptic information. Therefore we introduce randomness in synaptic weights of the interconnected network of seed and M1 neurons. With this, we study the closed-loop and the open-loop behaviors of the overall system in the absence of proprioceptive feedback. In the open-loop framework, i.e. in the absence of visual feedback, we find that the controller completely throws off the angular trajectory of the prosthetic finger from the desired path. We further study how the activities of M1 neurons are modulated by including the visual feedback which drives the final angular position of the prosthetic finger to the desired target space.

E. R. Kandel, J. H. Schwartz and T. M. Jessell, Principles of Neural Science, Fourth Edition, McGraw-Hill, 2000.

G. Kumar, V. Aggarwal, N. V. Thakor, M. H. Schieber and M. V. Kothare, An Optimal Control Problem in Closed-Loop Neuroprostheses, Submitted to the 50th IEEE Conference on Decision and Control and European Control Conference, 2011.

See more of this Session: Emerging Control Applications
See more of this Group/Topical: Computing and Systems Technology Division

(393a) Characterizing the Pore Structure of Biochars: A New Approach Based On Multiscale Pore Structure Models and Reactivity Measurements

Hao Sun[1], Caroline A. Masiello[2] and Kyriacos Zygourakis[1], (1)Chemical and Biomolecular Engineering, Rice University, Houston, TX, (2)Earth Sciences, Rice University, Houston, TX

Biochar is charcoal generated for intentional soil amendment by pyrolyzing sustainable biomass feedstocks. Properly "engineered" charcoals can increase the water holding and cation exchange capacities of soils, improving the ability of plants to survive under drought conditions and reducing fertilizer runoff into watersheds. Fertilizer runoff has become a serious problem, because as much as 70% of fertilizer applied to crop fields is leached into the groundwater or lost to streams and rivers, eventually leading to large hypoxic 'dead' zones in the world's oceans (including the Gulf of Mexico). The environmental performance of biochars depends on their ability to adsorb, retain and release water and nutrients. These biochar properties are controlled by their porosity and surface chemistry, which can vary widely depending on the composition of the biomass feedstocks and on the pyrolysis conditions employed during biochar production.

Biochars have a complicated pore structure consisting of multiple interconnected networks of micropores, mesopores and macropores that span multiple length scales: from sub-nanometer micropores to macropores with sizes of the order of 10 microns. Such pore structures cannot be characterized by a single analytical technique. Comprehensive studies will use a combination of analytical techniques to bridge the vastly different length scales: adsorption of multiple gases (like nitrogen, carbon dioxide and water) for the micropores, mercury porosimetry for the mesopores and sectioning with optical microscopy and 3-D reconstruction techniques for the macropores.

We report here the development of a novel approach to probe the multiscale pore structure of biochars. This approach uses thermogravimetric measurements of the transient reactivity of biochars in air over a wide range of temperatures. At low temperatures, combustion proceeds in the regime of kinetic control and the entire surface area attributed to micropores is completely accessible to the reactant. As the temperature rises, the reaction regimes shifts to diffusion control and strong diffusional resistances start appearing first in the micropores and subsequently in the mesopores. Thus, larger and larger fractions of the micropore and mesopore structure will become inaccessible to oxygen as the temperature is raised. At sufficiently high temperatures, combustion will take place only on the micropore and mesopores "mouths," where they open up into the large macropore cavities identified in SEM or optical microscopy images.

The novelty of our approach lies in the use of sophisticated discrete models that can accurately describe the temporal evolution of the surface area of solid reactants with complicated pore structure. The new models are significant extensions of the continuous [1-2] and discrete models [3] that have appeared in the literature. Simulations start by generating solids with the desired pore structure on three-dimensional computational grids. Pores of various shapes and sizes are distributed on the grid in a random or orderly fashion to match the experimentally determined micro, meso- and macroporosities, as well as any available information about the shape of the pores. The generated porous solids are then eroded using rules that simulate non-catalytic reactions between a gas and a solid reactant. Access of the gas reactant into pores of progressively increasing sizes is restricted to model pore diffusional limitations and parallel simulations are carried out to handle the multiple scales of this problem. Multiple realizations of a solid with the same pore structural properties are generated and reacted to estimate the average pore surface and to establish confidence intervals. For a given pore structure, this process will generate a family of curves that give the evolution of pore surface area with conversion as the intraparticle diffusional resistances increase. A comparison of these surface evolution patterns to the char reactivity patterns measured at different temperatures can provide important information about the complex pore structure of carbonaceous materials [1-4].

Simulation results were compared with experimental data from the combustion of different biochars to validate this technique. Chars produced from different biomass feedstocks (apple wood or corn stover) exhibited different reactivity patterns in the kinetic control regime. The simulations showed that the reactivity patterns of corn stover chars are consistent with a random distribution of the micropores. On the other hand, apple wood chars exhibited reactivity patterns that indicate the presence of a subpopulation of orderly distributed micropores. These micropores must be the slit pores formed between the graphitic-like layers of aromatic carbon clusters that are turbostratically arranged in nanometer-size crystallites, a structure that has been confirmed with NMR and XRD measurements. The same corn stover and apple wood chars exhibited similar reactivity patterns when combusted with oxygen in the regime of strong diffusional limitations, when oxygen cannot penetrate deeply into the micro- and mesopores and the reaction takes place at the pore "mouths" where they open up into the large macropore cavities. These patterns were consistent with a macropore structure that consists of many large cavities separated by walls of similar thickness.

These results demonstrate how TGA data and pore structure models can be used to gain significant insights into the pore structure of biochars. The new approach has the potential to overcome some of the major difficulties encountered in characterizing the complex pore structure of biochars and other carbonaceous solids.

REFERENCES:

1. G. Ballal and K. Zygourakis, *Ind. Eng. Chem. Res.* **26**, 911-921 (1987).
2. S. Sotirchos, *Chem. Eng. Science*, **42**, 1262-1265 (1987).
3. K. Zygourakis and C.W. Sandmann, Jr., *AIChE Journal*, **34**, 2030-2040 (1988).
4. M. Sahimi, G.R. Gavalas and T.T. Tsotsis, *Chem. Eng. Science*, **45**, 1443-1502 (1990).

(393b) Modeling and Simulation of Globular Protein Crystal Growth

Michael Nayhouse[1], Panagiotis D. Christofides[2] and G. Orkoulas[2], (1)Chemical and Biomolecular Engineering, UCLA, Los Angeles, CA, (2)Department of Chemical and Biomolecular Engineering, University of California, Los Angeles, Los Angeles, CA

The production of highly-ordered, high quality protein crystals through batch crystallization processes is vital in devising proteins for therapeutic purposes. Despite extensive experimental and theoretical work on understanding protein structure and function, there is lack of a systematic framework that relies on fundamental understanding of the nucleation and growth mechanisms of protein crystals at the microscopic level and utilizes such information to model and operate protein batch crystallization processes at the macroscopic level. The present work focuses on investigating the growth of protein crystals via molecular dynamics and kinetic Monte Carlo simulations and using such knowledge to model and control batch protein crystallizers through population, mass and energy balances. The modeling and simulation work is tested against available experimental data.

The crystal growth of globular proteins is a non-equilibrium process and it is thus simulated using kinetic Monte Carlo methods. As is common practice in simulations of crystal growth, the solid-on-solid model, in which molecules are deposited on the growing crystal without voids or overhangs, is adopted. In the simulations, only molecular attachment and detachment events are considered and surface diffusion effects are ignored. The implementation of the kinetic Monte Carlo methodology requires knowledge of the binding energies, the impingement rate, and the crystallization driving force. Previous work assigned a range of values to these parameters until satisfactory agreement between the calculated and the experimental growth rates was achieved. In the present work, the parameters required for the growth of protein crystals are calculated from molecular dynamics simulations of coarse-grained models for protein solutions.

The calculated growth rates are used in macroscopic models describing the evolution of entire crystal size distributions to achieve optimal crystallizer design and operation for large scale production. The population balance modeling framework is adopted to describe globular protein crystallization in batch systems from supersaturated solutions. The mathematical description of batch crystallization also requires conservation equations for the solute (protein) and the energy. The previous mathematical model is used in conjunction with stochastic model predictive control methodologies that account for model uncertainty to achieve a desired crystal size distribution at the end of the batch protein crystallization process. In the model predictive control formalism, the cost function typically maximizes the expected value of the volume-averaged crystal size subject to input and state constraints. Extensive simulation results covering the various aspects discussed above will be presented.

See more of this Session: Multiscale Modeling: Methods and Applications

(393c) Modeling Excitation Energy Transfer In Photosynthetic Systems: Application to Peridinin-Chlorophyll-Protein Complex In Dinoflagellates

William P. Bricker, Energy, Environmental & Chemical Engineering, Washington University in St. Louis, St. Louis, MO and Cynthia S. Lo, Energy, Environmental and Chemical Engineering, Washington University in St. Louis, St. Louis, MO

Photosynthetic light-harvesting complexes (LHC) have evolved over many millions of years to become extremely efficient at funneling excitation energy into their respective reaction centers. If these funneling processes are better understood, we can apply this knowledge to building an efficient bio-solar photovoltaic cell. Peridinin-chlorophyll-protein (PCP) complex is a LHC containing twenty-four peridinin and six chlorophyll-a pigment molecules, as shown by its x-ray crystallography structure.[1] We are studying the excitation energy transfer rates throughout the PCP complex using time-dependent density functional theory (TD-DFT) to calculate the pigment excited states, and Förster theory to calculate the transfer rates between pigment excited states. Since Förster theory does not take into account delocalized excitation states, we will extend our study to other quantum dynamics theories that allow delocalization. The excitation transfer pathway in PCP has been shown experimentally to go from the S2 to S1 state in peridinin, and from the S1 state in peridinin to the Qy state in chlorophyll-a.[2] Researchers have also been successful in reconstituting different species of chlorophyll molecules into the PCP complex array to tune the transfer efficiencies.[3] Our goal is to validate the transfer efficiencies of these known complexes using the computational methods outlined above and to extend these methods to study unknown LHC systems.

[1] Hofmann, E., et. al. *Science* **272**, 1788-1791 (1996).

[2] Damjanovic, A., Ritz, T. and Schulten, K. *Biophysical Journal* **79**, 1695-1705 (2000).

[3] Polivka, T., et. al. *Photosynthesis Research* **86**, 217-227 (2005).

See more of this Session: Multiscale Modeling: Methods and Applications
See more of this Group/Topical: Computing and Systems Technology Division

(393d) A Generalized Runge-Kutta Framework for Explicit Tau-Leaping Algorithms

Leonard A. Harris and James Faeder, Computational and Systems Biology, University of Pittsburgh School of Medicine, Pittsburgh, PA

Gillespie's tau-leaping algorithm has received considerable attention as a promising method for performing accelerated-stochastic simulations of multiscale chemical and biological systems. The simplest tau-leaping algorithm has been shown to be analogous to the simple forward Euler method for numerically integrating ordinary differential equations. A "midpoint" tau-leaping algorithm has also been proposed which is analogous to an explicit second-order Runge-Kutta integrator. In this vein, we develop a generalized Runge-Kutta formulation of tau-leaping. We focus exclusively on explicit methods, describing how reaction firing, postleap checking and tau-selection are accomplished within a Runge-Kutta context and contrasting with the simpler forward-Euler case. The advantages of the approach are demonstrated via illustrative example systems simulated within the open-source modeling and simulation software package BioNetGen.

See more of this Session: Multiscale Modeling: Methods and Applications
See more of this Group/Topical: Computing and Systems Technology Division

(393e) Multiphysics Model of Diesel Injector Deposit Formation

Richard H. West, Amrit Jalan and William H. Green, Department of Chemical Engineering, Massachusetts Institute of Technology, Cambridge, MA

Formation of deposits inside the narrow nozzles of diesel engine fuel injectors disturbs fuel flow and adversely affects engine efficiency and emissions. It is thought that the deposit is formed by oxidation of a thin film of diesel on the nozzle wall. Fuels and additives used to mitigate this problem are currently evaluated using very expensive tests in running engines. Modeling insights into this problem could therefore greatly reduce the cost of fuel and engine development.

The system comprises several different phenomena which span multiple timescales. The free radical chemistry responsible for the deposit precursors occurs on a microsecond timescale while the fuel injections every tens of milliseconds cause washing and replenishment of the fuel film. The deposit itself accumulates over thousands of injection events which occur over hundreds of hours. The challenge lies in addressing the many physical and chemical phenomena in a way that is computationally efficient and gives meaningful results.

We start with a detailed kinetic model for the autoxidation of a diesel fuel surrogate mixture, consisting of hundreds of intermediate species and thousands of elementary reactions, built using Reaction Mechanism Generator (RMG) software [1]. Thermochemical, transport, and solubility parameters are estimated for each of the intermediate species based on their molecular structure. The kinetic model is embedded in a coupled model of evaporation, chemical reaction, phase-separation, and washing, and solved for multiple injection cycles using PyDAS, a new Python wrapper for the stiff DAE solver DASSL.

A model such as this necessarily contains many uncertain and variable parameters. It is useful to identify which of these parameters are most important. To do this we use a modified version of the Morris method [2] to find the global sensitivity of the overall model with respect to each parameter. This has been developed into a Python framework that makes full use of a parallel computing environment and is simple to apply to other models written in the Python programming language.

As well as providing insight into the specific problem of deposit formation in diesel injectors, this work demonstrates an approach suitable for other multi-scale models of chemical engineering systems.

[1] W. H. Green et al. RMG reaction mechanism generator open-source software package, 2011. http://rmg.sourceforge.net/

[2] M. D. Morris. Factorial sampling plans for preliminary computational experiments. Technometrics, 33(2):161–174, 1991.

See more of this Session: Multiscale Modeling: Methods and Applications
See more of this Group/Topical: Computing and Systems Technology Division

(393f) Numerical Simulation of Fuel NOx and Thermal NOx Emissions from an Industrial Burner Using Biomass-Derived Producer Gas

Sujith Sukumaran and **Song-Charng Kong**, Mechanical Engineering, Iowa State University, Ames, IA

Most of the industrial burners in the market are designed for natural gas or synthesis gas resulting from coal gasification. These burners are optimized to reduce the thermal NOx emissions but often do not address the fuel NOx problem. On the other hand, producer gas derived from biomass gasification contains traces of ammonia. Combustion of such producer gas results in high emissions of NOx due to the significant amount of fuel NOx in addition to thermal NOx. In this study we performed modeling of an industrial burner to analyze NOx emissions resulting from combustion of biomass-derived producer gas. Multidimensional CFD modeling of the burner was performed initially using a detailed mechanism, and the results were validated using the experimental results. However, such a detailed mechanism is highly computationally intensive to be used with 3D modeling. Thus, the detailed mechanism was reduced to a skeletal mechanism, without loss of accuracy in NOx prediction. Further CFD modeling of the actual burner geometry were conducted and results were validated using experimental data. The relative ratios of thermal NOx and fuel NOx are characterized under different operating conditions. The resulting skeletal mechanism and modeling strategies can be used in the future as a tool to design low-NOx burner using producer gas containing ammonia.

See more of this Session: Multiscale Modeling: Methods and Applications
See more of this Group/Topical: Computing and Systems Technology Division

KINETIC MODELING OF SOLID-GAS REACTIONS AT REACTOR SCALE: A GENERAL APPROACH

Loïc FAVERGEON [a], Jacques MORANDINI [b], Michèle PIJOLAT [a], Michel SOUSTELLE [a]
[a] Centre SPIN, Ecole Nationale Supérieure des Mines, Saint-Etienne, FRANCE
[b] ASTEK Rhône-Alpes, Echirolles, FRANCE

Abstract

Understanding the industrial reactors behavior is a difficult task in the case of solid state reactions such as solid-gas reactions. Indeed the solid phase is a granular medium through which circulate gaseous reactants and products. The properties of such a medium are modified in space and time due to reactions occurring at a microscopic scale. The thermodynamic conditions are driven not only by the operating conditions but also by the heat and mass transfers in the reactor.

We propose to numerically resolve the thermohydraulic equations combined with kinetic laws which describe the heterogeneous reactions. The major advantage of this approach is due to the large variety of kinetic models of grains transformation (~40) compared to the usual approach, especially in the case of surface nucleation and growth processes which need to quantitatively describe the grain conversion kinetics at a microscopic scale due to nucleation frequency and growth rate laws obtained in separate isothermal and isobaric experiments.

The heat and mass transfers terms entering in the balance equations at a macroscopic scale depend on the kinetics evaluated at the microscopic scale. These equations give the temperature and partial pressure in the reactor, which in turn influence the microscopic kinetic behavior.

Keywords: kinetics, simulation, heat and mass transfers, multi-scale

1. Introduction

Solid-gas reactions are of great interest in many industrial fields such as nuclear, chemistry, metallurgy, CO_2 capture, etc... Industrial reactors where these reactions take place are difficult to understand. Indeed the solid phase is a granular medium through which circulate gaseous reactants and products. The properties of such a medium are modified in space and time due to reactions occurring at a microscopic scale. The thermodynamic conditions are driven not only by the operating conditions but also by the heat and mass transfers in the reactor. Several models have been developed to account for the complexity of these transformations such as the grain model [1] and the pore model [2] and all the related incremented models. However most of these models are based on the law of additive reaction times of Sohn [3] for which the order respective to the gas in the kinetic rate equation must be equal to 1. However such condition is scarcely encountered in many gas-solid reactions so that erroneous results may be obtained using this simplified approach.

Moreover, such approaches necessitate to describe the reaction kinetics using the shrinking core model or similar laws, that is to say using rate equations for which $d\alpha/dt$ is the product of an Arrhenius term by a f(α) function [1,2]. Such a behavior is unfortunately not always encountered for several reasons, especially in the case of reactions involving surface nucleation and growth processes [4].

In order to overcome such problems, we have developed a multi-physic approach based on the finite elements method which combines the resolution of the thermohydraulic equations with the kinetic laws describing the heterogeneous reactions at the scale of dense particles.

We propose [4] a large variety of models at the microscopic scale based on a general equation which accounts for the real influence of the partial pressures P_i of the relevant gases:

$$\frac{d\alpha}{dt} = \phi(T, P_i, ...) S_m(t, \phi, \gamma, r_0) \quad (1)$$

where $\phi(T,P_i,...)$ is the areic reactivity of growth, function of the thermodynamic conditions only (temperature, partial pressure, etc...) and $S_m(t,\phi,\gamma,r_0)$ is a function of the initial and current morphology of the sample and therefore of time, of ϕ, of the areic frequency of nucleation γ, and of the grains dimensions (e.g.: r_0 is the radius for spherical grain).

Using equation (1), about forty kinetic models may be derived based on various assumptions on the nucleation and growth processes in order to simulate heterogeneous reactions. These models can be simulated using CIN3 software (N° IDDN FR001130014.000.SP.2009.000.30625). Nevertheless CIN3 is able to simulate reactions where temperature and partial pressure of gases are uniform inside the granular medium. In the case of heterogeneous reactors where mass and heat transfers induce spatial and temporal variations of both temperature and partial pressures, it is necessary to couple kinetics of a grains population with the resolution of mass and heat transfers. This is the aim of the multi-physic software CIN4.

Section 2 presents the various scales approach and the main assumptions made in CIN4. Then section 3 summarizes a simplified presentation of the different mathematical models used, for kinetic modeling at grains scale as well as for heat and mass transfers at reactor scale. And finally section 4 presents the simulation of a thermobalance experiment, including both kinetic curves and mass and heat variations inside the system.

2. General presentation

2.1 The scales for modeling

Several geometrical scales characterize all phenomena which occur in a reactor. A rigorous description of our modeling process requires a precise definition of these different scales. We have deliberately set these scales from the objects: mesh, grain, grains population, agglomerate and reactor. In this paper, only three scales will be discussed: grain, grain population and reactor. At the grain scale, the nuclei appear at the grain surface[1] and grow at the expense of the initial solid phase: the kinetics is governed by the shape of the grain, the type of growth, the localization of the rate limiting-step of growth, the direction of development of the new phase. Then a microscopic population is a set of grains in the same thermodynamic conditions, i.e. the gradients of temperature and gas partial pressure are extremely small inside the population. Statistical considerations allow to define average behavior of this population from grain's one. And finally at the reactor scale, the chemical reaction is coupled with mass and heat transfers.

[1] Avrami's laws cannot be used for solid-gas heterogeneous reactions involving powders since they are based on bulk nucleation in an infinite volume, which is not physically acceptable.

2.2 The assumptions of the model

Six assumptions are done in this model:

A1: gas production and consummation by the reaction do not disturb the total pressure distribution,

A2: the gases are diluted in a carrier gas,

A3: gases flow rates are low (less than 5 $l.h^{-1}$ for a circular section of 2 cm),

A4: the gases are considered as ideal,

A5: the growth process can be described by a succession of elementary steps, one of them controls the growth kinetics (rate-determining step approximation),

A6: the grains shape and size distribution are known.

3. Mathematical modeling

3.1 Kinetic laws at grain and microscopic population scale

As previously recalled, in order to model reactions which involve surface nucleation and growth of the nuclei, one can use equation (1). For isothermal and isobaric conditions, the S_m function can take various expressions calculated from assumptions on the growth type (isotropic or anisotropic), the development direction of the new phase (inwards or outwards), the localization of the rate determining step of growth and the grains shape (spheres, cylinders, plates). Soustelle [5] has detailed the laws establishment and the expressions of S_m for each model.

The software CIN4 is able to solve forty-two kinetic models. Some of these models are derived at grain scale: in this case nucleation is considered as instantaneous compared to growth, so all the grains take the same kinetic law. Other models are derived at microscopic population scale: both nucleation and growth are taken into account, which leads to two kinds of models. According to the growth type (isotropic or anisotropic), two methods are used to calculate the fractional conversion of the reaction:

- in the case of isotropic growth of the nuclei, we used a statistical approach based on the works of Johnson and Mehl [6] and Mampel [7], then of Delmon [8],
- in the case of anisotropic growth, each grain starts its transformation at various times and the whole microscopic fractional conversion can be obtained considering the nuclei appearing on each grain of the microscopic population.

Moreover, CIN4 enables the calculation for a real grain size distribution.

3.2 Heat and mass transport at reactor scale

The macroscopic scale is the center of heat and mass exchanges that are described by the thermal and hydrodynamic models, as well as by a mass transfers equation.

3.2.1 Thermal model

Heat transfer is described by the following equation valid throughout each part of the reactor Ω^T:

$$\left[(1-\varepsilon)\rho_s c_{ps} + \varepsilon\rho_g c_{pg}\right]\frac{\partial T}{\partial t} + \varepsilon\rho_g c_{pg}\vec{u}\bullet\vec{\nabla}T + \vec{\nabla}\bullet\left(-k\vec{\nabla}T\right) = Q^T \qquad (2)$$

where s and g indexes refer to solids and gases respectively.

This equation is accompanied by two kinds of boundary conditions:

- Neumann conditions which impose the value of the heat flow on $\partial\Omega_N^T$ boundary:

$$-k\vec{\nabla}T\bullet\vec{n} = \phi_{imposed} + h(T-T_a) + \varepsilon_r\sigma\left(T^4 - T_a^4\right) \qquad (3)$$

- Dirichlet conditions which impose the value of the temperature on $\partial\Omega_D^T$ boundary:

$$T = T_D \tag{4}$$

These equations are valid for each part of the reactor: solid (ε=0), gas (ε=1), porous media (0<ε<1).

3.2.2 Hydrodynamic model

As the granular medium is static and as we consider solid-gas reactions, mass transfers are only due to the flowing of the gas produced or consumed inside the porous medium. The modeling of this flow inside the porous medium is based on notions of hydrodynamic charge and flow rate considerations. Hydrodynamic charge is calculated according to:

$$H = z + \frac{P}{\rho g} + \frac{1}{2g}u^2 \tag{5}$$

The flow rate is not the rate of fluidic particles inside the pores but an equivalent continuous rate. This flow rate is linked to the mass gradient according to the Darcy's law, which is expressed by:

$$\vec{u}_d = -\frac{K}{\eta}\vec{\nabla}(\rho g h) \approx -\frac{K}{\eta}\vec{\nabla}P = -k_p\vec{\nabla}P \tag{6}$$

The whole pressure field (or charge) is solution of a conservation equation. Neglecting source terms due to gas production and consummation (reaction is slow compared to the gas flow), hydrodynamic problem can be written as:

$$\vec{\nabla} \bullet \left(-k_p\vec{\nabla}P\right) = 0 \tag{7}$$

the boundary conditions being $-k_p\vec{\nabla}P \bullet \vec{n} = -u_n$ on $\partial\Omega_N^T$ and $P = P_D$ on $\partial\Omega_D^T$.

3.2.3 Mass transport

Gas phase is governed by the ideal gas law. Thus the total molar concentration inside the gas is equal to $C = \frac{P_r}{RT}$. The molar convective flow is due to the transport by the gas displacement and is expressed by:

$$\vec{J}_i^C = C\vec{u} = -\frac{P_i}{RT}\vec{u} \tag{8}$$

Diffusion in porous medium is a complex phenomenon including diffusion flow, Knudsen diffusion, Soret effect, which is well explained by Reid et al. [9]. Here, only diffusion flow is taking into account:

$$\vec{J}_i^D = -D_{ij}^D\vec{\nabla}C_j = -\frac{D_{ij}^D}{RT}\vec{\nabla}P_j \tag{9}$$

and finally each partial pressure is solution of the following equation:

$$\frac{\varepsilon}{RT}\frac{\partial P_i}{\partial t} + \frac{\varepsilon}{RT}\vec{u}_d \bullet \vec{\nabla}P_i - \vec{\nabla} \bullet \left(\frac{D_{ij}^D}{RT}\vec{\nabla}P_j\right) = Q_i^C \tag{10}$$

with the boundary conditions $-\frac{D_{ij}^D}{T}\vec{\nabla}P_j \bullet \vec{n} = R\phi_i$ on $\partial\Omega_N^T$ and $P_i = P_{iD}$ on $\partial\Omega_D^T$.

3.3 Coupling between the microscopic population scale and the reactor scale

At microscopic scale, the reaction fractional conversion is followed for a representative population of grains. By finite difference it is possible to evaluate the reaction rate $\frac{d\alpha}{dt}$. This rate allows to evaluate heat and mass sources produced by the reaction. Thus the heat source density is equal to:

$$Q^T = \sum_{i=1}^{n_g} v_i M_A \frac{1-\varepsilon}{V_{mA}} \frac{d\alpha}{dt} \Delta H_i \tag{11}$$

and the mass (partial pressure) source density for a gas i is equal to:

$$Q_i^C = v_i \frac{1-\varepsilon}{V_{mA}} \frac{d\alpha}{dt} \tag{12}$$

So, using these sources terms, microscopic reactions have an impact on the spatial and temporal evolution of the thermodynamic processes at the reactor scale. Inversely since it modifies the areic frequency of nucleation and the areic reactivity of growth, thermodynamic influences fractional conversion of the microscopic reaction. It is worthwhile to notice that CIN4 allows to simulate the behavior of reactions for which the kinetic rate at time t (at the microscopic population scale) depends on the local thermodynamic conditions that were established between t=0 and t. CIN4 highly improves the previous approaches which were limited to kinetic model based at grain scale only, excluding surface nucleation and growth models. The equations used in CIN4 require numerous physical properties (porosity, viscosity, density, heat capacity, thermal conductivity, diffusivity). CIN4's user should be able to calculate these properties using empirical laws or mathematical models like for example those given by Bird et al. [11].

4. Numerical modeling

At grain and microscopic population scales, two types of modeling are possible according to the type of growth: anisotropic or isotropic. In the first case, the method allows to calculate the evolution of the fractional conversion for all grains in each granulometric class. When the growth is isotropic, we used the Mampel's method [7] for three grain geometries: spheres, cylinders and plates. At reactor scale, for each equation (2), (7) and (10) (i.e. heat, hydrodynamic and mass transport problems), two finite element formulations were built, depending on the system symmetry (cylindrical or Cartesian coordinates). A classical way is followed to obtain weak formulations for each state variable: temperature, charge and partial pressures.

5. Results

Figures 1 to 3 present a thermobalance simulation achieved with CIN4. This simulation involves a model at the reactor scale coupled with a model of the reaction at the microscopic grains population scale. There, the system studied consists of a vertical cylindrical chamber in which a flow of reactive gas (G1) from the bottom up is maintained. In the middle position, crucible, whose base is porous, contains a granular medium initially composed of grains of A which are transformed into B according to the reaction:

$$A(s) + \frac{1}{2} G_1(g) \rightarrow 3B(s) + 2G_2(g) \tag{13}$$

The system being axi-symmetric, we have used weak forms written in cylindrical coordinates. Figure 1 shows the mesh used for solving heat and mass transfers' equations throughout the reactor by the finite elements method (the mesh was done by using Gmsh freeware [11]).

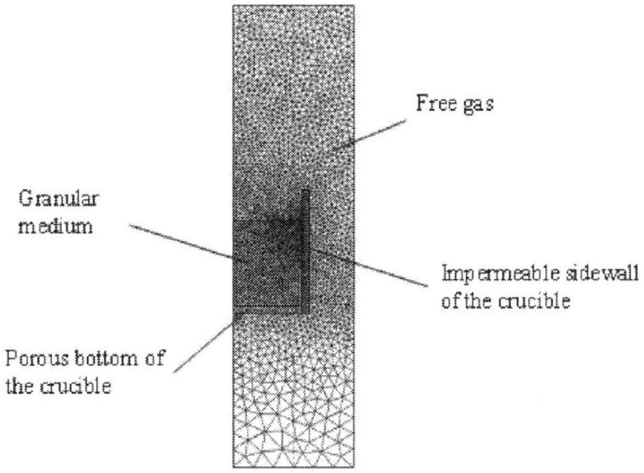

Figure 1: mesh of a thermobalance containing a crucible and a granular medium.

There are four areic regions in this mesh: the granular medium, the bottom of the crucible, the vertical wall of the crucible and the zone of free gas. We also the boundary conditions, which are:

- on the incoming boundary, the entering gaseous mixture is fixed: temperature, flow and composition (here 100% G1),
- on the outgoing boundary: free exit for the temperature, total pressure imposed at 1 atm,
- on the sidewall: the boundary is adiabatic and is impermeable for the mass flow,
- on the symmetry axis: the radial derivatives of all physical grandeurs are equal to 0.

In the following, the kinetic model is based on surface nucleation and anisotropic growth of the nuclei, the grains are spherical (r_0= 100μm), the rate determining-step of growth is located at the internal interface, with inward development.

The granular medium is the seat of heat and mass sources calculated by simulation of reaction phenomena at the microscopic level. Figure 2 shows the gas rate field obtained as well as some views of fields of temperature, partial pressure, fractional conversion and reaction rate for one instant. Figure 3 presents the whole fractional conversion versus time, and the whole rate of reaction versus fractional conversion.

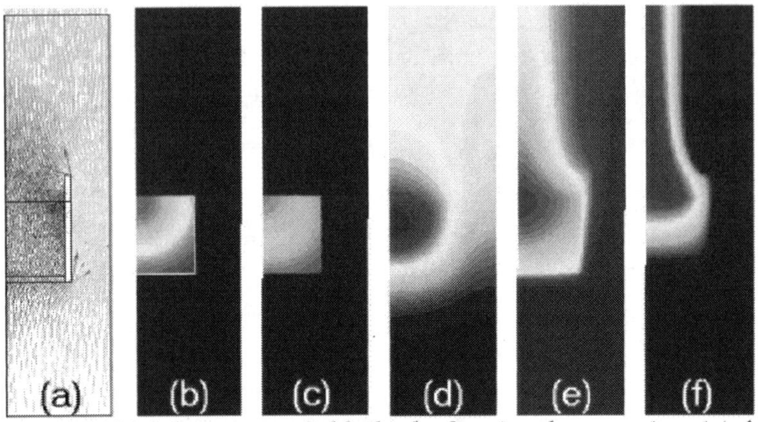

Figure 2: Representation of (a) the gas rate field, (b) the fractional conversion, (c) the rate of reaction, (d) the temperature, (e) G_1 partial pressure and (f) G_2 partial pressure in the thermobalance at a given time.

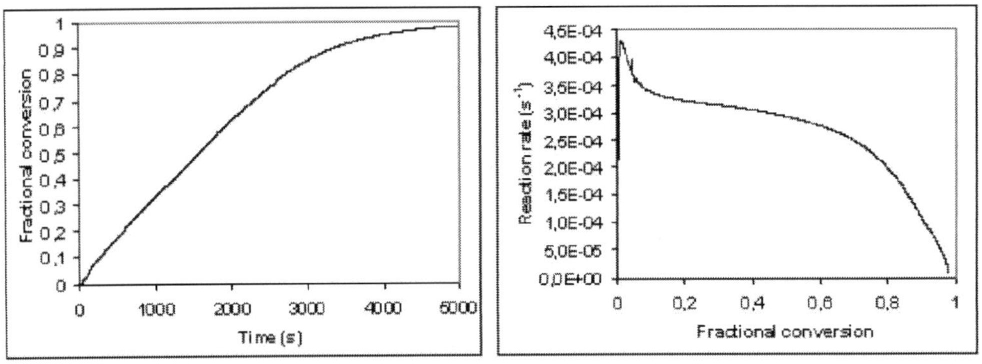

Figure 3: Whole fractional conversion vs. time, and rate of reaction vs. fractional conversion from the simulation.

6. Conclusions

We present a general approach for kinetic modeling of heterogeneous reactions by coupling kinetic models at grain and microscopic population scales with both heat and mass transfers by means of the finite element method. Kinetic models are based on a general method agreeing with physical reality, offering the possibility to simulate about forty models depending on the shape of the grains, the growth type (isotropic or anisotropic), the development direction of the new phase and the localization of the rate determining step of growth (internal interface, external interface or diffusion volume). Among these models, twenty-one include surface nucleation and growth processes. Heat and mass transfers are governed by differential equations and solved by a finite element method. This modeling process allows to simulate not only thermobalances but also industrial reactors.

Notations

α Fractional conversion of the powder
β Fractional conversion of a grain
C Concentration ($mol.m^{-3}$)
c_p Heat capacity at constant pressure ($J.kg^{-1}.K^{-1}$)
D Diffusivity ($m^2.s^{-1}$)
D_{ij} Multicomponent diffusion coefficient ($m^2.s^{-1}$)
S_m Space function ($m^2.mol^{-1}$)
ε Porous medium porosity
ε_0 Initial porous medium porosity
ε_r Emissivity
ϕ Areic reactivity of growth ($mol.m^{-2}.s^{-1}$)
g acceleration of gravity ($m.s^{-2}$)
γ Areic frequency of nucleation ($nuclei.m^{-2}.s^{-1}$)
h Heat transfer coefficient ($W.m^{-2}.K^{-1}$)
H Hydrodynamic charge (m)
J^D Diffusion flow vector ($mol.m^{-2}.s^{-1}$)
k Thermal conductivity ($W.m^{-1}.K^{-1}$)
k_p diffusion coefficient ($m^2.Pa^{-1}.S^{-1}$)

K Darcy's permeability (m^2)

M_i Molar mass for constituent i ($kg\ mol^{-1}$)
η Dynamic viscosity of the fluid (Pa.s)
P Total pressure (Pa)
P_i Partial pressure of reactive gas (Pa)
Q_T Heat sources density ($W.m^{-3}$)
Q_C Concentration sources density ($mol.m^{-3}.s^{-1}$)
R Ideal gas constant (= 8.314 $J.K^{-1}.mol^{-1}$)
r_0 Initial radius of the grain (m)
ρ Density ($kg.m^{-3}$)
σ Stefan Boltzmann constant ($W.m^{-2}.K^{-4}$)
T Temperature (K)
T_a Ambient temperature (K)
υ Rate vector ($m.s^{-1}$)
υ_d Flowing rate vector ($m^2.s^{-1}$)
u_n Incoming flow density ($m.s^{-1}$)
V_{mA} Molar volume of A ($m^3\ mol^{-1}$)

References

1. Szekely J. and Evans J.W. (1970) "A structural model for gas-solid reactions with a moving boundary", Chem. Eng. Sci., 25, pp 1091-1107.
2. Bhatia S. and Perlmutter D.D. (1980) "A random pore model for fluid-solid reactions: 1. Isothermal, kinetics control", AIChE J. 26, pp 379-386.
3. Sohn H.Y. (1978) "The law of additive reaction times in fluid-solid reactions", Metall. Trans., 9B, pp 89-96.
4. Pijolat M., Favergeon L. and Soustelle M. (2011) "From the drawbacks of "Arrhenius-f(α)" rate equation towards a more general formalism and new kinetic models for the kinetic analysis of solid-gas reactions", Thermochimica Acta, accepted on July 27th.
5. Soustelle M. (2010) "Handbook of Heterogeneous Chemistry", Edrs J. Wiley and ISTE Ltd.
6. Johnson W.A. and Mehl R.F. (1939) "Reaction kinetics in processes of nucleation and growth", Trans. Amer. Inst. Ming. Metal. Engrs. 135, pp 416-442.
7. Mampel, K.L. (1940) "Zeitumsatzformeln für heterogene Reaktionen an Phasengrenzen fester Körper", Z. Phys. Chem. A 187, pp 235-249.
8. Delmon B. (1969) "Introduction à la cinétique hétérogène", Eds. Technip, p. 397.
9. Reid R.C., Prausnitz J.M. and Poling B.E. (1987) "The properties of gases and liquids", MCGraw Hill, New-York.
10. Bird R.B., Stewart W.E. and Lightfoot E.N. (2002) "Transport phenomena", Wiley International Edition.
11. Geuzaine C. and Remacle J.F. (2009) "Gmsh : a three-dimensional finite element mesh generator with built-in pre- and post-processing facilities", Inter. J. for Numerical Methods in Engineering, 79, 11, pp 1309-1331.

(393h) Multiscale Methodology for Prototyping of Porous Catalysts

Milos A. Marek, Department of Chemical Engineering, Prague Institute of Chemical Technology, Prague, Czech Republic

The microstructure of the support determines a keyproperty of porous catalysts---effective diffusivity. Typically,supporting materials with bimodal pore size distribution are used thatinvolve both meso- and macro-pores. Spatial distribution of activemetal crystallites within the porous support then influences reactionrates and conversions. In order to optimize the catalyst supportmicrostructure and ultimately the whole catalyst, it is necessary torelate quantitatively the morphological features of the porousstructure both to its preparation conditions and to the final transportproperties and catalyst performance under reaction conditions.

In this paper we demonstratethe application of novel models based on the generalizedvolume-of-fluid method and 3D digital reconstruction of porousstructure. The procedure includes simulation of porous supportformation (virtual packing of primary particles of defined shapes andsizes), drying and crystallization of impregnated metal solution(growth of metal nano-particles), and solution of reaction andtransport within the final virtual catalyst structure to obtainvolume-averaged reaction rates that are then used in full-scale modelof catalytic monolith reactor. A parametric study is performed toinvestigate the effects of the sizes of primary particles (influencingthe meso- and macro-porosity and pore sizes) and active metalimpregnation conditions (influencing the distribution of activecatalytic surface area) on the macroscopic activity of a catalyticmonolith with Pt/g-Al2O3 washcoat used for automotive exhaust gasaftertreatment.

See more of this Session: Multiscale Modeling: Methods and Applications
See more of this Group/Topical: Computing and Systems Technology Division

(400a) A Mixed-Integer Linear Programming Model for Optimizing the Scheduling and Assignment of Tank Farm Operations

Sebastian Terrazas-Moreno, Chemical Engineering Department, Carnegie Mellon University, Pittsburgh, PA, **Ignacio E. Grossmann**, Department of Chemical Engineering, Carnegie Mellon University, Pittsburgh, PA and John Wassick, Engineering Sciences - Process Optimization, Dow Chemical Company, Midland, MI

Some chemical manufacturing sites ship finished products to customers using different modes of transportation (MOT) such as railcars, tank trucks, and pipelines. These MOTs are usually loaded from or connected to storage tanks. The set of storage tanks is known as a *Tank Farm*. All finished product has to be fed from the process into the storage tanks in the Tank Farm before being shipped to customers. This type of operation imposes the need for available storage space at all times in order to avoid unnecessary shut-downs of the upstream chemical process. When these shutdowns occur, they are said to be a result of storage tanks blocking the process. If the chemical process produces several products and each one requires dedicated tanks, the product-tank assignment and the processing schedule at the finishing lines determines how efficiently the storage space is used. An inefficient assignment of tanks or processing schedule can result in blocking the production of certain products even when there is plenty of available storage space in tanks assigned to other products. The problem outlined above is referred to in this work as the Tank Farm Operation Problem (TFOP).

This work presents a novel mixed-integer linear programming (MILP) formulation for the TFOP. The objective of the problem is to minimize blocking of the finished lines by obtaining an optimal schedule and an optimal allocation of storage resources. The scheduling part of the model is based on the Multi-operation Sequencing (MOS) model by Mouret et al. (2010). For scheduling purposes, there is a set of production orders that requires the processing of a certain amount of product in the finishing lines. Orders have a known release date and all are due at the end of the time horizon. The release date corresponds to the moment when a certain amount of an unfinished product becomes available for processing in the finishing lines. These lines can process the order immediately and feed the storage tanks, or delay the order for a while until there is available storage space. The main decisions in this problem are the tank-product assignment and the scheduling of processing orders.

The formulation is tested in three examples of different size and complexity. We solve a small example consisting of 2 lines, 3 products, and 5 tanks, spanning a two-week time horizon. This example illustrates the use of priority slots and the basic concept of multi-operation sequencing, where operations that can overlap are allowed to use the same priority slot. We solve two larger examples. In the first one, the system is extended to 8 products, 10 storage tanks, and 4 weeks; in the second one, the time horizon is further increased to cover 8 weeks. An important feature of the last example is that we consider a cyclic production schedule where inventory levels at the beginning and end of the time horizon have to be equal. The largest problem has 2,830 discrete variables, 36,011 continuous variables, and 87,745 constraints. Examples 1 and 2 are solved within optimality gap of 2% in less than 10 minutes of computations. However, Example 3 can only be solved within 6% of optimality after 7 CPU hours. An alternative for decreasing the computational complexity is considering unlimited storage in the cyclic schedule approach to determine the minimum storage space required to allocate all production.

The results point out to the possibility of incorporating the MILP model into a decision support system in combination a Discrete Event Simulation (DES) model of a tank farm, such as the one proposed by Sharda and Vazquez (2009).

References

Mouret S. I.E. Grossmann, and P. Pestiaux (2010).Time representations and mathematical models for process scheduling problems. *Computers and Chemical Engineering,* doi:10.1016/j. compchemeng.2010.07.007.

Sharda B., and A. Vazquez (2009). Evaluating Capacity and Expansion Opportunities at Tank Farm: A Decision Support System using Discrete Event Simulation. Winter Simulation Conference 2009, 2218 – 2224.

See more of this Session: Planning and Scheduling I
See more of this Group/Topical: Computing and Systems Technology Division

(400b) Integrated Operational Planning and Medium-Term Scheduling for Large-Scale Industrial Semicontinuous/Continuous Processes

Jie Li, Peter M. Verderame and Christodoulos A. Floudas, Department of Chemical and Biological Engineering, Princeton University, Princeton, NJ

The operationalplanning and the medium-term scheduling of industrial semicontinuous/ continuousmultiproduct processes are closely linked to each other, since both operationalplanning and scheduling involve the allocation of plant resources. The operationalplanning problem typically determines production targets for the plant to meet productdemands over a long time horizon from one to three months. The medium-termscheduling problem is defined over a shorter time horizon of between two tofour weeks and involves detailed decisions such as the sequence of processingunits, and detailed timings in which various products should be processed ineach unit to meet the production targets supplied by the operational planningproblem [1-2]. The lack of integration of planning and scheduling often makethe planning model to provide unrealistic production targets causing thescheduling model to allocate plant resources in a suboptimal manner. Theeffective integration of planning and scheduling can increase profits andreduce committed capital simultaneously [3]. However, because of theirdisparate time scales, the effective integration of planning and scheduling isa critically challenging task. Therefore, it is imperative to propose a novelframework that can integrate the operational planning and scheduling moreeffectively.

A simpleapproach for solving the integration of planning and scheduling problems is todevelop a single simultaneous planning and scheduling model over the entire theplanning horizon. However, this approach resulted in large problem size fortypical planning horizons, which is often computationally intractable. Therefore,the integration problem is often solved using bilevel decomposition approach [4], and rolling horizon algorithm [5]. Shah [6], Kallrath [7], Maravelias and Sung[8], and Verderame et al. [9] presented excellent reviews on the integration ofplanning and scheduling. However, most of the work related to integration ofplanning and scheduling has focused on batch processes. Moreover, most of the proposedoperational planning models provide the aggregate production targets, not the dailyproduction profile required by the scheduling model.

Recently, Verderameand Floudas [10-15] developed novel discrete-time mixed-integer linearprogramming (MILP) operational planning with production disaggregation models (PPDM)for a large-scale multipurpose and multiproduct batch plant with/withoutuncertainty. In their model, they disaggregated the production totals into afeasible distribution of daily production requirements and hence provided thedaily production profile. They also allowed any unutilized reactor time tocarry over from one day to the next and hence captured the continuous timenature of the plant. To integrate the operational planning and medium-term schedulingmodel developed by Janak et al. [16-17] effectively, they proposed a novelforward rolling horizon framework that allowed the feedback of schedulingproduction leading to additional operational planning model constraints. The computationalresults showed that the proposed framework yields greater aggregate productiontotals and a higher degree of daily demand satisfaction, compared to theapproach of isolated planning and scheduling, which did not allow the two-wayinteraction between planning and scheduling. However, all of their work is alsolimited to batch processes.

In thispresentation, we first introduce a novel mixed-integer linear programming(MILP) operational planning model based on discrete-time representation for large-scaleindustrial semicontinuous/ continuous multiproduct processes. In the model, we allowunused processing time to carry over from one day to the next and capture thecontinuous time nature of the plant. The production totals are disaggregatedinto a feasible distribution of daily production requirements and hence dailyproduction profiles are provided to meet the requirements of medium-termscheduling model developed by Shaik and Floudas [18]. The forward rollinghorizon framework proposed by Verderame and Floudas [10-15] is utilized to effectivelyaddress the integration of operational planning and medium-term schedulingmodel. The forward rolling horizon framework is applied to an industrial casestudy of a large-scale continuous multiproduct plant over a three-month timehorizon. The computational results show that the framework can lead to theoperational planning model providing a tighter upper bound on the productioncapacity and a higher degree of daily demand satisfaction compared to theapproach of isolated planning and scheduling.

References

[1] Floudas CA,Lin X. Continuous-time versus discrete-time approaches for scheduling ofchemical processes: A review. *Computers and Chemical Enginee*ring. 2004,28, 2109-2129.

[2] Floudas CA,Lin X. Mixed integer linear programming in process scheduling: Modeling,algorithms, and applications. *Annals of Operations Research*. 2005, 139,131-162.

[3] Shobrys DE,White DE. Planning, scheduling and control systems: Why cannot they worktogether. *Computers and Chemical Engineering*. 2002, 26, 149-160.

[4]Erdirik-Dogan M, Grossman IE. A decomposition method for the simultaneousplanning and scheduling of single-stage continuous multiproduct plants. *Industrialand Engineering Chemistry Research*. 2006, 45, 299-315.

[5] Li ZK,Ierapetritou MG. Rolling horizon based planning and scheduling integration withproduction capacity consideration. *Chemical Engineering Science*. 2010,65, 5887-5900.

[6] Shah N.Process industry supply chains: Advances and challenges. *Computers and*

ChemicalEngineering. 2005, 29, 1225-1235.

[7] Kallrath J.Planning and scheduling in the process industry. *OR Spectrum*. 2002, 24,219-250.

[8] Maravelias CT,Sung C. Integration of production planning and scheduling: Overview, challengesand opportunities. *Computers and Chemical Engineering*. 2009, 33,1919-1930.

[9] VerderamePM, Elia JA, Li J, Floudas CA. Planing and scheduling under uncertainty: Areview across multiple sections. *Industrial and Engineering ChemistryResearch*. 2010, 49, 3993-4017.

[10] Verderame PM,Floudas CA. Integrated operational planning and medium-term scheduling forlarge-scale industrial batch plants. *Industrial and Engineering ChemistryResearch*. 2008, 47, 4845-4860.

[11] Verderame PM,Floudas CA. Operational planning framework for multisite production anddistribution networks. *Computers and Chemical Engineering*. 2009, 33,1036-1050.

[12] Verderame PM,Floudas CA. Operational planning of large-scale industrial batch plants under demanddue date and amount uncertainty. I. Robust optimization frameowrk. *Industrialand Engineering Chemistry Research*. 2009, 48, 7214-7231.

[13] Verderame PM,Floudas CA. Integration of operational planning and medium-term scheduling forlarge-scale industrial batch plants under demand and processing time uncertainty.*Industrial and Engineering Chemistry Research*. 2010, 49, 4948-4965.

[14] Verderame PM,Floudas CA. Operational planning of large-scale industrial batch plants under demanddue date and amount uncertainty: II. Conditional value-at-risk framework. *Industrialand Engineering Chemistry Research*. 2010, 49, 260-275.

[15] Verderame PM,Floudas CA. Multisite planning under demand and transportation uncertainty: Robust optimization and conditional value at risk framework. *Industrial andEngineering Chemistry Research*. In press, 2011. **DOI:** 10.1021/ie101401k.

[16] Janak SL,Floudas CA, Vormbrock N. Production scheduling of a large-scale industrialbatch plant. I. Short-term and medium-term scheduling. *Industrial andEngineering Chemistry Research*. 2006, 25, 8234-8252.

[17] Janak SL,Floudas CA, Vormbrock N. Production scheduling of a large-scale industrial batchplant. II. Reactive scheduling. *Industrial and Engineering ChemistryResearch*. 2006, 45, 8253-8269.

[18] Shaik MA,Floudas CA, Kallrath J, Pitz HJ. Production scheduling of a large-scaleindustrial continuous-plant: Short-term and medium-term scheduling. *Computersand Chemical Engineering*. 2009, 33, 670-686.

(400c) Integrated Scheduling and Dynamic Optimization of Batch Processes Using State Equipment Networks

Yisu Nie and Lorenz T. Biegler, Chemical Engineering, Carnegie Mellon University, Pittsburgh, PA

Integrated Scheduling and Dynamic Optimization of Batch Processes Using State Equipment Networks

Yisu Nie and Lorenz T. Biegler

Department of Chemical Engineering, Carnegie Mellon University

Traditionally, scheduling and dynamic real-time optimization (DRTO) of a batch process are approached independently. However, an integrated decision-making approach is preferable for the sake of process profitability and flexibility. A foreseeable solution to the problem is to recover the dynamic characteristics of batch operations by directly incorporating dynamic models into scheduling formulations. The objective of this work is to propose an efficient structure based on state equipment network (SEN) that applies to the integration of general batch processes. This approach improves the overall batch process performance by simultaneously optimizing its production schedule and corresponding control profiles of each process unit.

As an alternative to conventional network structures for batch scheduling problems, such as state task network (STN) and resource task network (RTN), SEN addresses batch scheduling from an equipment-oriented perspective, in which complete connectivity of all process units is postulated and these units are described in the form of rigorous models without pre-assigned tasks. By virtue of that, SEN is able to provide a consistent modeling strategy and proper handling of units that operate under different operating modes. To start with, we adopt a unit-specific event-based continuous time representation for the scheduling problem, where time horizon is divided up into a finite number of asynchronous event time slots with regard to each unit. SEN can be efficiently applied to sequence, size and time all possible batch operations within the interest of time.

On the other hand, SEN regards different operating modes, which we also call operating states, as separate dynamic systems connected by exclusive disjunctions. Time-varying dynamic characteristics of batch units can therefore be captured by SEN, using binary variables with an additional temporal dimension. To facilitate DRTO in a batch process, we prefer to use first-principle models for predictions that suit the context of batch processing, employed in the form of differential-algebraic equations (DAE). Finally, SEN also recovers material quality information that has usually been discarded in STN and RTN, to account for time-varying operating strategies. The overall formulation falls into the category of mixed logical dynamic optimization (MLDO). We elaborate on a procedure to translate

MLDOs to mixed integer nonlinear programs (MINLP) coupled with a simultaneous collocation strategy. An illustrative example is studied, where a multiproduct batch plants are optimized to obtain the best revenue over a fixed time horizon, equipped with single or parallel units in each processing stage. We compare a traditional recipe-based approach, in which all controls are predefined, with an integrated approach. The comparison shows a significant increase of net income in the integrated scheme.

See more of this Session: Planning and Scheduling I
See more of this Group/Topical: Computing and Systems Technology Division

(400d) A Mathematical Programming Based Approach to Biologic Manufacturing Scheduling, Planning, Design, and Analysis

Spencer D. Miller, Steve Harding, George Applequist, Donald Miller and Joseph Pekny, Advanced Process Combinatorics, Inc., West Lafayette, IN

The proliferation of biologic based therapeutics and the complex nature of their manufacturing makes accurate detailed methods for generating production schedules valuable. The large capital costs and expenses in these plants, the convergent flow of buffer and media preparation to support mainline manufacturing, the need to satisfy regulations and medical demands, yet revise plans and designs in the face of changing knowledge, have motivated our work. Mathematical programming has proven to succeed when we layer it with a well-developed framework for model specification, algorithm customization, scenario analysis, and graphical results. We demonstrate an approach that can apply widely to biologics and allow for further custom detail. We discuss biologic processes in general and illustrate the methodology on perfusion based processes, which present some unique challenges to mathematical programming based approaches. The inputs take a natural, concise format to describe the plant and process steps. The model captures how perfusion reactors have varying yield over time with uncertainty. It also schedules supporting activities to supply growth media and buffers, clean in place, and satisfies restrictions on storage vessels and duration. The formulation ordinarily covers time scales from the daily variation of yield to many months for capacity analysis case studies. Scheduling has become rapid enough for large-scale scenarios to mine for data, helping to answer design questions and support routine rescheduling.

See more of this Session: Planning and Scheduling I
See more of this Group/Topical: Computing and Systems Technology Division

(400e) Scheduling As a Cornerstone of CPAS

Iiro Harjunkoski and Martin Hollender, Industrial Software and Applications, ABB AG, Corporate Research Germany, Ladenburg, Germany

Production sites face more and more intense world-wide competition. In the longterm only plants simultaneously managing to optimize quality, availability, flexibility and cost will remain competitive. Production must furthermore adhere to increasingly complex, all-embracing regulations.

A collaborative process automation system (CPAS) is often defined as a method to unify previously diverse systems in order to achieve operational excellence. Sharing the data, knowledge and functional views ensures that each functional group in the plant understands the operational situation, their interdependencies, and their role in improving it. CPAS thus provides a common platform for a plant's entire operations, from engineering to process optimization and asset management.

Control systems that automate and manage production are at the heart of process industries. These systems are networks of interconnected sensors, actuators, controllers and computers, often distributed across vast processing plants that help manufacturers run their operations safely and cost effectively, minimizing waste and ensuring consistent product quality.

Through a CPAS infrastructure that is functionally transparent, logically concise and standards based various functional components can be connected, as it efficiently and effectively supports vertical and horizontal integration of data access between various functional solution components. Thus, e.g. a control system can directly react to alarms in an intelligent way and planning and scheduling can base the performance assumptions on real time asset data. Having the possibility to cross-link various systems enables new level of applications that reaches far beyond current functional capabilities.

Being the coordinating element of a plant, scheduling can in fact through the utilization of existing knowledge reach far better and production plans, where both EWO and smart manufacturing objectives are included in a balanced way. With given data availability, it can be expected that e.g. energy and quality management become standard functional partners, which together drives the plant towards lower costs, better quality and optimal energy efficiency. In this presentation we will present two examples, where quality and energy information can create significant added value to the production facility - only through the availability of right data at the right place. The main focus can thus be put on developing and exploring new ideas of cross-collaboration between systems.

See more of this Session: Planning and Scheduling I
See more of this Group/Topical: Computing and Systems Technology Division

(400f) Real-Time and Rigorous Dynamic Hoist Scheduling

Chuanyu Zhao, Jie Fu and Qiang Xu, Dan F. Smith Department of Chemical Engineering, Lamar University, Beaumont, TX

Real-time and Rigorous Dynamic Hoist Scheduling†

Chuanyu Zhao, Jie Fu and Qiang Xu*

Dan F. Smith Department of Chemical Engineering

Lamar University, Beaumont, TX 77710, USA

Abstract

Dynamic hoist scheduling is often employed on-line for multi-stage material handling under consideration of simultaneously processing multiple types of jobs with different processing recipes. Unlike other scheduling problems, dynamic hoist scheduling requires multiple job processing time and changeover time to be precisely controlled by hoist movements. The objective is to maximize the job throughput of a production line. Dynamic hoist scheduling can be very complicated when various production uncertainties are considered, such as the random arrival of random types of jobs with different processing recipes. Furthermore, if stringent constraints such as multi-job processing capacity at some process stages and the real-time schedule implementation requirements are also taken into account, the mathematical-programming based solution identification for such a scheduling problem will be very challenging.

Hitherto, available methods are mostly heuristic based, and thus the solution optimality or even feasibility can not be guaranteed. In this paper, an MILP based modeling methodology is introduced to solve the real-time, multi-job capacity, and multi-recipe dynamic hoist scheduling, which has never been reported before. The MILP model covers all the major dynamic hoist scheduling issues. The online dynamic scheduling problem can be separated by a series of static scheduling problems. Each static hoist scheduling problem is triggered and solved reactively by every addition of jobs. For real-time application, the solving time for each static scheduling problem should be considered. When a job arrives, with the estimation of solving time of the reschedule, the expecting job and hoist information are collected, which will be used as the initial conditions of the rescheduling.

The efficacy of the proposed methodology is demonstrated by successfully tackling a real-time dynamic hoist scheduling problem, where 8 units with various capacities are employed in the production line to continuously produce three different types of jobs. Many case studies with various coming jobs and random arrival time are modeled and solved in GAMS version 23.3 with the solver CPLEX. The average solving time on an 8-Core Xeon 3.2GHz Dell server for the case studies is less than 0.6 seconds meanwhile, the scheduling results are also guaranteed as the global optimal.

Keywords: Hoist Scheduling, MILP, Dynamic scheduling, Real-time Optimization,

See more of this Session: Planning and Scheduling I
See more of this Group/Topical: Computing and Systems Technology Division

(404a) Use of Principal Components with Parallel Coordinates In Solid Oxide Fuel Cells Fault Diagnosis

Ricardo Dunia[1], Thomas F. Edgar[2] and Benjamin J. Spivey[2], (1)Chemical Engineering Department, The University of Texas at Austin, Austin, TX, (2)Chemical Engineering, The University of Texas at Austin, Austin, TX

The increase of electricity consumption and rising awareness of environmental conservation have expanded the use of fuel cell power generation systems [1]. Among the different type of fuel cells, the solid oxide fuel cells (SOFC) have demonstrated good fuel consumption adjustments and efficient operations during startup and shutdown conditions. Nevertheless, as all power systems, SOFC are prompt to fault operating conditions that need to be detected and identified promptly. The most typical SOFC faulty operations are:

- Micro-Cracks: High operating temperatures of SOFCs are advantageous from the efficient operating standpoint, but aggravate the mechanical stress in the fuel cell ceramic components. The results can be catastrophic if the stress cracks in the ceramic layers are not detected promptly [2].

- Catalyst Poisoning: Nickel particles in the anode material serve as Catalyst in the SOFC. The fact that SOFCs can use a great variety of fuels like biogas, Hydrogen and coal-derived syngas posts a potential risk of fast performance degradation for the anode nickel particles. In particular, syngas contains various impurities such as phosphorous, zinc and Sulphur that reacts with Nickel, reducing the energy generated in the fuel cells [3].

- Air and fuel Starvation: It has been pointed out that the performance of anode supported cells, like SOFC, is sharply reduced when the anode operates under alternatively oxidizing conditions at elevated temperatures [4]. Such conditions may occur during repeated startup and shutdown operations. Indeed, when the fuel supply is interrupted at high temperature, air may enter the anode compartment and partially oxidize the anode.

The prompt detection and identification of such type of faults can prevent catastrophic results during SOFC operations. Many techniques have been developed in the past decade to monitor and detect process faults. Among these methods are the use of a bank of Kalman Filters [5], Principal Component Analysis [6] and Parallel Coordinate Plots [7].

Parallel Coordinate Plots (PCP) is a tool that allows the visualization of multidimensional data into two dimensions. Its main function is to identify homogeneous patterns of data among many different variables [8]. It is a non-projective mapping where hypersurfaces are represented by their planar images.

In contrast, Principal Component Analysis (PCA) does project into a lower dimensional space in order to compact the operating information of the process. Although data projection permits to extract the most important process variations, the process may require more than three principal components. Plots that intend to illustrate more than three dimensions are not appropriate for monitoring visualization, detection and identification of process faults.

This work proposes the use of PCP with PCA for the detection and identification of the typical process faults in fuel cells systems. The proposed method of Principal Components in Parallel Coordinates (PCPC) brings the advantage of data projection and visualization in one technique. PCPC shows to be useful in cases where more than three principal components are needed to represent the main operating variations of a process system, as it is the case of SOFC.

References

[1] Wang, Caisheng and Nehrir, Hashem, Short-time overloading capability and distributed generation applications of solid oxide fuel cells, 2007 IEEE POWER ENGINEERING SOCIETY GENERAL MEETING, VOLS 1-10.

[2] Fischer, Katharina and Seume, Joerg R.,Impact of the Temperature Profile on Thermal Stress in a Tubular Solid Oxide Fuel Cell, JOURNAL OF FUEL CELL SCIENCE AND TECHNOLOGY, 2009, Volume 6, Number 1.

[3] Zhi, M. and Cayan, F. N. and Celik, I. and Gemmen, R. and Pakalapati, S. R. and Wu, N. Q., Temperature and Impurity Concentration Effects on Degradation of Nickel/Yttria-stabilised Zirconia Anode in PH3-Containing Coal Syngas, FUEL CELLS, 2010, Volume 10, Number 1.

[4] Laurencin, J. and Delette, G. and Morel, B. and Lefebvre-Joud, F. and Dupeux, M., Solid Oxide Fuel Cells damage mechanisms due to Ni-YSZ re-oxidation: Case of the Anode Supported Cell, JOURNAL OF POWER SOURCES, 2009, Volume 192, Number 2.

[5] Kobayashi, T and Simon, DL, Evaluation of an enhanced bank of Kalman filters for in-flight aircraft engine sensor fault diagnostics, JOURNAL OF ENGINEERING FOR GAS TURBINES AND POWER-TRANSACTIONS OF THE ASME, 2005, Volume 127, Number 3.

[6] Dunia, R and Qin, SJ and Edgar, TF and McAvoy, TJ, Identification of faulty sensors using principal component analysis, AICHE JOURNAL, 1996, Volume 42, Number 10.

[7] Inselberg, Alfred, Parallel Coordinates: Intelligent Multidimensional Visualization, INTELLIGENT COMPUTER GRAPHICS 2009, Studies in Computational Intelligence, 2009, Volume 240.

[8] Albazzaz, H and Wang, XZ, Historical data analysis based on plots of independent and parallel coordinates and statistical control limits, JOURNAL OF PROCESS CONTROL, 2006, Volume

16, Number 2.

See more of this Session: Process Monitoring, Fault Detection, and Diagnosis I
See more of this Group/Topical: Computing and Systems Technology Division

(404b) Fault Detection and Diagnosis in the Statistics Pattern Analysis Framework

Hector Galicia[1], Jin Wang[1] and Qinghua He[2], (1)Department of Chemical Engineering, Auburn University, Auburn, AL, (2)Department of Chemical Engineering, Tuskegee University, Tuskegee, AL

Statistics pattern analysis (SPA) is a new multivariate statistical monitoring framework proposed by the authors recently [1]. It addresses some challenges that cannot be readily addressed by the commonly used multivariate statistical methods such as principal component analysis (PCA) in monitoring batch processes in the semiconductor industry. It was later extended to the monitoring of continuous processes using a moving window based approach [2].

The major difference between the PCA-based and SPA-based fault detection methods is that PCA monitors process variables while SPA monitors the statistics of the process variables (e.g. mean, variance, autocorrelation, cross-correlation, etc.). In other words, in PCA singular value decomposition (SVD) is applied to the original process variables to build a model for normal operation, and the new measurements of the process variables are projected onto the PCA model to perform fault detection; while in SPA, SVD is applied to the computed statistics pattern matrix (i.e. various statistics of the process variables) of normal operating data to build the model, and the statistics calculated using the new measurements are projected onto the model to perform fault detection. In this way, different statistics that capture the different characteristics of the process can be selected to build the model for normal process operation, and higher order statistics can be utilized explicitly to cope with non-linearity and non-Gaussianity.

In this work, we explore the potential of our recently proposed SPA framework in fault diagnosis for continuous process. Specifically, the variable contributions are derived based on the fault detection indices used in SPA to generate contribution plots [3] for fault diagnosis. We use the Tennessee Eastman process [4,5] to evaluate the performance of the developed fault detection and diagnosis method based on the SPA framework, and compare with PCA, dynamic PCA (DPCA), kernel PCA (KPCA) and independent component analysis (ICA). The case study shows that in general all methods work well in detecting and diagnosing most of faults. However, for some faults that are difficult to detect and/or diagnose by PCA and DPCA based methods, SPA based method provides superior performance. In addition, because the SPA-based method breaks down the contribution of a fault to different variable statistics, they provide extra information in addition to identifying the major fault-contributing variable (s). For example, the SPA-based contribution plots tell us whether the fault is due to a change in variable mean or variance. It also should be noted that in general SPA requires more training data to build a reliable model due to the computation of variable statistics. However, this is not a big issue because most modern processes are equipped with DCS systems and therefore are data rich.

[1] He, Q. P. & Wang, J. Statistics pattern analysis: A new process monitoring framework and its

application to semiconductor batch processes. *AIChE J*, 2011, 57, 107-121

[2] Wang, J. & He, Q. P. Multivariate Statistical Process Monitoring Based on Statistics Pattern Analysis. *Industrial & Engineering Chemistry Research*, 2010, 49(17), 7858-7869.

[3] Miller, P.; Swanson, R. E. & Heckler, C. E. Contribution plots: A missing link in multivariate quality control. *Applied Mathemathics and Computer Science*, 1988, 8, 755-792.

[4] Downs, J. & Vogel, E. A plant-wide industrial process control problem. *Computers & Chemical Engineering, Industrial challenge problems in process control,* 1993, 17, 245-255.

[5] Lawrence Ricker, N. Decentralized control of the Tennessee Eastman Challenge Process. *Journal of Process Control*, 1996, 6, 205-221.

See more of this Session: Process Monitoring, Fault Detection, and Diagnosis I
See more of this Group/Topical: Computing and Systems Technology Division

ONLINE FAULT DIAGNOSIS DURING STARTUP OF DISTILLATION PROCESS USING ARTIFICIAL IMMUNE SYSTEM

Yiyang Dai, Tsinghua University, Beijing, China
Jinsong Zhao, Tsinghua University, Beijing, China

Abstract

The startup is one of the most important operations in a chemical process. Fault diagnosis is important for ensuring chemical processes stability and safety. The strong nonlinearity and complexity of startup make such diagnosis more difficult than for steady condition. Artificial immune system is a new artificial intelligence methodology which attempts to bridge the divide between immunology and engineering. Mathematical and computational methods were used to build models to solve engineering problem with the principle of immunology. In this paper an online fault diagnosis system is proposed for startup of distillation processes, based on an artificial immune system. The system generates diverse antibodies using known normal and fault samples. The antigens are generated using the testing data. The affinity of the antigen-antibody binding is calculated to detect and diagnose faults. A laboratory-scale distillation process has been designed and built for fault diagnosis study. Several faults introduced during the startup of distillation were detected and diagnosed using the proposed system. The result shows that the proposed system is efficient in online fault diagnosis during startup of distillation processes, especially when the number of historical samples is limited, and the self-learning ability ensures the system can remember and diagnose new faults immediately.

Introduction

The startup is one of the most important operations in a chemical process. It lasts a long period of time, leads to off-spec products, and costs much energy. Due to its nature of phase transition, large time delay and strong interaction between variables, it is prone to fault. In the nearly four decades, various process models and algorithms for fault diagnosis have been developed to eliminate any potential hazards to ensure that the chemical process runs safely (1, 2, 3, 4).

The establishment of most traditional methods needs a certain number of training samples. However, actually in the industrial processes, the samples of faults are quite limited. When there are only small amount of training samples, the traditional methods might not work very well. On the other hand, since the data during startup of chemical process may be much different between each time, the fault diagnosis models must have strong self-adaptive ability to retrain themselves with minimal human intervention.

In the recent years, many researchers have begun to use the concepts from immunology to solve engineering problems with strong self-adaptive ability. In this paper an online fault diagnosis system is proposed for startup of distillation processes, based on an artificial immune system. It is capable of recognizing identical fault types in different time during startup with varying magnitudes. A laboratory-scale distillation process is designed and built for study, and several faults were suggested to be introduced during the startup. The data read from real-time database combined with DCS system were given to the proposed system for online fault diagnosis. The result shows that the proposed system is efficient in online fault diagnosis during startup of distillation processes.

Artificial Immune System

Immune System

The immune system is a defense system that has evolved to protect its host from pathogens (harmful micro-organisms such as bacteria and viruses) (5). The immune system can protect the body against infections and has a very distributed and adaptive novel pattern recognition mechanism.

As the figure 1 shows, once pathogens enter the organism by breaching the physical barriers, the innate immune system provides an immediate response, such as phagocyte to destroy pathogens immediately. If pathogens successfully evade the innate response, the adaptive immune system is activated by the innate response. Some portions of the pathogen, called antigens are recognized by the adaptive immune system. Antibodies are generated by lymphocytes to defend the body against antigens. Adaptive immunity responds to a challenge with a high degree of specificity, as well as the remarkable property of "memory." Exposure to the same antigen sometime in the future results in a memory response: the immune response to the second challenge occurs more quickly than the first, is stronger, and is often more effective in neutralizing and clearing the pathogen.

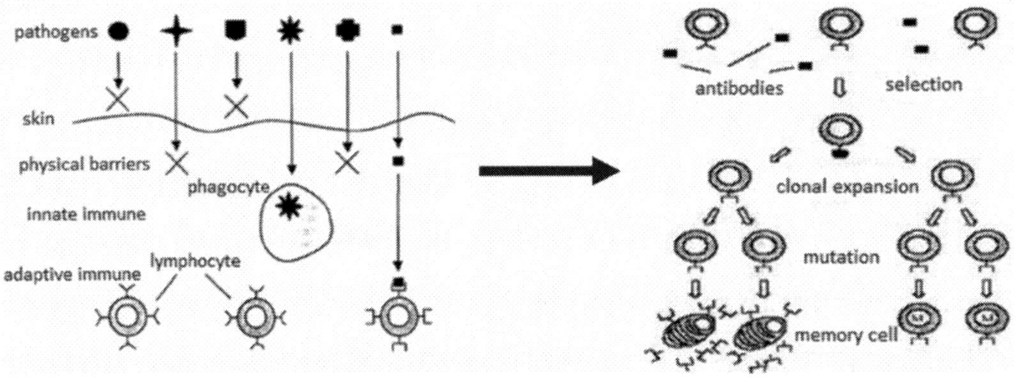

Figure 1. Mechanism of Immune Response

Artificial Immune System

Since many properties of the immune system are of great interest for engineers, the artificial immune system (AIS) was proposed to solve engineering problems. In general, AIS is a diverse area of research that attempts to bridge the gap between immunology and engineering and developed through the application of techniques such as mathematical and computational modeling of immunology, abstraction from those models into algorithm (and system) design and implementation in the context of engineering (6).

In 1990, Ishida used the principle of immune system to deal with the problem of fault diagnosis of sensor networks, supplied the immune system to engineering application for the first time [7]. Later, Forrest used immune system for computer security and virus detection, and this approach has been widely used soon (8). Then, AIS have been widely used in various diverse fields. Most researches can be found in the reviews of AIS by Timmis (9). Recently, some researches have used artificial immune system for fault diagnosis of chemical and petrol chemical process, such as in oil well (10), Tennessee Eastman (TE) process (11), CSTR process (12), batch chemical process (13).

Negative selection, clonal selection and immune networks are the most popular immune algorithms. No matter what algorithm is used in AIS, antigen and antibody are represented by vectors of data from a sample at a certain time. However, during startup of a chemical process, the data is not stable so a sample at a point of time cannot accurately reflect the dynamic working conditions of the process. Furthermore, once a fault is introduced into the process, data changes will be reflected in trends, and antigens and antibodies that are simply represented by vectors of data do not contain trend information. This is not conducive to the diagnosis of fault type or early fault detection. To deal with this problem, author have proposed dynamic time warping (DTW)-based AIS for fault diagnosis of batch processes, in which antibodies and antigens are represented by matrices of time-sampled data, instead of vectors of data (13). Figure 2 shows an antibody for one type of fault. The antibody in figure 2 contains a set of time-sampled data with 6 variables and 20 samples.

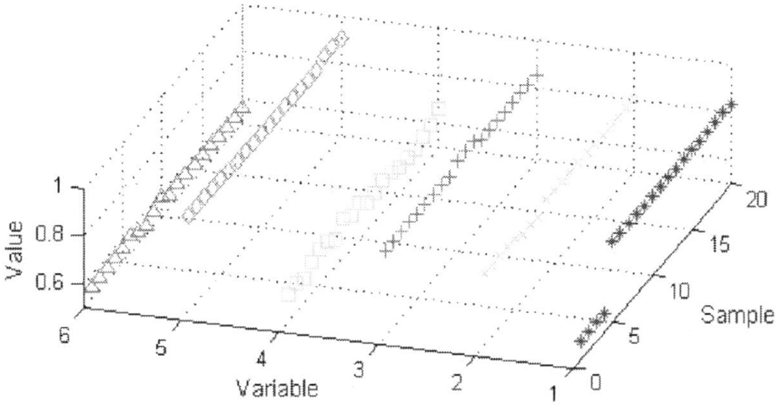

Figure 2. An Antibody of the Artificial Immune System

Online Fault Diagnosis System

Improved DTW-based AIS

In allusion to the fault diagnosis during startup of chemical process, we improved DTW-based AIS algorithm, such as dynamic selection of the length of antibodies and antigens, the clone mechanism of antibodies, etc. A scheme of the improved DTW-based AIS is shown in figure 3.

At first we divide the startup process into several periods by the operations. The original normal antibody is composed of the historical normal data between two operations will compose an antibody, as $Ab_{pi}^{m\times n}$=[$Ab_{pi}(1)$, $Ab_{pi}(2)$, …, $Ab_{pi}(k)$, ..., $Ab_{pi}(n)$], where $Ab_{pi}(k)$ indicates the data of variable k and m is the number of all samples in period i. The data between two near periods should have an intersection for transition, so the start sample of a normal antibody is not exactly at an operation time but some time before it.

The original fault antibody is composed of the historical data after the time that the fault was introduced, as $Ab_{fault}^{m\times n}$=[$Ab_{fault}(1)$, $Ab_{fault}(2)$, …, $Ab_{fault}(k)$, ..., $Ab_{fault}(n)$], where $Ab_{fault}(k)$ indicates the data of variable k and m represents the sample number of the fault antibody which is usually be taken from 15 to 20 in this paper.

All of the original normal and fault antibodies of different types will have to be cloned and mutated to create fault antibody libraries to enhance the capability of diagnosing faults varying in both of time and space scales. The cloning algorithm is proposed by integration of the DTW principle in reference (13). For the startup of distillation, in the cloning selection, normal antibody libraries should be generated in different periods.

When the antibody libraries have been constructed and the thresholds of the libraries have all been calculated, the system will have the capability for fault detection and diagnosis. As the online real-time data is introduced to the system, the period will be estimated at first. Some typical change of special variable can be used for criterion, such as a temperature rising from a steady state or a valve condition changed from off to on. Once an operation is taken, a period is end. And the data during this period will be used to generate a new original normal antibody of this period, and the normal antibody library will be updated.

When period is confirmed, the last l samples before the current time will compose an antigen, as $Ag^{m\times n}$=[$Ag(1)$, $Ag(2)$, …, $Ag(k)$, ..., $Ag(n)$], where Ag(k) indicates the data of variable k, and l is usually an integer taken from 10 to 15 in this paper.

The differences between the antigen and all the antibodies in the normal antibody library of same period are calculated by using DTW algorithm (14). If any of the differences is smaller than the threshold of the normal antibody library, the system is under the safe condition and the next sample is read in. Otherwise, the system will report a fault detected and initiate the fault diagnosis phase.

In the fault diagnosis phase, the differences between the antigen and all the antibodies in all of the fault antibody libraries are calculated by using DTW algorithms. If any of the differences between antigen and fault k antibodies is smaller the threshold of the fault k antibody library, then the fault type would be diagnosed as k. If all the differences are larger than the corresponding threshold, read the next sample and generate a new antigen to repeat the above steps. If the fault type still couldn't be diagnosed after N samples have been read in, the system would report that a new type of fault might have occurred. N is usually an integer taken from 3 to 5 in this paper.

Once a fault type k is diagnosed, the online data after the time that the fault was detected will be used to generate a new original fault antibody, and the fault antibody library k will be updated.

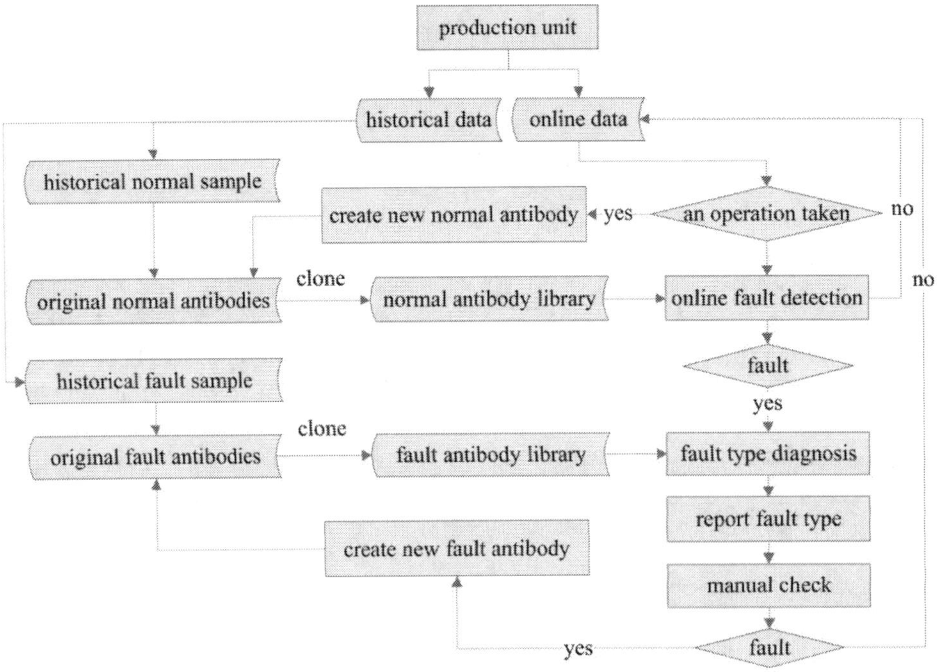

Figure 3. Flowchart of Improved DTW-based AIS

Online Fault Diagnosis System

Based on the improved DTW-based AIS algorithm, an online fault diagnosis system is developed for the industrial applications. The environment of online fault diagnosis system of distillation process is shown in figure 4.

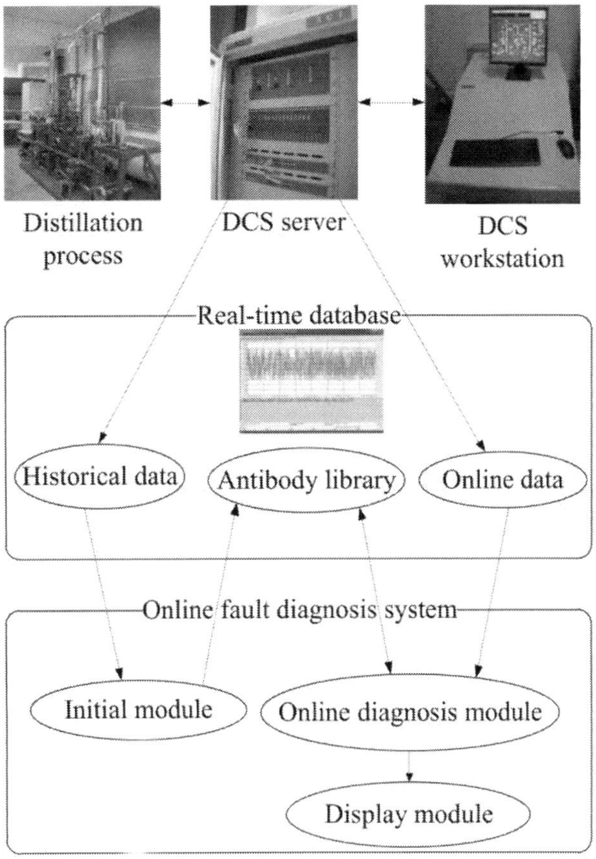

Figure 4. Environment of Online Fault Diagnosis System of Distillation Process

To study industrial fault diagnosis, a distillation process has been designed and installed. A DCS system is used to ensure the stable operation of the distillation process, and get the process data. A real-time database is combined with DCS system to get and store the online data. And the antibody libraries are also stored in the real-time database. The online fault diagnosis system can exchange information with real-time database. There are three different modules in the fault diagnosis system. Initial module can get the historical data from the real-time database to generate the antibody libraries and store these libraries back into the real-time database. The online diagnosis module can generate antigens using the online data read from real-time database. Antibodies will also be read from the real-time database to detect and diagnosis the fault. And when new antibodies can be generated, it will also be stored back into the real-time database. At last, display module will show the real time diagnosis result to the operator in a friendly interface.

Case Study

Process Description

The distillation process used in this case study is designed especially for fault diagnosis study. Several kinds of fault can be in traduced to the process during startup. The distillation column is of 2.2 meters height and 75mm width and has 15 trays, where the feed enters at tray 12 from top of the column. A DCS system is used to ensure the stable operation of the distillation process. In the process, condenser's level is controlled by the reflux rate, the reboiler's level is controlled by the rate of bottom residue, and the rate of feed, reflux and head product are all controlled by the metering pumps. 20 variables obtained by the DCS system are used for fault diagnosis, including 11 temperatures, 2 pressures, 4 rates, and 3 liquid levels. The process flow diagram (PFD) of the distillation process is shown in the figure 5, and all the measured variables can be found in it.

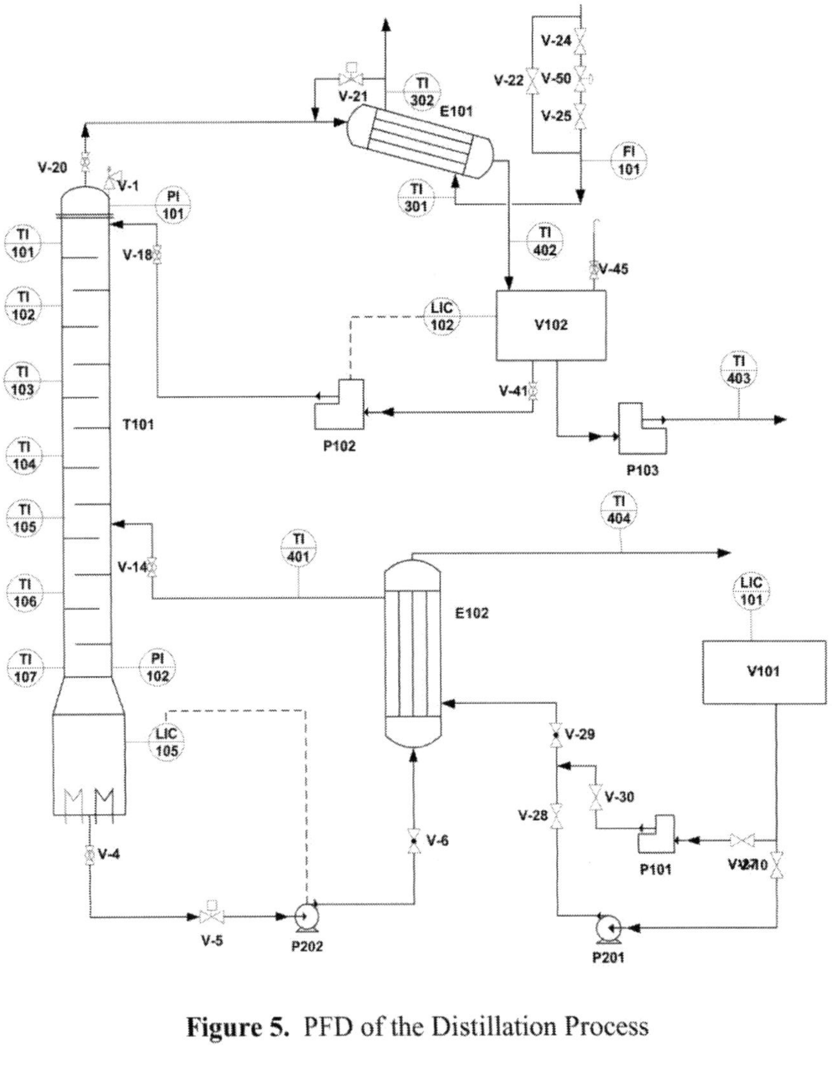

Figure 5. PFD of the Distillation Process

Cold startup of the distillation process is performed following the standard operating procedure (SOP) shown in table 1.

Table 1. Distillation Process Startup SOP

Unit	Operation
1	Fill reboiler with ethanol-water 30% v/v mixture in normal pressure and temperature using full-speed feed pump P201
2	When the reboiler's level get to 27.5cm, stop the full-speed feed pump and start reboiler heater with 100% power
3	When TI-102 get to 70℃, change the rebiler heater power to 60% and active reflux controller to control the condenser's level at 4cm
4	When the condenser's level and reflux rate are constant, start the metering pump of feed and head product, set the rate of them to be 1L/h and 10L/h. Active reboiler controller to pump bottom residue, and set the control level to be 25cm
5	Wait for all the variables to stabilize

The startup normally takes 30 to 40 minutes, and the variables are measured at 2-second intervals. Figure 6 shows temperatures of distillation column during normal startup.

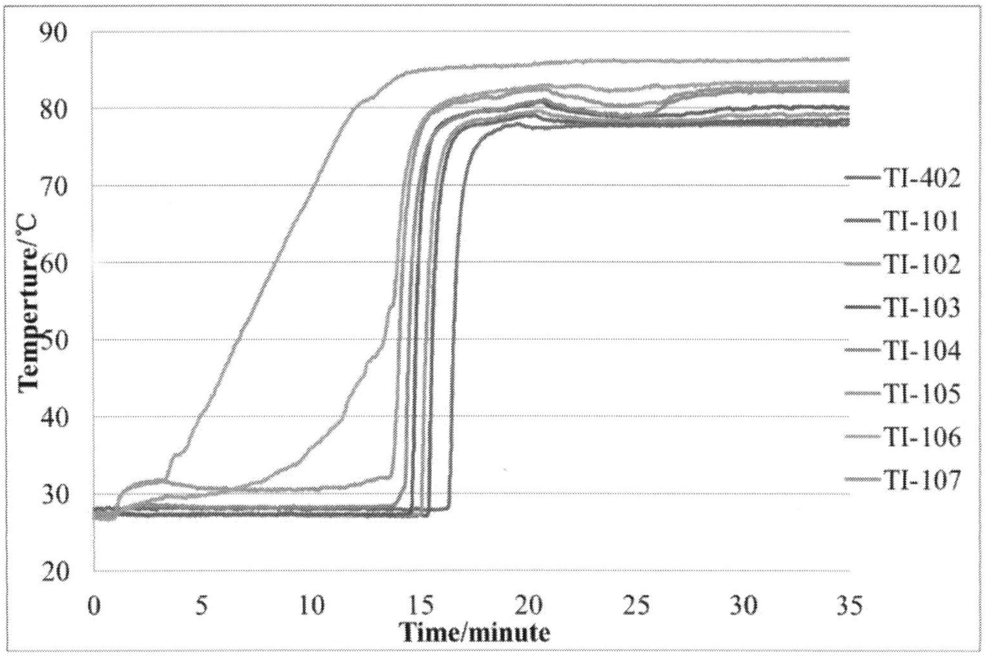

Figure 6. Temperatures of Distillation Column During Normal Startup

Fault Diagnosis

In this study, historical data of two normal samples and six fault samples were taken to generate the antibody libraries and initialize the system Six fault samples were of three fault types, with different introduced times and fault magnitudes, including reboiler heater off, cooling water rate decreasing, head valve V-20 off.

Eight fault samples were used to test the performance of the proposed online fault diagnosis system, two of which were running under the fault types of reboiler power increasing. that were not included in the historical fault sample.

Use the system described in last section to diagnose the test samples. The diagnosis result is shown in table 2.

Table 2. Example of a Sample Table and the Table Caption

Fault type	Introduced time/min	Detected time/min	Diagnosis Result
normal	-	-	normal
1 #: reboiler heater off	23	23.2	Fault 1
1 #: reboiler heater off	18	18.2	Fault 1
2 #: cooling water rate decreasing	25	25.4	Fault 1
2 #: cooling water rate decreasing	20	20.3	Fault 2
3 #: head valve V-20 off	0	16.3	Fault 2
3 #: head valve V-20 off	0	16.1	Fault 2
4#: reboiler power increasing	27	27.3	new fault
4#: reboiler power increasing	22	22.2	Fault 3

Conclusion

Result shows that the proposed system can detect and diagnose fault online during startup of distillation process. Compared with other methods, the proposed method has better capability when the number of historical fault samples is limited. When the system works, the antibody library is automatically updated in background when the startup was finished or a fault was diagnosed. So the system can learn online with an adaptive nature, and become more and more intelligent.

References

1. N.J. Scenna (2000) Some aspects of fault diagnosis in batch processes. Reliability Engineering and System Safety. 70. pp. 95–110.
2. R. Srinivasan and M.S. Qian (2006) Online fault diagnosis and state identification during process transitions using dynamic locus analysis. Chemical Engineering Science. 61. pp.6109-6132.

3. M.J Olanrewajua., B. Huang, and A. Afacana (2010) Onlinecomposition estimation and experiment validation of distillation processes with switching dynamics. Chemical Engineering Science. 65. pp. 1597-1608.
4. K. Ghosh and R. Srinivasan (2011). Immune-System-Inspired Approach to Process Monitoring and Fault Diagnosis. Industrial & Engineering Chemistry Research. 50. pp. 1637-1651.
5. Goldsby, R.; Kindt, T; Osborne, B. and Kuby J. (2003) *Immunology*, Fifth ed.: W. H. Freeman and Company, San Francisco.
6. Timmis, J.; Andrews, P.; Owens, N. and Clark, E. (2008) An interdisciplinary perspective on artificial immune systems, *Evolutionary Intelligence*. 1. pp. 5-26.
7. Ishida, Y. (1990) Fully distributed diagnosis by PDP learning algorithm: towards immune network PDP model, in *Proceedings of International Joint Conference on Neural Networks*, San Diego pp. 777-782.
8. Forrest, S.; Perelson, A.; Allen, L. and Cherukuri, R. (1994) Self-non-self discrimination in a computer, in *Proceedings of the 1994 IEEE Symposium on Research in Security and Privacy*, Oakland, USA, pp. 202-212.
9. Timmis, J. (2007) Artificial immune systems-today and tomorrow, *Natural Computing*. 6. pp.1-18.
10. Aguilar, J. (2004) An artificial immune system for fault detection, in *Proceedings of the 17th international conference on Innovations in applied artificial intelligence*, Ottawa, Canada, pp. 219-229.
11. Wang, C. and Zhao, Y. (2008) A new fault detection method based on artificial immune systems, *Asia-Pacific Journal of Chemical Engineering* 23. pp. 706–711.
12. Ghosh, K.; Srinivasan, R. (2011) Immune-system-inspired approach to process monitoring and fault diagnosis, *Industrial & Engineering Chemistry Research*. 50. pp. 1637-1651.
13. Dai, Y. and Zhao, J. (2011). Fault Diagnosis of Batch Chemical Processes Using a Dynamic Time Warping (DTW)-Based Artificial Immune System, *Industrial & Engineering Chemistry Research*. 50. pp. 4534-4544.
14. Kassidas, A.; Taylor, P. and MacGregor, J. (1998) Off-line diagnosis of deterministic faults in continuous dynamic multivariable processes using speech recognition methods, *Journal of Process Control*. 8. pp. 381-393.

(404d) A New Gaussian Mixture Model Based Bayesian Inferential Monitoring Framework for Complex Chemical Processes

Jie Yu, Chemical Engineering, McMaster University, Hamilton, ON, Canada

For complex chemical processes, the operating conditions often shift due to the changes of various factors such as feedstock, product specifications, set points and manufacturing strategies. Thus the plant operating data can be significantly away from multivariate Gaussian distribution so that the conventional statistical process monitoring methods like principal component analysis (PCA) and partial least squares (PLS) fail to detect the process abnormalities with the unreliable T^2 and SPE control limits. More recent effort has been reported to monitor such non-Gaussian multimode processes using multiple PCA/PLS models. However, a priori process knowledge is typically required to segment the historical operating data into multiple groups that correspond to different operating modes. In contrast, Gaussian mixture model (GMM) has not been explored for process monitoring until very recently and most of the literature study attempted to approach it in a deterministic way. That is, every monitored sample is classified into a particular Gaussian cluster instead of viewed as a random point that may comes from various modes with different posterior probabilities.

In this research, a finite Gaussian mixture model (FGMM) based Bayesian inferential framework is developed for fault detection and diagnosis of non-Gaussian processes with different operating modes. The normal process data at each individual operating mode are assumed to follow a multivariate Gaussian distribution and the operation status is treated as a random variable that can be at any of the possible modes with a prior probability. A FGMM is then built to characterize the multiple operating regions, each of which corresponds to a Gaussian component. With the estimated FGMM, a Bayesian inference strategy is used to compute the posterior probability of a monitored sample belonging to each Gaussian component. Further, the regularized Mahalanobis distance metrics are integrated across all different Gaussian modes through Bayesian inference strategy and the formed Bayesian inferential probability (BIP) index can be computed to detect abnormal process events. In addition to fault detection, a Bayesian inference based Gaussian mixture contribution (BIGMC) index is derived with posterior probabilities as weighting factors to isolate process faults and diagnose the root-cause variables. The Tennessee Eastman process (TEP) is used to demonstrate the utility of the Bayesian inferential monitoring approach and the comparison of monitoring results proves its superiority in terms of fault detection and diagnosis accuracy.

See more of this Session: Process Monitoring, Fault Detection, and Diagnosis I
See more of this Group/Topical: Computing and Systems Technology Division

(404e) Forecast Intervals In K-Steps-Ahead Prediction Modeling Under Continuous-Time Monitoring with Application to Blood Glucose Inference

Derrick K. Rollins Sr.[1], Lucas Beverlin[2], Kaylee Kotz[1], Nisarg Vyas[3] and Dave Andre[4], (1)Chemical and Biological Engineering, Iowa State University, Ame, IA, (2)Department of Statistics, Iowa State University, Ames, IA, (3)BodyMedia Inc., Pittsburgh, PA, (4)BodyMedia, Inc.

The ability to forecast the onset of hypo- or hyperglycemia is of critical importance in the management of glucose levels for insulin dependent diabetics. To achieve this objective, the individual must be under glucose monitoring. Recent years have seen the development of continuous glucose monitoring systems (CGMS) that report glucose measurements every five minutes. It is the purpose of this research to exploit this capability under modeling to forecast sufficiently early and with high certainty undesirable glucose levels so that corrective action can prevent these levels from occurring.

In the statistical literature a prediction interval gives a set of bounds that contain the value of the response given a value of the explanatory (i.e., input) variable (or vector for multiple explanatory variables), **x***, in the future, with a given level of confidence. In the context of this research, which is dynamic modeling, prediction intervals must address future input levels, **x***, as well as future time, *k*Dt, where *k* is an integer and Dt is the sampling time. To meet this requirement, forecast terminology from the time series literature is used. Thus, the goal of this work is to obtain a 100(1 - a)% forecast interval *k*-steps-ahead (KSA) (i.e., a distance of time *k*Dt into the future) for glucose response (*y*| *k*Dt, **x***) at a given input level of **x***.

This forecast interval is developed using classical linear regression as a basis. The expectation function is a static linear regression function in a Wiener block-oriented structure. In this structure each of the *p* inputs, x_i, $i = 1, \ldots, p$, is passed through a linear dynamic block producing outputs from these blocks v_i ($\mathbf{q}_i$), where $\mathbf{q}_i$ is a vector of dynamic parameters corresponding to the transfer of x_i to v_i. (In this work $p = 11$, consisting of 3 food variables, 7 activity variables from the BodyMedia® armband and time of day on a 24 hour clock.) The KSA forecast intervals are obtained under serially correlated noise from a normal distribution with mean 0 and a variance of s^2 for all time steps. In modeling real systems, one of the challenges is obtaining an estimator that maintains its accuracy in the presence of unmeasured disturbances. To insure this feature, this work uses a novel model development method based on the work of Rollins, et al. (2010), that maintains fitted correlation performance in training, validation and test sets. In addition, since the amount of data in training is very large, the $v_i(\mathbf{q}_i)$ can be assumed to be fixed, i.e., with standard errors of zero, making an errors-in-variable approach unnecessary. Moreover, for this same reason, the standard errors in the static and noise parameters can also be assumed negligible giving an approximate 100(1 - a)% forecast interval for (*y*| *k*Dt, **x***) as: , where is the 100(1 - a/2)th percentile of the standard normal distribution and is the estimate of s. Note that this interval does not depend on **x***. The lack of dependence on **x*** comes from the use of a large training set of data which occurs due to frequent sampling of CGM systems, that results in nearly zero estimation error for the parameters. Thus, essentially all the forecast errors come from the s that includes modeling lack of fit which is inversely related to r_{fit}, the correlation between the measured and estimated response.

The results of this work demonstrate high reliability of the proposed forecast interval using about 20 diabetic subjects over a validation period of two weeks or approximately 4,000 time intervals. For small values of *k* the forecast bands are narrow but grow as *k* increases since the forecast error, i.e., s, increases as *k* increases. The implications are that to obtain precise intervals far into the future, this will be primarily achieved by obtaining accurate fits for large values of *k*.

Rollins, D. K., N. Bhandari, J. Kleinedler, K. Kotz, A. Strohbehn, L. Boland, M. Murphy, D. Andre, N. Vyas, G.Welk and W. Franke, "Free-living inferential modeling of blood glucose level using only noninvasive inputs," Journal of Process Control **20** 95-107 (2010).

See more of this Session: Process Monitoring, Fault Detection, and Diagnosis I
See more of this Group/Topical: Computing and Systems Technology Division

(404f) Automatic Grading of TFT-LCD Glass Substrates Based On Machine Vision

J. Jay Liu, Department of Chemical Engineering, Pukyong National University, Busan, South Korea and **Seongkyu Yoon**, Chemical Engineering, University of Massashusetts, Lowell, MA

The outer appearance of a product is often one of important quality attributes for certain products that is mainly used for display or used as an exterior part of other products [1,2]. TFT-LCD (Thin Film Transistor – Liquid Crystal Display) glass substrates can be one of the representative cases. In such cases, visual quality (i.e., visual appearance) as well as the physical or mechanical quality attributes has to be controlled or maintained.

A main difficulty in this type of problems arises from the fact that prior information or knowledge is very limited: no class labels are often available nor is prior knowledge about important aspects of visual appearance often not available. This limitedness of prior information as well as the stochastic nature of the visual appearance of the products makes it much more difficult to apply machine vision to the problems.

This work presents an industrial application of a new machine vision methodology [3] to manufacturing of TFT-LCD glass substrates. Careful observation and interpretations of the results from the methodology enable one to automatically monitor the visual quality of the products. Other issues such as such as imbalanced training data that are frequently encountered in industrial applications will be addressed as well.

Literature cited

(1) Liu, J., MacGregor, J. F., "Modeling and Optimization of Product Appearance: Application to Injection-Molded Plastic Panels," Industrial & Engineering Chemistry Research, 44(13), pp. 4687-4696, 2005

(2) Liu, J., MacGregor, J. F., "Estimation and Monitoring of Product Aesthetics: Application to Manufacturing of Engineered Stone" Countertops," Machine Vision and Applications, 16(6), pp. 374-383, 2006

(3) Liu, J. and Han, C., "Wavelet texture anlaysis in process industries," Korean Journal of Chemical Engineering. Accepted, 2011

See more of this Session: Process Monitoring, Fault Detection, and Diagnosis I
See more of this Group/Topical: Computing and Systems Technology Division

(408a) Design of a Centrifugal Separator for Olive Oil Extraction Via Multiphase Flow CFD Simulation

Dimitrios I. Gerogiorgis, National Technical University of Athens (N.T.U.A.), Athens, Greece

Computational Fluid Dynamics (CFD) is a mature field comprising methodologies suitable not just for simulating complex fluid flows towards efficient food product design (forward problem), but also for deriving geometric and operational specifications towards robust process design (inverse problem).

Olive oil is the oil obtained solely from the fruit of the olive tree (*Olea Europaea*), a dominant tree crop of the Mediterranean Basin. Constituting a key Mediterranean diet component with outstanding nutritional quality, it contains a very high level of monounsaturated fats (most notably oleic acid) which epidemiological studies suggest it is strongly linked to a reduction in coronary heart disease risk. It is consumed raw or cooked and used in cosmetics, pharmaceuticals, soaps and traditional oil lamps. Quality assurance is a consideration of extreme importance in olive oil production, thereby excluding any oils obtained via solvents or re-esterification processes and of any mixture with oils of other kinds. Virgin olive oils' low acidity and supreme quality is ensured by exclusive use of mechanical/physical means under mild (particularly thermal) conditions, thereby avoiding any chemical alterations in the oil, and forbidding any treatment other than washing, decantation, centrifugation and filtration.

Modern, continuous industrial olive oil extraction relies on 4 processing stages: olive fruit cleaning and crushing towards a fine paste, malaxation towards olive droplet agglomeration and aroma generation, water-assisted olive paste decanting via centrifugation, and final wet olive oil purification via filtration. The centrifugal force induces rapid phase separation, due to density differences (solids > water > oil). A moving coil inside the decanter's rotating conical drum pushes the solid material out of the system. Oxygen and light adversely affect olive oil quality at all stages, resulting in severe quality deterioration. Oxidation begins immediately upon harvesting as active enzymes rapidly degrade the endogenous oil. If additional oxygen interaction with olive paste is allowed during extraction, acidity increases further.

Two-phase oil decanting centrifuges avoid excessive water use and oil polyphenol solubilization losses at the expense of a lower extraction capability; the olive paste is separated in 2 phases: oil and pomace. This type of decanter, instead of three separate exits (oil, water and solids), has only two outlet streams (water is discharged by the decanter coil together with the pomace, yielding a wetter residue stream).

The present paper focuses on the design problem for a continuous two-phase olive oil centrifugal separator. Addressing it by means of elaborate CFD simulations, we examine the effect of key geometric and operational parameters on efficiency and olive oil quality. Accurate thermodynamic and rheological property models have been employed and combined with a multiphase k-ε turbulent flow model to calculate phase, velocity and pressure distributions for various inlet flows and compositions, and in order to derive geometric specifications for optimal unit design.

See more of this Session: Separations Design
See more of this Group/Topical: Computing and Systems Technology Division

(408b) Optimal Design of Ionic Liquid Entrainers for Extractive Distillation of Azeotrope Systems

Brock C. Roughton, Bioengineering Graduate Program, University of Kansas, Lawrence, KS, John White, Department of Chemical and Petroleum Engineering, University of Kansas, Lawrence, KS, Raifqul Gani, Department of Chemical and Biochemical Engineering, Denmark Technical University, Lyngby, Denmark and Kyle V. Camarda, Chemical and Petroleum Engineering, University of Kansas, Lawrence, KS

Ionic liquids have shown great promise as alternative green solvents. The low vapor pressure exhibited by ionic liquids offers an environmental benefit over many traditional solvents used in industrial processes. However, the toxicity of ionic liquids can still be a concern and must be considered when selecting a solvent. Ionic liquids can also degrade due to thermal decomposition or hydrolysis. Such factors should be taken into account when designing or selecting an ionic liquid solvent. The cation, anion, and alkyl chain length on the cation can be altered to design an ionic liquid with desired properties. These tunable properties show the promise of designing ionic liquids as environmentally friendly entrainers for the separation of azeotropes. With an ionic liquid entrainer designed or selected for a specific azeotrope mixture, an extractive distillation process can be subsequently designed.

The design of ionic liquid entrainers for the separation of azeotropes has been investigated. The azeotropic systems considered were methanol-acetone, methanol-benzene, methanol-cyclohexane, water-ethanol, ethanol-benzene, and ethanol-cyclohexane. Several group contribution (GC) models were developed and used for the initial design and selection of candidate ionic liquids. Previously published experimental data was used in the development of the GC models. A solubility parameter model was used to optimally design cation, anion, and alkyl chain length combinations that were similar in value to the azeotrope mixture volume-fraction average solubility parameter value. Prediction intervals at 95% were used to represent the uncertainty in the predicted solubility parameter values and compare solutions to the design problem. Ionic liquid candidates were screened using a thermal decomposition temperature GC-model and several different GC-models for toxicity measures. Stable, non-toxic ionic liquids with optimal solubility parameter values were obtained.

A UNIFAC activity coefficient model for ionic liquids was developed using previously published experimental infinite dilution activity coefficient data. The UNIFAC model describes an ionic liquid as an anion group, a cation group, and alkyl groups for the chain on the cation. The UNIFAC model has been compared to several published experimental vapor-liquid equilibria (VLE) data sets and shows reasonable modeling of the VLE. The UNIFAC model was used to verify that a specified azeotrope was broken by the designed ionic liquid. The minimum amount of ionic liquid needed to break the azeotrope was determined for each system. The predicted VLE was tested for stability to determine if a liquid-liquid phase split occurred. After verification, an extractive distillation process was designed for each azeotrope mixture using the driving-force method. By designing the distillation column based upon the maximum driving force for separation, the energy input is minimized. For each azeotropic system considered, an optimal ionic liquid entrainer has been designed and an extractive distillation process has been designed using minimal energy inputs and ionic liquid concentrations.

See more of this Session: Separations Design
See more of this Group/Topical: Computing and Systems Technology Division

(408c) Design and Synthesis of Complex Distillation Networks with Hybrid Genetic Algorithm

Seon B. Kim, Chemical and Bioengineering, University of Illinois at Chicago, Chicago, IL and Andreas Linninger, Laboratory for Product and Process Design, University of Illinois at Chicago, Chicago, IL

We propose a novel computer-aided design for complex heat-integrated column networks to identify optimal energy efficient configurations. Since various complex column configurations including Petlyuk, Kaibel, divided wall columns, side-stripper and rectifiers have better energy efficiency than simple separation networks, we target to design complex column networks to separate multicomponent mixtures. However, complex column network synthesis is a challenging problem because the rigorous mass, equilibrium, summation and heat (MESH) equations lead to non-smooth fragmented search spaces. Therefore global optimization methods involving both structural and parametric degrees of freedom is still a challenge for existing math programming algorithms.

Moreover, combinatorial complexity grows rapidly with the number of components to be separated. Our design and synthesis approaches focus on solving these two main challenges. First new computer-aided design methodology deploys an inverse design methods based on temperature collocation which substantially reduces the problem dimension. Therefore it reduces the combinatorial complexity solving highly multi-dimensional mixture. Second, we developed hybrid genetic algorithm, in which stochastic GA elements can be combined with a local deterministic search. At first stage, GA iterates until individuals begin to coalesce and confines the region with niche technique. The gradient-based local optimizer follows to find a solution in this specified region. This hybrid scheme enables to find out all global and local solutions and thus improves accuracy of the solutions as well as the computation speed for finding solutions.

Our novel methods allow the global search for feasible, effective and optimal energy efficient designs of the complex distillation column network in a fully automatic fashion. To be verified as a design and synthesis solution in practice, we used the rigorous commercial flowsheet simulator AspenPlus and both results are prettily matched each other. Furthermore, the automatic and rigorous flowsheet synthesis is apt to systematically address industrial-size process design problems such as the synthesis of energy-efficient separation networks, layout of biorefineries with novel feedstocks or sustainable process for reduction of greenhouse gases emissions.

See more of this Session: Separations Design
See more of this Group/Topical: Computing and Systems Technology Division

A Systematic State-Space Superstructure Based Methodology for Batch Mass Exchange Network Design

Li-Juan Li[1], *Rui-Jie Zhou*[2] *and Hong-Guang Dong*[1]
[1]*School of Chemical Engineering, Dalian University of Technology, Dalian, 116024, PRC*
[2]*Center for Process Systems Engineering, Imperial College London, London, SW7 2AZ, UK*

Abstract

This paper aims at developing a systematic methodology to optimize operating cost and capital cost of batch mass exchange networks (MEN) in a simultaneous manner. Specifically, the traditional state space superstructure is divided into several interconnected subsystems through the property-based mixing/splitting rules, to capture the potential connecting and matching structures. The proposed superstructure can easily handle non-isocomposition mixing/splitting as well as flexible storage locations both in the component and time dimensions. Then, a mixed-integer nonlinear program (MINLP) is first formulated to minimize the total annual cost (TAC), in which the balance between operating cost (utility cost) and capital cost (investment on operating units and storage tanks) can be achieved in one step. Furthermore, the multi-purpose units and parallel operations are also introduced not only to enhance the flexibility of batch process production, but also to make a better use of all the equipment and stream resources in both space and time dimensions. Finally, the coke oven gases (COG) separating problem is presented to demonstrate the applicability and superiority the proposed approach for batch MEN design. It can be found that the TAC can be reduced by 5-70% compared with that of the previous works, and another over 10% can be decreased by using the multi-purpose units and parallel operations.

Keywords: Batch mass exchange networks; State Space superstructure; MINLP; Multi-purpose units; Parallel operations

Introduction

Optimization of mass exchange operations in chemical industries, including absorption, stripping, adsorption, extraction, leaching and ion exchange (Hallale, 2001), can indentify pollution prevention strategies that also satisfy the traditional process economics benefits. In a typical mass exchange network, the valuable or undesirable components in rich process streams can be reduced in countercurrent direct-contacted mass exchange units by contacting with lean process streams or external mass separating agents (MSAs). The main objectives on MEN design are minimum consumption of (external) MSAs or discharge of wastes, or the total cost including both utility cost and capital of these exchangers.

In the literature, studies on the MEN designs were mainly concerned with continuous processes. Methods to synthesis a MEN can be broadly classified into two general categories: the insight-based graphical pinch analysis (El-Halwagi & Manousiouthakis, 1989, 1990a, b; Hallale & Fraser, 2000a-d) and the mathematical-based optimization techniques (Papalexandri et al., 1994; Martin & Manousiouthakis, 2001, 2004; Chen & Hung, 2005a, b). Studies on batch mass exchange network design are relatively scarce and late, though batch productions own incomparable advantages in producing small-amount and high-value-added products. In batch MEN synthesizing, matches between rich and lean process streams are characterized not only by flow rate and concentration, but also time, so that storage strategies are always adopted for material-reusing purpose among different time periods. In the past, both the insight-based graphical pinch analysis (Foo et al., 2004, 2005) and the mathematical-based optimization techniques (Chen & Ciou, 2006, 2007) have been adopted in batch MEN studies, but none of these works can obtain the minimum total cost in one step. Furthermore, these works also failed to provide opportunities for non-isocomposition mixing/splitting and the locations for storage tanks are limited in the component dimension.

In this work, the state space superstructure is adopted to capture all the potential connecting and matching structures and also allows storage tanks to locate at any possible composition point and any time interval. Then, a general mathematical programming model based on the state-space superstructure, is formulated to minimize the total annual cost for the batch MEN. Finally, storage policies, multi-purpose units and parallel operations are introduced to further reduce the economical cost for the batch mass exchange processes. The coke oven gases (COG) separating problem is presented to demonstrate the applicability and superiority the proposed approach for batch MEN design.

Mathematical Model

The state-space superstructure (see Figure 1), proposed by Bagajewicz and Manousiouthakis (1992), has been improved to incorporate all the potential configurations in Batch MEN. The modified state-space superstructure is also viewed as a system of two interconnected blocks: the process operator (OP) and the distribution network (DN). In the distribution network (DN), all the mixers and splitters are embedded and the connections among them are determined through the property-based mixing rules. The process operator (OP), corresponding all the process units, can be further divided into two sub-blocks: mass exchangers (OP_ME) and storage tanks (OP_ST). For the sake of clearness and simplicity, the overall mathematical model on DN and OP will be introduced separately.

Distribution network

In the previous modeling approaches, system flows are specified by the identities of streams. However, ambiguities may arise while dealing with multi-stream mixing and splitting. To circumvent such problem, this work characterizes all flows with the splitting and/or mixing nodes (i.e., the splitters and/or mixers in DN) at both ends of the streams. Notice that, at every splitter and mixer, the flow rate and mass balances must all be satisfied, i.e.,

$$f_{sp,t}^{out} = \sum_{mx \in MX} fs_{sp,mx,t} \qquad \forall sp \in SP, t \in T \qquad (1)$$

$$f_{mx,t}^{in} = \sum_{sp \in SP} fs_{sp,mx,t} \qquad \forall mx \in MX, t \in T \qquad (2)$$

$$f_{mx,t}^{in} \cdot c_{mx,t}^{in} = \sum_{\forall sp \in SP} fs_{sp,mx,t} \cdot c_{sp,t}^{out} \qquad \forall mx \in MX, t \in T \qquad (3)$$

where $f_{sp,t}^{out}$ and $f_{mx,t}^{in}$ denote the total flowate out of the splitting node sp and the total flow rate into mixing node mx in the time interval t; $fs_{sp,mx,t}$ denotes the flowrate from node sp to mx in the time

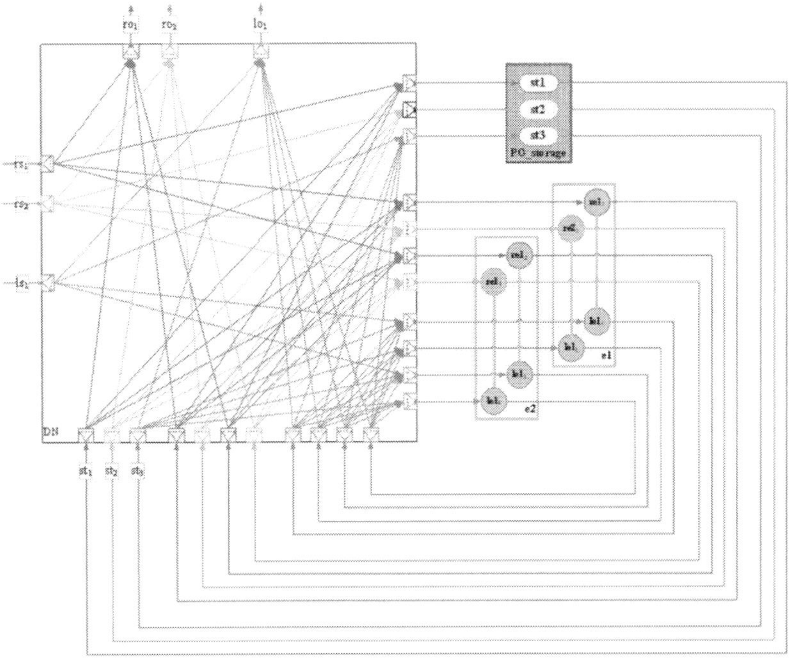

Figure 1 The improved State-Space Superstructure for batch MEN design

interval t; $c_{sp,t}^{out}$ and $c_{mx,t}^{in}$ stand for respectively the compositions of key component at the nodes sp and mx in the time interval t. Furthermore, since not all streams are allowed to mix at certain mixing points, the following constraints should be enforced based on property-based mixing rules:

$$fs_{sp,n_{sp},t} = 0 \qquad \forall sp \in SP, n_{sp} \in MX, t \in T \qquad (4)$$

$$nfs\left(sp, n_{sp}, t\right) = 0 \qquad \forall sp \in SP, n_{sp} \in MX, t \in T \qquad (5)$$

where set N_{SP} denotes all forbidden mixing nodes of the streams from node sp; binary variable $nfs\left(sp, n_{sp}, t\right)$ represents the existence/nonexistence of the stream between nodes sp and n_{sp} in the time interval t. Finally, a negligible amount of flow is not allowed in the optimal operating policy and such uneconomically amount can be eliminated by the addition of the following constraint:

$$Fs^{\min} \cdot nfs\left(sp, mx, t\right) \leq fs_{sp,mx,t} \leq Fs^{\max} \cdot nfs\left(sp, mx, t\right) \qquad \forall sp \in SP, mx \in MX, t \in T \qquad (6)$$

where $Fs^{\max}$ and $Fs^{\min}$ specify the upper and lower bounds of the flow rates in DN.

OP-ME

a. Flow Rate Balances for Each Mass Exchanger

The flow rate balances around unit in low component mass change processes can be written as the following linear equations:

$$f_{re,t}^{in} = f_{re,t}^{out} \qquad \forall re \in RE, t \in T \qquad (7)$$

$$f_{le,t}^{in} = f_{le,t}^{out} \quad \forall le \in LE, t \in T \quad (8)$$

where $f_{re,t}^{in}$, $f_{re,t}^{out}$, $f_{le,t}^{in}$ and $f_{le,t}^{out}$ are the flow rates of the inlet and outlet streams in the rich and lean side of the mass exchange unit in the time interval t.

b. Mass Balances in Each Mass Exchanger

To determine the mass load and the existence of each mass exchange unit, the following equations are needed:

$$m_{re,le,t} = f_{re,t}^{in} \cdot \left(c_{re,t}^{in} - c_{re,t}^{out}\right) \quad \forall re \in RE, le \in LE, t \in T \quad (9)$$

$$m_{re,le,t} = f_{le,t}^{in} \cdot \left(c_{le,t}^{out} - c_{le,t}^{in}\right) \quad \forall re \in RE, le \in LE, t \in T \quad (10)$$

$$M^{\min} \cdot w(re,le,t) \le m_{re,le,t} \le M^{\max} \cdot w(re,le,t) \quad \forall re \in RE, le \in LE, t \in T \quad (11)$$

where the binary variable $w(re,le,t)$ is activated when there is a mass load $m_{re,le,t}$ between $M^{\min}$ and $M^{\max}$, which corresponds to the lower and upper bounds of mass load.

c. Feasibility Constraints of the Equilibrium Relationships

The composition differences in PO-MSA are calculated as follows:

$$c_{re,t}^{out} - h_{re,le} \cdot \left(c_{le,t}^{in} + \Delta C_{re,le}^{\min}\right) - b_{re,le} + \Gamma \cdot \left[1 - w(re,le,t)\right] \ge 0 \quad \forall re \in RE, le \in LE, t \in T \quad (12)$$

$$c_{re,t}^{in} - h_{re,le} \cdot \left(c_{le,t}^{out} + \Delta C_{re,le}^{\min}\right) - b_{re,le} + \Gamma \cdot \left[1 - w(re,le,t)\right] \ge 0 \quad \forall re \in RE, le \in LE, t \in T \quad (13)$$

where $h_{re,le}$ and $b_{re,le}$ are equilibrium coefficients in the exchanger with rich stream *re* and lean stream *le*; $\Delta C_{re,le}^{\min}$ stands for the minimum composition difference for this exchanger. Note that if a match does not occur, the associated binary variable $w(re,le,t)$ equals zero and the large positive upper bound Γ can deem the equation redundant.

d. Equipment design for mass exchange units

In this work, the Kremser Equation can be adopted for determining the number of stages for tray columns if both operating and equilibrium lines are linear. To simplify calculation, the Kremser Equation for stage numbers is written as follows:

$$\Delta c_{re,le,t}^{out} = c_{re,t}^{out} - \left(h_{re,le} \cdot c_{le,t}^{in} + b_{re,le}\right) \quad \forall re \in RE, le \in LE, t \in T \quad (14)$$

$$\Delta c_{re,le,t}^{in} = c_{re,t}^{in} - \left(h_{re,le} \cdot c_{le,t}^{out} + b_{re,le}\right) \quad \forall re \in RE, le \in LE, t \in T \quad (15)$$

$$Ns_{re,le,t} = \left[\frac{\left(c_{re,t}^{in} - c_{re,t}^{out}\right)^{\frac{1}{3}} + \left(h_{re,le} \cdot c_{le,t}^{out} - h_{re,le} \cdot c_{le,t}^{in}\right)^{\frac{1}{3}}}{\left(\Delta c_{re,le,t}^{in}\right)^{\frac{1}{3}} + \left(\Delta_{re,le,t}^{cout}\right)^{\frac{1}{3}}}\right]^3 \quad \forall re \in RE, le \in LE, t \in T \quad (16)$$

where $\Delta c_{re,le,t}^{out}$ and $\Delta c_{re,le,t}^{in}$ denote the composition driving forces at both ends of the mass exchanger with rich stream *re* and lean stream *le* in the time interval t; $Ns_{re,le,t}$ is the number of trays for this exchanger in the time interval t. Furthermore, in practical designs, the optimal number of trays Ns_e should always satisfy the mass transferring needs in every time interval t.

$$Ns_e = \max_{(re,le)\in(RE,LE)_e} \left(Nst_{re,le,t} \right) \quad \forall e \in E, t \in T \quad (17)$$

$$w(re,le,t) \le z_e \le \sum_{t\in T} w(re,le,t) \quad \forall (re,le) \in (RE,LE)_e \quad (18)$$

where z_e denotes the existence/nonexistence of the mass exchanger e.

OP-ST

The mass and flowrate balances of storage tanks are summarized as follows:

$$v_{st,t} = v_{st,t-1} + \left(f_{st,t}^{in} - f_{st,t}^{out} \right) \cdot D(t) \quad \forall st \in ST, t \ge t_2, t \in T \quad (19)$$

$$v_{st,t} = v_{st}^0 + \left(f_{st,t}^{in} - f_{st,t}^{out} \right) \cdot D(t) \quad \forall st \in ST, t = t_1, t \in T \quad (20)$$

$$v_{st,t} \cdot c_{st,t}^{out} = v_{st}^0 \cdot c_{st}^0 + \left(f_{st,t}^{in} \cdot c_{st,t}^{in} - f_{st,t}^{out} \cdot c_{st,t}^{out} \right) \cdot D(t) \quad \forall st \in ST, t = t_1, t \in T \quad (21)$$

$$v_{st,t} \cdot c_{st,t}^{out} = v_{st,t-1} \cdot c_{st,t-1}^{out} + \left(f_{st,t}^{in} \cdot c_{st,t}^{in} - f_{st,t}^{out} \cdot c_{st,t}^{out} \right) \cdot D(t) \quad \forall st \in ST, t \ge t_2, t \in T \quad (22)$$

$$v_{st,t} = v_{st}^0 \quad \forall st \in ST, t = t_n, t \in T \quad (23)$$

$$c_{st,t}^{out} = c_{st}^0 \quad \forall st \in ST, t = t_n, t \in T \quad (24)$$

where v_{st}^0 and c_{st}^0 denote the initial volume and concentration in storage tank st; $v_{st,t}$ represents the volume in tank st in the time interval t; $c_{st,t}^{in}$ and $c_{st,t}^{out}$ represent respectively the inlet and outlet compositions of tank st in the time interval t; $D(t)$ it the length of time interval t. In addition to the constraints above, it is apparent that the optimal size of a tank should be larger than the volume in every time interval and less than the designing capacity $V^{\max}$.

$$v_{st,t} \le vf_{st} \le \sum_{t\in T} v_{st,t} \quad \forall st \in ST, t \in T \quad (25)$$

$$vf_{st} \le V^{\max} \quad \forall st \in ST \quad (26)$$

Objective Function

The objective function in this paper is to minimize the total annualized cost (TAC), including (1) the costs of mass separating agents, (2) the installation costs of mass exchange units and storage tanks. The objective function can be written as follows:

$$Obj = M \cdot \sum_{t\in T} \sum_{sls\in SLS} D(t) \cdot f_{sls,t}^{out} \cdot C_{ls} + \sum_{st\in ST} C^{st} \cdot vf_{st} + \sum_{e\in E} C^{fix} \cdot z_e + \sum_{e\in E} C^{var} \cdot Nst_e \quad (27)$$

where M represents the total numbers of production cycles in one year; set SLS denotes the splitting nodes of lean streams entering to systems; C^{st} is the annualized cost coefficient concerned with the installation of storage tank st, C^{fix} and C^{var} are the relevant annualized fixed and variable cost coefficients of mass exchange unit e.

Application Examples

Three examples are given to illustrate the advantages of the proposed approach for batch MEN deign. By introducing random initial values and perturbations, GAMS/DICOPT with CPLEX as the MILP solver and CONOPT as the NLP solver are used throughout the study.

All the case studies are focused on the process proposed by El-Halwagi & Manousiouthakis (1989), which involves the removal/recovery of H_2S from two rich streams (R1 and R2) with one process lean stream S1 and one external mass separating agent S2. The basic process data are provided in Table 1 and Figure 2. Capital cost for mass exchangers and storage tanks are set to 4552 $/yr per equilibrium stage and 10 $/yr/kg respectively. The minimum composition approach is chosen as 0.0001 and the equilibrium solubility data for H_2S in aqueous ammonia and methanol can be correlated by the following relations, respectively:

$$y=1.45*x1$$
$$y=0.26*x2$$

Table 1 Stream Data for examples 1-3

Stream	Description	Flowrate f(kg)	Mass fraction c^{in}	c^{out}	Operating cost $ s/kg
R1	COG	3240	0.0700	0.0003	
R2	Tail Gases	360	0.0510	0.0001	
S1	Aq. NH3	≤8280	0.0006	0.0310	0.004
S2	Methanol	∞	0.0002	0.0035	0.006

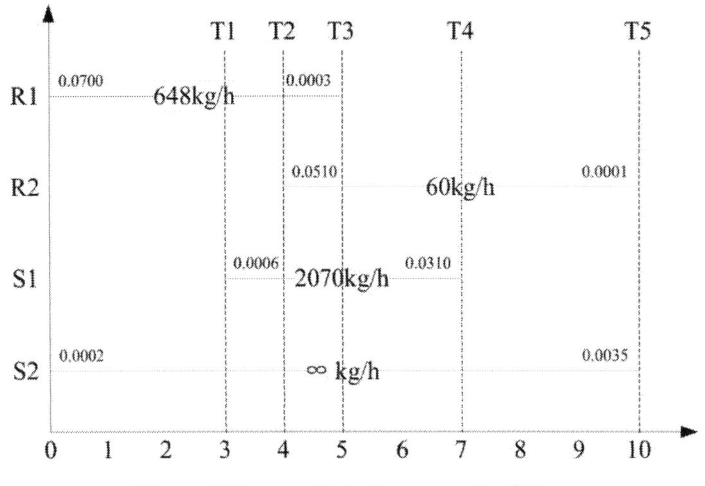

Figure 2 Process Data for examples 1-3

Example 1

The first example is addressed to demonstrate the effectiveness of our method in minimizing the total annual cost of MEN without storage tanks. The resulting optimal network and detailed design specifications of each mass exchanger are presented in Figure 3 and Table 2.

Table 2 Comparisons in Example 1 with the previous works

	TOC	TCC	TAC	Improvement
Foo et. al. (2005) Chen & Ciou (2006)	226523.84	262945.40	489469.24	42.54%
This work	247726.40	33511.45	281237.85	

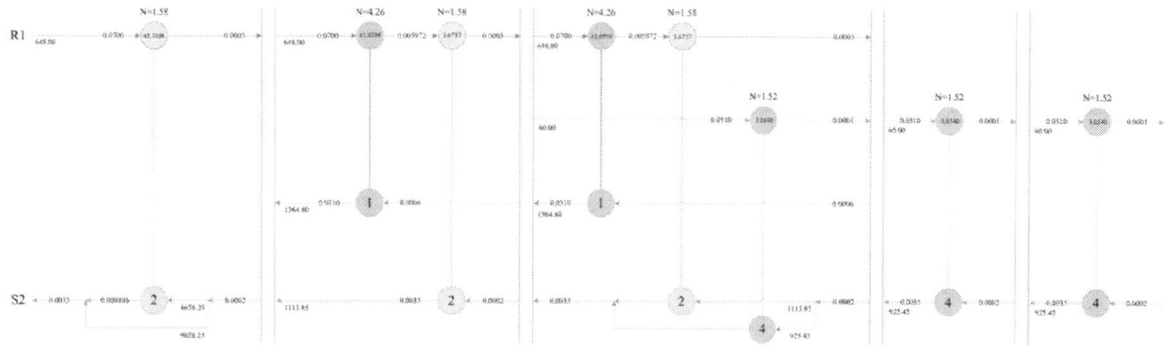

Figure 3 Optimal network configuration for example 1

Example 2

By introducing storage tanks, one can then generate the optimal structure in Figure 4 and the final results in Table 3.

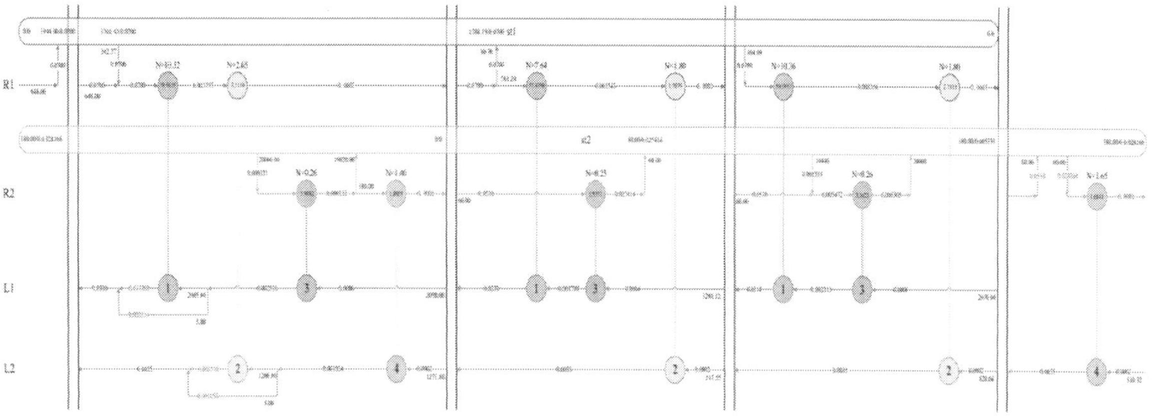

Figure 4 Optimal network configuration for example 3

Table 3 Comparisons in example 2 with the previous works

	TOC	TCC_ME	TCC_ST	TAC	Improvement
Chen & Ciou (2006)	42186.69	258325.47	20040.00	320552.15	56.98%
Chen & Ciou (2007)	48433.49	57045.47	40136.40	145615.36	5.32%
Example 2	48755.56	67900.34	21240.00	137895.90	

Example 3

In this example, Both multi-purpose units and parallel operations are introduced to further reduce the cost of the optimal batch MEN (see Figure 5 and table 4).

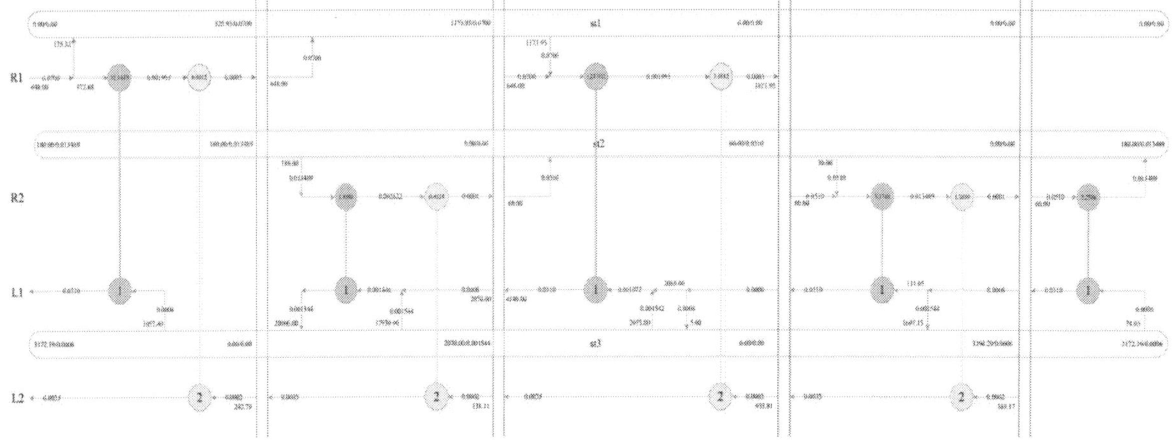

Figure 5 Optimal network configuration for example 4

Table 4 Comparisons in example 3 with the previous works

	TOC	TCC	TAC 1	Improvement 1	Storage	TAC 2	Improvement 2
Chen & Ciou (2006)	42186.69	258325.47	300512.16	65.88%	20040.00	320552.15	61.39%
Chen & Ciou (2007)	48433.49	57045.47	105478.96	2.79%	40136.40	145615.36	27.56%
Example 2	48755.56	67900.34	116655.90	12.10%	21240.00	137895.90	10.24%
Example 3	58574.59	43960.41	102535.00		21240.00	123775.00	

Conclusions

An MINLP model has been presented in this work for one-step optimization of Total Annual Cost of batch mass exchange networks. In this paper, non-isocomposition mixing/splitting scheme, flexible storage policy, multi-purpose units and parallel operations are introduced to enhance the benefits of batch MEN. Finally, it can be clearly observed from the optimal results in the illustrative examples that this approach is superior to all the conventional methods.

Acknowledgment

This work is supported by the National Natural Science Foundation of China, under Grant No. 20876020. The authors would also like to acknowledge the support from Center for Advanced Process Decision-making (CAPD) at Carnegie Mellon University.

References

1. Hallale, N. Mass Transfer Technology for Pollution Prevention. In Process Design Tools for the EnVironment; Sikdar, S. S., El-Halwagi, M. M., Eds.; Taylor & Francis: London, 2001; Chapter 7.
2. El-Halwagi, M. M.; Manousiouthakis, V. Synthesis of mass exchange networks. AIChE J. 1989, 35, 1233–1244.
3. El-Halwagi MM; Manousiouthakis V. Automatic synthesis of mass exchange networks with single component targets. Chemical Engineering Science. 1990a, 45: 2813-2831.
4. El-Halwagi, M. M., & Manousiouthakis, V. Simultaneous synthesis of mass exchange and regeneration networks. AIChE Journal, 1990b, 36, 1209–1219.
5. Hallale, N., Fraser, D.M., Capital and total cost targets for mass exchange networks. Part 1: Simple cost models. Computers and Chemical Engineering, 2000a, 23, 1661–1679.
6. Hallale, N., Fraser, D.M., Capital and total cost targets for mass exchange networks. Part 2: Detail capital cost models. Computers and Chemical Engineering, 2000b, 23, 1681–1699.
7. Hallale, N., Fraser, D.M., Supertargeting for mass exchange networks. Part 1: Targeting and design techniques. Transactions of the IChemE (Part A), 2000c, 78, 202–207.
8. Hallale, N., Fraser, D.M., Supertargeting for mass exchange networks. Part 2: Applications. Transactions of the IChemE (Part A), 2000d, 78, 208–216.
9. Papalexandri, K. P., Pistikopoulos, E. N., & Floudas, C. A. (). Mass exchange networks for waste minimization: A simultaneous approach. Transaction Institute of Chemical Engineering, 1994, 72, 279.
10. Lealon L. Martin, Vasilios Manousiouthakis, "Total annualized cost optimality properties of state space models for mass and heat exchanger networks" Original Research Article Chemical Engineering Science, October 2001, 56, 5835-5851.
11. Vasilios Manousiouthakis, Lealon L. Martin, "A minimum area (MA) targeting scheme for single component MEN and HEN synthesis" Computers & Chemical Engineering, 2004, 28, 1237-1247.
12. Chen, C. L.; Hung, P. S. Simultaneous synthesis of mass exchange networks for waste minimization. Comput. Chem. Eng. 2005, 29 (7), 1561.
13. Chen, C. L.; Hung, P. S. Retrofit of mass-exchange networks with superstructure-based MINLP formulation. Comput. Chem. Eng. 2005, 44(18), 7189.
14. Foo C. Y.; Manan Z. A.; Yunus R. M.; Aziz R. A. Synthesis of mass exchange network for batch processessPart I: Utility targeting. Chem. Eng. Sci., 2004, 59, 1009–1026.
15. Foo C. Y.; Manan Z. A.; Yunus R. M.; Aziz R. A. Synthesis of mass exchange network for batch processes Part II: Minimum units target and batch network design. Chem. Eng. Sci. 2005, 60, 1349–1362.
16. Chen, C. L.; Ciou, Y. J. Superstructure-Based MINLP Formulation for Synthesis of Semicontinuous Mass Exchanger Networks. Ind. Eng. Chem. Res. 2006, 45, 6728–6739.
17. Chen, C. L.; Ciou, Y. J. Synthesis of a Continuously Operated Mass-Exchanger Network for a Semicontinuous Process. Ind. Eng. Chem. Res. 2007, 46, 7136–7151.

(408e) Turndown Control Structures for Distillation Columns

William L. Luyben, Chemical Engineering, Lehigh University, Bethlehem, PA and Patrick J. Robinson, Chem. Eng., Lehigh University, Bethlehem, PA

Future chemical plants may be required to have much higher flexibility and agility than existing processes in order to be able to handle new hybrid combinations of power and chemical units. An important example is a gasification process producing synthesis gas that can either feed a combustion turbine for the generation of electricity or, during periods of lower power demand, feed a chemical plant. The chemical plant would be designed for the maximum capacity that is associated with periods of minimum electric power demand. But the plant would have to be able to turndown to low throughputs during periods of maximum electric power demand. The 24 hour power demand swings in many locations can be a factor of two or more from day to night.

The separations required in many chemical processes are achieved using distillation columns. If the process must operate over a wide range of throughputs, the columns must also have wide rangeability. There are several low-throughput limitations in distillation columns, usually involving hydraulic constraints. The most common is a low limit on vapor flowrate, below which weeping can adversely affect tray efficiency and separation. The vapor limit depends on tray design, with valve trays being the most rangeable. Even valve trays lose performance below about 50% of design vapor rates. If the plant throughput must be reduced to 25% of design capacity and the column vapor can only be reduced to 50% of design, a control structure that effectively handles this situation is required. The purpose of this paper is to explore three alternative control structures for columns with significant turndown requirements.

See more of this Session: Separations Design
See more of this Group/Topical: Computing and Systems Technology Division

(468a) Mathematical Modeling and Steady-State Analysis of a Hybrid Solid Oxide Fuel Cell

Mona Bavarian, Chemical and Biological Engineering Department, Drexel university, Philadelphia, PA and Masoud Soroush, Chemical and Biological Engineering, Drexel university, Philadelphia, PA

Electrolytes in solid oxide fuel cells (SOFCs) are typically based on stabilized zirconia, which has good oxygen-anion conductivity at high temperatures [1]. One disadvantage of this type of fuel cells is their high operation temperature. Many efforts are currently being made to develop novel lower temperature SOFCs that use electrolytes with high conductivity at intermediate temperatures. Among these electrolytes are the ones that have high proton conductivity at low temperatures between 600-800°C [2] or have mixed proton and oxygen ion conductivity in the temperature range of 600-1000°C [1, 3].

In this study, a mathematical model of a SOFC based on $BaCe_{1-x}Sm_xO_{3-\alpha}$ and two platinum electrodes are developed. This type of electrolytes exhibit both proton and oxygen anion conductivity [4]. In the modeling, heat transfer, mass transfer and electrochemical processes are taken into account. The model is then validated using the experimental data reported in [1]. The existence of steady-state multiplicity in this type of fuel cells is investigated under three modes of constant ohmic load, potentiostatic and galavanostatic operations. The cell shows up to three steady states under constant ohmic load and potentiostatic modes. Under galvanostatic mode, the cell exhibits a unique steady state, as in oxygen-ion conducting SOFCs [5].The effects of changes in the heat convection coefficient and the inlet fuel and air flow rates on the multiplicity are studied. The increase in the heat convection coefficient shifts the multiplicity region to lower values of load resistance and cell voltage in the constant ohmic load and potentiostatic modes, respectively. Interestingly, this study shows that thermal and concentration multiplicities can coexist. Ignition in the solid temperature is accompanied by extinction in the reactant concentrations and ignition in the product concentration.

References:

[1] H. Iwahara, T. Yajima, T. Hibino, H. Ushida, Journal of the Electrochemical Society, 140 (1993) 1687.

[2] H. Iwahara, Solid State Ion., 28 (1988) 573-578.

[3] N. Bonanos, Journal of Physics and Chemistry of Solids, 54 (1993) 867-870.

[4] H. Iwahara, H. Uchida, K. Morimoto, Journal of the Electrochemical Society, 137 (1990) 462.

[5] M. Mangold, M. Krasnyk, K. Sundmacher, J. of Applied Electrochemistry 36(3) (2006), 265-275.

(468b) Optimized Gas Loop Design for Fischer-Tropsch Process

Debanjan Chakrabarti[1], Lin Wang[1], Arun Cherian[1], Arno de Klerk[1] and Vinay Prasad[2], (1)Department of Chemical and Materials Engineering, University of Alberta, Edmonton, AB, Canada, (2)Department of Chemical and Materials Engineering, Univesity of Alberta, Edmonton, AB, Canada

Optimized gas loop design forFischer-Tropsch process

The Fischer-Tropsch (FT) process is usedto convert a feedstock such as coal or biomass to syngas ($CO + H_2$),and then to syncrude, which consists of a vast array of hydrocarbons rangingfrom methane to C35-C40 hydrocarbons. The syncrude is further refined to yieldgasoline, diesel or kerosene fractions. The process has been in use since theearly 1930s, when the Germans used a cobalt catalyst-based fixed bed FTsynthesis reactor to obtain transportation fuels from coal. The process hasundergone considerable developments since then, and is operated under a varietyof conditions.

Potential feeds to the FT processinclude natural gas, which would be reformed, or coal or biomass, which wouldbe gasified, to yield syngas in both cases. The syngas produced is thenprocessed in a Fischer-Tropsch reactor which yields the syncrude. The gas loopthus consists of the feed, the syngas generation technology and the FT reactor(Figure 1).

In general, it is desirable to generatea syncrude with a majority of its components in the diesel or gasolinefractions, while at the same time minimizing the production of methane. Thecomposition of the syncrude produced, however, depends on the gas loop, itscomponents and its design. The composition of the syngas is determined by thetype of feed and the type of syngas production technology used, and differentFT reactors favour specific syngas compositions to give the desirable output.In this work, we model the different technologies available in the gas loop andcarry out studies to select optimized gas loops for each of the followingmetrics: (1) minimizing methane generation, (2) increasing diesel cutselectivity in the syncrude, and (3) increasing conversion efficiency withrespect to the feed. The optimization is conducted at two levels, the selectionof the appropriate combination of technologies for the units in the gas loop,and in the optimization of operating conditions for each process unit. Theprocess technologies considered are steam reforming, partial oxidation andautothermal reforming for the reformer; fixed bed, fluidized bed and entrainedflow reactors for the gasifier; and slurry phase, fluidized bed, turbulentfluidized bed and fixed bed reactors for the Fischer-Tropsch synthesis reactor.Reaction kinetics for the FT reactor are obtained from Wang et al. [2] for ironcatalyst based processes, and from Yates and Satterfield [3] for cobaltcatalyst based processes.

References

[1] A. de Klerk, Fischer-Tropsch fuelsrefinery design' Energy Environ. Sci., 4 (2011) 1177-1205.

[2] Y-N. Wang, W-P. Ma, Y-J. Lu, J.Yang, Y-Y. Xu, H-W. Xiang, Y-W. Li, Y-L. Zhao and B-J. Zhang, Kineticsmodelling of FischerTropsch synthesis over an industrial FeCuK catalyst', Fuel,82 (2003) 195213.

[2] I.C. Yates and C.N. Satterfield, Intrinsickinetics of the Fischer-Tropsch synthesis on a cobalt catalyst', Energy andFuels, 5 (1991) 168-173.

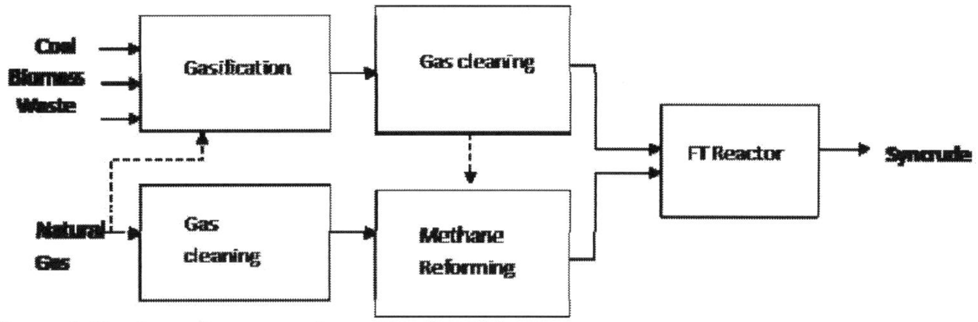

Figure 1.Gas loop for syncrude production [1]

See more of this Session: Modeling and Control of Energy Systems I
See more of this Group/Topical: Computing and Systems Technology Division

(468c) Decision Making for Unconventional Natural Gas Production: A Multivariate Analysis Approach

Srimoyee Bhattacharya, Chemical & Biomolecular Engineering, University of Houston, Houston, TX and Michael Nikolaou, Chemical and Biomolecular Engineering, University of Houston, Houston, TX

Recovering natural gas from unconventional resources, such as shales, tight gas, has recently increased the recoverable reserves in North America by a significant factor. Made possible through reservoir stimulation by massive hydraulic fracturing, the development of such resources poses engineering and environmental challenges that must be addressed. One such challenge is associated with making decisions, in a systematic way, about where to place and how to complete new production wells. The current practice relies on using a variety of available data, from early seismic studies and core analysis, to production from existing wells over a period of time. The entire process of data analysis requires significant human input, and can be quite laborious, with the possibility of not making optimal use of all available information. To address this need, we have undertaken a collaborative project, aiming to develop a suite of integrated software modules that facilitate the integrated use of all available data for systematic decision making. In this presentation we discuss our latest work on using multivariate data analysis and modeling methods that help the user analyze data from past production, so that factors affecting performance can be identified and subsequent decisions can be made to enforce factors that boost production while avoiding problematic ones.

The multivariate methods applied in this study are principal component analysis / partial least squares (PCA/PLS), Linear discriminant analysis (LDA), and symbolic regression (SR). The analysis was performed on real production data from the Holly Branch Field in Freestone County, Texas.

Production data from twelve existing wells in the field were fed to PCA pool. One big cluster of similar performance wells and three distinct behavior wells were obtained as a result of this analysis. Probing further into well completion data clearly demonstrated the reason for this distinction of the three wells from the rest of the nine-well cluster. It was also found that production data from all wells follow the stretched-exponential power-law decline curve ($q(t)=q_i\ (\exp(-D_i\ t^n))$) . Once these decline curve parameters were estimated, correlations and models were developed using PLS, LDA and SR with decline curve parameters q_i, D_i, and n as outputs and well properties such as porosity, permeability, fracture conductivity, fracture half-length, aspect ratio, drainage area, and cumulative gas in place in drainage area, as inputs. Adjusted R^2 of more than 90% for each of the decline curve parameters was obtained which proves good predictive ability of the models. Thus, these models obtained can be used confidently to predict production behavior of new wells in this field if other well planning parameters are known. Finally, the estimated ultimate recovery (EUR) can be evaluated once the production decline parameters are known.

Future extensions of this work include additional testing using real data, and exploration of other multivariate methods that might help identify patterns in available production data.

See more of this Session: Modeling and Control of Energy Systems I
See more of this Group/Topical: Computing and Systems Technology Division

(468d) Controlled Variables Selection for a Gas-to-Liquids Process

Mehdi Panahi, Chemical Engineering, Norwegian University of Science and Technology (NTNU), Trondheim, Norway and Sigurd Skogestad, Chemical Engineering, Norwegian Univeristy of Science and Technology, Trondheim, Norway

GTL (Gas To Liquid) is an important technology for converting natural gas to economical, clean and high quality liquid fuels. A GTL plant has 3 main parts (i) production of synthesis gas (syngas), (ii) Fischer-Tropsch (FT) reactor and (iii) upgrading units. Various technologies for producing synthesis gas and different types of FT reactor, catalyst and kinetics are available in literature.

In particular, Auto-Thermal Reforming (ATR) is claimed to be the best option for synthesis gas production [1]. In our work, we study in a detail; design, optimization and controlled variables selection for a GTL process based on ATR for synthesis gas production and a FT reactor with Cobalt catalyst in a Slurry Bubble Column Reactor (SBCR). The well-known kinetics of Iglesia [2] is used for the FT reactor; the remaining reactions are simulated assuming thermodynamic equilibrium. Selectivity to different hydrocarbons in FT reactions is described by ASF ideal model with a chain growth probability of hydrocarbons as function of the hydrogen to carbon monoxide ratio [3].

The UniSim process simulator is used to simulate and optimize one train of a GTL plant with capacity of around 34000 bbl/day, which is a current industrial scale capacity, similar to that of the Oryx (Qatar) and Nigeria plants.

The process includes heat integration, CO_2 removal, recycles and purge. Important process parameters, which are subject to optimization, include feed rates of water, oxygen, reactor temperatures, CO_2 removal percent and recycle flows. Nominally, the optimal ratios of oxygen and water to hydrocarbons (fresh natural gas and recycle hydrocarbons from FT reactor) were H_2O/hydrocarbons=0.6 and O_2/hydrocarbons=0.52, and the optimal ratio of H_2 to CO in fresh syngas is 2.1 . Optimization is used to find bottlenecks of the process while flowrate of natural gas is also a degree of freedom and it is tried to maximize the variable income of the plant respect to these bottlenecks.

In addition to determining the process and equipment design and finding the bottlenecks, optimization is used to find optimal operation (control) policies where an important issue is: what should we control? To answer this in a systematic manner we first need to define operational objective (which in this case is to maximize the variable income) and operational constraints. Next, we optimize the operation for various disturbances (such as feed composition and flowrate, pressure and temperature, constraint values, etc.). The first thing we need to control is the active constraints, for example, the outlet temperatures of the fired heater and ATR in syngas unit. Next, we look for "self-optimizing variables" [4], which are controlled variables which indirectly give close-to-optimal operation when held at constant setpoints, in spite of changes in the disturbances. Selection of the single and combinatorial measurements and its effect on the economy of the plant both in the nominal case and in the mode of maximizing throughput is discussed.

Acknowledgment

Discussions and suggestions from Dr. Dag Schanke at the Statoil R&D Centre in Trondheim are gratefully acknowledged.

References

[1] Bakkerud, P. K. (2005). "Update on synthesis gas production for GTL", Catalysis Today 106(1-4): 30-33.

[2] Iglesia, E., S. C. Reyes, et al. (1993). "Reaction-Transport selectivity models and the design of fischer-tropsch catalysts. Computer-aided design of catalysts", P. C. J. Becker E. Robert 199-257

[3] Song, H.-S., D. Ramkrishna, et al. (2004). "Operating strategies for Fischer-Tropsch reactors: A model-directed study", Korean Journal of Chemical Engineering 21(2): 308-317

[4] Skogestad, S. (2004). "Control Structure Design for Complete Chemical Plants", Computers and Chemical Engineering, 28, 219-234

See more of this Session: Modeling and Control of Energy Systems I
See more of this Group/Topical: Computing and Systems Technology Division

(468e) Modeling and Control of a Hybrid Renewable Energy System

Milana Trifkovic[1], Mehdi Sheikhzadeh[2], Khaled Nigim[2] and Prodromos Daoutidis[1], (1)Chemical Engineering and Materials Science, University of Minnesota, Minneapolis, MN, (2)Instrumentation and Control, Lambton College, Sarnia, ON, Canada

The utilization of renewable energy sources is becoming increasingly attractive and widely used as an alternative to oil-produced energy. Due to the widespread availability of renewable energy (RE) sources, power can be produced in a small scale and remote locations, as long as a sustainable integrated system for power production, storage and distribution exists. Energy storage is a key challenge, especially when the energy system is not connected to the grid, and is usually solved by incorporating an energy storage unit. For long-term storage, electrical energy can be converted into hydrogen, for later use in fuel cells to produce electricity.

This study deals with the development of a supervisory, optimization and power management tool for a 10kW hybrid energy system built in Lambton College, Sarnia, ON. The main rationale for such a tool lies in the fact that renewable energy sources are intermittent in nature and their availability may not coincide with the load demand or with the availability of energy carriers. In order to accomplish the task of developing the supervisory software, the main step is to develop a reliable plant model, which is then used for optimization and control purposes.

The wind, PV, and hydrogen production/storage/consumption components of the hybrid energy system are modeled individually. Process streams and electrical connections are applied for integration of the above-mentioned components. The RE sources and fuel cell are integrated either through an AC or DC bus. The hybrid energy system is modeled as stand-alone and grid-connected as well. When the hybrid system is without grid connection, the power demand (load) is the main criterion for power management. However, in the grid-connected system, power can be transferred from the grid to the hybrid system when power generated from RE sources does not meet the load demand. The integrated model was structured in a dynamic mode. In addition to main components, hydrogen storage tank, gas compressor and gas pressure regulator models are also developed.

The hybrid system is designed and modeled in a modular approach for all components, which allows proper sizing of equipment to achieve maximum efficiency. The control architecture consists of the low level local controllers for each component and a supervisory power controller, which manages the power flow. Excess RE power initiates the hydrogen production units and whenever the RE system cannot completely meet load demand, the FC system provides power to meet the remaining load. The supervisory controller deals with a complex control problem of handling binary and continuous variables, as well as constraints. The control strategy consists of tracking the power demand, producing sufficient amount of hydrogen, and switching on and off modular components. It is important to highlight the consideration of constraints such as minimum fuel cell, electrolyzer power threshold and time limitation between fuel cell and electrolyzer startups and shutdowns. A model predictive control strategy is developed to solve the optimal energy management problem.

See more of this Session: Modeling and Control of Energy Systems I
See more of this Group/Topical: Computing and Systems Technology Division

(468f) Offset-Free Model Predictive Control of Vapor Compression Cycle

Matt Wallace[1], Prashant Mhaskar[2], John House[3] and Tim Salsbury[3], (1)Department of Chemical Engineering, McMaster University, Hamilton, ON, Canada, (2)Chemical Engineering, McMaster University, Hamilton, ON, Canada, (3)Johnson Controls, Milwaukee, WI

The goal of energy efficiency is to make a device produce the same output by using less energy. This is clearly desirable in general, but must always be weighed carefully against any hidden costs required to achieve it. For example, improving the efficiency of one device in isolation might lead to more energy use somewhere else, thereby off-setting the overall benefit. There are also fundamental limits to how efficient energy conversion from one form to another can be. As we approach these limits, the potential for absolute improvements diminish accordingly.

One way to improve energy efficiency that has little or no hidden energy penalties is by modifying the way that a device is controlled. If no change is made to a device beyond improving the control algorithm, this is almost always a net benefit. Studies have shown that HVAC systems in buildings frequently operate above their theoretical minimal energy use level because of poor control design. Current practice fails to take into account interactions within a system and also handles constraints in a way that is separate and often damaging to control performance. Energy use is also not usually featured as an objective in the control performance and is handled instead by setting set-points in an ad-hoc way based on experience.

Model predictive control (MPC) is a control methodology that has reached a certain level of maturity in several industries but has not yet been applied to buildings at any substantial scale. MPC is designed to handle many of the shortfalls of existing control mentioned earlier including interaction between variables, constraints, and the incorporation of higher level objectives such as energy use. The work reported here involves applying MPC methods to vapor compression systems of the type used to provide cooling in buildings through equipment such as roof-top units. An MPC strategy is developed and tested for manipulating the electronic expansion valve and also the speed of a variable speed compressor. The control objective is to regulate the air temperature being blown over the evaporator subject to constraints on the superheat. An optimization objective is formulated to minimize the energy used by the compressor while meeting control objectives and constraints. A linear (identified) model is used for the air and superheat temperature predictions accounting for variations in the air and load conditions outside the vapor compression system.

An MPC formulation is devised to counter the inevitable plant model mismatch and achieve offset-free tracking. A detailed simulation model is used to evaluate and compare the performance of the MPC strategy with various combinations of traditional control strategies, including the use of MPC at a supervisory level. Results will be presented to demonstrate the improved temperature regulation and energy savings achievable using the MPC strategy.

See more of this Session: Modeling and Control of Energy Systems I
See more of this Group/Topical: Computing and Systems Technology Division

(468g) Networked Predictive Control of Distributed Energy Systems with Adaptive Communication

Yulei Sun and **Nael H. El-Farra**, Department of Chemical Engineering and Materials Science, University of California, Davis, Davis, CA

The convergence of competition in the power industry with the arrival of environmentally friendly micro-turbines, fuel cells, photovoltaics, small wind turbines, and other advanced distributed power technologies has sparked strong interest in distributed power generation. This convergence of policy and technology promises to transform the electric power system from one that relies primarily on central generation to one in which small Distributed Energy Resources (DERs) provide most of the power needed. The resulting major improvement in power reliability, quality and efficiency; and the greater flexibility to respond to changing energy needs could save billions of dollars now lost each year because of power disruptions. An important challenge with this evolution is the development of integrated control and communication systems that can handle the large number and diversity of DERs spread over the grid, dispatch these resources at the right time and account for the flow of energy correctly. This is significant given the fact that the distributed power market is primarily driven by the need for super-reliable, high-quality power and the fact that the impact of of even a brief communication disruption (e. g., due to local network congestion or server outage) can be substantial, particularly in sites where millisecond outages can cause costly computer crashes.

An effort to address this problem was initiated in [1] where a model-based networked control approach was developed for DERs that communicate with a supervisor over a shared resource-constrained communication network. The main idea was to reduce the susceptibility to communication outages by reducing the rate at which the data are exchanged between each DER and the supervisor as much as possible without sacrificing the desired stability and performance properties. A dynamic model that supplies the supervisor with the needed DER state information when communication is suspended over the network was embedded in the supervisor, and updated periodically using the actual state provided by the DER sensors at discrete time instances.

In this contribution, we extend this model-based networked control structure in two key directions. The first is to incorporate optimality considerations, as well as state and control constraints, in the design of the local control systems. The second is to design the networked control system in a way such that the necessary communication rate can be determined and adjusted dynamically based on the state of the DERs. Dynamic feedback-based communication policies have the advantage that they are more robust to unpredictable disturbances and allow the supervisor to respond quickly in an adaptive fashion to a DER that requires immediate attention to minimize power disruption. To address both problems, we initially synthesize for each DER a Lyapunov-based model predictive controller that enforces closed-loop stability and achieves the desired power output in the absence of communication suspensions. The controller minimizes a finite-horizon cost functional that imposes penalties on the deviation from the desired power reference and on the control action, subject to state, control and Lyapunov-based stability constraints. An adaptive communication policy in which the Lyapunov stability constraint is used as the basis for switching on and off the communication between the local sensor suite of each DER and the supervisor is then devised. The idea is to locally monitor the evolution of the Lyapunov function of each DER plant such that if it begins to breach a pre-specified performance threshold at any time, the local sensor suite is prompted to send its data over the network to update the corresponding model in the supervisor and provide corrective control action. The results are illustrated through an application to a multi-source hybrid DER micro-grid that includes a photo-voltaic panel, a PEM fuel cell stack, and a battery.

References:

[1] Sun, Y., S. Ghantasala and N. H. El-Farra, "Networked Control of Distributed Energy Resources: Application to Solid Oxide Fuel Cells," *Industrial and Engineering Chemistry Research*, **48**, 9590-9602, 2009.

See more of this Session: Modeling and Control of Energy Systems I
See more of this Group/Topical: Computing and Systems Technology Division

(473a) A Predictive Controller Design for Spatially Distributed Processes Using Adaptive Model Reduction

Sivakumar Pitchaiah, Department of Chemical Engineering, The Pennsylvania State University, University Park, PA and **Antonios Armaou**, Chemical Engineering, The Pennsylvania State University, University Park, PA

Over the last 20 years there has been increasing focus on analysis and control of processes that exhibit spatial variations. The research activity in this area has been motivated by a wealth of industrially important processes (e.g., chemical vapor deposition and etching, catalytic reaction and polymerization processes) which exhibit significant spatial variations due to the presence of strong diffusive and convective mechanisms. The key difficulty in designing control schemes for such PDE systems arises from the ``infinite-dimensional" nature of the distributed process model. Based on this property ideas for reducing the complexity of these models were considered through the formulation of the reduced order models (ROM). One of the widely used methodologies for developing ROM is proper orthogonal decomposition (POD). Initially, one collects the experimental data or data from detailed numerical simulations (snapshots) in a ``dataset". POD then extracts the characteristic basis (shape) functions from the collected data set. The computed basis functions are subsequently used in the framework of "method of weighted residuals" to compute the low-dimensional ROMs. The effectiveness of the POD methodology, however, is dependent on the quality of the collected dataset. Consequently, the ROMs are sufficiently accurate only in a restricted neighborhood around the state space regions where the system is sampled. On the other hand, no well defined methodology exists for constructing ROMs with a larger region of validity as this requires a ``representative" dataset which contains all the possible spatial modes (including those that might appear during closed-loop evolution of the PDE system). To address these issues we have developed an adaptive methodology called the adaptive proper orthogonal decomposition [1]. This methodology alleviates the requirement for the representative ensemble by recursively updating the basis functions in a computationally efficient way.

In this talk, we will present our recent results on using adaptive reduced order models (generated using APOD) in the design of model predictive controllers for stabilization of processes that are mathematically expressed as parabolic partial differential equation (PDE) systems. Initially, we construct a locally valid ROM of the PDE system employing the basis functions computed by applying the APOD methodology on a small data ensemble. This ROM is then utilized in the design of model predictive controllers (MPC) under constraints on the control action. As periodic closed-loop process data becomes available (during closed-loop operation under the constructed MPC), we recursively update the ROM by employing our computationally efficient adaptive model reduction methodology thus extending the validity of ROM over the current operating region. The effect of employing the adaptive methodology on performance of MPC will be discussed. The design of such MPC controllers is illustrated by employing the methodology on transport-reaction processes. The proposed controller successfully stabilizes the process at an open-loop unstable steady-state while minimal initial sampling is used.

Reference:

1. Pitchaiah, and A. Armaou, "Feedback control of dissipative distributed parameter systems using adaptive model reduction," Ind. & Eng. Chem. Res., **55**, 906-918, 2010.

See more of this Session: Optimization and Predictive Control
See more of this Group/Topical: Computing and Systems Technology Division

(473b) Constrained Multi-Rate State Estimation Using NLP Sensitivity-Based Moving Horizon Estimation

Rodrigo Lopez-Negrete and Lorenz T. Biegler, Chemical Engineering, Carnegie Mellon University, Pittsburgh, PA

Constrained Multi-Rate State Estimation using NLP Sensitivity-Based Moving Horizon Estimation

Model based control schemes, such as nonlinear model predictive control, assume that the full state vector is known for feedback control. However, in reality this is not always true. Most times only a set of noisy measurements are available, and thus, the unmeasured states need to be inferred from these measurements. Moreover, in most chemical processes, some state variables (e.g., concentrations, molecular weights, etc.) can be measured infrequently and, probably, with some time delay. The information provided by these infrequent measurements can aid the estimation process, and therefore multi-rate state estimators have been developed. Moving Horizon Estimation (MHE) provides a framework that allows to easily incorporate these infrequent observations because it uses a window of past measurements, where the slower ones can be introduced as they become available. Also, MHE is a preferred method for state estimation because it also allows the use of bounds and constraints on the state estimates that help the performance of the estimator.

Nonlinear Programming (NLP) sensitivity based MHE method has also been developed called advanced-step MHE (as-MHE). This method solves an MHE problem offline using approximated measurements, and when the new measurement becomes available a quick NLP sensitivity based correction is done to the offline solution. These sensitivity-based techniques also allow us to extract the reduced Hessian information from interior point NLP solvers that do not generally form them as part of the solution strategy. Furthermore, it is possible to show that the inverse of the reduced Hessian provides a good approximation of the covariance of the estimated states. Thus, this strategy also allows us to approximate the covariance of the state estimates at the beginning of the horizon. Therefore, it is possible to generate a fast approximation of the arrival cost covariance, as well. In this work we develop an as-MHE strategy to handle multi-rated measurements, and that takes advantage of NLP sensitivity to reduce the computational expense of the online estimation process. The proposed methodology is illustrated using benchmark examples from the literature.

Literature cited

Pirnay, H., Lopez-Negrete, R. , and Biegler, L.T. , Optimal Sensitivity Based on IPOPT, submitted to Math Prog Comp, 2011

Zavala, V. M.; Laird, C. D. & Biegler, L. T. A Fast Moving Horizon Estimation Algorithm Based on Nonlinear Programming Sensitivity. Journal of Process Control, 2008, 18, 876-884

Rao, C. V. & Rawlings, J. B. Constrained Process Monitoring: Moving-Horizon Approach. AIChE Journal, 2002, 48, 97-109

Krämer, S.; Gesthuisen, R. & Engell, S. Fixed Structure Multirate State Estimation. American Control Conference, 2005

Krämer, S. & Gesthuisen, R. Multirate State Estimation using Moving Horizon Estimation. 16th IFAC World Conference, 2005

See more of this Session: Optimization and Predictive Control
See more of this Group/Topical: Computing and Systems Technology Division

(473c) Resource-Aware Model Predictive Control Using Adaptive Sampling

Ye Hu and **Nael H. El-Farra**, Department of Chemical Engineering and Materials Science, University of California, Davis, Davis, CA

Model predictive control (MPC) has a long history in process control research as one of the few methods suited for handling constraints in an optimal control setting. In a typical MPC implementation, state measurements are received at the sampling time and the control action is obtained by solving, on-line, a model-based finite-horizon optimal control problem subject to state, control and stability constraints. Only the first part of the computed control trajectory is implemented until the next sampling time at which new state measurements are received to update the model state and re-solve the optimization problem. While many different MPC formulations currently exist in the literature, the MPC problem is often formulated and addressed within the classical feedback control paradigm in which the process output is passed at a fixed sampling rate to the controller which then generates the control input and in turn passes it back to the process.

The success of this paradigm notwithstanding, recent years have witnessed a number of calls for re-examining the conventional feedback control paradigm and expanding it in ways that account for the increasing complexity of the process/controller interface which features additional information-processing steps that need be accounted for. For example, with the emergence of applications involving wireless sensor and actuator networks, constraints on the information collection, processing and transmission capabilities of the measurement system and the communication medium are becoming commonplace and require attention in the MPC system design. In some of these applications, it is advantageous, if not necessary, to collect and transfer measurements at a variable (rather than a fixed) rate in order to optimize performance under limited resources.

In this contribution we present a framework for the design of resource-aware model predictive controllers for constrained nonlinear systems subject to sensor-controller communication constraints. The main objective is to enforce the desired closed-loop stability and optimality properties with minimal measurement sampling and/or sensor-controller communication. To this end, we initially design a Lyapunov-based model predictive controller that enforces closed-loop stability within a well-characterized stability region for a given sampling/communication rate. The controller minimizes an appropriate cost functional defined over the prediction horizon subject to state and control constraints, and includes a stability constraint that ensures the decay of the Lyapunov function during the sampling interval. The closed-loop stability properties are characterized in terms of a state-dependent bound on the decay of the Lyapunov function that can be tuned by proper selection of the sampling period and the controller design parameters. This characterization is then used to devise a dynamic sampling/ communication policy that adaptively adjusts the rate of information transfer from the sensors to the controller based on the evolution of the Lyapunov function within the stability region. The idea is to reduce the sampling/communication rate (or suspend communication altogether) for times when the prescribed stability bound is satisfied. During such periods, the control action computed from the model-based optimization is implemented without any model updates. At times when the state begins to breach the prescribed stability bound, the sampling/communication rate is increased (or communication is fully restored) to allow more frequent model updates. Switched system techniques are used to analyze closed-loop stability and derive precise conditions for the implementation of the proposed adaptive sampling policy. Finally, the theoretical results are illustrated using a chemical process example.

See more of this Session: Optimization and Predictive Control
See more of this Group/Topical: Computing and Systems Technology Division

LMI-Based Multi-Model Predictive Control of An Industrial C3/C4 Splitter

Bruno D. O. Capron and Darci Odloak, University of São Paulo, São Paulo, SP, Brazil

Model Predictive Control (MPC) is nowadays commonly implemented in chemical industries such as oil refineries, paper manufacturing, … The implementation of this control strategy consists in calculating at each time step an open-loop sequence of control moves that minimizes an objective function subject to constraints on the system inputs. Generally the objective function is composed by a term accounting for the difference between the predicted system outputs and their corresponding set points and a term penalizing the input moves. The first calculated control move is implemented and the optimization problem is solved again at the next step.

MPC is based on a linear model which may not be reliable if the real model of the plant is nonlinear, which is often the case in chemical industry. In that case, one unique model is uncertain and may not allow an efficient control of the plant in some operating zones. One way to overcome this disadvantage is to deal with the model uncertainties considering the Multi-plant uncertainty model (Porfírio et al., 2003), for which a discrete set of plants corresponding to different operating points of the system is considered. Defined an objective function for each model of the set, the multi-model predictive controller then results from minimizing the worst case objective function.

Such a controller has already been implemented by (Porfírio et al., 2003) on a real industrial C3/C4 splitter, where it actually showed to have a significantly better performance.

However, a disadvantage of the Multi-Model Predictive Control (MMPC) compared to the MPC is the computational burden involved in the optimization problem resolution, which can be prohibitive for high dimension systems. In this case, the LMI techniques that have been developed over the past decades are of a great interest as they allow to significantly reducing the computational complexity of the optimization problem (Kothare et al., 1996).

In this work, the classical MMPC problem developed for the industrial C3/C4 splitter system that can be found in (Porfírio et al., 2003) is re-casted as an LMI-based problem in order to compare the performances and computational costs of the classical NLP based MMPC and the LMI based MMPC. The simulations show that compared to the NLP based controller, the LMI based controller allows to significantly reducing the computational cost without penalizing the controller performance.

References

1. Kothare, M. V., Balakrishnan, V. and Morari, M. Robust constrained model predictive control using linear matrix inequalities. Automatica [S.I.], v. 32, n. 10, p. 1361-1379, 1996.

2. Porfirio, C. R; Almeida Neto, E. and Odloak, D. Multi-model predictive control of an industrial C3/C4 splitter. Control Engineering Practice [S.I.], v. 11, n. 7, p. 765-779, 2003.

(473e) Data-Based Modeling and Control of Nylon-6,6 Batch Polymerization

Siam Aumi[1], **Brandon Corbett**[1], Prashant Mhaskar[1] and Tracy L. Clarke-Pringle[2], (1)Chemical Engineering, McMaster University, Hamilton, ON, Canada, (2)Process Dynamics and Control, DuPont Engineering Research and Technology, Wilmington, DE

Batch processes are commonly used for the manufacture of high-value specialty products. As opposed to continuous reactors that are operated around a nominal steady-state, the end-point in a batch reactor is not necessarily an equilibrium/steady-state point, allowing for achieving a wide range of product specifications by changing the initial conditions and input trajectories. This flexibility is particularly important in the polymerization industry as polymerization reactions often require a wide range of operating conditions to yield polymers with a desirable end-use quality, which is the primary control objective. Batch polymerization systems, however, exhibit numerous numerous characteristics, such as nonlinear, time-varying, and coupled dynamics, that complicate the control problem. The biggest challenge in the control problem is that measurements related to the final end-use quality are often unavailable during the batch and only made (off-line) after the batch is complete. Thus, direct control of the end-use qualities is often impractical in practice, and the control objective is pursued indirectly via trajectory tracking methods. In trajectory tracking methods (e.g., see [1–3]), trajectories for a set of measurable secondary variables, which should be related to the end-use properties, are generated off-line (or re-calculated at specific time points during the batch (e.g, see [4,5])) and then subsequently tracked using local proportional-integral-derivative (PID) or predictive controllers. Classical control approaches, such as PID control, are typically inadequate due to the nonlinearities and constraints in the process combined with interactions among the control loops while the benefits from using a predictive controller are dictated by the underlying model's quality.

Polymerization models are identified either deterministically from first-principles or empirically from plant/process data. Deterministic polymerization models, however, have limited control applications. For instance, the number of system states is often excessive and/or the differential equations are overly complex (i.e., discontinuities, etc.) for use in any model-based or optimization-based control designs. Moreover, the model's predictive capabilities are subject to the accuracy of numerous model parameters, many of which may be inaccurate. Furthermore, many of the simplifying assumptions taken during model development are often violated in specific situations in practice. The increased availability of past process data, however, can be exploited to improve the achievable level of accuracy of polymerization models developed using simpler model structures (the empirical models). Identification experiments, such as those in which a pseudo-random binary signal (PRBS) is applied on the process, while suitable for identification at steady states, are often too expensive to justify for polymerization reactors since they result in wasted batches (and therefore wasted monomers and initiators and loss of time). Furthermore, within the range of operating conditions during a polymerization, the process behavior is highly nonlinear and characterized by stages with different dynamics, making conventional system identification approaches, where a single linear model is identified, ill-suited for identifying an accurate dynamic model. The high expenses associated with every batch dictate the need for developing dedicated modeling tools for polymerization processes that minimize wasted batches in the model-development process and yet provide a model that captures the essential nonlinear and complex nature of the process.

In our previously completed work [6], a multi-model approach was developed that exploits the simplicity of local linear models and uses an appropriate weighting scheme to capture the nonlinear nature of the process. However, in [6], full state measurements were assumed, which permitted identifying local linear state-space models in the form of linear discrete-time systems. In this work, we extend this modeling methodology and demonstrate its efficacy for a batch nylon-6,6 polymerization process where only limited measurements are available. The resulting data-based model is then used to formulate a trajectory tracking predictive controller. Through simulation studies, the modeling approach is shown to capture the major nonlinearities in the nylon-6,6 polymerization process, and closed-loop simulation results demonstrate the effectiveness of the proposed predictive controller and illustrate its advantages over existing trajectory tracking approaches such as conventional proportional-integral (PI) control and latent-variable MPC [3].

[1] Zhang GP, Rohani S. On-line optimal control of a seeded batch cooling crystallizer. Chem Eng Sci. 2003; 58:1887–1896.

[2] Soroush M, Kravaris C. Optimal-design and operation of batch reactors. 1. Theoretical framework. Ind & Eng Chem Res. 1993;32:866–881.

[3] Flores-Cerrillo J, MacGregor JF. Latent variable MPC for trajectory tracking in batch processes. J Process Control. 2005;15(6):651–663.

[4] Kozub DJ, MacGregor JF. Feedback control of polymer quality in semi-batch copolymerization reactors. Chem Eng Sci. 1992;47(4):929 – 942.

[5] Flores-Cerrillo J, MacGregor JF. Control of batch product quality by trajectory manipulation using latent variable models. J Process Control. 2004;14(5):539 – 553.

[6] Aumi S, Mhaskar P. Integrating Data-Based Modeling and Nonlinear Control Tools for Batch Process Control, (accepted). In: Proc. of the American Control Conference (ACC). 2011; .

See more of this Session: Optimization and Predictive Control
See more of this Group/Topical: Computing and Systems Technology Division

(473f) Lyapunov-Based Economic Model Predictive Control of Nonlinear Systems: Handling Asynchronous, Delayed Measurements and Distributed Implementation

Mohsen Heidarinejad[1], Jinfeng Liu[2] and Panagiotis D. Christofides[2], (1)Electric Engineering, University of California, Los Angeles, Los Angeles, CA, (2)Department of Chemical and Biomolecular Engineering, University of California, Los Angeles, Los Angeles, CA

Economic model predictive control (EMPC) refers to a class of model predictive control formulations in which the cost functional expresses directly economic optimization considerations of the plant under consideration, rather than penalizing the deviations of the plant states and of the manipulated inputs from desired steady-state values. As a result, EMPC may lead to the computation of time-varying optimal operating policies for the plant in contrast to MPC with traditional cost functionals which typically leads to stabilization of the plant at the desired steady state.

While there have been several calls, particularly within process control, for the integration of model predictive control (MPC) and economic optimization of processes as early as two decades ago, the subject of EMPC has received relatively little attention. Recently, in [1], MPC schemes using an economics-based cost function were proposed and the
stability properties were established using a suitable Lyapunov function. In a recent paper [2], the approach in [1] was extended to deal with cyclic process operation.
In a recent work [3], we presented an EMPC scheme for nonlinear systems that utilizes suitable Lyapunov-based stability constraints. The proposed EMPC is designed via Lyapunov-based techniques and has two different operation modes. The first operation mode corresponds to the periods in which the cost function should be optimized (e.g., normal production periods); and in this operation mode, the MPC maintains the closed-loop system state within a pre-defined stability region and optimizes the cost function to its maximum extent. The second operation mode corresponds to operation in which the system is driven by the MPC to an appropriate steady-state within the closed-loop system stability region.

In this work, we extend the results in [3] into two directions. First, we present an EMPC algorithm for nonlinear systems that efficiently handles asynchronous and delayed measurements using suitable Lyapunov-based constraints to ensure stability for a well-defined set of initial conditions, and demonstrate its application to a chemical process example. Second, we present a distributed EMPC (DEMPC) architecture for nonlinear systems. In this architecture, the distributed controllers communicate in a sequential fashion, optimize their inputs through maximizing a plant-wide (global) economic objective function and guarantee practical stability of the closed-loop system.

[1] M. Diehl, R. Amrit, and J. B. Rawlings, "A Lyapunov function for economic optimizing model predictive control," IEEE Transactions on Automatic Control, in press.
[2] R. Huang, E. Harinath, and L. T. Biegler, "Lyapunov stability of economically-oriented NMPC for cyclic processes," Journal of Process Control, in press.
[3] M. Heidarinejad, J. Liu, and P. D. Christofides, "Lyapunov-based economic model predictive control of nonlinear systems," in Proceedings of the American Control Conference, in press, 2011

See more of this Session: Optimization and Predictive Control
See more of this Group/Topical: Computing and Systems Technology Division

(473g) Economic MPC with Infinite Horizon

Donald J. Chmielewski and Benjamin P. Omell, Chemical and Biological Engineering, Illinois Institute of Technology, Chicago, IL

There is a growing interest in the development of MPC algorithms that replace the traditional quadratic objective function with a purely economic objective. However, implementation of such algorithms can result in unexpected and sometimes pathological closed-loop behavior, including inventory creep-down and bang-bang actuation during active constraint intervals. This paper will illustrate some of these issues. While an infinite horizon formulation seems to be a natural route to avoiding the above myopic behavior, implementation with an infinite horizon presents a number of technical and conceptual challenges. To address these issues, a number of subtle changes to the original problem formulation will be introduced and thus will allow for a computationally robust implementation of the infinite-horizon controller.

See more of this Session: Optimization and Predictive Control
See more of this Group/Topical: Computing and Systems Technology Division

(475a) Optimal Multi-Scale Capacity Planning Under Time-Sensitive Electricity Prices for Continuous Power-Intensive Processes

Sumit Mitra, Department of Chemical Engineering, Center of Advanced Process Decision-making, Carnegie Mellon University, Pittsburgh, PA, Ignacio Grossmann, Chemical Engineering, Carnegie Mellon University, Pittsburgh, PA, Jose M. Pinto, Praxair Inc., Danbury, CT and Nikhil Arora, Praxair, Inc., Tonawanda, NY

Electrical power costs are the major operating expense for many power-intensive processes, such as air separation plants and cement plants. Since electricity markets became deregulated in the 1990s, prices have been subject to hourly as well as seasonal variations. These issues are likely to become more acute as renewable sources of energy like wind and solar are introduced for power generation and expected to grow not least due to the disaster at the Fukushima nuclear plant. From an industrial consumer's point of view, this has introduced a considerable amount of uncertainty and variability in its short-term operating expenses, which in turn affects the long-term strategic capacity planning.

Multi-period capacity planning for continuous multi-product plants has been widely studied in literature. Sahinidis et al. (1989) propose one of the first deterministic MILP models. Liu and Sahinidis (1996) extend the model to account for demand uncertainty. Iyer and Grossmann (1998) develop a bi-level decomposition strategy to deal with the large number of scenarios that result from uncertain demand. All these papers share the idea that in order to cover a time horizon of multiple years, the problem is divided into a number of time periods. A time period is typically several months or several years long. This implies that model parameters, such as prices, demands are assumed to be constant over one time period. However, if prices fluctuate on an hourly and seasonal basis, there is need for a more detailed scheduling model and representation of the process' feasible region.

In recent work (Mitra and Grossmann (2011)), we studied the operational problem of minimizing production costs under time-sensitive electricity prices. In this work, we integrate the operational problem with the long-term strategic capacity planning problem, which involves decisions on installing or upgrading equipment, or increasing storage capacity. The resulting full-space mixed-integer linear program (MILP) model for a single plant is computationally hard to solve due to the multi-scale nature of the problem. Hence, we propose a tailored bi-level decomposition algorithm, similar in spirit to the work by Iyer and Grossmann (1998) and You et al. (2010). The decomposition relies on solving an aggregated master problem in which design decisions are optimized, and subproblems in which operations are optimized for that design. Furthermore, to effectively deal with the multi-scale nature of the problem we represent a yearly operation by a subset of cyclic schedules. The proposed strategy allows us to solve the MILP problem to optimality within a reasonable amount of time and much faster than a full-space solution. We illustrate the performance of the proposed methodology for an air separation plant based on industrial data for different operating capacities.

References:

Iyer, R.; Grossmann, I.E. (1998) A Bilevel Decomposition Algorithm for Long-Range Planning of Process Networks, Ind. Eng. Chem. Res.37, 474.

Liu, M. L.; Sahinidis, N. V. (1996) Optimization in process planning under uncertainty, Industrial & Engineering Chemistry Research35, 4154.

Mitra, S.; Grossmann, I.E. (2011) Optimal Production Planning under Time-sensitive Electricity Prices for Continuous Power-intensive Processes, to be submitted

Sahinidis, N.V.; Fornari, R.; Grossmann, I.E. (1989), Optimization model for long range Planning in the Chemical Industry, Comp. Chem. Eng.13, 1049-1063.

You, F.; Grossmann I.E.; Wassick, J. (2010), Multisite Capacity, Production and Distribution Planning with Reactor Modifications: MILP Model, Bilevel Decomposition Algorithm versus Lagrangean Decomposition Scheme, Ind. Eng. Chem. Res.,in press.

See more of this Session: Planning and Scheduling II
See more of this Group/Topical: Computing and Systems Technology Division

(475b) Integration of Scheduling and Control with Closed Loop Implementation

Jinjun Zhuge, Department of Chemical and Biochemical Engineering, Rutgers - The State University of New Jersey, Piscataway, NJ and Marianthi G. Ierapetritou, Department of Chemical and Biochemical Engineering, Rutgers, The State University of New Jersey, Piscataway, NJ

Traditionally production scheduling and process control problems are considered separately. Scheduling problem results in the optimal production sequence, production time and resources allocation while control problem focuses on dynamic behavior of transition periods between different products. However, the solutions obtained by considering scheduling are certainly suboptimal. Targeting better operating conditions in today's strict economic environment, a number of efforts have been made towards integration of scheduling and control problems. The integration of scheduling and control results in better modeling of the process operation since transitions are taking into consideration which are ignored in the traditional scheduling approach. With integrated modeling, information can be shared between scheduling and control without delay. Thus, a more economical process operation is achieved.

The existing approaches dealing with the integration of scheduling and control can be broadly categorized into simultaneous modeling approach and decomposition based methods. Using the simultaneous approach the process dynamic model is incorporated into the constraints of scheduling problem. Thus a Mixed Integer Dynamic Optimization (MIDO) problem is formed and then is discretized into Mixed Integer Nonlinear Programming (MINLP) using collocation point method ([1]). Using decomposition method, the control problem is modeled as dynamic optimization (primal problem) and the scheduling part as Mixed Integer Linear Programming (Master problem). The solution procedure proceeds by iterating between these two subproblems until convergence is achieved ([2]).

Most of the simultaneous based approaches however do not implement process control using close loop. In this study, we consider disturbance in real process and build a closed loop strategy for simultaneous scheduling and control, which can be regarded as real time scheduling and control. We assume that the state variables in both transition periods and production periods are subject to disturbance. We detect the disturbance and generate new solution for the integrated problem at the point of disturbance. As a result, process reacts quickly to eliminate the effects of disturbance. More specifically, we first solve the integrated problem off-line, and obtain the scheduling solution and control input. Then the solution is implemented in the process. If the real state track the reference (pre-calculated by solving the integrated problem off line) very well (i.e. their difference is within the tolerance), no feedback is needed. If significant disturbance occurs, the difference between state and reference is detected and feedback is activated. With the feedback information, the integrated problem is solved again for the remaining part of the production cycle. Thus both the scheduling solution and control input are updated, which ensure that the operation after the occurrence of disturbance is optimal. Case studies demonstrate that our approach is economically preferable compared to open loop strategy ([1]).

References:

[1] Tlacuahuac, A. Flores, & Grossmann, I.E.. (2006). Simultaneous Cyclic Scheduling and Control of a Multiproduct CSTR. *Industrial & Engineering Chemistry Research.* 45(20), 6698-6712.

[2] Nystrom, R.H., Franke, R, Harjunkoski, I., Kroll, A. (2005). Production campaign planning including grade transition sequencing and dynamic optimization. *Computers & Chemical Engineering.* 29(10) 2163-2179.

See more of this Session: Planning and Scheduling II
See more of this Group/Topical: Computing and Systems Technology Division

(475c) A New MILP/Discrete-Event Simulation Algorithm for Scheduling Automated Wet-Etching Stations

Pedro M. Castro[1], Adrin Aguirre[2], Luis Zeballos[1] and Carlos Mndez[2], (1)Modeling and Optimization of Energy Systems, LNEG, Lisbon, Portugal, (2)INTEC (Universidad Nacional del Litoral - CONICET), Santa Fe, Argentina

The efficient operation of complex semiconductor manufacturing facilities has attracted an increasing research interest in recent years. Industry is currently in a growth expansion, immersed in a series of highly competitive global markets that are characterized for being intensely technological and dynamic. This forces the wafer fabrication facilities to focus their efforts on providing to their customers, high-quality of affordable products with minimum delivery times and lower processing times. As a consequence, the development of efficient short-term scheduling strategies becomes a potential alternative to increase competitiveness, answering agilely to the requirements of highly demanding markets and customers.

The automated wet-etching station (AWS) is a key part of a modern semiconductor production system and has to deal with many complex constraints and limited resources. It consists of a series of chemical and water baths coupled with an automated lot transfer system, in which mixed intermediate storage policies must be strictly followed in order to avoid very expensive wafer contamination. The efficient operation of this stage will offer a substantial reduction of the processing time required to complete all the jobs, providing at the same time a better utilization of the critical transportation resources.

Previous works addressing this problem [1-3] have reported that the complexity lies with the sequencing of the transportation tasks that are performed by one or more robots and have proposed two-stage algorithms to reduce combinatorial complexity and hence allow the solution of larger problems. Unlimited robot availability was assumed in the first step and a relaxed MILP model was solved to find the optimal processing sequence. This was then fixed in the second step and a constrained MILP model was solved to find the optimal sequence of transportation tasks. Unfortunately, only problems featuring an extra few orders could still be tackled showing that other approaches are needed for sizes beyond 12 lots in 12 baths. Of the three conceptually different continuous-time models, the one by Castro et al. [1] exhibited the best performance.

In this work, we propose a new algorithm that combines mathematical programming and discrete-event simulation. The concern is no longer finding the optimal solution but very good solutions in a reasonable amount of time (e.g. 1 hour). The decomposition strategy involves three stages: (i) finding a near optimal processing sequence under unlimited robot availability by applying neighborhood search procedures [4-6] to an initial sequence, using the relaxed MILP model. Such rescheduling procedures use the ideas from our previous work [7-9] dealing with just a few lots per iteration but now lot selection is done randomly instead of using preordering heuristics. The best found sequence is then the input to (ii) the discrete-event simulation model that is able to generate a good feasible schedule to the one robot problem. This will be the starting point of the (iii) rescheduling procedure applied to the more complex MILP from [1], featuring the robot sequencing constraints. Note that the balance between solution quality and total computational effort can easily be shifted by changing the number of lots rescheduled per iteration.

Several examples are used to illustrate the capabilities of the proposed method, which is compared to the full-space model and to a constraint programming approach featuring a problem specific search strategy [10]. The results show significantly better solutions in about the same computational time. In particular, for the original 25 lots in 12 baths problem [2], the best found solution features a makespan that is 14.2% better than the A2' heuristic algorithm from [11] and 7.4% better than the constraint programming approach.

[1] Castro, P. M.; Zeballos, L. J.; Mndez, C. A. Hybrid Time Slots Sequencing Model for a Class of Scheduling Problems. *AIChE J.* doi: 10.1002/aic.12609.

[2] Bhushan S.; Karimi, I. A. An MILP Approach to Automated Wet-Etch Station Scheduling. *Ind. Eng. Chem. Res.* **2003**, *42*, 1391.

[3] Aguirre, A. M.; Mndez, C. A.; Castro, P. M. A Novel Optimization Method to Automated Wet-Etch Station Scheduling in Semiconductor Manufacturing Systems. *Comput. Chem. Eng.* **2011**, doi:10.1016/j. compchemeng.2011.02.14.

[4] Danna, E.; Rothberg, E.; Le Pape, C. Exploring relaxation induced neighborhoods to improve MIP solutions. *Mathematical Programming*, **2005**, *102*, 7.

[5] Rothberg, E. An evolutionary algorithm for polishing Mixed Integer Programming Solutions. *INFORMS Journal On Computing*, **2007**, *19*, 534.

[6] Fischetti, M.; Lodi, A. Local branching. *Mathematical Programming*, **2005**, *98*, 23.

[7] Mndez, C.; Cerdá, J. Dynamic scheduling in multiproduct batch plants. *Comp. Chem. Eng.* **2003**, *27*, 1247.

[8] Castro, P. M.; Harjunkoski, I.; Grossmann, I. E. Optimal Short-Term Scheduling of Large-Scale Multistage Batch Plants. *Ind. Eng. Chem. Res.* **2009**, *48*, 11002.

[9] Castro, P. M.; Harjunkoski, I.; Grossmann, I. E. Greedy Algorithm for Scheduling Batch Plants with Sequence-Dependent Changeovers. *AIChE J.* **2011**, *57*, 373.

[10] Zeballos, L. J.; Castro, P. M.; Mndez, C. A. Integrated Constraint Programming Scheduling Approach for Automated Wet-Etch Stations in Semiconductor Manufacturing. *Ind. Eng. Chem. Res.* **2011**, *50*, 1705.

[11] Bhushan, S.; Karimi, I. A. Heuristic Algorithms for Scheduling an Automated Wet-Etch Station. *Comput. Chem. Eng.* **2004**, *28*, 363.

See more of this Session: Planning and Scheduling II
See more of this Group/Topical: Computing and Systems Technology Division

(475d) In-Depth Study and Comparison of S-Graph Framework and Precedence Based MILP Formulations for Batch Process Scheduling

Mate Hegyhati, Tibor Holczinger and Ferenc Friedler, Department of Computer Science and Systems Technology, University of Pannonia, Veszprm, Hungary

Several review papers have been published on batch process scheduling (see, e.g., Floudas and Lin, 2004; Mendez *et. al.*, 2006; Hegyhati and Friedler, 2010), nevertheless, an in-depth study of specific approaches that can enhance each other's performance is yet to be done. It is especially interesting for the approaches based on task precedences: they are the so-called precedence based MILP formulations and the S-graph framework.

In the S-graph framework, batch process scheduling problems are modeled by a special class of directed graphs, where precedence of tasks is represented by an arc (Sanmarti *et. al.*, 1998). While it is natural for the MILP formulation based techniques that they need the complete mathematical programming formulation prior to solving the problem, it is not the case for the S-graph framework; the input is practically the graph representation of the recipe. The combinatorial algorithms extend this graph by so-called schedule-arcs as far as the allocation and sequencing decisions are made (Sanmarti *et. al.*, 2002). In the precedence based MILP models, binary variables are assigned to each pair of tasks that can be performed in the same unit (Mendez *et. al.*, 2001). The absence or existence of a schedule-arc on an S-graph is modeled by a binary variable in the MILP model.

The precedence based MILP approaches and the S-graph framework have some common features based on the one-to-one relation between a schedule-arc and a variable of sequencing. For example, limited wait storage policies, sequence dependent setup times can be addressed in a similar and straightforward way in both models (Hegyhati *et. al.*, 2011; Kopanos *et. al.*, 2009). The origin of the two mathematical representations are the same, nevertheless, the solution procedures are fundamentally different. While it is easier to avoid the so-called cross transfer in case of the S-graph framework (Hegyhati *et. al.*, 2009; Ferrer-Nadal *et. al.*, 2008), finite intermediate storage policy can be addressed more easily in the precedence based models (Mendez and Cerda, 2003; Romero *et. al.*, 2004). Despite the important difference between the applied optimization procedures, recent study has shown, that the two approaches have similar demands in terms of CPU time (Hegyhati and Friedler, 2011).

Present work summarizes the similarities and differences of these methods, and investigates, how they can contribute to each other's performance.

References

Ferrer-Nadal, S.; Capón-García, E.; Mndez, C. A. and Puigjaner, L., Material transfer operations in batch scheduling. A critical modeling issue, *Industrial & Engineering Chemistry Research,* 2008, 47, 7721-7732

Floudas, C. A. and Lin, X., Continuous-time versus discrete-time approaches for scheduling of chemical processes: a review, *Computers & Chemical Engineering,* 2004, 28, 2109-2129

Hegyhati, M. and Friedler, F., Overview of Industrial Batch Process Scheduling, *Chemical Engineering Transactions,* 2010, 21, 895-900

Hegyhati, M. and Friedler, F., Combinatorial Algorithms of the S-graph Framework for Batch Scheduling, *Industrial & Engineering Chemistry Research,* 2011, *50 (9)*, 5169-5174

Hegyhati, M.; Holczinger, T.; Szoldatics, A. and Friedler, F., Combinatorial Approach to Address Batch Scheduling Problems with Limited Storage Time, *Chemical Engineering Transactions*, 2011, in press.

Hegyhati, M.; Majozi, T.; Holczinger, T. and Friedler, F., Practical infeasibility of cross-transfer in batch plants with complex recipes: S-graph vs MILP methods, *Chemical Engineering Science,* 2009, 64, 605 – 610

Kopanos, G. M.; Lainez, J. M. and Puigjaner, L., An Efficient Mixed-Integer Linear Programming Scheduling Framework for Addressing Sequence-Dependent Setup Issues in Batch Plants, *Industrial & Engineering Chemistry Research,* 2009, 48, 6346-6357

Mendez, C. A. and Cerda, J., An MILP Continuous-Time Framework for Short-Term Scheduling of Multipurpose Batch Processes Under Different Operation Strategies, *Optimization and Engineering,* 2003, 4, 7-22

Mendez, C. A.; Cerda, J.; Grossmann, I. E.; Harjunkoski, I. and Fahl, M., State-of-the-art review of optimization methods for short-term scheduling of batch processes, *Computers & Chemical Engineering,* 2006, 30, 913-946

Mendez, C. A.; Henning, G. P. and Cerda, J., An MILP continuous-time approach to short-term scheduling of resource-constrained multistage flowshop batch facilities, *Computers & Chemical Engineering,* 2001, 25, 701-711

Romero, J.; Puigjaner, L.; Holczinger, T. and Friedler, F., Scheduling Intermediate Storage Multipurpose Batch Plants Using the S-Graph, *AIChE Journal,* 2004, 50(2), 403-417

Sanmarti, E.; Friedler, F. and Puigjaner, L., Combinatorial Technique for Short Term Scheduling of Multipurpose Batch Plants Based on Schedule-Graph Representation, *Computer Aided Chemical Engineering,* 1998, 22, S847-850

Sanmarti, E.; Holczinger, T.; Puigjaner, L. and Friedler, F., Combinatorial framework for effective scheduling of multipurpose batch plants, *AIChE Journal,* 2002, 48(11), 2557-2570

See more of this Session: Planning and Scheduling II
See more of this Group/Topical: Computing and Systems Technology Division

PLANNING AND SCHEDULING UNDER UNCERTAINTIES: DATA PROCESSING AND SOLUTION STRATEGY

Kailiang Tong, Yucheng Wu, Jiadong Xu, Yiping Feng, Gang Rong
Institute of Cyber-System and Control, Zhejiang University, Hangzhou, China

Introduction

Recent years, planning and scheduling are still great challenges in process optimization. Moreover, uncertainty is the most critical issue. Individual technique such as scheduling or planning is insufficient to deal with the problem and uncertainty. With the concept of Enterprise Wide Optimization[1], techniques should be integrated.

Planning usually determines the production profile as well as the raw material procurement and production distribution with the horizon from a week to a month. Scheduling gives task allocation and unit utilization with the information given by planning. Its time period is usually from an hour to a day. Integration of planning and scheduling saves cost or increases revenue. However, the problem size increases dramatically and the problem becomes intractable even a small time horizon is considered. In this article, a two level optimization framework for a petrochemical optimization problem, which is inherited from Luo[2] is presented. The stochastic programming planning model at upper level is integrated with the scheduling model at lower level using heuristic rules. This approach reduces the binary number caused by integration.

Many researchers focus on the external uncertainties such as demand uncertainty, supply uncertainty and price fluctuation in planning or scheduling optimization. Few focus on the process inherent uncertainty, such as product yield fluctuation, flow rate variation. Most people assume this kind of uncertainty obeying normal or uniform distribution. However, such assumption is neither accurate nor convincible compared with real world application. Lacking of information about process inherent uncertainty makes such optimization a big problem. Because of the process inherent uncertainty, the process data may be far from the one calculated by optimization model, especially when time goes on. Some treatment should be made to deal with such case, not only the optimization approach but also the data processing method. In real world application, process data depends on the accuracy of instruments. Moreover, in some cases, not all the process data are measureable. Process data with gross error or random error will distort some key constraints in the optimization model. As a result, data rectification approach should be executed before optimization.

The aim of this work is to provide a solution strategy of the optimization framework under both process inherent uncertainty and external uncertainty for a petrochemical plant. A real-time optimization framework is proposed. Planning and scheduling are integrated in the framework in a rolling horizon manner. The planning model is a MILP model with two-stage stochastic programming approach and the scheduling one is based on heuristic rules. The uncertainty of product yield is from both the uncertain raw material composition and the uncertain process. Data rectification and other statistical method are executed to obtain the distribution of uncertain yield. At the beginning of each optimization period, process data such as product yield and initial inventory is updated. The accuracy of the process data is guaranteed by an improved MT-NT method[3] for gross error detection and data reconciliation.

Solution Strategy

This subsection presents the solution strategy of planning and scheduling in real world petrochemical plant. The aim is to get an optimal or near-optimal solution of planning and scheduling under uncertainty. As the product yield fluctuates when tank feeding unit changes, regular optimization techniques are unable to solve such problem. A new optimization framework is proposed. In this framework, planning and scheduling are executed in rolling horizon manner. The product yield and other process data are updated at each beginning of the planning period. Moreover, the aim of this paper is not only to provide a solution strategy but also the procedure executing planning and scheduling in real world application.

Step 1. Data Processing: Before the planning and scheduling, we should analysis the historical data. After raw data received, use MT-NT approach to detect and rectify the gross error. Assume the distribution of product yield and use chi-square hypothesis testing to accept or reject the hypothesis.

Step 2. Planning and Scheduling: From historical data, we can see the product yield changes a little when one tank continues feeding the unit. As a result, we assume the distribution of product yield obeys normal distribution when one tank continues feeding the unit. And from a long time period of view, we assume the product yield obey normal of uniform distribution. These entire hypotheses should pass the hypothesis testing[4]. At the beginning of the planning period, a horizon of two week planning is built. Use the updated yield information to get the planning decision. Execute the scheduling with heuristic rules.

Step 3. Updating Planning Model and Replan: Move to next period, get the process data, and use data rectification method to rectify data. Update the initial data especially product yield. Redo the planning in the remaining planning period. To keep the production steady, it's not wise for us to execute the latest decision obtained by latest planning and scheduling every time. Only when the tank feeding unit changes or the uncertainty of process accumulates to large amount that the real situation differs a lot from the one in the original planning model, we should execute the new planning decision.

Step 4. New Planning: When all the 14 days planning and scheduling has finished, a new planning and scheduling in the next two weeks is executed in the same way.

Figure 1 shows the simple flowchart of the solution strategy. We conclude from case studies that the framework yields better results with lower inventory cost and less tank changeover.

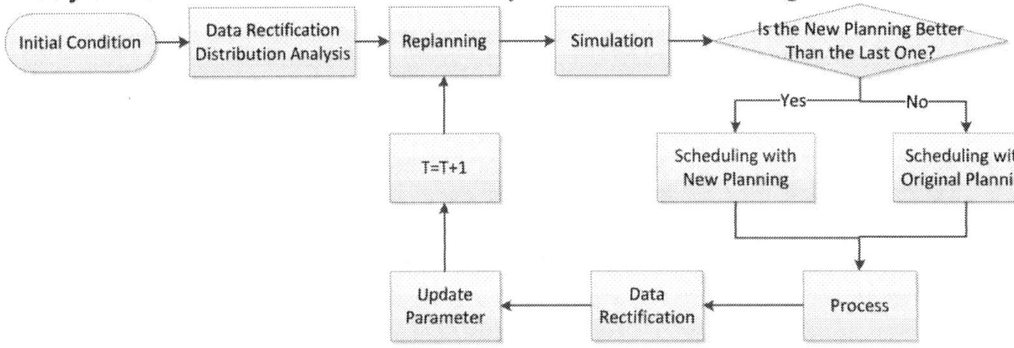

Figure 1 Solution Strategy including data processing and optimization

Reference

1. Grossmann, I., Enterprise-wide optimization: A new frontier in process systems engineering. *AIChE Journal* **2005,** 51, (7), 1846-1857.
2. Luo, C.; Rong, G., Hierarchical Approach for Short-Term Scheduling in Refineries. *Industrial & Engineering Chemistry Research* **2007**, 46, (11), 3656-3668.
3. Wang, F.; Jia, X.; Zheng, S.; Yue, J., An improved MT-NT method for gross error detection and data reconciliation. *Computers & Chemical Engineering* **2004,** 28, (11), 2189-2192.
4. Tamhane, A. C.; Dunlop, D. D., *Statistics and Data Analysis: From Elementary to Intermediate* Prentice Hall: 1999.

(475f) Advanced Optimization In Petroleum Refinery Planning

Omar J. Guerra, Ariel Uribe Rodriguez, Sandra Milena Montagut, Laura Andrea Duarte and Javier David Angarita, Colombian Petroleum Institute (ICP), ECOPETROL S.A., Piedecuesta, Colombia

Petroleum industry is nowadays a very complex and competitive business due to many changes in the dynamic global market place. These changes include most stringent environmental regulations, increasing in oil prices, increasing demand for petroleum products, etc [1]. For this reason, planning tools are very useful and essential for the economic success of petroleum companies. Despite this fact, the commercially available planning tools provide only an optimal solution point which is usually impossible to implement in the refinery operation. In other words, these planning tools do not give detailed information about the optimal operation region without degradation of the whole refinery profit [2].

This paper is focused on the implementation of rigorous analysis of the optimal solution to petroleum refinery planning problems by using advanced optimization techniques such as solution ranging, parametric analysis, and goal programming [2]. These optimization techniques provide an a-priori evaluation of solution robustness and flexibility of refinery operating plans. An ECOPETROL´s refinery planning tool (also called SIGMA-PLANNING) based on the nonlinear programming (NLP) technique was used for the implementation of all these advanced optimization techniques. Finally, many case studies for two ECOPETROL´s refineries were evaluated in order to show the applicability and usefulness of these advanced optimization techniques in the petroleum refinery planning.

[1] Shah, N.K, Zukui, L. and Ierapetritou, M. G, Petroleum Refining Operations: Key Issues, Advances, and Opportunities. *Industrial & Engineering Chemistry Research* **2011** *50* (3), 1161-1170.

[2] Varvarezos, D.K. (2008), Optimal Solution-Range Analysis in Production Planning: Refinery Feedstock Selection. *Industrial & Engineering Chemistry Research* **2008** *47* (21), 8282-8285.

See more of this Session: Planning and Scheduling II
See more of this Group/Topical: Computing and Systems Technology Division

(475g) Computational Performance of Big-M Formulations In Scheduling of Multipurpose Batch Plants

Esmael Reshid Seid and Thokozani Majozi, Chemical Engineering, University of Pretoria, Pretoria, South Africa

[1]E. Reshid and [1,2]T. Majozi

[1] *Department of Chemical Engineering, University of Pretoria, Lynnwood Road, Pretoria, 0002, South Africa*

[2] *Modeling and Digital Science, CSIR, Pretoria, South Africa*

Abstract

A large amount of research has gone into the development of optimization techniques for scheduling of batch plants. Scheduling of these batch plants becomes a challenging task which leads to several mathematical formulations to exist in the literature. The formulations that are based on continuous-time have gained significant attention for their advantage of requiring less number of time points, smaller problem size and less binary variables. The continuous-time formulations that use unit specific time points where tasks are allowed to start processing at different time with the same time point lead to big-M constraints. As the time horizon increases, the big M formulations require large computational time and difficult to close the relative gap. Due to this, cyclic scheduling is adopted for long time horizon by compromising objective value. If the formulation is beautiful both in structure and performance, there is no need to resort to cyclic scheduling as a way of approximating long term scheduling. In this presentation a novel scheduling model based on state sequence network (SSN) representation is presented. The formulation uses a continuous-time representation and a two index binary variable that result into mixed integer linear programming (MILP) problem. The model was tasted for a number of case studies taken from published literature. The results obtained showed that the model is computationally superior to existing model and capable of modelling and solving for long time horizon. The model is therefore advantageous over cyclic scheduling where the objective value is compromised by computational performance.

Key words: Scheduling; Multipurpose Batch Plant; Optimization; MILP; Big-M, continuous-time

(476a) Active Fault Isolation of Nonlinear Process Systems

Miao Du, Department of Chemical Engineering, McMaster University, Hamilton, ON, Canada and Prashant Mhaskar, Chemical Engineering, McMaster University, Hamilton, ON, Canada

Reliability and safety are of particular importance to process industries. Automatic control techniques have significantly improved the efficiency, reliability, and safety of chemical process operations. In a chemical plant, however, various faults are ubiquitously present, which can take place in control or processing equipment, due to reasons such as power failure, long-term use, or human-made mistakes. These abnormalities can jeopardize the profit of process operations or even lead to safety hazards to equipment and personnel. This realization has motivated significant research efforts to develop automated techniques that can detect the occurrence of faults, identify the failed equipment, and achieve fault tolerance. These techniques can be used to enhance safety and avoid potential economic losses in an intelligent way. The existing results on fault detect and isolation (FDI) can be categorized into data-based and model-based approaches.

In the data-based approach, faults are often detected through statistical process monitoring. From normal plant operating data, empirical correlation models can be built by using multivariate latent variable methods (e.g., [1]), such as principal component analysis (PCA) and partial least squares (PLS). These models are low dimensional and can capture the key information in normal process data. The current process data are compared with the normal variation contained in these low dimensional models, and the abnormal behavior is detected through statistical tests. Faults can then be isolated by using contribution plots, which are usually powerful to isolate simple faults, the effects of which are not propagated into other variables. The isolation of complex faults, the effects of which are propagated into many other variables (e.g., through feedback control), is improved by using additional data on past faults in [2]. The major benefits of this approach are that it can handle the case where there are a large number of measured variables and first principles models are unavailable. Due to the non-causal nature of the model built from normal plant operating data, however, the problem of fault isolation remains difficult when data on past faults are unavailable, which motivates the development of the model-based approach.

In the model-based approach, FDI is often achieved by generating residuals through the system model and input/output data (see [3] for a review). Under fault-free conditions, the magnitudes of these residuals are small. A fault is reported when a residual breaches the user-specified threshold. This approach has been studied extensively for linear (e.g., [4, 5]) and nonlinear (e.g., [6, 7]) systems. Due to the presence of plant-model mismatch, residuals that are sensitive to faults but insensitive to uncertainty and disturbances are desired. Unknown input observers are developed in [4] to decouple the effect of unknown inputs, such as disturbances, from that of the faults for linear systems. For nonlinear systems, the problem has been studied by using uniform thresholds in [6] (and adaptive thresholds in [7]), where fault isolation relies on the existence of a state variable which is directly and uniquely affected by the potential fault. Most of the existing FDI methods, however, are designed by using the measurement data, which are obtained from the process under the nominal control law. Therefore, these methods may not remain effective if the structure of the close-loop system inherently does not allow fault isolation under the nominal control law.

In comparison, there exist limited results on active fault isolation, namely utilizing the control action to facilitate fault isolation. Along this line, a method of decoupling the dependency between certain process state variables through feedback control (exploiting the system structure) has recently been proposed to enhance fault isolation under full state feedback control (e.g., [8]), where it is assumed that the state measurements are normally distributed. More recently, this approach has been extended to handle the case where only output measurements are available and studied with the use of model predictive control to optimize the input cost in [9]. Furthermore, the problem of distinguishing between faults that may directly affect the dynamics of the same process state (e.g., process disturbances and actuator faults) remains difficult. This problem is partly addressed in [10] for actuator faults by monitoring the outputs of the actuators, where it is assumed that the outputs of the (healthy or failed) actuators are constant between two consecutive discrete times and there exists a subsystem of the plant that satisfies a full rank condition.

Motivated by the above considerations, we in this work consider the problem of designing an active fault isolation scheme for nonlinear process systems. In particular, we consider the case where multiple faults may directly affect the dynamics of the same process state. To this end, we first develop a fault-tolerant control law by treating faults as disturbances, which can be used to achieve the desired control objective in FDI. We also develop a fault detection method that accounts for plant-model mismatch to provide timely and accurate information on the occurrence of faults. To utilize the control action to facilitate fault isolation, we design a mechanism to operate the process, upon fault detection, in a way such that for a given fault, the effect of other potential faults on the process dynamics can be reduced to an insignificant level. This allows the isolation of a particular fault by treating other potential faults as process disturbances. A simulation example of a chemical reactor is used to illustrate the effectiveness of the proposed method.

References

[1] J. V. Kresta, J. F. MacGregor, and T. E. Marlin. Multivariate statistical monitoring of process operating performance. Can. J. Chem. Eng., 69:35–47, 1991.

[2] S. Yoon and J. F. MacGregor. Fault diagnosis with multivariate statistical models part I: Using steady state fault signatures. J. Proc. Contr., 11:387–400, 2001.

[3] P. M. Frank. Fault diagnosis in dynamic systems using analytical and knowledge-based redundancy: A survey and some new results. Automatica, 26:459–474, 1990.

[4] J. Chen, R. J. Patton, and H.-Y. Zhang. Design of unknown input observers and robust fault detection

filters. Int. J. Contr., 63:85–105, 1996.

[5] F. Hamelin and D. Sauter. Robust fault detection in uncertain dynamic systems. Automatica, 36:1747–1754, 2000.

[6] P. Mhaskar, C. McFall, A. Gani, P. D. Christofides, and J. F. Davis. Isolation and handling of actuator faults in nonlinear systems. Automatica, 44:53–62, 2008.

[7] X. Zhang, M. M. Polycarpou, and T. Parisini. Fault diagnosis of a class of nonlinear uncertain systems with Lipschitz nonlinearities using adaptive estimation. Automatica, 46:290–299, 2010.

[8] B. J. Ohran, D. Munoz de la Pena, J. F. Davis, and P. D. Christofides. Enhancing data-based fault isolation through nonlinear control. AIChE J., 54:223–241, 2008.

[9] B. J. Ohran, J. Liu, D. Munoz de la Pena, P. D. Christofides, and J. F. Davis. Data-based fault detection and isolation using feedback control: Output feedback and optimality. Chem. Eng. Sci., 64:2370–2383, 2009.

[10] M. Du, J. Nease, and P. Mhaskar. An integrated fault diagnosis and safe-parking framework for fault-tolerant control of nonlinear systems. Int. J. Rob. & Non. Contr., Submitted.

See more of this Session: Process Monitoring, Fault Detection, and Diagnosis II
See more of this Group/Topical: Computing and Systems Technology Division

(476b) Kalman-Based Fault-Tolerant Control (FTC) for a Pilot-Scale Cooling Loop

Kris Villez[1], Humberto Garcia[2], Craig Rieger[2] and Venkat Venkatasubramanian[1], (1)School of Chemical Engineering, Purdue University, West Lafayette, IN, (2)Idaho National Laboratory, Idaho Falls, ID

TITLE:

Kalman-basedFault-Tolerant Control (FTC) for a pilot-scale cooling loop

AUTHORS:

Kris Villez, Humberto Garcia, Craig Rieger andVenkat Venkatasubramanian,

Listaddresses : Purdue and INL details

ABSTRACT:

We report on the real-life evaluation of afault-tolerant control strategy for a pilot-scale version of acooling loop of a nuclear plant [1]. This cooling loop is equippedwith two valves, a butterfly valve (BFV) and a glove valve (GLV)which both exhibit substantial hysteresis. This is the result of acoupling of the motor and vane axles design which allows for leewaybetween the valve motors and valve vanes, thereby leading to a safervalve subsystem. The whole system is used as a study object in thecontext of resilient system design and control [2].

In orderto design a supervisory control system allowing for fault-tolerance,the Fault Detection and Identification (FDI) method from [3] was usedas a first component. This method is based on the deployment of theKalman filter for FDI purposes as proposed earlier in literature[4,5]. The FDI component essentially consists of two steps, namelyfault detection and fault identification. The original method allowsfor a single type of fault only, namely the bias type, though in anyactuator or sensor in the studied system. In [3], it was specificallyextended to allow for a wider range of faults, such as drift, stuckand sticky faults in both actuators and sensors. With this newmethod, several types of faults can be identified correctly.

A secondcomponent consists of the supervisory controller which enablescorrective actions based on the diagnostic result. Whenever a faultis detected, the supervisory controller order the valves to open,leading to maximum cooling capacity, thus enabling the safest ofpossible operations. Following a fixed waiting time, collected datais used to identify the correct fault (second step of FDI). When thatis done, a final action is taken.

For a bias or drift fault, theregulatory control system can be corrected by means of a parametricadjustment (e.g. in the corresponding measurement or actuatorsignal). For other faults, like a stuck fault, one reconfigures thecontrol system so that the remaining working valve is used (only) forflow control. We call this a structural adjustment. In our earlierwork [1], we have applied this method successfully in simulationshowing that the hybrid behavior (discrete and continuous dynamicbehavior) can be handled well in a simulation based study. Indeed, itwas shown that several faults could be detected and diagnosedcorrectly. Figure 1 shows one simulated run in which the first valvegets stuck at 50 seconds in the simulation. Detection follows at 60seconds and at 80 seconds the fault is identified correctly.Following that, the supervisory controller decides to use the secondvalve instead of the first one.

With thiswork, we will report on the real-life implementation of the describedsupervisory controller. In particular, we expect to answer whether:

- The Kalman filter works well for the valves with hysteresis.
- The FDI strategy allows for successful detection and identification of faults in the real system
- To which extend the supervisory control system allows to mitigate the effect of introduced faults in the valve subsystems such as faults of the bias, drift and stuck type.

REFERENCES

1. Villez, K., Venkatasubramanian, V., Garcia, H., Rieger, C.,"Supervisory control of a pilot-scale cooling loop", submitted to the 4th International Symposium on Resilient Control Systems (ISRCS2011), 2011.
2. Rieger, C. G., Gertman, D. I., and McQueen, M. A., "Resilient control systems: Next generation design research." In 2nd IEEE Conf. on Human System Interaction, Catania, Italy, May 2009.
3. Villez, K., Srinivasan, B., Rengaswamy, R. and Narasimhan, S., "Kalman-based strategies for Fault Detection and Diagnosis: Extensions and critical evaluation for a simple benchmark system." Comput. Chem. Eng., 2011, In press.
4. Prakash, J., Narasimhan, S., and Patwardhan, S. C., "Integrating model based fault diagnosis with model predictive control." Ind. Eng. Chem. Res., 2005, 44 , 4344-4360.
5. Prakash, J., Patwardhan, S. C., and Narasimhan, S., "A supervisory approach to fault-tolerant control of linear multivariable systems." Ind. Eng. Chem. Res., 2002, 41 , 2270-2281.

Figure 1: Valve 1 stuck scenario. Top: Valve 1position; Middle: Valve 2 position; Bottom: Flow rate. Valve 1 getsstuck at 50 seconds. This is detected at 60 seconds and identified at80 seconds. Followingthat, the second valve is now used instead of the first.

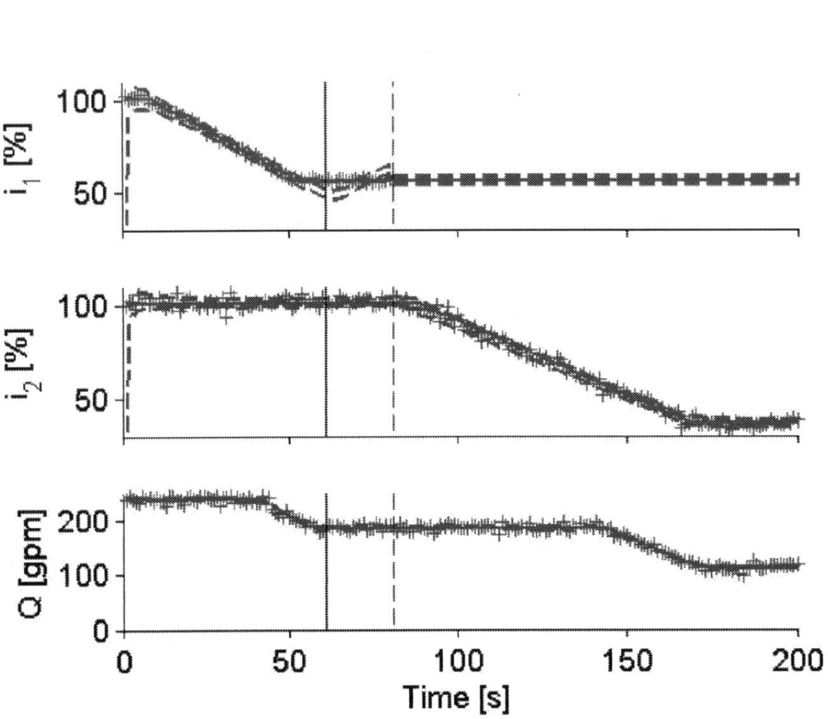

See more of this Session: Process Monitoring, Fault Detection, and Diagnosis II
See more of this Group/Topical: Computing and Systems Technology Division

(476c) Monitoring and Reconfiguration of Sampled-Data Nonlinear Hybrid Process Systems with Actuator Faults

Ye Hu and Nael H. El-Farra, Department of Chemical Engineering and Materials Science, University of California, Davis, Davis, CA

Compared with the substantial and growing body of research work on fault diagnosis and fault-tolerant control of continuous process systems, results for hybrid process systems that combine continuous dynamics and discrete events remain limited at present. Recently, we developed in [1] an integrated approach for fault detection and monitoring of a class of nonlinear hybrid processes with control actuator faults, uncertain continuous dynamics and uncertain mode transitions. A key idea was the design of a bank of dedicated mode observers using unknown input observer theory to identify the active mode at any given time and distinguish between faults, mode transitions and uncertainties. Beyond mode identification, measurement sampling due to the inherent limitations on the sensing devices or sensor-controller communication medium represents an important issue that needs to be accounted for in the design of the monitoring and control systems, since the discrete availability of measurements limits our ability to accurately monitor and precisely control the process. Another important aspect that requires attention is the design of the control system reconfiguration logic following fault detection. Unlike continuous processes, hybrid systems involve switching between multiple modes with different dynamics. In general, the different operating modes will have different fall-back control configurations, and this may limit the availability of certain control configurations for use as backup in the event of faults. Furthermore, the activation of a fall-back configuration in a hybrid system introduces new dynamics within the operating mode that should be taken into consideration to maintain the stability of the overall hybrid system.

In this work, we present a model-based framework of fault detection and control system reconfiguration for nonlinear hybrid process systems under output feedback control with measurement sampling rate constraints and control actuator faults. A dynamic model for each mode of the hybrid system is initially embedded within each controller so that an estimate of the process output between the sampling instants can be generated and used to calculate the control action; while at the sampling instants, the estimate is updated using the actual measurements from the sensors. The stability properties of each mode of the hybrid system are then analyzed, leading to an explicit characterization of the fault-free behavior and the maximum allowable sampling period for each mode. In order to detect actuator faults within each mode, the fault-free behavior is used as the basis for deriving a time-varying residual alarm threshold which is used to obtain the fault detection rules. Once a fault is detected in a given mode, the supervisor needs to determine which fall-back actuator configuration to activate. There are two key considerations for designing the reconfiguration logic. The first is the fact that the activation of a fall-back actuator configuration (that is not identical to the faulty one) alters the dynamics of the current operating mode and thus introduces a new "sub-mode" that may have a different stabilizing sampling period. Instability may ensue if the sampling period that is currently in use is larger than the maximum allowable sampling period for the new "sub-mode". The second consideration in designing the reconfiguration logic comes from the fact that a given actuator configuration that is suitable for use as backup in the current operating mode may not be available to future modes, and this requires that we take into account the availability of fall-back actuator configurations to future operating modes as well when performing the control system reconfiguration in any given mode. To address these two problems, one must identify all fall-back actuator configurations for each mode and characterize their maximum allowable sampling periods prior to process operation. Based on this information, the reconfiguration logic can then be implemented during process operation to ensure that the chosen fall-back configuration guarantees stability of the current mode as well as all future modes. Finally, the developed monitoring and reconfiguration strategies are illustrated using a simulated model of a hybrid chemical process.

References

[1] Y. Hu and N. H. El-Farra, "Robust fault detection and monitoring of hybrid process systems with uncertain mode transitions", *AIChE J.*, in press.

See more of this Session: Process Monitoring, Fault Detection, and Diagnosis II
See more of this Group/Topical: Computing and Systems Technology Division

(476d) The Square-Root Unscented Kalman Filter for Leak Detection and Location In Natural Gas Pipelines

Weiting Tang, Chemical Engineering, Texas Tech University, Lubbock, TX and M. Nazmul Karim, Department of Chemical Engineering, Texas Tech University, Lubbock, TX

Pipelines are the most efficient and economical way to transport and distribute natural gases all over the world. Leaks are the main problem of pipeline, which may cause excessive economical losses, or even worse, pose health hazards to communities. Therefore, designing a timely and reliable leak detection system is essential to maintain the normal operation of the pipeline company. In this work, a square-root unscented Kalman filter (SR-UKF) is proposed for leak detection and identifying location of the leak. The gas pipeline is modeled as a couple of quasi-linear hyperbolic partial differential equations (PDEs). Method of characteristics (MOC), which is suitable for large scale systems or systems with complex boundary conditions, is used to solve the quasi-linear hyperbolic PDEs. The proposed SR-UKF is applied to estimate the leak size and the location of the leak. The SR-UKF has the advantage of ease of implementation and has superior numerical stability. Comparative studies of extended Kalman filter (EKF) and the proposed SR-UKF are carried out to show the superiority of the proposed method.

See more of this Session: Process Monitoring, Fault Detection, and Diagnosis II
See more of this Group/Topical: Computing and Systems Technology Division

(476e) Thermodynamic Models for Leak Detection and Inventory Verification of Natural Gas In Salt Cavern Storage

Russell A. Ogle[1], D. "Trey" Morrison[1] and Ryan J. Hart[2], (1)Exponent, Lisle, IL, (2)Thermal Sciences, Exponent, Inc., Lisle, IL

Leak detection and inventory verification of natural gas storage is an important requirement for the safe and reliable operation of underground storage facilities. In depleted oil field and aquifer facilities the injection and withdrawal of gas is essentially an isothermal process: the gas temperature remains constant at the geothermal temperature of the storage medium. Natural gas exhibits more complex, transient behavior in salt caverns: the gas temperature increases during injection and decreases during withdrawal. During idle periods, the gas in the cavern either cools or warms towards the geothermal temperature. Depending on the gas storage operations, the cavern gas temperature can deviate considerably from the geothermal temperature. Furthermore, the gas under certain circumstances can temporarily cool or heat the cavern walls.

Sophisticated salt cavern storage models are available for simulating the thermodynamic and heat transfer processes that determine the gas pressure and temperature histories. These models, however, require a great deal of input information, can be cumbersome to operate, and may be prohibitively expensive for certain storage managers. They are also inadequate for leak detection and diagnosis.

This paper presents a simple thermodynamic analysis that permits the analyst to predict the gas storage conditions immediately following injection or withdrawal. The model uses the compressibility factor equation of state for the gas and treats the cavern as a perfectly insulated volume. Thus, the model should be used only to determine storage parameters either during or immediately after injection or withdrawal.

The success of the thermodynamic model is demonstrated by comparing its performance with both field measurements and with the predictions generated by the Salt Cavern Thermal Simulator, a software package used by the natural gas industry to predict storage behavior. Estimation of the gas state during injection/withdrawal or immediately thereafter is shown to be adequately predicted through thermodynamics alone; however, finite rate heat transfer effects dominate during idle periods. The thermodynamic model is also a powerful means for detecting leaks in the salt cavern facility. Criteria for leak detection are derived from the thermodynamic model and demonstrated with field data obtained from a salt cavern failure.

See more of this Session: Process Monitoring, Fault Detection, and Diagnosis II
See more of this Group/Topical: Computing and Systems Technology Division

HYPERSPECTRAL IMAGING BASED SENSING ARCHITECTURES FOR PARICULATE SOLIDS PROCESS AND MONTORING

Prof. Giuseppe Bonifazi Dr. Silvia Serranti
Sapienza – Università di Roma, Rome, Italy

Abstract

Many natural and/or artificial products are usually present on the market as **P***articulate* **S***olids* (PS). Their quality is defined according to the specific attributes of the single particles and/or the particles as a bulk system. PS characterization, inspection and quality control, inside a specific process and/or handling actions, are usually based on the collection of some specific particles physical-chemical attributes. These attributes represent the key factor to take into account when hardware and software (HW&SW) integrated architectures have to be developed to reach the previous mentioned goals. In this context the possibilities offered by specialized **H***yper*Spectral **I***maging* (HSI) based architectures, and related detection-control logics, able to play both *off-* and/or *on-line* are presented and critically evaluated in order to fulfill the previous mentioned goals. Some of the effects related to the adoption of different hardware configurations and different logics to perform the analysis and the selection of different algorithms according to the different characterization, inspection and quality control actions are presented

Introduction

The possibility to perform a fast and reliable identification of the different families of **P***articulate* **S***olids* (PS) resulting from different production processes and related to different materials, natural and/or synthetic, characterized by different compositional, and/or textural attributes, represents a key-issue in any industrial process addressed particulate solids handling and where control, selection and certification actions have to be carried out. To reach this goal specific particles physical-chemical attributes have to be collected, analyzed and processed in order to define specific classes of quality to assume as reference for the further particulate solids quality assessment and/or certification. The correct collection of these attributes thus represents the key factor to take into account when hardware and software (HW&SW) integrated architectures have to be developed to fulfill the previous mentioned targets. HW configurations, logics and algorithms are difficult to define because of

- the small range of variations, in many cases, of the physical-chemical detectable properties of many PS natural products and/or manufactured materials, belonging to the same family of products/materials,
- the difficulties to perform detection in respect of the handling and/or processing actions applied to particles,
- the constraints related to perform a correct inspection, minimizing, at the same time, any possible "negative" interference with the process,
- the processing speed of the collected data versus further possible *on-line* actions to apply,
- the need to define quality certification procedures for the investigated class of PS.

A platform embedding HW&SW architectures, allowing to fulfill the previous mentioned goals, primarily requires a sensing unit able to collect, in an affordable and reliable way, specific PS characteristics univocally discriminating them. The possibility to utilize **H***yper*Spectral **I***maging* (HSI) based sensing devices can thus represent a challenge topic to develop especially with reference to PS sector [1]. If the HSI sensing units represent the *eye* collecting particles information, physical devices and logics should be also designed, developed and implemented in order to allow:

- an optimal communication among the different hardware units constituting the platform,
- a simple installation, set up, calibration and handling of the inspection devices, according to different particulates solids families of products object of the investigations,
- a continuous synchronisation of the different acquired signals, as well as an efficient data-flow control,
- the possibility to define, modify and add multi-feature classifications algorithms finalized to identify the features to utilize for PS characterization, inspection and quality control and, finally,
- the utilization the platform to implement/test both *off-* and/or *on-line* sorting and/or certification procedures.

A primary role, in the realization of the previous mentioned recognition/sorting procedures, is played by the software engine utilized for data analysis, synchronization and control, for the man-machine interactions (user friendly interfaces and set up of customized easy-to-apply programming) and, finally, for sensors tuning. Furthermore the development of the previous mentioned integrated HD&SW architecture is affordable only if it is realized with high modularity and standardization. Starting from the developed HSI based platform, new applications can be thus implemented and/or optimized with minimum development and/or investments efforts, assuring, at the same time, flexibility, maintainability and a particulate solid products and/or materials full characterization. The HSI based sensing units represent an important class of devices [2] [3] [4] [5] allowing the logics integration inside PS processing.

In this paper an HSI based platform, to be utilized to develop recognition/classification logics on PS of different origin and characteristics is described. Examples of PS products quality testing, helping to define and implement innovative control strategies finalized to define quality classes of products and their certification are presented and critically evaluated.

H*yper*S*pectral* I*maging* (HIS) based platforms: hardware configurations

HW configurations, designed to realize HSI based platforms for particles analysis, have to take into account the possibility, not only to perform static, but also dynamic analysis, that is the possibility to carry out the analysis on limited number of non moving particles, as well as on a moving particles population. Results, in fact, are meaningful only if a large number of samples (particles) are investigated. For these reasons architectures have to mimic the typical characteristics of an *on-line* sorting unit, that is a device able to perform a continuous monitoring of a moving particulate solids flow stream (Figure 1).

The HSI based platform presented in this paper is constituted by a moving conveyor belt (width = 26 cm and length = 160 cm) with adjustable speed (variable between 0 and 50 mm/s). It is responsible of particles transport. Spectra acquisition is carried out continuously or at specific time intervals. Particular care was addressed to define and realize the proper energizing source (lighting devices) for the different hyperspectral devices installed on the platform (Figure 1). Diffused light cylinder sources were set up, embedding five halogen lamps producing an intense and continuous spectrum signal: a first one, $BaSO_4$ internal coated, acting in the 380-1500 nm wavelength range, properly designed for spectra acquisition in the wavelength range (400-1000) nm and a second one, optimized for the NIR, aluminum internal coating.

Sensing devices

Two different sensing devices able to acquire the reflectance spectra in the (400-1000) nm and (1000-1700) nm regions have been utilized. Both the device works as a push-broom-type line scan

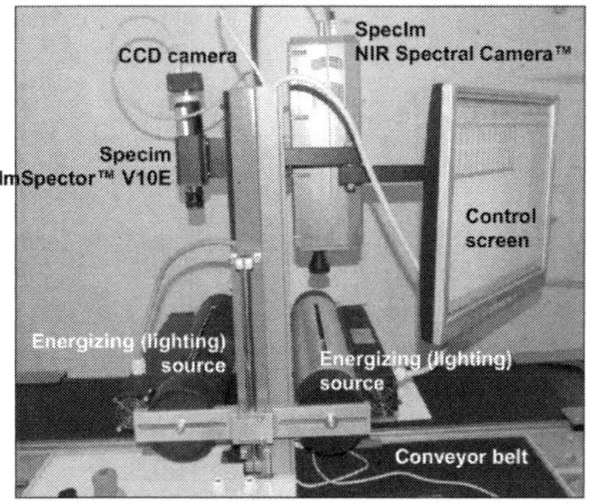

Figure 1: **H**yperSpectral **I**maging (HSI) based platform developed to perform particulate solids characterization, inspection and quality control.

camera providing full, contiguous spectral information for each pixel in the line. The transmission diffraction grating and optics provide high light throughput and high quality and distortion-less image for the devices. Their characteristics are synthetically reported in the following:

HyperSpectral device operating in the wavelength range: 400-1000 nm
An ImSpector™ V10E (SPECIM Ltd, Finland) with a spectral range (400-1000) nm, a spectral sampling/pixel of 2.8 nm, coupled with a CCD camera (780x580 pixels) with pixel resolution of 12 bits was utilized (Figure 1).

HyperSpectral device operating in the wavelength range: 1000-1700 nm
A NIR Spectral Camera™, embedding an ImSpector™ N17E (SPECIM Ltd, Finland) with a spectral range from 900 to 1700 nm, a spectral sampling/pixel of 2.6 nm, coupled with an TE-cooled InGaAs photodiode array sensor (320x240 pixels). A pixel resolution of 12 bits was utilized (Figure 1).

Acquisition set up

Sensing devices, as those described and utilized is this study, can produce "images" affected by geometrical distortion and/or wrong spectral acquisition.

Geometrical distortions, can negatively affect particles recognition/identification, in respect of their spatial position. They are usually related to mechanical problems due to the grating, prism and lens coupling. They can be minimized performing an adjustment by using the peaks within the spectrum of the wavelength calibration lamp. In the acquired image these peaks occur not as sharp lines but as relatively wide lines which are often parabolically bent by the spectrograph itself. Sharpening these lines means to minimize distortions. When the axis of the spectrograph and sensor array cannot be aligned by simple visual adjustment (i.e. the rotational difference is less than 1 pixel [5]), sub-pixeling techniques must be used [6].

Spectral calibration is required to correctly associate to each different material the related characterizing spectral signatures. With reference to the utilized devices, this means to perform a spectral, or wavelength, calibration to establish the correct assignment between pixel position and wavelength within each column. A gas extinction lamp with spectral bands covering the entire investigated spectral range is utilized to reach this goal [5].

The procedure was thus based on: i) spectral axis calibration with spectral lamps, ii) dark (*dark*) image acquisition, iii) measurement of white reference image (*white*). After these steps the image spectra (*sample*) is acquired and the reflectance (R_{ci}) (at wavelengths i and spatial pixels c of interest) is computed:

$$R_{ci} = [(sample)_{ci} - (dark)_{ci}]/[(white)_{ci} - (dark)_{ci}] \quad (1)$$

SW procedures for units control, spectra acquisition and storage, have been developed using Microsoft™ C++ in Windows XP™ environment.

HyperSpectral Imaging (HIS) based platforms: logics and algorithms

The software utilized to handle the devices and to acquire and analyze the spectra have to take into account several aspects:

- spectral sensor data and parameters control acquisition,
- conveyor belt control and synchronization with spectral data acquisition,
- reflectance data processing,
- hyperspectral image reconstruction, automatic and/or interactive spectral data plot,
- multi-feature classifications logics to identify and define the classes of features to utilize for particulate solids characterization, inspection and quality control,
- data export.

Spectra analysis

HSI based approach produces a large amount of information due to the large number of continuous spectral bands collected. Spectra analysis was mainly carried out:

- to verify the quality of the HW&SW HSI based procedures integrated in the platform/s,
- to analyze the proposed architecture flexibility, when other particulate solids materials have to be investigated,
- to build hyperspectral libraries to utilize for further recognition studies,
- to set up inspection or quality control logics, HSI based.

To reach the previous mentioned goals standard chemometrics methods [7] were used for data analysis [8]. More in detail, preliminary data processing, based on spectra manipulation utilizing baseline, derivative [9] and centering techniques, have been sequentially applied. Manipulated spectra have been then processed adopting linear multivariate PCA models. This approach was adopted to evaluate efficiency, robustness and reliability of the HSI based platform when particulates solids materials have to be characterized and particles related inspection and/or control actions, playing *off*- and/or *on*-line, have to be developed.

Logics and algorithms have been implemented using the PLS_Toolbox™ (Version 6.0.1, Eigenvector Research, Inc.) inside Matlab™ (Version 7.11.0, The Mathworks, Inc.).

Examples: particulate solids process monitoring and diagnosis

The HSI based sensing architectures was tested with reference to two completely different classes of products: i) dry fruits samples and ii) solid waste samples.

For the first class of samples (*dry fruits*) the main goal was to test the robustness and reliability of the proposed approach in order to utilize it as a "diagnostic tool" to evaluate the quality of the products. Furthermore the possibility to utilize this architecture to define, in an objective and automatic way, human-errors-free, different classes of quality of hazelnuts and/or almonds, was investigated.

For the second set of samples (*solid waste*) the main goal was to evaluate the potentialities of the HSI technique with reference to a specific and difficult to solve problem: the identification of polyethylene (PE) and polypropylene (PP) as fed and/or resulting from the processing of waste plastics flow streams coming from building and construction waste and/or automotive waste.

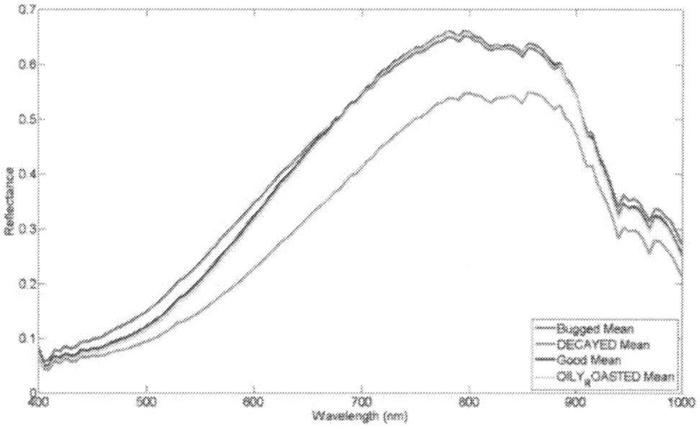

2a

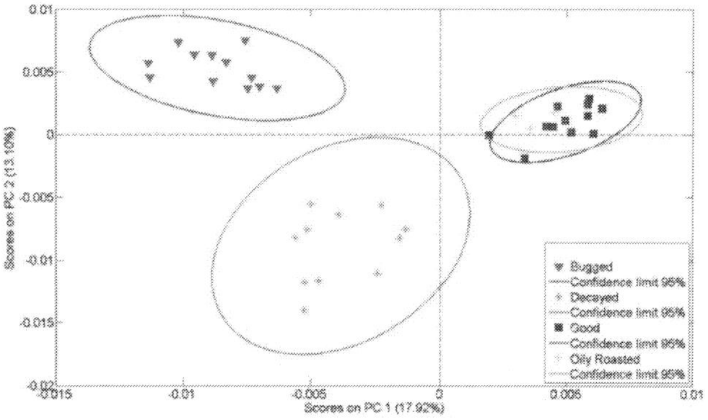

2b

Figure 2: Hazelnuts spectral (2a) and score (PC1-PC2) (2b) plots.
Average reflectance spectra (VIS-NIR field: 400-1000 nm), are strongly correlated to hazelnuts characteristics.
PCA (2b) clearly shows the optimal discrimination among quality classes.

Dry fruits samples

Different *hazelnuts* and *almonds* classes of products, characterized by different attributes (i.e. freshness, presence and entity of defects, mould and decays, etc.) influencing their organolectic properties, affecting their quality and their market value have been collected and analyzed.

Hazelnuts quality control

The analysis of the spectra shows as good quality hazelnuts are characterised by higher reflectance values, in comparison of those *bugged* and/or affected by *decays* (Figure 2). Spectra preliminary pre-processing and the further application of linear multivariate PCA models allows to perform a strong discrimination between the different classes of hazelnut products. Good ones (*good* and *oily roasted*) are well distinguished from *bugged* and *decayed* (Figure 2). Furthermore the last two classes belongs to two different regions of the PC1-PC2 plots, allowing this way a very detailed recognition, quite useful to develop control logics for hazelnuts quality certification.

Almonds quality control

As for hazelnuts, the almonds spectral and score plots (Figure 3), clearly show the high discriminating power, of the proposed architecture, for a robust and reliable quality assessment of such products. God quality almonds, hull, soft shells and stone are characterised by different spectral signatures, these attributes positively influence the further spectral pre-processing and the application PCA modelling.

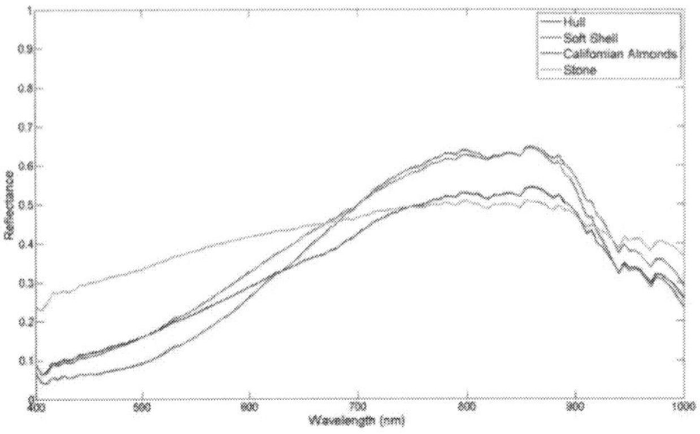

3a

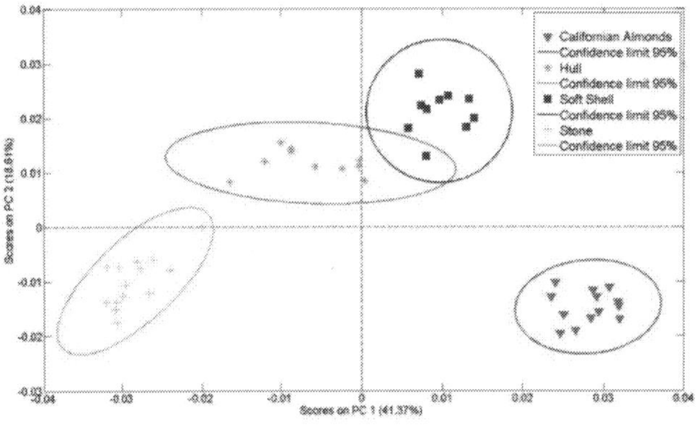

3b

Figure 3: Almonds spectral (3a) and score (PC1-PC2) (3b) plots.
Average reflectance spectra in the VIS-NIR field (400-1000) nm, are strongly correlated to almonds characteristics.
PCA (3a) clearly shows the optimal discrimination among quality classes.

The resulting score plot (PC1-PC2) clearly defines four regions where all the constitutive elements of the almonds sample are assigned. On these bases, as for hazelnuts, automatic quality control actions, HSI based, can be implemented.

Solid waste samples

Starting from plastic waste flow streams to recycle, polyolefins identification and discrimination was carried out. Polyethylene (PE) and polypropylene (PP) are, in fact, difficult to separate and chemically incompatible. PE and PP recovery, inside waste plastics flow streams, is a challenging issue actually not yet profitably solved [10].

Accurate and efficient separations exist, but they involve multiple processing stages, they are expensive, difficult to control and often do not allow the production of good concentrates.

Tests have been carried out to verify the possibility to develop efficient and low cost recognition logics to control the process and certify the products.

PE and PP recovery

Analyses show as PE and PP recognition is strongly influenced, in the visible region, by the colors of the samples. For these classes of particulate solids analyses have been thus performed in the near infrared region (NIR): (1000-1700) nm.

Analyses showed the high discriminating power of the NIR Spectral Camera™ to perform the identification of PE and PP spectral attributes (Figure 4). PE and PP spectra are different due

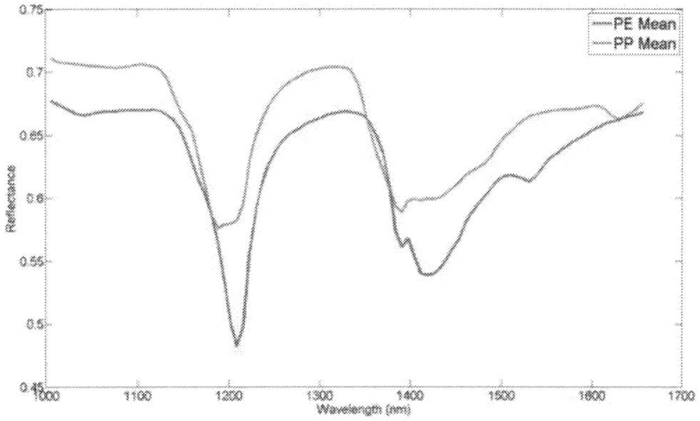

4a

Figure 4: PE and PP spectral plots. Average reflectance spectra in the NIR field (1000-1700) nm. Spectral signatures are strongly correlated to absorption of CH overtones regions (multiple of a vibration) and combinations of vibrations at specific wavelength intervals for both PE and PP.

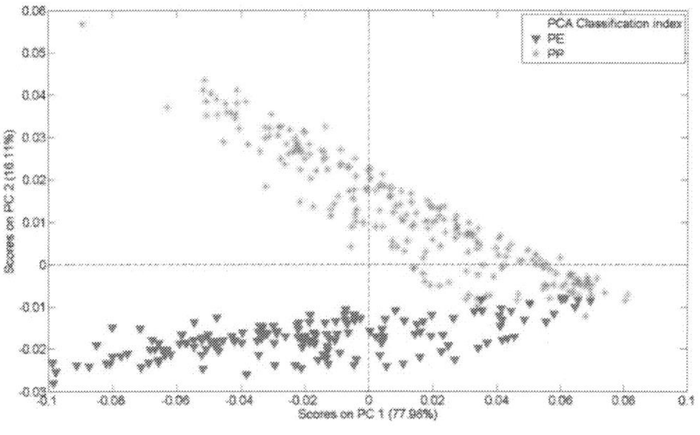

4b

Figure 5: PCA results (score plot: PC1-PC2) for PE and PP particles based on HSI average spectral signature in the NIR region: 1000-1700 nm.

to the different chemical formula and structure of the two polymers; spectral signatures are, in fact, strongly correlated to absorption of CH overtones regions (multiple of a vibration) and combinations of vibrations at specific wavelength intervals for both PE and PP (Figure 4).

Spectra pre-processing and the application of linear multivariate PCA models allow to perform a good discrimination of PE and PP (Figure 5).

Conclusions

The potentialities offered by an HSI based platform to perform the characterization, the inspection and the quality control of particulate solids was investigated. The platform was specifically designed and realized to be utilized, not only as a laboratory testing unit, but also as an analytical architecture to develop *on-line* strategies and/or sensing techniques to perform a continuous monitoring of a particle flow stream.

The examples reported, dry fruits (*hazelnuts* and *almonds*) and solid waste products (*polyolefins*), shows the flexibility and the potentialities of this platform in terms of collection and processing of the hyperspectral data related to materials physical-chemical attributes. Results demonstrated as, the proposed architecture can be utilized:

- to verify different logics and algorithms finalized to perform an HSI based *on-line*

detection of specific particle attributes conditioning their market value,
- to perform a continuous monitoring/certification of the different particles streams,
- to define innovative logics and procedures fro quality certification of the products.

According to the achieved results, investigations will be further addressed to:

- a systematic testing of the proposed HSI based platform on other particulate products and particle-like materials (i.e. minerals, chemicals, plants, etc.),
- the construction of affordable HSI based reference (i.e. standard) libraries for the different materials object of the investigations,
- the development of classification and recognition algorithms specifically designed for different sectors and for specific aims (sorting, quality control, certification).

References

1. Bonifazi G., Serranti S. and D'Aniello L (2008), "Hyperspectral Imaging Applied to Complex Particulate Solids Systems" (Invited Paper). *Proceedings of SPIE # 7003-14: Detector Technology and Imaging II*, Strasbourg, France.
2. Hyvarinen T., Herrala, E.E. and Dall'Ava A. (1998), "Direct sight imaging spectrograph: a unique add-on component brings spectral imaging to industrial applications". *Proceedings of SPIE # 3302: Electronic Imaging*, San Jose, CA, USA.
3. Gowen, A.A., O'Donnell, C.P., Cullen P.J. and Bellc S.E.J. (2008), "Recent applications of Chemical Imaging to pharmaceutical process monitoring and quality control". *European Journal of Pharmaceutics and Biopharmaceutics*, 69, 1, pp. 10-22.
4. Fortunato de Carvalho Rocha W., Post Sabin G., Março P.H. and Poppi R.J. (2010), "Quantitative analysis of piroxicam polymorphs pharmaceutical mixtures by hyperspectral imaging and chemometrics". *Chemomet. Intell. Lab. Syst.*, doi:10.1016/j.chemolab.2010.04.015.
5. Tatzer P., Wolf M. and Panner T. (2005), "Industrial application for inline material sorting using hyperspectral imaging in the NIR range". *Real-Time Imaging*, 11, pp. 99–107.
6. Rasmus N.J. (2002), "The VTTVIS line imaging spectrometer - principles, error, sources, and calibration", Denmark: Pitney Bowes Management Services, 2002.
7. Otto M. (1999), "Chemometrics, statistics and computer application in analytical chemistry". New York: Wiley-VCH.
8. Geladi P., Grahn H. and Burger J. (2007), "Multivariate Images, Hyperspectral Imaging: Background and Equipment". In: *Techniques and Applications of Hyperspectral Image Analysis*", 1-15, Grahn, H. and Geladi, P. Eds., John Wiley & Sons, West Sussex, England.
9. Savitzky A. and Golay M.J.E (1964), "Smoothing and Differentiation of Data by Simplified Least Squares Procedures". *Analytical Chemistry*, **36** (8): pp. 1627-1639.
10. Serranti S. (Guest Editor), Innovative Separation and Quality Control Strategies Applied to Mixed Polyolefins Waste. *The Open Waste Management Journal*, Special Issue, 3.

(476g) Soft Sensor Development of Batch Bioprocess Using Bayesian Inference Based Support Vector Regression

Jie Yu, Chemical Engineering, McMaster University, Hamilton, ON, Canada

The chemical and biochemical industry usually faces the challenge of unavailable on-line measurements of critical process and product variables, which are necessary to ensure successful control, monitoring and optimization of process operation. As an alternative solution, soft sensor can provide accurate estimates of unmeasured variables in a real-time fashion and has attracted increasing attention in advanced process control field. The commonly used data-driven models in building soft sensors require accurately recorded historical data and lab measurements. In industrial practice, however, the process input variables often have missing values and the time records of lab measurements may be inaccurate as well. Due to such uncertainty of modeling data, the traditional multivariate statistical or artificial intelligence methods can easily fail to construct reliable soft sensors. In this study, the Bayesian inference strategy is employed to handle the historical data uncertainty and estimate the posterior probabilities of input process variables and output lab measurements. Then the missing process values and misaligned lab data can be estimated and corrected within the probabilistic framework. Further, the support vector regression (SVR) is used to fit the predictive model between the process measurements and the lab based product quality variable. The proposed approach is tested on a fed-batch penicillin fermentation process and found to be effective in developing accurate soft sensors for key product variable predictions.

See more of this Session: Process Monitoring, Fault Detection, and Diagnosis II
See more of this Group/Topical: Computing and Systems Technology Division

(515a) Networked Control of Spatially Distributed Systems: Handling Communication Constraints and Delays

Zhiyuan Yao and Nael H. El-Farra, Department of Chemical Engineering and Materials Science, University of California, Davis, Davis, CA

The focus of this contribution is on the design of networked control systems for spatially distributed processes modeled by highly-dissipative partial differential equations (PDEs) with sensor-controller communication constraints and delays. This is an important problem given the increased reliance of process operation on networked sensors and controls, and the abundance of process systems characterized by spatial variations, such as transport-reaction processes and fluid flow systems. A key consideration in the design of the networked control system is the ability to enforce closed-loop stability with minimal sensor-controller information transfer to aggressively conserve limited network resources. In addition to resource constraints, sensor-controller communication delays represent an important issue that requires attention in the networked control system design to avoid potential instability or performance deterioration. To address these problems, we include within the networked control system: (1) an approximate finite-dimensional model that provides the controller with an estimate of the slow state of the infinite-dimensional system when communication is suspended to reduce information transfer, and (2) a propagation unit that compensates for the communication delay by using the model, together with the past control inputs, to calculate an estimate of the current output from the received delayed measurements. The model state is then updated using the latter estimate at discrete time instances. The networked closed-loop system is analyzed using a combination of tools from switched systems, infinite-dimensional systems and singular perturbations, leading to precise characterizations of the interplays between the minimum allowable sensor-controller communication rate, the finite-dimensional model order and accuracy, the delay size, and the selection of actuator and sensor configurations. Finally, the theoretical results are illustrated using a transport-reaction process example.

See more of this Session: Complex and Networked Systems
See more of this Group/Topical: Computing and Systems Technology Division

(515b) A Distributed Predictive Control Framework for Smart Grid Development

Wei Qi[1], Jinfeng Liu[2] and Panagiotis D. Christofides[2], (1)Chemical and Biomolecular Engineering, University of California, Los Angeles, Los Angeles, CA, (2)Department of Chemical and Biomolecular Engineering, University of California, Los Angeles, Los Angeles, CA

While the traditional electrical grid has been successful, in recent years there have been numerous calls for the development of the so-called "smart electrical grid" by expanding the traditional electrical grid with distributed, medium-scale renewables-based energy generation systems and digital technologies, for example, communications, computing, sensing and automation, to better meet the increasing energy demand and environmental regulations. Incorporated with two-way communication networks, digital devices and distributed optimization and control systems, the "smart grid" is expected to be more reliable, more secure, more energy efficient and more environmentally friendly. One important feature of the smart grid is its capability of integrating distributed energy resources and generation, for example, renewable energy resources, into the electrical
grid. Renewable energy resources, like wind and solar-based energy generation systems, are receiving national and worldwide attention owing to the rising rate of consumption of fossil fuels. In addition to the environmental benefits, solar and wind renewable energy generation systems also have a reduced investment risk and an increased energy efficiency. However, integrating renewable energy generation systems with the electrical grid requires addressing key fundamental challenges, for example, variable output, in the
operation of intermittent renewable resources like solar- and wind-based energy generation systems.

In this work, we propose a conceptual distributed control framework for electrical grid integrated with distributed renewable energy generation systems in order to enable the development of the so-called "smart electrical grid". First, we introduce the key elements and their interactions in the proposed control architecture and discuss the design of the distributed control systems which are able to coordinate their actions to account for optimization considerations on the system operation. Subsequently, we focus on a specific
wind/solar energy generation system connected to a reverse osmosis desalination system and the electrical grid and design two supervisory predictive controllers via model predictive control to operate the integrated system taking into account short-term and long-term optimal maintenance and operation considerations, respectively. Simulations are carried out to illustrate the applicability and effectiveness of the proposed approach.

See more of this Session: Complex and Networked Systems
See more of this Group/Topical: Computing and Systems Technology Division

(515c) Distributed Model Predictive Control of Switched Nonlinear Systems

Mohsen Heidarinejad[1], Jinfeng Liu[2] and **Panagiotis D. Christofides**[2], (1)Electric Engineering, University of California, Los Angeles, Los Angeles, CA, (2)Department of Chemical and Biomolecular Engineering, University of California, Los Angeles, Los Angeles, CA

Due to changes in raw materials, energy sources, product specifications and market demands, control of switched nonlinear systems with scheduled mode transitions has received considerable attention. From a controller design standpoint, in order
to achieve closed-loop stability, transition situations should be carefully accounted for in the control problem formulation and solution. In order to achieve scheduled mode transitions in an optimal setting and accomodate input/state constraints, model predictive control (MPC) framework can be employed. Motivated by these considerations, in previous work [1], we proposed a centralized Lyapunov-based MPC formulation for switched nonlinear systems with prescribed scheduled mode transitions. In particular, to account for the switched nature of the closed–loop system, we incorporated constraints, based on hybrid control theory, in the model predictive control formulation that preserve online implementation and enable preserving stability of the switched system. However, the computational complexity of a centralized MPC may significantly increase as the number of control inputs and states increases. Distributed MPC (DMPC) takes advantage of cooperation and communication of distributed predictive controllers, which communicate over a shared network, to reduce the computational burden of a centralized MPC solution at the cost of slight degradation of closed-loop performance.

In the present work, we extend the results in [3] in a distributed setting. We consider a nonlinear switched system consisting of several subsystems that follows a prescribed switching policy. A Lyapunov-based iterative DMPC scheme [2] is proposed with appropriate stability constraints to achieve stability of the switched closed-loop system subject to process disturbances and tracking of the prescribed switching policy. In terms of DMPC feasibility, the stability constraints make sure that at the end of each switching interval, the closed-loop system state enters the stability region of the subsequent mode while there is a reduction in the Lyapunov function value of the new mode compared to value that the Lyapunov function attained the last time that the closed-loop system had switched to that mode. The theoretical result are demonstrated with a chemical process example.

[1] Mhaskar, P., N. H. El-Farra and P. D. Christofides, "Predictive Control of Switched Nonlinear Systems with Scheduled Mode Transitions," IEEE Transactions On Automatic Control, 50, 1670-1680, 2005.

[2] Liu, J., X. Chen, D. Muñoz de la Peña and P. D. Christofides, "Sequential and Iterative Architectures for Distributed Model Predictive Control of Nonlinear Process Systems," AIChE Journal, 56, 2137-2149, 2010.

(515d) Coarse-Graining Dynamical Networks with Intrinsic Heterogeneities

Karthikeyan Rajendran[1], Andreas C. Tsoumanis[1], Andreas I. Reppas[2], Constantinos I. Siettos[2] and Ioannis G. Kevrekidis[1], (1)Chemical and Biological Engineering, Princeton University, Princeton, NJ, (2)School of Applied Mathematics and Physical Sciences, National Technical University of Athens, Athens, Greece

Networks play an important role in many fields ranging from the internet and social networks to biological neural systems and ecological food webs. Understanding the functional role of the *network structure* in the overall network dynamics is a critical research issue. On top of this, the nodes of many physical networks are inherently different from each other, and this intrinsic heterogeneity also affects the system dynamics. We consider coarse-graining approaches that can lead to reduced, more efficient network descriptions, and here we focus on taking the effect of network heterogeneity into consideration in such model reduction efforts.

We will present three distinct illustrative examples of networks of coupled *nonidentical* units connected by network structures. The first is a network of Kuramoto [1] oscillators; the eigenvectors of the graph Laplacian (corresponding to their network structure) can be used to coarse-grain the system [2], when the connectivity graph possesses a spectral gap (big jumps in the eigenspectrum of their graph Laplacian). We improve on this coarse model by taking into account the effect of the intrinsic oscillator heterogeneity; our approach is based on the observation of correlations between the states of the network nodes and their intrinsic heterogeneities. We also present two other network problems (one involving opinion propagation on a social network, the second involving epidemiological modeling) where network heterogeneities directly affect the system states. The examples highlight the generality of our approach, which links tools from the study of uncertainty quantification (polynomial chaos [3]) to the reduction of dynamical networks.

We computationally implement our coarse models using the equation-free framework [4], a modeling approach that designs and uses short bursts of detailed simulations to perform system-level computational tasks without requiring explicit coarse-grained equations. We discuss our selection of coarse variables and perform coarse-projective integration, coarse fixed point, coarse limit cycle as well as coarse stability computations.

References

[1] Y. Kuramoto, *Chemical oscillations, waves, and turbulence*, Berlin; New York: Springer-Verlag, 1984.

[2] K. Rajendran, A. C. Tsoumanis, I. G. Kevrekidis, *Coarse-Graining the Dynamics of (and on) Networks*, AIChE Annual Meeting, Salt Lake City (2010); Abstract 586d

[3] R. G. Ghanem and P. D. Spanos, *Stochastic Finite Elements: A Spectral Approach* (Springer-Verlag, New York, 1991).

[4] G. Kevrekidis, C. W. Gear, and G. Hummer, *Equation-free: The computer-aided analysis of complex multiscale systems*, AIChE Journal, vol. 50, no. 7, pp. 1346-1355, 2004.

See more of this Session: Complex and Networked Systems
See more of this Group/Topical: Computing and Systems Technology Division

(515e) Proactive Emission Source Detection and Evaluation with Air Quality Monitoring Network

Tianxing Cai, Dan F Smith Department of Chemical Engineering, Lamar University, Beaumont, TX and Qiang Xu, Dan F. Smith Department of Chemical Engineering, Lamar University, Beaumont, TX

Chemical facilities, where toxic or hazardous chemicals are manufactured and housed, present a high potential of dangerous accidents such as explosion or toxic gas leakage. These are usually caused by various factors of equipment failure, operation, or extreme conditions such as earthquake, tsunami, hurricane, transportation accidents, or terrorist attack. In certain accidental emission events, the emission source or severity are indeterminate and immeasurable due to event complexity or the simple fact that the accidental areas are too dangerous to approach. Lack of necessary information on emission source and severity may lead to insufficient decision makings that keep emergency rescue from timely and effective responses. Therefore the inverse determination of the emission source and evaluation of its severity is of great significance to all stake holders.

In this study, a methodology is developed for proactive emission source detection and evaluation in emergent situations. All the location of emission points have been assumed to be known because chemical facility positions are usually fixed and GPS technology is currently available for positioning chemical transportation vehicles. This method has included three stages. In the first stage of modeling constructing, the monitor data is collected from existent AQMN (air quality monitoring network). The changing trend of monitor result will be correlated with the normal emission condition under meteorological conditions. In the second stage of modeling optimization, the dispersion model and least square approximation are used for abnormal AQMN measurements diagnosis, which will help determine the possible emission positions. In the third stage of emission source evaluation, the air pollutant dispersion model is further refined with available abnormal emission measurements. The fine-tuned model is used to provide the dynamic prediction of future consequence to current or potential impacted areas so that timely response plan can be proactively taken.

The whole procedure is to setup a mechanism with risk early-warning and rapid response functions for dangerous chemical emissions from chemical plants, transportation vehicles, or accidental chemical releases. The developed methodology involves a complex inverse optimization problem. An effective solving strategy is also introduced.

See more of this Session: Complex and Networked Systems
See more of this Group/Topical: Computing and Systems Technology Division

(515f) Fate of Commensalistic Cultures In Identical Coupled Bioreactors

Satish J. Parulekar, Chemical and Biological Engineering, Illinois Institute of Technology, Chicago, IL

A comprehensive analysis of static and dynamic behavior of a mixed culture in two identical coupled bioreactors is presented considering anaerobic digestion involving acidogens (X) and methanogens (Y) as the example bioprocess. A single continuous culture may operate at up to seven steady states, including up to four coexistence steady states, with only one coexistence steady state being locally stable. The one-way interaction between X and Y allows for compartmentalization of the system for a stand-alone bioreactor and two coupled bioreactors into two subsystems, which facilitates the analysis of steady state types and stability characteristics of these and classification of dynamic behavior. The bioreactors in the two-reactor system are identical only in terms of feed composition and reactor space time. A two-reactor system may admit up to forty nine steady states, which are comprised of up to forty coexistence steady states, at least at very low interaction rate (R). The static and dynamic analysis of the two-reactor system is facilitated by appropriate grouping of large number of steady states arising for very low R into nine clusters. Numerical illustrations reveal the rich steady state structure of the bioprocess in coupled bioreactors. While a single bioreactor can operate at only one locally stable coexistence steady state, the coupled bioreactors can operate at up to five locally stable coexistence steady states over certain ranges of R. The two-reactor system is operationally more flexible and more robust vis-a-vis single reactor as concerns maintenance of mixed culture. Emergence of four additional steady state clusters and additional coexistence and partial washout steady states at intermediate R reveals that the coupled bioreactors are an example of a complex system.

See more of this Session: Complex and Networked Systems
See more of this Group/Topical: Computing and Systems Technology Division

(519a) Incorporating Sustainability Into the Design of Chemical Process: From Reaction Pathway Selection to Process Design

Kailiang Zheng and Helen Lou, Dan F. Smith Department of Chemical Engineering, Lamar University, Beaumont, TX

It is important to incorporate sustainability issues into the design of chemical process, as more and more social and environmental concerns arise, and natural resources become costly and less available. Sustainability is multi-dimensional and complex in nature, as economic, societal and environmental issues need to be considered together. Thus a structured approach is needed to assist in incorporating sustainability into the new process design, as well as the retrofit of an existing design. In this paper, a sequential approach from reaction pathway selection to process design is proposed, in light of sustainability. At the early design stage, the sustainability performance of each potential reaction pathway is evaluated in terms of possible profit, driving force of the pathway (Gibbs energy), inherent safety index, environmental analysis, mass efficiency, reaction selectivity and conversion, respectively. This can help the designers to screen out the inferior reaction alternatives systematically and efficiently, thus reduce the complexity and labor in the following design stage. Then in the detailed process design stage, the sustainability performance of each process is assessed sequentially considering economic, societal, environmental and efficiency issues.

Root cause analysis is a powerful tool to identify and illustrate the fundamental cause of a problem. Integrating root cause analysis into the presented sequential sustainable design approach can help identify the root cause and improvement opportunity for a specific design. The idea of this approach is illustrated via several case studies.

See more of this Session: Design and Operation for Sustainability
See more of this Group/Topical: Sustainable Engineering Forum

(519b) Optimization of CO_2 Capture Process with Aqueous Amines - a Comparison of Two Simulation-Optimization Approaches

Aroonsri Nuchitprasittichai, Chemical Engineering, The University of Tulsa, Tulsa, OK and Selen Cremaschi, Department of Chemical Engineering, The University of Tulsa, Tulsa, OK

Aqueous amine is a solvent considered for CO_2 recovery from the flue gas of a refinery gas turbine by chemical absorption/desorption process [1]. The performance and the economics of this process depends on the choice of amine absorbent and its concentration, the number of stages in absorber and stripper columns, and the operating conditions.

In this work, we used two different simulation-based optimization approaches to minimize the CO_2 capture cost for post-combustion CO_2 removal. The first simulation-based optimization technique uses local search space to estimate an appropriate direction to reduce the objective function, i.e., response surface methodologies. The optimal solution obtained using response surface methodologies will generally be a local optimum. In the second optimization approach, the simulation was used to build a surrogate model, i.e., an artificial neural network (ANN), of the objective function over the whole decision space, and the optimization was performed using the surrogate model. Depending on the accuracy of the surrogate models, the solutions obtained using this approach can be shown to be global within the bounds of the data used to generate the surrogate models.

The impact of different amines absorbents, design parameters and operating conditions on the CO_2 capture cost were analyzed using both optimization techniques. Monoethanolamine (MEA), diglycolamine (DGA), diethanolamine (DEA), methyl diethanolamine (MDEA), triethanolamine (TEA), and blended aqueous solutions of these amines have been considered in our analyses. The required specification for the CO_2 capture process was 96 mole% CO_2 purity at the product for all simulations. The results are the optimum number of stages for the absorber and stripper columns, and the optimum absorbent concentration and operating conditions, i.e., the ones that give the minimum cost for CO_2 capture process. In this presentation, the comparison of the results obtained from both simulation-optimization techniques will be discussed. This comparison will provide the accuracy and the validity of the response surface technique results for the CO_2 capture process optimization. It should be noted that response surface methodology is simpler to develop and requires less expert knowledge compared to the ANN approach. Therefore, the response surface methodology can be used for CO_2 capture cost optimization if the results are closer to the global optimum obtained by the ANN approach.

Reference:

[1] Mofarahi, M., Khojasteh, Y., Khaledi, H., & Farahnak, A. (2008). Design of CO_2 absorption plant for recovery of CO_2 from flue gases of gas turbine. Energy, 33, 1311-1319.

See more of this Session: Design and Operation for Sustainability
See more of this Group/Topical: Sustainable Engineering Forum

(519c) A New Process for TiO2 Production

David R. Corbin[1], Keith W. Hutchenson[1], Stephen E. Lyke[2], Eugene M. McCarron III[2], Mark B. Shiflett[1], Charlie C. Torardi[1] and Joseph J. Zaher[1], (1)Central Research and Development, DuPont Company, Wilmington, DE, (2)Titanium Technologies, DuPont Company, Wilmington, DE

Because of its high refractive index, stability and non-toxicity, TiO_2 has found application as the primary opacifying agent in paint, paper and plastics. Annual world consumption of TiO_2 is estimated to be about 10MMM lbs. With an annual growth rate of about 3%, significant new capacity is needed each year to meet the growing demand. This presentation will summarize our efforts at developing a novel next generation process for TiO_2 production based on ammonium binoxalate digestion of ilmenite ore.

See more of this Session: Design and Operation for Sustainability
See more of this Group/Topical: Sustainable Engineering Forum

(519d) Low Temperature Hydrothermal Crystallization for a New TiO2 Process

Keith W. Hutchenson[1], Sheng Li[2], David R. Corbin[1], Charles C. Torardi[1] and Eugene M. McCarron[1], (1)DuPont, Wilmington, DE, (2)Chemical and Biological Engineering, Princeton University, Princeton, NJ

This presentation summarizes the development and optimization of a novel low temperature hydrothermal route to recover base TiO_2 product from an intermediate titanyl hydroxide species formed in the TiO_2 process following digestion of the iron titanate ore feed material. This work was part of an effort to develop a new TiO_2 production process based on ammonium hydrogen oxalate digestion that can process low-grade ilmenite ores with lower energy and capital investment intensity and a smaller environmental footprint than conventional processes.

Statistically designed experiments have illustrated the feasibility of producing both pigment-grade and nano-size TiO_2 base materials at mild hydrothermal process conditions. Small scale screening experiments on synthetic intermediates were carried out to examine the effects of various reaction variables on the final TiO_2 properties. The screening study identified reaction time, hydrochloric acid to titanium ratio, and the concentration of a particular mineralizing salt as the three key factors most influential to the formation and growth of rutile TiO_2. Based on the results of the screening experiments, empirical process models, which express TiO_2 product properties as a function of the critical process variables, were developed using statistical response surface design techniques. The models were used to optimize the reaction conditions for the production of either pigmentary or nano rutile base products. The predicted reaction conditions were then validated at the same 10 mL scale.

The empirical model results identified a relatively narrow operating window for the production of pigmentary rutile TiO_2 at relatively low temperature (235-250 C) hydrothermal crystallization conditions. The process was successfully repeated on a 1L reactor scale and resulted in the production of pigmentary rutile base. This material was then successfully surface treated via the standard chloride wet treatment technology. Subsequent work validated the application of this process to an authentic ore-derived titanyl hydroxide intermediate and demonstrated the formation of pigmentary rutile TiO_2.

See more of this Session: Design and Operation for Sustainability
See more of this Group/Topical: Sustainable Engineering Forum

(519e) Integration of Electroplating Process Design and Operation for Simultaneous Productivity Maximization, Energy Saving, and Freshwater Minimization

Chaowei Liu, Chuanyu Zhao and Qiang Xu, Dan F. Smith Department of Chemical Engineering, Lamar University, Beaumont, TX

Electroplating annually produces numerous plated workpieces for U.S. pillar industries. However, it is also a major environment polluter in forms of wastewater, sludge, and spent solution. Thus, waste source reduction for the electroplating industry is a must. In practice, when productivity and energy use efficiency are also taken into account, the design and operation of an electroplating process become very complicated and need an in-depth study. In this paper, productivity maximization, energy saving, and freshwater/wastewater minimization are simultaneously addressed for the optimal design and operation of electroplating processes, which generates a triple-objective mixed-integer dynamic optimization (MIDO) model. The MIDO model is iteratively solved by a developed methodology to obtain the 3D Pareto frontier of the optimization problem, which provides important technical supports for the design and operation of electroplating processes. The efficacy has been demonstrated with a case study on an electroplating process.

Keywords: Hoist scheduling, Water network synthesis, Energy saving, Optimization

See more of this Session: Design and Operation for Sustainability
See more of this Group/Topical: Sustainable Engineering Forum

(519f) Design of Biofuel Additives Using Chemometric Modeling and Molecular Design Techniques

Subin Hada, Charles C. Solvason and Mario Richard Eden, Department of Chemical Engineering, Auburn University, Auburn, AL

Any type of feedstock that contains free fatty acids and/or triglycerides such as vegetable oil, waste oil, animal fat, and waste grease can be converted into biodiesel. Because of its biological origin, biodiesel has many advantages compared to its petroleum derived counterparts. However, it does exhibit several limitations like oxidative stability and low temperature operability that have impaired its wide use and commercialization. Simultaneous solution of these problems has proven difficult as improvements in one area tend to impair another. As a result, design of an optimal fuel for all environments can be a quite difficult task. In addition, feedstock origin will impact the final quality of the biodiesel product. Therefore, to produce fuel grade biodiesel, the characteristics of the feedstock are very important during the initial research and production stage.

Several approaches are possible for improving the fuel properties of biodiesel. Modification of the fatty acid composition through physical processes or uses of additives are the most prevalent. Fuel additives, such as antioxidants, cetane enhancers, or cold-flow improvers have become indispensable tools not only to alleviate the drawbacks described above, but also to assure that any fuel blend will meet international and regional standards regardless of origin. However, an additive solution to one problem often aggravates another problem. Furthermore, the questions of additive compatibility, required additive levels, the effect on other properties, and whether these additives function as designed for biodiesel fuels with differing fatty acid profiles still remain challenges and that need further investigation.

Therefore, it is desired to molecularly design biodiesel additives that simultaneously account for the unintended effect on other fuel properties in the neat and the blend fuel in order to achieve the performance properties of the petroleum based fuel. In this way, biofuels can be formulated that are adaptable to a range or blend of feedstocks and the desirable fuel characteristics like oxidative stability and wide operating temperature range.

In this work, we aim to identify all possible compounds which, when added to off spec biodiesel, results in a fuel that satisfies performance standards such as ASTM D6751 in the US and EN14214 in Europe. To meet this end, multivariate characterization data obtained from IR spectroscopy of common additives were combined with decomposition and property clustering techniques in a reverse problem formulation. In this approach, the fuel additive property targets are identified in the first reverse problem followed by molecular design to match the targets. The characterization data consists of multitude of properties of interest (such as cetane number, melting point, and kinematic viscosity) to ensure adequate performance. To facilitate an efficient design we consolidated these various properties into a latent property domain using principle component analysis (PCA) techniques. Finally, characterization based molecular design using group contribution parameters are then used to build novel additives that match the fuel specifications in the latent property space.

See more of this Session: Design and Operation for Sustainability
See more of this Group/Topical: Sustainable Engineering Forum

Rule of Thumb for Simulating Biomass Pyrolysis in Packed Bed Reactor

Widya Wijayanti, Ken-ichiro Tanoue Takahiro Suetomi, Tatsuo Nishimura,
Yamaguchi University, Ube, Japan

Miki Taniguchi and Ken-ichi Sasauchi,
Chugai Ro Co.ltd, Sakai, Japan

Introduction

Compared to other countries, Japan is particularly rich in forests; the percentage of Japan's landmass that is forested is 67%, which is the third highest in the world. Only 20% of all wood is effectively utilized as timber; the rest, including waste wood (4 million tons/year) and natural forests (10 million tons/year), is unused. In recent years, there has been increasing demand for quickly produced, environment-friendly energy resources; in particular, there is wide interest in the production of energy from woody biomass, using wood resources which are currently underutilized. The use of biomass to provide energy has many merits: it is renewable, abundant, easy to store, and carbon-neutral. As Japan has many mountains with steep slopes, it is difficult to transport the felled trees. So that, the development of a compact gasifier of high quality, which can cope with variations both in the amount of biomass collected and in energy demand, is necessary. Furthermore, in order to ensure that the system is used in the most efficient manner, it is necessary to increase our understanding of the reaction mechanisms involved in both decomposition of biomass and heat transfer in the biomass layer.

It has been well known that the woody biomass decomposed to char, gas with having the middle heating value and tar undergoing the pyrolysis. There are many chemical reaction models of pyrolysis[11)].

The heat transfer with including the chemical reactions undergoing the pyrolysis has been also investigated by Koufopanos et al [7)], Won et al[18)] and etc. The roughly mechanism of the heat transfer could be predicted by the numerical simulation [4)]. However, most of reports assumed that the volume didn't change undergoing the pyrolysis and there is no report how the thermal conduction in the bed changes with the effect of the volume reduction.

In this report, the numerical simulation of the unsteady thermal conduction without the chemical reaction has been conducted and compared with the experimental one. The yield of solid component undergoing the fast and slow pyrolysis in the packed bed of biomass particles has been also compared with the analyzed one by the different three kinetic models (model-1: Thurner et.al[16)], model-2: Mohan et.al[8)], model-3: Miller et.al[9)]). Furthermore, the relationship between the volume reduction process in the packed bed and the chemical reaction undergoing pyrolysis was elucidated by the formularization.

Numerical simulation for thermal conduction without considering chemical reaction

Two dimensional numerical simulation in the packed bed of the biomass particles for heat transfer has been conducted. The size of the calculation domain is r = 53mm and z = 90mm. The regular mesh is utilized. The number of the mesh along the radial component and along the axial component are 140 and 234, respectively. Specific heat in the packed bed can be expressed by using the specific heat of the biomass and air and volume fraction. Effective thermal conductivity in the packed bed is adapted by the equation, which was proposed by Yagi et al[19)]. Density in the packed bed is treated as experimental value.

The equation and the boundary conditions are descretized over a control volume using the finite difference method. The calculation program is made originally by using fortran language. The calculation has been conducted by using not only the dependence of the physical properties on the temperature bat also the change of the position of the top in the packed bed. The temperature profile in the packed bed and the temperature at the top of the packed bed are solved by the SOR method and Newton-Raphson method, respectively.

Experimental set up and procedure

The rig consists of the supplying of nitrogen gas, a tubular reactor, the cold traps for the tar and water, gas-sampling bag and gas flow meter[12-15]. The furnace has the maximum output of 1.5 kW and can set the wall temperature of the reactor up to about 1073 K. The cooled traps in the iced bath consist of two erlenmeyer flasks of 500cc that are filled with glass wool and solid of $CaCl_2$. Both the tar and water are adsorbed by the traps. Generated gas through the traps fills with the sampling bag (GL Science Co., Ltd) for measurement of gas component. Gas flow rate is measured by the wet type gas meter (Shinagawa Co., Ltd, W- NK-2B). O_2 monitor is set at the exit of the reactor. Air in the reactor is exhausted by the change of the nitrogen gas. The nitrogen gas introduces into the reactor until the concentration of O_2 at the exit of the reactor is less than 1 %. After that, the supplying of the nitrogen gas stops. The slow pyrolysis whose heating rate is 400 K/hr has been applied conducted to investigate the heat transfer.

Pyrolysis in the packed bed of biomass particles[13-15]

For the slow pyrolysis, woody biomass particles of $W_0 = 100$ g and $D_{p50} = 1.1$ mm are introduced into the tube reactor. The wall of the reactor is heated by the electric furnace with the heating rate at 400 K/hr up to 1073 K. The temperature of the wall is maintained at 1073 K for about 60 min. The temperatures of the packed bed of biomass particles and the wall of the reactor are measured by the K-type thermocouples (Yamari Co., Ltd., ϕ= 1mm) every 1 minute and recorded in a memory of the personal computer through the data logger (Keyence Co., Ltd, NR-1000). Heat flow rate can be calculated by the change of the average temperature in the biomass layer for 1 minute.

Shrinking process in the packed bed of biomass particles[14, 15]

Dry biomass is set in the quartz beaker (63mm in height and 50mm in internal diameter). The weight of the layer and height of the layer show about 9 g and about 30mm, respectively. The beaker with having three thermocouples is set in the test rig. After the switch on of the heater, visualization of the biomass layer and temperature measurement are started. In order to avoid the tar-adsorption on the visualized window, the nitrogen gas gets continuously into the reactor. *Pseudotsuga menziesii* and wood

Table 1. Properties of biomass

True density of wood[10] (g/cm^3)	ρ_T	1.5							
Kind of biomass	–	*Pseudotsuga menziesii*				Wood bark			
Size of biomass particles (mm)	Dp	1				0.5			
Component	–	Hemicellulose	Cellulose	Lignin	Ash	Hemicellulose	Cellulose	Lignin	Ash
Mass fraction[1] (wt%)	w_i	24.4	45.8	28.0	1.7	29.8	24.8	43.8	1.6

bark are used as the sample of biomass. The components of the sample are listed in table 1.

Numerical simulation for chemical reaction and shrinking process in the packed bed of biomass particles

Chemical reaction models undergoing pyrolysis of biomass

There are many models for pyrolysis of biomass. In this study, three models were chosen in these models. The model-1 shows the representative model undergoing the pyrolysis[16]. In the model-1, gas, tar and char are generated by three- parallel reactions. The parameters of the rate constant, which were reported by Di Blasi et al[5]., were used. The model-1 doesn't consider the effect of main components of biomass.

On the other hand, The model-2[10] and model-3[11] shows that Hemicellulose, Cellulose and Lignin decompose parallel to gas, tar and char. For the model-2, the parameters of the rate constant, which were reported by Varhegyi et al[17], Capart et al. [3] and Chan et al. [2], were used. For the model-3, the parameters of the rate constant, which were reported by Miller et al[9], were used.

The effect of the char formation due to the secondary decomposition by tar was ignored. The yield of solid component every model and every setting temperature were calculated by the Runge Kutta method. The mass fraction for Hemicellulose, Cellulose and Lignin in the table 1 was utilized as the initial weight of for the model-2 and –3.

Shrinking process in the packed bed of biomass particles[15]

The volume of the packed bed is given by

$$V_{\mathrm{PB}} = V_{\mathrm{G}} + V_{\mathrm{B}} \tag{1}$$

Where a V_{G} and a V_{B} show the volume in the gas phase and the volume in the solid phase, respectively. If the equation 1 is divided by $V_{\mathrm{PB,0}}$,

$$\frac{V_{\mathrm{PB}}}{V_{\mathrm{PB,0}}} = \frac{V_{\mathrm{G}}}{V_{\mathrm{PB,0}}} + \frac{V_{\mathrm{B}}}{V_{\mathrm{PB,0}}} \tag{2}$$

Firstly the V_{G} was decided experimentally using the volume fraction $\varepsilon_{\mathrm{PB}}$ and the equilibrium volume of the packed bed. The dependence of the V_{G} on the setting temperature were shown by the approximated lines[15]. Secondary the V_{B} was solved by the numerical simulation for the chemical reaction. A weight of the solid component of biomass particles $[S_{\mathrm{B}}]$ is given by

$$[S_{\mathrm{B}}] = \rho_{\mathrm{t}} V_{\mathrm{B}} \tag{3}$$

Where ρ_{t} shows the true density of woods and its value[10] is about 1.5 g/cm^3. If the true density doesn't change undergoing the pyrolysis, the relationship between the volume and the weight of the solid component of biomass is given by,

$$\frac{V_{\mathrm{B}}}{V_{\mathrm{B,0}}} = \frac{[S_{\mathrm{B}}]}{[W_{\mathrm{B,0}}]} \tag{4}$$

Where a V_{B0} and a $[W_{\mathrm{B,0}}]$ shows the initial volume and the initial weight of the biomass, respectively. From these procedures, the volume of the packed bed can be expressed by

For Pseudotsuga menziesii

$$\frac{V_{PB}}{V_{PB,0}} = \begin{cases} \dfrac{V_{B,0}}{V_{PB,0}} \dfrac{[S_B]}{[W_{B,0}]} + 0.899 & (T_S \le 532\,K) \\ \dfrac{V_{B,0}}{V_{PB,0}} \dfrac{[S_B]}{[W_{B,0}]} \\ + 0.344 + 12.6 * \exp\left(-\dfrac{T_S}{171}\right) & (T_S \le 532\,K) \end{cases} \tag{5}$$

For bark

$$\frac{V_{PB}}{V_{PB,0}} = \begin{cases} \dfrac{V_{B,0}}{V_{PB,0}} \dfrac{[S_B]}{[W_{B,0}]} + 0.833 & (T_S \le 665\,K) \\ \dfrac{V_{B,0}}{V_{PB,0}} \dfrac{[S_B]}{[W_{B,0}]} \\ + 0.672 + 108 * \exp\left(-\dfrac{T_S}{102}\right) & (T_S \le 665\,K) \end{cases}$$

The solid component $[S_B]$ was calculated by solving the equation of the pyrolysis using the Runge Kutta method. By substituting the measured temperature in the packed bed at $t = t$ $T_{PB}(t)$ to T_S, the volume of the packed bed was solved by equation 5.

Results and discussion

Experimental and calculated results for the heat transfer undergoing the pyrolysis [13, 14]

Fig. 1 shows the time course of the average temperature (T_{av}) in the packed bed of the biomass particles. Plots shows the experimental results while the solid line shows the calculated average temperature without the effect of chemical reaction. For T_{av} < 573K, the experimental results can be reproduced by the calculation. However, the experimental results for 573K < T_{av} didn't agree with the calculated results because the packed bed of biomass is shrinking dramatically due to the formation of tar from cellulose. The temperature beyond the biomass layer rather than the temperature in the biomass layer was measured by the thermocouples at z = 50mm. Therefore, in order to reproduce the heat transfer in the biomass layer, it is necessary to investigate the shrinking process in the packed bed of biomass particles.

Relationship between the Formation Rate of Solid Component and Volume Reduction of Packed Bed of Woody Biomass Particles during Pyrolysis [14,15]

Fig 2 shows the relationship between the yield of the solid component and the setting temperature. Fig. 2a shows the results for *Pseudotsuga menziesii*. ○ shows the experimental results while the lines shows the calculated results for three models. The dependence of the experimental yield on the setting temperature is divided by almost two zones. For the zone 1 at T_S < 623 K, the yield decreased

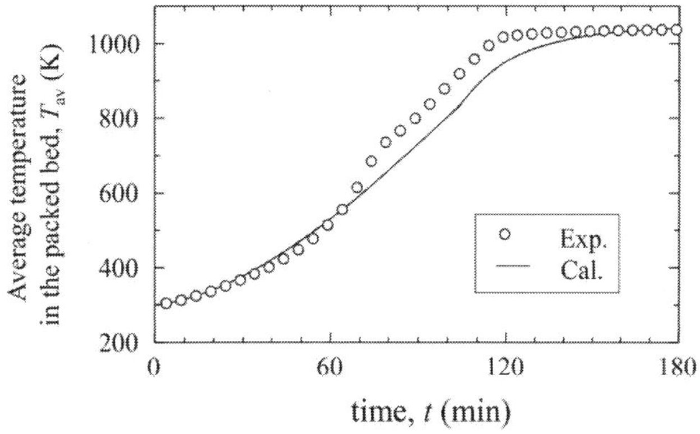

Fig.1 Comparison of the measured average temperature in the packed bed of biomass particles with the calculated one.

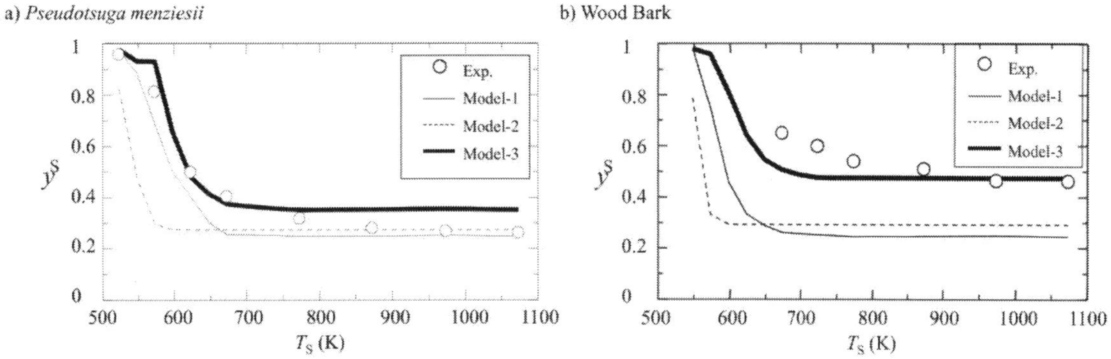

Fig. 2 Comparison of the experimental solid yield with the predicted one by three different kinetic models.

dramatically. In this zone, all components in the wood decomposed to tar and gas [6]. For zone 2 at $T_S >$ 623 K, the yield decreased slightly. In this zone, Lignin in the wood mainly decomposed to tar and gas[6]. The experimental yield for *Pseudotsuga menziesii* could be reproduced by the model-1 and model–3 except for the model-2. On the other hand, in fig 2b there was only 1 zone which the yield decreased slightly. As the experimental yield for Wood bark could be reproduced by the model–3, it was important to consider the lignin contents.

Fig.3 shows the time course of the volume in the packed bed for various setting temperatures. Fig. 3a shows the results for *Pseudotsuga menziesii*.○ shows the experimental results while the lines shows the calculated results. At $t <$ 50min, the average temperature in the packed bed of biomass was less than 523K and the volume of the layer changes hardly. At $t >$ 55min, the packed bed of biomass becomes carbonizing and then shrinking dramatically at $t >$ 65min. The experimental volume for the sawdust of *Pseudotsuga menziesii* could be reproduced by the model-1 and model–3. On the other hand, in fig 3b the volume decreased slightly with time and setting temperature. The experimental volume for Wood bark could be reproduced by the model–3.

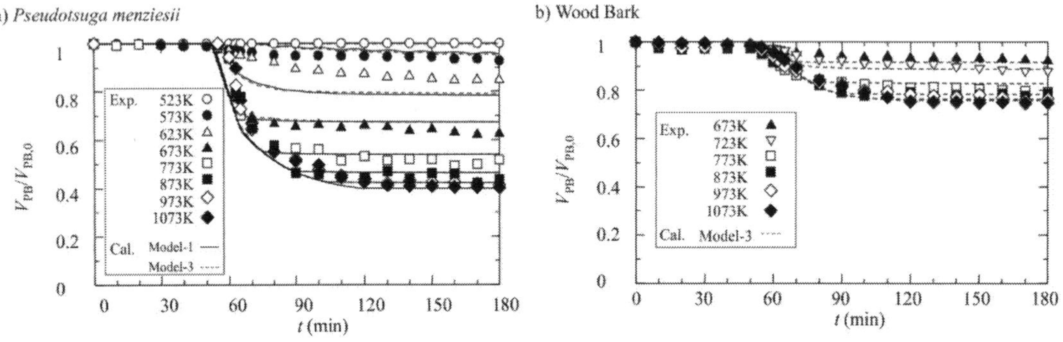

Fig. 3 Time course of the volume reduction process undergoing the pyrolysis.

Therefore the volume in the packed bed of the biomass undergoing the pyrolysis can be estimated by using the model-3 and the dependence of gas volume in the bed on the temperature.

The effect of the shrinking process on the heat transfer will be reported on the day of the presentation.

Conclusions

Transport phenomena in the pyrolysis of woody biomass has been investigated experimentally and numerically. Main conclusions are summarized as follows;

1) For $T_{av} < 573K$, the time course of the average temperature in the packed bed can be reproduced by the calculation. However, the experimental results for $573K < T_{av}$ didn't agree with the calculated results. The packed bed of biomass is shrinking dramatically due to the formation of tar from cellulose. The temperature beyond the biomass layer rather than the temperature in the biomass layer was measured by the thermocouples at $z = 50mm$. Therefore, in order to reproduce the heat transfer in the biomass layer, it is necessary to investigate the shrinking process in the packed bed of biomass particles.
2) The yield of solid component undergoing the fast and slow pyrolysis in the packed bed of biomass particles has been compared with the analyzed one by the different three kinetic models (model-1: Thurner et.al, model-2: Mohan et.al, model-3: Miller et.al). From the results, the yield of solid component could be reproduced by the analysis using the model-3 even if the heating rate and the lignin content changed. Furthermore, the relationship between the volume reduction process in the packed bed and the chemical reaction undergoing pyrolysis was elucidated by the formularization. From the result, the volume in the packed bed of the biomass undergoing the pyrolysis can be estimated by using the model-3 and the dependence of gas volume in the bed on the temperature.

Refernces

1. Ayhan D. (1997)," Calculation of higher heating values of biomass fuels," Fuel, **76-5**, 431-434.
2. Chan, R. W. C., Krieger, B. B. (1981), "Kinetics of dielectric loss microwave degradation of polymers: lignin," J. Appl. Poly. Sci., 26-5, 1533-1553.
3. Capart, R., Khezami, L., Burnham, A. K., (2004)," Assessment of various kinetic models for the pyrolysis of a microgranular cellulose," Thermo. ACTA, 417-1, 79-89.
4. Di Blasi, C. (2008), "Modeling chemical and physical processes of wood and biomass pyrolysis," Prog. Energy Comb. Sci., **34**, 47-90.
5. Di Blasi, C. and Branka, C. (2001)," Kinetics of Primary Product Formation from Wood

Pyrolysis," Ind. Eng. Chem. Res., **40-23**, 5547-5556.
6. Hasegawa, I., Fujisawa, H., Sunagawa, K., Mae, K. (2005)" Quantitative Prediction of Yield and Elemental Composition during Pyrolysis of Wood Biomass", J. Jpn. Inst. Energy, **84** (1), 46-52 (in Japanese).
7. Koufopanos, C. Papayannakos, A., N., Maschio, G. and Lucchesi, A.(1991), "Modelling of the Pyrolysis of Biomass Particles. Studies on Kinetics, Thermal and Heat Transfer Effects," Can. J. Chem. Eng., 69, pp.907-915.
8. Mohan, D., Pittman Jr., C. U. and Steele, P. H. (2006)," Pyrolysis of Wood/Biomass for Bio-oil: A Critical Review," Energy Fuels, **20-3**, 848-889.
9. Miller, R. S., Bellan, J. (1997)" A Generalized Biomass Pyrolysis Model Based on Superimposed Cellulose, Hemicelluloseand Liqnin Kinetics" , Combust. Sci. Technol., 126, 97-137.
10. Okano, K. and Sofue, N., Mokuzai kagaku handbook, Asakura shoten, 117(2006).
11. Okumura, Y., (2011)," Modeling of Pyrolysis and Gasification Reactions of Many Types of Biomass" , J. Jpn. Inst. Energy, **90** (2), 122-131 (in Japanese).
12. Tanoue, K., Nishimura, T., Hinauchi, T., Uchida, K., Taniguchi, M. and Sasauchi, K.(2007), "Analysis of pyrolysis of woody biomass with heating rate," Energy and Resources, 28-3, pp.186-192 (in Japanese).
13. Tanoue, K., Hinauchi, T., THAUNG, O., Nishimura, T., Taniguchi, M. and Sasauchi, K.(2007), "Modeling of heterogeneous chemical reactions caused in pyrolysis of biomass particles," Adv. Powder Technol., 18-6, pp.825-840.
14. Tanoue, K., Widya, W., Yamasaki, K., Kawanaka, T., Yoshida, A., Nishimura, T., Taniguchi, M., Sasauchi, K., (2010) "Numerical Simulation of the thermal conduction of packed bed of woody biomass particles accompanying volume reduction induced by pyrolysis", J. Jpn. Inst. Energy, **89** (10), 948 (in Japanese).
15. Tanoue, K., Yamasaki, K., Nishimura, T., Taniguchi, M., Sasauchi, K., "A Relationship between the Formation Rate of Solid Component and Volume Reduction of Packed Bed of Woody Biomass Particles during Pyrolysis", J. Jpn. Inst. Energy, in press (in Japanese).
16. Thurner, F., Mann U. (1981)" Kinetic investigation of wood pyrolysis" , Ind. Eng. Chem. Proc. Des. Dev., **20**, 482-488.
17. Varhegyi, G., Antal Jr., M.J., Szekely, T. and Szabo, P. (1989)"Kinetics of the thermal decomposition of cellulose ,hemicellulose, and sugar cane bagasse", Energy Fuels, 3-3, 329-335
18. Won, C. P, Arvind, A. and Howard, R. B. (2010)"Experimental and theoretical investigation of heat and mass transfer processes during wood pyrolysis" , Combust. Flame, **157**, 481-494.
19. Yagi,S and Kunii, D., International Deveropment in Heat Transfer, Part IV, 742&750 (1961).

Acknowledgements
This work was supported in part by a Grant-in Aid for Scientific Research C (No. 19560174) from the Japan Society for the Promotion of Science. Our special thanks are extended to Dr. Morihisa Yokota of Ube Industries, Ltd., for helpful discussions on the work.

(520a) Uncertainty Set Induced Robust Linear and Mixed Inter Linear Optimization and Their Probabilistic Guarantees: New Results and Comparative Study

Zukui Li and Christodoulos A. Floudas, Chemical and Biological Engineering, Princeton University, Princeton, NJ

In the vast majority of optimization applications, problem data are often subject to uncertainty due to their random nature, measurement errors or other reasons. The solution obtained without considering the data uncertainty may be suboptimal or even infeasible for practical applications. Thus, addressing uncertainty issues in optimization is of significant importance and has received a lot of attention in both academia and industry.

Robust optimization belongs to an important methodology dealing with optimization problems with data uncertainty. With a predefined set within the uncertain space, robust optimization techniques aim at finding the best solution which is feasible for any realization of the data uncertainty in the given set. The corresponding optimization problem is also called robust counterpart optimization problem.

Robust optimization has received lots of attention in the process operations applications [1], [2]. Lin et al., [3] and Janak et al., [4] extended the robust optimization formulation introduced for linear programming problems [5] to mixed integer linear optimization (MILP) problems and apply the method to process scheduling problem. Verderame and Floudas [6] studied both continuous (general, bounded, uniform, normal) and discrete (general, binomial, Poisson) uncertainty distributions and applied the robust optimization framework to operational planning problems. The work was further compared with the conditional-value risk based method [7].

In this work, we report our recent results on robust counterpart optimization techniques for linear optimization and mixed integer linear optimization problems, and perform a comparative study on the different robust formulations [8]. Specifically, different uncertainty sets, including those studied in literature (i.e., interval set [9]; combined interval and ellipsoidal set [5, 10]; combined interval and polyhedral set [11]) and new ones (i.e., adjustable box; pure ellipsoidal; pure polyhedral; combined interval, ellipsoidal, and polyhedral set) are studied in this work and their geometric relationship is discussed. For uncertainty in the left hand side, right hand side, and objective function of the optimization problems, robust counterpart optimization formulations induced by those different uncertainty sets are derived. Probabilistic guarantees on constraint violation are also derived for the above different uncertainty sets induced robust counterpart optimization models, for both bounded and unbounded uncertainty, with and without detailed probability distribution information.

Extensive numerical studies are performed to compare the solutions of the robust counterpart optimization models and the tightness of the different probability bounds, and also to assess the conservatism of several important robust counterpart optimization models. Applications in refinery production planning and batch process scheduling problems are presented. The usage of probability distribution information is introduced and tighter a priori probability bounds are proposed so as to find better robust solutions in these applications [8].

1. Verderame, P.M., et al., *Planning and Scheduling under Uncertainty: A Review Across Multiple Sectors.* Industrial & engineering chemistry research, 2010. **49**(9): p. 3993-4017.

2. Li, Z. and M.G. Ierapetritou, *Process scheduling under uncertainty: review and challenges.* Computers and Chemical Engineering, 2008. **32**: p. 715-727.

3. Lin, X., S.L. Janak, and C.A. Floudas, *A new robust optimization approach for scheduling under uncertainty: I. bounded uncertainty.* Computers and Chemical Engineering, 2004. **28**: p. 1069-1085.

4. Janak, S.L., X. Lin, and C.A. Floudas, *A new robust optimization approach for scheduling under uncertainty: II. Uncertainty with known probability distribution.* Computers and Chemical Engineering, 2007. **31**: p. 171-195.

5. Ben-Tal, A. and A. Nemirovski, *Robust solutions of Linear Programming problems contaminated with uncertain data.* Mathematical Programming, 2000. **88**: p. 411-424.

6. Verderame, P.M. and C.A. Floudas, *Operational Planning of Large-Scale Industrial Batch Plants under Demand Due Date and Amount Uncertainty. I. Robust Optimization Framework.* Industrial & engineering chemistry research, 2009. **48**(15): p. 7214-7231.

7. Verderame, P.M. and C.A. Floudas, *Operational Planning of Large-Scale Industrial Batch Plants under Demand Due Date and Amount Uncertainty: II. Conditional Value-at-Risk Framework.* Industrial & engineering chemistry research, 2010. **49**(1): p. 260-275.

8. Li, Z., R. Ding, and C.A. Floudas, *A Comparative Theoretical and Computational Study on Robust Counterpart Optimization: I. Robust Linear and Robust Mixed Integer Linear Optimization.* Ind. Eng. Chem. Res, 2011: p. In publication.

9. Soyster, A.L., *Convex programming with set-inclusive constraints and applications to inexact linear programming.* Operations Research, 1973. **21**: p. 1154-1157.

10. Ben-Tal, A. and A. Nemirovski, *Robust solutions of uncertain linear programs.* Operations Research Letters, 1999. **25**(1): p. 1-13.

11. Bertsimas, D. and M. Sim, *The price of robustness.* Operations Research, 2004. **52**(1): p. 35-53.

See more of this Session: Design and Operations Under Uncertainty I
See more of this Group/Topical: Computing and Systems Technology Division

NONCONVEX GENERALIZED BENDERS DECOMPOSITION FOR OPTIMAL PROCESS DESIGN AND SYNTHESIS UNDER UNCERTAINTY

Xiang Li
Process Systems Engineering Laboratory, Department of Chemical Engineering, Massachusetts Institute of Technology, Cambridge, MA

Asgeir Tomasgard
Department of Industrial Economics and Technology Management, Norwegian University of Science and Technology, Trondheim, Norway

Paul I. Barton
Process Systems Engineering Laboratory, Department of Chemical Engineering, Massachusetts Institute of Technology, Cambridge, MA

Mixed-integer nonlinear programming (MINLP) provides a powerful framework for the design and synthesis of process systems in an optimal and systematic manner. Traditionally, a deterministic process model is used in the MINLP framework, although real systems are almost always uncertain. Recently, more and more attention has been paid to including uncertainties into optimization models [1], especially when the uncertainties have a significant impact on the decision made, and stochastic programming with recourse [2] is a natural way to address uncertainties in various engineering problems.

Due to their special structures, stochastic programs with recourse have long been solved with duality-based decomposition strategies, typically the Benders decomposition/*L*-shaped method [3] [4] and generalized Benders decomposition [5]. However, these methods cannot solve nonconvex stochastic programs rigorously because of duality gap. This paper presents a rigorous, duality-based decomposition method, called nonconvex generalized Benders decomposition (NGBD), for two-stage stochastic nonconvex MINLP, which guarantees finding an ε-optimal solution within a finite number of iterations. The method follows the framework of concepts presented by Geoffrion [6] for the design of large-scale mathematical programming techniques and the notion of generalized Benders decomposition. By convexification of the nonconvex functions, the stochastic MINLP is relaxed into a lower bounding problem that is a potentially large-scale convex MINLP. Solution of this lower bounding problem is then decomposed into a sequence of relaxed master problems, which are mixed-integer linear programs (MILP) with much smaller sizes, and primal bounding problems, which are convex programs. The solutions of the relaxed master problems yield a sequence of nondecreasing lower bounds on the optimal objective value, and they also generate a sequence of integer realizations defining the primal problems which yield a sequence of nonincreasing upper bounds on the optimal objective value. It can be proved that NGBD terminates finitely when the lower and upper bounds coincide (or are close enough), or infeasibility of the problem is indicated.

Case studies of a literature problem (pump network configuration problem [7]) and a real industrial system (the Sarawak Gas Production System [8]) show the advantages of the stochastic programming model and the dramatic computational advantages of NGBD over the state-of-the-art branch-and-reduce global optimization method (BARON [9]). A nonconvex MINLP with almost 150,000 variables can be solved to global optimality with NGBD within 80 minutes.

References

[1] N.V. Sahinidis. Optimization under uncertainty: state-of-the-art and opportunities. *Computers and Chemical Engineering*, 28:971 – 983, 2004.

[2] J. R. Birge and F. Louveaux. Introduction to Stochastic Programming. Springer, New York, 1997.

[3] J. F. Benders. Partitioning procedures for solving mixed-variables programming problems. *Numerische Mathematik*, 4:238–252, 1962.

[4] R. M. Van Slyke and R. Wets. *L*-shaped linear programs with applications to optimal control and stochastic programming. *SIAM Journal on Applied Mathematics*, 17(4):638–663, 1969.

[5] A. M. Geoffrion. Generalized Benders decomposition. *Journal of Optimization Theory and Applications*, 10(4):237–260, 1972.

[6] A. M. Geoffrion. Elements of large-scale mathematical programming: Part I: Concepts. *Management Science*, 16(11):652–675, 1970.

[7] T. Westerlund, F. Pettersson and I. E. Grossmann. Optimization of pump configurations as a MINLP problem. *Computers and Chemical Engineering*, 18(9):845–858, 1994.

[8] A. Selot, L. K. Kuok, M. Robinson, T. L. Mason and P. I. Barton. A short-term operational planning model for natural gas production systems. *AIChE Journal*.54:495–515, 2008.

[9] N. V. Sahinidis. BARON: A general purpose global optimization software package. *Journal of Global Optimization*.8:201–205, 1996.

(520c) Development of Property Models with Uncertainty Estimate for Process Design Under Uncertainty

Amol Hukkerikar[1], Bent Sarup[2], Jens Abildskov[1], Grkan Sin[1] and Rafiqul Gani[1], (1)CAPEC, Department of Chemical and Biochemical Engineering, Technical University of Denmark, Kgs. Lyngby, Denmark, (2)Vegetable Oil Technology Business Unit, Alfa Laval Copenhagen A/S, Sborg, Denmark

Physical andthermodynamic properties of pure compounds and their mixtures play an importantrole in design, simulation and optimisation of chemical processes. The accuracy of process design and simulation largely depends onthe accuracy of the underlying physical and thermodynamic data and propertyprediction models. While use ofexperimentally measured values for the needed properties is desirable inprocess design, the experimental data for the compounds of interest may not beavailable in many cases. Therefore, development of efficient and reliableproperty prediction methods and tools that can also provide estimates ofuncertainties in predictions of properties and their effects on process designbecomes necessary. For instance, the accuracy of design of distillation columnto achieve a given product purity is dependent on many pure compoundproperties such as critical pressure, critical temperature, acentric factoretc. In such cases, accurate property values along with uncertainty estimatesare needed to perform sensitivity analysis and quantify the effects of these uncertaintieson the process design.

The objectiveof this work is to develop a systematic methodology to provide more reliablepredictions with a new and improved set of model parameters for GC (groupcontribution) based and CI (atom connectivity index) based models and to quantify theuncertainties in the estimated property values from a process designpoint-of-view. This includes: (i)parameter estimation using available GC based and CI based property predictionmodels and large training sets to determine improved group and atomcontributions; (ii) uncertainty analysis of property prediction models toestablish statistical information such as covariance, standard error andconfidence intervals; and (iii) use the results of uncertainty analysis topredict the uncertainties in process design. For parameter estimation, large data-sets of experimentallymeasured property values for a wide range of pure compounds are taken from theCAPEC database. Classical frequentistapproach i.e., least square method is adopted for the estimation of modelprediction uncertainties. The developedmethodology provides property values along with uncertainties for the following 20 properties: normalboiling point, critical temperature, critical pressure, critical volume, normalmelting point, standard Gibbs energy, standard enthalpy of formation, standardenthalpy of fusion, standard enthalpy of vaporization at 298 K and at thenormal boiling point, entropy of vaporization at the normal boiling point,surface tension at 298 K, viscosity at 300 K, flash point, auto ignitiontemperature, Hansen solubility parameters, Hildebrand solubility parameter,octanol/ water partition coefficient, aqueous solubility, acentric factor, andliquid molar volume at 298 K. Theperformance of property models for these properties with the revised set ofmodel parameters is highlighted through a set of compounds not considered inthe regression step. The comparison of model prediction uncertainties withreported range of measurement uncertainties is presented for the properties withrelated available data. The application of the developed methodology toquantify the effect of these uncertainties on the design of different unitoperations (distillation column, liquid-liquid extraction, heat exchanger,crystallizer, equilibrium reactor etc.) is presented. The results show thatdepending on the chemical systems involved and the operating conditions beingconsidered, some of the input uncertainties can result in significantuncertainties in design. The most sensitive properties for each unit operationare also identified. This analysis can be used to reduce the uncertainties inproperty estimates for the properties of critical importance (by performing additionalexperiments to get better experimental data and better model parameter values).Thus, the developed methodology can be usedto quantify the sensitivity of process design to uncertainties in property estimates; obtain rationally therisk/safety factors in process design; and identify additional experimentationneeds in order to reduce most critical uncertainties.

See more of this Session: Design and Operations Under Uncertainty I
See more of this Group/Topical: Computing and Systems Technology Division

(520d) A Chance-Constrained Framework for Optimization Under Uncertainty

Sebastian Werk, Tilman Barz, Gnter Wozny and Harvey Arellano-Garcia, Chair of Process Dynamics and Operation, Berlin Institute of Technology, Berlin, Germany

Nowadays, the process industry is highly competitive and has to fulfill high quality standards for their products. Therefore, the use of computer-aided optimization software became an integral part of today's planning in process industries. The increase of available computing power and important advancements in algorithms led to more and more complex models of real-world problems. In general those models are characterized by deterministic assumptions. Probabilities and uncertainties are only rarely considered as stochastic influences but as expected values. Anyway, since several years, uncertain parameters can be taken into account by implementing so called *chance constraints* [1]. Since chemical processes are subject to multiple uncertainties, the use of chance constraints can lead to more realistic models and give further useful information, as shown in several examples.

Even though chance constraints are known for several years, only few models using them were developed so far [2]. The major reason for the little use of chance constraints might be the absence of support of chance constraints by usual optimization software. Therefore the evaluation of chance constraints requires external processing, including multivariate numerical integration, root-finding and the use of further mathematical software. Those algorithms had to be individually implemented for each model so far. Furthermore, the evaluation of the chance constraints becomes often the most time consuming part of the optimization process, therefore a not-optimized implementation might lead to unacceptable high computation time.

This work presents a framework for implementing chance constraints in optimization models. It can be easily linked to numerous different solvers and provides interfaces to adopt further mathematical software or tools. The use of this framework should significantly reduce the effort to implement uncertain parameters into own optimization models.

Beside the usability, the improvement of the computing time was an essential part of the framework design. Therefore, the calculating algorithm was analyzed and key elements identified. Beside the provision of second derivatives for the solver, major computational time reduction could be realized by improving the integration procedure and the root-finding algorithm. The integration procedure was accelerated using fewer integration points without losing numerical precision [3]. The root-finding step could be improved by using higher derivatives for the Newton's method and implementing a data analysis for better starting points. Those improvements led to a significant lower computation time.

The talk will present the use of the framework on the basis of an example and give an insight in the algorithmic improvements.

[1] H. Arellano-Garcia, G. Wozny. (2009) Chance Constrained Optimization of Process Systems under Uncertainty: Strict Monotonicity. Computers and Chemical Engineering. 33 (1568–1583)

[2] T. Barz, H. Arellano-Garcia, G. Wozny. (2011) Robust Implementation of Optimal Decisions using a Two-Layer Chance Constrained Approach. Industrial and Engineering Chemistry Research. doi:10.1021/ie1014525

[3] T. Barz, G. Wozny, H. Arellano-Garcia. (2011) An Efficient Sparse Approach to Sensitivity Generation for Large-Scale Dynamic Optimization. Computers and Chemical Engineering. doi:10.1016/j.compchemeng.2010.10.008

See more of this Session: Design and Operations Under Uncertainty I
See more of this Group/Topical: Computing and Systems Technology Division

(520e) Scheduling of Crude Oil Operations Under Uncertainty: A Robust Optimization Framework Coupled with Global Optimization

Jie Li, Ruth Misener and Christodoulos A. Floudas, Department of Chemical and Biological Engineering, Princeton University, Princeton, NJ

Scheduling ofcrude oil operations is an important component of overall refinery operations,because crude oil costs account for about 80% of the refinery turnover. Optimal crudeoil scheduling can increase profits by exploiting cheaper but poor qualitycrudes, minimizing crude changeovers, avoiding ship demurrage, and managingcrude inventory optimally. The mathematical modeling of blending differentcrudes in storage tanks results in many bilinear terms, which transform theproblem into a challenging, nonconvex, mixed integer nonlinear programming(MINLP) optimization model.

The crudeoil scheduling problem has received considerable attention with researchersdeveloping different models based on discrete- and continuous-time representations[1]. In our previous work [1], we have developed a novel unit-specificevent-based continuous-time MINLP formulation, and have proposed abranch and bound global optimization algorithm with piecewise-linearunderestimation of the bilinear terms for this problem. The proposed model incorporatedmany realistic operational features such as single buoy mooring (SBM), multiplejetties, multi-parcel vessels, crude blending, brine settling, crudesegregation, and fifteen important volume-based or weight-based crude propertyindices, and significantly reduced the number of bilinear terms and problemsize compared to the discrete-time formulation of Reddy et al. [2] and Li etal. [3]. The proposed branch and bound global optimization algorithm resultedin better integer feasible solutions, which were guaranteed to be within 2% ofglobal optimality.

All ofthe aforementioned models assume that the parameters used in the models aredeterministic in nature. However, frequent uncertainties in practice areunavoidable such as demand fluctuations, ship arrival delays, crude qualityspecification variations, uncertainty on crude profit margin, demurrage cost, inventory cost, changeover cost and safety stock penalty, and equipmentmalfunction, and tank unavailability. In the presence of these uncertainties,an optimal crude schedule obtained using nominal parameter values may often besuboptimal or even become infeasible. In general, two approaches can be used toaddress those uncertainties: reactive scheduling and preventative scheduling [4].For detailed reviews on planning and scheduling under uncertainty, the readeris directed to Li and Ierapetritou [5], and Verderame et al. [4]

Robustoptimization focuses on developing preventive models to minimize the effects ofuncertainties on the performance measure such as profit, and operating cost.Its main objective is to ensure that the generated solutions are robust, whilemaintaining a high level of solution quality. Li et al. [6]

developedscenario-based models for demand and ship arrival uncertainties separately.However, the number of scenarios exponentially increases with the number ofuncertain parameters, and hence makes their model intractable for practical problemswith large number of uncertain parameters. Cao et al. [7] proposed anoptimization model based on chance constrained programming to generate robustschedules under demand uncertainty during scheduling of crude oil operations. However,their approach cannot be used to deal with uncertain parameters following adiscrete probability distribution [8]. More importantly, their approach resultsin composition discrepancy. Recently, Wang and Rong [8] developed a two-stagerobust optimization model for crude oil scheduling problem to address demandand ship arrival uncertainty separately. Although their model can cope with awide variety of uncertainties, the generated schedule from their model alsoresults in composition discrepancy. Most importantly, these approaches cannotensure the generated solution to be feasible for the nominal parameters.

In thispaper, we extend the robust optimization framework proposed by Lin et al. [9], Janaket al. [10], Verderame and Floudas [11-13] and Li et al. [14] to develop a deterministicrobust counterpart optimization model for demand and ship arrival uncertaintyduring crude oil scheduling operations. The robust solution from the robustoptimization framework is guaranteed to be feasible for the nominal parameters.The recently proposed branch and bound global optimization algorithm withpiecewise-linear underestimation of bilinear terms by Li et al. [1] is alsoextended to solve the non-convex MINLP deterministic robust counterpartoptimization model and generate robust schedules. Two examples are used toillustrate the capability of the proposed robust optimization approach and theextended branch and bound global optimization algorithm. The computationalresults demonstrate that the obtained schedules are robust in the presence ofdemand and ship arrival uncertainty.

References

[1] Li J,Misener R, Floudas CA. Continuous-time modeling and global optimizationapproach for scheduling of crude oil operations. *AIChE Journal*, Inpress, 2011.

[2] Reddy PCP,Karimi IA, Srinivasan R. Novel solution approach for optimization crude oil operations. *AIChE Journal*. 2004, 50: 1177-1197.

[3] Li J, Li WK,Karimi IA, Srinivasan R. Improving the robustness and efficiency of crudescheduling algorithms. *AIChE Journal*. 2007, 53: 2659-2680.

[4] Verderame PM,Elia JA, Li J, Floudas CA. Planning and scheduling under uncertainty: A reviewacross multiple sections. *Industrial and Engineering Chemistry Research*.2010; 49: 3993-4017.

[5] Li Z,Ierapetritou M. Processing scheduling under uncertainty: Review and challenges.*Computers and Chemical Engineering*. 2008; 32: 715-727.

[6] Li J, Karimi IA,Srinivasan R. Robust scheduling of crude oil operations under demand and shiparrival uncertainty. Presented at the AIChE Annual Meeting, San Francisco, CA,Nov. 12-17, 2006.

[7] Cao CW, GuXS, Xin Z. Chance constrained programming models for refinery short-term crudeoil scheduling problem. *Appl. Math. Model.* 2009; 33: 1696-1707.

[8] Wang JS,Rong G. Robust optimization model for crude oil scheduling under uncertainty. *Ind.Eng. Chem. Res.* 2010; 49: 1737-1748.

[9] Lin X, JanakSL, Floudas CA. A new robust optimization approach for scheduling underuncertainty: I. Bounded uncertainty. *Computers and Chemical Engineering*.2004; 28: 1069-1085.

[10] Janak SL,Lin X, Floudas CA. A new robust optimization approach for scheduling underuncertainty. II. Uncertainty with known probability distribution. *Computersand Chemical Engineering*. 2007: 31: 171-195.

[11] Verderame PM,Floudas CA. Operational planning of large-scale industrial batch plants underdemand due date and amount uncertainty. I. Robust optimization framework. *Industrialand Engineering Chemistry Research*. 2009; 48: 7214-7231.

[12] VerderamePM, Floudas CA. Integration of operational planning and medium-term schedulingfor large-scale industrial batch plants under demand and processing timeuncertainty. *Industrial and Engineering Chemistry Research*. 2010; 49:4948-4965.

[13] VerderamePM, Floudas CA. Multisite planning under demand and transportation uncertainty: Robust optimization and conditional value at risk framework. *Industrial andEngineering Chemistry Research*. In Press, 2011. **DOI:** 10.1021/ie101401k.

[14] Li ZK, Ding R, Floudas CA. Acomparative theoretical and computational study on robust counterpartoptimization: I. Robust linear optimization and robust mixed-integer linearoptimization. *Industrial and Engineering Chemistry Research*, Submittedfor Publication, 2011.

See more of this Session: Design and Operations Under Uncertainty I
See more of this Group/Topical: Computing and Systems Technology Division

(520f) Optimal Planning of Flexible Process Networks Under Uncertainty with Stochastic Inventory

Fengqi You, Department of Chemical and Biological Engineering, Northwestern University, Evanston, IL

Due to global competition and dynamic market environment, a major challenge in enterprise-wide optimization [1-3] has become to simultaneously optimize the production planning and inventory control decisions for chemical complexes under demand and supply uncertainty [4, 5]. Because chemical complexes are usually composed of many interconnected processes and various chemicals, it is an important but non-trivial task to integrate the production and inventory planning decisions with the uncertain supply and demand variations, and coordinate the activities of purchase, production, inventory and sale to minimize the total cost. The challenges arise not only from the modeling part about how to quantify the internal demand uncertainty and its propagation, but also from the computational perspective about how to effectively solve the resulting optimization problem that leads to a large-scale nonconvex mixed-integer nonlinear program (MINLP).

In this work, we address the mid-term planning under supply and demand uncertainty of flexible chemical process networks with dedicated and flexible processes. We propose a MINLP model that captures the stochastic nature of the supply and demand variations based on guaranteed service approach [6, 7], and integrates stochastic inventory control decisions with production planning and cyclic scheduling decisions. The model takes into account multiple tradeoffs and simultaneously determines the optimal purchases of feedstocks, production levels of processes, cyclic schedules of flexible processes, sales of final products and inventory levels of all chemicals in the process network. To solve the MINLP problems with modest computational times, we propose a tailored global optimization algorithm based on the transformation techniques [8] and successive piece-wise linear approximation [5]. The application of the model and the performance of the proposed algorithm are illustrated through two industrial-scale case studies.

References:

[1] Wassick JM. Enterprise-wide optimization in an integrated chemical complex. Computers & Chemical Engineering. 2009; 33:1950-1963.

[2] Grossmann IE. Enterprise-wide Optimization: A New Frontier in Process Systems Engineering. AIChE Journal. 2005; 51:1846-1857.

[3] Varma VA, Reklaitis GV, Blau GE, Pekny JF. Enterprise-wide modeling & optimization – An

overview of emerging research challenges and opportunities. Computers & Chemical Engineering. 2007; 31:692-711.

[4] Zipkin PH. Foundations of Inventory Management. McGraw-Hill: Boston, MA, 2000.

[5] You, F.; Grossmann, I. E., Stochastic inventory management for tactical process planning under uncertainties: MINLP model and algorithms. AIChE Journal 2011, 57, 1250-1277.

[6] Graves SC, Willems SP. Optimizing strategic safety stock placement in supply chains. Manufacturing & Service Operations Management. 2000; 2:68-83.

[7] Graves SC, Willems SP. Optimizing the supply chain configuration for new products. Management Science. 2005; 51:1165-1180.

[8] Lundell, A.; Westerlund, J.; Westerlund, T., Some transformation techniques with applications in global optimization. Journal of Global Optimization 2009, 43, 391–405.

See more of this Session: Design and Operations Under Uncertainty I
See more of this Group/Topical: Computing and Systems Technology Division

(520g) A Hybrid Approach for Process Scheduling Under Uncertainty

Martina Wittmann-Hohlbein and Efstratios N. Pistikopoulos, Department of Chemical Engineering, Centre for Process Systems Engineering, Imperial College, London, United Kingdom

In this work a novel hybrid methodology to address process scheduling problems under bounded uncertainty is proposed. A MILP formulation for short-term scheduling of batch processes is adopted [1]. We assume that the model is contaminated with uncertain data in the objective function (OFC), the right-hand side constraint vector (RHS) and in the constraint matrix (LHS), introduced by price, demand, and processing time or conversion rate uncertainty respectively.

Multi-parametric programming is a powerful tool to account for the presence of uncertainty in mathematical models and has previously contributed to proactive scheduling. Multi-parametric programming algorithms aim to derive the optimal solution as a function of the varying parameters without exhaustively enumerating the parameter space. However, not all types of uncertainty in MILP models can be addressed alike and, unlike OFC- and RHS-uncertainty, LHS-uncertainty still poses a major challenge.

In order to deal with all types of parameter variation in the scheduling model, we apply a combined robust optimization/multi-parametric programming procedure for its approximate solution [2]. In the first step a partial immunization of the model against LHS-uncertainty is performed embedding the worst-case oriented approach [3] and, alternatively, an approach that allows to control the degree of conservatism of the solutions [4], whereas in the second step explicit solutions of the partially robust counterpart are derived using a recently proposed state-of-the art mp-MILP algorithm [5].

We demonstrate that this hybrid approach is computationally efficient, and that it is an attractive alternative to the rigorous robust optimization approach employed in process scheduling under uncertainty ([6,7]) in terms of providing a tight estimate of the overall profit of the scheduling problem. We obtain piecewise affine partially robust scheduling policies and when the true values of the parameters are known the optimal policy is determined via simple function evaluation.

References

[1].Ierapetritou, M.G., Floudas, C.A. Effective continuous-time formulation for short-term scheduling. 1. Multipurpose Batch Processes. *Industrial & Engineering Chemistry Research,* 37, 4341-4359 (1998).

[2].Wittmann-Hohlbein, M., Pistikopoulos, E.N. A robust optimization based approach to the general solution of mp-MILP problems. *Accepted in Proceedings of 21st European Symposium on Computer Aided Process Engineering.*

[3].Ben-Tal, A., Nemirovski, A. Robust solutions of linear programming problems contaminated with uncertain data. *Mathematical Programming* 88, 411 (2000).

[4].Bertsimas, D., Sim, M. Robust discrete optimization and network flows. *Mathematical Programming,* 98, 49-71 (2003).

[5].Faísca, N.P., Kosmidis, V.D., Rustem, B., Pistikopoulos, E.N. Global optimization of multi-parametric MILP problems. *Journal of Global Optimization* 45, 131–151 (2009).
[6].Lin, X., Janak, S.L., Floudas, C.A. A new robust optimization approach for scheduling under uncertainty: I. Bounded uncertainty. *Computers & Chemical Engineering,* 28, 1069-1085 (2004).
[7].Li, Z., Ierapetritou, M.G. Robust Optimization for process scheduling under uncertainty. *Industrial & Engineering Chemistry Research,* 47, 4148-4157 (2008).
See more of this Session: Design and Operations Under Uncertainty I
See more of this Group/Topical: Computing and Systems Technology Division

SYNTHESIS OF SUSTAINABLE PROPERTY- BASED WATER NETWORKS

Luis Fernando Lira-Barragán,[a] José María Ortega-Ponce,[a] Medardo Serna-González,[a] Mahmoud M. El-Halwagi[c]

[a] *Universidad Michoacana de San Nicolás de Hidalgo, Morelia, Mich., México*

[b] *Chemical Engineering Department, Texas A&M University, College Station, TX, USA*

Introduction

Water is one of the most important natural resources that support the life in the world. The chemical and process industry is one of the biggest consumers of the available clean water, and at the same time the industry is one of the highest producers of polluted wastewater. In this regard, recently there have been proposed a set of techniques to economically reduce the clean water consumption in the process industry considering the possibility to recycle, reuse and regenerate the wastewater produced in the industrial facilities. The proposed techniques have addressed the problem from different perspectives (i.e., minimizing the fresh water consumption, minimizing the wastewater discharge, minimizing the regeneration costs, etc) and using diverse techniques (i.e., graphical, algorithmic and mathematical programming based approaches). The heuristic approaches allow to easily identify targets ahead of the design and to obtain a graphical representation of the problem (see for example Wang and Smith, 1994; El-Halwagi and Spriggs, 1998). All previous methodologies only have considered the processes that happen inside the industrial facilities, without taking into account the processes that occur outside the plant for the wastewater streams discharged to the environment.

Therefore, the goal of the proposed work is to overcome the drawbacks of the previously reported methods for the synthesis of water integration networks, considering the sustainability of the surrounding watershed that is based on properties for the cases that multiple pollutants are presented and that make difficult to characterize the streams in terms of the composition. The optimal quantity and quality for the wastewater discharged from the new plant depends on the location selected to install the plant because the impact in the watershed depends strongly on the location selected (considering the integration with other streams and the chemical and biochemical reactions that occur in the watershed system); and therefore, the recycle and reuse network is highly influenced by the location of the new plant.

Model Formulation

The main stream of the river is fed by several streams or tributaries ($FT_{r,t}$). As it can be seen in **Figure 1**, the watershed system exchanges water during its trajectory (i.e., water for agricultural use, wastewater discharged to the river with and without treatment, industrial and residential effluents, etc.) and finally the watershed discharges to the final disposal. Besides, natural phenomena like precipitation, filtration and vaporization are taken into account in the model. These processes modify significantly the characteristics of the river; therefore, to track adequately the mean properties, the river is sectioned in parts where the properties can be considered constant (these sections are represented in **Figure 1** with ovals). These parts o sections are called reaches and these identify zones where no big intermediate effluents are discharged and/or extracted. In this sense, the model also indentifies the main effluents discharged in each reach; these effluents are called tributaries. The tributaries can be channels or branches of the river, which may contain discharges with/without treatment, industrial discharges, etc., and their flowrates ($FT_{r,t}$) and properties ($P^{Trib}_{p,r,t}$) affect the reaches where they are discharged. The discharges to the watershed contain compounds that interact chemically and biochemically with the system thought the flora and fauna established in the watershed.

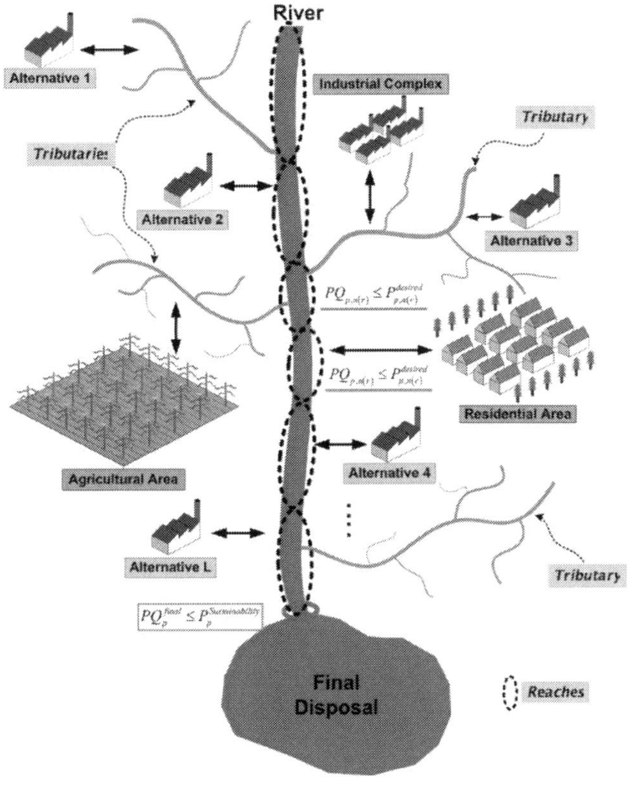

Figure. 1 Alternatives to locate the new industrial plant in a watershed system.

To ensure the sustainability of the surrounding watershed system, it is required to carry out an analysis prior to the optimization respect to the capacity of the final disposal to decompose chemically and biochemically the pollutants, and as consequence it is necessary to quantify these processes in terms of properties. The analysis yields the sustainability constraints which ensure that the pollutants will not be accumulated in the system and to maintain the properties of the watershed under reasonable limits.

In this sense, to consider the performance function of the interceptors, it is used a correction factor ($\gamma_p^{int^1,\ldots,int^N}$) that can be calculated prior to the optimization process, obtaining a linear function to model the interceptors. The correction factor must consider several aspects like the value of the property at the inlet conditions of the interceptor, the flowrate and design and operating parameters.

Thus, the correction factor is a very useful tool in this work, where the available technologies considered in the property interception network can be simulated (i.e., using ASPEN) to get the efficiency factors. It is worth of notice that the model includes a set of fictitious interceptors for each property to be intercepted with conversion factors and costs equal to zero. Besides, the property operators values for the inlet condition of the process sources are also known prior to the optimization process, $\psi_p\left(P_{i,p}^{InSource}\right)$, then the property values at the exit conditions of the property interception network can be calculated prior to the optimization process as follows:

$$\psi_p\left(P_{i,p}^{OutPIN,\mathrm{int}^1,\ldots,\mathrm{int}^N}\right)=\gamma_p^{\mathrm{int}^1,\ldots,\mathrm{int}^N}\psi_p\left(P_{i,p}^{InSource}\right),\quad i\in I, p\in P, \mathrm{int}^1\in INT^1,\ldots,\mathrm{int}^N\in INT^N$$

MFA Model based on properties

$$Q_r=Q_{r-1}+P_r+D_r+H_r+\sum_{t=1}^{N_{t(r)}}FT_{r,t}+QPNEW_{r(l)}-L_r-U_r,\quad \forall r\in R \tag{1}$$

$$\begin{aligned}\psi_p\left(P_{p,r}^{\mathrm{Reach}}\right)Q_r&=\psi_p\left(P_{p,r-1}^{In\mathrm{Reach}}\right)Q_{r-1}+\psi_p\left(P_{p,r}^{\mathrm{Prec}}\right)P_r+\psi_p\left(P_{p,r}^{Ind}\right)D_r+\psi_p\left(P_{p,r}^{\mathrm{Resid}}\right)H_r+\sum_{t=1}^{N_{t(r)}}\psi_p\left(P_{p,r,t}^{Trib}\right)FT_{r,t}\\&+\psi_p\left(P_{p,r(l)}^{PNEW}\right)QPNEW_{r(l)}-\psi_p\left(P_{p,r}^{Loses}\right)L_r-\psi_p\left(P_{p,r}^{Uses}\right)U_r-\int_{V=0}^{V_r}r_{p,r}dV_r,\ \forall p\in P, r\in R\end{aligned} \tag{2}$$

$$FT_{r,t} = S_{r,t}^{Untreat} + S_{r,t}^{Treat} + I_{r,t} + P_{r,t} + D_{r,t} + QPNEW_{r,t(l)} - L_{r,t} - U_{r,t}, \qquad \forall r \in R, t \in T \tag{3}$$

$$\begin{aligned} \psi_p\left(P_{p,r,t}^{Trib}\right)FT_{r,t} &= \psi_p\left(P_{p,r,t}^{Untreat}\right)S_{r,t}^{Untreat} + \psi_p\left(P_{p,r,t}^{Treat}\right)S_{r,t}^{Treat} + \psi_p\left(P_{p,r,t}^{Ind}\right)I_{r,t} + \psi_p\left(P_{p,r,t}^{Prec}\right)P_{r,t} + \psi_p\left(P_{p,r,t}^{Direct}\right)D_{r,t} \\ &+ \psi_p\left(P_{p,r,t(l)}^{PNEW}\right)QPNEW_{r,t(l)} - \psi_p\left(P_{p,r,t}^{Loses}\right)L_{r,t} - \psi_p\left(P_{p,r,t}^{Uses}\right)U_{r,t} - \int_{V=0}^{V_{r,t}} r_{p,r,t}\, dV_{r,t}, \qquad \forall p \in P, r \in R, t \in T \end{aligned} \tag{4}$$

$$\int_{V=0}^{V_r} r_{p,r}\, dV_r = k_p\left[\psi_p\left(P_{p,r}^{Reach}\right)\right]^{\sigma_p} V_r, \qquad \forall p \in P, r \in R \tag{5}$$

$$\int_{V=0}^{V_{r,t}} r_{p,r,t}\, dV_{r,t} = k_p\left[\psi_p\left(P_{p,r,t}^{Trib}\right)\right]^{\sigma_p} V_{r,t}, \qquad \forall p \in P, r \in R, t \in T \tag{6}$$

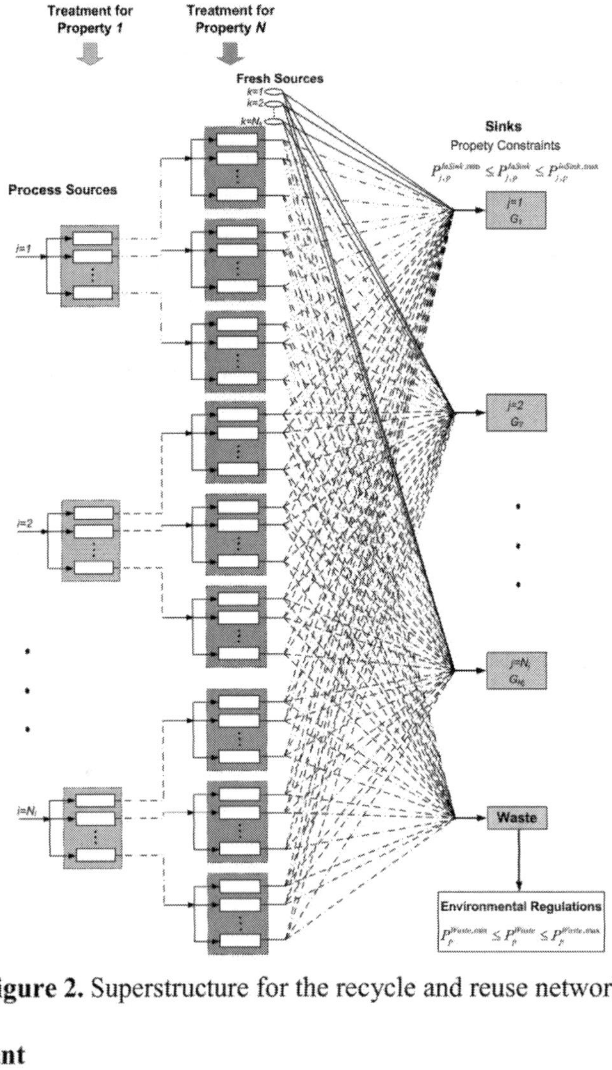

Figure 2. Superstructure for the recycle and reuse network

Location of the new plant

This work considers a set of alternatives to locate the new industrial facility (L), these options are generated prior to the optimization process. Then, for each alternative, it is necessary to consider the possible discharge of the new wastewater flowrate ($QPNEW_l$) and their property operators ($\psi_p\left(P_{p,l}^{PNEW}\right)$), thus, if the new industrial plant is located in the alternative 1, the wastewater flowrate and their property operators associated to this site must be bigger than zero; otherwise, if the new plant is not located in the position 1, the wastewater flowrate and their property operators discharged would be zero. This is included in the following disjunction:

$$\begin{bmatrix} Y_1 \\ \psi_p\left(P_{p,1}^{PNEW}\right) \geq 0, QPNEW_1 \geq 0 \\ \psi_p\left(P_{p,2}^{PNEW}\right) = 0, QPNEW_2 = 0 \\ \vdots \\ \psi_p\left(P_{p,l}^{PNEW}\right) = 0, QPNEW_l = 0 \end{bmatrix} \vee \begin{bmatrix} Y_2 \\ \psi_p\left(P_{p,1}^{PNEW}\right) = 0, QPNEW_1 = 0 \\ \psi_p\left(P_{p,2}^{PNEW}\right) \geq 0, QPNEW_2 \geq 0 \\ \vdots \\ \psi_p\left(P_{p,l}^{PNEW}\right) = 0, QPNEW_l = 0 \end{bmatrix} \vee \cdots \vee \begin{bmatrix} Y_l \\ \psi_p\left(P_{p,1}^{PNEW}\right) = 0, QPNEW_1 = 0 \\ \psi_p\left(P_{p,2}^{PNEW}\right) = 0, QPNEW_2 = 0 \\ \vdots \\ \psi_p\left(P_{p,l}^{PNEW}\right) \geq 0, QPNEW_l \geq 0 \end{bmatrix}, \quad \forall p \in P$$

The reformulation includes the transformation of the Boolean variables (Y_l) into a set of binary variables (y_l), when the Boolean variables are true, the associated binary variables must be set as one; otherwise, when the Boolean variables are false, the associated binary variables are zero. Therefore, the following relationship is used to select only one location to install the plant:

$$\sum_{l \in L} y_l = 1 \tag{7}$$

$$QPNEW_l \leq QPNEW^{Up} y_l, \quad \forall l \in L \tag{8}$$

$$\psi_p\left(P_{p,l}^{PNEW}\right) \leq \psi_p\left(P^{PNEW}\right)^{Up} y_l, \quad \forall p \in P, l \in L \tag{9}$$

where $QPNEW^{Up}$ and $\psi_p\left(P_{p,l}^{PNEW}\right)^{Up}$ are the upper limits for $QPNEW_l$ and $\psi_p\left(P_{p,l}^{PNEW}\right)$, respectively. Notice that $\psi_p\left(P_{p,l}^{PNEW}\right)^{Up}$ is an upper bound for all properties p. These constraints are required to ensure that $QPNEW_l$ and $\psi_p\left(P_{p,l}^{PNEW}\right)$ take values bigger than zero only when the new plant is located in a position l in particular.

$$\sum_{l \in L} QPNEW_l = Waste \tag{10}$$

$$\sum_{l \in L} \psi_p\left(P_{p,l}^{PNEW}\right) = \psi_p\left(P_p^{disc}\right), \quad \forall l \in L \tag{11}$$

Superstructure for the water network

$$F_k = \sum_{j \in J} f_{k,j}, \quad \forall k \in K \tag{12}$$

$$W_i = \sum_{int^1 \in INT^1} w_i^{int^1}, \quad \forall i \in I \tag{13}$$

$$w_i^{int^1} = \sum_{int^2 \in INT^2} w_i^{int^1, int^2}, \quad \forall i \in I, int^1 \in INT^1 \tag{14}$$

$$w_i^{int^1, int^2, \ldots, int^{N-1}} = \sum_{int^N \in INT^N} w_i^{int^1, int^2, \ldots, int^N}, \quad \forall i \in I, int^1 \in INT^1, \ldots, int^{N-1} \in INT^{N-1} \tag{15}$$

$$w_i^{int^1, \ldots, int^N} = \sum_{j \in J} g_{i,j}^{int^1, \ldots, int^N} + g_{i,Waste}^{int^1, \ldots, int^N}, \quad \forall i \in I, int^1 \in INT^1, \ldots, int^N \in INT^N \tag{16}$$

$$G_j = \sum_{i \in I} \sum_{int^1 \in INT^1} \cdots \sum_{\text{int}^N \in INT^N} g_{i,j}^{int^1,\ldots,\text{int}^N} + \sum_{k \in K} f_{k,j}, \quad \forall j \in J \tag{17}$$

$$\psi_p\left(P_{j,p}^{InSink}\right) G_j = \sum_{i \in I} \sum_{\text{int}^1 \in INT^1} \cdots \sum_{\text{int}^N \in INT^N} \left[\psi_p\left(P_{i,p}^{OutPIN,int^1,\ldots,\text{int}^N}\right) g_{i,j}^{int^1,\ldots,\text{int}^N}\right] + \sum_{k \in K}\left[\psi_p\left(P_{k,p}^{Fresh}\right) f_{k,j}\right], \quad \forall j \in J, p \in P \tag{18}$$

$$Waste = \sum_{i \in I} \sum_{\text{int}^1 \in INT^1} \cdots \sum_{\text{int}^N \in INT^N} g_{i,Waste}^{int^1,\ldots,\text{int}^N} \tag{19}$$

$$\psi_p\left(P_p^{Waste}\right) Waste = \sum_{i \in I} \sum_{\text{int}^1 \in INT^1} \cdots \sum_{\text{int}^N \in INT^N} \left(\psi_p\left(P_{i,p}^{OutPIN,int^1,\ldots,\text{int}^N}\right) g_{i,Waste}^{int^1,\ldots,\text{int}^N}\right), \quad \forall p \in P \tag{20}$$

$$\psi_p\left(P_{j,p}^{InSink,\min}\right) \le \psi_p\left(P_{j,p}^{InSink}\right) \le \psi_p\left(P_{j,p}^{InSink,\max}\right), \quad j \in J, P \in P \tag{21}$$

Environmental constraints

$$\psi_p\left(P_p^{Waste,\min}\right) \le \psi_p\left(P_p^{Waste}\right) \le \psi_p\left(P_p^{Waste,\max}\right), \quad \forall p \in P \tag{22}$$

Constraints for the quality of specific reaches

$$\psi_p\left(P_{p,m(r)}^{DesiredMIN}\right) \le \psi_p\left(P_{p,m(r)}^{Reach}\right) \le \psi_p\left(P_{p,m(r)}^{DesiredMAX}\right), \quad \forall p \in P, m(r) \in M(R) \tag{23}$$

Sustainability constraints

$$\psi_p\left(P_p^{SustainableMIN}\right) \le \psi_p\left(P_{p,final}^{Reach}\right) \le \psi_p\left(P_p^{SustainableMAX}\right), \quad \forall p \in P \tag{24}$$

Objective Function

The objective function consists in the minimization of the total annual cost (*TAC*), which is constituted by the installation costs for the new plant, the fresh sources costs, the treatment cost (including operational cost, variable and fixed capital cost) and the piping costs. The installation cost ($Cland_l$) takes into account the land cost, the transportation costs for raw materials, products and services. Fresh sources cost ($FreC_k$) is caused by the required flowrate of fresh sources to satisfy the inlet flowrate in the process sinks as well as the properties constraints imposed on them. Also, process sources need to be treated prior to be sent to the process sinks, as well as prior to be discharged to the environment, in this sense; a set of technologies must be installed to treat the process sources yielding capital and operational costs for each one of the interceptors required. The operational cost ($VarC_{op}^{\text{int}^N}$) for the interceptors depends on the treated flowrate, while the capital cost contains a variable cost ($VarC_{cap}^{\text{int}^N}$), which also depends on the treated flowrate and a fixed component ($FixC_{cap}^{\text{int}^N}$). Finally, the piping costs for the four sections showed in the superstructure are considered. The first section are the pipes requires when the process sources are sent to treatment ($TrPipC$), once the process sources are treated they can be sent to wastewater stream ($WaPipC$) or process sinks ($EqPipC$), the next part covers the pipes to connect the fresh sources with process sinks ($FrPipC$).

$$
\begin{aligned}
&Min\, TAC = \sum_{l\in L} Cland_l y_l + H_Y \sum_{k\in K} FreC_k F_k \\
&+H_Y \sum_{i\in I}\left[\sum_{int^1\in INT^1} VarC_{op}^{int^1} w_i^{int^1} + \cdots \sum_{int^N\in INT^N} \left(VarC_{op}^{int^N} w_i^{int^1,\ldots,int^N} \right) \right] \\
&+H_Y \sum_{i\in I}\left[\sum_{int^1\in INT^1} VarC_{cap}^{int^1} w_i^{int^1} + \cdots \sum_{int^N\in INT^N} \left(VarC_{cap}^{int^N} w_i^{int^1,\ldots,int^N} \right) \right] \\
&+k_f \sum_{i\in I}\left[\sum_{int^1\in INT^1} FixC_{cap}^{int^1} z_i^{int^1} + \cdots \sum_{int^N\in INT^N} \left(FixC_{cap}^{int^N} z_i^{int^1,\ldots,int^N} \right) \right] \\
&+H_Y \left[\begin{array}{l} TrPipC \sum_{i\in I} \left(\sum_{int\in INT} w_i^{int^1} + \cdots \sum_{int^N\in INT^N} w_i^{int^1,\ldots,int^N} \right) + WaPipC \sum_{i\in I} \sum_{int^1\in INT^1} \cdots \sum_{int^N\in INT^N} g_{i,Waste}^{int^1,\ldots,int^N} \\ +EqPipC \sum_{i\in I} \sum_{int^1\in INT^1} \cdots \sum_{int^N\in INT^N} \sum_{j\in J} g_{i,j}^{int^1,\ldots,int^N} + FrPipC \sum_{k\in K} \sum_{j\in J} f_{k,j} \end{array} \right]
\end{aligned}
\tag{25}
$$

Results and discussion

Case of study

This case considers the Balsas watershed that is one of the biggest systems in Mexico. This system transports several types of discharges until the Pacific Ocean. CONAGUA 20010a, 2010b reported the information for the system sowed on **Figure 6**. In this case, the properties considered were the composition, toxicity, density, COD, odor and color; however, there is a set of properties restricted in the environment and other one restricted in the process sinks. Thus, the properties restricted for the stream discharged to the environment are composition, toxicity, COD, odor and color, while the process sinks limit the composition and density. The system offers twenty possible locations to install the new industrial facility and their installation costs are presented in **Table 6**

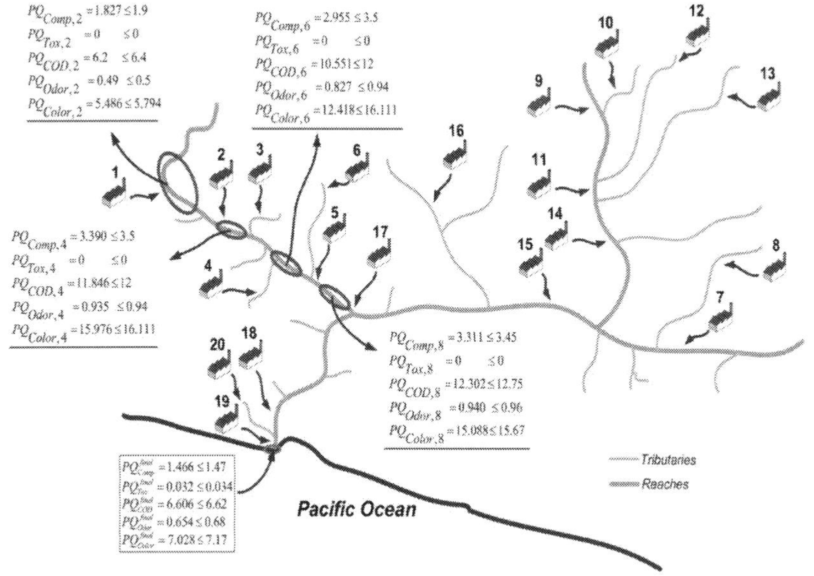

Figure 6. Balsas watershed system (possible locations and the constraints on the river)

Table 6. Installation costs for Example 2

Site	Annualized installation cost, $/year
1	8,000,000
2	8,500,000
3	9,000,000
4	10,500,000
5	9,500,000
6	9,000,000
7	15,500,000
8	14,500,000
9	17,700,000
10	22,000,000
11	18,000,000
12	17,500,000
13	24,000,000
14	28,000,000
15	23,500,000
16	17,000,000
17	20,000,000
18	19,000,000
19	16,000,000
20	17,000,000

Table 7. Fresh and process sources for Example 2

Sources	*Flowrate, m^3/s*	*Composition, ppm*	*Toxicity, %*	*COD, mgO_2/lt*	*Odor*	*Color*	*Density, g/ml*
			Process				
1	0.363	64.6	0.11	149	6.3	250	0.94
2	0.241	58	0.04	88	7.7	234	2.41
3	0.046	45.3	0	108	5.1	298	1.14
4	0.447	60	0.13	185	4.8	202	1.20
5	0.196	52	0.21	120	7.6	194	2.70
6	0.204	62	0.25	76	8.2	164	1.64
7	0.281	54.5	0	243	3.7	179	1.85
8	0.329	47.8	0.13	164	7.4	346	1.75
9	0.282	46	0.09	297	4.8	264	2.45
10	0.266	82.1	0.7	326	8.5	394	0.89
11	0.215	26.4	0.3	146	5.7	195	2.15
			Fresh				
1	-	0	---	---	---	---	2.3
2	-	1	---	---	---	---	2.45
3	-	2	---	---	---	---	1.8

Table 7 shows the characteristics for the fresh and process sources. **Table 8** shows the requirements for the flowrate, maximum composition as well as the minimum and maximum density for each process sink. The cost for fresh source 1 is $0.128/m^3, for fresh source 2 is $0.113/m^3, for fresh source 3 is $0.097/m^3. **Table 9** shows the characteristics for the property interceptors. For this case, the variables costs are included in just one term.

Table 8. Requirements for process sinks for Example 2.

Process Sink	*Flowrate, m^3/s*	*$Comp^{max}$*	*$Density^{min}$*	*$Density^{max}$*
1	0.348	16	2.2	2.4
2	0.428	19.4	1.95	2.1
3	0.406	23.3	2.35	2.55
4	0.312	28.5	1.7	1.9
5	0.336	18	2.2	2.3
6	0.294	14	1.8	2
7	0.196	22.5	2.45	2.55

Table 9. Available interceptors for Example 2.

Interceptor	Efficiency Factor	Variable Cost, $/m^3
	Composition	
Rec^1	0.89	0.026
Rec^2	0.77	0.021
	Toxicity	
Tox^1	1	0.054
Tox^2	0.92	0.049
	COD	
Aer^1	0.82	0.022
Aer^2	0.57	0.018
	Odor	
Ads^1	0.86	0.013
Ads^2	0.61	0.011
	Color	
Dec^1	0.90	0.019
Dec^2	0.80	0.015

On the other hand, the environmental legislation imposes the next constraints for the wastewater stream: $P_{Comp}^{Sustainable}=50$, $P_{Tox}^{Sustainable}=0.1$, $P_{COD}^{Sustainable}=75$, $P_{Odor}^{Sustainable}=2.7$, and $P_{Comp}^{Sustainable}=100$. While the piping cost for each section in the superstructure are: $TrPipC=0.001\,\$/m^3$, $WaPipC=0.002\,\$/m^3$, $EqPipC=0.005\,\$/m^3$, and $FrPipC=0.003\,\$/m^3$.

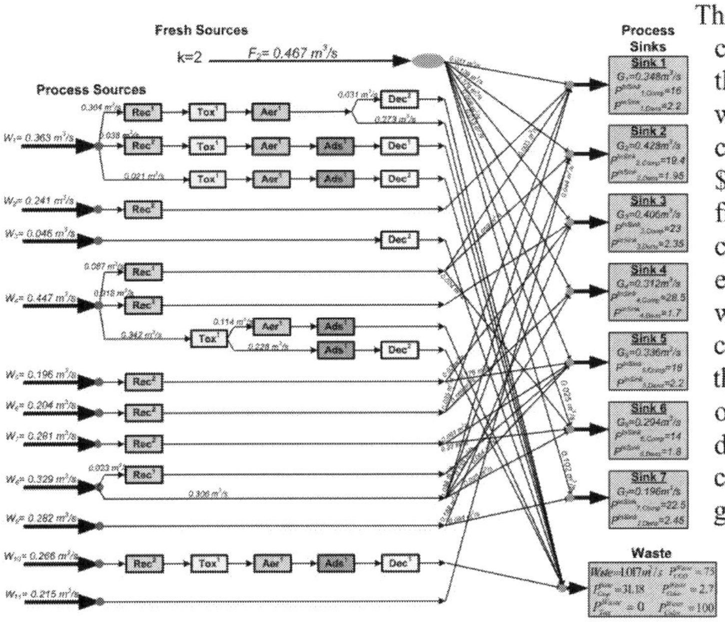

Figure 7. Mass integration for Example 2

Then, the proposed model was applied for this case of study, giving an optimal solution where the new facility must be placed in alternative 6 with a total annual cost of $14.9x10^6/yr, composed by $9x10^6/yr of installation cost, $3.6x10^6/yr of treatment cost, $1.5203x10^6/yr of fresh source cost and $779,040/yr of piping cost. **Figure 7** shows the water network this example, notice that the value of toxicity in the wastewater stream is zero, satisfying the constraint in the reach 8, where is not allowed the existence of this property. Notice that most of the properties in the wastewater streams are discharged in the limit of the environmental constraints; however, the composition requires a greater reduction to satisfy the constraints imposed over the reach 8 ensuring the quality of the water extracted for human activities.

Conclusions

This paper proposes a new water integration approach to consider simultaneously the integration inside the plant and the sustainability of the surrounded watershed. The model is based on the properties of the streams, which is suited for the wastewater streams that are constituted by several pollutants. The proposed approach considers the integration of the wastewater discharged by the industrial facility to the other discharges and uses through the optimal selection of the location of the industrial facility. In addition, the model considers the natural degradation of the pollutants through the watershed and the effect to the specific properties that are constrained in the system. The proposed model is applied to one case from Mexico. The results show that the traditional water integration schemes that do not consider the surrounding watershed yield inappropriate solutions, because the environmental regulations are not enough to satisfy the sustainability of the system.

References

CONAGUA (2009a). Mexican National Water Commission. Water Statistics 2008. http://www.conagua.gob.mx/OCB07/Temas/EstadisticasBALSAS-FINALago09.pdf

CONAGUA (2009b). Mexican National Water Commission. Water Atlas in Mexico 2009.

El-Halwagi, M. M., and Spriggs, H. D. (1998). Solve design puzzles with mass integration. *Chemical Engineering Progress, 94* (8), 25-44.

Wang, Y. P., and Smith, R. (1994). Wastewater minimization. *Chemical Engineering Science, 49* (7), 981-1006.

(521b) Comparing Continuous and Batch Process Design Under Uncertainty for Biodiesel Production

Pahola T. Benavides[1], Urmila Diwekar[2] and Juan M. Salazar[2], (1)VRI- Custom, Chicago, IL, (2) Vishwamitra Research Institute, Center for Uncertain Systems: Tools for Optimization and Management, Clarendon Hills, IL

Abstract:

The continuing depletion of fossil fuel reserves and the increasing environmental concerns has encouraged engineers and scientists to look for an alternative, clean and renewable fuel that can reduce the negative environmental impact. Biodiesel has been considered as one of the best candidate of one of these renewable fuels. One of the pathways to biodiesel production is the transesterification reaction of triglycerides from vegetable oils and short-chain alcohols. A batch reactor is employed for the production of biodiesel. However, most of the other unit operations like decanters and distillation columns are operated continuously. The success of biofuel introduction in economy depends on the economics of these fuels. One of the factors which are problematic in this is the seasonal variation of feedstock, and transportation and storage costs of material for continuous processing. In this work, we compare batch and continuous processing options for biodiesel production under feedstock variation. These models will allow us to choose an optimal process configuration, depending on the size of the market, distance to market, and feedstock availability,

Key words: *batch process, continuous process, biodiesel, stochastic simulation, economical analysis.*

See more of this Session: Design of Sustainable Processes
See more of this Group/Topical: Computing and Systems Technology Division

(521c) Simultaneous Optimization of Flowsheet, Heat Recovery, and Water Network

Linlin Yang and Ignacio Grossmann, Chemical Engineering, Carnegie Mellon University, Pittsburgh, PA

Significant research work has been done in the area of heat integration in the past thirty years due to the incentive of improving energy recovery. Also, stricter environmental regulations have caused an increasing demand for minimization of water consumption and effluent generation for industrial processes. Various techniques have been developed in response to this need for process integration and they are presented in these reviews[1][3][4]. However, most of this research has been focused on minimization of energy and water separately. Furthermore, traditionally, process synthesis design is performed sequentially - the flowsheet is optimized first, then the resulting operating parameters are used in heat exchanger network synthesis (HENS) and water network (WN).

This presentation describes a novel and effective approach for the simultaneous optimization of process flowsheet subject to heat and water integration. Through simultaneous optimization approach, the idea is to represent the HENS and the WN through design targets such as energy consumption and water consumption to avoid solving a very large MINLP problem. The advantage of performing simultaneous optimization is that proper trade-offs of operation cost and investment cost can be taken into account.

While effective NLP targeting formulation for simultaneous process optimization with heat integration has already been presented by Duran & Grossmann[2], this is not the case for targeting minimum water use. Typical WN problems are formulated using NLP or MINLP models using a detailed superstructure as reported in many references[5]. Freshwater is first used for process units, and then wastewater generated from these processes is treated in various treatment units. Process units are defined by their allowable contaminant concentration level, whereas wastewater treatment units are defined based on their contaminant removal ratio. The superstructures consider systematic alternatives for water recycle and water reuse with the objective of minimizing the freshwater consumption, or more generally the total cost. In general, single-contaminant systems can be modeled using LP formulations, while multi-contaminant systems take the form of NLP models.

In order to incorporate water integration in the simultaneous optimization framework, a new LP formulation is presented to determine the target for a multi-contaminant water network with process units only. In particular, the bilinear equations of the mixer are replaced by a set of inequalities that are evaluated at the limits of inlets and outlet concentrations of the units. This relaxation, which follows the optimality conditions proved by Bagajewicz & Savelski[6], yields an LP that can be shown to predict the exact minimum amount of freshwater that is required by the process units. The LP targeting model is incorporated in the simultaneous optimization of a methanol synthesis process flowsheet with heat and water integration and resulted in an improvement of the economical objective function. The important observation here is that in the simultaneous approach, heating and cooling are integrated as part of the optimization so that improved operating conditions can be achieved. It is shown for a specific instance that the resulting flowsheet from the simultaneous method improves the profit by 17%.

For the case of multi-contaminant water network problems including wastewater treatment units with process units, the proposed LP targeting formulation is shown to predict a rigorous minimum freshwater requirement for special cases of the WN problem. For the general case, the lack of exact upper bound of contaminant concentration in the treatment units complicates the interactions in the mixer units. To circumvent this problem, a number of linear relaxations are used to obtain approximate but tight water targets for general WN problems. The proposed targeting model yields an MILP problem. The application of this model is also applied in the simultaneous optimization, heat and water integration of the methanol process yielding further improvements in the profit.

References:

[1] Bagajewicz, M., (2000). A review of recent design procedures for water networks in re0neries and process plants. *Computers & Chemical Engineering*, 24 (9), 2093-2115.

[2] Duran, M.A. & Grossmann, I.E. (1986). Simultaneous optimization and heat integration of chemical processes. *AIChE Journal*, 32, 123-138.

[3] Foo, D.C.Y. (2009). State-of-the-art review of pinch analysis techniques for water network synthesis. *Ind. Eng. Chem. Res.* 48, 5125–5159.

[4] Furman, K.C. & Sahinidis, N.V. (2002). A critical review and annotated bibliography for heat exchanger network synthesis in the 20th Century. *Ind. Eng. Chem. Res.*, 41, 2335-2370.

[5] Karuppiah, R. & Grossmann, I. E. (2008). Global optimization of multi-scenario mixed integer nonlinear programming models arising in the synthesis of integrated water networks under uncertainty. *Computers & Chemical Engineering*, 32(1-2), 145-160.

[6] Savelski, M. & Bagajewicz, M. (2003). On the necessary conditions of optimality of water utilization systems in process plants with multiple contaminants. *Chemical Engineering Science*, 58(23-24), 5349-5362.

See more of this Session: Design of Sustainable Processes
See more of this Group/Topical: Computing and Systems Technology Division

(521d) Incorporating Safety and Efficiency Considerations In Sustainability Assessment of Poly-Generation Systems

Preeti Gangadharan[1], Kailiang Zheng[2], Abhishek V. Jayswal[3], Anand Zanwar[1] and Helen Lou[1], (1) Dan F. Smith Department of Chemical Engineering, Lamar University, Beaumont, TX, (2)Department of Chemical Engineering, Lamar University, Beaumont, TX, (3)Dan F. smith Department of Chemical Engineering, Lamar University, Beaumont, TX

Sustainability assessment determines the most sustainable pathway, taking into consideration the economic, environmental, efficiency and societal aspects. Embedding exergy analysis and novel safety evaluation approach to complement the widely-accepted economic assessment and environmental impacts assessment in a decision tree evaluation framework provides a more comprehensive, yet faster methodology to compare related processes in terms of sustainability. Considering their promise to provide high material and energy conversion from natural resources, poly-generation systems are recognized as promising technologies for future chemical and power industries.

The commonly used inherent safety approach considers chemical and process safety, but does not give due emphasis to complexity of the process flow sheet. This becomes important in the case of poly-generation systems that accept multiple feeds and produce multiple products through many pathways for chemical and energy conversion. Since the simultaneous production of chemicals and electricity offers both economical viability and thermodynamic efficiency, exergy efficiency analysis is employed for thermodynamic comparison of proposed processes.

In this paper three different poly-generation systems, which use coal and natural gas as feed to produce DME and power are compared using a comprehensive sustainability assessment methodology, which takes into account the economic, environmental, societal and efficiency factors.

See more of this Session: Design of Sustainable Processes
See more of this Group/Topical: Computing and Systems Technology Division

(521e) Technology Evaluation and Decision Making for Sustainability Enhancement of Industrial Systems Under Uncertainty

Zheng Liu, Department of Chemical Engineering and Materials Science, Wayne State University, Detroit, MI and Yinlun Huang, Dept. of Chemical Engineering & Material Science, Wayne State University, Detroit, MI

Depletion of natural resources, environmental pressure, economic globalization, etc., demand seriously industrial organizations to ensure that their development is sustainable. Technology based sustainability enhancement is one of key approaches for industrial sustainability realization. However, identification of effective technologies for a given industrial system could be a combinatorial problem; this is especially true when multiple technologies are considered for adoption together. On the other hand, if the available data and information about the industrial system and the known technologies are incomplete, imprecise, and uncertain, the technology identification could be even more difficult.

In this paper, we introduce a simple approach for systematic sustainability assessment of industrial systems and technologies, and effective system sustainability enhancement under uncertainty. An industrial system can be an industrial zone, corporate, company, or plant of any complexity. We assume that the industrial organization has set a strategic plan that includes detailed development goals of economic, environmental, and social sustainability with a budget set aside for goal achievement. It is known that a number of technologies are seemingly promising for sustainability enhancement. A challenge is how to derive efficiently the most suitable solutions for identification of superior sustainability technologies.

The technology-assisted sustainability enhancement methodology to be introduced will consist of a sustainability assessment method for evaluating the performance of industrial systems as well as technologies, and a sustainability decision making and analysis method for identifying superior technology combinations for achieving the sustainability goals of industrial systems. The interval parameter based information processing and decision making permit the acceptance of uncertain information that could be available at the best. The logically designed solution identification procedure can make the combinatorial problem solved efficiently through reducing the solution space stage-wisely using different criteria set by the strategic plan. The identified solutions are also sufficiently detailed which can greatly facilitate the industrial organization in final decision making. The methodology is general that can be applied to the sustainability enhancement problems of any size and scope, and the efficacy of the approach is illustrated through a sustainability development problem about regional biodiesel manufacturing improvement.

See more of this Session: Design of Sustainable Processes

(521f) Integration of Production Processes, Waste Management and Energy Utility Generation In Chemical Batch Plants

Stavros Papadokonstantakis, Claude Rerat and Konrad Hungerbuhler, Chemistry and Applied Biosciences, Institute for Chemical- and Bioengineering, Swiss Federal Institute of Technology, Zurich (ETHZ), Zurich, Switzerland

The production of chemicals on industrial scale heavily relies on the availability of utilities such as energy in different forms e.g., steam, electricity, cooling water, brine and others. These utilities represent a significant share of the production cost of chemicals and define to a major degree the environmental impact resulting from chemical production. The efficiency of utility usage in chemical industry is primarily defined by two major factors, the efficiency of the utility generation (e.g. steam production) and the efficiency in using the generated utilities in achieving a productivity of chemicals as high as possible.This description evokes the idea of two separated problems and unfortunately nowadays the two fields are often treated like this and consequently substantial integration potentials are overlooked resulting in increased cost and environmental impact. However, the generation and the usage of utilities in chemical production are highly interlinked and integration potentials must be identified during early design stages of both utility generation systems and chemical production processes. Major aspects in this context are waste treatment in general and solvent management in particular. Increased process integration, e.g., recycling of material flows, reduces the amount of waste to be treated and therefore the additional utility input required by many waste treatment technologies (e.g., stripping, sewage treatment). However, waste can also be a source of utility generation, e.g., by generating steam through waste incineration. This is particularly an issue for the huge amount of waste solvents that are generated in pharmaceutical and fine chemical production typically operating in batch mode.

Chemical batch plants are characterized by frequent changes in production portfolio which implies dynamic patterns in energy demand and waste generation. Since waste treatment plays an important role in the energy system of a chemical plant, these dynamic patterns can increase energy demand and simultaneously offer opportunities for primary energy savings. In constrast, utility generation systems cannot be changed frequently because high investments are associated and high efficiency can only be achieved for a well-defined utility mix (e.g. the proportion of electricity to steam in Combined Heat and Power devices).

In order to study the potential of an integrated design in this three-component system (i.e., production processes, waste management and energy utility generation) suitable models are needed. In the present study the energy demand of batch chemical production processes was based on bottom-up unit operation models adjusted to real process measurements using calibrated control valve openings. The difference between theoretical and real utility consumption was used to develop empirical parametric equations describing thermal losses due to inefficient heating/cooling system operations. Validation of these bottom-up models was performed at different aggregation levels, including equipment specific unit operations, production lines and overall batch plant energy consumption. The results indicated that for all types of energy utility, equipment and unit operation the consumption could be estimated with a relative error between 5% and 35% depending on the aggregation level.

Typical batch plant waste treatment facilities were considered like incineration, wastewater treatment and wet air oxidation, while solvent recovery was treated with continuous and batch distillation units. This waste management system can make full use of the dual properties of waste solvents which can be an energy source as combustible for incineration or energy sink as material for recycling in distillation units. In order to include life cycle analysis aspects in the investigation, multi-input allocation models were used for the waste treatment units to include energy requirements and pollutant emissions of waste handling. These models assess emissions and utility consumption with few waste stream data using linear transfer coefficients derived by extensive industrial studies of the respective operations.

The energy utility generation system includes steam boilers and the associated steam network, gas turbines, a refrigeration cycle producing brine and of course the waste incineration plant. The heat load provided by utilities corresponds to the heat demand of the process taking into account an average efficiency for heat exchangers. Concerning the process streams temperatures, it is possible to assume that they are ΔT_{min} lower (or higher, if the utility is a cold stream) than the utility lower (higher) temperature. The size of the corresponding heat exchanger network and the amount of necessary primary energy sources can be optimized using Pinch analysis.

For the optimization of the integrated system in a modular form, three nested optimization problems were solved: the production planning of the batch processes (MILP), the mixing problem of the generated waste streams (MINLP), and the heat integration problem for energy utilities geneartion (LP). The operational costs and diverse life-cycle impact oriented environmental indicators were used as objective functions. A synthetic case study plant consisting of a mono-product and a multi-product building was used for investigating the efficiency of the integrated design. The case study plant reflects the complexity of a real batch plant that provided the data for the development of the bottom-up energy consumption models, the desired production capacities, the waste management constraints and the utility generation system. In this way, it was possible to define a base case corresponding to the current plant operation, to which the optimization results were compared. It is shown that an optimal waste mixing together with a suitable production planning (i.e., more homogeneously distributed production over the time horizon) can result in a significant reduction of ancillaries in waste treatment operations and primary sources for energy utility generation (approx. up to 50% less energy consumption related operational costs and 20% less environmental impacts expressed in Eco-Indicator-99 points).

See more of this Session: Design of Sustainable Processes
See more of this Group/Topical: Computing and Systems Technology Division

(521g) Economic Analysis of Municipal Wastewater Utilization for Thermoelectric Power Production

Iman Safari[1], Michael E. Walker[1], Javad Abbasian[1], Hamid Arastoopour[2], Ming-Kai Hsieh[3], Ranjani B. Theregowda[3], David A. Dzombak[3] and David C. Miller[4], (1)Chemical and Biological Engineering, Illinois Institute of Technology, Chicago, IL, (2)Department of Chemical and Biological Engineering, Wanger Institute for Sustainable Energy Research (WISER), Illinois Institute of Technology, Chicago, IL, (3)Department of Civil and Environmental Engineering, Carnegie Mellon University, Pittsburgh, PA, (4)U.S. Department Of Energy, National Energy Technology Laboratory, Morgantown, WV

The thermoelectric power industry in the U.S. uses a large amount of freshwater. The large water demand is increasingly a problem, especially for new power plant development, as availability of freshwater for new uses diminishes in the United States. Reusing non-traditional water sources, such as treated municipal wastewater, provides one option to mitigate freshwater usage in the thermoelectric power industry. The amount of freshwater withdrawal that can be displaced with non-traditional water sources at a particular location requires evaluation of the water management and treatment requirements, considering the quality and abundance of the non-traditional water sources.

This paper presents the development of an integrated costing model to assess the impact of degraded water treatment, as well as the implications of increased tube scaling in the main condenser. The model developed herein is used to perform case studies of various treatment, condenser cleaning and condenser configurations to provide insight into the ramifications of degraded water use in the cooling loops of thermoelectric power plants. Further, this paper lays the groundwork for the integration of relationships between degraded water quality, scaling characteristics and volatile emission within a recirculating cooling loop model.

See more of this Session: Design of Sustainable Processes
See more of this Group/Topical: Computing and Systems Technology Division

(537a) Modeling of Dynamic Hindered Diffusion of Drugs From Biodegradable PLGA Microspheres with Evolving Porous Structure

Ashlee N. Ford Versypt, Department of Chemical and Biomolecular Engineering, University of Illinois at Urbana-Champaign, Urbana, IL, Daniel W. Pack, Chemical and Biomolecular Engineering, University of Illinois Urbana Champaign, Urbana, IL and Richard D. Braatz, Department of Chemical Engineering, Massachusetts Institute of Technology, Cambridge, MA

Controlled-release drug delivery systems provide alternatives to conventional medical drug therapy regimens which require frequent dosages due to short pharmaceutical in vivo half-life and poor oral bioavailability. Controlled-release systems have the potential to enhance control of drug concentrations, reduce side effects, and improve compliance as compared to conventional regimens.

Poly(D,L-lactic-co-glycolic acid) (PLGA) microspheres are the controlled-release drug delivery system modeled in this work. Biodegradable PLGA microspheres have been extensively studied as drug delivery devices [1-3]. Modeling of the drug release from the devices must consider the interdependent phenomena that contribute to drug release. The microspheres encapsulate drug molecules dispersed throughout the polymer. The polymer undergoes autocatalytic hydrolysis, breaking the polymer bonds and generating smaller polymer chains with acidic end groups that catalyze further hydrolysis of the degradation products. Autocatalytic hydrolysis is more significant in the interior of large microspheres where the diffusion of degradation products is more limited [4-6]. Sufficiently small oligomers produced by the degradation are water-soluble and can diffuse out of the polymer microspheres through water-filled pores. The resulting polymer mass loss increases the pore volume in the microspheres. Encapsulated drug molecules diffuse through the aqueous pores in the polymer by hindered diffusion [7]. As the pore network in the microspheres evolves, the effective diffusivity of the drug increases.

A reaction-diffusion model has been developed to investigate the degradation heterogeneities observed in PLGA microparticles of different sizes due to autocatalytic polymer hydrolysis and polymer erosion. The model considers the coupling between the hydrolysis reactions, the diffusion of polymer oligomers, and the dynamic effective diffusivity of the drug. The unique contribution of the modeling work is that it combines autocatalytic PLGA degradation mechanisms [8-9] with hindered diffusion in aqueous pores [10] having variable pore sizes. Simulations using the model capture microsphere-size-dependent release phenomena observed experimentally and show the dependence of drug release profiles on polymer morphology.

References

[1] J. M. Anderson and M. S. Shive. Biodegradation and Biocompatibility of PLA and PLGA

Microspheres. *Advanced Drug Delivery Reviews*, 28(1):5-24, 1997.

[2] N. K. Varde and D. W. Pack. Microspheres for Controlled Release Drug Delivery. *Expert Opinions on Biological Therapy*, 4(1):35-51, 2004.

[3] S. Freiberg and X. X. Zhu. Polymer Microspheres for Controlled Drug Release. *International Journal of Pharmaceutics*, 282(1-2):1-18, 2004.

[4] J. Siepmann, K. Elkharraz, F. Siepmann, and D. Klose. How Autocatalysis Accelerates Drug Release from PLGA-Based Microparticles: A Quantitative Treatment. *Biomacromolecules*, 6(4):2312-2319, 2005.

[5] C. Berkland, E. Pollauf, C. Raman, R. Silverman, K. Kim, and D. W. Pack. Macromolecule Release from Monodisperse PLG Microspheres: Control of Release Rates and Investigation of Release Mechanism. *Journal of Pharmaceutical Sciences*, 96(5):1176-1191, 2007.

[6] D. Klose, F. Siepmann, K. Elkharraz, and J. Siepmann. PLGA-Based Drug Delivery Systems: Importance of the Type of Drug and Device Geometry. *International Journal of Pharmaceutics*, 354(1-2): 95-103, 2008.

[7] A. Zhao, S. K. Hunter, and V. G. J. Rodgers. Theoretical Prediction of Induction Period from Transient Pore Evolvement in Polyester-Based Microparticles. *Journal of Pharmaceutical Sciences*, 99 (11):4477-4487, 2010.

[8] H. Antheunis, J. C. van der Meer, M. de Geus, W. Kingma, and C. E. Koning. Improved Mathematical Model for the Hydrolytic Degradation of Aliphatic Polyesters. *Macromolecules*, 42 (7):2462-2471, 2009.

[9] S. Lyu, R. Sparer, and D. Untereker. Analytical Solutions to Mathematical Models of the Surface and Bulk Erosion of Solid Polymers. *Journal of Polymer Science Part B: Polymer Physics*, 43(4):383-397, 2005.

[10] P. Dechadilok and W. M. Deen. Hindrance Factors for Diffusion and Convection in Pores. *Industrial and Engineering Chemistry Research*, 45(21):6953-6959, 2006.

See more of this Session: Mathematical Modeling In Transport Processes
See more of this Group/Topical: Engineering Sciences and Fundamentals

(537b) Applications of Fractional Calculus to Anomalous Mass and Heat Transfer Phenomena In Complex Media

Ruben D. Vargas and Watson L. Vargas, Chemical Engineering, Universidad de los Andes, Bogota, D. C, Colombia

Many problems of mass and heat transfer in complex media exhibit what is called anomalous transport phenomena. Anomalous phenomena are all those process that don't obey the classical models for transport (e.g., Fick and Fourier's laws for mass and heat, respectively), describing this kind of processes with anomalous diffusion is a problem that has led investigators to formulate corrections of the classical laws. In this challenge of making successful corrections authors have commented on the necessity of including memory as an important factor to be considered when working in complex media [1]. Two of such approaches include the introduction of Levy- like processes and the Continuous-Time Random Walks (CTRW) theory, both of which arrive to an Integro-differential equation called space-time fractional diffusion equation (STFDE) (which is applicable for both mass and heat transfer).[2]

In this work analytic and numerical methods are shown for solving specific boundary value problems on the time fractional diffusion equation (TFDE) compared with experimental data for mass and heat transport in complex media – solute diffusion in a gel matrix and heat conduction in granular media, respectively. The analytical treatment uses a combined transform method were both Laplace and Fourier Transforms are applied. Evaluation of the analytical solution is made by numerical Laplace inversion and the deformation of contours method developed by Luchko.[3] The experimental parameters for the TFDE are found by procedures similar to those implemented by Kosztołowicz *et al.* [4] using a penetration distance function.

For the numerical solution of the TFDE a finite differences method has been developed that evaluates fractional derivates of the Caputo kind, computational implementations for both mass and heat are developed and are applied to different boundary conditions.

Finally this work makes a comparison of classical and fractional models for transport phenomena in complex media, observing a significant advantage of using the fractional approach, especially when predicting outside of the time interval used for determination of the parameters (Diffusivity for classical models and fractional derivative and Diffusivity for the TFDE).

References:

1. W. L. Vargas, J. J. McCarthy, Stress effects on the conductivity of particulate beds. Chemical

Engineering Science E 57, (2002).
2. D. Hernandez, C. Varea, and R. A. Barrio. Dynamics of reaction-diffusion systems in a subdiffusive regime. Physical Review E 79, 026109 (2009).
3. Y. Luchko. Algorithms for evaluation of the Wright function for the real arguments' values. Fractional Calculus & Applied Analysis V 11, N 1 (2008).
4. T. Kosztołowicz, K. Dworecki, and St. Mrówczyñski, Measuring subdiffusion parameters. Physical Review E **71**, 041105 (2005).

See more of this Session: Mathematical Modeling In Transport Processes
See more of this Group/Topical: Engineering Sciences and Fundamentals

(537c) Stress Distribution In Flow Through Porous Media

Naga Rajesh Tummala, School of Chemical, Biological and Materials Engineering, The University of Oklahoma, Norman, OK, Roman S. Voronov, School of Chemical, Biological, and Materials Engineering, University of Oklahoma, Norman, OK and Dimitrios V. Papavassiliou, School of Chemical Biological and Materials Engineering, The University of Oklahoma, Norman, OK

Velocity and stress fields within both structured and random porous media were computed using the Lattice Boltzmann method (LBM).[1-2] The structured porous media consisted of spheres packed in simple lattice, body centered cubic lattice, and face centered cubic lattice, respectively. Random porous media consisted of random close packed spheres with 3 different porosities and micro-computer tomography of the Boise sandstone core.[3-4] The permeability values computed from LBM were in agreement with the values found in the literature.[5] The ability to predict stress distribution within a porous medium *a priori* will not only help in the better design of catalyst particles and packing material in reactors, but also to effectively design nano-particles that can propagate through the open pore space in enhanced oil recovery applications. The normalized flow-induced stresses observed within various porous media follow a common probability density function, except for the length of the distribution tail. However, the parameters of the functions used to fit the normalized stress distributions are related to the porosity and the crystallinity of the porous media. We quantify such behavior by comparing the distributions and corresponding parameters to the bond-order parameter (to quantify crystallinity) and porosity. Results of such comparisons are presented and compared with the stress distribution in high porosity bio-scaffolds.[6]

References

[1] Voronov, R.; VanGordon, S.; Sikavitsas, V. I.; Papavassiliou, D. V., J. Biomech. 43 (2010) 1279.

[2] Voronov, R. S.; VanGordon, S. B.; Sikavitsas, V. I.; Papavassiliou, D. V., International Journal for Numerical Methods in Fluids (2010) 10.1002/fld.2369.

[3] Martys, N. S.; Torquato, S.; Bentz, D. P., Phys. Rev. E 50 (1994) 403.

[4] Torquato, S.; Truskett, T. M.; Debenedetti, P. G., Phys. Rev. Lett. 84 (2000) 2064.

[5] Chapman, A. M.; Higdon, J. J. L., Physics of Fluids A: Fluid Dynamics 4 (1992) 2099.

[6] Voronov, R. S.; VanGordon, S. B.; Sikavitsas, V. I.; Papavassiliou, D. V., Appl. Phys. Lett. 97

(2010) 024101.

See more of this Session: Mathematical Modeling In Transport Processes
See more of this Group/Topical: Engineering Sciences and Fundamentals

(537d) Modeling Cellular Metabolism and U(VI) Immobilization In Shewanella Oneidensis MR-1 Biofilms

Bulbul Ahmed, Ryan S. Renslow and Haluk Beyenal, The Gene and Linda Voiland School of Chemical Engineering and Bioengineering, Washington State University, Pullman, WA

Uranium (U) contamination of the environment due to nuclear processing activities causes significant economic and health problems due to its presence in terrestrial and aquatic ecosystems. Understanding the mobility of U at the microscale is critical for predicting its macroscale fate and for optimizing remediation strategies. Mathematical modeling can provide a non-intrusive and low-cost alternative to difficult and expensive laboratory and field experiments. In this study, we developed a 2-dimentionsal mathematical model to monitor U mobility inside *Shewanella oneidensis* MR-1 biofilms. Biofilms are comprised of a matrix of cells and abundant extracellular polymeric substances (EPS), both loosely-associated and bound EPS. Specifically, our model focused on (1) the contribution of cells and EPS to immobilize U, (2) the effect of U on cellular metabolism and EPS production, and (3) the relative role of adsorption and reduction on overall U immobilization. The model was developed and solved using COMSOL Multiphysics software. Our model demonstrated that U preferentially binds to specific locations within the biofilm, which U detrimentally affects cellular metabolism, and that loosely associated and bound EPS interact differently with U.

See more of this Session: Mathematical Modeling In Transport Processes
See more of this Group/Topical: Engineering Sciences and Fundamentals

Retro-Convection Enhanced Drug Delivery: Computational Analysis and Pharmacodynamic Modeling

Daniel Lepek Ph.D. and Michael Cerro
The Cooper Union for the Advancement of Science and Art

Introduction

Purpose

In the year 2004, approximately 124,000 persons in the United States were living with a diagnosis of a primary malignant brain and central nervous system (CNS) tumor. Despite the advancements in cancer research, 12,920 deaths were attributed to this form of cancer in the United States in the year 2009 [1]. CNS malignancies represent the fastest growing area of untreatable diseases in medicine. Finding protocols which deliver therapeutic agents to the CNS in an effective and safe manner is in high demand. Difficulties in finding administration routes to the CNS are primarily due to the physiological properties of the blood-brain barrier (BBB).

Blood-brain barrier

The blood-brain barrier (BBB) aids in separating blood and cerebrospinal fluid (CSF) in the central nervous system (CNS). The tight junctions between endothelial cells which guard the capillary walls prohibit the diffusion of large hydrophilic molecules into the CSF, while allowing small hydrophobic molecules to pass. Specifically, the BBB only allows highly lipid-soluble molecules under 400-600 Daltons to penetrate [2]. Only two percent of low-molecular-weight drugs cross the BBB, while 100% of large-molecular-weight molecules are stopped [3].

Substances are able to cross the BBB by different pathways; for example, transport proteins and receptor-mediated transcytosis. However, the biopharmaceuticals necessary to combat malignant tumors in the CNS are too large in size to make use of these pathways. Because of their inability to cross the BBB, circumventing this barrier entirely remains the only option. Morrison and coworkers utilized a high-flow microinfusion probe which introduced therapeutic agents directly into the interstitial area [4]. The delivery of drugs to malignant tissue is highly dependent upon convective fluxes from the source, in addition to losses due to metabolism and reactions.

Convection enhanced drug delivery

Convection enhanced drug delivery (CED) is a method by which a drug is administered by direct intracerebral infusion using a microdialysis probe. The concept of microdialysis began in the 1960s, when probes were inserted into animal tissue to study transmitter release in rodents. Microdialysis works by inserting a probe with a membrane attached at the tip to the desired area. The inflow or

[1]CBTRUS [2004-2006]
[2]Pardridge [2005]
[3]Pardridge [2002]
[4]F.Morrison, Laske, Bobo, Oldfield, and Dedrick [1994]

perfusate is introduced to the probe by a source while the outflow or dialysate is removed.

The direct contact with the brain and CNS bypasses the problems of the blood brain barrier (BBB) when trying to administer drugs systemically. The benefits of CED are that: (1) the high infusion rate causes convective fluxes to dominate diffusion at the injection site; compared to diffusion, convection has been shown to increase the distribution of the drug [5]; and (2) the BBB remains unaffected from the procedure, reducing the side-effects that may occur with other methods. However, there are some drawbacks with CED. Reflux, or "backflow", can occur at the injection site. Due to the lack of adhesion between the probe and brain tissue, the brain tissue can move during infusion. Reflux is the result of this motion because the perfusate is not allowed to move freely. This phenomenon has been widely studied and is a main impediment to a wider clinical use of CED. Many complications in CED are caused by reflux; particularly, the poor control over the drugs being infused. The therapeutic drugs are prepared to fight malignant tissue, however, the reflux may cause them to target healthy cells [6]. A recent alternative drug delivery method, retro-convection enhanced drug delivery (R-CED), helps to avoid the reflux problem.

Retro-convection enhanced drug delivery

R-CED reverses the direction of fluid velocity by sucking interstitial fluid (ISF) towards the microdialysis probe. Because of this, reflux does not occur. R-CED helps administer drugs which otherwise may be blocked by the BBB by reducing the tumor pressure to increase the pressure gradient across the endothelial cells. This allows the blood to drive fluid from the capillary wall into the tissue more easily [7]. DiResta and coworkers began their attempts at R-CED by using a needle connected to a vacuum to remove ISF, in effect lowering the tumor pressure. Later, Huynh *et al.* replaced the needle with an implanted microdialysis probe which utilizes a polymeric osmolyte [8].

While there is much literature on the macromolecular transport in brain tissue using the CED method, there is currently very little known about the effects of R-CED. The most extensive study on the mathematical modeling of R-CED was done by Wang and Olbricht in early 2010 [9]. Wang and Olbricht performed a thorough sensitivity analysis on the various parameters associated with macromolecular transport, including differentiations of the membrane used with the microdialysis probe. It was concluded that an impermeable membrance is best for drug delivery, and that R-CED does not provide a competitive therapeutic method compared to CED. However, there were many uncertainties in the values of their parameters. Most values were retrieved from CED protocols, which can alter the numbers due to the backflow phenomenon. Specifically, they mention that hydraulic conductivity and the reflection coefficient (both of which are presented in the mathematical model section) may vary due to bulk flow.

The strengths of each case was shown in a recent study by Motion *et al.* [10]. They used a dual-catheter setup to infuse the drugs past the blood-brain barrier by utilizing a CED catheter, in

[5] Wang and Olbricht [2010]
[6] Krauze [2005]
[7] DiResta, Lee, Healey, Larson, and Arbit [2004-2006]
[8] Huynh, Ozawa, Deen, Tihan, and Jr. [2007]
[9] Wang and Olbricht [2010]
[10] Motion, Huynh, Jr., and Siegel [2010]

addition to a R-CED catheter in order to pinpoint where they wanted to localize the drugs. Similarly, they also used COMSOL Multiphysics as a tool to predict the mass transfer processes of the drug molecules, and then utilized a wet lab setup to varify their findings. In addition, they conclude that using a spherical geometry is a safe approximation in modeling both cases.

This study elaborates on their attempts and will aim to determine which geometry is more accurate when modeling the R-CED protocol, cylindrical or spherical coordinates. Furthermore, analyzing drug molecules during the time after the procedure has ended, or the post-extraction period, will yield a complete comprehension of drug behavior. In addition to the above, a pharmacodynamics analysis, specifically, an area under the curve metric, was performed to better gauge where the drugs are being localized in the interstitial region. And finally, this study will conclude as to whether R-CED is a viable alternative to the more understood CED procedure.

Mathematical Model

Pressure and velocity profiles

To develop the mathematical model for the pressure and velocity profiles, the brain tissue is assumed to be a homogenous rigid porous medium. Because of this assumption, Darcy's law, which is a constitutive equation describing fluid flow through porous media, to compare the interstitial velocity and pressure gradient, was used. Darcy's law is commonly expressed as

$$\varepsilon v = -\kappa \nabla p_i \tag{1}$$

where ε is the brain tissue porosity, κ is the local hydraulic conducivity, and p_i is the interstitial pressure.

We can develop an expression for the pressure profile that develops in brain tissue (equation 4) by combining Darcy's law with the equation for the conservation of water (or Starling's equation)

$$\nabla \cdot (\varepsilon v) = L_p s(p_e - p_i) \tag{2}$$

where the net water transport into the interstitial volume element equals the water losses along the capillary walls in that element; L_p is the capillary hydraulic conductivity, s is the capillary area per tissue volume, and p_e is the effective Starling pressure

$$p_e = p_v + \sigma(\pi_i - \pi_v) \tag{3}$$

where p_v is the venous vascular pressure, σ is the refelction coefficient, and π_i and π_v are the interstitial and vascular osmotic pressures. For this study, it is assumed that the effective Starling pressure is essentially equal to the cerebrospinal fluid pressure.

$$\nabla \cdot (-\kappa \nabla p_i) = L_p s(p_e - p_i) \tag{4}$$

Starling's equation of transcapillary exchange defines the forces driving the bulk movement of water between the vascular and interstitial regions. This motion is due to an imbalance between filtrative and absorptive forces.

The boundary conditions for equation 4 are as follows:

$$r \to \infty,\ p_i \to p_e \tag{5}$$

$$r = R,\ S\varepsilon v = Q \tag{6}$$

where S is the surface area of the probe, and Q is the volumeric flow rate. For CED, Q is positive, but negative for R-CED. The first boundary condition is derived from the second assumption stated previously, and the second boundary condition is simply the flow rate at the tip of the probe.

Due to the geometry of the microdialysis probe, it is best to utilize a one-dimensional cylindrical coordinate system. The equivalent form of equation 4 and respective boundary conditions are as follows (taking ε and κ to be constants)

$$\frac{d^2p_i}{dr^2} + \frac{1}{r}\frac{dp_i}{dr} = \frac{L_p s}{\kappa}(p_i - p_e) \tag{7}$$

$$r \to \infty,\ p_i \to p_e \tag{8}$$

$$r = R,\ 2\pi R\varepsilon v = Q \tag{9}$$

where R is the radius of the microdialysis probe.

If we define the dimensionless parameter, $\alpha^2 = L_p s R^2/\kappa$, which is the ratio of interstitial to vascular resistences of fluid flow, then the explicit solution for the interstitial pressure in equation 7 is

$$p_i = p_e + \frac{Q}{2\pi\kappa\alpha \cdot K(1,\alpha)} \cdot K\left(1, \frac{\alpha}{R}r\right) \tag{10}$$

where $K(a,r)$ is the modified Bessel function of the second kind of order a. Furthermore, because equations 1 and 2 form a closed set with respect to v and p_i, an expression for the fluid velocity can also be obtained.

$$v = \frac{Q}{2\pi\varepsilon R \cdot K(1,\alpha)} \cdot K\left(1, \frac{\alpha}{R}r\right) \tag{11}$$

Conservation of mass

The general equation of continuity for a species C_i is

$$\frac{DC_i}{dt} = \nabla \cdot (\mathcal{D}\nabla C_i) + r_{C_i} \tag{12}$$

where $\frac{DC_i}{dt}$ is the substantial derivative and r_{C_i} is the reaction term. By expanding the substantial derivative one obtains

$$\frac{dC_i}{dt} + \nabla \cdot (v_i C_i) = \nabla \cdot (\mathcal{D}\nabla C_i) + r_{C_i} \tag{13}$$

If we split the reaction term into the sum of a source term and a degradation term, the governing equation for the local conservation of mass of the macromolecule is [11]

[11] Baxter and Jain [1989]

$$\frac{dC_i}{dt} + \nabla \cdot (r_F v_i C_i) = \nabla \cdot (\mathcal{D} \nabla C_i) + \phi_s - \phi_r \tag{14}$$

where r_F is the retardation factor (solute velocity over interstitial velocity, assumed to be 1 for tumor tissue), $\mathcal{D}$ is the interstitial diffusion coefficient, C_i is the interstitial concentration, ϕ_s is the source term for the solute, and ϕ_r is the solute reaction term. Equation 14 states that the time rate of change of interstitial concentration (first left-hand term) plus net gain due to bulk flow (second left-hand term) equals the net diffusion (first right-hand term) plus gains by pore diffusion and bulk flow across the vasculature (second right-hand term) less losses due to irreversible metabolism (third right-hand term).

The source term is obtained from the pore model for transcapillary exchange given by *Curry et al.* [12]

$$\phi_s(r) = \frac{J_v(1-\sigma)}{V} \cdot \frac{C_0 \exp[\mathrm{Pe}_{mv}] - C_i}{\exp[\mathrm{Pe}_{mv}] - 1} - \frac{J_L}{V} C_i \tag{15}$$

where J_v is the transcapillary fluid flux, J_L is the fluid flux into the lymphatics (we will disregard the lympathic pathway, following the guidelines of Morrison *et al.*), C_0 is the plasma concentration of the macromolecule (constant), V is unit volume for transport in the tumor, and Pe_{mv} is the microvasculature Péclet number which is defined as

$$\mathrm{Pe}_{mv} = \frac{J_v(1-\sigma)}{Vps} = \frac{L_p s(p_e - p_i)(1-\sigma)}{ps} \tag{16}$$

where p is the capillary permeability and σ is the reflection coefficient. The Péclet number is a comparison of the convective to diffusive flux across the capillaries.

By combining equations 14, 15, and 16, a more specific form of the mass conservation equation, which includes the pore model for the source term, as well as a first-order reaction for elimination, ϕ_r, can be developed

$$\varepsilon \frac{dC_i}{dt} = \nabla \cdot (\mathcal{D} \varepsilon \nabla C_i) - \nabla \cdot (\varepsilon v_i C_i) + ps\mathrm{Pe}_{mv} \cdot \frac{C_0 \exp[\mathrm{Pe}_{mv}] - C_i}{\exp[\mathrm{Pe}_{mv}] - 1} - k_d C_i \tag{17}$$

where k_d is the irreversible metabolism constant. Rearrangement of equation 17 produces a partial differential equation in general form

$$\frac{d\overline{C}}{dt} + \nabla \cdot \left(-\mathcal{D} \nabla \overline{C} + v_i \overline{C}\right) = \frac{ps\mathrm{Pe}_{mv}}{\varepsilon} \cdot \frac{\exp[\mathrm{Pe}_{mv}] - \overline{C}}{\exp[\mathrm{Pe}_{mv}] - 1} - \frac{k_d \overline{C}}{\varepsilon} \tag{18}$$

where $\overline{C} = \frac{C_i}{C_0}$. Because we are only concerned with the behavioral trends of the interstitial concentration, eliminating the initial concentration of the macromolecule, C_0, in the governing PDE is not detrimental to the significance of our results. In fact, this only helps to simplify the model by reducing the number of required parameters by one.

The mass conservation equation can be represented in the one-dimensional cylindrical coordinate system as

[12] Curry [1984]

$$\frac{d\overline{C}}{dt} + \frac{1}{r}\frac{d}{dr} \cdot \left[r \cdot \left(-\mathcal{D}\frac{d\overline{C}}{dr} + v_i\overline{C} \right) \right] = \frac{ps\text{Pe}_{mv}}{\varepsilon} \cdot \frac{\exp[\text{Pe}_{mv}] - \overline{C}}{\exp[\text{Pe}_{mv}] - 1} - \frac{k_d\overline{C}}{\varepsilon} \tag{19}$$

Constraints

According to assumptions by Morrison *et. al*, the infused molecule is not produced by the organism. Hence, before retro-convection enhanced drug delivery begins, the concentration of the molecule in the interstitial region is zero.

$$\text{At } t \leq 0, \ \overline{C} = 0 \tag{20}$$

Due to the geometry for R-CED, we can assume that as the radial coordinate approaches infinity, i.e. as one moves further away from the microdialysis probe, there will be negligible variation in dimensionless concentration.

$$\text{As } r \to \infty, \ \frac{d\overline{C}}{dr} \to 0 \tag{21}$$

Several cases were analyzed when interpreting the effects of the membrane of the microdialysis probe on the system: (1) If the membrane is completely permeable to the drug, the flux through the membrane will be equal to the convective flux due to the high flow rate, resulting in a zero diffusive flux approximation; (2) If the membrane is impermeable to the drug, the total flux is zero; (3) If the membrane is semipermeable to the drug, the boundary will be described by a combination of the two previous cases.

$$\text{Permeable: At } r = R, \ \mathcal{D}\frac{d\overline{C}}{dr} = 0 \tag{22}$$

$$\text{Impermeable: At } r = R, \ -\mathcal{D}\frac{d\overline{C}}{dr} + v_i\overline{C} = 0 \tag{23}$$

$$\text{Semipermeable: At } r = R, \ -\mathcal{D}\frac{d\overline{C}}{dr} + v_i\overline{C}(1-\eta) = 0 \tag{24}$$

where η gives the property of the membrane, ranging from zero to one. Essentially, η is an artificially imposed reflection coefficient for the membrane of the microdialysis probe. Any value for η in between zero and one will present a semipermeable membrane to the model. If η equals zero, meaning no drugs are allowed to pass, the membrane will be modeled as impermeable. If η equals one, allowing all drugs to pass, a permeable membrane will be modeled.

COMSOL Implementation

Mathematical preparation

The equation of interest is presented in COMSOL as follows

$$e_a\frac{d^2u}{dt^2} + d_a\frac{du}{dt} + \nabla \cdot \Gamma = f \tag{25}$$

where u is the dependent variable, e_a is the mass coefficient, d_a is the damping coefficient, Γ represents total flux, and f is the source term. COMSOL assumes Cartesian coordinates for the del operator (∇). In order to utilize equation 19 in COMSOL, the governing PDE must rearranged.

$$r\frac{d\overline{C}}{dt} + \frac{d}{dr} \cdot \left[r \cdot \left(-\mathcal{D}\frac{d\overline{C}}{dr} + v_i\overline{C} \right) \right] = \left(\frac{ps\text{Pe}_{mv}}{\varepsilon} \cdot \frac{\exp[\text{Pe}_{mv}] - \overline{C}}{\exp[\text{Pe}_{mv}] - 1} - \frac{k_d\overline{C}}{\varepsilon} \right) \cdot r \tag{26}$$

Parameter values

The value of the parameters used in the model are listed in Table 1.

Parameters	Symbol	COMSOL Name	Value	Reference
Capillary hydraulic conductivity	L_p	Lp	7.1×10^{-13} cm^3/dyn/s	13
Tissue hydraulic conductivity	κ	kappa	2×10^{-9} cm^4/dyn/s	14
Capillary permeability	p	perm	1.1×10^{-9} cm/s	15
Capillary area per tissue volume	s	s	100 cm^{-1}	16
Reflection coefficient	σ	sigma	0.95	17
Tissue porosity	ε	eps	0.26	18
Microdialysis probe radius	R	R	2.5×10^{-2} cm	19
Diffusion coefficient	$\mathcal{D}$	D	10^{-7} cm^2/s	20
Flow rate[a]	Q	Q	-8×10^{-6} cm^2/s	21
Degradation rate constant	k_d	kd	0	22
Membrane property	η	Z	0, 0.5, 1	Artificial

Table 1: Parameters for COMSOL. [a]Wang and Olbricht utilized a 0.1μL/min withdrawal rate and a 0.2 cm microdialysis membrane length

Conclusion and Recommendations

This study focused on the mathematical modeling of mass transfer processes in R-CED utilizing COMSOL Multiphysics. It was shown that using a membrane for the microdialysis probe that is impermeable to the drug molecules is beneficial to drug delivery, as this increases the amount of drug accumulating at the boundary of the probe and interstitial region, which is where a malignant tumor would be located. As the membrane property parameter gravitates from a permeable to an impermeable membrane, the value at $r = R$ increases.

By comparing spherical and cylindrical coordinate geometries, the spherical coordinate geometry resulted in higher overall values for drug concentration than the axisymmetrical case. Overall, however, the trends in concentration profile for both geometries are identical, with the absolute values for dimensionless concentration being within one order of magnitude of one another. Both coordinate systems are a safe approximation in modeling either CED or R-CED.

In addition to quantifying the behavior of drug molecules during the experimental time, it was important to analyze drug behavior after the protocol ended, or, the post-extraction phase. After the procedure stops, the dimensionless concentration profile remains the same, but begins with a lower value at $r = R$ as time progresses, until the interstitial region is completely removed of drug molecules. These results concluded that CED has a greater penetration depth than R-CED, due to the concentration profile precipitously decaying at distances much farther away from the probe than the R-CED case. This discrepancy may be attributed to the fact that CED has a much longer experimental time, resulting in a greater amount of drug molecules being infused into the interstitial region than the R-CED procedure. R-CED, however, cannot occur for long durations, as the removal of interstitial fluid is limited in order to avoid physiological ramifications.

Our pharmacodynamic study utilizing the area under the curve metric resulted in a fairly flattened concentration profile for the CED case. Naturally, the value of concentration was lower as one moved away from the surface of the microdialysis probe, but there was no precipitous drop as shown with the R-CED protocol. For R-CED, there is a sharp drop as one moves into the interstitial region, showing that most of the drugs are localized near the surface of the catheter. This tells us that CED has greater penetration depth, but if we are concerned with targeting a specific spot within the interstitial region, then the R-CED protocol is advantageous.

Overall, the positive aspects of retro-convection enhanced drug delivery are not great enough to outweigh its drawbacks. Avoiding the reflux problem associated with CED is one of its advantages, but, the inability to infuse a large amount of drugs into the interstitial region is detrimental to the future use of the R-CED protocol. The strength of the R-CED procedure lies in its ability to pinpoint where the drug molecules will accumulate, near the surface of the microdialysis probe. It does not, however, show great promise in aiding biopharmaceutics to circumvent the blood-brain barrier. Because of this, retro-convection enhanced drug delivery cannot stand on its own as a way to treat malignancies in the CNS. However, Motion and coworkers optimized the use of R-CED by using it solely to isolate drugs in a confined area, and in its current stage, this is the only benefit of the procedure.

We have shown through a sensitivity analysis for each parameter that there are certain values which will benefit drug delivery. Further investigation should be conducted in how to manipulate these parameters in a physical setting to promote drug delivery. For example, vascular endothelial growth factor, a protein which stimulates the growth of new blood vessels, could help alter capillary permeability to a desired value.

When comparing the R-CED and CED procedures with mathematical modeling, one should avoid non-dimensionalizing the drug concentration variable. The two procedures are non-dimensionalized with respect to different values of initial concentration, and it may be misleading visually when trying to determine which protocol is best for drug delivery.

Finally, one should elaborate on the simplicity of the geometry of this study. Approximating the R-CED procedure as a one-dimensional geometry is sufficient for this preliminary analysis. However, the complexities when implementing a two-dimensional or three-dimensional geometry garners attention.

References

N. K. B., C. T. Lo, C. P. Foley, W. M. Saltzman, and W. L. Olbricht. Fabrication and Characterization of Microfluidic Probes for Convection Enhanced Drug Delivery. *Journal of Controlled Release*, 111:252–262, 2006.

L. T. Baxter and R. K. Jain. Transport of Fluid and Macromolecules in Tumors. *Microvascular Research*, 37:77–104, 1989.

CBTRUS. Statistical Report. *Central Brain Tumor Registry of the United States*, 2004-2006.

F. E. Curry. Mechanics and Thermodynamics of Transcapillary Exchange. *Handbook of Physiology, Section 2: The Cardiovascular System*, pages 309–374, 1984.

G. R. DiResta, J. Lee, J. H. Healey, S. M. Larson, and E. Arbit. Enhancing the Uptake of Chemotherapeutic Drugs into Tumors Using an Artificial Lymphatic System. *Annals of Biomedical Engineering*, 28:556–564, 2004-2006.

P. F.Morrison, D. W. Laske, H. Bobo, E. H. Oldfield, and R. L. Dedrick. Highflow Microinfusion: Tissue Penetration and Pharmacodynamics. *American Journal of Physiology*, 266:R292–R305, 1994.

G. H. Huynh, T. Ozawa, D. F. Deen, T. Tihan, and F. C. S. Jr. Retro-Convection Enhanced Delivery to Increase Blood to Brain Transfer of Macromolecules. *Brain Research*, 1128:181–190, 2007.

Krauze. Reflux-Free Cannula for Convection-Enhanced High-Speed Delivery of Therapeutic Agents. *J Neurosurg*, 103:923–929, 2005.

J. P. M. Motion, G. H. Huynh, F. C. S. Jr., and R. A. Siegel. Convection and Retro-Convection Enhanced Delivery: Some Theoretical Considerations Related to Drug Targeting. *Pharm Res*, DOI 10.1007/(s11095-010-0296-2), 2010.

W. Pardridge. Drug and Gene Delivery to the Brain: The Vascular Route. *Neuron.*, 36:555558, 2002.

W. Pardridge. The Blood-Brain Barrier: Bottleneck in Brain Drug Development. *NeuroRx*, 2:3–14, 2005.

S. Rapoport. Osmotic Opening of the Blood-Brain Barrrier: Principles, Mechanism, and Therapeutic Applications. *Cellular and Molecular Neurobiology*, 20:217–230, 2000.

J. Z. Sarntinoranont M., X. Chen and T. H. Mareci. Computational Model of Interstitial Transport in the Spinal Cord Using Diffusion Tensor Imaging. *Annals of Biomedical Engineering*, 34:1304–1321, 2006.

J. Smith and J. A. C. Humphrey. Interstitial Transport and Trasnvascular Fluid Exchange During Infusion into Brain and Tumor Tissue. *Microvascular Research*, 73:58–73, 2007.

P. Wang and W. L. Olbricht. Retro-Convection Enhance Drug Delivery: A Computational Study. *Annals of Biomedical Engineering*, 38(8):2512–2519, 2010.

(537f) An Area-Averaging Approach for Electrochemical Systems and It's Validity

Rajavardhan [Raja] Reddy Nagolu[1], Vinten Diwakar[1] and Pedro E. Arce[2], (1)Department of Chemical Engineering, Tennessee Technological University, Cookeville, TN, (2)Chemical Engineering, Tennessee Technological University, Cookeville, TN

The mathematical analysis of mass transport in chemical systems usually involves the method of averaging to obtain system level information from the microscopic domain. Reactions in such systems are usually assumed to be occurring in the bulk (homogeneous reactions). However, in electrochemical systems (Batteries or Fuel Cells), the reactions are heterogeneous and they usually occur on catalytic surfaces. A diffusion-reaction model is developed to study concentration distributions and the behavior of the system within the porous anode catalyst layer of Direct Formic Acid Fuel Cells (DFAFCs). And at the same time an area-average approach is then implemented to obtain macroscopic concentration distributions within the domain, i.e., porous cavity. To analyze the role of averaging, this work is organized based on different cases of electrochemical kinetics. Firstly, the linearized form of the Butler-Volmer (concentration independent) kinetics is analyzed followed by classic Butler-Volmer (concentration dependent) type of kinetics. The system parameters varied are geometric (aspect ratio) and kinetic parameters. From this paper it is inferred that the averaging technique is valid for linearized form of Butler-Volmer kinetics at all conditions. But for the case of general Butler-Volmer kinetics, the area-averaging technique has certain limitations and it valid only for long pores. Also, the percentage of errors which is calculated from two methods is reported for different cases of Butler-Volmer kinetics to know the limitations of the system.

See more of this Session: Mathematical Modeling In Transport Processes
See more of this Group/Topical: Engineering Sciences and Fundamentals

(537g) Mixing Improvement of Fluid Flow Using Lagrangian Coherent Structures

Mojtaba Izadi, Chemical & Materials Engineering, University of Alberta, Edmonton, AB, Canada and Stevan Dubljevic, Chemical and Materials Engineering, University of Alberta, Edmonton, AB, Canada

Abstract

Considering fluid flow problem as a dynamical system, there are emergent patterns that influence the transport phenomena in the flow. These patterns are known as coherent structures [1]. When studied in terms of the velocity field of a fluid flow in a Lagrangian framework, these structures are called Lagrangian Coherent Structures (LCSs). Relating LCSs to the ridges of finite-time Lyapunov exponent (FTLE) (or Direct Lyapunov Exponent (DLE) [2]), Shadden et al. developed the theory and computation procedure of LCSs [3,4]. These structures are known as dynamically driven barriers to transport processes in a fluid flow and they can help reveal the mixing properties and transport mechanisms in the system. Today, identification of these structures have become an important tool for studying transport phenomena in fluid mechanics.

Moving LCSs can improve mixing by inducing transport mechanisms causing material stretching and folding [5]. In this study, it is intended to move the LCSs present in the fluid domain of a rectangular geometry to enhance mixing properties of the flow. To this end, Navier-Stockes equations with Boussinesq approximation are considered along with continuity as governing equations of fluid flow. These equations can be projected to a divergence-free space to obtain the input-output behaviour of the fluid flow. By Boussimesq approximation, the velocity field of fluid domain can be manipulated by the boundary actuation in terms of temperature distribution or heat flux to the boundary.

Finally, we are going to provide the input function (temperature distribution or heat flux on the boundary) to bring the velocity field to a desired state for which the mixing properties of the flow has been enhanced.

References

1. S. Wiggins, Chaotic transport in dynamical systems, Springer-Verlak, 1992.
2. G. Haller, Distinguished material surfaces and coherent structures in 3d fluid flows, Physica D 149 (2001) 248-277.
3. S.C. Shadden, F. Lekien, J.E. Marsden, Definition and properties of Lagrangian coherent structures from finite-time Lyapunov exponents in two-dimensional aperiodic flows, Physica D 212 (2005) 271-304.
4. F. Lekien, S.C. Shadden, J.E. Marsden, Lagrangian coherent structures in n-dimensional systems, Journal of Mathematical Physics 48 (2007) 065404.

5. Y. Wang, G. Haller, A. Banaszuk, G. Tadmor, Closed-loop Lagrangian separation control in a bluff body shear flow model, Physics of Fluids 15 (2003) 2251-2266.

See more of this Session: Mathematical Modeling In Transport Processes
See more of this Group/Topical: Engineering Sciences and Fundamentals

(537h) Suffocating On a Plastic Bag? Modeling Hypoxia In Porous Polymer-Cell Constructs

Roger L. York, Monica Neugebauer, Minglin Ma, Robert Langer and Daniel Anderson, Koch Institute for Integrative Cancer Research, Massachusetts Institute of Technology, Cambridge, MA

This talk will describe modeling the simultaneous reaction (consumption) and diffusion of oxygen through porous polymer-cell constructs (encapsulated cells), based on reaction-diffusion models first used in the context of chemical reaction engineering. Encapsulated cells are used for xenotransplantation in the context of diabetes treatment; they allow for the transport of chemicals into and out of the construct while maintaining surface properties that prevent the foreign body response from the host's immune system. We will discuss under what conditions hypoxia-defined as a region of tissue which is deprived of oxygen-could occur in these constructs. We will elucidate the oxygen concentration profile for the zeroth and first order reaction kinetics in slabs, cylinders and spheres, using the Robin set of boundary conditions, which take external resistance to diffusion into account. The concentration profiles require two parameters: the Thiele Modulus, which is analogous to a ratio of a reaction rate to a diffusion rate, and the Biot number for mass transfer, which is analogous the ratio of mass transfer limitations inside of the cells to the mass transfer limitations outside of the cells (i.e. the polymer). We find that under certain sets of conditions, "dead zones" or hypoxic regions can be formed. We will conclude by making recommendations on the prevention of hypoxia in these constructs.

See more of this Session: Mathematical Modeling In Transport Processes
See more of this Group/Topical: Engineering Sciences and Fundamentals

(579a) Understanding the Evolution of by-Product Synergy Networks by Network Analysis

Shweta Singh and Bhavik R. Bakshi, William G. Lowrie Department of Chemical and Biomolecular Engineering, The Ohio State University, Columbus, OH

By-Product synergy (BPS) networks aim to mimic nature by using waste from one process as a resource in another. They are increasingly popular due to their ability to reduce environmental impact while saving money. Exploiting this win-win opportunity is a primary driver in the development of BPS networks, as evidenced by the many successes of the networks developed by the US Business Council for Sustainable Development and similar efforts elsewhere in the world. Development of a BPS network enhances the economic and environmental efficiency, but like most complex networks, it is likely that BPS networks also exhibit a trade-off between efficiency and robustness. If such a trade-off exists then it would make sense to rely on criteria other than just maximizing efficiency for the design of BPS networks. This may reduce the chance of developing fragile networks that are vulnerable to collapse.

This work aims to understand the trade-off between efficiency and resilience in BPS networks and use this insight to design more robust networks. . The field of network analysis provides different studies that look at resilience of networks such as service oriented networks , supply chain logistics for military purposes and many others. Resilience is an essential property for the sustainability of networks, and indicates the ability of a network to maintain their function in case of perturbation. Several metrics like average path length, network disconnection, and others have been used to understand network structure. Such metrics will be used in this work to understand the evolution of BPS networks. Examples from real networks in the US and toy networks will be used to illustrate the approach. Besides BPS networks, this work will also consider the behavior of techno-ecological networks. These are networks that include ecological systems such as forests and wetlands that provide essential goods and services such as oxygen, water and carbon sequestration for the sustenance of the technological networks. It is also desired that BPS networks must be such that the functioning of industries in the network is not severely affected by the collapse of one or two industries. The insight from such work be explored for guiding the development of current and future BPS and techno-ecological networks.

See more of this Session: Complexity, Networks, Dynamics and Sustainability
See more of this Group/Topical: Sustainable Engineering Forum

(579b) Optimal Rerouting of Traffic Flows for Resilient Management of Recharging Station Networks

Kris Villez[1], Craig Rieger[2] and Venkat Venkatasubramanian[1], (1)School of Chemical Engineering, Purdue University, West Lafayette, IN, (2)Idaho National Laboratory, Idaho Falls, ID

One of the most important challenges to be met inthe transition towards an ecologically friendly society is theimplementation of better ways of transportation, which reduce theamounts of energy used and which permit the switch to "greener"sources of energy. One part of the solution is likely to be met bymeans of electrical vehicles. Because electrical vehicles are limitedin range and their "refueling" (i.e. recharging) takestime, a system in which batteries can be replaced quickly has beenproposed [1]. Still, for such a system to work, it is of criticalimportance that electric vehicles can efficiently reach a servicestation. We have studied this problem in the past by optimizing thelocation of service stations for a given traffic flow in the form ofan Integer Linear Program (ILP). Originally, this was solved for afailure-free setting [2]. More recently, we have extended this forscenarios in which parts of the infrastructure (roads, road junctionsand service stations) can fail [3].

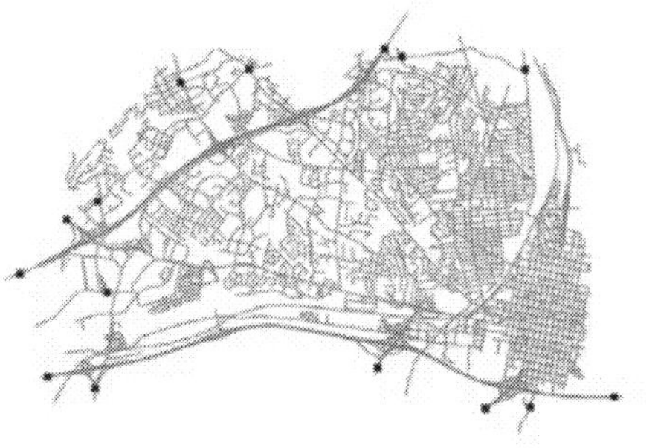

In ourvision, the previously adopted strategy to optimize the location ofservice stations is only one necessary element. Next to this, oneshould account for the fact that the route of electrical vehicles canbe modified. Indeed, the routes of electrical vehicles can be guidedand manipulated so to make sure that they can reach service stationsefficiently under varying conditions. We explore this option in thiswork by evaluation of the following algorithm.

Considerfirst that the set of service stations is fixed and the normaltraffic flow of conventional cars is known in the form of routes(paths) over a city network. In such case, we reroute the cars in agreedy fashion as follows:

The paths which contain a service station remain the same

The paths which do not contain a service station are modified as follows:

Select the most busy path (highest car flows)

For the selected path, find the shortest path from start location to destination which contain one of the service stations. Do this for all available service stations. To this end, we apply Dijkstra's algorithm to find (1) the shortest path between the start location and the selected service station and (2) the shortest path between the selected service station and the destination location.

If no paths could be found which include a service station, exclude the traffic flow associated with this path in subsequent steps and go to 7. Otherwise, continue to 4.

Following the above enumeration of shortest paths, sort the obtained paths from shortest to longest.

Now the flow in the original path is distributed over the paths starting with the shortest paths while not exceeding the capacity of any of the roads. If all computed paths are used to their maximum capacity, retain the excessive amount of cars on the original path.

If any of the roads in the city network become saturated, i.e., reach their maximum capacity, exclude these roads as available roads in the network for repeated optimization steps.

Stop the algorithm if any of the following conditions is met:

All edges in the network are saturated

All car flows have been excluded from optimization (step 3) or have been rerouted successfully (step 5).

We apply the above algorithm to the case studyused before, which is that of the city of Alexandria, VA. To computethe normal flow of conventional cars, we use the model as implementedin TRANSIMS.

REFERENCES

K. Galbraith, "Better Place unveils battery swap station," http://green.blogs.nytimes.com/2009/05/13/better-place-unveils-battery-swap-station/ , 2009

A. Shukla, J. Pekny, and V. Venkatasubramanian, "An optimization framework for cost effective design of refueling station infrastructure for alternative fuel vehicles," Computers & Chemical Engineering, In Press, Available online, 2011.

Kris Villez, Akshya Gupta, Venkat Venkatasubramanian, Craig Rieger. Resilient design of recharging station networks for electric transportation vehicles. Submitted to the 4th International Symposium on Resilient Control Systems (ISRCS2011).

(579c) Scheduling of Multiple Chemical Plant Start-Ups for Minimizing Regional Air Quality Impacts

Tianxing Cai, Dan F Smith Department of Chemical Engineering, Lamar University, Beaumont, TX and Qiang Xu, Dan F. Smith Department of Chemical Engineering, Lamar University, Beaumont, TX

Chemical industrial regions concentrate many chemical plants, whose emissions may case highly localized and transient air pollution events violating national ambient air quality standards (NAAQS). Flaring emissions, especially the intensive start-up flaring, emits huge amounts of pollutants, which has potentially significant impacts on local air quality. Give the situations that many chemical plants are spatially located in an industrial region and their startup flare emissions exhibit significant temporal variability. Such variability may cause an unexpected coincidence of flare emissions casing local air quality problems. Therefore, from the emission control point of view, the best practices for improving regional air quality should integrate both efforts on plant-wide emission minimization at every industrial point source, and regional-wide multi-plant emission variability control.

In this paper, a scheduling methodology for coordinating multiple chemical plant start-ups is developed to minimize regional air quality impacts. This methodology includes two stages of work. In the first stage, geometric locations of plants and air quality monitor stations are both taken into account. Each plant proposes its startup time window, emission profile, and expected start-up time. In the second stage, a scheduling model for regional chemical plants' startup is developed. The objective of the model is to reschedule multi-plant start-up sequence to achieve the minimum sum of delay time compared to the expected start-up time of each plant. Certainly, the major constraints include the atmosphere dispersion model constraints and predicted air quality indexes from air quality monitor stations satisfying NAAQS. The development provide quantitative decision support for multiple stake holders, including government environment agencies, chemical industry, and local communities.

See more of this Session: Complexity, Networks, Dynamics and Sustainability
See more of this Group/Topical: Sustainable Engineering Forum

(579d) Assessing Risks Due to Loss of Natural Capital: The Case of Pollination Services

Vikas Khanna, Department of Civil and Environmental Engineering, University of Pittsburgh, Pittsburgh, PA and Bhavik R. Bakshi, William G. Lowrie Department of Chemical and Biomolecular Engineering, The Ohio State University, Columbus, OH

Several recent events in the U.S. have highlighted the criticality and vulnerability of systems to sudden shocks such as natural disasters and deliberate attacks. Proper understanding of such disruptive scenarios and their impact using holistic and integrated systems modeling techniques is crucial for effective resource allocation and disaster management. Attempts to study the economic-environmental impact of sudden shocks or disruptive scenarios pose both methodological and empirical challenges for engineers and policy makers alike. This is because of the complexity and the highly interconnected nature of the industrial-ecological networks and the associated systems. Such complex systems are observed to have high efficiency, performance, and robustness, but are often fragile and vulnerable to unanticipated perturbations. Proper understanding of such disruptive scenarios and their impact is pivotal for guiding design of complex systems and their long-term sustainability.

The present work focuses on exploring the utility of Input-Output (IO) models for studying the effect of sudden shocks and quantifying the associated risks to complex industrial systems [1]. We are using the IO model to understand the impact of changes in the availability of natural resources including natural capital. This includes understanding the potential impact of loss of services such as pollination, water scarcities, and soil fertility or loss of certain key resources such as fossil fuels. Such information is used to determine sectors that are likely to face maximum risk due to sudden disruptions. Since the EIO model is based on empirically available data and considers a static and linear state of the economy, it is not able to simulate long-term effects of such disruptions. However, it is appropriate for gaining insight into the short-term effects of disruptions before any adaptation due to market forces or policies. Such simulation is relevant to understanding the effect of environmental changes as well as human-induced changes such as terrorism and natural disasters.

We illustrate the utility of our framework by studying the effect of loss of pollination services attributable to managed honeybees on the various industrial sectors. Natural systems provide a number of ecosystem goods and services to mankind. One such ecosystem service is the pollination provided by wild insects and managed honeybees. We restrict our focus to the critical ecological pollination service provided by domesticated or managed honeybees thus excluding the role of wild insects. Concerns have been expressed over the disappearance of 50-90 percent of the honeybee colonies across the U.S. as a result of the colony collapse disorder. Honeybees are responsible for pollinating more than 14 billion worth of crops in the United States [2]. There have been several attempts to quantify the direct economic value of pollination services provided by wild insects and honeybees. However, no studies exist in the literature to the best of our knowledge that try to study and quantify the effect of the loss of critical pollination services on complex industrial systems in a systematic manner. This is the first comprehensive attempt to quantify the direct and indirect effects of the loss of pollination services provided by domesticated honeybees on the complex industrial systems using an IO model. Besides quantifying the economic impact, the methodology can identify industrial sectors with greatest sensitivity under a given perturbation. It is observed that sectors that suffer the highest economic loss under agiven perturbation are not always the ones with the highest production degradation. In addition, the IO model is also able to identify sectors experiencing the inoperability and the economic losses as a result of mostly the indirect effect of the perturbation. The IO model is also attractive in this case as it is based on readily available empirical data for a specified time period and can simulate the short term effect of shocks before any adaptation due to market forces or policies. The implications of the results for guiding conservation policies will also be discussed.

References

1. R. E. Miller and P. D. Blair. Input-Output analysis. Foundations and extensions. Prentice-Hall, Inc., Englewood Cliffs, New Jersey, 1985.
2. M.E. Watanabe. Pollination worries rise as honey bees decline. Science, 265(5176):1170–1170, 1994.

See more of this Session: Complexity, Networks, Dynamics and Sustainability
See more of this Group/Topical: Sustainable Engineering Forum

(579e) Supply Chain Planning In Oil & Gas Industry: Energy and Clean Technologies Under Uncertainty

Ivan Ordoez Sr., **Ariel Uribe** and Andrs Joaqun Calderon, Colombian Petroleum Institute, ECOPETROL S.A., Piedecuesta, Colombia

Conventional planning strategies are complicated to develop because the future events are uncertain. For a supply chain of oil & gas industry is difficult to planning new production and transport networks because oil & gas reserves, energy generation technologies and availability of electric public net is very uncertain at long term. This work shows then behavior future of decision process in investments of new supplier energy systems when three elements act: 1) expansion in oil crude and gas production 2) dynamic of growing in electric public network 3) imperative use of clean technologies to generation of energy. In this work the topological connectivity among generation and consume point is a manipulated variable and this effect is evaluated on taking decisions process. For this propose a representative chain supply of oil & gas industry composed by around 300 oil fields and 33 transport terminals is represented by a MILP model in GAMS (General Algebraic Modeling System). This model is used to an optimization of several future scenarios. The analysis of results shows that decisions in new investments of clean technologies are too sensible with dynamics of expansion in oil crude and gas production and growing in electric public network. Several solutions can obtain according to supply chain connectivity where clean technologies to energy generation is economically viable if the supply chain will have generation and consumes points poorly connected.

See more of this Session: Complexity, Networks, Dynamics and Sustainability
See more of this Group/Topical: Sustainable Engineering Forum

(581a) Designing Subsea Production Facilities for the Worst Case Using Semi-Infinite Programming

Matthew D. Stuber and Paul I. Barton, Process Systems Engineering Laboratory, Dept. of Chemical Engineering, Massachusetts Institute of Technology, Cambridge, MA

Diminishing petroleum supply from traditional sources alongwith the desire for reduced dependence on petroleum from foreign sources hasmotivated exploration into increasingly more hostile environments. Onepromising frontier is ultra deep-sea oil and gas in the Gulf of Mexico. Theapplication of traditional floating platform technologies to ultra deep-seaproduction has been considered to be a very risky choice. Alternatively,remote compact subsea production facilities are considered to be the enablingtechnology for ultra deep-sea oil and gas production. However, as recent eventshave demonstrated, operational failures in this environment are extremelycostly in terms of economic and environmental damages. In this case, the costassociated with over-designing the process does not outweigh the costassociated with operational failure. Therefore, remote compact subsea productionfacilities must be designed for the worst-case scenario, however improbable.

In order to assess the robustness of a proposed design, arigorous model-based approach must be taken. In such an approach, uncertaintyin the input disturbances to the process as well as uncertainties that areinherent in the modeling parameters must be considered. The question that mustbe answered is for every realization of uncertainty, does there exist acontrol setting such that all performance and safety specifications are neverviolated?

In [3], the authors attempt to address the robustnessproblem by formulating it as a bilevel optimization problem with the modelequations taken as equality constraints. Since there are no algorithms currentlyavailable for solving equality constrained bilevel optimization problems ingeneral, this technique is not applicable to the complex case of modeling a subseaproduction facility. Alternatively, in [5], the authors propose the idea ofreformulating the bilevel optimization program into an equivalent semi-infiniteprogram (SIP). They suggest that, due to developments in globally solving SIPs[1,2], the SIP reformulation is more tractable. However, there is one majorcaveat with this reformulation: an implicit function is embedded in thesemi-infinite constraint from solving the equality constraints (modelequations) for the state variables as an implicit function of the uncertainvariables and the controls [5]. Therefore, it does not have a known closedform and can only be approximated using a procedure such as a fixed-pointiteration [5].

New developments in the construction of convexunderestimators and concave overestimators (or relaxations) of nonconvexfunctions [4,6] have enabled the authors of this paper to extend the algorithmof [1,2] to SIPs with embedded implicit functions. Utilizing the most recentdevelopments [6], the SIP reformulation can be solved globally in the generalcase.

In this paper the authors discuss a zeroth-order model ofa remote compact subsea production facility. The novel ideas of relaxingimplicit functions [6] were applied within the new SIP algorithm to solve theSIP reformulation and answer the question of robust feasibility. The result isa working proof-of-concept that the robust SIP formulation is tractable fornontrivial examples.

[1] Bhattacharjee, B., Green Jr. W. H., and P. I. Barton. Interval Methods forSemi-Infinite Programs. *Computational Optimization andApplications*, 30:63-93, 2005.

[2] Bhattacharjee, B., Lemonidis, P., Green Jr. W. H., and P. I. Barton. GlobalSolution of Semi-Infinite Programs. *Math. Program.*, Ser. B103:283-307, 2005.

[3] Halemane, K.P. and Grossmann, I.E. OptimalProcess Design Under Uncertainty. *AIChE Journal*, 29 (3):425-433, 1983.

[4] Scott, J.K., Stuber, M.D., and P.I. Barton.Generalized McCormick Relaxations. *Journal of Global Optimization*, DOI:10.1007/x10898-011-9664-7, In press 2011.

[5] Stuber, M.D. and P.I. Barton. Robust Simulationand Design Using Semi-Infinite Programs with Implicit Functions. *Int. J. ofReliability and Safety*, In press 2011.

[6] Stuber, M.D., Scott, J.K., and P.I. Barton.Global Optimization of Implicit Functions. In preparation 2011.

See more of this Session: Design and Operations Under Uncertainty II
See more of this Group/Topical: Computing and Systems Technology Division

(581b) Rigorous Approach for Robust Design of Nonlinear Dynamic Systems

Yao Zhao and Mark A. Stadtherr, Department of Chemical and Biomolecular Engineering, University of Notre Dame, Notre Dame, IN

Robust design problems involving dynamic processes arise in many areas of engineering. In chemical engineering, processes and control systems often must be designed to satisfy certain safety and/or quality control constraints, and to do so in a way that is robust to uncertainties in process parameters and to possible disturbances. To address this problem, dynamic processes can be modeled as continuous-time hybrid systems [1] subject to appropriate terminal constraints and path constraints. The design problem considered here is then to rigorously determine robust subregions of the design space; that is, subregions in which all constraints will be satisfied over a finite time horizon, and this will remain true for all possible parameter values and disturbances. Most existing methods [2-4] for addressing this problem are restricted to linear models. Huang et al. [5] have proposed a region-transition-model (RTM) framework for uncertain nonlinear systems. However, they used difference equations to approximate the evolution of the continuous states, which may not realistically represent the true dynamics of a nonlinear system. Recently, Lin and Stadtherr [6] extended the RTM framework to continuous-time systems and used it to develop a rigorous approach for safety analysis. In general, their approach was not designed to handle uncertainties, although the approach could be used to address robust design problems by treating uncertain parameters as additional design variables, which introduces unnecessary computational cost.

In this study, we present a new, rigorous approach for the robust design of nonlinear, continuous-time dynamic systems. This method is based on constraint propagation using Taylor-models [7-9] under a double-layer RTM framework. In the inner layer, a possible subregion of the design space is tested over different subregions of uncertainty. In the outer layer, the results from the inner layer are collected and summarized to determine overall feasibility and subsequent action. A key feature of the method is the use of a validated solver (VSPODE) for parametric ODEs [10], which is used to produce guaranteed bounds on the solutions of dynamic systems with interval-valued parameters and/or initial states. VSPODE consists of two phases applied at each integration step. In the first phase, existence and uniqueness of the solution are proven, and a coarse enclosure of the solution for the entire integration step is computed. In the second phase, a tighter enclosure of the solution is computed and bounded by Taylor models, which are symbolic (algebraic) functions of the uncertain parameters and initial states.

Our strategy provides four main advantages in this application. First, the approach can be directly applied to the original nonlinear dynamic model without introducing any linearization or discrete-time approximation. Second, the path constraint functions can be easily formulated using additional state variables and then tested for feasibility over the entire time interval of an integration step in the first phase of VSPODE, thus providing rigorous assurance that the path constraints are satisfied over the entire time horizon, and not just at discrete points. Third, the use of Taylor models to represent bounds provides the power to obtain rigorous and very tight path enclosures with a lower computational cost, in comparison to other commonly used bounding methods. Fourth, because the Taylor model method used provides an explicit algebraic representation of the state trajectories in terms of the design variables, constraint propagation techniques can be exploited to efficiently distinguish feasible subregions and infeasible ones. The computational aspects of this new approach will be demonstrated through application to several example problems.

References

[1] Barton, P.; Pantelides, C. Modeling of Combined Discrete-continuous Process. *AIChE* J. **1994**, 40 , 966.

[2] Srinivasan, R. et al. Safety Verification using a Hybrid Knowledge-based Mathematical Programming Framework. *AIChE* J. **1998**, 44 , 361.

[3] Park, T.; Barton, P. Implicit Model Checking of Logic-based Control Systems.

AIChE J. **1997**, 43 , 2246.

[4] Dimitriadis, V. et al. A Case Study in Hybrid Process Safety Verification.

Comput. Chem. Eng. **1996**, 20 , S503.

[5] Huang, H et al. Quantitative Framework for Reliable Safety Analysis. *AIChE* J. **2002**, 48 , 78.

[6] Lin, Y.; Stadtherr, M.A. Rigorous Model-based Safety Analysis for Nonlinear Continuous Time Systems. *Comput. Chem. Eng.* **2009**, 32. 493.

[7] Lin, Y.; Stadtherr, M. A. Deterministic Global Optimization of Nonlinear Dynamic Systems. *AIChE J.* **2007**, 53 , 866.

[8] Lin, Y.; Stadtherr, M. A. Deterministic Global Optimization for Parameter Estimation of Dynamic Systems. *Ind. Eng. Chem. Res.* **2006**, 45, 8438.

[9] Lin, Y.; Stadtherr, M. A. Enclosing All Solutions of Two-Point Boundary Value Problems for ODEs. *Comput. Chem. Eng.* **2008**, 32, 1714.

[10] Lin, Y.; Stadtherr, M. A. Validated Solutions of Initial Value Problems for Parametric ODEs. *Appl. Num. Math.* **2007**, 57, 1145.

(581c) An Optimization-Based Framework for Process Planning Under Uncertainty with Risk Management

Cheng Seong Khor[1], Sara Giarola[2], **Benoit Chachuat**[1] and Nilay Shah[1], (1)Centre for Process Systems Engineering, Department of Chemical Engineering, Imperial College London, London, United Kingdom, (2)CAPE-Lab - Computer-Aided Process Engineering Laboratory, Dipartimento di Principi e Impianti di Ingegneria Chimica, Universit di Padova, Padova, Italy

The process systems engineering (PSE) community has been instrumental in carrying out a key role in extending the systems engineering boundaries from a sole focus on process systems to the incorporation of important business issues. The latter involves the inevitable consideration for uncertainty in decision-making that gives rise to a need for risk management in enhancing the robustness of process planning activities under numerous possible operating scenarios. In current challenging and volatile political and economic environment, the process industry is exposed to a high degree of uncertainty that renders the production planning task to be a risky and complex optimization problem that typically requires high computational expense.

This work proposes a computationally-tractable, optimization-based framework for risk management in mid-term process planning under uncertainty. In general, the two major methodologies for addressing optimization problems under uncertainty are recourse-based two-stage or multi-stage stochastic programming [1,2] and robust optimization [3,4,5]. In this work, the latter is employed with a scenario-based approach adopted to represent uncertainties in the stochastic parameters. The problem is formulated as a recourse-based two-stage stochastic program that incorporates a mean-risk structure in the objective function. Two risk measures, with origins in the insurance and finance industries, are considered, namely Value-at-Risk (VaR) [6] and Conditional Value-at-Risk (CVaR) [7]. We account for uncertainty in prices of crude oil and commercial products, market demands of products, and production yields. However, since a large number of scenarios are often required to capture the stochasticity of the problem, the model suffers from a curse of dimensionality since the model size scales exponentially with the number of scenarios and the uncertain parameters. To circumvent this problem, we propose a framework with relatively low computational burden that involves the following two major steps. First, a linear programming (LP) approximation of the risk-inclined version of the planning model is solved for a number of randomly generated scenarios. Subsequently, the VaR parameters of the model are simulated and incorporated into an LP approximation of the risk-averse version of the planning model, which is a mean-CVaR stochastic program [8].

The proposed approach is illustrated through a petroleum refinery planning case study to demonstrate how solutions with relatively affordable computational expense can be attained in a risk-averse model in the face of uncertainty.

Bibliography

1. Birge, J. R., and F. Louveaux (1997). *Introduction to Stochastic Programming*. New York: Springer-Verlag.
2. Kall, P., and S. W. Wallace (1994). *Stochastic Programming*. New York: John Wiley & Sons.
3. Ben-Tal, A., and A. Nemirovski (1998). Robust convex optimization. *Mathematics of Operations Research* **23**(4):769-805.
4. Ben-Tal, A., and A. Nemirovski (1999). Robust solutions of uncertain linear programs. *Operations Research Letters* **25**(1):1-13.
5. El Ghaoui, L., F. Oustry, and H. Lebret (1998). Robust solutions to uncertain semidefinite programs. *SIAM Journal on Optimization* **9**(1):33-52.
6. Tsang, K. H., N. J. Samsatli, and N. Shah (2007). Capacity investment planning for multiple vaccines under uncertainty: 2. Financial risk analysis. *Food and Bioproducts Processing* **85**(C2):129-140.
7. Verderame, P. M. and C. A. Floudas (2010). Operational planning of large-scale industrial batch plants under demand due date and amount uncertainty: II. Conditional Value-at-Risk framework. *Industrial & Engineering Chemistry Research* **49**(1):260-275.
8. Rockafellar, R. T. and S. Uryasev (2000). Optimization of conditional value-at-risk. *Journal of Risk* **2** (3):21-41.

See more of this Session: Design and Operations Under Uncertainty II
See more of this Group/Topical: Computing and Systems Technology Division

(581d) Robust Optimization of the Capacitated Vehicle Routing Problem Under Demand Uncertainty

Chrysanthos E. Gounaris[1], Wolfram Wiesemann[2] and Christodoulos A. Floudas[1], (1)Department of Chemical and Biological Engineering, Princeton University, Princeton, NJ, (2)Department of Computing, Imperial College London, London, England

Vehicle Routing has evolved as one of the core operations research and combinatorial optimization problem classes and covers a wide range of supply chain and production scheduling applications. In this presentation, we focus on the well-known "Capacitated Vehicle Routing Problem" (CVRP) [1,2]. The CVRP considers a set of geographically scattered customers that should be served by a fleet of identical finite-capacity vehicles. All vehicles are stationed at a depot, and the objective is to select a set of round-trip routes that serve all customer demands at minimum cost. The routes must be designed such that each customer is visited only once by exactly one vehicle, and the cumulative demand serviced by any given vehicle must not exceed the vehicle capacity.

In the deterministic version of the CVRP, the demand of each customer is known a priori. Instances of this problem with up to a few hundred customers can be solved efficiently by branch-and-cut [3,4] or branch-and-cut-and-price [5,6,7] methods. Here, we consider uncertainty in the customer demands, that is, the demand realizations are unknown at the time the vehicle routes are selected. In this setting, we aim at choosing robust vehicle routes, i.e., routes that are feasible for any possible realization of the demand vector within a prespecified uncertainty set [8,9,10,11]. Previous efforts to address demand uncertainty in the CVRP [12,13] have focused on problems where each customer demand can attain its worst-case individually, independent of the other customers. Problems of this type can be solved via deterministic CVRP methods, but their solutions may be very conservative.

We address the more general case where the vector of all customer demands lies in a prespecified (typically non-rectangular) uncertainty set [14]. Suitable uncertainty sets can be derived from statistical estimation procedures, and they produce significantly less conservative solutions in practice. We derive four families of robust CVRP formulations, which are based on Generalized Subtour Elimination constraints, Miller-Tucker-Zemlin constraints, commodity-flow models, and a novel vehicle-to-customer assignment model. All of these formulations are exact, that is, they do not exhibit any over-conservatism beyond the specification of the uncertaint set. We present a branch-and-cut procedure that allows us to solve instances with up to 50 customers to optimality, and we provide computational results that illustrate our approach.

[1] Laporte G. "Fifty years of vehicle routing", Transportation Science, 43(4):408-416 (2009).

[2] Baldacci R., P. Toth and D. Vigo. "Exact algorithms for routing problems under vehicle capacity constraints", Annals of Operations Research, 175(1):213-245 (2010).

[3] Lysgaard J., A.N. Letchford and R.W. Eglese. "A new branch-and-cut algorithm for the capacitated vehicle routing problem". Mathematical Programming, Ser.A 100:423-445 (2004).

[4] Gounaris C.E., P.P. Repoussis, C.D. Tarantilis, and C.A. Floudas "A Hybrid Branch-and-Cut Approach for the Capacitated Vehicle Routing Problem". In: E.N. Pistikopoulos, M.C. Georgiadis and A. Kokossis (Eds.) 21st European Symposium on Computer Aided Process Engineering – ESCAPE 21 (2011).

[5] Fukasawa R., H. Longo, J. Lysgaard, M.P. de Aragao, M. Reis, E. Uchoa and R.F. Werneck. "Robust branch-and-cut-and-price for the capacitated vehicle routing problem". Mathematical Programming, Ser.A 106:491-511 (2006).

[6] Baldacci R., N. Chrstofides and A. Mingozzi. "An exact algorithm for the vehicle routing problem based on the set partitioning formulation with additional cuts". Mathematical Programming, 115:351-385 (2008).

[7] Baldacci R., E. Bartolini, A. Mingozzi and R. Roberti. "An exact solution framework for a broad class of vehicle routing problems". Computational Management Science 7:229-268(2010).

[8] Ben-Tal A. and A. Nemirovski. "Robust convex optimization". Mathematics of Operations Research, 23(4):769-805 (1998).

[9] Bertsimas D. and M. Sim. "The price of robustness". Operations Research, 52(1):35-53 (2004).

[10] Lin X., S.L. Janak and C.A. Floudas. "A new robust optimization approach for scheduling under uncertainty: I. Bounded uncertainty". Computers & Chemical Engineering, 28(6-7):1069-1085 (2004).

[11] Janak S.L., X. Lin and C.A. Floudas. "A new robust optimization approach for scheduling under uncertainty: II. Uncertainty with known probability distribution". Computers & Chemical Engineering, 31(3):171-195 (2007).

[12] Sungur I., F. Ordonez and M. Dessouky. "A robust optimization approach for the capacitated vehicle routing problem with demand uncertainty". IIE Transactions, 40(5):509-523 (2008).

[13] Sungur I., Y. Ren, F. Ordonez, M. Dessouky and H. Zhong. "A Model and Algorithm for the Courier Delivery Problem with Uncertainty". Transportation Science, 44(2):193-205 (2010).

[14] Gounaris C.E., W. Wiesemann and C.A. Floudas. "Robust Vehicle Routing". In preparation (2011).

(581e) Managing Large-Scale Uncertainty In Process Flow Sheet Synthesis and Design

Mihael Kasa, Zdravko Kravanja and Zorka Novak Pintari•, Faculty of Chemistry and Chemical Engineering, University of Maribor, Maribor, Slovenia

Introduction

Design and synthesis of processes in chemical, biochemical and pharmaceutical industry involve a great deal of uncertainty in the external, measured and model data. Including the uncertainty into design approaches brings the flexibility to process flow sheets, however, it requires transformation of deterministic mathematical formulation into the stochastic problem, and further, into its deterministic equivalent. The latter represents much larger mathematical model than the deterministic, therefore, the computational effort to find its optimal solution could be extremely high or even impossible. This is particularly problematic in large-scale process systems with several tens or even hundreds of uncertain parameters. Various approaches have been proposed for reducing the number of scenarios in stochastic programming, e.g. Karuppiah et al. 2010. Authors of this contribution have also proposed a few approaches for MINLP synthesis of flexible processes flow sheets (Novak Pintari• and Kravanja, 2004). The emphasize of this contribution is on the approaches for solving flow sheet problems with large number of uncertain parameters (up to 100) that have not yet been managed satisfactorily.

Outline of the approach for managing large-scale uncertainty

The proposed approach is based on a two-stage stochastic programming formulation with fixed recourse. Its main feature is the approximation of stochastic result by using a small set of discrete points while ensuring flexibility of flow sheet design within the pre-specified variations of uncertain parameters. The main effort is on the transformation of the flow sheet's stochastic formulation into its minimum deterministic equivalent. The latter is defined over a considerably reduced set of scenarios (points) which are determined in advance either for fixed process topology in the case of process design, or for topological alternatives in process synthesis by using a mixed integer nonlinear programming (MINLP) approach. The reduced set of scenarios is composed of the nominal point and the minimum set of vertex points, called critical vertices. The nominal point could be a mean or a mode value for skewed distributions. The critical vertices define the values of the first-stage (design) variables in the flow sheet model, while the expected cost of the second-stage (operating) variables is approximated at the nominal point.

Identification of critical vertices

Identification of critical vertices is the most difficult step of the proposed approach. First, a sensitivity analysis of the influence of uncertain parameters on design variables is carried out. The parameters are classified into three groups based on the results of the sensitivity analysis: a) parameters with no influence, b) parameters with monotonic influence, and c) parameters with non-monotonic influence on design variables. The non-influential parameters could be fixed at any arbitrary value in the deterministic equivalent model. Those parameters with monotonic influence are fixed to the worst-case values, i.e. either the lower or upper bound.

Determination of critical vertices is performed only for those parameters that show non-monotonic influence during the sensitivity analysis. In previous work, authors have proposed several approaches for identification of critical points ranging from the exact formulation based on the Karush-Kuhn-Tucker conditions to the heuristic ones (Novak Pintaric and Kravanja, 2008). The common idea of all these methods was to identify those critical values of uncertain parameters that determine the largest values of first stage-variables at minimum cost. This requires the solution of min-max problem which could not be solved straightforwardly.

While not all of the proposed methods are suitable to manage large-scale processes with large number of uncertain parameters, in recent research, two methods for finding the critical vertices have been further investigating: approximate one-level method (Novak Pintaric and Kravanja, 2008), and a novel one, a two-stage method using the external functions in GAMS.

a) Approximate one-level method

In this method, a simple deterministic model is solved for each first-stage variable. Both objectives, minimization of the cost function and maximization of the design variable, are accomplished simultaneously in this model, while uncertain parameters are treated as optimization variables. Approximate min-max solution is obtained by a composite objective function, which consists of the minimum cost function from which the design variable is subtracted multiplied by a large positive constant (big-M). The results of the model are the critical values of uncertain parameters that maximize design variable under study. The maximum number of critical points would be as high as the number of the first-stage variables, but is usually considerably lower. The identified vertices can be further reduced by using a set-covering procedure. A drawback of one-level method is its sensitiveness to the value of big-M constant which could influence the selection of critical vertices. This challenge is studied further in this research.

b) Two-level method with external functions

This method starts with the solution of the deterministic problem at a single, i.e. the nominal point. As the solution obtained is inflexible, the objective value represents a lower bound for minimization problem. In the next iterations, critical vertices are added to the model one by one, and dimensionality of the problem gradually grows from iteration to iteration. The additional critical point at every major iteration is calculated simultaneously as the optimization variable by using the external functions in GAMS program, which enable to solve min-max problem iteratively. Adding the critical points into the model forces the objective value to rise. The complete minimum set of critical vertices is found when the objective value stops rising. The case-studied examples indicate fast convergence of this method and precise identification of critical vertices, however, a termination criterion still needs to be improved. The advantage of this method is also that the iteration at which all critical vertices are identified represents a solution of deterministic equivalent problem whose result is the approximate flexible flow sheet design.

Conclusions

This contribution presents an approach for approximating a stochastic model of process flow sheet under uncertainty by a minimum deterministic equivalent. The main goal was to improve and upgrade those approaches developed by authors in the past in order to be able of managing industrial flow sheet problems with high level of uncertainty.

Minimum deterministic equivalent of stochastic flow sheet model is defined over significantly reduced set of scenarios composed of the nominal point and critical vertices. Prescreening approach was introduced in order to identify those uncertain parameters that influence design variables. Two methods are proposed for efficient and quick identification of critical vertices in large-scale problems with large number of uncertain parameters. Some illustrating examples will be presented in order to illustrate the main characteristics of the proposed approach, as well as a synthesis of large-scale bio-ethanol process via MINLP. By managing large-scale uncertainty in process flow sheet synthesis and design, mathematical programming evolves towards the efficient supporting tool for investment decision-making in process industries.

References

Karuppiah R., Martín M., Grossmann I. E. A simple heuristic for reducing the number of scenarios in two-stage stochastic programming. Comput. Chem. Eng. 2010, 34, 1246–1255.

Novak Pintari• Z., Kravanja Z. A strategy for MINLP synthesis of flexible and operable processes.

Comput. Chem. Eng. 2004, 28, 1105–1119.

Novak Pintari• Z., Kravanja Z. Identification of critical points for the design and synthesis of flexible processes. Comput. Chem. Eng. 2008, 32, 1603–1624.

See more of this Session: Design and Operations Under Uncertainty II
See more of this Group/Topical: Computing and Systems Technology Division

Improving Supply Chain Management in a Competitive Environment under Uncertainty

M. Zamarripa and A. Espuña, Universidad Politécnica de Cataluña, Barcelona, Spain.

Abstract

The consideration of uncertainty is highly important in nowadays and there are lots of academic and industrial based studies that deal with uncertainty in the parameters of the problem to make decisions in production planning problems. This work aims to optimize the planning of Supply Chains (SCs) in competitive environments under demand and market competition uncertainty. The proposed model approach considers a two stage stochastic programming formulation to solve a MILP (Mixed integer linear programming) or a MINLP (Mixed Integer Non-Linear Programming) problem to improve the decision making on the operation of different Supply Chains; different optimization criteria are considered to deal with cooperative and non cooperative behavior between SCs.

Introduction

The problem of decision making in the chemical process industry becomes more complex as the scope covered by decisions is extended. This increasing complexity is additionally complicated by the need to consider a certain degree of uncertainty in the models used to forecast the events that should be considered in this decision making. But although some of the published works in this area explicitly address the problem associated with the uncertainty in the available data, very few of them put the issue in a realistic operational environment, taking into account that the SC of interest will have to compete with other Supply Chains which, logically, would also like to work as efficiently as possible to cover the same demands on the basis of raw materials obtained in the same (global) markets. But even an efficient behavior should be assumed for these third parties (competitors), their specific way to face demands, suppliers, etc. is, from the point of view of the own interest, a new uncertain element to be considered.

Within this scope, nowadays it is not enough to study the deterministic problems that the industry deals with. There are some works (Lababidi et al (2004), Gupta & Maranas (2003), You and Grossmann (2010), Pistikopoulos (2004), Wellons & Reklaitis (1989), etc.) that solve the SC planning under uncertainty in terms of demand and market prices, using tools like Model Predictive Control, Multi-parametric programming and stochastic programming. In order to deal with this more realistic problem (introduce uncertainty as market competition behavior this work proposes to introduce into the model information, the expected performance of the competing supply chains (including the SC of interest); such information is entered in form of scenarios to manage the uncertainty in the optimization model. Based on this information, it is possible to construct a mathematical model to solve the problem using stochastic programming considering different sources of uncertainty.

This paper addresses the SC planning problem (inventory, production and distribution tasks) for Supply Chains that work in a common scenario. As mention before it is necessary consider a greater degree of uncertainty, and develop more complex models to solve the same supply chain problems. In these sense, this work aims to develop a model that considers several scenarios for the demand satisfaction using stochastic programming to solve the problem. It is also important to consider a higher complexity in the models; e. g. considers uncertainty, non linear models, etc.

Problem Statement

Supply Chain Planning

The typical scope of a SC planning problem is to determine the optimal production, inventory and distribution levels in the organization network (production centers, distribution equipments, storage centers, and market places), taking care of the constraints of the raw materials, production and distribution limits. These problems are typically formulated as MILP problems and are solved using mathematical programming tools, heuristics rules, metaheuristics, etc.

The model originally proposed by Zamarripa et al 2011 (this model introduce the use of game theory as a decision technique to determine the optimal SC production, inventory and distribution levels in a competitive planning scenario, and model the competition behavior of several SC's as an uncertainty source) has been adopted as a basis for the formulation presented in this paper, and consists in a two stage stochastic programming. Also this work considers as organization network; several SC's (production sites, storage centers and distribution options) that should face the same market competition for multiple time periods, there are some availability of machine and man hours to limit the production and there is a fixed capacity to distribute the products to the nodes of the network, also there are some fixed cost associated to the production, distribution and the storage of the products.

SC Planning under Uncertainty

Since 1980 there is an important increase of optimization applications in the study of the uncertainty behavior in the chemical systems, production planning and scheduling systems. Dolgui et al 2002, identify different sources of uncertainties along the supply chain (supplying reliability, assembly and manufacturing random lead times; random level and customers demand), the oldest fashion solutions for this problems was the use of safety stocks. Due to this, the academy focuses its efforts to solve the demand uncertainty in the production problems.

Sahinidis 2004, makes an exhaustive revision of the state of the art of the optimization under uncertainty setting as a principal recourse the two stage stochastic programming; the first stage variables are those that have to be decided before known the uncertainty parameters, subsequently, the random scenarios have presented the policy of the model selects the values of the second stage variables.

This model approach sets the production as first stage variable, being the second stage variables the storage and distribution levels, to be fixed/optimized once the uncertain parameters mentioned before are revealed.

The problem consists in two criteria to improve the decision making:

- Minimize the total cost of the Supply Chain under consideration (production, inventory, distribution and backorder costs)

$$
\begin{aligned}
Min\ z1(g) = & \sum_{i \in I\_G(i,g)} \sum_{n} \sum_{h} a_{in} Q_{inh} (1+e_b)^h \\
& + \sum_{sc} \frac{1}{SC} \sum_{i \in I\_G(i,g)} \sum_{n} \sum_{h} c_{in} W_{scinh} (1+e_b)^h + \sum_{i \in I\_G(i,g)} \sum_{n} \sum_{h} d_{in} E_{scinh} (1+e_b)^h \\
& + \sum_{i \in I\_G(i,g)} \sum_{n} \sum_{h} \sum_{j} k_{inj} T_{scinhj} (1+e_b)^h
\end{aligned}
$$

- Minimize the expenses of the buyers.

$$
Min\ CST(g) = \sum_{i \in I\_G(i,g)} \sum_{n} \sum_{h} \sum_{j} Ps_{inj} T_{scinhj}\, Prate + \sum_{sc} \frac{1}{SC} \sum_{i \in I\_G(i,g)} \sum_{n} \sum_{h} \sum_{j} Ps_{inj} T_{scinhj}\, Disc_{sc}
$$

Several Constraints are introduced to manage:

- Material balances.
- Production capacity.
- Storage limit.
- Demand satisfaction.
- Law of demand elasticity.
- Distribution capacity.
- Budget capacity.

Parameters under uncertainty

- **Demand.** As mentioned before, different scenarios (sc) for the satisfaction of the consumers are usually defined to take into account the overall effect of all exogenous sources of uncertainty. This is usually modeled as a normal probability distribution curve in the products demand. Since all the exogenous sources of uncertainty are considered to be included in the "Demand uncertainty" model, only the Total Cost of the Supply Chain under consideration (*z1*) is needed to optimize the SC behavior.
- **Competence behavior.** Actually, the demand to be covered is the result of the uncertain market demand and the demand covered by the competitor SCs, which will depend on the uncertain competitors' behavior. One way to consider this competition is to assume that the markets get the products from the cheaper SC; then, the second criteria above ("minimize the expenses of the buyers") can be included in the model as a second objective to be optimized.
  In order to manage this specific source of uncertainty, the eventual discount in the price of the products has been introduced as a new "first stage" variable to be optimized.
  Obviously a new challenge appears, related to the evaluation of uncertain behavior of the competitor SCs.

Case study

These concepts have been applied to a supply chain case study adapted from Liang (2008). The factory's strategy is to maintain a constant work force level over the planning horizon, and supply as much product as possible (demanded), playing with inventories and backorders. Two products are considered (P1 and P2) with a market demand of 3 months horizon demanded from 4 distribution centers (Distr1 to Distr4). The information about the considered scenarios, production, etc. and the rest of problem conditions (initial storage levels, transport capacities, etc.) can be found at http://cepima.upc.edu/papers/Competitive_SCs.pdf (Tables 3-6).

The cooperative problem is solved as a MILP that minimize the total cost z1 (sum of transportation, production and inventory costs). Also, the benefit is calculated (difference between sales and total cost) of each SC is estimated to highlight the results (cooperative/competitive). The competitive problem is solved as a MINLP that minimize the expense of the buyers CST. Figure 1 shows the considered SCs basic configuration, composed by 2 SCs (2+2 plants, Plant1/Plant2 and Plant3/Plant4) which collaborate or compete to fulfill the global demand from the 4 distribution centers.

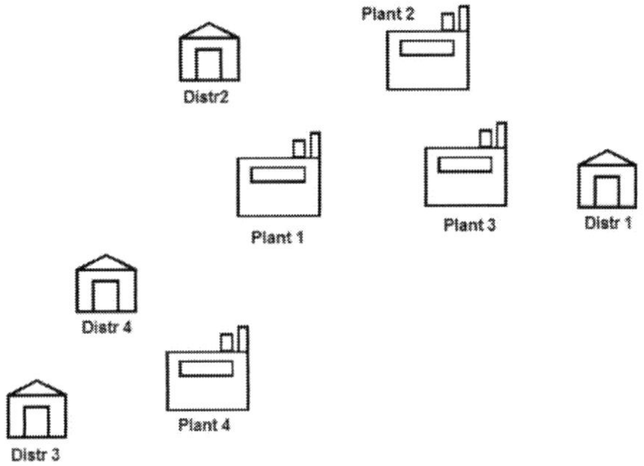

Figure 1 Description of the problem. Plant1-4 serves D1-4.

Results

Demand Uncertainty

This paper considers 3 different scenarios for the product demand (table 1). The deterministic solution for each scenario are presented in table 2.

From figure 2 it can be observed the inventory levels when the deterministic problem is solved for each scenario considered in the stochastic problem, after solving the deterministic cases proceed to resolve the stochastic programming model.

Table 1. Products demand for each distribution centers (D1-D4) for each scenario (sc1-3).

Destiny-Scenario		D1			D2			D3			D4		
Product	time	sc1	sc2	sc3	sc1	sc2	sc3	sc1	sc2	sc3	sc1	sc2	sc3
P1	t1	750	1000	1250	615	820	1025	375	500	625	923	1230	1538
	t2	2250	3000	3750	173	230	288	900	1200	1500	2550	3400	4250
	t3	375	500	625	3000	4000	5000	1800	2400	3000	3975	5300	6625
P2	t1	488	650	813	375	500	625	225	300	375	533	710	888
	t2	683	910	1138	540	720	900	300	400	500	5288	7050	8813
	t3	2250	3000	3750	1800	2400	3000	863	1150	1438	2325	3100	3875

Table 2. Solution for the deterministic cases

	Scenario 1		Scenario 2		Scenario 3	
	SC1	SC2	SC1	SC2	SC1	SC2
Cost	382081	210663	532941	269572	649097	363342
Total cost	592744		802513		1012439	
Benefit	1744418	1036129	2384058	1311079	2894902	1714612

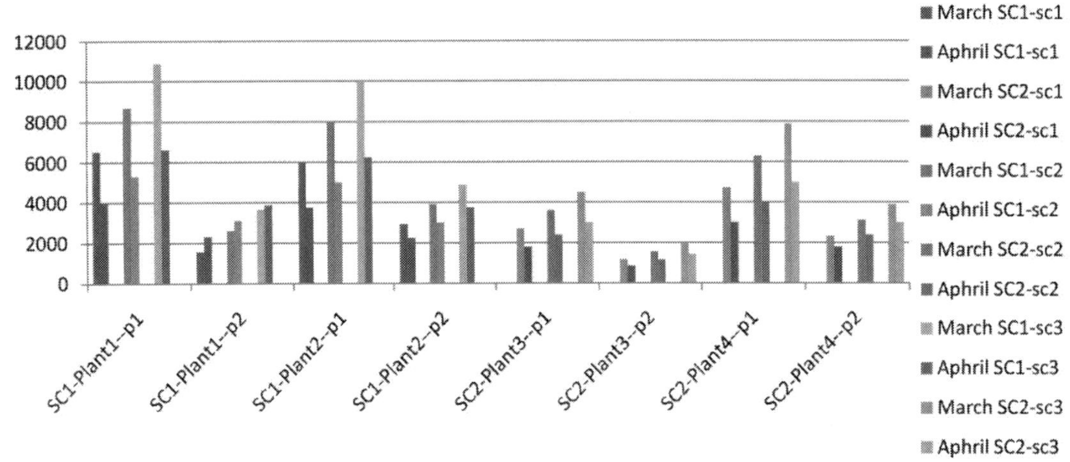

Figure 2. Optimal Inventory levels for the different Supply Chains (SC1 and SC2) at the different scenarios

Using the same case study, the proposed approach intends to obtain the production level necessary to achieve the different demands for the products (3 scenarios). The two stage stochastic programming model consists: the first stage variables which are the production and raw materials, and the second stage variables are the inventory and distribution tasks (equation 1). Some considerations are made in the problem: since the consumers in this case are the distribution centers of the Supply Chain suppose that the production centers can deliver more than the demanded in each time; because the problem needs to produce to satisfy the higher demand of the scenarios and can't storage the product at the end of the time horizon.

The solution for the stochastic model is given in tables 3-5; in the table 3 can be found the distribution tasks for the different scenarios of the problem, the table 4 contains the different inventory levels, and the production level is shown in Table 5. The solution of production levels is the same for all three scenarios, due to the need to set production levels before knowing the demand.

Table 3. Distribution tasks for each scenario.

Plant/product	Destiny		D1			D2			D3			D4		
	time\scenario		sc1	sc2	sc3	sc1	sc2	sc3	sc1	sc2	sc3	sc1	sc2	sc3
Plant 1	P1	April	0	0	0	4975	2493	10	0	0	0	923	1230	1538
		March	0	0	0	0	0	0	0	0	0	2550	3400	4250
		May	0	0	0	0	0	0	0	0	0	3975	5300	6625
	P2	April	0	0	0	2440	1226	10	0	0	0	533	710	888
		March	0	0	0	0	0	0	0	0	0	788	1050	1313
		May	0	0	0	0	0	0	0	0	0	2325	3100	3875
Plant 2	P1	April	5250	3250	1250	0	0	0	0	0	0	0	0	0
		March	2250	3000	3750	0	0	0	0	0	0	0	0	0
		May	3750	5000	6250	0	0	0	0	0	0	0	0	0
	P2	April	2749	1772	794	0	0	0	0	0	0	0	0	0
		March	683	910	1138	0	0	0	0	0	0	0	0	0
		May	2250	3000	3750	0	0	0	0	0	0	0	0	0
Plant 3	P1	April	0	0	0	0	0	0	2425	1525	625	0	0	0
		March	0	0	0	0	0	0	900	1200	1500	0	0	0
		May	0	0	0	0	0	0	1800	2400	3000	0	0	0
	P2	April	0	0	0	0	0	0	1140	753	365	0	0	0
		March	0	0	0	0	0	0	300	400	500	0	0	0
		May	0	0	0	0	0	0	863	1150	1438	0	0	0
Plant 4	P1	April	0	0	0	4165	2590	1015	0	0	0	0	0	0
		March	0	0	0	1725	2300	2875	0	0	0	0	0	0
		May	0	0	0	3000	4000	5000	0	0	0	0	0	0
	P2	April	10	10	19	1567	1095	615	627	319	10	0	0	0
		March	0	0	0	540	720	900	0	0	0	0	0	0
		May	0	0	0	1800	2400	3000	0	0	0	0	0	0

Table 4. Inventory levels.

Plant/product		sc1 (scenario 1)			sc2 (scenario 2)			sc3 (scenario 3)		
		April	March	May	April	March	May	April	March	May
PLANT 1	P1	6523	3975	0	8698	5300	0	10873	6625	0
	P2	1575	2325	0	2612	3100	0	3650	3875	0
PLANT 2	P1	6000	3750	0	8000	5000	0	10000	6250	0
	P2	2933	3350	0	3910	3000	0	4888	3750	0
PLANT 3	P1	2700	1800	0	3600	2400	0	4500	3000	0

	P2	1163	863	0	1550	1150	0	1938	1438	0
PLANT 4	P1	4725	3000	0	6300	4000	0	7875	5000	0
	P2	2340	1800	0	3120	2400	0	3900	3000	0

Table 5. Production level

PRODUCTION		April	March	May
PLANT 1	P1	120221	2	0
	P2	4348	1538	0
PLANT 2	P1	10950	0	0
	P2	5482	0	0
PLANT 3	P1	4725	0	0
	P2	2103	0	0
PLANT 4	P1	8590	0	0
	P2	4344	0	0

Market Competition Uncertainty

The proposed approach to model the SCs competition has been used to solve the same case study presented before, considering a fixed demand. This work looks for the best discount rate in the price for the products of Supply Chain 1 (Plant 1 and Plant 2), considering 4 scenarios of discount (percent) of the prices for the Supply Chain 2 (Plant 3 and Plant 4) sc1=0.1, sc2=0.2, sc3=0.3, sc4=0.4. As mentioned above this work uses the solution of the deterministic cases when the competition chooses the optimum discount rates. The best solution for the SC1is summarized by the discount rates shown in Table 6. The corresponding payoff matrix can be found in: http://cepima.upc.edu/papers/Competitive_SCs.pdf.

Table 6. Deterministic solutions for the different scenarios of the discount rate.

	Scenario SC2			
	sc1	sc2	sc3	sc4
Best solution of SC1	0.3	0.4	0.4	0.3

The solutions for the stochastic non linear model are given considering the different scenarios (tables 7-9), also this model choose the discount in the price for the Supply Chain 1 (discount of 0.3%).

Table 7. Production levels

		sc1 (scenario 1)			sc2 (scenario 2)			sc3 (scenario 3)			sc4 (scenario 4)		
Production Plant/product		April	March	May	April	March	May	April	March	May	April	March	May
PLANT 1	P1	14852	1498	0	12650	0	0	10350	0	0	9530	0	0
	P2	810	7270	0	3562	4518	0	5160	0	0	4660	0	0
PLANT 2	P1	8700	0	0	8700	0	0	8700	0	0	8700	0	0
	P2	4360	0	0	4360	0	0	4360	0	0	4360	0	0
PLANT 3	P1	3700	0	0	3700	0	0	3700	0	0	3700	0	0
	P2	1650	0	0	1650	0	0	1650	0	0	1650	0	0

PLANT 4	P1	0	0	0	3700	0	0	6000	0	0	6820	0	0
	P2	0	0	0	0	0	0	2920	0	0	3420	0	0

Table 8. Distribution tasks.

Plant/product		Destiny	D1				D2				D3				D4			
		time\scenario	sc1	sc2	sc3	sc4	sc1	sc2	sc3	sc4	sc1	sc2	sc3	sc4	sc1	sc2	sc3	sc4
Plant 1	P1	april	0	0	0	0	820	820	820	0	0	0	0	0	1230	1230	1230	1230
		March	0	0	0	0	2300	2300	0	0	0	0	0	0	3400	3400	3400	3400
		May	0	0	0	0	3700	0	0	0	0	0	0	0	5300	5300	5300	5300
	P2	april	0	0	0	0	300	500	500	0	0	0	0	0	710	710	710	710
		March	0	0	0	0	720	720	0	0	0	0	0	0	1050	1050	1050	1050
		May	0	0	0	0	2400	2200	0	0	0	0	0	0	3100	3100	3100	3100
Plant 2	P1	april	1000	1000	1000	0	0	0	0	0	0	0	0	0	0	0	0	0
		March	3000	3000	3000	0	0	0	0	0	0	0	0	0	0	0	0	0
		May	5000	5000	5000	0	0	0	0	0	0	0	0	0	0	0	0	0
	P2	april	650	650	650	0	0	0	0	0	0	0	0	0	0	0	0	0
		March	910	910	910	0	0	0	0	0	0	0	0	0	0	0	0	0
		May	3000	3000	3000	0	0	0	0	0	0	0	0	0	0	0	0	0
Plant 3	P1	april	0	0	0	0	0	0	0	0	500	500	500	500	0	0	0	0
		March	0	0	0	0	0	0	0	0	1200	1200	1200	1200	0	0	0	0
		May	0	0	0	0	0	0	0	0	2400	2400	2400	2400	0	0	0	0
	P2	april	0	0	0	0	0	0	0	0	300	300	300	300	0	0	0	0
		March	0	0	0	0	0	0	0	0	400	400	400	400	0	0	0	0
		May	0	0	0	0	0	0	0	0	1150	1150	1150	1150	0	0	0	0
Plant 4	P1	april	0	0	0	0	0	0	0	820	0	0	0	0	0	0	0	0
		March	0	0	0	0	0	0	2300	2300	0	0	0	0	0	0	0	0
		May	0	0	0	0	300	4000	4000	4000	0	0	0	0	0	0	0	0
	P2	april	0	0	0	0	200	0	0	500	0	0	0	0	0	0	0	0
		March	0	0	0	0	0	0	720	720	0	0	0	0	0	0	0	0
		May	0	0	0	0	0	200	2400	2400	0	0	0	0	0	0	0	0

The table 6 shows that in 2 deterministic cases the discount rate will be 0.4% but the global solution of the nonlinear stochastic model determines that there is better to set the discount rate in 0.3%.

The optimal solution of the SC1 (0.3 % of discount) for several scenarios of SC2 is driven by its capacity to adapt the prices, the model takes into account the global demand, and the budget capacity of the production plants. This budget capacity acts as the bottleneck of the problem, since there is a point in which don't care how much discount in the price there isn't more production capacity.

Table 9. Inventory levels.

Inventory Plant/product		sc1 (scenario 1)			sc2 (scenario 2)			sc3 (scenario 3)			sc4 (scenario 4)		
		April	March	May	April	March	May	April	March	May	April	March	May
PLANT 1	P1	13202	9000	0	11000	53000	0	8700	5300	0	8700	5300	0
	P2	0	5500	0	2552	5300	0	4150	3100	0	4150	3100	0
PLANT 2	P1	8000	5000	0	8000	5000	0	8000	5000	0	8000	5000	0
	P2	3910	3000	0	3910	3000	0	3910	3000	0	3910	3000	0
PLANT 3	P1	3600	2400	0	3600	2400	0	3600	2400	0	3600	2400	0
	P2	1550	1150	0	1550	1150	0	1550	1150	0	1550	1150	0
PLANT 4	P1	300	300	0	4000	4000	0	6300	4000	0	6300	4000	0
	P2	0	0	0	200	200	0	3120	2400	0	3120	2400	0

Finally the total cost and the benefit for the competitive case for several scenarios are shown in the next table (table 10). The preferences of the consumers have been modeled as just based on service (due date's maintenance) and customers cost, these elements have been introduced in the final objective function as previously indicated. A nominal selling price has been also introduced to maintain data integrity.

Table 10. Costs summary.

	Scenario 1		Scenario 2		Scenario 3		Scenario 4	
	SC1	SC2	SC1	SC2	SC1	SC2	SC1	SC2
Total Cost	702559	100734	621159	181684	538423	264089	515516	286997
Benefit	3144863	544265	2857373	831285	2419675	1268299	2310978	1375326

CONCLUSIONS

This work introduces the use of the SC competition behavior as a specific source of uncertainty in order to optimize the management of a certain SC, proposing a stochastic non-linear programming model that jointly determines the price discount rates and the production, inventory and distribution decisions to be made in order to optimize the expected performance of this SC.

The obtained solution can be compared with the one resulting from other approaches considering that the effect of the competition can be model as part of the "SC demand uncertainty". Even this approach results in a more easy to solve two stage MILP stochastic model, the obtained results are clearly different, since the approach proposed in this contribution allows the SC manager to consider a new "control parameter" (prices or discount rates) to directly manage this source of uncertainty, obtaining improved solutions at the tactical decision level for real Supply Chain industrial problems.

References

Alexandre Dolgui and Mohamed Aly Ould Louly, 2002, Optimization of supply chain planning under uncertainty, Int. J. Production Economics 78, 145-152.

Haitham M. S. Lababidi, Mohamed A. Ahmed, Imad M. Alatiqi and Adel F. Al-Enzi, 2004, Optimizing the Supply Chain of a Petrochemical Company under Uncertain Operating and Economic Conditions. Ind & Eng. Chem. Res., 43, 63-73.

T. Liang, 2008, Fuzzy multi-objective production/distribution planning decisions with multi product and multitime period in a supply chain, Computers and Industrial Engineering, 55, 678-694.

Gupta, A. & Maranas (2000). A two-stage modeling and solution framework for multisite midterm planning under demand uncertainty. Industrial and Engineering Chemistry Research 39, 3799_/3813

Gupta A. Maranas 2003, Managing demand uncertainty in supply chain planning, Computers and Chemical Engineering 27, 1219-1227.

Nikolaos V. Sahinidis, 2004, Optimization under uncertainty: state of the art and opportunities, Computers and Chemical Engineering, 28, 971-983.

Wellons, H. S., & Reklaitis, G. V. 1989. The design of multiproduct batch plants under uncertainty with staged expansion. Computers and Chemical Engineering 13, 115.

Fengqi You and Ignacio E. Grossmann, 2010, Stochastic inventory management for tactical process planning under uncertainties: MINLP Models and Algorithms, 57,1250-1277.

Miguel Zamarripa, Adrian Aguirre, Carlos Mendez, Antonio Espuña, 2011, Improving supply chain management in a competitive environment, 21 European Symposium on Computer Aided Process Engineering- ESCAPE21, Elsevier, 1000-1005.

(581g) Stochastic Programming and Uncertainty Management In Electricity System Operation

Bri-Mathias S. Hodge, Erik Ela and Michael Milligan, National Renewable Energy Laboratory, Golden, CO

The electricity grid is one of the most complex systems designed by man and its continued reliable operation is essential for modern life. The system is currently undergoing substantial changes through the integration of large amounts of variable and uncertain renewable electricity generators, that must be properly managed in order to maintain current levels of system reliability. The scheduling of generating units to meet forecast demand is a key process in the management of the electricity system in the United States, known as unit commitment and economic dispatch. The addition of large amounts of variable generation such as wind and solar power into the process make tools for adequately handling the increased uncertainty involved in system operations a critical component of integration strategies. One such tool is the use of stochastic programming techniques in the unit commitment process.

Wind power forecasting is another important tool for mitigating the uncertainty involved in integrating large amounts of renewable electricity generation into the electricity grid. The wind power forecast error distribution assumed can have a large impact on the confidence intervals produced in wind power forecasting. We have examined the shape of the error distributions produced using different forecasting techniques for a number of wind plants in United States interconnections over multiple timescales [1]. Comparisons are made between the experimental distribution shapes and that of some commonly assumed distributions, including the normal [2-5], beta [6] and Weibull [7] distributions. The shape of the distribution is found to change significantly with the length of the forecasting timescale and the forecasting technique used. The Cauchy distribution is proposed as a model distribution for the forecast errors associated with both the persistence model and numerical weather prediction models commonly used in system operations. Parameters for the Cauchy distribution have been fitted for a number of different wind plants and timescales.

The differences in confidence intervals obtained using the newly characterized distributions and the commonly assumed normal distribution are compared. The practical implications of these different confidence intervals are then examined through their use in a stochastic unit commitment and economic dispatch model. The advantages of utilizing stochastic programming in the unit commitment process for systems with high wind penetration levels have been previously demonstrated [8]. In this work we demonstrate the additional advantages that can be realized from a more accurate characterization of the wind forecasting errors.

References:

[1] B.-M. Hodge and M. Milligan, "Wind Power Forecasting Error Distributions over Multiple

TImescales," in *IEEE Power & Energy Society General Meeting*, Detroit, MI, 2011.

[2] K. Methaprayoon, W. J. Lee, C. Yingvivatanapong, and J. Liao, "An integration of ANN wind power estimation into UC considering the forecasting uncertainty," in *IEEE Industrial and Commercial Power Systems Technical Conference*, Saratoga Springs, NY, 2005.

[3] V. Pappala, I. Erlich, K. Rohrig, and J. Dobschinski, "A Stochastic Model for the Optimal Operation of a Wind-Thermal Power System," *IEEE Transactions on Power Systems,* vol. 24, pp. 940 - 950, May 2009.

[4] E. Castronuovo and J. Lopes, "On the Optimization of the Daily Operation of a Wind-Hydro Power Plant," *IEEE Transactions on Power Systems,* vol. 19, pp. 1599 - 1606, August 2004.

[5] R. Doherty and M. O'Malley, "A New Approach to Quantify Reserve Demand in Systems With Significant Installed Wind Capacity," *IEEE Transactions on Power Systems,* vol. 20, pp. 587 - 595, May 2005.

[6] K. Dietrich, J. Latorre, L. Olmos, A. Ramos, and I. Perez-Arriaga, "Stochastic unit commitment considering uncertain wind production in an isolated system," in *4th Conference on Energy Economics and Technology*, Dresden, Germany, 2009.

[7] H. Bludzuweit, J. A. Dominguez-Navarro, and A. Llombart, "Statistical Analysis of Wind Power Forecast Error," *IEEE Transactions on Power Systems,* vol. 23, pp. 983 - 991, August 2008.

[8] A. Tuohy, P. Meibom, E. Denny, and M. O'Malley, "Unit Commitment for Systems with Significant Wind Penetration," *IEEE Transactions on Power Systems,* pp. 592-601, May 2009.

See more of this Session: Design and Operations Under Uncertainty II
See more of this Group/Topical: Computing and Systems Technology Division

(595a) In-Silico Design of Experiments As a Tool for Nonlinear Sensitivity Analysis of Detailed Models

Alexandros Kiparissides[1], A. Mantalaris[1], Stratos Pistikopoulos[1] and **Christos Georgakis**[2], (1)Centre for Process Systems Engineering, Department of Chemical Engineering, Imperial College, London, United Kingdom, (2)Systems Research Institute and Chemical and Biological Engineering, Tufts University, Medford, MA

Mathematical modellingprovides a systematic way to study quantitatively the behaviour of complexsystems composed of numerous components and some of which might be described atdifferent levels of detail. Engineers and scientists who postulate such modelsinvariably chose to include more rather than less detail is their descriptionof the physical, chemical or/and biological system of interest. The inclusionof more and more details in such models increases the number of parameters thatneed be estimated from experimental data. The estimation of an increased numberof parameters increases the number of experimental data needed. However, evenif this requirement is met, the estimation of parameters associated withinsignificant model details is extremely difficult because of their minimalimpact on the model predictions being matched to the experimental data. Thisdifficulty cannot be easily removed because we do not know *a priori*which model components are insignificant in the region of present interest.Even in the case where the values of the model parameters are known, it is ofinterest to investigate the sensitivity of the model predictions on the valuesof the input variables of the process. Exhaustive simulations, especially ofvery complex models, do not provide an effective answer to the problem becauseof the computational cost. Knowing which parameters and to what extenthave a significant effect on the model output allows for their accurateestimation through tailor-made experiments (Kiparissides et al 2011).Parameters with negligible sensitivity, quantified by a Sensitivity Index (SI),can be fixed at their literature or their initially assumed value as they donot affect the model predictions

Existing model sensitivity methodsare commonly classified in three broad categories: screening, local and globalmethods (Saltelli, 2000). For the general case of non-linear ODE models,commonly met in engineering applications, global methods and variance basedglobal methods in particular have the edge over their counterparts as discussedpreviously (Chan et al 1997, Saltelli et al 2000, Kiparissides et al 2009). Briefly, the ability to estimate higher-level indices, which quantify theeffect on the output of parameter-parameter interactions, gives global methodsthe edge over their local counterparts. Alas, global methods usually requirethe estimation of a large, often prohibitive, number of model evaluations.

Here we propose touse, in a different context, a methodology developed more than fifty years ago.It is called "Design of Experiments" and has become a classic experimental toolin understanding how different operating conditions affect the performance of aprocess (see for example Box and Draper

(1987)). Herein, we will apply this techniqueon the mathematical model of the process instead of on an experimental set-up, hencethe use of the term *in silico*. Similar to its original use in extractingthe maximum amount of information from the process with the minimum number ofproperly designed experiments, we will use it here so that we can obtain the maximuminformation on the sensitivity of the model to its parameters with the minimum numberof simulations of the full model. This parallel use of the design ofexperiments concepts and tools has also been used to understand the operabilitycharacteristics of continuous chemical processes by Li and Georgakis (2010),where it was called design of selective calculations.

We utilise a DAEmodel that describes in substantial detail hybridoma cell growth and proliferationin batch cultures (Bang and Jarford 2000). We show that the use of DoE for thegeneration of a response surface model (RSM) approximation of the full model canlead to the same quality of sensitivity information as the established variancebased GSA methods at a significantly lower computational cost. We show that theexplicit information in the RSM model provides the needed sensitivityinformation with respect to the parameters of the detailed model. Besides theirrelative absolute magnitude, the RSM model parameters provide additionalinformation that previous methods did not avail. The results of the proposedapproach are in full agreement with those of the Sobol' method, one of the mostwidely used previously, and this is done at a fraction of the computationalcost.

The proposedapproach offers directional and interactional information about the full model ata minimal computational cost. Moreover, we show that by estimating an approximateRSM model, minimal information is lost in terms of parameter significance and sensitivity.This is highlighted by comparing the results of the Sobol' method when appliedto the full model to the results of the Sobol' method when applied to the estimatedRSM model.

References:

1. Box G.E.P., Draper N.R. (1987) Empirical Model Building and Response Surfaces. Wiley Press.
2. Chan K., Saltelli A., Tarantola S. (1997) Sensitivity Analysis of the model output: variance-based methods make the difference. Proceedings of the 1997 Winter Simulation Conference.
3. Jang J.D., Barford J.P. (2000) An unstructured kinetic model of macromolecular metabolism in batch and fed-batch cultures of hybridoma cells producing monoclonal antibodies. Biochem. Eng. Journal, 4, 153-168.
4. Kiparissides A., Kucherenko S., Mantalaris A., Pisitikopoulos E.N. (2009) Global Sensitivity Analysis Challenges in biological systems modeling. Ind. Eng. Chem. Res., 48 (15), pp 7168-7180.
5. Kiparissides A., Koutinas M., Kontoravdi C., Mantalaris A., Pistikopoulos E.N. (2011) °Closing the loop in biological systems modeling From the in silico to the in vitro. Automatica, doi:10.1016/j.automatica.2011.01.013.
6. Li, L., Georgakis, C. (2010), On the Calculation of Operability Sets of Nonlinear High-Dimensional Processes Ind. & Eng. Chem. Res. 49, (17) pp.8035-8047
7. Saltelli A., Chan K., Scott E.M. (2000) Sensitivity Analysis, Wiley Press.

(595b) In Silico Prediction of Cancer GI50

E. A. Whitebay, J. D. Ramsey, B. J. Neely and K. A. M. Gasem, School of Chemical Engineering, Oklahoma State University, Stillwater, OK

Current methods of cancer treatment have improved vastly the survivability of humans from many forms of cancer. Traditional chemotherapy remains a key treatment method, offering distinct advantages over other treatment options, especially in the treatment of metastasized tumors. Much effort has been devoted to identifying new, more effective chemotherapeutic agents. Currently, the National Cancer Institute (NCI) analyzes the potential anticancer activity of over 3,000 compounds per year. However, less than one percent of these drugs will be marketed, mostly due to poor anticancer activity and toxicity issues.

To provide a faster and more accurate methodology for identification of potential anticancer agents, we have developed *in silico* QSAR (quantitative structure-activity relationship) models to predict the concentration of chemical required to reduce the cancer growth rate by 50% (GI_{50}) for the breast cancer cell line MCF-7. These screening models use nonlinear neural networks to provide *a priori* predictions of GI_{50} for potential anticancer agents with an overall root-mean square error of 0.49 Log (GI_{50}). As a screening tool capable of greatly reducing time and expense of new drug development, the model is employed to remove unqualified candidate anticancer compounds, while identifying and retaining highly qualified candidates for experimental trials. Further, this methodology facilitates a greater understanding of the relationship between molecular structure and growth inhibition potential.

See more of this Session: Mathematical and Computational Biosystems Engineering
See more of this Group/Topical: Computing and Systems Technology Division

(595c) A Mesoscopic Membrane Model for the Investigation of Mechanisms Underlying Caveolae-Mediated Endocytosis

Belinda S. Akpa[1], Richard D. Minshall[2], Lewis E Wedgewood[3] and Ludwig C Nitsche[3], (1)Chemical Engineering and Bioengineering, University of Illinois at Chicago, Chicago, IL, (2)Pharmacology and Anesthesiology, University of Illinois at Chicago, Chicago, IL, (3)Chemical Engineering, University of Illinois at Chicago, Chicago, IL

Caveolae, small, flask-shaped invaginations of mammalian plasma membranes, are ubiquitous features of endothelial cells. They comprise about 15% of the cell volume and account for >95% of the plasmalemmal vesicles within the cells. Caveolae are implicated as essential vesicle carriers mediating endocytosis in endothelial cells. In this work, we report the development of a computational model which can address how changes in caveolin-1 oligomerization and key signaling reactions impact upon membrane invagination, budding, and internalization by (i) controlling the stability of caveolin polymers in the plasma membrane and (ii) their ability to facilitate changes in membrane curvature.

A novel mesoscopic, particulate model is developed for cell membranes in which local directionality (surface normal vector) is extracted from the relative arrangement of isotropic particles rather than being embedded in the more complex (i.e., rod-like or multi-bead) internal structure of lipid particles frequently employed in coarse-grained or atomistic simulations. Based on the normal vector, an anisotropic force law is applied having three parts: (1) A pairwise additive, two-dimensional Lennard-Jones force preserves in-plane cohesion. (2) A normal restoring force penalizes excursions from the plane and maintains a spontaneous curvature, thereby imparting bending elasticity. (3) An anisotropic Brownian force confers in-plane fluidity. This approach is contrasted with previous models in the dependence of the normal restoring force on the mean and Gaussian curvatures.

The membrane model is to be combined with a previously developed model of the Cav-1 protein coat characteristic of caveolae in endothelial cells. This Brownian dynamics simulation, adapted from polymer kinetic theory, represents the protein chains as bead-spring assemblies and tracks their stochastic motion as forces that are transmitted to the membrane. By combining Brownian dynamics of Cav-1 oligomers anchored to the membrane with the elastic membrane response, a model of caveolin-mediated membrane deformation will be obtained.

See more of this Session: Mathematical and Computational Biosystems Engineering
See more of this Group/Topical: Computing and Systems Technology Division

(595d) Modeling Heterogeneity In Populations of Self-Renewing Human Embryonic Stem Cells

Jincheng Wu and Emmanuel (Manolis) S. Tzanakakis, Chemical and Biological Engineering, State University of New York at Buffalo, Buffalo, NY

Despite the great potential of human pluripotent stem cells as a source of therapeutics for regenerative medicine, their efficient expansion and differentiation remains challenging. To that end, cell population modeling can contribute to the optimization and standardization of conditions for the efficient maintenance and differentiation of stem cells. In this work, we developed a population balance equation (PBE) model to describe and predict the dynamics of human embryonic stem cell (hESC) self-renewal. Intrinsic and extrinsic sources of stem cell population heterogeneity were considered. The model is based on experimental data obtained in our laboratory.

A PBE model was constructed for the heterogeneous self-renewal of hESCs with a physiological state vector comprising elements pertinent to hESC proliferation and the expression level of pluripotency markers. These variables were measured experimentally for a variety of hESC maintenance conditions and their rates of change were represented by ordinary and stochastic differential equations. Solutions to the model were obtained primarily by Monte Carlo methods.

Model parameters pertinent to cell growth, division and marker expression were calculated by comparing simulation results with experimental data. Sensitivity analysis was also performed to gauge the effect of parameter perturbation on hESC heterogeneity. Additional experiments for the self-renewal of hESCs under different conditions were carried out to validate the model. The results correlate well with the flow cytometry data for hESC populations. The contribution of intrinsic sources considered here to hESC heterogeneity was found to be almost 40%. Finally, the PBE model successfully depicts both the proliferation of hESCs and maintenance of their pluripotency in a heterogeneous population. Current efforts concentrate on expanding the model to include additional sources of heterogeneity stemming from hESC differentiation.

The quantitative framework developed can complement and accelerate investigations utilizing stem cell populations by reducing costly and labor-intensive experimental procedures. Moreover, the model's predictive power can be useful in the design and operation of processes for the large-scale expansion and differentiation of human pluripotent stem cells to therapeutically useful cells.

See more of this Session: Mathematical and Computational Biosystems Engineering
See more of this Group/Topical: Computing and Systems Technology Division

(595e) Systematic Engineering Approach to Development and Identification of Physiologically-Based Pharmacokinetic Models

Cierra Hall[1], Martina Heitzig[2], Grkan Sin[2], Rafiqul Gani[2] and Andreas Linninger[1], (1)Laboratory for Product and Process Design, University of Illinois at Chicago, Chicago, IL, (2)Department of Chemical and Biochemical Engineering, Technical University of Denmark, Kgs. Lyngby, Denmark

Whole body physiologically-based pharmacokinetic (PBPK) modeling attempts to model the uptake and distribution of a drug inside the body. Traditionally, pharmacokinetic models have been developed by simple curve fitting of experimental data without considering the mechanistic and physics of the system. The drawback of such models is their lack of physical insights and the pure extrapolation features to untested conditions, to other drugs as well as to individuals of the same species or different species. Lately, PBPK models, which are based on first principles, have received increasing attention. In this work, a first principles engineering approach to PBPK modeling that is based on conservation equations and advanced scaling laws is proposed. The developed global pharmacokinetic models are based on biochemical, anatomical and physiological data.

The resulting PBPK models are able to extrapolate and scale between species and individuals because they are based on actual occurring physical mechanisms and allow the application of more advanced scaling laws than just considering the differing masses of the individuals. Due to their improved scaling qualities, first principles PBPK models have the potential to reduce the number of undesirable as well as time- and cost-intensive animal experiments during the drug development phase. Further, the development of first principle-based PBPK models leads to a gain of insight in the occurring processes during drug uptake by an organism. Having these models at hand moreover allows the determination of accurate and optimal drug dosage for desired therapeutic effects.

However, the development of first principles PBPK models, their identification, and discrimination between different candidate models is a non-trivial task inherent with a number of challenges which are related to identifying the occurring mechanisms inside the body and the appropriate degrees of detail of the models with respect to the modeling goal and the available experimental data:

Firstly, a large number of alternative PBPK models can be generated depending on the degree of detail considered and the assumptions made about the different considered mechanisms in the body. Further, first principle PBPK models are usually complex models which contain a large number of unknown parameters that are not always all identifiable by the limited experimental data accessible. A strategy needs to be found to address the problem of discriminating between the different candidate models and identifying their parameters in order to find the model that is best supported by the experimental data available. In this context, design of experiments for model discrimination and parameter estimation plays an important role. If the models cannot be discriminated based on the available data, or model parameters that are important for the desired model outputs are not identifiable by the available data, methods for optimal design of experiments need to be applied. Design of experiments is especially important in pharmacokinetic modeling because in order to obtain the required datapoints animals need to be sacrificed.

In this work, a systematic methodology for the development, analysis, identification (including sensitivity analysis, identifiability analysis and parameter estimation) and discrimination between different pharmacokinetic candidate models is proposed. The methodology is implemented into a computer-aided modeling framework. The benefit of the proposed systematic approach is an improvement of model quality and an increase of efficiency of the modeler as well as the experimentalist and thereby a reduction of time and resources required for model development, identification, discrimination and application. This is achieved by designing the structure of the computer-aided modeling framework such that it can handle the work-flows and data-flows associated with the different modeling tasks, combining state-of-the-art modeling techniques for the different work-flow steps as well as supporting model-documentation and model reuse.

The proposed modeling framework for PBPK modeling is highlighted through a case study which deals with the development of a PBPK model for a rat based on available experimental data.

See more of this Session: Mathematical and Computational Biosystems Engineering
See more of this Group/Topical: Computing and Systems Technology Division

(595f) A Compartmentalized Sepsis Model Based On Neutrophil Kinetics As a Decision Supporting Tool of Therapeutic Intervention

Sang Ok Song[1], Justin S. Hogg[2], Robert S. Parker[3] and Gilles Clermont[1], (1)Critical Care Medicine, University of Pittsburgh, Pittsburgh, PA, (2)Computational Biology, University of Pittsburgh, Pittsburgh, PA, (3)Chemical and Petroleum Engineering, University of Pittsburgh, Pittsburgh, PA

The complex pathophysiological interaction network of sepsis and its systemic nature involve many inflammatory mediators and a number of cell types, tissues and organs. Mechanisms driving the high mortality and morbidity observed in sepsis are therefore difficult to intuit and elucidate. Consequently, diagnostic, prognostic and therapeutic options have been limited due to the high variability and non-specificity of symptoms. As a broad-spectrum immune modulating therapy, an extracorporeal blood purification by hemoadsorption (HA) device using CytoSorb^TM beads (CytoSorbents, Monmouth Junction, NJ) has shown the capacity to alter circulating cytokine levels and improve survival in experimental endotoxemia and cecal ligation and puncture (CLP) induced sepsis in animals. However, the exact mechanisms by which the HA treatment resulted in reduced organ injury and improved outcome remain unclear.

Balancing the complexity of the inflammatory response, the insufficient knowledge of mechanistic details, and the relative paucity of data, we developed a mathematical model having a conceptual abstraction of the system behaviors across multiple tissue compartments. We seek to account for available experimental evidence and provide useful perspective on the pathophysiologic mechanisms of sepsis. The model focuses on the systemic behavior of neutrophils in response to infectious challenges by phenotyping circulatory neutrophils as they respond to inflammatory signals. The network components and interactions were assembled based on qualitative domain knowledge of the acute inflammatory response, including multiple phenotypes of neutrophils and major molecular effectors in distinct compartments (blood, peritoneum, and lung). In order to capture the impaired recruitment of neutrophils, which is one of the key pathophysiologic features of severe sepsis, several lines of evidence about the mechanisms influencing neutrophil migration in sepsis were incorporated in the model. Based on prior and concurrent experimental data, and our previous simulation results, we hypothesized that the HA device adsorbs circulating neutrophils as well as pro- and anti-inflammatory mediators (e.g., IL6, IL10, TNF, IL1, IL8) from the circulation.

Serial cytokine/chemokine measurements and neutrophil counts in distinct compartments were measured in rats that underwent CLP with sham (no HA treatment) or an HA intervention of 4 hour duration 18 hours after the induction of CLP. The experimental data sets were categorized into four subgroups (HA survival, HA death, Sham survival, Sham death) and the resulting four objective functions were constructed to constrain the model behaviors separately.

We computed quantitative predictions over an ensemble of parameter sets statistically drawn from all sets that are consistent with the available experimental data. An adaptive Markov Chain Monte Carlo where the step scales are periodically optimized based on the Hessian matrix of the cost function was used to efficiently sample the plausible parameter sets. Analysis of the parameter ensembles sampled from different target distributions provided useful insights on the pathophysiologic mechanisms represented by parameter subsets strongly associated with an adverse outcome and hypothesized the effects of HA on the dynamics of the system. In addition, model simulations predict a subset of animals (in their parametric description) most likely to benefit from the HA intervention. Other factors such as treatment timing and duration are discussed based on different plausible hypotheses. The mathematical model constructed here provides a platform for generating and testing hypotheses in silico, as well as motivating further translational (experimental) studies to advance our understanding of the complex biological response to inflammation and sepsis.

See more of this Session: Mathematical and Computational Biosystems Engineering
See more of this Group/Topical: Computing and Systems Technology Division

(595g) A Lattice-Based Computational Model of Microscopic Bone Resorption In Cortical Bone Multicellular Units

Junhwan Jeon[1], Pascal R. Buenzli[2], Peter Pivonka[2], David W. Smith[2] and Peter T. Cummings[1], (1) Deptartment of Chemical and Biomolecular Engineering, Vanderbilt University, Nashville, TN, (2) Engineering Computational Biology Group, The University of Western Australia, Australia

Bone resorption by osteoclasts is an essential phase in the bone remodelling cycle as it creates the physiological conditions for subsequent bone formation. While several properties of osteoclastic bone resorption in cortical Bone Multicellular Units (BMU) have been assessed experimentally, the precise spatio-temporal dynamics, movement pattern, apoptotic state (single nuclei vs. whole cell) of the osteoclasts remain to be elucidated. Furthermore, the individual effects that these behaviours confer on the shape and extent of the resorption cavity are unclear. In this work, we develop a lattice-based computational model focused on bone resorption in cortical BMUs to address these questions. Our model takes into account the interaction of osteoclasts with the bone matrix, the interaction of osteoclasts with each other, the production of osteoclasts from the tip of a growing blood vessel, apoptosis time of osteoclasts, and the renewal of osteoclasts' nuclei by cell fusion. All these features are shown to strongly affect the geometrical properties of the developing resorption cavity including shape, size, and progression rate, but also the resorption pattern, lifespan and average resorption rate of individual osteoclasts, as well as their spatial distribution in the BMU. In particular, our numerical simulation results indicate that the growth of a blood vessel and the continual renewal of the osteoclasts' nuclei are essential physiological features for functional resorption by cortical BMUs, without which balanced bone renewal may be compromised.

See more of this Session: Mathematical and Computational Biosystems Engineering
See more of this Group/Topical: Computing and Systems Technology Division

(595h) Optimal Control of Fed-Batch Fermentation Process Using Modified Iterative Dynamic Programming

Zheng Li and **M. Nazmul Karim**, Department of Chemical Engineering, Texas Tech University, Lubbock, TX

This article proposes a feeding strategy based on a metabolic networks model to optimize ethanol production using *Escherichia coli* KO11. In metabolic engineering, growth behavior and product secretion from microorganisms can be predicted using constraint-based analysis of metabolic networks. In this study, a new approach of using stoichiometric model combined with kinetic model allows for precise prediction of ethanol production and formation of other byproducts under specific fermentation environment. An optimal feeding strategy and minimization of total fermentation time are accomplished by iterative Dynamic Programming approach, which maximizes ethanol production and also minimizes toxic byproduct formation. The switching time from batch to fed- batch is also taken as variable to be optimized. All variables are computed by using iterative Dynamic Programming utilizing a stoichiometric model. Iterative dynamic programming is modified to increase computational efficiency. Simulation results are promising.

See more of this Session: Mathematical and Computational Biosystems Engineering
See more of this Group/Topical: Computing and Systems Technology Division

Dynamical analysis of sucker-rod string in artificial lift systems for control applications

James Ng and Stevan Dubljevic

2011 AIChE Annual Meeting: Minneapolis, MN

The use of sucker-rod pumping systems continues to be the most common method of artificial lift in the oil-well industry. Although the basic design of the pump-jack systems utilized today remains based on centuries old principles, the advancements and implementation of modern equipment, materials, and control systems technology have all contributed to the improvement in well-site extraction efficiency and an overall increase in oil production from existing deposits, [4, 5]. One of the critical components of the pump-jack which has benefitted from the use of computer aided design is the sucker-rod string which connects the above ground mechanical lift system to the downhole pump apparatus, [7, 6]. However, the depth and underground well conditions prohibit the placement of sensory equipment, and together with the unavailability of output production measurements, prevent the collection of data which would enable the direct use of parametric modelling techniques. As a result, dynamic analysis of the sucker-rod is carried out using the available surface-card measurements of position and rod-load as inputs to a mathematical model with primary parameters of damping and rod elasticity to describe the overall rod-string behaviour at various well-depths, [1, 2, 3]. There exists a large amount literature on the subject matter and potential models of the rod-string behaviour have been proposed including systems of coupled ordinary differential equations and variations of the damped wave-equation. The primary focus of this work is the comparison of various models which are suitable for the purpose of controller design applications. The models must be reliable, sufficiently capture the rod-string dynamics, predict the downhole pump-behaviour, facilitate diagnostic and predictive well-analysis, and also allow for the design of low order controllers for practical implementations in individual well-sites and for further use in plant wide production optimization strategies.

References

[1] S.G. Gibbs. Predicting the behavior of sucker rod pumping systems. *Journal of Petroleum Technology (July)*, pages 769–778, 1963.

[2] S.G. Gibbs. A general method for predicting rod pumping system performance. *Society of Petroleum Engineers*, 6850:9–12, 1977.

[3] S.G. Gibbs. A review of methods for design and analysis of rod pumping installation. *Society of Petroleum Engineers*, 9980:49–78, 1982.

[4] B. Guo, W.C. Lyons, and A. Ghalambor. *Petroleum Production Engineering*. Gulf Professional Publishing, 2007.

[5] S. Miska, A. Sharaki, and J. M. Rajtar. A simple model for computer-aided optimization and design of sucker-rod pumping systems. *Journal of Petroleum Science and Engineering*, 17:303–312, 1997.

[6] Dawood Momeni, George V. Chilingarian, and Axelson W.B. Hatcher. An alternative approach tot he analysis of sucker-rod dynamics in vertical and deviated wells. *Developments in Petroleum Science*, 19(1):531–634, 1987.

[7] G. Takacs. *Sucker-rod pumping manual*. PennWell, Tulsa, OK, U.S.A., 2002.

(596b) Plantwide Control of a Hybrid IGCC/ Methanol Plant

William L. Luyben, Chemical Engineering, Lehigh University, Bethlehem, PA and Patrick J. Robinson, Chem. Eng., Lehigh University, Bethlehem, PA

The coupling of an Integrated Gasification Combined Cycle (IGCC) electric power plant with a hydrogen-consuming chemical (methanol) plant can handle swings in electric power demand. Hydrogen gas from the combustion turbine and synthesis gas from the gasifier can be diverted to a methanol plant for the production of an easily-stored, hydrogen-consuming liquid product.

This paper extends previous work on dynamic studies of a gasifier and downstream units of an IGCC to explore the steady-state economic design, control and successful turndown of the methanol plant. The plantwide control structure and interaction among units are also shown.

The methanol plant is sized to reduce the power generation from an IGCC by 50%, producing a high-purity methanol stream of 99.5 mol%. Regulatory control structures are designed and play a significant role for the successful turndown of the methanol plant to 20% capacity. The exit temperature of the cooled tubular methanol reactor is controlled instead of a peak temperature within the reactor. During times of low capacity and minimum vapor rate within the distillation column, tray temperature is controlled by recycling a portion of the distillate and bottoms back upstream so that temperature can be effectively controlled by manipulating feed flowrate.

The gasifier feed is held constant. The product hydrogen from the IGCC is fed to the combustion turbine as required by electric power demand. Synthesis gas fed into the methanol plant maintains pressure of the hydrogen stream. Make-up hydrogen is also fed to the methanol plant to maintain reaction stoichiometry by controlling carbon monoxide composition of the recycle gas in the methanol plant.

See more of this Session: Modeling and Control of Energy Systems II
See more of this Group/Topical: Computing and Systems Technology Division

(596c) Optimization of Hydraulic Fracturing In Horizontal Wells for Unconventional Gas Production

Srimoyee Bhattacharya, Chemical & Biomolecular Engineering, University of Houston, Houston, TX and Michael Nikolaou, Chemical and Biomolecular Engineering, University of Houston, Houston, TX

Massive hydraulic fracturing of horizontal wells has made the recovery of natural gas from unconventional gas reservoirs both technically and economically feasible. However, the design of hydraulic fracturing relies on extensive what-if analysis using commercial simulators. The entire design process is fairly demanding, and may lead to designs that are far from optimal. In this work, we present a computer-aided approach that facilitates engineers to efficiently converge to solutions that are at or close to the optimum.

The proposed optimization objective is the net present value (NPV) for horizontal well drilled in the direction of the minimum horizontal stress, which allows multiple transverse fractures. Primary decision variables are the number of transverse fractures in a horizontal well; horizontal well length; fracturing fluid viscosity; proppant concentration; injection rate of fracturing fluid; and injection time. Key constraints are fracture growth control, operational constraints, and geometric constraints. Critical reservoir properties (e.g. permeability, porosity, and stress orientation) needed for optimization is assumed to be available from a variety of sources (e.g., seismic, prior production logs). The LMN model for hydraulic fracturing propagation by Lietard et. al.[1] which addresses the transverse case is used. The productivity of multifractured horizontal wells is computed by modifying the analytical method presented by Guo et. al.[2] and compared with Soliman et. al.[3] for consistency. The optimization algorithm used is NLPQLY developed by Schittkowski[4]. Probabilistic optimization algorithms are also used for comparison.

The proposed approach is used to design horizontal well geometry and stimulations for a number of test cases, based on numerical simulations. Results are compared with other available case studies in literature. The proposed approach demonstrates significant increase in NPV over other available methods. The simulations indicate good numerical efficiency and ability of the approach to be used as a decision making tool for unconventional gas field development. Work is currently underway to integrate this tool with vertical well hydraulic fracture optimization.

References:

1. Lietard, O., Maniere J., and Norris M.: Modeling of Transverse Hydraulic Fracturing, paper SPE 106251, Hydraulic Fracturing Technology Conference, College Station (January 29-31, 2007).
2. Guo, B., and Schechter, D.S. (1997), A Simple and Rigorous Mathematical Model for Estimating

Inflow Performance of Wells Intersecting Long Fractures., *SPE 38104 presented at the 1997 SPE Asia Pacific Oil and Gas Conference and Exhibition, Kuala Lumpur, Malaysia.*
3. Soliman, M. Y., Hunt, J. L., and Azari, M (1996): Fracturing Horizontal Wells in Gas reservoirs, in SPE 35260, presented at the Gas Technology Symposium, edited, Calgary, Canada.
4. Schittkowski, K. (2009), NLPQLY: An Easy-To-Use Fortran Implementation of a Sequential Quadratic Programming Algorithm - User's Guide, Report, Department of Computer Science, University of Bayreuth.

See more of this Session: Modeling and Control of Energy Systems II
See more of this Group/Topical: Computing and Systems Technology Division

(596d) Dynamic Modeling and Analysis of a Post-Combustion Carbon Dioxide Capture System

ZhiXin Wang[1], JianHong Lu[2], **B. Wayne Bequette**[1] and TieJun Zhang[3], (1)Chemical and Biological Engineering, Rensselaer Polytechnic Institute, Troy, NY, (2)Southeast University, School of Energy & Environment, Nanjing, China, (3)Masdar Institute of Science and Technology, Abu Dhabi, United Arab Emirates

While there is a desire for the use of sustainable/renewable energy to reduce net carbon dioxide emissions, fossil fuels will continue to serve as the dominant energy source in the near future, because of their availability and high energy density. To reduce carbon dioxide emissions from hydrocarbon energy sources, many carbon capture methods are being explored; post-combustion carbon dioxide chemical absorption technology is the most feasible near-term solution. Since MEA absorption or chilled ammonia based carbon dioxide capture processes are energy-intensive and expensive to construct and operate, model-based design and operation of the carbon capture process is useful to reduce capital cost and energy consumption.

In this work a rate-based steady-state model of an integrated MEA-based carbon dioxide capture system, consisting of an absorber and stripper, is developed using the widely-accepted commercial platform-Aspen Plus. The rate-based model has better predictive capability than the equilibrium-based model, and agrees well with experimental data from the pilot plant at the University of Texas at Austin (Zhang et al, 2009). Based on this steady-state model, comprehensive flowsheet parametric studies have been done to investigate the integrated system characteristics.

The steady-state model is converted into a dynamic model in Aspen Dynamics; because of software limitations, the equilibrium reactions in Aspen Dynamics must be converted into kinetic expressions with the equivalent equilibrium behavior. The resulting dynamic model is used to analyze the dynamic characteristics of the carbon dioxide capture process. Model-based control and optimization is shown to be a valuable tool to improve overall energy efficiency and therefore reduce the operational cost of clean coal-fired power generation system with carbon capture.

Literature Cited

Zhang, Y., H. Chen, C.-C. Chen, J.M. Plaza, R. Dugas and G.T. Rochelle "Rate-based process modeling study of CO2 capture with aqueous monoethanolamine solutions," Industrial & Engineering Chemistry Research, 48(20), 9233-9246 (2009).

See more of this Session: Modeling and Control of Energy Systems II
See more of this Group/Topical: Computing and Systems Technology Division

(596e) Dynamic Modeling and Model-Based Control Strategy for Dimethyl-Ether (DME) Production Reactor

Yu Kyung Lim, En Sup Yoon, Jong Min Lee, Shin Je Lee and Sung Ho Kim, School of Chemical and Biological Engineering, Seoul National University, Seoul, South Korea

Dimethyl ether (DME) is considered as a clean multi-source energy resource because it can be synthesized from low-grade coal, natural gas, or biomass without producing side products such as sulfur compounds and aromatic impurities. It has the potential to replace liquefied natural gas (LNG) or diesel fuel in the near future. DME production process consists of two major steps; the first step is methanol synthesis from syngas and the second is methanol dehydration to produce DME. Significant obstacles exist in commercialization of the direct DME synthesis since combining two highly exothermic equilibrium reactions makes separation and online measurements more complex. Therefore better understanding on the reactor is requisite for performance improvement by optimal operation and catalyst design.

Due to the complexity in the DME synthesis models, most studies have been mainly focused on the yield estimation and optimization at steady state. In this work, we present a dynamic, distributed parameter model for synthesis DME reactor, a parameter estimation scheme, and a model-based optimal control strategy by adopting existing steady-state reaction rate models into the mass balance equation.

We combine the Vanden Bussche model for methanol formation reaction [1] and the Bercic model for methanol dehydration reaction [2]. The Bercic model is employed to include the effect of concentrations of non-methanol components in calculation of methanol dehydration rate. The model equations have highly complex polynomial structures because of the assumptions of Langmuir adsorption isotherm and existence of reversible reactions. To reduce the complexities and stiffness of the combined model, we simplify the modeling equations as an isothermal fixed-bed shell-and-tube reactor model. The resulting model equation is reduced to a first-order partial differential equation. The finite difference method is used to conduct sensitivity analysis between the manipulated inputs and the DME yield. The inputs are reactor temperature and feed flowrate. A model-based optimal control strategy will also be presented to propose an operation strategy to improve the DME yield while maintaining the process stability. This study is conducted with the support of Samsung Heavy Industries.

References:

[1] K. Bussche and G. Froment, "A Steady-State Kinetic Model for Methanol Synthesis and the Water Gas Shift Reaction on a Commercial Cu/ZnO/Al2O3Catalyst," *Journal of Catalysis,* vol. 161, pp. 1-10, 1996.

[2] G. Bercic and J. Levec, "Intrinsic and global reaction rate of methanol dehydration over. gamma.-alumina pellets," *Industrial & engineering chemistry research,* vol. 31, pp. 1035-1040, 1992.

See more of this Session: Modeling and Control of Energy Systems II
See more of this Group/Topical: Computing and Systems Technology Division

(596f) Studying Various Optimal Control Problems In Biodiesel Production In a Batch Reactor

Pahola T. Benavides, VRI- Custom, Chicago, IL and Urmila Diwekar, Vishwamitra Research Institute, Center for Uncertain Systems: Tools for Optimization and Management, Clarendon Hills, IL

The optimal control problem encountered in biodiesel production can be formulated using various performance indices like maximum concentration, minimum time, and maximum profit. The problems involve determining optimal temperature profile so as to maximize these performance indices. This paper presents these three formulations and analyzes the solutions. Determining time-varying or space-varying profiles through dynamic optimization are common examples of optimal control problems. These problems involve the solution of partial and second order differential equation depending on the method used for their solution, resulting in difficult tasks to solve due to their mathematical representation. It has been shown that the application of maximum principle in optimal control problems provides the same results but its formulation avoids the solution of second order and partial differential equations. In this work, we use the maximum principle to solve these three problems.

See more of this Session: Modeling and Control of Energy Systems II
See more of this Group/Topical: Computing and Systems Technology Division

(596g) Model Predictive Control for Load-Following of An Integrated Gasification Combined Cycle (IGCC) Plant with CO2 Capture

Debangsu Bhattacharyya[1], Richard Turton[1] and Stephen E. Zitney[2], (1)Department of Chemical Engg., West Virginia University, Morgantown, WV, (2)Collaboratory for Process and Dynamic Systems Research, National Energy Technology Laboratory, U.S. Department of Energy, Morgantown, WV

Future integrated gasification combined cycle (IGCC) plants with CO_2 capture may have to adjust their power output as demand for electricity from the grid fluctuates over time. Such load-following requirements will become far more challenging as power produced by renewable energy is connected to the grid and where seasonal and diurnal change in the load is expected. Considering multiple single-loop controllers for load-following, the preferred control strategy from the perspective of the power producers is gas turbine (GT) lead with gasifier follow. In this strategy, the GT controls the load by manipulating its firing rate while the slurry feed flow to the gasifier is manipulated to control the syngas pressure at the GT inlet. However, the syngas pressure control is an integrating process with significant time-delay mainly because of the large piping and equipment volumes between the gasifier and the GT inlet. In addition, with growing concern over environmental emissions, the emission limits must be satisfied in IGCC plants along with the operational constraints.

While operational and environmental constraints are difficult to satisfy with decentralized proportional–integral–derivative (PID) controllers, model predictive controllers (MPC) are very good candidates for such problems. In this work, a linear model predictive controller (LMPC) is implemented in Matlab while the nonlinear IGCC process is simulated in Aspen Plus Dynamics. A multiple-input multiple-output (MIMO) linear model is identified considering various information-theoretic criteria and by comparing various linear polynomial models. The performance of the LMPC is compared with a modified PID controller where the integrating process is converted to an open-loop stable process by implementing an internal feedback loop.

To study the load-following control strategies described above, a plant-wide dynamic simulation of an IGCC plant with CO_2 capture has been developed. The coal slurry-fed gasifier is a single-stage, downward-fired, oxygen-blown, entrained-flow type with a radiant syngas cooler (RSC). The syngas from the outlet of the RSC goes to a scrubber followed by a two-stage sour shift process with inter-stage cooling. The acid gas removal (AGR) process is a dual-stage physical solvent-based process for selective removal of H_2S in the first stage and CO_2 in the second stage. Sulfur is recovered using a Claus unit with tail gas recycle to the AGR. The recovered CO_2 is compressed by a split-shaft multistage compressor and sent for sequestration after being treated in an absorber with triethylene glycol for dehydration. The clean syngas is sent to two advanced "F" class GT's partially integrated with an elevated-pressure air separation unit. A subcritical steam cycle is used for heat recovery steam generation. A treatment unit for the sour water strips off the acid gases for utilization in the Claus unit. A three-phase, top-down, optimization-based approach is taken for designing the IGCC plant with CO_2 capture in the steady-state process simulator, Aspen Plus®. The steady-state model is converted to an Aspen Plus Dynamics® simulation with appropriate modifications in the plant configuration for a pressure-driven dynamic simulation and by providing the dynamic data such as equipment sizes, equipment heat transfer, etc.

See more of this Session: Modeling and Control of Energy Systems II
See more of this Group/Topical: Computing and Systems Technology Division

APPLYING THE ANNULAR FLOW MODEL TO COMPLEX FLOW SYSTEMS

Ardson dos Santos Vianna Jr. *
University of São Paulo, São Paulo, Brazil
Department of Chemical Engineering, ardson@usp.br

José Carlos G. Peres
University of São Paulo, São Paulo, Brazil
Department of Chemical Engineering, josecarlos.gperes@gmail.com

Abstract

A stochastic model that was proposed to annular flow in a tubular reactor is used here to other complex fluid dynamics. The model was constructed from basic diffusion equations with a stochastic term. Numerical simulation was utilized to generate sample paths fitting the residence time distributions (RTD) of a circulating fluidized bed riser. The model encompassed the two deterministic parameters, mean residence time and Peclet number, as well as three stochastic parameters: stochastic relevance (b), updated time (ΔT) and the seed that begins the Wiener process. The satisfactory results indicate that the model may be applied to multiphase flows, and can be renamed as stochastic axial dispersion model.

Keywords: Stochastic model, RTD, multiphase flow, Fluid mechanics, Mathematical modeling, Simulation

Introduction

The Residence Time Distribution (RTD) can point out how to model the complex flow patterns, such as polymers, suspensions and multiphase systems. Indeed, many chemical processes have used RTD analysis and characterization, for instance, fluidized bed risers (Harris et al, 2002) and polymerization reactors (Vianna et al, 2006, 2007).

Vianna Jr. and Nichele (2010) have evaluated the fluid dynamics of three polymerization tubular reactors. They proposed an annular flow model that is, in reality, a stochastic axial dispersion model. It can generate several sample paths that representing tracer concentration responses which exhibit stochastic behavior. The model is based on a stochastic differential equation (SDE), requiring specific fundamentals and numerical routines such as the Itô integration formula. The SDE model was numerically solved by considering the Wiener process, and using the Euler-Maruyama semi-implicit method of integration. RTD sample paths were generated to reproduce experimental results similar to those in the numerical experiments undertaken by Kloeden, Platen and Schwurz (1991). The annular flow reactor model was able to represent the available experimental data fairly well for the polymerization tubular flow reactor.

Harris el at (2002) had simulated the residence time distribution of solids in the riser of a circulating fluidized bed. The models simulate the motion of single particles moving up and down the riser, and they solve for the RTD by assigning probabilities for the transition of single particles between states. The arrangements of the states that describe a particular model depend upon the hydrodynamics, and in particular, the macroscopic axial flow structure of the riser. A typical flow structure in a circulating fluidized bed had regions, an exit region, a dilute region with clusters, a dense transition region and a dense churn turbulent, bottom zone.

In the present work, the multiphase flow of a circulating fluidised bed riser was evaluated using the annular flow reactor model (Vianna Jr. and Nichele, 2010). It can be viewed as a stochastic axial dispersion model, since it can generate several sample paths that representing tracer concentration responses which exhibit stochastic behavior. The model was able to represent the available experimental data fairly well for the circulating fluidised bed riser.

Fundamentals

The model used here is a stochastic model generated from an axial dispersion model (Vianna Jr. and Nichele, 2010). However, some points are discussed here.

RTD

When a fluid flows trough a vessel, each volume element spends different time intervals in the vessel, so that a distribution of frequencies is established (Danckwerts, 1953; 1958; Zwietering, 1959). This concept would be related to experimental data. Thus, the characterization of fluid dynamics in equipment can be done experimentally by stimulus-response techniques. A known quantity of inert substance, the tracer, is introduced into the fluid entering the equipment, and the tracer concentration of the exiting fluid, $C(t)$, is monitored. The response of a pulse input is related to the residence time distribution (RTD), i.e., the normalized time distribution of all volume elements in the system (Eq. 1)

$$E(t) = \frac{C(t)}{\int_0^\infty C(t)dt} \tag{1}$$

Annular Flow tubular reactor

Vianna Jr. and Nichele (2010) proposed a stochastic model that represents the annular flow in a tubular polymerization reactor. They generated residence time distributions that can imitate the random behavior of volume elements as have been observed elsewhere (Vianna et al, 2006; 2007). The random flow is composed of two layers; one is considered light and the other heavy. This heavy layer is eventually dragged by the central layer (light), which flows much faster. The process can be seen as an erosion process, where the creeping flow around the slow layer results in its movement. The slow layer exerts a force on the light layer but it is a fluctuating force. However, the model may be extended to other stratified flows and other boundary conditions, since the equation is a diffusion-type with a stochastic term.

The model can represent complex fluid dynamics that are associated with macrofluids behavior. The RTD fluctuations are transport phenomena results, and they are not random outcomes obtained by experiment. Since the random responses are phenomenological, the source of randomness is a global term that encompasses the various contributions to randomness. The model equation is similar to the axial dispersion model without reaction, with the addition of a Wiener process, and is presented in Eq. 2:

$$\frac{\partial C_A}{\partial t} + v_z \frac{\partial C_A}{\partial z} = D_{axial} \frac{\partial^2 C_A}{\partial z^2} + \sigma W(t) \tag{2}$$

where σ is the variance of concentration of A and $W(t)$ is a noise that is dependent only on time.

Using the variables $y = (C_A - C_{Af}) / (C_{A0} - C_{Af})$; $x = z/L$ and $t' = t.v_z/L$, the dimensionless form is expressed as:

$$\frac{\partial y}{\partial t'} + \frac{\partial y}{\partial x} = \frac{1}{Pe}\frac{\partial^2 y}{\partial x^2} + bW(t) \qquad b = \frac{\sigma}{C_{A0} - C_{Af}}\frac{L}{v_z} \tag{3}$$

The three parameters of the model describe the fluid dynamics of the equipment. The Peclet number is related to the slope of the $F(t)$ curve, which indicates the proximity to plug flow. The stochastic relevance b is associated with the amplitude of the oscillations, because molecules of tracer were retained in the external stagnant layer. The thickness of the stagnant layer could be connected to this parameter. Since the update time can be linked to frequency of random perturbations, returning to Brownian movement, the update time is associated with the time in which volume elements do not interact with others, and changes are not observed.

Numerical solution of SDE

The stochastic partial differential equation (SPDE) must be integrated over time and space. As a partial equation, it has to be solved by a double integration process like the Crank-Nicolson algorithm. Assuming a parabolic heat equation with noise that is dependent only on time, the numerical solution is given by Gaines (1995):

$$du_i(t) = \frac{1}{(\Delta x)^2}\left[\frac{u_{i-1} - 2u_i + u_{i+1}}{h^2}\right] dt + \frac{1}{\sqrt{\Delta x}} dW_i(t) \qquad (4)$$

where $u(x,t)$ is the dimensionless temperature with Brownian noise, x is space and t is time. This is a starting point for solving another SDPE, in which the final discretized formula will depend on the driving noise.

Methodology

The model verification and validation was done by fitting available experimental data to the simulation model (Aris, 1994). The experimental data were collected from circulating fluidized bed in which solids flow in the riser (Harris el at, 2002). The motion of single particles was analyzed, and a typical flow structure with several regions was observed.

The fluid dynamics were evaluated using a pulse perturbation, and the experimental conditions indicated a closed boundary condition should be assumed. The responses of the pulse perturbations are the $E(t)$ curves.

The model used here was composed of mean residence time and Peclet number, which are deterministic parameters. The other two parameters, b and ΔT, are random parameters. The parameter estimation process must follow this structure, so a two-level procedure was done. In the first deterministic step, mean residence time and Peclet number are evaluated to represent the mean path by using parameter estimation software. The model used here was the deterministic axial dispersion model. The parameters are correlated, thus a maximum likelihood procedure was used, the ESTIMA software (Santos and Pinto, 1998). At this step, no oscillations were considered, therefore the model described the pulse response.

The second step is stochastic therefore a trial-and-error process was necessary to represent the experimental data set. The trial-and-error procedure was carried out by assigning a seed to start the pseudo-random generator, so each seed generated a sample path. At this step, the model proposed here was used (i.e., Eq. 3). The parameters estimated were stochastic relevance, b, updated time, ΔT and the seed which began the Wiener process. This part can be seen as a fine-tuning, since the macroscopic behavior was represented by mean residence time and Peclet number, and then the oscillations had to be simulated.

The ultimate model is a stochastic partial differential equation solved by the Euler-Maruyama semi-implicit method. The Wiener process used here is the Marsaglia algorithm (Kloeden et al, 1991) which is a modification of the Box-Muller procedure that doesn't use trigonometric functions. The seed is inserted into the randomly generated algorithm, and this initiates the Marsaglia procedure. The method is numerically integrated over the space by a semi-implicit algorithm (Crank-Nicolson of one-half order) and the time by the Itô integration formula (Gaines, 1995). The tridiagonal system is presented in Eq. 5:

$$u_{i-1}^{j+1}\left[-\frac{\Delta t}{2Pe\,\Delta z^2}-\frac{\Delta t}{4\Delta z}\right]+u_i^{j+1}\left[1+\frac{\Delta t}{Pe\,\Delta z^2}\right]+u_{i+1}^{j+1}\left[-\frac{\Delta t}{2Pe\,\Delta z^2}+\frac{\Delta t}{4\Delta z}\right]=u_{i-1}^{j}\left[\frac{\Delta t}{2Pe\,\Delta z^2}+\frac{\Delta t}{4\Delta z}\right]$$

$$+u_i^{j}\left[1-\frac{\Delta t}{Pe\,\Delta z^2}\right]+u_{i+1}^{j}\left[\frac{\Delta t}{2Pe\,\Delta z^2}-\frac{\Delta t}{4\Delta z}\right]+b\,\Delta W_i(t) \qquad (5)$$

where u_i^j is the numeric evaluation of concentration at i Δx position and j Δt time.

The numerical pulse perturbation can be carried out by using an imperfect pulse. The perfect pulse is mathematically represented by a delta function, i.e., infinitely high zero width but with area equal o one (Levenspiel, 1998). Since infinity value is an abstraction, the input can be viewed as rectangular shaped, which the width is equal to Δt_{pulse} and the length is equal to 1 (because E(t) is a normalized distribution, see eq. 1):

The initial condition that represents the step input is:

$$u_0^0=\frac{1}{\Delta t_{pulse}} \quad and \quad u_0^j=0 \qquad (6)$$

The closed and open boundary conditions of Danckwerts can be assumed. The discretizations of these conditions are:

$$1\,B.C.\quad 1=\frac{1}{Pe}\frac{u_1^j-u_{-1}^j}{\Delta x}+u_0^j$$

$$2\,B.C.\quad u_{n+1}^j-u_n^j=0 \qquad (7)$$

Results and discussion

The result presented in Figure 1 is obtained to pulse input perturbation that has applied to fluidised bed riser, and an open-closed boundary condition. Harris et al (2002)

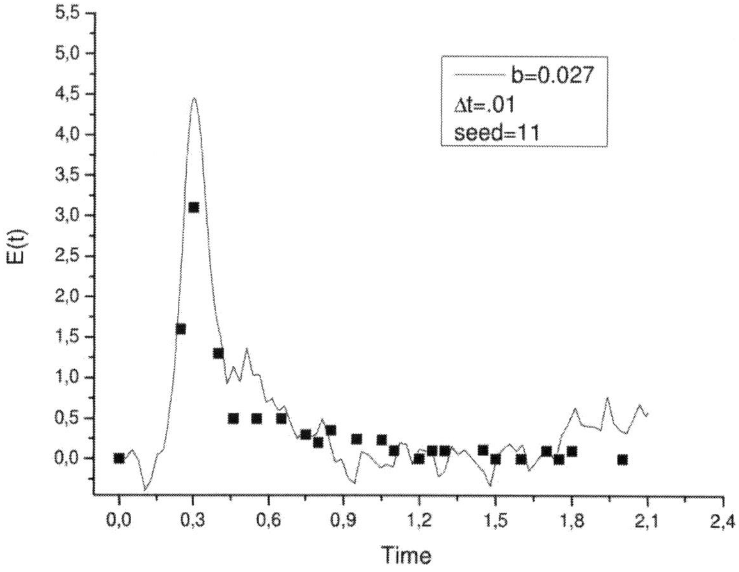

Figure 1: E(t) macroscopic axial flow of the circulating fluidized bed riser, tr= 0.3 min, Pe = 81.7. (Harris et al, 2002).

The model can fit the available data fairly well for the circulating fluidized bed riser, that is, the deterministic tendency of E(t) is allowed and the oscillations are well represent by the sample paths. The sample paths are very sensitive to seed input, in order negative values of E(t) and values greater than one could be obtained. Therefore the model shouldn't be use to estimate flow systems, instead of this it could be use to represent available data set.

However, these fluctuations cannot be credited to measurement experimental errors at all, but it is caused by the complex macrofluid-dynamic behavior of the flowing mixtures. This is already discussed to tubular polymerization reactor where three parameters are required, the thickness of the stagnant layer, the characteristic period of the random fluctuations of the flow velocity and the maximum allowed flow velocity of the external layer (Vianna Jr. et al, 2007).

The model could be a generalization to any segregation and slug flow. The new parameter b stochastic relevance can be view from a generalization of the thickness of the polymer stagnant layer, and so it can be associated to a characteristic dimension of stagnant layer. If multiphase flow with solids is being studied, the b parameter is associated to the element of volume segregated or slugged. The characteristic period of the random fluctuations has the same interpretation, i.e., the slug element of volume is periodically (and randomly) dragged by the tensions induced by the bulk flowing. The stochastic relevance and updated time are global parameters since many different factors are being considered.

5. Conclusions

The results are satisfactory indicating that one more step was given in the complex fluid dynamics comprehension of system flows.

The model encompasses deterministic parameters, mean residence time and Pelcet number, and other three stochastic parameters, stochastic relevance - b, updated time - Δt, and the seed which begins the Wiener process. The model could fit quite well the experiment data set of the flow systems of a tubular polymerization reactor and a fluidised bed.

The Euler-Maruyama procedure was efficient to solve the SPDE, since there are not additional problems to find the sample paths.

Acknowledgements

The authors would like to acknowledge the support received from CNPq, Grant number 81330/2004-6, and Faperj, Grant number E-26/111895/2008.

References

Aris, R. (1994), "Mathematical Modeling Techniques", Dover, New York.

Danckwerts, P.V. (1953), "Continuous flow systems. Distribution of residence times," Chemical Engineering Science, 2, pp. 1-15.

Danckwerts, P.V. (1958), "The effect of incomplete mixing on homogeneous reactions", Chemical Engineering Science, 8, 93.

Gaines, J.G. Numerical Experiments with S(P)DE's, in Stochastic Partial Differential Equations, A.M. Etheridge. ed., London Mathematical Society Lecture Note Series 216, Cambridge Univ. Press, 1995, 55-71.

Harris, A.T., Thorpe, R.B., Davidson, J.F. (2002), "Stochastic modelling of the particle residence time distribution in circulating fluidised bed risers", Chemical Engineering Science, 57, pp. 4779-4796.

Kloeden, P.E; Platen, E. and Schurz, H. (1991). "Numerical Solution of SDE through Computer Experiments," Springer Verlag.

Levenspiel, O. (1998). Chemical Reaction Engineering, 3rd Edition, Wiley, New York.

Santos, T.J., Pinto, J.C., (1998), "Taking Variable Correlation into Consideration during Parameter Estimation," Brazilian Journal of Chemical Engineering, 15, 1, PP. 1-15.

Vianna Jr., A. S, Biscaia Jr., E.C., Pinto, J.C. (2006), "Flutuações estocásticas para a DTR de um Reator Tubular de Polimerização", Polímeros: Ciência e Tecnologia, 16, 1, PP. 19-25.

Vianna Jr., A. S, Biscaia Jr., E.C., Pinto, J.C. (2007), "A Stochastic Flow Model for a Tubular Solution Polymerization Reactor", Polymer Engineering and Science, 47, pp. 1839-1846.

Vianna Jr., A. S, Nichele, J., (2010), "Modeling an annular flow tubular reactor," Chemical Engineering Science, 65, pp. 4261-4269.

Zwietering, Th.N., (1959), "The degree of mixing in continuous flow systems," Chemical Engineering Science, 11, 1, pp. 1-15.

Nomenclature

$a(t)$	stochastic parameter
b	stochastic relevance
C_A	concentration of the specie A
D_{axial}	axial dispersion parameter
$E(t)$	response of pulse input
Pe	Peclet number
t	time
t'	dimensionless time
t_r	residence time
$u(x,t)$	numeric evaluation at x position and t time
u_i^j	numeric evaluation at i Δx position and j Δt time
v_z	superficial velocity
$v(x,t)$	convective velocity
x	dimensionless axial position
X	random variable
y	dimensionless concentration
z	axial position
$W(t)$	Wiener process

Greek letters

ΔT	updated time
σ	variance

(602b) Modeling of a Bubble-Train Flow and Interphase Mass Transfer In a Long Microchannel

Dmitry Eskin, Farshid Mostowfi and Shahnawaz Molla, DBR Technology Center, Schlumberger, Edmonton, AB, Canada

A model of a bubbletrain flow, accompanied with mass transfer, in a lengthy capillary tube isdeveloped. In contrast to models presented in literature, our modeling approachaccounts for expansion of gas bubbles and flow velocity increase along thechannel due to the pressure drop caused by friction losses. The modelperformance is illustrated by a number of computational examples. Thedistributions of the bubble velocity and the volumetric mass transfercoefficient along the channels of different diameters are presented. Thedeviation of the dissolved gas concentration from the saturation concentrationalong the channel is used as a characteristic of closeness of a fluid system tothe phase equilibrium. The calculations explicitly demonstrate that thedeviations of gas-liquid mixture flows from equilibrium in lengthy capillarychannels of small diameters are small (see Figure below). The effect of thechannel diameter, the channel length and the bubble nucleation frequency on thedeviation of the system from equilibrium is also studied.

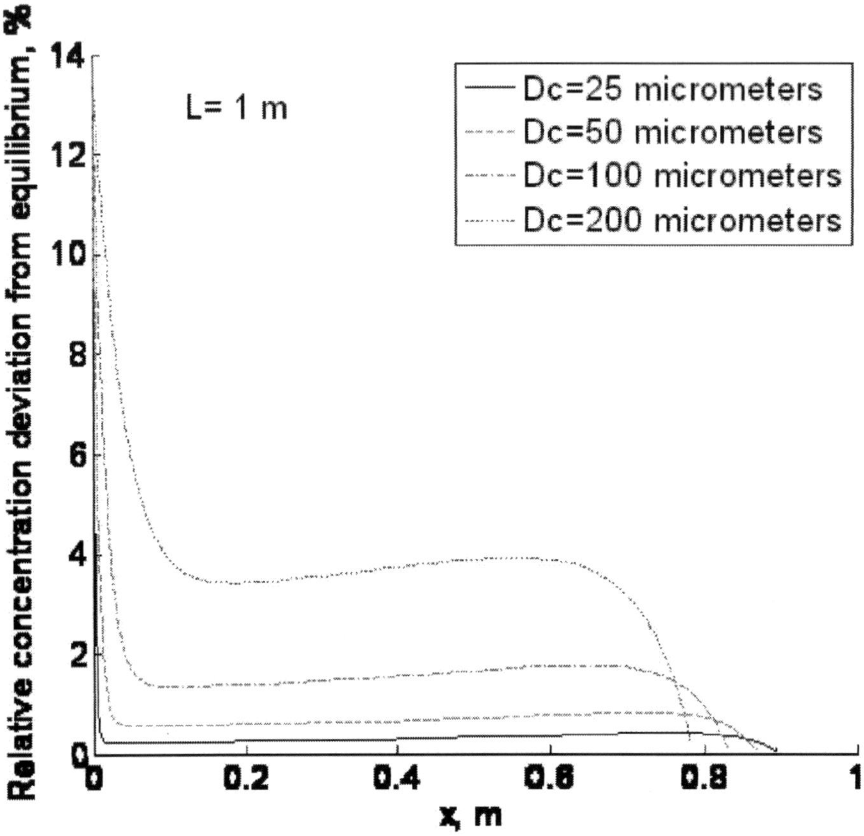

See more of this Session: Novel Numerical Methods In Fluids
See more of this Group/Topical: Engineering Sciences and Fundamentals

Numerical Investigation of Mixing inside the Dispersed Phase in Segmented Flows in Microchannel

Yuehao Li[1], Rupesh Reddy[1], Krishnaswamy Nandakumar[1] and Challa S.S.R. Kumar[2]

[1]Cain Department of Chemical Engineering, Louisiana State University, Baton Rouge 70803

[2]Center for Advanced Microstructures and Devices (CAMD), Louisiana State University, 6980 Jefferson Hwy, Baton Rouge 70809

Introduction

Multiphase segmented flows offers unique environment for chemical synthesis at micron scale.[1] The dispersed phase provides controlled environment for chemical reactions with minimal contamination and dispersion effects while intensifying mass transfer due to internal circulations.[2, 3] The numerical simulations of mixing inside one slug[4] and between two adjacent slugs[5] based on moving frame method have been studied by Sarrazin et al. and Kashid et al. respectively. The influence of parameters such as flow rate, viscosity and geometry on mixing inside the dispersed phase has not been extensively studied and is not well – understood.

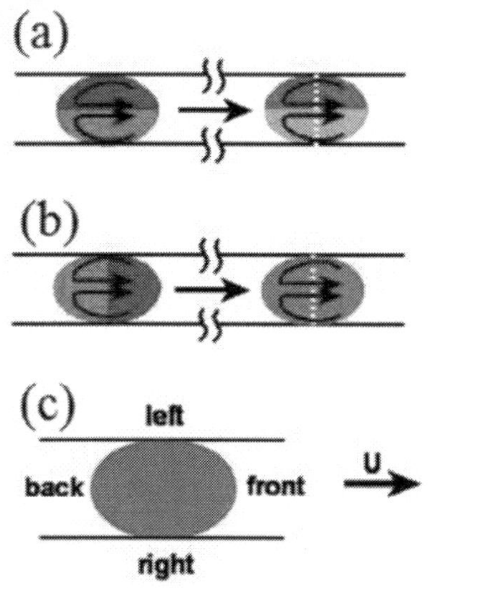

Fig. 1 Effect of recirculation on mixing (a) the reagents are initially localized at left and right half (b) the reagents are localized at back and front half. (c) Notations used to identify different regions of the slug. (reproduced from Tice et al)

When slugs move through microchannels, the shear between the wall surface and slug axis generates internal recirculation. The recirculating flows are localized in the left and right half of the slugs as shown in Fig. 1. If the reagents are initially distributed on left and right half, the recirculation does not mix the reagents efficiently as shown in Fig. 1(a). In contrast, if the reagents are localized on the back and front half of the slug, the recirculations mix the reagents well as shown in Fig.1 (b). The initial distribution in a given slug depends on the slug formation mechanics. Therefore, in present study, the effect of slug formation process on mixing is taken into account.

In this work, we study mixing inside dispersed phase in a T – junction microfluidic device through extensive 3D numerical simulations. The Volume of Fluid (VOF) method is used to study the hydrodynamics of water drop / slug formation in perfluorodecaline (PFD), and species transport equation is applied to the dispersed phase to study mixing inside slugs. The results are compared to experimental work done by Tice et al[6].

Numerical Method

In this study, the VOF method is used to simulate the dynamics of slug formation. The flow is operated in laminar region (Re<10) and surface tension force dominates the slug formation process (Ca <0.01). All fluids are modeled as incompressible Newtonian fluids.

VOF method

A single set of continuity (1) and momentum equation (2) are solved continuously throughout the computational domain,

$$\frac{\partial \rho}{\partial t} + \nabla \cdot (\rho \vec{v}) = 0 \quad (1)$$

$$\frac{\partial (\rho \vec{v})}{\partial t} + \nabla \cdot (\rho \vec{v}\vec{v}) = -\nabla P + \nabla \cdot [\mu(\nabla \vec{v} + \nabla \vec{v}^T)] + \vec{F} \quad (2)$$

Of which ρ and μ are the volume – averaged density and viscosity, which are defined as (3) and (4) respectively where α_q is introduced to identify the underlying fluid volume inside each computational cell

$$\rho = \alpha_q \rho_q + (1 - \alpha_q)\rho_p \quad (3)$$

$$\mu = \alpha_q \mu_q + (1 - \alpha_q)\mu_p \quad (4)$$

The body force term in Eq. (2) is computed in Fluent by the continuum surface force (CSF) model[7] as shown in Eq. (5),

$$\vec{F} = \sigma \left[\frac{\rho \kappa \hat{n}}{\frac{1}{2}(\rho_p + \rho_q)} \right] \quad (5)$$

where σ is the surface tension coefficient, and $\hat{n}$ is the unit surface normal vector, and κ is the local surface curvature, which are defined by Eq. (6) to Eq. (8).

$$n = \nabla \cdot \alpha_q \quad (6)$$

$$\hat{n} = \frac{n}{|n|} \quad (7)$$

$$\kappa = \nabla \cdot \hat{n} \quad (8)$$

The VOF method uses the geometric reconstruction scheme which based on piecewise linear interpolation[8] to obtain the reconstructed interface. The interface is tracked by solving an additional advection equation (9), which yields the volume fraction of each fluid.

$$\frac{\partial (\alpha_q \rho_q)}{\partial t} + \nabla \cdot (\alpha_q \rho_q \vec{v}_q) = 0 \quad (9)$$

When the liquid – liquid interface contacts the wall, wall adhesion force exists and affects the liquid – liquid – wall interaction. Thus the wall adhesion force must be taken into account in the calculation. If the three phase contact angle is defined as θ_w, then the interface normal near in the cells near the wall is calculated by

$$\hat{n} = \hat{n}_w \cos\theta_w + \hat{t}_w \sin\theta_w \quad (10)$$

where $\hat{n}_w$ and $\hat{t}_w$ are the unit vectors normal and tangential to the wall. The CSF model combines the contact angle with the normally surface normal calculated from one cell away from the wall to determine the local curvature κ, and this curvature is used to calculate the body force term in Eq. (5).

In this study, we assume that the contact line maintains a constant value as the static contact angle and is independent of the velocity and the direction of contact line.

Species transport model

The species transport model is used to study the mixing of two fluids inside slugs. We assume that two aqueous streams are injected from the side inlet. One stream has concentration as 1 which we model as dye in our work while the other one has concentration of 0 as pure water. The species transport equation (11) is solved.

$$\frac{\partial C_i}{\partial t} + v \cdot \nabla C_i = D\nabla^2 C_i \quad (11)$$

The diffusivity of dye in water is set as 10^{-11} m^2/s, and dye does not diffuse into the carrier fluid.

Solution methods

The commercial CFD software, FLUENT, is used to solve the equations. A transient explicit solver is used along with the implicit body force formulation. The PISO (pressure – implicit with splitting of operators) scheme is used for pressure – velocity coupling of the momentum equation and continuity equation. For spatial discretization, Green – Gauss node – based method is used for gradients, and the PRESTO! (pressure staggering options) scheme is used for pressure interpolation, and the second – order up – wind differencing scheme is used for the momentum and species transport equation.

The 3D computational mesh is generated by ICEM software (ANSYS INc.). The liquids are water solution ($\rho_{water} = 1000\ kg/m^3$, $\mu_{water} = 0.001$ Pa · s) and PFD ($\rho_{PFD} = 1893\ kg/m^3$, $\mu_{PFD} = 0.0051$ Pa · s). The interfacial tension between two phased are assumed to have a constant value of 0.014 N/m. Velocity inlets are specified for the two inlets and the outlet is modeled as outflow. No – slip boundary condition is imposed on the channel walls, and the contact angle between the water and PFD is set as 140 degree.

Results and Discussion

After performing mesh dependence study, a computational domain containing 593,750 hexahedral elements is used for parametric study. All these simulations are performed on the Louisiana State University HPC facility.

The effect of water fraction on slug length

Previous studies[6, 9-11] have shown that the slug length is sensitive to the volumetric ratio of dispersed phase to continuous phase. Tice et al.[6] demonstrated in their work that the slug length could be easily controlled by varying the relative volumetric flow rates of aqueous phase and oil phase. The comparison of slug length and period (the center – to –center distance between adjacent slugs) that obtained from Tice's experimental observations and our numerical studies are shown in Fig.2. Water fraction is defined as the ratio of volumetric flow rate of aqueous phase to total flow rate, and the relative size is defined as the ratio of lengths to channel width (50 μm). The total flow rate is maintained at 50 mm/s, and water fraction varies to produce slugs with different lengths are shown in Fig.2.

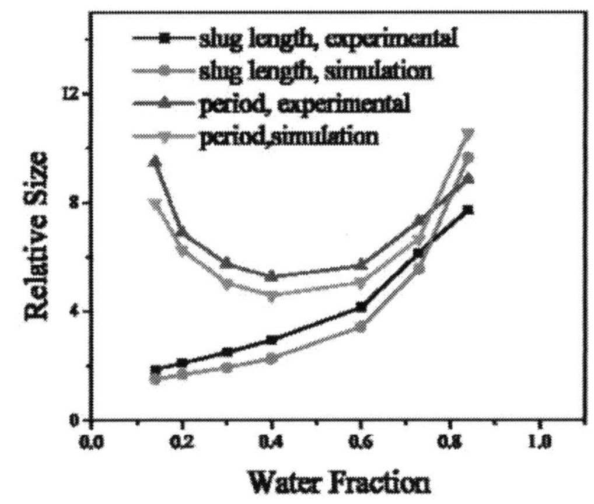

Fig. 2. The spatial period and the lengths of slugs as a function of water fraction.

If the flow rate of aqueous phase is lower than PFD phase, small slugs are produced; in contrast, long slugs are formed when the flow rate of aqueous phase is higher than PFD phase. The period, depending on slug length itself, passes through a minimal at the water fraction around 0.5.

The slug lengths and periods predicted by numerical simulations agree well with the experimental observations except the point when the water fraction is 0.84. A possible reason for the deviation might be the failure capturing wettability in this condition. The wettability, which is considered in the model in terms of contact angle, is a function of water fraction. When the water fraction increases, the interaction of water, PFD and wall surface changes resulting to a different contact angle from that at the lower water fraction. However, the contact angle is set as a constant in all the simulations. The deviation of contact angle might cause the failure on predicting the slug length.

Despite of water fraction at 0.84, a good agreement of simulation results with experimental observation validates the adequacy of our computational model to study mixing inside slugs.

Mixing inside slugs

The mixing of dye and water inside slugs with different lengths are shown in Fig. 2. In this study, the total velocity is fixed at 50 mm/s, and the water fraction of aqueous phase is varied.

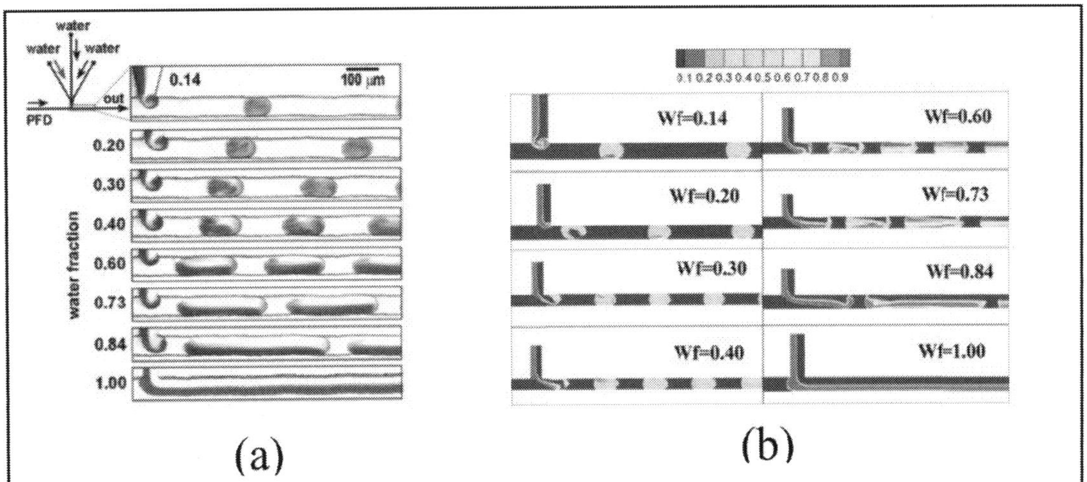

Fig. 2 Mixing inside slugs. The results are demonstrated by (a) experimental observation (reproduced from Tice et al[6]) (b) numerical simulation

As the aqueous phase is injected into the PDF phase, the shearing interaction between two phases induced an eddy, which is referred as "twirling" by Tice et al[6]. Twirling is observed at all velocities, and it redistributes dye to different regions of the slug, thus enhance the efficiency of recirculating.

Twirling shows different effect on slugs with different sizes. Recirculation does not significantly enhance mixing in long slugs (wf=0.73 and wf=0.84) because not enough dye is transferred into the front side of the slug. Mixing between left and right half of the slug is dominated by molecular diffusion as mixing between two parallel laminar flows without carrier fluid (wf=1.00). In contrast, mixing inside slugs of medium sizes (wf=0.30 and wf=0.40) is accelerated by recirculation as twirling brings dye into the left half of the slug. For short slugs (wf=0.14 and wf=0.20), "overtwirling" occurs. In this case, the twirling effect is so strong that too much dye is brought into the left half. The mixing is expected to be less efficient than slugs of medium sizes because initial condition is resembled as long slugs.

Conclusion

In this study, we employ numerical simulation to study the slug formation and mixing performance inside slugs. The numerical results show good agreement with experimental results. The twirling effect, resulting from the shear between carrier fluid and aqueous phase, redistribute the dye inside the slug and accelerating mixing inside small slugs. The twirling effect is less efficient in long slugs, thus recirculation efficiency on mixing is comparatively lower. Mixing efficiency index will be employed in the later studies and give quantitative results on mixing efficiency of slugs with different size and residence time.

Acknowledgment

We thank Louisiana State University HPC and Louisiana Optical Network Initiative (LONI) for providing access to computational facility and timely support.

References

1. S. Duraiswamy and S. A. Khan, *Small*, 2009, **5**, 2828-2834.
2. J. R. Burns and C. Ramshaw, *Lab on a Chip*, 2001, **1**, 10-15.
3. G. Dummann, U. Quittmann, L. Groschel, D. W. Agar, O. Worz and K. Morgenschweis, *Catal. Today*, 2003, **79**, 433-439.
4. F. Sarrazin, K. Loubiere, L. Prat, C. Gourdon, T. Bonometti and J. Magnaudet, *Aiche J.*, 2006, **52**, 4061-4070.
5. M. N. Kashid, D. W. Agar and S. Turek, *Chem Eng Sci*, 2007, **62**, 5102-5109.
6. J. D. Tice, H. Song, A. D. Lyon and R. F. Ismagilov, *Langmuir*, 2003, **19**, 9127-9133.
7. J. U. Brackbill, D. B. Kothe and C. Zemach, *J. Comput. Phys.*, 1992, **100**, 335-354.
8. W. J. Rider and D. B. Kothe, *J. Comput. Phys.*, 1998, **141**, 112-152.
9. Y. C. Zhao, G. W. Chen and Q. Yuan, *Aiche J.*, 2006, **52**, 4052-4060.
10. P. Garstecki, M. J. Fuerstman, H. A. Stone and G. M. Whitesides, *Lab on a Chip*, 2006, **6**, 437-446.
11. S. K. R. Cherlo, S. Kariveti and S. Pushpavanam, *Ind. Eng. Chem. Res.*, 2010, **49**, 893-899.

(602d) Pseudo-Analytical Solution of the Power LAW MODEL for for NON-NEWTONIAN CARRIER FLUIDS IN FREE CONVECTIVE CELLS

Mario Oyanader, Chemical Engineering, Tenneessee Tech University, Cookeville, TN, Patricio Trigo, Ingenieria, Universidad Arturo Prat, Iquique, Chile and Pedro E. Arce, Chemical Engineering, Tennessee Technological University, Cookeville, TN

The present contribution focus on the analysis of the Joule heating effects for the case of non-Newtonian fluids, i.e. carrier fluid flowing in a batch electrophoretic cell. In particular, this study proposes and develops a more efficient and economical procedure to obtain a formal analytical solution for the momentum equation based on the *Ostwald*-de Waele model. In the solution of the differential equation this approach uses a Lagrange polynomial that outperforms previous efforts based on Taylor's series. The outcomes of this contribution have potential applications on the pharmaceutical, food, and polymers industries as well as in biological and environmental operations, to name a few, where the non-Newtonian characteristics of carrier fluids could significantly impact the efficiency of mixing or separations. The proposed strategy and the subsequent solution of the non-Newtonian momentum equation are alternative tools for process scaling and research. Although the primary interest of this contribution deals with the velocity field only, the important role that this variable plays in the determination of the effective parameter for this type of systems cannot be over looked.

See more of this Session: Novel Numerical Methods In Fluids
See more of this Group/Topical: Engineering Sciences and Fundamentals

(602e) Interaction Between An Elastic Filament and the Vesicle Membrane

Yuan-Nan Young, Department of Mathematical Sciences, New Jersey Institute of Technology, Newark, NJ and Christopher Jacobs, Biomedical Engineering, Columbia University, New York, NY

The primary cilium is found for all non-dividing mammalian cells. Since its discover a century ago, only recently has more understanding of the biological role of primary cilia been gained. In this work slender-body formulation is utilized to describe the dynamics of the primary cilium, modeled as an elastic filament attached to a solid wall or membrane. Comparison with the experimental data will be provided. Coupling between the filament/membrane system and the mechanosensitive channel (MscL) show how the primary cilium functions as a probe of the extracellular flow.

See more of this Session: Novel Numerical Methods In Fluids
See more of this Group/Topical: Engineering Sciences and Fundamentals

(612a) Implementing Chem-e-Car Competition Into the Curriculum: Ten Year Experience

Sundararajan. V. Madihally, School of Chemical Engineering, Oklahoma State University, Stillwater, OK

Chem-e-car activity has emerged as a signature event at the regional and national conferences since its inception in 1999. The creation of these cars is fun, and the competition is entertaining. Most schools take part in the contest by thriving on voluntary participation of students in the AIChE student chapter. At Oklahoma State University, the chem-e-car competition has been administered for the past ten years as part of multiple courses in the chemical engineering curriculum. In this presentation, general methodology adapted is described. Based on those experiences, the advantages and challenges of this approach are discussed. In addition, some of the deciding factors to be considered for implementation of this approach are discussed.

See more of this Session: Software, Technology and Computation In Chemical Engineering Education
See more of this Group/Topical: Education

(612b) Group Contribution Methods In Undergraduate Chemical Engineering Thermodynamics

Rebecca K. Toghiani, Dave C. Swalm School of Chemical Engineering, Mississippi State University, MS State, MS

Thermodynamics at the undergraduate level in a chemical engineering curriculum traditionally covers topics ranging from the first law of thermodynamics, the second law of thermodynamics, properties of pure components and mixtures as well as phase equilibria and chemical equilibria. The challenge is often to provide the necessary depth in these important areas while still preparing the student for the activities/endeavors that require thermodynamics in subsequent design courses or in the workplace. Thus, design of out-of-class assignments that allow the student to explore some of the concepts not covered in depth in the classroom is a mechanism that allows newer topics relevant to chemical engineering practice/design to be included in the course, yet not at the expense of essential topics in the curriculum.

One concept/topic that often is not covered in depth in the undergraduate chemical engineering thermodynamics sequence is the use of group contribution methods for estimation of physical properties and for estimation of vapor-liquid equilibria. For many compounds, necessary physical property data (critical properties, acentric factor, vapor pressure, heat capacity, phase change enthalpies) may be readily available in textbooks [1-4] or in data compilations [5-7]. However, upon graduation, chemical engineers may develop and design new chemical processes and/or products involving compounds for which these data are not available. Experimental measurement of these types of data is expensive, requiring precise and specialized equipment, and often is undertaken only after a process is sufficiently developed to indicate a favorable economic return on investment. At MSU, Chemcad (Chemstations, Inc.) is used as the process simulation software package. However, for compounds not included in its database, the user must provide a minimum set of data, including critical temperature, critical pressure and acentric factor among others. An assignment employing the student version of Cranium [8] was developed to allow students to explore group contributions methods with respect to estimation of these properties for organic chemicals, including isomers. Exposure to group contribution methods for physical property estimation at the undergraduate level is beneficial. Not only does the student become conversant in the use of these methods, they also are made aware of the limitations inherent in certain methods through use of properly designed assignments.

The expanded use of process simulation software in the senior design courses also warrants inclusion of group contribution methods at an earlier point in the undergraduate chemical engineering curriculum. Chemcad is also equipped with a 'Thermodynamics Wizard' the program will identify an appropriate model for estimation of K-values (phase equilibria) based on the specified components and upper/lower bounds on temperature and pressure provided by the user. When experimental data for a given chemical system are lacking, 'UNIFAC' may be identified as the appropriate model for phase equilibria estimation. A project focused on use of UNIFAC for estimation of activity coefficients for a binary system taken from the literature allowed students to gain an appreciation for this method and for its implementation in Chemcad.

Incorporation of group contribution models into undergraduate thermodynamics instruction for both pure component properties as well as phase equilibria is beneficial. Such inclusion results in students having the necessary background in estimation techniques, allowing them to readily employ these tools in future endeavors.

References

[1] Sandler, Stanley I. Chemical, Biochemical and Engineering Thermodynamics, (4th ed., Wiley & Sons: 2006).

[2] Smith, J.M.; Van Ness, .C.; Abbott, M.M. Introduction to Chemical Engineering Thermodynamics. (7th edition, McGraw-Hill: 2005).

[3] Elliott, J.R.; Lira, C.T. Introductory Chemical Engineering Thermodynamics, (Prentice Hall: 1999).

[4] Koretsky,M. D. Engineering and Chemical Thermodynamics (Wiley & Sons: 2003).

[5] Poling, B.E.; Prausnitz, J.M.; O'Connell, J.P. The Properties of Gases and Liquids. (5th ed., McGraw-Hill Professional: 2000).

[6] Green, D.; Perry, R. Perry's Chemical Engineers Handbook. (8th ed., McGraw-Hill: 2008).

[7] Yaws, C. Chemical Properties Handbook: Physical, Thermodynamics, Engironmental Transport, Safety & Health Related Properties fo Organic & Inorganic Chemicals. (McGraw Hill: 1999).

[8] Molecular Knowledge Systems, Inc., CRANIUM - Property Estimation Software, <http://www.molknow.com/Cranium/cranium.htm>.

See more of this Session: Software, Technology and Computation In Chemical Engineering Education
See more of this Group/Topical: Education

(612c) Browser-Based Simulations for the Illustration of Chemical Engineering Concepts

Anthony Butterfield, Chemical Engineering, University of Utah, Salt Lake City, UT

Interactive online components have been found to be effective in science and engineering education. Such components may allow students to work at their own pace and in their own time, and allow for interactive projects without the consumption of lab resources. In general, these gains have been shown in the literature to come at no educational cost or even with improved student learning when compared to traditional approaches. However, for chemical engineers, such tools are relatively rare and often require proprietary software or software installation, which limits their reach and impact.

We have developed the tools to build interactive and visually dynamic simulations to illustrate important chemical engineering concepts and processes using JavaScript. Use of JavaScript presents some distinct difficulties, due to its limited native library of mathematical operations and limited means of visualization. We have, however, assembled a toolbox of JavaScript functions to overcome those restrictions. The result is a set of simulations that require only a browser to operate, and, unlike Java or Flash-based simulations, these simulations will function on the majority of smart phones and all major operating systems with no need to install additional software, proprietary or otherwise.

Many simulations have been created and, due to their usefulness and accessibility as pedagogical tools for our students, others are in development. For example, a CSTR simulation has been created to demonstrate concepts of reaction kinetics, mass and energy balances, and control. A 2-D heat conduction simulation has been used to demonstrate concepts of heat transfer and numerical methods. We have also created browser-based simulations of existing lab equipment and used these simulations in conjunction with hands-on work to better illustrate the interplay between theory and practice.

These web tools allow our students (or any student with internet access) to quickly develop an intuition for complicated engineering concepts by literally interacting with them from the device in the palm of their hands. Furthermore, as compared to the traditional use of static example problems which we may find in texts, the interactivity and visual feedback emphasized in these simulations have improved our ability to quickly cover multiple scenarios with a single system and educate a more diverse set of learners.

See more of this Session: Software, Technology and Computation In Chemical Engineering Education
See more of this Group/Topical: Education

(612d) Shifts In Student Attitudes to a Technology-Based Active Learning Pedagogy

Milo D. Koretsky and Bill J. Brooks, Chemical, Biological, and Environmental Engineering, Oregon State University, Corvallis, OR

Manyengineering classes emphasize student problem-solving skills almost to theexclusion of the understanding of underlying concepts. However, with this typeof instruction, students are better rewarded by rote learning than byconceptual understanding. It has also been shown that the lack of conceptualunderstanding severely restricts the students' ability to solve a new problemsince they do not have the functional understanding to use their knowledge innew situations. Alternatively, classes that use active learning pedagogies havebeen shown to allow students to integrate new concepts with their priorknowledge and develop richer conceptual understanding.

Active learning pedagogies havebecome enabled by technology-based classroom tools. For example, the use of personalresponse systems (clickers) has increased substantially. Clicker technologiesenable students to provide instantaneous feedback to instructor questions via ahandheld device. Each clicker unit has a unique signal so that the answer fromeach individual student can be identified and recorded. However, most clickersare limited to multiple choice questions.

This study uses an alternative,technology-based tool, the Web-based Interactive Science and Engineering (WISE)Learning Tool. Its use of computer technology permits a significantly widerrange of learning activities than clickers allow. Specific to this study is theability to ask students to provide short-answer, written explanations followingmultiple choice questions. Pedagogically, the short answers provide studentsopportunities for metacognition through reflection.

Changesin student perceptions to this novel technology-based active learning pedagogywere studied over the first five years it was used in a chemical engineeringthermodynamics course. Students tended to view active learning more favorablyover time, particularly in regards to statements which required them to beinterpretive of their own learning. They also perceived that being required towrite explanations prompted them to think more deeply about multiple choiceconceptual questions.

Implicationsfor adopting active learning techniques are discussed.

See more of this Session: Software, Technology and Computation In Chemical Engineering Education
See more of this Group/Topical: Education

(612e) Applying Reaction Engineering In a Virtual Chemical Company

Marcel A. Liauw[1], Steffen Heddrich[1] and Volker L. Deringer[2], (1)Itmc, RWTH Aachen University, Aachen, Germany, (2)RWTH Aachen University, Aachen, Germany

A computer game has been developed that approaches the students as a (virtual) chemical company where problems must be solved by applying reaction engineering concepts. The immersive character of the game is supposed to enhance retention of skills learned.

See more of this Session: Software, Technology and Computation In Chemical Engineering Education
See more of this Group/Topical: Education

(612f) The Benefits of Using Computational Modeling In the Classroom to Complement Experiment

David Gallagher, CAChe Research, Beaverton, OR

The latest computational modeling tools enable students to quickly and easily perform virtual experiments in reaction chemistry or thermodynamics, etc. on their own laptop computers. As well as kinetics and thermodynamics, the latest tools can predict a wide range of physical properties of compounds and mixtures, including vapor pressure, boiling points, activity coefficients, azeotropes, partition coefficients, liquid-liquid equilibria, etc.

However, more than just numbers, visualization of the computational models offers an intuitive understanding of the fundamental phenomena that influence the chemistry or physical properties. For example, visualization of electrostatic potentials by color on the solvent-accessible surface and sigma profiles, etc., illustrate the potential for hydrogen bonding, dipole-dipole, dipole-induced-dipole and Van der Waals interactions, providing clear insight into the molecular interactions that determine the physical properties.

Software tools that incorporate quantum chemistry and statistical thermodynamics methods do not require empirical data, enabling the student to experiment with entirely novel chemicals. This releases the student from the constraints of their own limited experience and the chemicals available in the lab.

Finally, these new tools enable the student to explore their own creativity and experiment with off-the-wall ideas, with the safety and convenience of a laptop computer anywhere: in the lab, the classroom, or at home.

This talk will be illustrated with examples of experiments performed with COSMOtherm and other software tools that are available free for educational use, along with a brief overview of the underlying computational methodologies.

See more of this Session: Software, Technology and Computation In Chemical Engineering Education
See more of this Group/Topical: Education

(612g) Energy Modules for Hydrogen and Fuel Cells In the Chemical Engineering Curriculum

Jason Keith[1], Daniel Lopez Gaxiola[1], Daniel A. Crowl[1], Dave Caspary[1], Abhijit Mukherjee[2], Dennis DeSheng Meng[2], Jeff Naber[2], Jeff Allen[2], John Lukowski[3], Barry Solomon[4], Jay Meldrum[5] and Thomas F. Edgar[6], (1)Department of Chemical Engineering, Michigan Technological University, Houghton, MI, (2)Department of Mechanical Engineering - Engineering Mechanics, Michigan Technological University, Houghton, MI, (3)Department of Electrical and Computer Engineering, Michigan Technological University, Houghton, MI, (4)Department of Social Sciences, Michigan Technological University, Houghton, MI, (5)Keweenaw Research Center, Michigan Technological University, Houghton, MI, (6)Chemical Engineering, The University of Texas at Austin, Austin, TX

In this project, we report on the development of energy modules for use in the chemical, mechanical, or electrical engineering curriculum. This project follows on our successful development of hydrogen and fuel cell related modules. We base the module design in a format which enables students to apply principles from their core classes to learn about traditional and alternative energy technology. Each module contains an introduction, problem motivation and background, example problem statement, example problem solution, and a homework problem statement. Instructors can obtain the solutions from the lead author by email.

Both the fuel cell and energy modules have been used at Michigan Technological University in the following manner:

- Short lecture on a chemical engineering (or energy) topic
- Handout of a module to serve as an in-class problem
- Instructor helps students work through the example problem and apply to the homework problem
- The homework problem is due at a future class meeting

In this paper we report on the development, testing, and assessment of these modules and report future directions.

See more of this Session: Software, Technology and Computation In Chemical Engineering Education
See more of this Group/Topical: Education

(614a) Optimal Operation and Design of Pooling and Other Bilinear Networks

Keith Zorn and Nick Sahinidis, Chemical Engineering, Carnegie Mellon University, Pittsburgh, PA

Pooling and blending problems typically involve maximizing profit subject to supply, demand, storage capacity, and product quality constraints. Applications of the pooling problem include: petroleum refining, wastewater treatment, communications, and supply-chain operations [1]. Bilinear network flow problems typically involve maximizing profit subject to network topology constraints. Applications of the bilinear network flow problem include: superstructure optimization, reactor network design, and heat and water integration [2].

Pooling problems and network flow problems can be formulated as non-convex mixed-integer non-linear programs and generally have many local solutions. Large-scale problem instances lead to challenging global optimization problems. Algorithmically, one is interested in solving optimization problems with mostly bilinear nonlinearities and possibly integer variables to model fixed charges. These types of problems are very challenging and continue to constitute a very difficult and important class of problems for the modeling and optimization of chemical supply chains.

Tight convex relaxations are essential when solving these difficult problems to global optimality with a branch-and-bound algorithm [3]. It is now understood that the successful application of convexification techniques often requires unconventional mathematical reformulations of the problem constraints. We combine reformulation linearization techniques [4] and advanced convex envelope construction techniques [5] to produce tight subproblem formulations for these bilinear programming applications. Extensive computational results will be provided.

[1] Misener, R. and Floudas, C.A.. "Advances for the pooling problem: modeling, global optimization, and computational studies." Appl. Comput.. Math. 8, 3-22, 2009.

[2] Ruiz, J.P. and Grossmann, I.E., "Exploiting vector space properties to strengthen the relaxation of bilinear programs arising in the global optimization of process networks." Optimization Letters, 5, 1-11, 2011.

[3] Tawarmalani, M. and Sahinidis, N.V.. "Global optimization of mixed-integer nonlinear programs: A theoretical and computational study." Mathematical Programming. 99, 563-591, 2004.

[4] Sherali, H.D. and Adams, W.P.. "A reformulation-linearization technique for solving discrete and continuous nonconvex problems." Norwell: Kluwer, 1999.

[5] Bao, X., Sahinidis, N.V., and Tawarmalani, M.. "Multiterm polyhedral relaxtions for nonconvex, quadratically-constrained quadratic programs." Optimization Methods and Software. 24, 485-504, 2009.

See more of this Session: Supply Chain Optimization
See more of this Group/Topical: Computing and Systems Technology Division

(614b) Sizing Safety Stock for Supply Chain Risk Management: The Sole Source Disruption Risk

Russell A. Ogle[1], Mark J. Viz[2] and Dennis M. McCann[2], (1)Exponent, Lisle, IL, (2)Exponent, Wood Dale, IL

Relying on a sole source supplier for raw material used at process facilities (such as petrochemical facilities, integrated coke ovens, blast furnaces at steel production facilities, or coal-fired power plants) may be strategically chosen for economic or other reasons, but such a strategy is not without significant risks if safety stock / inventory is not properly determined and maintained. Consider the following from current research in this area: "Single sourcing, where one supplier is responsible for the supply of a specific item or service may be advantageous from a cost and quality management perspective, but is dangerous in terms of resilience. While it may be desirable to have a lead supplier, wherever possible alternative sources should be available."[1] However, the availability of alternative sources may not mean that alternative or multiple supply transport routes are available, leading to a "sole source supplier" problem even if multiple suppliers are available.

Properly determining the size of an inventory stockpile presents numerous challenges even without the additional constraint of sole sourcing.[2],[3],[4] In addition, the "decision maker must balance the cost of holding inventory against the risk of running out of fuel,"[5] that is, the classic trade-off between inventory holding cost and potential shortage cost. The option of continually increasing stockpile size without knowledge of what size is sufficient given anticipated risks might protect against certain supply disruptions and is easy to implement,[6] but it can be cost prohibitive. Reducing the number of suppliers for a significant raw material (such as coal for a utility) can itself "breed vulnerability in the supply chain."[7]

We define "sole source supplier" in a general sense to not only include single commodity suppliers but also the situation that might involve multiple suppliers, all of whom rely on the same individual transport route for shipping the commodity to its point of use. Examples of this situation are numerous. The Joint Line in the Powder River Basin (Wyoming) is a "single-path-to-destination" transport route that is relied upon by a number of utilities across the United States for a steady and regular supply of low-sulfur coal. The Ohio River with its multiple transloading facilities (from rail to barge) is another primary example. Commodity stockpiles have been seen in certain circumstances as a "backstop"[8] against unreliable rail or waterway service that might be caused by a significant disruption (e.g., derailment or flooding, respectively) in these transportation routes.

When relying on a sole source supplier or multiple suppliers that rely on a single transportation route (such as a single captive rail line or waterway), how much raw material should be held as safety stock? The answer depends on many factors, including the manufacturer's awareness of impending disruptions [9] ("warning time") and its contingency plan to procure emergency inventory from other sources. In this paper we focus on protecting a continuous manufacturing operation from supply disruptions due to transportation failures involving single-path-to-destination routes. Specifically, we consider how to determine the optimal size of a coal stockpile for a coal-burning power plant subject to disruptions in single-path rail and/or barge transport. The frequency distributions of transportation disruptions involving single-path-to-destination or captive routing are drawn from historical data collected in the United States from 1990 to 2010. The data reveal a broad statistical distribution for frequency and duration of disruption events.

It is assumed that demand is constant and that plant operation responses such as reduced burn are not viable options. The size of the safety stock is determined using discrete event simulation guided by ideas drawn from queueing theory and performed in the context of existing research in this area. The economic factors contributing to the objective function are drawn from published data. The study objective is to determine the safety stock level required to cope with the median (50%), 90th percentile, and 95th percentile disruptions based on the selected data sets.

[1] Christopher, M. and Peck, H., "Building the resilient supply chain," *International Journal of Logistics Management*, v. 15, n. 2, p. 15, 2004.

[2] Morris, P.A., Sandling, M.J., Fancher, R.B., Kohn, M.A., Chao, H.P., Chapel, S.W., "A utility fuel inventory model," *Operations Research*, Operations Research Society of America: v. 35, n. 2, p. 2, 1987.

[3] Brady, T.F. and Pfitzer, C.M., "A prescriptive analysis of the Indiana coal transportation infrastructure," report of the *Center for Coal Technology Research*, Purdue University: p. 23, 2007.

[4] Liberatore, M.J., "Using MRP [material requirements planning] and EOQ/safety stock [economic order quantity] for raw materials inventory control: Discussion and case study," *Interfaces*, Institute of Management Sciences: v.9, n. 2, pp. 4, 6, 1979.

[5] Morris, loc. cit., p. 1.

[6] Tomlin, B. and Wang, Y., "Operational strategies for managing supply chain disruption risk," pre-publication draft for *Handbook of Integrated Risk Management in Global Supply Chains*, Wiley: p. 3, 2009.

[7] Stecke, K.E. and Kumar, S., "Sources of supply chain disruptions, factors that breed vulnerability, and mitigating strategies," *University of Texas Dallas School of Management*, p. 12, 2006.

[8] Kaplan, S.M., "Rail transportation of coal to power plants: Reliability issues," *CRS Report for Congress*, Congressional Research Service: CRS report no. RL34186, p. 2, 2007.

[9] Morris, loc. cit., p. 12.

(614c) A Holistic Framework for Drug Development and Capacity Planning In Novel Pharmaceutical Supply Chains

Arul Sundaramoorthy[1], James M.B. Evans[1] and Paul I. Barton[2], (1)Department of Chemical Engineering, Massachusetts Institute of Technology, Cambridge, MA, (2)Process Systems Engineering Laboratory, Dept. of Chemical Engineering, Massachusetts Institute of Technology, Cambridge, MA

Upon successful completion of clinical trials and the FDA process, a new drug is introduced commercially (launched) into the market under patent protection. The launch phase includes capacity planning for meeting customer demands, and optimizing the production and inventory from launch to peak demands. Since time to market is the major driver in the pharmaceutical industry, decisions pertaining to the launch phase are usually made in parallel with the development process. Clearly, these decisions have to be made under both technical and market uncertainties. Furthermore, the risk in the investment decisions must be properly taken into account. Thus, a holistic framework to optimize the above strategic decisions under uncertainty in an integrated manner is highly desired (Shah, 2004).

Existing approaches for simultaneous drug development and capacity planning were developed for traditional pharmaceutical supply chains in which the primary and secondary production facilities are operated in geographically different locations (Maravelias and Grossmann, 2001; Shah, 2004). Furthermore, the batch mode of manufacturing has dominated the pharmaceutical industry for decades. Recently, the Novartis-MIT Center for Continuous Manufacturing has embarked on an innovative project to shift from the batch mode of manufacturing to the continuous mode, where raw materials through the APIs to the finished products are produced seamlessly in an integrated facility. Such an integrated end-to-end production scheme is very promising for several reasons: a) low costs of inventory, logistics, and production, b) short supply-chain cycle times, c) less exposure to supply-chain disruptions, and d) resilient supply chain. Thus, the option of continuous manufacturing in the pharmaceutical industry brings in additional challenges and opportunities in the context of drug development and investment strategy.

In this work, we develop a holistic optimization under uncertainty framework to address drug development and capacity planning for pharmaceutical supply chains that employ novel integrated production schemes. Since the outcomes of clinical trials are uncertain, the problem naturally leads to a stochastic programming problem, which is formulated as a multi-period mixed-integer linear programming (MILP) model. However, the model size can grow dramatically with the number of scenarios, resulting in intractable large-scale MILP models. The underlying problem structure motivates a novel solution method to solve the above large-scale models, and we illustrate the application of the proposed framework using several problem instances.

Reference:

1. Shah, N. Pharmaceutical supply chains: key issues and strategies for optimization. *Comput. Chem. Eng.*, **2004**, 28, 929-941.
2. Maravelias, C. T.; Grossmann, I. E. Simultaneous planning for new product development and batch manufacturing facilities. *Ind. Eng. Chem. Res.*, **2001**, 40, 6147-6164.

See more of this Session: Supply Chain Optimization
See more of this Group/Topical: Computing and Systems Technology Division

(614d) Integrated Planning and Scheduling of Multisite Production and Distribution Facilities

Nikisha Shah and Marianthi Ierapetritou, Chemical and Biochemical Engineering, Rutgers University, Piscataway, NJ

Modern process industries operate as a large integrated complex that involve multiproduct, multipurpose, and multisite production facilities serving a global market. The process industries supply chain can be defined to be composed of production facilities and distribution centers, where final products produced at production facilities are transported to distribution center to satisfy the customers demand. The multisite plants produce a number of products driven by market demand under operating conditions such as sequence dependent switchovers and resource constraints. Each plant within the enterprise may have different production capacity and costs, different product recipes, and different transportation costs to the markets according to the location of the plants. To maintain economic competitiveness in a global market, interdependences between the different plants, including intermediate products and shared resources need to be taken into consideration when making planning decisions. Furthermore, the planning model should consider not only individual production facilities constraints but also transportation constraints because in addition to minimizing the production cost, it's important to minimize the costs of products transportation from production facilities to the distribution centers.

Thus, simultaneous planning of all activities from production to distribution stage is important in a multisite process industry supply chain [1]. To achieve enterprise wide optimization (EWO) in spatially distributed production facilities and distribution centers, interactions between different complexes should be taken into consideration and their optimization should be tackled simultaneously. In the recent years, multisite production and distribution planning problem has received lot of attention in the literature [2-4].

The planning problem covers a time horizon of few months to a year and is concerned with decisions such as production, inventory, and distribution whereas the scheduling problem deals with issues such as assignment of tasks to units and sequencing of tasks in each unit which covers time horizon of few days to few weeks. Since there is a significant overlap between different decisions levels, it is necessary to integrate planning and scheduling problems to achieve global optimal solutions for the entire supply chain [5]. For multisite facilities, the size and level of interdependences between these sites present unique challenges to the integrated tactical production planning and day-to-day scheduling problem and these challenges are highlighted by Kallrath, 2002 [6].

In this work, we focus on the integration of planning (medium-term) and scheduling (short-term) problems for the multiproduct plants that are located in different sites and supply different markets. In the recent years, the area of integrated planning and scheduling for single site has received much attention [7-9]. Even though most companies operate in a multi-site production manner, very limited attention has been paid on integrating planning and scheduling decisions for multi-site facilities.

In this work, we first propose an integrated planning and scheduling model for multisite production and distribution facilities that takes into consideration shared resources and intermediates between production facilities, transportation time between production facilities, between production site and distribution center, and in some rare case, between distribution centers. To account for the situations when due to production capacity limitations or raw material availability limitations, industry cannot satisfy the demand; we consider the option of hiring external contractors. The full-scale integrated planning and scheduling optimization model spans the entire planning horizon of interest and includes decisions regarding all the production sites, distribution centers, and transportation between them. Since the production planning and scheduling levels deal with different time scales, the major challenge for the integration using mathematical programming methods lies in addressing large scale optimization models. When typical planning horizon is considered, the integrated problem becomes intractable and a mathematical decomposition solution approaches are necessary. To effectively deal with complexity issues of the integrated problem, the block angular structure of the constraints matrix is exploited by relaxing the inventory constraints between adjoining time periods using the augmented lagrangian decomposition method. To resolve the issues of non-separable cross-product terms in the augmented lagrangian function, we apply diagonal approximation method. This decomposition then results in separable planning and scheduling problems for each planning period and for each production site. To illustrate the effectiveness of the proposed model and decomposition approach, we apply them to different sizes case studies.

References:

(1) Shah, N., *Single and multisite planning and scheduling: current status and future challenges.* AIChE Symposium Series, 1998. 94(320): p. 75.

(2) Verderame, P.M. and C.A. Floudas, *Operational planning framework for multisite production and distribution networks.* Computers & Chemical Engineering, 2009. 33(5): p. 1036-1050.

(3) Jackson, J.R. and I.E. Grossmann, *Temporal Decomposition Scheme for Nonlinear Multisite Production Planning and Distribution Models.* Industrial & Engineering Chemistry Research, 2003. 42 (13): p. 3045-3055.

(4) Timpe, C.H. and J. Kallrath, *Optimal planning in large multi-site production networks.* European Journal of Operational Research, 2000. 126(2): p. 422-435.

(5) Maravelias, C.T. and C. Sung, *Integration of production planning and scheduling: Overview, challenges and opportunities.* Computers & Chemical Engineering, 2009. 33(12): p. 1919-1930.

(6) Kallrath, J., *Planning and scheduling in the process industry.* OR Spectrum, 2002. 24(3): p. 219-250.

(7) Li, Z. and M.G. Ierapetritou, *Rolling horizon based planning and scheduling integration with production capacity consideration.* Chemical Engineering Science, 2010. 65(22): p. 5887-5900.

(8) Li, Z. and M.G. Ierapetritou, *Production planning and scheduling integration through augmented Lagrangian optimization.* Computers & Chemical Engineering, 2010. 34(6): p. 996-1006.

(9) Verderame, P.M. and C.A. Floudas, *Integrated Operational Planning and Medium-Term Scheduling for Large-Scale Industrial Batch Plants.* Industrial & Engineering Chemistry Research, 2008. 47(14): p. 4845-4860.

See more of this Session: Supply Chain Optimization
See more of this Group/Topical: Computing and Systems Technology Division

(614e) Operability Considerations In Process Supply Chain Design for Forest Industry Transformations

Richard Mastragostino and Christopher L. E. Swartz, Chemical Engineering, McMaster University, Hamilton, ON, Canada

The forest products industry (FPI) is primarily commodity based and facing challenges as a result of global competition and a sustained reduction in demand of key sector products. This has led to proposals for a shift towards revenue diversification of existing pulp and paper mills, through the production of biofuels or high-value specialty products alongside conventional commodity products. The strategy will result in the transformation of existing pulp and paper mills into integrated forest biorefineries (IFBR), which implies an increase in capital expenditure compensated for by the ability for plants to manufacture high value products. A key consideration within this new business paradigm is that the process and supply chain have sufficient capability to be resilient against market volatilities such as demand, supply or price fluctuations. The present research draws on concepts of operability analysis associated with process plant operation and design.

Process plants operate in a continually changing environment and are required to operate satisfactorily under sustained changes as well as short-term fluctuations. Process plant operability reflects the ability of a system to function satisfactorily under the expected range of process conditions. Two notions of operability are flexibility and dynamic operability. Flexibility reflects a system's ability to remain feasible in the face of sustained changes, while dynamic operability refers to a system's ability to respond rapidly to short-term fluctuations. Systematic, quantitative approaches for flexibility and dynamic operability analysis of process plants have been proposed (Grossmann and Morari, 1983), as well as the incorporation of operability considerations in optimization-based plant design (Mohideen et al., 1996; Baker and Swartz, 2004; Sakizlis et al., 2004).

The above considerations have motivated an investigation into an optimization based framework for the design of operable supply chains. Notions of operability that are commonly used within a supply chain context include lead time, order fill rate, on-time deliveries and delivery flexibility. A comprehensive review of supply chain performance metrics is given in Beamon (1999). Of note is that agile supply chains (Christopher and Towill, 2001) include elements of both static and dynamic operability. In a recent study, You and Grossmann (2008) developed a framework which allows for network design, within a bi-criterion optimization, that considers both net present value and lead time of the supply chain design. In an earlier study, Sabri and Beamon (2000) developed a multi-criterion optimization framework for supply chain planning which considers net present value, customer service level, and distribution and plant flexibility.

The present study considers three key objectives: (i) development of an optimization based framework to quantify the operability of a process supply chain design, (ii) utilization of the operability metrics in an optimization framework for the design of operable process supply chain systems, and (iii) development of appropriate case studies which are relevant to the forest products industry. The literature on biorefinery supply chain optimization is primarily concerned with high level strategic decisions. However, of relevance to the present study is the work of Eksioglu et al. (2009) who developed an optimization framework that captures deterioration, supply seasonality and supply availability of the biomass-to-biorefinery supply chain. The supply chain model developed in our study can be characterized by time discretization at the operational level to capture response dynamics, flexible manufacturing plants, manufacturing and transportation delays and uncertain demand patterns. A goal of the current study is to develop a tool that can be used to investigate the long term sustainability of new business paradigms within the forest products industry. The paper will present the context of the study in relation to developments within the forest products industry, draw parallels between process plant and supply chain operability, and present an optimization-based framework for operability analysis of process supply chains. Preliminary results will be presented, and insights regarding the relationship between effective revenue diversification and operability will be discussed. In addition, key challenges and future research directions will be identified.

References:

Baker, R. and Swartz, C.L.E. (2004). Simultaneous solution strategies for inclusion of input saturation in the optimal design of dynamically operable plants. Optimization and Engineering, 5, 5-24.

Beamon, B.M. (1999). Measuring supply chain performance. International Journal of Operations and Production Management, 19, 275-292.

Christopher, M. and Towill, D. (2001). An integrated model for the design of agile supply chains. International Journal of Physical Distribution and Logistics Management, 31, 235-246

Eksioglu, S.D., Acharya, A., Leightley, L.E. and Arora, S. (2009). Analyzing the design and management of biomass-to-biorefinery supply chains. Computers and Industrial Engineering, 57,1342–1352.

Grossmann. I.E. and Morari, M. (1983). Operability, resiliency and flexibility – Process design objectives for a changing world. Proc. Int. Conf. on Foundations of Computer-Aided Process Design, A. W. Westerberg and H.H. Chien, eds., 931-1030.

Mohideen, M.J., Perkins, J.D. and Pistikopoulos, E.N. (1996). Optimal design of dynamic systems under uncertainty. AIChE Journal, 42, 2251-2272.

Sabri, E.H. and Beamon, B.M. (2000). A multi-objective approach to simultaneous strategic and operational planning in supply chain design. Omega, 28, 581-598

Sakizlis, V., Perkins, J.D. and Pistikopoulos, E.N. (2004). Recent advances in optimization-based simultaneous process and control design. Comp. Chem. Eng., 28, 2069-2086.

You, F. and Grossmann, I.E. (2008). Design of responsive supply chains under demand uncertainty. Comp. Chem. Eng., 32, 3090–3111.

See more of this Session: Supply Chain Optimization
See more of this Group/Topical: Computing and Systems Technology Division

(614f) Planning and Long-Term Scheduling of Single-Stage Multi-Product Continuous Lines with a Complex Recycling Structure

Ricardo M. Lima, Department of Chemical Engineering, Carnegie Mellon University, Pittsburgh, PA, Ignacio E. Grossmann, Chemical Engineering Department, Carnegie Mellon University, Pittsburgh, PA and Yu Jiao, PPG Inc, Glass Business and Discovery Center, Pittsburgh, PA

In this presentation we address the planning and scheduling of a real-world process for manufacturing tinted glass in a single-stage multi-product process with two continuous lines. The goal of this work is the development of a decision making tool to support supply chain decisions at the level of short and medium term scheduling, and also strategic decisions such as product portfolio composition. The core element of this tool is a Mixed-Integer Linear Programming (MILP) model involving production planning and scheduling decisions. An important feature of this model is the integration of the consumption, production and storage of waste glass (cullet) with the production schedule. This tool has been used by the supply chain group in PPG Glass to address strategic decisions.

The scheduling of multiproduct continuous processes in the processing industries has not received as much attention as batch scheduling. However, multi-product continuous lines are common in the paper, glass, steel, and specialty chemicals industries, and their optimal scheduling has been studied by some authors. Sahinidis and Grossmann (1991) addressed the optimal scheduling of continuous parallel lines with a cyclical operation and constant demand. Latter Pinto and Grossmann (1994) extended the former work considering multiple production stages. In both works the problems were formulated as Mixed-Integer NonLinear Programming (MINLP) models using slots to represent the time activities, and specific solution methods were proposed. Cooke and Rohdler (2006) studied the scheduling of a multi-product continuous reactor for manufacturing polymers with different grades, with also constant demand. These authors propose a MINLP lot-size model based on an extended basic period approach, and a specific decomposition algorithm with three levels. An interesting feature of their model is the integration of the optimal scheduling with safety stocks calculation under demand and production uncertainty. Miegeville (2005) developed a customized MILP lot-sizing model for the scheduling of tinted glass production. The model involves the aggregation of products into families and different levels of time resolution, in order to handle the hierarchical production scheduling and inventory control. At the planning level, Liu and Papageorgiou (2008) proposed a TSP based MILP model for continuous lines, while Erdirik-Dogan and Grossmann (2008) have addressed the integration of planning and scheduling using a bi-level decomposition algorithm, involving a TSP based MILP model in the upper level, and a continuous time slot based model in the lower level.

The model proposed in this work was developed based on the continuous time slot MILP model proposed by Erdirik-Dogan and Grossmann (2008). Recently, Lima et al. (2011) have extended the model of Erdirik-Dogan and Grossmann by including the following features: a) carry-over changeovers across the due dates; b) minimum run lengths across the due dates; c) a rigorous aggregation of the products based on the type of changeovers; d) definition of minimum inventory levels at the end of the time horizon; and e) some specific constraints for glass production. Meanwhile, the model was further developed to capture the cullet operations integrated with the production, which represent real constraints of operation, and have a significant economic value. These constraints were not addressed by Miegeville (2005) in his glass production planning model.

From the process point of view, we address the optimal scheduling of the production of specialty glass characterized by different colors. The color of each glass depends on the color of the substrate and the color of a coating if added. The changeover from one substrate to another substrate implies a transition time in the order of days, while the changeover from a specific substrate color to a product with the same substrate with a coating requires a comparatively short changeover. During changeovers, the process produces cullet at the same conditions (energy, raw materials, and production rate). The process scheduling is restricted by: 1) sequence-dependent changeovers between a subset of the products and no changeovers between another subset of products; 2) impositions on the sequence of production and production times between products without changeovers; 3) minimum run lengths due to process control and stability; and 4) by a complex compatibility and recycling profile of cullet. The market demand for each product is not constant during the time horizon, and therefore we do not consider a cyclic operation. Each product has a minimum requirement for cullet feed to the process in order to achieve glass quality standards, and to minimize energy costs. An interesting feature is the bidirectional relation between the scheduling, and the cullet generated and consumed, since a given sequence of production will generate specific amounts of cullet, but also the quality and quantity of cullet available in storage can restrict the products to be produced. The cullet is generated due to the process yield and during the transitions, creating two major types of cullet, transition cullet and product cullet. The recycling of cullet is characterized by a compatibility matrix between cullet recycled and product to produce, and by specific rates of cullet consumption as a function of the production run. The cullet recycling is modeled using mass balances for the transition cullet and product cullet that is produced and consumed, and inventory balances, since there is a maximum storage capacity. Cullet generated over the storage capacity has to be sold. These balances are applied at the end of each slot because the dilution cullet generated in slot l of a specific changeover may be used for a different product in slot $l+1$.

In this work we consider two parallel production lines operating in parallel. One line produces most of the products, while the other produces only a subset of products plus a dedicated product. Two valid assumptions in this process are the following: 1) both lines will not produce the same product in the same time period; and 2) one production line does not consume cullet from produced during changeovers. These two features of the process prevent the need to synchronize the start and end of the slots of both lines to account for the simultaneous consumption of the same type of cullet. Therefore, the balances for the cullet consumption and generation are made over each slot, and synchronized at the end of each the time period.

To solve the current MILP models, the hybrid rolling horizon strategy described in the Lima et al. (2011) is used. However, in order to cope with the increased size of the models, the time horizons for the scheduling model and planning model have been reduced to 3 and 6 months, respectively. Different case studies are presented to demonstrate a) the impact of considering the cullet model in the scheduling in terms of number of transitions and length of the production runs; b) the impact of different final inventory level conditions or constraints at the end of the planning horizon. Variations of the rolling horizon algorithms, and different inventory features are compared in terms of the size of the models, and computational efficiency in terms of integrality gaps and CPU time.

The model developed is the core of a software tool, where the MILP model developed in GAMS is integrated with a spreadsheet software for data input and display the results. This model has already been used to study strategic decisions at the level of product portfolio composition.

References

Cooke, D.L., Rohleder T.R., (2006). *Inventory evaluation and product slate management in large-scale continuous process industries.* Journal of Operations Management. 24, 235-249.

Erdirik-Dogan, M., Grossmann, I.E., (2008) *Simultaneous planning and scheduling of single-stage multi-product continuous plants with parallel lines*, Computers and Chemical Engineering, 32, 11, 2664-2683.

Lima, R.M., Grossmann, I.E., Jiao, Y. (2011) *Long-term scheduling of a single-unit multi-product continuous process to manufacture high performance glass.* Computers and Chemical Engineering. 35 (2011) 554–574

Liu, S., Pinto, J. M., & Papageorgiou, L. G. (2008). *TSP-based MILP model for medium term planning of single-stage continuous multiproduct plants.* Industrial & Engineering Chemistry Research, 47(20), 7733–7743.

Miegeville N. (2005), *Supply chain optimization in the process industry: Methods and case study of the glass industry.* PhD thesis. École Centrale Paris

Pinto, J.M., Grossmann, I.E. (1994), *Optimal cyclic scheduling of multistage continuous multiproduct plants.* Computers and chemical Engineering, 18(9), 797-816.

Sahinidis, N.V., Grossmann, I.E., (1991), *MINLP model for cyclic multiproduct scheduling on continuous parallel lines.* Computers and Chemical Engineering. 15 (2), 85-103.

See more of this Session: Supply Chain Optimization
See more of this Group/Topical: Computing and Systems Technology Division

AGENT-BASED CHEMICAL SUPPLY CHAIN MODELS ASSESSING DYNAMIC DISRUPTIONS

Mark Pepple, Sandia National Laboratories, Albuquerque, NM
Amy Sun, Sandia National Laboratories, Albuquerque, NM
Mark A. Ehlen, Sandia National Laboratories, Albuquerque, NM
Brian S. Jones, Sandia National Laboratories, Albuquerque, NM

Abstract

The chemical industry is one of the largest industries in the United States and a vital contributor to global chemical supply chains. The U.S. Department of Homeland Security (DHS) Science and Technology Directorate has tasked Sandia National Laboratories (Sandia) with developing an analytical capability to assess interdependencies and complexities of the nation's critical infrastructures on and with the chemical sector. This work is being performed to expand the infrastructure analytical capabilities of the National Infrastructure Simulation and Analysis Center (NISAC). To address this need, Sandia has focused on development of an agent-based methodology towards simulating the domestic chemical supply chain and determining economic impacts resulting from large-scale disruptions to the chemical sector.

Modeling the chemical supply chain is unique because the flow of goods and services are guided by process thermodynamics and reaction kinetics. Sandia has integrated an agent-based microeconomic simulation tool N-ABLE™ with various chemical industry datasets to abstract the chemical supply chain behavior. An enterprise design within N-ABLE™ consists of a collection of firms within a supply chain network; each firm interacts with others through chemical reactions, markets, and physical infrastructure. The supply and demand within each simulated network must be consistent with respect to mass balances of every chemical within the network. Production decisions at every time step are a set of constrained linear program (LP) solutions that minimize the difference between desired and actual outputs.

We illustrate the methodology with examples of modeled petrochemical supply chains under an earthquake event. The supply chain impacts of upstream and downstream chemicals associated with organic intermediates after a short-term shutdown in the affected area are discussed.

Introduction

Domestic events such as hurricanes (Katrina, Ike) and acts of terrorism can potentially have significant impacts on the U.S. chemical industry. Chemical plants, pipelines, rail, and road transportation routes can be damaged indefinitely during these events, thus impacting the ability of chemical facilities to produce, transact, and deliver chemicals. There has been a growing national interest in understanding and estimating the impact of chemical supply chains under natural and man-made events. In doing so, the insight can enhance industry's preparation towards responding to these events. This paper describes an agent-based model of chemical supply chains that is used to estimate supply chain impacts caused by homeland security-related events, including brief descriptions of the mathematical foundation, the chemical sector database, and an illustrated example of supply chains under a natural disruptive event.

Agent-Based Model

As a model, chemical supply chains behave as control-and-adaptation systems in which some components perform as traditional "state-and-control" systems (e.g., in-plant operations, cross-plant logistics networks) while others are more adaptive in nature (e.g., merchant markets with many, heterogeneous buyers and sellers; road transportation systems with many independent vehicles). Agent-based modeling (ABM) is one tool for modeling chemical supply chains as control-and-adaptation systems. ABM has been used widely to capture the structured, complex, and adaptive behavior of collections of economic entities. Examples of seminal ABM work include Palmer et al. (1994), Axelrod (1997), Arthur et al. (2007), and Epstein and Axtell (1996). For homeland security applications, Downes et al. (2005) developed a 3000-firm chlorine supply chain control-and-adaptation model and used it to estimate the impacts to chlorine firms of a rail transport disruption and private policies that reduce impact times by expediting shipments.

ABM has some important benefits over other modeling techniques, including that it can assemble disparate mathematical and network modeling approaches, some basic and others complex, into a single model and simulation environment. In our ABM, individual chemical plants are modeled as synthetic firms that carry out the purchasing, production, scheduling, storage, and sale of chemicals, as illustrated in Fig. 1. Each buyer is responsible for the purchase and inventory management of one chemical. These buyers

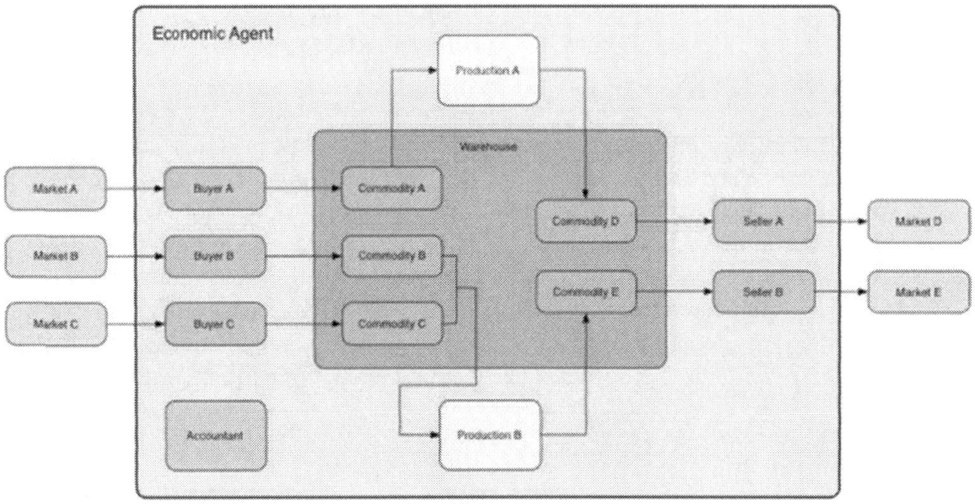

Figure 1. An agent-based chemical plant composed of buyers, productions, warehouses, production scheduler, and sellers.

purchase the chemicals in merchant markets, alongside buyers from other chemical plants, thereby competing for resources. Analogously, the sellers in these plants sell in one or more markets, attempting to set prices that maximize plant profits. Each chemical plant contains one-to-many production units that are managed by a production scheduler that follows the linear program described below, managing inventories of each commodity within the constraints of the warehouse and direct connections between production units.

Chemical Data

Robust and up-to-date information regarding chemical names, chemical manufacturers, economic statistics, and chemical reactions are foundational to conducting a static or dynamic analysis of the chemical supply chain. As such, Sandia has acquired and modified disparate sets of commercial databases over the years to create an in-house suite of application-ready database servers that analysts and software developers can use for software development, fast analysis, and chemical sector statistics. To establish a modeled production market, different layers of data are created to represent different components within a chemical supply chain. Table 1 outlines the different data categories used to represent chemical supply chains.

Table 1. Categories of chemical supply chain data used in constructing supply chain models.

Category	Detailed Information
Plant Information	Parent company; units; geo-location; chemicals; capacities captive production
Infrastructure	Transportation; pipelines;
Chemistry	Process; technologies; stoichiometry
End use	Wholesalers; geo-locations; quantities
Economics	Employees; imports/exports

The fundamental building blocks of chemical data consist of the most current chemical manufacturers information available from SRIC.[1] The Environmental Protection Agency also provides a chemical storage dataset that supplement our knowledge of how the chemicals are stored throughout the U.S.[2] Detailed data sources are described in the next section. The most unique feature of the chemical data is the chemistry information housed within Sandia. Such information relies on years of experience from chemical subject matter experts and hours of data recording to create a stoichiometric table that enables the analysts to link facility-level production data with their perspective end consumers.

---

[1] SRI Consulting: Directory of Chemical Producers, World Petrochemicals, and Chemical Economic Handbook

[2] Environmental Production Agency, E-Plan Chemical and Facility Hazards Data https://erplan.net/eplan/login.htm

Building a supply chain not only relies on accurate chemical producer information but also technology information that links the manufacturers to their perspective consumers. Each producing facility must transport its chemical product to another chemical user, and so on. The chemical user at the end of the chemical network chain can either be a chemical producer or an end consumer (such as a distributor or a wholesaler or a consumer product manufacturer). Figure 2 illustrates a simple example that centers around 1,4-butanediol. The upstream step shows the four different reaction pathways representative of how 1,4-butanediol is manufactured from different chemical raw materials. Likewise, three different processes then use 1,4-butanediol to manufacture other chemicals, mostly polyurethane and tetrahydrofuran. Overall, eleven chemicals are immediately tied to 1, 4-butanediol while approximately fifty-three domestic facilities are directly impacted by the flow of these eleven chemicals. The collection of chemical data, known as the Chemical Data Model (CDM) allows different application software to access the chemical manufacturing data and their respective technology information for analysis at different granularities from individual facilities to global chemical supply chain networks.

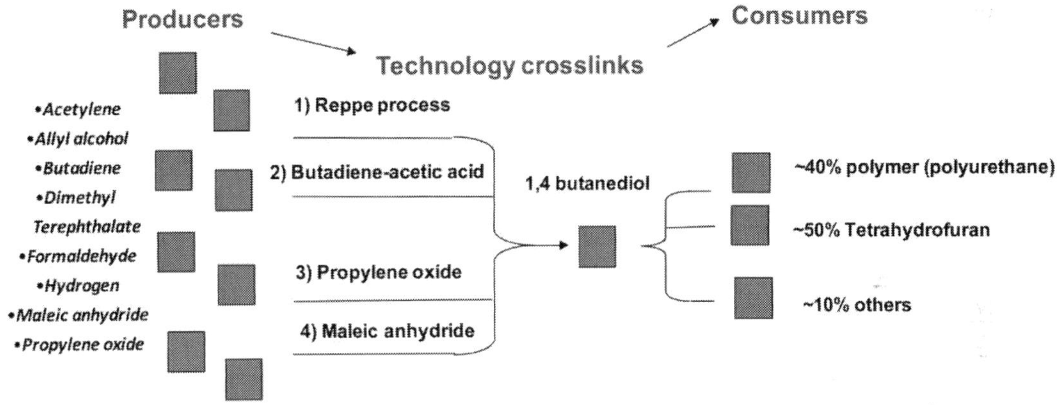

Figure 2. Chemicals directly impacted by the production of 1,4-butanediol.

Earthquake Scenario

This capability demonstration study analyzes the impacts on the chemical industry of an earthquake scenario centered around the New Madrid Seismic Zone (NMSZ). Figure 3 shows the impact area in different magnitudes.

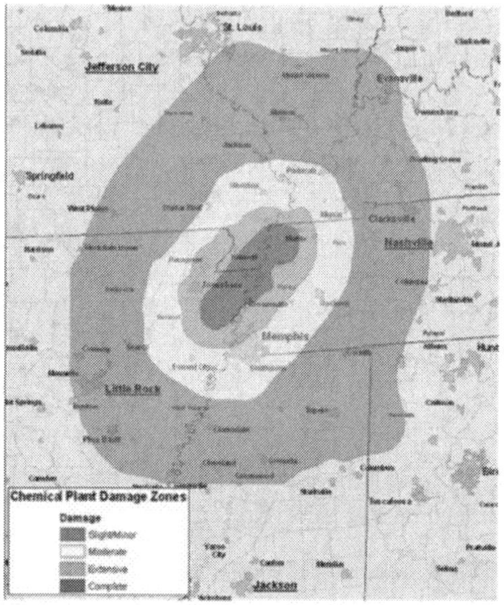

Figure 3. Impacted zones in different magnitudes for New Madrid Seismic zone.

N-ABLE Results

The N-ABLE model was built on 80 chemicals and 900 manufacturing facilities; a subset of those contained in the CDM. The 900 chemical facilities were comprised of both CDM plants and synthetic firms that act as "sources" and "sinks" at the boundaries of the supply chain. The analysis was focused on changes in supply chain structure and performance caused by the disruption. The length of the disruption is the primary factor determining supply chain-wide impacts. Changes in structure do not automatically result in changes in performance due to supply chain inventories and market liquidity. The analysis includes that of an individual chemical plant, a particular chemical in the supply chain, and across the entire supply chain.

Sample Plant

An individual chemical plant is linked to the supply chain through the markets by which it purchases and sells chemical. When the plant is unable to operate, its ability to purchase and produce is reduced. Internally, a plant's response to a disruption may include shutdown, restart, and accumulation or depletion of inventories. Figure 4 shows the mass of input chemical shipments over a period of time that includes the disruption. Each chemical this facility uses is represented by a color band in Figure 4. Day 1 is the beginning of the disruption and it takes approximately 50 days to return to baseline shipment levels. The decline in shipments is not immediate upon the disruption because in-transit shipments continue to arrive. Once the disruption is over, shipments levels rise above the previous baseline to make up for those lost during the disruption.

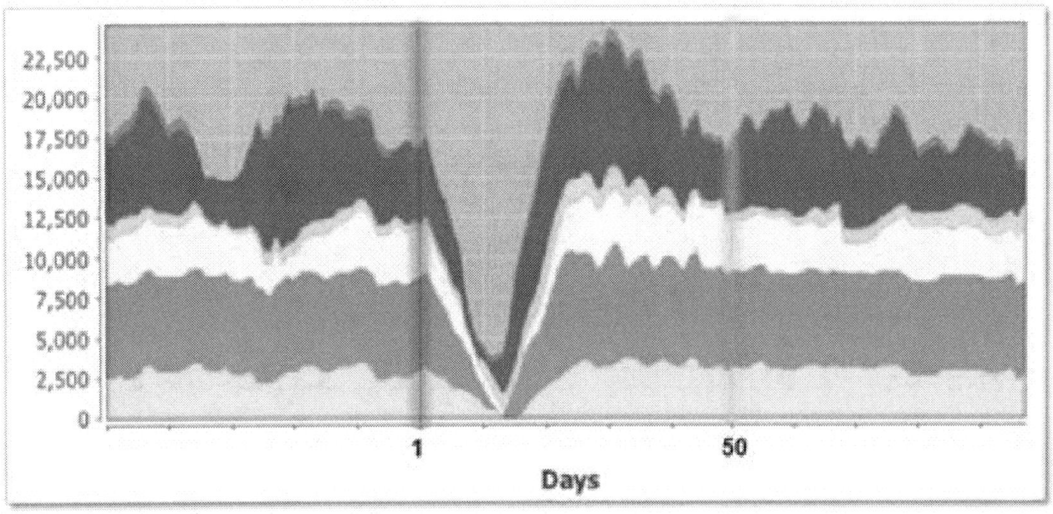

Figure 4. Input shipments by chemical

Sample Commodity

N-ABLE provides analysis of individual commodities (chemicals) by aggregating the production of this chemical across all the facilities where it is manufactured. For example, chemical 'A' is produced by approximately 30 plants, most of whom experience a direct or indirect impact from the earthquake, as seen in Figure 5. During the disruption, aggregate supply decreased by approximately 50%. Although most plants experienced significant decreases in production, in a few cases, plants increased production during the disruption to compensate for an overall decreased supply.

The nature of the chemical 'A' market during the disruption also changed. Buyer-seller relationships before the earthquake were changed by the disruption and new relationships were created during the disruption. Figure 6 shows the flow of chemical 'A' on A) a base line day B) day1 of the disruption and C) on day 200 of the disruption. The gray cylinders are buyers and sellers, with their height and diameter representative of production level and inventory, respectively. The turquoise arcs show transactions; the lighter end is the seller and the darker end the buyer. N-ABLE shows that some firms impacted by the earthquake do not return to their pre-disruption status. Furthermore, during the disruption several cross-country arcs are established as firms seek buyers and sellers outside the damage zone. Finally, N-ABLE also shows that in some cases buyer-seller relationships established during the disruption continue post disruption.

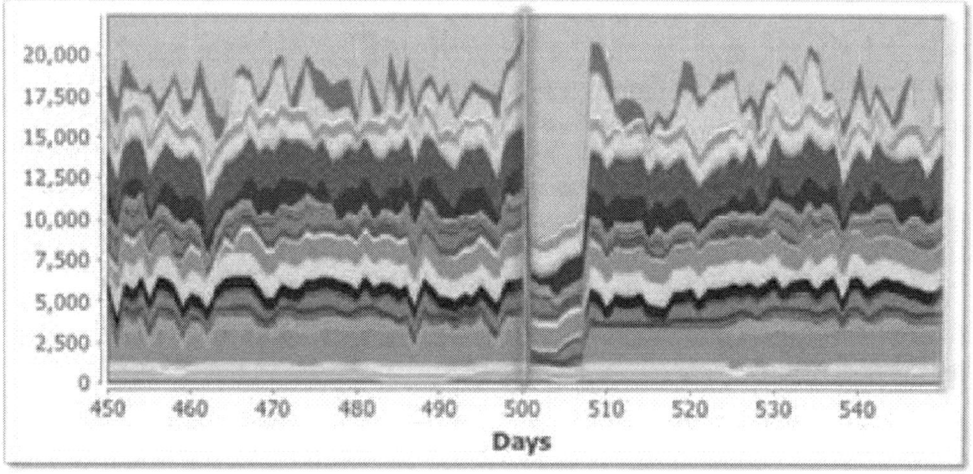

Figure 5. Chemical 'A' production by plant

Supply Chain-Wide Impacts

The disruption impacted chemicals across the product network, which indicates potentially wide-spread changes in market conditions. Figure 7 shows graphically how the chemicals are networked. During the first week of disruption, the supply chain experiences a 30 percent decline in production by mass. The overall supply-chain production levels return to normal after approximately 30 days, though some individual chemicals take much longer, see Figure 8. Sales experienced a 15 percent decline and take approximately 20 days to recover. The reduction in sales is less than in production due to sales from inventory for those markets that are not supply constrained during baseline and short duration conditions.

Aggregate unmet demand stabilizes after an estimated 80 days, although there is unmet demand for certain chemicals that remain. Longer term unmet demand for certain chemicals is expected due to the lack of production of particular chemicals even after the overall supply-chain production levels return to normal. Finally, shipment levels experienced a reduction during the disruption, followed by a surge immediately after, and return to baseline conditions after about 80 days. N-ABLE indicated that some firms in the impacted areas will have production processes out of service for six months or more. However, beyond the core damage area, most production flows return to normal status after 25-50 days.

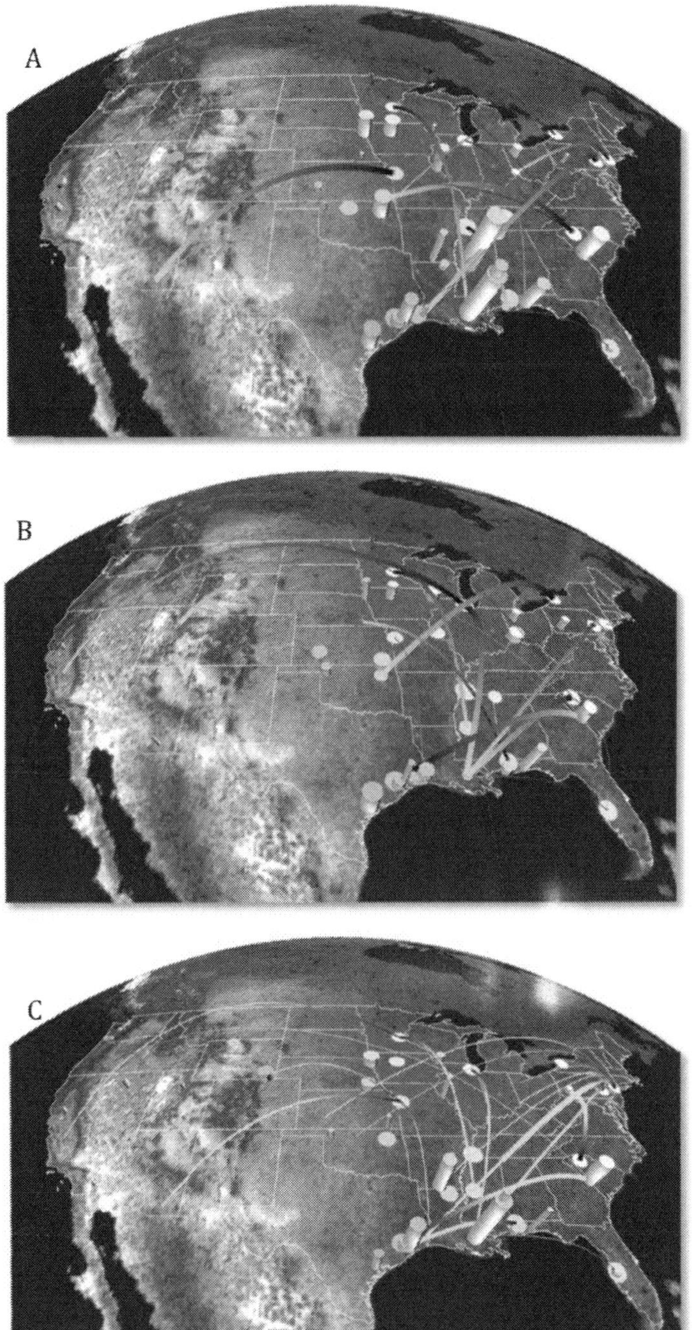

Figure 6. Chemical 'A' market A) baseline day, B) day 1 of disruption, and C) day 200 of disruption

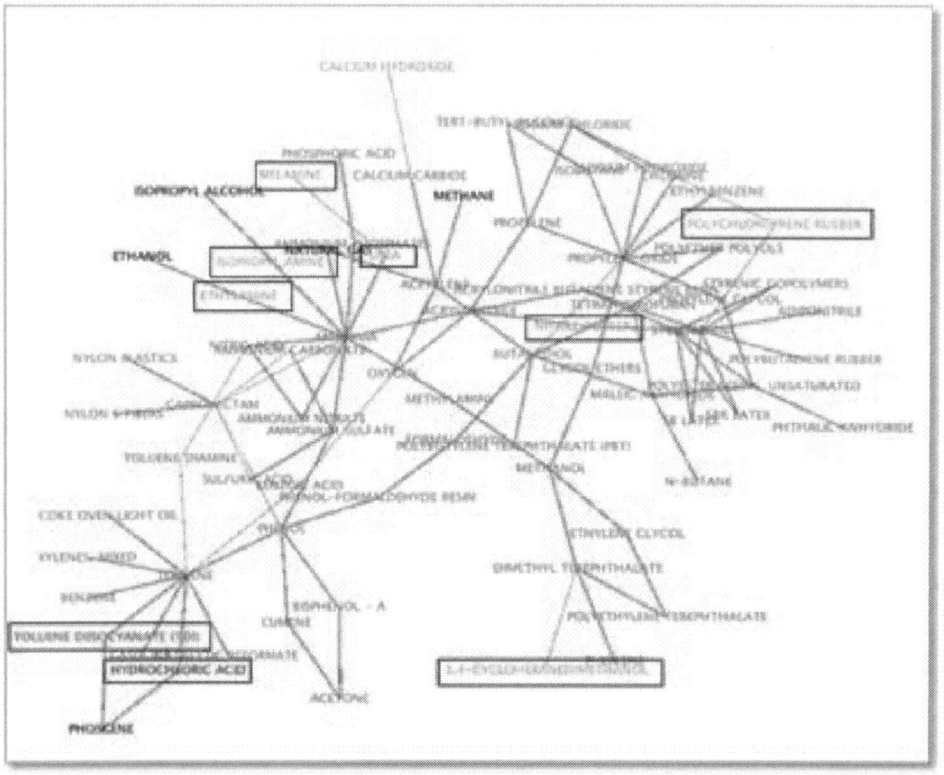

Figure 7. Chemical markets

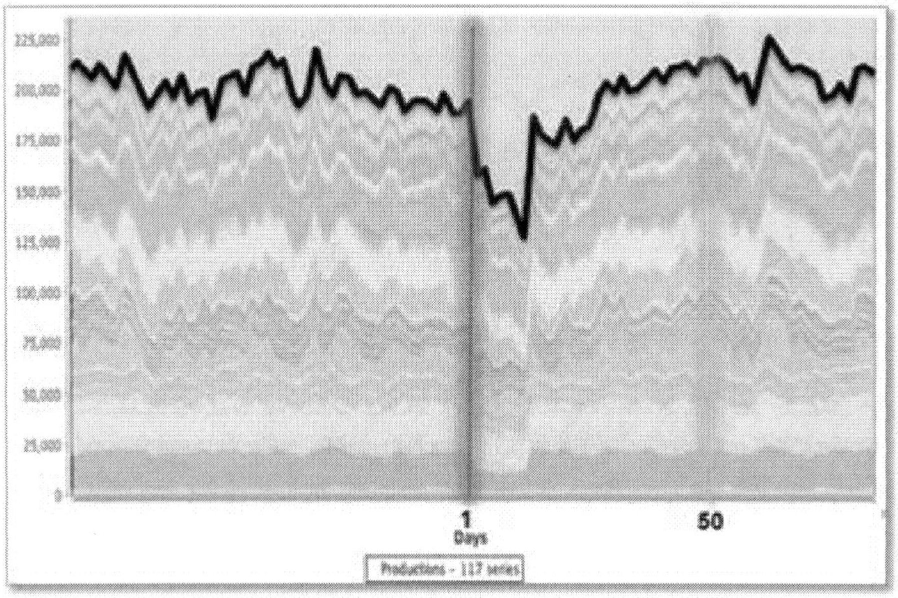

Figure 8. Chemical production

Summary

Sandia has developed an agent-based tool, N-ABLE, capable of simulating the domestic chemical supply chain that can be sued to estimate economic impacts resulting from large-scale disruptions to the chemical sector. N-ABLE draws upon various chemical industry datasets to simulate firms within a supply chain network where each firm interacts with others through chemical reactions, markets, and physical infrastructure. The supply and demand within each simulated network is consistent with respect to mass balances of every chemical within the network. Production decisions at each time step are a set of constrained linear program (LP) solutions that minimize the difference between desired and actual outputs.

The methodology is demonstrated through the simulation and analysis of an 80 chemical supply chain disrupted by an earthquake in the New Madrid Seismic Zone. It was found that disruptions impact chemicals across the product network, suggesting potentially wide-spread changes in chemical plants' operations, aggregate output declines by 30 percent and recovers in about 2 months, and sufficient capacity exists in the system to adapt and recover from lost capacity, despite the assumed total loss of some facilities.

Acknowledgement

Sandia National Laboratories is a multi-program laboratory managed and operated by Sandia Corporation, a wholly-owned subsidiary of Lockheed Martin Company, for the United States Department of Energy National Nuclear Security Administration under contract DE-AC04-94AL85000.

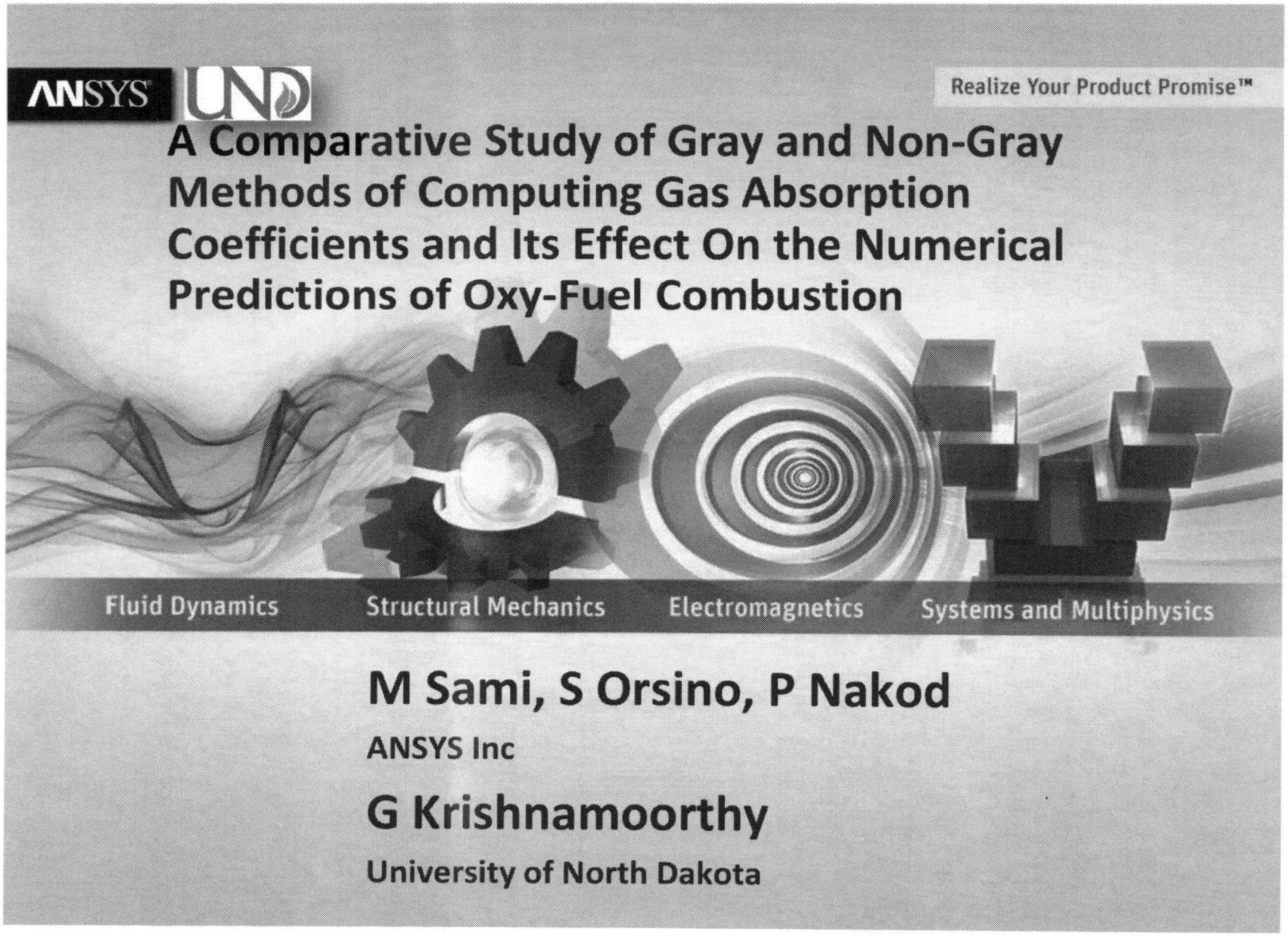
ANSYS
UND
Realize Your Product Promise™
A Comparative Study of Gray and Non-Gray Methods of Computing Gas Absorption Coefficients and Its Effect On the Numerical Predictions of Oxy-Fuel Combustion
Fluid Dynamics
Structural Mechanics
Electromagnetics
Systems and Multiphysics
M Sami, S Orsino, P Nakod
ANSYS Inc
G Krishnamoorthy
University of North Dakota

ANSYS

Introduction

- Oxy-fuel technology helps in Clean Coal usage with CO2 recycle
  - Fuel reacts with pure oxygen in CO2-rich environment.
  - At high CO2 concentrations, the default radiative property method in FLUENT can not be used due to its limited range of applicability (H2O/CO2 =0.5 to 2.5).
  - New models are proposed that cover a larger range of H_2O/CO_2 ratio.
    - EWBM can cover entire range of H_2O/CO_2 ratio
      - Computational cost is high (~4 times)
    - Extension of the range of applicability for Perry's model is an alternative to EWBM
- Several papers are available that cover these new radiative property models
  - For example, see International Journal of Computational Fluid Dynamics-Vol. 24 (2010)

ANSYS H_2O/CO_2 ratios in BERL Furnace – an example

H_2O/CO_2 ratios

- **depart from the single step stoichiometric value of 2** (methane combustion)
- **can have a wide range of variation in air-fired case (1 to 5).**
- **are very low (max=0.7) in case of oxy-fired furnace, due to CO_2-rich oxidizer stream.**

  - **Therefore, radiative property models that are accurate over the entire range of H_2O/CO_2 ratios need to be employed.**

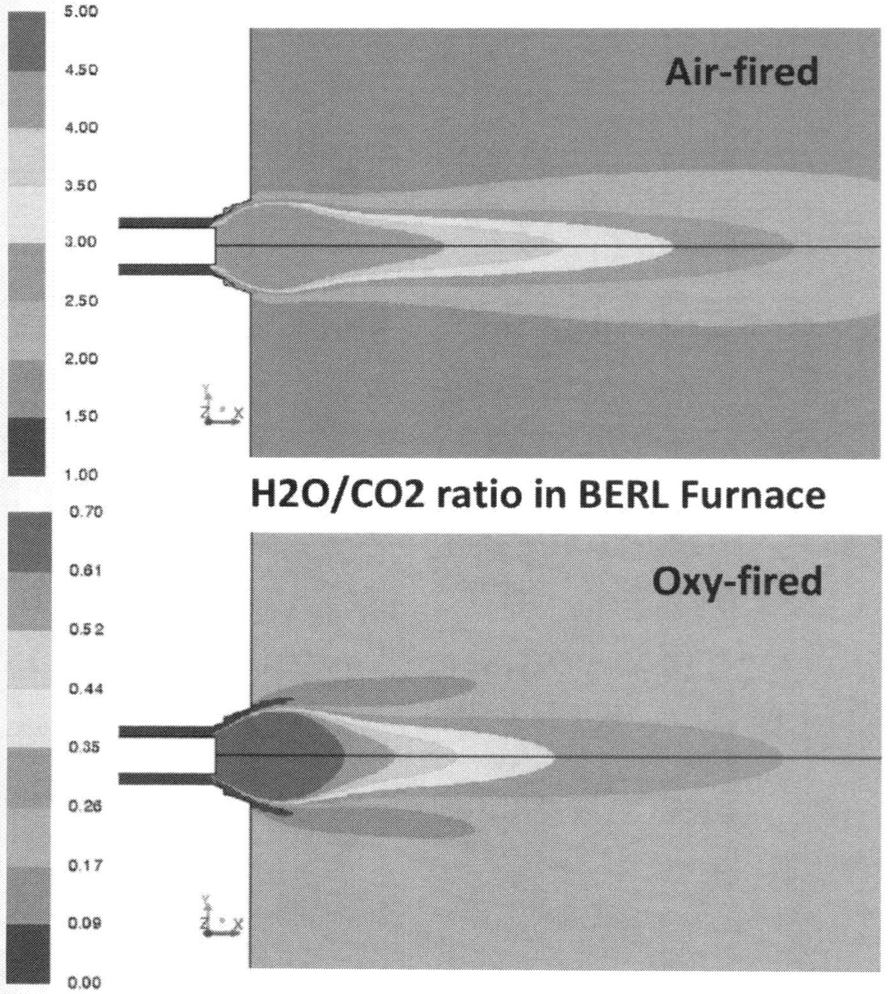

  August 25, 2011

ANSYS

Current Study

- Six different radiative property models were tested using the industrial strength case of the Costa furnace*.
- Two cases were setup for each model: air-fired and oxy-fired
- The radiative property models were collected from literature and previous work from the same authors.
- An example[1] of computing total gas emissivity, using the Perry gray model, in the form of a polynomial is listed below:

$$\log_{10} \varepsilon_g T_g = a_0 + a_1 \log pL + a_2 \log pL^2 + a_3 \log pL^3$$

*Costa, M. and Azevedo, JL., "Experimental Characterization of an Industrial Pulverized Coal-Fired Furnace Under Deep Staging Conditions", Combust. Sci. and Tech., 179: 1923–1935, 2007

[1]Hottel, H.C., Noble, J.J., and Sarofim, A.F., 2007. Heat and mass transfer, Chapter 5. In: D.W. Green and R.H. Perry, eds. Perry's chemical engineers' handbook. 8th ed. New York: McGraw-Hill.

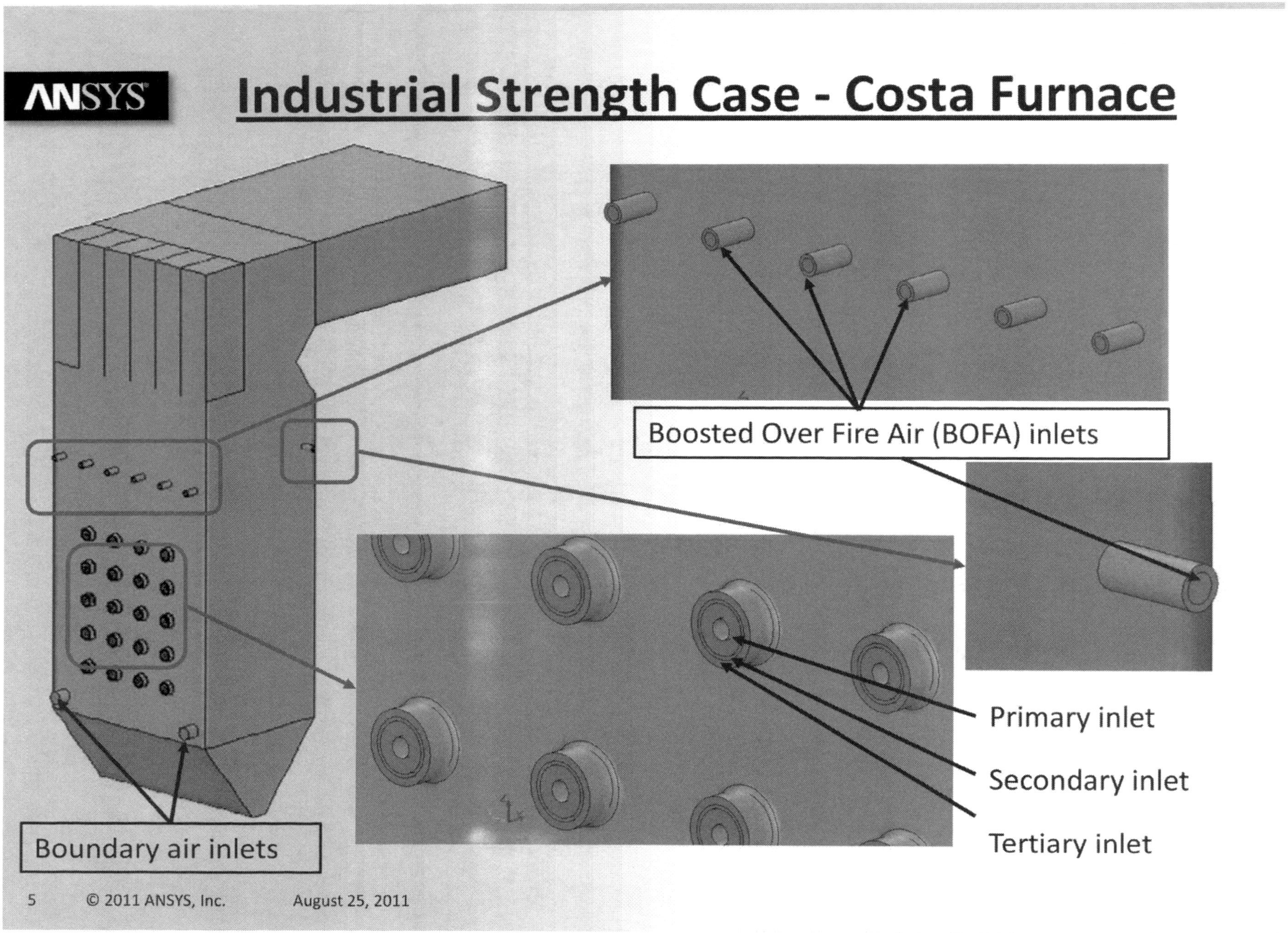
ANSYS
Industrial Strength Case - Costa Furnace
Boosted Over Fire Air (BOFA) inlets
Primary inlet
Secondary inlet
Tertiary inlet
Boundary air inlets
5 © 2011 ANSYS, Inc. August 25, 2011

ANSYS

Setup in ANSYS FLUENT

Cases

Air-fuel

- Inlet compositions
  - $O_2 \rightarrow$ 23.3%
  - $N_2 \rightarrow$ 76.7%

Oxy-fuel

- Inlet compositions
  - $O_2 \rightarrow$ 23.3%
  - $CO_2 \rightarrow$ 76.7%

Models

- Turbulence model
  - Standard k-ε
- Reaction model
  - Eddy dissipation with 2 step reaction mechanism
- Radiative property models
  - Smith, gray model (Fluent's default) [Smith Gray]
  - Perry, gray model (UDF) [Perry Gray]
  - Smith, non gray: 4band (UDF) [Smith: 4GG]
  - Perry, non gray: 4band (UDF) [Perry: 4GG]
  - Perry, non gray: 5band (UDF) [Perry: 5GG]
  - Chalmers ,non gray: 5band (UDF) [Chalmers: 5GG]

ANSYS

Operating Conditions

Primary air inlet
- Flow rate = 305.2 ton/h
- Temperature = 75°C

Secondary + Tertiary air inlets
- Total flow rate = 574.8 ton/h
- Secondary to tertiary mass flow rate distribution = 15% - 85%
- Temperature = 377°C

Boundary air inlet
- Flow rate = 50 ton/h
- Temperature = 377°C

BOFA air inlet
- Flow rate = 257 ton/h
- Temperature = 377°C

Wall
- Temperature = 452 K
- Emissivity = 0.6

Coal
- Feed rate = 110.2 ton/h
- Temperature = 75°C

Proximate analysis (dry wt %)
- Ash = 13.83
- Volatiles = 33.07
- Fixed carbon = 53.10

Ultimate analysis (dry wt %)
- Carbon = 70.91
- Hydrogen = 4.52
- Nitrogen = 1.47
- Sulfur = 1.39
- Oxygen = 7.89

High heating value
- 28.9 MJ/kg

Temperature Contours (air-fired)

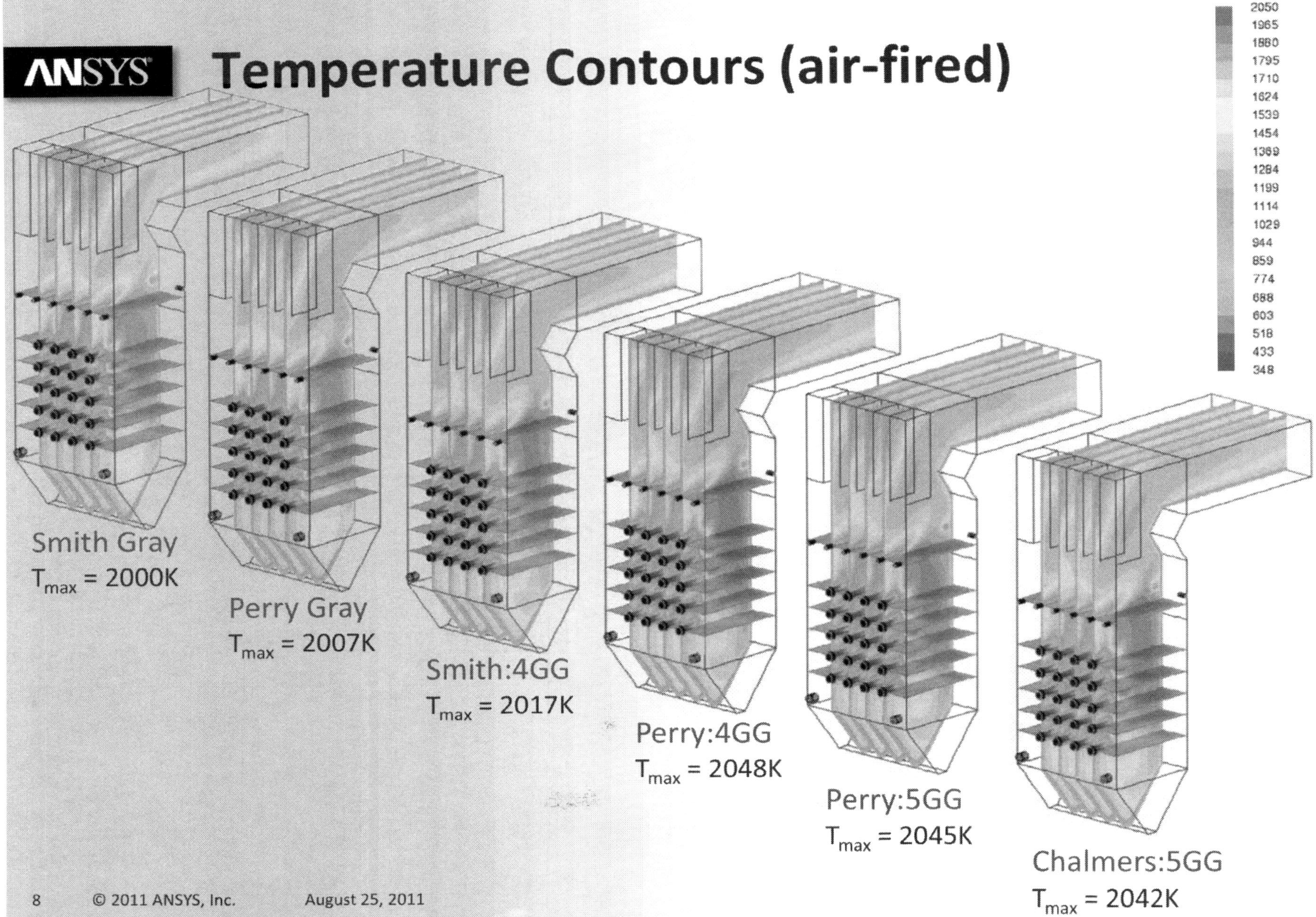

  August 25, 2011

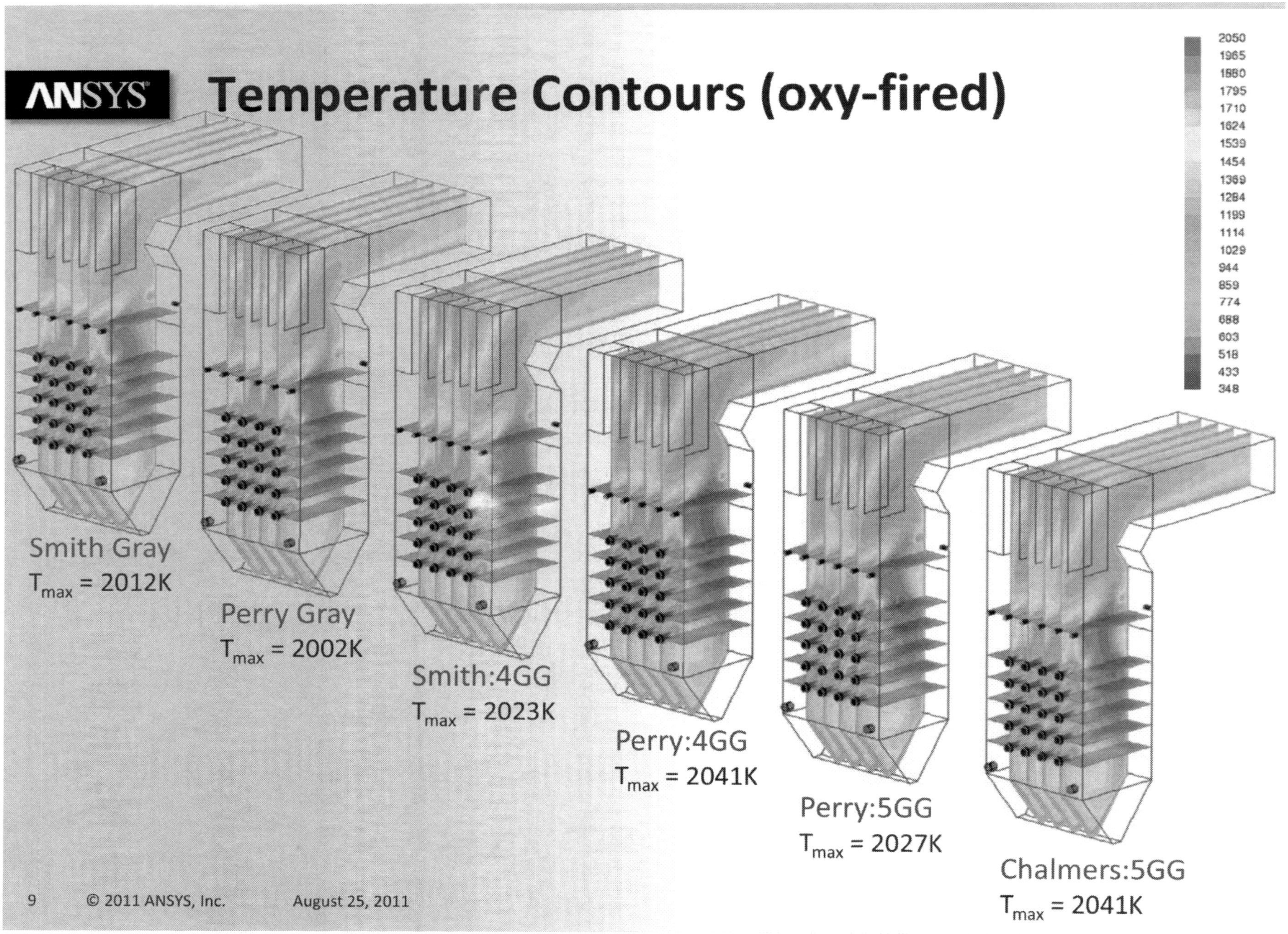
ANSYS
Temperature Contours (oxy-fired)
2050
1965
1880
1795
1710
1624
1539
1454
1369
1284
1199
1114
1029
944
859
774
688
603
518
433
348
Smith Gray
T_{max} = 2012K
Perry Gray
T_{max} = 2002K
Smith:4GG
T_{max} = 2023K
Perry:4GG
T_{max} = 2041K
Perry:5GG
T_{max} = 2027K
Chalmers:5GG
T_{max} = 2041K
9
© 2011 ANSYS, Inc.
August 25, 2011

ANSYS

H_2O/CO_2 Distribution (Perry gray)

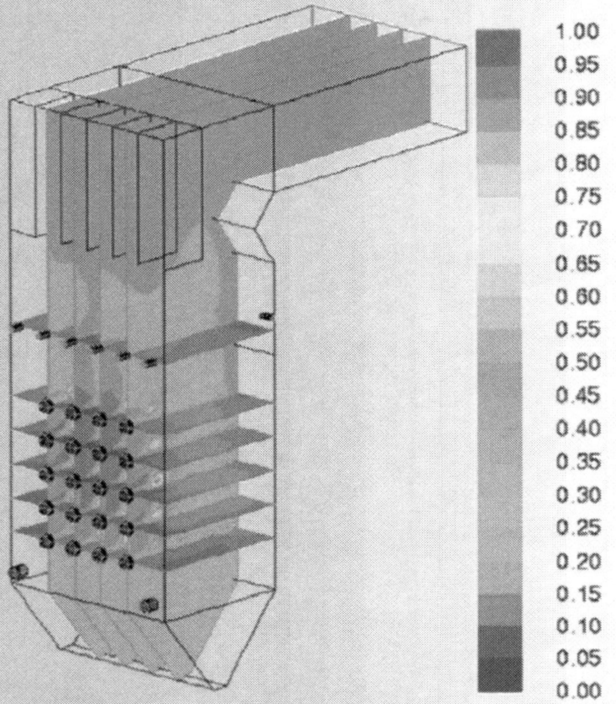

Air-fired
Max ratio =1.0

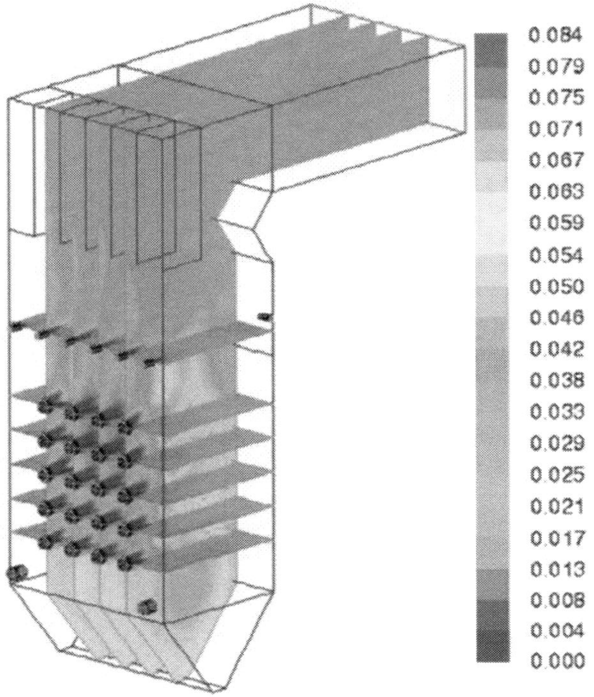

Oxy-fired
Max ratio=0.084

10 August 25, 2011

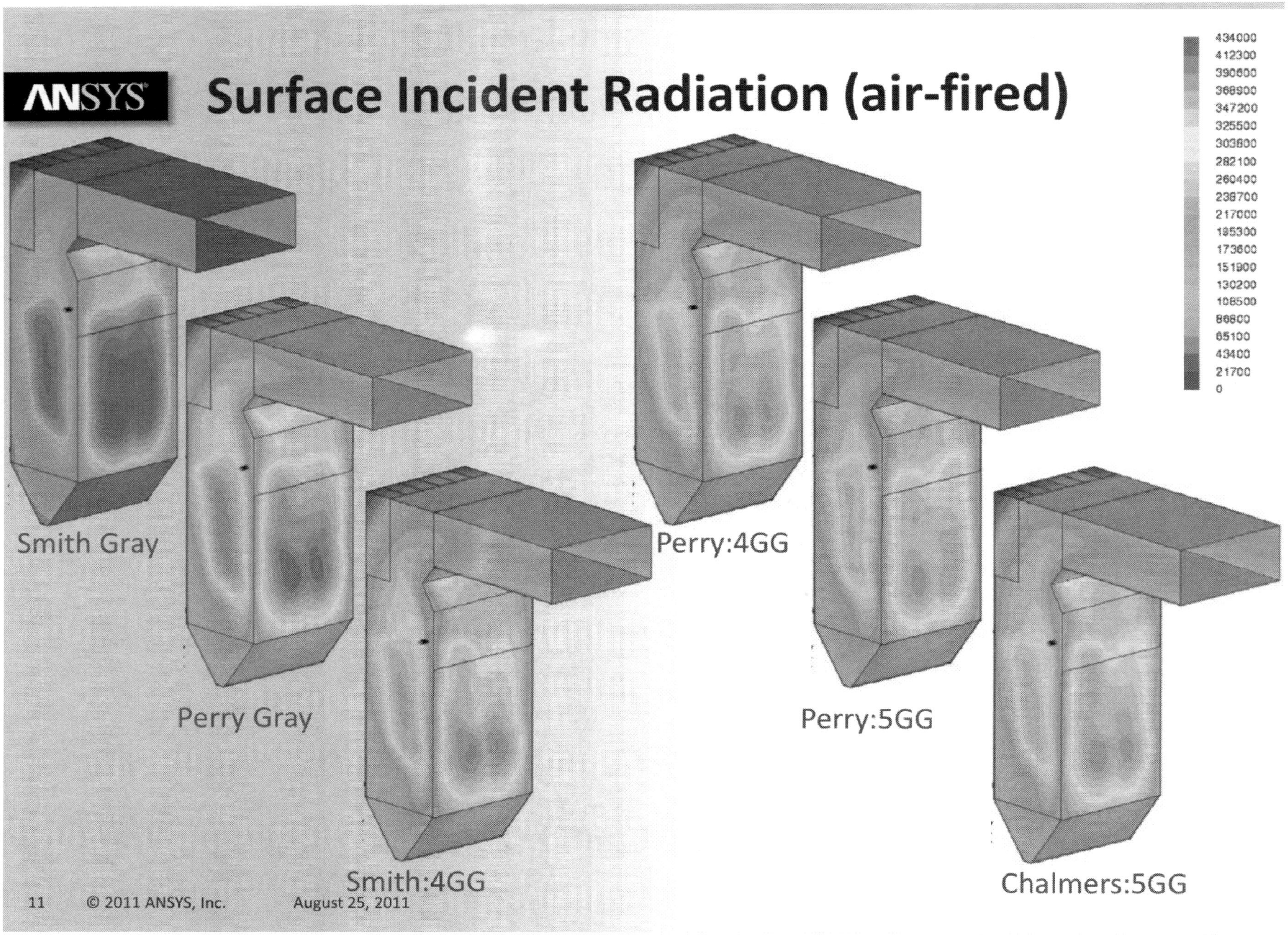
ANSYS
Surface Incident Radiation (air-fired)
434000
412300
390600
368900
347200
325500
303800
282100
260400
238700
217000
195300
173600
151900
130200
108500
86800
65100
43400
21700
0
Smith Gray
Perry Gray
Smith:4GG
Perry:4GG
Perry:5GG
Chalmers:5GG
11 © 2011 ANSYS, Inc. August 25, 2011

ANSYS Surface Incident Radiation (oxy-fired)

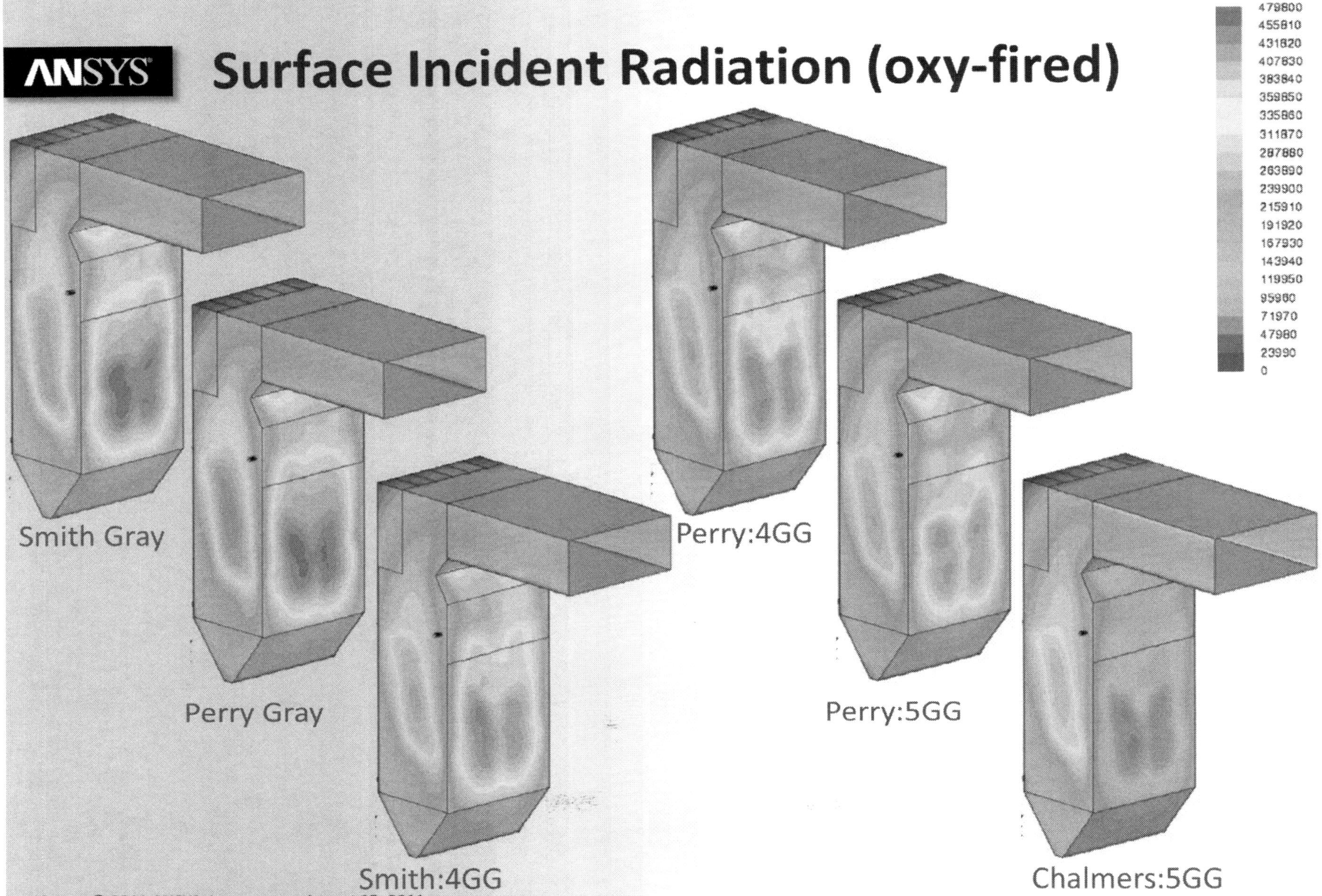

12 August 25, 2011

Temperature Profiles

ANSYS

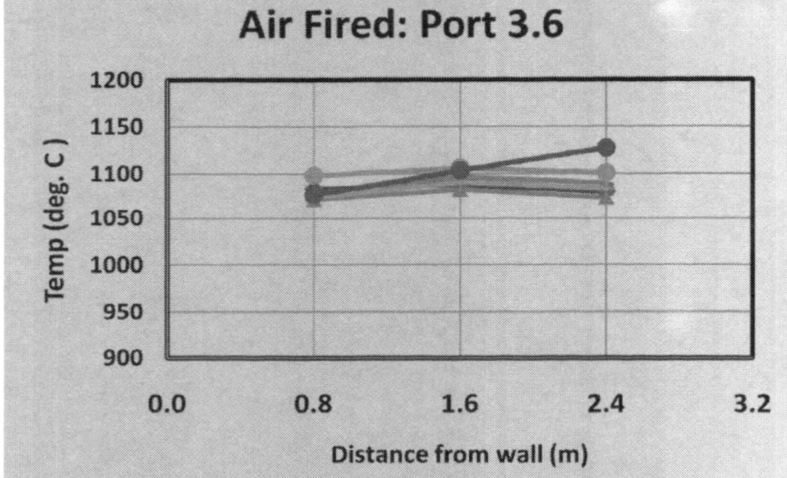

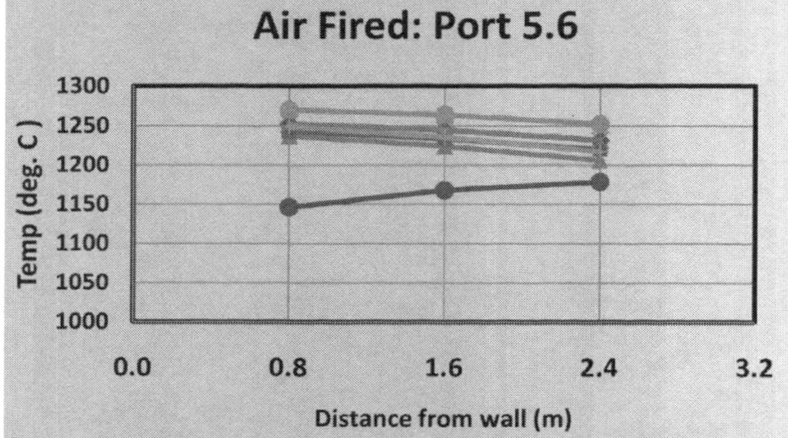

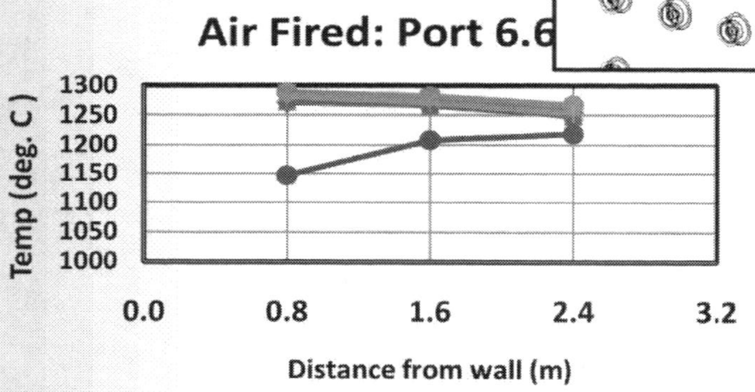

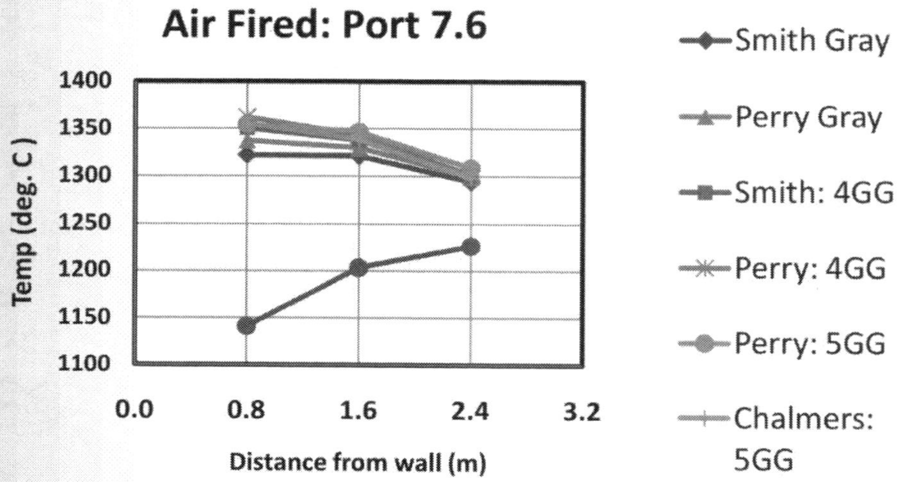

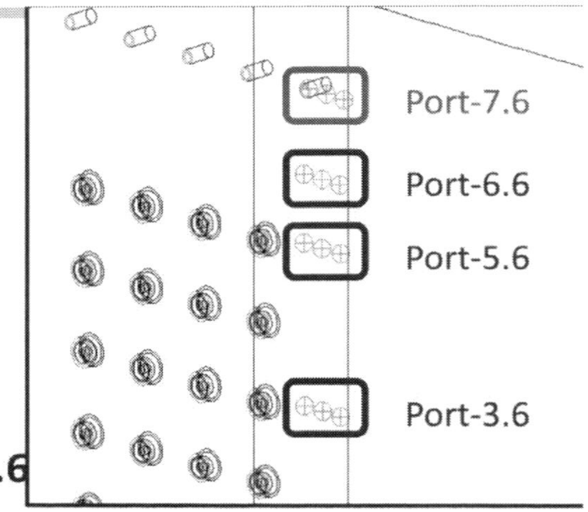

  August 25, 2011

ANSYS

Summary

- Temperature profiles predicted using gray and non gray radiation models are similar(max difference = 40K).
- Particles dominate the absorption process and hence the difference between gray/non-gray models is small.
- Non gray models predict 9-10% lower surface radiative fluxes on the furnace wall and the pendant plates than gray models.
- Choice of the radiative property models and the number of bands does not significantly affect the radiative flux predictions.
- Choice of the radiative property models does not affect the species distribution

(620aa) Energy Systems Integration Using Reduced Order Methods

Parag Jain[1], Lorenz T. Biegler[2] and Myung S. Jhon[2], (1)UOP LLC - A Honeywell Company, Des Plaines, IL, (2)Chemical Engineering, Carnegie Mellon University, Pittsburgh, PA

In integrating multi-component system models, such as polymer electrolyte fuel cells (PEFCs), situations often arise where information for one or more sub-components is not fully available; nor is direct integration among the disparate sub-components (especially in multi-scale systems) trivial. In these cases, one novel methodology for achieving system integration is to capture very complex sub-system information via simple parametric representations, yet keeping the essence of the physics. In this study, we explore whether such compact mathematical representations of the missing information exist.

We propose a generalized strategy for integration of two systems A & B, which can be represented via spatially distributed models, and for which direct model-integration is nontrivial. The essence of our methodology can be stated as the following:

Assuming a parametric representation of the adjoining boundary between systems A & B (which has to be determined), we repeatedly solve for system A *only* by spanning the entire parametric space, and generate snapshots for the output flux from system A towards system B. Through these snapshots, we construct a compact reduced order model (ROM) for the output flux. Finally, system B and the unknown boundary are determined using the above generated ROM for the flux as the input. Via this methodology we determine spatially distributed state information within systems A & B such that conditions of state and flux continuity at the adjoining boundary are satisfied.

For illustration, we present a benchmark case study for integration of catalyst layer (CL) – gas diffusion layer (GDL) sub-component models for our PEFC system [1]. Here, the GDL sub-component information is projected onto the adjoining boundary between GDL and CL, with manageable number of parameters and approximate integrated system solution is obtained. We obtained promising computational times & accuracy-levels for both simulation and optimization cases, when approximate solution is compared with previously obtained rigorous solution. The integration methodology described here can be applied to several unknown systems; using these ideas two different sub-systems can be easily linked with reduced number of parameters in conjunction with system optimization. Further, extension of integrating multi-scale systems will be given.

1. Jain P., Biegler L.T., and Jhon M.S., "Optimization of Polymer Electrolyte Fuel Cell Cathodes," *Electrochem Solid-State Lett*, **11** (10), B193 (2008).

See more of this Session: Poster Session

Distillation Column Conceptual Design in the Presence of Uncertain Parameters using Bayesian Network

F. Jalali, Professor, Chemical Eng. Dept., University of Tehran, Iran

Z. Amini, M.Sc. students of Chemical Eng. Dept., University of Tehran, Iran

M. Nili, Associated Professor, Electric and Computer Eng. Dept., University of Tehran, Iran

Abstract

In recent years, uncertain parameters have been affecting the final design and operation results in most science and engineering applications. Uncertain parameters have to be implemented in modelling and evaluating the results. Uncertainties along with inputs enter in the system, are extended in the model and, ultimately, appear with the outputs. Therefore, to minimize the error of system, they should be considered in the calculation. Two viewpoints are used to determine uncertain parameters which include deterministic and stochastic. In this research, a comparison was made between deterministic and stochastic methods. Distillation column is one of the key equipment and is usually present in most industrial plants. In this paper, the debutanizer column in olefin plant was conceptually designed. The external reflux flow, pressure deference between top and bottom of column, temperature of the top tray, feed temperature and steam flow rate were selected as uncertain parameters in the design of the debutanizer. Uncertain parameters were the functions of probability and Bayesian Network (BN) was used for calculating the most probable conditions. The nodes were variables and the edges represented probabilistic dependencies among the corresponding variables, constructed by the BN structure. Using industrial data formed by the BN structure. FUG method was applied for modelling column while the specified parameters were replaced by probability functions. The objective function was considered as the re-boiler steam flow. The investment cost decreased and the energy and size of column were minimized as the result of this calculation. The new design can improve the flexibility of the column operating condition using minimum energy.

Introduction

Nowadays, experts are interested in considering uncertain parameters in science and engineering applications. If uncertainties are not implemented in design process or considered as deterministic values, precision reduces and, therefore, operation of values deviates from the target ones. Uncertainties are in different types and sources. After defining and classifying uncertain parameters, those which have an insignificant effect on the final design and operation results are ignored. This simplification reduces the calculation load. Two viewpoints are used to implement uncertain parameters in design calculations which are deterministic and stochastic ones. The main purpose of this paper is to compare these two views.

Stochastic method can be presented using the probability function. BN calculates probability functions and conditional probability between the parameters in the network.

Expert systems are achieved by joining process technologies, expert knowledge and present operation data. Uncertain parameters are considered in the expert system for improving design and operation results.

There are similar units in petrochemical plants and their data can be used for efficient design and optimum operating conditions. This data determines the probability network structure and parameters. BN is one of the useful tools for probability networks which is based on the Baye's Theory. BN structure includes nodes and edges that represent parameters and their relations. Every parameter has a probability distribution.

Distillation column is the key equipment that is more applied in industrial plants. Therefore, it is the goal of engineers to achieve a suitable design for column and optimum operation conditions. In this paper, uncertain parameters that influence the precision design are considered and the probability functions and column sizes are calculated. And at the end, deterministic and stochastic viewpoints are compared.

Theory

Uncertainty

Every processing has two inherent properties, uncertainty and variability. Descriptive phenomena as well as processing systems include uncertainty as well. Uncertainties must be imported in the model and effective parameters selected.

Uncertain Parameters in Distillation Column Design

Nominal parameters, determined by extensive studies in sources, are needed for the column design. Uncertainties in distillation column design are represented in feed, model and cost parameters.

In this work, uncertain parameters were selected according to the available data. Although more recorded data were necessary for improving the results, the study was conducted by studying the debutanizer of olefin plant in a petrochemical plant in Iran. In this work, the external reflux flow, pressure deference between top and bottom of column, the temperature of the top tray, the feed temperature and steam flow rate were the recorded parameters of this unit. Refining data and their value-ranges were considered and probability functions were determined using the BN.

Bayesian Theory

The Bayesian interpretation of probability is an extension of logic that enables reasoning with uncertain statements. The Bayesian probabilistic specifies some prior probability in order to evaluate the probability of hypothesis which is then updated in new relevant data. Bayesian inference was provided in the 18th century by Thomas Baye's (1702-1761).

Bayesian Probabilities

The posterior probability was calculated by the observed data, prior probability and Baye's theorem. Thus:

$$P(H \mid D) = \frac{P(D \mid H).P(H)}{P(D)} \qquad \text{(Eq.1)}$$

where:
P(H) is the prior probability of H.
P(D|H) is the conditional probability of seeing the data, D, given that the hypothesis H is true.
P(D) is the marginal probability of D.
P(H|D) is the posterior probability.

Bayesian Network

Bayesian Network (BN) is one of the probabilistic graphic models (GMs) that is used for representing knowledge about uncertain domains. Each node represents a variable and an edge between two nodes is the probabilistic dependency among the corresponding variables. Conditional dependencies were estimated by statistical and computational methods. BN combines the principles from graph theory, probability theory, computer science and statistics.

Discussion

In this work, debutanizer column in Olefin plant of a petrochemical plant was studied. The schematic diagram of this column is shown in Figure 1:

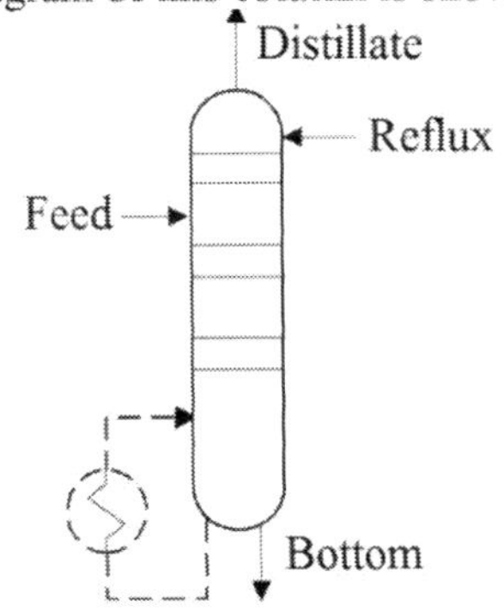

Fig 1- Diagram of the debutanizer column

Network Formation using Industrial Data

The network was formed from industrial data which were recorded for five months in the petrochemical plant. 402 data sets were available. Values of the variables were recorded every 8 hours.

Data distribution, design and operation ranges were implemented for division of bounds. Numbers of discrete values affected precision and, for wide ranges, more divisions were necessary. Of course, a large number of divisions increased the run time; therefore, suitable divisions guaranteed the precision and simplicity of calculations.

Primarily, every parameter was plotted. Next, design and operation ranges of parameters were applied and, finally, the number and sequence of divisions were determined.

Parameter values must have been normalized in order to reduce errors in the system; therefore, every value was set in the [0, 1] range, which was done using Eq.2:

$$New_{Data} = \frac{Data - \min Data}{\max Data - \min Data} \qquad \text{(Eq.2)}$$

Bayesian Network Structure Prediction

After data became discrete, the BN shape could be formed. In this step, the sequences of variables were more important; therefore, their arrangement was necessary. Impressive variables were the parents which had a priority over the children.

Finally in this system, Figure 2 was chosen as the BN shape. Reflux flow and feed temperature were selected as parent nodes:

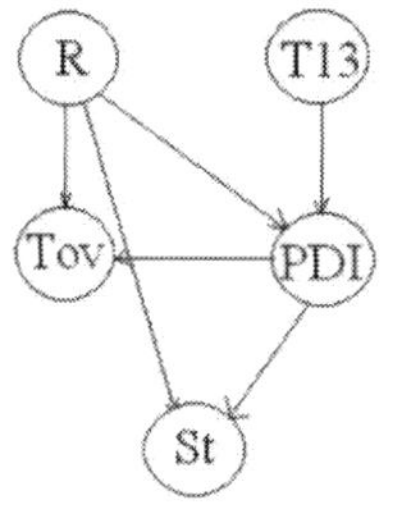

Fig. 2- Selected BN

In this network, feed temperature depended on the prior unit and did not have controllability. Reflux flow was the main operation parameter which had an effect on other parameters. Feed temperature and reflux flow were considered parent nodes. Pressure deference between top and bottom and top tray temperature were operation conditions which could be controlled in acceptable ranges; they were middle nodes. The final node was the steam flow rate which was influenced by other nodes.

The objective is to find a minimum value for the steam flow rate because the size of the column is a function of this parameter.

Preliminary Distillation Column Design

Distillation column is a key piece of equipment in industrial plants. Suitable design and adjustable operation of these columns are always desirable for engineers. In this section, the BN is used for redesigning debutanizer column of the olefin plant.

Two methods were applied for distillation column design: shortcut and rigorous. The preliminary design was done using shortcut methods and FUG was one of the common methods of this work, which was chosen for studying the BN effect on the final column sizing. FUG was coded using Matlab and was connected to the BN by means of an optimization program. One algorithm that is suitable for optimization under uncertainty is the EM algorithm; but the existing BN was simple and the optimization code that was written included several loops for this purpose.

Results

According to the above arguments, considering uncertain parameters is necessary for achieving the best results in the design and operation. The equipment must be flexible in all states and confidence coefficients can guarantee flexibility. One of the reasons of deviation between design and operation is the existence of uncertainty.

Sometimes, uncertain parameters affecting the design and operation results are not known; in such cases, if they are imported in calculations, the precision will increase. In other cases, uncertainties are imported as deterministic values into the model; in such cases, if uncertain parameters are considered as stochastic values, optimum design and operation results can be determined.

In this work, five uncertain parameters were considered as stochastic and the design result was compared with the deterministic state. Three cases were studied:

1- Uncertain parameters were considered as deterministic values and just the FUG model was used,

2- Uncertain parameters were considered as stochastic values; the FUG model and BN were connected using the optimization program, and

3- Only the BN was used.

In all cases, steam flow rate was calculated and; therefore, diameter of the column was determined. In cases 1 and 2, the height of column was calculated using the model.

Conclusion

Reboiler duty was calculated in three cases and the results are shown in Table 1:

Table 1. Reboiler duty in three cases

Cases	Reboiler
1- Model	$9/1156 \times 10^{8}$
2- Model + BN	$1/3072 \times 10^{9}$
3- BN	$4/2 \times 10^{9}$

In case 1, the amount of reboiler duty was in minimum because all parameters were approximated with usual values. In case 2, uncertainties were considered as stochastic values and the results were close to the actual values which were calculated as BN results. Case 3 was higher than other cases because industrial data were used for learning BN and actual reboiler duty was calculated considering confidence coefficients.

In case 2, the optimum results were reported as the below vector shows:

[R, TF, PDI, Tov] =[517/62; 176; 5/145; 122]

This vector was used for sizing the column. The distance between trays was 400 mm and; therefore, the height of column could be calculated. This column was a pressurized column and, using Table 2, the velocity in the column was obtained and used in the diameter calculation:

Table 2. Gas velocity in the columns

Operation	velocity (ft/s)
Atmospheri	۳
Vacuum	۸-۶
Pressurized	۱

Finally, column size was defined and these results are shown in Table 3:

Table 3. Column size results

Case	Diameter (ft)	height (ft)
1- Model	0/3466	۲۳/٦۲۱٤
2- Model + BN	0/4135	۲٤/۹۳۳۷
3- BN	0.7412	-

Case 2 had better design results compared with that of case 1 because they were close to the actual unit. Therefore, to have the best preliminary design, available uncertainties should be considered as stochastic.

References

1. Oberkampf W.L., Helton J.C., Joslyn C.A., Wojkiewicz S.F. and Ferson S., 2001, Challenge problems: uncertainty in system response given uncertain parameters
2. Ierapetritou M.G., Acevedo J. and Pistikopoulos E.N., 1996, An optimization approach for process engineering problems under uncertainty, Computer chem. Engng Vol 20, No. 6/7, pp. 703-709
3. Dua V. and Pistikopoulos E.N., 1998, Optimization techniques for process synthesis and material design under uncertainty, Trans IChem E, Vol 76, part A
4. Dua V. and Pistikopoulos E.N., 1997, A reduce space branch and bound algorithm for global optimization
5. Pistikopoulos E.N., 1995, Uncertainty in process design and optimization, Computer chem. Engng Vol. 19, suppl., pp. s553-s563
6. Hoch P.M. and Eliceche A.M., 1996, Flexibility analysis lead to sizing strategy in distillation columns
7. Edwin T., 2003,Cambridge University Press, Probability theory: the logic of science
8. Stich T., 2004, Bayesian networks and structure learning, Diploma Thesis, Computer Science and Engineering, University of Mannheim

9. Pearl J. and Kaufmann M., 1988, Probabilistic reasoning in intelligent systems, San Francisco
10. Pearl J. and Russel S., 2000, in Handbook of Brain Theory and Neural Networks
11. Friedman N., Geiger D. and Goldszmidt M., 1997, Bayesian network classifiers, Machine Learning 29, 131–163
12. Murphy K., 1998, A brief introduction to graphical models and Bayesian networks
13. Seader J.D. and Henley E. J., 1998, Separation process principles, John Wiley & Sons, Inc.
14. Ben-Gal I., Ruggeri F., Faltin F. and Kenett R., 2007, Encyclopedia of Statistics in Quality & Reliability, Wiley & Sons, Bayesian networks
15. Jordan M.I., 1999, Cambridge, Learning in graphical models, MIT Press

A MATHEMATICAL PROGRAMMING FORMULATION FOR THE SYNTHESIS OF PROPERTY-BASED BATCH WATER NETWORKS

José Antonio Vázquez-Castillo [a], José María Ponce-Ortega [b], Juan Gabriel Segovia-Hernández [a], and Mahmoud M. El-Halwagi [c]*

[a] Chemical Engineering Department, Universidad de Guanajuato,Campus Guanajuato, Gto., México, 36005.
[b] Chemical Engineering Department, Universidad Michoacana de San Nicolás de Hidalgo, Morelia, Mich., México, 58060.
[c] Chemical Engineering Department, Texas A&M University, College Station, TX, USA.

1. Introduction

Mass integration strategies aim to optimize the allocation, transformation, and separation of species and streams. An important class of mass integration problems deals with the synthesis of networks that allow the recycle and reuse of process sources to minimize the consumption of fresh sources and waste discharges, the mass integration techniques for recycle/reuse can be classified into two categories, methodologies that use the principles of pinch analysis and methods based on mathematical programming , it should be noted that in many of the reported methods based on mathematical programming, the model formulation for the synthesis of recycle/reuse networks have ignored the fact that environmental and process constraints do not depend only on compositions and flows of the streams, but also on properties such as: pH, density, viscosity, toxicity, color, and theoretical oxygen demand. Recently, there have been some works in which the model formulation is based on mass and property integration[1,2]. However, these have been developed for processes that operate under a continuous regime leaving aside processes that can be classified as batch.

This work presents a mathematical programming model for the synthesis of a property-based recycle and reuse batch network. The batch network model formulation for mass balances is based on a state-task network representation[3], with a stated horizon of time over which states and tasks are allocated, states are represented by buffer tanks whereas property interception units are tasks. The model formulation includes mass and property balances in process sinks, property treatment units, waste disposal, and these mass and property balances are based on mass and property integration. In addition, mass balances in states throughout the horizon of time are formulated, which take into account mass retention in current and previous periods of time, the model is also formulated in such a way that both process and environmental constraints imposed by process sinks and environmental regulations respectively are satisfied, properties constrained by sinks include: Composition, density, viscosity, pH. On the other hand properties constrained by the environment include: Composition, toxicity, and theoretical oxygen demand. The objective function consists in the simultaneous minimization of the total annual cost and the total retention of the batch network, this has been done in order to include a innovative concept of process intensification[4] to the synthesis of the batch network moreover this inclusion is enforced with mass and property balances at the exit of the property interception units, such a balances allow the discharge of treated process streams and they are no longer stored. The formulated model yields a moMINLP, (multi objective Mixed-Integer Non-Linear Problem). In order to ensure an optimal solution in the objective function, the constraint method was used and the model was programmed in the General Algebraic Modeling System (GAMS)[5], SBB MINLP GAMS solver was used to solve the model.

2. Problem statement

The problem can be expressed as follows: Given is a process with a set of process sources or process streams, $NSOURCES=\{i/i=1,2,...,N_{sources}\}$, from which the values of their properties ($P_{i,p}^{In\ Source}$) are known, such properties include: *pH, density, viscosity, composition, etc.*, it is also known the period of time in which every one of these process sources is available, in addition the value of the mass flow rates, ($W_{i,t}$), is known. Here, the process sources can be either sent to the network or sent to the waste, the latter only if the values of the properties satisfy a set of environmental restrictions. A set of units for the treatment of properties is also available; each unit is able to treat only one property *P*, the processing time, and the efficiency for each property interception unit is known. A set of buffer tanks for the storage of intermediate treated process sources is also available; a class of these tanks is associated with every process sources and another with each treatment unit. There is also a set of process sinks, $NSINKS=\{j/j=1,2,\ldots,N_{sinks}\}$, with known demands of mass flow rates and periods of time in which these flows are required $(G_{j,t})$, the sinks also restrict the values of properties $(P_{j,p}^{In\ Sink})$. Also, a set of fresh sources is allowed in order to feed any of the process sinks, $NFRESH=\{r/r=1,2,\ldots,N_{fresh}\}$, there are also known the property values $(P_{r,p}^{Fresh})$ for fresh sources. Finally, the mass flow rates and the period of time in which any of these fresh sources send mass flow rates to any sink are determined after solving the problem $(f_{r,j,t})$, all of the operations of the process are carried out in a known horizon of time (H), which is divided in periods of time of *1* hour of length.

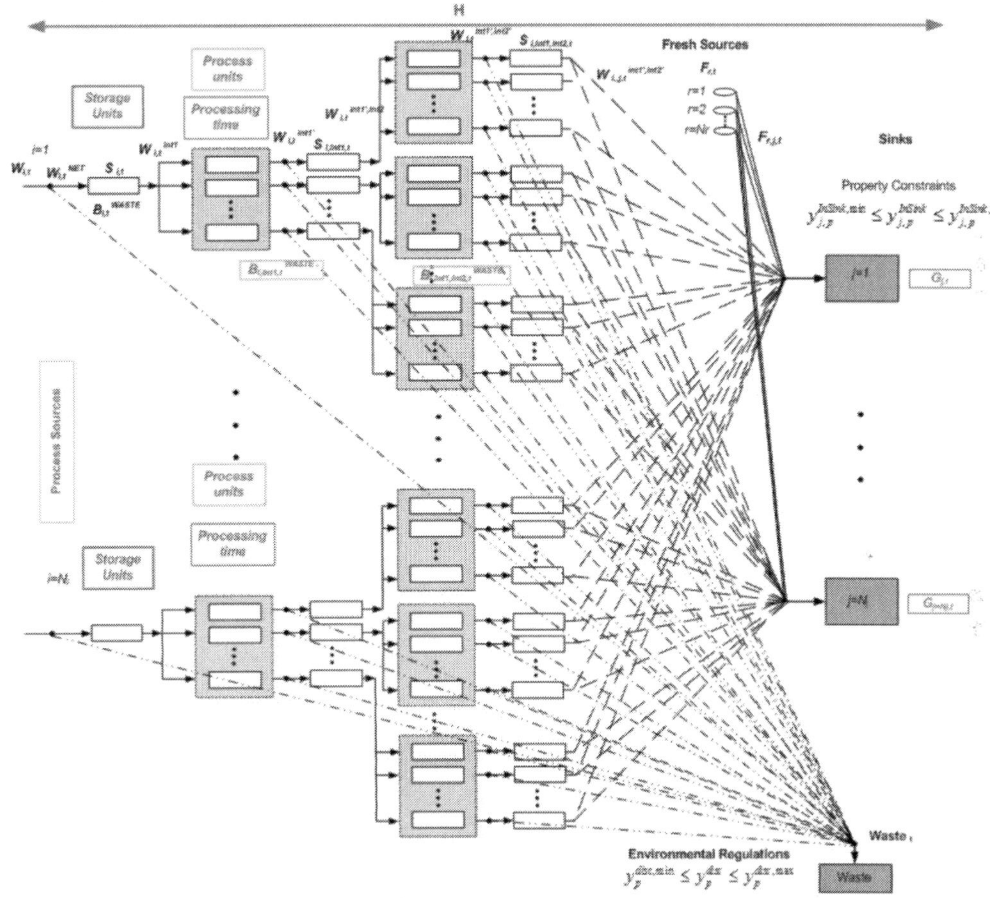

Figure 1. Intensified batch network (Note that mass flow rates are allowed to be send to the waste after processing units or property interceptors). Not all the sets of interceptors have been included because of limitation in size.

3. Model formulation

The mathematical model formulation is based on the intensified batch network shown in Figure 1. The model equations are listed hereafter.

3.1 . Splitting of the fresh sources

Every fresh source is segregated and directed to all process sinks:

$$F_{r,t} = \sum_{j\in NSINKS} f_{r,j,t}, \ \forall r,j,t \quad r\in NFRESH \tag{1}$$

In Equation (1), $F_{r,t}$ is the total mass flowrate for the fresh source r and $f_{r,j,t}$ is the mass flow rate sent to any sink j at any time period t.

3.2. Mass balance for the division of the process streams

Each one of the process streams may be directed either to the network or to the waste:

$$W_{i,t} = W_{i,t}^{Net} + B_{i,t}^{Waste}, \quad \forall\, i,t \;\; i \in NSOURCES \tag{2}$$

Where, $W_{i,t}$ is the total mass flow rate of process stream i at time t, $W_{i,t}^{Net}$ is the mass flow rate of process stream i sent to the network, and $B_{i,t}^{Waste}$ is the mass flow rate sent to the waste. Notice that this allows for the inclusion of intensification to the batch network as a stream with possible hazardous material is not stored in the network but it is discharged to the waste.

3.3. Mass balance in tank i before treatment of property 1

This is the first class of tanks; every one of these is associated with one i process source which does not allow for mixing of different process streams:

$$S_{i,t-1} + W_{i,t-1}^{Net} = S_{i,t} + \sum_{\text{int}1 \in INT1} w_{i,t}^{in\,\text{int}^1}, \quad \forall\, i, \text{int}^1, t \geq 2 \tag{3}$$

Where, $W_{i,t-1}^{Net}$ is the total mass flow rate of process stream i entering tank i at time $t-1$, $S_{i,t-1}$ is the accumulation of mass in tank i at time $t-1$, $S_{i,t}$ is the accumulation of mass in tank i during period of time t, finally, $w_{i,t}^{in\,\text{int}^1}$ is the segregated mass flow rate sent to all interceptors of property 1 *(int^1)*, if the property is not intercepted it is used a fictitious unit for such a property with effectiveness, cost and processing time equal to zero. This interceptor is placed at the end of the set of the corresponding interceptors.

3.4. Mass balance at the exit of interceptor for property 1

This is the mass balance at the exit of interceptor *int^1* utilized for the treatment of property 1:

$$w_{i,t-time\text{int}^1}^{in\,\text{int}^1} = B_{i,\text{int}^1,t}^{Waste} + w_{i,t}^{in\,\text{int}^{1'}}, \quad \forall\, i, \text{int}^1, t > time\text{int}^1 \tag{4}$$

In this equation, $w_{i,t-time\text{int}^1}^{in\,\text{int}^1}$ is the total mass flow rate entering to interceptor *int^1*, $B_{i,\text{int}^1,t}^{Waste}$ is the mass flow rate sent to the waste in the period of time t and, $w_{i,t}^{in\,\text{int}^{1'}}$ is the mass flow rate sent to tank *int^1*. Notice that the time of processing of *int^1* for treatment of property 1 has been included in this balance as $time\text{int}^1$.

3.5. Mass balance in tank int^1 before treatment of property 2

This is the second class of tanks; every one of these is associated with one *int^1* interceptor, and the second class of tanks also includes tanks associated with each property interceptor for the treatment of different properties:

$$S_{i,\text{int}^1,t-1} + w_{i,t-1}^{in\,\text{int}^{1'}} = S_{i,\text{int}^1,t} + \sum_{\text{int}^2 \in INT^2} w_{i,t}^{in\,\text{int}^{1'},\text{int}^2}, \quad \forall\, i, \text{int}^1, \text{int}^2, t > time\text{int}^1 \tag{5}$$

In this equation, $S_{i,\text{int}^1,t-1}$ is the accumulated mass in tank *int^1* at time $t-1$, $w_{i,t-1}^{in\,\text{int}^{1'}}$ is the mass flow rate sent to tank *int^1* at time $t-1$, $S_{i,\text{int}^1,t}$ is the accumulated mass in tank *int^1* in period of time t. Whereas, $w_{i,t}^{in\,\text{int}^{1'},\text{int}^2}$ is the segregated mass flow rate sent to all the interceptors of property 2.

3.6. Mass balance at the exit of interceptor for property 2

$$w_{i,t-time\text{int}^2}^{in\,\text{int}^{1'},\text{int}^2} = B_{i,\text{int}^2,t}^{Waste} + w_{i,t}^{in\,\text{int}^{1'},\text{int}^{2'}}, \quad \forall\, i, \text{int}^1, \text{int}^2, t > time\text{int}^2 \tag{6}$$

In this equation, $w_{i,t-time\text{int}^2}^{in\,\text{int}^{1'},\text{int}^2}$ is the total mass flow rate entering to interceptor *int^2*, $B_{i,\text{int}^2,t}^{Waste}$ is the mass flow rate sent to the waste in the period of time t and, $w_{i,t}^{in\,\text{int}^{1'},\text{int}^{2'}}$ is the mass flow rate sent to tank *int^2*.

Notice that it is only in the tanks that the mass flow rate is segregated to all the interceptors of the next property, e.g. from tank int^1 to all interceptors int^2.

3.7. Mass balance at the exit of interceptor for property N

It should also be noted that the above equations, excluding Equations 1 and 2, can be extended for all the properties to be intercepted as follows:

$$w_{i,t-time\,int^N}^{in\,int^1,int^2,...,int^{(N-1)},int^N} = B_{i,int^N,t}^{Waste} + w_{i,t}^{in\,int^1,int^2,...,int^{(N-1)},int^N} \quad (7)$$

$$\forall\, i, int^1, int^2, ..., int^{N-1}, int^N, t > time\,int^N$$

Where, $w_{i,t-time\,int^N}^{in\,int^1,int^2,...,int^{(N-1)},int^N}$ is the total mass flow rate entering to interceptor int^N, $B_{i,int^N,t}^{Waste}$ is the mass flow rate sent to the waste in the period of time t and, $w_{i,t}^{in\,int^1,int^2,...,int^{(N-1)},int^N}$ is the mass flow rate sent to tank int^N.

3.8. Mass balance in the tank int^N at the exit of the batch network

Once that all the properties have been treated, the treated process sources are sent to the process sinks.

$$S_{i,int^1,int^2,...,int^{N-1},int^N,t-1} + w_{i,t-1}^{in\,int^1,int^2,...,int^{(N-1)},int^N} = S_{i,int^1,int^2,...,int^{N-1},int^N,t} + \sum_{j\in SINKS} w_{i,j,t}^{int^1,int^2,...,int^{N-1},int^N} \quad (8)$$

$$\forall\, i, int^1, int^2, ..., int^{N-1}, int^N, j, t > time\,int^N$$

Where, $S_{i,int^1,int^2,...,int^{N-1},int^N,t-1}$ is the accumulated mass in tank int^N at time $t-1$, $w_{i,t-1}^{in\,int^1,int^2,...,int^{(N-1)},int^N}$ is the mass flow rate sent to tank int^N at time $t-1$, $S_{i,int^1,int^2,...,int^N,t}$ is the accumulated mass in tank int^N in period of time t. Whereas, $w_{i,j,t}^{int^1,int^2,...,int^{N-1},int^N}$ is the segregated mass flow rate sent to all the process sinks.

3.9. Overall mass balance at the mixing point prior to any sink

The total mass flow rate entering any sink consists of the sum of the mass flow rates of the treated process sources plus the fresh sources mass flow rates:

$$G_{j,t} = \sum_{i\in NSOURCES}\ \sum_{int^1\in INT^1}, ..., \sum_{int^{N-1}\in INT^{N-1}}\ \sum_{int^N\in INT^N} w_{i,j,t-1}^{int^1,int^2,...,int^{N-1},int^N} + \sum_{j\in NSINKS} f_{r,j,t-1} \quad (9)$$

$$r \in NFRESH, \forall\, i, int^1, int^2, ..., int^{N-1}, int^N, r, j, t > 1$$

3.10. Property balance at the mixing point prior to any sink

A property balance for the properties constrained by the sinks is developed in order to determine the values of the properties at the inlet conditions of the sinks:

$$\psi_p\left(P_{j,p}^{in\,Sink}\right)G_{j,t} = \sum_{i\in NSOURCES}\ \sum_{int^1\in INT^1}, ..., \sum_{int^{N-1}\in INT^{N-1}}\ \sum_{int^N\in INT^N}\left[\psi_p\left(P_{i,p}^{out\,PIN\,int^1,int^2,...,int^{N-1},int^N}\right)\cdot w_{i,j,t-1}^{int^1,int^2,...,int^{N-1},int^N}\right] + \sum_{j\in SINKS}\left[\psi_p\left(P_{r,p}^{in\,Fresh}\right)f_{r,j,t-1}\right] \quad (10)$$

$$j \in NSINKS, p\in P, r \in NFRESH, \forall\, p, i, int^1, int^2, ..., int^{N-1}, int^N, r, j, t > 1$$

In Equation (10) there have been used the mixing rules for properties[6,7]. Some mixing rules are linear whereas others are not, in order to avoid numerical complications with these nonlinearities, in this work it is proposed to utilize the quantity given by the property operator $\psi_p\left(P_{j,p}^{in\,Sink}\right)$ as an optimization variable without the need to know these values of the properties at the inlet to the sinks conditions, $P_{j,p}^{in\,Sink}$, to do so, the sink constraints must also be given in terms of property operators $\psi_p\left(P_{j,p}^{in\,Sink}\right)$ as shown in Equation (13).

3.11. Overall mass balance for the waste

The total mass flow rate of the waste stream is given by the sum of the treated process streams in all the interceptors.

$$Waste_t = \sum_{i \in NSOURCES} B^{Waste}_{i,t-1} + \sum_{int^1 \in INT^1} B^{Waste}_{i,int^1,t-1} + \sum_{int^2 \in INT^2} B^{Waste}_{i,int^2,t-1} + \sum_{int^3 \in INT^3} B^{Waste}_{i,int^3\,t-1}$$
$$+,\ldots,+ \sum_{int^{N-1} \in INT^{N-1}} B^{Waste}_{i,int^{N-1}\,t-1} + \sum_{int^{N} \in INT^{N}} B^{Waste}_{i,int^{N}\,t-1} \quad (11)$$
$$\forall\, i, int^1, int^2, int^3, \ldots, int^{N-1}, int^N, t > 1$$

3.12. Property balance for the waste

Because of the intensified nature of the batch network it has been necessary to include property balances at the exit of each set of property interceptors for the property balance in the waste, e.g. int^1, int^2, int^3,…,int^{N-1}, int^N . A global property balance which involves the property operator at the exit of the *batch network* is also included:

$$\psi_p\left(P_p^{\,in\,Waste}\right) Waste_t = \sum_{i \in NSOURCES}\left[\psi_p\left(P_{i,p}^{\,in\,Source}\right) B^{Waste}_{i,t-1}\right] + \sum_{i \in NSOURCES}\sum_{int^1 \in INT^1}\left[\psi_p\left(P_{i,p}^{\,out\,PIN\,int^1}\right) B^{Waste}_{i,int^1\,t-1}\right] +$$
$$\sum_{i \in NSOURCES}\sum_{int^1 \in INT^1}\sum_{int^2 \in INT^2}\left[\psi_p\left(P_{i,p}^{\,out\,PIN\,int^1,int^2}\right) B^{Waste}_{i,int^2,t-1}\right] +$$
$$\sum_{i \in NSOURCES}\sum_{int^1 \in INT^1}\sum_{int^2 \in INT^2}\sum_{int^3 \in INT^3}\left[\psi_p\left(P_{i,p}^{\,out\,PIN\,int^1,int^2,int^3}\right) B^{Waste}_{i,int^3\,t-1}\right] +,\ldots,+ \quad (12)$$
$$\sum_{i \in NSOURCES}\sum_{int^1 \in INT^1}\sum_{int^2 \in INT^2}\sum_{int^3 \in INT^3},\ldots,\sum_{int^{N-1} \in INT^{N-1}}\left[\psi_p\left(P_{i,p}^{\,out\,PIN\,int^1,int^2,int^3,\ldots,int^{N-1}}\right) B^{Waste}_{i,int^{N-1}\,t-1}\right] +$$
$$\sum_{i \in NSOURCES}\sum_{int^1 \in INT^1}\sum_{int^2 \in INT^2}\sum_{int^3 \in INT^3},\ldots,\sum_{int^{N-1} \in INT^{N-1}}\sum_{int^{N} \in INT^{N}}\left[\psi_p\left(P_{i,p}^{\,out\,PIN\,int^1,int^2,int^3,\ldots,int^{N-1},int^N}\right) B^{Waste}_{i,int^{N}\,t-1}\right]$$
$$p \in P, \forall\, i, p, int^1, int^2, int^3, \ldots, int^{N-1}, int^N, t > 1$$

3.13. Sink constraints

Process sinks impose limits on the properties. Each process sink includes a set of specific process constraints. Regularly, process sinks constraint the composition of specific chemical compounds, density, pH, viscosity, etc.:

$$\psi_p\left(P_{j,p}^{\,min\,Sink}\right) \le \psi_p\left(P_{j,p}^{\,in\,Sink}\right) \le \psi_p\left(P_{j,p}^{\,max\,Sink}\right), \quad j \in NSINKS, p \in P \quad (13)$$

Notice that the process sinks limits are imposed in terms of the property operators to avoid nonlinear terms in Equation (10).

3.14. Environmental constraints

The environmental regulations impose a set of constraints in terms of composition of hazardous materials and specific properties like toxicity, chemical oxygen demand, pH, color, and odor, among others:

$$\psi_p\left(P_{i,p}^{\,min\,Env}\right) \le \psi_p\left(P_p^{\,in\,Waste}\right) \le \psi_p\left(P_{i,p}^{\,max\,Env}\right), \quad p \in P \quad (14)$$

3.15. Existence or inexistence of property interception units (Logical constraints).

To determine the existence or inexistence of a property interceptor unit, a binary variable is needed; this variable is activated through a simple logical relationship utilizing minimum and maximum known values of mass flow rates entering to any interceptor:

$$w_i^{in\,min\,int^1} \cdot y_{i,int^1,t} \le w_{i,t}^{in\,int^1} \le w_i^{in\,max\,int^1} \cdot y_{i,int^1,t}, \qquad \forall\, i, int^1, t \quad (15)$$

However, as this binary variable, $y_{i,int^1,t}$ could be activated or inactivated many times during the iteration process, it is necessary a new binary variable z_{i,int^1} to specify that the interception unit associated with this first binary variable should only be taken into account just one time for purchase as it is the case of a fixed cost. Therefore, a logical relationship between both binary variables is formulated:

$$1 - y_{i,int^1,t} + z_{i,int^1} \ge 1, \qquad \forall\, i, int^1, t \quad (16)$$

Equation (16) allows to the binary variable z_{i,int^1} to take the value of *1* if the values for $y_{i,\text{int}^1,t}$ are *0* or *1*, which should be understood as the existence of the interception unit int^1 in different periods of time where its capacity is not surpassed. Consequently, a unique value of *1* for z_{i,int^1} is registered and applied on the Total Annual Cost objective function. See Equation (27). Equations (15 and 16) can be extended for the interceptors of all properties in this way:

$$w^{in\ \min \text{int}^N}_{i,\text{int}^1,\text{int}^2,...,\text{int}^{N-1},\text{int}^N} \cdot y_{i,\text{int}^1,\text{int}^2,...,\text{int}^{N-1},\text{int}^N,t} \le w^{in\,\text{int}^1,\text{int}^{2},...,\text{int}^{(N-1)},\text{int}^N}_{i,t} \le w^{in\ \max \text{int}^N}_{i,\text{int}^1,\text{int}^2,...,\text{int}^{N-1},\text{int}^N} \cdot y_{i,\text{int}^1,\text{int}^2,...,\text{int}^{N-1},\text{int}^N,t}$$
$$\forall\, i,\text{int}^1,\text{int}^2,...,\text{int}^{N-1},\text{int}^N,t \tag{17}$$

$$1 - y_{i,\text{int}^1,\text{int}^2,...,\text{int}^{N-1},\text{int}^N,t} + z_{i,\text{int}^1,\text{int}^2,...,\text{int}^{N-1},\text{int}^N} \ge 1,$$
$$\forall\, i,\text{int}^1,\text{int}^2,...,\text{int}^{N-1},\text{int}^N,t \tag{18}$$

3.16. Existence or inexistence of storage tanks.

Similarly to the interception units, the existence or inexistence of storage tanks is determined through logical relationships of binary variables. For tanks associated with process sources

$$S_i^{\min} \cdot x_{i,t} \le S_{i,t} \le S_i^{\max} \cdot x_{i,t}, \qquad \forall\, i,t \tag{19}$$

$$1 - x_{i,t} + w_i \ge 1, \qquad \forall\, i,t \tag{20}$$

For tanks associated with interception units:

$$S^{\min}_{i,\text{int}^1} \cdot x_{i,\text{int}^1,t} \le S_{i,\text{int}^1,t} \le S^{\max}_{i,\text{int}^1} \cdot x_{i,\text{int}^1,t}, \qquad \forall\, i,\text{int}^1,t \tag{21}$$

$$1 - x_{i,\text{int}^1,t} + w_{i,\text{int}^1} \ge 1, \qquad \forall\, i,\text{int}^1,t \tag{22}$$

A generalized formulation:

$$S^{\min \text{int}^N}_{i,\text{int}^1,\text{int}^2,...,\text{int}^{N-1},\text{int}^N} \cdot x_{i,\text{int}^1,\text{int}^2,...,\text{int}^{N-1},\text{int}^N,t} \le S_{i,\text{int}^1,\text{int}^2,...,\text{int}^{N-1},\text{int}^N,t} \le S^{\max \text{int}^N}_{i,\text{int}^1,\text{int}^2,...,\text{int}^{N-1},\text{int}^N} \cdot x_{i,\text{int}^1,\text{int}^2,...,\text{int}^{N-1},\text{int}^N,t}$$
$$\forall\, i,\text{int}^1,\text{int}^2,...,\text{int}^{N-1},\text{int}^N,t \tag{23}$$

$$1 - x_{i,\text{int}^1,\text{int}^2,...,\text{int}^{N-1},\text{int}^N,t} + w_{i,\text{int}^1,\text{int}^2,...,\text{int}^{N-1},\text{int}^N} \ge 1,$$
$$\forall\, i,\text{int}^1,\text{int}^2,...,\text{int}^{N-1},\text{int}^N,t \tag{24}$$

3.17. Objective function

The objective function consists in the simultaneous minimization of the Total Annual Cost of the Batch Network, and the total mass retention of the batch network, these two objectives are stated as follows:

$$\min Z = \{TAC; TRET\} \tag{25}$$

Generally the Total Annual Cost is the unique objective function in many problems in literature[1,2], but in this paper the Total Retention of the batch network is also included as an objective function which is aimed to take into account the broader definition of process intensification[4], especially in reference to the minimization of storage of hazardous substances.

A). **Total Retention**: This objective function consists of the sum of the accumulated mass of all the tanks in all the periods of time. This objective function can be mathematically expressed in this way:

$$TRET = \sum_i \sum_t S_{i,t} + \sum_i \sum_{\text{int}^1} \sum_t S_{i,\text{int}^1,t} + \sum_i \sum_{\text{int}^1} \sum_{\text{int}^2} \sum_t S_{i,\text{int}^1,\text{int}^2,t} +,...,+$$
$$\sum_i \sum_{\text{int}^1} \sum_{\text{int}^2},...,\sum_{\text{int}^N} \sum_t S_{i,\text{int}^1,\text{int}^2,...,\text{int}^N,t} \qquad ,\forall\, i,\text{int}^1,\text{int}^2,...,\text{int}^N,t \tag{26}$$

B). **Total Annual Cost Objective function**: It consists of the costs of fresh sources, the operational, variable and fixed costs of property treatment units, the costs of tanks to store mass of process streams, and the cost of tanks associated with interceptors for the treatment of properties. This objective function can be mathematically expressed as shown in Equation (27).

Finally, the optimization formulation consists in minimizing Equation (26) and (27) subject to restrictions given by Equations (1) to (24), which yields a moMINLP, (multi objective Mixed-Integer Non-Linear Problem).

4. *Solution strategy*

In this work the constraint method was utilized to solve the moMINLP, so that one of the objectives is transformed as a limit, and the problem consists in minimizing the other objective yielding a single objective problem. Therefore, multiples values for TAC and TRET were obtained and with these values a Pareto curve was obtained. For a further explanation on the constraint method, the reader is referred to other paper[8].

$$
\begin{aligned}
TAC = & \sum_{r\in NFRESH}\sum_{t\in T} c_r^{Unit}\cdot F_{r,t}\cdot \Delta t\cdot N_B + \sum_{i\in NSOURCES}\left(C_i^{Fix.ST}\cdot y_i + C_i^{Var.ST}\cdot\left(V_i^{\max}\right)^{\beta}\right)\cdot A_f + \\
& \sum_{i\in NSOURCES}\sum_{int^1\in INT^1}\left(C_{i,int^1}^{Fix.ST}\cdot y_{i,int^1} + C_{i,int^1}^{Var.ST}\cdot\left(V_{i,int^1}^{\max}\right)^{\beta}\right)\cdot A_f + \\
& \sum_{i\in NSOURCES}\sum_{int^1\in INT^1}\sum_{int^2\in INT^2}\left(C_{i,int^1,int^2}^{Fix.ST}\cdot y_{i,int^1,int^2} + C_{i,int^1,int^2}^{Var.ST}\cdot\left(V_{i,int^1,int^2}^{\max}\right)^{\beta}\right)\cdot A_f + \\
& \sum_{i\in NSOURCES}\sum_{int^1\in INT^1}\sum_{int^2\in INT^2}\sum_{int^3\in INT^3}\left(C_{i,int^1,int^2,int^3}^{Fix.ST}\cdot y_{i,int^1,int^2,int^3} + C_{i,int^1,int^2,int^3}^{Var.ST}\cdot\left(V_{i,int^1,int^2,int^3}^{\max}\right)^{\beta}\right)\cdot A_f + \\
& \sum_{i\in NSOURCES}\sum_{int^1\in INT^1}\sum_{int^2\in INT^2}\sum_{int^3\in INT^3},\dots,\sum_{int^N\in INT^N}\left(\begin{array}{l} C_{i,int^1,int^2,int^3,\dots,int^N}^{Fix.ST}\cdot y_{i,int^1,int^2,int^3,\dots,int^N} \\ + C_{i,int^1,int^2,int^3,\dots,int^N}^{Var.ST}\cdot\left(V_{i,int^1,int^2,int^3,\dots,int^N}^{\max}\right)^{\beta}\end{array}\right)\cdot A_f + \\
& \sum_{i\in NSOURCES}\sum_{int^1\in INT^1}\sum_{t\in T} C_{int^1}^{oper\, int^1}\cdot w_{i,t}^{in\, int^1}\cdot L_B\cdot N_B + \sum_{i\in NSOURCES}\sum_{int^1\in INT^1}\left(C_{Fix.}^{int^1}\cdot z_{i,int^1} + C_{Var.}^{int^1}\cdot\left(w_{i,int^1}^{in\max int^1}\right)^{\alpha}\right)\cdot A_f + \\
& \sum_{i\in NSOURCES}\sum_{int^1\in INT^1}\sum_{int^2\in INT^2}\sum_{t\in T} C_{int^2}^{oper\, int^2}\cdot w_{i,t}^{in\, int^1,int^2}\cdot L_B\cdot N_B + \\
& \sum_{i\in NSOURCES}\sum_{int^1\in INT^1}\sum_{int^2\in INT^2}\left(C_{Fix.}^{int^2}\cdot z_{i,int^1,int^2} + C_{Var.}^{int^2}\cdot\left(w_{i,int^1,int^2}^{in\max int^2}\right)^{\alpha}\right)\cdot A_f + \\
& \sum_{i\in NSOURCES}\sum_{int^1\in INT^1}\sum_{int^2\in INT^2}\sum_{int^3\in INT^3}\sum_{t\in T} C_{int^3}^{oper\, int^3}\cdot w_{i,t}^{in\, int^{1'},int^{2'},int^3}\cdot L_B\cdot N_B + \\
& \sum_{i\in NSOURCES}\sum_{int^1\in INT^1}\sum_{int^2\in INT^2}\sum_{int^3\in INT^3}\left(C_{Fix.}^{int^3}\cdot z_{i,int^1,int^2,int^3} + C_{Var.}^{int^3}\cdot\left(w_{i,int^1,int^2,int^3}^{in\max int^3}\right)^{\alpha}\right)\cdot A_f + \\
& \sum_{i\in NSOURCES}\sum_{int^1\in INT^1}\sum_{int^2\in INT^2}\sum_{int^3\in INT^3},\dots,\sum_{int^N\in INT^N}\sum_{t\in T} C_{int^N}^{oper\, int^N}\cdot w_{i,t}^{in\, int^{1'},int^{2'},int^{3'},\dots,int^{(N-1)'},int^N}\cdot L_B\cdot N_B + \\
& \sum_{i\in NSOURCES}\sum_{int^1\in INT^1}\sum_{int^2\in INT^2}\sum_{int^3\in INT^3},\dots,\sum_{int^N\in INT^N}\left(C_{Fix.}^{int^N}\cdot z_{i,int^1,int^2,int^3,\dots,int^N} + C_{Var.}^{int^N}\cdot\left(w_{i,int^1,int^2,int^3,\dots,int^N}^{in\max int^N}\right)^{\alpha}\right)\cdot A_f
\end{aligned} \qquad (27)
$$

For a further explanation on the terms listed in this equation, the reader is referred to the notation section.

5. Results and discussion

A case of study to show the applicability of the model formulation and the discussion of results are offered in this section. Three process sources and two fresh sources were considered as shown in Table 1, a total of six properties were taken into account but only four were considered to be treated by property interceptors. Table 2 offers detailed information on these process units. Table 3 displays data on process and environmental constraints.

Table 1. Mass flow rates, periods, and properties characteristics for process and fresh sources.

Sources and time	Flow rate	Composition	Tox(%)	pH	THOD(mgO_2/l)	ρ(lb/l)	μ (cP)

of availability *(i,t)*	$W_{i,t}$(lb/hr)	(ppm)					
(1,3)	8083	0.016	0.30	5.4	0.187	2.000	1.256
(2,4)	3900	0.024	0.50	4.8	92.10	2.208	1.220
(3,5)	3279	0.220	1.50	5.1	48.85	2.305	1.261
Fresh sources *r*	-	-	-	-	-	-	
1	-	0.00	0.00	7.0	0.00	2.204	1.002
2	-	0.01	0.10	6.8	0.01	2.209	0.992

Table 2. Characteristics of process interceptors.

Property	Interceptor	Efficiency, α	Processing time (h)	Operative cost $/lb	Fixed cost $(USD)
Composition Set (INT^1)	REC1	0.985	2	0.0065	8000
	REC2	0.852	2	0.0033	7000
Toxicity Set (INT^2)	TOX1	1.000	2	0.0075	9000
	TOX2	0.900	2	0.0063	8200
THOD Set (INT^3)	AER1	0.800	1	0.0055	6800
	AER2	0.700	1	0.0043	5200
pH Set (INT^4)	PH^1	0.500	1	0.0085	5200
	PH^2	0.300	1	0.0093	4800
	POH^1	-0.500	1	0.0055	3700
	POH^2	-0.300	1	0.0065	2200
None	NONE	0	0	0	0

Table 3. Process and Environmental property constraints for the case of study.

Sinks and time of required flow *(j,t)*	Flow rate $G_{j,t}$(lb/hr)	Composition (ppm)	Tox(%)	pH	THOD (mgO_2/l)	ρ(lb/l)	μ (cP)
Maximum							
(1,10)	6000	0.013	-	8.0	-	1.8	1.202
(2,15)	4400	0.011	-	7.8	-	1.7	2.230
(3,21)	2490	0.100	-	8.2	-	1.85	1.260
Waste	-	0.050	0.0	9.0	70.00	-	-
Minimum							
(1,10)	6000	-	-	5.3	-	2.8	0.900
(2,15)	4400	-	-	5.4	-	2.5	0.905
(3,21)	2490	-	-	5.2	-	2.9	0.903
Waste	-	-	-	5.5	-	-	-

A horizon of operation of 24 hours as the length of batch was fixed, 333 batches per year and 7992 hours of yearly operation were considered to solve the moMINLP problem. The costs of fresh sources were *(r=1)*, \$0.009/lb and *(r=2)* \$0.006/lb. The four properties treated were: composition, toxicity, THOD and pH, the main features of the interceptors to treat these properties are given in Table 2. SBB MINLP solver of GAMS[6] was used to solve the moMINLP all the runs were performed on an Intel Core i7, 2.6 GHz and 9 GB RAM. The solution that correspond to the minimum TAC delivered a solution with \$ 9,319,922.97/yr and a Total Retention of 50823.109 lb/batch, on the other extreme the solution for the minimum Total Retention provided a TAC of \$ 10,947,980/year and a Total Retention of 0 lb/batch. The Pareto curve between these two objectives is shown in Figure 2, intermediate values of Total Retention between the minimum and maximum values of this objective were fixed so that different values of Total Annual Cost were obtained. It should clearly be noted from Figure 2, that the point for minimum retention corresponds to the absence of tanks and the process tends to fall in a continuous regime. Another important feature of the behavior of the batch network observed in Figure 2 is that as the system approaches to the minimum retention there is a remarkable increase in the Total Annual Cost contrary to the increase in the Total Retention. An explanation of this behavior is attributed to the absence of tanks, which are by far less expensive that treatment units. Figure 3 offers a full description of total number of units including tanks and property interceptors units, it is worth to observe that the number of units that correspond to the minimum TAC is 15 from which 9 are property interceptors and 6 are tanks.

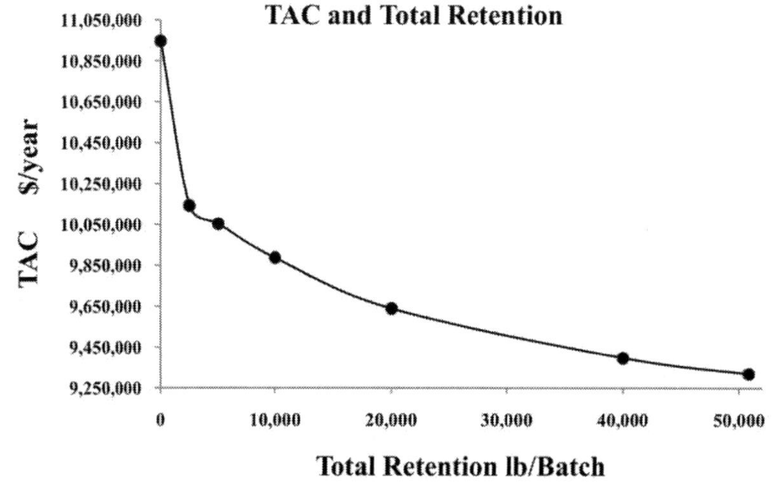

Figure 2. Pareto curve between the two objectives, Total Annual Cost and Total Retention.

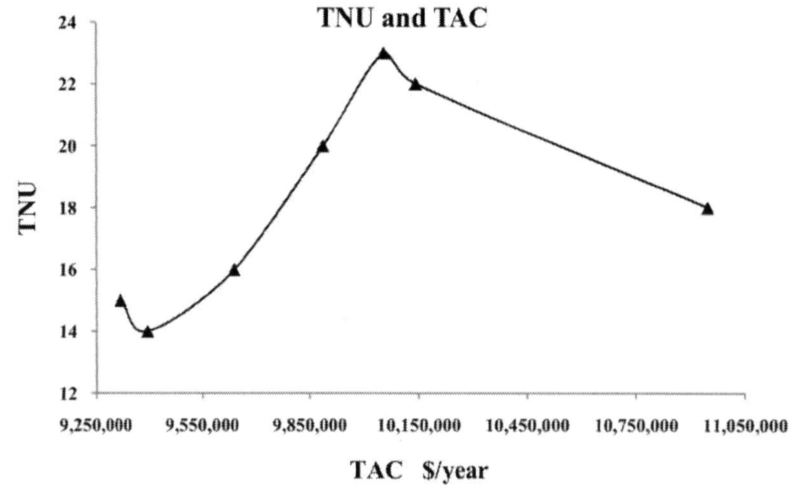

Figure 3. Total number of units (TNU, tanks and interceptors), and Total Annual Cost.

On the other hand, the point for the maximum TAC, and minimum retention corresponds to 18 units all of them are property interceptors, it is also interesting notice that there is a remarkable increase in the total number of units reaching the highest point as 23 units (3 tanks and 20 property interceptors) for a total retention of 5000 lb/batch.

6. Conclusions

A MINLP formulation based on properties for the synthesis of a batch network was presented in this paper; the formulation allows the inclusion of the concept of process intensification in an extended way, in particular in reference to the minimization of storage in a process. A total of six properties were considered in this study and four of them were treated. According to the results the batch network tends to favor the existence of storage tanks and a high mass retention in order to provide solutions with total annual costs lower than those solutions in which there is a decrease in both the number of tanks and the total retention. It seems to suggest that as the batch network approaches a continuous operation an increase in the total annual cost is observed. Furthermore, the increase in the total annual cost is a direct consequence of the increase in the number of treatment units which are more expensive that storage tanks.

Notation

c_r^{Unit}	Unitary cost of fresh sources	$C_{i,\text{int}^1}^{Fix.ST}$	Fixed Cost of Storage tank INT^1
Δt	Length of time period (1 hour)	$C_{i,\text{int}^1}^{Var.ST}$	Variable Cost of Storage tank INT^1
N_B	Number of batches per year	$V_{i,\text{int}^1}^{\max}$	Maximum volume of tank INT^1
$C_i^{Fix.ST}$	Fixed Cost of Storage tank i	$C_{int^1}^{oper\ \text{int}^1}$	Operational cost of interceptor INT^1
$C_i^{Var.ST}$	Variable cost of storage tank i	L_B	Length of batch (24 hours)
$V_i^{\max}$	Maximum volume of tank i	$C_{Fix.}^{\text{int}^1}$	Fixed Cost of interceptor INT^1
β	Constant equal to 1,	$C_{Var.}^{\text{int}^1}$	Variable Cost of interceptor INT^1
α	Constant equal to 1		
A_f	Anualization factor		

References

1. Ponce-Ortega José María, Carolina Hortua Ana, El-Halwagi MM, Jiménez-Gutiérrez Arturo. A property-based optimization of direct recycle networks and wastewater treatment processes. *AIChE Journal*. 2009; 55;9: 2329-2344.
2. Ponce-Ortega José María, El-Halwagi MM, Jiménez-Gutiérrez Arturo. Global optimization for the synthesis of property-based recycle and reuse networks including environmental constraints. *Comput Chem Eng*. 2010; 34: 318-330.
3. E. Kondili, C. C. Pantelides , R. W. H. Sargent. A general algorithm for short-term scheduling of batch operations-I. MILP formulation. *Comp. Chem. Eng.,*1993: 17:2:211-227.
4. Ponce-Ortega José María, Al-Thubaiti Musaed, El-Halwagi MM. Process Intensification: New Understanding and Systematic Approach. Chemical and engineering processing journal. In press.
5. Brooke A, Kendrick D, Meeruas A, Raman R. 2006. *GAMS-Language guide*. Washington, D.C.: GAMS Development Corporation.
6. Shelley MD, El Halwagi MM. Componentless design of recovery and allocation systems: a functionality-based clustering approach. *Comput Chem Eng*. 2000;24:2081-2091.
7. El Halwagi MM, M.M., Gabriel, F., & Harell, D. Rigurous graphical targeting for resource conservation via material recycle/reuse networks. *Industrial and Engineering Chemistry Research*,.2003 42(19), 4319-4328.
8. López-Maldonado Lizbeth Anabel, Ponce-Ortega José María Segovia-Hernández Juan Gabriel. Multiobjective Synthesis of Heat Exchanger Networks Minimizing the Total Annual Cost and the Environmental Impact. *Applied Thermal Engineering*.2011; 31(6-7),1099-1113.

Optimization of the Biofouling Control In Integrated Desalination/Power Plants

Fabricio Nápoles-Rivera,[1] *Abdullah Bin Mahfouz,*[2] *Arturo Jiménez-Gutiérrez*[1], *Mahmoud M. El-Halwagi,*[2] *José María Ponce-Ortega*[3]

(1) *Chemical Engineering Department, Instituto Tecnológico de Celaya, Celaya, Gto., México,*
(2) *Chemical Engineering Department, Texas A&M University, College Station, TX, 77843, USA*
(3) *Chemical Engineering Department, Universidad Michoacana de San Nicolás de Hidalgo, Morelia, Mich., 58060, México.*

Introduction

The use of seawater as a cooling agent in chemical plants is a common practice in several countries, especially those with severe water scarcity. Although it prevents the use of low salinity water, this practice represents a challenge due to its operational problems, such as biofouling and consequently reduction in the thermal efficiency of the heat exchangers. To alleviate these problems the use of biocides and periodical maintenance is commonly used. This paper presents a model for the optimal planning of biocide dosing and maintenance scheduling in a seawater cooled power/desalination plant. The model includes the use of disjunctive programming to avoid bilinear terms between continuous and binary variables and to add flexibility to the maintenance schedule by allowing different levels of biofilm thickness, depending on the plant schedule. The model considers the biocide concentration decay in the network, the biofilm thickness dynamics as a function of temperature and biocide concentration, and the heat reduction as a function of the biofilm thickness. It includes constraints over the maximum allowed biofilm thickness and heat reduction. When using biocides (usually chlorine due to its availability and cost) the residual concentration in the discharge is also an important issue because of its negative environmental impact; thus, an end of pipe treatment system is also considered along with a constraint over the residual concentration. The resulting model is a mixed integer nonlinear programming problem (MINLP), which is solved using the GAMS software. A practical case study with a planning span of a year is presented, in which the implementation of several dosing policies is tested.

Model

The considered network is shown in Figure 1. The network consist of 27 process segments (tanks, pipeline segments and heat exchangers). The initial concentration in the basin tank is

$$C_t^{basin} = \frac{F_t^B C^{Feed}}{F_{basin,t}}, \qquad \forall t \in NT$$

In order to calculate the residual concentration in each of the segments, a first order decay kinetic for the biocide is assumed

$$C = C_0 e^{-kt}$$

Thus the residual concentration in any process segment is:

$$C_{i,j,t} = C_{i-p,j,t} e^{-k_{i,t}\tau_{i,t}}, \qquad \forall i \in NI, \forall j \in NC, \forall t \in NT$$

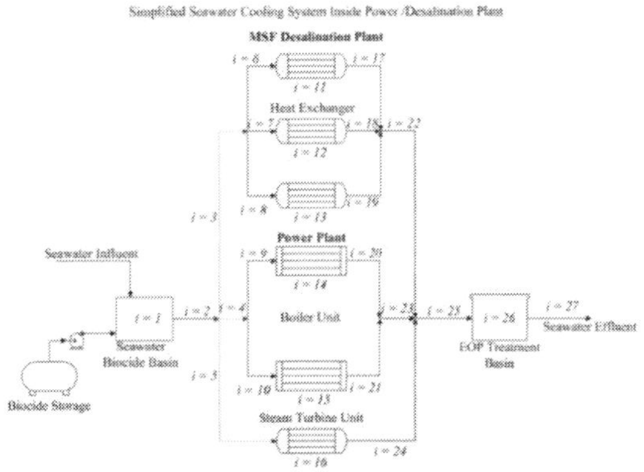

Figure 1. Cooling system for a power/desalination facility (Bin-Mahfouz et al. 2011)

$C_{i,j,t}$ and $C_{i-p,j,t}$ are the concentrations in process segments i and i-p at time t, (mg/l), $\tau_{i,t}$ is the residence time in process segment i at time t, (min), and $k_{i,t}$ is the decay constant in process segment i at a time t, (min^{-1}). The decay constant is proportional to the temperature $T_{i,t}$

$$k_{i,t} = 6.8\times10^{-4}\frac{1}{min^{\circ}C}\cdot T_{i,t}, \qquad \forall i\in NI, \forall t\in NT$$

To calculate the residual concentration in the EOP tank, the effect of the EOP chemical (dechlorination agent) has to be taken into account, thus

$$C_{i,j,t} = \frac{C_{i-p,j,t}F_{i-p,t} - C_t^{Chem}F_t^{Chem} + \psi_t C_t^{Chem}}{F_{i,t}}e^{-k_{i,t}\tau_{i,t}}, \qquad \forall i = EOP, \forall j\in NC, \forall t\in NT$$

C_t^{Chem} is the concentration of the EOP chemical, F_t^{Chem} is the flow of chemical used in the basin, and ψ_t is the excess of chemical used.

The percentage of heat transfer reduction (due to biofilm growth) is considered to be proportional to the biofilm thickness:

$$\%HTR_{i,t} = 0.2156\cdot d_{i,t}, \qquad \forall i\in NHE, \forall t\in NT$$

$\%HTR_{i,t}$ is the percentage of heat transfer reduction in process segment i at time t, $d_{i,t}$ is the biofilm thickness in process segment i at time t (μm). Here $NHE\subset NI$ (NHE is a subset of NI that contains all the process segments that are heat exchangers). The actual heat transferred is

$$Q_{i,t} = Q_{i,t}^{Req}\left(1-\frac{\%HTR_{i,t}}{100}\right), \qquad \forall i\in NHE, \forall t\in NT$$

$Q_{i,t}^{Req}$ is the required heat in process segment i at time t (W), and $Q_{i,t}$ is the actual heat transferred in process segment i at time t (W). The temperature at the entrance of any process segment i is equal to the temperature at the exit of process segment i-p, which is the segment that precedes i (see Figure 2).

The energy balance in process segment i at time t states that:

$$Q_{i,t} = F_{i,t}\rho^{SW}Cp^{SW}(\Delta T) = F_{i,t}\rho^{SW}Cp^{SW}(T_{i,t} - T_{i-p,t}), \quad \forall i\in NI, \forall t\in NT$$

Rearranging

$$T_{i,t} = \frac{Q_{i,t}}{F_{i,t}\rho^{SW}Cp^{SW}} + T_{i-p,t}, \qquad \forall i\in NI, \forall t\in NT$$

For the energy balance it is assumed that the density and Cp are independent of temperature in the range of operation of the process and equal to the values for seawater. This is a mild assumption since the changes of these properties are not very significant in the range 20-40°C (Millero and Huang, 2009; Millero, 1973).

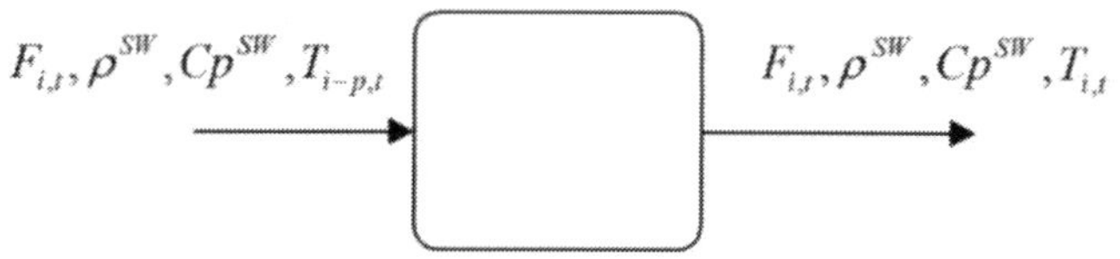

Figure 2. Schematic representation of a segment *i*.

$F_{i,t}$ is the total flow in process segment *i* at time *t* (m^3/hr), ρ^{SW} is the density of seawater (kg/m^3), and Cp^{SW} is the heat capacity of seawater (*J/Kg°C*). The biofilm thickness over the time span can be described as:

$$d_{i,t} = d^{\infty}\frac{\left[1-e^{-\beta_{i,t}\cdot\sigma_{i,t}}\right]}{e^{\gamma_{i,t}\cdot\omega}}, \qquad \forall i\in NHE, \forall t\in NT$$

With $d^{\infty} = 250\mu m$. The biofilm growth rate constant $\beta_{i,t}$ for process segment *i* at time interval *t* (min^{-1}) is:

$$\beta_{i,t} = 2.909*10^{-7}\frac{1}{min°C}\cdot T_{i,t}, \qquad \forall i\in NHE, \forall t\in NT$$

$\sigma_{i,t}$ is a mathematical representation of the period over which the biofilm has grown in process segment *i* and at time interval *t* (*min*) and it can be described with the following disjunction

$$\begin{bmatrix} Y_t \\ \sigma_{i,t+1} = \omega \end{bmatrix} \vee \begin{bmatrix} \neg Y_t \\ \sigma_{i,t+1} = \omega + \left(\frac{-1}{\beta_{i,t}}\right)\ln\left(1-\frac{d_{i,t}}{d^{\infty}}\right) \end{bmatrix}$$

$d_{i,t}$ is the biofilm thickness in process segment *i* and time interval *t* (μm), and ω is the length of a single time interval (*min*). If the binary variable (y_t) is set to one, maintenance is scheduled in time *t*. In this scenario $\sigma_{i,t+1}$ is equal to a single time interval; if the binary variable is equal to zero, no maintenance is scheduled and $\sigma_{i,t+1}$ increases as a function of the biofilm growth rate constant and the biofilm thickness. The disjunction is reformulated using a big-M (Vecchietti et al., 2003) technique as follows:

$$\sigma_{i,t+1} - \omega \le M^U(1-y_t), \qquad \forall i \in NHE, \forall t \in NT$$

$$\sigma_{i,t+1} - \omega \ge -M^U(1-y_t), \qquad \forall i \in NHE, \forall t \in NT$$

$$(\sigma_{i,t+1} - \omega)\beta_{i,t} + \ln\left(1 - \frac{d_{i,t}}{d^\infty}\right) \le M^U(y_t), \qquad \forall i \in NHE, \forall t \in NT$$

$$(\sigma_{i,t+1} - \omega)\beta_{i,t} + \ln\left(1 - \frac{d_{i,t}}{d^\infty}\right) \ge -M^U(y_t), \qquad \forall i \in NHE, \forall t \in NT$$

This representation avoids the bilinear terms of the original equation presented by Bin-Mahfouz et al. (2011).

The biofilm destruction rate $\gamma_{i,t}$ for process segment i and time interval t depends on the concentration of the key component j in time interval t-1 (min^{-1}):

$$\gamma_{i,t} = 2.829\times10^{-4}\frac{1}{\min\cdot(mg/l)}\cdot C_{j,i,t-1}, \qquad j = KC, \forall i \in NHE, \forall t \in NT$$

To select the month in which maintenance is to be scheduled, the following disjunction is used:

$$\begin{bmatrix} Y_M \\ d_{i,M} \le 70\mu m, \ \forall i \in NHE \\ d_{i,M-1} \le 60\mu m, \forall i \in NHE \end{bmatrix} \vee \begin{bmatrix} Y'_M \\ d_{i,M} \le 50\mu m, \ \forall i \in NHE \\ \ \end{bmatrix}, \quad \forall M \in NM$$

This disjunction is also reformulated using a Big-M technique to yield the following relationships,

$$d_{i,M} - 70 \le M(1-y_M), \qquad \forall i \in NHE, \forall M \in NM$$

$$d_{i,M-1} - 60 \le M(1-y_M), \qquad \forall i \in NHE, \forall M \in NM$$

$$d_{i,M} - 50 \le M(1-y'_M), \qquad \forall i \in NHE, \forall M \in NM$$

If the binary variable y_M is equal to one, maintenance is scheduled in month M. On the other hand, if the binary variable y'_M is equal to one, then no maintenance is scheduled in months M and M+1. The relationship between y'_M, y_M and y_{M+1} can be described as follows:

$$y'_M = 1 - y_M - y_{M+1}, \quad \forall M \in NM$$

Substituting y'_M into the last equation:

$$d_{i,M} - 50 \le M(y_M + y_{M+1}), \qquad \forall i \in NHE, \forall M \in NM$$

These relationships imply that the maximum allowed biofilm thickness is $50\mu m$ if no maintenance is scheduled in two consecutive months; on the other hand, the maximum thickness is $70\mu m$ for month M if maintenance is scheduled in that month and $60\mu m$ for the month before maintenance is scheduled (i.e., month M-1)

If a maximum of one treatment per month is allowed, the following constraint is used:

$$y_M \le 1, \qquad \forall M \in NM$$

The number of treatments per year is limited by a maximum of n allowed treatments,

$$\sum_M y_M \le n$$

If t is the time interval in days and M is the time interval in months, the relation between these variables is:

$$\sum_{t \in M} y_t = y_M, \qquad \forall M \in NM$$

Finally, the objective function consists in minimizing the total annual cost divided in biocide cost, EOP chemical cost and cost of maintenances (including the loss of revenue caused for plant shutdowns):

$$\min TAC = \sum_{t=1}^{Nt} \left[F_t^B \cdot \frac{\omega}{60 \frac{\min}{hr}} \cdot BiocideCost + F_t^{Chem} \cdot \frac{\omega}{60 \frac{\min}{hr}} \cdot ChemicalsCost + MaintenanceCost \cdot y_t \right]$$

Case study

The requirements of cooling seawater for a MSF power/desalination plant during a year are presented in Table 1 (Bin-Mahfouz et al., 2011). The planning span of the project is one year. The maximum allowed concentrations in the discharge are 0.002 mg/L for the biocide and the EOP chemical.

Table 1. Requirements of cooling seawater for a MSF power/desalination plant

		Cooling water (m^3/hr)			
Month	(%) of Cap.	MSF unit	Boiler	Steam Turbine	Total
Jan	75	27750	1356	9496	38602
Feb	70	25615	1252	8765	35632
Mar	81	29884	1461	10226	41571
Abr	84	30952	1513	10591	43056
May	93	34154	1669	11687	47510
Jun	93	34154	1669	11687	47510
Jul	99	36288	1774	12417	50479
Aug	100	36822	1800	12600	51222
Sep	94	34687	1695	11869	48252
Oct	93	34154	1669	11687	47510
Nov	90	33086	1617	11322	46025
Dec	81	29884	1461	10226	41571
Average	88	32286	1578	11048	44912

Results

Two different dosing policies were tested (monthly and bimonthly). The maximum number of treatments per year range from 1 to 4 treatments. Table 2 shows the results obtained.

Table 2. Results for the case study

Policy	Allowed number of treatments	TAC (MM$/year)	Real number of treatments	January	February	March	April	May	June	July	August	Septembe	October	November	December
Monthly	1	4.0460	1	0	0	0	0	0	0	0	1	0	0	0	0
Monthly	2	4.0169	2	0	0	0	0	0	0	0	1	0	1	0	0
Monthly	3	3.9989	3	0	0	0	0	0	1	0	1	0	1	0	0
Monthly	4	3.9962	4	0	0	0	0	1	0	1	0	1	0	1	0
Bimonthly	1	4.2722	1	0	0	0	0	0	0	0	0	0	1	0	0
Bimonthly	2	4.2104	2	0	0	0	0	0	0	0	1	0	1	0	0
Bimonthly	3	4.1753	3	0	0	0	0	0	1	0	1	0	1	0	0
Bimonthly	4	4.1432	4	0	0	0	1	0	1	0	1	0	1	0	0

It can be seen that the cheapest option corresponds to a monthly dosing, with the maximum allowed number of treatments per year (four treatments). It can be noted that the difference for the monthly dosing between the three and four treatments per year is negligible. Thus to avoid technical difficulties associated to plant shutdowns, it seems recommendable to use a monthly dosing with three treatments at the end of June, August and October. Figure 3 shows the detailed results for this scenario. Figures 3a and 3b show the linear relation between the biocide dosing and the chemical for end of pipe treatment. A gradual increment in the biocide dosing during the first four months of the year is observed, until it reaches a maximum of 0.09 m^3/h during the month of April. In May a step down in dosing is observed because the constraint for the maximum allowed biofilm thickness increases to 60μm. Something similar is observed for the next month (when the constraint becomes 70μm), and the first scheduled maintenance is performed at the end of June. During the next two months (July and August), despite the fact that maintenance is planned, the dosing is maintained almost constant to compensate for the amount of water processed during these months (the plant operates at 99 and 100% respectively). In September, a new step down is observed caused by the same effect from the constraints. Figure 3c shows the behavior of the biofilm thickness during the year. It can be seen that during the first months the biofilm thickness is maintained almost constant with a value around 43μm for process segments 11, 12, 13 and 16. And 48μm for process segments 14 and 15. In the month prior to the scheduled maintenances, an increment in the biofilm thickness is observed, with heat exchangers 14 and 15 dictating the operation of the plant because they reach the thickness limit faster in every case. Finally, Figure 3d shows the biocide concentration dynamics in the network; the environmental constraint is always satisfied with average values of 0.0013 mg/L for the biocide and 0.0007 mg/L for the EOP chemical.

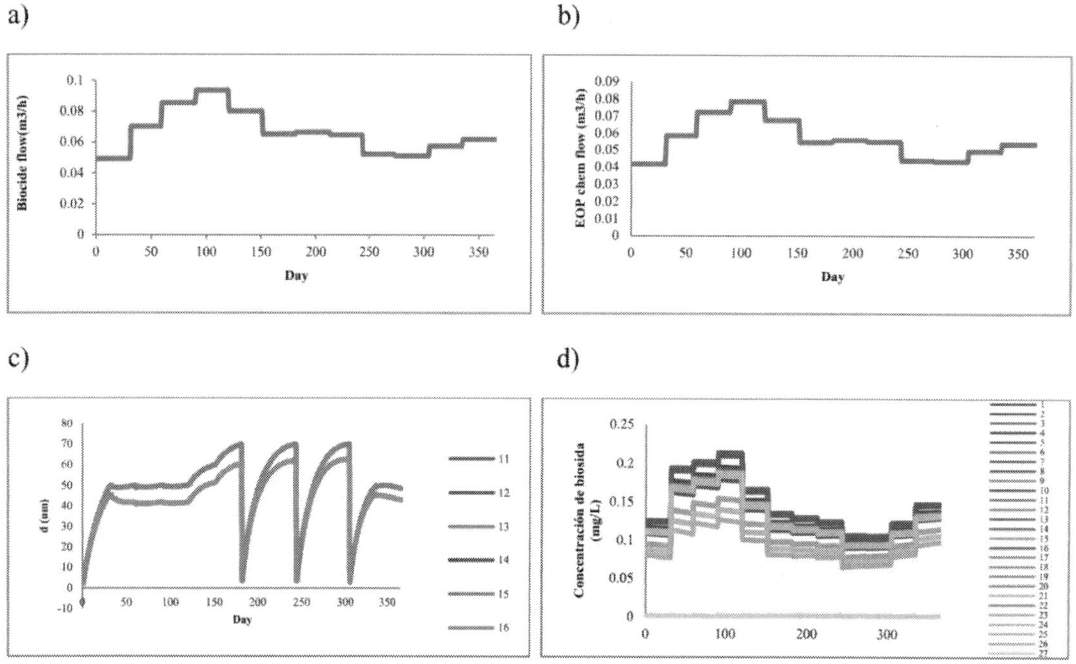

Figure 3. a) Biocide dosing b) EOP chemical dosing c) Biofilm thickness d) Biocide concentration in the network

Conclusions

A mixed integer nonlinear programming model for the optimal planning of biocide dosing and maintenance scheduling for seawater cooled plants was presented. A case study for a MSF power/desalination plant was solved. The results show that the best policies correspond to a monthly dosing with three treatments per year. Although this is not the cheapest option, the small difference between the four treatments scheme and the three treatments scheme does not seem to justify the scheduling of an additional plant shut down. For the selected scheme the biocide dosing gradually increases during the first months and a sudden decrease is seen for the months when maintenance is planned. The results also show that the boilers are the units that build up the biofilm faster and thus determine the operation schedule of the plant.

References

Bin Mahfouz, A., Atilhan, S., Batchelor, B., Linke, P., Abdel-Wahab, A., & El-Halwagi, M. (2011). Optimal scheduling of biocide dosing for seawater-cooled power and desalination plants. Clean Technologies and Environmental Policy, 1-14.

Millero, F.J., Perron, G., & Desnoyers, J.E. (1973). Heat Capacity of Seawater Solutions from 5° to 35°C and 0.5 to 22%. Chlorinity Journal of Geophysical Research, 78, 4499-4507.

Millero, F.J., & Huang, F. (2009). The density of seawater as a function of salinity (5 to 70 g kg− 1) and temperature (273.15 to 363.15 K). Ocean Science Discussions, 6, 153-169.

Vecchietti, A., Lee, S., & Grossmann, I.E. (2003). Modeling of discrete/continuous optimization problems: characterization and formulation of disjunctions and their relaxations. Computers & Chemical Engineering, 27, 433-448.

OPTIMAL DESING OF DISTRIBUTED TREATMENT SYSTEMS FOR THE EFFLUENTS DISCHARGED TO THE RIVERS

Oscar Burgara-Montero[1], José María Ponce-Ortega[1], Medardo Serna-González, Mahmoud M. El-Halwagi[2]

[1]*Chemical Engineering Department, Universidad Michoacana de San Nicolás de Hidalgo, Morelia, Michoacán, México, 58060*

[2]*Chemical Engineering Department, Texas A&M University, College Station, TX, USA, 77843-3122*

Introduction

Nowadays most of the rivers around the world are highly polluted because many wastewater streams are discharged without an adequate treatment. Usually central treatment facilities are considered to clean the polluted rivers, where treatment technologies are simply installed at the end of the rivers (Kennish,[1] Best et al.[2]). Recently, the concept of distributed treatment systems has been investigated systematically, and several methodologies for the distributed treatment of the effluents in the industry have been reported and implemented with excellent results to reduce the treatment cost (see, for example, Savelski and Bagajewicz,[3-5] Wang and Smith,[6] Alva-Argaez et al.,[7,8] Benko et al.,[9] Teles et al.,[10] Gabriel and El-Halwagi,[11] Kuo and Smith,[12] Doyle and Smith,[13] Galan and Grossmann,[14] Hernandez-Suarez at al.,[15] Gunaratnam et al.,[16] Karuppiah and Grossmann,[17]; Putra and Amminudin,[18] Ponce-Ortega et al.,[19] Nápoles-Rivera et al.,[20] Ng et al.[21]). The methodologies aforementioned have demonstrated that the distributed treatment systems are better than centralized treatment systems; however, such methodologies only consider the process and environmental constraints on the industrial wastewater streams for treating them before discharge, without evaluating how the resulting discharge affects the sustainability of the surrounding watershed even if meets the environmental regulations imposed by authorities.

The rivers are affected by several polluted inputs such as effluents from surrounding agricultural, residential and industrial systems. These effluents have a great impact on the biological, physical, health, social and surrounding environments. To understand the characteristics of watersheds, how they are affected by their environment, and how they affect their surroundings it is important to develop quantitative models that track the flows and compositions of different pollutants through the watershed. In this regard, it is necessary to use models to assess the quality of the water bodies and rivers at each point and throughout the period of interest, when the modified conditions and the initial state are known. Cooper[22] employed the Streeper-Phelps model, which takes the river as a plug flow reactor for the oxygen dynamics in the system. Drolc and Koncan[23] used the water quality model QUAL2E to evaluate the impact of wastewater discharges on dissolved oxygen concentration in the river Sava nearby to Ljubljana. The material flow analysis technique (MFA) is particularly useful for tracking chemical species that cause pollution in rivers and to analyze the causes of the problems of pollution in a region considering all relevant activities, water supplies, water use, discharges of polluted streams, physical phenomena, and chemical and biological processes that impact the rivers; in this context, Baccini and Brunner[24] developed an MFA model to analyze ecosystems with human activities which exchange mass, energy and information around. Also, Lampert and Brunner[25] proposed an MFA model to track the major nutrients discharged, processed and flowing into the Danube River. Drolc and Koncan[26] made a study using the strategy of material flow analysis on the Krka river in Slovenia, where they determined that in addition to the wastewater treatment plants, agriculture contributes significantly to the total annual phosphorus load in the macroscopic system. El-Baz et al.[27] combined a simulation tool with the material flow analysis technique to provide a complete analysis and useful information on the transport and destination of the key component. This methodology was used in the Egyptian drainage system Bar-El Baqar, where targeted species were ammonium ions. Then, El- Baz et al.[28] developed a systematic approach to provide solution strategies joining an MFA model with a synthesis tools to detect and find alternative optimal solutions. Lovelady[29] inverted the environmental quality model (material flow analysis) and included it into an optimization formulation that seeks to determine the maximum allowable target for the process discharges while meeting the overall environmental requirements of the watershed.

The proposed model considers the optimal selection of the tributaries to be treated, the location for the new treatment units and the type of technology used. The model is based on the material flow analysis technique to consider all discharges into the rivers, their development trough the river and their final disposal. In addition, the model considers independent constraints for the water quality in the watershed regions where it is modified by new discharges; these restrictions are because in some sections of the river, water can be used for agricultural, industrial and human uses.

Methodology.

The equations for the model are described as follows:

Overall mass balance for each reach.

The exit flowrate from each reach i (Q_i) is equal to the inlet flowrate to the reach (Q_{i-1}) plus the industrial discharges (IND_i), direct discharges (D_i), the average precipitation (P_i), the sum of all effluents inlet to the reach $\sum_{j=1}^{N_{Efl,i}} T_{i,j}$, minus the extractions for filtration and evaporation (L_i) and the use of water in that section of the river (U_i).

$$Q_i = Q_{i-1} + P_i + D_i + IND_i + \sum_{j=1}^{N_{Efl,i}} T_{i,j} - L_i - U_i \qquad \forall i \in I \qquad (1)$$

Component balance for each reach.

The exit mass from each reach for the hazardous compound $\left(Q_i * CQ_{i,l}\right)$ is equal to the inlet mass to the reach $\left(Q_{i-1} * CQ_{i-1,l}\right)$, plus the precipitation $\left(P_i * CP_{i,l}\right)$, direct discharges $\left(D_i * CD_{i,l}\right)$, industrial discharges $\left(IND_i * CIND_{i,l}\right)$, effluents $\sum_{i=1}^{N_{Efl,i}} T_{i,j} * CT_{i,j,l}$, minus the loss $\left(L_i * CL_{i,l}\right)$ and the use $\left(U_i * CU_{i,l}\right)$, and minus the degradation for the component in the reach ($\int_{V=0}^{V_i} r_i dV_i$). Therefore, the component balance for each reach is stated as follows:

$$Q_i * CQ_{i,l} = Q_{i-1} * CQ_{i-1,l} + P_i * CP_{i,l} + D_i * CD_{i,l} + IND_i * CIND_{i,l} + \sum_{i=1}^{N_{Efl,i}} T_{i,j} * CTR_{i,j,l} - L_i * CL_{i,l} - U_i * CU_{i,l} - \int_{V=0}^{V_i} r_i dV_i$$

$$\forall i \in I \qquad (2)$$

The last term of equation (2) corresponds to the reaction for each reach, and it can be expressed through a first order kinetic reaction as follows:

$$\int_{V=0}^{V_i} r_{i,l} dV_i = K_{i,l} * CQ_{i,l} * V_i$$

where $K_{i,l}$ is the rate constant of Arrhenius that must be determined experimentally, and V_i is the volume for the section i of the river. The reactive term $\int_{V_i=0}^{V_i} r_{i,l} dV_i$ is very interesting because it considers the chemical and biochemical reactions that take place in the river by the flora and fauna inside the system, which can produce or eliminate the component.

Mass balances for tributaries.

The mass balance for the tributary j that discharges to the reach i is described as follows:

$$T_{i,j} = S_{i,j}^{untreated} + S_{i,j}^{treated} + IND_{i,j} + P_{i,j} + D_{i,j} - L_{i,j} - U_{i,j} \quad \forall i \in I, j \in J \qquad (3)$$

The total flowrate that leaves the tributary j and enters to the reach i $\left(T_{i,j}\right)$ is equal to the sum of the discharge with or without treatment ($S_{i,j}^{treated}, S_{i,j}^{untreated}$), industrial wastewater ($IND_{i,j}$), the flow of precipitations ($P_{i,j}$), the industrial discharge ($D_{i,j}$) minus the loss by evaporation and filtration ($L_{i,j}$) and the use or extraction streams ($U_{i,j}$).

Component balances for each tributary.

The mass for each component leaving each tributary j and directed to the reach i is calculated as follows:

$$T_{i,j} * CT_{i,j} = S_{i,j}^{untreated} * CS_{i,j,l}^{untreated} + S_{i,j}^{treated} * CS_{i,j,l}^{treated} + IND_{i,j} CIND_{i,j,l} + P_{i,j} * CP_{i,j,l} + D_{i,j} * CD_{i,j,l} - L_{i,j} * CL_{i,j,l}$$
$$-U_{i,j} * CU_{i,j,l} - \int_{V=0}^{V_{i,j}} r_{i,j} dV_{i,j}, \qquad \forall i \in I, j \in J \qquad (4)$$

It is important to note that the component balances for each tributary include the reaction term $\int_{V=0}^{V_{i,j}} r_{i,j} dV_{i,j}$ to consider the natural degradation of the component in the system. To determine this reactive term, usually first order kinetic reactions are considered.

Agricultural discharges and uses.

The agricultural discharges $D_{i,j}$ and uses $U_{i,j}$ are proportional to the agricultural areas. These discharges and uses are calculated as follows:

$$D_{i,j} = \lambda_{i,j} * A_{i,j} \qquad \forall i \in I, j \in J \qquad (5)$$
$$U_{i,j} = \beta_{i,j} * A_{i,j} \qquad \forall i \in I, j \in J \qquad (6)$$

Where $\lambda_{i,j}$ is the flowrate per unit of area in m^3/acre*s, $\beta_{i,j}$ is the agricultural use of water per unit of area for the tributary t in m^3/acre*s and $A_{i,j}$ is the area that covers the tributary j in acres.

Design for the distributed treatment system.

There are several regeneration technologies that can be used to treat the different effluents discharged to the main river (see **Figure 1**). For a macroscopic system, there are a big number of tributaries; many of them can be treated to satisfy the environmental constraints and ensure the sustainability of the system. Therefore, the problem consists in determining the flowrate for the tributaries that must be treated and the type of technologies that must be used for accomplishing these constraints at the minimum cost. To design this distributed treatment system, the following disjunction is used for each tributary j that discharges to the river i:

$$\begin{bmatrix} Y_{i,j} \\ T_{i,j} = \sum_{x \in X} fs_{i,j,x} \\ \sum_{x \in X} fs_{i,j,x} * (1-\alpha_{x,l}) * CT_{i,j,l} = T_{i,j} * CTR_{i,j,l}, \forall l \in L \\ \begin{bmatrix} Z_{i,j,x} \\ fs_{i,j,x} \geq \Omega_{i,j,x}^{\min} \\ fs_{i,j,x} \leq \Omega_{i,j,x}^{\max} \end{bmatrix} \vee \begin{bmatrix} \neg Z_{i,j,x} \\ fs_{i,j,x} = 0 \end{bmatrix}, \forall x \in X \end{bmatrix} \vee \begin{bmatrix} \neg Y_{i,j} \\ CTR_{i,j,l} = CT_{i,j,l}, \forall l \in L \\ fs_{i,j,x} = 0, \forall x \in X \end{bmatrix}, \forall i \in I, \forall j \in J$$

To reformulate previous disjunction as a set of algebraic equations, the Boolean variables are transformed into binary variables. This way, when the Boolean variable is true, the associated binary variable is equal to one, and when the Boolean variable is false, the associated binary variable is equal to zero. Finally, previous disjunction is modeled using the convex hull reformulation (see Raman and Grossmann,[30] Ponce-Ortega et al.[31]).

Constraints for the quality of specific reaches.

There are some constraints required for guaranteeing the water quality in specific places. These constraints depend on the kind of water using operation or activity, for example, agricultural (including the type of agriculture), industrial or domestic uses. The proposed model is able to consider these constraints through the following relationships:

$$CQ_{n(i)} \leq CQ_{n(i)}^{desired}, \qquad \forall n(i) \in N(I) \tag{7}$$

where, $N(I)$ represents a subset of reaches n from all the reaches r that require a specific water quality.

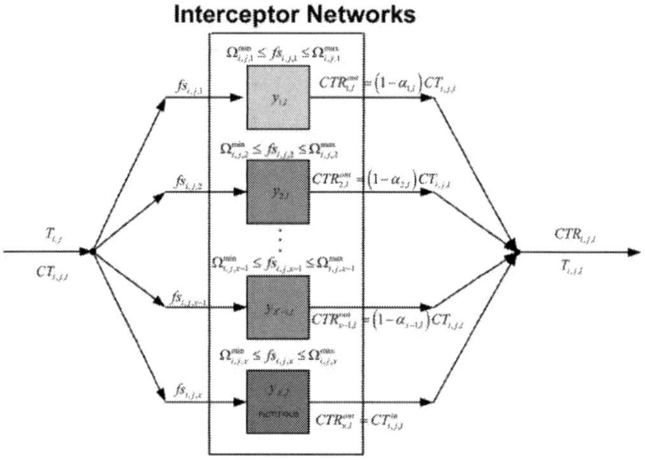

Figure 1. Superstructure for n interceptors for any effluent.

Objective function.

The model is formulated as a multi-objective problem for the simultaneous minimization of the total annual cost and the final concentration for the hazardous compound at the final disposal for the watershed as follows:

$$OF = \{\min TAC \ ; \ \min CQ^{dis}\} \tag{8}$$

Notice that these two objectives contradict each other; in other words, when the concentration in the final disposal decreases, the total annual cost to treat the effluents increases; whereas, when the concentration at the final disposal is relaxed, the total annual cost decreases.

The total annual cost function objective function is expressed as follows:

$$TAC = k_f \sum_i \sum_j \sum_x \left[FC_x * z_{i,j,X} + VC_x * fs_{i,j,x}^{\gamma} \right] + H_Y \sum_i \sum_j \sum_x Cu_x^{op} * fs_{i,j,x} \tag{9}$$

Result and Discussions.

The Bahr El-Baqar drainage is the longest drainage in the east of the Nile Delta in Egypt. The starting point originates from the intersection of drainage Qalyoubia and drainage Belbies beginning in the northern part of Cairo. The end point is the entrance on the south side of Lake Manzala, which has an area of 1000 km^2 and an average depth of 1 m. The drain receives several types of discharges including agricultural drainage, domestic waste water treated and untreated, and industrial wastewater streams. This river is polluted mainly by phosphorus and sulfur. The distributed treatment system can be placed along the river to assist the removal of phosphorus and thereby to decrease the final concentration in the catchment area. Originally, without a distributed treatment system, the concentrations that exit the system are 1.3 mg/L for phosphorus and 1.132 mg/L for sulfur.

Figure 2 shows the optimal solutions for various constraints placed on the final disposition for the key components; this set of solutions is called Pareto curve.

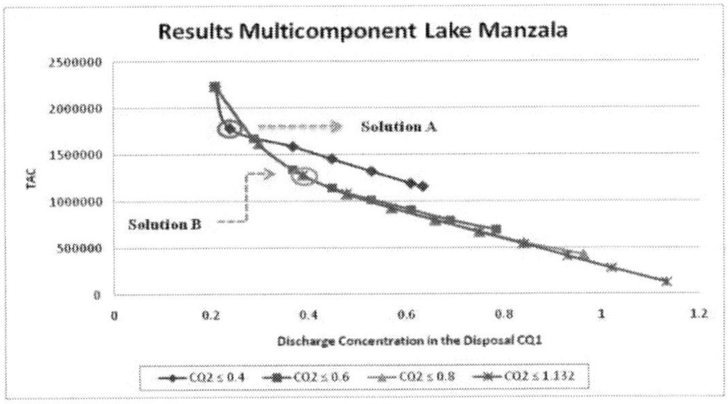

Figure 2. Pareto curves.

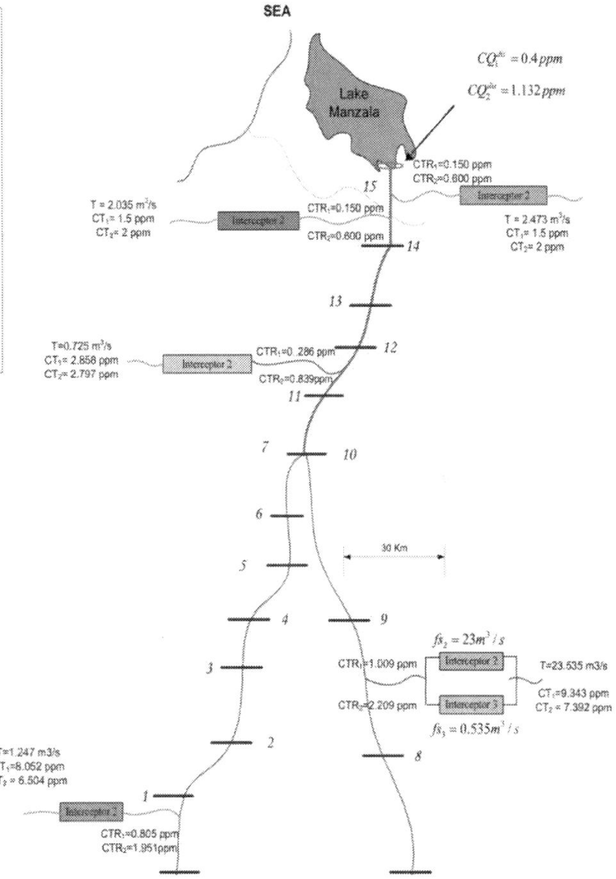

Figure 3. Optimal configuration for the Case B scenario.

Figure 3 shows the optimal solution for the concentration discharges $CQ_1 \leq$ 0.4 ppm for the key component 1 and $CQ_2 \leq$ 1.132 ppm for the key component 2. The optimal solution indicates that 6 treatment units are required to satisfy the restriction imposed in the catchment area with a total annual cost of $1,250,000.

Conclusions.

This work presents a multi-objective mixed integer non linear programming problem for the optimal design of distributed treatment systems of watersheds to satisfy environmental constraints that are able to guaranty sustainable process. The model is based on a disjunctive programming formulations and it considers all discharges and extractions for the watershed through an MFA model. The model is able to identify the distributed treatment system to satisfy the environmental constraints for the polluted rivers, results shows significant savings for the distributed treatments respect to the central systems.

Literature Cited

(1) Kennish, M. J. Estuary restoration and maintenance. CRC Press, 1999, Boca Raton, Florida, USA.

(2) Best, G.; Bogacka, T.; Niemirycz, E. International river water quality: Pollution and restoration. Spon Press, 1998, London, UK.

(3) Savelski, M. J.; Bagajewicz, M. J. On the optimality conditions of water utilization systems in process plants with single contaminants. *Chem. Eng. Sci.* **2000,** *55*, 5035-5048.

(4) Savelski, M. J.; Bagajewicz, M. J. Algorithmic procedure to design water utilization systems featuring a single contaminant in process plants. *Chem. Eng. Sci.* **2001,** *56*, 1897-1911.

(5) Savelski, M.; Bagajewicz, M. On the necessary conditions of optimality of water utilization systems in process plants with multiple contaminants. *Chem. Eng. Sci.* **2003,** *58*, 5349-5362

(6) Wang, Y. P.; Smith, R. Wastewater minimization. *Chem. Eng. Sci.* **1994,** *49*, 981-1006.

(7) Alva-Argaez, A.; Kokossis, A. C.; Smith, R. Wastewater minimization of industrial systems using an integrated approach. *Comput. Chem. Eng.* **1998,** *22*, S741-S744.

(8) Alva-Argaez, A.; Vallianatos, A.; Kokossis, A. A multi-contaminant transshipment model for mass exchange networks and wastewater minimization problems. *Comput. Chem. Eng.* **1999,** *23*, 1439-1453.

(9) Benko, N.; Rev, E.; Fonyo, Z. The use of nonlinear programming to optimal water allocation. *Chem. Eng. Commun.* **2000,** *178*, 67-101.

(10) Teles, J.; Castro, P. M.; Novals, A. Q. LP-based solution strategies for the optimal design of industrial water networks with multiple contaminants. *Chem. Eng. Sci.* **2008,** *63*, 376-394.

(11) Gabriel, F. B.; El-Halwagi, M. M. Simultaneous synthesis of waste interception and material reuse networks: Problem reformulation for global optimization. *Environ. Prog.* **2005,** *24*, 171-180.

(12) Kuo, W. C. J.; Smith, R. Effluent treatment system design. *Chem. Eng. Sci.* **1997,** *52*, 4273-4290.

(13) Doyle, S. J.; Smith, R. Targeting water reuse with multiple contaminants. *Process Saf. Environ. Prot.* **1997,** *75*, 181-189.

(14) Galan, B.; Grossmann, I. E. Optimal design of distributed wastewater treatment networks. *Ind. Eng. Chem. Res.* **1998,** *37*, 4036-4048.

(15) Hernandez-Suarez, R.; Castellanos-Fernandez, J.; Zamora, J. M. Superstructure decomposition and parametric optimization approach for the synthesis of distributed wastewater treatment networks. *Ind. Eng. Chem. Res.* **2004,** *43*, 2175-2191.

(16) Gunaratnam, M.; Alva-Argaez, A.; Kokossis, A.; Kim, J. K.; Smith, R. Automated design of total water systems. *Ind. Eng. Chem. Res.* **2005,** *44*, 588-599.

(17) Karuppiah, R.; Grossmann, I. E. Global optimization for the synthesis of integrated water systems in chemical processes. *Comput. Chem. Eng.* **2006,** *30*, 650-673.

(18) Putra, Z. A.; Amminudin, K. A. Two-step optimization approach for design a total water system. *Ind. Eng. Chem. Res.* **2008,** *47*, 6045-6057.

(19) Ponce-Ortega, J. M.; El-Halwagi, M. M.; Jiménez-Gutiérrez, A. Global optimization of property-based recycle and reuse networks including environmental constraints. *Comput. Chem. Eng.* **2010,** *34*, 318-330.

(20) Nápoles-Rivera, F.; Ponce-Ortega, J. M.; El-Halwagi, M. M.; Jiménez-Gutiérrez, A. Global Optimization of Mass and Property Integration Networks with In-Plant Property Interceptors. *Chem. Eng. Sci.* **2010,** *65*, 4363-4377.

(21) Sum Ng-Kok, D.; Chwan-Yee Foo, D.; R. Tan R. Automated Targeting Technique for Single-Impurity Resource Conservation Networks. Part 2: Single-Pass and Partition Waste-Interception Systems. *Ind. Eng. Chem. Res.* **2009,** *48,* 7647–7661.

(22) Cooper B. Developing Management Guidelines for River Nitrogenous Oxygen Demand. *Water Pollution Control Federation*. **1986,** 58, 845-852.

(23) Drolc A. and Zagorc-Koncan J. Water Quality Modeling of the river Sava, Slovenia. Elsevier Science Ltd. **1996**, 30, 2587-2592.

(24) Baccini P.; Brunner P. *Metabolism of the anthroposphere*. Springer-Verlag, Berlin. **1991**.

(25) Lampert, C.; Brunner P. H. Material accounting as a policy tool for nutrient management in the Danube Basin. *Water Sci. Technol.* **1999**, *40*, 43-49.

(26) Drolc A. and Zagorc-Koncan J. Estimation of source of total phosphorus in a river basin and assessment of alternatives for river pollution reduction. *Environment International*. **2002**, 28, 393-400.

(27) El-Baz, A. A.; Ewida K. T.; Shouman, M. A.; El-Halwagi M. M. Material flow analysis and integration of watersheds and drainage systems: I. Simulation and application to ammonium management in Bahr El-Baqar drainage system. *Clean Technol. Environ. Policy* **2005** *7*, 51–61.

(28) El-Baz, A. A.; Ewida K. T.; Shouman, M. A.; El-Halwagi M. M. Material flow analysis and integration of watersheds and drain systems: II. Integration and solution strategies with application to ammonium management in Bahr El-Baqar drain system, *Clean Techn. Environ Policy*. **2005**, 78–86.

(29) Lovelady, E. M.; El-Baz, A. A.; El-Monayeri, D.; EL-Halwagi, M. M. Reverse problem formulation for integrated process discharges with wastewater and drainage systems: Managing phosphorus in lake Manzala. *J. Ind. Ecol.* **2009,** *13*, 14-927.

(30) Raman, R.; Grossmann, I.E. Modeling and computational techniques for logic based integer programming. *Comput. Chem. Eng.* **1994,** *18*, 563-578.

(31) Ponce-Ortega, J. M.; Serna-González, M.; Jiménez-Gutiérrez A. A disjunctive programming model for simultaneous synthesis and detailed design of cooling networks. *Ind. Eng. Chem. Res.* **2009,** *48*, 2991-3003.

(620f) Property-BASED MASS INTEGRATION Considering Efficient Property Operators

Ma Guadalupe Rojas-Torres[1], Jos Mara Ponce-Ortega[2], Medardo Serna-Gonzlez[2] and M.M. El-Halwagi[3], (1)Chemical Engineering, Universidad Michoacana de San Nicols de Hidalgo, Morelia, Mexico, (2)Chemical Engineering Department, Universidad Michoacana de San Nicols de Hidalgo, Morelia, Michoacan, Mexico, (3)The Artie McFerrin Department of Chemical Engineering, Texas A&M University, College Station, TX

The proper use of the resources is an ongoing need in all processes because of the global competitiveness and the environmental deterioration. One of the most used strategies to have an efficient use of resources is the mass integration, which traditionally has been carried out based on the composition of waste streams; however, the waste streams characterization is highly challenging because the large number of components presented. Recently have been reported a series of methodologies for the synthesis of mass integration networks based on properties considering the recycle, reuse and regeneration of waste streams, these methodologies consider property balances instead of component balances in the mixing points; however, several of the property operators considered in the property balances produce significant errors in some specific cases. Therefore, this paper proposes a disjunctive model for a better prediction of mixing properties, within a recycle, reuse and regeneration strategy, considering simultaneously the effects of temperature that drastically affect the value of some properties of the streams (depreciated in most of the previous works). The model considers to synthesize the water network that meets the process and environmental constraints based on properties at the minimum total annual cost. The methodology proposed has been applied to several cases of study, and the results show that the model accurately predicts better mixing properties than the models previously reported, without numerical complications because the model avoids nonconvex terms.

See more of this Session: Poster Session
See more of this Group/Topical: Computing and Systems Technology Division

(620g) Development of Moving Bed Simulation Model for Carbon Capture From Fossil Energy Systems

Hosoo Kim, Computational Science Division, National Energy Technology Laboratory, Morgantown, WV and David C. Miller, U.S. Department Of Energy, National Energy Technology Laboratory, Morgantown, WV

The capture and separation of carbon dioxide (CO2) has been identified as a high-priority topic to cope with global climate change. Fossil fuels currently supply the most of the world's energy needs, and their utilization is the major source of the anthropogenic CO2 emission [1]. Particularly, the existing coal-fired power plants annually emit about 2 billion tons of CO2 which is equivalent to two-thirds of the total emissions from U.S. power sector [2]. Therefore, it is critical to develop the cost-effective technologies to mitigate this problem. There are three options for capture for capturing CO2 from fossil energy system: post-combustion capture, pre-combustion capture, and oxy-combustion. Among them, post-combustion capture has the greatest near-term potential for reducing CO2 emission, because it can be applied to the existing coal-fired power plant with relative ease through a retrofit. The current commercially available solvent-based processes have advantages of fast kinetics and strong reactions, however only at a significant cost and efficiency penalty. Recently, various solid sorbents are being explored for one of promising CO2 capture technology, which are expected to reduce energy requirement and water usage with the approaches of fluidized or moving bed. However, solids are inherently more difficult to work with than liquids and no large scale system has yet been commercialized. In this study, we developed the rigorous 1-D PDE model for moving beds in Aspen Custom Modeler; the entire system consists of adsorbers, regenerators, and auxiliary equipments. The simulation result will be expected to compare with those of other post-combustion processes. We will deal with not only advantages of lower capital costs and power requirements but also problems associated with pressure drop and heat transfer.

[1] Figueroa JD, Fout T, Plasynski S, McIlvried H, Srivastava RD. Advances in CO2 capture technology--The U.S. Department of Energy's Carbon Sequestration Program. *International Journal of Greenhouse Gas Control*. 2008;2(1):9-20.

[2] Ciferno JP, Fout T,E Jones AP, Murphy JT. Capturing Carbon from Existing Coal-Fired Power Plants. *Chemical Engineering Progress*, 2009;April:33-41

See more of this Session: Poster Session
See more of this Group/Topical: Computing and Systems Technology Division

AUTHOR INDEX

AUTHOR INDEX

AUTHOR INDEX

AUTHOR INDEX

AUTHOR INDEX

AUTHOR INDEX

AUTHOR INDEX

AUTHOR INDEX

AUTHOR INDEX

Computing and Systems Technology Division

Core Programming Topic at the 2011 AIChE Annual Meeting

Minneapolis, Minnesota, USA
16-21 October 2011

Volume 2 of 2

ISBN: 978-1-61839-731-7

Printed from e-media with permission by:

Curran Associates, Inc.
57 Morehouse Lane
Red Hook, NY 12571

Some format issues inherent in the e-media version may also appear in this print version.

Printed by Curran Associates, Inc. (2012)

AIChE
3 Park Avenue
New York, NY 10016-5991

Phone: (203) 702-7660
Fax: (203) 775-5177

www.aiche.org

Additional copies of this publication are available from:

Curran Associates, Inc.
57 Morehouse Lane
Red Hook, NY 12571 USA
Phone: 845-758-0400
Fax: 845-758-2634
Email: curran@proceedings.com
Web: www.proceedings.com

TABLE OF CONTENTS

OPTIMAL PLANNING AND SCHEDULING OF A BIOMASS CONVERSION SYSTEM CONSIDERING ECONOMIC AND ENVIRONMENTAL ASPECTS

José Ezequiel Santibañez-Aguilar[1], Janett Betzabe González-Campos[1], José María Ponce-Ortega[1], Medardo Serna-González[1] and Mahmoud M. El Halwagi[2]
[1]Universidad Michoacana de San Nicolás de Hidalgo, Morelia, Michoacán, México.
[2]Texas A&M University, College Station, Texas, USA.

Introduction

Currently, extensive research efforts have been made to achieve energy sustainability by replacing fossil fuels with renewable sources of energy. In this context, biomass is emerging as an option to phase out fossil fuel dependence, as well as to reduce environmental impact through the reduction of greenhouse gas emissions; furthermore, the biomass is a renewable resource that can be guaranteed through the time.

The available biomass mainly comes from the agro-industrial sector; such is the case of animal and vegetable residues. However, the production, cost and availability for the biomass is not constant through the year. This way, it is necessary to develop a new methodology for planning the production of a biomass conversion system considering the year-round availability of the bioresources to satisfy the demand with the maximum profit and at the same time with the lowest environmental impact.

This work presents a multiobjective optimization model for the optimal planning and scheduling of a biomass conversion system considering the selection of feedstocks, processing technologies, and products. It also takes into account seasonality aspects, since the availability and demand of bioresources and products are different for each season, central supply zone, storage and market. It is important to note, that this work uses network based models of states and tasks to make a good scheduling, in which the states are referred to different types of materials, for example bioresources, intermediate products and final products, and on the other hand, the tasks are the activities for the treatment of each state and they are time dependent.

The proposed multiobjective problem considers simultaneously the maximization and minimization of the economic and environmental objectives, respectively. The economic objective function involves the availability of bioresources, demand of products and processing limits, as well as the costs associated to the raw materials, products and processing, besides transport and storage costs. Whereas, the environmental objective function is measured through the life cycle analysis methodology by the assessment of the *Eco-Indicator 99,* in order to obtain a more objective measure of the environmental impact. Since both objectives contradict each other, a multiobjetive optimization strategy is proposed to identify the *Pareto* curve that describes the set of solutions that simultaneously satisfies the economic and environmental objectives. The proposed methodology is applied to a case of study for a biomass conversion system in Mexico, which allows to identify the technologies and the location as well as the best feedstocks to yield economical and environmental attractive solutions.

Model Formulation

The addressed problem is described as follows: Given a set of M available feedstocks that can be converted to K products using R routes available with α as a conversion factor, it considers a horizon of time H, this horizon is divided into several periods time T, since, the biomass and product are required in different amount depending on the time. The optimization process must select the feedstocks, products and processing routes to yield the solution with the maximum profit and the minimum environmental impact. The model is based on the representation shown in Figure 1.

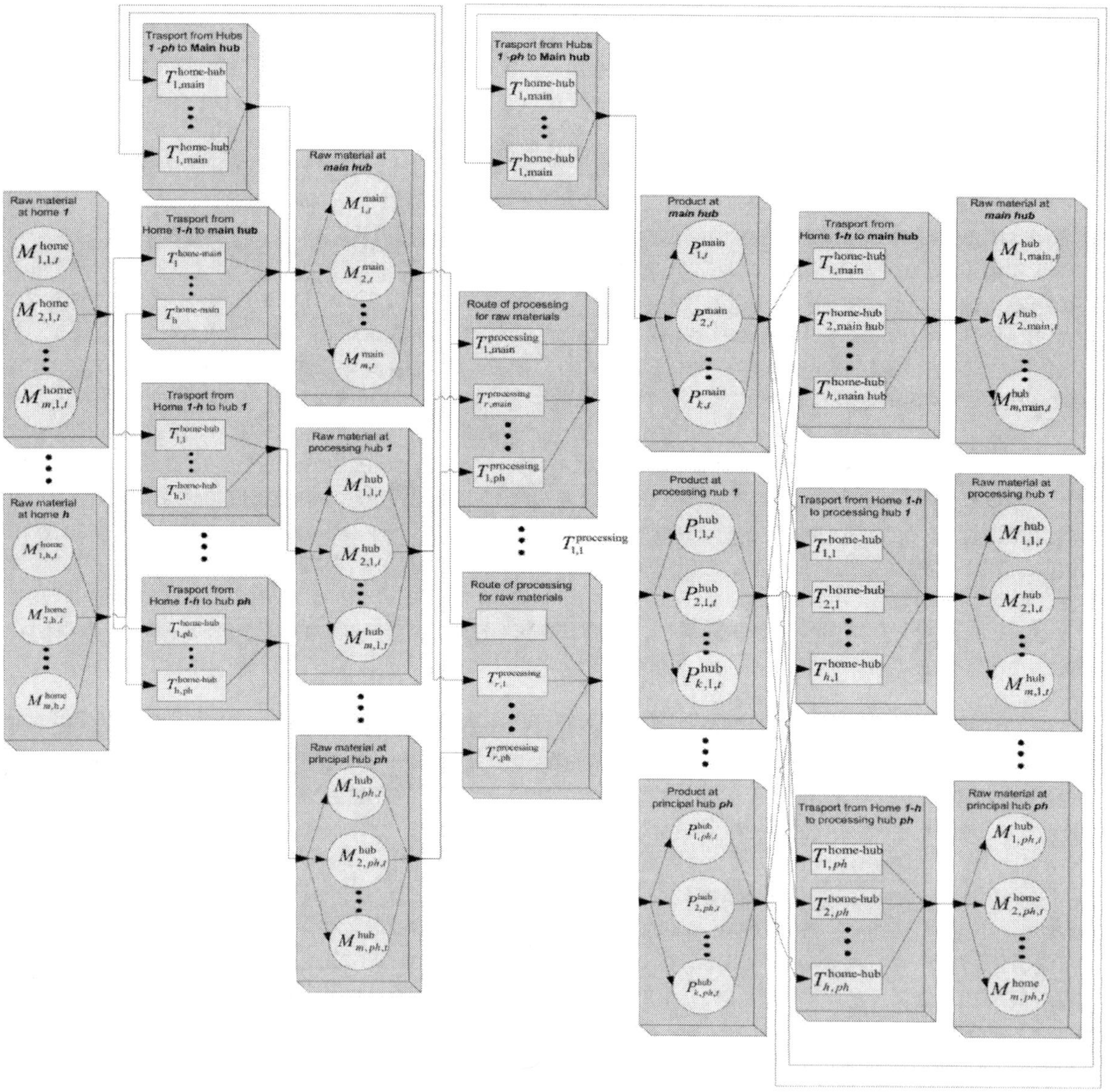

Figure 1. Superstructure for planning and scheduling of a biomass conversion system.

The model formulation can be stated as follows:

Mass balances

Mass balances are made for each place of the chain supply; they consider every activity done at the end of each time period. Each mass balance is based on the following general equation.

$$S^{l}_{s,l,t} = S^{prodrucida}_{s,l,t} + \sum_{l2} S^{l2-l}_{s,l,l2,t} + Se^{principal-l}_{s,l,t} - \sum_{l3} S^{l-l3}_{s,l,l3,t} - S^{l-principal}_{s,l,t} - S^{procesamiento}_{s,l,t} - S^{venta}_{s,l,t} + S^{l}_{s,l,t-1} \quad \forall s,l,t \tag{1}$$

Where $S^{l}_{s,l,t}$ is the amount of material *s* in the place *l* at the end of period time *t*, therefore, $S^{prodrucida}_{s,l,t}$ is the amount of material *s* yielded in the place *l* at the end of time *t*. On the other hand, $S^{l2-l}_{s,l,l2,t}$ and $Se^{principal-l}_{s,l,t}$ are the amounts of material *s* that go to/from *l2* and the main processing plant to *l* respectively, therefore, $S^{l-l3}_{s,l,l3,t}$ and $S^{l-principal}_{s,l,t}$ are the amounts of resource *s* that go to/from *l3* or the main processing plant to *l*. $S^{procesamiento}_{s,l,t}$ and $S^{venta}_{s,l,t}$ are the amounts of material *s* that is processed or sold.

Raw material balance at place of production

$$M^{origen}_{m,h,t} = M^{prodrucida}_{m,h,t} - \sum_{ph} M^{origen-planta}_{m,h,ph,t} - M^{origen-principal}_{m,h,t} + M^{origen}_{m,h,t-1} \quad \forall m,h,t \tag{2}$$

Raw material balance at processing hub

$$M^{planta}_{m,ph,t} = \sum_{h} Me^{origen-planta}_{m,h,ph,t} - \sum_{k}\sum_{r} M^{procesamiento}_{m,k,r,ph,t} - M^{planta-principal}_{m,ph,t} + M^{planta}_{m,ph,t-1} \quad \forall m,ph,t \tag{3}$$

Raw material balance at main processing plant

$$M^{principal}_{m,t} = \sum_{h} Me^{origen-principal}_{m,h,t} + \sum_{ph} Me^{planta-principal}_{m,ph,t} - \sum_{k}\sum_{r} M^{procesamiento-principal}_{m,k,r,t} + M^{principal}_{m,t-1} \quad \forall m,t \tag{4}$$

Product balance at processing hub

$$P^{planta}_{k,ph,t} = \sum_{m}\sum_{r} P^{producido}_{m,k,r,ph,t} - P^{planta-principal}_{k,ph,t} - \sum_{mk} P^{planta-mercado}_{k,ph,mk,t} + P^{planta}_{k,ph,t-1} \quad \forall k,ph,t \tag{5}$$

Product balance at main processing plant

$$P^{principal}_{k,t} = \sum_{m}\sum_{r} P^{producido-principal}_{m,k,r,t} + \sum_{ph} Pe^{planta-principal}_{k,ph,t} - \sum_{mk} P^{principal-mercado}_{k,mk,t} + P^{principal}_{k,t-1} \quad \forall k,t \tag{6}$$

Product balance at market

$$P^{mercado}_{k,mk,t} = \sum_{ph} Pe^{planta-mercado}_{k,ph,mk,t} + Pe^{principal-mercado}_{k,mk,t} - P^{vendido}_{k,mk,t} + P^{mercado}_{k,mk,t-1} \quad \forall k,mk,t \tag{7}$$

Storage

The model considers that it is possible to store material in different areas and at different periods of time, for example, raw material can be stored in the site of origin but not the product, the raw material and product can be stored in the processing plants and, finally, the final product can be stored in the market. These storages have fixed and variable costs, because the amount of the material stored in a period of time. If the storage is required in at least one period, it involves the capital cost for the purchase. The equations for the storages are stated in terms of disjunctions to consider the possibility of storing or not. The general disjunctions and the logical relationships for the selection of storage are shown as follows:

$$\begin{bmatrix} Ys_{s,l}^{\overset{l}{necesario}} \\ Cm_{s,l}^{l\;\overset{almacenamiento}{}} = Csm_{s,l}^{l} Sm_{s,l}^{l} + \sum_{t} Csm_{s} M_{s,l,t}^{l} \end{bmatrix} \vee \begin{bmatrix} \neg Ys_{s,l}^{\overset{l}{necesario}} \\ Cm_{s,l}^{l\;\overset{almacenamiento}{}} = 0 \end{bmatrix} \quad \forall s,l \tag{8}$$

$$\begin{bmatrix} Ys_{s,l,t}^{l} \\ S_{s,l,t}^{l} \le Sm_{s,l}^{l} \end{bmatrix} \vee \begin{bmatrix} \neg Ys_{s,l,t}^{l} \\ S_{s,l,t}^{l} = 0 \end{bmatrix} \quad \forall s,l,t \tag{9}$$

$$\sum_{t} ys_{s,l,t}^{l} \ge ys_{s,l}^{\overset{l}{necesario}} \tag{10}$$

$$(Max\ periods)\, ys_{s,l}^{\overset{l}{necesario}} - \sum_{t} ys_{s,l,t}^{l} \ge 0 \;\; \Rightarrow \;\; 12\, ys_{s,l}^{\overset{l}{necesario}} - \sum_{t} ys_{s,l,t}^{l} \ge 0 \tag{11}$$

Transport

Transport modeling was performed similarly to the storage, i.e., disjunctions are used to select the transport from one place to another place when it is needed. Thus, when material transport between two points is required, this involves a cost of transport as well as maximum and minimum limits. This material is received in the target site at a later time.

Processing

The processing involves additional disjunctions, which includes the type of processing that takes place in both, in the main processing plant and in the secondary processing plants. These two types of processing differ in the amount of raw material that can be treated as well as the processing costs, which are generally cheaper in the main processing plant. When the binary variables are activated, then processing occurs with a certain amount processed, which is bounded by upper and lower limits, thus obtaining a certain amount of product associated with a conversion factor of raw materials that depends on the processing route.

Objective Functions

This problem has two objective functions, which oppose each other, therefore a method to show the compromise between these objectives is required. First, the economic objective function is given by the annual net profit, which involves the economic benefit generated by product sales minus the cost of raw materials, processing cost of raw materials for conversion to products, transportation costs for raw material and products from the place of origin to processing plants, secondary plants to the main processing plant, or processing plants to the markets. Also, the objective function considers the material storage costs throughout the supply chain, for example from the place of extraction of raw materials to storage in the product market.

$$\begin{aligned} Profit = {} & \sum_{k}\sum_{mk}\sum_{t} Ck_{k,mk,t}^{vendido} P_{k,mk,t}^{vendido} - \sum_{m}\sum_{h}\sum_{t} Cm_{m,h,t}^{producida} M_{m,h,t}^{producida} - \sum_{m}\sum_{h} Cm_{m,h}^{\overset{almacenamiento}{origen}} - \sum_{m}\sum_{ph} Cm_{m,ph}^{\overset{almacenamiento}{planta}} - \sum_{m} Cm_{m}^{\overset{almacenamiento}{principal}} - \sum_{k}\sum_{ph} Ck_{k,ph}^{\overset{almacenamiento}{planta}} \\ & - \sum_{k} Ck_{k}^{\overset{almacenamiento}{principal}} - \sum_{k}\sum_{mk} Ck_{k,mk}^{\overset{almacenamiento}{mercado}} - \sum_{m}\sum_{h}\sum_{ph}\sum_{t} Ctm_{m,h,ph,t}^{\overset{origen-}{planta}} M_{m,h,ph,t}^{\overset{origen-}{planta}} - \sum_{m}\sum_{h}\sum_{t} Ctm_{m,h,t}^{\overset{origen-}{principal}} M_{m,h,t}^{\overset{origen-}{principal}} - \sum_{k}\sum_{ph}\sum_{mk}\sum_{t} Ctk_{k,ph,mk,t}^{\overset{planta-}{mercado}} P_{k,ph,mk,t}^{\overset{planta-}{mercado}} \\ & - \sum_{k}\sum_{ph}\sum_{t} Ctk_{k,ph,t}^{\overset{planta-}{principal}} P_{k,ph,t}^{\overset{planta-}{principal}} - \sum_{k}\sum_{mk}\sum_{t} Ctk_{k,mk,t}^{\overset{principal-}{mercado}} P_{k,mk,t}^{\overset{principal-}{mercado}} - \sum_{m}\sum_{k}\sum_{r}\sum_{ph}\sum_{t} C_{m,k,r,ph,t}^{procesamiento} M_{m,k,r,ph,t}^{procesamiento} - \sum_{m}\sum_{k}\sum_{r}\sum_{t} C_{m,k,r,t}^{\overset{procesamiento}{principal}} M_{m,k,r,t}^{\overset{procesamiento}{principal}} \end{aligned} \tag{12}$$

On the other hand, the environmental objective function is given by the overall environmental impact, measured by the Eco-indicator 99, which is based on the methodology of life cycle analysis. This approach considers the Eco-indicator 99 for the extraction of raw materials, for the disposal of products, for the raw material processing and for each type of transport.

$$
\begin{aligned}
EI = & \sum_{k}\sum_{mk}\sum_{t} EcoInd_{k}^{producto} P_{k,mk,t}^{vendido} + \sum_{m}\sum_{h}\sum_{t} EcoInd_{m}^{materiaprima} M_{m,h,t}^{producida} + \sum_{m}\sum_{h}\sum_{ph}\sum_{t} EcoIndm_{m,h,ph}^{\substack{origen-\\planta}} M_{m,h,ph,t}^{\substack{origen-\\planta}} + \sum_{m}\sum_{h}\sum_{t} EcoIndm_{m,h}^{\substack{origen-\\principal}} M_{m,h,t}^{\substack{origen-\\principal}} \\
& + \sum_{k}\sum_{ph}\sum_{mk}\sum_{t} EcoIndk_{k,ph,mk}^{\substack{planta-\\mercado}} P_{k,ph,mk,t}^{\substack{planta-\\mercado}} + \sum_{k}\sum_{ph}\sum_{t} EcoIndk_{k,ph}^{\substack{planta-\\principal}} P_{k,ph,t}^{\substack{planta-\\principal}} + \sum_{k}\sum_{mk}\sum_{t} EcoIndk_{k,mk}^{\substack{principal-\\mercado}} P_{k,mk,t}^{\substack{principal-\\mercado}} \\
& + \sum_{m}\sum_{k}\sum_{r}\sum_{ph}\sum_{t} EcoInd_{m,k,r,ph,t}^{procesamiento} M_{m,k,r,ph,t}^{procesamiento} + \sum_{m}\sum_{k}\sum_{r}\sum_{t} EcoInd_{m,k,r,t}^{\substack{procesamiento\\principal}} M_{m,k,r,t}^{\substack{procesamiento\\principal}}
\end{aligned} \quad (13)
$$

For the solution of the multiobjective optimization problem, the constraint method is applied to generate the set of Pareto optimal solutions that simultaneously compensate both objectives.

Example

To test the proposed methodology, a case of study to establish a biorefinery system in Mexico is used. This case involves 8 bioresources available in Mexico which can be used as feedstocks to obtain 2 products; furthermore, there are 5 different processing routes available. It is worth of notice that this problem considers the geographic situation of the central part of Mexico. On the other hand, this case of study considers 6 places of supply, 5 secondary processing hubs and 1 main processing plant, also, 5 markets are considered. These places were determined for the infrastructure in each place and the demand of products. The model formulation can be applied to any region anywhere. The efficiency factors for all processing technologies where taken from previous results shown by Santibañez-Aguilar et al. (2011). Additional required data such as feedstock, products and by-products costs, availability and demand were taken from governmental institutions of Mexico (i.e., SAGARPA, SENER, SE, SEMARNAT, SIAP). The schematic representation for the addressed problem can be seen in Figure 2.

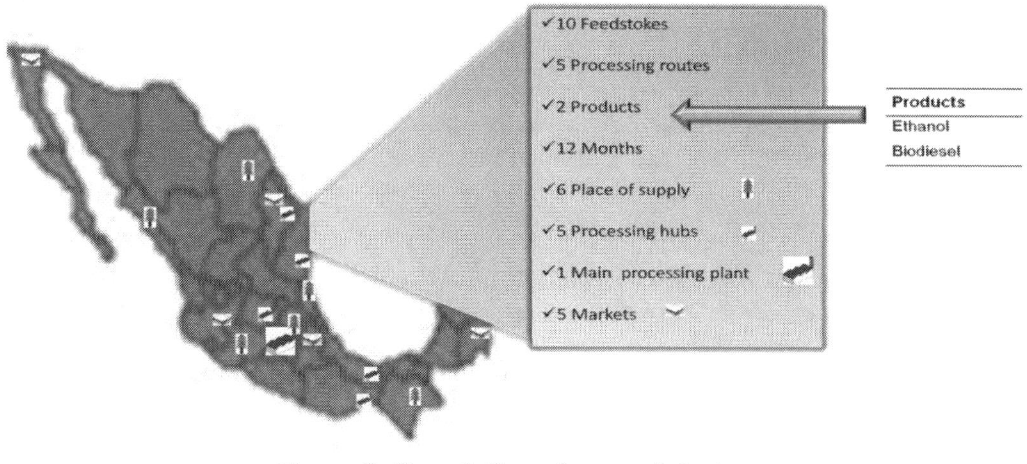

Figure 2. Description of case of study

The set of optimal solutions can be identified in Figure 3. Notice the significant gap between the minimum and maximum *EI* identified in the Pareto curve. It is worth of notice that the Pareto curve represents a set of optimal solutions that compensate both objectives simultaneously and from which the decision maker can take the one that best fits the specific requirements. In this case, we identify solution B (see Figure 3) that corresponds to about 50% of the maximum environmental impact and about 80% of the maximum profit.

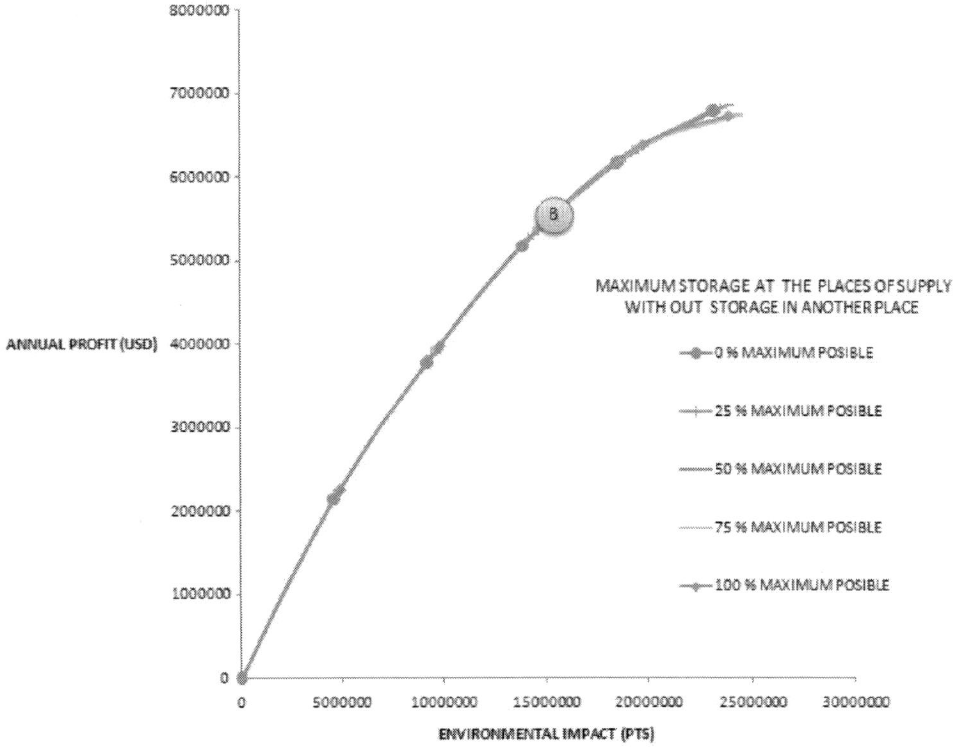

Figure 3. Pareto solution for the case studied

The point B represents a set of raw material, products, processing routes and places chosen for the horizon of one year. It is important to note that raw materials such as corn, wheat or sugar cane are not chosen, also, the processing hubs in Guanajuato and Salina Cruz are not selected. An important aspect is that the transport defines the amounts processed.

Conclusions

This paper presents a new mathematical formulation for the optimal planning of a biorefinery considering simultaneously the maximization of the total net profit and the minimization of the total environmental impact. The environmental impact is measured through the eco indicator-99 that is based on the life cycle analysis for the raw materials, processing and products. This paper presents an efficient method to adequately consider the objective function since these two objectives contradict each other. The resulting model is an

MILP problem that can be easily solved without numerical complications. The model is applied to a case of study for the planning and scheduling of the production of a biorefinery system in Mexico accounting for the specific bioresources available in this region. Our results show that the proposed model may be really helpful to solve the problem about biofuels production from food-grade feedstocks, since it does not consider this type of crops as part of the optimal solutions. Finally, the model is general and it can be applied to other different cases.

References

1. Aksoy B.; Cullinan H. T.; Sammons Jr. N. E.; Eden M.R. (2008), "Identification of optimal poultry litter biorefinery location in Alabama through minimization of feedstock transportation cost." *Environmental Progress.* 27(4), 515-523.

2. Sammons, Jr. N.; Eden, R. M.; Yuan, W.; Cullinan, H.; Aksoy, B. (2008), "A flexible framework for optimal biorefinery Product allocation." *Environmental Progress,* 26(4):349-354.

3. Cherubini, F.; Bird, N. D.; Cowie, A.; Jungmeier, G.; Schlamadinger, B.; Woess-Gallasch, S. (2009), "Energy and greenhouse gas-based LCA of biofuel and bioenergy systems: Key issues, ranges and recommendations." *Resources Conservation and Recycling.* 53(8), 434-447.

4. Van Dyken, S.; Bakken, B. H.; Skjelbred, H. I. (2010), "Linear mixed-integer models for biomass supply chains con transport, storage and processing." Energy. 35(3), 1338-1350.

5. Sammons, Jr. N. E.; Yuan, W.; Eden, M. R.; Aksoy, B.; Cullinan, H. T. (2008) "Optimal biorefinery product allocation by combining process and economic modeling." *Chemical Engineering Research and Design.* 86(7), 800-808.

6. Hugo, A.; Pistikopoulos, E. N. (2005), "Environmentally conscious long-range planning and design of supply chain networks." *Journal of Cleaner Production.* 13(15), 1471-1491.

7. Elms, R. D.; El-Halwagi, M. M. (2009), "Optimal scheduling and operation of biodiesel plants con multiple feedstokes." *International Journal of Process Systems Engineering.* 1(1), 1-28.

8. Castro, P. M.; Grossmann, I. E. (2006), "An efficient MILP model for the short-term scheduling of single stage batch plants." *Computers and Chemical Engineering.* 1003-1018.

9. Méndez, C. A.; Cerdá, J.; Grossmann, I. E.;Harjunkoski, I.; Fahl, M. (2006), *Computers and Chemical Engineering.* 913-946

10. Santibañez-Aguilar, J. E.; González-Campos, J. B.; Ponce-Ortega, J. M.; Serna-González, M.; El-Halwagi, M. M. (2011), "Optimal planning of a biomass conversion system considering economic and environmental aspects." *Industrial and Engineering Chemistry Research.* 50 (14): 8558-8570.

11. Dogan, M. E.; Grossmann, I. E.; Wassick, J. (2008) "Short-Term Scheduling of Batch Plants with Parallel Reactors Forming Mobile, Workgroups." *Industrial and Engineering Chemistry Research.*

12. Maravelias, C. T.; Grossmann, I. E. (2008), "New General Continuous-Time State-Task Network Formulation for Short-Term Scheduling of Multipurpose Batch Plants." *Industrial and Engineering Chemistry Research.* 3056-3074.

13. Wei, L.; Pordesimo, L. O.; Igathinathane, C.; Batchelor, W. D. (2009), "Process engineering evaluation of Ethanol production from wood through bioprocessing and chemical catalysis." *Biomass and Bioenergy.* 33(2), 255-266.

28350

Process Design: Principles and Methods

Process Design Project

Group 11

Kate Harboe, s072208

Alessandra Pennati, s101275

Yu Jiang, s102309

May 24, 2011

DTU Chemical Engineering

Department of Chemical and Biochemical Engineering

Preface

This design project has been conducted in the MSc course 28350 Process Design: Principles and Methods under the supervision of Professor Rafiqul Gani in the spring semester 2011 at DTU. The design project covers a conceptual design of a chemical process of ethanol production and is divided into 12 sequential tasks that include all stages from preliminary calculations to detailed process simulations, equipment sizing, costing and economic evaluation. The design project also includes heat and process optimization and environmental impact evaluation.

Kate Harboe, s072208

Alessandra Pennati, s101275

Yu Jiang, s102309

Abstract

Ethanol is a very important chemical as it is a very common solvent used in industrial processes and widely used as an alternative fuel. It is expected that the importance of ethanol will increase and that ethanol consumptions will grow significantly in the coming years because of its energy and environmental advantages compared to gasoline.

A plant for producing 190 proof ethanol from ethylene is designed by applying a systematic method. The process design work is divided into 12 sequential tasks that include selection of process alternatives, development of a base case followed by economic and environmental evaluations and improvements of the base case through heat integration and optimization. This procedure can be applied to design or analysis of any other chemical process.

The base case selected is direct catalytic hydration of ethylene at high temperature and pressure, which is currently the most widely used process in industry. Generation of diethylether from ethanol also occurs as a side reaction. The ethylene feed contains impurities of methane and propylene. Both the byproduct and the impurities make the downstream separation of the product quite complex. Separation is done by a sequence of distillation units and ethanol at azeotropic composition is obtained as the product. First information about the product and process was obtained in tasks 1-2. A preliminary process flow sheet was obtained in task 3 using a modified Douglas-approach for flow sheet synthesis. Next, design decisions related to separation factors, reactor operating conditions, product purity, etc., were made and further refined in moving from tasks 4-7. In tasks 4 mass balance is performed. After operating conditions are set in task 5, energy balances are performed in tasks 6-7. All simulations are made with PRO/II. Tasks 8-9 make the sizing and economic evaluation calculations. At this point, the base case design is obtained, which is then further refined and improved with respect to heat integration and process optimization (tasks 10-11). In the final task 12, the environmental impact of the process design is evaluated together with some of the key sustainability measures. In addition to PRO/II, the following software are used: ICAS (for property prediction, environmental analysis) and ECON (cost and economic analysis).

It is estimated that the annual profit of base case is 36 million US$ with a payback time of 3 years for a production rate of 72,360 metric tons/yr. The capital and operating costs are divided into the individual cost items. It is found that compressors are the most expensive equipment (59.3% of the total purchased equipment cost) and that raw material accounts for 65.1% of the total operating cost, while steam represents the 39.5% of utility cost. This information is then used as target for process improvement by heat integration and process optimization, which then further increases the annual profit to 42 million US$. The environmental impact analysis identifies impact due to the release of chemicals. The 12-tasks design method helped to perform the process design related work systematically and efficiently, where first design decisions were made based on collected information and then verified through simulations.

Contents

List of Figures

List of Tables

Introduction

In this project the objective is the conceptual design of a process of production of ethanol from ethylene. There is 75 million kg/year of ethylene available as a raw material. This ethylene is not 100% pure since it contains 96.1% ethylene, 3.0% propylene and 0.9% methane by mole. The use of this raw material is evaluated with respect to production of the most common products obtained from it to prove that production of ethanol is profitable and competitive with other process alternatives. If this is proven to be true, the process is designed.

Twelve tasks in a specified sequence are involved in the design, from the consideration of qualitative aspects of the process and preliminary calculations to detailed process simulation, equipment sizing, and evaluation analysis of the process in terms of economy, efficiency operational and environmental impact. Improvements to the base case are then suggested in the form of heat integration and process optimization.

1 Task 1

In this section information about the product, ethanol, is collected, including applications, production, market and properties.

1.1 Applications

Ethanol has been consumed for many years. Depending on its water content, preparation, and final use, several ethanol products exist in the market. They can be classified into three categories: fuel use, industrial use and beverages ethanol [7, 8].

Fuel Use The largest use of ethanol is as a motor fuel and fuel additive used to operate vehicle engines. The market for this sector has the highest growth potential. Many countries have attempted to replace some of their gasoline consumption by mixing it with ethanol in a low-level blend (E10, 10% ethanol and 90% gasoline) or in higher percentages as an alternative fuel (E85, 85% ethanol and 15% gasoline). The ethanol mixed with gasoline must be anhydrous. Hydrous ethanol (about 95% ethanol and 5% water) can be used as fuel if it is used alone. The vast majority of ethanol for use as fuel is produced from starchy materials in a biological process. The largest national fuel ethanol industries exist in Brazil and U.S.

Industrial Use For industrial use, ethanol is mostly used as chemical in industrial processes for other organic compounds as can be seen in Figure 1. These include ethyl halides, ethyl esters, diethylether, acetic acid, ethyl amines and to a lesser extent butadiene. In addition, 99.9% alcohol (often referred to as absolute alcohol) is used extensively for tinctures and pharmaceutical preparations, as a solvent and preservative, as an antiseptic, and in perfumes.

Beverages ethanol The market of ethanol for beverage is the smallest and has the lowest growth among the three because alcohol beverages can be produced from fruits such as grapes and malt as well as from ethanol. The ethanol content of a beverage is usually measured in terms of the volume fraction of ethanol in the beverage, expressed either as a percentage or in alcoholic proof units. China is the main market for ethanol beverages followed by Russia, U.S., Brazil and Japan.

1.2 Production and market

Ethanol is industrially produced mainly in two variants depending on the water content, namely 190 proof and 200 proof, which contain 95.6 wt% and 99.9 wt% ethanol, respectively. In this project we use the U.S. definition of proof: the ethyl alcohol content of a liquid at 60 degrees Fahrenheit, stated as twice the percent of ethanol by volume [9]. Ethanol is synthesized by two main industrial processes: a biological process, which involves the fermentation of carbohydrates contained in biomass, and a petrochemical process, specifically

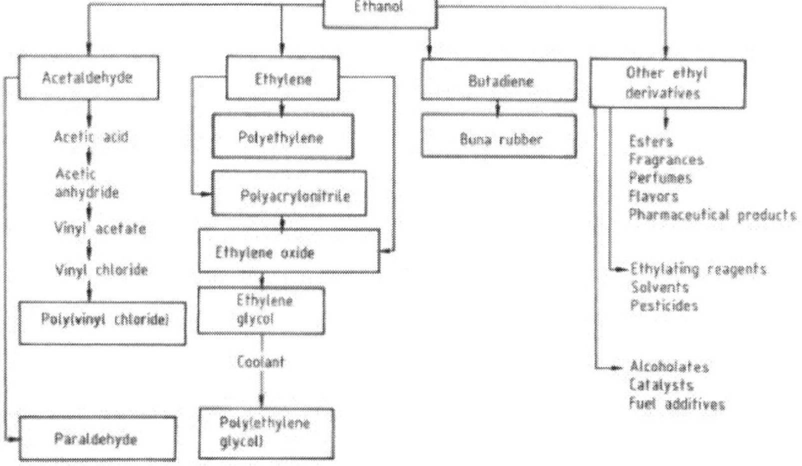

Figure 1 – Chemicals produced from ethanol [1].

by ethylene hydration. The larger portion of ethanol world production is based on biological processes. Most of the ethanol produced is 190 proof, while a purer product can be obtained by further refining it depending on the market demand. The ethanol market is wide and in continuous growth. In 2009 world ethanol production has reached a volume of around 20 billions gallons; U.S. and Brazil are the main producers and they account for around 90% of the total world production. Figures of the global world production volumes and of the most important producing countries in 2009 are listed in Table 1.

Production and consumption of ethanol has grown significantly in the past years all over the world. Between 2000 and 2007 the number of ethanol plants more than doubled and production capacity tripled in the U.S. The main factors driving the ethanol market growth are environmental and energy concerns. Studies have estimated that ethanol and other biofuels could replace 30% or more of U.S. gasoline demand by 2030. The use of ethanol as a fuel increases energy security and availability because domestically production leads to a reduction in petroleum imports, protecting the economy from unstable markets and growing prices and reducing the dependency from foreign countries. Furthermore, if obtained from biomass, ethanol production supports the agricultural sector because ending feedstock can be employed in its production. Thus, government policies encourage the production and the consumption of ethanol, especially in the U.S. [3].

Market researches show that ethanol production will continue to grow in the future [10, 2]. The trend forecast according to the recent OECD-FAO Agricultural Outlook 2010–2019 [2] is reported in Figure 2, which also reports the biological raw materials from which ethanol is obtained. Thus, ethanol market is huge and very dynamic and promising.

Table 1 – World ethanol production by country [3].

Country	Production (Mill. U.S. Gal)		
	2007	2008	2009
U.S.	6,499	9,000	10,600
Brazil	5,019	6,472	6,578
Europe	570	734	1,040
China	486	502	542
Canada	211	238	291
Thailand	79	90	435
Colombia	75	79	83
India	53	66	92
Australia	26	26	57
Other	82	128	247
World	13,101	17,335	19,965

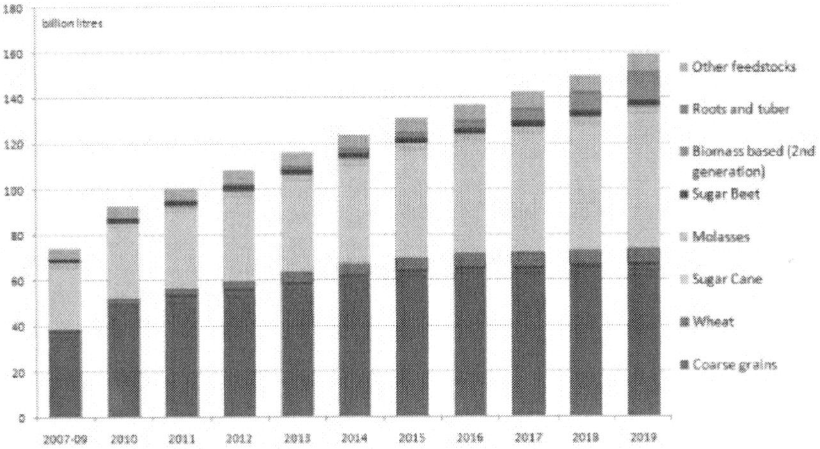

Figure 2 – World ethanol production forecast and sources [2].

1.3 Physical properties

For the design of the process of production of ethanol it is important to know information about the physical properties to determine phase conditions. The properties of anhydrous ethanol are listed in Table 2.

Table 2 – Properties for anhydrous ethanol.

Property	Value	Unit	Reference
CAS no.	64-15-5	-	[11]
Molar weight (MW)	46.069	g/mol	[6]
Melting point (T_m)	159.05	K	[6]
Boiling point (T_b)	351.44	K	[6]
Critical temperature (T_c)	513.19	K	[6]
Critical pressure (P_c)	60.68	atm	[6]
Density at 25°C (ρ)	17.06	kmol/m^3	[6]
Viscosity at 25°C (η)	1.074	mPa s	[12]
Enthalpy of formation of gas ($\Delta H^\circ_{\text{f,gas}}$)	-234 ± 2	kJ/mol	[13]
Enthalpy of formation of liquid ($\Delta H^\circ_{\text{f,liquid}}$)	-276 ± 2	kJ/mol	[13]
Enthalpy of vaporization (ΔH_{vap})	38.56	kJ/mol	[13]

In its pure form ethanol is a colorless liquid and is miscible with water. However, ethanol and water form an azeotrope which has a minimum boiling point at 78.2 °C at atmospheric pressure and whose mass and molar compositions are shown in Table 3.

Table 3 – Ethanol-water azeotrope mass and molar compositions [4].

	EA	W
Weight composition	0.956	0.044
Molar composition	0.895	0.105

2 Task 2

The characteristics and the feasibility of the possible process alternatives are investigated. Specifically, the production of ethanol from ethylene (base case) is compared with the production of a different product from the same raw material and with another method of production based on a different raw material. Approximate economical evaluations are performed in order to discard the less profitable alternatives on the basis of the maximum potential profit.

2.1 Synthesis of ethanol from ethylene

Ethanol can be produced from hydration of ethylene. Commercial production of synthetic ethanol on an industrial scale was first accomplished in 1930 by the Carbide and Chemical Corporation (now Union Carbide). The process used involved absorption of ethylene in sulfuric acid and subsequent hydrolytic cleavage of the resulting ethyl sulfates [1]. This indirect hydration of ethylene is now almost entirely obsolete. Its main drawbacks are connected with corrosion problems created by the use of sulfuric acid and the high cost of reconcentrating the acid, which counts for the most significant part of the capital expenditure [14]. The indirect process has been superseded by direct catalytic hydration of ethylene, introduced by the Shell Chemical Company in 1948. Here we focus on the direct hydration.

2.1.1 Process description

Chemistry In the temperature range at which industrial catalysts operate, the direct hydration of ethylene to ethanol is a reversible reaction controlled by equilibrium:

$$C_2H_4(g) + H_2O(g) \rightleftharpoons C_2H_5OH(g) \quad \Delta H = -43.4\,kJ \tag{2.1}$$

Only about 5-7% of the ethylene is converted into ethanol at each pass through the reactor. The high exothermicity of this reaction, which is accompanied by a decrease in the number of moles, would justify operation at low temperature and high pressure. In practice, however, temperatures between 250-300°C and pressures in the range 5-8 MPa are required to reach acceptable reaction rates [1].

The reaction is accompanied by reversible formation of diethylether according to reaction (2.2).

$$2\,C_2H_5OH \rightleftharpoons C_2H_5OC_2H_5 + H_2O \tag{2.2}$$

A lower temperature favors higher ethylene conversion but enhances reaction (2.2). At higher pressures, ethylene polymerizes to yield butenes and higher olefins, which are converted to higher alcohols by hydration.

Both the desired reaction and the combination of ethanol are equilibrium controlled, thus Le Chatelier's principle applies. The most obvious choice would be to operate in excess of water in order to shift the equilibrium of the reactions in the desired direction, that is towards ethanol. However, excess of steam would cause damage to the catalyst, because too much steam would dilute the catalyst and can even wash it off the support [14, 15]. Therefore, in practice an excess of ethylene is used and the molar ratio between water and ethylene is between 0.6 and 1 [14, 16].

Catalysts The technical and patent literatures describe numerous catalysts for ethylene hydration. Most of them are acidic because the reaction involves carbocations. However, only phosphoric acid catalysts supported by diatomaceous earth (kieselguhr, Celite), montmorillonite, bentonite, silica gel, or opoka (Volga sandstone) are of industrial importance [1].

Production process The production process of direct hydration involves reaction and several steps of separation. The basic flow sheet is shown in Figure 3.

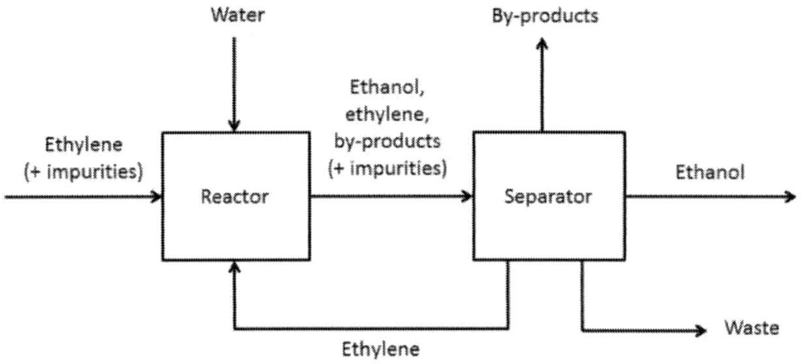

Figure 3 – Basic flow sheet of direct hydration of ethanol.

According to the process description in literature the ethanol-water feed reacts in a packed bed reactor [1, 14]. The vapor leaving the reactor is partly condensed by heat exchange with the feed, and scrubbed by a dilute caustic solution to neutralize all traces of phosphoric acid. The gaseous effluent is again cooled in a second heat exchanger and a water/ethanol fraction is thus condensed. After scrubbing with water and purge, the off-gases are compressed and recycled. Crude product is collected at the bottom and contains ca. 10-25 wt% ethanol. It is then decompressed to recover the dissolved ethylene, which is recycled to the reactor. The ethanol is then purified, usually by extractive distillation followed by rectification, to obtain a 95 wt% ethanol-water azeotrope. The mixture can be further dehydrated by azeotropic distillation to give anhydrous ethanol.

2.2 Alternative product from ethylene

The main alternative use of ethylene is the manufacturing of polymers, where polyethylene makes up the largest share and is the most widely used plastic

[17]. Another use of ethylene is the production of ethylene oxide which is mainly employed to make ethylene glycol, useful to make polyester fibers for textiles applications. Table 4 shows the world consumption of ethylene in 2000 with the relevant prices.

Table 4 – World consumption of ethylene in 2000 and product prices.

Use	World [1] [%]	Price [18] [$/lb]
High density polyethylene	26	0.63-0.89
Low density polyethylene	18	0.79-0.97
Linear low density polyethylene	14	0.64-0.81
Ethylene oxide/Ethylene glycol	19	0.74-0.79 / 0.65
Ethylene dichloride	14	0.16
Others	9	-

From an economic perspective only, it would be most reasonable to produce polymers since the highest price is achieved here [18]. However, the price ranges are close to each other, but as it is seen low density polyethylene can be sold to the highest price. Based on this, it is decided to look at the production of low density polyethylene as an alternative product to ethanol.

2.2.1 Process description

The polymerization of ethylene leads to polyethylenes of different molecular weights. Low density polyethylene is produced by free-radical bulk polymerization using traces of oxygen and peroxide as initiators [19]. The overall reaction scheme for polymerization of ethylene is shown below.

$$\mathrm{n\,CH_2{=}CH_2 \longrightarrow -[CH_2{-}CH_2]_n-} \tag{2.3}$$

The production of polyethylene takes place in a tubular reactor. The polymerization is started by adding initiators to the monomer. The radical polymerization reaction is highly exothermic and different cooling cycles are used to remove heat. The conversion achieved is often only about 25-30 % and hence, the unreacted monomer is separated and recycled [20]. A flow sheet of the overall process is seen in Figure 4.

2.3 Alternative raw material for production of ethanol

Ethanol can be produced from a variety of raw materials. Considering chemicals of petroleum derivation, the main production route consists in the catalytic hydration of ethylene, but ethanol can also be obtained via ethylene or acetylene, from calcium carbide, coal, oil gas and other sources. However, the other main industrial process for large scale production which has proven to be viable is a biological process involving fermentation.

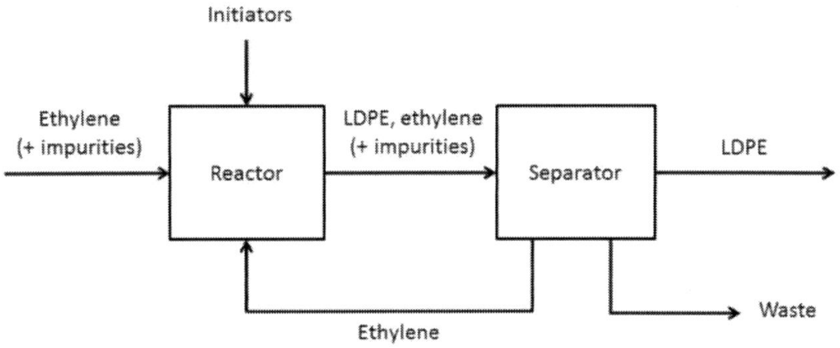

Figure 4 – Flow sheet of the overall polymerization process.

2.3.1 Process description

Fermentation is a metabolic process consisting of a series of chemical reactions that convert sugars into ethanol and carbon dioxide thanks to yeasts. The simplified fermentation reaction equation is written for glucose as follows:

$$C_6H_{12}O_6 \longrightarrow 2\,CH_3CH_2OH + 2\,CO_2 \tag{2.4}$$

Prior to fermentation, glucose must be obtained from biomass materials which contain high levels of glucose or precursors that can be converted into it with hydrolysis (carbohydrates such as cellulose and starch). After the fermentation the product is distilled and eventually dehydrated. A simple flow sheet of the overall process is shown in Figure 5.

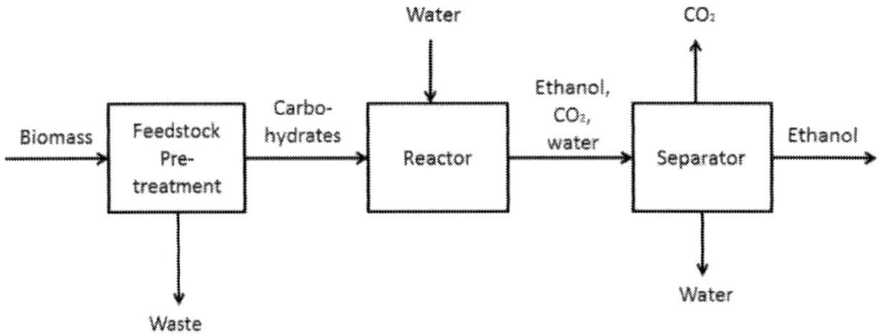

Figure 5 – Flow sheet of the overall fermentation process.

Almost any plant-based material or waste can be feedstock for the process. Some feedstocks are easier to process into ethanol than others. The treatment of highly cellulosic feedstocks involves more difficulties and the process is still under development. Thus, the feedstocks mainly used are biomasses rich in sugars and starch, such as sugar cane or corn and wheat. Corn is the feedstock for more than 90% of current U.S. ethanol production while Brazil mainly uses sugar cane [21].

The content of carbohydrates is highly variable from sample to sample requiring a high flexibility of the process and leading to uncertainties in the revenue. A number of database exists which report the composition of a wide range of biomass products [21, 22]. Ash and lignin cannot be converted into ethanol by fermentation while the other carbohydrates can. Thus, based on the data collected in database [21], for both corn stover and sugarcane the percentage of useful mass is roughly around 65% of the total. Since cellulose is harder to covert completely into simple sugar and some of it is used by organisms, a reasonable estimation of the mass converted into glucose for rough computations could be around 50% of the dry feedstock mass.

2.4 Economic evaluation

A rough economic evaluation is performed for the processes presented above. The target product is chosen to be 190 proof ethanol. The chosen product itself has a big market and furthermore it can be further purified to 200 proof ethanol by adding an extra separation unit to the original process or selling it to companies making 200 proof ethanol. The maximum potential profit is computed as the difference between the revenue from the sale of the product and the cost paid for the raw material. Issues concerning the fixed and operating costs of the process are not yet considered.

2.4.1 Base case

For simplification, total conversion of ethylene in ethanol is assumed and the effect of impurities is neglected. Once the composition and the quantity of the feed are known, it is possible to compute the quantity of the product. The results are shown in Table 5.

Table 5 – Economic potential of ethanol production by ethylene hydration.

Description	Price [18] [$/kg]	Quantity [kg/yr]	Price x Quantity [$/yr]
96.1% mol ethylene	0.4	75,000,000	30,000,000
190 proof ethanol	1.14	124,875,590	142,156,374

Thus, the economic potential (EP) is given by:

EP=142,156,374 $/yr - 30,000,000 $/yr =112,156,374 $/yr

2.4.2 Alternative product from ethylene

For simplification, total conversion of ethylene into LDPE is assumed. The molecular weight of low density polyethylene is typically between 6,000 and 40,000. For the economic evaluation it is assumed that polyethylene with a molecular weight of 20,000 is produced. The effect of impurities is neglected. Once the composition and the quantity of the feed are known, it is possible to compute the quantity of the product. The chain length n is computed as $n = \frac{MW_{\text{LDPE}}}{MW_{\text{ethylene}}}$; thus, the stoichiometric relationship between reagent and product is provided. The results are shown in Table 6.

Table 6 – Economic potential of ethylene polymerization into LDPE.

Description	Price [18] [$/kg]	Quantity [kg/yr]	Price x Quantity [$/yr]
96.1% mol ethylene	0.4	75,000,000	30,000,000
LDPE	1.93	71,280,048	137,378,948

Thus, the economic potential is given by:

EP=137,378,948 $/yr-30,000,000 $/yr =107,378,948 $/yr

2.4.3 Alternative raw material for production of ethanol

For simplification, 50% conversion of the feedstock biomass into glucose mass is assumed, and conversion of glucose into ethanol and carbon dioxide is assumed to be 100%. The effect of impurities is neglected. Assuming that the aim is to produce the same amount of 190 proof ethanol that could be obtained in the base case process, the necessary feed can be computed by stoichiometric computation. The results are shown in Table 7.

Table 7 – Economic potential of ethanol production by fermentation.

Description	Price [$/kg]	Quantity [kg/yr]	Price x Quantity [$/yr]
Corn	0.284 [23]	457,818,663	130,047,969
190 proof ethanol	1.14 [18]	124,875,590	142,156,374

Thus, the economic potential is given by:

EP=142,156,374 $/yr-130,047,969 $/yr =12,108,405 $/yr

As it is evident from the comparison with the result obtained for the base case, this way of producing ethanol would be much less profitable. The main reason why the process is unprofitable is the high quantity of biomass necessary to produce ethanol. Even if a conversion of 65% is assumed, the revenue does not improve significantly. A solution of this type would be profitable when coupled with an agricultural complex, where raw material is not to be bought but is available at no expense.

2.4.4 Conclusion

On the basis of this economic evaluation, the production of ethanol from ethylene has the highest economical potential both compared with an alternative use of the feed stream (the production of LDPE) and with a competitive production route(producing ethanol from biomass). The reason why the production from biomass is unprofitable is the high quantity of biomass necessary to produce ethanol. Even if a conversion of 65% is assumed, the revenue does not improve significantly. A solution of this type would be profitable when coupled with an agricultural complex, where raw material is not to be bought but is available at no expense. From the economic point of view, the process selected as a base case seems to be better than the other alternatives.

3 Task 3

The selected process is now analyzed in more detail. The process conditions are examined and all the relevant substances involved in the process are identified along with their physical properties. On this basis, decisions are made following hierarchical levels in order to better define the base case.

3.1 Process conditions and physical property

3.1.1 Process conditions

The core of the process is the reactor in which the hydration of ethylene into ethanol takes place. The reaction conditions and the chemicals reacting influence the subsequent process development and conditions, therefore it is the starting point of the analysis. We choose to operate the reactor according to temperature and pressure in agreement with what is reported in literature, namely T=300°C and P=68 atm. Likewise, the molar ratio between ethylene and water is set to 0.6. The desired and side reactions that take place in the reactor are:

$$C_2H_4(g) + H_2O(g) \leftrightharpoons C_2H_5OH(g) \tag{3.1}$$

$$2\,C_2H_5OH \leftrightharpoons C_2H_5OC_2H_5 + H_2O \tag{3.2}$$

The stream of ethylene fed to the reactor is impure; its composition is given in Table 8.

Table 8 – Ethylene feed composition (mole %).

Ethylene	Propylene	Methane
96.1	3.0	0.9

Thus, the substances present in the reactor at this stage would be ethylene, propylene, methane, water, ethanol and diethylether. All these chemicals can react with each other except methane that is inert. Some information about expected undesired and side reactions are reported [24]. Isopropylalcohol is formed by oxidation of propylene with water according to the following reaction:

$$C_3H_6 + H_2O \leftrightharpoons C_3H_7OH \tag{3.3}$$

It is also known that crotonaldehyde is formed in trace amount. Considering the chemistry of alcohols and ethers, it seems reasonable to assume that other compounds such as aldehydes and ketones can be formed only in trace amount, thus they are neglected. Therefore, the main reactions and chemical species present in the reactor are taken into account.

3.1.2 Physical properties

The physical properties of the chemical species in the feed and product streams are shown in Table 9, where the constants for the Antoine equation are also listed.

Table 9 – Properties for substances present in the process [5, 6].

Property	EA Ethyl-alcohol	W Water	EL Ethylene	DEE Diethyl-ether	M Methane	PL Propylene	IPA Iso-propanol	CA Croton aldehyde
Formula	CH_3CH_2OH	H_2O	$CH_2{=}CH_2$	$(C_2H_5)_2O$	CH_4	$CH_3CH{=}CH_2$	$CH_3CHOHCH_3$	$CH_3CH{=}CHCHO$
MW[g/mol]	46.07	18.02	28.05	74.12	16.04	42.08	61.10	70.09
T_m[K]	159.05	273.15	104	156.85	90.694	87.89	185.28	196.65
T_b[K]	351.44	373.15	169.41	307.58	111.66	225.46	355.41	375.15
T_c[K]	513.19	647.13	282.34	466.7	190.564	365.57	508.3	569
P_c [atm]	60.68	217.67	49.741	35.924	45.389	46.04	46.997	41.944
ρ^1[kg/m^3]	790.49	1000	1.178	714	0.68	514.4	768	846
η^1[mPa] s	1.20	1.00	-	0.22	0.20	-	1.96	-
ΔH_{vap}[kJ/mol]	38.56	40.65	13.544	-	8.17	14.39	43.40	-
VP^2 A	8.04	8.11	6.75	7.40	6.61	6.82	6.66	7.32
VP B	1554.30	1750.29	585.00	1391.40	389.93	785.00	813.06	1480.16
VP C	222.65	235.00	255.00	273.16	266.00	247.00	132.93	230.81

[1] at 20°C

[2] VP(mm Hg) $= 10^{A-B/(C+T(^\circ C))}$, VP is vapor pressure and T is temperature.

3.2 Decisions

The information about the reactor conditions and the substance present in the process are used to refine the flow sheet of the process. The procedure to select process alternatives is based on the guidelines suggested by Douglas and reported in "*Systematic Methods of Chemical Process Design*" by Biegler et al. [24].

3.2.1 Level 1: Batch vs. continuous

The quantity of raw material available, impure ethylene, is relatively high (75,000,000 kg/year) and can be all used to obtain the profitable product, ethanol. The product is likely to be profitable for a long time in the future, so the production will be in operation for several years. The process is already established in the industry, therefore there is no need to set it up quickly to take advantage of market factors. For the same reason, there is also no need of setting up a pilot plant to obtain the optimal operation conditions, since they are well known from common practice. Thus, a continuous process instead of a batch process is selected.

3.2.2 Level 2: Input–Output structure

The stream of ethylene to feed to the reactor is not pure, but propylene and methane are present as impurities in a small percentage as seen in Table 8. Their presence leads to the production of waste products so we could consider the option of separating them before the stream is fed to the reactor. In order to separate the impurities by distillation a cryogenic column would be necessary because of the low boiling points of the substances (see Table 9). A even lower temperature would be required for crystallization, so this alternative is discarded. We can think about a membrane separation for removing methane, but this is not worth because methane is inert and its amount is very low. However, since both impurities are in low quantity, we can feed them to the reactor.

The outlet stream of the reactor contains a number of compounds, namely the reactants (ethylene, water), the impurities present in the feed (propylene, methane), the desired product (ethanol), the by-product (diethylether) and the waste products (isopropanol).

The reactants must be recycled because the conversion in the reactor is very low. Since water is cheap and can be employed in high quantities without much expense, the possibility of recycling it is not taken into consideration at the moment. The impurities need to be purged possibly without recycling them to the reactor, or at least not in large amount, in order to avoid their build-up and also because the molar fraction of methane must not exceed 10% to prevent coking. The desired product must be recovered. There are two options regarding the by-product: it can be recovered if considered valuable for the market or it can be recycled to the reactor in order to shift the equilibrium towards ethanol. In any case, the separation of this chemical must be performed. We decide to recycle it to the reactor, so that at steady state reaction (3.2) would be at equilibrium. The waste products should be removed. Then, at first the flow sheet is as shown in Figure 6.

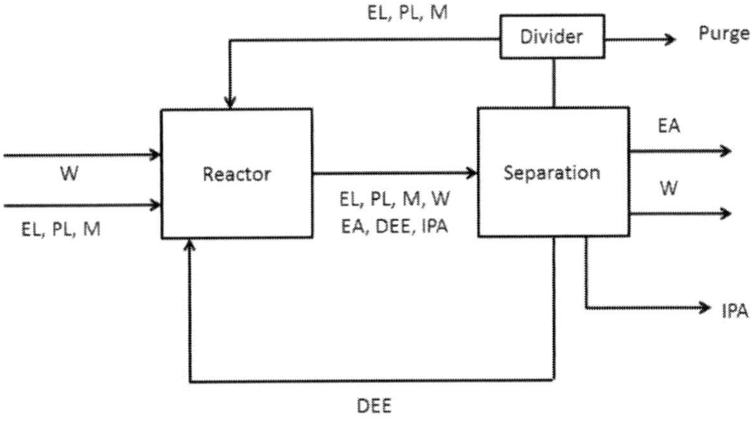

Figure 6 – Basic flow sheet.

3.2.3 Level 3: Reactor and recycles structure

Since we have only one desired reaction, we will have only one reactor. All the other reactions take place simultaneously at the same conditions.

The desired reaction is exothermic ($\Delta H_R = -43.4\,\text{kJ/mol}$); the effect of the other reactions on the energy release or consumption can be neglected because the concentration of the reactants involved is low compared to that of ethylene. The reaction is not highly exothermic and the conversion is low, so the temperature rise would not be very significant.

For this reason, in principle an adiabatic reactor could be used. However, an isothermal reactor seems more reasonable because in this way we can have controlled conditions that assure constant process parameters and minimize the possibility of parasite reactions that can start at high temperatures. Furthermore, the heat released from the reactor can be used for heating flows of the plant, for example for pre-heating the feed.

The number of components to recycle was already discussed in the previous section. Since the boiling point of diethylether is much higher than the one of ethylene and the impurities, we assume that there will be two different recycles paths.

3.2.4 Level 4: Separation system

The reactor outlet will be at the same temperature and pressure at which it is operated. In order to separate the components we need to lower the temperature. According to their boiling point, the components can be divided into two groups: condensible components, whose boiling point is higher than 273 K, and incondensible, whose boiling point is lower than 273 K. Since the boiling points of the two groups are very different, they can be separated using a flash; condensible components including ethylene, propylene and methane are vaporized while the incondensibles are found in the liquid phase. However, flash separation is not sharp and the vapor stream may contain some condensible species, thus it is better to add an absorber unit after the flash to remove these compounds from the gas stream before sending it to the recycle and purge. In this way, we recover ethanol and avoid recycling it to the reactor. Water is a very good solvent to be use in the absorber, since it is able to dissolve polar compound such as alcohols and ethers and it is cheap and environmentally friendly. The liquid stream from the absorber is then sent with the condensible stream obtained from the flash to condensible purification.

We would like to recycle the reactants to the reactor ethylene but not propylene and methane because we want to avoid the build-up of these impurities. Again, since both propylene and methane are in low quantity, it is not worth to separate them using membrane, adsorption or distillation. So, we decide to remove a fraction of the stream with a purge. The purge is combustible and can be used as a fuel if its amount is considerable or can be flared if it is small and environmentally safe. At this point, the flow sheet is as shown in Figure 7.

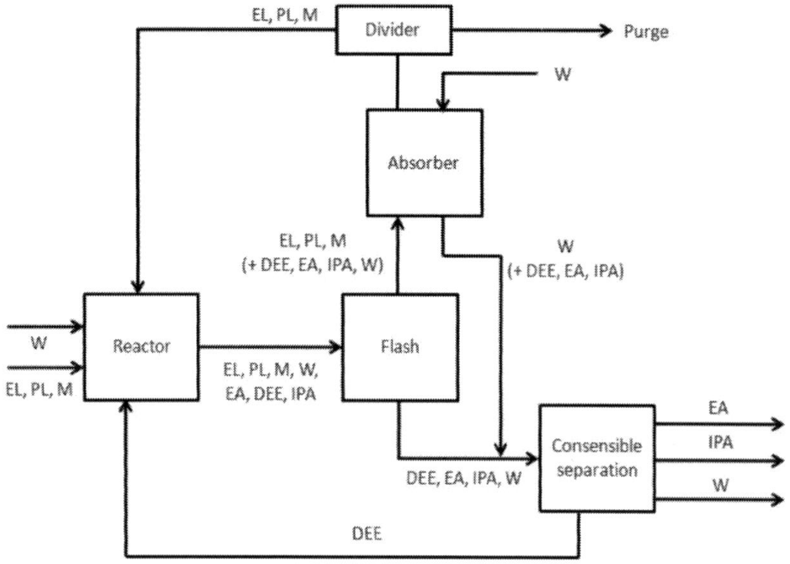

Figure 7 – Intermediate flow sheet.

In order to perform the condensible separation it is useful to arrange the compounds accordingly to their volatility, as shown in Table 10. Some organic compounds form azeotropes with water; the azeotropic temperature is also reported and is the one that we have to consider when selecting a separation method. The most effective separation is obtained if the split between the compounds is made between the two which have the highest difference in boiling point. However, in this case we prefer to make the first cut between ethanol

Table 10 – Normal boiling temperatures of compounds and their water azeotropes.

Compound	T_{NB} [K]	$T_{NB,az}$ [K] [12]
DEE	307.58	307.35
EA	351.44	351.25
IPA	355.41	353.55
W	373.15	-

and isopropanol, so that isopropanol can be partly discarded with water. Furthermore, water accounts for the biggest part of the stream and if we remove it at first we will have to deal with a smaller stream afterwards, with consequent lower costs for the separation units. The top product is sent to further separation units. The criteria to select the steps of the separation is that the easiest separation is made first, so we first separate DEE from EA and then EA from the remaining water; however the product stream will contain a certain amount of water because ethanol will be extracted in a concentration near to its azeotrope. The ethanol stream is sent to storage, while the diethylether stream is recycled to the reactor. A flow sheet summarizing the decisions made so far is presented in Figure 8.

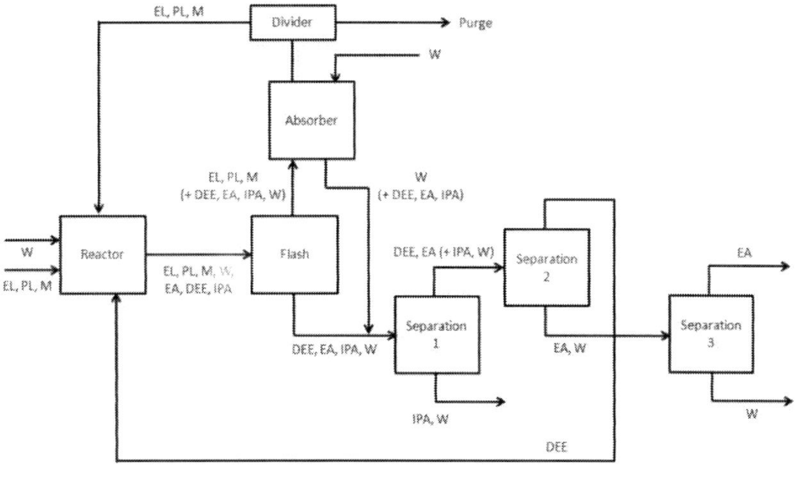

Figure 8 – Final flow sheet.

4 Task 4

In this task the mass balance for the process designed in Task 3 is carried out. The flow sheet developed and the information collected in the previous tasks will be applied to estimate the performance of the process.

The process analyzed has a number of unit operations, which can be described properly by complex and detailed models. However, we need some simple and fast methods to evaluate the performance of our selected flow sheet without too much expense of time and computation. In order to do this, linear mass balances are developed and solved. Both hand calculations and process simulations are used.

4.1 Linear mass balance equations

The flow sheet used for mass balance computations is given in Figure 9. A progressive number is associated with each unit operation. When more than one stream enters one unit, a mixer is introduced, which is also included in the progressive numeration. The streams are identified as μ_{ij}^k, which represents the molar flow rate of component k in the jth stream out of unit i. In order to have linear mass balance equations, all the separation units (flash, distillation columns, absorber) are substituted with a splitter with two outlet streams representing the vapor and the liquid outlets; ξ_{jk} is the recovery of component k in top stream of unit j. In this way, the outlet flows are linearly related to inlet flows.

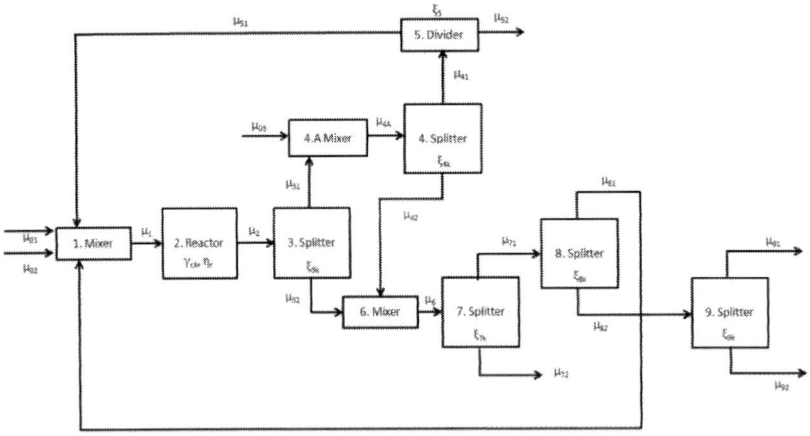

Figure 9 – Flow sheet for calculating mass balance.

The linear model approximations of the units in the process are:

1. Mixer

$$\mu_1^k = \mu_{01}^k + \mu_{02}^k + \mu_{51}^k + \mu_{81}^k \tag{4.1}$$

2. Reactor

$$\mu_2^k = \mu_1^k + \sum_{r=1}^{NR} \gamma_{r,k}\eta_r\mu_1^{l(r)} \tag{4.2}$$

where $\gamma_{r,k}$ represents the normalized coefficient of component k in reaction r, η_r is the conversion of the reference reactant $l(r)$ in reaction r and NR is the total number of reactions taking place in the reactor.

3. Flash

$$\mu_{31}^k = \xi_{3k}\mu_2^k \tag{4.3}$$

$$\mu_{32}^k = (1 - \xi_{3k})\mu_2^k \tag{4.4}$$

4A. Mixer

$$\mu_{42}^k = \mu_{03}^k + \mu_{31}^k \tag{4.5}$$

4. Absorber

$$\mu_{41}^k = \xi_{4k}\mu_{4A}^k \tag{4.6}$$

$$\mu_{42}^k = (1 - \xi_{4k})\mu_{4A}^k \tag{4.7}$$

5. Divider

$$\mu_{52} = \xi_5\mu_{41} \tag{4.8}$$

$$\mu_{51} = (1 - \xi_5)\mu_{41} \tag{4.9}$$

6. Mixer

$$\mu_6^k = \mu_{42}^k + \mu_{32}^k \tag{4.10}$$

7. Column 1

$$\mu_{71}^k = \xi_{7k}\mu_6^k \tag{4.11}$$

$$\mu_{72}^k = (1 - \xi_{7k})\mu_6^k \tag{4.12}$$

8. Column 2

$$\mu_{81}^k = \xi_{8k}\mu_{71}^k \tag{4.13}$$

$$\mu_{82}^k = (1 - \xi_{8k})\mu_{71}^k \tag{4.14}$$

9. Column 3

$$\mu_{91}^k = \xi_{9k}\mu_{82}^k \tag{4.15}$$

$$\mu_{92}^k = (1 - \xi_{9k})\mu_{82}^k \tag{4.16}$$

4.2 Degree of freedom analysis

The problem can be solved only if the system is completely and uniquely specified, that is, only if a number of parameters equal to its degrees of freedom are specified. To address this problem, we compare the number of equations of the model with the number of variables and define the number and the type of specifications to give.

Table 11 shows the total number of variables of the system and Table 12 shows the total number of equations available, with reference to the previous section. NC is the total number of components and NR is the total number of reactions.

Table 11 – Number of variables.

Variables	Number
μ_{ij}^{k}	17NC+2
$\gamma_{r,k}$, η_r	NC·NR+NR
ξ_{jk}	5NC
ξ_5	1
Total	22NC+NC·NR+NR+3

Table 12 – Number of equations.

Equations	Number
Eq.4.1	NC
Eq.4.2	NC
Eq.4.3	NC
Eq.4.4	NC
Eq.4.5	NC
Eq.4.6	NC
Eq.4.7	NC
Eq.4.8	1
Eq.4.9	1
Eq.4.10	NC
Eq.4.11	NC
Eq.4.12	NC
Eq.4.13	NC
Eq.4.14	NC
Eq.4.15	NC
Eq.4.16	NC
Total	14NC+2

If we define NV as the total number of variables and NE as the total number of equations, the degrees of freedom of the system are given by:

$$\text{DOF} = \text{NV–NE} = 8\text{NC}+\text{NC}\cdot\text{NR}+\text{NR}+1$$

Thus, in order to solve the mass balance, we need to specify 8NC+NC·NR+NR+1 variables, which in our specific case, given NC=7 and NR=3, is 81 variables. Table 13 shows how the degrees of freedom are saturated.

Table 13 – Number of variables specified.

Variables to specify	Number
Reactor parameters ($\gamma_{r,k}$, η_r)	NC·NR+NR
Stream calculator parameters (ξ_{jk})	5NC
Divider parameter (ξ_5)	1
Input flow rate (μ_{0j}^k)	3NC
Total	8NC+NC·NR+NR+1

The reactor parameters are chosen on the basis of what is known of the reactions (see section 4.3.1). The stream calculator parameters are computed on the basis of simplified models of unit operations (see section 4.3.2, 4.3.3, 4.3.5). The divider parameter is associated with the purge fraction, and it is chosen in order to satisfy the constraints on the inert concentration in the reactor (see section 4.3.4). The input flow rates are given, since μ_{01}^k is our ethylene feed stream (see Table 8), μ_{02}^k is the water make-up which can be calculated on the basis of μ_{01}^{EL} and μ_{03}^k is the absorbing water, known once a model for the absorber is defined.

4.3 Unit models

In order to perform mass balance computations, the knowledge of the split of each component in the different outlet streams of a unit is required. To obtain them, a model for the considered unit is needed. The models used are described in detail in Appendix A and B. The recoveries obtained from the models are shown in this section. Here the basis of calculation is 100 kmol/hr of ethylene feed.

4.3.1 Reactor

The composition of stream μ_2 can be calculated from stream μ_1 on the basis of the reactions taking place in the reactor, which are shown below.

$$\mathrm{EL} + \mathrm{W} \leftrightharpoons \mathrm{EA} \tag{4.17}$$

$$\mathrm{PL} + \mathrm{W} \leftrightharpoons \mathrm{IPA} \tag{4.18}$$

$$2\mathrm{EA} \leftrightharpoons \mathrm{DEE} + \mathrm{W} \tag{4.19}$$

For the first two reactions the extents of conversion are known for the reference reactant EA in the first and for PL in the second. The conversions are chosen to be $\eta_1 = 7\%$ for the first reaction and $\eta_2 = 0.7\%$ for the second reaction which are values in agreement with literature [24]. The normalized coefficients $\gamma_{r,k}$ are given in Table 14.

Table 14 – Coefficients of the reactions.

	M	EL	PL	DEE	EA	IPA	W
Reaction 1 (Eq. (4.17))	0	-1	0	0	1	0	-1
Reaction 2 (Eq. (4.18))	0	0	-1	0	0	1	-1

The third reaction is governed by equilibrium. The equilibrium constant is given by:

$$K_p = \sum_{i=1}^{NC} a_i^{\gamma_i} = \frac{a_{\mathrm{DEE}}\, a_{\mathrm{W}}}{a_{\mathrm{EA}}^2} \tag{4.20}$$

where a_i is the activity and γ_i is the stoichiometric coefficient of component i. If an ideal gas mixture is assumed, the activity of each compound is equal to its partial pressure and the equilibrium constant can be calculated as follows:

$$K_p = \frac{p_{\mathrm{DEE}}\, p_{\mathrm{W}}}{p_{\mathrm{EA}}^2} = \frac{y_{\mathrm{DEE}}\, y_{\mathrm{W}}}{y_{\mathrm{EA}}^2} \tag{4.21}$$

Since the reaction is equimolar, the equilibrium is not influenced on the total pressure and depends only on temperature and reactants/products concentrations. Different values are reported in literature [25, 26, 24]. However, this values differ a lot and this is due to the different experimental catalyst used. Thus, we use the value for the equilibrium constant computed from thermodynamic data, which is calculated to be

$$K_p = 4.04 \tag{4.22}$$

The calculation is described in Appendix A. This value is higher than all the values computed with literature relationships ($K_p = 3.58$ [25], $K_p = 0.009$ [26], $K_p = 0.2$ [24]). We believe that this is acceptable and set us in a conservative situation: the worst case is considered when the production of DEE is higher than what it could be in reality.

At first conversion data are considered and the stream of each component k in μ_2 is defined by:

$$\mu_2^k = \mu_1^k + \sum_{r=1}^{2} \gamma_{r,k}\eta_r\mu_1^k \tag{4.23}$$

After the desired conversion on the reactants has been imposed, the equilibrium on the third reaction is imposed and the flows of DEE, EA and W are computed accordingly, using equation (4.21) and the equilibrium constant of 4.04. So new values for μ_2^{DEE}, μ_2^{EA} and μ_2^{W} are obtained.

4.3.2 Flash

According to the model of the flash (see Appendix B.1) there are two degrees of freedom. They are saturated choosing pressure and temperature. The pressure is set equal to the pressure in the reactor (68 atm) and the temperature is chosen to be just above the temperature of cooling water ($T_{\text{top}} = T_{\text{CW}} + \Delta T_{\text{approach}}$, being $T_{CW} \cong 25°\text{C} \cong 300$ K the temperature of cooling water and $\Delta T_{\text{approach}} \cong 10$ K the approach). In order to compute the recoveries a key component and its recovery must be chosen. The key component is selected to be DEE, because this is the one that has intermediate volatility compared to the two lumps that we want to separate. We choose its recovery to be $\xi_n = 0.5$. The recoveries obtained for each component are then listed in Table 15. The stream entering the flash is the outlet stream from the reactor, μ_2^k.

Table 15 – Flash recoveries.

	M	EL	PL	DEE	EA	IPA	W
$P_k^0(310)$ [atm]	277.648	72.828	14.902	1.079	0.149	0.098	0.061
$\alpha_{k/DEE}$	257.304	67.491	13.810	1.000	0.139	0.091	0.057
ξ_{3k}	0.996	0.985	0.932	0.500	0.122	0.083	0.054

4.3.3 Absorber

According to the model of the absorber (see Appendix B.2), the absorber has four degrees of freedom. They are saturated choosing the recovery of the key component, the pressure and the flow and the temperature of the liquid solvent. The adsorption process is favored at high pressures and low temperatures. The pressure is set equal to the pressure in the reactor taking into account a reasonable pressure drop of 0.5 atm and the pressure therefore becomes 67 atm. The temperature is set just above the one of the commonly available water which is used as a solvent (310 K). The operation is assumed to be isothermal, while in reality it is exothermic, thus the selected temperature takes into account a small increase in temperature in the absorption process with the inlet water which is available at 300 K. Since we are interested in recovering ethanol into the liquid phase, we select EA as our key component and we set its recovery to be $r_n = 0.99$. Instead of selecting directly the flow rate of the liquid flow, we select the effective absorption factor relying on the heuristic value $A_E = 1.4$, so that $L_0 = A_E K_{EA} V_{N+1}$. The vapor phase entering the absorber is the stream from the flash, so $V_{N+1} = \mu_{31}$, and the liquid phase is fed from outside, $L_0 = \mu_{03}$. Theoretically, in order to compute the split factors the flow rates of the different components in the inlet mixture must be known because we need to use the Kremser equation(see Appendix B.2). If we know that the key component flow rate in the inlet liquid stream is zero, $l_0^{EA} = 0$, the Kremser equation is simplifies as:

$$N = \ln\left(\frac{r v_{N+1}^n - A_E v_{N+1}^n}{-A_n(1-r)v_{N+1}^n}\right) / \ln(A_E) = \ln\left(\frac{r - A_E}{-A_n(1-r)}\right) / \ln(A_E) \qquad (4.24)$$

so that the number of ideal trays can be computed without knowledge of the composition of the previous streams. Also the recoveries for all the component not present in the solvent stream are independent on the inlet vapor stream composition as $l_0^k = 0$ and:

$$\xi_k = \frac{v_1^k}{v_{N+1}^k} = \frac{\frac{v_{N+1}^k}{\beta_N^k}}{v_{N+1}^k} = \frac{1}{\beta_N^k} \qquad (4.25)$$

Thus, we need to make assumption for v_{N+1}^W in order to compute ξ_W as:

$$\xi_k = \frac{v_1^k}{v_{N+1}^k + l_0^k} = \frac{\frac{v_{N+1k}}{\beta_N^k} + \frac{\beta_{N-1}^k}{\beta_N^k} l_0^k}{v_{N+1}^k + l_0^k} \qquad (4.26)$$

where β_N^k is a coefficient defined in Appendix B.2. Afterwards we need to compute a new ξ_W with $v_{N+1}^W = \mu_{31}^W$ obtained from simulation, and check if it is almost the same as the previous one; if not, the procedure must be iterated with the new value computed. The first guess for μ_{31} is made assuming no recycles and a flow rate of inlet water of 0.6 times the one of feed ethylene (which means, assuming μ_{01} and μ_{02}) and following the path of the streams in the reactors and then in the flash computing the following streams, with the specified conversions, equilibria and the specified split factors.

Before accepting the values of the recoveries thus obtained, we need to check that not much water solvent is evaporated and leaves the absorber in gas phase, because, as mentioned before, the absorption process in exothermic and causes a raise of temperature. We can estimate the fraction of solvent water leaving as a vapor as:

$$\frac{\beta_{N-1}^{W}}{\beta_{N}^{W}} \tag{4.27}$$

With $A_E = 1.4$ we obtain 29.4%, which is too much if we are in the assumption of isothermal process. This leads us to reconsider our operating parameters. It is not worth changing temperature and pressure; higher pressures and lower temperatures would lead to higher operation costs. So we consider the option of incrementing the solvent flow rate increasing the absorption factor. The heat released in the absorption process is the same, so the amount of water vaporized would be the same but its fraction over the total water flow rate inlet would be lower. If we set $A_E = 10$, we obtain a percentage of evaporated water of 4.1%. This is acceptable, so we proceed to the computation of new recoveries with this new value, which are shown in Table 16. With $A_E = 10$, the number of stages in the absorber is calculated to be $N = 2$.

Table 16 – Absorber recoveries at $A_E = 10$.

	M	EL	PL	DEE	EA	IPA	W
$P_k^0(310)$ [atm]	277.648	72.828	14.902	1.079	0.149	0.098	0.061
$\alpha_{k/EA}$	1857.635	487.260	99.703	7.220	1.000	0.655	0.411
A_k	0.005	0.021	0.100	1.385	10	15.270	24.329
β_N	1.005	1.021	1.110	4.304	111	249.420	617.218
β_{N-1}	1.005	1.021	1.100	2.385	11	16.269	25.329
ξ_{4k}	0.995	0.979	0.901	0.232	0.009	0.004	0.019

4.3.4 Purge

The purge is introduced to minimize the methane build-up into the reactor which could be a consequence of the recycle of the unreacted reactant mixture. The minimum value of the purge is found with the constraint that the feed to the reactor contains less than 10 mole% methane. We define b as the fraction of the vapor recovery stream extracted as the purge stream. Its minimum value can be found by mass balance computations on the units dealing with the recycle loop. Then we can perform the material balance to compute the size of the recycle stream and the loss of ethylene as a function of the purge fraction. Here the basis of calculation is 1 kmol/hr of ethylene feed. The result of purge stream analysis is listed in Table 17. See Appendix B.3 for details of the recycle loop materials balance.

From Table 17 we see that for $b > 0.003$ the constraint on the maximum methane concentration is satisfied. To be conservative, we set $\xi_5 = b = 0.005$. If we then take into account also the recycle of DEE, the molar fraction of methane would be even lower and we can consider to be totally protected from the problem. Since the aim of the purge is to limit the loss of EL, we would be lead to minimize the purge split to the lowest value allowable. However, the smaller the purge split the larger the flow of the recycle and the higher the costs. The optimal purge fraction must be chosen; however, we do not have enough information to make such a decision at the moment.

Table 17 – Flows for purge stream analysis.

b	μ_1^M	μ_1^{EL}	μ_1^{PL}	$\frac{\mu_1^M}{\mu_1}$ %	μ_{52}^{EL}	μ_{52}^{PL}	μ_{51} (total recycle)
0.001	9.00	13.55	3.75	26.14	0.0129	0.0033	25.17
0.002	4.50	13.37	3.34	15.39	0.0254	0.0059	20.12
0.003	3.00	13.20	3.01	11.06	0.0376	0.0080	18.15
0.004	2.25	13.04	2.73	8.71	0.0495	0.0097	16.99
0.005	1.80	12.87	2.51	7.23	0.0611	0.0111	16.17
0.006	1.50	12.72	2.32	6.21	0.0725	0.0123	15.53
0.007	1.29	12.56	2.15	5.46	0.0835	0.0133	15.02
0.008	1.13	12.41	2.01	4.89	0.0943	0.0142	14.57
0.009	1.00	12.26	1.88	4.44	0.1048	0.0150	14.19
0.01	0.90	12.12	1.77	4.08	0.1151	0.0157	13.84

4.3.5 Distillation columns

To set up the mass balance for a distillation column it is necessary to identify the light and heavy key components as well as the temperature of the column. The split fractions of the light and heavy key components, ξ_{lk} and ξ_{hk}, are then specified. The minimum number of stages can be calculated using the Fenske equation:

$$N_m = \ln\left[\frac{\xi_{lk}(1-\xi_{hk})}{\xi_{hk}(1-\xi_{lk})}\right] / \ln(\alpha_{lk/hk}) \tag{4.28}$$

The split fractions can now be calculated as:

$$\xi_k = \frac{\alpha_k^{Nm}\xi_{hk}}{1+(\alpha_k^{Nm}-1)\xi_{hk}} \tag{4.29}$$

Each standard distillation column has three degrees of freedom. They are saturated specifying the pressure and the recovery of selected light and heavy component. In the first distillation column the purpose is to recover ethanol and remove water, hence the light key component is ethanol and the heavy key component is water. We would be interested to have as much isopropanol as possible in the bottom stream so that it will not contaminate the streams separated afterwards, which will be recycled or extracted as a product. However, making a cut between ethanol and isopropanol is costly because their relative volatility is quite near to 1, consequently as a first separation we choose water to be the heavy key component. In the second distillation column we are interested in separating diethylether to recycle it to the reactor and hence this becomes the light key component, while we want that ethanol goes at the bottom to be further refined, so it would be the heavy component. The purpose of the last distillation is to extract ethanol in azeotropic composition separating it from the bulk of water, and at the same time we would like to purify the product as much as possible from isopropanol, so isopropanol will be the heavy key component and ethanol the light. The split fractions for each key component and for each distillation column have to be chosen. They must be as high as possible to give satisfactory performance, but complete recovery is impossible to obtain in practice. Table 18 lists the key components for each of the distillation columns and the chosen split fractions. In order to make the computations, the knowledge of temperature is necessary. However, since we are interested in obtaining a general mass balance, we select a temperature just above the one of cooling water (310 K) to perform our calculation. The recoveries obtained for each column are seen in Table 19.

Table 18 – Key components and split fractions for distillation columns.

	Light key	ξ_{lk}	Heavy key	ξ_{hk}
Column 1	EA	0.995	W	0.100
Column 2	DEE	0.995	EA	0.100
Column 3	EA	0.995	IPA	0.050

Table 19 – Column recoveries.

		M	EL	PL	DEE	EA	IPA	W
	$P^\circ(310K)[atm]$	277.648	72.828	14.902	1.079	0.149	0.098	0.061
Column 1	$\alpha_{lk/hk}$	4519.397	1185.445	242.566	17.564	2.433	1.593	1.000
	$\xi_{7,k}$	1.000	1.000	1.000	1.000	0.995	0.849	0.100
Column 2	$\alpha_{lk/hk}$	1857.635	487.260	99.703	7.220	1.000	0.655	0.411
	$\xi_{8,k}$	1.000	1.000	1.000	0.995	0.005	0.001	0.000
Column 3	$\alpha_{lk/hk}$	4519.397	1185.445	242.566	17.564	2.433	1.593	1.000
	$\xi_{9,k}$	1.000	1.000	1.000	1.000	0.995	0.050	0.146

The recoveries for the key components of the last column are selected so that 99.5% of the azeotrope is recovered and that isopropanol does not exceed a 0.01% molar fraction in the top product. Thus, the recovery of ethanol is directly fixed to $\xi_{9EA} = 0.995$ and the recovery of water is determined by imposing that the top product has water in azeotropic composition:

$$\mu_{91}^{W} = \frac{x_{az,W}}{x_{az,EA}} \mu_{91}^{EA}$$

$$\xi_{9W} = \frac{\mu_{91}^{W}}{\mu_{82}^{W}}$$

where the molar composition of the azeotrope is $x_{az,W}, x_{az,EA}$, as reported in Table 3. The recovery of isopropanol is set to be $\xi_{9IPA} = 0.05$, as we see that the constraint of its maximum composition is satisfied.

4.4 Tear stream

The equation presented in section 4.1 cannot be solved in the sequence in which they are presented, because of the presence of recycle loops, in total 2, which cause some of the inlet flows to be unknown at the beginning of the computations. For example, equation (4.1) cannot be solved before we know μ_{51}^k and μ_{81}^k which are outlets of later units and thus require the knowledge of all the flows upstream. Therefore, the solution of the mass balance is not straightforward and an iterative procedure is to be followed. In order to do this, a tear stream where the computation should be started is selected. The value of this stream is guessed and all the computations are made, then the values of the same stream are recomputed and the first guess is verified. If the resulting values are the same as the previous iteration, the procedure stops and convergence has been reached; if not, the procedure is reiterated until convergence.

The identification of the tear stream is performed using the signal flowgraph based technique. On the basis of the flow sheet in Figure 9 a table of nodes and precursors is created (Table 20); the input and output streams are not taken into account. Nodes with single precursors are eliminated and a new list of precursors is made in sequence substituting each stream number with its precursors, until a short circuit is identified meaning that a stream has itself as a precursor. The streams thus identified are candidate tear streams. From the results in Table 20 it is seen that stream μ_1, right before the reactor, is our tear stream. The computations will then start and converge on this stream, and its value will initially be guessed.

Table 20 – Signal flowgraph based technique to identify tear stream.

Node	Precursers				
μ_1	(81),(51)	(71),(41)	(6),(31)	(6),(2)	(6),(1)
μ_2	(1)				
μ_{31}	(2)				
μ_{32}	(2)				
μ_{41}	(31)				
μ_{42}	(31)				
μ_{51}	(41)				
μ_6	(32),(42)	(2),(31)	(1),(2)	(1),(1)	
μ_{71}	(6)				
μ_{81}	(71)				
μ_{82}	(71)				

4.5 Incidence matrix

After the tear stream has been selected, the sequence in which the equations are solved must be defined. The equations are solved starting from the tear stream, that is, at the reactor. Thus, equation (4.2) is the first one to be solved. Then, the equations are solved in a sequence so that all the input variables are known and the output can be computed. The incidence matrix in Table 21 shows how the variables are interlinked through the equations. The equations are set in the rows and the variables in the columns. For each equation, the full dot stands for the variables that we need to know in order to use the equation (input), while the empty dot indicates the variable that the equation computes (output). The variables are listed going downstream after the tear stream. In this way if any sign is seen above the diagonal it means that in order to solve the equation it is necessary to know information downstream. If equation (4.1) is set after equation (4.14), this is true for equation (4.2) only and the problem can be solved once μ_1 is known. The star represents the variables which needs a first guess and convergence.

Table 21 – Analysis of incidence matrix of equations.

	μ_2	μ_{31}	μ_{32}	μ_{4A}	μ_{41}	μ_{42}	μ_{51}	μ_{52}	μ_6	μ_{71}	μ_{72}	μ_{81}	μ_{82}	μ_1	μ_{91}	μ_{92}
Eq.(4.2)	○													⋆		
Eq.(4.3)	•	○														
Eq.(4.4)	•		○													
Eq.(4.5)		•		○												
Eq.(4.6)				•	○											
Eq.(4.7)				•		○										
Eq.(4.8)					•		○									
Eq.(4.9)					•			○								
Eq.(4.10)			•			•			○							
Eq.(4.11)									•	○						
Eq.(4.12)									•		○					
Eq.(4.13)										•		○				
Eq.(4.14)										•			○			
Eq.(4.1)							•					•		○		
Eq.(4.15)													•		○	
Eq.(4.16)													•			○

In Table 22 the equations are replaced with the physical units to which they correspond, so that the sequence in which the units are to be solved is clearer. The process is then divided into two blocks or partitions: the first consists in all the equipment before the last column, and the second one in the whole process. The first partition is lead to convergence before the second partition is solved. A scheme of the solution strategy is shown in the flowchart below.

Table 22 – Analysis of incidence matrix of units.

	μ_2	μ_{31}	μ_{32}	μ_{4A}	μ_{41}	μ_{42}	μ_{51}	μ_{52}	μ_6	μ_{71}	μ_{72}	μ_{81}	μ_{82}	μ_1	μ_{91}	μ_{92}
2. Reactor	○													⋆		
3. Flash	•	○	○													
4A. Mixer		•		○												
4B. Absorber				•	○	○										
5. Divider					•		○	○								
6. Mixer			•			•			○							
7. Column 1									•	○	○					
8 . Column 2										•		○	○			
1. Mixer							•					•		○		
9. Column 3													•		○	○

Solution strategy

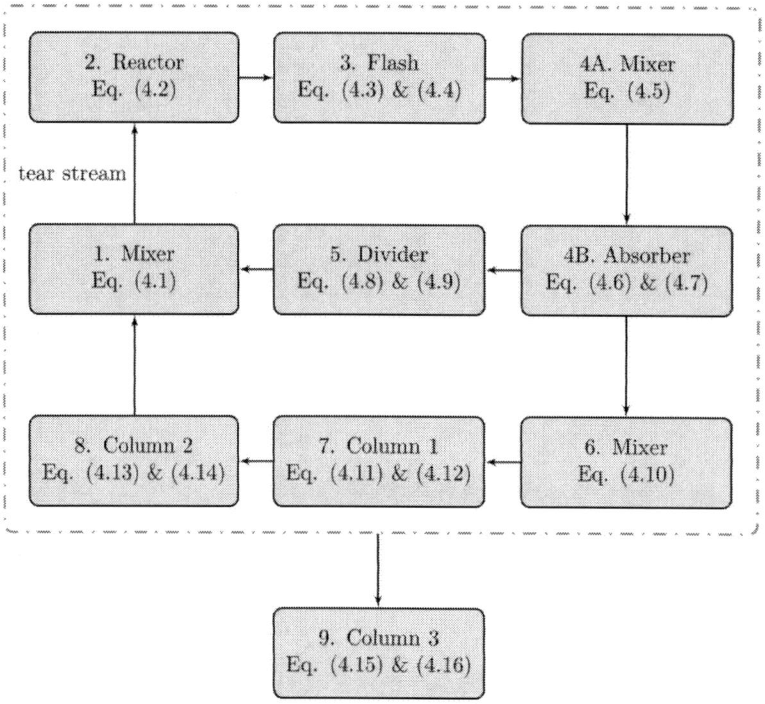

4.6 Iteration in PRO/II

The iterations for the mass balance are performed with PRO/II. There are only four types of units in the preliminary Pro/II flow sheet: reactor, mixer, stream calculator and divider. Stream calculators are used in place of all separation units like flash and distillation column. Since all unit operations are simulated as splitters, we do not need to set their temperature and pressure yet. Also, we select an ideal thermodynamic method because no equilibrium computations are needed. The first guess for each recovery ξ_{ik} in the i stream calculator is made by the short-cut method presented for each unit operation, assuming the various specifications as described in the relevant sections. The computations are based on 100 kg-mol/hr of impure ethylene feed stream.

Step 1. Building the flow sheet without recycle
First, we build the flow sheet unit by unit without recycles in PRO/II. Each time a unit is added, the simulation is run and the stream results given by PRO/II are compared with our hand calculations. This way helps us to check if there is any problem with each unit and to make sure that we set the right parameters for every unit. The flow sheet without recycles built in PRO/II is shown in Figure 10. The single reactor is simulated with two reactors, R2A and R2B: R2A is a conversion reactor where the reaction of ethylene to ethanol and reaction of propylene are defined according to their conversions, while R2B is an equilibrium reactor where the reaction of ethanol to diethylether is defined according to its equilibrium constant. In front of the reactor, the feed streams are mixed with mixer M1. F3 is the flash and M4A and M4 together acts as the absorber; M4A is a virtual mixer which does not exist in reality, it is just added so that the absorber can be simulated. C7, C8 and C9 are column 1, 2 and 3, respectively. The purge unit is represented by the divider D5. One controller is used to control the flow rate of water feed S02 to keep the ratio of water/ethylene in the feed of reactor S1. Another controller is used to control the flow rate of water feed S03 according to the solvent requirements of the absorber.

Step 2. Iteration with recycles
Once we are sure that there are no problems in all the units, we close the recycle loops. Iterations are done automatically in PRO/II. The final flow sheet is shown in Figure 11. We define the calculation sequence of units in PRO/II according to our previous analysis. In order to make it easier to get the system converge, we choose to converge only tear streams in PRO/II - Problem Recycle Convergence and Acceleration Options, and also set a large stream tolerance. After the system converges, we reduce the stream tolerance and run again. When the system converges with a small tolerance, we change to converge all streams. Finally, the system converges with an acceptable tolerance of 0.0001.

The recovery of water obtained for the absorber in the previous section is dependent on the inlet molar flow rates of gaseous water, $v_{N+1}^W = v_{31}^W$. After convergence has been obtained, we must check if the new values of the recoveries match with the old ones. If the difference is not significant we keep the same

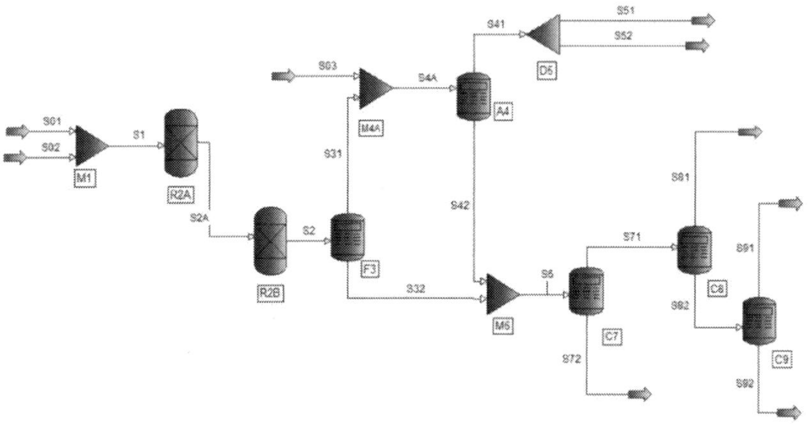

Figure 10 – PRO/II flow sheet without recycles.

values, if it is we must use the new recomputed recoveries to perform the mass balance until convergence on the recoveries of the absorber is reached. The final value obtained is $\xi_{3W} = 0.0215$.

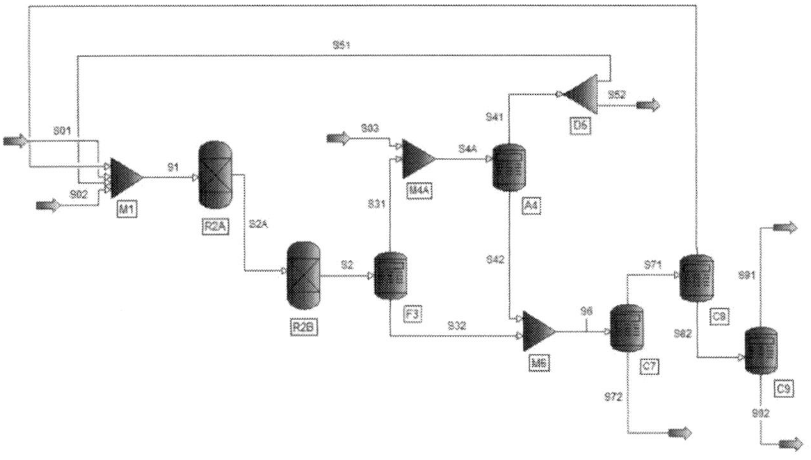

Figure 11 – PRO/II flow sheet with recycles.

The final mass balance obtained is presented in Table 23. On its basis, we can check if the process leads to satisfactory performance. We notice that the mole fraction of methane in the feed of reactor is about 7.2%, which is lower than 10%, and 5.7% of ethylene raw material is wasted in the purge stream. This is considered satisfactory and the selected purge fraction is kept.

From the results obtained in Table 23 we can verify if the total material balance is conserved. Based on the total molar rates we obtain:

$$\begin{aligned}
\text{IN} + \text{PRODUCED} &= S01 + S02 + S03 + (S2 - S1) \\
&= 100 + 772.67 + 37.46 + (2467.64 - 2559.851) \\
&= 817.92 \\
\text{OUT} &= S52 + S72 + S91 + S92 \\
&= 7.83 + 647.09 + 101.00 + 62.00 \\
&= 817.92
\end{aligned} \tag{4.30}$$

Relative error $= 0.0$

The total flow in plus the produced equals the total flow out. The balance is also made with respect to mass rate and the following is obtained:

$$\begin{aligned}
\text{IN} &= S01 + S02 + S03 \\
&= 2836.65 + 13919.81 + 674.89 \\
&= 17431.35 \\
\text{OUT} &= S52 + S72 + S91 + S92 \\
&= 225.97 + 11682.15 + 4384.90 + 1138.03 \\
&= 17431.05
\end{aligned} \tag{4.31}$$

Relative error $= (17431.35 - 17431.05)/17431.35 = 1.7 \cdot 10^{-5}$

Both ways show that the total material balance of the system is conserved.

Table 23 – Mass balance for ethanol process flow sheet.

Stream	UOM	S01	S02	S03	S1	S2	S2A	S31	S32	S41	S42
Total Molar Component Rates	kg-mol / hr										
METHANE		0.90	0.00	0.00	178.28	178.26	178.26	177.57	0.69	176.62	0.96
ETHYLENE		96.10	0.00	0.00	1290.14	1199.83	1199.83	1182.31	17.52	1158.06	24.25
PROPENE		3.00	0.00	0.00	268.51	266.63	266.63	248.63	18.00	223.92	24.71
DEE		0.00	0.00	0.00	48.12	48.36	48.12	24.18	24.18	5.62	18.56
ETHANOL		0.00	0.00	0.00	0.55	90.37	90.86	10.99	79.38	0.10	10.90
IPROPANOL		0.00	0.00	0.00	0.00	1.88	1.88	0.16	1.72	0.00	0.16
WATER		0.00	772.67	37.46	774.25	682.31	682.06	36.76	645.55	1.59	72.62
Total Molar Rate	kg-mol / hr	100.00	772.67	37.46	2559.85	2467.64	2467.64	1680.61	787.04	1565.91	152.15
Total Mass Rate	kg / hr	2836.65	13919.81	674.89	67892.86	67892.56	67892.56	49450.10	18442.45	45194.01	4930.99
Stream	UOM	S4A	S51	S52	S6	S71	S72	S81	S82	S91	S92
Total Molar Component Rates	kg-mol / hr										
METHANE		177.57	175.74	0.88	1.65	1.65	0.00	1.65	0.00	0.00	0.00
ETHYLENE		1182.31	1152.27	5.79	41.77	41.77	0.00	41.77	0.00	0.00	0.00
PROPENE		248.63	222.80	1.12	42.71	42.71	0.00	42.71	0.00	0.00	0.00
DEE		24.18	5.59	0.03	42.74	42.74	0.00	42.53	0.21	0.21	0.00
ETHANOL		10.99	0.10	0.00	90.27	89.82	0.45	0.45	89.37	88.93	0.45
IPROPANOL		0.16	0.00	0.00	1.88	1.60	0.28	0.00	1.60	0.08	1.52
WATER		74.22	1.59	0.01	718.17	71.82	646.36	0.00	71.82	10.46	61.35
Total Molar Rate	kg-mol / hr	1718.07	1558.09	7.83	939.19	292.10	647.09	129.10	163.00	99.68	63.32
Total Mass Rate	kg / hr	50125.00	44968.04	225.97	23373.44	11691.30	11682.15	6168.37	5522.93	4305.94	1216.99

5 Task 5

In the previous Task 4, temperature and pressures were assigned to each unit in order to make it possible to perform the computations, but they may not necessarily match with the actual value needed to obtain the desired performance once more complex models are considered, which are more similar to reality. In Task 5 the real operating conditions of each unit operation are specified. Temperature and pressure are set according to the results obtained from the mass balance in order to obtain the same recoveries specified.

The same flow sheet used in Task 4 is kept. The separation units are still represented as stream calculators apart from the flash: the flash operation is critical because it determines the first split between condensible and incondensible species, which is a key point for designing the following units. Thus, a new mass balance must be performed with the actual recoveries obtained with the detailed model for the flash unit. After this, temperature and pressures are computed for all the streams and the units in the flow sheet.

5.1 Thermodynamic model

The mass balance in Task 4 was performed using an ideal thermodynamic model only. The ideal model gives us a good estimate of the parameters involved in the process, but it is not sufficient for a more accurate estimation. Furthermore, the mixtures treated are highly non-ideal because the operating pressures are often very high and binary and ternary azeotropes are formed, consequently complex calculations are required in order to compute the equilibrium. Here we take the non-ideality of the system into account with selection of a proper model of liquid-vapor equilibrium.

The most generic equilibrium expression takes the form of the following equation:

$$\bar{\phi}_k y_k P = \gamma_k x_k f_k^0 \tag{5.1}$$

where $\bar{\phi}_k$ is the vapor fugacity coefficient of component k in the gas mixture, γ_k is the liquid activity coefficient and f_k^0 is the pure component liquid fugacity. We choose to take into account the non-ideality of the mixture with the activity coefficient of the liquid phase only, while the vapor phase is considered nearly ideal; this leads to an equilibrium constant in the form:

$$K_k = \frac{P_k^0 \gamma_k}{P} \tag{5.2}$$

$$y_k = K_k x_k \tag{5.3}$$

Equilibrium between vapor and liquid is concerned in the separation units, thus it is important to take into account the non-ideality of the mixture here, and for these units the NRTL model is chosen [27]. A different choice is made for the reactors: all the components involved there are vapor phase and most of them are hydrocarbons, therefore it is reasonable to assume that the behavior of the mixture is nearly ideal and the ideal model is kept for it.

5.2 Flash

The pressure in the flash is specified so that it is almost the same as the pressure in the reactor, apart from the pressure drop. We are interested in maintaining a high pressure so that the top vapor phase can be sent to the absorber unit at high pressure, so that the efficiency of the absorption is high. Furthermore, a high pressure in the vapor phase reduces the work necessary to recompress the recycle at the exit of the absorber before sending it to the reactor. Thus, assuming a pressure drop of ΔP=0.5 atm, the flash operates at 67.5 atm. The operating temperature should lie between the dew point and the bubble point of the inlet stream μ_2, whose composition is known from the mass balance performed in the previous task. The bubble and dew point of the inlet stream are shown in Table 24. The flash should operate above the temperature of cooling water (300 K $\cong$ 25°C) so that water can be used as a coolant, also considering a reasonable approach. The flash temperature must also be lower than the critical temperature of the feed.

The simple model for the flash is not suitable to define the temperature with a satisfactory precision. Thus, the temperature of the flash is computed with the help of the simulator and of the NRTL model. The stream calculator corresponding to the flash is substituted with a flash simulator. The bubble point and the dew point of the inlet stream are calculated and then the temperature is varied to check the recoveries obtained. There are several factors that we need to consider when choosing the temperature. When a non-ideal model is considered the flash separation is not that effective, so to obtain satisfactory recoveries of the incondensible we should select a high temperature. However, a high temperature causes a bigger amount of water as a solvent in the absorber, leading to higher costs for water supply and also for the following columns, because a bigger flow stream needs to be treated. If a lower temperature is selected, the recovery of the incondensible is not as high, but the absorption process is more effective and part of the incondensibles would be recycled to the reactor in any case with the top stream of the second column. Consequently, we should keep the temperature reasonably low so that we can obtain satisfactory recoveries of incondensible. Thus, a trade-off is needed. We select 120°C, which we believe being reasonably low for sending the vapor phase to be treated in the absorber and high enough to obtain satisfactory recoveries. These recoveries are different from those specified in Task 4 and the system must be taken again to convergence and this will result in a new mass balance. Table 24 reports the bubble point and the dew point of the feed, as well as the selected operating temperature, which obviously lies between the two. Recoveries obtained under the selected operating condition of T=120°C and P=67.5 atm are presented in Table 25.

5.3 Absorber

The pressure in the absorber is selected to be as near as possible to the one in the reactor; this is desirable to have a more effective absorption and a smaller work supply to the gas compressor. If we assume a pressure drop of ΔP=0.5

Table 24 – Feed dew and bubble point and operating conditions of flash.

Pressure	67.5 atm
Operation temperature	120.00 ℃
Dew point	215.23 ℃
Bubble point	33.46 ℃

Table 25 – Recoveries in flash at T=120 ℃ and P=67.5 atm.

Component	ξ_k
M	0.930
EL	0.806
PL	0.518
DEE	0.482
EA	0.121
IPA	0.123
W	0.036

atm, the operating pressure will then be $P = 67.0$ atm.

We are interested in keeping the temperature as low as possible, because the absorption process is favored at low temperatures in order not to vaporize too much solvent. Since water is the absorbing solvent and it is commonly available at ambient temperature, the temperature of the inlet liquid is set to be equal to T≅25℃≅300 K. No cooling of absorbing water is taken into consideration.

For the gas-liquid absorption it is assumed that vapor-liquid equilibrium exists at every stage of the absorber. At the bottom of the absorber the liquid outlet will therefore be in equilibrium with the vapor phase and the temperature of the liquid stream μ_{42} is its bubble point. Similarly, the vapor outlet μ_{41} at the top of the absorber will be in equilibrium with the liquid and will have the temperature of its dew point. From the stream calculator in PRO/II it is found that the dew point of vapor stream μ_{41} is $T_{41} = 64.84$℃ and for the liquid stream the bubble point is $T_{42} = 141.45$ ℃. Similar temperatures were obtained by hand calculations.

Another constraint is that the bubble point should be lower than the critical temperature. The pseudo-critical temperature of the mixture liquid stream, T_c^m, is calculated by

$$T_c^m = \sum T_c^i x_i \tag{5.4}$$

where x_i is the molar fraction of component i. It is found that the critical temperature for stream μ_{42} is $T_c^m = 508.65$ K $= 235.5$ ℃ which is well above the bubble point of stream μ_{42} so the constraint is satisfied.

5.4 Distillation columns

For the distillation columns the pressure is set first. The value of pressure is chosen considering several factors:

• The top temperature should not be lower than 310 K in order to make it possible to use water as a coolant fluid, whose temperature is around ambient temperature ($T_{top} = T_{CW} + \Delta T_{approach}$, being $T_{CW} \cong 25$°C$\cong$ 300 K the temperature of cooling water and $\Delta T_{approach} \cong 10$ K the approach).

• The bottom temperature should not be higher than the one allowed for using pressurized steam as a heating fluid (because of its cost, pressurized steam is almost never used at higher pressure than 20 atm, which corresponds to a hot utility temperature of around 480 K $\cong$ 205 °C).

• The bottom temperature should not be higher than the critical temperature of the mixture.

For simplicity, the pressure drop in all the columns is assumed independent on the number of trays and is set to be $\Delta P = 1.0$ atm.

After the pressure has been set, the top and bottom temperatures are found as the saturation temperature of the outlet streams. The temperatures of the outlet top and bottom streams depend on the type of condenser and reboiler chosen. The condenser can be total or partial. In a total condenser all the fluid is condensed and part of it is sent to the column as a reflux and the rest is taken out as outlet liquid top product, so the temperature of the top outlet stream is the bubble point of the mixture. In a partial condenser only part of the fluid is condensed: the condensed part is sent to the column as a reflux and the remaining part is sent out as a vapor top stream, so the temperature of the top outlet stream is at the dew point of the mixture. In case a partial condenser is selected, it counts as an additional equilibrium stage. The choice of one or the other alternative is determined by the temperature necessary to condense the top mixture completely (if it is too low, a partial condenser is preferred) and by the next treatment of the top product (if it must be used in vapor phase, there is no need to waste energy to condense it). Analogous considerations can be made for a total and partial reboiler. In our case, the bottom product is liquid for all the columns, so the temperature of the bottom stream is equal to its bubble point and the reboiler always counts as an equilibrium stage. The top product can be liquid or vapor depending if a total or partial condenser is chosen, so each case is considered individually.

5.4.1 Column 1

The function of column 1 is to remove water from μ_6. The pressure at the top is computed so that the vapor phase can be condensed with water.

First, the possibility of using a total condenser is investigated. If cooling water is used, the minimum top temperature is T_b=310 K. If this value is set to be the bubble point of the top product, the minimum pressure is obtained. Any pressure higher than this value can be acceptable. The bubble point pressure is computed by the short-cut algorithm for the computation of the bubble point pressure of a mixture of composition like μ_{71} with the bubble point temperature set to be T_b=310 K. Based on the composition of stream μ_{71} and $n = EL$ (the most abundant component), the minimum top pressure to have a total condenser is:

$$P_{\text{top, min}} = P_n^0(310K)\bar{\alpha} = 38.3\,\text{atm} \tag{5.5}$$

This pressure is quite high. We would be interested in keeping the column pressure above atmospheric, so that also in the second column the operating pressure would be higher and the diethylether recycled would be at higher pressure, thus reducing the recompressing costs. However, if the pressure is too high the column would be expensive to manufacture and difficult to operate: better materials are required to build the column to avoid leaks; the difference in the relative volatilities of the different components is lower, so the separation becomes more difficult and a high number of trays would be required. Additionally, if the following column operates at lower pressure the stream must be expanded before it is fed to the column, so μ_{71} will become vapor again and therefore there is no need to use a total condenser. Thus, it is better to choose a lower pressure and a partial condenser.

If we want to use a partial condenser, the minimum top pressure is provided by the short-cut algorithm for the computation of the dew point pressure of a mixture of composition like μ_{71} with the dew point temperature set to be T_d=310 K. Based on the composition of stream μ_{71} and being $n = EL$ (the most abundant component), we compute the minimum top pressure:

$$P_{\text{top, min}} = P_n^0(310K)/(\sum y_k/\alpha_{k/n}) = 0.22\,\text{atm} \tag{5.6}$$

This is the lower limit of the pressure at the top. If a higher pressure is selected, the equilibrium temperature would be higher than 310 K and condensation with cooling water is still possible. If a lower pressure is selected, the equilibrium temperature would be lower than 310 K and the stream can not be condensed with cooling water. We are not interested in operating the column in vacuum, so any pressure above 1 atm would be required.

Since we do not have any special requirement on the pressure of this first column a part from the extreme values discussed above, a proper pressure should be determined after the pressure of the second column is specified so as to make sure that the second column can work well with the feed from the first column. The top pressure of the first column will be determined as being slightly higher than the pressure in the second one, so that the flow of the feed stream is allowed.

5.4.2 Column 2

In the second column light components are separated from ethanol product. Since the overhead stream goes back to the vapor recycle loop, there is no need of considering the option of a total condenser and we choose a partial condenser with saturated vapor product. The pressure at the top of the second column is computed so that the vapor phase at the top can be condensed with water. The first guess is provided by the short-cut algorithm for the computation of the dew point pressure of a mixture of composition like μ_{81} with the dew point set to be T_d=310 K. Using the composition of stream μ_{81} and $n = EL$ (the most abundant component), the pressure at the top is computed as:

$$P_{\text{top, min}} = P_n^0(310K)/(\sum y_k/\alpha_{k/n}) = 8.15\,\text{atm} \tag{5.7}$$

This initial estimate is the lowest pressure to make sure that the vapor can be condensed with cooling water. A higher pressure is preferred to save the cost of compressing this recycle vapor stream. However, it is easier and cheaper to operate with a column at low pressure. Thus, a pressure of 8.15 atm is selected for the top. Consequently, the bottom pressure is obtained considering a pressure drop of 1.0 atm as $P_{\text{bottom}} = 9.15$ atm. The temperature at the bottom can then be computed as the bubble point temperature of the mixture at the given bottom pressure and is found to be $T_{\text{bottom}} = 147.0$ °C. The temperature of the bottom outlet stream is the same as the bottom temperature in the column, because the bottom product is extracted as a liquid. Thus, we have $T_{82} = 147.0$°C. It is lower than its critical temperature $T^m_{c,82} = 298.5$°C

Now that the conditions in the second column are set, we go back to set the conditions in the first one. As the feed into the second column, the optimal pressure of μ_{71} is in the middle of the top and bottom pressures of the second column, so it should be around $P_{\text{feed}} = 8.5$ atm; thus, this is the minimum pressure at the top of the first column. If the top pressure of the first column is specified as 8.5 atm there is no need of a valve before the top product flows into the second column. Thus, this value is selected. Since the top product is partially condensed, its temperature is the dew point of the mixture, which is computed with PRO/II and is found to be $T_{71} = 106.1$°C, higher than the temperature of cooling water so that condensation with this fluid is possible. Taking into account the pressure drop, the pressure at the bottom becomes $P_{\text{bottom}} = 9.5$ atm. The temperature at the bottom can then be computed as the bubble point temperature of the mixture at the given bottom pressure and is found to be $T_{\text{bottom}} = 177.7$ °C. The temperature of the bottom outlet stream is the same as the bottom temperature in the column, because the bottom product is extracted as a liquid. Thus, we have $T_{72} = 177.7$°C, which is lower than the critical temperature of the mixture $T^m_{c,72} = 373.8$°C.

5.4.3 Column 3

In the last column water is removed from the 89 mol% ethanol-water azeotrope plus impurities. The top product now is our final product, which we desire to have liquid and at atmospheric pressure. For this reason, the column works near to atmospheric pressure and a total condenser is used. The boiling point of the azeotrope at 1 atm is $T_{\text{top}} = 78.2$°C. Due to the presence of impurities, the temperature of the overhead liquid product is slightly lower than that and it is found to be $T_{91} = 77.6$°C.

With the assumed pressure drop of 1.0 atm, the bottom pressure is $P_{\text{bottom}} =$ 2.0 atm. The temperature at the bottom can be then computed as the bubble point temperature of the mixture at the bottom pressure and is found to be $T_{\text{bottom}} = 109.4$°C. The temperature of the bottom outlet stream is the same as the bottom temperature in the column, because the bottom product is extracted as a liquid. Thus, we have $T_{92} = 109.4$°C, which is lower than its critical temperature $T^{m}_{c,92} = 369.1$°C.

5.5 Conclusion

In the previous sections, the pressure and temperature of all streams and unit operations are determined. A summary of the operating conditions of the three columns is seen in Table 26. Table 27 presents all the streams with their temperature and pressure specifications.

Table 26 – Operating conditions for the columns.

	P_{bottom} [atm]	P_{top} [atm]	T_{bottom} [°C]	T_{top} [°C]
Column 1	8.50	9.50	106.1	177.7
Column 2	8.15	9.15	35.5	147.0
Column 3	1.00	2.00	77.6	109.4

Table 27 – Mass balance with T and P specifications for Task 5.

Stream	UOM	S01	S02	S03	S1	S2	S2A	S31	S32	S41	S42
Total Molar Component Rates	kg-mol / hr										
METHANE		0.90	0.00	0.00	192.20	192.19	192.19	179.39	12.80	178.43	0.97
ETHYLENE		96.10	0.00	0.00	1303.84	1212.58	1212.58	986.78	225.80	966.54	20.24
PROPENE		3.00	0.00	0.00	317.87	315.62	315.62	167.45	148.17	150.81	16.64
DEE		0.00	0.00	0.00	48.67	48.91	48.67	24.31	24.61	5.65	18.66
ETHANOL		0.00	0.00	0.00	0.56	91.34	91.83	11.57	79.77	0.10	11.47
IPROPANOL		0.00	0.00	0.00	0.00	2.23	2.23	0.29	1.94	0.00	0.29
WATER		0.00	780.98	31.12	782.29	689.04	688.80	25.24	663.80	1.32	55.04
Total Molar Rate	kg-mol / hr	100.00	780.98	31.12	2645.43	2551.91	2551.91	1395.03	1156.88	1302.85	123.30
Total Mass Rate	kg / hr	2836.65	14069.57	560.56	70763.73	70762.78	70762.78	40413.97	30348.81	36770.86	4203.68
Temperature	°C	300.00	300.00	26.85	195.92	300.00	300.00	120.00	120.00	64.84	141.45
Pressure	atm	68.00	68.00	68.00	68.00	68.00	68.00	67.50	67.50	67.00	67.00
Stream	**UOM**	**S4A**	**S51**	**S52**	**S6**	**S71**	**S72**	**S81**	**S82**	**S91**	**S92**
Total Molar Component Rates	kg-mol / hr										
METHANE		179.39	177.53	0.89	13.76	13.76	0.00	13.76	0.00	0.00	0.00
ETHYLENE		986.78	961.71	4.83	246.03	246.03	0.00	246.03	0.00	0.00	0.00
PROPENE		167.45	150.05	0.75	164.81	164.81	0.00	164.81	0.00	0.00	0.00
DEE		24.31	5.62	0.03	43.27	43.27	0.00	43.05	0.22	0.22	0.00
ETHANOL		11.57	0.10	0.00	91.23	90.78	0.46	0.45	90.32	89.87	0.45
IPROPANOL		0.29	0.00	0.00	2.23	1.89	0.34	0.00	1.89	0.09	1.79
WATER		56.36	1.31	0.01	718.84	71.88	646.96	0.00	71.88	10.57	61.31
Total Molar Rate	kg-mol / hr	1426.15	1296.33	6.51	1280.18	632.43	647.75	468.12	164.31	100.76	63.56
Total Mass Rate	kg / hr	40974.53	36587.00	183.85	34552.49	22856.19	11696.30	17270.51	5585.68	4352.52	1233.16
Temperature	°C	121.70	64.84	64.84	122.22	106.10	177.71	35.55	146.99	77.60	109.44
Pressure	atm	67.50	67.00	67.00	67.00	8.50	9.50	8.15	9.15	1.00	2.00

6 Task 6

In Task 5 the temperatures and pressures were specified for each stream and unit operation. Now the auxiliary equipment needed to obtain the desired conditions must be placed; this includes the positioning of heat exchangers, pumps, compressors and valves. Then the energy balance is performed, also computing the duties and the power required to run the plant at steady state conditions.

6.1 Equipment

The feed of the reactor should be at a pressure of 68 atm and a temperature of 300 °C while the raw materials and recycle streams are at lower pressure and temperature. Compressors or pumps and heat exchangers are therefore needed to increase the pressure and temperature of these streams before they are fed to the reactor. First the streams are compressed and mixed, then the mixture will be heated with a heat exchanger to the reaction temperature 300 °C. The stream is heated up to the reactor temperature after compression so that the gaseous volume rate to treat is smaller with consequent reduction cost, even if the problems related to operating the heat exchanger under pressure must be considered. It is assumed that there is a pressure drop of 0.5 atm in the heat exchanger while the pressure drop in the mixer can be neglected. The various streams are mixed after they have been separately compressed because the degree to which their pressure must be increased is different. Additionally, there would be instability when streams of different pressures are mixed, which is not desired. The raw material ethylene is at 1 atm and 25°C in vapor phase. Considering the pressure drop in the heat exchanger, it is compressed to 68.5 atm. Also, the vapor recycle streams μ_{51} and μ_{81} are compressed to 68.5 atm from 67.0 atm and 12.0 atm, respectively. The raw material water is at 1 atm and 25°C in liquid phase, and it must be converted to vapor before it is mixed with other vapor streams. Since it is easier to pump a liquid than to compress a vapor, water is pumped to 69 atm and then vaporized in a heat exchanger with pressure drop of 0.5 atm. Thus, μ_{01}, μ_{02}, μ_{51} and μ_{81} are all at 68.5 atm when they finally go into the mixer.

Also the water sent to the absorber must be compressed. The solvent water is originally at 1 atm, much lower than the pressure in the absorber, to whose top water is inserted is at 67.0 atm. Therefore a pump is added to increase the pressure of the solvent water to 67.0 atm.

The regulation of the pressure of the column is done expanding the feed to the desired pressure. If we consider the feed to the first columns, the pressure of the mixture of μ_{32} and μ_{42} is 67.5 atm. However, the optimal pressure for the feed is 9 atm, in the middle of the top and bottom pressure. Thus, a valve should be equipped in μ_6 to decrease its pressure. Also, the bottom product of the second column should go through a valve and expand to 1.5 atm before it is fed into the last column.

The stream coming out from the reactor, μ_2, is at 300°C and 68 atm. In the flash its temperature decreases to 120°C, while the pressure becomes 67.5

atm for the only effect of pressure drop. No heat exchanger is added for this part because it is assumed that the flash owns a proper cooling jacket which is able to maintain the desired temperature in it. Also no heat exchanger is added to cool the stream at the top of the flash to the temperature of the absorber, because this stream is at its dew point temperature and cooling it down would cause part of the stream to become liquid.

After all the necessary units have been added, the flow sheet looks like Figure 12.

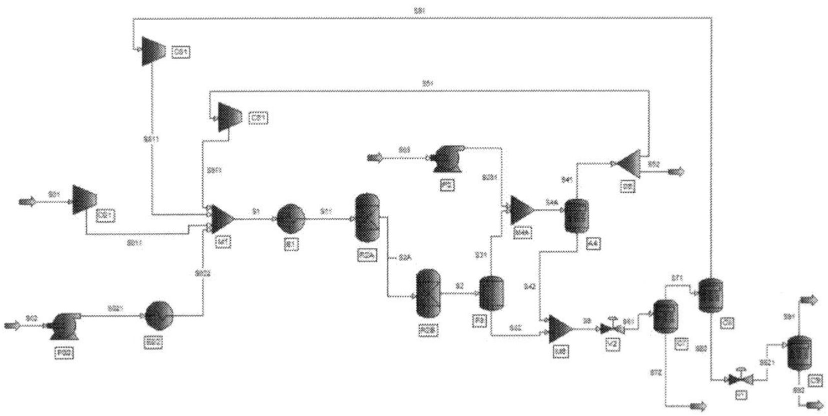

Figure 12 – Flow sheet for Task 6.

6.2 Solution strategy

As explained in the previous section, the pressure drop in each heat exchanger is assumed to be 0.5 atm while no pressure drop is set for the mixer. As assumed in Task 5, the pressure drop in each piece of equipment is also 0.5 atm. The compressors' efficiency is set to 0.8 and the pumps' efficiency is 0.5.

For each piece of equipment, the proper thermodynamic model is chosen. In addition to what was discussed in the previous task about the model for each separation unit, thermodynamic models are selected to make the newly added units of compressors, pumps and valves work properly. For the pumps the NRTL model is chosen, so that the non-ideality of the liquid water is properly considered. The same is true for the heat exchanger in which water is vaporized. For all the compressors, one of the equation of state (EOS) models, SRK, is used. It is because that the compressor model in PRO/II requires methods for vapor entropy and liquid entropy to be included in its thermodynamic method set, while EOS models are better at providing entropy than activity coefficient models (NRTL is one of them).Since the mixer in front of the reactor is treating gases, we use ideal model for the same reasons why it was chosen for the reactors. For the mixer in front of the absorber instead the NRTL model is used because both a liquid and a vapor stream are treated.

6.3 Results

In Table 29 a complete version of the properties of each stream is shown. The difference observed in the flow rates of each stream with respect to the previous task is due to the tolerance with which the solver is asked to converge, which in our case is 0.01%. In Table 28 the duty of each unit operation is seen. The convention adopted is that a positive Q represents energy given to the system, while a negative Q represents energy subtracted.

Table 28 – Unit duties for basis of 100 kmol/hr ethylene feed.

	Unit	Q (MJ/hr)
Compressor	C01	1862.62
	C51	4984.62
	C81	5787.88
Reactor	R2A	-2388.66
	R2B	6.18
Flash	F3	-52381.66
Heat Exchanger	E02	36874.73
	E1	15978.22
Pump	P02	194.75
	P2	7.57
Absorber	A4	-4791.23
Column	C7	8921.40
	C8	-6091.33
	C9	-1232.90

The overall energy balance is verified: the energy entering the system must be equal to the energy exiting it. The energy entering in the system is the sum of the enthalpies of the inlet streams and all the positive duties for the equipment, while the energy exiting it is the sum of the enthalpies of the outlet streams and the absolute value of all the negative duties for the equipment. The total enthalpy of a selected stream m_j is computed as the specific mass enthalpy times the total mass flow rate:

$$\Delta H_j = \Delta h_j m_j \tag{6.1}$$

From the results obtained in Tables 28 and 29 we can verify the total energy balance (units are MJ/hr):

$$\begin{aligned}
\text{IN} &= \Delta H_{S01} + \Delta H_{S02} + \Delta H_{S03} + Q_{C01} \\
&\quad + Q_{C51} + Q_{C81} + Q_{R2B} + Q_{E02} + Q_{E1} + Q_{P02} + Q_{P2} + Q_{C7} \\
&= 1004.55 + 1477.20 + 58.73 + 1862.62 + \\
&\quad 4984.62 + 5787.88 + 6.18 + 36874.73 + 15978.22 + 194.75 + 7.57 + 8921.40 \\
&= 77158.45 \\
\text{OUT} &= \Delta H_{S52} + \Delta H_{S72} + \Delta H_{S91} + \Delta H_{S92} + \\
&\quad Q_{R2A} + Q_{F3} + Q_{A4} + Q_{C8} + Q_{C9} \\
&= 38.74 + 8789.70 + 895.80 + 548.21 + \\
&\quad 2388.66 + 52381.66 + 4791.23 + 6091.33 + 1232.90 \\
&= 77158.23
\end{aligned}$$

$$\text{Relative error} = 0.0 \qquad (6.2)$$

The total flow in plus the produced equals the total flow out. The total energy balance of the system is conserved.

Table 29 – Mass and energy balance for Task 6.

Stream	UOM	S01	S011	S02	S021	S022	S03	S031	S1	S11	S2
Phase		Vapor	Vapor	Liquid	Liquid	Vapor	Liquid	Liquid	Mixed	Vapor	Vapor
Total Molar Component Rates	kg-mol / hr										
METHANE		0.90	0.90	0.00	0.00	0.00	0.00	0.00	190.57	190.57	190.57
ETHYLENE		96.10	96.10	0.00	0.00	0.00	0.00	0.00	1303.84	1303.84	1212.57
PROPENE		3.00	3.00	0.00	0.00	0.00	0.00	0.00	319.20	319.20	316.97
DEE		0.00	0.00	0.00	0.00	0.00	0.00	0.00	48.70	48.70	48.95
ETHANOL		0.00	0.00	0.00	0.00	0.00	0.00	0.00	0.56	0.56	91.34
IPROPANOL		0.00	0.00	0.00	0.00	0.00	0.00	0.00	0.00	0.00	2.24
WATER		0.00	0.00	780.52	780.52	780.52	31.03	31.03	781.83	781.83	688.57
Total Molar Rate	kg-mol / hr	100.00	100.00	780.52	780.52	780.52	31.03	31.03	2644.71	2644.71	2551.20
Total Mass Rate	kg / hr	2836.65	2836.65	14061.34	14061.34	14061.34	559.01	559.01	70787.74	70787.81	70787.74
Temperature	°C	25.00	350.13	25.00	28.31	295.24	25.00	28.24	213.32	300.00	300.00
Pressure	atm	1.00	68.50	1.00	69.00	68.50	1.00	67.50	68.50	68.00	68.00
Total Specific Enthalpy	kJ / kg	354.13	1010.76	105.05	118.90	2741.32	105.05	118.60	915.18	1140.90	1107.25
Total Enthalpy	MJ/hr	1004.55	2867.16	1477.20	1671.95	38546.68	58.73	66.30	64783.83	80762.06	78379.58
Stream	UOM	S2A	S3	S31	S32	S41	S42	S4A	S51	S511	S52
Phase		Vapor	Mixed	Vapor	Liquid	Vapor	Liquid	Mixed	Vapor	Vapor	Vapor
Total Molar Component Rates	kg-mol / hr										
METHANE		190.57	190.57	177.87	12.71	176.91	0.96	177.87	176.02	176.02	0.88
ETHYLENE		1212.57	1212.57	986.55	226.03	966.31	20.23	986.55	961.48	961.48	4.83
PROPENE		316.97	316.97	168.06	148.90	151.36	16.70	168.06	150.60	150.60	0.76
DEE		48.70	48.95	24.28	24.67	5.64	18.64	24.28	5.61	5.61	0.03
ETHANOL		91.83	91.34	11.55	79.79	0.10	11.44	11.55	0.10	0.10	0.00
IPROPANOL		2.24	2.24	0.29	1.95	0.00	0.29	0.29	0.00	0.00	0.00
WATER		688.33	688.57	25.20	663.38	1.32	54.91	56.23	1.31	1.31	0.01
Total Molar Rate	kg-mol / hr	2551.20	2551.20	1393.79	1157.42	1301.64	123.17	1424.82	1295.13	1295.13	6.51
Total Mass Rate	kg / hr	70787.74	70787.80	40404.84	30382.90	36762.83	4201.03	40963.86	36579.01	36579.01	183.81
Temperature	°C	300.00	120.00	120.00	120.00	64.89	141.38	119.21	64.89	76.67	64.89
Pressure	atm	68.00	67.50	67.50	67.50	67.00	67.00	67.50	67.00	68.50	67.00
Total Specific Enthalpy	kJ / kg	1107.16	359.16	353.27	385.87	210.76	428.72	350.07	210.76	347.03	210.76
Total Enthalpy	MJ/hr	78373.40	25424.08	14274.02	11723.90	7748.03	1801.06	14340.32	7709.29	12693.91	38.74
Stream	UOM	S6	S61	S71	S72	S81	S811	S82	S821	S91	S92
Phase		Mixed	Mixed	Vapor	Liquid	Vapor	Vapor	Liquid	Mixed	Liquid	Liquid
Total Molar Component Rates	kg-mol / hr										
METHANE		13.66	13.66	13.66	0.00	13.66	13.66	0.00	0.00	0.00	0.00
ETHYLENE		246.26	246.26	246.26	0.00	246.26	246.26	0.00	0.00	0.00	0.00
PROPENE		165.61	165.61	165.61	0.00	165.61	165.61	0.00	0.00	0.00	0.00
DEE		43.31	43.31	43.31	0.00	43.09	43.09	0.22	0.22	0.22	0.00
ETHANOL		91.23	91.23	90.78	0.46	0.45	0.45	90.32	90.32	89.87	0.45
IPROPANOL		2.24	2.24	1.90	0.34	0.00	0.00	1.90	1.90	0.09	1.80
WATER		718.29	718.29	71.83	646.46	0.00	0.00	71.83	71.83	10.57	61.26
Total Molar Rate	kg-mol / hr	1280.59	1280.59	633.34	647.25	469.07	469.07	164.26	164.26	100.75	63.52
Total Mass Rate	kg / hr	34583.93	34583.93	22896.49	11687.44	17311.35	17311.35	5585.14	5585.14	4352.37	1232.77
Temperature	°C	122.21	102.05	106.05	177.71	35.54	182.69	146.99	89.76	77.60	109.41
Pressure	atm	67.00	9.00	8.50	9.50	8.15	68.50	9.15	1.50	1.00	2.00
Total Specific Enthalpy	kJ / kg	391.08	391.08	596.45	752.06	282.38	616.72	479.29	479.29	205.82	444.70
Total Enthalpy	MJ/hr	13524.96	13524.96	13656.66	8789.70	4888.42	10676.30	2676.91	2676.91	895.80	548.21

7 Task 7

In Task 7 the stream calculators put in place of the columns are substituted with rigorous models. In this way a more realistic estimation of the performance of the separation units, and thus of the whole process, can be done. First, the operating parameters of the columns like the minimum reflux ratio and the minimum number of trays are computed with short-cut methods embedded in the process simulator, to which follow the actual reflux and number of trays as well as the feed tray. Then, these parameters are used in the rigorous distillation columns.

7.1 Short-cut methods

The short-cut method is used to compute the relevant parameters. The UNIQUAC thermodynamic method is used since we have non-ideal vapour-liquid equilibrium at pressures of lower than 12 bar[27]. The calculations are performed on the basis of the specifications required for each column and using as feed stream the stream computed in the non-rigorous mass balance. In the embedded short-cut methods in PRO/II, the minimum number of trays is computed with the Fenske equation and the minimum reflux ratio is computed with the Underwood equation. Then a proper combination of number of trays and reflux is chosen. We make simulations with $\frac{R}{R_{min}} \leq 1.5$ and the results given allow to see the effect of a variation of reflux ratio around this value. If we choose a high value for the reflux ratio more top product is recycled to the column and retreated, so we will have a lower number of trays with consequent lower operating costs; however, we will have a high duty to the condenser with consequent higher operation costs. The opposite is true if we select a lower reflux ratio. We prefer to use a low reflux ratio and have a higher number of trays in order to keep the operating costs low.

7.1.1 Column 1

For the first column we specify the recoveries of the light component being ethanol and of the heavy component being water as selected in Task 4 (see Table 18) and we select a partial condenser at the top. The minimum number of stages is N_{min}=3.94 and the minimum reflux ratio is R_{min}= 0.171. PRO/II returns the results for the variation of the operating parameters as showed in Appendix C. As discussed above, it can be seen that the duties at the condenser and the reboiler decrease as the number of trays increases. We choose to operate the column with N=10, in order to have reasonable duties at the condenser and the reboiler. Thus, R=0.256 and the feed is fed around the second tray. This positioning of the feed is just approximative.

7.1.2 Column 2

For the second column we specify the recoveries of the light component being diethylether and of the heavy component being ethanol as selected in Task 4 (see Table 18) and we select a partial condenser at the top. The minimum number of stages is N_{min}=4.32 and the minimum reflux ratio is R_{min}=0.459. The variation of the operating parameters are shown in Appendix C. We choose to operate the column with N=11, in order to have reasonable duties at the condenser and the reboiler. Thus, R=0.573 and the feed is fed around the sixth tray.

7.1.3 Column 3

For the third column, problems arise if the recoveries used in the previous tasks are specified. This is because of the high complexity of the system of components involved: water, ethanol and isopropanol form binary and ternary azeotropes between each other, as noted before. The composition of the top product would be determined by these interactions between the components, which are not taken into account by the simple model used to compute the recoveries in the previous sections, but they are considered by the model used to simulate the short-cut column. Thus, we are forced to select different recoveries in order to make the separation feasible. We choose water as a heavy key component and specify its recovery so that it is in azeotropic composition with respect to ethanol, as we did in Task 4; however, we do not specify any recovery for isopropanol and we let it vary according to the equilibrium model selected for the system. Thus, the two specification required are given as top recoveries of ethanol and water. Then, in this column we will have a total condenser. The minimum number of stages is N_{min}=9.22 and the minimum reflux ratio is R_{min}=1.253. The variation of the operating parameters are shown in Appendix C. We choose to operate the column with N=18, in order to have low duties at the condenser and the reboiler. Thus, R=1.723 and the feed is fed to the fifth tray. In this case we obtain a molar fraction of isopropanol in the top product of around 2%. This high percentage of impurity in the product is not acceptable. This issue is addressed after the stream calculators are replaced by the rigorous columns, as discussed in the following section.

The operating parameters for all three columns are seen in Table 30.

Table 30 – Operating parameters for columns.

	N	N_{feed}	R/$R_{\min}$	R	Duty [10^3MJ/hr] Condenser	Reboiler
Column 1	10	1.850	1.500	0.256	-6.90	15.42
Column 2	11	6.210	1.250	0.573	-7.581	1.547
Column 3	18	4.650	1.375	1.723	-10.950	4.002

7.2 Rigorous methods and results

The stream calculators are then replaced with rigorous column models. Each column is first converged independently and then inserted in the complete flow sheet. This operation is done sequentially for the first, the second and then the third column. The specifications given to each column are the recoveries of the light and key components, the number of trays, and the pressure drop. The number of trays and the feed tray are selected on the basis of the range indicated by the short-cut computations and sometimes with a slight adjustment by one tray in order to make the column converge. The feed tray number for the three columns is seen in Table 31.

Table 31 – Number of trays and feed tray number for the columns.

	N	$N_{feedtray}$
Column 1	10	2
Column 2	11	5
Column 3	18	6

The final product contains about 2 mol% isopropanol while it is desired to have less than 0.1 mol % to assure the purity of the product. Since the separation of isopropanol from the other components is not feasible in the third column, it must be performed in the first one, where it should be removed with water at the bottom. According to the simulation performed, in the first column most isopropanol goes up to the top stream instead of the bottom stream. However, most isopropanol should go out of the system from the bottom, which is the fact in practice. The problem is that the thermodynamic models provided in PRO/II are not good at predicting the behavior of ternary azeotrope of ethanol, water and isopropanol. To solve this problem, further research of a thermodynamic model suitable for the ternary azeotrope is conducted as an external study. Before such a model is found, a steam calculator is placed in the top stream of the first column to make 95% isopropanol in it flow into the bottom stream, thus the actual first column is the combination of the first column in the simulation and this stream calculator. Once the model is found for the first column, the modification in the simulation is only within the first column and other parts do not need to be changed. The modification of the quantity of isopropanol in the top stream of the first column causes the inlet streams of the second and the third column to be different; however, this does not affect the short-cut calculations because the specifications were not related to isopropanol, which is also in very low quantity compared to the streams of the other components. So there is no need to perform the short-cut computations again. The flow sheet is showed in Figure 13.

In the rigorous simulation it turns out to be impossible to make the last column converge with the specification that the top pressure is 1 atm and the recovery of water corresponds to the composition of ethanol-water azeotrope. The reason is easily found in the equilibrium diagram obtained with the chosen thermodynamic model, which is shown in Figure 14. The irregularity of the equilibrium curve is such that there is no possibility of obtaining a product exactly at azeotropic composition if the top pressure is 1 atm, because even if an infinite number of stages are used it is not possible to obtain product in azeotropic composition without intersecting the equilibrium line. The slope of the rectification curve in the equilibrium diagram is in fact limited by the

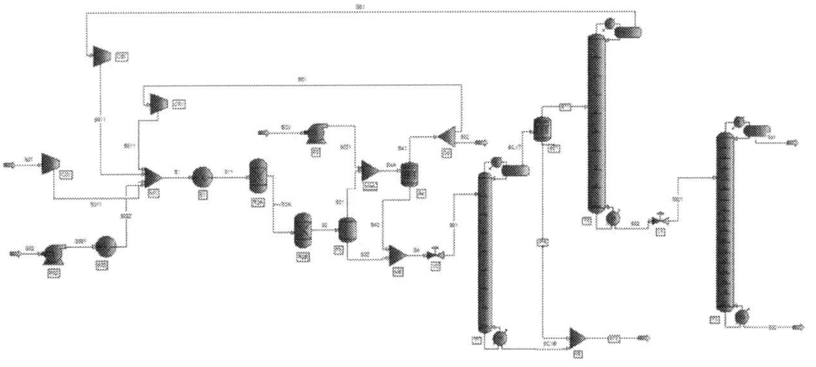

Figure 13 – Flow sheet for Task 7.

tangent to the equilibrium line. If one aims at a product of exactly azeotrope composition, the column must be operated under vacuum conditions, since at lower pressure the equilibrium curve shifts so to allow to reach concentrations of ethanol-water nearer to the azeotrope; however, this is practical only for small amount of production. Since our production is high and must be done continuously, it is not worth using a vacuum column which is quite difficult to operate. We therefore choose to keep the last column operating with the top pressure of 1 atm and obtain a little lower concentration of ethanol than the azeotrope composition in the product. With the given specifications, the composition of the product obtained is 86.1% ethanol and 13.6% water, with 0.3% of impurities, mainly constituted by diethilether; the azeotropic composition instead is 89.5% ethanol and 10.5% water. We believe that this small difference is acceptable.

The mass balance, the energy balance and the operating parameters obtained for the complete flow sheets with the rigorous units is seen in Table 33. The pressure remains specified for each unit, so the temperature is regulated as the equilibrium temperature of the streams. If the mass balance is compared with the one obtained in the previous task, we see that the flow rates are a bit different: this is because of the effect of non-ideality of the mixture which is now taken into account into the columns, and which affects also the previous flow rates because of the recycle.

The duties of the units are seen in Table 32. From the results in Table 32 and Table 33 we can verify the total material balance and the total energy balance.

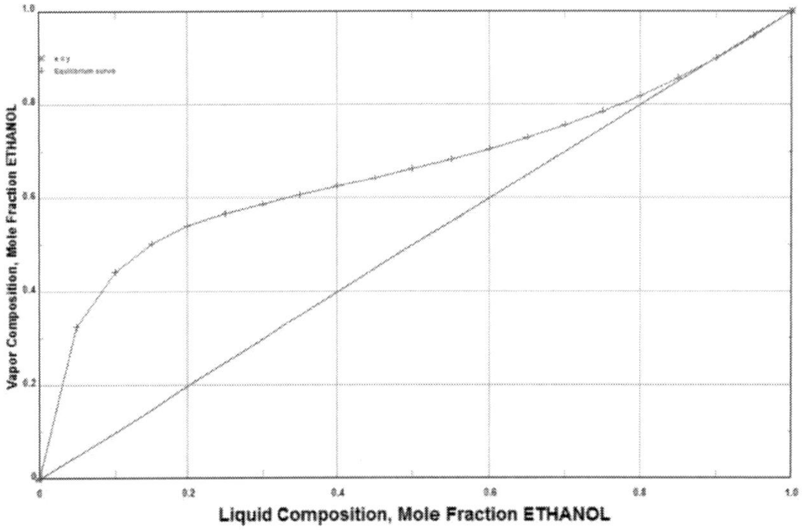

Figure 14 – X-Y plot for ethanol and water at 1 atm.

For the mass balance (units: kg/hr):

$$\begin{aligned}
\text{IN} &= S01 + S02 + S03 \\
&= 2836.65 + 14067.82 + 560.51 \\
&= 17464.98 \\
\text{OUT} &= S52 + S72 + S91 + S92 \\
&= 183.74 + 11823.09 + 4417.24 + 1139.11 \\
&= 17463.18
\end{aligned} \tag{7.1}$$

Relative error = 0.00

For the energy balance (units: MJ/hr):

$$
\begin{aligned}
\text{IN} &= \Delta H_{S01} + \Delta H_{S02} + \Delta H_{S03} + Q_{C01} + Q_{C51} + \\
&\quad Q_{C81} + Q_{R2B} + Q_{E02} + Q_{E1} + Q_{P02} + Q_{P2} + Q_{SCI} + Q_{T1} \\
&= 215.04 + 1477.88 + 58.88 + 2849.49 + 4995.26 + \\
&\quad 5824.69 + 6.17 + 36891.73 + 15690.11 + 194.75 + 7.59 + 0.04 + 8922.35 \\
&= 77134.08 \\
\text{OUT} &= \Delta H_{S52} + \Delta H_{S72} + \Delta H_{S91} + \Delta H_{S92} + \\
&\quad Q_{R2A} + Q_{F3} + Q_{A4} + Q_{T2} + Q_{T3} \\
&= 38.63 + 8855.71 + 915.56 + 512.22 + \\
&\quad 2388.33 + 52417.64 + 4815.94 + 5996.40 + 1192.71 \\
&= 77133.13
\end{aligned}
$$

$$\text{Relative error} = 0.00 \tag{7.2}$$

Table 32 – Duties for units in Task 7.

	Unit	Q (MJ/hr)
Compressor	C01	2849.49
	C51	4995.26
	C81	5824.69
Reactor	R2A	-2388.33
	R2B	6.17
Flash	F3	-52417.64
Heat exchanger	E02	36891.73
	E1	15690.11
Pump	P02	194.84
	P2	7.59
Absorber	A4	-4815.94
Stream calc.	SC1	0.04
Columns	T1	8922.35
	T2	-5996.40
	T3	-1192.71

Table 33 – Mass and energy balance for Task 7.

Stream	UOM	S01	S011	S02	S021	S022	S03	S031	S1	S11	S2	S2A
Phase		Vapor	Vapor	Liquid	Liquid	Vapor	Liquid	Liquid	Mixed	Vapor	Vapor	Vapor
Total Molar Component Rates	kg-mol / hr											
METHANE		0.90	0.90	0.00	0.00	0.00	0.00	0.00	193.32	193.32	193.32	193.32
ETHYLENE		96.10	96.10	0.00	0.00	0.00	0.00	0.00	1303.83	1303.83	1212.56	1212.56
PROPENE		3.00	3.00	0.00	0.00	0.00	0.00	0.00	316.13	316.13	313.88	313.88
DEE		0.00	0.00	0.00	0.00	0.00	0.00	0.00	48.58	48.58	48.82	48.58
ETHANOL		0.00	0.00	0.00	0.00	0.00	0.00	0.00	0.56	0.56	91.34	91.83
IPROPNOL		0.00	0.00	0.00	0.00	0.00	0.00	0.00	0.00	0.00	2.21	2.21
WATER		0.00	0.00	780.88	780.88	780.88	31.11	31.11	783.56	783.56	690.32	690.08
Total Molar Rate	kg-mol / hr	100.00	100.00	780.88	780.88	780.88	31.11	31.11	2645.97	2645.97	2552.46	2552.46
Total Mass Rate	kg / hr	2836.65	2836.65	14067.82	14067.82	14067.82	560.51	560.51	70724.00	70724.00	70722.72	70722.72
Temperature	°C	25.00	375.64	25.00	28.31	295.24	25.00	28.24	212.59	300.00	300.00	300.00
Pressure	atm	1.00	68.50	1.00	69.00	68.50	1.00	67.50	68.50	68.00	68.00	68.00
Total Specific Enthalpy	kJ / kg	75.81	1080.34	105.05	118.90	2741.32	105.05	118.60	920.54	1142.39	1108.72	1108.63
Total Enthalpy	MJ / hr	215.04	3064.54	1477.88	1672.72	38564.45	58.88	66.48	65104.33	80794.44	78411.37	78405.20

Stream	UOM	S3	S31	S32	S41	S42	S4A	S51	S511	S52	S6	S61
Phase		Mixed	Vapor	Liquid	Vapor	Liquid	Mixed	Vapor	Vapor	Vapor	Mixed	Mixed
Total Molar Component Rates	kg-mol / hr											
METHANE		193.32	180.44	12.88	179.47	0.97	180.44	178.57	178.57	0.90	13.85	13.85
ETHYLENE		1212.56	986.68	225.88	966.44	20.24	986.68	961.61	961.61	4.83	246.12	246.12
PROPENE		313.88	166.49	147.39	149.94	16.55	166.49	149.19	149.19	0.75	163.94	163.94
DEE		48.82	24.29	24.53	5.64	18.65	24.29	5.62	5.62	0.03	43.18	43.18
ETHANOL		91.34	11.58	79.76	0.10	11.47	11.58	0.10	0.10	0.00	91.23	91.23
IPROPNOL		2.21	0.29	1.93	0.00	0.29	0.29	0.00	0.00	0.00	2.21	2.21
WATER		690.32	25.27	665.05	1.32	55.07	56.39	1.31	1.31	0.01	720.11	720.11
Total Molar Rate	kg-mol / hr	2552.46	1395.04	1157.42	1302.92	123.23	1426.15	1296.41	1296.41	6.51	1280.65	1280.65
Total Mass Rate	kg / hr	70722.72	40387.17	30335.55	36748.12	4199.56	40947.68	36564.38	36564.38	183.74	34535.11	34535.11
Temperature	°C	120.00	120.00	120.00	64.78	141.46	119.21	64.78	76.59	64.78	122.22	94.92
Pressure	atm	67.50	67.50	67.50	67.00	67.00	67.50	67.00	68.50	67.00	67.00	9.00
Total Specific Enthalpy	kJ / kg	359.41	353.53	386.20	210.25	429.20	350.31	210.25	346.86	210.25	391.43	391.43
Total Enthalpy	MJ / hr	25418.24	14278.06	11715.66	7726.16	1802.44	14344.54	7687.53	12682.78	38.63	13518.11	13518.12

Stream	UOM	IPA	SC1B	SC1T	S71	S72	S81	S811	S82	S821	S91	S92
Phase		Liquid	Liquid	Vapor	Vapor	Mixed	Vapor	Vapor	Liquid	Mixed	Liquid	Liquid
Total Molar Component Rates	kg-mol / hr											
METHANE		0.00	0.00	13.85	13.85	0.00	13.85	13.85	0.00	0.00	0.00	0.00
ETHYLENE		0.00	0.00	246.12	246.12	0.00	246.12	246.12	0.00	0.00	0.00	0.00
PROPENE		0.00	0.00	163.94	163.94	0.00	163.93	163.93	0.01	0.01	0.01	0.00
DEE		0.00	0.00	43.18	43.18	0.00	42.96	42.96	0.22	0.22	0.22	0.00
ETHANOL		0.00	0.46	90.78	90.78	0.46	0.45	0.45	90.32	90.32	89.85	0.45
IPROPNOL		2.10	0.00	2.21	0.11	2.10	0.00	0.00	0.11	0.11	0.11	0.00
WATER		0.00	648.10	72.01	72.01	648.10	1.36	1.36	70.65	70.65	14.15	56.52
Total Molar Rate	kg-mol / hr	2.10	648.56	632.09	629.99	650.66	468.68	468.68	161.31	161.31	104.33	56.97
Total Mass Rate	kg / hr	126.30	11696.79	22838.32	22712.02	11823.09	17255.15	17255.15	5456.87	5456.87	4417.24	1039.11
Temperature	°C	106.98	178.02	105.32	106.98	171.45	37.47	184.85	147.34	89.84	77.51	118.20
Pressure	atm	8.50	9.50	8.50	8.50	8.50	8.15	68.50	9.15	1.50	1.00	2.00
Total Specific Enthalpy	kJ / kg	317.95	753.67	596.58	598.13	749.02	287.91	625.47	480.23	480.23	207.27	492.94
Total Enthalpy	MJ / hr	40.16	8815.52	13624.94	13584.82	8855.71	4967.87	10792.56	2620.56	2620.55	915.56	512.22

8 Task 8

The objective of this task is to perform the sizing for all equipment in the process. Previously, the basis of ethylene feed has been of 100 kmol/h. When sizing the equipment, the actual feed of ethylene processed has to be considered, which is 75 million kg/year. The actual flow rate to be considered is estimated on the basis of the average running hours of a chemical plant, which is estimated to be between 7,000 and 8,000 hours per year, since shutdowns, maintenance and repair will occur. The feed flow rate is now based on the average of this, namely 7,500 hr/yr, which corresponds to 312.5 days/yr. This gives a flow rate of:

$$\frac{75,000,000\,\text{kg/yr}}{46.07\,\text{kg/kmol}}\frac{1}{7,500\,\text{hr/yr}} = 217\,\text{kmol/hr} \tag{8.1}$$

The flow sheet is now based on the new flow rate for sizing of all the equipments.

8.1 Vessels

Vessels include the reactor R2, the flash drum F3 and the mixer after the flash and the absorber M6. Only this mixer is sized as a vessel, because M1 deals with gas phase only, and thus it is an inline mixer, and M4A is a virtual mixer. The volumes of these vessels are computed in the following sections respectively. After their volume is obtained, their length L and diameter D can be determined with the assumption that the aspect ratio L/D is equal to 4, which is the optimal ratio if the bottom and top caps are four times more expensive than the sides [24].

8.1.1 Reactor

The volume of the reactor is occupied by the vapor reactants and the solid particles of catalyst. The void volume is equal to the inlet gas volume, which can be calculated from the inlet volumetric flow rate F and the residence time τ as $V_{\text{void}} = F \cdot \tau$. Then the total volume can be computed knowing the void fraction, ϵ, as:

$$V = \frac{F\tau}{1-\epsilon} \tag{8.2}$$

The volumetric flow rate is $F = 66.2\,\text{m}^3/\text{min}$. The residence time τ is assumed to be 0.5 minutes, which is a reasonable value for gas phase, and the void fraction ϵ is assumed equal to 0.5. Then the total volume is found to be $V = 66.2$ m^3. With this value of volume, the dimensions of the flash are $D = 2.8$ m and $L = 11.0$ m.

8.1.2 Flash

The volume of the flash drum is computed based on two assumptions: first, the liquid hold-up time is set to be a typical value of 5 minutes, which is dictated by maintaining a liquid buffer for on/off switching times for pumps. Second, the volume for vapor flow is the same as that for liquid. Thus the volume of the flash is:

$$V = 2F_{\mathrm{L}}\tau \tag{8.3}$$

where F_L is the liquid volume flow rate leaving the vessel and τ is the liquid hold-up time.

The volumetric flow rate of the liquid stream from the bottom of the flash is $F_{\mathrm{L}} = 116.74\ \mathrm{m}^3/\mathrm{hr}$. The volume of the flash drum is then calculated to be $V = 27.79\ \mathrm{m}^3$. With this value of volume, the dimensions of the flash are $D = 2.1$ m and $L = 8.3$ m.

In Task 6, when the additional equipment for heat exchange and pressure adjustment was settled, no heat exchanger was considered for the flash because it was assumed that it was provided with internal cooling jacket. However, considering its dimensions now, it would be difficult to maintain the flash isothermal with this type of system: it is difficult to cool quickly and effectively such a big hold-up and furthermore the temperature of the inlet streams is hundreds of degrees higher than the temperature to be maintained in the flash (it must be decreased from 300°C to 120°C). Thus, the operation of the flash may be compromised by this. Consequently a new piece of equipment is added to the flow sheet in the form of an heat exchanger in front of the flash right after the reactor. This heat exchanger takes the process stream from 300°C to 120°C.

8.1.3 Mixer

The liquid streams from the flash and the absorber come into the mixer M6 and get mixed before going to the first column. Again, the liquid hold-up time is set to be 5 minutes for safety consideration. It is assumed that the mixing vessel is 80% filled with liquid. Thus, the volume of the mixer is computed as

$$V = \frac{F_{\mathrm{L}}\tau}{0.8} \tag{8.4}$$

The total volume flow rate entering the mixer is $F_{\mathrm{L}} = 3.1\ \mathrm{m}^3/\mathrm{min}$. The volume of the mixer M6 is found to be $V = 19.6\ \mathrm{m}^3$. With this value of volume, the dimensions of the flash are $D = 1.8$ m and $L = 7.4$ m.

The sizing results for the vessels are listed in Table 34.

Table 34 – Sizing of vessels.

Unit	No.	V [m^3]	D [m]	L [m]
Reactor	R2	66.2	2.8	11.0
Flash	F3	27.8	2.1	8.3
Mixer	M6	19.6	1.8	7.4

8.2 Absorber

The number of theoretical stages was previously derived from the Kremser equation, and was found to be $N = 2$. However, in absorption operation the assumption of equilibrium on a tray is quite poor because the trays have a quite low efficiency. If an efficiency of 20% is assumed, the number of stages becomes $N = 10$. In the absorber, all the equilibrium stages are constituted by trays, so $N_{trays} = N$. Since it is a staged operation, the absorber can be dimensioned as a column. The height of the absorber can then be computed as in equation (8.15).

As for the column, it is assumed that the tray spacing is d=0.6 m. The total height occupied by the trays is computed as $h_{\text{trays}} = (N-1) \cdot d$. Then, the height of the skirt is $h_{\text{skirt}} = 1.5$ m, the extra feed space is $h_{\text{feed}} = 1.5$ m and the disengagement is $h_{\text{disengagement}} = 1.5$ m. With these values, the height of the absorber is found to be H=12.0 m.

The principal factor that determines the column diameter is the vapor flowrate. According to "*Chemical Engineering Design*" [28], the diameter of the absorber can be calculated as:

$$D = \sqrt{\frac{4W_v}{\pi \rho_v u_v}} \tag{8.5}$$

where W_v is the maximum vapor mass rate, which in our case is the mass rate of the inlet vapor stream S31, $W_v = 24.4$ kg/s, ρ_v is the density of the vapor stream of maximum mass rate, $\rho_v = 60.6$ kg/m^3, and u_v is the maximum allowable vapor velocity. The value of u_v can be estimated by the following equation:

$$u_v = (-0.171 l_t^2 + 0.27 l_t - 0.047)\left[\frac{\rho_{Lo} - \rho_{vo}}{\rho_{vo}}\right]^2 \tag{8.6}$$

where l_t is the tray space, $l_t = d = 0.6$ m, ρ_{Lo} is the density of the bottom liquid stream, which in our case is the stream S42, $\rho_L = 419.6$ kg/m^3, and ρ_{vo} is the density of the top vapor stream, which in our case is the stream S41, $\rho_{vo} = 68.2$ kg/m^3. Using these correlations, the maximum allowable vapor velocity is found to be $u_v = 0.12$ m/s and the diameter of the absorber is $D = 2.1$ m. The pressure drop is still considered to be $\Delta P = 0.5$ atm, as assumed before, and the maximum pressure for which the tower should be designed is $P_{max} = 67.5$ atm. Table 35 summarizes the parameters necessary to size the absorber. The type of trays for the absorber is chosen to be sieve tray which is commonly used for absorbers.

Table 35 – Absorber sizing parameters.

	N_{trays}	D [m]	H [m]	ΔP [atm]	P_{max} [atm]
Absorber	10	2.05	11.4	0.5	67.5

8.3 Heat exchanger

The heat exchanger used in the process are countercurrent shell and tube heat exchangers. Two types of heat exchangers are distinguished: condensers and reboilers of the columns and independent heat exchangers. Both are treated here.

8.3.1 Condensers and reboilers

The condensers and reboilers are sized by PRO/II connecting shell and tube heat exchangers to each end of every column. Thus, we have CT1, CT2, CT3 and RT1, RT2 and RT3.

At the condensers the process stream needs to be cooled down and in order to do this cooling water is used. This cold utility is used as it is the cheapest refrigerant available and the ranges of temperatures allow its use. We assume the inlet cooling water temperature to be $T_{\mathrm{W,in}} = 25$ °C and a temperature increase of $\Delta T = 15$ °C, as seen in Table 36. If the initial water temperature and the temperature increase are specified in PRO/II, the simulator computes the amount of water necessary to cool the process stream down in the condenser. Also a value for the global heat exchange coefficient must be specified. The global heat exchange coefficient U depends on type of fluid, structure of the heat exchanger, flow properties, temperature,etc. For simplicity, it is assumed that the global heat exchange coefficient is a constant and only depends on type of flows. It can be estimated based on the typical values reported in the literature. In all the columns, the hot fluid is composed by light organics and the cold fluid by water, so U=375-750 W/K·m^2 [29]; consequently, an average value of U=563 W/K·m^2 is selected. At the reboiler the process stream needs to be heated up and the hot side of the heat exchanger is chosen to be water steam. If T_B is the bottom temperature of the column, the temperature of the steam necessary to warm the process stream up is:

$$T_{\mathrm{steam}} = T_B + \Delta T_{\mathrm{approach}} \tag{8.7}$$

being $\Delta T_{\mathrm{approach}} \cong 10$°C. The corresponding pressure of the steam is obtained from saturated water tables reported in literature [30] and is seen in Table 37. If the steam pressure is specified in PRO/II the simulator computes the necessary amount of steam to heat the process stream up in the reboiler. A different value of global heat exchange coefficient is used depending on the nature of the cold fluid to treat. In the first and the third column the cold side is composed almost totally by water, for which U=1000-3500 W/K·m^2 [29], so U=2250 W/K·m^2 is selected; in the second column the cold side is composed totally by light organics, for which U=500-1000 W/K·m^2 [29], so U=750 W/K·m^2 is selected.

The results returned by the sizing made in PRO/II are acceptable if the area of the heat exchanger does not exceed A $\geq$ 1000 m^2 and if the duty to be supplied is not enormous. Area greater than the value specified would require a heat exchanger which would be too big in dimensions, and higher duties would imply much auxiliary fluid and too high operation costs. If the duties result to be unreasonably high, the reflux and the vapor reboiled must be decreased. Thus, previous to compute the effective water and steam supply necessary to operate the columns, the duties at the reboilers and condensers obtained according to the design decisions specified in the previous tasks are checked. The duties required at the reboiler and condenser of the first and second columns are not as huge as what would be the duty in the third, which with the old specifications would be $Q \geq 200{,}000$ MJ/hr in absolute value at both the reboiler and condenser. In order to reduce it, one option is to increase the number of trays, because it is preferable to have higher installation costs rather than operational costs. The number of trays is therefore increased by 1 and becomes 18, but the duties are still high. As it is not good to have a very high column, instead of further increasing the number of trays, another way to reduce the duties is used, which is adjusting the feed stage. If the feed stage is changed to a more suitable position, the vapor and liquid streams change along the column and it is possible to obtain a significant reduction of the duties required. It is found that feeding the feed stream to tray 16 instead of tray 6, the duty obtained is $Q \cong 40{,}000$ MJ/hr at both the reboiler at the condenser. This change in the feed does not alter significantly the product compositions, so this change is acceptable. Table 36 and Table 37 show the features of the reboiler and condenser for each column, as they were obtained on the basis of the considerations above.

Table 36 – Condensers.

	Q [MJ/hr]	U [W/(m^2 K)]	A [m^2]	$T_{CW,in}$ [°C]	$T_{CW,out}$ [°C]	m_W [kg/hr]
Column 1	7,544.81	563	48.02	25	40	120,035.99
Column 2	18,087.02	563	602.05	25	40	287,759.87
Column 3	43,698.20	563	483.91	25	40	695,227.07

Table 37 – Reboilers.

	Q [MJ/hr]	U [W/(m^{2}K)]	A [m^2]	P_{steam} [atm]	m [kg/hr]
Column 1	26,384.24	2250	222.92	11.71	13,261.71
Column 2	4,841.58	750	117.06	5.70	2,313.93
Column 3	41,115.96	2250	316.70	1.52	18,473.61

8.3.2 Independent heat exchangers

The heat exchangers independent from distillation units in the plant are E02, E1 and E2. In E02 and E1 the process stream constitutes the cold side and needs to be warmed up, while in E2 the process stream is to be cooled down.

The sizing for E2 is done in PRO/II, as seen for the condensers. Cooling water is assumed as utility fluid and we just need to specify a proper global heat exchange coefficient on the basis of the fluid treated as hot side. On the hot side we have mostly light hydrocarbons with some water, so the global heat exchange coefficient can be selected as U=80-200 Btu/ft^2·hr·F [30], which correspond to an average value of U=795 W/K·m^2. The inlet cooling water temperature is 25°C and a temperature increase of 15°C is specified.

A different situation is verified for E1 and E02. At first, the possible use of water steam is considered. The parameters given are the inlet and outlet process stream temperature, as specified in Task 5. If T_1 and T_2 are the inlet and outlet process stream temperature respectively, the temperature of the steam necessary to warm the process stream up is:

$$T_{\text{steam}} = T_2 + \Delta T_{\text{approach}} \tag{8.8}$$

being $\Delta T_{\text{approach}} \cong 10°C$. The corresponding pressure of the steam is obtained from saturated water tables reported in literature [30]. If the steam pressure is specified in PRO/II the simulator computes the amount of steam necessary to heat the process stream up to the desired final temperature. However, the highest temperature on the process side of these heat exchangers (which is the cold side) is $T_2 \cong 300°C$ (for E1: T_2=300°C, for E02: T_2=295°C). Thus, the temperature of the steam needed is around T_{steam}=310°C≅ 583 K; in the steam table, the value of pressure corresponding to T_{steam}=580 K is P=94.51 atm. This pressure is too high. Thus, an alternative heating fluid must be considered and hand-calculations are required to size the heat exchangers, since PRO/II does not provide such an option.

The procedure followed is reported below. First, an alternative source of heat other than steam is required. This is chosen to be a heavy oil commonly used for heat exchangers, Dowtherm A.

The design equation for a general countercurrent heat exchanger is:

$$Q = UA\Delta T_{ln} \tag{8.9}$$

where Q is the heat duty, U is the overall heat exchange coefficient, A is the total area of the heat exchanger and ΔT_{ln} is the logarithmic mean temperature which is defined as:

$$\Delta T_{ln} = \frac{(T_1 - t_2) - (T_2 - t_1)}{\ln(\frac{T_1 - t_2}{T_2 - t_1})} \tag{8.10}$$

where T_i is the hot fluid temperature and t_i is the cold fluid temperature; $i = 1$ stands for inlet and $i = 2$ stands for outlet. Thus, $(T_1 - t_2)$ would be the temperature difference at one end of the heat exchanger and at the other end we have $(T_2 - t_1)$. From the flow sheet results, we know the duty Q and the cold side inlet and outlet temperatures, t_1 and t_2, since both heat exchangers need to heat up the process stream. If we assume a certain $\Delta T_{\text{approach}}$ between the inlet hot fluid temperature and the outlet cold fluid temperature (approach on the hot side), we have T_1 set as $T_1 = t_2 + \Delta T_{\text{approach}}$; if we also set a certain temperature decrease in the hot side ΔT, we also have T_2 set as $T_2 = T_1 - \Delta T$. We choose reasonable values such as $\Delta T_{\text{approach}}$=25°C and ΔT=25°C. Then, U is found from empirical data and ΔT_{ln} can be computed; thus, the only unknown A is determined by:

$$A = \frac{Q}{U \cdot \Delta T_{ln}} \tag{8.11}$$

If the area is found to be too high ($A > 1000$ m^2), the single heat exchanger must be split into two or more heat exchangers in series.

If the heat exchanger involves phase changing fluid, this must be taken into account in the sizing, because the global heat exchange coefficient is affected by the presence of a phase-changing fluid. According to the procedure illustrated in [24], when phase changes take place in the heat exchanger, as it happens in E02, the single heat exchanger can be split into two sections, one dealing with the non-phase changing fluid and the other one dealing with the phase changing fluid. Then, the same method shown above can be used for both sections, provided that the relevant fraction of heat Q and the proper global heat exchange coefficient is chosen for each section; in the phase change section, for the mean logarithmic temperature we have $t_1 = t_2 = t_3$, being t_3 the temperature at which the phase change takes place. If 1 is the non-phase changing section and 2 is the phase changing section and m is the amount of fluid to be treated, we have $Q_1 = mc_P\Delta t$ and $Q_2 = m\Delta h_{vap}$, being Δt the temperature variation from the inlet temperature to the phase change temperature and Δh_{vap} the specific latent heat of phase change (evaporation or condensation). The total area of the heat exchanger is then given by the sum of the two areas:

$$A_{\text{tot}} = A_1 + A_2 \tag{8.12}$$

For condensers and reboilers, this was not considered since the sizing was automatically done by PRO/II once the proper global heat exchanger coefficient U was specified, thanks to the prediction of the non-ideal thermodynamic model used. Here instead, this distinction needs to be made clear to compute the correct value of the surface area.

Finally, the amount of utility necessary can be computed from the equation:

$$Q = m_u c_{P,u}(T_1 - T_2) \Rightarrow m_u = \frac{Q}{c_{P,u}(T_1 - T_2)} \tag{8.13}$$

The temperature variations of the hot fluid utility were specified before, while its heat capacity is known from literature. Considering that heavy oil is used, and assuming no phase change in the utility fluid, which is used as a liquid, the heat capacity is found as a function of temperature from experimental data as [31]:

$$c_{P,u} = 2.808 + 250.844 \cdot T[K] \tag{8.14}$$

In order to perform the computations, its heat capacity is evaluated at the mean temperature between inlet and outlet.

The global heat exchange coefficient depends also on the type of cold fluid treated. In E1 the cold fluid is composed mainly of water and light hydrocarbons, while in E02 the cold fluid is pure water, first liquid and then evaporating. The most suitable values reported in literature for the specified situation are in the range U= 20-200 W/K·m^2 for E1 [29], while for E02 U=15-50 Btu/ft^2·hr·F in the non evaporating part [30] and U=30-300 W/K·m^2 in the evaporating part [29]. Thus, the average values of U=184.5 W/K·m^2 and U=165 W/K·m^2 are selected for E02 and U=110 W/K·m^2 is selected for E1.

Heat exchanger E1 The sizing of heat exchanger E1 is quite straightforward, since no phase changing fluid is involved, and it is done as described from the procedure above. The total area computed in this way is way too high, namely 1721.48 m^2, so the heat exchanger is split into two parts of 860.74 m^2 each.

Heat exchanger E02 In E02 phase change is involved, and the sizing is done as described in the procedure above. The heat exchanger is split into two parts, one in which water is heated and one in which water is evaporated and taken 10°C above its dew point. The duty of each part is specified as: $Q_1 = mc_P\Delta t$, being c_P the specific heat capacity of water evaluated at its mean temperature [5], and $Q_2 = Q_{tot} - Q_1$. It is assumed that the same flow of heating fluid flows first in the phase-changing part and then in the non-phase changing part. The temperature approach between cold and hot fluid is imposed at the separation between the two, and in this way the minimum temperature approach is assured in any part of the heat exchanger; this is obtained if the inlet temperature of Dowtherm A is set 35°C above the outlet temperature of water. Once the inlet and outlet temperature of the utility is known, as well as the duty, we can compute the mass flow. The utility flow rate is the same that flows in the non-phase changing section of the heat exchanger. In this section, knowing the duty, the outlet temperature of the utility is determined and thus the computations of the areas are possible in each part. For the first part an acceptable value for the area is obtained, namely 460.23 m^2, while for the second one it is necessary to split the heat exchanger of 2609.05 m^2 in three parts of 869.68 m^2 each. The results are summarized in Table 38.

Table 38 – Independent heat exchangers.

	Q [MJ/hr]	U [W/m$^2 \cdot K$]	A [m^2]	N	P [atm]	Utility fluid	m [kg/s]
E2	113,468.54	563	347.83	1	68.0	Cooling water	1547.72
E1	34,000.75	110	860.74	2	68.5	Dowtherm A	157.69
E02-1	79,937.30	200	460.23	1	69.0	Dowtherm A	264.62
E02-2	12,794,302.63	165	869.68	3	69.0	Dowtherm A	264.62

The use of Dowtherm A as a utility fluid is translated in additional costs. While steam can be bought and returned as it is, this is not true for Dowtherm A. A circuit to heat it up and recycle it must be present, and this is translated in additional costs for a furnace and piping. The heat duty of the furnace will be equal to the duty to be supplied to the heat exchangers.

8.4 Columns

The sizing of the distillation columns is done by PRO/II. A common tray column is used. The tray type must be selected. We can choose between a sieve tray, a valve tray or a bubble cap tray. The sieve tray is cheap and offers a low pressure drop, but has the disadvantage of allowing some fluid to go down from the holes instead from the downcomers. The valve tray does not have this problem, but it is more expensive and has a higher pressure drop. A compromise between the two is the bubble cap tray, which is consequently selected. The inputs to be specified are the distance between two trays and the flooding factor. The tray distance is assumed to be d=0.6 m according to common practice. The nearer the system is to flooding the more effective is the mass transfer between vapor and liquid, but the drawback may be instability of the column; so we are interested in keeping it as high as reasonable and we select it to be 80%.

For each column, the simulator computes the number of flow paths on the tray, the suggested diameter for each tray and the pressure drop between one tray and the other. The flow path is one for every tray. For each tray, we have the suggested diameter dimensions computed according to the liquid and vapor streams: the higher the tray, the bigger the diameter because the vapor volume flow rate is higher. In order to have the simplest and cheapest design possible, a constant diameter is chosen as the maximum required. Also the pressure drop varies from tray to tray. We select an average value for all the trays and if the total pressure drop thus obtained is different from what was specified in the simulation before (1 atm) the new value is selected as a new specification for the column. A change in the global pressure drop would change the pressure of the bottom, so also the pressure of the feed stream is modified according to the pressure drop found. When the total pressure drop is not significant ($\Delta P \leq 0.1atm$), the pressure of the feed stream is selected equal to the top pressure, in the other case is selected as an average between the top and the bottom pressure. Thus, also the outlet valve pressure and the saturation temperatures of the bottom streams are modified accordingly. The height of the column is computed as the sum of the space occupied between the trays and summing the height of the skirt, the extra space needed for the feed and the disengagement space both at bottom and top as:

$$H = h_{\text{trays}} + h_{\text{skirt}} + h_{\text{feed}} + 2h_{\text{disengagement}} \tag{8.15}$$

The total height occupied by the trays is computed as $h_{\text{trays}} = (N_{trays} - 1) \cdot d$. Then, with N being the number of stages, for a column with total condenser, it is $N_{trays} = N - 1$ so $h_{\text{trays}} = (N - 2) \cdot d$, while for a column with partial condenser another stage accounts for the condenser and $N_{trays} = N - 2$, so $h_{\text{trays}} = (N - 3) \cdot d$. Then, the height of the skirt is $h_{\text{skirt}} = 1.5$ m, the extra feed space is $h_{\text{feed}} = 1.5$ m and the disengagement is $h_{\text{disengagement}} = 1.5$ m.

Table 39 shows the necessary parameters for sizing the column: the number of trays, the diameter and the total height along with the average pressure drop for each tray and the maximum pressure, which is of importance in the design stage.

Table 39 – Column sizing parameters.

	N_{trays}	D [m]	H [m]	ΔP_{tray} [atm]	$P_{\max}$ [atm]
Column 1	8	1.87	18.7	0.006	8.55
Column 2	9	2.29	20.3	0.004	8.19
Column 3	18	2.60	33.7	0.007	1.11

8.5 Pumps

Two pumps are needed to pump liquid water into the system. Pump *P02* is used to feed the raw material water and the pump *P2* is used for the solvent water. For the two pumps the brake horsepower is calculated by:

$$W_b = \mu(P_{\text{out}} - P_{\text{in}})/(\rho\eta_p\eta_m) \tag{8.16}$$

where μ is the flow rate, P_{in} and P_{out} are the inlet and outlet pressures, respectively, and ρ is the density. The pump efficiency η_p is assumed to be 0.5 and the motor efficiency η_m is assumed to be 0.9. PRO/II also calculates the work based on the η_p specified, but does not take the motor efficiency into account; consequently, the values obtained must be corrected. The results are seen in Table 40.

Table 40 – Work of pumps.

	P_{in} [atm]	P_{out} [atm]	Flow rate [kg/hr]	W_b [kW]
P02	1	69.0	30,482	130.3
P2	1	67.7	1,215	5.1

8.6 Compressors

Three compressors are used for the streams to the mixer before the reactor. The work of the compressors is calculated by PRO/II where the efficiency is chosen to be $\eta_c = 0.8$. The actual horsepower for a compressor is given by:

$$W_b = W/(\eta_m\eta_c) \tag{8.17}$$

The work given by PRO/II takes into account η_m only, so the value obtained is divided by $\eta_m = 0.9$ to account for the motor efficiency. The work of the three compressors is seen in Table 41. All the pressures are below 70 atm (1000 psig) so it is acceptable to use a single compressor [24].

8.7 Materials of construction

Two main concerns in selection of materials of construction are their mechanical properties and corrosion resistance. A material should be easy to work and weld so that it can be processed into a desired equipment. It should have good

Table 41 – Work of compressors.

	P_{in} [atm]	P_{out} [atm]	Flow rate [kg/hr]	W_b [kW]
C01	1.00	68.5	6,155.5	1,908.5
C51	67.00	68.5	79,189.4	3,350.9
C81	8.15	68.5	37,174.5	3,881.0

tensile strength to hold the pressure, which is important especially for high pressure units. Corrosion resistance is also very important as equipment failure would happen due to corrosion. For static equipment, corrosion is of higher concern, while for equipment with moving parts, like pump and compressors, the mechanical strength.

Materials of construction in our process are selected among steels since they are relatively cheap and able to meet our requirements. Carbon steel is the most commonly used engineering material. It is cheap, available in a wild range of standard forms and sizes and it can be easily worked and welded and it has good tensile strength. It can be used for units where process fluid is not highly corrosive and mechanical strength is not required to be very high. If carbon steel is not suitable, stainless steel or a higher alloy should be considered.

The reactor can not be constructed with carbon steel, as corrosion within the reactor can be very strong due to the phosphoric acid content of the catalyst and to the high temperature and pressure. To avoid corrosion of the inside of the reactor, Incoloy 825 is recommended by Perry's Chemical Engineers' Handbook (1997) [30]. This alloy has a high nickel content and therefore is highly resistant to many aggressive and corrosive environments. This material constitutes a thin layer only to protect the internal surface from corrosion, while the bulk vessel is built of cheaper material. However, there is no need of using such expensive material for the cooling jacket outside the reactor since it only contains cooling water. Carbon steel is a better choice of the cooling jacket. According to Perry's Chemical Engineers' Handbook (1997) [30], the most common used material of construction for shell of heat exchangers is carbon steel while the tubes are usually constructed with alloys. A good choice for tubes may be Stainless Steel 304.

For compressors and pumps, carbon steel can be used because they have moving parts that require a better mechanical strength, which is related to the carbon content of the steel. Carbon steel therefore selected as the material to construct compressors. However, for pumps Stainless Steel 304 is used because pumps continuously deal with water, which is quite corrosive compared to hydrocarbons.

All the other pieces of equipment such as pressure vessels, columns, absorbers are constructed with carbon steel according to common practice in absence of strong corrosive fluid. Materials of construction selected for each unit is shown in Table 42.

Table 42 – Materials of construction.

Unit	Fluid Phase	Main components(> 1mol%)	Material
Absorber			
A4	Vapor-liquid	M,EL,PL,DEE,W	Carbon steel
Column			
T1	Vapor-liquid	M,EL,PL,DEE, EA, W	Carbon steel
T2	Vapor-liquid	M,EL,PL,DEE,EA,W	Carbon steel
T3	Vapor-liquid	EA,W	Carbon steel
Compressor			
C01	Vapor	EL,PL	Carbon steel
C51	Vapor	M,EL,PL	Carbon steel
C81	Vapor	M,EL,PL,DEE	Carbon steel
Flash			
F3	Vapor-liquid	M,EL,PL, DEE, EA, W	Carbon steel
Heat exchanger			
CT1-shell	Vapor-liquid	M,EL,PL, DEE,EA,W	Carbon steel
CT1-tube	Liquid	cooling water	Stainless steel 304
CT2-shell	Vapor-liquid	M,EL,PL,DEE	Carbon steel
CT2-tube	Liquid	cooling water	Stainless steel 304
CT3-shell	Vapor-liquid	EA,W	Carbon steel
CT3-tube	Liquid	cooling water	Stainless steel 304
E02-shell	Steam	W	Carbon steel
E02-tube	Vapor-liquid	W	Stainless steel 304
E1-shell	Steam	stream	Carbon steel
E1-tube	Vapor	EL, PL, W	Stainless steel 304
RT1-shell	Steam	W	Carbon steel
RT1-tube	Vapor-liquid	W	Stainless steel 304
RT2-shell	Steam	W	Carbon steel
RT3-tube	Vapor-liquid	EA,W	Stainless steel 304
RT3-shell	Steam	W	Carbon steel
RT3-tube	Vapor-liquid	W	Stainless steel 304
Mixer			
M1	Vapor	M,EL,PL,DEE,W	Carbon steel
M6	Liquid	M,EL,PL,DEE, EA, W	Carbon steel
Reactor			
R2-inside	Vapor	H_3PO_4,M,EL,PL, DEE, EA, W	Incoloy 825
R2-cooling jacket	Liquid	cooling water	Carbon steel
Splitter			
D5	Vapor	M,EL,PL	Carbon steel
Pump			
P02	Liquid	W	Stainless steel 304
P2	Liquid	W	Stainless steel 304

9 Task 9

The economic analysis of the design process is executed in this task. First the cost of building the plant and manufacture the product are computed and then on the basis of the revenues the profitability of the process is assessed. Some economic figures are presented, as well as profitability indexes and sensitivity analysis.

The costs of raw material, purchase cost of equipment and utility costs have to be estimated and also other factors concurring to the total operation and capital costs are taken into account. The majority of the equipment and utility costs is calculated bases on the model reported in "*Plant Design and Economics for Chemical Engineers*" [29] which is also used in the software ECON. The manufacturing costs vary significantly depending on where the chemical plant is build. Since many entries of the operational costs include labor work, it is necessary to decide where to build the plant. Since we are selling a bulk chemical the price per kg will be relatively low so we are interested in locating the chemical plant in a place where wages are low and expenses will be decreased. For this a developing country like China is a good option. Additionally, this location also provides a very promising market due to its dynamic economy, which can help to reduce the costs associated with transportation of the final product to the sales. Thus, the following economic analysis is based on China's price level for water and labor work.

9.1 Cost of raw materials

The prices of the raw materials were given in Table 5 and the amount of ethylene entering the process is 75,000,000 kg/yr. The water supplied to the reactor, stream S02, and to the absorber, stream S03, is in total 31,698 kg/hr, which can be translated into annual quantity on the basis of the estimated operating hours. The price of the impure ethylene stream is reported as 0.4 $/yr and the price of water is highly dependent on the location of the plant. The price of water for a chemical plant in China is reported as 0.003 $/kg [32]. The reactor contains catalysts, which may be considered as a raw material. However, the price of catalyst is neglected in the following analysis. The cost of the raw materials can be seen in Table 43.

Table 43 – Cost of raw materials.

Raw material	Quantity [kg/yr]	Price [$/kg]	Annual price [$/yr]
Ethylene	75,000,000	0.4	30,000,000.00
Process water	237,738,417	0.003	713,215.25
Total			30,713,215.25

9.2 Equipment purchasing cost

In the following, the price of the equipment in the plant is calculated. All equipment purchase costs except the vessels are calculated from the method of the ECON software. The price of the vessels is calculated based on the method suggested by Biegler et al. [24]. The calculation methods are presented in Appendix D.

9.2.1 Vessels

In Table 44 the relevant sizing parameter and the costs are reported. For the vessels, the significant sizing parameters are the diameter, the length, the pressure and the material, as they were selected in the previous task. If the diameter is greater than 1.2 m, the vessel is designed as horizontal. For safety reasons, the pressure for the economic evaluation is selected as 50% higher than the design operating pressure. Appendix D.1 gives the parameters used in the cost calculations.

Table 44 – Vessels costing.

	Flash	Mixer	Reactor
Orientation	Horizontal	Horizontal	Horizontal
Material	Carbon Steel	Carbon Steel	Inconel 375
D [m]	2.1	1.8	2.8
L [m]	8.3	7.4	11
P [atm]	101.25	12.75	102
Cost [$]	100,575.14	36,371.44	646,078.95

9.2.2 Distillation columns and absorber

The input parameters for the column costing are reported in Table 45. The absorber is treated as a column as it is a staged operation and it is indicated as T4. The diameter of the body of the column was chosen to be the nearest bigger integer to the tray diameter identified. For the tray diameter, instead, the values reported in Table 45 are used. The pressure is specified to be the nearest bigger value available; on the basis of the maximum pressure specified in Table 35 and Table 39 in the previous task, for all the columns the pressure specified is P_{max}=1035 kPa while for the absorber P_{max}=10000 kPa. The cost estimation for the columns is divided into cost calculations of the columns, the trays, the reboilers and the condensers. The reboilers and the condensers are treated in the heat exchanger section.

Columns The parameters and the details for the computation of the cost of the columns are seen in Appendix D.2. The cost of the columns is seen in Table 46.

Table 45 – Sizing parameters for the columns, absorber and trays.

	N_{tray}	Tray type	D_c [m]	D_{tray} [m]	H [m]	P_{max} [kPa]	Material (body)
T1	8	Bubble cap tray (CS)	2	1.9	18.7	1035	Carbon steel
T2	9	Bubble cap tray (CS)	3	2.3	20.3	1035	Carbon steel
T3	18	Bubble cap tray (CS)	3	2.6	33.7	1035	Carbon steel
T4	10	Sieve tray (CS)	3	2.1	11.4	10000	Carbon steel

Table 46 – Data and cost of columns and absorber.

	D_c [m]	H [m]	PF	MF	Cost [$]
T1	2	18.7	1.6	1.0	182,374.98
T2	3	20.3	1.6	1.0	286,566.56
T3	3	33.7	1.6	1.0	427,055.47
T4	3	11.4	4.6	1.0	518,987.80

Trays The cost of the trays is calculated as seen in Appendix D.3, where the cost for one tray is calculated and multiplied by the total number of trays for each column. To calculate the cost of the trays the price must be multiplied with a quantity factor (QF) that depends on the number of trays. The total tray price for each column can be seen in Table 47.

Table 47 – Data and cost of trays.

	N_{tray}	D_{tray} [m]	QF	Cost [$]
T1-tray	8	1.9	1.80	40,053.52
T2-tray	9	2.3	1.65	52,434.04
T3-tray	18	2.6	1.10	81,473.28
T4-tray	10	2.1	1.50	22,669.64

9.2.3 Heat exchangers

The condensers and the reboilers of all the columns are also treated here, since they are heat exchangers at all effects. The procedure for computing the cost of heat exchangers is shown in Appendix D.4.

The area of the heat exchangers depends on the utility fluid available, so the two are strictly dependent on each other. In the previous Task 8, the amount of steam necessary for the reboiler was computed assuming that steam at any pressure was available. However, the price of utilities are known from ECON for steam at low pressure (500 kPa, 160 °C), medium pressure (1,000 kPa, 184 °C) and high pressure (4,100 kPa, 254 °C). If one of these hot utilities needs to be used, the reboilers area will change if compared to the one computed previously, since the pressure and temperature at which heat is released are different. Thus, a new simulation in PRO/II is performed to obtain the area relevant for the utility selected so that the heat exchanger is dimensioned correspondingly to these utilities. Of course, also the previous areas were acceptable to be used with these utilities, but the amount of utility needed would be different. In this way, the reboilers are dimensioned to be used with exactly the specified type of utility. This is only relevant for the reboilers since the utility fluid of the other heat exchangers are either cooling water or Dowtherm A. All heat exchangers are of the type 'Shell and tube, fixed tubes'. The pressure was selected to be the nearest value available according to the pressure on the tube side. For simplicity, for the costing calculations it is assumed that the heat exchangers are built completely with carbon steel.

The cost of the heat exchanger in front of the flash, E2, is given in Table 48. The pressure here is 68 atm, equal to 6890 kPa, and the nearest value available is 10,000 kPa. For heat exchanger E1 and E02 the results are also seen in in Table 48. The maximum area of the heat exchangers when using the method of ECON is 635 m^2 [29]. Since the areas of E1 and E02 are above this more heat exchangers with smaller area are used. As reported in section 8.3 the total area of E1 is 1721 m^2 so 3 heat exchangers each with an area of 574 m^2 are used. Heat exchanger E02 is divided into one heat exchanger with an area of 460 m^2 where heating of water takes place, and then 5 other heat exchangers each with an area of 521 m^2 where the phase change takes place.

Table 48 – Data and cost of heat exchangers.

	A [m^2]	N	Pressure [kPa]	Duty [MJ/hr]	Utility fluid	Material	Cost [$]
E2	347.83	1	10,000	113,468.54	Cooling water	CS	46,483.00
E1	573.83	3	1035	34,000.75	Dowtherm A	CS	195,624.23
E02-1	460.23	1	1035	79,937.30	Dowtherm A	CS	53,671.93
E02-2	521.81	5	1035	12,794,303.63	Dowtherm A	CS	299,627.81

The sizing parameters and the costs of the columns and the reboilers are reported in Table 49.

Table 49 – Data and cost for condensers.

	A [m^2]	Pressure [kPa]	Duty [MJ/hr]	Utility fluid	Material	Cost [$]
CT1	47.96	1035	7,537.34	Cooling water	CS	11,804.54
CT2	600.82	1035	18,091.52	Cooling water	CS	67,949.23
CT3	480.98	1035	43,432.00	Cooling water	CS	55,778.98
RT1	461.91	1035	26,392.43	Medium Pres. Steam	CS	53,842.47
RT2	182.69	1035	4,838.19	Low Pres. Steam	CS	25,486.57
RT3	89.92	1035	40,848.65	Low Pres. Steam	CS	16,065.17

9.2.4 Furnace

For heating of the Dowtherm A to operate the heat exchangers E1 and E02, a furnace has to be used. The Dowtherm A is at low pressure and the type of furnace available is a carbon steel tube at 3450 kPa. The heat duty will be equal to the sum of the duties of the heat exchangers E1 and E02. The parameters for the cost calculations are seen in Appendix D.5, and the data and cost of the furnace are seen in Table 50.

Table 50 – Data and cost for furnace.

	Type	Heat duty [kW]	Cost [$]
Furnace	CS tube 3450 kPa	31642.22	2,683,686.91

9.2.5 Pumps

The data for the pumps are seen in Table 51 where also the costs are given. The calculations are seen in Appendix D.6.

Table 51 – Data and cost for pumps.

	Type	Pressure [kPa]	Capacity [m^3/hr]	Material	Cost [$]
P02	Centrifugal	10000	30.48	Stainless steel 304	26,124.50
P2	Centrifugal	10000	1.22	Stainless steel 304	12,848.70

9.2.6 Compressors

For the compressors the type and material is specified and Table 52 shows the data used as well as the cost calculated. The computation of the costs can be found in Appendix D.7.

Table 52 – Data and cost for compressors.

	Type	Duty [kW]	ΔP [atm]	Material	Cost [$]
C01	Centrifugal-motor	1,908.46	67.50	CS	1,860,881.43
C51	Centrifugal-motor	3,350.95	1.50	CS	3,165,119.86
C81	Centrifugal-motor	3,881.00	60.35	CS	3,635,487.68

9.2.7 Summary of cost of equipments

In total the cost of equipment is

$$\text{Cost}_{\text{total}} = \$14,605,123.81 \quad (9.1)$$

The pie charts in Figure 15 shows how the cost of the equipment is distributed among the units. The pie chart in Figure 16 shows the price of the equipment lumped by type. It can be seen that compressors have the highest impact, followed by the furnace and then by heat exchangers.

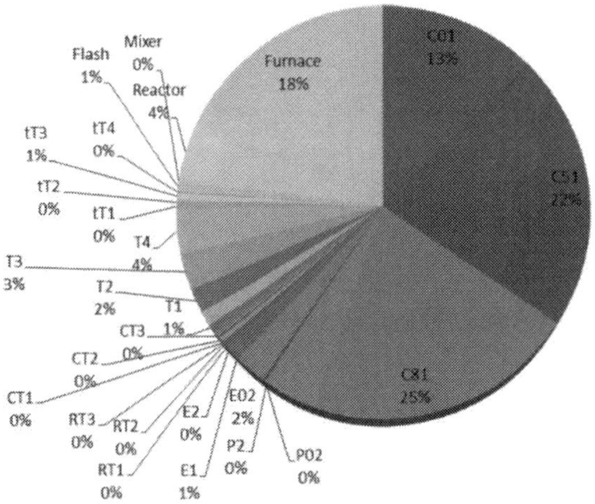

Figure 15 – Equipment purchase costs.

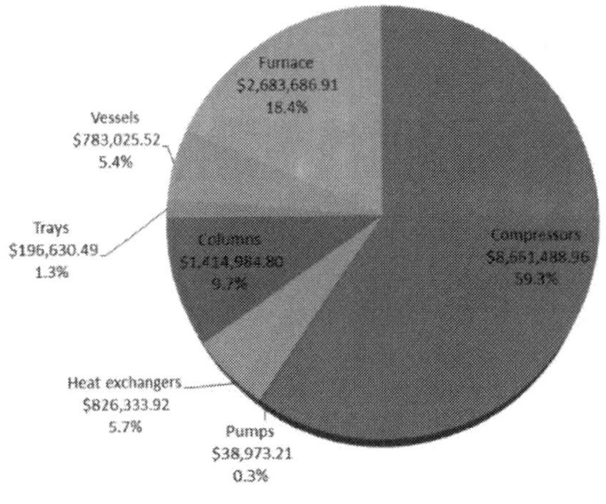

Figure 16 – Equipment purchase costs - lumped by type.

9.3 Cost of utilities

The cost of the utilities includes the electricity for the compressors, the steam, Dowtherm A and cooling water for the heat exchangers and the fuel for the furnace.

Electricity, steam and cooling water The cost of the previous utilities is computed on the basis of the input data specified and with prices given by Turton [33].

Dowtherm A The price of Dowtherm A is computed considering the flow rate needed for each heat exchanger reported in Table 38 and a reasonable design for the recycle of the utility stream. It is assumed that all the Dowtherm A is treated in the same furnace and it travels along a pipe of 100 m of length with a velocity of 1.5 m/s. With these assumptions, the amount of mass flowing in the cycle can be computed and it is found to be m_D=28,153.2 kg for both heat exchangers, which equals to a total volume to purchase of 34,814.5 L (data on density are available from the supplier [31]). The price per unit volume of Dowtherm A is 93.90 €/L [34], which leads to an expense of \$ 4,737,599.63. However, since the Dowtherm A is continuously recycled, this price has to be divided over the lifetime of the utility purchase, which can be reasonably assumed to be 5 years.

Furnace fuel Coal was chosen as the fuel for the furnace since this is most available fuel in China, where the plant is placed.

The list of the utility cost per year for each equipment is seen Table 53 and the contribution of each equipment is seen in the pie chart in figure 17. The contribution to the cost of each type of utility instead is seen in Figure 18. It can be seen that the most significant expense goes for electricity and steam.

Table 53 – Cost of utilities.

	Type	Utility	Size [kW]	Price [$/GJ]	Utility cost [$/yr]
C01	Compressor	Electricity	1,908	16.80	865,676.53
C51	Compressor	Electricity	3,351	16.80	1,519,989.18
C81	Compressor	Electricity	3,881	16.80	1,760,420.22
P02	Pump	Electricity	130	16.80	59,106.65
P2	Pump	Electricity	5	16.80	2,304.86
RT1	Reboiler	Medium Pres. Steam	7,331	8.22	1,627,093.20
RT2	Reboiler	Low Press. Steam	1,344	7.78	282,308.39
RT3	Reboiler	Low Press. Steam	11,347	7.78	2,383,518.67
CT1	Condenser	Cooling water	2,096	0.35	19,805.13
CT2	Condenser	Cooling water	5,024	0.35	47,478.44
CT3	Condenser	Cooling water	12,132	0.35	114,645.03
E2	Heat exchanger	Cooling water	31,519	0.35	297,854.91
Reactor	Heat exchanger	Cooling water	1,435	0.35	13,562.33
Dowtherm A[1]	Heating fluid	-	-	-	947,519.93
Furnace	Furnace	Coal	31,642	1.07	914,143.79
Total					10,854,783.40

[1]The cost of Dowtherm A is $4,737,599.63 for 28,153 kg that is used in 5 years. The utility is calculated per year.

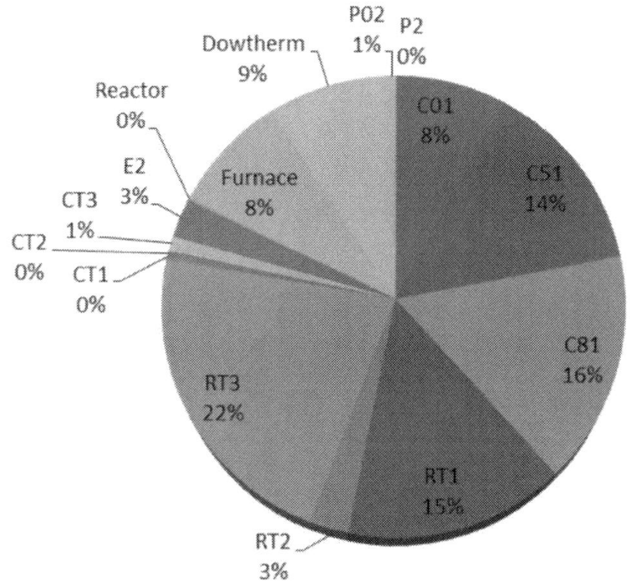

Figure 17 – Cost of utilities.

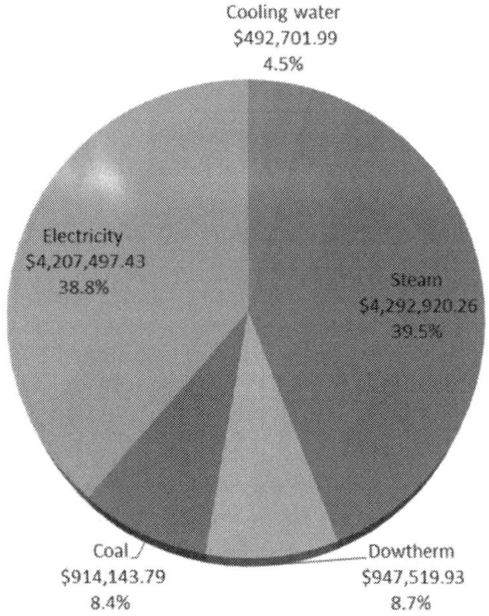

Figure 18 – Cost of utilities - lumped by type.

9.4 Economic evaluation

When the cost of raw materials, equipment and utilities has been calculated, other factors relevant when starting a plant are considered. When all of these are computed, economic considerations are made to see if the base case design is economically feasible.

First of all, we should distinguish between capital costs and operating costs. The former account for the expenses to be sustained at the beginning of the plant life, and include the costs to build and start up the process; the latter account for the costs to operate the plant. Once both have been calculated, the total revenue of the plant can be computed.

9.4.1 Capital costs

The price of the equipment calculated above did not take the installation, piping, instrumentation etc. into account. These are fixed costs that must be considered at the beginning. Their price is estimated as a percentage of the purchased equipment cost. The factors for the expenses are based on "*Plant Design and Economics for Chemical Engineers*" [29].

The capital costs are divided into fixed-capital investment and working capital. Fixed-capital investment is subdivided into direct and indirect costs. Direct cost is the capital that is necessary for installation of the process equipment. The delivery cost of the equipment also has to be included in the equipment price and can be estimated to be 10% of the purchased equipment cost. Obviously it depends on the distance from source to plant and the method of transportation. The price for installation of the equipment includes the cost for labor, foundations, construction expenses etc. and the cost depends on the size and the type of the equipment. The cost is estimated to be between 25-50% of the delivered purchased equipment. For a plant like ours the installation cost is estimated to be 47% of the delivered purchased equipment cost. The instrumentation and control systems and their installation are also considered and are estimated to account for 36% of the delivered purchased equipment cost. Previously, in the estimation of the equipment cost, the pipes were not taken into account. The cost of the piping includes the materials of construction, fittings, valves, supports and installation. Piping is a major expense and is estimated to account for 68% of the delivered equipment cost. Another 11% of the delivered purchased equipment cost goes to install the electrical systems required to run the plant. The capital cost also includes services and land, estimated to be 40% of the purchased equipment cost, and buildings, estimated to be in total 18% of the delivered equipment cost.

The indirect cost is non-manufacturing fixed-capital investment, meaning that it is not directly related to the process operation but it is necessary to make it run. This includes engineering and supervision of the construction design, cost of software, travel and communication, in total estimated to be 30% of the delivered purchased equipment cost. Another cost is the construction expenses which include everything during the construction of the plant such as temporary constructions, construction tools, insurance etc. The construction expenses are estimated to be around 30% of the purchased equipment cost. The cost of the necessary contracts for construction and compliance to the government and environmental requirements are taken into account and in total the legal expenses are assumed to be 4% of the delivered purchased equipment cost. Finally, a contingency amount is included for unexpected expenses that will increase the cost of the project and is approximately 25% of the purchased equipment cost [29].

The working capital consists of the total amount of money required to operate the plant due to delay in payment, because the credit term for the customers is often based on 1 month payment period. The working capital also represents the money needed to carry the raw materials in stock and the initial payments. This expense account for around 19% of the fixed capital investment [24]. The different entries can be seen in Table 54.

9.4.2 Operating costs

The operating costs are estimated on a yearly basis and include manufacturing costs and general expenses associated with the run of the process.

The manufacturing costs are subdivided into variable production costs and fixed charges. The variable production costs include the cost of the raw materials, utilities, labor and maintenance. The price of raw materials was discussed in section 9.1 and the costs of the utilities was calculated in section 9.3. The labor work necessary to operate a plant like ours is assumed to include 20 workers, 2 managers and 2 laboratory technicians [35]. Since we have decided to build the plant in China, the wage is based on Chinese price level [36]. Additionally a certain amount of operating supervision is also required and accounts for about 15% of the cost of operating labor. Another cost is the maintenance and repair of the equipment, which is estimated to be 2% of the fixed capital investment for a chemical plant like ours, and 15% of the cost of maintenance and repair is estimated to account for the operating supplies, such as spare parts and test chemicals that cannot be considered as raw materials.

The fixed charges include property taxes and insurance. The tax level depends on the region that the plant is located in, but usually the property taxes are about 2% of the fixed-capital investment. The insurance depends on the type of process and the available protection facilities and the rate is estimated to be 1% of the fixed-capital investment per year.

The general expenses include the cost of administration and the distribution and marketing costs. The administrative costs include salaries to administrators, secretaries, accountants etc. and can be approximated to account for 15% of the operating labor. Distribution and marketing costs are mainly associated with the sales of the product and for a bulk chemical the costs are relatively low and are estimated to represent 2% of the total product cost [29]. The different entries can be seen in Table 54.

Table 54 – Schematic representation of the costs.

Capital cost

Direct costs		% of total
Purchased equipment delivered (DPEC)	16,065,636.20	24.81%
Purchased-equipment installation	7,550,849.01	11.66%
Instrumentation and controls (installed)	5,783,629.03	8.93%
Piping (installed)	10,924,632.61	16.87%
Electrical systems (installed)	1,767,219.98	2.73%
Building (including service)	2,891,814.52	4.47%
Yard improvements	1,460,512.38	2.26%
Service facilities (installed)	4,381,537.14	6.77%
Total direct plant cost	50,825,830.87	78.48%
Indirect cost		
Engineering and supervision	4,819,690.86	7.44%
Construction expenses	4,819,690.86	7.44%
Legal expenses	642,625.45	0.99%
Contigency	3,651,280.95	5.64%
Total indirect plant cost	13,933,288.12	21.52%
Fixed capital investment	64,759,118.99	
Working capital investment	12,304,232.61	
Total capital investment	77,063,351.60	

Operating cost

Item		[$/yr]	% of total cost
Raw materials	Ethylene	30,000,000.00	65.11%
	Water	713,215.25	1.55%
Operating labor	20 workers	91,800.00	0.20%
Manager	2 managers	36,720.00	0.08%
Operating supervision		13,770.00	0.03%
Utilities		10,854,783.40	23.56%
Maintanance and Repair		1,295,182.38	2.81%
Operating suppliers		194,277.36	0.42%
Laboratory analysis	2 lab technicians	18,360.00	0.04%
Variable cost		43,218,108.38	93.79%
Property taxes		1,295,182.38	2.81%
Insurance		647,591.19	1.41%
Fixed charges		1,942,773.57	4.22%
Manufacturing cost		45,160,881.95	98.01%
Administration costs		13,770.00	0.03%
Distribution and marketing costs		903,493.04	1.96%
General expense		917,263.04	1.99%
Total product cost		46,078,144.99	
Revenue		82,490,752.25	
Operating profit		36,412,607.25	

9.4.3 Total revenue

The total revenue comes from the sales of the product, 190 proof ethanol. Table 55 shows the annual revenues.

Table 55 – Revenue from product.

Product	Quantity [kg/yr]	Price [$/kg]	Revenue [$/yr]
190 proof Ethanol	72,360,308.99	1.14	82,490,752.25

9.4.4 Performance indexes

In order to estimate the process economic performance, some indexes exist. It is desirable to have a process with large added value, short time of regaining the initial investment and high return of investment. These can be estimated with the Net Present Value (NPV), the payback time and the return on investment (ROI). These indexes can be calculated on the basis of the investment and the annual profit and they are discussed in the following. The future horizon investigated is limited to the estimated process lifetime, the time after which the process is assumed to be dismantled for low profitability or too high costs for maintenance. A project lifetime of 15 years is assumed.

Net Present Value The Net Present Value allows for an estimation of the value that the firm gains or losses if the investment is done on the basis of the cash flows forecast associated with the project. This is the easiest way to judge if the investment is profitable or not. Once a project lifetime is assumed, it can be verified if the investment leads to an increase of the value of the company, NPV>0, in which case the investment is profitable, or to a decrease of the value of the company, NPV<0, in which case the investment is not profitable. In the hypothesis that all the fixed capital is handed out at the very beginning of the investment and that the investment has no residual value after its lifetime of N years, the value of the investment is:

$$NPV = -C_0 + \sum_{n=1}^{N} \frac{P}{(1+i)^n} \tag{9.2}$$

where C_0 is the initial total capital investment and P is the net annual profit, computed as the difference between the revenues and the operating costs and actualized over time with the interest coefficient i. If the investment is considered at the end of the project lifetime (year N) the global earning associated with the project is obtained, and if NPV>0 it means that the total capital investment has been more than repaid, also taking into account depreciation and risk, which are accounted for with i. So, the greater NPV the more profitable the investment is. On the basis of the data in Table 54, we have:

$$NPV = -\$77,063,351.60 + \sum_{n=1}^{15} \frac{\$36,412,607.25}{(1+0.1)^n} = \$199,893,834.19 \quad (9.3)$$

Figure 19 shows the discounted cumulative cash flow diagram associated to the process and it can be seen that the value of the investment increases until it reaches the expected final value.

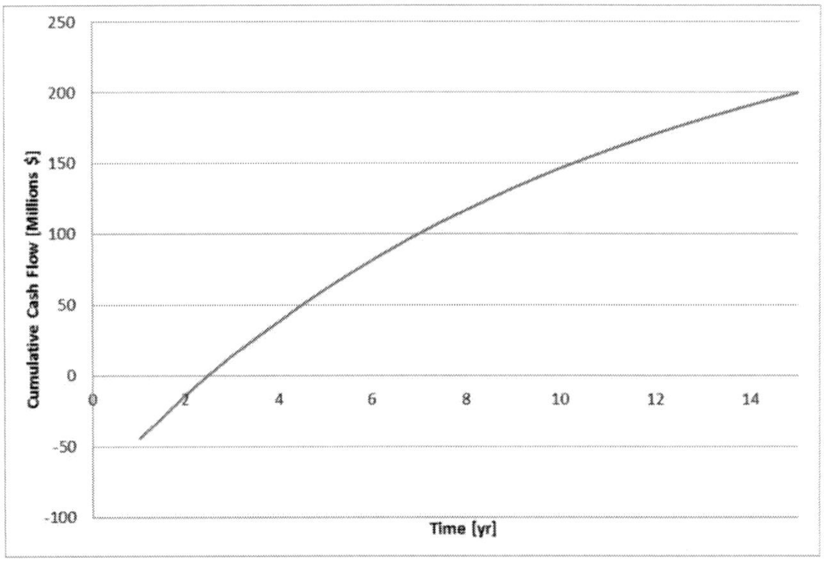

Figure 19 – Discounted cumulative cash flow diagram.

Payout time The payout time is the time in which the money spent in the initial investment is regained. Of course, the lower the payout time the better. The payout time is associated with the NPV: when NPV=0, the cumulative actualized revenues are equal to the sum of money invested initially. In Figure 19, this is seen to be 3 years. This is a very good payout time value, especially considering that further improvement can be achieved after process optimization.

Return on investment Another significant index is the return on investment (ROI), defined as:

$$ROI = \frac{\text{Net annual profit}}{\text{Total capital investment}} \quad (9.4)$$

This index tells how much money is earned for 1 \$ invested in the process; however, it does not take into account the discount rate with which money loses values with the years. The higher it is, the most profitable the process is.

On the basis of the data in Table 54, we have:

$$ROI = \frac{\$36,424,336.35}{\$76,805,164.00} = 47.3\% \tag{9.5}$$

A more realistic estimation is obtained if the effect of taxation on the net profit is considered. However, such kind of estimation involves factors which are highly dependent on the location and the government policies that are very difficult to predict. If the net profit is roughly cut to -50% because of taxation, the ROI would be halved to 23.6%. This is considered a satisfactory result: it means that for each 1 $ invested in the project 23.6 $ is obtained.

9.4.5 Sensitivity analysis

The effect of a variation of any of the factors affecting the expected profit of the process is evaluated.

First, a variation in the operating costs is evaluated. The main variation in the operating costs can happen to the price of raw material, labor, utility, product cost and purchase equipment cost. Figure 20 shows the sensitivity of NPV to the variation of the five. It can be seen that the most influencing factor is the price of the product, and if it drops by more than 30% the NPV is negative. However, such a big variation in the product price is not expected, also according to what evidenced from the market researches considered in section 1.2. Considering the operative costs in itself, the highest influence is to be attributed to the cost of raw material, which in fact accounts for the highest percentage of the operating costs, namely 65.1%, as seen in Table 54. For the same reason, utility costs have a lower influence, since they account for 23.6% of the total operating cost, but their influence is still quite significant. Some improvements on the NPV can be obtained reducing the utility use. The same cannot be said for labor cost, which does not affect the NPV even if it varies of ±50% since it accounts for 0.32% only of the total operating costs. Also if the plant was located in Europe the influence would not be significant, since the European wages would cause the percentage of labour work to raise to 2.5%, being them eight times higher than the Chinese ones. Figure 20 also shows how the capital costs influence the NPV; the variable chosen is the purchased equipment cost, since all the other element of the capital cost have been calculated as a function of it. The purchased equipment cost overlap with the utility cost in the figure. It can be seen that even if purchased equipment costs are raised of +50% the NPV still remains positive. However, if the capital cost is increased too much the NPV becomes more sensitive to variation in the operating costs.

Figure 21 shows the effect of the same variation of the operating costs over ROI. The trend is similar to the one observed for NPV. Figure 22 instead shows the effect of the most significant variables on the payback time. The costs of raw material and capital cost are the most significant and comparable, while the effect of utility cost is less significant.

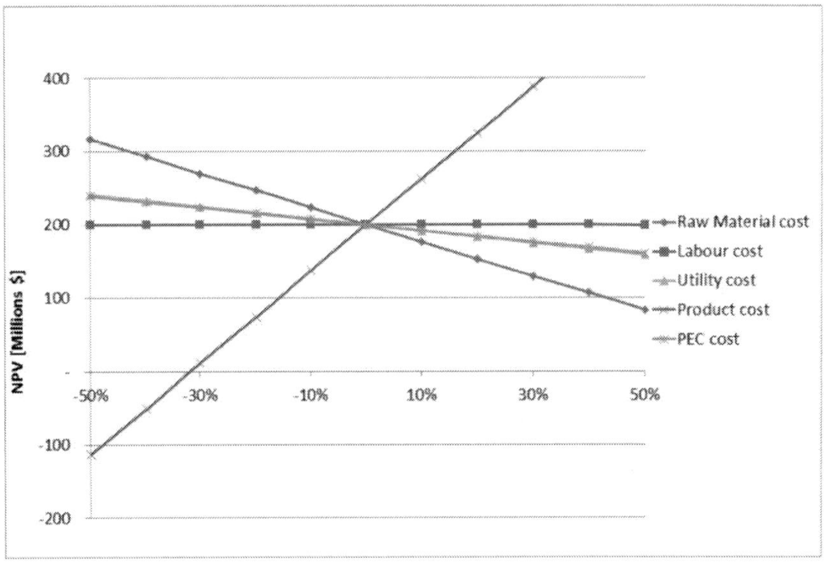

Figure 20 – Sensitivity of NPV to operating and capital costs.

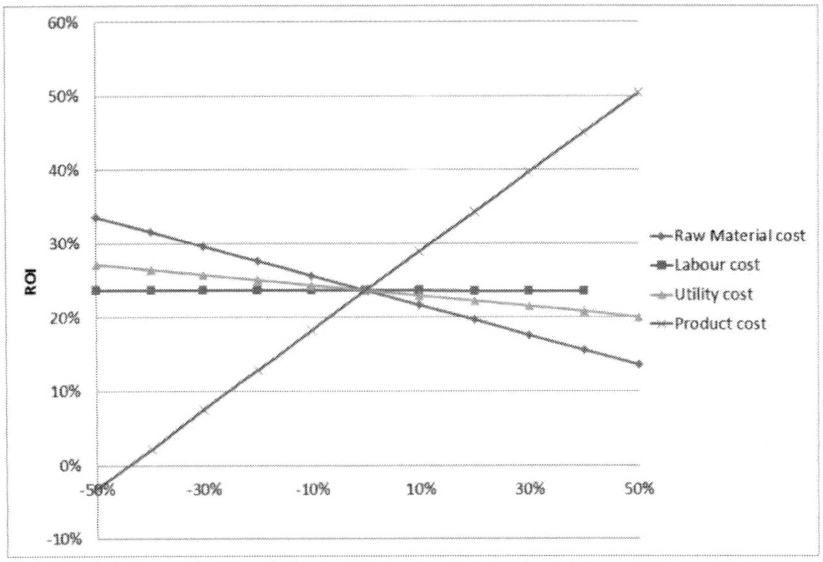

Figure 21 – Sensitivity of ROI to operating and capital costs.

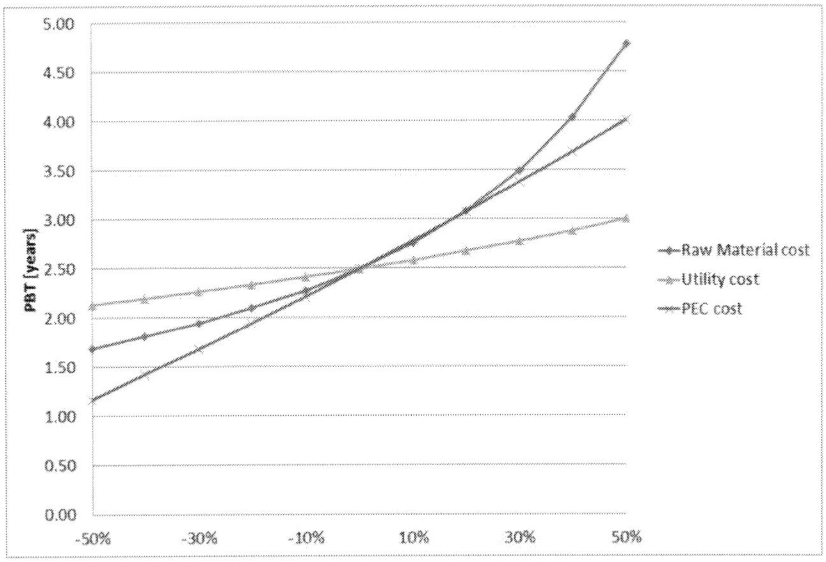

Figure 22 – Sensitivity of payback time to some significant cost variables.

9.5 Conclusion

On the basis of the economic analysis performed, it is evident that the process is profitable and thus it is worth investing in it. With an initial investment of $ 77,063,351.60, an annual revenue of $ 36,412,607.25 is expected (non-actualized value); the payback time is 3 years. The expectation for this process are to increase the value of the firm of $ 199,893,834.19 , with a ROI of 23.6%. It should be remembered that these simple indexes should be handled with care, because they represent rough estimations only. They just indicate if the process is worth to be considered for a more detailed study or not. For more reliable economic figures, a more detailed analysis is required. The economic gain from the process may be improved with heat integration and optimization, which will be explored in the following tasks. In addition to these economic figures, it is significant to report some key performance indexes which are useful to evaluate the performance of the process in terms of raw material usage and energy consumption. These are reported in Table 56.

Table 56 – Key performance indexes of base case.

		Base Case
Yield	kg P / kg RM	0.263
Material loss	kg EL lost in purge / kg EL feed	5.0%
	kg EA lost / kg EA out of reactor	10.2%
Water use	kg W (feed+absorber) / kg EL feed	5.149
Energy use	Energy used [MJ / kg RM]	7.963
	Cooling water used [MJ CW / kg RM]	4.981

10 Task 10

The possibility of heat and mass integration is explored in this task. By heat and mass integration energy and material consumption can be reduced, which helps to achieve sustainable development. From the economic point of view, higher profit will become possible with reduced energy and material consumption. From the environmental point of view, the environmental impact of the plant will be less significant. However, no mass integration is possible in this case because solvent streams cannot be used for more than one absorption process.

10.1 Heat integration

There are ten units involving heat exchange in the base case process: the heater of the raw material water E02, the heater before the reactor E1, the cooling jacket of the reactor R2, the heat exchanger between reactor and flash E2 and the condensers and the reboilers of the three columns, CT1, CT2, CT3 and RT1, RT2, RT3. It is found from the economic evaluation that RT3, RT1 and E02 have the three highest utility cost, which is 15%, 17% and 7% respectively. Therefore they become the main targets of reducing utility consumption by heat integration.

The first consideration when making decisions for heat integration is the distance between hot and cold streams. It is not feasible to integrate hot and cold streams that are located far away from each other. Therefore, we choose to integrate the hot and cold streams around the reactor while reboilers and condensers are not integrated with them. Thus, the heat integration is done first in the reactor area and then heat integration for the columns is considered.

10.1.1 Heat integration in the reactor area

In the area round the reactor, there is one hot stream, the outlet stream of the reactor S2, and two cold steams, the raw material water S021 and the inlet of reactor S1. The heat contents of each stream in each interval are obtained in PRO/II and listed in Figure 23, details about this procedure can be found in Appendix E. The minimum temperature approach between hot and cold streams is assumed to be 10°C. The heat cascade diagram (Figure 24) is built based on the flows of heat contents, with the heat contents of the hot stream introduced into the corresponding intervals while the heat contents of the cold streams are extracted from their corresponding intervals. In the following, R_i represents the heat residual of interval i, while Q_s and Q_w represent the heating and cooling loads respectively.

The problem can be treated as an optimization problem with the objective of minimizing utility consumption. In order to formulate the minimum utility

Temperature						Heat Contents(GJ/hr)		
H1(S2)	Hot		Cold	C1(S1)	C2(S021)	H1(S2)	C1(S1)	C2(S021)
	(310.0)		300.0					
		1					1.080	
	(305.2)		295.2					
		2					1.217	0.329
	300.0		(290.0)					
		3				18.635	31.686	56.277
	(222.6)		212.6					
		4				94.833		13.078
	120.0		(110.0)					
		5						10.414
	(38.3)		28.3					

Figure 23 – Temperature intervals and heat contents.

consumption, heat balance is first performed in each interval.

$$
\begin{aligned}
Q_s &= R_1 + 1.080 \\
R_1 &= R_2 - 1.546 \\
R_2 &= R_3 - 69.328 \\
R_3 &= R_4 + 81.755 \\
R_4 &= Q_w - 10.414
\end{aligned} \tag{10.1}
$$

Then the optimization problem can be written as

$$
\begin{aligned}
\min Z &= Q_s + Q_w \\
s.t.\ R_1 &= Q_s - 1.080 \\
R_2 &= Q_s - 2.627 \\
R_3 &= Q_s - 71.954 \\
R_4 &= Q_s + 9.801 \\
Q_w &= Q_s - 0.614 \\
Q_s, Q_w, & R_1, R_2, R_3, R_4 \geq 0
\end{aligned} \tag{10.2}
$$

The solution to the optimization problem is $Q_s = 71.95, Q_w = 71.34, R_1 = 70.87, R_2 = 69.33, R_3 = 0, R_4 = 81.76$. The minimum heating load is 71.95 GJ/hr and the minimum cooling load is 71.34 GJ/hr. Since $R_3 = 0$, the pinch point is 222.6-212.6℃.

A feasible solution of the heat integration in the reactor area is integrating the hot stream S2 and the raw material water stream S02. The inlet of reactor S1 is not involved in the network as the hot stream S2 is not able to heat both cold streams. 42.128 GJ/hr energy can be supplied from S2 to S02, shown in Figure 25. It means that 42.1 GJ/hr energy is saved from heating and that the same amount is saved from cooling.

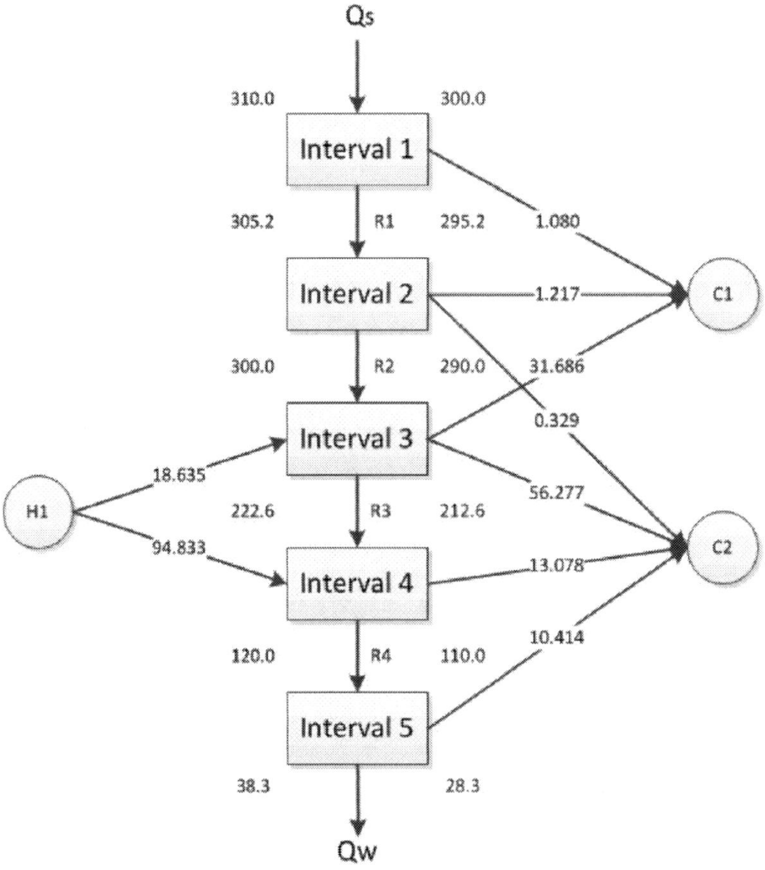

Figure 24 – Heat cascade diagram.

10.1.2 Heat integration of columns

One way to reduce the energy consumption of distillation columns is heat integration of the columns. The idea is to use the heat released from one column's condenser to heat the reboiler of another column. In oder to do this, the temperature in the condenser should be higher than the temperature at the reboiler so that the former can give heat to the latter. Table 57 shows the temperatures at the top and the bottom of the columns in the current flow sheet, which represent the heat sources and the heat sinks, respectively. It is seen that in the state of the process heat integration between columns is not feasible, because the temperatures at the top are lower than temperatures at the bottom of every column, except if we consider the condenser of column 1 and the reboiler of column 3, but the temperature approach between the two is not enough to assure a feasible heat integration. Thus, the pressure of some columns must be increased to make the equilibrium temperature shifted towards higher values at both the condenser and the reboiler, making it possible to give heat from one column's condenser to another column's reboiler. However, if operated at a higher pressure, a column would cost more to construct and operate, thus this option is not explored.

Table 57 – Condensers' and reboilers' temperatures.

Heat sources	CT1	CT2	CT3
T [°C]	107.1	37.4	76.9
Heat sinks	RT1	RT2	RT3
T [°C]	173.5	143	100.9

A more immediate solution to reduce the energy consumption of distillation columns is to make steam for reboilers using the waste heat in other parts of the process. It is noticed that the stream out of the reactor is still able to give out a large amount of heat after it is integrated with the stream of raw material water. Also, the heat of reaction can be used to generate steam.

The temperature of the steam used in a reboiler is determined by adding 10°C to the temperature of the bottom product. The steam temperature for the three reboilers are listed in Table 58.

The total heat released in the reaction is 5.18 GJ/hr at 300°C. The heat that stream S2 can provide when cooled from 201.6°C to 120°C is 71.34GJ/hr. Comparing these values and the duty of reboilers (see Table 58), a feasible solution is to generate steam for RT2 with the heat of reaction and to generate steam with S2 for RT1 first and then for RT3, shown in Figure 25. Thus, all the utility cost of the reboilers can be saved. Only 4.0 GJ/hr heat from the steam S2 and 0.4 GJ/hr from the reactor needs to be removed by cooling water.

Table 58 – Steam temperature and duty for column reboilers.

	RT1	RT2	RT3
Steam temperature[°C]	183.5	153	110.9
Duty[GJ/hr]	26.4	4.8	41.0

After heat integration, the utility cost in the plant is reduced from 10,854,783 $/year to 5,921,893 $/year meaning that 54.6% utility cost is reduced by heat integration. The cost of heat exchanger network may be effected by heat integration, but this is not further investigated here.

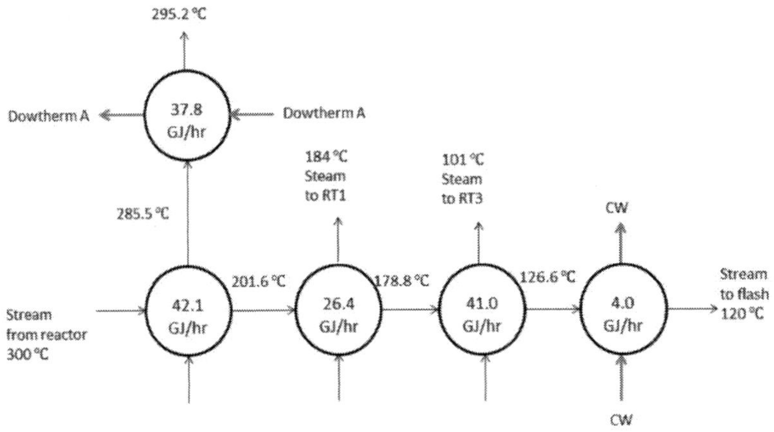

Figure 25 – Network of heat exchangers.

11 Task 11

In this task attempts are made to optimize the process. Some of the design decisions made so far can be reconsidered in order to obtain a better use of raw material, lower energy consumption and an increase in the profit. This can be achieved with slight modifications of the design. In order to accomplish this, optimization problems are formulated to improve the base case and other options of improvements are considered.

The focus of the optimization is on the factors that influence the profit the most. From the economic analysis done in Task 9 it was evident that the operating costs that have the highest impact are the price of the product, the cost of raw material and the utility costs. It is not possible to act on the first, which is determined by the market, but the other two can be reduced with a better use of raw material and lower energy consumption. A reduction in the utility cost was already obtained with heat integration, as seen in Task 10. All the steam needed can be produced by the heat available in the plant and some cooling water can be saved as well. It is seen that the main expense for the utilities is electricity, which is primarily associated with the work of the compressors. The compressors also represent the biggest cost for equipment purchase. Thus, we would be interested in reducing the load of the compressors so that their energy consumption and their purchase cost are lower. This can be done reducing the amount of the gas stream to compress, which means increasing the purge fraction. However, increasing the purge fraction would cause a higher loss of raw material. In the following section, an optimization problem is formulated to find a trade-off between these factors.

Another way of reducing the utility costs without much impact on the design is optimizing the feed tray of the columns. This is done on the column with the highest energy duty, being column 3 where azeotropic ethanol is separated from water. Also the possibility of burning the purge in order to provide heat to the furnace is investigated. Finally, the recycle of water obtained at the bottom of the third column is considered.

11.1 Optimal purge fraction

The optimal purge fraction can be obtained setting up an optimization problem in which the objective function to minimize is the sum of the cost of the electricity used for compressor C51 and the cost of raw material lost. Thus, the objective function is:

$$f(x):\ c_e \cdot W_c + c_{RM} \cdot \mu_{52}^{EL} \cdot MW_{EL} \tag{11.1}$$

For simplicity the raw material cost is assumed to be the one for pure ethylene. The constraint that must be accomplished is that the build-up of methane in the reactor feed does not exceed 10%:

$$g(x): \mu_{11}^{M} \leq 0.1 \cdot \mu_{11} \tag{11.2}$$

The optimization problem is solved varying the purge fraction and obtaining values for the objective function so that the minimum within the constraints is

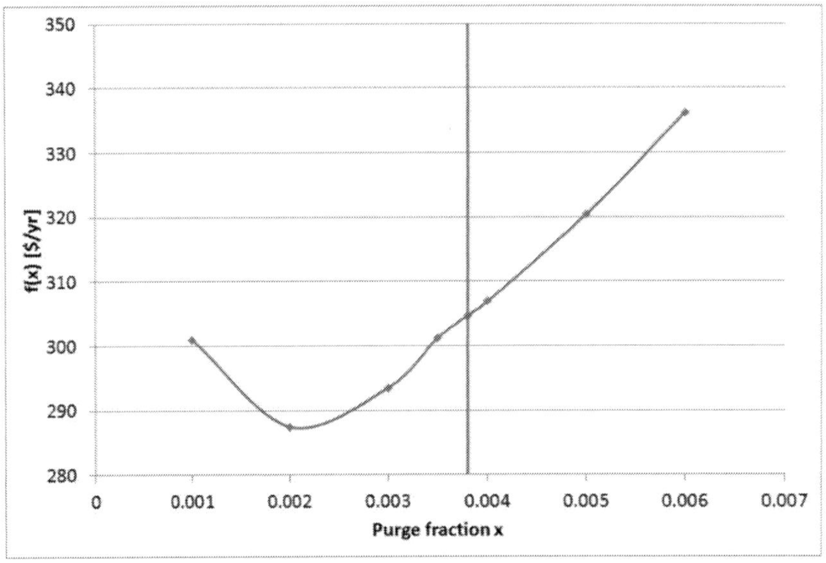

Figure 26 – Optimal purge fraction.

found. Figure 26 shows how the objective function varies as the purge fraction varies; also the constraint is shown, which separates the region in which the solutions are feasible on the right to the one in which the solutions are not feasible to the left. At the beginning, the purge fraction was assigned 0.005. It can be seen that it can be pushed down to 0.0038 still keeping the methane build-up within the constraints. In the region of acceptance, the effect of the loss of raw material prevails on the cost of compressing the whole gas stream at the reactor pressure, so the function is descending. It can be seen that the absolute minimum of the objective function is reached in the region of non-acceptance, thus the feasible minimum is found at the boundary of the region of acceptance and the constraint is active. With a purge fraction of 0.0038, the methane concentration in the inlet reactor stream is 9.51% and the work of the compressor is incremented from 3351.48 kW to 3521.33 kW.

11.2 Optimal feed tray

The position of the feed tray can be optimized in order to reduce the heat duty to the condenser and reboiler of the column which involves the highest utility consumption, namely the third. This was already considered in Task 7, however here this is reconsidered with respect to the new optimal purge fraction. The objective function to minimize is the sum of the duty to the reboiler and the condenser, which use steam and cooling water. The price of utility is not chosen as an objective function because steam is produced in the plant itself and the minimum price of cooling water is obtained when the minimum utility in the condenser is used. Minimizing the sum amount of duty at the condenser and reboiler allows for having more steam to be sold and less water to be bought.

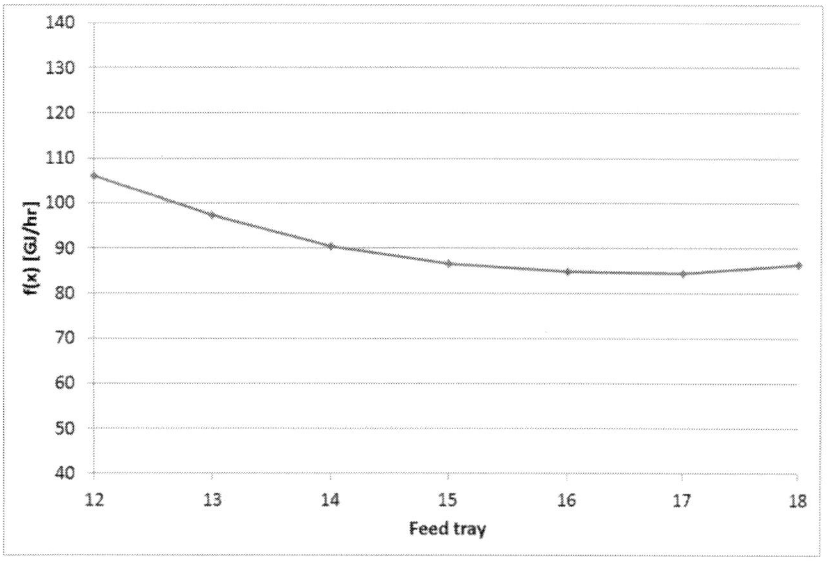

Figure 27 – Optimal feed tray for column 3.

Thus, the objective function is:

$$f(x) : Q_R + Q_C \tag{11.3}$$

Figure 27 shows how the total amount of the utilities needed for the column varies as a function of the feed tray. The optimal solution is to place the feed at tray 17. This result is different from what was obtained before, since now the purge fraction has decreased.

11.3 Purge combustion

The possibility of burning the purge to provide energy to the furnace is investigated. In this case, it is not possible to formulate an optimization problem but the advantage of saving energy is considered. The energy obtained from the combustion of the purge corresponds to the heat of combustion of the most abundant hydrocarbons found in the gas stream purged, methane, ethylene and propylene. If the energy of combustion is assumed to be released at 25°C with an efficiency of 60%, the heat available from the purge is 4,436.80 kW. Thus, the energy to be provided to the furnace passes from 31,642.22 kW to 27,205.43 kW, with savings on the coal needed to provide this utility.

11.4 Water recycle

It is noticed that the bottom stream of the last column consists of 99.2 mol% water and 0.8 mol% ethanol, which is almost pure water. Therefore it is feasible to recycle this stream as feed water. Since this water contains impurities it would be better to recycle it to the absorber, since they can be recovered and do not influence the degree of conversion in the reactor. However, the high temperature of the water stream (≅ 100 °C) and its low flow rate compared with the raw material water allows it to be recycled as water feed without any significant problem. By water recycling, the feed water can be saved by 2200 kg/hr. It means that 16,500,000 kg water can be saved every year.

11.5 Conclusion

The effect of the improvements introduced by heat integration presented in the previous section and process optimization discussed in this section is evaluated with respect to the profit.

The effect of the purge fraction optimization is on both the capital and operation costs. If the purge fraction is diminished, the cost of purchase the compressor and the cost of electricity increases but the amount of ethanol produced increases with a positive effect on the revenues. The other improvements affect the operating costs only.

After the purge fraction is optimized, the stream flow rates are slightly changed. Therefore heat integration is updated with new flow rates of hot and cold streams with the same procedure in Task 10. Then a final improved design is obtained. The key performance and the operating cost of the improved case are summarized in Table 59 and Table 60. Compared to the base case, the improved case has better performance and costs less. The new operating profit is $ 42,457,179.34, which is an increase of 16.6% compared to the base case.

Table 59 – Key performance indexes of base case and improved case design.

		Base Case	Improved Case	Difference
Yield	kg P / kg RM	0.263	0.280	6.4%
Material loss	kg EL lost in purge / kg EL feed	5.0%	3.9%	-22.2%
	kg EA lost / kg EA out of reactor	10.2%	10.3%	0.1%
Water use	kg W (feed+absorber) / kg EL feed	5.149	4.859	-5.6%
Energy use	Energy used [MJ / kg RM]	7.963	1.776	-77.7%
	Cooling water used [MJ CW / kg RM]	4.981	2.118	-57.5%

Table 60 – Effect of improvements on operating costs.

	Base case [$/yr]	Optimized case [$/yr]	Difference [%]
Steam	4,292,920.26	-	-100
Coal	914,143.79	488,008.45	-47
Cooling water	492,701.99	193,787.68	-60.7
Water	713,215.25	672,956.75	-6
Electricity	1,519,989.18	1,597,276.73	5

12 Task 12

In this last task the environmental impact is evaluated for the base case and for the new improved design. For the evaluation the program ICAS Process Analysis is used which is based on the WAR algorithm explained below.

For the following environmental analysis all the inlet and outlet streams are considered except the product stream. This is done in order to make the comparison between plant producing different chemicals possible. The boundary is set only around the plant, so the power consumption is not considered.

12.1 The WAR algorithm

To determine the potential environmental impacts (PEI) the waste reduction (WAR) algorithm is used. The idea behind the WAR algorithm is to evaluate not only the quantity of pollution but also the environmental impact of the pollution. The PEI is defined as the effect that a given material and energy would have on the environment if they were to be emitted into the environment.

The WAR algorithm uses eight PEI categories in its evaluation, four local toxicological and four global atmospheric. The two first categories are Human toxicity potential by ingestion (HTPI) and Human toxicity potential by exposure both dermal and inhalation (HTPE). If the chemicals are liquid or solid at temperature of 60°C and atmospheric pressure HTPI is calculated, which typically is based on the lethal-dose that produces death for 50% of rats by oral ingestion (LD_{50}). High LD_{50} are typical of chemicals with low toxicity. HTPE is calculated if the chemical exists as a gas at 60°C and atmospheric pressure and the value is obtained from National Institutes of Health and Safety. The category Aquatic toxicity potential (ATP) is estimated by using the representative species of fish, Pimephales promelas again using the method of LD_{50} and the Terrestrial toxicity potential (TTP) is also estimated by LD_{50} data from rats. The Global warming potential (GWP) is also estimated by measuring the extent to which a unit mass of a chemical absorbs infrared radiation over its chemical lifetime and comparing this values with the extent that CO_2 absorbs infrared radiation over its respective lifetimes. Another category is the Ozone depletion potential (ODP) where the chemical must stay long enough in the atmosphere to reach the stratosphere and must also contain a chlorine or bromine atom to obtain a value. The Photochemical oxidation potential (PCOP) is determined by comparing the rate at which a unit of mass of the chemical reacts with a hydroxyl radical to the rate at which a unit of mass of ethylene reacts with hydroxyl radical. The last category is Acidification potential (AP), which is measured by comparing the rate of release of H^+ in the atmosphere as promoted by the chemical to same rate but promoted by SO_2 [37, 38].

12.2 Base case

The results of the environmental impact for the base case design are seen in Table 61. The clean water does not contribute to the PEI, but the PCOP from the ethylene feed gives the primary contribution to the total PEI of the inputs. For the output streams the PCOP from the purge has the largest effect on the total PEI since it contains ethylene. It is seen that the impact generated is negative meaning that input PEI is larger than the output PEI.

12.3 Improved case

For the improved design the environmental impact is also analyzed as seen in Table 61. Now the downstream from column 3 is recycled, so one output stream is removed. The output now only consists of the downstream from column 1 and the purge, whose impact is lowered compared to the base case design. For the improved case the purge is burned so it is assumed that complete combustion is taken place with air at a temperature of 2000℃, where all the carbon is converted into CO_2. Again the total PEI impact generated is negative, but further it is lower than the result of the base case. This analysis means that the improved design is not only better in term of lowered costs, but also in terms of reduced environmental impact.

Table 61 – Potential Environmental impact.

Stream No	Total PEI	HTPI	HTPE	ATP	TTP	GWP	ODP	PCOP	AP
Base case									
Ethylene feed	12718.80	3.91	0.07	93.42	3.91	0	0	12617.40	0
Process water	0	0	0	0	0	0	0	0	0
Absorber water	0	0	0	0	0	0	0	0	0
Input sum	12718.80	3.91	0.07	93.42	3.91	0	0	12617.40	0
Downstream C1	67.30	21.03	0.09	0.03	21.03	0	0	25.11	0
Downstream C3	29.43	2.26	0.01	0	2.26	0	0	24.91	0
Purge	778.79	5.24	0.07	5.38	5.24	0	0	762.86	0
Output sum	875.52	28.53	0.17	5.41	28.53	0	0	812.88	0
Impact generated	-11843.20	24.61	0.10	-88.01	24.61	0	0	-11804.60	0
Improved case									
Ethylene feed	12718.80	3.91	0	93	3.91	0	0	12617.40	0
Process water	0	0	0	0	0	0	0	0	0
Absorber water	0	0	0	0	0	0	0	0	0
Input sum	12718.80	3.91	0.07	93.42	3.91	0	0	12617.40	0
Downstream C1	67.30	21.03	0.09	0.03	21.03	0	0	25.11	0
Purge	37.32	0	0.03	36.98	0	0.31	0	0	0
Output sum	104.617	21.03	0.12	37.00	21.03	0.31	0	25.11	0
Impact generated	-12614.1	17.12	0.05	-56.42	17.12	0.31	0	-12592.30	0
Difference between the improved and the base cases									
Impact generated	-770.90	-7.49	-0.05	31.59	-7.49	0.31	0	-787.70	0

13 Conclusion

The design of the process of production of ethanol from ethylene was designed. This was done with the help of a systematic methodology which guides the building of the process flow sheet step by step. The process design project was divided into 12 tasks: the first 7 tasks encompass the selection of process alternatives and development of the base case, with its sizing and economic evaluation following in tasks 8 and 9, while the remaining tasks deal with possible improvements of the base case.

From a rough economic evaluation it was found that production of ethanol from ethylene was the most profitable process compared to production of polyethylene and ethanol produced from biomass. The flow sheet for the ethanol production was developed based on the physical properties of the chemicals involved and the reaction conditions. First linear mass balance equations were carried out and models for each operating unit were used to find the recoveries. The flow sheet was built in PRO/II step by step using stream calculators. Later when the operating conditions of each unit were assigned in order to obtain the desired performance, additional equipment like heat exchangers, compressors, pumps and valves were included in the process simulations and the energy balance was conducted. After substituting the stream calculators of the columns with rigorous column methods, the base case design was obtained. The process designed is able to produce 72,360 metric tons per year. The yield obtained was 0.263 kg per each kg of raw material. The amount of water used for the feed and the absorber was 5.15 kg for each kg of ethylene feed. The energy supplied to the process was 7.96 MJ per kg raw material and the energy removed by cooling water was 4.98 MJ per kg raw material.

Task 8-9 included the sizing of equipment and the economic evaluation. It was decided to build the plant in China and the payback time was found to be 3 years with an initial investment of \$77,063,351.60 and an annual revenue of \$36,412,607.25.

In Task 10-11 the base case was improved by heat integration as well as optimization of the purge fraction and the feed tray of the column with the highest duty. Also the possibility of burning the purge to produce energy and the recycle of water was considered. After these improvements, the utility cost was reduced by 68% and the water consumption was reduced by 6%. The annual production of ethanol was increased to 73,245 metric tons per year. Further, the annual revenue was increased to \$42,457,179.34.

In Task 12 the environmental impact of the process was evaluated with the WAR algorithm and the improved design was compared with the base case design. The environmental impact of both the base case design and the improved design was acceptable and furthermore the impact of the improved design was reduced by 770 PEI.

Nomenclature

$\alpha_{k/n}$	Relative volatility of component k w.r.t the key component n	-
$\bar{\alpha}$	Average relative volatility	-
$\bar{\phi}_k$	Vapor figacity coefficient of component k	-
ΔG_R	Reaction standard Gibbs energy	kJ/mol
Δg_{fi}	Standard Gibbs energy of formation for component i	kJ/mol
ΔH	Enthalpy	kJ/mol
$\Delta H^\circ_{\text{f,gas}}$	Enthalpy of formation of gas	kJ/mol
$\Delta H^\circ_{\text{f,liquid}}$	Enthalpy of formation of liquid	kJ/mol
ΔH_R	Enthalpy of reaction	kJ/mol
ΔH_{vap}	Enthalpy of evaporation	kJ/mol
Δh_{fi}	Standard enthalpy of formation for component i	kJ/mol
ΔT_{ln}	Logarithmic mean temperature	K, ℃
η	Viscosity	mPa s
η_c	Compressor efficiency	-
η_m	Motor efficiency	-
η_p	Pump efficiency	-
η_r	Conversion of the reference reactant $l(r)$ in reaction r	-
γ_i	Stoichiometric coefficient of component i	-
γ_k	Liquid activity coefficient of component k	-
$\gamma_{r,k}$	Normalized coefficient of component k in reaction r	-
μ^k_{ij}	Molar flow rate of component k of jth stream out of unit i	kmol/hr
ρ	Density	kg/m^3, $kmol/m^3$
ρ_L	Liquid density	kg/m^3, $kmol/m^3$
τ	Residence time	s
ξ_n	Recovery of component n in the vapor phase	-
ξ_{jk}	Recovery of component k in top stream of unit j	-
A	Area	m^2
A_E	Absorption factor	-

a_i	Activity of component i	-
b	Purge fraction	-
c_P	Specific heat capacity	J/(mol K)
D, d	Diameter	m
F	Volumetric flow rate	m^3/min
f_k^0	Fugacity of component k	-
F_L	Liquid volumetric flow rate	m^3/hr
H, h	Height	m
K	Equilibrium ratio for vapor-liquid equilibria	-
K_p	Equilibrium constant	-
L	Length	m
L	Liquid flow rate	kmol/hr
l	Component flow rate in vapor	kmol/hr
l_t	Tray space	m
m	Mass flow rate	kg/hr
MW	Molecular weight	g/mol
N	Number of equilibrium trays	-
NC	Number of components	-
NE	Number of equations	-
NR	Number of reactions	-
NV	Number of variables	-
P	Pressure	atm, bar, Pa
p	Partial pressure	Pa, atm
P_k^0	Vapor pressure of component k	-
P_c	Critical pressure	atm, bar
Q	Heat duty	MJ/hr
Q_s	Heating loads	GJ/hr
Q_w	Cooling loads	GJ/hr
R	Gas constant	8.31447 J/(mol K)

R	Reflux ratio	-
r	Recovery	-
R_i	Heat residual of interval i	GJ/hr
R_{min}	Minimum reflux ratio	-
T	Temperature	K, °C
T_b	Boiling temperature	K, °C
T_c	Critical temperature	K, °C
T_c^i	Pseudo-critical temperature of component i	K, °C
T_c^m	Pseudo-critical temperature of the mixture liquid stream	K, °C
T_m	Melting temperature	K, °C
U	Overall heat transfer coefficient	$W/K \cdot m^2$, $Btu/ft^2 \cdot hr \cdot F$
u_v	Maximum allowable vapor velocity	m/s
V	Vapor flow rate	kmol/hr
V	Volume	m^3
v	Component flow rate in vapor	kmol/hr
W_b	Work	kW
W_v	Maximum vapor mass rate	kg/s
x_i	Molar fraction in liquid phase of component i	-
y_i	Molar fraction in vapor phase of component i	-
z_i	Molar fraction in feed of component i	-

References

[1] Heinz Zimmermann and Roland Walzi. *Ullmann's Encyclopedia of Industrial Chemistry.* Wiley Online Library, 2009.

[2] OECD-FAO Agricultural Outlook 2010-2019. http://www.agri-outlook.org, February 2011.

[3] U.S. Department of Energy. http://www.afdc.energy.gov/, February 2011.

[4] Ronald W. Rousseau and James R. Fair. *Handbook of separation process technology.* Wiley-IEEE, 1987.

[5] National Institute of Standards and Technology. NIST chemical webbook. http://webbook.nist.gov/chemistry/, May 2011.

[6] CAPEC Department of Chemical and Biochemical Engineering. ICAS.

[7] IEC Business Partners. Ethanol application. http://www.iecbp.com/product_ethanolKnowEtaApp.php, February 2011.

[8] Kim Ha Viet. http://www.kimhaviet.com/ethanol/ethanol-applications.aspx?lang=en, March 2011.

[9] Firearms U.S. Bureau of Alcohol, Tobacco and Explosives. Federal Register, August 1996. Volume 61, Number 151.

[10] Market Research Analyst. World's ethanol production forecast 2008 - 2012. http://www.marketresearchanalyst.com/2008/01/26/world-ethanol-production-forecast-2008-2012/, February 2011.

[11] Technological University of Denmark. Kemibrug. http://kemibrug.dk/KBA/CAS/109573/, February 2011.

[12] W.M. Haynes, editor. *CRC Handbook of Chemistry and Physics.* CRC, 2010.

[13] National Institute of Standards and Technology. Ethanol data. http://webbook.nist.gov/cgi/cbook.cgi?ID=64175, February 2011.

[14] Alain Chauvel and Gilles Lefebvre. *Petrochemical Processes: Major Oxy-Genated, Chlorinated And Nitrated.* Editions Technip, 2. edition, 1989.

[15] Chemical Guide. The manufacture of ethanol: http://www.chemguide.co.uk/physical/equilibria/ethanol.html.

[16] Naonobu Katada, Yohei Iseki, Akira Shichi, Nobuo Fujita, Iwao Ishino, Kazuhiro Osaki, Tatsuya Torikai, and Miki Niwa. Production of ethanol by vapor phase hydration of ethene over tungsta monolayer catalyst loaded on titania. *Applied Catalysis A: General*, 349:55–61, 2008.

[17] ICIS Chemical Intelligence. Ethylene: http://www.icis.com/v2/chemicals/9075777/ethylene/uses.html, February 2011.

[18] ICIS Chemical Intelligence. http://www.icis.com/StaticPages/a-e.htm, February 2006.

[19] Joel R. Fried. *Polymer science and technology.* Prentice Hall, 2003.

[20] M. Häfele, A. Kienle, M. Boll, and C.-U. Schmidt. Modeling and analysis of a plant for the production of low density polyethylene. Master's thesis, Otto-von-Guericke Universität Magdeburg, 2006.

[21] U.S. Department of Energy Biomass Department. Biomass program. http://www1.eere.energy.gov/biomass/index.html, February 2011.

[22] Energy Research Center of the Nederlands. http://www.ecn.nl/phyllis/, February 2011.

[23] Index Mundi. Corn price. http://www.indexmundi.com/commodities/?commodity=corn, February 2011.

[24] Lorenz T. Biegler, Ignacio E. Grossmann, and Arthur W. Westerberg. *Systematic Methods of Chemical Process Design.* Prentice Hall Internatonal Series, 1997.

[25] J. E. Connett. Chemical equilibria. Enthalpy of dehydration of ethanol to diethyl ether by measurement of equilibrium constants in ethanol + ether + water by a vapour flow technique. *J. Chem. Thermodynamics*, 4:135–138, 1972.

[26] H. Momose, K. Kusumoto, Y. Izumi, and Y. Mizutani. Vapor-phase direct hydration of ethylene over zirconium tungstate catalyst i. Catalytic behavior and kinetics at atmospheric pressure. *Journal of Catalysis*, 77:23–31, 1982.

[27] Prof. Rafiqul Gani. Lecture 4: Mass and energy balance. Lecture notes, 2011. Slide no. 14.

[28] R. K. Sinnott, John Metcalfe Coulson, and John Francis Richardson. *Chemical Engineering Design*, volume 6. Elsevier, 4. edition, 2005.

[29] M. S. Peters, K. D. Timmerhaus, and R. E. West. *Plant Design and Economics for Chemical Engineers.* McGraw-Hill, 2003.

[30] Don W. Green and Robert H. Perry. *Perry's Chemical Engineers' Handbook.* McGraw-Hill, 2008.

[31] Dow Chemical Company. Dowtherm A, heat transfer fluid - product technical data. http://www.dow.com/heattrans/prod/synthetic/dowtherm.htm, May 2011.

[32] Shanghai City South Water Company. http://www.water-sh.com, May 2011.

[33] Richard Turton. *Analysis, Synthesis and Design of Chemical Process.* Prentice Hall, 3. edition, 2008.

[34] Sigma-Aldrich. `http://www.sigmaaldrich.com/denmark.html`, April 2011.

[35] Prof. Rafiqul Gani. Lecture 7: Sizing, costing and economic evaluation. Lecture notes, 2011.

[36] Average salary in Shanghai. `http://www.jobyun.com/salary.html`, May 2011.

[37] Douglas M. Young, Richard Scharp, and Heriberto Cabezas. The waste reduction (WAR) algorithm: environmental impacts, energy consumption and engineering economics. *Waste management*, 20:605–615, 2000.

[38] Douglas M. Young and Heriberto Cabezas. Designing sustainable processes with simulation: the waste reduction (WAR) algorithm. *Computers and chemical engineering*, 23:1477–1491, 1999.

[39] Preeyaporn Saengwirum. *Manual for ECON: a software for cost evaluation and economic analysis.* CAPEC.

This page intentionally left blank.

Appendix A Equilibrium constant calculation

The following calculation is done to obtain the equilibrium constant of the reaction of ethanol to form DEE and water (Reaction 4.19)from thermodynamic data. Under the hypothesis that ΔH_R does not vary much with temperature, the equilibrium constant is computed as:

$$K_p(1\text{ atm, T K}) = K_{p0} \cdot \exp\left(\frac{\Delta H_R(1\text{ atm, 298 K})}{R}\left(\frac{1}{T} - \frac{1}{298\,\text{K}}\right)\right) \quad \text{(A.1)}$$

with

$$K_{p0} = K_p(1\text{ atm, 298 K}) = \exp\left(\frac{\Delta G_R(1\text{ atm, 298 K})}{R \cdot 298\,\text{K}}\right) \quad \text{(A.2)}$$

where $\Delta H_R(1\text{ atm, 298 K})$ and $\Delta G_R(1\text{ atm, 298 K})$ can be computed from thermodynamics data in Table 62 as:

$$\Delta H_R(1\text{ atm, 298 K}) = \sum_{i=1}^{NC} \gamma_i \Delta h_{fi}(1\text{ atm, 298 K}) \quad \text{(A.3)}$$

$$\Delta G_R(1\text{ atm, 298 K}) = \sum_{i=1}^{NC} \gamma_i \Delta g_{fi}(1\text{ atm, 298 K}) \quad \text{(A.4)}$$

Table 62 – Thermodynamics data [30].

	EL	W	EA	DEE	PL	IPA
Δg_f(1 atm, 298 K)[J/mol]	68440	-228590	-167850	-122100	62150	-173470
Δh_f(1 atm, 298 K)[J/mol]	52510	-241814	-234950	-252100	19710	-272700

The equilibrium constant computed in this way is

$$K_p = 4.04 \quad \text{(A.5)}$$

Appendix B Linear unit models

B.1 Flash

The flash unit splits the inlet stream f of composition z_k into a top product of vapor of composition y_k, v, and a bottom product of liquid of composition x_k, l; the overhead split fraction of component k is $\xi_k = v_k/f_k$. From the general expression of vapor/liquid equilibrium we have:

$$\bar{\phi}_k y_k P = \gamma_k x_k f_k^0 \tag{B.1}$$

where $\bar{\phi}_k$ is the vapor fugacity coefficient of component k in the gas mixture, γ_k is the liquid activity coefficient and f_k^0 is the pure component liquid fugacity. If ideal behavior is assumed, $\bar{\phi}_k = 1$, $\gamma_k = 1$ and $f_k^0 = P_k^0$, being P_k^0 the vapor pressure of component k at the temperature considered. The assumptions lead to Raoult's law:

$$y_k/x_k = P_k^0/P = K_k \tag{B.2}$$

A key component n is selected and its recovery in the vapor phase ξ_n is specified. The relative volatility of each component k with respect to the key component n is defined as:

$$\alpha_{k/n} = K_k/K_n = P_k^0/P_n^0 \tag{B.3}$$

Then, substituting ξ_k, ξ_n and $\alpha_{k/n}$ in the equilibrium expression for component k and rearranging it, the recoveries of the all the k components can be found as:

$$\xi_k = \frac{\alpha_{k/n}\xi_n}{1 + (\alpha_{k/n} - 1)\xi_n} \tag{B.4}$$

In order to perform the calculations, the operating temperature is needed. If the temperature of the flash is not given as a specification, it should be computed. This can be done solving the following equation:

$$\frac{P_k^0(T)}{P} = \frac{\alpha_{k/n}}{\bar{\alpha}} \tag{B.5}$$

where the index k relates to the most abundant component in the liquid phase and $\bar{\alpha}$ being an average relative volatility, defined as

$$1/K_n = \sum (K_i/K_n) x_i = \sum \alpha_{i/n} x_i = \bar{\alpha} \tag{B.6}$$

The computed temperature should be between the bubble and the dew point of the inlet mixture. At the bubble point of the inlet mixture the following is true:

$$\sum z_i = \sum y_i = \sum K_i x_i = 1 \tag{B.7}$$

while at dew point:

$$\sum z_i = \sum x_i = \sum \frac{y_i}{K_i} = 1 \quad \text{(B.8)}$$

Thus the bubble and dew point can be computed solving numerically the previous equations, for T unknown and P known or vice versa.

B.2 Absorber

The absorber unit purifies the vapor phase V_{N+1} by absorbing the desired heavy components in the liquid phase L_0. This is possible thanks to the higher tendency of heavy components to pass into the liquid phase. The temperature and the pressure are set and the recovery r of a desired component n is selected. For simplicity the absorption factor is assumed to be constant all over the column and can be computed as:

$$A_E = \frac{L_0}{V_{N+1}K_n} \quad \text{(B.9)}$$

Once the absorption coefficient has been defined, the flow rate of the absorbing liquid stream L_0 is found rearranging the expression into:

$$L_0 = V_{N+1}K_n A_E \quad \text{(B.10)}$$

The number of stages necessary to obtain the desired recovery r of n is calculated with the Kremser equation:

$$N = \ln\left(\frac{rv_{N+1}^n + l_0^n - A_E v_{N+1}^n}{l_0^n - A_{EA}(1-r)v_{N+1}^n}\right) / \ln(A_E) \quad \text{(B.11)}$$

where v_{N+1}^n and l_0^n are the flow rates of component n in the inlet vapor and liquid stream, respectively. This number must be rounded to the higher integer in order to have an indication of the number of stages necessary. The number of stages N is used to compute the absorption factors A_k for each component k:

$$A_k = \frac{A_E}{\alpha_{k/n}} \quad \text{(B.12)}$$

where again $\alpha_{k/n}$ is the relative volatility of component k with respect to the key component n. These are then used to define the coefficients β_{N-1}^k and β_N^k:

$$\beta_N^k = \frac{1 - A_k^{N+1}}{1 - A_k}; \quad \beta_{N-1}^k = \frac{1 - A_k^N}{1 - A_k} \quad \text{(B.13)}$$

These coefficients are needed to find the outlet vapor and liquid flows for each component, which can be computed as:

$$v_1^k = \frac{v_{N+1^k}}{\beta_N^k} + \frac{\beta_{N-1}^k}{\beta_N^k} l_0^k \quad \text{(B.14)}$$

$$l_N^k = \left(1 - \frac{\beta_{N-1}^k}{\beta_N^k}\right) l_0^k + \left(1 - \frac{1}{\beta_N^k}\right) v_{N+1}^k \tag{B.15}$$

The split fraction for each of component k is defined as the amount that goes out in the vapor phase over the total in the inlet, as:

$$\xi_k = \frac{v_1^k}{v_{N+1}^k + l_0^k} \tag{B.16}$$

B.3 Purge

The recycle loop material balance is set up as the following.

The methane molar fraction in the inlet stream of the reactor is:

$$x_1^M = \frac{\mu_1^M}{\sum_{k=1}^{NC} \mu_1^k} = \frac{\mu_1^M}{\mu_1^M + \mu_1^{EL} + \mu_1^{PL} + \mu_1^W}$$

having assumed that the only compounds present in significant quantity are methane, ethylene, propylene and water. Here we make the conservative assumption of total recovery of the incondensibles in the flash and none in the absorber, so that $\mu_2^k = \mu_{31}^k = \mu_{41}^k$ for M, EL and PL. The methane present in the stream is the sum of the methane of the feed and of the one recycled:

$$\mu_1^M = \mu_{01}^M + \mu_{51}^M = \mu_{01}^M + (1-b)\mu_1^M \tag{B.17}$$

which can be rearranged to give:

$$\mu_1^M = \frac{\mu_{01}^M}{b}$$

The same is true for ethylene and propylene:

$$\mu_1^{EL} = \mu_{01}^{EL} + \mu_{51}^{EL} = \mu_{01}^{EL} + (1-b)(1-\eta_1)\mu_1^{EL}$$

$$\mu_1^{EL} = \frac{\mu_{01}^{EL}}{1-(1-b)(1-\eta_1)}$$

$$\mu_1^{PL} = \mu_{01}^{PL} + \mu_{51}^{PL} = \mu_{01}^{PL} + (1-b)(1-\eta_2)\mu_1^{PL}$$

$$\mu_1^{PL} = \frac{\mu_{01}^{PL}}{1-(1-b)(1-\eta_2)}$$

The flow rate of water in the feed is determined by the ratio of water of ethylene, that is 0.6:

$$\mu_1^W = 0.6\mu_1^{EL} = 0.6\frac{\mu_{01}^{EL}}{1-(1-b)(1-\eta_1)}$$

Thus, if the previous equalities are substituted in (B.17) we obtain an expression function of the feed composition μ_{01}^k and the purge fraction b only, which can be solved for different values of b.

Appendix C Column operating parameters

Table 63 – Column 1 operating parameters.

N	N_{feed}	R/R_{min}	R	Duty [10^3MJ/hr] Condenser	Reboiler
12	2.080	1.250	0.214	-5.78	14.29
11	1.900	1.375	0.235	-6.34	14.86
10	1.850	1.500	0.256	-6.90	15.42
10	1.800	1.625	0.278	-7.46	15.98
10	1.760	1.750	0.299	-8.02	16.53

Table 64 – Column 2 operating parameters.

N	N_{feed}	R/R_{min}	R	Duty [10^3MJ/hr] Condenser	Reboiler
11	6.210	1.250	0.573	-7.581	1.547
10	5.820	1.375	0.631	-8.316	2.283
10	5.500	1.500	0.688	-9.051	3.018
9	5.230	1.625	0.746	-9.785	3.752
9	5.000	1.750	0.803	-10.520	4.485

Table 65 – Column 3 operating parameters.

N	N_{feed}	R/R_{min}	R	Duty [10^3MJ/hr] Condenser	Reboiler
20	5.060	1.250	1.567	-10.320	3.373
18	4.650	1.375	1.723	-10.950	4.002
17	4.340	1.500	1.880	-11.580	4.632
16	4.110	1.625	2.037	-12.210	5.262
15	3.980	1.750	2.193	-12.840	5.892

Appendix D Method for computing equipment costs

The purchase cost is calculated based on the method of the software ECON using following calculation method:

$$E_i' = UF \cdot E_i \tag{D.1}$$

In the previous expression, E_i' is the purchase cost of equipment i updated to the current year and E_i is the reference purchase cost. The factor UF is the update factor, which is defined using the Chemical cost index (CPI) as:

$$UF = \frac{\text{CPI Index at present year}}{\text{CPI Index} = 390.4} \tag{D.2}$$

where 390.4 is the CPI for the year at which the reference price E_i is known (1999) [39]. The CPI for year 2011 has been estimated to be $CPI = 664.9$, so the update factor becomes $UF = 1.703$.

The reference purchased price E_i is defined as $E_i = \text{cost}_i \cdot AF$, AF being an adjustment factor and cost_i is calculated by either one of the two following equations:

$$\text{cost}_i = \alpha \cdot S^{\beta} \tag{D.3}$$

or

$$\text{cost}_i = aS^6 + bS^5 + cS^4 + dS^3 + eS^2 + fS + C \tag{D.4}$$

where α, β, a, b, c, d, e, f and C are parameters dependent on the equipment type and S is the variable for the given equipment. Whether equation (D.3) or (D.4) should be used depends on the type of equipment. For further details and for the value of the various factors used, please refer to [29]. This method allows for cost calculations for heat exchangers, columns, compressors, pumps and furnace, but not for vessels.

The price for the vessels are found accordingly to the method suggested by Biegler et al. [24]. The procedure is based on Guthrie method, with which the equipment cost is computed as a combination of factors which take into account the size, the material and the installation costs, as well as the inflation. The general formula is:

$$Cost = UF \cdot BC \cdot (MPF + MF - 1) \tag{D.5}$$

where UF is the update factor, BC is the bare cost, MPF is the material and pressure correction factor and MF is the module factor. For each type of equipment, a reference cost at a reference year is available. The bare cost BC represents the cost of a similar type of equipment taking into account the exact size of the desired equipment. The factor MPF is a function of the type of material through F_m and the operating pressure F_p, as MPF=$F_m \cdot F_p$. The uninstalled equipment cost is thus $BC \cdot MPF$. The module factor MF is a function of BC and consider the cost associated with installation, which amount to $BC \cdot (MF-1)$. Thus, the total cost of the equipment is $BC \cdot (MPF+MF-1)$.

The price computed in this way is not updated to the present day, but it is computed for the date for which the reference cost is given. The update factor UF relates prices of equipment known in a certain date with the current date of the analysis, thus taking into account the effect of inflation. This factor is defined as in D.2; however, the reference price are known for 1967, for which the CPI is 115 [24], thus for vessels UF=5.782. In our work, the price of the vessels is estimated without considering installation, which is considered as an entry in itself in the capital costs, thus the price of the vessels is barely $Cost = UF \cdot BC \cdot MPF$. This method is very general and can be applied to any type of equipment. However, in our work it was applied to vessels only to maintain continuity between the equipment costing and the relevant utilities computed with ECON. For further details, for the criteria of selection of the various factors used in the following and for their value, please refer to Biegler [24].

D.1 Vessel

The parameters for computing the price of the vessels are given here. Equation (D.5) has been used for the calculations.

Table 66 – Vessels costing.

	Flash	Mixer	Reactor
Orientation	Horizontal	Horizontal	Horizontal
Material	Carbon Steel	Carbon Steel	Inconel 375
D [m]	2.1	1.8	2.8
L [m]	8.3	7.4	11
P [atm]	101.25	12.75	102
BC [$]	6957.50	5469.72	11489.42
MF2	3.18	3.18	3.18
Fp	2.5	1.15	2.5
Fm	1	1	3.89
MPF	2.5	1.15	9.725
UF	5.782	5.782	5.782
Cost [$]	100,575.14	36,371.44	646,078.95

D.2 Columns

The base cost of the columns is calculated by equation (D.4) where S is the column height and the parameters are given depending on the column diameter and the height of the column, see Table 67. The cost is multiplied with a material factor MF as well as a pressure factor PF depending on P_{max}, so that $E_i = cost \cdot MF \cdot PF$, and finally the update factor UF.

Table 67 – Parameters for column costs.

	e	f	C
D_c=2 m and H= 4-45 m	3.011	3139.4	7166.9
D_c=3 and H= 6-50 m	-23.555	5119.4	10945

D.3 Trays

For the cost of the trays equation (D.4) is used where S is the tray diameter and with the following parameters depending on the type of tray.

Table 68 – Parameters for trays.

	d	e	f	C
Bubble cap tray (CS)	-84.874	638.2	-454.1	774.21
Sieve tray (CS)	-32.7	234.91	-66.321	293.53

The total cost_i of the trays for each column is the price of one tray multiplied with the number of trays and also multiplied with a quantity factor QF which is depended on the total number of trays. The purchase cost is again multiplied with UF.

D.4 Heat exchangers

The cost is calculated based on equation (D.4) with the parameters showed in Table 69, where S is the area of the heat exchangers.

Table 69 – Parameters for heat exchangers.

	f	C
Heat exchanger (CS)	59.628	4071.2

D.5 Furnace

Equation (D.3) is used to calculate the cost with the parameters in Table 70.

Table 70 – Parameters for heat exchangers.

	α	β
Furnace	199.85	0.8659

D.6 Pump

The cost is calculated based on equation (D.4) with the parameters from Table 71. The material factor and the pressure factor is also seen.

Table 71 – Parameters for heat exchangers.

	e	f	C	MF	PF
Pump	$-2 \cdot 10^7$	318781	1017.3	2.4	2.8

D.7 Compressor

Equation (D.3) is used to calculated the cost of the compressors where S is the duty given in kW and with the parameters in Table 72.

Table 72 – Parameters for heat exchangers.

	α	β
Furnace	877.3	0.9435

Appendix E Heat contents from PRO/II

In order to do heat integration in Task 10, heat contents of streams in each interval must be provided. A quick way to get the heat contents of streams is setting up a series of heat exchangers for each stream presenting each interval in PRO/II, as seen in Figure 28. The outlet temperature of each heat exchanger in Figure 28 is set according to the temperature intervals. The inlet streams are set to be the exactly same as the streams in the flow sheet. For example, stream H1 is referred to stream S2 in the flow sheet. Then after the simulation is run, the duties for these heat exchangers are the heat contents of streams in each interval.

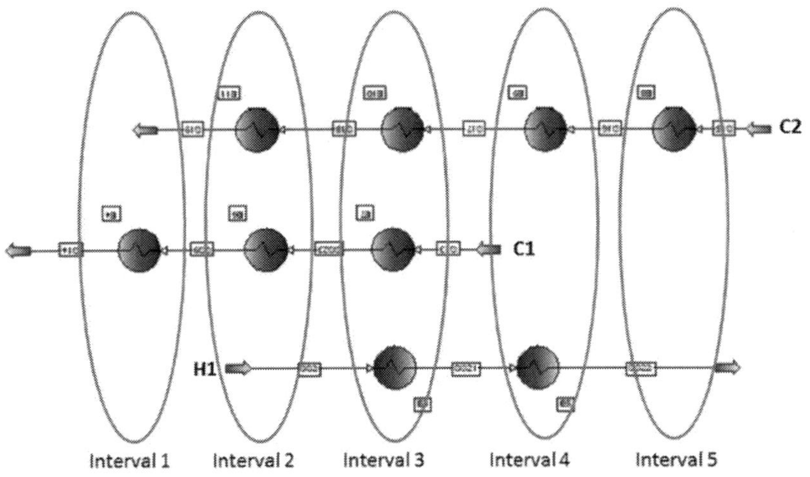

Figure 28 – Heat exchangers presenting temperature intervals.

Sustainable Integration of Industrial Gaseous Emissions

Pascual Eduardo Murillo-Alvarado,[1] José María Ponce-Ortega,[2] Mahmoud M. El-Halwagi,[3] Juan Gabriel Segovia-Hernandez[1]

[(1)] Chemical Engineering Department, Universidad de Guanajuato, Guanajuato, Gto., México.
[(2)] Chemical Engineering Department, Universidad Michoacana de San Nicolás de Hidalgo, Morelia, Mich., México.
[(2)] Chemical Engineering Department, Texas A&M University, College Station TX, USA.

1. Introduction

The process and chemical industries generate large amounts of liquid and gaseous emissions. These emissions represent severe adverse impacts in the environment. This way, several environmental constraints have been imposed to the industrial emissions in order to promote sustainable processes. For the design of a new industry, the environmental constraints only impose limits to the concentration and discharges of the hazardous pollutants; and when the effect of the emissions is observed after the design, the industries have to install treatment units to recover the hazardous pollutants and to satisfy the environmental constraints, which represents an additional cost. In addition, in the industry generally there is a set of process units that require streams with specific characteristics. In this context, there has been reported several methodologies for the recycle, reuse and regeneration of the process streams to satisfy the process units requirements, and at the same time to reduce the fresh sources consumption, the total waste discharged to the environment and the associated cost (see the papers review by Bagajewicz, 2000; Dunn and El-Halwagi, 2003; Foo, 2009); however, most of the reported methodologies consider only liquid streams. A lower number of methods have been reported for the reuse of the gaseous process stream in the industry (see for example the works by Shonnard and Hiew, 2000; Parthasarathy and El-Halwagi, 1999; Hamad and Fayed, 2004; El-Halwagi et al., 1996; Dunn and El-Halwagi, 1994a; b; Gassner and Marechal, 2010; Parthasarathy and Dunn, 2003). **Figure 1** shows a schematic representation of these recycle, reuse and interception schemes for the gaseous streams. Notice that previous formulations only have considered the effects that happen inside the industrial facilities; this means that previous methodologies do not consider the effect of the gaseous emission in the surroundings neither the interaction with other emissions. High concentration of the gaseous pollutants affects the air quality of the surroundings, causing environmental and public health problems, these pollutants are dispersed into the air that may be inhaled by people from nearby communities to the point of emission. To reduce the concentration of the pollutants emitted is necessary to implement certain environmental constrains to ensure the air quality and to implement a treatment system to the gaseous streams inside the plant before they are discharged to the environment. Recently, Lira-Barragán et al. (2011a; 2011b) have reported approaches to include the water integration inside the industrial facilities together with the sustainability of the surrounding watershed, these approaches show that the consideration of the surroundings is required to yield sustainable process, in addition, these works show that the effect of the wastewater discharged can be considered using a material flow analysis model (MFA) (Baccini and Brunner, 1991; Brunner and Rechberg, 2004; Lovelady et al., 2009) based on the mean composition in the segments of the watershed to avoid numerical complications. Therefore, to overcome the limitations of previous works, this paper proposes a new mathematical programming model for the synthesis of recycle, reuse and interception networks for gaseous streams since the installation of a new industrial facility considering the interaction with their surroundings; the model considers the possibility to direct recycle of the process streams, the interception of the streams, in addition to the interaction of the gaseous emissions with their surroundings through the selection of the best location to install the new industrial facility, the model is able to account for constraints for the gaseous emissions to the neighboring cities and the interaction with other sources of emissions, this way the Pasquill-Gifford (Turne, 1994; Barratt, 2001) model is used to consider the effect of the gaseous emissions in the surroundings. This Pasquill-Gifford model accounts for the dispersion of the emissions considering the properties, air quality, speed and direction of the wind; this way, the proposed model is based in the calculation of the dispersion parameters to calculate the concentration of the pollutants at a certain distance from the point of emission to determine the optimal location of the industrial facility accounting for the surroundings.

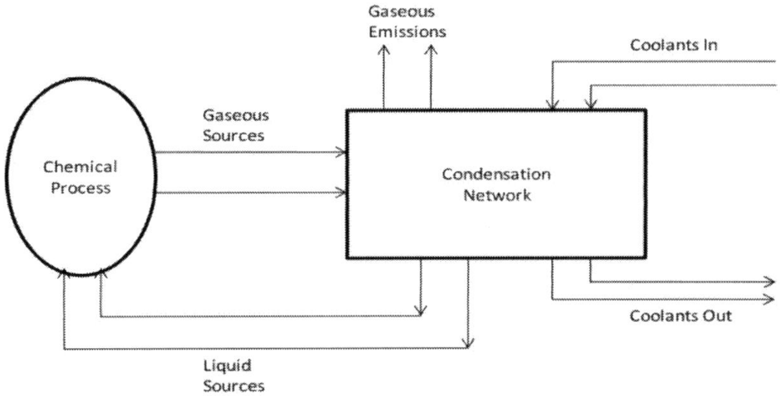

Figure 1. Integration of inside the plant

2. Outline of the proposed model

The problem addressed in this paper can be described as follows. Given a set of possible locations to install a new industrial plant (L={$l | l=1,2,...,N_l$}), in each possible location l is assigned an installation cost ($CLand_l$), this cost includes the land cost, transportation of the raw material cost and product and services costs. There are several nearby cities that can be impacted by the emissions from the new industrial plant (P={$p | p=1,2,...,N_p$}), in each nearby city p is evaluated the effects for the emissions using the dispersion model of Pasquill-Gifford, this model determine the concentration $Conc_{p,c}$ of the each toxic compound (C={$c | c=1,2,...,N_c$}) in each city taking into account the speed and direction of the wind. The Pasquill-Gifford model is based in the calculations of the concentrations like the MFA model for the liquid streams. The calculation of the concentration by an emission point is given by equation (1),

$$C(x,y,z)=\frac{Q_m}{2\pi\sigma_y\sigma_z u}\exp\left[-\frac{1}{2}\left(\frac{y}{\sigma_y}\right)^2\right]\times\left\{\exp\left[-\frac{1}{2}\left(\frac{z-H_r}{\sigma_z}\right)^2\right]+\exp\left[-\frac{1}{2}\left(\frac{z+H_r}{\sigma_z}\right)^2\right]\right\} \quad (1)$$

were H_r is the height of the source, Q_m is the amount of the gases emitted, u is the wind velocity, x is the wind direction, y is the ratio of dispersion, z is the height of dispersion and σ_y and σ_z are the dispersion parameters for the directions y and z, respectively. If the emissions source is found at $z=0$, the equation (2) is obtained and it depends only on the ratio of dispersion.

$$C(x,y,0)=\frac{Q_m}{\pi\sigma_y\sigma_z u}\exp\left[-\frac{1}{2}\left(\frac{y}{\sigma_y}\right)^2-\frac{1}{2}\left(\frac{H_r}{\sigma_z}\right)^2\right] \quad (2)$$

The concentration calculated by the Pasquill-Gifford model is used to determine the location of the plant considering the emissions of the new plant (Em_c). Also, the values obtained by the Pasquill-Gifford model are grouped in the Pasquill-Gifford constant ($CPG_{i,p,c}$) in a model reformulation to obtain one location of a new plant and also to consider what happens inside the plant. Inside of the plant is given a set of gaseous process sources ($I = \{i | i = 1,2,...,N_i\}$) with composition $z_{i,c}^{In}$ and flowrate W_i, which can be segregated to be recycled and reused in the process sinks ($J = \{j | j = 1,2,...,N_j\}$) with composition $z_{j,c}^{InSink}$ and flowrate G_j fixed by the process. To satisfy the requirements of the process sinks, there is available a set of fresh sources ($R = \{r | r = 1,2,...,N_r\}$) with composition given $z_{r,c}^{InFresh}$ and the flowrate F_r needs to be determined. Also, there is available a set of interceptor for the treatment of gaseous streams (INT). The composition at the exit of each treatment unit for each component is calculated using an efficiency factor for the removal of pollutant by unit of treatment as follows:

$$z_{i,c}^{Out,\text{int}} = \left(1 - \gamma_c^{\text{int}}\right) z_{i,c}^{In}, \forall i \in I, c \in C, \text{int} \in INT$$

To satisfy the environmental regulation after the installation of the new plant, a set of environmental constrains are stated to comply with the permissible limits for the amount of the emissions for the new plant and for the maximum permissible concentration in each nearby city.

Therefore, the problem consists in finding the optimal recycle, reuse and interception of gaseous streams inside the industrial facilities considering the effect of the emissions in the surrounded cities, to consider the effects of the emissions on the environment a new MFA model based on the Pasquill-Gifford approach is used, this way the optimal solution must consider the location of the new industrial facility to integrate the process gaseous streams with the environment. The objective function consists in minimizing the total annual cost (TAC), which includes the installation cost for the new plant, the treatment for the gaseous process sources, the fresh sources cost, the cost for the interceptor cost and piping cost. **Figure 2** shows a schematic representation to integrate the industrial emissions with the surroundings, where there are L possible locations to install the new plant and there are P neighboring cities that are affected by the emissions of pollutants from the new plant.

The dispersion parameters for the Pasquill-Gifford model are used to determine the influence by the dispersion of the contaminants discharged from the new plant in the surrounding air quality, where it is necessary to know the wind direction and speed, because this form the model locates the new plant in best option according to the air quality in each city. To satisfy the process, environmental and sustainability constraints, a recycle, reuse and interception scheme is used. In this paper, the superstructure shown in **Figure 3** is used. This superstructure avoids mixing of different process streams to yield convex relationships, and then there is a set of treatment units to treat different pollutants with different efficiencies and costs. There is included a fictitious interceptor to model the bypass streams (the last one for each set of interceptors). The optimization model must select the flowrate and the treatment technologies used to satisfy the process and environmental constraints.

3. Model Formulation

The emissions discharged from the new plant that satisfy the constraint for the different surrounding cities depend on the location of the plant. Then, the model must consider the location of the new plant, the treatment network inside the plant to satisfy the process, environmental and sustainability constraints (notice that the sustainability constraints depend on the location of the new plant) as well as the relationships required to interconnect these models, minimizing the total annual cost.

3.1 Location of a new plant

To locate the new plant, there are l alternatives (see **Figure 2**), and the Pasquill-Gifford model is used to evaluate the effects of the emissions in the p surrounded cities. For example, if the new plant is located in the position 1, the concentration of the pollutants and the emissions associated to this site must be bigger than zero, on the other hand if the new plant is not located in the position 1, the values for the emissions must be zero. This is modeled through the following disjunction:

$$\begin{bmatrix} Y_{l=1} \\ Conc_{p,c} = CPG_{l,p,c} Em_c \end{bmatrix} \vee \begin{bmatrix} Y_{l=2} \\ Conc_{p,c} = CPG_{l,p,c} Em_c \end{bmatrix} \vee \ldots \vee \begin{bmatrix} Y_L \\ Conc_{P,C} = CPG_{L,P,C} Em_c \end{bmatrix}, \forall c \in C, p \in P$$

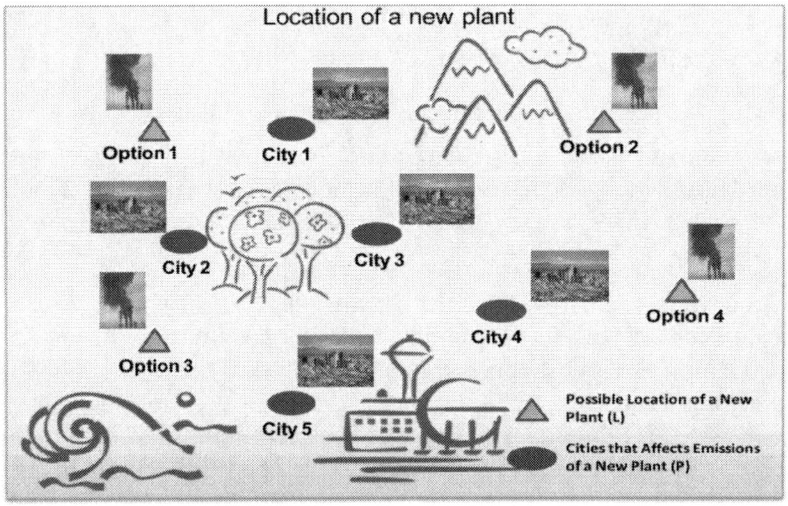

Figure 2. Schematic representation for the integration of the industrial emissions and the surroundings

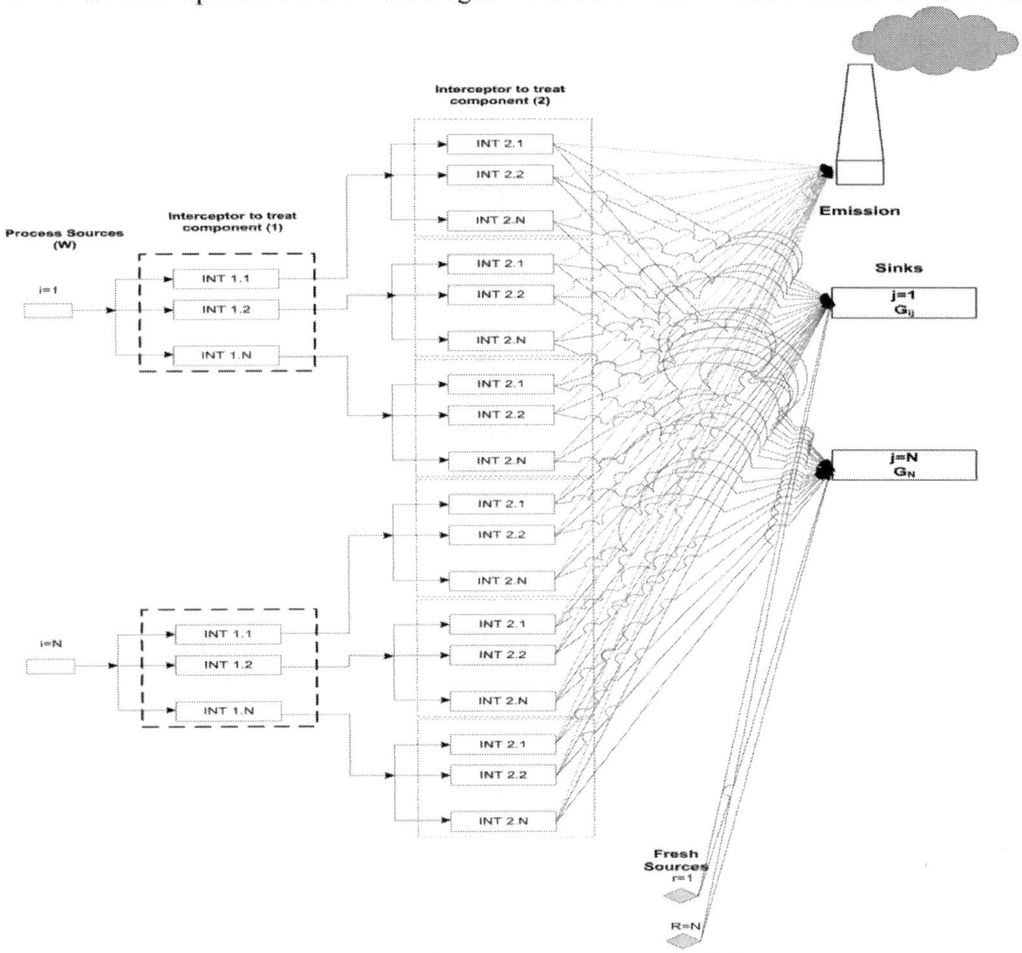

Figure 3. Network to treat the gaseous emissions.

The Boolean variable Y_l is associated with the possible location of a new plant, $Conc_{p,c}$ is the concentration for the component in the emission discharged from the new plant to the neighboring city p and Em_c is the total flowrate of the emission of the new plant.

The Boolean variables Y_l are transformed into binary variables y_l to reformulate the disjunction as an algebraic problem (Raman and Grossmann, 1994). The binary variables determine only one location for the new plant as follows:

$$\sum_{l \in L} y_l = 1 \tag{3}$$

Previous disjunction is modeled using the convex hull reformulation and the following algebraic relationships are obtained.

3.1.1 Disaggregate the continuous variables. The continuous variables that depend on the location of the new plan are disaggregated as follows:

$$Conc_{p,c} = \sum_{l \in L} dConc_{p,c,l}, \forall c \in C, p \in P$$
$$Em_c = \sum_{l \in L} dEm_{c,l}, \forall c \in C \tag{4}$$

3.1.2 Relationships in term of disaggregated variables. The concentration of a given component in a given position depends on the location of the new industrial facility, and this is calculated using the disaggregated variables as follows:

$$dConc_{p,c,l} = CPG_{p,c,l} \cdot dEm_{c,l}, \forall p \in P, c \in C, l \in L \tag{5}$$

3.1.3 Upper limits to disaggregated variables. There are required upper limits for the disaggregated variables to active these variables only when the corresponding location is selected:

$$dConc_{p,c,l} \leq Conc_c^{\max} \cdot y_l, \forall p \in P, c \in C, l \in L$$
$$dEm_{c,l} \leq Em_c^{\max} \cdot y_l, \forall c \in C, l \in L \tag{6}$$

3.2 Recycle and interception model inside the plant

The proposed model is based on the superstructure shown in **Figure 3**, which shows the configuration for the interceptors and recycle network for the gaseous process streams. This superstructure is based on the ones previously reported by Gabriel and El-Halwagi (2005) and Ponce-Ortega et al. (2010). The model for this superstructure is presented as follows.

3.2.1 Splitting of the fresh sources. First, to satisfy the process sink constraints fresh streams can be used, then different type of fresh sources can be used with different costs and compositions, and the model must select the amount of each type of fresh source selected to satisfy the process and environmental constraints at the minimum cost. Therefore, the superstructure allows splitting the fresh sources and sent them to the process sinks, which is modeled as follows:

$$F_r = \sum_{j \in NSINKS} f_{r,j}, r \in FRESH \tag{7}$$

where F_r is the total flowrate for the fresh source r and $f_{r,j}$ is the flowrate segregated from the fresh source r and directed to the process sink j. Notice that the fresh sources only are directed to the process sinks and these are not discharged to the environment.

3.2.2 Splitting of the process sources to the interceptors. The process sources in the interception network are segregated avoiding mix different process streams, the mixing of different process streams is allowed only after the treatment and prior to inlet to the process sinks. If a gaseous process source does not require any treatment, a fictitious unit with effectiveness and cost equal to zero is used to model the bypass of this stream.

The flowrates for the gaseous process sources are segregated $w_i^{\text{int}^1}$ and sent to the different interceptors to treat the first pollutant as follows:

$$W_i = \sum_{\text{int}^1 \in INT^1} w_i^{\text{int}^1}, i \in NSOURCES \tag{8}$$

To treat the second pollutant, the flowrates from each interceptor of the first stage are segregated $w_i^{\text{int}^1,\text{int}^2}$ and sent to the interceptors of the stage two.

$$w_1^{\text{int}^1} = \sum_{\text{int}^2 \in INT^2} w_i^{\text{int}^1,\text{int}^2}, i \in NSOURCES, \text{int}^1 \in INT^1 \tag{9}$$

Similar balances are required for the other stages required to treat different pollutants.

3.2.3 Splitting of the sources at the exit of the interceptors. Once the pollutants of the gaseous process sources are intercepted, they are directed to the process sinks $g_{i,j}^{\text{int}^1,\ldots,\text{int}^N}$ and to the emissions discharged to the environment $g_{i,emission}^{\text{int}^1,\ldots,\text{int}^N}$:

$$w_i^{\text{int}^1,\ldots,\text{int}^N} = \sum_{j \in SINKS} g_{i,j}^{\text{int}^1,\ldots,\text{int}^N} + g_{i,emission}^{\text{int}^1,\ldots,\text{int}^N}, i \in NSOURCES, \text{int}^1 \in INT^1, \ldots, \text{int}^N \in INT^N \tag{10}$$

3.2.4 Overall mass balance at the mixing point prior to any sink. The total flowrate inlet to any process sink G_j is given by the sum of the flowrates from the gaseous process source intercepted $g_{i,j}^{\text{int}^1,\ldots,\text{int}^N}$ and the flowrates from the fresh sources $f_{r,j}$:

$$G_j = \sum_{i \in NSOURCES} \sum_{\text{int}^1} \ldots \sum_{\text{int}^N} g_{i,j}^{\text{int}^1,\ldots,\text{int}^N} + \sum_{r \in FRESH} f_{r,j}, j \in NSINKS \tag{11}$$

3.2.5 Component balance at the mixing point prior to any sink. This balance is use to determine the composition of the compounds at inlet conditions of the process sinks $z_{j,c}^{InSink}$:

$$G_j z_{j,c}^{InSink} = \sum_{i \in NSOURCES} \sum_{\text{int}^1} \ldots \sum_{\text{int}^N} g_{i,j}^{\text{int}^1,\ldots,\text{int}^N} z_{i,c}^{Out,\text{int}} + \sum_{r \in FRESH} f_{r,j} z_{r,c}^{InFresh}, j \in NSINKS, c \in COMPONENT \tag{12}$$

3.2.6 Overall mass balance for gaseous emissions. The total flowrate of the gaseous emission discharged to the environment *emission* is given by the sum of the gaseous process sources from the interceptors emitted to the atmosphere $g_{i,emission}^{\text{int}^1,\ldots,\text{int}^N}$:

$$emission = \sum_{i \in NSOURCES} \sum_{\text{int}^1} \sum_{\text{int}^N} g_{i,emission}^{\text{int}^1,\ldots,\text{int}^N} \tag{13}$$

3.2.7 Component balance for the emission streams: This balance is needed to determine the amount for the gaseous stream emitted to the environment for each hazardous compound Em_c:

$$Em_c = \sum_{i \in NSOURCES} \sum_{\text{int}^1} \ldots \sum_{\text{int}^N} g_{i,emission}^{\text{int}^1,\ldots,\text{int}^N} z_{i,c}^{Out,\text{int}^1,\ldots,\text{int}^N}, \forall c \in C \tag{14}$$

The term Em_c is defined as the total amount of the emission by the composition of each hazardous compound *c*.

3.2.8 Equations to active the fixed cost for the interceptors: The following relationships are required to determine the existence of the interceptor that depends of the maximum and minimum flowrate necessary to operate the treatment unit:

$$w_i^{\text{int}} \le w_i^{\text{int}MAX} \cdot z_i^{\text{int}}, \forall i \in I, \text{int} \in INT$$
$$w_i^{\text{int}} \ge w_i^{\text{int}MIN} \cdot z_i^{\text{int}}, \forall i \in I, \text{int} \in INT \tag{15}$$

3.2.9 Sink constraints: Each process sink includes a set of specific process constraints given on terms of the maximum $z_{j,c}^{InSink,UP}$ and minimum $z_{j,c}^{InSink,LO}$ compositions for specific compound required to work properly:

$$z_{j,c}^{InSink,LO} = z_{j,c}^{\min}, \forall j \in J, c \in C$$
$$z_{j,c}^{InSink,UP} = z_{j,c}^{\max}, \forall j \in J, c \in C \qquad (16)$$

3.2.10 Environmental constraints for the emission of the plant: The following constraint is required to specific that the plant is not allowed to discharge more emissions than the maximum allowed by the environmental constraints for the hazardous compounds $Em_c^{\max}$:

$$Em_c \le Em_c^{\max}, \forall c \in C \qquad (17)$$

3.2.11 Environmental constraints for the location of a new plant: This constrains determine the maximum permissible concentration for the hazardous compounds in each possible location $Conc_{p,c}^{\max}$.

$$Conc_{p,c} \le Conc_{p,c}^{\max}, \forall p \in P, c \in P \qquad (18)$$

3.2.12 Objective function. The objective function consists in the minimization of the total annual cost (TAC), including the installation cost for the new plant ($CLand_l$), the cost for the fresh sources ($FreC_r$), and cost for the treatment required ($FixC_{cap}^{\text{int}}, VaC_{cap}^{\text{int}}$), and the piping cost for the different sections including the treatment ($TPipingC_i^{\text{int}}$), emission ($EmPipingC_i^{\text{int}}$), equipment ($EqPipingC_{i,j}^{\text{int}}$) and fresh sources ($FshPipingC_{r,j}^{\text{int}}$). This way, the objective function is stated as follows:

$$\begin{aligned}
\min TAC = & \sum_{l \in L} CLand_l y_l + H_Y \sum_{r \in R} FreC_r F_r + H_Y \sum_{i \in I} \sum_{\text{int} \in INT} C_{op}^{\text{int}} w_i^{\text{int}} \\
& + k_f \left[\sum_{i \in I} \sum_{\text{int} \in INT} FixC_{cap}^{\text{int}} z_i^{\text{int}} + \sum_{i \in I} \sum_{\text{int} \in INT} VaC_{cap}^{\text{int}} w_i^{\text{int}} \right] \\
& + H_Y \left[\begin{array}{l} \sum_{i \in I} \sum_{\text{int} \in INT} \left(TPipingC_i^{\text{int}} + EmPipingC_i^{\text{int}} g_{i,emmission}^{\text{int}} \right) \\ + \sum_{i \in I} \sum_{\text{int} \in INT} \sum_{j \in J} EqPipingC_{i,j}^{\text{int}} g_{i,j}^{\text{int}} + \sum_{r \in R} \sum_{j \in J} FshPipingC_{r,j} f_{r,j} \end{array} \right]
\end{aligned} \qquad (19)$$

where H_Y is the hours per year that operates the new plant, and k_f is a factor used to annualize the capital costs.

3.2.12 Interceptors performance. The interceptors performance is modeled through a conversion factor that is determined prior to the optimization process based on the experimental data or simulation of given treatment units (see for example Gabriel and El-Halwagi, 2005; Ponce-Ortega et al., 2010). This way, the outlet concentration for the interceptors depend on the design and operating conditions for these units and these are fixed prior to the optimization process; therefore, the conversion factor $\alpha_p^{\text{int}^1,\ldots,\text{int}^N}$ is used to determine the concentration for the streams at the exit of the interception network as follows:

$$z_{i,c}^{Out,\text{int}^1,\ldots,\text{int}^N} = \alpha_c^{\text{int}^1,\ldots,\text{int}^N} z_{i,c}^{In} \qquad (20)$$

Therefore the model must determine the flowrate that is intercepted in each available treatment unit (including the last one that is fictitious to model the bypass) to satisfy the process and environmental constraints. The overall optimization model consists in minimizing (19) subject to (1-18), and this is a mixed integer non linear programming problem (MINLP).

4. Case Study

This example considers the VOC recovery from adhesive tapes plant, the contaminants in this process are methanol (MeOH) and methyl-iso-butyl ketone (MIBK) due to the solvents used in the process of the adhesive tape manufacturing. The recovery of VOC is using condensation systems. In this case study only the treatment of the MeOH is considered. The mass integration for the VOC recovery was previously reported by Parthasarathy and El-Halwagi (1999 and Hamad and Fayed (2004); however, only the integration inside the industrial facility was considered previously, this means that the environmental problem for the pollutants emitted in the surrounding cities have not been considered properly. **Figure 4** shows a schematic representation for the adhesive tape manufacturing process, where three process sources and two process sinks for the mass integration are identifies. The data for the sources and sinks are shown in **Table 1**, which presents the characteristics of flowrates and compositions of the process sources. **Table 1** also shows the conditions for the flowrates and compositions necessaries for the process sinks. There is available one fresh source to satisfy the process requirements and the unitary cost is $0.009/kg.

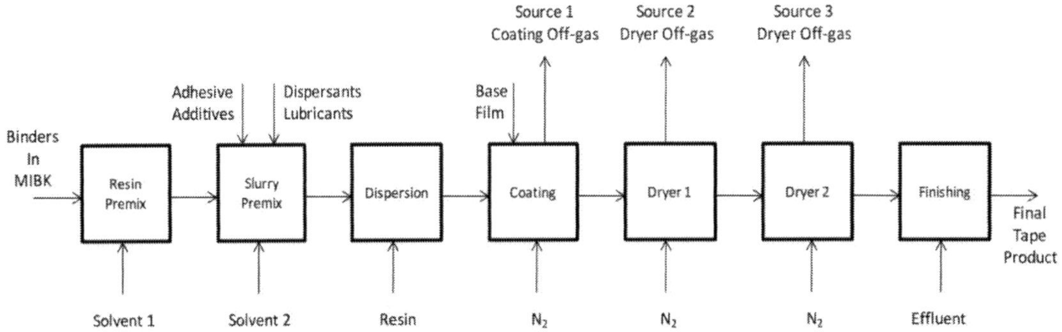

Figure 4. Flowshet for the process addressed in the Example 1

Table 1. Data for gaseous process sources and the process sinks for Example 1

Source	W_i *(Kg/hr)*	$z^{In}_{i,MeOH}$ *(ppm)*	*Sink*	G_j *(Kg/hr)*	$z^{InSink}_{j,MeOH}$ *(ppm)*
1	200	14	1	231	0.310
2	200	10	2	105	0.252
3	200	6			

In this example, the central part of Mexico has been considered for the installation of the new industrial facility. **Figure 5** identify four possible options to install the new industrial plant and nine surrounded cities that may be affected by the gaseous emissions for the new plant. To consider the effect of the emissions in the surrounding cities, the conditions of the weather, direction and speed of wind are taken into account in each possible location. **Table 2** shows the values for the constants of the Pasquill-Gifford model for each city depending on the option to locate the plant. These values are necessary to calculate the dispersion parameters of the emissions and to know the concentrations in the nearby cities. To consider properly the interactions between the processes inside the plant and the surroundings, it is necessary to include the environmental constraints for the emissions allowed in each city as are indicated in **Table 3** for this case of study. These environmental constraints are different for each city, and these depend on the environmental, geographic and weather conditions as well as the interactions with other emissions in the specific zone, therefore, these constraints are used to ensure a sustainable process. For each option to locate the new plant there is a given installation cost that depends on the land cost, cost of transportation of raw materials and products, services, etc. **Table 4** shows the installation cost for the locations identified in this example. The treatment system has different interceptors allowed to treat the gaseous process streams, **Table 5** shows the operation and fixed costs and the efficiency to remove the hazardous compound for the interceptors allowed in this example.

Table 2. Pascall-Giffort constants for the different locations for Example 1

City	Option 1	Option 2	Option 3	Option 4
Guadalajara	3.19×10^{-8}	8.79×10^{-9}	3.52×10^{-8}	3.94×10^{-9}
Zacatecas	5.22×10^{-8}	1.20×10^{-8}	1.00×10^{-8}	3.28×10^{-9}
Aguascalientes	1.30×10^{-7}	3.04×10^{-8}	1.88×10^{-8}	4.69×10^{-9}
León	2.75×10^{-8}	9.77×10^{-8}	3.52×10^{-8}	7.77×10^{-9}
Querétaro	8.25×10^{-9}	2.64×10^{-8}	2.68×10^{-8}	2.45×10^{-8}
Morelia	7.98×10^{-9}	1.23×10^{-8}	1.26×10^{-8}	1.57×10^{-8}
Lázaro Cárdenas	3.98×10^{-9}	3.36×10^{-9}	4.82×10^{-8}	5.64×10^{-9}
Chilpancingo	2.28×10^{-9}	2.57×10^{-9}	1.27×10^{-8}	1.24×10^{-8}
Toluca	3.93×10^{-9}	6.10×10^{-9}	2.19×10^{-8}	1.45×10^{-7}
Distrito Federal	3.44×10^{-9}	5.38×10^{-9}	1.54×10^{-8}	7.19×10^{-7}

Table 3. Constraints for the emissions in each surrounded city for Example 1

City	Environmental Concentration
Guadalajara	0.00014
Zacatecas	0.00017
Aguascalientes	0.00016
Leon	0.00016
Queretaro	0.00015
Morelia	0.00018
Cd. Lazaro Cardenas	0.00019
Chilpancingo	0.00020
Toluca	0.00010
Distrito Federal	0.00005

Table 4. Installation cost for the locations of Example 1

Option	Annualized installation cost ($/year)
1. Tabasco, Zacatecas	12×10^{-6}
2. San Felipe, Guanajuato	11×10^{-6}
3. Uruapan, Michoacán	13×10^{-6}
4. Cuatitlan Izcalli, Mexico	14×10^{-6}

Table 5. Efficiency and costs for the interceptors for Example 1.

Interceptor	*Efficiency α*	*Operation Unit Cost ($/lb)*	*Fixed Interceptor Cost*
INT^1	0.98	0.0065	1500
INT^2	0.85	0.0033	1300
INT^3	0.00	0.0000	0000

This example consists of 13 binary variables, 168 continuous variables, and 20 constraints, and it was solved in 0.094 seconds of CPU time. The solution of the optimization problem proposed in this paper yields the optimal location for the new plant in the option 2 corresponding to San Felipe, Gto (see **Figure 5**) with a total annual cost of $11,353,772/year. The in-plant treatment network is shown in **Figure 6**, notice that there are required several interceptors to eliminate the pollutant to satisfy the process and environmental constraints. **Table 6** shows the solutions of the problem for the cases when the location for the new industrial facility is fixed (i.e., not optimized). This information is very useful to identify the governmental incentives required to install the new industrial facility in a location different to the optimal to bring to the location social benefit obtained for the generated jobs.

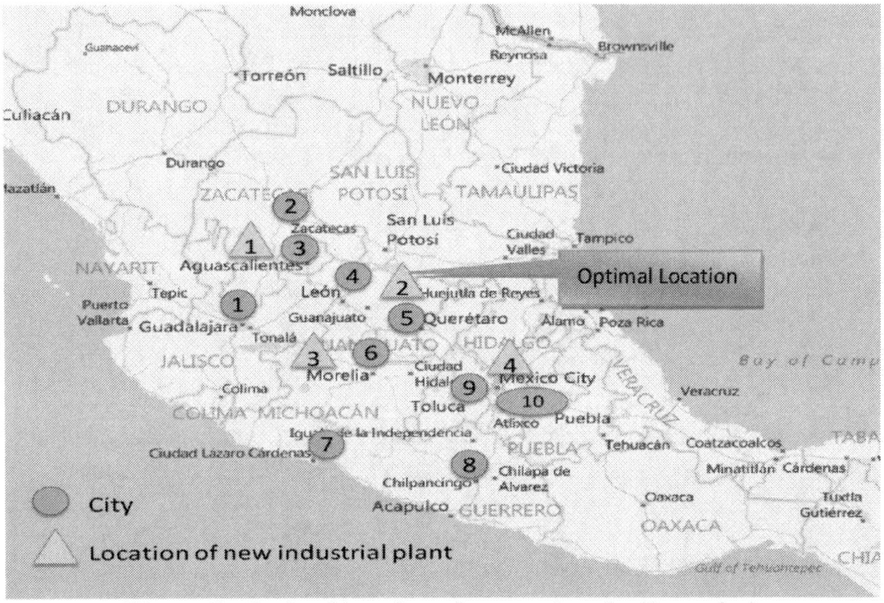

Figure 5. Optimal location of a new plant for Example 1

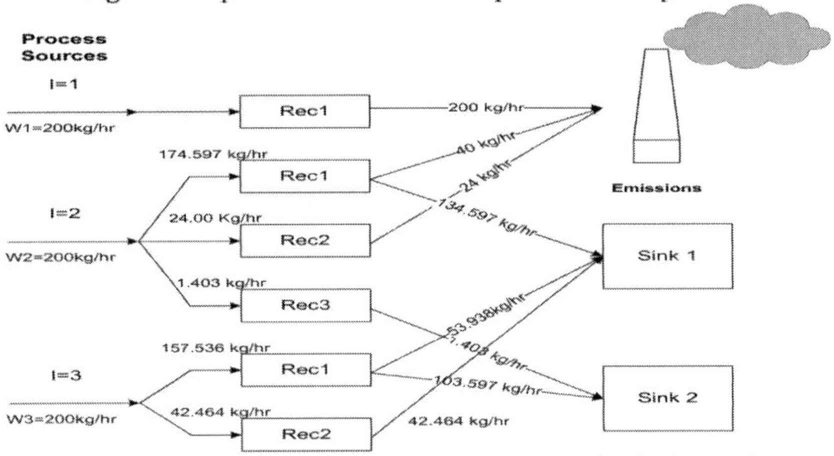

Figure 6. Optimal solution for the treatment system for the Example 1

Table 6. Economic results for the location of the new plant for Example 1

Location	Total annual cost	Economic incentives
Option 1	12'353,770	$ 999,998
Option 2	11'353,772	Optimal
Option 3	13'353,770	$ 1'999,998
Option 4	14'353,770	$ 2'999,998

5. Conclusions

This paper presents a new mathematical programming model for the in-plant integration for the gaseous emissions in addition to the phenomena that happen in the surrounded areas. The proposed model allows tracking the pollutant in the surrounded areas and this way determines the best location for the industrial facility integrating the gaseous emission inside and outside the plant. The model consists in the minimization of the total annual costs including the plant installation cost, the fresh sources costs, and the interception costs to satisfy the process and environmental constraints. The proposed model allows to identify the governmental incentives required to yield a given location economically attractive satisfying the environmental constraints, this situation is very important for development countries (like Mexico) because the social benefit obtained for the number of jobs generated. The proposed methodology also allows to identify different options for the inside and outside gaseous emissions integration for different levels of environmental constraints, the major advantage of the proposed methodology is to identify these objectives prior the design. The proposed methodology is general, and it can be applied to different plants with gaseous emissions and different locations that can be affected by the industrial gaseous emissions. Finally, no numerical complications were observer in the application of the proposed methodology.

References

Baccini, P., Brunner, P. (1991). Metabolism of the anthroposphere. Berlin: Springer-Verlag.

Bagajewicz, M., (2000). A review of recent design procedures for water networks in refineries and process plants. *Computers and Chemical Engineering* 24 (9-10), 2093-2113.

Barratt, R. (2001). Atmospheric dispersion modeling. Earthscan.

Brunner, P. H., Rechberg, H. (2004). Practical handbook of material flow analysis. CRC.

Dunn, R. F., El-Halwagi, M. M. (1994a). Selection of optimal VOC-condensation systems. *Waste Management*, 14 (2): 103-113.

Dunn, R. F., El-Halwagi, M. M. (1994b). Optimal design of multicomponent VOC-condensation systems. *Journal of Hazardous Materials*, 38 (1): 187-206.

Dunn, R.F., El-Halwagi, M.M., (2003). Process integration technology review: background and applications in the chemical process industry. *Chemical Technology and Biotechnology* 78 (9), 1011-1021.

El-Halwagi, M. M., Hamad, A. A., Garrison, G. W. (1996). Synthesis of waste interception and allocation networks. *AIChE Journal*, 42 (11): 3087-3101.

Foo, D.C.Y., (2009). State-of-the-art review of pinch analysis techniques for water network synthesis. *Industrial and Engineering Chemistry Research* 48 (11), 5125-5159.

Gabriel, F. B., and El-Halwagi, M. M. (2005). Simultaneous synthesis of waste interception and material reuse networks: Problem reformulation for global optimization. *Environmental Progress*, 24 (2), 171-180.

Gassner M., Marechal F. (2010). Combined mass and energy integration in process design at the example of membrane-based gas separation systems. *Computers and Chemical Engineering*, 34 (12): 2033-2042 .

Lira-Barragán, L. F., Ponce-Ortega, J. M., Serna-Gonzales, M. and El-Halwagi, M.M. (2011a). An MINLP model for the optimal location of a new industrial plant with simultaneous consideration of economic and environmental criteria. *Industrial and Engineering Chemistry Research*, 52 (2): 953-964.

Lira-Barragán, L.F., Ponce-Ortega, J.M., Serna-González, M., El-Halwagi, M. M. (2011b). Synthesis of water networks considering the sustainability of the surrounding watershed. Accepted in *Computers and Chemical Engineering*, doi:10.1016/j.compchemeng.2011.03.021

Lovelady, E. M., El-Baz, A. A., El-Monayeri, D., EL-Halwagi, M. M. (2009). Reverse problem formulation for integrated process discharges with wastewater and drainage systems: Managing phosphorus in lake Manzala. *Journal of Industrial Ecology*, 13(6), 914–927

Parthasarathy G., Dunn R. F. (2003). Graphical strategies for design of evaporation crystallization networks for environmental wastewater applications. *Advances in Environmental Research*. 8: 247-265.

Parthasarathy G., El-Halwagi, M. M., Optimum mass integration strategies for condensation and allocation of multicomponent VOCs. *Chemical Engineering Science*. 55: 881-895.

Ponce-Ortega, J. M., El-Halwagi, M. M., and Jiménez-Gutiérrez, A. (2010a). Global optimization of property-based recycle and reuse networks including environmental constraints. *Computers and Chemical Engineering*. 34 (3): 318-330.

Shonnar, D. R., Hiew, D. S. (2000). Comparative environmental assessments of VOC recovery and recycle design alternatives for a gaseous waste stream. *Environmental Science and Technology*, 34 (24): 5222-5228.

Turner, D. B. (1994). Workbook of atmospheric dispersion estimates: Second Edition. CRC-Press.

(620k) Sustainable Optimization of Steam Power Plants

Csar Giovani Gutirrez-Arriaga[1], Medardo Serna-Gonzlez[2] and Jos Mara Ponce-Ortega[2], (1)Chemical Engineering Department, Universidad Michoacana de San Nicols de Hidalgo, Morelia, Mexico, (2) Chemical Engineering Department, Universidad Michoacana de San Nicols de Hidalgo, Morelia, Michoacan, Mexico

This work presents a new methodology for the multi-objective optimization of steam power plants determining simultaneously the configuration and operational conditions. The proposed methodology is based on search strategy through genetic algorithms because the large sets of nonconvex terms present and to avoid get trapped in a very bad local solution. The proposed methodology considers the optimal selection for the type of fuel used, because this affects drastically the economic and the environmental objective functions; also is considered the selection of a solar collector to provide the required energy because this can be a good option to reduce the green house gas emissions; however the economic objective function is drastically impacted when the solar collector is used. The economic objective function considers the minimization of the total annual cost, including the heating costs (fossil, biofuels and solar), the capital costs for the turbine, condenser, pump, reboiler, deareator, and the solar collector, as well as the operational costs associated to the function of these units. On the other hand, the environmental objective function is based in the eco indicator 99 that measures the overall environmental impact considering the life cycle analysis. These two objectives contradict each other, and the constraint method is used to determine the set of optimal solutions that compensate these two objectives. The proposed methodology allows identifying the incentives required to yield economically attractive a solution that reduces significantly the green house gas emissions.

See more of this Session: Poster Session
See more of this Group/Topical: Computing and Systems Technology Division

(620l) Simulation of a Typical and An Advanced Fischer Tropsch Reactor Technology

Marwan Abbas[1], **Fadwa T. Eljack**[2], Elfatih E. Elmalik[3], Aswani K. Mogalicherla[3] and Nimir O. Elbashir[3], (1)Chemical Engineering, Qatar University, Doha, Qatar, (2)Department of Chemical Engineering, Qatar University, Doha, Qatar, (3)Chemical Engineering Program, Texas A&M University at Qatar, Doha, Qatar

Simulation of a Typical and an Advanced Fischer Tropsch Reactor Technology

Marwan Abbas[1], FadwaT. El jack[1], Aswani Mogalicherla[2], Elfatih Elmalik[2]and Nimir O. Elbashir[2]

(1)Departmentof Chemical Engineering , Qatar University, Doha, Qatar

(2)ChemicalEngineer Department, Texas A&M, Doha, Qatar

10A – 00 Systems & Process Design

During the past twenty years, tremendous investments from the majorplayers in the energy market have been directed towards Gas to Liquid (GTL)technology. The advantage of GTL technology in energy production is that itfurnishes a broad range of environmentally clean fuels, additives andvalue-added. While the GTL market continues to grow, there are still manyhurdles to overcome in order to optimize the use of this important technology– all of which are related to technology limitations. Our proposedproject is focused on the design ofadvanced Fischer-Tropsch synthesis (FTS) reactortechnology to facilitate operating the reaction on both conventional media (gasphase) and non-conventional media (supercritical phase). The latter assumed toprovide unique reaction medium to leverage certain advantages over thecurrent commercial technologies (slurry reactor and multi-tubular reactor) andat the same time overcome their limitations[1]

The main aim of this study isto develop simulation models of Fischer Tropsch Synthesis (FTS) reactortechnology. The approach todesign this novel reactor technology has been described in a previous study[2]. The modeledreactor configurations operating at gas phase , slurryphase , near and super critical phase conditions (SCF) will be compared to existingFTS commercial reactors (Fixed Bed Reactors , Shell Bintulu Plant in Malaysia)and experimental reactors (Supercritical , Roberts and Elbashir [1]).Simulating FTS reactors have always been a challenge. This due to their abnormalchemistry behavior that led available design tools to not best represent the FTand SCF-FT reactions. The main reactor design focus will be on SCF-FT. SCF-FThas been strongly considered to yield better fuel products. The SCF media usedhas the capability of improving the diffusivity (gas like) and heat transfer (liquidlike) [3]. Our study will be consisting of two phases. Phase one will mainlyfocus on using the existing academia tools such as (ASPEN Plus) to modeldifferent FT reactors and then compare it to available commercial andexperimental results. This will help conclude the advantages and the disadvantagesof the existing solutions used to model the FT reactors. The second phase willbe focusing on creating our own FT and SCF-FT reactor model configuration and compareit to the experimental results. This step will be achieved using otheravailable tools such as (ASPEN Simulation Workbook (ASW) and ComputationalFluid Dynamics model (CFD)). Our final results of the reactor design is to closelyfollow the results of SCF-FT reactors previously reported .The study will alsoexamine the economic aspect of SCF-FT reactors to account for the additionalcost accompanied with the use of technology.

References

[1]. Elbashir N. O., Bukur D. B.,Durham E., Roberts C. B. (2010) "Advancement in Fischer-Tropsch Synthesis viaUtilization of Supercritical Fluids as Reaction Media" AIChE J; 56 (4)997-1015.

[2]. Elbashir, N. O., and Eljack, F.T. (2010) "Models forDesigning Advanced Fischer Tropsch ReactorTechnology" Proceedings of the 2nd Annual Gas Processing Symposium;Elsevier.

[3] Bao B., El-Halwagi M. M.,Elbashir, N. O. (2010) "Simulation, Integration, and Economic Analysis ofGas-to-Liquid Processes" Fuel Proc. Techn.; in press.

See more of this Session: Poster Session
See more of this Group/Topical: Computing and Systems Technology Division

Design Project

Production of Ethanol from Excess Ethylene

Group 2

Adam S. Kadhim (s072192)

Kim B. Carlsen (s072202)

Thomas Bisgaard (s072213)

Department of Chemical and Biochemical Engineering
Technical University of Denmark
Course: 28350 Process Design: Principles and Methods
Date: 24. May 2011

Preface

This report has been produced to satisfy the course graduation criteria for the MSc-level course 28350 Process Design: Principles and Methods at the Department of Chemical and Biochemical Engineering, Technical University of Denmark (DTU). The project were supervised by Professor Rafiqul Gani.

The work has been evenly distributed among the group participants:

Adam Samir Kadhim (s072192)

Kim Braad Carlsen (s072202)

Thomas Bisgaard (s072213)

Abstract

Ethyl alcohol is one of the most important and used chemicals. Two common routes exist for the production: synthetic route typically based on petroleum feedstock and a fermentation route. The fermentation route comprises the majority of the produced ethyl alcohol. In this work, however, we will focus on the synthetic method, which employs direct hydration of ethylene.

A conceptual process design of an ethyl alcohol producing plant is performed in a MSc-level course on Process Design at the Department of Chemical and Biochemical Engineering at DTU. In the designed process, 190 proof ethyl alcohol (azeotropic mixture) is produced from excess ethylene containing propylene and methane as impurities. The design work is based on a systematic approach consisting of 12 tasks performed in a specified hierarchy. According to this 12-tasks design procedure, information about the product and process is obtained in tasks 1-2. A preliminary process flowsheet is obtained in task 3 using a modified Douglas hierarchical process synthesis method. The next tasks involves making design decisions and then further refining them in tasks 4-7 related to separation factors, reactor operating conditions, product purity, etc. In tasks 4 mass balances is are performed while in tasks 6-7 energy balances are performed. All simulations are made with PRO/II. Tasks 8-9 make the sizing and economic evaluation calculations. At this point, the base case design is obtained, which is then further refined and improved with respect to heat integration and process optimization (tasks 10-11). In the final task-12, the environmental impact of the process design is evaluated together with some of the key sustainability measures. In addition to PRO/II, the following software is used: ICAS (for property prediction, analysis) and ECON (cost and economic analysis). This design therefore covers all stages of conceptual design, starting from the consideration of qualitative aspects of the process flowsheet and preliminary calculations to detailed process simulations, equipment sizing, costing and an economic evaluation of the designed process.

The resulting design utilizes 75 million kg/year ethylene feed in order to obtain an ethyl alcohol production of 90.5 million kg/year. The total capital investment has been estimated to 43 million USD and the total product cost without depreciation estimated to 58.5 million USD. Furthermore, computer aided economic analysis method has been applied to investigate the potential economic improvements. This analysis helps to define targets for improvement, which are then achieved through heat and mass integration as well as mathematical optimization. In the final step, the environmental impact of the process is analyzed and key sustainability factors such as energy used per kg of product, water used per kg of product, etc., are also determined. A sizable reduction of the operating costs has been possible through heat integration and process optimization yielding a 20 % reduction in total capital investment and a 15 % reduction in product cost without depreciation.

Contents

1 Introduction

Ethanol is a bulk chemical that has a wide range of applications, for example it is used as solvent, cosmetics, food and beverage, fuel, chemical feedstock etc. Due to increasing focus on sustainability bioethanol takes up the majority of the worldwide ethanol market.

In this project the preliminary design of a process to produce 190 proof ethanol from excess ethylene containing propylene and methane as impurities will be described. The project is formulated at the stage of preliminary (conceptual) process design which typically requires involving of multiple partners e.g. construction companies, operating company and process engineers. The process design is made from a process engineer's scope using a systematically predefined work-flow defined by set of 12 tasks. Within each task, a set of decisions need to be made and calculations are made to verify the decisions. By this work-flow we are able to work systematic throughout the conceptual design and therefore avoid trial-and-error simulations. Furthermore the systematic approach is organized in a way such changes performed in a task only affects the task itself and the proceeding tasks. This enables a time efficient, iterative, systematic procedure which is highly valued when screening process design alternatives in the chemical industry. The final design in this report covers all stages of conceptual design, starting from the consideration of qualitative aspects of the process flowsheet and preliminary calculations to detailed process simulations, equipment sizing, costing and an economic evaluation of the designed process. Various commercial softwares have been used in this work including PRO/II (process simulator), ICAS (for property prediction, analysis) and ECON (cost and economic analysis).

2 Task 1: Product Information

The main goal of this task is to gather information about the desired product, which is ethanol in our case, where the importance, uses, physical properties and biggest produces of this chemical will be in focus.

The reason behind selecting ethanol CH_3CH_2OH or ethyl alcohol as a product in the first place is due to that it is one of the most applied and used organic chemicals in the world. Ethanol is also referred to as alcohol, spirit and ethyl hydrate. Depending on its purity in regard to water content, preparation and final use, ethanol consist in plenty of products in the market.

2.1 Importance and Use

Ethanol is the most important solvent after water. Its major commercial applications are in the manufacture of cosmetics, detergents, disinfectants, pharmaceuticals, surface coatings and in food and drug processing. Both synthetic and fermentation ethanol can be used for these purposes. The 99% pure ethanol, often referred to as absolute alcohol, is used extensively as a solvent in the manufacturing of perfumes, pharmaceuticals, preservatives and tinctures. Meanwhile will less pure ethanol be used as an important functional component of alcoholic beverages, which are produced by fermentation. The fermentation broth itself may constitute, after processing and ageing, a beverage in the case of beer or wine. Nevertheless the alcohol can be concentrated from the broth to produce high-alcohol-containing spirits. If the alcohol is used for purposes other than as a beverage, it is denatured by the addition of substances such as methanol, pyridine, formaldehyde, or sublimate and sometimes colored to prevent its use as a beverage. Since a beverage tax is not charged on such alcohol its cost is quite low. The denatured alcohol is then used by industry and commerce, principally as a solvent, as a raw material for manufacturing chemicals, or as a fuel [1].

Chemically synthesized ethanol is usually derived from petroleum sources by hydration of ethylene and has found wide application as industrial alcohol. Since the objective of the design project focuses on the production of ethanol from excess ethylene, the industrial chemical reactions for producing synthetic ethanol will be mentioned in chapter 3.

The grades of ethanol can be seen in table 1, where the most common types of ethanol and its applications are listed.

Table 1: Grades of ethanol in the market [1]

Type	Purity	Application
Industrial ethanol	96.5 vol.%	Industrial and technical purposes as a solvent and a fuel. Can be converted into many other products. Usually denaturated and colored.
Denaturated spirit	88 vol.%	Generally used for heating and lightning. Usually denaturated and colored.
Fine alcohol	96-96.5 vol.%	Purer type of ethanol used mainly for pharmaceuticals, cosmetics and human consumption.
Absolute/Anhydrous ethanol	99.7-99.8 vol.%	Very pure extremely dry ethanol. Used in food and pharmaceutical industry and If mixed with 70-80 vol.% of gasoline it can be used as fuel for internal combustion engines.
Motor fuel ethanol		Refers to fermentation ethanol that is used as a fuel or as blending agent to improve the octane number of gasoline.

2.2 Properties

Ethanol is at room temperature a colorless liquid that is miscible with water in all proportions and is separated from water only with difficulty. As seen on figure 1, ethanol and water forms an azeotropic mixture that contains 85 mole% ethanol. This shows that the highest concentration of ethanol obtained by distillation from ethanol-water mixture can only be op to 85-86 mole% ethanol due to the presence of the azeotrope. However, the water can be removed by adding benzene to the mixture and distilling it off, and thereby getting absolute alcohol (99 mole% ethanol) [1]. The data for ethanols physical properties can be seen on Table 2.

The chemical properties of ethanol are dominated by the functional -OH group, which can undergo many industrially important chemical reactions, such as dehydration, halogenation, ester formation, and oxidation. Ethanol reacts with active metals to form the metal ethoxide and hydrogen. It reacts with certain acids to form esters and it can also be oxidized to form acetic acid and acetaldehyde. It can also be dehydrated to form diethyl ether or, at higher temperatures, ethylene. Because ethanol can be produced efficiently not only by chemical synthesis from petroleum and coal-based feedstock, but also by fermentation of abundantly available organic materials, its commercial role as a raw material for various chemicals is of increasing importance.

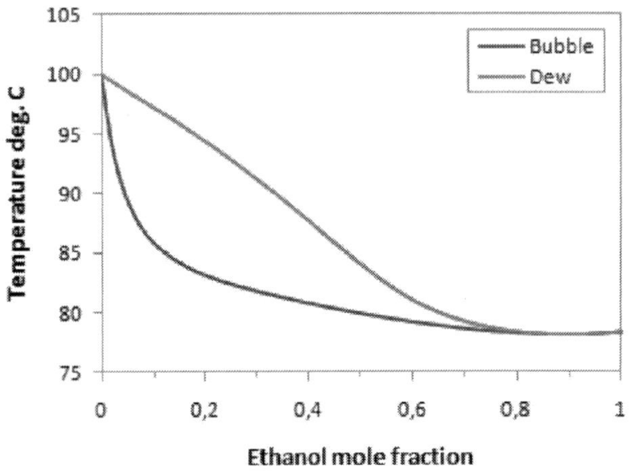

Figure 1: Txy - diagram for Ethanol-water in mole fraction. Created by [2]

2.3 Biggest Producers of Synthetic Ethanol

Since our major concern is about producing ethanol synthetically, we would like to know who the biggest producers are to get an idea about how well established this production is and who we are going to compete with in the ethanol market. The total worldwide production capacity for synthetic ethanol is approximately $8.5 \cdot 10^6$ L/day. Most chemical plants that produce synthetic ethanol are located in industrialized countries, where the developing countries produce mainly fermentation ethanol. In Italy, Germany, UK, and Canada, 50% of the ethanol is produced by ethylene hydration, while the US produce 35% by this method [1].

The total worldwide production capacity for fermentation ethanol far exceeds that for synthetic ethanol and amounts to more than $32 \cdot 10^6$ L/day. The reason is that fermentation plants in most of the developing countries are supplied with raw materials they produce and are usually tax-free as an support from the governments of these countries to make profit. Fermentation ethanol producers in the industrialized nations import some of their raw materials in the form of molasses from the sugar-producing industries of developing countries and don't get the same kind of support from their governments, which make it unprofitable [1].

Synthetic alcohol is concentrated in the hands of a couple of mostly multi-national companies such as Sasol with operations in South Africa and Germany, SADAF of Saudi Arabia, a 50:50 joint venture between Shell of the UK and Netherlands and the Saudi Arabian Basic Industries Corporation (SABIC), and BP of the UK as well as Equistar in the US [3].

3 Task 2: Literature Study

In this task will relevant literature about the manufacture of ethanol from ethylene be investigated, especially in regard to the chemical reactions since knowledge about those are essential and will influence the design decisions that have to be taken in this work. Another uses of ethylene and other raw materials to produce ethanol will also be examined as alternative cases. To compare those cases an economic potential (EP) will be done for each case.

3.1 Direct hydration of ethylene

The most common method for producing ethanol from ethylene is done by direct hydration, where the chemical reaction can be seen in equation 1:

$$CH_2{=}CH_2 + H_2O \longrightarrow CH_3CH_2OH \tag{1}$$

During the past few decades deionized water is usually mixed with ethylene in a 1:0.3 - 1:0.8 feed ratio (ethylene:water) and then passed to a reactor containing an acidic catalyst (H_3PO_4/SiO_2-based catalysts are commonly used). An exothermic gas phase hydration of ethylene takes place in the reactor which operates continuous at about 250 - 300 °C and 6 - 8 MPa [1]. However, the phosphoric acid catalyst still poses a problem of vaporization of acid from catalyst support during operation, which causes a decline in catalytic activity and corrosion of equipment. Tungtosilicic acid on silica gel as an other alternative with quite high activity at lower temperatures has also been suggested but its activity still falls rapidly owing to acid depletion. $ZnO{-}TiO_2$ is a binary metal oxide catalyst that has good activity too, though its problem is poor reproducibility in its preparation. Zirconium Tungstate $ZrW_{2-8}O_{0.5-3.5}$ has shown high catalytic activity, but it is still an acidic binary-metal oxide [4]. To prevent corrosion and the concern about catalysis efficiency we choose to use a high-temperature, high-pressure homogeneous noncatalytic reactor. The reactor temperature lies around 535 - 575 K with a pressure about 69 bar [5]. Around 5-7 % of the ethylene is converted to ethanol in the reactor and thus recirculation of ethylene is required. Lower temperatures favours ethanol production but also the formation of diethyl ether:

$$2\,CH_3CH_2OH \rightleftharpoons (CH_3CH_2)_2O + H_2O \tag{2}$$

Diethyl ether is a byproduct which typically consumes what corresponds to ca. 2 % of the converted ethylene. If recycled, diethyl ether can be reversibly converted to ethanol in the reactor.

In the presence of impurities in the feed stream (available feed contains 95 % ethylene,

3 % propylene and 2 % methane on mole basis) propylene hydrates similar to ethylene:

$$CH_2{=}CHCH_3 + H_2O \longrightarrow (CH_3)_2CHOH \tag{3}$$

Croton aldehyde is also one of the byproducts of this method, but since this waste occurs at low amounts we assume that it is negligible and therefore will not get any further attention in this work.

Table 2 show the physical property data for the process, of which they will be useful in the analysis of it. The table shows the species in the feed and possible products for the reactor. It is noticeable that methane, ethylene and propylene have low boiling points. Thus we can see that it is not possible to condense methane and ethylene at room temperature.

Table 2: Physical property data for species [5].

Species	W water	EA ethylalcohol	EL ethylene	DEE diethylether	M methane	PL propylene	IPA isopropylalcohol
Formula	H_2O	CH_3CH_2OH	$CH_2{=}CH_2$	$(C_2H_5)_2O$	CH_4	$CH_3CH{=}CH_2$	$CH_3CHOHCH_3$
MW	18.02	46.07	28.05	74.12	16.04	42.08	60.10
Sp. Gr.	1.0	0.789	0.56	0.708	N/A	0.609	0.785
Melt Pt, °C	0	-114.5	-169.2	116.3 (α)	-182.5	-185.3	-89.5
BP, °C	100	78.4	-103.7	34.6	-161.5	-47.7	82.4
$\triangle H_v(Kcal/mol)$	539.55	204.3	115.4		121.9	104.6	159.4
VP A	8.10765	8.04494	6.74756	7.4021	6.61184	6.81960	6.66040
VP B	1750.286	1554.3	585	1391.4	389.93	785	813.055
VP C	235	222.65	255	273.16	266	247	132.93
t_c, °C	374.14	243.5	9.6	193.8	-82.1	91.4	235.16
P_c, atm	217.6	63.1	50.7	35.5	45.8	45.4	47.02

3.2 Alternative Routes for Ethylene Utilization

Besides that ethanol can be produced from ethylene other reactions are more propagated. In the United States and Europe, approximately 90 % of ethylene is used to produce three chemical compounds ethylene oxide, ethylene dichloride, and ethylbenzene and a variety of kinds of polyethylene [6]. In figure 2 the reactions routes from ethylene are schematically listed. As about 95 % of the world production of vinyl chloride monomer (VCM) is used for the production of polyvinyl chloride (PVC) we will take a closer look at this production process.

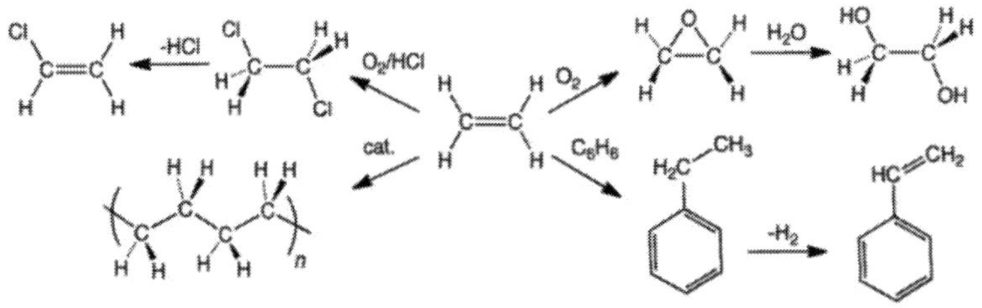

Figure 2: Another possible products from ethylene instead of Ethanol.

There are three manufacturing steps for producing VCM: Direct chlorination, oxychlorination and thermal cracking. In short terms it is possible to react ethylene directly with chlorine in the present of a iron(III)chloride as catalyst, these compounds react exothermically and gives 1,2-dichloroethane (EDC):

$$CH_2{=}CH_2 + Cl_2 \longrightarrow ClCH_2CH_2Cl \tag{4}$$

Modern VCM plants use recycled HCl to produce more EDC via oxychlorination, which entails the reaction of ethylene, oxygen and hydrogen chloride over a copper(II)chloride catalyst to produce more EDC:

$$CH_2{=}CH_2 + 2\,HCl + \tfrac{1}{2}O_2 \longrightarrow ClCH_2CH_2Cl + H_2O \tag{5}$$

The reaction is highly exothermic. One useful by-product of the oxychlorination is ethyl chloride, used as anesthetic. EDC is cracked in thermal reactor heated to 500 °C at 15 to 30 atm pressure, EDC vapor decomposes to produce vinyl chloride and HCl (g).

$$ClCH_2CH_2Cl \longrightarrow CH_2{=}CHCl + HCl \tag{6}$$

The thermal cracking reaction is highly endothermic. Even though residence time and temperature are carefully controlled, it produces significant quantities of chlorinated hydrocarbon side products. In practice, EDC conversion is relatively low (50 to 60 percent). The resulting vapor-liquid mixture then goes to a purification system. In the purification process an absorber-stripper system can be used to separate HCl from the chlorinated hydrocarbons, but it should also be possible to use a continuous distillation system.

A simple flowsheet summarizing the manufacturing producing step are shown in figure 3.

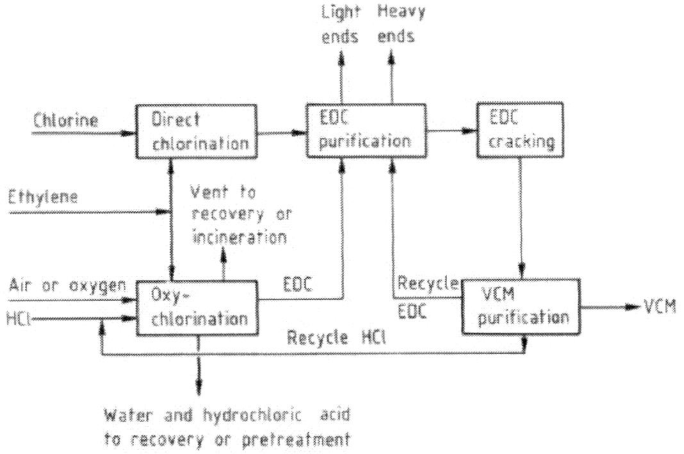

Figure 3: Ethylene-based process for the production of vinyl chloride [7].

3.3 Alternative Routes for Ethanol Production

A common way of producing ethanol in thousands of years has been through fermentation. In the 1930s a synthetic route for ethanol manufacturing was introduced referred to as indirect hydration of ethylene. This route has later been superseded by a direct hydration method described earlier. Furthermore a synthesis method has been developed which homologate (addition of $-CH_2-$ group) methanol using industrial synthesis gas consisting of CO and H_2. The ethanol production is thus distinguished in agricultural (sugar, corn etc.) and synthetic (petrochemical feedstock). Two alternative routes (fermentation and homologation) will shortly be described in terms of reactions, byproducts and process solutions.

3.3.1 Homologation

This process is also referred to as hydrocarbonylation where $-CH_2-$ groups are added using a catalyst. The desired reaction is the gas phase reaction:

$$CH_3OH + CO + 2\,H_2 \longrightarrow CH_3CH_2OH + H_2O \tag{7}$$

But side reactions such as formation of CO_2, ethanal, dimethyl ether and even more hydrocarbonylated alchohols does also occur. The alcohol formation is increased at high pressures. The reactions takes place over a solid catalysts where high selectivity to ethanol is obtained by catalysts based on rhodium. These are however expensive and no feasible alternatives have yet been discovered. Production of ethanol using homologation has not yet been used commercially and the currently status on this technology is still on pilot-scale stage.

3.3.2 Fermentation

Another way of producing ethanol is by fermentation. Using cells to produce ethanol has been known for ages and today this industry accounts for the majority of the ethanol production world wide (around 95 %). Typically glucose is fermented using yeast (common used: Saccharomyces cerevisiae) under anaerobic conditions according to the Gay-Lussac equation:

$$C_6H_{12}O_6 \rightarrow 2\,CH_3CH_2OH + 2\,CO_2 \tag{8}$$

Glucose is frequently extracted directly from various sugar crops such as sugarcanes or converted from starch rich raw materials like corn using various enzymes. Thus latter requires a pretreatment (extraction and hydrolysis) which is performed by the enzymes.

The typical operating conditions for ethanol production by fermentation is at pH between 4 and 6 and around 35 °C for the Saccharomyces cerevisiae yeast type. In general yeasts are sensitive to the operation conditions.

Most fermentation systems today are continuous and a typical process configuration can be seen in figure 4.

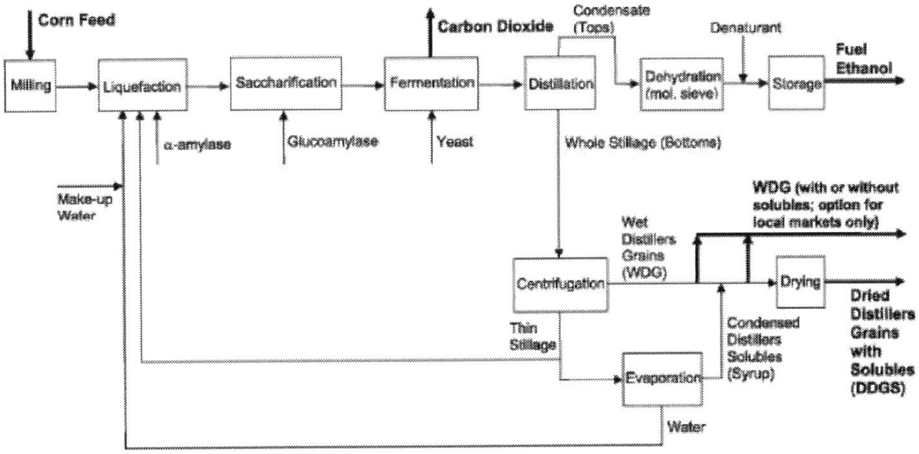

Figure 4: Flowsheet for ethanol production using continuous operation.

The typical configurations is preprocessing followed by fermentation which produces ethanol. The ethanol produced is relatively diluted since it inhibits both growth and production of yeast cells. This requires ultimately water removal by distillation.

3.4 Economic Potential

It is important to determine the economic potential, EP, of the conversion of ethylene to ethyl alcohol in order to assess the marketability of a future production. The EP values are rough estimates and do not take operation cost into account and should thus be used relative to each other (assuming same order of magnitude of the operation costs). EP values are determined for an alternative route for ethyl alcohol (compare competitiveness) and for alternative uses of ethylene (compare benefit). Indicative chemical prices have been found in [8] and are listed in table 3. Following chemical routes are considered:

1. Ethanol production from ethylene (equation 1)
2. Vinyl Chloride monomer (VCM) production from ethylene (equation 4)
3. Alcohol production by fermentation (equation 8)

The maximum economic potential calculations are based on reaction stoichiometry with 100% conversion (except fermentation route). The results of the calculations are summarized in table 4.

Table 3: Indicative chemical prices from American marked. Reference: [8]

Feedstock	Price (US Dollar/kg)
Feed	0.4
Water	0.00083
190 Proof ethyl alcohol	0.9
Chlorine gas	0.3
Vinyl Chloride monomer (VCM)	0.375
Feedstock*	Price (mill. US Dollar/yr)
Fermentation medium	3.93
Corn	30.3

*Scaled to stoichiometric amounts of ethyl alcohol from ethylene for a 119.1 mill. kg/year capacity plant [9]

Table 4: EP for selected routes concerning ethylene.

Route	Feedstock mill. US Dollar/yr	Products mill. US Dollar/yr	EP mill. US Dollar/yr
Route 1	30.0	110	80.1
Route 2	83.7	93.6	9.97
Route 3	34.2	110	75.9

It appears that route 1 is competitive to other ethanol producers from the fermentation industry. Furthermore route 1 appears superior with respect to EP to route 3. If we would like to know predictively how much our production based on reaction stoichiometry will satisfy the ethanol market in percent, we can by calculating how much our plant are capable of producing per day in theory ($4.5 \cdot 10^5$ L/day) and compare it to the daily worldwide production ($8.5 \cdot 10^6$ L/day) see that we will potentially be able to cover approximately 4.5 % of the market.

So it has thus been shown that it is worth to continue considering the conversion of excess ethylene to ethyl alcohol route which will be done in following chapters.

4 Task 3: Process Flowsheet

In this task a feasible preliminary process flowsheet has to be chosen for our production of ethanol from 75 kg/year excess ethylene. We have chosen to develop our flowsheet with Hierarchical Decomposition. In this chapter we will show how we base our decision making.

Since our ethylene source is not absolute pure (95 mole% ethylene, 3 mole% propylene, 2 mole% methane) we should also keep in mind the consequences of these impurities, and how we can get rid of them.

A simple representation of a common used flowsheet for ethanol production from ethylene is displayed in 5. This will be the basis for the process synthesis and the objective is thus to verify it qualitatively. This verification is based on a five-level hierarchical decomposition which will be discussed in this chapter.

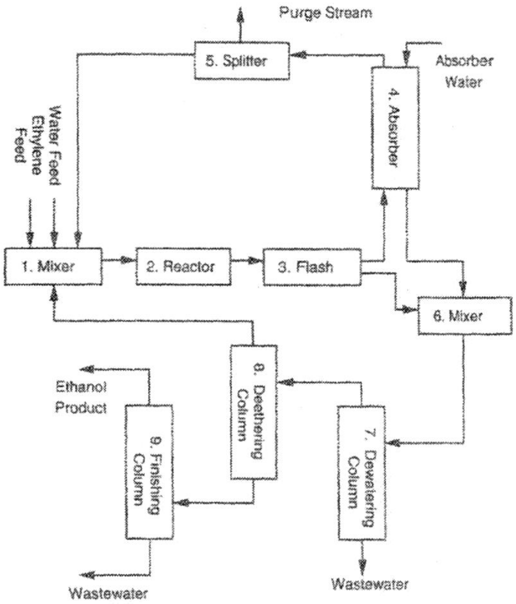

Figure 5: Preliminary Ethanol flowsheet.

All the chemicals within the systems are in this work referred to as the abbreviations ethylene (EL), propylene (PL), methane (M), water (W), ethyl alcohol (EA), isopropyl alcohol (IPA) and diethyl ether (DEE).

4.1 Level 1: Batch versus Continuous

In the first level of the hierarchical flowsheet, we have to consider our decision on whether we will use a batch or a continuous process synthesis. Based on the information gathered in task 2 about the technical information, production rate, product life time and value of product, we have chosen to consider a continuous process.

Since the process behind 200 proof ethyl alcohol is more complicated and therefore more expensive we have decided to produce 190 proof ethyl alcohol. This product is as known a commodity chemical that have a long product lifetime in an already established market. So we are going to be added as one more producer of the same product, which means that we have to be competitive to get a share of the market by being cost effective.

We have also to be aware of that we have to convert a continuously flowing supply of ethylene throughout the year, which will be almost impossible to do so in few days by batch process. It is also cheaper to produce large amounts of ethanol through long expecting operating time by using a continuous process, which confirms the reason behind our decision. Last but not least the continuous process gives the most satisfactory EP.

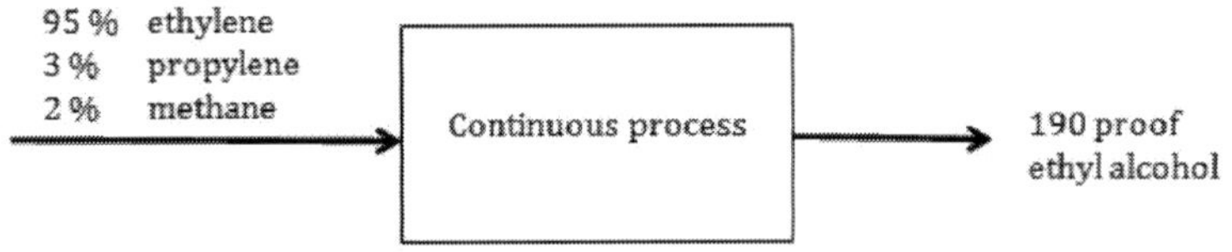

Figure 6: Flowsheet for ethanol production by converting ethylene in continuous operation for level 1.

4.2 Level 2: Input-Output Structure

In this level we have to consider what to do with the impurities in our raw material, how many product streams and recycle streams we have to use. Considerations of selectivity versus cost should also be considered.

Our ethylene feed contains as mentioned some impurities and inert components (all of them can be seen on Table 2). Moreover, the conversion per pass of ethylene to ethanol is about 7% which indications points out the need of process recycle. There are some several guidelines that can be used to decide what to do with the raw materials. If the impurities are inert, they can be removed after the reaction if they are valuable. If they are presented in large amounts and can be easily separated, then they can be removed before the reactor. If they have boiling points lower than reactans and products and are also inert, then it is possible to remove them before the reactor if necessary by using a purge unit. If the impurities are also products from the reactor, then the feed stream has to be placed before the unit that will remove the impurities.

In regard to the product streams there are some several rules and guidelines for recycle and removal. The rules are based recycling unreacted reactants and intermediate reactants, while removing/recovering the main products and valuable byproducts. Things that can be recycled and/removed are azeotropes with reactants and byproducts that are not valuable depending on the circumstances for the process. The guidelines that are based on the rules lie in e.g. for less than 100% conversion of reactants and reaction in the gas phase, then a recycle of gases will be necessary while if impurities are present purge units will be necessary. Compared to the liquid phase then the recycle of liquids of the reactants will be necessary and a purge units will also be used if impurities are present.

So if we assume that methane and propylene in the feed stream and the impurities in the outlet stream are invaluable, then they will eventually need to be removed from the process. To begin with methane and propylene can either be removed before the ethylene enters the reactor, or we can let either or both of them enter the reactor and remove them (and their possible products) after the reactor. However because of the difficulties and expenses that lies behind separating both methane and propylene in the feed, we choose to let both components enter the reactor as shown in figure 7. As we refine the flowsheet and continuing with the tasks, we may choose to return to the other alternatives and evaluate them as part of our bonding strategy.

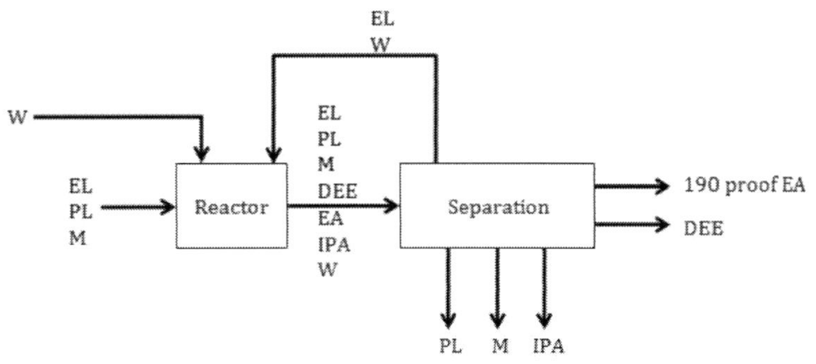

Figure 7: Flowsheet for ethanol production by converting ethylene in continuous operation for level 2.

4.3 Level 3: Recycle Structure

In this level we care about the reactor equilibriums, components and conditions. Our interest is to increase the gain of the reactor by recycle strategies. Therefore many considerations have to been taken, such as number of reactors and recycle streams, components phases, reactor type, desired and undesired reactions kinetics and equilibrium, and capital and operation cost of the reactor.

It is only necessary with one reactor as ethanol is formed directly from ethylene and water, due to the main reaction. All the non-reacted ethylene is of course recycled to the reactor again to increase the effectiveness. In principle we can convert all the ethylene to ethanol.

As the reaction is exothermic the reactor should not been operated adiabatically to avoid accidents. Instead it seems logical to operate it as isothermal do to operations conditions mention in section 3.1.

We would like to prevent these undesired reactions to take place, even though that the by-product can be sold, because it leads to selective losses of ethanol. One way to prevent the undesired reaction is by recycling ethyl ether to force/shift the undesired reaction against ethanol. The reaction equilibrium is limited and therefore this method works.

As the feed stream contain impurities of methane and propylene we also have the opportunity to recycled these components. As methane is an inert component and expensive to separate off we choose so far to purge it. Regarding propylene we must also consider its undesired reaction to isopropylalcohol.

Due to PL and IPA boiling point (see table 5) it more easily to separate IPA from the other components and therefore we decide to reflux PL to the reactor to compete the above mentioned reaction.

In this level (Level 3) we have discussed which components we want to recycle and the decisions behind those. In the next section (Level 4) we will discuss the separation decisions and due to the conclusion of this we end up with reflux streams to the reactor, one vapor and one liquid stream.

4.4 Level 4: Separation System

Separation system is necessary in order to divide products, by-products and raw material from each other. Products are removed while raw materials are recycled to the reactor. In liquid recovery distillation are the preferred unit operation, but still a lot of strategies are to be considered. In vapor recovery purge streams are often used as they are no-cost separation, but still the recycle of components can have a huge effect on the reactor.

In order to design the separation operations needed the species are listed in table 5 ordered in boiling points and information about recycling. It is preferred and easiest to separate the condensible and the noncondensible components in an early stage. Therefore a flash unit is placed after the reactor (see figure 8). It was decided to recycle methane and propylene in task 3 which requires a purge stream to avoid accumulation of impu-

Table 5: Species ordered in boiling point.

Species	BP, °C	Recycled	Condensible
M	-161.5	×	
EL	-103.7	×	
PL	-47.7	×	
DEE	34.6	×	×
EA	78.4		×
IPA	82.4		×
W	100	×	×

rities in the plant. In literature it is found that a sharp separation cannot be obtained

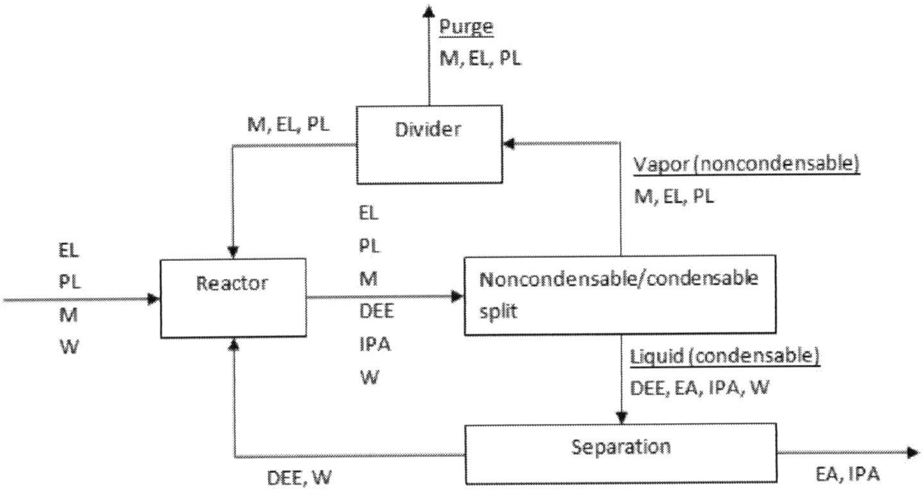

Figure 8: Separation of condensible and noncondensible components.

by a flash unit which will causes some diethyl ether and ethyl alcohol to escape with the noncondensibles. Due to high water solubility of both components these compounds can be recovered from the noncondensibles using an absorber. The remaining issue is now the condensible/liquid recovery section. At this preliminary stage distillations columns are sufficient to obtain the necessary separations. It is preferred to operate distillation columns at ambient conditions (typically 1 atm) to prevent high operation costs. Often it is proposed to remove the lightest component (diethyl ether) as distillate first. However diethyl ether is expected to be present in very small amounts relative to water and hence it is not considered feasible for a distillation column. Instead it is suggested to add a dewatering/impurity removal column where a fraction of water with impurities (isopropanol) is drained from the bottom as a waste stream. After the dewatering column is it preferred to use a deethering column which purpose is to recover and recycle diethyl ether and eventual noncondensibles since reaction (2) is limited by equilibrium, and thus the reaction eventually will reach equilibrium when diethyl ether is not removed from the plant.

Finally the remaining water/ethyl alcohol mixture is distilled to obtained an azeotropic composition as distillate and a waste water stream at the bottom (see figure 9).

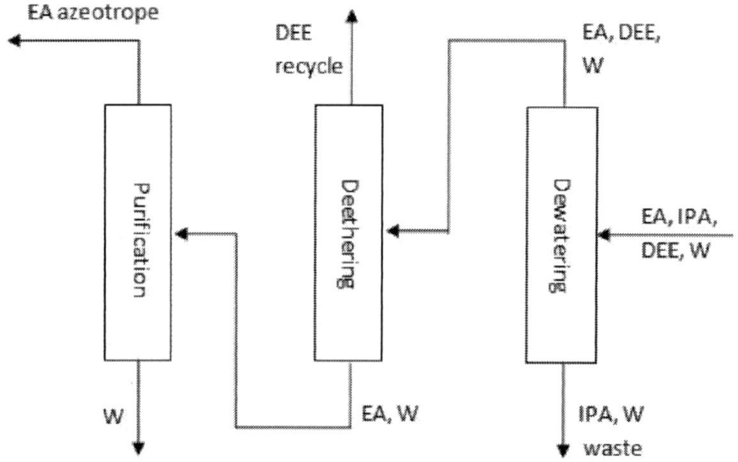

Figure 9: Distillation sequence.

The entire preliminary flowsheet can be seen in figure 10.

4.5 Level 5: Heat Recovery Network

Heat recovery networks are becoming increasingly popular. Typically they can reduce the required heat significantly. It is however almost impossible to synthesize a network based on pure qualitative considerations. At this point we can expect that heat integration has a great potential for lowering the heat requirements e.g. due to high reactor operating temperature.

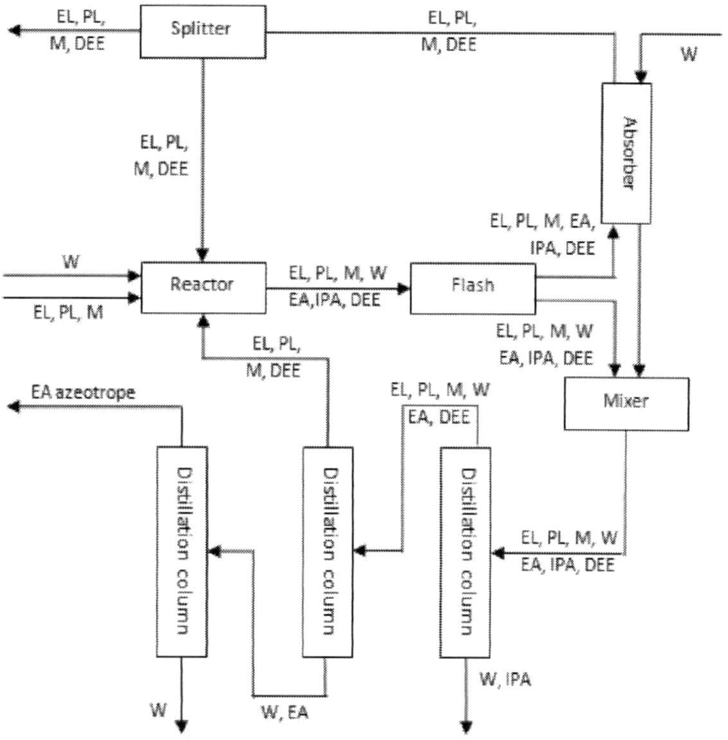

Figure 10: Preliminary flowsheet

5 Task 4: Linear Mass Balance Model

At this point the flowsheet can be solved with respect to mass balances. Therefore each operation in the flowsheet is decomposed into three different elementary operations:

- Reaction: Single input, single output, single reaction
- Mixing: Multiple input, single output
- Splitting (and dividing): Single input, multiple output

The decomposed flowsheet can be seen in figure 11. As it can be seen the decomposed flowsheet consists of two partitions and two recycle streams. Furthermore there are in the entire system three input streams and four output streams.

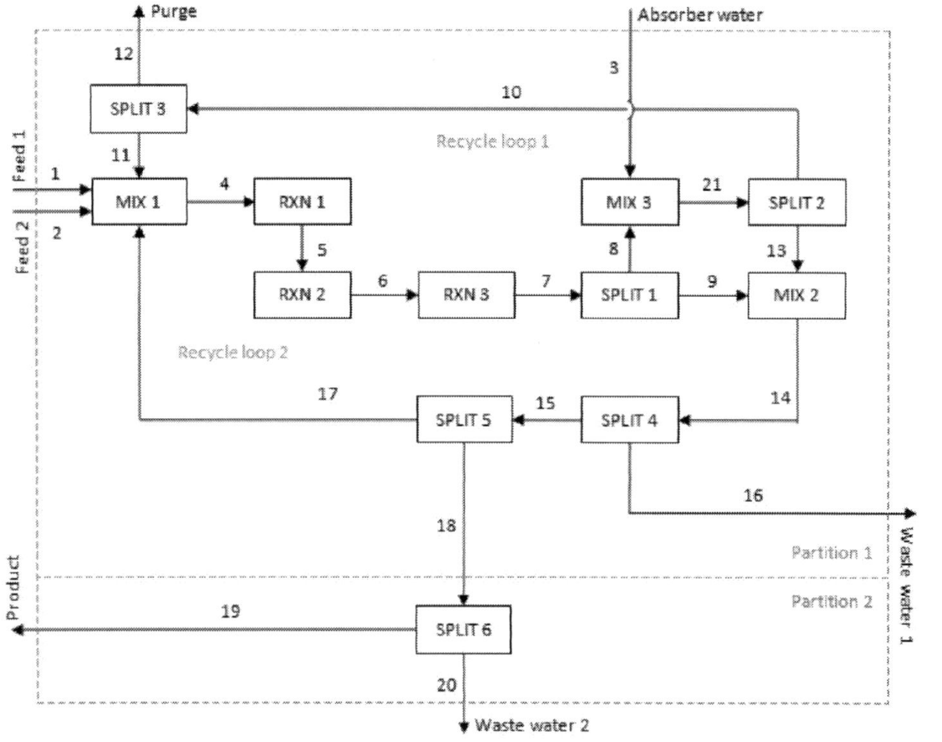

Figure 11: Flow sheet decomposed into elementary operations.

5.1 Degree of Freedom Analysis

We want to determine the number of degrees of freedom for partition 1. The result of the analysis is summarized in table 6.

Table 6: Determination of degrees of freedom for partition 1.

Type	Number of equations
Reactor eq. (9)	$3 \cdot N_C$
Mixer eq. (13)	$3 \cdot N_C$
Splitter eq. (9a-b)	$5 \cdot 2 \cdot N_C$
Total	$16N_C$

Type	Number of variables
Streams	$19 \cdot N_C$
Reaction parameters	$3 \cdot (1 + N_C)$
Split fractions	$5 \cdot N_C$
Total	$27N_C + 3$

Degrees of freedom	$11N_C + 3$

It is shown that partition 1 contains $11N_C + 3$ degrees of freedom. The following sections will explain how these are eliminated which ultimately should lead to a solution strategy. The variables that need to be specified are given in table 7.

Table 7: Determination of variables that need to be specified.

Variable type	Variables	Degrees of freedom
Process variables	$f_{1,k}, f_{2,k}, f_{3,k}$	$3N_C$
Equipment parameters	$\eta_1, \eta_2, \eta_3, \gamma_{1,k}, \gamma_{2,k}, \gamma_{3,k}$	$3 + 3N_C$
Split fractions	$\xi_{8,k}, \xi_{10,k}, \xi_{12,k}, \xi_{15,k}, \xi_{17,k}$	$5N_C$
Total		$11N_C + 3$

5.2 Reaction

Reactors where a reaction takes place are modelled as (see figure 12):

$$f_{j+1,k} = f_{j,k} + \eta_r f_{j,l(r)} \gamma_{r,i} \tag{9}$$

Where $f_{j,k}$ is the molar flow rate of component $k = EL, M, PL, W, IPA, EA, DEE$ in stream j. η_r is the molar conversion for a reactor pass belonging to reaction r using the limiting component $l(r)$ as basis. $\gamma_{r,k}$ is the normalised stoichiometric coefficient belonging to reaction r. The stoichiometric coefficient for products is positive, for reactants negative and zero for inerts.

Three reactions take place in the reactor in the process. In this decomposition procedure the physical reactor is divided in three theoretical reactors: RXN 1, RXN 2 and RXN 3 (see figure 12).

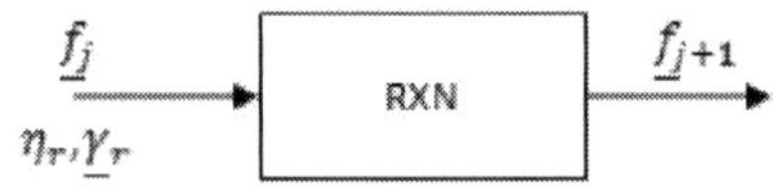

Figure 12: Reaction.

5.2.1 RXN 1

Here the reaction (1), referred to as reaction $r = 1$, is modelled. The conversion is $\eta_1 = 0.07$ and the limiting reactant is defined to be ethylene. Thus the mass balance becomes:

k	EL	M	PL	W	IPA	EA	DEE
γ_1	-1	0	0	-1	0	1	0

$$f_{5,k} = f_{4,k} + \eta_1 \gamma_{1,k} f_{4,EA} \tag{10}$$

5.2.2 RXN 2

Here the reaction (3), referred to as reaction $r = 2$, is modelled. The conversion is $\eta_2 = 0.007$ and the limiting reactant is defined to be propylene. Thus the mass balance

k	EL	M	PL	W	IPA	EA	DEE
γ_1	0	0	-1	-1	1	0	0

becomes:

$$f_{6,k} = f_{5,k} + \eta_2 \gamma_{2,k} f_{5,EA} \tag{11}$$

5.2.3 RXN 3

Here the reaction (2), referred to as reaction $r = 3$, is modelled. This reaction is equilibrium based and thus the conversion depends on the upstream composition. In steady state the conversion is zero as diethyl ether is recycled, $\eta_3 = 0$.

Instead, due to equilibrium, the diethyl ether composition is given by the expression if vapour phase is assumed ideal:

$$f_{7,DEE} = \frac{K_P (f_{7,EA})^2}{f_{7,W}} \tag{12}$$

Where $K_P = 0.2$ is the equilibrium constant.

5.3 Mixing

The mixing operation can be formulated as (see figure 13):

$$f_{j+N_M+1,k} = \sum_{m=j}^{j+N_M} f_{m,k} \tag{13}$$

Where N_M is the total number of inlet streams.

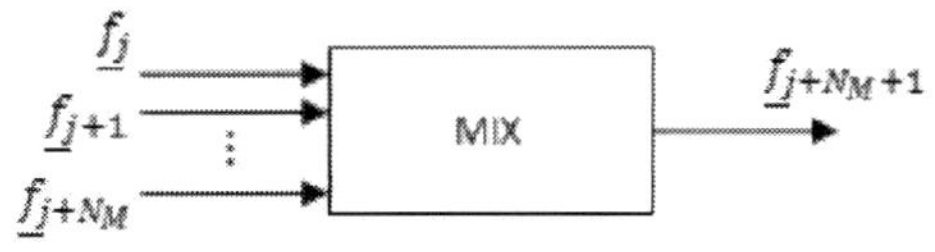

Figure 13: Mixing.

Four mixers have been chosen, and the mass balances are:

MIX 1: $$f_{1,k} + f_{2,k} + f_{11,k} + f_{17,k} - f_{4,k} = 0 \tag{14}$$

MIX 2: $$f_{9,k} + f_{13,k} - f_{14,k} = 0 \tag{15}$$

MIX 3: $$f_{3,k} + f_{8,k} - f_{21,k} = 0 \tag{16}$$

5.4 Splitting

The splitting (and dividing) operation can be formulated as (see figure 14):

$$f_{j+u,k} = \xi_{j+u,k} f_{j,k} \,, \quad u = 1, 2, ..., N_S \tag{17a}$$

$$\xi_{N_S,k} = 1 - \sum_{u=1}^{N_S-1} \xi_{u,k} \tag{17b}$$

N_S is the total number in outlet streams (split streams). $\xi_{u,k}$ is the split fraction between 0 and 1 for outlet stream u. The component split fractions for the N_S'th outlet stream are fixed.

This operation covers both separations like flash distillations, rectification and absorption (splitting) and stream dividing where the outlet streams have equal compositions.

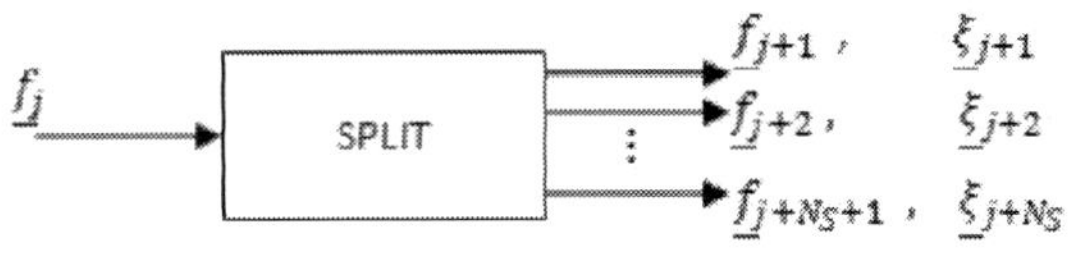

Figure 14: Splitting/dividing.

5.4.1 SPLIT 1 - Flash

We want to determine the split fractions for the flash unit. At this early design stage ideal vapour phase and ideal liquid phase is assumed which enables a simple flash calculation where the commonly used Antoine equation is used to determine pure component vapour pressures. All component split fractions can thus be determined when fixing T and a key component split fraction $\xi_{j,n}$. Assuming a cooling water temperature of $T = 310K$ and a overhead split fraction for the medium boiling compound diethyl ether ($n = DEE$) of $\xi_{6,n} = 0.50$ a mass balance yields:

$$\xi_k = \frac{\alpha_{k/n}\xi_n}{1 + (\alpha_{k/n} - 1)\xi_n}, \quad \alpha_{k/n} = \frac{P_k^0(T)}{P_n^0(T)} \tag{18}$$

Where $\alpha_{k/n}$ is the relative volatility and $P_k^0(T)$ is the temperature dependant pure component liquid vapour pressure.

k	EL	M	PL	W	IPA	EA	DEE=n
P_k^0 [mmHg]	55348	211011	11325	45.7	74.4	113.6	820.1
$\alpha_{k/n}$	67.5	257.3	13.8	0.057	0.091	0.139	1
$\xi_{8,k}$	0.985	0.996	0.932	0.054	0.083	0.122	0.500

The SPLIT 1 mass balances are:

$$f_{8,k} = \xi_{8,k} f_{7,k} \tag{19}$$

$$f_{9,k} = (1 - \xi_{8,k}) f_{7,k} \tag{20}$$

5.4.2 SPLIT 2 - Absorber

Physically the absorber consists of both MIX 2 and SPLIT 2. In the absorber liquid water (f_3) is passed through the vapour stream (f_8) in order to absorb noncondensed ethyl alcohol. The absorber split fractions are determined on the basis of a plate absorber with constant tray absorption factor $A_E = L/(VK_i)$ where L is the total liquid flow rate and V the total vapour flow rate. We set $n = EA$ and assume $A_n = 10$, recovery $1 - \xi_{10,n} = 0.99$ and $P = 68bar$. The Kremser equition yields the number of equilibrium stages:

$$N = \ln\left(\frac{\xi_{13,n} - A_n}{-A_n(1 - \xi_{13,n})}\right) / \ln A_n \tag{21}$$

While mass balances yield:

$$A_k = \frac{A_n}{\alpha_{k/n}}; \; \beta_N^k = \frac{1 - (A_k)^{N+1}}{1 - A_k}; \; \beta_{N-1}^k = \frac{1 - (A_k)^N}{1 - A_k} \tag{22}$$

From the absorber calculations we can get an estimate of the water stream needed to

k	EL	M	PL	W	IPA	EA=n	DEE
P_k^0 [mmHg]	55348	211011	11325	45.7	74.4	113.6	820.1
$\alpha_{k/n}$	487.3	1857.6	99.7	0.41	0.66	1	7.22
A_k	0.021	0.005	0.100	24.3	15.3	10	1.39
$\beta_{N,k}$	1.021	1.005	1.110	534.2	220.4	100	4.20
$\beta_{N-1,k}$	1.020	1.005	1.099	21.92	14.37	9.90	2.31
$\xi_{10,k}$	0.979	0.995	0.901	0.002	0.005	0.010	0.238
$N = 1.96$, $K_n = 0.022$ and $\beta_{N-1,W}/\beta_{N,W} = 0.041$ solvent loss.							

obtain the wanted ethyl alcohol recovery as:

$$L = A_n V K_n \Rightarrow F_3 = f_{3,W} = 0.223F_8 \tag{23}$$

Where F_j are the total molar flow rates of stream j.

The SPLIT 2 mass balances are:

$$f_{10,k} = \xi_{10,k} f_{21,k} \tag{24}$$

$$f_{13,k} = (1 - \xi_{13,k}) f_{21,k} \tag{25}$$

5.4.3 SPLIT 3 - Divider

The split fractions for a stream divider are identical and will be called the purge fraction, p. We have to specify the purge fraction such that the constraint of max. 10% methanol in the reactor is satisfied. Assuming all ethylene, propylene and methane in stream 14 is recycled we can write:

$$f_{1,k} + \xi_{8,k}\xi_{10,k}(1-\eta_r)(1-p)f_{4,k} + (1-\xi_{10,k})(1-\eta_r)f_{4,k} \\ - \xi_{8,k}(1-\xi_{10,k})(1-\eta_r)f_{4,k} - f_{4,k} = 0 \tag{26}$$

For the cases $k = EL, r = 1$, $k = PL, r = 2$ and $k = M, \eta_r = 0$. Equation (26) can be used to investigate the purge fractions influence on methanol concentration in the reactor feed. Assuming water is present in 0.6 times the amount of ethylene and no other components are present the methane content can be plotted versus purge fraction which can be seen in figure 15. A purge fraction around 0.0085 corresponding to 0.85% is the lower limit. The disadvantage of a low purge fraction is a higher recycle flow rate which requires equipment with larger capacities.

At this stage we choose $p = \xi_{12,k} = 0.0085$ which gives the SPLIT 3 mass balances:

$$f_{12,k} = \xi_{12,k} f_{10,k} \tag{27}$$

$$f_{11,k} = (1 - \xi_{12,k}) f_{10,k} \tag{28}$$

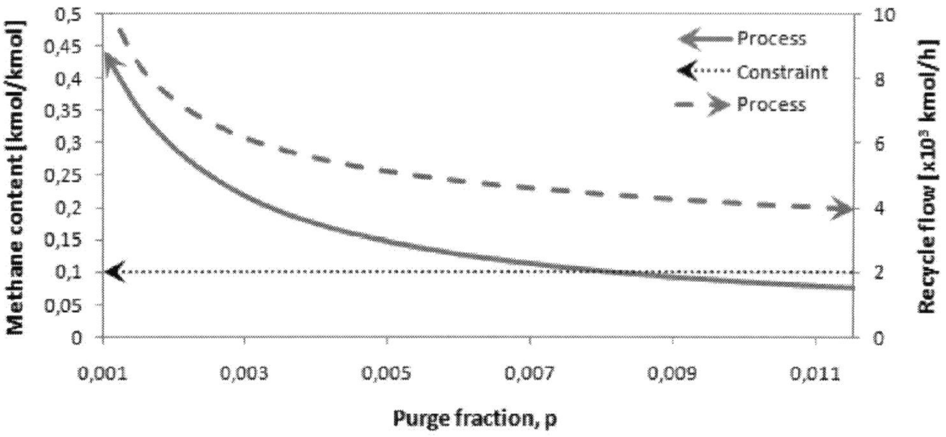

Figure 15: Invenstigation of influence of purge fraction on methane content in reactor feed.

5.4.4 SPLIT 4 - Distillation

Assuming vapour boilup, liquid flow in each stage, ideal mixtures and total reflux, we have to specify cooling temperature, recovery of a light component (lk) and a heavy component (hk). The minimum number of stages is given by the Fenske equation:

$$N_m = \ln\left(\frac{\xi_{lk}(1-\xi_{hk})}{\xi_{hk}(1-\xi lk)} / \ln \alpha_{lk/hk}\right) \tag{29}$$

The component recovery of components lighter than lk is 1 and the recovery heavier than hk is 0. The intermediate recoveries are given as:

$$\xi_{j,k} = \frac{(\alpha_{k/hk})^{N_m}\xi_{j,hk}}{1+[(\alpha_{k/hk})^{N_m}-1]\xi_{j,hk}} \tag{30}$$

We choose $T = 310K$, $\xi_{lk} = 0.995$ and $\xi_{hk} = 0.1$ with $lk = EA$ and $hk = W$.

k	EL	M	PL	W=hk	IPA	EA=lk	DEE
$P_k^0(T)$ [mmHg]	55348	211011	11325	45.7	74.4	113.6	820.1
$\alpha_{k/hk}$	1185.4	4519.4	242.6	1	1.59	2.43	17.6
$\xi_{15,k}$	1	1	1	0.1	0.849	0.995	1
$N_m = 8.4$							

The SPLIT 4 mass balances are:

$$f_{15,k} = \xi_{15,k} f_{14,k} \tag{31}$$

$$f_{16,k} = (1-\xi_{16,k}) f_{14,k} \tag{32}$$

5.4.5 SPLIT 5 - Distillation

The equations are the same as in SPLIT 4. Here we choose $T = 310K$, $\xi_{lk} = 0.995$ and $\xi_{hk} = 0.005$ with $lk = DEE$ and $hk = EA$.

k	EL	M	PL	W=hk	IPA	EA=lk	DEE
$P_k^0(T)$ [mmHg]	55348	211011	11325	45.7	74.4	113.6	820.1
$\alpha_{k/hk}$	487.3	1857.6	99.7	0.411	0.655	1	7.22
$\xi_{17,k}$	1	1	1	0	0	0.005	0.995
$N_m = 5.4$							

The SPLIT 4 mass balances are:

$$f_{17,k} = \xi_{17,k} f_{15,k} \tag{33}$$

$$f_{18,k} = (1 - \xi_{17,k}) f_{15,k} \tag{34}$$

5.4.6 SPLIT 6 - Distillation

The SPLIT 6 unit is not a part of partition 1 and thus it does not have to be considered in the linear mass balance model. In any case we have to specify the product requirements. These are 99.5% of the azeotrope (corresponding to 85.4 mol% ethyl alcohol at exit composition) should be recovered and the product stream should contain less then 0.1% isopropyl alcohol.

5.5 Feed Stream Summary

There are three feed streams. Stream 1 containing ethylene, propylene and methane is fixed by the supplier(s). The amount of water for the reactor that must be fed is based on a design decision. It was mentioned earlier that ethylene was present in excess and thus a water to ethylene feed ratio of 0.6 has been chosen which gives:

$$f_{2,W} = 0.6 f_{4,EL} \tag{35}$$

$$f_{2,k} = 0,\ k \neq W \tag{36}$$

In section the section dealing with SPLIT 2 (absorber) a relation for the required water stream for the absorption was suggested. Recapitulating equation (23):

$$f_{3,W} = 0.223(f_{8,EL} + f_{8,M} + f_{8,PL} + f_{8,W} + f_{8,IPA} + f_{8,EA} + f_{8,DEE}) \tag{37}$$

$$f_{3,k} = 0,\ k \neq W \tag{38}$$

5.6 Solution Strategy and Results

There are two approaches to solve the mass balances: 1. simultaneously using linear algebra or 2. sequentially using a tear stream approach (iterative).

In this work the tear stream approach has been applied. Using a signal flowgraph based technique (see table 30 in Appendix A) it has been found that only one tear stream is needed which is stream 4. Thus the calculation sequence is:
Guess all component flow rates for $f_{4,k}$,RXN 1, RXN 2, RXN 3, SPLIT 1, MIX 3, SPLIT 2, SPLIT 3, MIX 2, SPLIT 4, SPLIT 5, MIX 1, check newly calculated $f_{4,k}$ with guess - if within tolerance continue else change start guess and start from RXN 1, SPLIT 6.
This method has been used in PRO/II (see figure 16 for process flowsheet) and excel (using add-in "problem solver").

If the simultaneous method was to be applied one has to consider the component order since special constraints like equation (12) where the balances for water and ethylene must be solved before diethyl ether and (23) where ethylene must be solved before water.

The stream summary is given in table 31.

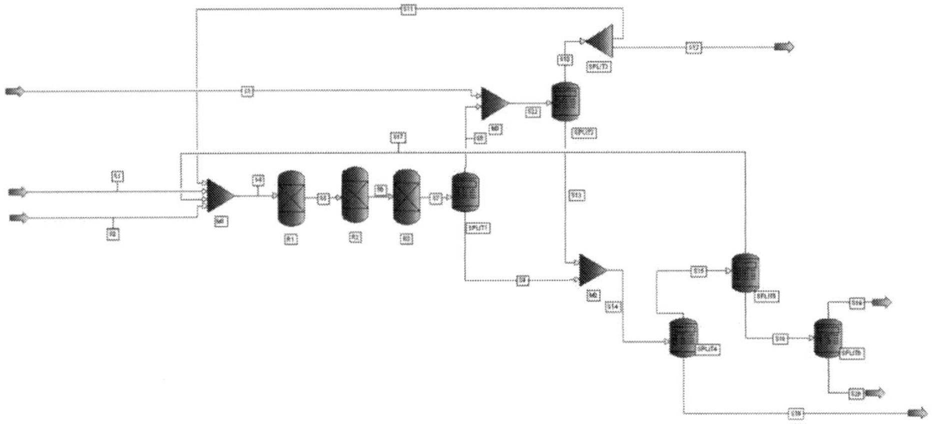

Figure 16: The simple flowsheet build in PROII containing stream calculators, reactors, mixers and splitters.

6 Task 5: Stream/process conditions

To perform energy balances it is necessary to determine the temperature and pressure for all streams. Therefore we will in the following perform calculations of pressure and temperatures for every process unit.

6.1 Reactor

In regard to the reactor conditions we have not performed any calculations, as we have implemented the conditions found in the literature search (see section 3.1).

6.2 SPLIT 1 - Flash

From the mass balance provided in Chapter 5 we choose DEE as the intermediate key component with a split factor of $\zeta_{DEE} = 0.5$. From that we could determine the mole fractions of feed and effluent streams and furthermore we can find the critical temperature T_c^m for the flash unit with the given component composition:

$$T_c^m = \sum x_k T_c^k = 619K \tag{39}$$

When we determine the flash temperature we assume that the cooling water has a temperature of 310 K. From this temperature we base our calculations on relative volatilities, α and therefore we assume that α is independent of temperature. The relative volatilities of each component is based on the ratio of partial pressure between every component and the key component, which for this unit is DEE. Partial pressure is found according to Antoine.

$$\alpha_{k/DEE} = \frac{P_k^0}{P_{DEE}^0} \tag{40}$$

And from the these calculations the average relative volatilities, $\overline{\alpha}_{DEE}$ with respect to DEE can be expressed as:

$$\overline{\alpha}_{DEE} = \sum \alpha_{k/DEE}\, x_k = 2.2 \tag{41}$$

Basing the flash calculation on the most abundant component which is water, leads to:

$$P_W^0(T) = P\frac{\alpha_{k/DEE}}{\overline{\alpha}_{DEE}} \tag{42}$$

Based on this pressure determination of the temperature can also be done by Antoine. This results in a temperature of $T_8 = T_9$ =389 K for both vapor and liquid phase. As the temperature is between the cooling water temperature (310 K) and the critical estimate (620 K) and therefore we can conclude that it is an acceptable temperature. The flash unit operate isobaric as the pressure in and out of the unit is the same, only a pressure drop can influence this situation. We will assume that we have pressure drop of 0.5 bar for every unit operations we have in the plant and therefore the out going pressure is $P_8 = P_9$ =68.5 bar.

6.3 SPLIT 2 - Absorber

For the absorber we determine the critical estimate temperature to $T_c^m = 516$K as we then can check if the operation temperature is in-between this value and the cooling water. Since water again is the most abundant component, we determine the bubble point with respect to DEE as the intermediate key component from the following equation. The relative volatilities is evaluated at T=310 K

$$P_W^0(T) = P\frac{\alpha_{k/DEE}}{\overline{\alpha}_{DEE}} \tag{43}$$

From this calculation we can determine the temperature $T_{13} = 339$K.
For the gas stream 10, we evaluate the dew point for the vapor mixture giving in the table from the absorber in previous section. Again the relative volatilities is found for the reference temperature (310 K) and the most abundant component for the vapor stream is Ethylene (EL). Therefore we can write the dew point equation:

$$P_{EL}^0(T) = P\sum\frac{y_k}{\alpha_{k/DEE}} \tag{44}$$

where P=68 bar. Here we find the $P_{EL}^0(T) = 138230\,mmHg$ which correspond to $T_{10} =$ 382 K. For both T_{10} and T_{13} we see the temperature is between the cooling water temperature (310 K) and the critical estimate (620 K) and therefore we can conclude that it is a acceptable temperature.

6.4 Distillation Units

For a distillation column the idea is to separate the light and heaviest key components from each other whose recoveries are degrees of freedom. By choosing the correct operation conditions for the separation column there will be a vapor and liquid phase. In the bottom end of the column there is a reboiler which deliver the necessary heating and after the vapor leaves the column it has to be condensed and tab off/sent further around in the system, while a fraction is recycled to the column again. For the condenser part there are 2 options as it is possible to have a partial or total condenser. The idea of a total condenser is that the vapor stream will be 100 % condensed and then a liquid stream will be obtained. A partial condenser will only cool a degree of the stream and therefore a stream of liquid at bubble point will be reflux in the reactor and a stream of saturated vapor product will be obtained. The choice of condenser depends on the lowest energy equipments, cascade of pressure between a series of distillation units and e.g. if a stream should be recycled what phase is the desired.

The decision of condenser type will effect the column and reboiler conditions as well as shaft work needed. There are potential for saving energy with separation columns using partial condenser, but pressure cascade have to be taken into consideration. Therefore we will investigate both types condenser for split 4 and 5, and estimate the temperatures and pressures for each case, as only the energy balance and price list of equipment in the end can tell us what favours the economy. For split 6 it needs to be a total condenser to obtain the ethanol product as liquid.

6.4.1 SPLIT 4 - Distillation

The purpose of this unit is to separate the light key components from water. We can consider two options, a total condenser and a partial condenser. We assume that the condenser is operated with cooling water and therefore $T_{con} = 310$ K. The critical temperature for this composition is $T_c^m = 647$ K.

Total condenser

From stream 15 and basing the calculation on the most plentiful component, EA, we can determine the total pressure when total condenser is used:

$$P_{15} = P_{EA}^0(310) \cdot \overline{\alpha} = 17.5 bar \tag{45}$$

Where the pressure and temperature for the reboiler can be found as we assume a pressure drop of 0.5 bar and therefore the pressure in the reboiler is $P_{16} = 18$ bar and from Antoine we determine the temperature as if almost every thing was water at $T_{16} = 478$ K at the given pressure.

Partial condenser

For a partial condenser we calculate the pressure from the dew point equation:

$$P_{15} = \frac{P_{EA}^0(310\,K)}{\sum \frac{y_{EA}}{\alpha_{k/EA}}} = 0.14\,bar \tag{46}$$

From this result we must be skeptical as a vacuum distillation is cost-full due to safety from leaks etc. The pressure for the reboiler would be $P_{16} = 0.6$ bar according to pressure drop. The temperature would then be $T_{16} = 361$ K much less than the temperature for the reboiler in connection with the total condenser.

Both temperatures are acceptable as they are between the cooling water temperature and the critical estimate.

6.4.2 SPLIT 5 - Distillation

This unit is purpose to separate the light key components from ethanol. From this unit we also want to recycle the EL and DEE to the reactor to maximize/shift the equilibrium reaction against EA and therefore will all the light key components remained get recycled. We can consider to options, a total condenser and a partial condenser. We assume that the condenser is operated with cooling water and therefore T_{con} = 310 K. The critical temperature for this composition is $T_c^m = 574$ K.

Total condenser

From stream 17 and basing the calculation on the most plentiful component, EL, we can determine the total pressure:

$$P_{17} = P_{EL}^0(310) \cdot \overline{\alpha} = 51.9bar \tag{47}$$

and the pressure and temperature for the reboiler can be found as we assume a pressure drop of 0.5 bar and therefore the pressure in the reboiler is P_{18} = 52.4 bar and from Antoine we determine the temperature to T_{18} = 501 K.

Partial condenser

For a partial condenser we calculate the pressure from the dew point equation:

$$P_{17} = \frac{P_{EL}^0(310)}{\sum \frac{y_{EL}}{\alpha_{k/EL}}} = 10.4bar \tag{48}$$

We see that the pressure is much less than the total condenser. The pressure for the reboiler would be P_{18} = 10.9 bar according to pressure drop. The temperature would then be T_{18} = 427 K less than the temperature for the reboiler in connection with the total condenser. However it is always a possibility to operate with a higher pressure as long as it is within a feasible range.

Both temperatures are acceptable as they are between the cooling water temperature and the critical estimate. Still partial condensers will maintain to be the preferred ones in this situation.

6.4.3 SPLIT 6 - Distillation

For the last split unit we will only consider the total condenser as we need to tab off the azeotropic composition of EA/W as liquid and therefore the overhead stream is transferred to a storage tank with commercial product. As the product is stored after this unit operation the condenser is operated at atmospheric pressure, P_{19} = 1 bar. The pressure at the reboiler is therefore P_{20} = 1.5 bar. The composition in the bottom stream is 97 %

water and the boiling point is determined to $T_{20} = 361$ K.

6.4.4 Summary

If we reflect on the design of the distillation units with respect to partial and total condenser, we would say that the optimal solution must be where split 4 and 5 are partial condensers and split 6 must be a total condenser. However to obtain a cascade pressure for the distillations units we however increase the pressure more in split 4 and also to avoid that the temperature of the recycle stream is lower than the cooling water the pressure is also increased in split 5 and the temperature in the flash unit is also increased from 390 to 420 K. These changes are okay as long as we operate within a feasible temperature/pressure range.

7 Task 6: Energy Balance

In section 5 the simple linear mass balances assuming ideality were calculated. In this task energy balance models are included using the obtained process conditions from 6. Furthermore the SPLIT 1 unit is replaced by a corresponding flash unit. For checking, the energy balances are both calculated by hand as described in section 7.1 and by PRO/II.

Appendix A contain a complete summary of streams for hand-calculations (verification) and PRO/II calculations.

7.1 Mass and Energy Balances

When the flow rates, compositions, states and temperatures of the streams are known, energy balances can be performed. The energy balances are based on the first law of thermodynamics: the principle of conversion of energy. The steady-state energy balances for an open system is involving the heat flow to the system, Q, minus the work, W, done by the system on the surroundings, W plus the energy added to the system by mass flow $\dot{E}_{in}$, minus the energy leaving the system by mass flow $\dot{E}_{out}$, given by equation:

$$0 = Q - W + \dot{E}_{in} - \dot{E}_{out} \tag{49}$$

If we specify $T_{ref} = 298.15K$ and assume we only want vapor heat capacities we have the following relations for stream enthalpies:

$$\Delta H_v(T, \mathbf{y}) = \sum_{k=1}^{7} y_k \Delta_f H_k(T_{ref} + \sum_{k=1}^{7} \int_{T_{ref}}^{T} C_{p,k}^{0}(T)dT \tag{50}$$

$$\Delta H_l(T, \mathbf{x}) = \sum_{k=1}^{7} x_k \Delta_f H_k(T_{ref}) + \sum_{k=1}^{7} \int_{T_{ref}}^{T} C_{p,k}^{0}(T)dT - \sum_{k=1}^{7} x_k \Delta_{vap} H_k(T) \tag{51}$$

Since enthalpy is a state function the liquid enthalpy can be obtained by adjusting the temperature of a vapor to the stream temperature and then condensate it. This will require a correlation between heat of vaporization and temperature. In this work the Watson correlations has been used:

$$\Delta_{vap} H_k = \Delta_{vap} H_k(T_b^k) \left[\frac{T_c^k - T}{T_c^k - T_b^k} \right] \tag{52}$$

8 Task 7: The Total Flowsheet

In this task the assumption of ideality is rejected which requires rigorous modelling in PRO/II. This includes following major changes:

- Suitable activity coefficient models and/or equation of states must be applied.
- Stream calculators must be replaced by corresponding rigorous, multicomponent equilibrium models.
- Iterative adjustments (fine tuning) with respect to operation conditions to minimize deviations from streams in the simple flowsheet arising from rejection of ideality. These adjustments can also be performed by changing the process equipment.

8.1 Replacement of Stream Calculators

The stream calculators SPLIT 4, SPLIT 5 and SPLIT 6 (appear in figure 16) are replaced by rigorous distillation models. The absorber, SPLIT 2, has not been changed due to model parameter uncertainty. Instead the corresponding rigorous model is included for column sizing using reference streams. Following procedure was applied for stream calculator replacements: Previous design decisions concerning key component recoveries and operation conditions were supplied to units called "Shortcut Columns" (in PRO/II) whose input streams were referenced to the actual flowsheet. These shortcut columns gave estimates for feed stages, minimum number of stages with corresponding condenser and reboiler duties. Reasonable values were supplied to rigorous distillation models where the recoveries of the heavy and the light components were supplied fix the duties. The number of stages and feed stages were slightly changed to investigate sensitivity. For the finishing distillation column a driving force based technique was used to determine feed stage: 26 equilibrium stages giving $26(1 - 0.12) \approx 23$ i.e. feed stage located at 23 for minimum energy consumption (see figure (17) for driving force diagram). Inside-Out solution algorithm has been applied as default convergence algorithm for VLE column systems.

Furthermore a stream calculator has been added to remove existing isopropanol which is acceptable at current design stage. This action does only affects the real design of the dewatering column.

8.2 Thermodynamic Models

Since the majority of the process streams existing in the gas phase exceeds a pressure that normally can be considered as ideal and since existing liquid phases can not be considered ideal, the well-known gamma-phi approach has been applied. To calculate liquid activity coefficients the group contribution model UNIFAC has been chosen since no detailed investigations on model parameters have been performed. For vapor phase properties an

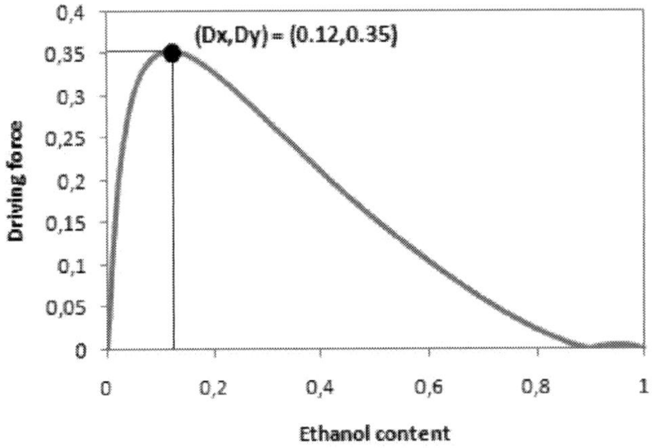

Figure 17: Driving force diagram for finishing column at 1 bar. All tray compositions are approximated as the binary mixture water-ethyl alcohol.

equation of state has been use to account for the non-ideality due to high pressure. The used equation of state is Soave-Redlich-Kwong model. This model is also applied to predict vapor phase properties.

As a special case it was found necessary to include liquid-liquid equilibrium for the deethering column due to immiscibility of water and diethyl ether. Even though water is the heavy component and diethyl ether is the light component it is presumed that this immiscible region is achieved since convergence can not be obtained by Inside-Out VLE (water is saturated at 0.5 mol% diethyl ether at e.g. 300 K). As a result the gamma-phi approach is extended to cover LLE using UNIFAC as well for the corresponding column. The immiscible region is displayed in figure 18.

8.3 Flowsheet Completion

After replacing the stream calculators with the designed distillation columns and checked that they worked properly we add pumps, compressors, heat exchangers and throttling valves to complete the flowsheet. Pumps are placed to increase pressure in liquids and compressors for vapors. Opposite valves are placed to reduce the pressure. All inlet streams entering unit operations have equal pressures. Furthermore heat exchangers are placed to control stream phases, e.g. one phase in mixers etc. Furthermore heat exchangers has been placed to simulate reactor and flash heat jackets. Heuristics about pressure drops in various unit operations from [10] have been applied as guidelines. E.g. pressure drops around heat exchangers are assumed to be 0.2 bar whereas tray pressure drops are assumed values of 0.007 bar per tray.

A major drawback of the existing flowsheet is the distillate temperature at the deether-

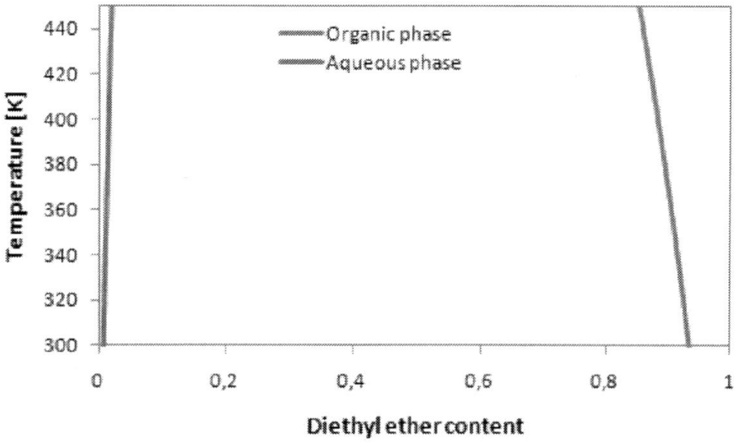

Figure 18: LLE diagram for the binary system Water-Diethyl ether at 19.5 bar.

ing column which is around 320 K. This temperature is low in terms of sizing the heat exchanger equipment since typical cooling water temperatures are assumed to be 310 K. The temperature has been maximized by increasing the column operating pressure to be near the previous column.

8.4 Results

The fully implemented flowsheet can be seen in figure 19. When all these process specifications are implemented into the PRO/II simulator, the program provide a stream summary, which includes the result of the mass and energy balances for each stream. The total stream summaries are given in Appendix B together with the simple and advanced flowsheet.

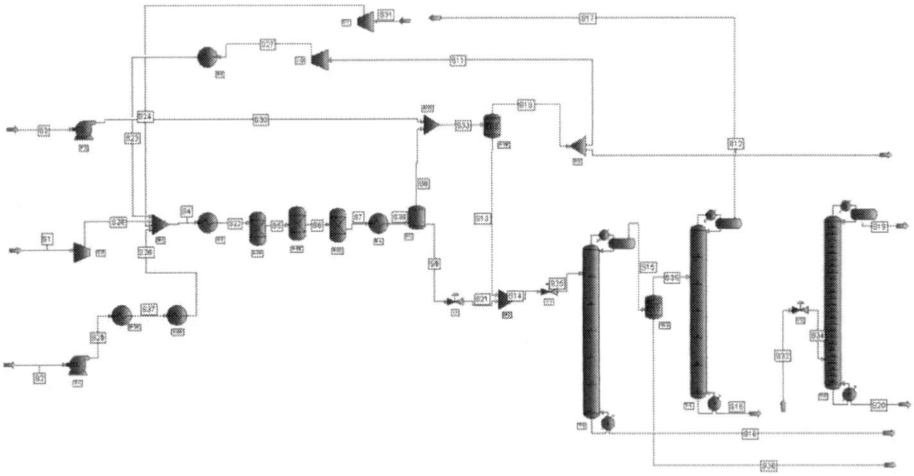

Figure 19: The advanced flowsheet build in PRO/II.

9 Task 8: Sizing and Costing

Sizing is conducted on each piece of equipment incorporated in the base-case design. The basis for the sizing calculations are the design decisions stated in the previous chapters which include mass balance and heat balance results. The resulting equipment cost estimates will be used to estimate the cost of buildings, piping etc. (other direct costs) and ultimately a total economic evaluation in the subsequent chapter (task 9).

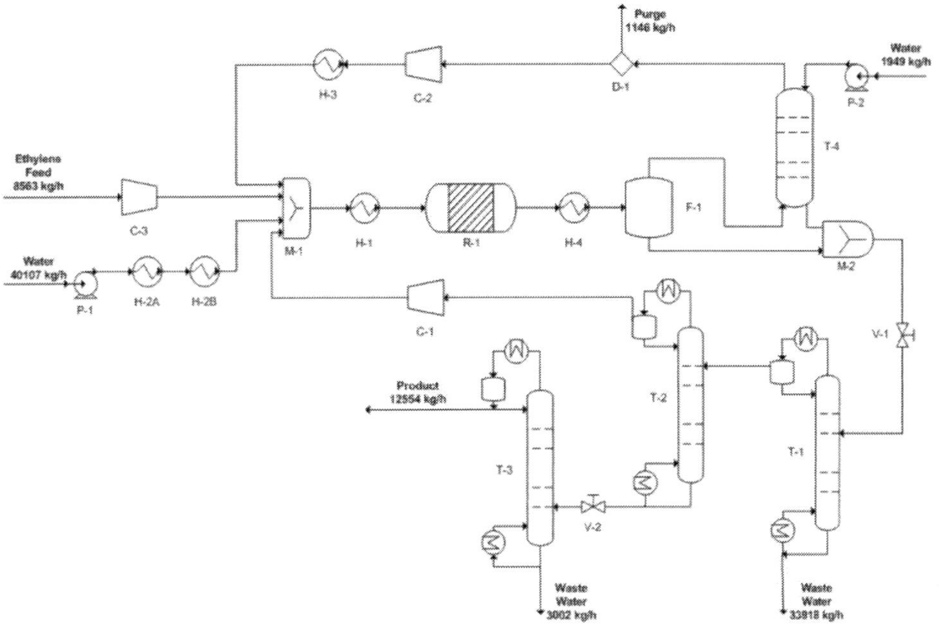

Figure 20: Flowsheet for base-case design with equipment ID used in this chapter.

In the following subsections will go through the complete range of equipment types with respect to dimensioning. At the end of this chapter a complete list of equipment with estimated costs are summarized in a table. The layout is set so the following dimensioning sections will work as input to the software ECON which has been used.

9.1 Pumps and Compressors

Pumps and compressors are installed to increase pressures of liquids and gasses respectively. Centrifugal pumps are used where a pump efficiency of 50% and a motor efficiency of 90% are assumed which gives a combined efficiency of 45%. Design information of pumps is given in table 8. The shaft works are determined by PRO/II.

In regard to compressor sizing we will perform some assumptions on gas compression: We assume centrifugal compressors are sufficient to provide the required compression ratios since we have relative high capacities. We choose to run with adiabatic compressors.

Table 8: Design information of pumps.

Pump ID	Work [HP]	P_{in} [bar]	P_{out} [bar]	Type	Material
P-1	227.72	1	69.2	Centrifugal	Cast Steel
P-2	10.96	1	68.3	Centrifugal	Cast Steel

Typically staged compressors with intercooling are used to reduce the required compression work when a high pressure increase is needed. Since heating is required (stream handled by C-3) a compressor working in the entire pressure range is used. A combined adiabatic efficiency of 72% is used resulting from 80% compressor efficiency and 90% driver efficiency. Design information of compressors is given in table 9. The compression works are determined by PRO/II.

Table 9: Design information of compressors.

Compressor ID	Work [HP]	P_{in} [bar]	P_{out} [bar]	Type	Material
C-1	259.97	16	69	Centrifugal	Carbon Steel
C-2	600.85	65	69.2	Centrifugal	Carbon Steel
C-3	2930.05	1	69	Centrifugal	Carbon Steel

9.2 Heat Exchangers

A total number of twelve heat exchangers are incorporated in the base-case design. These heat exchangers consist of conventional heat exchangers, condensers, cooling jacket for the reactor, cooling jacket for the flash unit and final reboilers and condensers for the distillation columns. The heat exchanger duties is given by PRO/II which are used to calculate the heat transfer area, A, from the following equation:

$$A = \frac{Q}{U\Delta T_{lm}} \tag{53}$$

Where Q is the heat duty, U is the overall heat transfer coefficient and ΔT_{lm} is the log-mean temperature difference for a countercurrent heat exchanger that is given by:

$$\Delta T_{lm} = \frac{(T_1 - t_2) - (T_2 - t_1)}{\ln\left(\dfrac{T_2 - t_2}{T_2 - t_1}\right)} \tag{54}$$

Here T_1 and T_2 denote the hot stream while t_1 and t_2 denote the cold stream. The simple relationship given in equation (53) is a simplification which, however, is adequate for the sizing and costing the this preliminary base-case design [5]. In order to perform the calculations we estimate the overall transfer coefficients, U, from [5] table 4.3 based on the states and components in the passing streams. Furthermore we assume that a temperature difference between the exit streams of ΔT between cold and hot streams of 10 K can be achieved. This scenario is dependant on the utility stream flowrate as illustrated in figure

21. We assume scenario (a) from the figure is achievable and thus the outlet temperature of the utility stream.

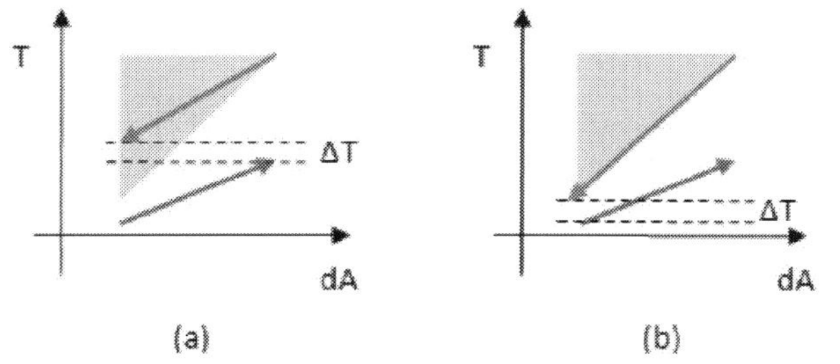

Figure 21: Example illustrating two possible configurations for a cold stream and a hot stream having different log-mean temperatures.

The heat exchanger used are counter-current shell and tube heat exchanger, because counter-current flow will utilize the capacity of the heat exchanger best, and a shell and tube heat exchanger has a good application for high pressures. If a heat transfer area exceeds 1000 m^2 it has to be divided and placed in series or parallel configurations.

Table 10: Design information of heat exchangers. Special heat exchangers are reactor cooling jacket (R-1), flash jacket (F-1) and distillation column condensers (T-1C, T-2C and T-3C) and reboilers (T-1R, T-2R and T-3R). Cooling water available at 310 K and MP steam at 1000 kPa and 457 K and HP steam at 4100 kPa ad 527 K.

HEX ID	U $[BTU/°F/ft^2]$	A $[m]$	ΔT_{lm} $[°F]$	P $[bar]$	Type	Material
H-1	40	445	127	69	Fixed Tubes	Carbon Steel
H-2A	100	120	277	69.2	Fixed Tubes	Carbon Steel
H-2B	100	708	75	69	Floating Heat	Carbon Steel
H-3	40	382	305	69.2	Fixed Tubes	Carbon Steel
H-4	40	255	269	68.5	Fixed Tubes	Carbon Steel
R-1	40	189	177	69	Fixed Tubes	Carbon Steel
F-1	100	439	167	68.5	Fixed Tubes	Carbon Steel
T-1C	100	102	104	20	Fixed Tubes	Carbon Steel
T-1R	250	219	75	20.042	Fixed Tubes	Carbon Steel
T-2C	100	1000	30	19	Floating Heat	Carbon Steel
T-2R	250	45	132	19.042	Fixed Tubes	Carbon Steel
T-3C	100	641	53	1	Floating Heat	Carbon Steel
T-3R	250	49	131	1.175	Fixed Tubes	Carbon Steel

High pressure (HP) steam is used for T-1R and T-2R while medium pressure (MP) steam is used for T-3R. For heating of H-1, H-2A, H-2B and H-3 a heating liquid is applied (DOWTHERM G) which will be discussed in the proceeding section, Furnace, since the external heating of the DOWTHERM requires special equipment. An operating temperature of the DOWTHERM is assumed to be 633 K.

The utility prices are listed in table 11.

Table 11: Utility Prices.

	MP Steam	HP Steam	Cooling Water	Electricity
Price [$/GJ]	8.22	9.83	0.354	16.8

9.3 Furnace

Heat supplied at H-1, H-2A, H-2B and H-3 to the process is supplied by a heat transfer liquid. A solution of an external heat system has been chosen for improved flexibility. The heat transfer liquid is available through The Dow Chemical Company [11]. Specifications to the DOWTHERM can be found in the reference. It has been estimated that a sum of 51,111 kW must be supplied to the furnace using natural gas available at 6 $/GJ.

9.4 Tower Units

The process has four tower units: Three distillation columns and one absorber. The tower units are sized according to the number of trays (theoretical trays in case of the absorber). A flooding factor of 80% and a tray spacing of 0.60 m are used for sizing the tower units. Physically the distillation towers consist of trays (valve trays for all distillation columns and a pressure drop across each tray of 0.007 bars as suggested by [10]). A total pressure drop across the absorber of 0.5 bar is assumed. The tray stack height is obtained be subtracting 1 from the number of stages, N, since the condenser is not considered a tray. Extra heights have to be added to get the actual height of the columns: The feed tray needs extra feed spacing of 1.5 m, $H_{feed\,space}$, a vapor disengagement spacing in the top and bottom of a total of 3 m for liquid level and reboiler return, $H_{top\,\&\,bottom}$, and there needs to be a skirt height of 1.5 m [10]. The total column height is:

$$H_{total} = H_{skirt} + H_{feed\,space} + H_{trays} + H_{top\,\&\,bottom} \tag{55}$$
$$H_{trays} = 0.60m \cdot (N-1)$$

For the absorber a height equivalent to a theoretical plate (HETP) of 0.9144 m for 2"Pall Rings is used as suggested by [10], two extra feed spacing (top and bottom) and a skirt height is used:

$$H_{total} = H_{skirt} + 2H_{feed\,space} + H_{trays} \tag{56}$$
$$H_{trays} = N \cdot HETP$$

From a supplier of column packing material, data is given that 6500 pieces per cubic meter is need and with bulk density of 395 $\frac{kg}{m^3}$, which means that 17 kg is needed. The price per kilo i 5.15$\frac{USD}{kg}$ and with shipping fee of 390 USD. This gives in total 478 USD, which must be considered as a very small cost.

The design information is summarized in table 12.

Table 12: Design information of tower units. T-1, T-2 and T-3 denotes distillation columns and T-4 is an absorber. The column called type refers to tray type for distillation towers and packing material for absorber.

Tower ID	N	H_{total} [m]	D [m]	P [bar]	Type	Material
T-1	8	10.8	1.5	20	Valve tray	Carbon Steel
T-2	8	10.8	1.4	19	Valve tray	Carbon Steel
T-3	27	22.2	2.2	1	Valve tray	Carbon Steel
T-4	10	13.6	1.5	65	2" Pall Rings	Carbon Steel

9.5 Vessels

Vessels include mixers (M-1 and M-2), reflux drums (coupled to distillation columns, T-1C, T-2C and T-3C), the flash unit (F-1) and the reactor (R-1). The geometry of all vessels is cylindrical with a length-to-width ratio of $L/D = 4$ which is considered as the optimal as the top and bottom cap are a lot more expensive than the sides [5]. Residence/hold-up times, τ of 5 min for liquids and 10 seconds for gasses are used (20 seconds residence time for reactor). A 20 seconds residence time in the reactor is chosen because it is operating without catalysis. The volume are based on the giving formula:

$$V = 2\frac{F\ M\ \tau}{\rho} \tag{57}$$

Where F is the total molar flowrate leaving the vessel, M the molar mass of the stream and ρ is the density. A buffer volume of +10% is included to avoid overfilling as varying throughput of pumps and compressors can be expected (typically resulted form process controlling). If the diameter of the vessel exceed 1.2 m then a horizontal design would be provided otherwise vertical. This would requires more space but less cost of structural support. As a safety precaution we assume that the vessels have a pressure relief valve of 50 % of the calculated operation pressure. The sizing of vessels are given in Table 13 where it can be seen the all vessels should be horizontal.

The reason behind selecting stainless steel as a reactor material is due to be able to adopt changes in the process such as running a catalyst based hydration of ethylene if required. Over-sizing has has also been done through this task to be more flexible regarding e.g. to produce more product.

9.6 Costing Summary

All expected equipment with utility cost and purchase cost is listed in table 14.

Table 13: Design information of vessels. All vessels are horizontal and cylindrical.

Vessel ID	Volume [m^3]	Length [m]	Res. time [min]	Pressure [bar]	Material
R-1	27.3	8.2	0.33	69	Stainless Steel
F-1	10.1	5.9	5	68.5	Carbon Steel
M-1	23.9	7.9	0.17	69	Carbon Steel
M-2*	14.2	6.6	5	65.5	Carbon Steel
T-1C	7.6	1.3	5	20	Carbon Steel
T-2C	12.4	1.6	5	19	Carbon Steel
T-3C	10.1	1.5	5	1	Carbon Steel

* Ignoring present, minor condensation.

Table 14: Equipment purchase and utility cost for the base-case design of the plant.

ID	Equipment Type	Material	Purchase Cost	Utility Cost
P-1	Pump	Cast Steel	$ 16,481	$ 73,945
P-2	Pump	Cast Steel	$ 5,969	$ 3,571
C-1	Compressor	Carbon Steel	$ 176,555	$ 84,427
C-2	Compressor	Carbon Steel	$ 379,098	$ 194,831
C-3	Compressor	Carbon Steel	$ 2,257,312	$ 1,372,750
H-1	Heat Exchanger	CS shell & tube	$ 70,998	$ -
H-2A	Heat Exchanger	CS shell & tube	$ 16,720	$ -
H-2B	Heat Exchanger	CS shell & tube	$ 89,653	$ -
H-3	Heat Exchanger	CS shell & tube	$ 62,782	$ -
H-4	Heat Exchanger	CS shell & tube	$ 23,878	$ 79,012
F-1	Flash Vessel	Carbon Steel	$ 22,160	$ -
	Heat Exchanger	CS shell & tube	$ 37,469	$ 212,353
M-1	Mixer	Carbon Steel	$ 34,839	$ -
M-2	Mixer	Carbon Steel	$ 26,503	$ -
T-1	Tower Unit	Carbon Steel	$ 132,420	$ -
	Valve tray	Carbon Steel	$ 10,365	$ -
	Condenser	Carbon Steel	$ 14,116	$ 30,850
	Reboiler	CS shell & tube	$ 45,207	$ 3,302,456
	Reflux Drum	Carbon Steel	$ 19,085	$ -
T-2	Tower Unit	Carbon Steel	$ 132,420	$ -
	Valve tray	Carbon Steel	$ 9,650	$ -
	Condenser	CS shell & tube	$ 104,012	$ 87,976
	Reboiler	Carbon Steel	$ 12,088	$ 1,201,273
	Reflux Drum	Carbon Steel	$ 24,682	$ -
T-3	Tower Unit	Carbon Steel	$ 125,224	$ -
	Valve tray	Carbon Steel	$ 31,364	$ -
	Condenser	CS shell & tube	$ 57,780	$ 97,468
	Reboiler	CS shell & tube	$ 7,616	$ 1,069,360
	Reflux Drum	Carbon Steel	$ 22,160	$ -
T-4	Tower Unit	Carbon Steel	$ 161,178	$ -
	Packing Material		$ 500	$ -
R-1	Reactor Vessel	Stainless Steel	$ 56,957	$ -
	Cooling-jacket		$ 19,003	$ 38,790
FX-1	Furnace	Carbon Steel	$ 2,384,303	$ 7,948,783
		Total	$ 6,590,547	$ 15,797,838

10 Task 9: Economic Evaluation

A crucial task in the process design discipline is to estimate and evaluate the plant economics in order to assess its economic potential. This evaluation is broken down to two parts which are the capital expenditures (CAPEX) and operation expenditures (OPEX). The CAPEX covers following expenditures:

- Manufacturing Fixed-capital Investment (Direct Cost) which involves sizing of the equipment
- Nonmanufacturing Fixed-capital Investment (Indirect Cost) which are estimates from heuristic percentage of the equipment cost
- Selecting plant location to determine land cost

Whereas the OPEX covers:

- Raw materials
- Utilities (e.g. steam, cooling water, shaft work)
- Taxes, labor, insurance etc.

From the sizing data and knowledge about the market price it is possible to perform an economic potential of the plant. To this evaluation we use the software ECON. Calculations, data source and utility costs are based on Peters and Timmerhaus work [12] and Turton et al. [13].

10.1 CAPEX

In capital cost also working capital investment (WCI) and fixed-capital investment (FCI) such as piping, equipment delivered, service facilities, engineering, construction expenses etc. are included. All capital expenditures are assumed to be proportional to the total equipment cost obtained in chapter 9. We assume the plant is operated in China. A land cost for 17 acres (740,520 square feet) in China is available for 1.5 million US dollar [14]. A summary is given in table 15.

10.2 OPEX

Operating cost such as raw materials, operating labor, administration, insurance, taxes etc. Raw material and product flows are listed in table 16 while a complete summary is given in table 17. We assume a sum of 44 employees (operation labor 44,000 $/year) are working on the plant of which four are managers (operation labor 90,000 $/year) taking up roughly 6% of the fixed capital investment (estimates of Chinese labor cost).

Table 15: Summary of base-case Capital Expenditures (CAPEX). Basis: Purchase Equipment Cost.

Manufacturing Fixed-capital Investment (Direct Cost)	Percent of Delivered Equipment	Result
Purchased Equipment Delivered	1.1	$ 7,249,601.70
Purchased Equipment Installation	0.47	$ 3,407,312.80
Instrumentation and Controls	0.36	$ 2,609,856.61
Piping	0.68	$ 4,929,729.16
Electrical Systems	0.11	$ 797,456.19
Buildings (Including Services)	0.18	$ 1,304,928.31
Yard Improvement	0,1	$724,960.17
Service Facilities	0.7	$ 5,074,721.19
Total Direct Cost		**$ 26,098,566.12**
Nonmanufacturing Fixed-capital Investment (Indirect Cost)	Percent of Delivered Equipment	Result
Engineering and Supervision	0.33	$ 2,392,368.56
Construction Expenses	0.41	$ 2,972,336.70
Legal Expenses	0.04	$ 289,984.07
Contractor's Fees	0.22	$ 1,594,912.37
Contingency	0.44	$ 3,189,824.75
Total Indirect Cost		**$ 10,439,426.45**
Fixed-capital Investment (FCI)		$ 36,537,992.57
Working Capital Investment	Percent of Delivered-equipment	Result
Working Capital Investment (WC)	0.89	$ 6,452,145.51
Total Capital Investment (TCI)		**$ 42,990,138.08**

Table 16: Raw Materials and Products.

	Ethylene Feed	Water	190 Proof Ethanol
Price [$/kg]	0.4	0.00083	0.9
Annual Flow [mill. kg/year]	75	28.9	90.5
	Raw Materials		Products
Annual Cash Flow [mill. $/year]	31.3		81.5

Table 17: Summary of base-case Operation Expenditures (OPEX).

Items of Operating Cost	Factor	Basis	Cost per year
Raw Materials			\$ 30,251,334.00
Operating Labor	0.06		\$ 2,192,279.50
Operating Supervision	0.15	Operating Labor	\$ 328,841.93
Utilities	-		15,797,838.00
Maintenance and Repairs	0.05	FCI	\$ 1,826,899.60
Operating Supplies	0.15	Maintenance and Repairs	\$ 274,034.94
Laboratory Charges	0.15	Operating Labor	\$ 328,841.93
Royalties	0.01	Total Product Cost	\$ 585,334.32
Catalysts and Solvents	0		\$ 0
Variable Cost			**\$ 51,585,404.21**
Property Taxes	0.02	FCI	\$ 730,759.84
Financing (interest)	0	FCI	\$ 0
Insurance	0.02	FCI	\$ 730,759.84
Rent	0	FCI	\$ 0
Fixed Charges			**\$ 1,461,519.68**
Plant Overhead	0.6	Labor, Supervision and Maintenance	\$ 2,608,812.60
Manufacturing Cost			**\$ 53,046,923.89**
Administration	0.2	Labor, Supervision and Maintenance	\$ 869,604.20
Distribution and selling	0.04	Total Product Cost	\$ 2,341,337.28
Research and Development	0.04	Total Product Cost	\$ 2,341,337.28
General Expenses			**\$ 5,552,278.76**
Total Product Cost without Depreciation			**\$ 58,533,432.00**

10.3 Evaluation

Key numbers for evaluating the base-case profitability are given in table 18.

Table 18: Base-case Profitability.

Without Time Value of Money	
Rate of Return	47%
Pay Back Period	1.8
Net Return	$ 14,649,889.91
With Time Value of Money	
Annual Ene of Year cash flows and discounting	
Net Present Worth	$ 10,321,652.00
DCFR	0.15
Continuous Cash Flows and Discounting	
Net Present Worth	$ 11,077,582.00
DCFR	0.14

We expect that we can sell our product at 0.9 $/kg which result in an operating profit of roughly 23 mill. US dollar per year which is a high operating profit (see table 19 and figure 22). From the time the money are loaned/withdrawn from the bank and construction has been initiated a total period of 4.5 year have to pass before profit is made. If the money should be loaned the bank could consider the project as a risky investment and refuse to do so due to the long period. To see how a decrease in product price affects the operating profit, it has been found that the plant still make money until a lower value of approximately 0.7 USD/kg ethanol is reached as shown in figure 34 in Appendix C. For the situation with a product price on 0.7 USD/kg ethanol the plant have a return rate of 8% and a pay back period at 6.5 years.

Table 19: Economic potential dependent on the product sells price.

CAPEX	
Total Capital Investment	$ 42,990,138.08
OPEX	
Total Product Cost Without Depreciation	$ 58,533,432.00
Sale of Products	$ 81,459,898
Operating profit	$ 22,926,466

Numbers that comprises a basis of the economic evaluation are summarized in table 20.

As it appears in table 19 the plant economical potential is strongly dependent of the product price. Therefore a sensitivity analysis of the plant was made to find out where are the critical elements with respect to the economic and get and idea of where improvements and optimizations have to been done. From Figure 23 there are no doubt of that the product sales price is the way most sensitive with respect to the economic. Thereafter the

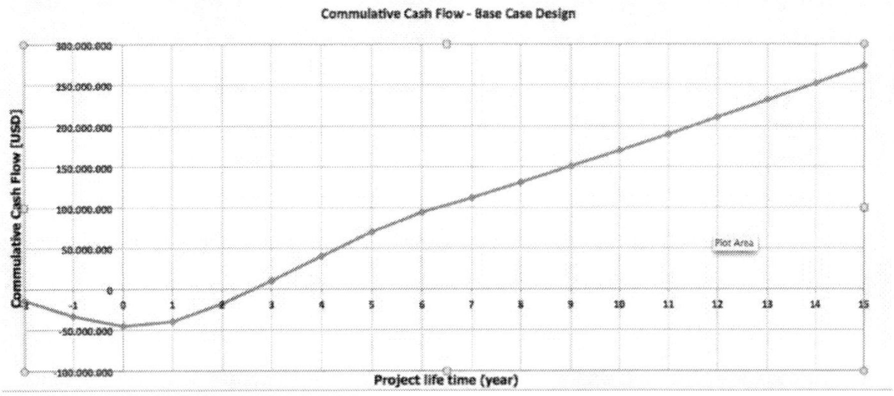

Figure 22: Commulative cash flow for a product price at 0.9 USD/kg

Table 20: Economic potential dependent on the product sells price.

Construction Inflation Rate	2 [%/year]
Product Price Inflation Rate	2 [%/year]
TPC Inflation Rate	2 [%/year]
Taxes	25%
Depreciation time (linear)	5 [years]
Project Lifetime	15 [years]
Production Capacity Sequence	50%, 80%, 90%, 100%, 100% ...

purchase price of raw material is the second sensitive. Further from the sensitivity analysis we see that there are also money to spare by lowering the utility cost and the equipment purchase cost. To find out which utility cost there is the most dominant we use the values from table 14 and illustrate them by pie charts. Some critical increase percentages for the major expenses are -20% product value, +40% raw material price, +80% utility cost and +90% equipment cost which all will cause zero net present values (NPV).

In figure 24 we see that total equipment purchase cost is 6.6 million USD and here the compressors and the furnace is quite expensive as they constitute 77% of the total amount.

Figure 25 shows that the total utility cost is 15.8 million USD and here 50% of the cost is spent on heating the DOWTHERM G with natural gas of which the reboilers and compressors are next dominants values.

We have seen in the sensitivity analysis that product and raw material prices have most influence on the profitability of the plant. As the world petroleum resources are becoming increasingly scarce [15] the prices of petrochemical derived feedstock are expected to rise. Furthermore the demand of ethanol is rising due to increasing application in fuel e.g. some countries do have regulations for ethanol content in car gasoline. Both scenarios suggest a decrease in the plant's profitability (base-case design) and therefore a

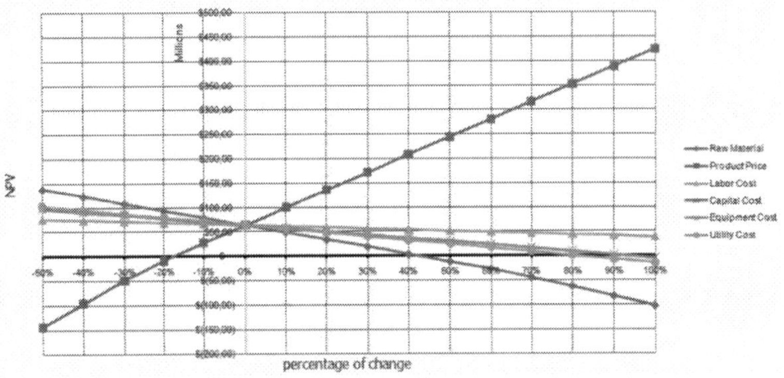

Figure 23: Sensitivity analysis for base-case design. NPV is Net Present Value.

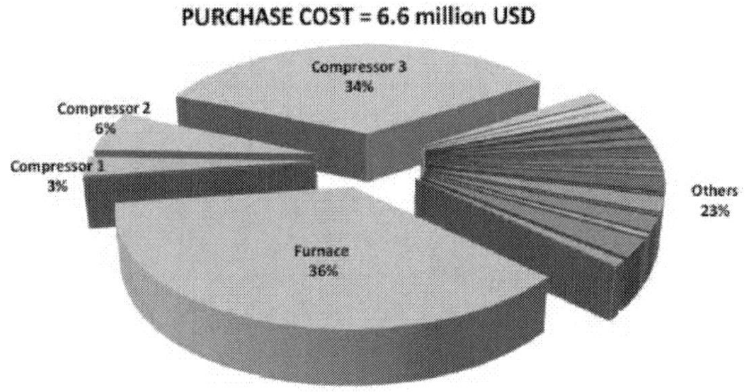

Figure 24: The equipment purchase cost for the base-case design

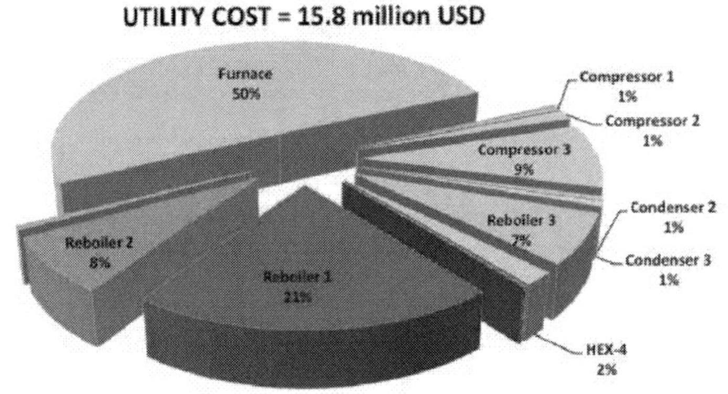

Figure 25: The Utility cost of the base-case design

large project life time can not be expected for the base-case. The sensitivity analysis do not shown significant deviations from equipment cost changes. Therefore room is left for substituting material from carbon steel to stainless steel for the crucial parts of the plant if needed. Further analysis should be performed for this purpose but as water is present (in a hydrocarbon mixture) corrosion could turn out to be a problem which can be avoided by using stainless steel equipment and piping. Other expenses that have not been taken into account are storage facilities, miscellaneous buildings, purchase of heat transfer liquid etc. As earlier stated a +90% equipment cost increase will still make a profitable plant.

The above considerations are still based on the base-case design. There is still potential for improvements with respect to operation economics. The proceeding chapters will handle improvements that can be obtained through heat and mass integration and flowsheet optimization.

It should be kept in mind that what we are evaluate here is based on a conceptual design. If this plant should be build spare equipment such as spare pumps, compressors, heat exchangers, etc. shall be included in regard to build a safe and functional plant that shall work with minimum probability of failure or hazard at any time. That means that safety barriers and similar shall be implemented in the plant design and a risk assessment shall be done from time to time, of which also has some expenses.

11 Task 10: Heat and Mass Integration

At the current stage a base case design is completed. As the first step of improving the design with the ultimate to maximize the profit, possibilities for heat and mass integration are investigated. Both mass and heat integrations typically have huge potentials for significantly reducing the utility cost at a cost of increasing operation complexity. For example the amount of working time that must be spent on designing appropriate control structures for highly integrated networks and furthermore operation start-up and shut-down can become challenging.

This chapter contain reflections, results and discussions related to both heat integration and mass integration.

11.1 Heat Integration

It was discussed in chapter 9 that a potential for decreasing the plant's operation expenditures exists by considering the heat transfer systems in the base case design which take up a significant part of the utility cost (see figure 25). In the base case design cooling water is used excessively to remove heat from the system which leads to a increased demand for e.g. supply of heat for the heat transfer liquid. In order to exploit heat integration following streams have been taken into account (Stream labels are illustrated in figure 26):

- C1: The combined stream that is fed to the reactor
- CS1: Water feed indicating phase change for evaporator
- C2: Water feed
- C3: Recycle stream leaving the absorber
- H1: Reactor effluent cooling
- HS1: Flashed stream which is condensing
- HS2: Heat of reaction from reactor
- H2: Waste water from dewatering column
- H3: Waste water from finishing column
- H4: Product stream

The necessary stream data are summarized in table 33 in Appendix D with heat flows (kW/K) and inlet and outlet temperatures. The heat integration and heat exchanger network synthesis problem is carried out in the following subsections. The approach is based on the Pinch Design Approach given in [5] that overall consists of two steps: First identifying a pinch point followed by determination of minimum utilities (hot and cold)

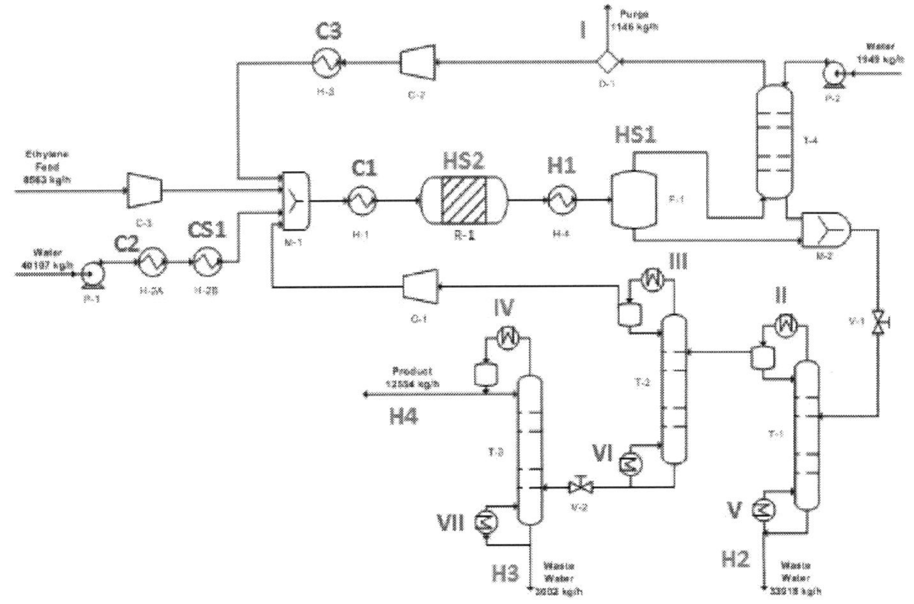

Figure 26: All streams that are considered in heat integration. Streams that are not considered but have potential are indicated by roman numbers: I) Heat of combustion; II-IV) Condensers (hot streams); V-VII) Reboilers (cold streams).

using a grand composite curve (GCC) and then synthesizing the heat exchanger network (HEN) based a partitioning strategy and process insight.

11.1.1 Pinch Point Identification

We define a minimum recovery approach temperature of $\Delta T = 20\ K$. From this temperature all relevant temperatures for all combinations of inlet, outlet, hot and cold streams can be ordered to obtain certain temperature intervals (NB: CS1 and HS2 not included). From these intervals all heat required and heat available can be calculated and cascaded resulting in the table given in table 34 in Appendix D. The grand composite curve is then constructed using the total cascaded heat which can be seen in figure 27 from which the pinch point can be identified as the temperature of the lowest cascaded heat.

Using the GCC we can determine the minimum hot utility as 7363 kW, the minimum cold utility as 10465 kW and the pinch point at 480/460 K. These values are in this case only used as guidances since it is assumed that waste streams must be cooled to 310 K (which is assumed to be unnecessary) and the heat of reaction (HS2) and water evaporation (CS1) are not included. Adjustments of the minimum utilities can be carried

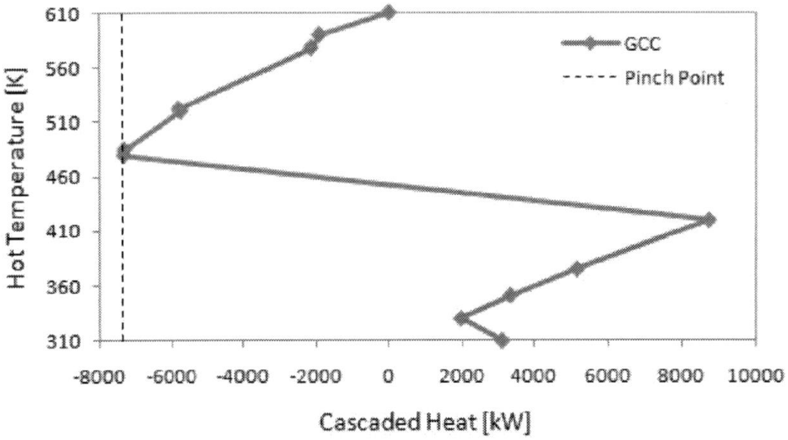

Figure 27: Grand composite curve used for identifying pinch point.

out as:

$$Q_{hot}^{min} = 7363kW - 4222kW + 16778kW = 19918kW \quad (58)$$

$$Q_{cold}^{min} = 10465kW - (42.85 + 3.51 + 10.53)(460 - 310)kW = 1932kW \quad (59)$$

Here the latent heat for vaporization of CS1 will not be integrated and the heat of reaction is considered as a hot utility.

11.1.2 Invention of Network

In this subsection the heat integration network will be developed based on the knowledge of the pinch point location. The pinch point is used to divide the problem in two partitions: One requiring only hot utility and one requiring only cold utility - rest can be integrated. Following equation is used to determine temperature of outlet streams when integrating two streams:

$$[FC_p]_{hot}(T_{in} - T_{out}) = [FC_p]_{cold}(t_{out} - t_{in}) \quad (60)$$

The result is illustrated in figure 28 indicating the amount of heat transferred for each heat exchanger and the final network is given in figure 29.

11.1.3 Sizing and Costing

As the last step in the heat integration task the new heat exchangers must be sized and cost estimated. The result is displayed in table 21.

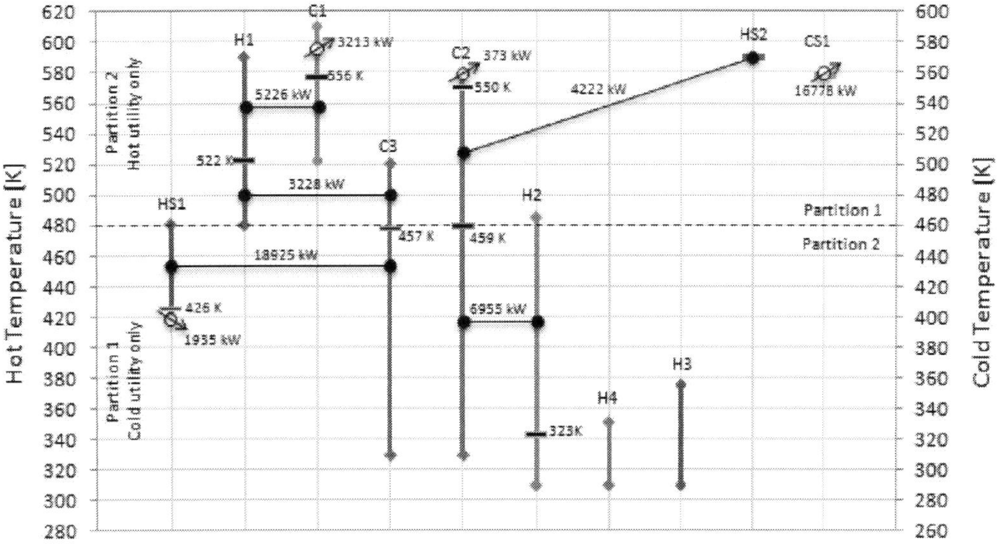

Figure 28: Synthesizing of heat exchanger network.

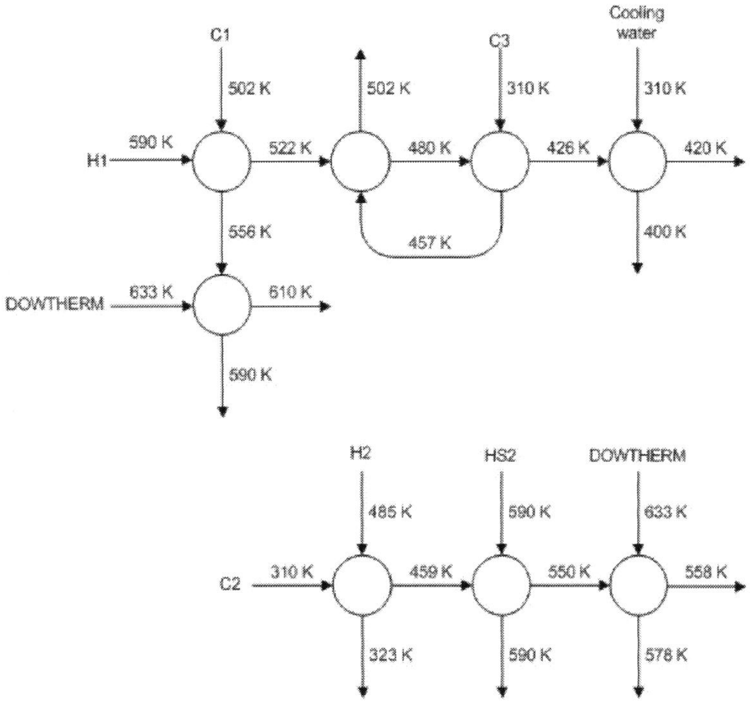

Figure 29: Heat Exchanger Network.

Table 21: Sizing and costing for heat exchangers for synthesized network.

HEX ID	Hot-Cold	U $(BTU/(ft^2h^\circ F))$	A (m^2)	Cost ($US)	Type
HN1	DT-C1	40	294	26,759	Fixed Tubes
HN2	H1-C1	40	879	130,986	Floating Head
HN3	H1-C3	40	637	95,881	Floating Head
HN4	HS1-C3	100	585	48,253	Fixed Tubes
HN5	HS1-CW	100	59	9401	Fixed Tubes
HN6	DT-(C2+CS1)	120	528	44,043	Fixed Tubes
HN7	HS2-C2	40	243	22,992	Fixed Tubes
HN8	H2-C2	200	332	29,566	Fixed Tubes

DT: Hot utility (DOWTHERM), CW: Cold utility (cooling water)

11.1.4 Purge Heat of Combustion

The purged stream contains mainly ethylene, methane and propylene which also are the most abundant components in available natural gas used for combustion. Natural gas has been chosen as the fuel for the installed furnace for two reasons: First the plant is expected to be operated in a petrochemical complex thereby assuming natural gas is readily available and second the opportunity of integrating the purge stream as a fuel in the furnace. The typical natural gas composition is compared with the purge composition in figure 30. The disadvantage of using the purge as fuel is the relative high water content on 267 mg/m^3 compared to typical natural gas which is below $80mg/m^3$. The influence of the water content should be investigated further before integrating the stream. It has been estimated that a sum of 15940 kW heat of combustion is available from the purge stream which has a huge potential for covering up to 80% of the required energy to drive the furnace after heat integration and roughly 30% before heat integration.

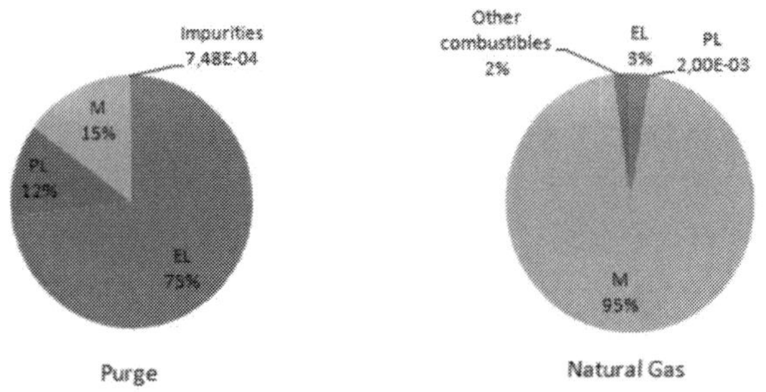

Figure 30: Comparison of purge gas and natural gas compositions. Natural gas reference: [16].

11.2 Process Impact

It can be seen in table 22 that a significant gain can be achieved through heat integration. As expected we see that the purchase cost is increased due to extensive heat exchanger network resulting in increased number of heat exchangers. On the other hand the furnace purchase purchase and utility cost have more than halved. Also heat exchanger utilities have been reduced significantly.

Table 22: Summary of base-case improvement by heat integration. Column heat exchangers not included. HEX is abbreviation for heat exchanger.

Expense Type	Base-case	With HEN	With HEN and fired purge
HEX Purchase Cost [$]	320,503	407,881	407,881
HEX Utility Cost [$/year]	330,155	17,755	17,755
Furnace Purchase Cost [$]	2,384,303	1,074,740	1,074,740
Furnace Utility Cost [$/year]	7,948,783	3,167,009	688,021

11.3 Mass Integration

Advanced mass integration will not be performed but instead simple considerations of recycling process exit streams will be discussed by qualitative considerations. As indicated in figure 20 streams available for mass integration are waste streams consisting mainly of water and the purge stream. Two recycle loops have already been implemented in the base-case design during the flowsheet synthesis in Task 3 (chapter 4). In addition the purpose of the purge stream is to bleed off methane otherwise accumulation would happen and thus the purge stream can not be mass integrated unless methane is removed. As discussed during the heat integration investigation the purge stream can be combusted to replace natural gas in the furnace. The disadvantage of purging the methane in a ethylene rich stream results in a lowered atomic conversion, i.e. some ethylene that could have been converted to product is lost. However, browsing through the literature for different available technologies for separating methane from ethylene reveals that it is most likely not worth to perform any separation. Wozny et al. [17] states that separation of ethylene from light gas mixture is one of the most energy intensive separations in petrochemical processes. For this reason purge recycle is precluded. Waste water mass integration is a likely candidate for mass integration (in the sense of minimum fresh-water and water reuse). Two waster water streams exist (see compositions in table 23):

- WW1: Water with isopropanol from the dewatering column
- WW2: Water stream with minor impurities from the finishing column.

Like methane isopropanol is an inert specie if the loop of the stream leaving from the dewatering columnwas closed so similar to methane it will cause a snowball-like effect by increasing the inventory within the plant.

Possibilities for mass integration are recycling WW2 which can improve the atomic conversion slightly. It is not suggested to use WW2 as feed for absorber due to lowering of ethanol driving force because of the ethanol content. As discussed above WW1 should

Table 23: Waste water stream compositions.

Impurity	WW1 flow [kg/h]	WW2 flow [kg/h]
Water	33605	2943
Isopropanol	253	0
Ethanol	60	59

not be recycled unless purified. A recycling of stream WW2 will result in a minor 11,647 US dollar saving per year.

12 Task 11: Optimization

In this task we would like to investigate if it is possible to further improve the base case design. Earlier the optimization approach have been used with heat integration, where it was seen that operation cost could be reduced. There are though still place for improvement, of which this section will provide details on how this will be done.

As mentioned the heat integration has reduced a big part of the utility cost calculated for the base case design. To figure out what else could be done we take a look at figure 23 showing the sensitivity analysis achieved in chapter 9. Hence one can deduce that the product price has the largest effect on the economics followed by the price of the raw material. Nevertheless, the third factor that plays a role is the utility cost. So the focus will then be targeting how to formulate an objective function that takes that into account, and how to perform that in practise.

To begin with an optimization model can be written in the general form:

$$\textbf{Objective function: } min\ Cost\ [\$/h] = product\ + raw\ material\ + utility \tag{61}$$

$$\textbf{Subject to: } Methane\ in\ reactor\ < 10\% \tag{62}$$

Now we have to determine the logical constraints. We see that there is a room for improvements regarding the purge stream. As mentioned the purge stream is used to keep the mole fraction of methane in the reactor feed less than 10% to prevent coking in the reactor under the extreme conditions the process faces. But the more that we purge the less we send to the recycle stream, which means that the flow to the compressor (C-2) will be reduced (see figure 20 for flowsheet illustration). A reduced flow to the compressor will result in a less consumption of energy in terms of that the compressor duty will be reduced, which in other words means savings in the utility cost. Although this from the first thought may sound favourable, we shall not forget that some raw material and product get lost with the purge, and since the system is more sensitive to that we shall try to find a good balance between those criteria.

So by changing the purge stream and monitor the work of compressor C-2, the content of EL, EA and DEE purged, the content of EA in the product stream and the concentration of methane in the reactor inlet in PRO II, the plots seen on figure 31 and 32 are created. The objective function as seen in the equations above minimizes the cost of EL, EA and C-2 work in US $/h. The dashed line in both figures illustrate the limit/constraint for methane.

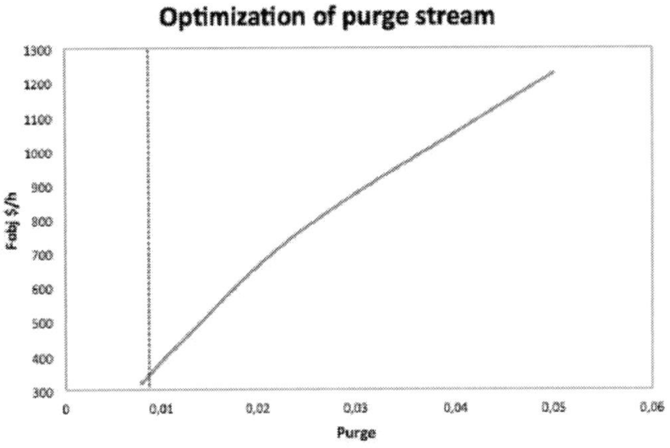

Figure 31: Plot showing the the optimization of the purge stream subject to keeping the methane mole fraction below 10% in the process (illustrated by dashed line).

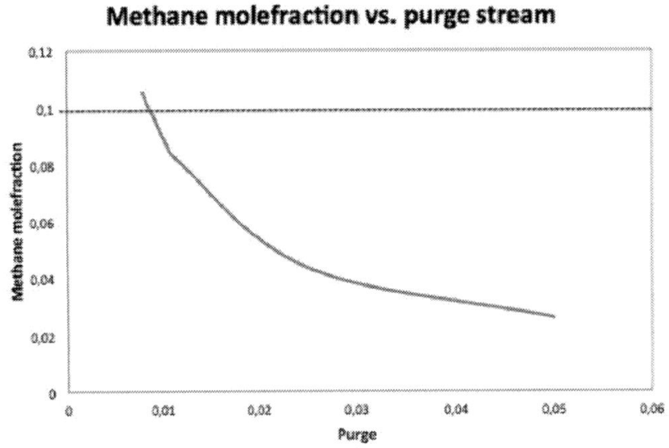

Figure 32: Optimization of the purge stream subject to keeping the methane molefraction below 10% in the process. The dashed line show the limit for reaching the constraints.

Thereby it can be derived from the objective function and the plots seen on figure 31 and 32 that the purge fraction should be maintained at 0.90%, which means that no significant improvement or changes will be favourable to make for the process flowsheet, since we already are very close to the optimum purge fraction value. This is also confirmed by the optimization regarding minimizing cost based on the sensitivity analysis

A more extensive optimization problem is suggested as future work where all major design decisions should be considered as adjustable e.g. water recovery in dewatering column, flash temperature, purge simultaneously subject to mass balances and the remaining design decisions, with the ultimate goal to minimize a measure for the total operation cost.

This measure typically involves annualizations of the capital cost such that a final NLP formulation can be obtained. These types of optimization problems do tend to be work intensively which is why this have not been performed.

13 Task 12: Environmental Impact and Sustainability

It is familiar that the chemical industry represents an important economic activity to the modern world, not least it is central to it in a lot of aspects especially regarding the conversion of raw materials into products that people consume on a daily basis, keeping the living standard as it known nowadays.

In the past decades there has been a remarkable change in the perception of the environment and consuming the world natural resources, which increases the concern and focus on waste reduction and pollution prevention. Since there will always be a request for higher profits in order to be competitive on the market, it is necessary to include the environmental impact factors in the considerations while generating a sustainable process alternative with respect to new process design as well as retrofit design.

The environmental impact for our design is in this chapter determined in order to be able to run our business by balancing economic, environmental and social responsibilities, where the important aspects of environmental responsibility are the utilization of natural resources and disposal of wastes.

Table 24: The important parameters used in the WAR algorithm used to indicate the total PEI (potential environmental impact).

Parameter	Definition
HTPI	Human toxicity potential by ingestion
HTPE	Human toxicity potential by dermal/inhalation exposure
TTP	Terrestrial toxicity potential
ATP	Aquatic toxicity potential
GWP	Global warming potential
ODP	Ozone depletion potential
PCOP	Photochemical oxidation potential
AP	Acidification potential

So to evaluate how sustainable our process' environmental impact factors, the waste reduction algorithm (WAR) calculated by the ICAS software is used, which defines indexes that characterize the generation and output of potential environmental impact (PEI) from a process. The PEI is a relative measure of the potential for a chemical to have an adverse affect on human health and the environment. The potential environmental impacts are calculated from stream mass flow rates, stream compositions and a relative potential environmental impact score for each chemical present. There are several categories of impact that can be subdivided into physical potential impacts given as acidification, greenhouse enhancement, ozone depletion and photochemical oxidant depletion. Human toxicity effects by air, water and soil plus aquatic and terrestrial eco-toxicity effects does also account in those categories. These are classified in the parameters seen in table 24, where all of these are gathered in one total factor, the total PEI.

First we define our system boundary, which is the whole system also called the traditional system boundary. Here we have all the inlet and outlet streams for the base-case design (see figure 33). In the inlet we have the ethylene feed and the two water feed streams, while we in the outlet have the product and purge stream plus the two waste water streams from the distillation column T-1 and T-3 (see figure 20). The generated WAR algorithm for the base-case design can be seen in table 25. From the table it can be seen that there are given either positive and negative numbers under each category. If the value is positive it is equal to that the environmental impact has become worse for this parameter and vice versa.

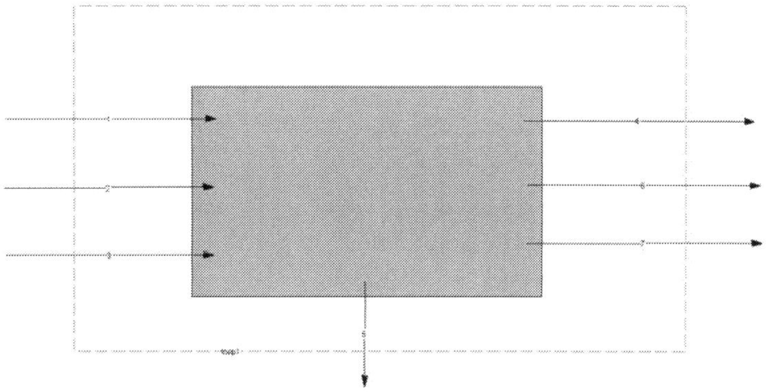

Figure 33: The base-case design system build in ICAS showing the input and output streams in the process (the grey box), where the red dashed box shows where the boundary is set. There are three input streams, four output of which one of them is the product (pointing down)

Even though the PEI is negative as seen in the tables (table 25, 26 and 27), it should not be interpreted as if the process is sustainable and have a good environmental impact. The goal of the methodology used in the WAR algorithm is to minimize the PEI for a process instead of minimizing the amount of waste (pollutants) generated by a process, which distort the picture. So each parameter has to be evaluated separately and the interpretation of the PEI value will not be weighed highly in this work.

In table 25 we have the WAR calculated for the base-case design. The boundary could have been chosen such that the product stream was not classified as an outlet stream but as a part of the system since it will be tapped and sold. Although this will give better results in the WAR algorithm, we choose not to do that since the product is a conversion

Table 25: Results of WAR algorithm for the base-case design.

Stream	Total PEI	HTPI	HTPE	ATP	TTP	GWP	ODP	PCOP	AP
Ethylene (feed)	17552.5	0	0	131.384	0	0.34	0	17420.8	0
Water (feed)	0	0	0	0	0	0	0	0	0
Water (absorber)	0	0	0	0	0	0	0	0	0
Input sum	17552.5	0	0	131.4	0	0.34	0	17420.8	0
Purge	2195.43	0.35	3.95E-4	178.55	0.35	0.34	0	2176.53	0
Product	7657.42	587.14	18.29	0.79	587.14	4.77E-4	0	6480.53	0
Waste water (T-1)	742.20	206.07	0.09	0.03	206.07	3.18E-4	0	328.93	0
Waste water (T-3)	38.48	294.71	0.01	0.004	294.71	7.07E-16	0	325.69	0
Output sum	9965.54	611.04	192.37	186.75	611.04	0.34	0	8722.52	0
Impact generated	-7586.98	611.043	192.368	-112.71	611.04	-0.004	0	-8698.28	0

of an other material that our plant is responsible for and it will get sold and thereby gets out of the system at the end. It is seen that it is only the ethylene feed that has an affect on the parameters in the inlet, since water is assumed to be neutral. It should be noticed that the waste water even though it contains mainly water it does still contain impurities and therefore has an influence in the calculated WAR. The base-case design does according to the results obtained show that the this design has almost the same negative impact on HTPI and TTP, and a less unfavourable impact in HTPE. It does also have almost no influence on GWP and a good impact at ATP and PCOP. There are though no outputs generated from ODP and AP for all three tables.

Table 26: Results of WAR algorithm after mass integration and thereby with water recycle.

Stream	Total PEI	HTPI	HTPE	ATP	TTP	GWP	ODP	PCOP	AP
Ethylene (feed)	17552.5	0	0	131.38	0	0.34	0	17420.8	0
Waste water (T-3)	38.48	294.72	0.01	0.004	294.72	7.06E-12	0	3.26E+5	0
IWater (feed)	0	0	0	0	0	0	0	0	0
Input sum	17591	294.72	0.009	131.39	294.72	0.34	0	17453.4	0
Purge	2195.43	0.35	3.95E-4	178.55	0.35	0.34	0	2176.53	0
Product	7657.42	587.14	18.29	0.79	587.14	4.77E-4	0	6480.53	0
Waste water (T-1)	742.2	206.07	0.09	0.03	206.07	3.18E-4	0	328.93	0
Output sum	9927.06	608.1	191.45	186.72	608.1	0.34	0	8689.95	0
Impact generated	-7663.93	605.15	190.53	-112.72	605.15	-0.004	0	-8763.41	0

Now we will try to calculate a new WAR algorithm for an alternative case design inspired by the mass integration (see 11.3), where the water stream is recycled from (T-3) since it almost contain water with a very low concentration of product and impurities. The results can be seen in table 26. Compared to the table obtained before it can be seen that there are some improvements on HTPI, HTPE, TTP and PCOP, where the rest of the parameters still maintain same level. So all in all it can be seen that this alternative has a better environmetal impact than the base-case design.

Table 27: Results of WAR algorithm after mass integration and burning of purge with water recycle.

Stream	Total PEI	HTPI	HTPE	ATP	TTP	GWP	ODP	PCOP	AP
Ethylene (feed)	17552.5	0	0	131.38	0	0.34	0	17420.8	0
Waste water (T-3)	38.48	294.72	0.009	0.004	294.72	7.06E-12	0	325.69	0
Water (feed)	0	0	0	0	0	0	0	0	0
Input sum	17591	294.72	0.009	131.39	294.72	0.34	0	17453.4	0
Product	7657.42	587.14	18.29	0.79	587.14	4.77E-4	0	6480.53	0
Waste water (T-1)	742.2	206.07	0.09	0.03	206.07	3.18E-4	0	328.93	0
Burned purge	134.94	0	0.12	133.69	0	112.85	0	0	0
Output sum	7866.58	607.74	203.06	134.51	607.74	112.85	0	6513.42	0
Impact generated	-9724.42	604.79	202.15	312.49	604.79	0.79	0	-10939.9	0

Table 28: Chemical decomposition reactions for burning the purge stream where the quantities of needed oxygen for absoulte combustion are listed

Reactants		Products	Amount of O_2 in $kmol/h$
$CH_4 + 2\,O_2$	$\longrightarrow$	$CO_2 + 2\,H_2O$	12.24
$C_2H_4 + 3\,O_2$	$\longrightarrow$	$2\,CO_2 + 2\,H_2O$	89.9952
$(C_2H_5)O + 6\,O_2$	$\longrightarrow$	$4\,CO_2 + 5\,H_2O$	0.0876
$C_2H_6O + 3\,O_2$	$\longrightarrow$	$2\,CO_2 + 3\,H_2O$	0.0528
$C_3H_6 + 4{\cdot}5\,O_2$	$\longrightarrow$	$3\,CO_2 + 3\,H_2O$	21.84
$C_3H_8o + 4{\cdot}5\,O_2$	$\longrightarrow$	$3\,CO_2 + 4\,H_2O$	0.0528

A third alternative will be if we burn the purge stream and recycle the water as done before. The impact of that can be seen in table 27, where the burned purge has resulted in that the PCOP is improved where HTPI and TTP has slightly improved. Otherwise HTPE, ATP and GWP has got worse, which is explained by the burning process itself where more CO_2 will be vented to the atmosphere. Even though this will have an influence on global warming as seen on GWP and human toxicity by inhalation, it will still have lower quotes compared to sending the purge as waste containing e.g. methane that has a larger environmental emissions. The purge stream will be burned in the furnace as mentioned in chapter 11.1, also be used to heat the system up and thereby minimizing the utility cost for buying natural gas which is more profitable when considering sustainability. But when burning the purge one shall not forget that under realistic circumstances to try to have an absolute combustion to prevent harmful health and environmental substances such as NO_x and CO that are more harmful than CO_2. The chemical reactions can be seen below where the amounts of needed oxygen. The quantities of oxygen can be used to ensure that the right amount of air is sent to the furnace so an absolute combustion in the furnace can take place in further work if this shall be realized.

Because of the conflicting effects of operational costs, operability, sustainability, environmental impact, safety and hazards on the process performance, it is difficult to determine a process design that can simultaneously satisfy all these criteria. Although the last mentioned alternative with water recycle and purge burning seems to be the most sustainable one with a fine balance between the specified criteria determined by the systematic approach of chemical process design.

14 Conclusion and Future Work

A preliminary process design has in this work been developed using a systematic process synthesis approach. The final design utilizes 75 million kg/year ethylene feed and in order to obtain an ethanol production of 90.5 million kg/year (190 Proof 0.8545 %(mol/mol)). The plant have been designed and cost estimated for a possible at 10 % increase in production which enables an adjustable production rate depending on the demand. A complete equipment costing and economic evaluation have been performed indicating a profitable design already for the base-case design. A sizable reduction of the operating costs has been achieved through heat integration and process optimization yielding a 20 % reduction in total capital investment and a 15 % reduction in product cost without depreciation giving a rate of return of 77 % resulting in a total profit of 400 million US dollars during the 15 year project life time. A sensibility analysis show however high sensibility towards product and feed prices. A decrease in ethanol price of more 30 % will turn the plant non-profitable. Therefore a comprehensive market analysis considering ethanol price forecast must be performed before further project progress. At the bottom line the conclusion based in this work shows that the project is an investment worthy which enables the company to take up 4.5 % of the synthetic ethanol marked. The total capital investment is expected to be 35 million US dollars and the expected duration between investment and positive commulative cash flow is four years. Key performance indicators are displayed in table 29 revealing the improvements of the base-case design throughout the work.

Table 29: Key Performance Indicators.

Indicator	Base-case	Final
kg product/kg RM	0.25	0.25
kg purge/kg RM	0.02	0.02
kJ energy need/kg RM	6415	3295
kg water used/kg RM	0.72	0.72
kg recycle/kg RM	2.62	2.68
mol EA/mol EL	0.88	0.88

Attention for further work should be put on increasing the atomic efficiency since the plant profits are sensitive toward feed and product price changes. Potential improvements of the current molar conversion of 88 % meaning this percentage of the carbon atoms in ethylene are converted to ethanol through the reactor. Conversions above 90 % are often achieved in the chemical industry.

References

[1] *Ullmann's Encyclopedia of Industrial Chemistry*, chapter Ethanol. Wiley-VCH Verlag GmbH & Co. KGaA, 2007.

[2] Rafiqul Gani and Partners. Icas.

[3] Industrial ethanol, February 2011. http://www.ethanolmarket.com/industrialethanol.html.

[4] Hiroo Momose, Koshi Kusumoto, Yusuki Izumi, and Yukio Mizutani. Vapor-phase direct hydration of ethylene over zirconium tungstate catalyst. *Journal of Catalysis*, 77:23–31, 1982.

[5] L. T. Biegler, I. E. Grossmann, and A. W. Westerberg. *Systematic Methods of Chemical Process Design*. Prentice Hall PTR, 1997. ISBN 0-13-492422-3.

[6] Oecd sids initial assessment profile - ethylene. Technical report, inchem.org, 2008.

[7] *Ullmann's Encyclopedia of Industrial Chemistry*, chapter Chlorinated Hydrocarbons. 2006.

[8] ICIS. Ethanol pricing, April 2011. www.ICIS.com.

[9] Jason R. Kwiatkowski, Andrew J. McAloon, Frank Taylor, and David B. Johnston. Modeling the process and costs of fuel ethanol production by the corn dry-grind process. *Industrial Crops and Products*, 23:288–296, 2006.

[10] *Chemical Process Equipment: Selection and Designs*. Butterworth-Heinemann, 1990.

[11] Dowtherm g, March 2011.

[12] *Plant Design and Economic for Chemical Engineering*. McGraw-Hill Book Co., 1991.

[13] *Analysis, Synthesis, and Design of Chemical Process*. Prentice Hall, 1998.

[14] Industrial land cost in china, April 2011. http://www.made-in-china.com/showroom/doherty1598/product-detailVTzxRPUYFnqk/China-Industrial-Land-for-Sale.html.

[15] Tsoskounoglo M.; Ayerides G.; Tritopoulou E. The end of cheap oil: Current status and prospects. *Energy Policy*, 36:3797–3806, 2008.

[16] Union Gas Limited. Natural gas properties, May 2011. http://www.uniongas.com/aboutus/aboutng/composition.asp.

[17] Daniel Salerno, Harvey Arellano-Garcia, and Günter Wozny. Ethylene separation by feed-splitting from light gases. *10th International Symposium on Process Systems Engineering*, 2009.

15 Appendices

Appendix A: Task 4-6

Signal Flowgraph Based Technique

Table 30: Signal Flowgraph for identification of tear stream(s).

Recycle loops: 2 Partitions: 2 Tear streams: 1 (=stream 4)			
Nodes	Precursor (1)	Precursor (2)	Precursor (3)
1			
2			
3			
4	11,17	4	
5	4		
6	5		
7	6		
8	7		
9	7		
10	21		
11	10		
12	10		
13	21		
14	9,13	4	
15	14		
16	14		
17	15		
18	15		
21	8		

Linear Mass and Energy Balances

Table 31: Complete list of stream data for task 4-6 calculated by hand. $T_{ref} = 298.15K$.

Stream name	S1	S2	S3	S4	S5	S6	S7	S8	S9	S10
Component flow rate (kmol/h)										
Ethylene	288.111	0.000	0.000	3711.117	3451.338	3451.338	3451.338	3400.947	50.391	3331.184
Methane	6.065	0.000	0.000	720.179	720.179	720.179	720.179	717.390	2.788	713.529
Propylene	9.098	0.000	0.000	645.755	645.755	641.235	641.235	597.938	43.297	538.569
Water	0.000	2226.280	108.184	2226.677	1966.899	1962.379	1962.417	105.709	1856.708	0.400
Isopropyl alcohol	0.000	0.000	0.000	0.002	0.002	4.522	4.522	0.376	4.146	0.002
Ethyl alcohol	0.000	0.000	0.000	1.614	261.392	261.392	261.316	31.792	229.524	0.318
Diethyl ether	0.000	0.000	0.000	6.922	6.922	6.922	6.959	3.480	3.480	0.828
Total flow rate (kmol/h)	303.274	2226.280	108.184	7312.265	7052.487	7047.967	7047.967	4857.632	2190.334	4584.831
Temperature (K)	300.000	300.000	310.000	590.000	590.000	590.000	590.000	389.091	389.091	331.587
Pressure (bar)	1.000	1.000	68.000	69.000	69.000	69.000	69.000	68.500	68.500	68.000
Vapor fraction	1.000	0.000	0.000	1.000	1.000	1.000	1.000	1.000	0.000	1.000
Enthalpy (kcal/s)	984.217	-42309.173	-2050.029	-18322.265	-19132.613	-19148.005	-19148.059	8251.876	-37679.978	9198.869

S11	S12	S13	S14	S15	S16	S17	S18	S19	S20	S21
3302.869	28.315	69.763	120.154	120.154	0.000	120.154	0.000	0.000	0.000	3400.947
707.464	6.065	3.862	6.650	6.650	0.000	6.650	0.000	0.000	0.000	717.390
533.992	4.578	59.369	102.666	102.666	0.000	102.666	0.000	0.000	0.000	597.938
0.397	0.003	213.492	2070.200	207.020	1863.180	0.000	207.020	43.954	163.066	213.893
0.002	0.000	0.374	4.520	3.838	0.682	0.000	3.838	0.288	3.550	0.376
0.315	0.003	31.474	260.998	259.693	1.305	1.298	258.395	257.103	1.292	31.792
0.821	0.007	2.652	6.131	6.131	0.000	6.101	0.031	0.031	0.000	3.480
4545.860	38.971	380.986	2571.320	706.152	1865.168	236.869	469.284	301.376	167.908	4965.816
331.587	331.587	335.847	505.118	310.000	478.179	310.000	439.345	310.000	361.012	300.000
67.500	67.500	68.000	68.000	17.484	17.984	10.479	10.979	1.000	1.500	68.000
1.000	1.000	0.000	0.000	1.000	0.000	1.000	0.000	0.000	0.000	1.000
9120.678	78.190	-4228.334	-39824.894	-6991.445	-33368.225	411.144	-8175.339	-5554.250	-3137.529	5086.774

Table 32: Complete list of stream data for task 4-6 calculated by PRO/II. $T_{ref} = 0K$.

Name	S1	S2	S3	S4	S5	S6	S7	S8	S9	S10
Component flow rate (kmol/h)										
Ethylene	288.11	0	0	3709.47	3449.81	3449.81	3449.81	3398.06	51.7471	3326.7
Methane	6.06548	0	0	389.401	389.401	389.401	389.401	387.843	1.55761	385.904
Propylene	9.09822	0	0	470.745	470.745	467.45	467.45	435.664	31.7866	392.533
Water	0	2226.28	108.184	2226.71	1967.04	1963.75	1963.79	106.044	1857.74	0.42846
Isopryl alcohol	0	0	0	0.00135	0.00135	3.29657	3.29657	0.27362	3.02295	0.00137
Ethyl alcohol	0	0	0	1.61376	261.277	261.277	261.2	31.8664	229.334	0.31866
Diethyl ether	0	0	0	6.91018	6.91018	6.91018	6.94862	3.47431	3.47431	0.82689
Total flow rate (kmol/h)	303.274	2226.28	108.184	6804.85	6545.19	6541.89	6541.89	4363.23	2178.66	4106.72
Temperature (K)	310	310	310	332.313	590	590	590	401.931	524.975	331.587
Pressure (bar)	1	1	68	69	69	69	69	68.5	68.5	68
Phase	Vapor	Liquid	Liquid	Mixed	Vapor	Vapor	Vapor	Vapor	Liquid	Vapor
Enthalpy (kcal/s)	59.0267	411.358	19.9896	1989.47	13478.6	13476.2	13476.3	3016.33	3126.59	1411.51
S11	S12	S13	S14	S15	S16	S17	S18	S19	S20	S22
3298.43	28.277	71.3594	123.106	123.106	0	123.106	0	0	0	3398.06
382.624	3.28019	1.93922	3.49682	3.49682	0	3.49682	0	0	0	387.843
389.196	3.33653	43.1307	74.9173	74.9173	0	74.9173	0	0	0	435.664
0.42481	0.00364	213.8	2071.54	207.154	1864.39	0	207.154	43.9167	163.237	214.228
0.00136	1.2E-05	0.27225	3.2952	2.79762	0.49758	0	2.79762	0.20982	2.5878	0.27362
0.31596	0.00271	31.5477	260.881	259.577	1.30441	1.29788	258.279	256.988	1.29139	31.8664
0.81986	0.00703	2.64742	6.12173	6.12173	0	6.09113	0.03061	0.03061	0	3.47431
4071.81	34.9071	364.697	2543.36	677.171	1866.19	208.91	468.261	301.145	167.117	4471.41
331.587	331.587	331.587	498.426	310	478.18	310	439.345	310	361.012	400.697
68	68	68	68	17.484	17.984	10.479	10.979	1	1.5	68
Vapor	Vapor	Liquid	Mixed	Mixed	Liquid	Mixed	Mixed	Liquid	Liquid	Mixed
1399.52	11.9979	116.161	3242.75	154.366	1951.99	123.035	808.182	77.3691	75.3162	3036.32

Appendix B: PROII

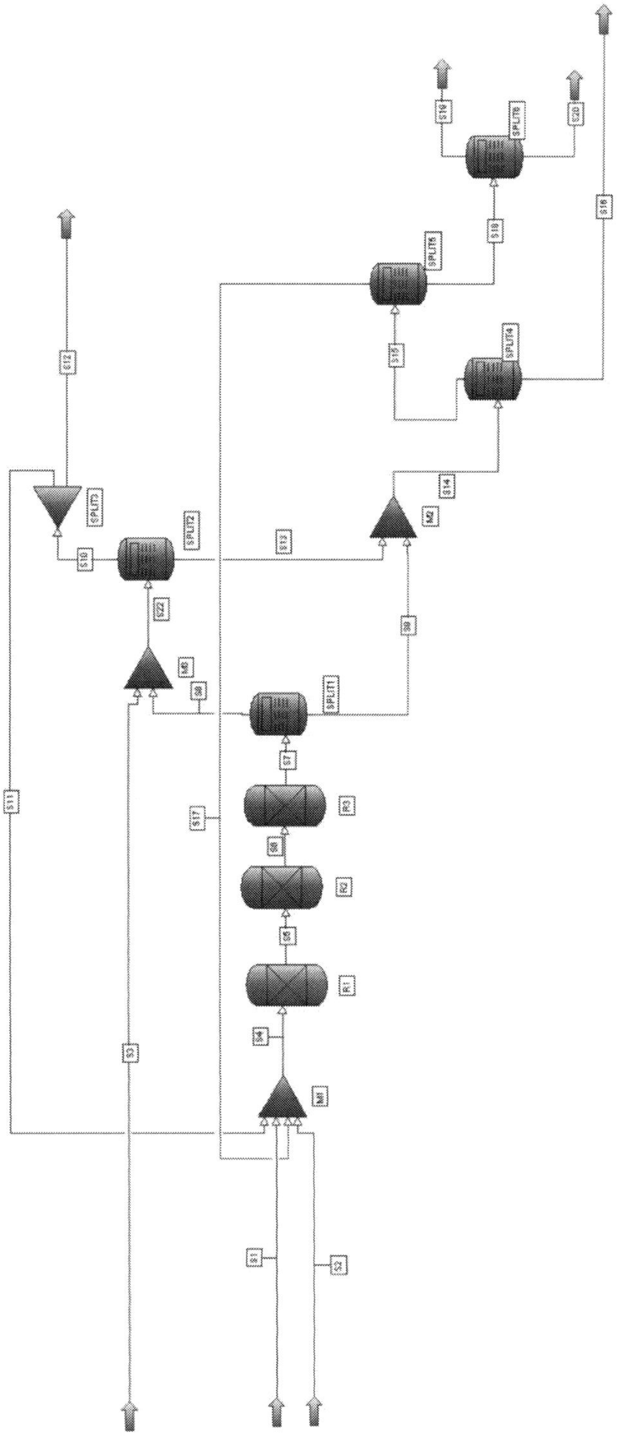
S1
S2
S3
S4
R1
R2
R3
M1
M2
S7
SPLIT1
SPLIT2
SPLIT3
SPLIT4
SPLIT5
S10
S11
S12
S13
S14
S15
S16
S17
S20
S22

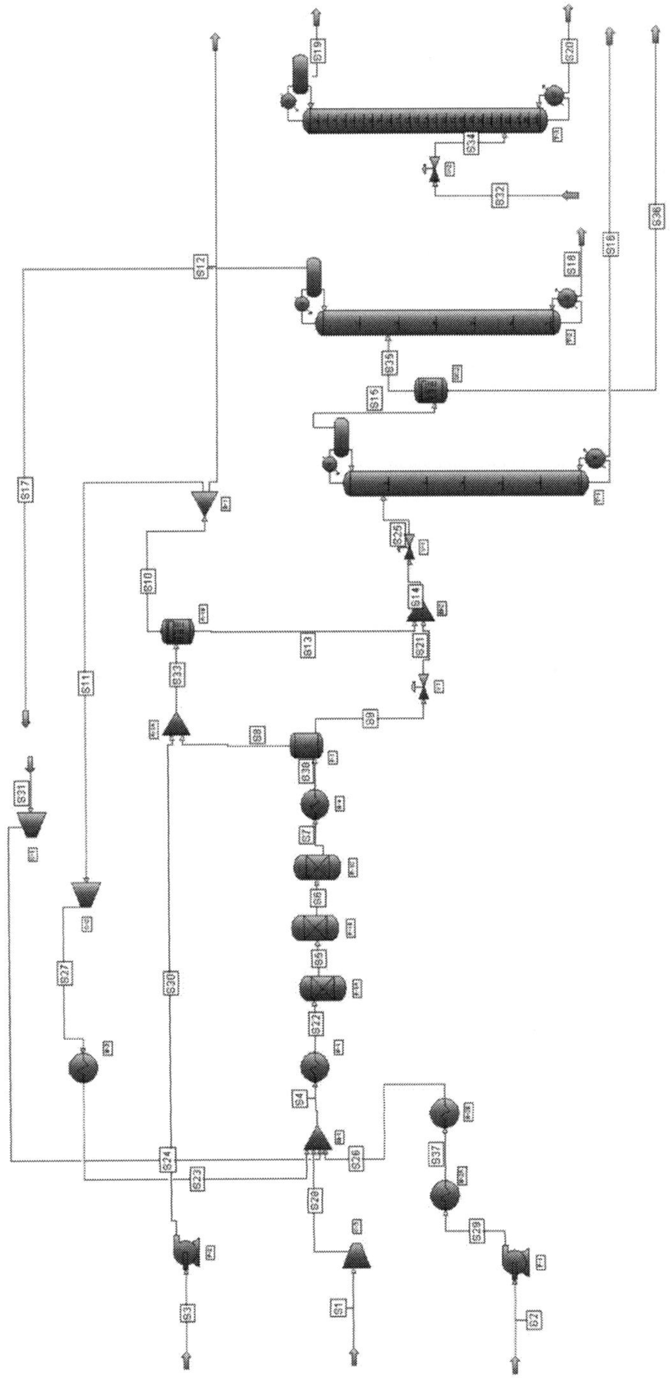

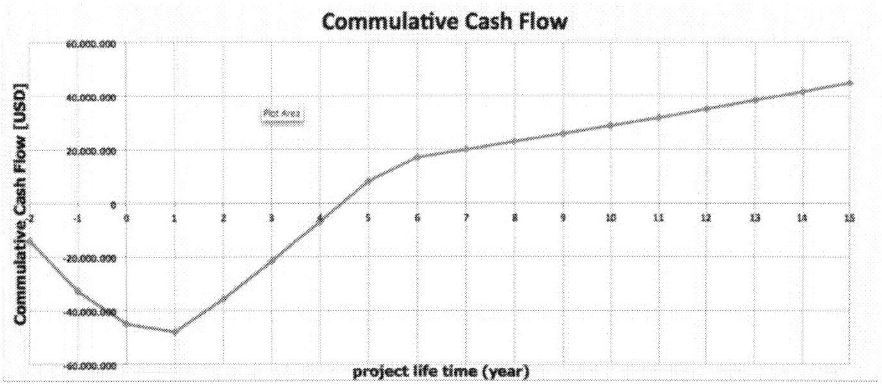

Figure 34: A commulative cash flow for the plant with 0.7 $/kg product sales price.

Appendix C: Economic Potential

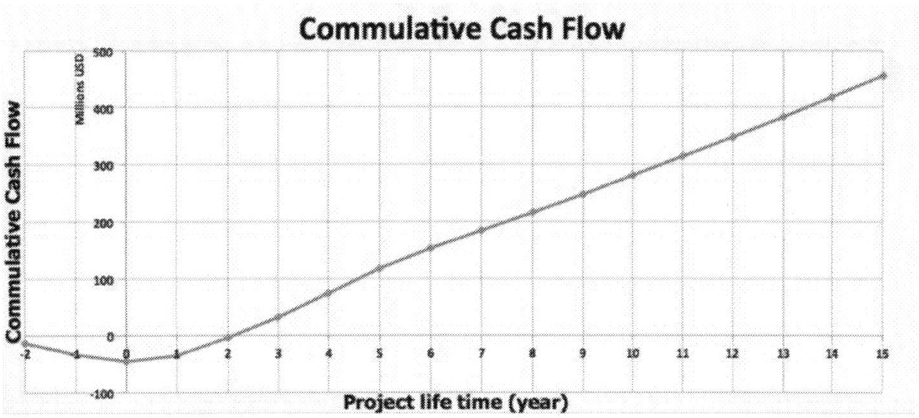

Figure 35: Commulative cash flow for a product price at 1.1 USD/kg

Appendix D: Heat Integration

Table 33: Streams including in the heat integration considerations.

Stream	T_{in} (K)	T_{out} (K)	FC_p (kW/K)
H1	590	480	76.85
HS1	480	420	348
H2	485	300	43
H3	375	300	3.5
H4	351	300	10.5
C1	502	590	95.9
C2	310	558	46.6
C3	310	500	75.7
Stream	T_{in} (K)	T_{out} (K)	Latent Heat (kW)
HS2	590	590	4222
CS1	558	558	16777

Values for FC_p for the multicomponent mixture condensing in the flash unit is estimated by simulating duties arising from stepwise decreasing the temperature in PRO II. The result is illustrated in figure 36.

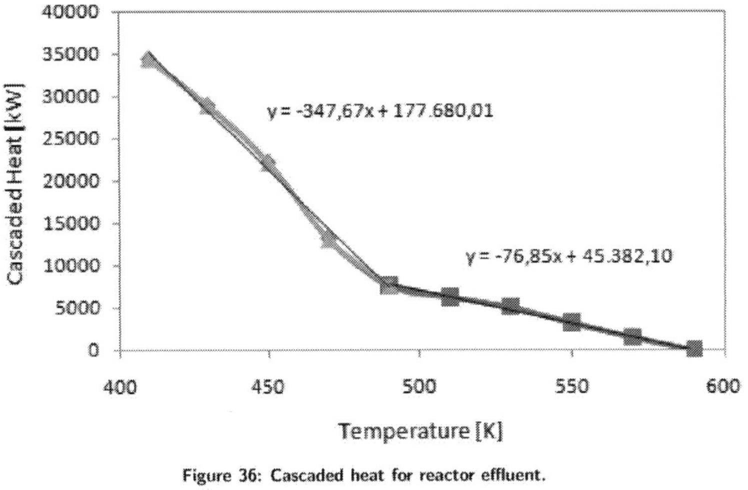

Figure 36: Cascaded heat for reactor effluent.

Table 34: Heat Diagram for Heat Integration with $\Delta T = 20$. All heat values denote FC_p in kW. Minimum hot utility 7363 kW, minimum cold utility 10465 kW and pinch point at 480/460 K (highlighted).

Heat Available	Cascade Heat	HOT (K)	COLD (K)	Heat Required	Cascade Heat	Net Heat	Cascade Net Heat	Final
								7363
	0	610	590		0		0	7363
0				1918		-1918		
	0	590	570		1918		-1918	5445
922				1151		-229		
	922	578	558		3069		-2147	5216
4304				7978		-3675		
	5226	522	502		11047		-5821	1542
154				93		61		
	5380	520	500		11140		-5761	1602
2690				4279		-1589		
	8069	485	465		15420		-7350	13
599				611		-13		
	8668	480	460		16031		-7363	**0**
23431				7336		16095		
	32099	420	400		23367		8732	16095
1928				5502		-3574		
	34027	375	355		28869		5159	12522
1113				2934		-1822		
	35140	351	331		31803		3337	10700
1195				2568		-1373		
	36334	330	310		34370		1964	9327
1138				0		1138		
	37472	310	290		34370		3102	***10465***

(620n) Design for a Medium to Small-Scale Hydrogen Liquefaction Plant

Young Kim, Junseok Ko, Yong-Ju Hong and Kong Hoon Lee, Energy Plant Research Division, Korea Institute of Machinery and Materials, Daejeon, South Korea

Hydrogen is currently drawing attention as a clean fuel producing less carbon dioxide. The transport and storage, however, require more energy and rather complex infrastructure than fossil fuels. Hydrogen may be stored as liquid, compressed gas, or hydrides.

Liquid hydrogen has relatively small density of 50~70 kg/m3, i.e., the volume of hydrogen shrinks less than other gas when it is transformed to a liquid. Moreover, it has low boiling point, and its spin conversion during liquefaction consumes a significant amount of energy. 20~30% of its energy content is required for transport and storage of hydrogen in the liquid state, while compression of hydrogen gas requires 15%. Although such features of hydrogen makes it more complicated to be liquefied, compression of hydrogen gas to 350~700 bar also has technological barriers to be realized. Compression and high pressure storage require larger facilities and heavier storage vessels. It also involves safety concerns, and the compression technology has not yet been matured enough. In this regard, the most economic state of hydrogen for transport and storage is likely to be liquid although the decision may vary according to technological developments.

A precooled Claude cycle and a He refrigerated cycle are investigated to develop a pilot scale hydrogen liquefaction system. The simulation results provide the specifications for component development, as well as the advantages of each type of liquefaction cycles.

See more of this Session: Poster Session
See more of this Group/Topical: Computing and Systems Technology Division

USE OF SURROGATE MODELS IN WATER NETWORK SYNTHESIS

José Eduardo A. Graciano, Galo A. C. Le Roux,

Introduction

The water is used in large quantities by the chemical industries, for being such a strategically resource it is the aim of severe restrictions imposed by environmental surveillance and control organisms, which should be followed strictly, causing a significant increase of its cost.

Several studies were developed with the aim of reducing water consumption through segregation and integration of wastewater streams. The water network problem was initially formulated by Takama [1]. After the initial problem formulation there were no major changes in the models for wastewater treatment processes synthesis. The research in this area focused mainly in the development of new methods for solving the NLP (nonlinear programming) or MINLP (mixer integer nonlinear programming) generated in the superstructure model formulated by Takama. Although it remained dormant, research in this area has raised up in the mid 90s, with the several works [2-5], and continues to increase steadily from 2000.

According to Jezówski [6] "Just a few studies considered the most detailed treatment process models, however these approaches are often limited to a single technology for treating and of a fixed type." In the opinion of Jezówski "the crux involves the application of more rigorous processes models: water use units, as well as, effluent process treatment." However, the use of phenomenological models in superstructure process synthesis could be unfeasible due to its high computational time.

Oil refineries are large consumers of water, producing wastewater with high concentrations of H2S, oils and salts, because of the direct contact processes between water and oil (e.g., desalting, stream stripping and many washing operations) [7].

The most used process for such wastewater treatment is the steam stripper, where the volatile compounds are removed by distillation of the heavy component, water. The working temperature of this type of equipment is usually high, close to the water bubble point. Due to the temperature, the steam stripping allows the removal of compounds heavier than air stripping processes do.

The steam stripper is a device that consists of a column with previously heated wastewater feed at the top, and with a steam feed at the bottom. The phase contact takes place down the column while the wastewater becomes leaner and the steam more enriched with contaminants respect [8].

The objective of this contribution is the synthesis of a water treatment system at petroleum refineries using a steam stripper surrogate model for this. This surrogate model includes equipment construction variables (number of stages and steam flow) and process variables (feed conditions: flow, temperature and contaminant concentrations). The data for the adjustment of the surrogate model are generated from process simulations in, ASPEN PLUS.

Methodology

The mixture to be treated contains H_2S, oils and salts. As a simplifying assumption, we adopted the n-hexane to represent the oils and used NaCl as the contaminant salt of the process. The electrolytes dissociation equations in aqueous medium, where the equilibrium constants are temperature function, are considered.

There are several thermodynamic models capable of predicting the thermodynamic behavior of liquid vapor mixtures, but the mixture to be treated is formed by a combination of electrolytes, which generate non-idealities in the system, requiring a thermodynamic model that takes into account their behavior. The model selected for this system was the electrolyte-NRTL [9].

A stripper can use direct steam at the plant boiler or a column bottom reboiler. Moreover, we can choose whether or not to integrate energetically the feed and output streams and to preheat the feed stream. Eight equipment variations are considered for the simulations.

It is usual to use the treatment cost as the objective function in the wastewater treatment synthesis so, the fixed and operating equipment cost are calculated using cost graphics updated by the Marshall & Swift's cost index in the third quarter of 2010 [10]. In the objective function, TAC (total annualized cost), the fixed cost is multiplied by an annualized factor, that divides the equipment fixed cost during its life cycle, taking into account the interest rate.

The data are obtained from process simulators with operating conditions varying. The input variables that are changed in the stripper operating ranges are the flow rate, the temperature and contaminants concentration in the feed stream, reflux ration, number of theoretical stages and the flow of steam. The data obtained from the simulations are normalized, and then used in the regression of a polynomial model linear (with respect to the parameters) for the following output variables: TAC, stripping factor for each component, temperature and flow of output stream.

The regression is performed using the least squares method. This complete model is then reduced using the PLS method (partial least squares regression). The validation of the surrogate model is made by cross-validation.

The model for the stripper is a MINLP composed by the mass balance of a mixer, stripper surrogate model and splitter. Binary variables are used to represent decision variables for the stripper type and their sum is imposed to be 1.

Conclusions

A steam stripper surrogate model was formulated which decreases the number of variables of the system, reduces the great sensitivity to the initial guess and its complexity, without affecting much its prediction in the specific range. In comparison to the usage of a phenomenological model there is a large economy in computing time in the water network synthesis, beyond possibility the implementation in an optimization platform such as GAMS.

References

[1] Takama N., "Optimal water allocation in a petroleum refinery", *Computers & Chemical Engineering* 4, n. 4 (1980): 251-258.

[2] Alvaargaez A., "Wastewater minimization of industrial systems using an integrated approach", *Computers & Chemical Engineering* 22 (1998): S741-S744.

[3] Galan B. e Grossmann I. E., "Optimal Design of Distributed Wastewater Treatment Networks", *Industrial & Engineering Chemistry Research* 37, n. 10 (1998): 4036-4048.

[4] Huang C. H. et al., "A Mathematical Programming Model for Water Usage and Treatment Network Design", *Industrial & Engineering Chemistry Research* 38, n. 7 (1999): 2666-2679.

[5] Bagajewicz M., "A review of recent design procedures for water networks in refineries and process plants*1", *Computers & Chemical Engineering* 24, n. 9-10 (2000): 2093-2113.

[6] Jeżowski J., "Review of Water Network Design Methods with Literature Annotations", *Industrial & Engineering Chemistry Research* 49, n. 10 (2010): 4475-4516.

[7] Wang Y.P. and Smith R., "Wastewater minimization", *Chemical Engineering Science* 49, n. 7 (1994): 981-1006.

[8] Zygula T. M., "Designing steam stripping columns for wastewater", *Periodic* 115, n. 5, Chemical engineering (May 2008): 52-57.

[9] Lee S. Y. , J. M. Lee, D. Lee, I. B. Lee, "Improvement in Steam Stripping of Sour water through an Industrial-Scale Simulation", *Korean J. of Chemical Engineering*, vol. 21, issue 3 (2004): 549-555.

[10] Peters M. S.. Plant design and economics for chemical engineers. 5 ed. New York: McGraw-Hill, 2003, 988p.

(620q) Simultaneous Environmental and Economic Optimization of Reverse Osmosis Plants Coupled with Solar Rankine Cycles

Gonzalo Guilln-Goslbez[1], Raquel Salcedo-Daz[2], Ekaterina Antipova[1], Dieter Boer[1] and Laureano Jimnez[1], (1)Department of Chemical Engineering, University Rovira i Virgili, Tarragona, Spain, (2) Department of Chemical Engineering, University of Alicante, Alicante

This work addresses the design of seawater desalination plants that integrate reverse osmosis, Rankine cycles and solar collectors. The design task is formulated as a multiobjective mixed-integer nonlinear programming problem (MINLP) that considers the simultaneous minimization of cost and environmental impact given a certain water demand to be fulfilled. The environmental performance is quantified following life cycle assessment (LCA) principles. A case study that considers weather data of Tarragona (Spain) is used to illustrate the capabilities of the approach. By applying the epsilon constraint method, a set of Pareto solutions representing the optimal trade-off between both conflicting objectives is generated. We show that coupling seawater desalination with solar collectors leads to both, environmental and economical benefits. The systematic tool presented herein is intended to guide decision-makers towards the adoption of more sustainable technological alternatives for seawater desalination.

See more of this Session: Poster Session
See more of this Group/Topical: Computing and Systems Technology Division

(620r) Combined Use of Simulation Packages, Multi-Objective Optimization and Statistical Tools for the Environmentally Conscious Design of Thermodynamic Cycles

Robert Brunet[1], Daniel Corts[1], Dieter Boer[2], **Gonzalo Guilln-Goslbez**[1] and Laureano Jimnez[1], (1) Departament d'Enginyeria Qumica, Universitat Rovira i Virgili, Tarragona, Spain, (2)Departament d'Enginyeria Mecnica, Universitat Rovira i Virgili, Tarragona, Spain

Energy consumption is one of the main factors affecting global warming. In the last years the installation of air conditioning systems has increased worldwide (Balaras et al. 2007, Henning 2007), leading to higher electricity consumption. Energy is usually produced in power plants, using mainly Rankine cycles, which consume fossil fuels. -On the other hand, conventional air conditioning systems are based on compression cycles requiring mechanical energy for its operation. Considering all these issues, there is a clear need to develop systematic tools to assist in their design according to environmental and economic concerns.

This work introduces a novel framework for the optimal design of sustainable thermodynamic cycles that integrates process simulation packages (i.e. Aspen Hysys and Aspen Plus), multi-objective optimization, life cycle assessment (LCA) and principal component analysis (PCA). The optimization problems are formulated using multi-objective mixed-integer linear programming (MINLP) techniques in which the LCA metrics are described through standard algebraic equations. A solution strategy is proposed to tackle these problems based on decomposing them into two levels: a nonlinear programming (NLP) subproblem, where the binary variables are fixed, and a master mixed-integer linear programming (MILP) problem that predicts new values for the binaries. The NLP subproblems are solved by combining an external NLP solver with process simulation packages (i.e., Aspen Plus), while the master problem is solved with standard MILP solvers (i.e., CPLEX). PCA is used in this context to identify redundant LCA metrics that can be omitted, thereby reducing the complexity of the model.

Two different thermodynamic cycles are used to highlight the capabilities of the approach presented: a 10 MW Rankine power cycle, and an ammonia-water absorption cooling cycle with 100 kW of cooling demand operating at cooling and refrigeration conditions. Our approach provides valuable insight into the design problem, and a set of alternatives for reducing the environmental impact in energy systems.

See more of this Session: Poster Session
See more of this Group/Topical: Computing and Systems Technology Division

(620s) Effect of Salt Addition On Extractive Distillation Process

Russell Cruz[1], **Nelly Ramrez**[1], Luis Ros[1] and Arturo Jimnez[2], (1)Departamento de Ingeniera Qumica, Alimentos y Ambiental, Universidad de las Amricas Puebla, Cholula,Puebla, Mexico, (2)Departamento de Ingeniera Qumica, Instituto Tecnolgico de Celaya, Celaya, Gto., Mexico

In the last decade, anhydrous ethanol production has been acquiring an increasing industrial importance due to its growing demand as a renewable fuel. Most bioethanol industrial production is based on the fermentation process, from which a maximal yield is about 15% (v/v), a reason why the product purification is an important stage for the process. The ethanol-water separation is the energetic bottleneck for bioethanol production because of the presence of the azeotrope, which makes it necessary the use of alternative techniques to common distillation to achieve the required purity. Extractive distillation using liquid entrainers for the separation of homogeneous azeotropic mixtures is a widely accepted technique in the industrial environment; however, in most cases its implementation requires a high demand on solvent feed flow. On the other hand, from several experimental researches it is known that salt addition can modify activity coefficients, even breaking the azeotropic points, but their industrial implementation has been rejected because of problems associate to the potential salt precipitation. In this work we explore the implementation of a hybrid extractive process combining a liquid entrainer (ethylene glycol) and a salt, in this case $CaCl_2$. The analysis was conducted using the Aspen Plus process simulator. In order to identify the convenience of using an ethanol pre-concentrator column in this process, different feed compositions for the ionic-extractive distillation were analyzed. According to the obtained results, in most cases there is not an important reduction on energy consumption (3 to 5%), but the combination of both entrainers allows reducing significantly the use of liquid solvent, even more than 50% in comparison with the conventional process. As in conventional extractive distillation, in this hybrid process the minimum solvent and energy requirements are reached when the mixture is fed close to the azeotropic composition.

See more of this Session: Poster Session
See more of this Group/Topical: Computing and Systems Technology Division

(620t) Dynamic Behavior of a Thermally Coupled Distillation Column Implemented On a Reactor-Separation-Recycle System

Daniel Mascote[1], **Nelly Ramrez**[1] and Arturo Jimnez[2], (1)Departamento de Ingeniera Qumica, Alimentos y Ambiental, Universidad de las Amricas Puebla, Cholula,Puebla, Mexico, (2)Departamento de Ingeniera Qumica, Instituto Tecnolgico de Celaya, Celaya, Gto., Mexico

Nowadays, the installation and operation of systems with minimum energy and material consumption is a primary objective in the industrial environment; this goal can be achieved through the implementation of recycle streams to recover raw materials and the use of energy integration techniques to minimize the utilities requirements. However, It is well known that processes with recycle streams can present several control problems when the control structure is not properly selected, because of loops interactions and potential snowball effects. Several researchers have analyzed the dynamic and control properties for reactor-separator-recycle systems using flash units and conventional distillation columns in the separation stage. This work presents the dynamic behavior of a thermally coupled distillation sequence implemented in RSR system. In particular, the implementation of the coupled system with side rectifier is considered. The analysis was conducted using the Aspen Dynamics process simulator and the dynamic responses of the thermally coupled system are compared to those generated by the use of a conventional distillation sequence. The plantwide control strategy was developed using the Luyben's rule for the recycle streams, and the control configuration in the distillation columns was implemented taking into account the flow magnitude in the product streams; the L-B configuration was used for the thermally coupled arrangement, and the L-V configuration for the conventional sequence. A simplified process for ethylbenzene production was taken as a basis. The results obtained showed that selected control structure for the thermally coupled arrangement can ensure the process stability for inventory control against disturbances on feed streams in the RSR system. Despite that dynamic responses presented by this complex arrangement are not as smooth as those presented by the conventional scheme, it is shown that their implementation in a process with recycles streams can be feasible.

See more of this Session: Poster Session
See more of this Group/Topical: Computing and Systems Technology Division

(620u) Model Based Optimization of Bioethanol Production From Lignocellulosic Biomass

Federico Andersen, Susana Moreno and **Maria Soledad Diaz**, Chemical Engineering, Planta Piloto de Ingenieria Quimica (PLAPIQUI), Universidad Nacional del Sur - CONICET, Bahia Blanca, Argentina

Lignocellulosic biomass is largely available for the production of bioethanol; it is relatively inexpensive and does not compete with food. Much effort is being spent on the development of process schemes that can be economically and environmentally attractive. The production of bioethanol from lignocellulosic biomass through a biochemical route, requires a pretreatment stage, whose goal is to alter lignin and hemicellulose structure to make cellulose more accessible to hydrolysis. This step includes reduction of the sample size, breakdown of hemicellulose to sugars, and disclosure of cellulose to be in turn hydrolyzed by enzymes into glucose using microbial cellulases (saccharification). Following pretreatment and hydrolysis, the sugars extracted from cellulose and hemicellulose materials can be converted to ethanol through microbial fermentation.

In this work, we propose a model based optimization strategy for the production of bioethanol from lignocelllulosic biomass, through a continuous process. The objective is to determine optimal operation and design variables that maximizes ethanol production. A unique feature in this work is the integration of kinetic models for the pre-treatment stage (dilute acid hydrolysis) to saccharification (enzymatic hydrolysis) and simultaneous fermentation (co-fermentation) models. Although several pretreatment techniques exist,[1,2] the dilute acid hydrolysis is selected due to its simplicity, effectiveness, and economic convenience, as compared to other pretreatment methods. Kinetic equations describing hemicellulose polymers (xylan, arabian) and a small fraction of glucan decomposition have been included. At this stage, the main sugar decomposition products are furfural and hydroxymetil furfural. Furthermore, an additional operation of detoxification is required after pretreatment because fermentation inhibitory compounds, mainly furfural and hydroximetyl furfural, are produced in the acid hydrolysis. A kinetic model for detoxification of acid hydrolyzates by the addition of $Ca(OH)_2$ (overliming)[3,4] is included. As in this latter stage not only furans degradation takes place but also sugar consumption[3], a minimal sugar recovery has been imposed, while maximizing loss of inhibitors.

The problem has been formulated as a non-linear programming (NLP) model. Key variables are: equipment volumes, acid and enzyme concentrations, operating temperatures, and water consumption. Detailed kinetic equations[3,5,6,7] and mass balances have been formulated in GAMS. Numerical results show the need for integrated models development for the optimization of the entire biochemical route to optimize bioethanol production and to be able to later minimize energy consumption.

1- Kumar et al., 2009. Methods for pretreatment of lignocellulosic biomass for efficient hydrolysis and biofuel production. Ind. Eng. Chem. Res. 48, 3713 – 3729.

2- Taherzadeh and Karimi, 2008. Pretreatment of lignocellulosic wastes to improve ethanol and biogas production: A review. Int. J. Mol. Sci. 9, 1621 – 1651.

3- Purwadi et al., 2004. Kinetic study of detoxification of dilute-acid hydrolyzates by $Ca(OH)_2$. Journal of Biotechnology 114, 187 – 198.

4- Aden et al., 2002. Lignocellulosic Biomass to Ethanol Process Design and Economics Utilizing Co-Current Dilute Acid Prehydrolysis and Enzymatic Hydrolysis for Corn Stover. NREL / TP-510-32438.

5- Lavarack et al., 2002. The acid hydrolysis of sugarcane bagasse hemicelluloses to produce xylose, arabinose, glucose and other products. Biomass and Bioenergy 23, 367 – 380.

6- Kadam et al., 2004. Development and validation of a kinetic model for enzymatic saccharification of lignocellulosic Biomass. Biotechnology Progr. 20, 698 – 705.

7- Leksawasdi et al., 2001. Mathematical modeling of ethanol production from glucose/xylose mixtures by recombinant *Zymomonas mobilis*. Biotechnology Letters 23, 1087 – 1093.

See more of this Session: Poster Session
See more of this Group/Topical: Computing and Systems Technology Division

(620v) Eutrophication Control In Lakes and Reservoirs Through Integrated 3D Ecological and Hydrodynamic Models

Sabrina Belen Rodriguez Reartes, Vanina Estrada and **Maria Soledad Diaz**, Chemical Engineering, Planta Piloto de Ingenieria Quimica (PLAPIQUI), Universidad Nacional del Sur - CONICET, Bahia Blanca, Argentina

Eutrophication in most water bodies constitutes an important environmental problem. It is caused by increasing nutrient loading associated to urban, industrial and agricultural activities. Eutrophication is associated to accelerated growth of phytoplankton (algal blooms). The presence of certain species of cyanobacteria within the blooms is of special care, as they may produce hepato- and neurotoxins that can severely compromise human and animal health. To address nutrient point sources, such as urban and industrial, much effort has been devoted to the development of wastewater treatment processes and facilities (Gernaey et al., 2004; Karuppiah and Grossmann, 2008). However, nonpoint nutrient sources, such as those associated to agricultural activities, have not received much attention (Estrada et al., 2009). In particular, increased eutrophication can be a key feature associated to the large-scale production of biofuels from energy crops when compared to fossil fuels. The life cycle wide emissions of nutrients depend on the application and losses of fertilisers during the agricultural production of biofuel feedstocks. External restoration techniques for nonpoint sources include the construction of artificial wetlands nearby the water bodies to remove nutrients and, in this way, decrease nutrient loading. But the application of external restoration techniques is not enough to control algae growth in the middle term. Within in-lake restoration strategies, biomanipulation is based on the trophic chain theory and it has been applied to control phytoplankton growth in lakes and reservoirs. The basic idea is to keep a high grazing pressure on the phytoplankton community by the herbivore zooplankton by performing zooplanktivorous fish removal (Sondegaard et al., 2007). Another in-lake restoration strategy is hypolimniom aeration, so as to precipitate phosphorous compounds to the sediment, not becoming available for phytoplankton production. Three dimensional eutrophication models have been proposed for large lakes with horizontal uneven distributions (Hu et al., 2006) and for costal systems (Moll and Radach, 2003).

In this work, we formulate a three-dimensional eutrophication model, based on a previous one-dimensional model (Estrada et al.; 2010) for a reservoir providing drinking water to two cities. Continuity and momentum equations have been formulated for the lake, considering a stretching vertical coordinate that allows for spatially and temporally varying depth. The model is based on first principles, with ecological parameters that have been tuned with collected data from the specific reservoir under study (Estrada et al., 2009). Dynamic mass balances for three phytoplankton groups and main nutrients have also been included. The distributed parameter model is formulated within a control vector parametrization dynamic optimization framework. Numerical results provide quantitative information for planning restoration strategies over a one year horizon, as well as its effect on algae growth and nutrient concentration.

References

V. Estrada, E. Parodi, M.S. Diaz, 2009, Determination of biogeochemical parameters in eutrophication models as large scale dynamic parameter estimation problems, Computers and Chemical Engineering, 33, 1760-1769.

V. Estrada, M.S. Diaz, 2010, Global sensitivity analysis and dynamic parameter estimation in eutrophication models, Environmental Modelling and Software, 2010, 25,1539-1551.

K.V. Gernaey, M. van Loosdrecht, M. Henze, M. Lind, S. B. Jørgensen, 2004, Activated sludge wastewater treatment plant modelling and simulation: state of the art, Environmental Modelling & Software 19, 763–783.

W. Hu, Jørgensen, S. E. and Zhang, F., 2006. A vertical-compressed three-dimensional ecological model in Lake Taihu, China. Ecol. Model. 190, 367-398

R. Karuppiah, I.E. Grossmann, 2008, Global optimization of multiscenario mixed integer nonlinear programming models arising in the synthesis of integratedwater networks under uncertainty. Computers&Chemical Engineering, 32, 145–216.

A. Moll, G. Radach, 2003, Review of three-dimensional ecological modelling related to the North Sea shelf system. Part 1: models and their results, Progress in Oceanography 57, 175–217.

E. Parodi, V. Estrada, N. Trobbiani, G. Argañaraz Bonini, 2004, Análisis del estado trófico del Embalse Paso de las Piedras. Ecología en tiempos de Cambio. 178..

PSEnterprise, 2009, gPROMS User guide.

M. Søndergaard, E. Jeppesen, T. L. Lauridsen, C. Skov, E. H. van Nes, R. Roijackers, E. Lammens, and R. Portielje, 2007, Lake restoration: successes, failures and long-term effects. J. Appl. Ecol., 44, 1095–1105.

See more of this Session: Poster Session
See more of this Group/Topical: Computing and Systems Technology Division

(620w) Quantum-Level Descriptors In Computational Molecular Design

Qi Chen[1], Karen D. Camarda[2], Kenneth Bishop[1] and Kyle V. Camarda[3], (1)Chemical and Petroleum Engineering, The University of Kansas, Lawrence, KS, (2)Physics and Astronomy, Washburn University, Topeka, KS, (3)Chemical and Petroleum Engineering, University of Kansas, Lawrence, KS

A fundamental need in chemical engineering is the ability to predict the properties of small molecules *a priori*, purely from 2-D structure. While group contributions are effective for some properties, other properties require knowledge of the topology of the molecule, as well as a quantum-level description of the bonding environments around each atom. However, current quantum-level simulations require a great deal of time to arrive at acceptable accuracies. By developing links between quantum-level simulation information and linearized property prediction models, quantum-level molecular information can be incorporated into optimization-based design algorithms for novel molecules. In this project, parallel quantum simulations were performed for a representative set of small molecules, which span the combinatorial space of interest for a solvent design problem. The results of the quantum computations were then used compute numerical descriptors, which were next correlated with sets of experimentally measured property values for properties such as solubility, toxicity, viscosity, and melting point. The correlations were then combined with a set of structural constraints to form an optimization framework for molecular design. This optimization problem, an MINLP with some black-box constraints, can be solved to near-optimality by stochastic methods, including the Tabu search algorithm applied in this work. Examples are shown which present successful designs for both solvent and ionic liquid design tasks.

See more of this Session: Poster Session
See more of this Group/Topical: Computing and Systems Technology Division

(620x) Characterization Based Molecular Design of Biofuel Additives for Feedstock Flexibility

Subin Hada, Charles C. Solvason and Mario Richard Eden, Department of Chemical Engineering, Auburn University, Auburn, AL

Any type of feedstock that contains free fatty acids and/or triglycerides such as vegetable oil, waste oil, animal fat, and waste grease can be converted into biodiesel. Because of its biological origin, biodiesel has many advantages compared to its petroleum derived counterparts. However, it does exhibit several limitations like oxidative stability and low temperature operability that have impaired its wide use and commercialization. Simultaneous solution of these problems has proven difficult as improvements in one area tend to impair another. As a result, design of an optimal fuel for all environments can be a quite difficult task. In addition, feedstock origin will impact the final quality of the biodiesel product. Therefore, to produce fuel grade biodiesel, the characteristics of the feedstock are very important during the initial research and production stage.

Several approaches are possible for improving the fuel properties of biodiesel. Modification of the fatty acid composition through physical processes or uses of additives are the most prevalent. Fuel additives, such as antioxidants, cetane enhancers, or cold-flow improvers have become indispensable tools not only to alleviate the drawbacks described above, but also to assure that any fuel blend will meet international and regional standards regardless of origin. However, an additive solution to one problem often aggravates another problem. Furthermore, the questions of additive compatibility, required additive levels, the effect on other properties, and whether these additives function as designed for biodiesel fuels with differing fatty acid profiles still remain challenges and that need further investigation.

Therefore, it is desired to molecularly design biodiesel additives that simultaneously account for the unintended effect on other fuel properties in the neat and the blend fuel in order to achieve the performance properties of the petroleum based fuel. In this way, biofuels can be formulated that are adaptable to a range or blend of feedstocks and the desirable fuel characteristics like oxidative stability and wide operating temperature range.

In this work, we aim to identify all possible compounds which, when added to off spec biodiesel, results in a fuel that satisfies performance standards such as ASTM D6751 in the US and EN14214 in Europe. To meet this end, multivariate characterization data obtained from IR spectroscopy of common additives were combined with decomposition and property clustering techniques in a reverse problem formulation. In this approach, the fuel additive property targets are identified in the first reverse problem followed by molecular design to match the targets. The characterization data consists of multitude of properties of interest (such as cetane number, melting point, and kinematic viscosity) to ensure adequate performance. To facilitate an efficient design we consolidated these various properties into a latent property domain using principle component analysis (PCA) techniques. Finally, characterization based molecular design using group contribution parameters are then used to build novel additives that match the fuel specifications in the latent property space.

See more of this Session: Poster Session
See more of this Group/Topical: Computing and Systems Technology Division

(620y) Group Contribution Based Process Synthesis and Design

Susilpa Bommareddy and Mario Richard Eden, Department of Chemical Engineering, Auburn University, Auburn, AL

There are many different approaches to process synthesis including expert systems, optimization or algorithmic methods, and conceptual methods based on physical insights. This paper highlights a novel hybrid method for Computer Aided Flowsheet Design (CAFD) that combines physical insights with algorithmic reverse design approaches to enable systematic identification of feasible flowsheets at significantly reduced computational expense.

The framework presented in this paper is based on the process group (PG) contribution approach developed by d'Anterroches and Gani (2005). The CAFD approach was inspired by the group contribution based methods for Computer Aided Molecular Design (CAMD), which includes building blocks (atoms and functional groups) to generate and represent molecules; group contribution (GC) based property models to predict target properties; a standard molecular structure notation system (such as SMILES) to store and visualize the molecular structure information; and a synthesis method to generate and screen molecules that match the target (design) properties. Analogous to CAMD, in the CAFD approach, flowsheets are generated and represented by functional process groups; process group contribution based property models are employed to predict flowsheet properties; a notation system (called SFILES) is used for storing the flowsheet structural information; and a synthesis method is used to generate and identify the feasible flowsheets.

Like functional groups in molecules that are characterized by atoms and their molecular weight, each process group is characterized by the type of unit operation/process and their corresponding driving force. Each PG contributes to the flowsheet (performance) properties, which can be calculated once a feasible configuration has been identified. The candidate flowsheets are ranked based on performance criteria like energy consumption, amount (mass) of external agents used and/or cost/profit. Once a set of near optimal flowsheet alternatives have been identified, rigorous simulation is used to verify the predicted performance and select the best option.

See more of this Session: Poster Session
See more of this Group/Topical: Computing and Systems Technology Division

(620z) Modeling and Optimization of a Multiple Tube Solar Receiver for High Temperature Solar-Thermal Processes

Janna Martinek[1], Carl Bingham[2] and Alan W. Weimer[1], (1)Department of Chemical and Biological Engineering, University of Colorado, Boulder, CO, (2)National Renewable Energy Laboratory, Golden, CO

A three-dimensional steady state computational fluid dynamics model is developed for a solar receiver consisting of a reflective cavity with a windowed aperture enclosing an array of five tubes. The computational model couples radiative transfer with fluid flow, heat transfer, mass transfer, and chemical reaction kinetics. Radiation heat transfer is included via a combination of ray tracing, Monte Carlo, and finite volume methods. Ray trace modeling of the concentrating system provides the magnitude and direction of the solar energy incident on the receiver window surface. Transport of solar radiation in the cavity is decoupled from all other transport processes occurring in the receiver and profiles of the absorbed solar radiation are determined via a Monte Carlo model using only the receiver geometry, magnitude/direction of radiation incident on the window, and spectral/directional optical properties. A finite volume model is implemented in conjunction with the overall CFD model to account for thermal radiation emitted by heated surfaces in the receiver. Absorption coefficients, scattering coefficients, and the scattering phase function for the cloud of particles in each tube are determined via Mie theory. Transport of the aerosol particles is modeled using a single fluid mixture model with a population balance including transport by convection, Brownian motion, and thermophoretic diffusion. Gasification of 40nm carbon particles is used as a test reaction with Arrhenius kinetic parameters from the literature. Temperature distributions and, correspondingly, reaction conversions among the five tubes are found to be considerably non-uniform, with large temperature gradients developing between the front and back surfaces of each tube. Validation of the base heat transfer and reaction models is accomplished using experimental data taken on-sun at the High Flux Solar Furnace (HFSF) at the National Renewable Energy Laboratory (NREL) with both inert and reacting materials reaching temperatures up to 1400°C.

This model is used to investigate the impact of factors describing receiver configuration and operating conditions on overall receiver performance. Thirteen individual factors are considered with four factors describing operating conditions (gas flow rate, carbon feed rate, solar power input, and steam to carbon ratio) and nine two-level factors describing receiver geometry. The geometric factors are defined relative to a base design in which all tubes are arranged in a semicircle around the back cavity wall and allow for variations in cavity size, and number, radius, and arrangement of tubes with both staggered and non-staggered arrangements considered along with factors allowing for offset of all or part of the tube semicircle toward the window. Designs are evaluated in the framework of a 2^{13-8} fractional factorial design on the basis of receiver solar to chemical efficiency and reaction conversion for both absorbing and reflective cavity walls.

See more of this Session: Poster Session
See more of this Group/Topical: Computing and Systems Technology Division

(621a) Simultaneous Linear and Nonlinear Approach to Solving Complex Mathematical Models In Chemical Engineering

Sreenivas Vemulapalli, Cell for Industrial Safety and Risk Analysis, Chemical Engineering Department, Chennai, India

Solving complex mathematical models and optimization functions during design, modeling and simulation of chemical engineering problems routinely involves finding solutions to nonlinear equations and adopting linear programming techniques. Nonlinear algebaric equation solvers attempt to find solutions to sets of nonlinear equations by linearizing the function using gradient functions and making some simplifying assumptions. This approach often results in compromising on the robustness or efficiency of the algorithm and other computational problems. Therefore, an alternative approach is being proposed which transforms the original mathematical functions into its logarthmic linear (log-linear) representation. This technique can be applied while trying to solve both linear as well as nonlinear equations that are encountered in chemical engineering mathematics.

See more of this Session: Poster Session Applied Mathematics and Numerical Analysis
See more of this Group/Topical: Computing and Systems Technology Division

THREE-DIMENSIONAL MICRO-SCALE MODEL FOR CONVECTIVE DRYING

Abdolreza Kharaghani[1], Christoph Kirsch[2], Thomas Metzger[1] and Evangelos Tsotsas[1]

[1]Otto von Guericke University, Magdeburg, Germany
[2]University of North Carolina at Chapel Hill, USA

Abstract

A micro-scale model for convective drying is presented and used to simulate the liquid distribution over time in a model aggregate made from glass beads. For the same model system, experimental phase distributions during drying are obtained from X-ray microtomography measurements. The simulated liquid phase distributions are qualitatively in good agreement with the experimental result. An extended version of this drying model shall help to understand the micro-scale phenomena during the convective drying of gels, such as cracks and shrinkage.

Introduction

The sol-gel route can produce highly porous solid aggregates with large surface area, which are usually dried under supercritical conditions in order to preserve the structure. These aerogels can be used, for example, as lightweight insulators or in supercapacitors. Convective drying is not yet suited to produce such dry gels: the liquid-gas interface causes high capillary stress in the small pores, which results in mechanical damage of the commonly very fragile solid. However, a better understanding of the micro-scale phenomena inducing shrinkage and cracks during convective drying is expected to extend the range of applications for this inexpensive and safe process [1].

Several attempts have been made so far in order to understand drying characteristics of porous aggregates bringing forward various modeling approaches at different length scales. At the effective medium length scale, models based on volume averaging have been developed to describe the macroscopic behavior of porous media during drying by spatially averaged quantities [2]. Although macro-scale information such as the moisture distribution and temperature within the material can be resolved by these models, they do not provide spatially local information about the micro-scale phenomena in the medium.

In an attempt to retain this information, some authors have investigated pore-level effects arising during convective drying of particle aggregates, and gels in particular. For example, Brinker and Scherer [3] have done extensive studies on the driving forces for undesired effects such as shrinkage and cracks. In [3], it is well understood that the liquid phase distribution in a medium is responsible for the damage, but it is not modeled at the pore level. A well established approach that takes into account the pore-level phenomena in simulations is the use of pore network models [4]. In these models, the void space of a porous medium is represented by a network of pores, and all relevant transport phenomena are modeled directly at the pore level. Recently, a 3D pore network model has been combined with a discrete element method (DEM) to describe the micro-mechanical behavior of numerically generated dense particle aggregates [5]. These network models are based on cylindrical pores with liquid flow in the axial direction. Therefore, they are not suited to approximate the void space of highly porous particle aggregates, where this assumption cannot be justified.

Several numerical techniques have been developed, which describe the evolution of interfaces in pores of arbitrary shape and size. For example, the volume-of-fluid (VOF) method [6] is widely used. This method employs a stationary discretization of the domain; the spatial distribution of the phases is described by time-dependent volume fractions f defined in each mesh cell. As the mesh size $h \to 0$, the cell volume fractions for each phase converge to the characteristic function of the phase domain. The interface between two immiscible fluids is obviously located at the discontinuity of this characteristic function; therefore, it is located for $h>0$ in the set of mesh cells, in which the phase volume fractions satisfy $0<f<1$. This yields an implicit representation of the interface, and therefore topological changes over time, such as splitting or merging, are easily handled by the VOF method.

The shape and exact location of the interfaces are determined from the phase volume fractions using interface reconstruction methods such as SLIC (simple line interface calculation), PLIC (piecewise linear interface calculation), or Young's PLIC method with varying degree of interface smearing (see [7], for example).

In the present work, a 3D micro-scale model for convective drying is proposed, which resolves the three phases (solid, liquid, and gas) as well as the interfaces between them. The solid phase is assumed to be given in a discrete way as an aggregate of rigid primary particles sintered at the mutual contacts. The diameter of the primary particles also determines the characteristic length scale of the problem, which is small enough such that gravitational forces can be neglected as compared to surface tension forces (~10^{-8} m for the solid matrix in a gel). A continuum description is used for the fluid phases. The spatial distribution of all three phases is described by volume fractions defined on a stationary grid of cubic cells, in a VOF-like approach.

For a slow drying process, the scale separation in time justifies the use of a quasi-steady state approach, where the solid and liquid phases are allowed to relax to a hydrostatic equilibrium during evaporation:

- *large time scale:* liquid volume fractions in liquid-gas interface cells are reduced according to local evaporation rates, which may be determined from the solution of an auxiliary vapor diffusion problem in the gas phase.
- *small time scale:* liquid redistribution governed by volume conserving mean curvature flow, with an additional contact angle condition at three-phase contact lines. This is different from the classical VOF method, where the interface velocity is usually determined from the solution of the incompressible Navier-Stokes equations in the bulk liquid. The presumably computationally cheaper approach of mean curvature flow was chosen because the fluid dynamics at the small time scale is not of interest here, but

only the liquid distribution after relaxation. For any given liquid distribution, the capillary forces acting on the primary particles of the solid phase can be computed by numerical integration of the liquid pressure and surface tension. The discrete element method shall be used to simulate the motion of the solid phase as the primary particles are accelerated by these forces.

Aside from numerical simulations, non-invasive experimental techniques such as laser induced fluorescence [8], magnetic resonance imaging [9], and X-ray microtomography [10] have been used to investigate drying characteristics of porous media.

In the following sections, numerical results of the liquid distribution over time shall be presented, together with experimental data from X-ray microtomography measurements performed on highly porous glass bead assemblies.

X-ray microtomography (XMT) for particle aggregates

Highly porous model aggregates were produced by sintering glass beads in a graphite mold. The sintering was done in an oven at 600 °C for 7 minutes. The sintered particle structures were used to create highly porous packings in a cylindrical tube made of PMMA, with a diameter of 8 mm.

XMT measurements were done with the cone beam X-ray tomography scanner CT-ALPHA (manufactured by ProCon, Germany). The X-ray source of this device operates at up to 160 kV with up to 15 W power. Its detector is a 2300×2300 pixel sensor with a 16 bit grayscale resolution. The maximum spatial resolution is < 1 μm. For the measurement, the sample is placed on a rotation holder between the X-ray source and the detector. X-rays are attenuated as they pass through the sample; the extent of this attenuation is proportional to the local density, and this yields a gray-level transmission image on the detector side. Images obtained from different angles are used for the reconstruction of a three-dimensional map of the attenuation coefficient within the sample, which then provides insight into its microstructure.

Image acquisition and processing

For the image acquisition by XMT, X-ray parameters such as voltage, filter parameters and exposure time were adjusted with reference to the systems being analyzed, in order to obtain the best possible image contrast. The images were acquired at 90 kV and 160 mA, and 800 different projections were considered in a full scan. The tube was covered by a layer of aluminum of 1 mm thickness in order to reduce beam hardening effects.

In addition to a suitable selection of parameters during image acquisition, extensive corrections during image reconstruction were also necessary so as to minimize a variety of artifacts, including misalignment of the sample rotation axis, beam hardening, motion and ring artifacts. A histogram of normalized gray values together with a slice of the reconstructed 3D grayscale image for a model glass bead aggregate is shown in Fig. 1.

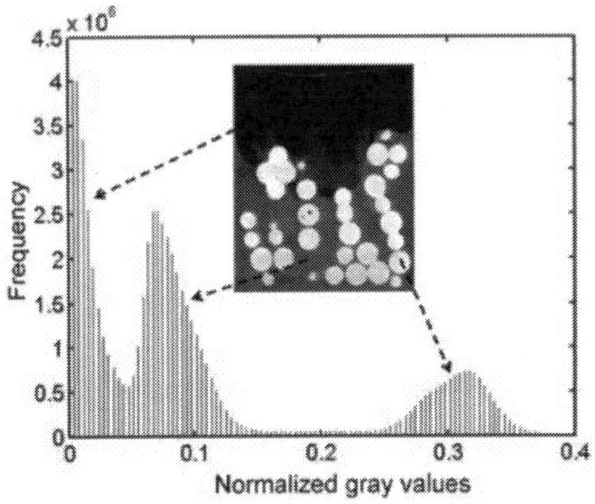

Figure 1: A histogram of normalized gray values together with a slice of the reconstructed 3D grayscale image of a model glass bead aggregate. The peaks in the frequency distribution correspond to gas, liquid and solid (from left to right).

The three phases (gas, liquid, and solid) can be discriminated from the contrast in the gray values: for the scan shown in Fig. 1, we assign voxels with gray values < 0.05 to the gas phase, voxels with gray values > 0.2 to the solid phase, and voxels with gray values between these two thresholds to the liquid phase. The thresholds can be determined from the histogram of gray values. The thresholding process yields a segmentation of the image into three regions corresponding to the three phases, so that the phase distribution in the sample can be analyzed.

After segmentation of the images, median filter and morphological operations were performed to eliminate the artifacts. Air bubbles inside the particles were removed by a sequence of erosion and dilation morphological operations. The center coordinates and radii of primary particles were obtained using watershed segmentation of the solid phase. All image processing and analysis was done using the MAVI software package developed at Fraunhofer [11].

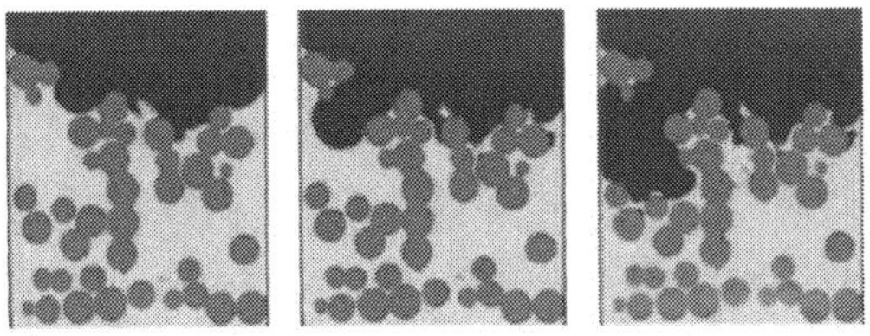

Figure 2: Cross sections of 3D tomograms for a packing of sintered glass beads at various liquid saturations (saturation decreases from left to right).

Figure 2 shows a cross section of the phase distributions in a drying glass bead structure. The microstructure geometry of the solid phase and the evolution of the liquid phase during drying are clearly visible. Corresponding 3D visualizations are given in Fig. 3: the three phases have been successfully separated. The major physical effect of capillary flow from large pores into small pores is easily recognized: large pores dry out first while smaller regions of the void space stay saturated with liquid.

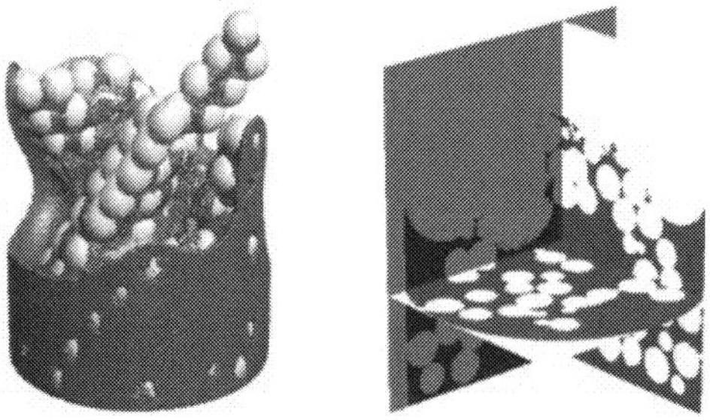

Figure 3: 3D visualization of a sintered particle aggregate partially saturated with liquid.

Description of the modeling approach

At the beginning of the drying process, the void space in the particle aggregate is completely filled with liquid. The simulation of drying over time involves two alternate basic processes, which are computed in an iterative scheme: (1) evaporation from the liquid-gas (*lg*) interface and (2) liquid relaxation to capillary equilibrium.

Local evaporation rates on the *lg* interface are computed from the gradient of the vapor pressure, which is obtained from a simulation of the vapor diffusion in the gas phase, with no-flux boundary condition at the dry solid surface and with equilibrium vapor pressure imposed on the *lg* interface. For simplicity, we have used a uniform and constant evaporation rate for all interface cells in the simulation presented here.

The equilibrium liquid distribution must fulfill the contact angle condition $\mathbf{n}_w \cdot \mathbf{n}_s = \cos\theta_{eq}$ at all three-phase boundaries, where $\mathbf{n}_w$ and $\mathbf{n}_s$ denote the outward unit normal vectors on the liquid and solid phase, respectively. Furthermore, the mean curvature κ of the *lg* interface must be uniform in each liquid cluster. Normal vectors and mean curvature are computed for each phase's volume fractions by the incorporation of a Gaussian smoothing step.

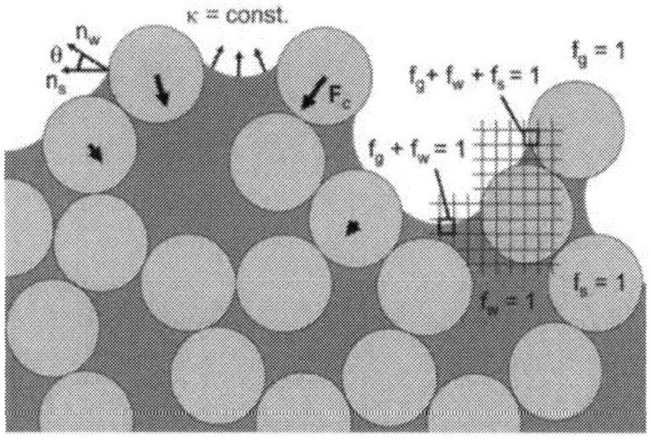

Figure 4: Illustration of the equilibrium liquid distribution in a particle aggregate, indicating the spatial discretization, phase volume fractions and capillary forces.

The geometrical redistribution of liquid towards capillary equilibrium is described by volume-preserving mean curvature flow: The normal velocity of the *lg* interface in each liquid cluster is given by the deviation of the local mean curvature κ from the cluster average $\overline{\kappa}$. This method has been shown to work successfully for two-phase problems: a liquid cube transforms into a liquid sphere of equal volume, for example. Topological changes in the liquid domain such as splitting and merging of the liquid clusters, are conveniently handled by the implicit description of the *lg* interface via cell volume fractions.

In the presence of a third phase (solid), new rules for the liquid redistribution had to be developed: the equilibrium contact angle condition does not provide any insight into the dynamics of contact line motion and, moreover, the micro-scale phenomena in the vicinity of a moving contact line are not well understood from a physical point of view. Therefore, we assume that the direction of the liquid velocity at the three-phase boundary is tangential to the solid surface, orthogonal to the contact line, and proportional to the deviation from the equilibrium contact angle. The velocity in three-phase cells is thus proportional to $(\cos\theta_{eq} - \mathbf{n}_w \cdot \mathbf{n}_s)(\mathbf{n}_w - (\mathbf{n}_w \cdot \mathbf{n}_s)\mathbf{n}_s)$. This vector points outward of the liquid phase if and only if the current contact angle is greater than the equilibrium contact angle, indicating that the computed velocity has the correct sign. Two weighting parameters *a*, *b* > *0* have been introduced in the code to balance the two competing dynamics in the *lg* and *slg* cells, respectively: the liquid volume fractions f_w are updated over time according to

$$f_w^{new} = f_w^{old} + a(\kappa - \kappa_\Sigma) \qquad \text{in } lg \text{ cells,} \qquad (1)$$

$$f_w^{new} = f_w^{old} + b(\cos\theta_{eq} - \mathbf{n}_w \cdot \mathbf{n}_s) \qquad \text{in } slg \text{ cells.} \qquad (2)$$

Additionally, volume conservation needs to be satisfied, which is not trivial in the three-phase case (in two phase problems, Gauss's divergence theorem can be used). Since the sum of all "desired" liquid volume fraction changes $f_w^{old} - f_w^{new}$ is not zero in general, the individual changes have to be corrected. We designed the corresponding routine to ensure volume conservation, while sign changes of $f_w^{old} - f_w^{new}$ in the correction of three-phase cell volume fraction updates are prevented. This means that the correction cannot lead to a local change in direction of the three-phase contact line.

The updating (1) and (2) is applied iteratively, and liquid normal vectors and local mean curvatures are recomputed in each step, until the magnitudes of the interface velocities are below a prescribed threshold. A similar approach has been used by Štěpánek et al. to simulate, e.g., the wetting of particles with a rough surface [12].

The local liquid redistributions described above leave the set of interface cells unchanged, which would only allow for very small liquid displacements. Larger changes are reflected in unphysical liquid volume fractions, such as $f_w > 1 - f_s$ or $f_w < 0$, which require *passing liquid to neighboring cells* or taking it from them. This is done by another local redistribution routine, which is less computationally expensive than a global routine. The basic idea is to pass through the neighbors in a certain order and to check for each of them whether they may accept/donate liquid [13]. If yes, liquid is passed to/taken from this cell and the over-/underflow in the center cell is reduced accordingly. This procedure is repeated for further neighbor cells if necessary; remaining over-/underflow after a certain number of correction steps is discarded. We notice that this last step violates the volume conservation constraint, but the scaling parameters a, b allow us to bound the maximum over-/underflow and make liquid volume loss/gain in the redistribution step arbitrarily small.

Simulated liquid phase distribution

The initial configuration of the solid phase comprised of 192 glass-bead particles in a cylindrical container has been obtained from the tomography measurement as described above. The solid phase was assumed to be static and the porosity is ~ 0.83. The whole domain including the sintered primary particles and the cylindrical container are discretized with a mesh of $90 \times 90 \times 94$ cells. Initially, the aggregate is fully saturated with liquid; evaporation starts from the top face.

The liquid phase distribution over time obtained from the numerical simulation is shown in Fig. 5: At the onset of evaporation, a curved meniscus develops in the cylinder and recedes. Upon further evaporation, capillary effects become clearly visible: large pores dry out earlier, while small pores stay saturated for a longer time.

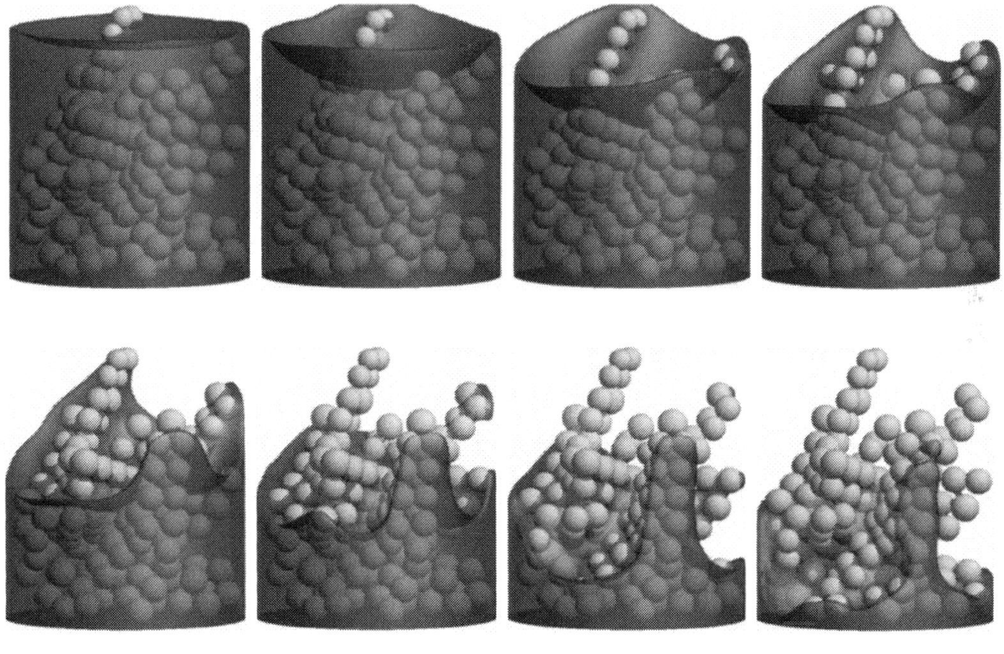

Figure 5: The liquid distribution during drying in a sintered particle aggregate for a small equilibrium contact angle: $\cos\theta_{eq} = 0.95$. The mesh resolution used for the computation is 10 cells per particle diameter.

Conclusions

In this work, important advances in both simulation and experiment have been presented towards a quantitative understanding of pore-scale phenomena, which occur during the convective drying of highly porous particle aggregates, and which cause undesired damage and shrinkage.

The experimental results from X-ray microtomography (XMT) measurements resolve the pore-scale phenomena in highly porous model particle aggregates made of sintered glass beads. Simulation results performed for the corresponding solid phase data exhibit a similar qualitative behavior. A quantitative comparison of experimental and simulation results shall evaluate the accuracy of our computational tool.

Future work on the simulation part will include full integration of the capillary force computation into the simulation, as well as the coupling with a micro-mechanical simulation of the solid phase. This should be sufficient to simulate the damage and shrinkage during the drying process. XMT measurements of real gel systems shall be carried out under various drying conditions and for different material and structural properties, in order to assess the influence of these parameters on the micro-mechanical behavior of particle aggregates during convective drying.

References

1. Metzger, T., A. Léonard, W. Jomaa, and H. Tamon (2011), "Understanding and Preventing Structural Changes during Drying of Gels," in Modern Drying Technology, Vol. 3, E. Tsotsas and A. S. Mujumdar, ed. John Wiley & Sons.
2. Perré, P. and I. W. Turner (1999), "A 3-D version of TransPore: a comprehensive heat and mass transfer computational model for simulating the drying of porous media," Int. J. Heat Mass Transfer. 42, pp. 4501–4521.
3. Brinker, C. J. and G. W. Scherer (1990), "Sol-Gel Science," New York, Academic Press.
4. Metzger T., A. Irawan A, and E. Tsotsas (2007), "Influence of pore structure on drying kinetics: a pore network study," AIChE J. 53, pp. 3029–3041.
5. Kharaghani, A., T. Metzger, and E. Tsotsas (2011), "A proposal for discrete modeling of mechanical effects during drying, combining pore networks with DEM," AIChE J. 57(4), pp. 872-885.
6. Hirt, C. W. and B. D. Nichols (1981), "Volume of fluid (VOF) method for the dynamics of free boundaries," J. Comput. Phys. 39, pp. 201-225.
7. Ranade, V. V. (2002), Computational flow modeling for chemical reactor engineering," London, Academic Press.
8. Fontenot, M. M. and R. D. Vigil (2002), "Pore-scale study of nonaqueous phase liquid dissolution in porous media using laser-induced fluorescence," J. Colloid Interface Sci. 247(2), pp. 481-489.

9. Faure, P. and P. Coussot (2010), "Drying of a model soil," Phys. Rev. E. 82, pp. 036303.

10. Léonard, A., S. Blacher, P. Marchot, J. P. Pirard, and M. Crine (2004), "Measurement of shrinkage and cracks associated to convective drying of soft materials by X-ray microtomography," Drying Technol. 22, pp. 1695-1708.

11. Fraunhofer ITWM (2005), "MAVI – Modular Algorithms for Volume Images".

12. Štěpánek, F. and P. Rajniak (2006), "Droplet morphologies on particles with macroscopic surface roughness," Langmuir. 22, pp. 917-923.

13. Kirsch, C., T. Metzger, and E. Tsotsas (2010), "Micro-scale model for drying of highly porous particle aggregates," 17th IDS, Magdeburg, Germany, October 3-6, 2010, Vol. A, pp. 407-414.

(621c) Multivariate Analysis and Reduced Order Modeling Based On Discrete Element Method (DEM) Simulations for a Powder Blender

Yijie Gao, **Fani Boukouvala**, Fernando. J Muzzio and Marianthi G. Ierapetritou, Chemical and Biochemical Engineering, Rutgers University, Piscataway, NJ

In this work, a set of Discrete Element Method (DEM) simulations are performed based on a computer experimental design for the qualitative and quantitative characterization of a periodic section of a powder blender. Mixing of powders is a crucial operation in areas such as pharmaceuticals, cosmetics and food manufacturing since poor blend homogeneity can affect the quality of the final products. It is therefore extremely important to have a good understanding of mixing mechanisms and the effect of parameters such as operating conditions and mixer design on blend uniformity [1-2]. The available modeling approaches for powder blending range from purely first-principle based, (DEM modeling) [3] to simply data-driven based (Response Surface Methodology or Kriging modeling) [4], or hybrid-based models which combine first-principle and empirical correlations (Population Balance modeling) [5]. Choosing between the available modeling approaches depends on the desired level of detail as well as the preferred computational cost. DEM is an expensive modeling approach that has been applied to study many different kinds of complex solid-based processes including powder mixing, since it can provide detailed information about every particle involved in the process. However, DEM is very computationally intensive tool which cannot be used for online purposes, optimization and control. The specific aims of this work include (a) the investigation of the effects and significance of different operating variables and design parameters on the performance of a blender and (b) the development of a fast reduced order model (ROM) using the information obtained by the set of DEM simulations which can predict the performance of the blender at unexplored operating regions.

Design of computer experiments for expensive simulations has attracted a lot of attention in the optimization literature [6-8]. Specifically, a lot of work has focused on designing a set of experiments that can lead to an accurate surrogate response surface which is less computationally expensive and more appropriate for process optimization. However, the minimization of sampling cost in the process of attaining the required information about the underlying process is a very important aspect. For this purpose, Latin Hypercube sampling has been considered in this work, due to its ability to produce a sampling set which is shown to be representative of the entire experimental region but smaller than a full factorial design [9]. The six significant input variables considered in this work are the blade speed, blade angle, blade width, fill level, weir height and shaft angle. By applying the periodic section modeling developed in our previous study [10], effects of these variables on cross-sectional mixing and axial motion are distinguished. As a result, the variance decay rate of the batch-like mixing in the cross-sectional directions (k_b), and the mean particle velocity in the axial direction (v_x), which are key indices of mixing performance, are considered as output variables.

The first goal of this study is to understand the effect of different operating conditions on the mixing performance of continuous mixing system. To achieve this goal, Projection to Latent Structure (PLS) is applied to the data set of a total of 64 samples. The cross-validation method recommended by Wold *et al.* [11] is applied, which is used to derive the confidence intervals of loading scores of these variables in order to assess the sufficiency of the obtained data. The results reveal that increase of blade speed and shaft angle, or decrease of fill level, weir height and blade width, lead to significant increase of v_x; increase of blade speed and blade angle, or decrease of fill level, blade width and shaft angle, lead to significant increase of k_b. Based on the loading scores of different variables, simultaneously increasing blade speed and decreasing shaft angle is the optimal strategy for the improvement of mixing performance. This conclusion can be used for further design of continuous mixing system. Results from this initial qualitative study agree with experimental observations as well as expected conclusions based on theoretical knowledge of the mixing system. In addition, the bootstrapping approach followed increases the reliability of the database sufficiency for describing the process performance. As a result, the next step of this work is the development of a more quantitative predictive model based on this database.

The second aspect of this work is the development of a Reduced Order Model (ROM) based solely on the data obtained by the performed simulations. ROM modeling based on Principal Component Analysis (PCA) or Proper Orthogonal Decomposition (POD) has attracted a lot of attention in fluid dynamics where processes are modeled by expensive CFD simulations [12-13]. In this work, the collected data is comprised of the steady-state magnitudes of certain state variables inside the periodic section, such as average concentration, average velocity components in x, y and z direction, and the average particle kinetic energy. PCA is employed in order to compress the available data and Kriging models are built for the input-output mappings as well as input-PC loading mappings. Using this approach, the process outputs (k_b and v_x), and the values of the state variables at different locations of the periodic section can be predicted at unsampled operating conditions and design configurations of the mixer. Results show that the developed model can predict the contours of the desired state variables as well as the outputs at different input variable combinations with good accuracy at a very low computational cost.

Summarizing, this work addresses three challenges associated with expensive DEM simulations for modeling powder blending. Firstly, critical operating conditions and design parameters for the mixing performance of a continuous blender are identified through a PLS model. Secondly, the sufficiency of the data obtained based on a Latin Hypercube sampling design in the case of a high dimensional input variable space is assessed through a cross-validation approach in terms of the accuracy of the PLS loadings values. Finally, a reduced order model is built based on a PCA decomposition of the obtained data and the use of Kriging algorithm. DEM simulations are gaining popularity in modeling powder-based processes in the pharmaceutical industry, however, their high computational requirements prohibits their use for optimization, control and process flowsheet design. To our knowledge, this is the first attempt of developing a ROM model for DEM modeling and thus the proposed methodology will have significant applications in all areas of powder based processing.

References:

1. Portillo, P.M., M.G. Ierapetritou, and F.J. Muzzio, *Characterization of continuous convective powder mixing processes.* Powder Technology, 2008. **182**(3): p. 368-378.

2. Portillo, P.M., M.G. Ierapetritou, and F.J. Muzzio, *Effects of rotation rate, mixing angle, and cohesion in two continuous powder mixers--A statistical approach.* Powder Technology, 2009. **194**(3): p. 217-227.

3. Dubey, A., et al., *Computational Approaches for the study of granular dynamics of continuous blending process- I : DEM based methods.* Macromolecular Materials and Engineering, 2010(Accepted for Publication).

4. Jia, Z., et al., *Predictive Modeling for Pharmaceutical Processes Using Kriging and Response Surface.* Journal of Pharmaceutical Innovation, 2009. **4**(4): p. 174-186.

5. Ramachandran, R. and P.I. Barton, *Effective parameter estimation within a multi-dimensional population balance model framework.* Chemical Engineering Science, 2010. **65**(16): p. 4884-4893.

6. Kleijnen, J.P.C., et al., *State-of-the-Art Review: A User's Guide to the Brave New World of Designing Simulation Experiments.* INFORMS JOURNAL ON COMPUTING, 2005. **17**(3): p. 263-289.

7. Pistone, G. and G. Vicario, *Design for Computer Experiments: Comparing and Generating Designs in Kriging Models*, in *Statistics for Innovation*, P. Erto, Editor. 2009, Springer Milan. p. 91-102.

8. Sacks, J., et al., *Design and Analysis of Computer Experiments.* Statistical Science, 1989. **4**(4): p. 409-423.

9. Simpson, T., D. Lin, and W. Chen, *Sampling strategies for computer experiments: design and analysis.* International Journal of Reliability and Safety (IJRS), 2001. **2**(3): p. 209-240.

10. Gao, Y., M.G. Ierapetritou, and F.J. Muzzio, *Periodic section modeling of convective continuous powder mixing processes.* AIChE Journal, 2011. **In Press**: p. DOI: 10.1002/aic.12348.

11. Wold, S., M. Sjöström, and L. Eriksson, *PLS-regression: a basic tool of chemometrics.* Chemometrics and Intelligent Laboratory Systems, 2001. **58**(2): p. 109-130.

12. Alonso, D., A. Velazquez, and J.M. Vega, *A method to generate computationally efficient reduced order models.* Computer Methods in Applied Mechanics and Engineering, 2009. **198**(33-36): p. 2683-2691.

13. Lang, Y.-d., et al., *Reduced Order Model Based on Principal Component Analysis for Process Simulation and Optimization†*. Energy & Fuels, 2009. **23**(3): p. 1695-1706.

See more of this Session: Poster Session Applied Mathematics and Numerical Analysis
See more of this Group/Topical: Computing and Systems Technology Division

(621d) A Quasi-Continuum Approach for the Computational Modeling of the Compaction of Bi-Layer Pharmaceutical Tablets Prediction of Axial Tensile Strength

Athanas A. Koynov, Rutgers University, Piscataway, NJ, Ilgaz Akseli, R&D Pharmaceutical Development, Boehringer Ingelheim, Ridgefield, CT and Alberto Cuitino, MAE, Rutgers University, Piscataway, NJ

Lack of sufficient bonding and adhesion at interfaces are key sources of product performance and compliance failures in the pharmaceutical industry. In this study, a hybrid quasi-continuum approach has been implemented for the prediction of the tensile strength of bi-layer tablets manufactured under different process conditions. For the purposes of validation, bi-layer tablets made from typical pharmaceutical excipients have been subjected to axial tensile loads to the point of fracture and axial tensile strength values have been determined. The relationship between fracture and compaction forces was obtained from the experiments and compared with the model. Good agreement between the experimental data and model results is observed, which demonstrates that quasi-continuum approach can capture the major features of the bi-layer powder compaction.

See more of this Session: Poster Session Applied Mathematics and Numerical Analysis
See more of this Group/Topical: Computing and Systems Technology Division

(621e) Mathematical Modeling of Intravascular Drug Delivery In Drug-Eluting Stents with Biodegradable Coating

Xiaoxiang Zhu and Richard Braatz, Chemical Engineering, Massachusetts Institute of Technology, Cambridge, MA

Drug-eluting stents (DES) are commonly used in the coronary angioplasty procedure for preventing vessel remodeling and reducing in-stent restenosis. Typical DES consists of a metallic mesh coated with a polymeric thin film, within which the drug is embedded and released for a prolonged time at the implanted coronary artery site. While both biodurable and biodegradable polymers can be used for the stent coating, biodurable coatings exclusively produce very slow release after a fast initial burst [1]. In contrast, a coating of biodegradable polymer such as poly(D,L-lactic-co-glycolic acid) (PLGA) exhibits multiphase drug release characteristics due to polymer degradation and erosion [2], and is more versatile in the design and control of drug release profiles.

This work presents a mathematical model that describes the coupled process of polymer degradation, erosion, and drug transport in a PLGA-coated DES. The delivery of a hydrophobic drug (sirolimus) is considered. A combined mechanism of bulk random-chain scission and catalyzed end-chain scission was used for the degradation model [3], which was able to capture the polymer molecular weight change and erosion rates. The drug transport is modeled by an evolving effective diffusivity that is dependent on the current porosity, diffusivity in the bulk polymer, and partitioning between the polymer and liquid phase (in the pores).

The model is first simulated for an *in vitro* drug release study and compared with experimental data in the literature [4]. The model correctly predicts the *in vitro* release of sirolimus from the PLGA-coated DES. Following that, the model is extended to study *in vivo* drug delivery from the DES implanted in a coronary artery [5]. Drug in the arterial wall is assumed to undergo a first-order reversible binding and an internalization mechanism [6]. Different configurations of the stent strut in the arterial wall (contacting, half-embedment, and full-embedment) were studied. The full-embedment configuration is observed to have the longest duration of release as well as the highest drug concentration profiles in the arterial wall. The porous structure generated by degradation and erosion facilitates a significant amount of drug dissipation into the lumen for the half-embedment (or contacting) configuration, resulting in shorter release time and lower arterial drug concentration. This finding potentially distinguishes biodegradable coating greatly from biodurable coating in terms of *in vivo* DES performance.

Reference

[1] G. Acharya, K. Park, Mechanisms of controlled drug release from drug-eluting stents, Adv Drug

Deliver Rev, 58 (2006) 387-401.

[2] A. Finkelstein, D. McClean, S. Kar, K. Takizawa, K. Varghese, N. Baek, K. Park, M.C. Fishbein, R. Makkar, F. Litvack, N.L. Eigler, Local drug delivery via a coronary stent with programmable release pharmacokinetics, Circulation, 107 (2003) 777-784.

[3] R.P. Batycky, J. Hanes, R. Langer, D.A. Edwards, A theoretical model of erosion and macromolecular drug release from biodegrading microspheres, Journal of Pharmaceutical Sciences, 86 (1997) 1464-1477.

[4] X.T. Wang, S.S. Venkatraman, F.Y.C. Boey, J.S.C. Loo, L.P. Tan, Controlled release of sirolimus from a multilayered PLGA stent matrix, Biomaterials, 27 (2006) 5588-5595.

[5] X. Zhu, D.W. Pack, R.D. Braatz, Modeling intravascular delivery from drug-eluting stents: effect of anisotropic vascular diffusivity and drug loading on arterial drug distribution (submitted).

[6] M.A. Lovick, E.R. Edelman, Computational simulations of local vascular heparin deposition and distribution, Am J Physiol-Heart C, 40 (1996) H2014-H2024.

See more of this Session: Poster Session Applied Mathematics and Numerical Analysis
See more of this Group/Topical: Computing and Systems Technology Division

(621f) Derivation of Equations of State by Discrete Modelling of Entropy

Martin Pfleger, Institute of Chemical Engineering and Environmental Technology, Graz University of Technology, Graz, Austria and Thomas Wallek, Institute of Chemical Engineering and Environmental Technology, Graz University of Technology, 8010 Graz, Austria

Background

Entropy is the basic thermodynamic state variable in order to understand the behaviour of systems consisting of a huge number of particles like gases and liquids. In conjunction with Gibbs' thermodynamics it can be used to deduce equations of state, one of the building blocks in chemical engineering. They provide the theoretical fundamentals for specifying and estimating the properties of gases and liquids. There are several classical approaches in order to deduce an equation for correlating dependencies between pressure and temperature of a gas and the volume it needs: empirical deductions (ideal gas), mathematical constructs (viral equation), and statistical approaches (Boltzmann statistics).

The method

The proposed algorithm consists in a simple approach which combines the Gibbs' fundamental equation with discrete modelling of the state variables S, U, V and N. For modelling the entropy term we use information theory. Essential in this approach is that the four state variables are modelled using the same inner variables. The Gibbs' fundamental equation builds the glue between the state variables and will lead to constraints that finally lead to the Maxwell-Boltzmann distribution of energies, and additionally to the ideal gas law.

Application

In this paper it will be demonstrated how modelling is done, especially for the inner energy and the entropy for which we need information theory. Following the mathematical constraints given by the Gibbs' fundamental equation this leads in a first approach to the ideal gas law and at the same time to a distribution of energy comparable to the well-known Maxwell-Boltzmann distribution. These results build important criteria for the quality of approach. Finally approach can be extended to modelling of real gas.

See more of this Session: Poster Session Applied Mathematics and Numerical Analysis
See more of this Group/Topical: Computing and Systems Technology Division

(621g) Gap Gene Regulation In Drosophila: A Study In Multi-Objective Optimization of Spatiotemporal Models

James Hengenius[1], Michael Gribskov[1], Ann Rundell[2] and David Umulis[3], (1)Biology, Purdue University, West Lafayette, IN, (2)Biomedical Engineering, Purdue University, West Lafayette, IN, (3) Agricultural and Biological Engineering, Purdue University, West Lafayette, IN

The axial bodyplan of *Drosophila melanogaster* is specified during the process of embryonic morphogenesis. Maternally deposited *bicoid* mRNA is translated into Bicoid. This protein establishes a spatially graded morphogen distribution along the anterior-posterior (AP) axis. Bicoid in turn triggers gap gene expression in a spatially graded manner. The gap gene regulatory network (GRN) further refines AP patterning. Reaction-diffusion models of gap gene expression on 1D and 3D domains have been used to infer complex GRN interactions by optimizing model parameters with respect to gap gene expression data. Previous studies have approached this optimization with a variety of algorithms (e.g., GA, SA), but have primarily relied on scalar cost functions (LSE, RMSE) to quantify concentration errors. Moreover, these functions are ill-suited for identification of parameter sets which produce qualitatively correct (proper AP placement of protein bands) because they consider only quantitative agreement. We found that the optimization problem is highly dependent on the specific choice of objective function to be minimized and that there is no clear best choice. We evaluated spatially-extended cost functions incorporating spatial metrics (cosine similarity, spatial gradients) in 1D and 3D models. We also decomposed the objective into concentration error and spatial contributions and explore tradeoffs in each via multi-objective optimization. To appropriately optimize spatiotemporal systems such as the gap gene network, we found that compromise between multiple objectives produced superior GRN inference.

See more of this Session: Poster Session Applied Mathematics and Numerical Analysis
See more of this Group/Topical: Computing and Systems Technology Division

(621h) Numerical Methods for a Case of Fractional Isoperimetric Problem

Annie X.W. Tangpong, Department of Mechanical Engineering, North Dakota State University, Fargo, ND and Md. Mehedi Hasan, Mechanical Engineering, North Dakota State University, Fargo, ND

An isoperimetric problem is a problem of constrained optimization and it is one case of study in Calculus of Variations. When the objective function or the constraint equations or both contain fractional order derivatives, the problem becomes a fractional isoperimetric problem. Unlike integer order isoperimetric problems that can be solved analytically, analytical solutions for fractional isoperimetric problems either do not exist or are extremely complex. This work proposes numerical methods for a case of fractional isoperimetric problem. The fractional derivatives are defined in term of Riemann-Liouville derivatives and with zero terminal conditions, these are the same as Caputo derivatives. An iterative numerical scheme is developed in which the Grünwald–Letnikov (GL) approach is used to approximate the fractional derivatives. The spatial domain and the objective function are both approximated by three types of GL definitions. These three methods all show good convergence as the integration step sizes decrease. The convergence errors of these methods are obtained and compared. As the order of the derivatives approaches the integer value of 1, the numerical results recover the analytical solution of the corresponding integer order isoperimetric problem.

See more of this Session: Poster Session Applied Mathematics and Numerical Analysis
See more of this Group/Topical: Computing and Systems Technology Division

A NEW GLOBAL OPTIMIZATION ALGORITHM AND ITS APPLICATION TO PHASE STABILITY PROBLEMS

Stanislaw K. Wasylkiewicz
Computer Modelling Group Ltd., Calgary, Alberta, Canada
Monika J. Wasylkiewicz
The Schulich School of Engineering, University of Calgary, Calgary, Alberta, Canada

Finding all solutions of mathematical models is still a challenging problem in many areas of science and engineering. There are various methods for finding one or more solutions of the models but they do not guarantee that the solution is a global one or that all solutions have been found, which can be of paramount importance for some modeling problems of physical systems. E.g. in liquid phase stability analysis, we have to find all stationary points of a tangent plane to the Gibbs free energy surface to be sure whether the examined liquid is stable or not.

In this paper, we describe a general approach to finding all singular points of mathematical models by exploring the natural connectedness that exists between singular points of an objective function. The idea of following ridges and valleys using information gathered along the way is not new and was used in many of the hill climbing or direct search optimization techniques developed in the 1960s. However, these optimization techniques do not use derivative information to determine the search direction, have difficulties tracking strong curvature and are not useful for large-scale problems. They are generally considered to be less efficient than derivative-based methods.

In 1984, Van Dongen and Doherty (1) used ideas from differential geometry and the theory of differential equations to describe dynamics of distillation processes. They defined a ridge on a topographic map that contains contour lines for different elevations as a location where the contour lines exhibit maximum curvature. They gave also another definition for ridges (valleys) as maxima (minima) of an objective function $F(\mathbf{x})$ taken in the constraint plane normal to the ridge-top locus. The second definition, together with some properties of the gradient system, was used by Wasylkiewicz et al. (2) to develop a robust global algorithm for implementation of the Gibbs tangent plane analysis for complex liquid mixtures that involves locating all the stationary points of the tangent plane distance function by tracking its ridges and valleys. They conjectured that the ridges and valleys of $F(\mathbf{x})$ are all connected to each other and that this connected domain contains all the stationary points. This was proved later by Cheng-Dong and Xiao-Xia (3).

Valleys and ridges of the least-squares landscape are specific integral curves of the gradient vector field and have been defined by Lucia and Feng (4) as a collection of constrained extrema over a set of level curves. Based on this definition they developed a so called "global terrain method", which consists of a series of downhill, equation-solving computations and uphill, predictor–corrector calculations. They noticed that the nonlinear optimization problems used to define valleys and ridges can be quite difficult to solve because they are inherently poorly scaled. Mathematically, these definitions can be expressed as the following constrained optimization problem (5)

$$\underset{x}{\mathrm{opt}}[\nabla F(\boldsymbol{x})^{\mathrm{T}}\,\nabla F(\boldsymbol{x})]$$
$$s.t.\quad F(\boldsymbol{x}) = c$$

where *c* is any given value (or level, or elevation) of the potential function $F(\boldsymbol{x})$. For any given contour line (specified value of *c*), we can find a point on the line that corresponds to a local optimum in the gradient norm by solving the above constrained optimization problem.

By applying the Kuhn-Tucker conditions to the Lagrangian function $\boldsymbol{L}(\boldsymbol{x},\lambda)$ for the optimization problem we can get the following set of equations

$$\boldsymbol{L}_x(\boldsymbol{x},\lambda) = 2\,\boldsymbol{H}[F(\boldsymbol{x})]^{\mathrm{T}}\,\nabla F(\boldsymbol{x}) + \lambda\,\nabla F(\boldsymbol{x}) = \boldsymbol{0}$$

$$L_\lambda(\boldsymbol{x},\lambda) = F(\boldsymbol{x}) - c = 0$$

The set of equations contains *n+1* unknowns $(x_1, x_2, \ldots, x_n, \lambda)$ and one parameter - contour line value *(c)*. A current branch that follows valley-bottoms or ridge-tops can be tracked by gradually increasing or decreasing parameter *c* and solving the set of equations. We have chosen the contour line value (*c*) as a natural continuation parameter ($t = c$) for this global optimization problem.

For efficient tracking of continuation branches, we use a modified arc length continuation method (6), where an additional equation $g(\boldsymbol{x}, \lambda, t, s)$ for the arc length has been added to the original equations to form the following augmented system

$$\boldsymbol{G}(\boldsymbol{x},\lambda,t,s) = \begin{bmatrix} \boldsymbol{L}_x(\boldsymbol{x},\lambda,t) \\ L_\lambda(\boldsymbol{x},\lambda,t) \\ g(\boldsymbol{x},\lambda,t,s) \end{bmatrix} = \boldsymbol{0}$$

that contains *n+2* unknowns $(x_1, x_2, \ldots, x_n, \lambda, t)$ and one parameter - arc length *(s)*. The current branch can be tracked by gradually increasing parameter *s* and solving the augmented system.

While tracking the continuation branches, we monitor the determinant of the Jacobian matrix of the set of equations. If it changes sign, we calculate a new bifurcation point, from which additional branches are later started in eigendirections, using the following approximation of the algebraic bifurcation equation

$$\begin{bmatrix} \boldsymbol{L}_{xx}(\boldsymbol{x},\lambda,t) & \nabla F(\boldsymbol{x}) & \boldsymbol{0} \\ \nabla F(\boldsymbol{x})^T & 0 & -1 \\ \frac{d\boldsymbol{x}_b^T}{ds} & \frac{d\lambda_b}{ds} & \frac{dt_b}{ds} \end{bmatrix} \begin{bmatrix} \frac{d\boldsymbol{x}_0}{ds} \\ \frac{d\lambda_0}{ds} \\ \frac{dt_0}{ds} \end{bmatrix} = \boldsymbol{0}$$

where the subscript *b* indicates derivatives computed at the bifurcation point for the old branch and the subscript *0* for the new branch, and the top left part of the above matrix is calculated from the following equation (5)

$$\boldsymbol{L}_{xx}(\boldsymbol{x},\lambda,t) = 2\,\boldsymbol{J}\{\boldsymbol{H}[F(\boldsymbol{x})]^{\mathrm{T}}\,\nabla F(\boldsymbol{x})\} + \lambda\,\boldsymbol{H}[F(\boldsymbol{x})]$$

The Jacobian matrix in the above equation is calculated numerically using a routine specially design to give smooth derivatives close to the composition boundaries and the component vertices.

As an example, we apply this new global optimization algorithm to the Gibbs tangent plane stability test for multiphase liquid mixtures. The algorithm relies on a new arc length continuation method that gives an efficient and robust scheme for locating all stationary points of the tangent plane distance function predicted by a thermodynamic model. The algorithm is self-starting and significantly improves reliability and robustness of multiphase equilibrium calculations.

References

1. Van Dongen, D. B. and M. F. Doherty (1984), "On the Dynamics of Distillation Processess V. The Topology of the Boiling Temperature Surface and its Relation to Azeotropic Distillation", *Chem. Eng. Sci.*, 39, pp. 883-892.
2. Wasylkiewicz, S. K., L. N. Sridhar, M. F. Malone and M. F. Doherty (1996), "Gibbs Stability Analysis and Calculation of Liquid-Liquid Equilibrium in Multicomponent Mixtures", *Ind. Eng. Chem. Res.*, 35, pp. 1395-1408.
3. Cheng-Dong, Z. And W. Xiao-Xia (2001), "The existence and uniqueness of trajectories joining critical points for differential equations in R^3 ", *Chaos, Solitons and Fractals*, 12, pp. 153-158.
4. Lucia, A. and Y. Feng (2003), "Multivarible terrain methods", *AIChE J.*, 49, pp. 2553–2563.
5. Wasylkiewicz, S. K. (2009), "Search for all Stationary Points", Private Communication, Calgary, 2009.03.01.
6. Wasylkiewicz, S. K., M. F. Malone and M. F. Doherty (1999), "Computing All Homogeneous and Heterogeneous Azeotropes in Multicomponent Mixtures", *Ind. Eng. Chem. Res.*, 38, pp. 4901–4912.

(621j) Optimal Control of Fed-Batch Fermentation Process Using Iterative Dynamic Programming

Zheng Li, Department of Chemical Engineering, Texas Tech University, Lubbock, TX and M. Nazmul Karim, Chemical Engineering Department, Texas Tech University, Lubbock, TX

This article proposes a feeding strategy based on a metabolic networks model to optimize ethanol production using *Escherichia coli* KO11. In metabolic engineering, growth behavior and product secretion from microorganisms can be predicted using constraint-based analysis of metabolic networks. In this study, a new approach of using stoichiometric model combined with kinetic model allows for precise prediction of ethanol production and formation of other byproducts under specific fermentation environment. An optimal feeding strategy and minimization of total fermentation time are accomplished by iterative Dynamic Programming approach, which maximizes ethanol production and also minimizes toxic byproduct formation. The switching time from batch to fed- batch is also taken as variable to be optimized. All variables are computed by using iterative Dynamic Programming utilizing a stoichiometric model. Iterative dynamic programming is modified to increase computational efficiency. Simulation results are promising.

See more of this Session: Poster Session Applied Mathematics and Numerical Analysis
See more of this Group/Topical: Computing and Systems Technology Division

(621k) Tuning of Tabu Search Method for Optimal Scheduling Problems

William Bryant, Department of Chemical and Petroleum Engineering, The University of Kansas, Lawrence, KS and Kyle V. Camarda, Chemical and Petroleum Engineering, University of Kansas, Lawrence, KS

Tabu search is a stochastic optimization method which is appropriate for large nonconvex optimization problems. While it has been applied to many problems including some in the area of scheduling, there have previously been no systematic studies on the selection of algorithm parameters. Tabu search has many such parameters, including the length of the Tabu list and the number of instances of the algorithm to run simultaneously. These parameters are functions of problem size and structure, and currently are set via a trial-and-error method. This work seeks to develop simple heuristics which allow the parameters for Tabu search to be set automatically. The heuristics also must be varied based on the serial or parallel computer architecture available. These heuristcs are evaluated for the solution of a set of batch scheduling problems, including the resource-constrained, continuous-time scheduling problem of Jain and Grossmann [1].. Results are provided which compare the novel tuning method with the use of fixed values for Tabu search parameters, and computation times are compared to those obtained with deterministic methods as well. The tuning of a parallel Tabu search implementation was also evaluated, in terms of both speedup and robustness on the same set of test problems.

[1] Jain, V. and I.E. Grossmann, Resource-Constrained Scheduling of Tests in New Product Development, *Ind. Eng. Chem. Res.* 1999, **38,** 3013-3026.

See more of this Session: Poster Session Applied Mathematics and Numerical Analysis
See more of this Group/Topical: Computing and Systems Technology Division

(621l) An Improved Kinetic Model of Mint Essential Oil Biosynthesis That Accounts for Transport and the Early Monoterpene Metabolic Steps

Daniel A. Carmona-Alvarez[1], B. M. Lange[2], Rigoberto Rios-Estepa[1] and Rigoberto Rios-Estepa[3], (1) Chemical Engineering, University of Antioquia, Medellin, Colombia, (2)Institute of Biological Chemistry, Washington State University, Pullman, WA, (3)School of Bioscience, National University of Colombia, Medellin, Colombia

We have previously reported a tested kinetic model for essential oil biosynthesis in peppermint [Rios-Estepa et al., 2008, Rios-Estepa et al., 2010]. The mathematical model used, as input variables, the diverse profiles of gene expression, enzyme concentration/activity, levels of metabolites and the number of secretory cells involved in oil biosynthesis, and the kinetic properties as parameters. A second generation model described changes in essential oil profiles under various environmental conditions and in different transgenic lines. It was also included variables accounting for variations in the number of glandular trichomes (specialized essential oil-producing structures on leaf surfaces), the amount of essential oil stored per trichome, the distribution of glandular trichomes at different stages of leaf development, and modulation of gene expression patterns in transgenic lines with modified essential oil composition. The refined model was used to identify the key factors controlling monoterpenoid essential oil biosynthesis under adverse environmental conditions. It also was used for investigating determinants of essential oil biosynthesis in transgenic peppermint lines with modulated essential oil profiles. A computational perturbation analysis, which was implemented to identify the variables that exerted prominent control over the outputs of the model, indicated that the essential oil composition should be highly dependent on certain biosynthetic enzyme concentrations [(+)-pulegone reductase and (+)-menthofuran synthase], whereas oil yield should be particularly sensitive to the density and/or distribution of leaf glandular trichomes, the specialized anatomical structures responsible for the synthesis and storage of essential oils.

Here we present an improved modeling approach to account for transport in the secretory cell wall, the endoplasmatic reticulum and the mitochondria where some of most important metabolic catalysts are present. Also, the enhanced model did consider the early metabolic steps in monoterpene biosynthesis.

The implications of these model modifications for wild type and transgenic approaches aimed at improving essential oil yield and composition are discussed. [Rios-Estepa et al., 2008. PNAS Feb 26;105 (8):2818-23; Rios-Estepa et al., 2010. April 152: 2105-2119]

See more of this Session: Poster Session Applied Mathematics and Numerical Analysis

(622a) A Fuzzy Sliding Mode Control Approach for Nonlinear Chemical Processes

Azar Shahraz, Chemical Engineering, Pennsylvania State University, University Park, PA

Fuzzy sliding mode control (FSMC) as a robust and intelligent nonlinear control technique is proposed to control processes with severe nonlinearity and unknown models. The performance of the proposed method has been evaluated for both single input single output (SISO) and MIMO nonlinear systems through its application in three severely nonlinear processes that are frequently used as benchmarks of nonlinear process control strategies. The evaluation shows that, despite its lack of dependence on the process model, the proposed method performs almost the same as conventional sliding mode control alternatives that utilize all the information that exists in the mathematical model of the process.

See more of this Session: Poster Session - Area 10B: Systems and Process Control
See more of this Group/Topical: Computing and Systems Technology Division

(622b) Mathematical Formulation of Carbon Capture and Storage Under the Presence of Uncertainty

Jun-hyung Ryu[1], Jeehoon Han[2] and In-Beum Lee[2], (1)Nuclear and Energy System Engineering, Dongguk University, Gyeongju, South Korea, (2)Chemical Engineering, POSTECH, Pohang, South Korea

A large number of research works were undertaken for the planning of CO_2 Capture and Storage (CCS). CO_2 emissions are considerably difficult to be estimated clearly because CO_2 is emitted from various sources according to changing environments. In this study, a two-stage stochastic programming model is developed for planning CO_2 emissions disposing under uncertainty. The proposed CO_2 disposal network model allows us to determine where and how much the captured CO_2 to be held for storage and where to sequester the given amount of CO_2 among multiple potential candidates for the purpose of minimizing the total cost of handling demand uncertainty of CO_2. The proposed model is applied to a case study of examining the robustness in terms of handling changing environments in the context of CO_2 emissions and disposal targets of Korea in 2030. The gained results aid determining policy to plan in the budget of disposing CO_2.

See more of this Session: Poster Session - Area 10B: Systems and Process Control
See more of this Group/Topical: Computing and Systems Technology Division

(622c) Optimal Design of Acid Gas Removal Unit In LNG Plant

Kiwook Song[1], Chul-Jin Lee[2], Jeongwoo Jeon[1] and Chonghun Han[3], (1)School of Chemical and Biological Engineering, Seoul National University, Seoul, South Korea, (2)Chemical and biological engineering, Seoul National University, Seoul, South Korea, (3)Chemical and Biological Engineering, Seoul National University, Seoul, South Korea

Acid gases such as carbon dioxide and hydrogen sulfide in raw natural gas need to be removed before the main liquefaction process of natural gas to avoid corrosion, solidification and plugging problems. Amine based chemical solvents are the most generally accepted and widely used method for removal of acid gases. Amine processes including gas absorber and solvent regenerator require large energy consumption and therefore optimization in design of such processes are essential.

First, conceptual design and simulation of gas sweetening process using aqueous amines are performed in Aspen HYSYS. AMINES property package is selected for the thermodynamic properties. The goal is to lower carbon dioxide and hydrogen sulfide concentration to 50 ppmv and 4 ppmv, respectively.

Then, optimization formulation of the sweetening process is developed to minimize total capital and operational cost. HYSYS simulation is linked with MATLAB and GAMS to find the optimal design. Main design variables are liquid solvent to gas ratio (L/G ratio) and lean amine solvent loading value. Total trays of absorber and regenerator including the feed location of rich solvent in the regenerator are integer variables. The optimization formulation is a mixed-integer nonlinear programming (MINLP) problem. Design alternatives such as split-flow scheme are also analyzed and compared.

This research was supported by a grant from the LNG Plant R&D Center funded by the Ministry of Land, Transportation and Maritime Affairs(MLTM) of the Korean government.

See more of this Session: Poster Session - Area 10B: Systems and Process Control
See more of this Group/Topical: Computing and Systems Technology Division

(622d) Design and Control of a Reactive Distillation Process for Naphtha Hydrodesulfurization

Ankit Sharma and **Nitin Kaistha**, Chemical Engineering, IIT Kanpur, Kanpur, India

Design and Control of a Reactive Distillation Process for Naphtha Hydrodesulfurization

Ankit Sharma and Nitin Kaistha

Department of Chemical Engineering, Indian Institute of Technology Kanpur 208016

Reactive distillation (RD) based hydrodesulfurization (HDS) has been proposed as an alternative to conventional trickle-bed reactor technologies towards meeting the increasingly stringent fuel sulfur content specification [1]. In this work the steady state optimum design and control of a naphtha HDS process is studied. The process consists of an RD column followed by a closed loop H_2S absorption system using aqueous MDEA. Fresh naphtha and H_2 are fed to an RD column which consists of a reactive tray section with a stripping section below it. Naphtha is fed into the reactive zone while H_2 is fed towards the lower portion of the column. In addition to hydrodesulfurization in the RD column, the naphtha gets fractionated to recover desulfurized C_7 free light naphtha as the distillate and desulfurized C_6 free heavy naphtha as the bottoms. The excess hydrogen, H_2S and other non-condensables leave the column along with some C_4 - C_5 hydrocarbons as vapor distillate. This vapor stream is sent to an H_2S absorber to recover a desulfurized fuel gas stream which is recycled to the RD column. Some of the fuel gas is purged to prevent a build-up of inerts in the gas recycle loop. The H_2S rich MDEA stream from the absorber is stripped to release the H_2S and recirculate lean MDEA back to the absorber.

Exploratory simulations show that the process can be designed to produce ppb level ultra low sulfur heavy naphtha for blending with gasoline. The RD column specifications are chosen as 1.6 wt% C_7 in the liquid distillate, 2.4 wt% C_6 in the bottoms and a condenser temperature of 50 °C with cooling water as the coolant. The absorber is designed for 4 ppm H_2S in the fuel gas while the stripper is designed for 99% H_2S recovery. The dominant design variables for the process are the RD column pressure, total H_2 to thiophene ratio in the RD column feed and the fuel gas purge fraction. The other "local" design variables such as number of trays are obtained using well established heuristics. The dominant design variables are optimized to obtain the steady state economic optimum design of the RD HDS process. Classic economic tradeoffs are observed in the optimization, which are described in detail. A comparison with a conventional HDS process shows that the RD process is about 30% cheaper than the conventional process in terms of the total annualized cost.

A simple decentralized plantwide control system with a two point temperature controller in the RD column is synthesized for the process. It is found to effectively handle large throughput changes and changes in the naphtha / H_2 fresh feed composition. The RD HDS process is thus a promising alternative to conventional HDS processes.

References

1. A.D. Estrada-Villagrana, G.B. Quiroz-Sosaa, M.L. Jimnez-Alarcóna, L.O. Alemán-Vázqueza and J.L. Cano-Domíngueza, Comparison between a conventional process and reactive distillation for naphtha hydrodesulfurization, Chemical Engineering and Processing, Volume 45, Issue 12, December 2006, Pages 1036-1040

See more of this Session: Poster Session - Area 10B: Systems and Process Control
See more of this Group/Topical: Computing and Systems Technology Division

ROBUST AVERAGING LEVEL CONTROL

Peter Rosander[1], * **Alf J. Isaksson**, *,** **Johan Löfberg**, *
Krister Forsman ***

* *Department of Electrical Engineering, Linköping University, SE-581-83 Linköping Sweden (email: {rosander,alf,johanl}@isy.liu.se)*
** *ABB AB Corporate Research, SE-721-78 Västerås Sweden (email: alf.isaksson@se.abb.com)*
*** *Perstorp AB, SE-284-80 Perstorp Sweden (email: krister.forsman@perstorp.com)*

Abstract

Frequent inlet flow changes typically cause problems for averaging level controllers. For a frequently changing inlet flow the upsets do not occur when the system is in steady state and the tank level at its set-point. For this reason the tuning of the level controller gets quite complicated, since not only the size of the upsets but also the time in between them relative to the hold up of the tank have to be considered.

One way to obtain optimal flow filtering while directly accounting for future inlet flow upsets is to use closed-loop robust MPC, as proposed here. The behavior of the robust MPC controller differs from earlier proposed level controllers as it does not return the tank level to a fixed set-point following an inlet flow upset. Guidelines on the tuning of the controller is presented and its performance is compared to that of a previously proposed MPC approach.

Keywords

Averaging level control, Robust MPC, Surge tanks

Introduction

Surge tanks are widely used within the process industry to prevent upstream flow variations from upsetting downstream processes. The outlet flow is kept smooth by using the capacity of the tank to average out any inlet flow variations. Averaging level control thus differs from most other control problems in the sense that the objective is not to obtain tight reference following but rather to keep the control signal (the outlet flow) smooth.

Over the years many approaches to control surge tanks have been suggested. Luyben and Buckley (1977) proposed to control the tank by a proportional part and a feed forward compensation from the inlet flow. Gain scheduled PI controllers which filter small upsets slowly and large ones more rapidly were presented in (Shunta and Fehervari, 1976). Design charts for the tuning of PI controllers were given by Cheung and Luyben (1979) to achieve optimal filtering for inlet flow step changes. A recent contribution to the tuning of PI controllers is the optimization approach taken by Shin et al. (2008). A two-degree of freedom controller which permits designing the disturbance attenuation and reference following separately was proposed by Wu et al. (2001). The continuous-time optimal level controller, which minimizes the maximum rate of change of the outlet flow, was derived in (McDonald et al., 1986). Its discrete-time counterpart as well as a formulation of the control problem in the model predictive control (MPC) framework, was presented in (Campo and Morari, 1989).

The aforementioned controllers achieve averaging control by, following an inlet flow change, allowing the tank level to initially deviate from its steady state set-point of (typically) 50% while smoothly adapting the outlet flow to match the new inlet flow and eventually bringing back the tank level to the set-point. The various approaches do, of course, differ in exactly how this is performed and also in what tuning knobs that are available. We have an application in one of Perstorp AB's factories where the inlet flow upsets are rather frequent, see Figure 1. This gives that, in general, the system might not have reached steady state when an inlet flow upset occurs. Therefore the guidelines of previous contributions are not directly applicable (since the steady state assumption is violated). Using historical data and some extra safety margin for future uncertainty it is possible to find a controller tuning which will perform well for a specific surge tank. How to generalize the result or to give general guidelines is, however, not obvious.

By using the framework of robust model predictive control it is possible to obtain optimal flow filtering while directly accounting for future inlet flow changes. This paper presents the controller design, some early results and discusses how the insight gained from the design can be used to modify other control structures to improve their performance for frequent inlet flows.

System Description

We assume a cylindrical buffer tank with level x and inlet and outlet flows q_{in} and q_{out} respectively. Furthermore we assume that we manipulate the outlet flow directly, $u = q_{out}$. Using mass balance we obtain the continuous time model

[1] To whom all correspondence should be addressed.

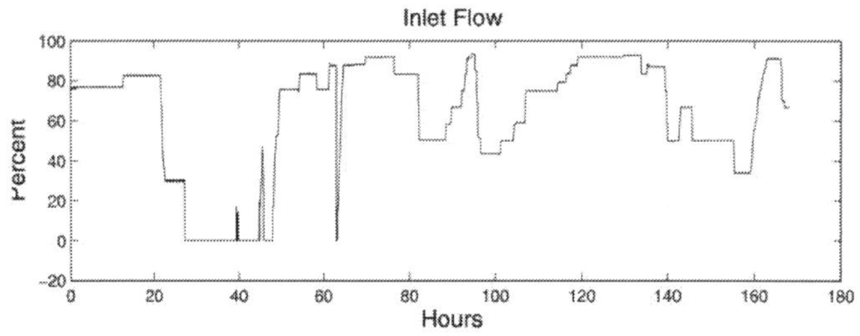

Figure 1. Inlet flow data from Perstorp AB second week of March 2011.

$$\dot{x}(t) = k_v\left(q_{in}(t) - u(t)\right) \tag{1}$$

Under the assumption that the inlet and outlet flow is constant over a sampling instance of length T_s the discrete time model is

$$x(k+1) = x(k) + k_v T_s\left(q_{in}(k) - u(k)\right) \tag{2}$$

Furthermore we assume that all quantities are given in percent and that the outlet and inlet flow have equal range

$$\begin{aligned} x_{min} &\le x \le x_{max} \\ q_{min} &\le q_{in},\, u \le q_{max} \end{aligned} \tag{3}$$

where $q_{min} = x_{min} = 0\%$ and $q_{max} = x_{max} = 100\%$ if no extra security limitations are put on the tank level. The parameter k_v describes how one percent of flow imbalance per time unit relates to change in tank level. With the outlet being the manipulated flow we could allow it to have greater range than the inlet flow, but for the scope of this paper that would only serve to obstruct the view and complicate the notation. For the same reason we assume that the inlet flow is directly measurable. In any case, the linear dynamics of the system allow for a straightforward estimation of the inlet flow using the Kalman filter as shown in (Khanbaghi et al., 2001).

Model Predictive Control

Model predictive control use a prediction of the future response of the system to find the optimal control signal. In every sample a control sequence that minimizes the optimization problem

$$\begin{aligned} &\min_{u(0),\,u(1),\ldots,\,u(N)} \quad J(x(k), u(k)) \\ &\text{subject to} \\ &x(k+1) = Fx(k) + Gu(k) \\ &x(0),\, u(-1) \text{ known} \\ &\text{constraints on } x(k) \text{ and } u(k) \end{aligned} \tag{4}$$

is found. The function $J(x(k), u(k))$ quantifies what we want to achieve and the initial conditions $x(0)$, $u(-1)$ are the current state and the last control signal respectively. The first control signal, $u(0)$, of the sequence $u(0), u(1), \ldots, u(N)$, is then actuated and at the next measurement instant a new optimization problem with new initial conditions is solved. At every sample instant the control policy employed is thus an open-loop one, but since the procedure is repeated, using fresh knowledge of the state, the controller is indeed working in closed loop. A comprehensive description of MPC can be found in (Maciejowski, 2002).

The averaging level control problem can be formulated in the MPC framework as first reported in (Campo and Morari, 1989). The presentation here follows their framework. To obtain a "smooth outlet flow" we minimize the maximum change of the outlet flow

$$\min_{u(0)),\ldots,\,u(N)} \max_k |u(k) - u(k-1)| \tag{5}$$

To ensure that the constraints (3) are fulfilled for the whole prediction horizon we note that the dynamics of the system yields

$$x_{min} \le x(k) \le x_{max} \Leftrightarrow x_{min} \le x(0) + k_v T_s \sum_{i=0}^{k} (q_{in}(i) - u(i)) \le x_{max} \tag{6}$$

where $u(i)$, $i = 0, \ldots, N$ are the optimization variables and $x(0)$ is the current state. Future inlet flows q_{in} are assumed to be constant and equal to the last known inlet flow value. If the inlet flow is directly measurable the last known value is $q_{in}(0)$ while if it is estimated a delay of one sample is needed. The observer (the Kalman filter) is, however, not particularly computationally demanding and is thus typically run at higher execution frequency than the MPC controller. It is thus a reasonable approximation that $q_{in}(0)$ is available. This gives the optimization problem

$$\min_{u(0),\ldots,u(N)} \max_k |u(k) - u(k-1)|$$
$$\text{subject to}$$
$$x(k+1) = x(k) + T_s k_v (q_{in}(0) - u(k)) \tag{7}$$
$$x(0),\ q_{in}(0),\ u(-1) \text{ known}$$
$$x(k) \in [x_{min},\ x_{max}],\ k = 0, 1, \ldots, N$$
$$u(k) \in [q_{min},\ q_{max}],\ k = 0, 1, \ldots, N$$

which, as shown in (Campo and Morari, 1989), can be reformulated as a linear program. The behavior of this formulation is shown in Figure 2 where an inlet flow step from 20% to 50% occurring at $t = 1$ is simulated. This simulation as well as the succeeding ones were performed using YALMIP, (Löfberg, 2004) and the Gurobi Mex interface, (Yin, 2009-2011). The optimal

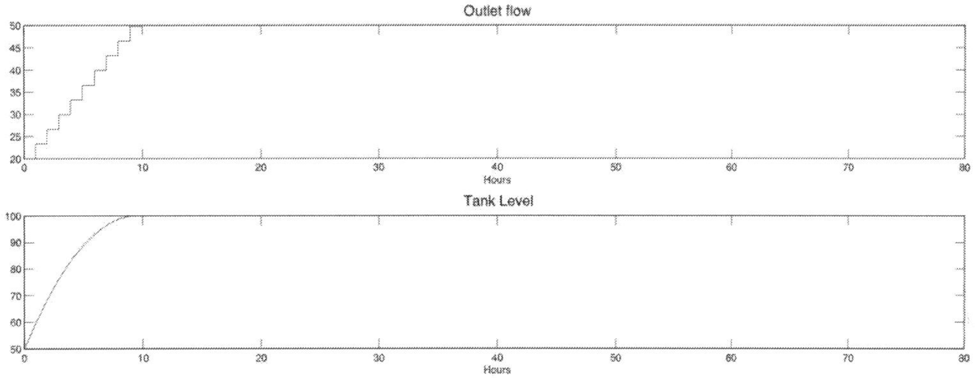

Figure 2. Simulation of model predictive control using $k_v = \frac{1}{3}$ h^{-1}, a sampling interval of 1 hour and a prediction horizon of 20 hours.

control strategy is, as pointed out by (Campo and Morari, 1989) and (McDonald et al., 1986), a ramp that removes the flow imbalance just as the tank reaches its limit. The performance criterion for the above formulation evaluates to 3.3.

The drawback with the above behavior should be rather obvious as a positive inlet flow change after $t = 10$, will have to be directly transferred to the outlet flow to avoid tank overflow. That the tank level stays on the tank boundary is on the other hand not surprising, since the controller is "told" that future inlet flows are constant and equal to $q_{in}(0)$. We know, however, that this will not be the case and consequently must augment the controller with a mechanism that returns the tank level to, e.g., 50% to prepare the system for the next upset. In the MPC formulation this can be achieved by augmenting (7) with the constraint

$$x(k+N) = 50\% \tag{8}$$

as suggested by Campo and Morari (1989). The control sequence $u(0),\ u(1), \ldots,\ u(N)$ found in every sample will thus bring back the tank level to 50% in the open-loop prediction. To illustrate the effect of the end-value constraint we simulate the system for three different prediction horizons 10, 20 and 60. The system response is shown in Figure 3 and the obtained performance criteria are shown in Table 1. Note, that since only $u(0)$ is actuated, the resulting settling time is greater than N and, according to Campo

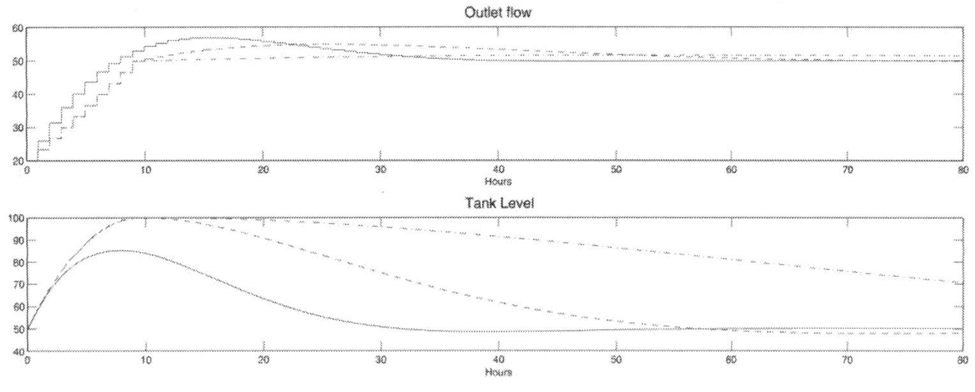

Figure 3. Simulation of model predictive control using $k_v = \frac{1}{3}$ h^{-1}, a sampling interval of 1 hour. The different prediction horizon are 10, 20 and 60 hours and drawn with solid, dashed and dash-dotted respectively.

and Morari (1989), often in the region between $2N$ and $2.5N$ sampling intervals. An effect of the end-value constraint is that the prediction horizon's importance as a tuning parameter is further increased compared to other MPC problems. In "normal" MPC

Table 1. Performance criteria for different prediction horizons for a step from 20% to 50%, $k_v = \frac{1}{3}$ and using end-value constraint.

Pred. Horizon	Objective
10	6.0
20	3.3
60	3.3

typically N should be chosen to capture the essential dynamic behavior (the transient) of the system and thus there is typically no obvious disadvantage (apart from an increased computational burden) nor any major advantage in using large N-values. When using (7) and (8) the selection of N does, however, become a delicate business since it affects the settling time of the system. We see, for example, in Table 1 that both 20 and 60 give the same performance criterion, 3.3, (which is optimal) but from Figure 3 it is clear that $N = 60$ gives a considerably longer settling time. A long prediction horizon does, however, filter more upsets (smaller ones) optimally than what a short horizon does, as shown in (Campo and Morari, 1989). This connection gives the user a way of tuning the controller. By specifying the smallest inlet flow upset, Δq_{in}, that should be filtered optimally the prediction horizon can be calculated according to

$$N \geq N_{crit} = \frac{1}{2|u^\star|}\left(2|\Delta q_{in}| + \sqrt{\alpha}\right) \tag{9a}$$

where

$$\alpha = \left(|u^\star| - 2|\Delta q_{in}|\right)^2 \tag{9b}$$

$$u^\star = \frac{2|\Delta q_{in}|}{k^\star + 1} - \frac{100}{k_v T_s k^\star (k^\star + 1)} \tag{9c}$$

$$k^\star = \left\lceil \frac{100}{k_v T_s |\Delta q_{in}|} \right\rceil \tag{9d}$$

as derived by (Campo and Morari, 1989). The notation $\lceil\,\rceil$ means rounding to the nearest larger integer. The above equations assume that the system is in steady state. Another tuning procedure would be to specify a desired settling time and then select T_s and N according to

$$NT_s = \frac{\text{Settling time}}{2} \tag{10}$$

In Figure 3 one of the drawbacks that might arise with this approach is visible. Even though the tank level eventually is returned to 50% the tank level deviation is still rather large. If, for example, a second positive inlet flow change occurs at $t \approx 10$ it would, more or less, have to be directly transferred to the outlet flow since little tank volume remains to surge the upset.

Robust MPC

The MPC controller in the previous section shares one feature with other previously suggested level controllers. Following an inlet flow step, there is a time interval when the available tank capacity to filter a subsequent upset is reduced. For example we see that for $t \in [5, 25]$ the tuning $N = 20$ will have less than 20% tank capacity available to surge a positive inlet flow step. To improve the handling of frequently changing inlet flows the controller should typically be tuned more tightly. An example of this can be seen in Figure 2 where $N = 10$ has considerably much more tank capacity available than $N = 20$.

By alternatively formulating the control problem in a robust MPC framework we can obtain a controller that achieves optimal filtering while accounting for rapidly varying inlet flows. Analogously to the previous approach, flow smoothing is obtained by minimizing $\max|u(k) - u(k-1)|$ but future inlet flows $q_{in}(1), q_{in}(2), \ldots, q_{in}(N)$ are treated as unknown but bounded $q_{in} \in [q_{min}, q_{max}]$, instead of being assumed to be constant. The robust MPC framework then permits to find an open-loop sequence $u(0), u(1), \ldots, u(N)$ that is robustly optimal. The control law which fulfills the aforementioned description is the solution to the following optimization problem

$$\begin{aligned}
&\min_{u(0),\ldots,u(N)} \; \max_{q_{in}(1),\ldots,q_{in}(N)} \; \max_{k} |u(k) - u(k-1)| \\
&\text{subject to} \\
&x(k+1) = x(k) + T_s k_v \left(q_{in}(k) - u(k)\right) \\
&x(0),\, q_{in}(0),\, u(-1) \text{ known} \\
&x(k) \in [x_{min}, x_{max}] \;\forall\, q_{in}(k) \in [q_{min}, q_{max}] \\
&q_{in}(k),\, u(k) \in [q_{min}, q_{max}]
\end{aligned} \tag{11}$$

Note that the above formulation does not include any mechanism that promotes the return of the tank level to 50% but only assures that the tank level stays within its limits.

The above formulation suffers, however, from severe conservativity which can even lead to lack of feasibility. The reason for this is that the controller tries to find an open-loop solution $u(0), u(1), \ldots, u(N)$ that will keep the system feasible for every possible inlet flow sequences $q_{in}(0), q_{in}(1), \ldots, q_{in}(N)$. This problem arises since the MPC formulation only uses information available at the present time when deciding the input signals N samples into the future. We know that in the "real world" new information (future measurements) will become available gradually. To mitigate the conservativity, (11) can be augmented with an artificial feedback mechanism, according to which, future control signals are decided. One simple policy is to assume that future control signals are given by the affine control law $u(k) = l_k x(k) + v(k)$. The drawback with this policy is that the optimization problem solved in every sample becomes bi-linear in the decision variables l_k and $v(k)$ and thus typically difficult to solve. An alternative policy, proposed by Löfberg (2003), which gives a convex optimization problem is to allow every future input to be a linear feedback of past disturbances

$$u(k) = \sum_{i \leq k-1} l_{k,i} q_{in}(i) + v(k) = L_k \mathbf{q_{in}} + v(k) \tag{12}$$

where $L_k = (l_{k,0}, l_{k,1}, \dots, l_{k,k-1}, 0, \dots, 0)$ and $\mathbf{q_{in}} = (q_{in}(0), q_{in}(1), \dots, q_{in}(N))^T$. The set of admissible states using (12) is in fact equivalent to the one using $u(k) = l_k x(k) + v(k)$, as shown in (Goulart et al., 2006). We will, however, still assume that $q_{in}(k)$ is available when $u(k)$ is decided and thus use $i \leq k$ in (12). This yields the optimization problem

$$\begin{aligned}
&\min_{L_{0:N},\, v(0:N)} \max_{q_{in}(1:N)} \max_k |(L_k - L_{k-1})\mathbf{q_{in}} + v(k) - v(k-1)| \\
&\text{subject to} \\
&x(k+1) = x(k) + T_s k_v (q_{in}(k) - u(k)) \\
&x(0),\ q_{in}(0),\ v(-1) \text{ known} \\
&x(k) \in [x_{min}, x_{max}]\ \forall q_{in}(k) \in [q_{min}, q_{max}] \\
&q_{in}(k),\ u(k) \in [q_{min}, q_{max}]
\end{aligned} \tag{13}$$

where the initial condition should be interpreted as $v(-1)$ equal to $u(0)$ from the last sample. Note that the control law used is still receding horizon control and not the mapping (12), i.e., only $u(0)$ is actually applied to the system . The mapping does, however, influence the MPC controller since it is part of the optimization problem. The optimization problem (13) can be formulated as a linear program as shown in (Löfberg, 2003) and solved by a standard LP-solver.

To illustrate the behavior of the robust MPC controller it is simulated on an artificial inlet flow which illuminates the overall behavior of the controller. The inlet and outlet flows as well as the resulting tank level are shown in Figure 4. The controller was initialized with $x(0) = 50\%$ and $v(-1) = q_{in}(0)$.

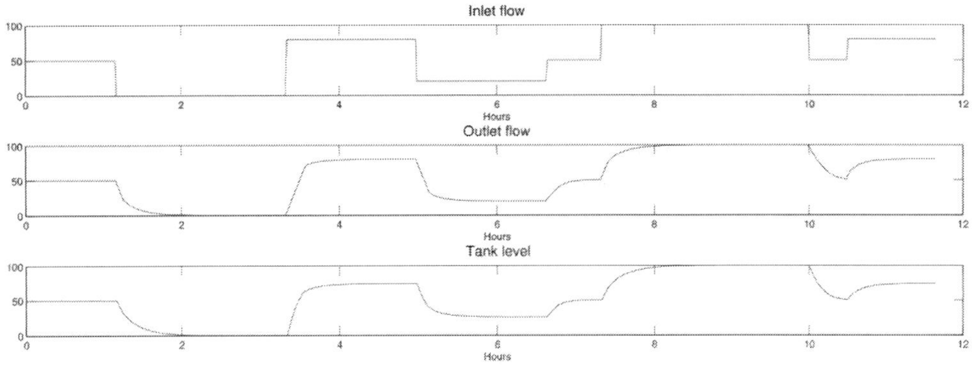

Figure 4. Simulation of closed loop robust MPC with a sampling period of one minute, a prediction horizon of 60 minutes and $k_v = \frac{1}{10}$.

The controller does not return the tank level to 50% following an inlet flow upset but instead adapts the steady state tank level to that of the inlet flow. For the case of $q_{in} = 0\%$ and $q_{in} = 100\%$ the steady state tank level is on the lower and upper boundary respectively. This behavior is an effect of the assumption that the inlet and outlet flows having equal and known capacities. For a real application any uncertainty of the capacity estimates could be handled by adding a safety margin to the tank level limits. That $q_{in} = 100\%$ does give a completely filled tank is, however, rather intuitive from filtering perspective. By assumption we know that the next upset must be downwards and by having a completely filled tank, maximum capacity is available to filter the upset.

Implementation Issues

Up to a prediction horizon of approximately 40 the computational burden of the robust controller is not prohibitive. The optimization is done in less than one second using YALMIP (Löfberg, 2004) and the Gurobi Mex interface of (Yin, 2009-2011) on a 2.67 GHz processor. The parameter space is also of low dimension, $x(0)$, $q_{in}(0)$ and $v(-1) = u(-1)$, and thus it should also be possible to obtain a very rapid explicit formulation, (Bemporad et al., 2002). The latter has, however, not yet been tested. All in all it should be possible to use the controller in a real time implementation even with modest computational power.

Tuning Knobs

The controller has two tuning knobs, the prediction horizon, N and the sampling interval, T_s. The combination of these, NT_s (the actual prediction time), should, as with most MPC controllers, be chosen to capture the overall dynamic behavior of the system. A small as possible sampling interval is of course preferred. With T_s fixed to 1, the effect N has on the mapping has been investigated by using an inlet flow that does steps of size 2% and observing the steady state tank level. The result is shown in

Figure 5. The choice $N = 10$ is too short to capture the transient of the system which is in the region of 15 hours. Using $N = 20$ (enough to capture the essential step response) and $N = 40$ (more than twice as long the transient) does not give rise to any major changes of the mapping. The controller is thus easy to tune, just choose NT_s so that the transient of the system is captured.

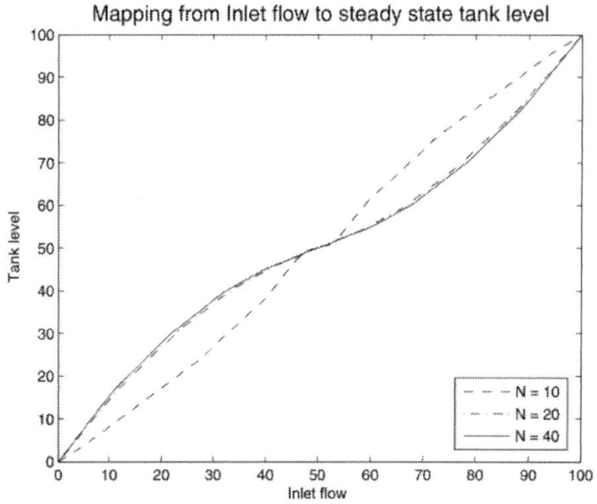

Figure 5. How the steady state tank level depends on the inlet flow for different prediction horizons using $T_s = 1$ and $k_v = \frac{1}{3}$.

Which controller to use

The robustly optimal control strategy is not to return the tank level to 50% but rather to adapt the new steady state tank level to the new inlet flow level. Previous approaches assume the inlet flow to be constant or at least to change very infrequently. Our approach, on the other hand, assumes that the inlet flow can change in every sample. A real inlet flow typically ends up somewhere in between these two extremas and one can expect that the inlet flow description which fits the considered process best will also achieve the best filtering. One advantage with our approach is, however, that the controller will never run into problems even if inlet flow changes are frequent.

Simulation Results

To give an indication of how the controller might work for Perstorp AB's application we show the results of using the real inlet flow data from Perstorp shown in Figure 1. This is, however, only data for one week and, admittedly, amongst the ones with frequent upsets. The data is sampled every minute and the tank has roughly a hold up time of 3 hours. The MPC controllers (both (13) and (7)) are executed every 30 minutes as it was a good compromise between limiting the simulation time and still gave fair filtering results. For a real-time application it would probably be wise to execute the controllers more frequently. The robust MPC controller was tuned with $N = 40$ to give a prediction time of 20 hours which captures the transient response. The tuning of the controller by Campo and Morari (1989), hereafter denoted CM MPC, is more involved. The inlet flow performs steps in the range $[5, 80]$ and some upsets occur shortly after the previous one. Obviously we can not filter all upsets optimally since that would give a controller with a very long settling time. One possibility is to say that upsets of magnitude $\Delta q_{in} = 20$ should be filtered optimally. That means that steps like the ones at $t \approx 89, 107, 143, 146$ and 155 will be filtered optimally. Inserting $\Delta q_{in} = 20$ and $T_s = 30$ into (9) gives the prediction horizon $N = 62$. Another approach on how to select N would be to use (10) and specify a desired settling time. To have returned the tank level to 50% when the upsets occur the settling time must be short. A settling time of 15 hours is probably too long to handle all upsets but should be enough for the major upsets at $t \approx 20, 50, 80, 140$ and 159. With $T_s = 30$ we obtain $N = 15$. Both configurations have been tested and compared to the robust MPC controller.

The resulting filtering performances are shown in Table 2 while the simulated outlet flow and tank level is shown in Figure 6. For comparison reasons we display both the maximum change of the outlet flow as well as the sum of the squared outlet flow changes. Note that the tuning of the CM MPC controllers are customized for this specific inlet flow sequence. Their performance on other inlet flows has not been investigated.

Table 2. Tuning and performance criteria for the different controllers.

Controller	Tuning	$\max \lvert u(k) - u(k-1) \rvert$	$\sum_{k=0}^{N} (u(k) - u(k-1))^2$
CM MPC[1]	$T_s = 30$ min, $N = 62$	47.4	3795
CM MPC[2]	$T_s = 30$ min, $N = 15$	33.3	3318
Robust MPC	$T_s = 30$ min, $N = 40$	7.7	1817

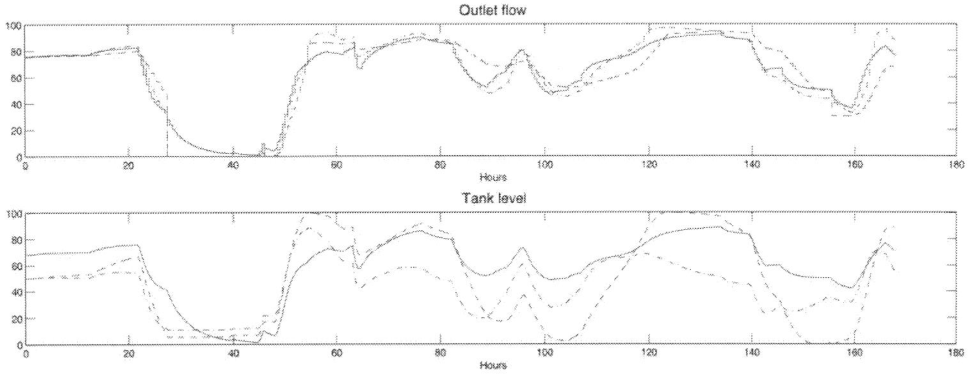

Figure 6. Outlet flow and tank level for the robust MPC (solid), CM MPC[1] (dashed) and CM MPC[2] (dash-dotted).

Conclusions

By approaching the averaging level control problem via robust MPC a controller that can handle frequently changing inlet flows was obtained. The behavior of the controller resembles that of the proportional only controller in that it does not return the tank level to 50%. The steady state tank level is instead adapted to the current inlet flow level.

The robust MPC controller depends on accurate estimates of the minimum and maximum inlet and outlet flow levels. This could be obtained by process knowledge and by investigating historical data. It was shown that the steady state tank level depends on the prediction horizon, but this dependency is almost constant for horizons that capture the essential transient behavior of the system. This makes the controller easy to tune compared to many other level control approaches.

One issue that remains to investigate is if it is possible to give guidelines for when it is optimal to use the robust MPC controller and when it is indeed optimal to use a controller that returns the level to 50%. Given a typical time interval between upsets, the size of the upsets and the hold-up of the tank it should be possible to give guidelines on which controller that is optimal to use.

The concept of not having a fixed set-point but rather letting the tank level depend on the inlet flow could be incorporated into other control structures. One example would be to use a standard PI controller whose set-point is given by a mapping from the inlet flow. Another possibility would be to use the optimal controller and replace the end-value constraint with one depends on the current inlet flow level.

Acknowledgments

This work was funded by the Swedish Foundation for Stategic Research as part of Process Industry Center Linköping (PIC-LI).

References

A Bemporad, F Borrelli, and M Morari. Model Predictive Control Based on Linear Programming – The Explicit Solution. *IEEE Transactions on Automatic Control*, 47:1974–1985, December 2002.

P.J. Campo and M. Morari. Model predictive optimal averaging level control. *AIChE Journal*, 35(4):579–591, Apr 1989.

T. Cheung and W.L. Luyben. Liquid-level control in single tanks and cascade of tanks with proportional-only and proportional-integral feedback control. *Industrial and Engineering Chemistry Fundamentals*, 18(1):15–21, Jan 1979.

P.J. Goulart, E.C. Kerrigan, and J.M. Maciejowsky. Optimization over state feedback policies for robust control with constraints. *Automatica*, 42(4):523–533, Apr 2006.

M. Khanbaghi, B. Allison, and R. Harper. Optimal averaging level control for surge tanks. In *IEEE Conference on Advanced Process Control Applications for Industry*, Vancouver, Canada, 2001.

J. Löfberg. Approximations of closed-loop minimax MPC. In *Proceedings of the 42nd Conference on Decicion and Control*, Maui, USA, Dec 2003.

J. Löfberg. Yalmip : A toolbox for modeling and optimization in MATLAB. In *Proceedings of the CACSD Conference*, Taipei, Taiwan, 2004. URL `http://users.isy.liu.se/johanl/yalmip`.

W.L. Luyben and P.S. Buckley. A proportional-lag level controller. *Instrumentation Technology*, 18(1):65–68, 1977.

J.M. Maciejowski. *Predictive Control with Constraints*. Prentice Hall, 1st edition, 2002. ISBN 0201398230.

K.A. McDonald, T.J. McAvoy, and A. Tits. Optimal averaging level control. *AIChE Journal*, 32(1):75–86, Jan 1986.

J. Shin, J. Lee, S. Park, K. Koo, and M. Lee. Analytical design of a proportional-integral controller for constrained optimal regulatory control of inventory loop. *Control Engineering Practice*, 16(11):1391–1397, Nov 2008.

J.P. Shunta and W. Fehervari. Nonlinear control of liquid-level. *Instrumentation Technology*, 23(1):43–48, Jan 1976.

K.L. Wu, C.C. Yu, and Y.C. Cheng. A two degree of freedom level control. *Journal of Process Control*, 11(3):311–319, Jun 2001.

W. Yin. Gurobi Mex: A MATLAB interface for Gurobi, 2009-2011. URL `http://convexoptimization.com/wikimization/index.php/gurobi_mex`.

(622f) Effect of Uncertainty In Time Delay On Multi-Step-Ahead Predication Performance of Some Black Box Models

Lemma D. Tufa, Chemical Engineering Department, Universiti Teknologi PETRONAS, Tronoh, Malaysia and Marappagounder Ramasamy, Chemical Engineering, Universiti Teknologi PETRONAS,, Tronoh, Malaysia

Abstract

Multi-step-ahead prediction is part of several model based technologies. One of the most illustrative example is in model predictive control (MPC). In MPC, in many cases, the value of the control signal is calculated by penalizing the future deviation of the trajectory of the system output from a desired output trajectory. The future trajectory of the plant is predicted using the plant model. The most commonly used predictive models are the black box models. For linear time invariant systems, step response models, ARX and FIR models are commonly used. Orthonormal Basis Filter Based MPC is also becoming popular in literature.

It is believed that the prediction performance of models depend on several factors, like the quality of the excitation signals used during identification test, model structure selection and model parameter estimation. While there are several possible model structures to select from, understanding the strengths and weaknesses of the various model structures greatly helps in selecting a structure for a specific application. One of the issues related to model structures is time delay. It is known that model structures like ARX, ARMAX and Box Jenkins need the time delay to be separately estimated and included in the model during the parameter estimation. Nevertheless, time delay estimation itself is not a trivial task and generally involves uncertainty in its accuracy. Therefore, it is essential to understand how uncertainties in time delay affect the multi-step-ahead prediction performance of the models. On the other hand, model structures like OBF, ARX-OBF, has inherent capability to estimate the time delay of the system because of their Pade' like structures. They do not necessarily need a separate estimate of the time delay during model development.

In this paper, the effect of uncertainty in time delay on the prediction performance of various model structures is studied and compared to OBF based structures. The study includes both open-loop and closed-loop identification simulation case studies. The result shows that one-step-ahead prediction is not affected by the uncertainty for all types of structures. While as the number of steps-ahead and the uncertainty increases, the prediction performance of structures like ARX and ARMAX deteriorates much more than that of the OBF-Based model structures.

See more of this Session: Poster Session - Area 10B: Systems and Process Control

(622g) Energy Network Dispatch Optimization Under Emergent Events of Local Energy Shortage

Tianxing Cai, Dan F Smith Department of Chemical Engineering, Lamar University, Beaumont, TX and Qiang Xu, Dan F. Smith Department of Chemical Engineering, Lamar University, Beaumont, TX

Agile energy dispatching due to emergent local energy shortage through a complex energy network is one of the critical tasks in the world, especially in certain emergent situations. In recent several years, there are frequent occurrences of extreme events such as U.S. Southern Storm and Tornado (2011), Japan's earthquake and tsunami (2011), India power shortage (2010), and U.S. Atlantic Hurricane (2005). Such kind of events always causes problems of local energy shortage, which make local areas suffer from rescue delay, large-scale blackout, economic losses, and even social safety problems. Therefore, the reduction of energy recovery time should be the first priority of the emergency rescue plan.

In his study, a new methodology is developed for energy network dispatch under emergent event of local energy shortage. This methodology has included two stages. In the first stage, the energy network has been structured and characterized. The capacity, quantity, and availability of different energy sources have been determined, where multi-dimensional "Energy Vector" and "Energy Source Vector" are proposed. In the second stage of modeling optimization, the discussion of local energy shortage has included the conditions of both energy source shortage and transmission network interruption. The optimization problem can be formulated as the sequence scheduling of energy source dispatch, energy conversion technology and energy transportation with the introduction and adoption of energy conversion efficiency metrics, energy conversion speed metrics and energy transportation speed vector. The objective of the optimization problem is to minimize the restoration time of local energy supply. Due to uncertainty in demand quantity of energy shortage, parametric programming is utilized for the comprehensive consideration.

See more of this Session: Poster Session - Area 10B: Systems and Process Control
See more of this Group/Topical: Computing and Systems Technology Division

Modeling and Control of the Solar Powered Membrane Distillation System

Jiann-Shiun Lin[a], *Hsuan Chang*[b], and *Gow-Bin Wang*[a*]

[a] *Department of Chemical and Materials Engineering, Chang Gung University, Tao-Yuan 33302, Taiwan*

[b] *Department of Chemical and Materials Engineering / Energy and Opto-Electronic Materials Research Center, Tamkang University, Taipei 25137 Taiwan*

Abstract

In many arid and semiarid regions desalination of sea and brackish water has become a necessity. It is commonly reported that the lack of fresh water will increase dramatically in the near future. It is known solar radiation can be transferred to thermal energy by collectors and used to solar thermal energy systems. There are many potentially commercial advantages to use seawater desalination systems driven by solar power.

Membrane distillation (MD) is a thermal process in which only vapor molecules, driven by a difference in temperature, transport through porous hydrophobic membranes and condense on the opposite side. This work studies the dynamic modeling and process control of the solar powered membrane distillation system. Direct contact membrane distillation (DCMD) is considered in the proposed solar powered MD system. The mathematical models of solar collector, thermal storage tanks, heat exchanger, MD module and fresh water storage tank are built on MATLAB platform. The simulated dynamic characteristics of these units are compared with those shown in the literature.

This study also presents two temperature control loops and determines the feasible PI controller parameters for the solar powered MD systems. According to surrounding climate data of Taipei and Kaohsiung cities of Taiwan reported in the literature, the simulation results demonstrate effective performance of the proposed control loops for the solar powered MD systems.

Introduction

The water desalination was started early in 20th century. True expansion and spread of the water desalination industry occurred during the 1960s. It is well known that multistage flash (MSF) and reverse osmosis (RO) technologies are the main standards in seawater desalination industry. However, RO desalination gradually becomes the leading desalination technology due to low energy requirement, low operating temperature, improved materials and low production cost. Recently it has been reported the cost-efficient DCMD is apparently advantageous over RO in some real applications (Bui et al., 2010).

Figure 1 shows the schematic diagram of the solar powered membrane distillation system proposed in this work. Main units of this system include a solar collector, two thermal storage tanks, a heat exchanger, a MD module and a fresh water storage tank. Here, direct contact membrane distillation (DCMD) is considered as the MD module because it is best suited for applications in which the major fluxing component is water. The model equations of each unit of the proposed solar powered MD system are built and simulated on MATLAB platform (Lin, 2010).

DCMD Performance Analysis

For the DCMD module shown in Figure 1, hot liquid and cold liquid are in cross flow layout and allocated in shell side and tube side, respectively. It is known various theoretical mathematical models have been developed for different MD configurations (Lawson and Lloyd, 1997). In this work, the mathematical models for DCMD module are according to those proposed by Chang and Liau (2009). In the models, correlations for mass and heat transfer coefficients are used for the liquid side, cold liquid side and the condensing water. For the porous membrane, Knudsen diffusion and molecular diffusion are taken into account for mass transfer, while vapor conduction and membrane material conduction are considered for heat transfer. Figure 2 illustrates the effects of feed temperature on overall DCMD performance. It shows the responses of experimental mass flux and simulated mass flux are very close. The simulated dynamic characteristics of other units shown in Figure 1 are also compared with those shown in the literature.

Dynamic Simulation for the Solar Powered DCMD System

Recently, Oh et al. (2008) have taken islandwide solar radiation data from Central Weather Bureau (CWB) of Taiwan as proof to analyze the climatic characteristics of Taiwan. By analyzing the relationship between solar radiation amount and related climate data, it is able to calculate reliable solar radiation data of each weather station by statistic methods. Figure 3 shows the average solar radiation energy and environmental temperature in June for the Taipei and Kaohsiung cities of Taiwan (Oh et al., 2008).

The overall control loops of the solar powered DCMD system are shown as Figure 4. Controller C1 is used to switch the flow direction of the working fluid from outlet of the solar collector. Two temperature control loops are also set for thermal storage tank S2 and fresh water storage tank S3, respectively. The feasible PI controller parameters for these two temperature control loops are then determined. According to surrounding climate data of Taipei and Kaohsiung cities of Taiwan shown as Figure3, the temperature control performances for the solar powered MD system are illustrated in Figure 5. The simulation results demonstrate effective performance of the proposed control loops for the solar powered MD systems.

Conclusions

Membrane distillation (MD) is a thermal process in which only vapor molecules, driven by a difference in temperature, transport through porous hydrophobic membranes and condense on the opposite side. This work studies the dynamic modeling and process control of the solar powered membrane distillation system. Direct contact membrane distillation (DCMD) is considered in the proposed solar powered MD system. The mathematical models of solar collector, thermal storage tanks, heat exchanger, MD module and fresh water storage tank are built on MATLAB platform. The simulated dynamic characteristics of these units are compared with those shown in the literature.

This study also presents two temperature control loops and determines the feasible PI controller parameters for the solar powered MD systems. According to surrounding climate data of Taipei and

Kaohsiung cities of Taiwan reported in the literature, the simulation results demonstrate effective performance of the proposed control loops for the solar powered MD systems.

Acknowledgements

Financial support from Chang Gung University (UERPD 280261) is gratefully acknowledged.

References

1. Bui, V. A., L. T. T. Vu and M. H. Nguyen (2010), "Simulation and Optimisation of Direct Contact Membrane Distillation for Energy Efficiency," Desalination, 259, 29-37.
2. Chang, H. and J. S. Liau (2009), "Simulation of Membrane Distillation Modules for Desalination by Developing User's Model on Aspen Plus Platform," Desalination, 249, 380-387.
3. Lawson, K. W. and D. R. Lloyd (1996), "Membrane Distillation II. Direct Contact MD," Journal of Membrane Science, 120, 123-133.
4. Lawson, K. W. and D. R. Lloyd (1997), "Membrane Distillation," Journal of Membrane Science, 124, 1-25.
5. Lin, J. S. (2010), Modeling and Control of the Solar Powered Membrane Distillation System, M.S. Thesis, Dept. of Chemical and Materials Eng., Chang Gung University, Tao-Yuan, Taiwan. (in Chinese)
6. Oh, W. S., M. C. Ho, J. L. Chen, J. F. Chen and S. C. Lo (2008), "The Study on the Typical Radiation for Solar Architecture Design of Taiwan," Journal of Architecture, 64, 103-118. (in Chinese)

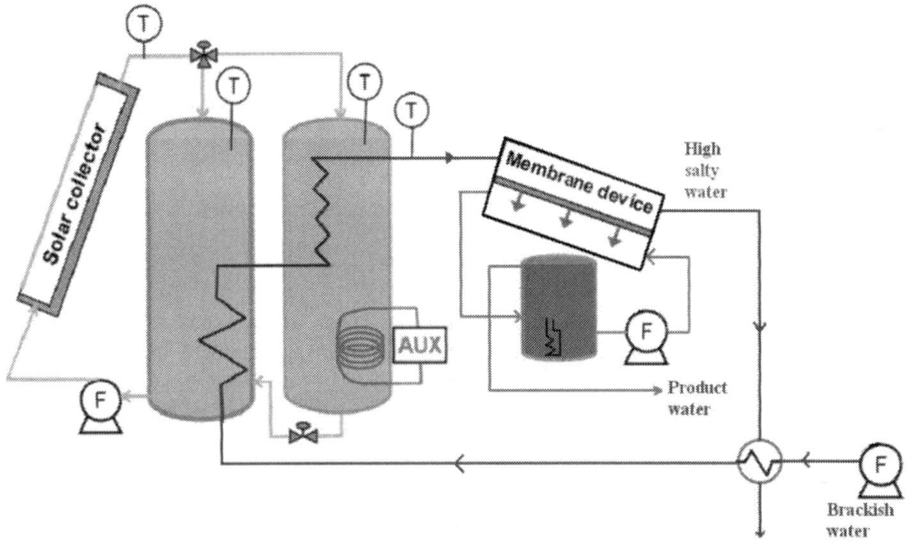

Figure 1. Schematic diagram of the solar powered MD system (Lin, 2010).

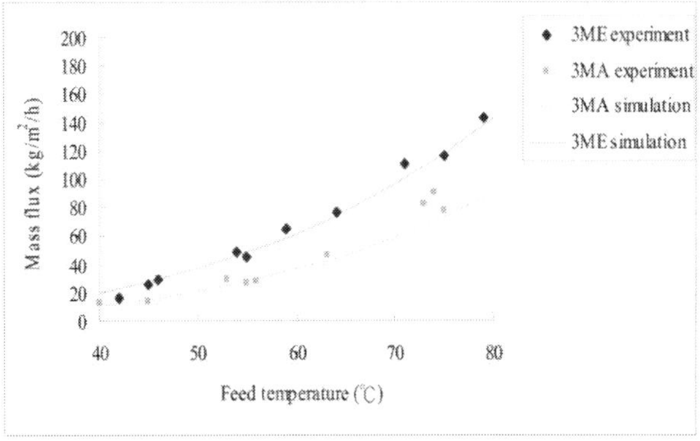

Figure 2. Comparison of experimental and simulated mass flux (Lawson and Lloyd, 1996).

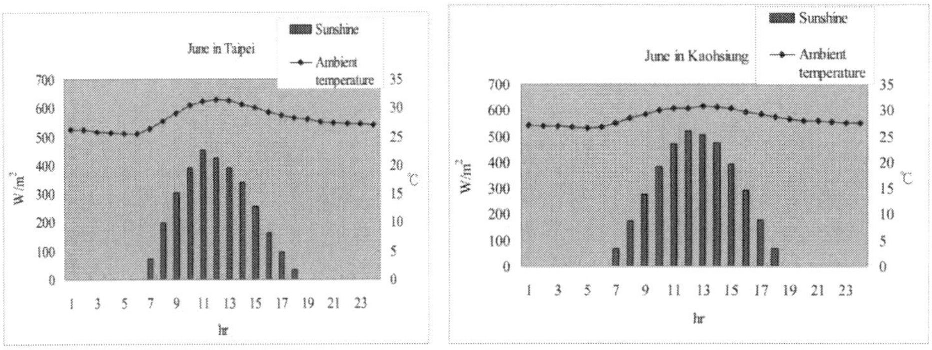

Figure 3. Surrounding climate data of Taipei and Kaohsiung cities of Taiwan (Oh et al., 2008).

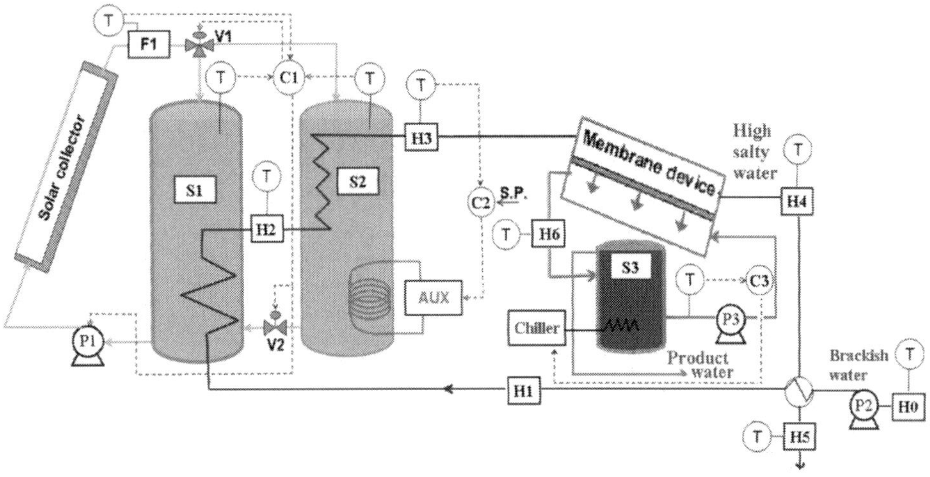

Figure 4. Solar powered MD control system (Lin, 2010).

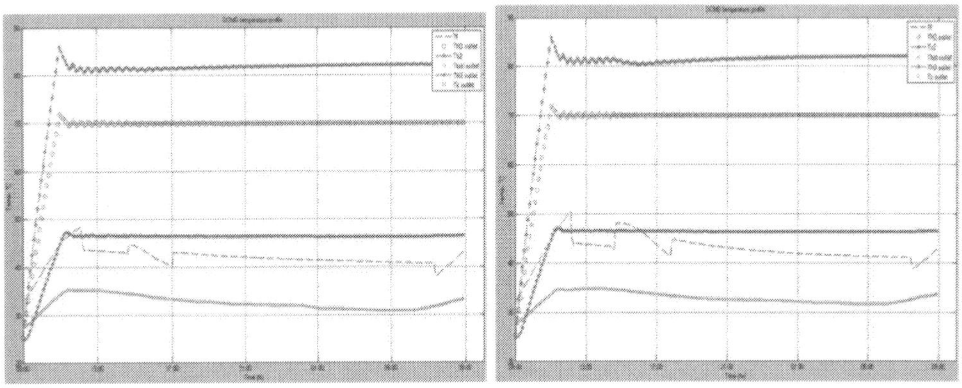

Figure 5. Temperature control performances for the solar powered MD system.

(622i) The Square-Root Unscented Kalman Filter for Leak Detection and Identifying Location of Leaks In Natural Gas Pipelines

Weiting Tang, Chemical Engineering, Texas Tech University, Lubbock, TX and M. Nazmul Karim, Department of Chemical Engineering, Texas Tech University, Lubbock, TX

Pipelines are the most efficient and economical way to transport and distribute natural gases all over the world. Leaks are the main problem of pipeline, which may cause excessive economical losses, or even worse, pose health hazards to communities. Therefore, designing a timely and reliable leak detection system is essential to maintain the normal operation of the pipeline company. In this work, a square-root unscented Kalman filter (SR-UKF) is proposed for leak detection and identifying location of the leak. The gas pipeline is modeled as a couple of quasi-linear hyperbolic partial differential equations (PDEs). Method of characteristics (MOC), which is suitable for large scale systems or systems with complex boundary conditions, is used to solve the quasi-linear hyperbolic PDEs. The proposed SR-UKF is applied to estimate the leak size and the location of the leak. The SR-UKF has the advantage of ease of implementation and has superior numerical stability. Comparative studies of extended Kalman filter (EKF) and the proposed SR-UKF are carried out to show the superiority of the proposed method.

See more of this Session: Poster Session - Area 10B: Systems and Process Control
See more of this Group/Topical: Computing and Systems Technology Division

(622j) Constrained Nonlinear State Estimation of An Acid Gas Removal Process As Part of An Integrated Gasification Combined Cycle (IGCC) Power Plant with CO2 Capture

Debangsu Bhattacharyya[1], **Prokash Paul**[1], Richard Turton[1] and Stephen E. Zitney[2], (1)Department of Chemical Engg., West Virginia University, Morgantown, WV, (2)U.S. Department of Energy, National Energy Technology Laboratory, Morgantown, WV

In future IGCC plants with CO_2 capture, the acid gas removal unit (AGR) needs to satisfy strict emission standards for CO_2 and H_2S. With the current target of capture of CO_2 and the sulfur species in the AGR, CO_2 concentration in the clean syngas is expected to be 2-5% (vol) while the H_2S concentration (in ppmv) is expected to be in the single digits. Since in a full-scale IGCC plant, the amount of syngas is large, small deviations from the target values can result in significant amounts being vented to the atmosphere. The reliability and precision of the typical sensors measuring gas compositions are usually low. In addition, the typical sampling periods for these sensors are in the orders of several minutes. A large noise-to-signal ratio further complicates the situation. In addition, a number of states in this system is very important for effective control of the plant. To address this problem, a constrained nonlinear state estimator is developed in this work.

The AGR plant, modeled in this work, is a two-stage unit for selective removal of H_2S and CO_2 using a physical solvent SELEXOL®. In the first stage, H_2S is captured in an absorber and then recovered in a stripper before sending it to the Claus unit for sulfur recovery. An intermediate stripper column, known as the H_2S concentrator, ensures that the concentration of H_2S in the Claus feed satisfies certain limits. In the second stage, CO_2 is captured in an absorber and then recovered by multi-stage flash blocks before being sent for compression. An intermediate flash vessel ensures that the H_2 slip in the stripped-off CO_2 does not exceed some stipulated guide-lines. In the chilled solvent process, the solvent is cooled to about 4°C by using a NH_3 refrigeration process. For calculating accurately the thermodynamic properties of the system, the simulation is implemented using the perturbed chain statistical associating fluid theory (PC-SAFT) EOS where the model parameters are regressed from experimental data.

For linear systems with linear constraints, various algorithms can be considered that can lead to the same optimal constrained linear state estimates. However, constrained state estimation in a nonlinear system is usually not optimal and largely depends upon the algorithm being considered for a particular system.

The estimation problem becomes complicated due to the presence of a multimodal probability density function. For the AGR process under consideration, state constraints naturally arise due to non-negative concentrations of the chemical species. In addition, this system has more than one thousand states. So an efficient algorithm is also needed for solving the problem with reasonable computational effort. In this work, the nonlinear constrained state estimation problem is solved by using a variant of the unscented Kalman filter (UKF) that uses a Jacobian free algorithm. A nonlinear programming problem (NLP) is considered in the update step considering constraints on individual updated sigma points and updated states. With certain simplifications, the NLP problem is reduced to a quadratic programming (QP) problem. Computational performance, estimation errors, and constraint errors of this constrained UKF are compared with the results from the unconstrained Kalman filter and unconstrained UKF.

See more of this Session: Poster Session - Area 10B: Systems and Process Control
See more of this Group/Topical: Computing and Systems Technology Division

(622k) Nonlinear Control of Dynamic Processes Using Models Based On Granular Runge-Kutta Methods

Tomas Co, Chemical Engineering, Michigan Technological University, Houghton, MI

Granular computing have previously been applied to process modeling of dynamic processes using several approaches including fuzzy logic and neural networks. They allow for different levels of granular (coarse to fine) description of a process. As such, different control strategies can be used to arrive at the target, as well as different compensators for disturbance rejection can be implemented based on the local topology of the states. However, these models are predominantly based on discrete-time models and approaches. In this paper, we combine the well-known method of Runge-Kutta with a granular domain description of the states and external inputs to obtain continuous-time process description. This approach allows us to obtain critical nonlinearity of the continuous-time dynamics as well as the multiplicity of steady states of the process. Moreover, the granular description has the advantage of allowing smooth switches among different control modes via hybrid control methods. One proposed strategy is to use the steady state granular maps for feedforward control, while using the granular dynamic model to design and implement nonlinear feedback strategies. Another approach is to use the granular models for predictive control. The paper focuses on the former appoach. We will include analytical and simulation results based on the proposed approach. The results are very encouraging and point to a generic and effective approach to nonlinear control that also provide a practical generalization of other approaches including gain-scheduled and other hybrid control methods.

See more of this Session: Poster Session - Area 10B: Systems and Process Control
See more of this Group/Topical: Computing and Systems Technology Division

Separating the Identifications of Nonlinear and Linear Parts in Block-Oriented Nonlinear Models Using Subspace Identification Approach

Jyh-Cheng Jeng*

Department of Chemical Engineering and Biotechnology
National Taipei University of Technology, Taipei, Taiwan
*(Tel : +886-2-27712171 ext. 2540; E-mail: jcjeng@ntut.edu.tw)

Abstract: In this work, novel noniterative algorithms for the identification of (multivariable) block-oriented nonlinear models (Hammerstein and Wiener models), consisting of a cascade combination of a linear dynamic block and a nonlinear static one, are presented. The proposed algorithms enable the identifications of the nonlinear and linear blocks fully separated without any special design for input test signals. Therefore, both the experimental and computational complexities during the identification procedures can be significantly reduced. The proposed method represents the linear dynamics as finite impulse response (FIR) model and approximates the static nonlinearity by expansions of basis functions. The key identification algorithms are derived from the subspace identification method.

Keywords: Block-oriented nonlinear models; Nonlinear identification; Subspace identification method; Process control.

1. INTRODUCTION

In the last decades, a considerable amount of research has been carried out on modeling, identification, and control of nonlinear systems. Most dynamical systems can be better represented by nonlinear models, which are able to describe the global behavior of the system over wide ranges of operating conditions, rather than by linear ones that are only able to approximate the system around a given operating point. One of the most frequently studied classes of nonlinear models are the so-called block-oriented nonlinear models [1], which involve a cascade combination of a linear dynamic block and a nonlinear static (memoryless) one. Such models are related very closely to linear ones and can be easily adapted to linear control techniques. Two typical block-oriented model structures are the Hammerstein and Wiener models, as shown in Fig. 1. In the Hammerstein structure, the linear dynamic element is preceded by the nonlinear static element. The order of connection is reversed in the Wiener structure. These model structures have been successfully used to describe nonlinear systems in a number of practical applications in the areas of chemical processes, biological processes, signal processing, communications, and control [2-4].

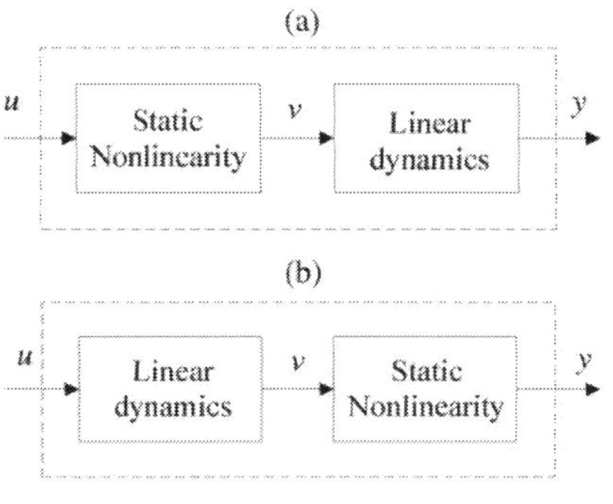

Fig. 1 Schematic diagrams of block-oriented models: (a) Hammerstein model (b) Wiener model.

Many approaches have been proposed in the literature for the identification of Hammerstein and/or Wiener models, and they fall into two main groups. The first group is simultaneous parameter identification of the linear and nonlinear elements, which usually leads to iterative procedures or nonlinear optimization problems [5-7]. The major difficulty is that the convergence of parameter estimates to the actual values is not always warranted. The second group is elaborated to separate the identification of the nonlinear static part from that of the linear dynamic part. This often involves a special design of the test input signal that enables the decoupling of the linear and nonlinear parts [8-11]. The requirement on the test signal would cause extra implementation complexity during the identification test.

To overcome the aforementioned drawbacks in the existing methods, this work develops a novel method for the identification of block-oriented nonlinear models (Hammerstein and Wiener models). The linear dynamic element is represented as its finite impulse response (FIR) model, and the static nonlinearity is approximated by expansions of basis functions. The identification problem is to estimate the unknown FIR model and expansion coefficients characterizing the linear and the nonlinear parts, respectively, from a data set of observed input-output (u and y) measurements. Notice that the intermediate variable v, as shown in Fig. 1, is non-accessible. The algorithm is derived from the subspace identification method (SIM). In comparison with other works, the proposed method has three advantages: (1) special design of the test input signal is not required; (2) parameter estimations of the nonlinear and linear parts are decoupled; (3) iterative and/or nonlinear optimization procedures are avoided. These features significantly reduce the identification complexities in process test and computation, which makes the proposed method more suitable for practical applications.

2. SUBSPACE IDENTIFICATION METHOD

Subspace identification method (SIM) is originally developed for the estimation of system matrices in a linear state-space model, and offers an attractive alternative to input-output methods due to its simple and general parameterization for multivariable systems. Recently, SIM has also been used for the identification of FIR model and dynamic matrix for model predictive control (MPC) application [12, 13].

A brief description for part of the subspace identification method, which is useful for FIR estimation, is first summarized as follows. The innovation form of state-space representation of linear systems is given by:

$$\begin{aligned} x_{k+1} &= Ax_k + Bu_k + Ke_k \\ y_k &= Cx_k + Du_k + e_k \end{aligned} \qquad (1)$$

where $x_k \in \mathbb{R}^n$, $u_k \in \mathbb{R}^{n_u}$, $y_k \in \mathbb{R}^{n_y}$, and $e_k \in \mathbb{R}^{n_y}$ are the system state, input, output, and white noise (innovations), respectively. The symbols n, n_u, and n_y denote the number of state, input, and output, respectively. Matrices A, B, C, D are system matrices with appropriate dimensions, and K is the steady-state Kalman gain. Based on the innovation form in Eq.(1), an extended state-space model can be formulated as [14]:

$$Y_f = \Gamma_N X_f + H_N^d U_f + H_N^s E_f \qquad (2)$$

where the subscript f denotes future horizon. In Eq.(2), the future input, output, and innovation block-Hankel matrices U_f, Y_f, and E_f, respectively, are defined as:

$$U_f \triangleq U_{N|2N-1} \triangleq \begin{pmatrix} u_N & u_{N+1} & \cdots & u_{N+j-1} \\ u_{N+1} & u_{N+2} & \cdots & u_{N+j} \\ \vdots & \vdots & \ddots & \vdots \\ u_{2N-1} & u_{2N} & \cdots & u_{2N+j-2} \end{pmatrix} \qquad (3)$$

$$Y_f \triangleq Y_{N|2N-1} \triangleq \begin{pmatrix} y_N & y_{N+1} & \cdots & y_{N+j-1} \\ y_{N+1} & y_{N+2} & \cdots & y_{N+j} \\ \vdots & \vdots & \ddots & \vdots \\ y_{2N-1} & y_{2N} & \cdots & y_{2N+j-2} \end{pmatrix} \qquad (4)$$

$$E_f \triangleq E_{N|2N-1} \triangleq \begin{pmatrix} e_N & e_{N+1} & \cdots & e_{N+j-1} \\ e_{N+1} & e_{N+2} & \cdots & e_{N+j} \\ \vdots & \vdots & \ddots & \vdots \\ e_{2N-1} & e_{2N} & \cdots & e_{2N+j-2} \end{pmatrix} \qquad (5)$$

where the subscript $k_1|k_2$ indicates that the first column of the matrix starts from time k_1 and ends by time k_2. The state matrix, X_f, is defined as:

$$X_f \triangleq X_N = \left(x_N \; x_{N+1} \; \cdots \; x_{N+j-1} \right) \qquad (6)$$

The extended observability matrix Γ_N is given as:

$$\Gamma_N = \begin{pmatrix} C \\ CA \\ \vdots \\ CA^{N-1} \end{pmatrix} \qquad (7)$$

The lower triangular block-Toeplitz matrices H_N^d and H_N^s are given by:

$$H_N^d = \begin{pmatrix} D & 0 & \cdots & 0 \\ CB & D & \cdots & 0 \\ \vdots & \vdots & \ddots & \vdots \\ CA^{N-2}B & CA^{N-3}B & \cdots & D \end{pmatrix} \qquad (8)$$

$$H_N^s = \begin{pmatrix} I & 0 & \cdots & 0 \\ CK & I & \cdots & 0 \\ \vdots & \vdots & \ddots & \vdots \\ CA^{N-2}K & CA^{N-3}K & \cdots & I \end{pmatrix} \qquad (9)$$

The Kalman state X_f is unknown, but it can be estimated from past input and output data as [13]:

$$X_f = L_p \begin{pmatrix} Y_p \\ U_p \end{pmatrix} \triangleq L_p W_p \qquad (10)$$

where the subscript p denotes past horizon, L_p is the subspace matrix corresponding to past data, and the past input and output block-Hankel matrices U_p and Y_p, respectively, are defined as:

$$U_p \triangleq U_{0|N-1} \triangleq \begin{pmatrix} u_0 & u_1 & \cdots & u_{j-1} \\ u_1 & u_2 & \cdots & u_j \\ \vdots & \vdots & \ddots & \vdots \\ u_{N-1} & u_N & \cdots & u_{N+j-2} \end{pmatrix} \qquad (11)$$

$$Y_p \triangleq Y_{0|N-1} \triangleq \begin{pmatrix} y_0 & y_1 & \cdots & y_{j-1} \\ y_1 & y_2 & \cdots & y_j \\ \vdots & \vdots & \ddots & \vdots \\ y_{N-1} & y_N & \cdots & y_{N+j-2} \end{pmatrix} \qquad (12)$$

Substituting Eq. (10) into Eq. (2), one can write

$$Y_f = L_w \begin{pmatrix} Y_p \\ U_p \end{pmatrix} + H_N^d U_f + H_N^s E_f \quad (13)$$

where $L_w = \Gamma_N L_p$. Thus, the subspace matrices L_w and H_N^d can be identified from data Hankel matrices using regression techniques such as least-squares method, provided the following conditions satisfied [13]:

- The input u_k is uncorrelated with e_k.
- The input u_k is persistently exciting.
- The number of measurement is sufficiently large (i.e. j is sufficiently large).

With the input-output data, the solution of this least-squares problem is given by:

$$\begin{pmatrix} L_w & H_N^d \end{pmatrix} = Y_f \begin{pmatrix} W_p^T & U_f^T \end{pmatrix} \left[\begin{pmatrix} W_p \\ U_f \end{pmatrix} \begin{pmatrix} W_p^T & U_f^T \end{pmatrix} \right]^{-1} \quad (14)$$

The matrix L_w is useful for the estimation of system matrices (A, B, C, D, K), but it is not needed here because our goal is to identify the FIR model which can be extracted from the subspace matrix H_N^d. The relationship between FIR sequences, $\mathbf{h} = \{h_i\}_{i=1,2,3,\cdots}$, and state-space matrices is given as:

$$h_i = CA^{i-1}B, \quad i = 1, 2, 3, \cdots \quad (15)$$

From Eq. (8), it is found that the first column of matrix H_N^d contains the first N sequences of FIR. In other words, the first N sequences of FIR (i.e. h_i, i=0,1,...,N-1) can be obtained from the solution of H_N^d in Eq. (14).

It is noted that the future and past horizons, N, can be arbitrarily chosen in this algorithm, which is the key of the proposed method to enable decoupling of the identification of linear and nonlinear parts in block-oriented nonlinear models.

3. IDENTIFICATION OF BLOCK-ORIENTED NONLINEAR MODELS

In the proposed method, the nonlinear static part is first identified. Generally, the system matrix D in state-space models of most (chemical) dynamic systems is zero (i.e., the first FIR parameter is zero). Therefore, by choosing the future horizon as N =1, the matrix H_N^d given in Eq. (8) reduces to a zero matrix.

3.1 Hammerstein model

By choosing N =1, the subspace matrix equations of Eq. (13) for the linear dynamic element in Hammerstein model can be written as:

$$Y_f = L_w \begin{pmatrix} Y_p \\ V_p \end{pmatrix} + H_N^s E_f \triangleq \begin{bmatrix} L_{w,y} & L_{w,v} \end{bmatrix} \begin{pmatrix} Y_p \\ V_p \end{pmatrix} + H_N^s E_f \quad (16)$$

where

$$Y_f \triangleq Y_{1|1} \triangleq \begin{bmatrix} y_1 & y_2 & \cdots & y_j \end{bmatrix} \quad (17)$$

$$Y_p \triangleq Y_{0|0} \triangleq \begin{bmatrix} y_0 & y_1 & \cdots & y_{j-1} \end{bmatrix} \quad (18)$$

$$V_p \triangleq V_{0|0} \triangleq \begin{bmatrix} v_0 & v_1 & \cdots & v_{j-1} \end{bmatrix} \quad (19)$$

It is assumed that the static nonlinearity, $f(\bullet)$, can be represented by expansions of basis functions as:

$$f(u_k) = v_k = \sum_{i=1}^{r} a_i g_i(u_k) \quad (20)$$

where a_i are unknown parameters to be identified, and $g_i(\bullet)$, i = 1,2,..., r, are known basis functions.

The parameters of blocked-oriented models are not unique, because the process gain can be arbitrarily distributed in the linear block and nonlinear block. To obtain the unique parameters, it is assumed, without loss of generality, that a_1 = 1. As a result, v_k in Eq. (20) can be expressed as:

$$v_k = g_1(u_k) + \sum_{i=2}^{r} a_i g_i(u_k) = g_1(u_k) + \mathbf{a}\mathbf{g}_k^u \quad (21)$$

where

$$\mathbf{a} \triangleq \begin{bmatrix} a_2 & a_3 & \cdots & a_r \end{bmatrix} \quad (22)$$

$$\mathbf{g}_k^u \triangleq \begin{bmatrix} g_2(u_k) & g_3(u_k) & \cdots & g_r(u_k) \end{bmatrix}^T \quad (23)$$

Then V_p in Eq. (19) can be written as:

$$V_p = \mathbf{g}_{p,1}^u + \mathbf{a}\mathbf{G}_p^u \quad (24)$$

where

$$\mathbf{g}_{p,1}^u \triangleq \begin{bmatrix} g_1(u_0) & g_1(u_1) & \cdots & g_1(u_{j-1}) \end{bmatrix} \quad (25)$$

$$\mathbf{G}_p^u \triangleq \begin{bmatrix} \mathbf{g}_0^u & \mathbf{g}_1^u & \cdots & \mathbf{g}_{j-1}^u \end{bmatrix} \quad (26)$$

Substituting Eq. (24) into Eq. (16), regression equations for the unknown parameters a_i can be set up as:

$$Y_f = L_{w,y} Y_p + L_{w,v} \mathbf{g}_{p,1}^u + L_{w,v} \mathbf{a} \mathbf{G}_p^u + H_N^s E_f \quad (27)$$

The least-squares estimation for the unknown parameter matrices is given by:

$$\begin{bmatrix} L_{w,y} & L_{w,v} & (L_{w,v}\mathbf{a}) \end{bmatrix} = Y_f \begin{bmatrix} Y_p^T & \mathbf{g}_{p,1}^{u\,T} & \mathbf{G}_p^{u\,T} \end{bmatrix} \left[\begin{pmatrix} Y_p \\ \mathbf{g}_{p,1}^u \\ \mathbf{G}_p^u \end{pmatrix} \begin{pmatrix} Y_p^T & \mathbf{g}_{p,1}^{u\,T} & \mathbf{G}_p^{u\,T} \end{pmatrix} \right]^{-1} \quad (28)$$

Then, the unknown expansion coefficients can be calculated as:

$$\hat{\mathbf{a}} = (L_{w,v})^{-1} (L_{w,v}\mathbf{a}) \quad (29)$$

3.2 Wiener model

By choosing $N=1$, the subspace matrix equations of Eq. (13) for the linear dynamic element in Hammerstein model can be written as:

$$V_f = L_w \begin{pmatrix} V_p \\ U_p \end{pmatrix} + H_N^s E_f \triangleq \begin{bmatrix} L_{w,v} & L_{w,u} \end{bmatrix} \begin{pmatrix} V_p \\ U_p \end{pmatrix} + H_N^s E_f \quad (30)$$

where

$$V_f \triangleq V_{1|1} \triangleq \begin{bmatrix} v_1 & v_2 & \cdots & v_j \end{bmatrix} \quad (31)$$

$$U_p \triangleq U_{0|0} \triangleq \begin{bmatrix} u_0 & u_1 & \cdots & u_{j-1} \end{bmatrix} \quad (32)$$

and the definition of V_p is the same as given in Eq. (19). The static nonlinearity in Wiener models will be assumed to be invertible, and that its inverse, $f^{-1}(\bullet)$, can be described as:

$$f^{-1}(y_k) = v_k = \sum_{i=1}^{r} a_i g_i(y_k) \quad (33)$$

It is also assumed, without loss of generality, that $a_1 = 1$. As a result, v_k in Eq. (33) can be expressed as:

$$v_k = g_1(y_k) + \sum_{i=2}^{r} a_i g_i(y_k) = g_1(y_k) + \mathbf{a}\mathbf{g}_k^y \quad (34)$$

where

$$\mathbf{g}_k^y \triangleq \begin{bmatrix} g_2(y_k) & g_3(y_k) & \cdots & g_r(y_k) \end{bmatrix}^T \quad (35)$$

Then, V_p in Eq. (19) and V_f in Eq. (31) can be written as:

$$V_p = \mathbf{g}_{p,1}^y + \mathbf{a}\mathbf{G}_p^y \quad (36)$$

$$V_f = \mathbf{g}_{f,1}^y + \mathbf{a}\mathbf{G}_f^y \quad (37)$$

where

$$\mathbf{g}_{p,1}^y \triangleq \begin{bmatrix} g_1(y_0) & g_1(y_1) & \cdots & g_1(y_{j-1}) \end{bmatrix} \quad (38)$$

$$\mathbf{g}_{f,1}^y \triangleq \begin{bmatrix} g_1(y_1) & g_1(y_2) & \cdots & g_1(y_j) \end{bmatrix} \quad (39)$$

$$\mathbf{G}_p^y \triangleq \begin{bmatrix} \mathbf{g}_0^y & \mathbf{g}_1^y & \cdots & \mathbf{g}_{j-1}^y \end{bmatrix} \quad (40)$$

$$\mathbf{G}_f^y \triangleq \begin{bmatrix} \mathbf{g}_1^y & \mathbf{g}_2^y & \cdots & \mathbf{g}_j^y \end{bmatrix} \quad (41)$$

Substituting Eqs. (36) and (37) into Eq. (30), regression equations for the unknown parameters a_i can be set up as:

$$\mathbf{g}_{f,1}^y = -\mathbf{a}\mathbf{G}_f^y + L_w \begin{pmatrix} \mathbf{g}_{p,1}^y \\ U_p \end{pmatrix} + L_{w,v}\mathbf{a}\mathbf{G}_p^y + H_N^s E_f \quad (42)$$

The least-squares estimation for the unknown parameter matrices is given by:

$$\begin{bmatrix} \mathbf{a} & L_w & (L_{w,v}\mathbf{a}) \end{bmatrix} = \mathbf{g}_{f,1}^y \begin{bmatrix} -\mathbf{G}_f^{y\,T} & \mathbf{g}_{p,1}^{y\,T} & U_p^T & \mathbf{G}_p^{y\,T} \end{bmatrix} \left[\begin{pmatrix} -\mathbf{G}_f^y \\ \mathbf{g}_{p,1}^y \\ U_p \\ \mathbf{G}_p^y \end{pmatrix} \begin{pmatrix} -\mathbf{G}_f^{y\,T} & \mathbf{g}_{p,1}^{y\,T} & U_p^T & \mathbf{G}_p^{y\,T} \end{pmatrix} \right]^{-1} \quad (43)$$

The unknown expansion coefficients is directly obtained from Eq. (43).

3.3 Identification of linear dynamic subsystem

Once the static nonlinearity (i.e., expansion coefficients) of the block-oriented nonlinear models has been estimated, the unknown intermediate variable, v_k, can be reconstructed. Consequently, the SIM presented in Section 2 can be applied straightforwardly to identify the linear dynamic element, because the input-output data of linear block are now available.

REFERENCES

[1] R. Haber, H. Unbehauen, "Structure Identification of Nonlinear Dynamic Systems - A Survey on Input/Output Approaches", *Automatica*, Vol. 26, No. 4, pp. 651-677, 1990.

[2] H. Bloemen, C. Chou, T. van den Boon, V. Verdult, M. Verhaegen, T. Backx, "Wiener Model Identification and Predictive Control for Dual Composition Control of a Distillation Column", *J. Process Control*, Vol. 11, pp. 601-620, 2001.

[3] I. Hunter, M. Korenberg, "The Identification of Nonlinear Biological Systems: Wiener and Hammerstein Cascade Models", *Biological Cybernetics*, Vol. 55, pp. 135-144, 1986.

[4] J. Stapleton, S. Bass, "Adaptive Noise Cancellation for a Class of Nonlinear Dynamic Reference Channels", *IEEE Trans. Circuits Syst.*, CAS-32, pp. 143-150, 1985.

[5] T. Wigren, "Recursive Prediction Error Identification Using the Nonlinear Wiener Model", *Automatica*, Vol. 29, pp. 1011-1025, 1993.

[6] J. Voros, "Iterative Algorithm for Parameter Identification of Hammerstein System with Two-Segment Nonlinearities", *IEEE Trans. Autom. Control*, Vol. 44, pp. 2145-2149, 1999.

[7] H. Chen, S. Hwang, C. Chang, "Iterative Identification of Continuous-Time Hammerstein and Wiener Processes Using a Two-Stage Estimation Algorithm", *Ind. Eng. Chem. Res.*, Vol. 48, pp. 1495-1510, 2009.

[8] E. Bai, "Frequency Domain Identification of Wiener Models", *Automatica*, Vol. 39, pp. 1521-1530, 2003.

[9] H. Park, J. Lee, "Step and Pulse Response Methods for Identification of Wiener Processes", *AIChE J.*, Vol. 52, pp. 668-677, 2006.

[10] H. Park, S. Sung; J. Lee, "Modeling of Hammerstein-Wiener Processes with Special Input

Test Signals", *Ind. Eng. Chem. Res.*, Vol. 45, pp. 1029-1038, 2006.

[11] J. C. Jeng, H. P. Huang, "Nonparametric Identification for Control of MIMO Hammerstein Systems", *Ind. Eng. Chem. Res.*, Vol. 47, pp. 6640-6647, 2008.

[12] R. Kadali, B. Huang, "Estimation of the Dynamic Matrix and Noise Model for Model Predictive Control Using Closed-Loop Data", *Ind. Eng. Chem. Res.*, Vol. 41, pp. 842-852, 2002.

[13] J. C. Jeng, Y. Lin, "Closed-Loop Identification of Dynamic Models for Multivariable Systems with Applications to Monitoring and Redesign of Controllers", *Ind. Eng. Chem. Res.*, Vol. 50, pp. 1460-1472, 2011.

[14] B. Huang, R. Kadali, *Dynamic Modeling, Predictive Control and Performance Monitoring*, Springer, London, 2008.

MODEL PREDICTIVE CONTROL WITH GRADIENT OF A FUNCTION OF THE OPTIMIZING TARGETS

Luz Adriana Alvarez, Darci Odloak, University of Sao Paulo, Sao Paulo, Brazil

The industrial processes usually operate under a hierarchical structure, where the Real Time Optimization (RTO) and the Model Predictive Control (MPC) are executed in separated layers, resulting in a two or three-layer architecture of plant-operation strategy [1]. The RTO stage determines the operating values of the outputs and inputs that produce the maximum economic profit or the lowest operating costs. In the context of the hierarchical structure, the objective of the MPC controller is to follow the economic targets in presence of disturbances and manipulating directly the process inputs or the set-points of local PID controllers in a cascade configuration. In a recent work [2], the gradient of the economic function is incorporated in the objective function of the MPC controller in order to solve the RTO and MPC problem in one single layer. The approach was successfully simulated in a FCC unit. The strategy consisted in minimizing the gradient of the economic function in a quadratic term so as to maximize the economic profit, and therefore the method requires the convexity of the economic function.

In this work we propose to use the gradient of a convex function that links the RTO targets to the MPC controller, resulting in a two-layer approach. The convex function is a $nu+1$ dimensional paraboloid of the process inputs, with the minimum corresponding to the input targets resulting from the RTO routine. This controller has infinite prediction horizon and terminal state constraint [3] and operates according to the zone control strategy [4]. The inclusion of slack variables in the MPC optimization problem makes it implementable in practice. In addition, we provide a stability analysis of the closed-loop system for the nominal case.

The proposed MPC was initially tested in a linear system and then, the real time optimization of a continuous polymerization process under the proposed structure was developed. The simulation results of the MPC in a polymerization process showed that the production rate can be maximized in presence of disturbances, preserving the polymer quality and respecting the allowed limits for the controlled outputs.

References

1. C.M. Ying and B. Joseph. Performance and Stability Analysis of LP-MPC and QP-MPC Cascade Control Systems. AIChE J. 45 (1999) 1521-1534.
2. G. De Souza, D. Odloak and A. Zanin. Real Time Optimization (RTO) with Model Predictive Control (MPC). Comp. Chem. Eng. 34 (2010) 1999-2006.
3. D. Odloak, Extended Robust Model Predictive Control. AIChE J. 50 (2004) 1824-1836.
4. A. González and D. Odloak. A stable model predictive control with zone control. J. Proc. Cont. 19 (2009) 110-122.

(622n) Model-Based Sensor Placement for Component Condition Monitoring and Fault Diagnosis In Fossil Energy Systems

Jeevan Maddala[1], Babji Srinivasan[2], Debangsu Bhattacharyya[3], Richard Turton[3] and Raghunathan Rengasamy[4], (1)Chemical Engineering, Texas Tech University, Lubbock, TX, (2)Department Chemical Engineering, Texas Tech University, Lubbock, TX, (3)Department of Chemical Engg., West Virginia University, Morgantown, WV, (4)Department of Chemical Engineering, Texas Tech University, Lubbock, TX

With the current energy related concerns, it is imperative that the efficiency and effectiveness of fossil-fuel-based power plants are improved tremendously. One of the reasons for the loss of efficiency and effectiveness of such plants is due to low availability caused by the shutdown of a section of the plant or the whole plant after short span of operation. This shutdown of equipment/plant may occur immediately or over a period of time. The root cause for productivity losses, the "faults", can be anticipated or unanticipated. Even though many of the faults are anticipated, the true evolution of faults depends upon the current dynamical state of the process and hence, cannot be properly predicted based on past operational experiences or even using high-fidelity steady-state models. If the faults can be identified in a timely manner then it may be possible to avoid serious damages to the equipment/plant. It is also possible to plan for the shutdown in advance. This improves the plant efficiency by improving the plant availability and reducing economic losses that result due to damage to valuable equipment. In this work, this problem is addressed by optimal placement of sensors. In this paper, we propose an innovative two-tier framework for the identification of locations, types, and optimum number of sensors for detection of system-level and component level faults in a coal-based IGCC (Integrated Gasification Combined Cycle) plant. This is a challenging problem because of the significant interaction between the various components and nonlinearity in the process. Several illustrative examples will be discussed to demonstrate the power of the proposed approach in identifying maximally informative sensor locations.

See more of this Session: Poster Session - Area 10B: Systems and Process Control
See more of this Group/Topical: Computing and Systems Technology Division

(622o) Mathematical Modeling of a Dye-Sensitized Solar Cell

Mona Bavarian, Siamak Nejati, Kenneth K.S. Lau and Masoud Soroush, Chemical and Biological Engineering, Drexel University, Philadelphia, PA

Dye sensitized solar cells (DSSCs) have drawn extensive attention since the seminal report of O'Regan and Grätzel [1]. DSSCs have been able to achieve photoconversion efficiencies of up to 11.1% under 1-sun AM1.5 illumination [2]. This method of harvesting sunlight is a promising alternative to the silicon-based solar cells due to the lower manufacturing cost and sufficiently high energy conversion efficiency. To be commercially feasible, the cells need to show stable performance and long durability of at least 10 years [3]. It has been reported that photovoltaic characteristics, i.e., open circuit voltage, short circuit current, fill factor and efficiency, decrease at elevated temperatures [4]. The cell temperature can be increased to as high as 72°C under severe environmental conditions [5]. A dynamic model of the cell that accurately predicts the response of the cell to different changes will allow one to optimize the cell design and operation.

In a DSSC, a photoactive sensitizer (dye) adsorbed within a thin mesoporous TiO_2 film harvests the sunlight. The excited dye injects an electron into the TiO_2 conduction band which diffuses through the interconnected network of the semiconductor and eventually is collected at the transparent conductive glass and moves through the external circuit to the counter electrode, where it reduces the tri-iodide ions to iodide ions. The iodide ions diffuse through the electrolyte and reach the TiO_2/dye interface, where they regenerate the oxidized dye molecules [6].

This paper presents mathematical modeling of a DSSC and investigates steady-state and dynamic behaviors of the cell. To develop the model, the equations of continuity and transport for all species in the cell including electrons, iodide, tri-iodide and cation are considered. The Butler-Volmer kinetics is applied at the platinum counter electrode.

Under steady state conditions, the developed model is validated against the current-voltage characteristics of the cell. Parameter estimation is conducted to determine the unknown parameters of the model. The dynamic responses of the cell to changes in solar irradiance, ambient temperature and load resistance are simulated.

References

[1] B. O'Regan, M. Grätzel, Nature, 353 (1991) 737-740.

[2] M. Grätzel, Accounts of Chemical Research, 42 (2009) 1788-1798.

[3] E. Figgemeier, A. Hagfeldt, International Journal of Photoenergy, 6 (2004) 127-140.

[4] H.S. Lee, S.H. Bae, Y. Jo, K.J. Kim, Y. Jun, C.H. Han, Electrochimica Acta, 55(24), (2010) 7159-7165.

[5] N. Kato, Y. Takeda, K. Higuchi, A. Takeichi, E. Sudo, H. Tanaka, T. Motohiro, T. Sano, T. Toyoda, Solar Energy Materials and Solar Cells, 93 (2009) 893-897.

[6] S. Huang, G. Schlichthörl, A. Nozik, M. Grätzel, A. Frank, Journal of Physical Chemistry B, 101 (1997) 2576-2582.

See more of this Session: Poster Session - Area 10B: Systems and Process Control
See more of this Group/Topical: Computing and Systems Technology Division

(622p) The Use of a Time Variant IMC Controller for Robust Performance Improvement

Jonathan Whitlow and Latifat Adebayo, Chemical Engineering, Florida Institute of Technology, Melbourne, FL

The design of the IMC controller requires that the filter time constant be specified as a single tunable parameter with an increase resulting generally in lower system performance but increased system robustness. In this work the effect of using a variable filter time constant for implementation in an Internal Model Control (IMC) system is presented. The basic premise is that during an initial closed loop response a high performance filter constant can be employed with it changing during the course of the response to a final value resulting in a robust filter constant. This work will demonstrate the potential performance improvement that can be obtained by this approach for various types of plant model structures with the internal models used evaluated with and without model error. Implementation details and rules for how to adjust the filter constant over time will also be presented.

See more of this Session: Poster Session - Area 10B: Systems and Process Control
See more of this Group/Topical: Computing and Systems Technology Division

(622q) Energy Integration In Solid Oxide Fuel Cell Systems Under Different Steam Reformer Conditions: a Design and Control Approach

Dimitrios Georgis[1], Sujit S. Jogwar[1], Ali Almansoori[2] and Prodromos Daoutidis[1], (1)Chemical Engineering and Materials Science, University of Minnesota, Minneapolis, MN, (2)Department of Chemical Engineering, The Petroleum Institute, Abu Dhabi, United Arab Emirates

Solid oxide fuel cell (SOFC) systems are considered promising alternative to conventional energy systems for both stationary and stand-alone applications. Their fuel flexibility provides the ability for coupling with other well known technologies, such as, natural gas steam reforming [1], coal [2] and biomass [3] gasification units. The SOFC high operating temperature shows the potential for energy integration, especially with highly endothermic processes, such as steam reforming, leading to a higher overall system's efficiency.

In our previous work [4] we proposed and analyzed two different energy integrated SOFC systems (incorporating an external steam reformer). In the first configuration, the hot effluent streams leaving the SOFC were utilized directly for energy integration, while in the second configuration, the effluent streams leaving the SOFC were catalytically oxidized in a burner to produce more energy. The open- and closed-loop analysis revealed interesting features and challenges for each configuration.

In this study, we analyze the effect of the reformer design parameters (steam-to-carbon (S/C) ratio and operating temperature) on the steady-state performance of the first configuration described above. Our previous work showed that this configuration is more challenging in terms of open-loop dynamics (instability) and the requirement for an external utility (furnace). Additionally, the effect of the S/C ratio on the open-loop behavior is also investigated. Lastly, in order to compensate for the process nonlinearities, a nonlinear controller for the fuel cell temperature is derived and implemented.

The simulations show that the S/C ratio has a greater impact on the steady state design rather than on dynamics and control. Also, a comparative evaluation between the nonlinear and linear controllers indicates a superior performance of the nonlinear controller even in the presence of modeling errors.

[1] T.A. Adams II, P.I. Barton, High-efficiency power production from natural gas with carbon capture, Journal of Power Sources. 195 (2010) 1971-1983.

[2] M.Li, J. Brouwer, A.D. Rao, G.S. Samuelsen, Application of a detailed dimensional solid oxide fuel cell model in integrated gasification fuel cell system design and analysis, Journal of Power Sources. (2011) doi:10.1016/j.jpowsour.2011.02.080.

[3] A.O. Omosuna, A.Bauenc, N.P. Brandon, C.S. Adjimana, D. Hart, Modelling system efficiencies and costs of two biomass-fuelled SOFC systems, Journal of Power Sources. 131 (2004) 96-106.

[4] D.Georgis, S.S. Jogwar, A.S. Almansoori, P. Daoutidis, Design and control of energy integrated SOFC systems for in situ hydrogen production and power generation, Computers & Chemical Engineering. (2011) doi:10.1016/j.compchemeng.2011.02.006.

See more of this Session: Poster Session - Area 10B: Systems and Process Control
See more of this Group/Topical: Computing and Systems Technology Division

(624a) A New Method for the Global Optimization of Nonlinear Problems with Bilinear Terms

Joo Teles[1], **Pedro M. Castro**[1] and Henrique A. S. Matos[2], (1)Modeling and Optimization of Energy Systems, LNEG, Lisbon, Portugal, (2)Chemical Engineering, Instituto Superior T°Cnico, Lisbon, Portugal

Nonlinear programming problems with bilinear terms can be found in a variety of design problems (e.g. water networks). Existing solution methods include gradient-based algorithms and stochastic approaches (e.g. genetic algorithms, particle swarm optimization) that while capable of finding good feasible solutions with modest computational resources, often fail to find the global optimal solutions to the problem. More reliable methods employ spatial branch-and-bound, where the bilinear terms are replaced by convex over and underestimators (e.g. McCormick linear envelopes). The drawback of such global optimizations solvers is that it may be difficult to prove global optimality, meaning that the lower bound (when minimizing) from the relaxed problem at termination may still be far from the best solution, which naturally increases the chances of failure with respect to finding the global optimal solution.

In this paper, we evaluate the performance of a recently proposed multiparametric disaggregation technique for generalized signomial optimization problems [1], on the optimal design of water-using [2] and wastewater treatment networks [3]. Besides applying it to a problem of industrial relevance instead of performing a proof-of-concept with toy problems [1], the novelty is the use of a binary system for the parameterized variables instead of a decimal numeric system. In such problems, bilinear terms appear in the contaminant mass balances resulting from the product of flows by concentrations.

The proposed approach is to approximate the original NLP by a MILP, where the flowrate variables are disaggregated and the concentration variables are parameterized. Any desired accuracy level can be set for the approximation being the MILP exact in the limit of an infinite number of significant digits. Fortunately, the value of the objective function rapidly approaches the global optimal solution following small increments in the number of significant digits. A given MILP can also be used as an effective initialization method prior to the solution of the original NLP with a local optimization solver, provided that the discretized solution space is not empty.

The results on a set of test problems taken from the literature, for different accuracy levels, show that the method is better suited for wastewater treatment design problems, for which it is more difficult for BARON to reduce the optimality gap to acceptable values. Through the consecutive solution of the MILP for a single increment in accuracy, our new method estimated the optimality gap to almost zero, thus increasing our confidence that the best solutions found are indeed global optima. Regarding the comparison between the two numeric representation systems, the binary had a slight advantage since it could still find good feasible solutions in two cases where the decimal system failed.

[1] J. Teles, P. M. Castro and H. A. Matos. Multi-parametric disaggregation technique for global optimization of nonlinear signomial programming problems. Submitted to Journal of Global Optimization.

[2] J. Teles, P. M. Castro and A. Q. Novais. LP-based solution strategies for the optimal design of industrial water networks with multiple contaminants. Chemical Engineering Science 63 (2008) 376.

[3] P. M. Castro, J. P. Teles, A. Q. Novais. Linear program-based algorithm for the optimal design of wastewater treatment systems. Clean Techn Environ Policy 11 (2009) 83.

See more of this Session: Poster Session: Computers In Operations and Information Processing
See more of this Group/Topical: Computing and Systems Technology Division

(624b) CFD Simulation of Large-Scale Bioreactor with Side-Entry Agitator

Minhua Zhang, Yonghui Li, Zhiqiang Zhang and Jing Ma, R & D Center for Petrochemical Technology, Tianjin University, Tianjin, China

The scale-up and design of commercial scale stirred bioreactors from lab data is still a big challenge. The performance of a bioreactor largely depends on the efficient and homogeneous dispersion of gas-liquid and suspension of solid particles. It is difficult for engineers to rely on rules-of-thumb alone to deal with the uncertainties encountered in the design, scale-up and optimization of this complex three phase system. Mathematical modeling has been suggested to be a useful tool in the scale-up process.

In the present work, a detailed computational fluid dynamics (CFD) simulation was conducted to predict the flow pattern in a large-scale bioreactor (3400 m^3) with side-entry agitators. The CFD analysis was firstly carried out in a single-phase reactor. With 4 side agitators installed in the reactor, the horizontal continuous circular flow was indentified, which would facilitate the mixing of solid particles. Subsequently, the CFD simulation was extended to solid-liquid two-phase. The uniformity of solid suspension and the power input were used as the objectives to assess the effects of the shaft angle, stirring speed as well as the agitator quantity and an optimum installation mode was obtained. Finally, the bubbles generation and distribution in gas-liquid two-phase flow were examined by using Population Balance Model (PBM). The modeling results suggest that the formation and escape of different size bubbles could change the flow distribution at the upper tank, which would benefit the suspension of solid particles.

See more of this Session: Poster Session: Computers In Operations and Information Processing
See more of this Group/Topical: Computing and Systems Technology Division

(624c) Supply Chain Management Challenges In Continuous Pharmaceutical Manufacturing (CPM)

Dimitrios I. Gerogiorgis, National Technical University of Athens (N.T.U.A.), Athens, Greece

Continuous Pharmaceutical Manufacturing (CPM) is a concept with a demonstrated potential to revolutionize the pharmaceutical industry and achieve Quality by Design (QbD) via implementation of knowledge-driven process modeling and online Process Analytical Technology (PAT) data acquisition. The main advantages of CPM over batch pharmaceutical processes include process intensification via continuous flow reactions and separations, lean manufacturing and elimination of long (idle) batch times. Nevertheless, product quality and uniformity must strictly adhere to regulatory specifications and may well be affected by feedstock quality disturbances, which are currently addressed by offline lot analyses. Considering that full-scale CPM production lines are already under construction and may streamline and facilitate high-volume manufacturing of some blockbuster drugs which enjoy ample patent protection, it is useful to investigate the potential impact of CPM practices on traditional pharmaceutical supply chains.

The pharmaceutical supply chain is characterized by the incessant quest for more efficacious formulations: the latter relies on a highly regulated R&D pipeline for novel API screening and new product development, and on the elaborate standardization of manufacturing, prescription marketing and distribution operations. Rapid licensing and patent longevity is a *sine qua non* prerequisite towards ensuring economic success. Launching "first to market" often relies on technically robust (rather than optimal) manufacturing recipes, which have a clear potential to be licensed quickly and replicated worldwide without any reconfiguration.

Batch pharmaceutical manufacturing processes are the overwhelming majority in modern pharmacopoieia: Multinational pharmaceutical firms possess several strategically located production facilities worldwide and many major distribution centers in order to effectively meet the demand within each country or region. The vast variability of legal, technical and economic regulations in the host countries further perplexes importing, storage, exporting and disposal constraints on feedstock, key intermediate and API compounds. Concurrently, the past proliferation and current prevalence of batch pharma production plants (tank farms) is a consequence of the century-long established paradigm of iterative pharmaceutical chemistry research, reinforced by the wide versatility of these facilities, where myriads of synthetic chemical routes have been conceived, explored, documented and scaled up, resulting in generations of structurally similar compounds. Most importantly, batch processing times, lot manufacturing and product changeovers have been studied exhaustively, thus enabling systematic production planning which efficiently address demand fluctuations.

Continuous manufacturing lines, in contrast to batch-oriented tank farms, eliminate idle times between processing units by definition; however, a deeper understanding of the process is required (Plumb,

2005). Demand variability is harder to tackle, but an entire CPM line can fit in a lab room (Roberge et al., 2008). Multiproduct CPM plants can thus be constructed, provided that the combination of target products and the geographic location are judiciously selected, so as to achieve high demand and close proximity to markets.

The present paper focuses on a comparative quantitative analysis of several key technical and economic performance indicators for two different (batch vs. CPM) strategic pharmaceutical manufacturing scenaria. The geographic region, production/distribution basis, drug portfolio, patent protection and market share data are assumed given and constant, with reasonable technical assumptions made with respect to the comparative technical efficiency and economic viability of process alternatives (Gerogiorgis et al., 2010). The focus of this investigation is to explore the conditions under which a rapid CPM transition is advisable.

Keywords: Continuous Pharmaceutical Manufacturing (CPM), Supply Chain Management (SCM).

REFERENCES

Gerogiorgis, D.I., Barton, P.I., et al. (2010). Comparative economic analysis of continuous and batch pharmaceutical manufacturing: A case study (submitted).

Gerogiorgis, D.I., Barton, P.I., et al. (2009). Steady state optimization of a continuous pharmaceutical process. *Computer-Aided Chemical Engineering (Proceedings of PSE 2009)* **27**(1): 927-932, Elsevier, Amsterdam.

Plumb, K. (2005). Continuous processing in the pharmaceutical industry – Changing the mind set. *Chem. Eng. Res. Des.* **83**(A6): 730-738.

Roberge, D.M., Zimmermann, B., Rainone, F., Gottsponer, M., Eyholzer, M., Kockmann, N. (2008). Microreactor technology and continuous processes in the fine chemical and pharmaceutical industry: Is the revolution underway? *Org. Proc. Res. Dev.* **12**(5): 905-910.

See more of this Session: Poster Session: Computers In Operations and Information Processing
See more of this Group/Topical: Computing and Systems Technology Division

(624d) Simulating the Impact of Distillation Operating Parameters On Energy Requirement for Methanol Separation From Biodiesel

Bipro Ranjan Dhar, Chemical & Biochemical Engineering, The University of Western Ontario, London, ON, Canada and Kawnish Kirtania, Chemical Engineering, Bangladesh University of Engineering and Technolog, Dhaka, Bangladesh

Biodiesel is a wonderful replacement to conventional petro-diesel fuel, which can be produced from a renewable domestic resource. In biodiesel production process, oil or animal fat (Triglyceride) react with a primary alcohol in presence of a catalyst to give the corresponding alkyl esters of the fatty acid mixture that is found in the parent vegetable oil or animal fat. Methanol is a widely used primary alcohol for biodiesel production. After the biodiesel process is complete, a lot of methanol is available for recovery and reuse. To meet ASTM D6751 or EN 14214 standards the removal of excess methanol becomes a vital step. Residual methanol in the biodiesel fuel is a major environmental and health safety issue. Methanol is toxic, and the emission of excess methanol from the use of biodiesel can be hazardous for our life and environment. Excess methanol can also make the fuel flammable and more dangerous to handle and store. Besides, excess methanol may corrode metal components of engine. For these reasons, most conventional biodiesel manufacturers waste a lot of unused methanol through washing the final product. A simulation study has been done using ASPEN PLUS^{TM} for excess methanol separation in continuous biodiesel production process using a distillation column. The feedstock used for biodiesel production was triolein containing 15% free fatty acid (as oleic acid). The special attention was devoted to the effect of different alcohol to oil ratio and important design and operating parameters (reflux ratio, operating pressure, number of stages etc.) of distillation column on excess methanol separation from the product. Analysis of simulation results shows that for a certain distillation operating condition and reaction parameters it is possible to separate around 95-98% of excess methanol before phase separation of biodiesel and glycerol, although for high alcohol to oil ratio the energy requirement increases exponentially.

See more of this Session: Poster Session: Computers In Operations and Information Processing
See more of this Group/Topical: Computing and Systems Technology Division

A Novel Technique for Prediction of Time Points for Scheduling of Multipurpose Batch Plants

[1]E. Reshid and [1,2]T.Majozi
[1] *Department of Chemical Engineering, University of Pretoria, Lynnwood Road, Pretoria, 0002, South Africa*
[2] *Modeling and Digital Science, CSIR, Pretoria, South Africa*

Abstract

This presentation addresses a mathematical technique for prediction of the optimal number of time points in short-term scheduling of multipurpose batch plants. The mathematical formulation is based on state sequence network (SSN) representation. The developed method is based on the principle that the optimal number of time points depends on how frequent the critical unit is used throughout the time horizon. In the context of this work, a critical unit refers to a unit that is most frequently used and it is active for most of the time points when it is compared to other units. The statistical R-squared value obtained between the predicted and actual optimal number of time points in all the problems considered was 0.998, which suggests that the developed method is accurate in determining optimal number of time points. Consequently this avoids costly computational times due to iterations.
Key words: Scheduling; Multipurpose Batch Plant; Optimization; MILP; Optimal time point, continuous-time

Introduction

Batch processes differ from the continuous processes in many ways, the main of which is that time is inherent in batch processes. In batch process every task has a definite duration, with starting and finishing times, whereas in continuous processes time is important during non-steady state operation. As a result of this, the scheduling of batch processes is vital to the operation of any batch facility. In batch plants, detailed requirements for the various products may be specified on a day-to-day basis. A production schedule must indicate the sequence and manner in which the products are to be produced and specify the times at which the process operations are to be carried out. It is clear that the overall productivities and economic effectiveness of batch plants depend critically on the production schedule as it harmonizes the entire plant operation to attain production goals. While flexibility of batch plants improves productivity, it also makes plant scheduling a challenging task. Much research has going in the past developing mathematical optimization models for scheduling of batch plants targeting in obtaining a better optimal objective value and computational time.

In all models that are based on continuous-time representation for scheduling of multipurpose batch plants, the optimal number of event points is obtained by iteration. Little work has been done in the past to determine the optimal number of event points. Li and Floudas (2010) presented optimal time point determination based on the model of Shaik and Floudas (2009). They used an iterative procedure to obtain the maximum number of event points and determine the critical intermediate states. The minimum number of event points is determined by the approach proposed by Janak and Floudas (2008). A branch and bound strategy is then used based on the minimum and maximum number of event points to determine the optimal number of event points.

The developed method to determine the optimal number of time points is based on the branch and bound strategy on the predicted optimal number of time points obtained by solving linear programming maximization (LP MAX) and linear programming minimization (LP MIN) of the scheduling problem. LP MAX and LP MIN are solved only once to predict the optimal number of time points.

Mathematical model using state sequence network (SSN)

Below we presented the condensed mathematical formulation. The notations for this mathematical model are found in the paper an advanced scheduling technique for multipurpose batch plant by the same Author.

Allocation constraints

$$\sum_{s_{in,j} \in S^*_{in,j}} y(s_{in,j},p) \leq 1, \ \forall\, j \in J, \ \ p \in P \tag{1}$$

Capacity constraints

$$V^L_{s_{in,j}}\, y(s_{in,j},p) \leq mu(s_{in,j},p) \leq V^U_{s_{in,j}}\, y(s_{in,j},p), \ \forall\ p \in P,\ j \in J,\ s_{in,j} \in S_{in,J} \tag{2}$$

Material balance for storage

$$q_s(s,p) = q_s(s,p-1) \ - \sum_{s_{in,j} \in S^{sc}_{in,J}} \rho^{sc}_{s_{in,j}}\, mu(s_{in,j},p) \ + \sum_{s_{in,j} \in S^{sp}_{in,J}} \rho^{sp}_{s_{in,j}}\, mu(s_{in,j},p-1)\ \forall\ p \in P,\ s \in S \tag{3}$$

$$q_s(s^p,p) = q_s(s^p,p-1) + \sum_{s_{in,j} \in S^{s^p p}_{in,J}} \rho^{sp}_{s_{in,j}}\, mu(s_{in,j},p),\ \forall\ p \in P,\ \ s^p \in S^p \tag{4}$$

Duration constraints (batch time as a function of batch size)

$$t_p(s_{in,j'},p) \ \geq\ t_u(s_{in,j'},p) + \tau(s_{in,j'})\, y(s_{in,j'},p),\ \forall\ j \in J,\ p \in P,\ s_{in,j} \in S^{sp}_{in,J},\ s_{in,j'} \in S^{sc}_{in,J} \tag{5}$$

Sequence constraints

The two subsections address the proper allocation of tasks in a given unit that ensures the starting time of a new task to be later than the finishing time of the previous task.

Same task in same unit

$$t_u(s_{in,j},p) \ \geq\ t_p(s_{in,j},p-1),\ \ \forall\ j \in J,\ p \in P,\ s_{in,j} \in S^*_{in,j} \tag{6}$$

Different tasks in same unit

$$t_u(s_{in,j},p) \geq t_p(s'_{in,j},p),\ \forall\ j \in J,\ p \in P,\ s_{in,j} \neq s'_{in,j},\ \ s_{in,j}, s'_{in,j} \in S^*_{in,j} \tag{7}$$

$$t_p(s^{usp}_{in,j},p-1) \geq t_u(s^{usc}_{in,j},p) - H(1-y(s^{usp}_{in,j},p-1)),\ \forall\ j \in J,\ p \in P,\ s^{usc}_{in,j} \in S^{usc}_{in,j},\ s^{usp}_{in,j} \in S^{usp}_{in,j} \tag{8}$$

Different tasks in different units

If an intermediate state s is produced from one unit.

$$\rho(s^{sp}_{in,j})\, mu(s_{in,j},p-1) \leq qs(s,p) + V^U_j\, t(j,p)\ \forall\ j \in J, p \in P, s_{in,j} \in S^{sp}_{in,j} \tag{9}$$

$$tu(s_{in,j'},p) \geq tp(s_{in,j},p) - H((2 - y\ (s_{in,j},p-1) - t\ (j,p))), \quad \forall\ j \in J,\ p \in P,\ s_{in,j} \in S^{sp}_{in,J},\ s_{in,j'} \in S^{sc}_{in,J} \tag{10}$$

If an intermediate state is produced from more than one unit

$$\sum_{s_{in,j} \in S^{sc}_{in,J}} \rho^{sc}_{s_{in,j}}\, mu\ (s_{in,j},p) \leq qs\,(s,p-1) \ + \sum_{s_{in,j} \in S^{sp}_{in,J}} \rho^{sp}_{s_{in,j}}\, mu\ (s_{in,j},p-1)\ t(j,p),\ \forall\ j \in J,\ p \in P \tag{11}$$

$$t_u(s_{in,j'},p) \geq t_p(s_{in,j},p-2) - H(1-y(s_{in,j},p-2)),\ \forall\ j \in J,\ p \in P,\ \ s_{in,j} \in S^{sp}_{in,J},\ \ s_{in,j'} \in S^{sc}_{in,J} \tag{12}$$

Sequence constraint for fixed intermediate storage (FIS) policy

$$\sum_{s_{in,j}\in S^{sp}_{in,J}} \rho^{ps}_{s_{in,j}} mu\left(s_{in,j}, p-1\right) + qs(s, p-1) \le QS^U + \sum_{j\in J_s} V^U_j\left(1 - x(s,p)\right) \ \forall\ j \in J, p \in P, s \in S \quad (13)$$

$$tp\left(s_{in,j'}, p-1\right) \le tp\left(s_{in,j}, p-1\right) + H\left(2 - y\left(s_{in,j'}, p\right) - y\left(s_{in,j}, p-1\right)\right) + H\left(x(s,p)\right)$$
$$\forall\ j \in J,\ p \in P,\ s_{in,j} \in S^{sp}_{in,J},\ s_{in,j'} \in S^{sc}_{in,J} \quad (14)$$

Storage constraints

$$q_s(s,p) \le QS^U + \sum_{s_{in,j}\in S^{sp}_{in,J}} u\left(s_{in,j}, p\right) \quad \forall s \in S, p \in P, j \in J \quad (15)$$

$$u\left(s_{in,j}, p\right) \le \rho^{sp}_{s_{in,j}} mu\left(s_{in,j}, p-1\right) + u\left(s_{in,j}, p-1\right) \quad \forall p \in P,\ j \in J \quad (16)$$

$$u\left(s_{in,j}, p\right) \le V^U_j\left(1 - \sum_{s_{in,j}\in S^{*}_{in,J}} y\left(s_{in,j}, p\right)\right) \quad \forall p \in P,\ j \in J \quad (17)$$

Time horizon constraints

$$t_u\left(s_{in,j}, p\right) \le H,\ \forall\ s_{in,j} \in S_{in,J}, p \in P, j \in J \quad (18)$$

$$t_p\left(s_{in,j}, p\right) \le H,\ \forall\ s_{in,j} \in S_{in,J}, p \in P,\ j \in J \quad (19)$$

Calculation of the total duration a unit is active within the time horizon

The knowledge of this value, i.e. the total duration a unit is active over a time horizon, is fundamental in the prediction of the optimal number of time points. The following two sections show the calculation of this value in the case of fixed and variable duration constraints, respectively.

Duration constraints are fixed

Constraint (20) is an important constraint which is used to predict optimal number of time points. The equation is used in the model when the model is solved as a linear programming maximization, (LP MAX) and a linear programming minimization, (LP MIN). In solving LP MAX and LP MIN, the binary variable $y(s_{in,j},p)$ is considered as a continuous variable with a value between 0 and 1. Constraint (20) expresses that the total duration a unit is active is limited by the time horizon, the sum of mean processing times of previous tasks and the sum of mean processing times of all subsequent tasks leading to a product as prescribed by the recipe.

$$\sum_{p}\sum_{s_{in,j}} \tau\left(s_{in,j}\right) y\left(s_{in,j}, p\right) \le H - \sum_{s_{in,j'}\in S^{PT}_{in,J}} \tau\left(s_{in,j'}\right) - \sum_{s_{in,j'}\in S^{ST}_{in,J}} \tau\left(s_{in,j'}\right),\ \forall\ j \in J, p \in P, S_{in,j} \ne s_{in,j'} \quad (20)$$

In the beginning, the parameter $\sum_{s_{in,j'}\in S^{PT}_{in,J}} \tau(s_{in,j'})$ caters for the situation where the unit j must wait for all the previous tasks to finish before it can start processing. If there is an intermediate at the beginning of the time horizon so that unit j can start its operation, then the parameter becomes 0. The parameter $\sum_{s_{in,j'}\in S^{ST}_{in,J}} \tau(s_{in,j'})$ states that at the end of its operation unit j gives its intermediate to the subsequent tasks which lead to a product (the shortest path of subsequent tasks lead to a product), so that the unit is not active at this time. If the unit is one producing a product then the term becomes 0, this assumption is used not to underestimate time points because of underutilization of a unit. Constraint (20) tightens the model better than one simply stating that the sum of the durations of all tasks suitable in each unit should be less than the scheduling time horizon without loss of any generality.

Duration constraints dependent on batch size

Constraint (21) is similar to (20), but for the situation where duration is a function of batch size.

$$\sum_{p}\sum_{s_{in,j}}\tau\left(s_{in,j}\right)y\left(s_{in,j},p\right)+\beta\left(s_{in,j}\right)mu\left(s_{in,j},p\right)\leq H-\sum_{s_{in,j'}\in S_{in,J}^{PT}}\tau\left(s_{in,j'}\right)-\sum_{s_{in,j'}\in S_{in,J}^{ST}}\tau\left(s_{in,j'}\right),\ \forall\ j\in J,p\in P,s_{in,j}\neq s_{in,j'} \quad (21)$$

Objective function

Constraints (1) – (8), (10), (13) - (15), (18) - (21) together with the objective function given in Constraint (22) constitute LP MAX and LP MIN. Sequence Constraint (10) and (15) for LP MAX and LP MIN can be expressed as:

$$tu\left(s_{in,j'},p\right)\geq tp\left(s_{in,j},p-1\right)-H\left(1-y\left(s_{in,j},p-1\right)\right)\ \forall\ j\in J,p\in P,s_{in,j}\in S_{in,J}^{sp},s_{in,j'}\in S_{in,J}^{sc} \quad (10)$$

$$q_s(s,p)\leq QS^{\max}\quad \forall\ s\in S,p\in P \quad (15)$$

Model ML2 constitutes Constraints (1) – (22). For the model ML2 Constraint (20) and (21) is expressed as:

$$\sum_{p}\sum_{s_{in,j}\in S_{in,j}^{*}}\tau\left(s_{in,j}\right)y\left(s_{in,j},p\right)\leq H,\ \forall\ j\in J,\ p\in P \quad (20)$$

$$\sum_{p}\sum_{s_{in,j}\in S_{in,j}^{*}}\tau\left(s_{in,j}\right)y\left(s_{in,j},p\right)+\beta\left(s_{in,j}\right)mu\left(s_{in,j},p\right)\leq H,\ \forall\ j\in J,\ p\in P \quad (21)$$

The objective of the scheduling problem is to maximize the product throughput or to minimize the makespan, as given in Constraint (22).

$$\text{maximize}\sum_{s}price\left(s^{p}\right)qs\left(s^{p},p\right),\ \forall\ p=P,s^{p}\in S^{p}$$

or

$$\text{minimize } H \quad (22)$$

Prediction of optimal number of time points

The first step is to solve the model as a LP MAX problem using the maximum number of time points calculated with $\left\lceil \frac{H}{\tau^{L}}\right\rceil$. In this step, the maximum achievable throughput that is very close to the dual objective value at the optimal number of time points is determined. The number of time points each unit can be active in the time horizon is also determined. Equation (23) is used to determine the total time points required using the frequency of utilization of unit j obtained from the LP MAX problem. The unit which gives the maximum number of time points is the unit most frequently used in the time horizon. This becomes the *critical unit*; therefore, the optimal number of time points is dependent on the frequency of utilization of this unit. In essence, this particular observation is the cornerstone of this contribution.

The first term on the right of Equation (23), $\sum_{s_{in,j}\in S_{in,j}^{*}}\sum_{p}y(s_{in,j},p)$ is used to determine how many times a unit j starts processing throughout the time horizon of interest.

$$p_j=\left\lceil \sum_{s_{in,j}\in S_{in,j}^{*}}\sum_{p}y(s_{in,j},p)+\sum_{i\in I^{PT}}i+\sum_{i\in I^{ST}}i\right\rceil,\ j\in J,\ p\in P \quad (23)$$

The term $\sum_{i\in I^{PT}}i$ is to cater for those time points not shared by a unit because at the beginning, a unit must wait until all the previous tasks have been processed. If there is an intermediate initially in storage to start a task at the beginning of the time horizon, then $\sum_{i\in I^{PT}}i$ = 0. Unit j at the end of its operation passes its intermediate to the subsequent units (that lead to a product). The time points associated with the subsequent tasks can be calculated with $\sum_{i \in I^{ST}} i$; these time points are not shared by unit j. If the unit j is a unit producing a product then the term $\sum_{i \in I^{ST}} i$ becomes 0, this assumption prevents overestimation of the predicted optimal number of time points.

The second step is to minimize the number of time points the critical unit is active as given in Constraint (24). This is done by fixing the objective value and number of time points for each unit other than the critical unit to the values obtained from LP MAX. The model used in this step, i.e. LP MIN, comprises of Constraints (1) - (8), (13) - (15), (18) - (24). The term j^c and z_j represents the critical unit and the frequency each unit is active obtained from LP MAX problem respectively. The predicted optimal number of time points which is very close to the actual optimal number of time points is then given in Equation (25). Equation (23) for LP MIN can be expressed as

$$z_j = \left\lceil \sum_{s_{in,j} \in S^*_{in,j}} \sum_{p} y(s_{in,j}, p) \right\rceil, p \in P, \ j \in J \tag{23}$$

$$\text{Minimize} \sum_{s_{in,j^c}} \sum_{p} y(s_{in,j^c}, p) \tag{24}$$

$$p^{pre} = \left\lceil \sum_{s_{in,j^c}} \sum_{p} y(s_{in,j^c}, p) + \sum_{i \in I^{PT}} i + \sum_{i \in I^{ST}} i \right\rceil \tag{25}$$

The third step is to solve the exact MILP problem using the branch and bound strategy around the predicted optimal number of time points. Solving the MILP problem starts at the root node p^{pre}, the left node is a node that is smaller than p^{pre} and the right node is a node that is larger than p^{pre}. After solving the MILP problem at the predicted optimal number of time points, the iteration starts at the immediate node $p^{pre} - 1$. If the optimal objective value at $p^{pre} - 1$ is worse than that of p^{pre}, it is not necessary to continue any time point lower than $p^{pre} - 1$. Otherwise, continue solving until the objective value becomes worse. If the objective value at $p^{pre} - 1$ is worse than at p^{pre}, then start solving the right node at $p^{pre} + 1$. If the objective value at $p^{pre} + 1$ is better continue solving $p^{pre} + 2$... until the solution converges. Otherwise stop the iteration.

Case study

This case study has been studied by Ierapetritou and Floudas (1998) and Majozi and Zhu, (2001). The STN and SSN representations are shown in Figure 1. The parameters for Case I are given in Table 1.

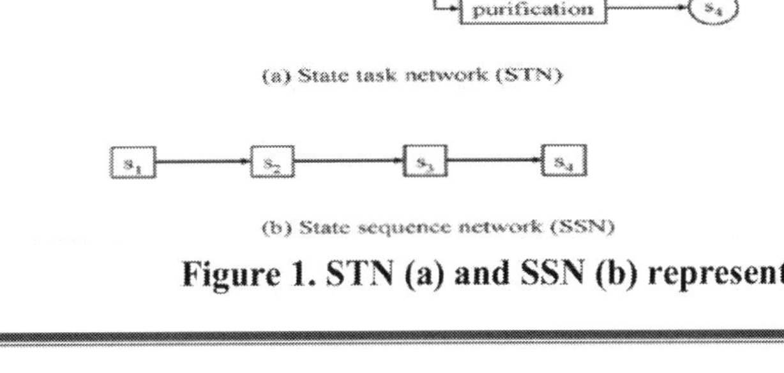

Figure 1. STN (a) and SSN (b) representation of Case I.

Table 1. Given parameters for Case I.

Unit	Label j	Capacity	Suitability	$\tau(s_{in,j})$	$\beta(s_{in,j})$
Heater	1	100	Mixing	3 for batch dependent 4.5 for fixed duration	0.03
Reactor	2	75	Reaction	2 for batch dependent 3 for fixed duration	0.0266
Separator	3	50	Purification	1 for batch dependent 1.5 for fixed duration	0.02
	State	**Storage capacity**	**Initial amount**	**Price**	
	1	unlimited	Unlimited	0	
	2	100	0	0	
	3	100	0	0	
	4	unlimited	0	1	

Results and discussions

The computational statistics that show the optimal objective value, number of constraints required and the total number of continuous and binary variables required by the model for different time horizons are given in Table 2. The duration constraints in the model are fixed. In the iteration method a lower time point is chosen first and then increased by one until no change in the objective value is observed. For a time horizon of 12 h, the iteration was started at time points 2. The objective value was 0. The time points were increased by one to time points 3 and a better objective value of 50 was recorded. This process of increasing the number of time points by one was continued and the objective value of 100 did not change after time points 4. Hence, the optimal objective value was 100. The CPU time required to solve the problem consists not only of the CPU time required at optimal number of time points 4, but is the sum of the CPU times of each iteration from time points 2 to 5 which is (0.07+ 0.036 + 0.086 + 0.051) s.

For a time horizon of 24 h, a similar procedure was followed until the solution converged. The optimal objective value of 350 was obtained at time points 9 (after which there was no improvement in value). For the time horizon of 24 h, the model consisted of 351 constraints, 215 continuous variables and 61 binary variables at the optimal number of time points. The model was solved to 0% relative gap and required a CPU time of 0.08 s. The model was also solved for a time horizon of 168 h to see how the computational time increased with this time horizon. The problem required 75 iterations to solve. The optimal number of time points was 74 with an objective value of 3550. The relative gap does not close in the specified CPU time of 40,000 s. It can be observed, as is expected, that the CPU time required in solving the problem increases exponentially as the number of time points increases. This is due to an increasing number of binary variables. Evidently, the iteration method for obtaining the optimal objective value is computationally costly which renders it impractical as the time horizon for scheduling problems increases. This necessitates prediction of the optimal number of time points beforehand.

Table 2. Results found for Case I under maximization of profit (fixed duration).

Model	P	CPU time (s)	Nodes	RMILP	MILP	B.V	C.V	Constraints	Non-zeros	Relative Gap (%)
				(H = 12)						
ML2	2	0.07	0	0	0	12	47	78	163	
ML2	3	0.036	0	50	50	19	71	117	267	-
ML2	4	0.086	2	100	100	26	95	156	365	-
ML2	5	0.051	10	150	100	33	119	195	467	
				(H = 24)						
ML2	8	0.01	0	300	300	54	191	312	787	-
ML2	9	0.08	10	350	350	61	215	351	875	-
ML2	10	0.5	1301	400	350	68	239	390	977	
				(H = 36)						
ML2	14	0.12	25	600	600	96	335	546	1385	-
ML2	15	42.5	92,785	650	625	103	359	585	1487	-
ML2	16	246.7	45,4229	700	625	110	383	624	1589	-
				(H = 168)						
ML2	73	40,000	2,054,460	3550	3525	509	1751	2847	7403	0.7
ML2	74	40,000	2,436,601	3600	3550	516	1775	2886	7607	1.4
ML2	75	40,000	1,903,820	3650	3550	523	1799	2925	7705	2.8

Table 3. Computational results for LP MAX and LP MIN for Case I under maximization of profit (duration constraints are fixed).

H	Obj. value	Heater	Reactor	Separator	CPU time (s) for LP max	Separator After LP min	CPU time (s) for LP min	p^{pre}
12	150	1.5	2	3	0.015	3	0.015	5
24	433.3	4.33	5.8	8.7	0.015	8.7	0.031	11
36	700	7	9.33	14	0.031	14	0.046	16
168	3633.3	36.3	48.4	72.7	0.33	72.7	0.34	75

The results for this literature example obtained from the prediction method are explained here in detail. For the first scenario, which is a time horizon of 12 h, the first step is to solve a LP MAX problem to maximize the throughput. The time points for the LP model can be calculated with $\left\lceil \frac{H}{\tau^L} \right\rceil$ (12 h/1.5 h) to be 8. From Table 3, it can be seen that the LP MAX problem for the time horizon of 12 h resulted in an objective value of 150. The value of 1.5 for heater implies that the number of times the heater can be utilized is two (rounded up to the nearest integer). The number of times the reactor can be used is two and the number of times the separator can be used is three. Equation (23) was used to determine the time points required by using the frequency of utilization of each unit obtained from the LP MAX problem. For the mixer it becomes $p_1 = \left\lceil \sum_p y(s_1,p) + \sum_{i \in I^{PT}} i + \sum_{i \in I^{ST}} i = 1.5 + 0 + 2 \right\rceil = 4$. The parameter $\sum_{i \in I^{PT}} i$ in the above equation is zero because there is no task before the mixer that is producing an intermediate. The parameter $\sum_{i \in I^{ST}} i$ in the above equation is two because at the end of the process mixing, the mixer passes its intermediate to the reactor and separator to be converted into product. There are two time points associated with the reactor and separator after the end of process by the mixer. Applying the same thought as that of p_1, the time points required by the reactor and separator were calculated.

$p_2 = \left\lceil \sum_p y(s_2,p) + \sum_{i \in I^{PT}} i + \sum_{i \in I^{ST}} i = 2 + 1 + 1 \right\rceil = 4$ and $p_3 = \left\lceil \sum_p y(s_3,p) + \sum_{i \in I^{PT}} i + \sum_{i \in I^{ST}} i = 3 + 2 + 0 \right\rceil = 5$

The maximum number of time points obtained from the LP MAX problem is, therefore, 5. Consequently, the separator becomes the critical unit.

The second step is to solve the LP MIN problem by setting the objective value to 150, $\sum_p y(s_1,p) = 1.5$, $\sum_p y(s_2,p) = 2.0$ and minimizing $\sum_p y(s_3,p)$, to obtain a tighter solution for the number of time points the critical unit is used throughout the time horizon. From the minimization problem it was found that the critical unit is used 3 times. Equation (25) was used to predict the optimal number of time points.

$p^{pre} = \sum_p y(s_3,p) + \sum_{i \in I^{PT}} i + \sum_{i \in I^{ST}} i + 1 = 3 + 2 + 0 = 5$, which is close to the actual optimal number of time points, i.e. 4.

For the time horizon of 36 h, the first step of the LP MAX problem gave an objective value of 700. Using Equation (23) and $\sum_p y(s_1,p) = 7$, $\sum_p y(s_2,p) = 9.33$, $\sum_p y(s_3,p) = 14$, the value of p_1 was calculated to be 9 (7 + 0 + 2), p_2 was calculated to be 12 = $\lceil 9.33 + 1 + 1 \rceil$ and p_3 was calculated to be 16

(14 + 2 + 0). From these values, the maximum of p_j (highest integer value) is 16; hence the separator is still the critical unit.

The second step is to solve the LP MIN using the objective value (700) and the number of time points each unit can be active, obtained from the LP MAX problem to minimize $\sum_p y(s_3,p)$. $\sum_p y(s_3,p)$ was found to be 14 implying that the critical unit is used 14 times. Using Equation (25), the predicted optimal number of time points, p^{pre} was 16 (14 + 2 + 0) which is very close to the actual optimal number of time points, 15.

For the time horizon of 168 h, the first step of LP MAX problem gave an objective value of 3633.3. Using Equation (23) and $\sum_p y(s_1,p)=36.3$, $\sum_p y(s_2,p)=48.4$ and $\sum_p y(s_3,p)=72.7$, the value of p_1 was calculated to be 39 $\lceil 36.3+0+2 \rceil$, p_2 was calculated to be 51 $\lceil 48.4+1+1 \rceil$, and p_3 was calculated to be 75 $\lceil 72.7+2+0 \rceil$. From these values, the maximum of p_j (highest integer value) is 75; hence the separator is the critical unit.

The second step is to solve the LP MIN using the objective value (3633.3) and the number of time points each unit can be active, obtained from the LP MAX problem to minimize $\sum_p y(s_3,p)$. $\sum_p y(s_3,p)$ was found to be 72.7 implying that the critical unit is used 73 times (rounded up to the nearest integer). Using Equation (25), the predicted optimal number of time points, p^{pre} was $\lceil 72.7+2+0 \rceil = 75$ which is very close to the actual optimal number of time points, 74. It is also interesting to see that the CPU time in solving both LP MAX and LP MIN was 0.67 (0.33+0.34) which is less than 1 s. This implies that the method is also accurate in predicting over long time horizon in few seconds of processing time, which makes the prediction method more valuable in avoiding computational costly iterations especially as the time horizon increases.

Conclusions

A method for predicting the optimal number of time points has been presented in this paper. Linear programming maximization and minimization models were used to identify the critical unit and the number of times this critical unit was used. Based on the principle that the optimal number of time points depends on how many times this critical unit is used, an equation was developed to predict the optimal number of time points. The model was tested for a number of case studies taken from published literature. A variance of 0.998 was obtained when comparing predicted optimal number of time points with the actual optimal number of time points. This indicates that the method is accurate. Application of the proposed method will avoid costly computational iteration as the time horizon increases.

References

1 Shaik, M., Floudas, C., 2009. Novel unified modeling approach for short term scheduling. Industrial Engineering Chemistry Research 48, 2947–2964.

2 Li.J, Floudas, C.A., 2010. Optimal event point determination for short-term scheduling of multipurpose batch plants via unit-specific event-based continuous-time approaches. Industrial and Engineering Chemistry Research 49, 7446–7469

3 Janak, S.L., Floudas, C.A., 2008. Improving unit-specific event based continuous time approaches for batch processes: Integrality gap and task splitting. Computer and Chemical Engineering 32, 913-955.

(624f) SCONTO: A Supply Chain ONTOlogy That Extends and Formalizes the SCOR Model

Alicia C. Bhm[1], Horacio P. Leone[1] and **Gabriela P. Henning**[2], (1)INGAR (CONICET - UTN), Santa Fe, Argentina, (2)Universidad Nacional del Litoral, CONICET, Santa Fe, Argentina

SCONTO: A Supply Chain ONTOlogy that extends and formalizes the SCOR model

Alicia C. Böhm, Horacio P. Leone

INGAR, Instituto de Desarrollo y Diseño, UTN – CONICET, Santa Fe, Argentina.

{abohm, hleone}@santafe-conicet.gov.ar

Gabriela P. Henning

INTEC, Instituto de Desarrollo Tecnológico para la Industria Química, UNL–CONICET, Santa Fe, Argentina

ghenning@intec.unl.edu.ar

Abstract. Nowadays, industrial supply chains (SCs), involving different actors (e.g. manufacturers, suppliers, logistics providers, etc.), are striving to boost efficiency and responsiveness. In the face of highly competitive and global markets there are big pressures to reduce lead-times, minimize logistic costs and improve customer service, by increasing agility, flexibility and responsiveness. Thus, enterprises consider supply chain management (SCM) to be a key area where improvements in the previous issues can actually be made. In this hostile environment, both an efficient management and a fluent communication among the various players are mandatory. The development of a common representation of the SC business processes, and their associated concepts, facilitates the appropriate interrelation of the various stakeholders, as well as supports correct decision making and improvement activities.

The Supply Chain Operations Reference (SCOR) model captures the Supply Chain Council (SCC) consensus view of supply chain management (Supply Chain Council, 2011) from a business process perspective. It has been widely used to identify, measure, reorganize, and improve supply chain (SC) business processes. Much of its underlying content has been employed by practitioners of all over the world for many years, thus becoming a de facto standard. The SCOR model has helped managing SC related business problems through a standardized language, pre-established metrics, and common business practices, which aim at accelerating business change, improving performance, and reducing costs. Its mains goals are to support communication among supply chain partners, as well as to improve the effectiveness of SC management and its related supply chain enhancement activities.

Reference models – such as SCOR – are abstract frameworks for identifying and defining concepts, as well as significant relationships among the entities of a domain. These frameworks, which are independent of specific technologies, implementations, or other concrete details, have several purposes. A reference model can be used as a standard, to improve the communication among people. Also, it can be employed by a community (e.g., a set of interacting companies), to define clear roles and responsibilities, and to allow the comparison of different things. Reference models can have various degrees of formalization. As these models become more formal, they turn out to be less ambiguous, more robust, contain fewer errors, and serve better to the purpose of promoting knowledge sharing, effective communication and comparison.

The SCOR model has evolved and improved very much along its lifespan, until reaching the current SCOR 10.0 version, which includes online browsing facilities throughout the framework. Navigation through the SCOR model is now much easier, and linkages between business processes, metrics, practices, etc., are made more explicit. Despite these improvements, the model is so far informal; it is mostly based on textual and graphical semi-formal definitions, which may lead to some confusing interpretations. In addition, certain aspects of the model (e.g. temporal relationships among activities that are part of a workflow, the temporal perspective of metric calculations) seem to be still underspecified or omitted. Moreover, associations between the SC business processes and the performance evaluation and assessment issues are until now rather weak.

This presentation will provide an overview of SCONTO, a Supply Chain ONTOlogy that takes the SCOR model as a foundation and attempts to overcome some of the limitations pointed out previously.

An ontology is a formal, explicit specification of a shared conceptualization (Studer, 1998). In the last 15 years, ontologies of various types have been proposed in several domains, including the supply chain one. In the last decade, the ontology field has matured and ontologies have been recognized as key elements to reach semantic integration, becoming one of the pillars of the Semantic Web stack. Thus, the benefits of adopting an ontology not only reside in its intrinsic capacity to promote share, reuse and communicate information, but also in its ability to establish a solid foundation for interoperability and seamless data integration among distinct applications. This last capacity turns out to be of enormous importance in the context of extended supply chains, where several partners, having different information systems, need to interplay and carry out together the associated SC business processes.

SCONTO is a domain ontology that has been developed during the last five years by our group. It formalizes the SC business process representation and extends the SCOR model by formally capturing the SC physical structure, its participating organizations, as well as the resources that partake in the SC business processes, their roles and flows. Besides, SCONTO includes an explicit, comprehensive, and formal representation of SC assessment related concepts (metrics and best practices). The ontology has been developed following the methodological guidelines devised by Methontology (Fernández-López et al., 1999). As the main result of its conceptualization phase, a semiformal conceptual model was built using the UML language and OCL constraints. Later, this model has been formalized and implemented in OWL. To test the ontology, different case studies were carried out, and are still being addressed.

The presentation will describe the main characteristics of SCONTO´s conceptual model and its application to a particular case-study. This example will illustrate that SCONTO can unambiguously describe SC business processes and activities, their scope, associated organizational units and resources, their interrelationships, etc. For instance, it will show how the relationships between two somehow "overlapping" processes, *Source* and *Deliver,* which are carried out by different enterprises or distinct organizational units, can be explicitly captured by the model and later analyzed for improvement. The presentation will also show how performance related concepts are formally captured within the framework of an assessment system, which explicitly models domain notions (performance dimensions and concepts, performance attributes, metrics, measures, appraisal activities, etc.) and their interrelationships. The example being discussed will also cover these aspects and will demonstrate how measures can be clearly specified (i.e. what is measured, when, where, and how something is evaluated).

The talk will conclude summarizing the benefits of turning the SCOR model into an ontology by extending and formalizing it. It will also point the advantages of this approach over the ones that are based on the use of generic business process description languages, such as BPMN (OMG, 2011).

References:

Fernández-López, M., A. Gómez-Prez, J.P. Sierra and A.P. Sierra. Building a Chemical Ontology Using Methontology and the Ontology Design Environment. Intelligent Systems, 14, 37-46, 1999.

OMG, Object Management Group. Business Process Model and Notation (BPMN) Version 2.0. http://www.omg.org/spec/BPMN/2.0/, Accessed April 2011.

Supply Chain Council. SCOR Supply Chain Operations Reference Model Version 10.0. http://supply-chain.org/f/SCOR-Overview-Web.pdf, Accessed April 2011.

Studer, R., V.R. Benjamins and D. Fensel. Knowledge Engineering: Principles and Methods. IEEE Transactions on Data and Knowledge Engineering, 25, 161-197, 1998.

See more of this Session: Poster Session: Computers In Operations and Information Processing
See more of this Group/Topical: Computing and Systems Technology Division

An Advanced Scheduling Technique for Multipurpose Batch Plants

[1]E. Reshid and [1,2]T. Majozi
[1]*Department of Chemical Engineering, University of Pretoria, Lynnwood Road, Pretoria, 0002, South Africa*
[2] *Modelling and Digital Science, CSIR, Pretoria, South Africa*

Abstract

Several scheduling techniques exist in literature based on continuous time representation. The models based on unit specific time points have shown better solution efficiency by reducing the number of time points and problem size. In this presentation novel scheduling techniques based on unit specific time point continuous time representation are presented. The proposed models allow nonsimultaneous material transfer into a unit like in the models of Susarla et al. (2010). This approach gives a better schedule as compared to most published models. The developed scheduling models are based on state sequence network representation that has proven to inherently result in smaller problems in terms of binary variables. The models require a smaller number of time points as compared to single-grid and multi-grid continuous time models. Consequently, they exhibit much better computational performance. Numerical evaluation using literature examples indicate in some of the complex examples, that the proposed models give a better objective value as compared to other scheduling models. An added feature of the proposed models are their ability to exactly handle fixed intermediate storage operational philosophy, which has proven to be a subtle drawback in most published scheduling techniques.
Key words: Scheduling; Multipurpose Batch Plant; Optimization; MILP; Continuous-time

Introduction

In recent years, batch processes have been getting more attention due to their suitability for the production of small volume, high value added products which are becoming increasingly important with fast market change. The flexibility of batch plants allows the production of different products within the same facility. This makes plant scheduling a challenging task. Much research has focused on developing optimization techniques for scheduling batch plants with the aim of reducing the time required to attain an optimal objective value within a reasonable computer processing (CPU) time.

Méndez et al. (2006) and Shaik and Floudas (2006) presented excellent reviews of the current scheduling techniques based on different time representations and associated challenges. In their review the different models are classified as slot based, event based and precedence based (sequence - based) time representation.

In the slot based models (Liu and Karimi, 2008) the time horizon is divided into "nonuniform unknown slots" and "tasks start and finish in the same slot". On the other hand, there exist slot models that use nonuniform unknown slots where tasks are allowed to continue to the next slots (Sundaramoorthy and Karimi, 2005). The event based models can also be categorized into those that use uniform unknown events where the time associated with the events is common across all units (Maravelias and Grossmann, 2003) and those that use unit specific events where the time associated to the events can be different across the units (Ierapetritou and Floudas, 1998; Majozi and Zhu, 2001; Shaik and Floudas, 2009). Shaik et al.(2006) and Shaik and Floudas (2008) concluded that due to heterogeneous location of events across the unit, the unit specific event based models require less events and perform better as compared to both the global event based and slot based models. The sequence-based or precedence-based representation uses either direct precedence (Liu and Karimi, 2007) or indirect precedence sequencing of pairs of tasks on units (Ferrer-Nadal et al., 2008). The models do not require pre postulation of events and slots.

Shaik and Floudas (2009) proposed a MILP formulation that used three index binary variables. Their formulation used unit specific event based time representation, where tasks were allowed to continue processing on multiple event points. Susarla et al. (2010) presented models that use unit specific slots that allowed tasks to span over multiple slots. Their models also allow nonsimultaneous transfer of material into a unit to get a better schedule.

In this paper, novel scheduling techniques that use unit specific time points are presented. The proposed models can handle the different storage operational policies that do not violate real time storage constraints. Paramount among these is the fixed intermediate storage (FIS) operational policy, which has been inadvertently violated in most published models. The developed models are compared to the different models in the literature based on bench mark examples. Through literature examples, it is demonstrated that the proposed formulations are simpler, tighter and give less number of time points as compared to other formulations in the literature. Moreover, the resulting formulations give better optimal objective value.

Mathematical model using state sequence network (SSN)

The mathematical model presented in this paper entails the following constraints.

Allocation constraints

$$\sum_{s_{in,j} \in S^*_{in,j}} y(s_{in,j}, p) \leq 1, \ \forall j \in J, \quad p \in P \tag{1}$$

Capacity constraints

$$V^L_{s_{in,j}} y(s_{in,j}, p) \leq mu(s_{in,j}, p) \leq V^U_{s_{in,j}} y(s_{in,j}, p), \ \forall \ p \in P, \ j \in J, \ s_{in,j} \in S_{in,J} \tag{2}$$

Material balance for storage

$$q_s(s,p) = q_s(s,p-1) - \sum_{s_{in,j} \in S^{sc}_{in,J}} \rho^{sc}_{s_{in,j}} mu(s_{in,j}, p) + \sum_{s_{in,j} \in S^{sp}_{in,J}} \rho^{sp}_{s_{in,j}} mu(s_{in,j}, p-1) \tag{3}$$

$$\forall \ p \in P, \quad s \in S$$

$$q_s(s^P, p) = q_s(s^P, p-1) + \sum_{s_{in,j} \in s^{s^P p}_{in,J}} \rho^{sp}_{s_{in,j}} mu(s_{in,j}, p), \ \forall \ p \in P, \quad s^P \in S^P \tag{4}$$

Duration constraints (batch time as a function of batch size)

$$t_p(s_{in,j}, p) \geq t_u(s_{in,j}, p) + \tau(s_{in,j}) y(s_{in,j}, p) + \beta(s_{in,j}) mu(s_{in,j}, p), \ \forall \ j \in J, \ p \in P, \ s_{in,j} \in S_{in,J} \tag{5}$$

Sequence constraints

Same task in same unit

$$t_u(s_{in,j}, p) \geq t_p(s_{in,j}, p-1), \ \forall \ j \in J, \ p \in P, \ s_{in,j} \in S^*_{in,j} \tag{6}$$

Different tasks in same unit

$$t_u(s_{in,j}, p) \geq t_p(s'_{in,j}, p), \ \forall \ j \in J, \ p \in P, \ s_{in,j} \neq s'_{in,j}, \ s_{in,j}, s'_{in,j} \in S^*_{in,j} \tag{7}$$

If the state is consumed and produced in the same unit, where the produced state is unstable then in addition to Constraint (7), Constraints (8) is used.

$$t_p\left(s_{in,j}^{usp},p-1\right) \geq t_u\left(s_{in,j}^{usc},p\right) - H\left(1-y\left(s_{in,j}^{usp},p-1\right)\right),\ \forall\ j \in J,\ p \in P,\ s_{in,j}^{usc} \in S_{in,j}^{usc},\ s_{in,j}^{usp} \in S_{in,j}^{usp} \quad (8)$$

Sequence constraints different tasks in different units.

These constraints state that for different tasks that consume and produce the same state, the starting time of the consuming task at time point p must be later than the finishing time of any task at the previous time point p-1 provided that the state is used.

If an intermediate state s is produced from one unit.

Constraints (9) and (10) work together in the following manner:

$$\rho\left(s_{in,j}^{sp}\right) mu\left(s_{in,j}, p-1\right) \leq q_s(s,p) + V_j^U t(j,p),\ \forall\ j \in J,\ p \in P,\ s_{in,j} \in S_{in,J}^{sp} \quad (9)$$

$$tu\left(s_{in,j'}, p\right) \geq tp\left(s_{in,j}, p-1\right) - H\left(\left(2 - y\left(s_{in,j}, p-1\right) - t(j,p)\right)\right),$$
$$\forall\ j \in J,\ p \in P,\ s_{in,j} \in S_{in,J}^{sp},\ s_{in,j'} \in S_{in,J}^{sc} \quad (10)$$

Constraint (9) states that if the state s is produced from unit j at time point p-1 but is not consumed at time point p by another unit j', i.e. $t(j,p) = 0$, then the amount produced cannot exceed allowed storage, i.e. $q_s(s,p)$. On the other hand, if state s produced from unit j at time point p-1 is used by another unit j' then the amount of state s stored at time point p, i.e.$q_s(s,p)$ is less than the amount of state s produced at time point p-1. The outcome is that the binary variable $t(j,p)$ becomes 1 in order for Constraint (9) to hold. If the unit performs tasks like separation, distillation and other tasks that produce more than one intermediate at time point p then the binary variable $t(j,p)$ becomes $t(j,s,p)$. This allows us at the same time point for Constraint (10) to be relaxed for the unit that is not using the state produced by unit j at time point p. Simultaneously, for the other unit that uses the state produced by unit j at time point p the sequence Constraint (10) holds.

Constraint (10) states that the starting time of a task consuming state s at time point p must be later than the finishing time of a task that produces state s at the previous time point p-1, provided that state s is used. Otherwise the sequence constraint is relaxed.

If an intermediate state is produced from more than one unit

Constraint (11) states that the amount of state s used at time point p can either come from storage, or from other units that produce the same state depending on the binary variable $t(j,p)$. If the binary variable $t(j,p)$ is 0, which means that state s produced from unit j at time point p-1 is not used at time point p, then Constraint (10) is relaxed. If $t(j,p)$ is 1, state s produced from unit j at time point p-1 is used, as a result Constraint (10) holds. Although Constraint (11) is nonlinear it can be linearized exactly using Glover transformation developed by Glover (1975). It is highly imperative to realize that Constraint (10) plays a pivotal role in both instances when a state is produced from one unit and when a state is produced from many units.

$$\sum_{s_{in,j} \in S_{in,J}^{sc}} \rho_{s_{in,j}}^{sc}\, mu\left(s_{in,j}, p\right) \leq qs(s,p-1) + \sum_{s_{in,j} \in S_{in,J}^{sp}} \rho_{s_{in,j}}^{sp}\, mu\left(s_{in,j}, p-1\right) t(j,p)\ \forall\ j \in J,\ p \in P \quad (11)$$

Constraint (12) states that a consuming task can start after the completion of the previous task. Constraint (12) taken care of proper sequencing time when a unit uses material which is previously stored that is when the producing task is active at time point p-2 later produce and transfer the material to the storage at time point p-1. This available material in the storage at time point p-1 is then used by the consuming task in the next time points this necessitate the starting time of the consuming task must be later than the finishing time of the producing task at time point p-2.

$$t_u\left(s_{in,j'},p\right) \geq t_p\left(s_{in,j}, p-2\right) - H(1-y(s_{in,j}, p-2)),\ \forall\ j \in J,\ p \in P,\ s_{in,j} \in S_{in,J}^{sp},\ s_{in,j'} \in S_{in,j'}^{sc} \quad (12)$$

Sequence constraint for FIS policy

According to Constraint (13), any state s can only be stored if the capacity of available storage will not be exceeded. Otherwise, state s will either be produced and consumed immediately or not produced at all. Constraint (14) enforces this condition.

$$\sum_{s_{in,j} \in S_{in,j}^{sp}} \rho_{s_{in,j}}^{ps} mu\left(s_{in,j}, p-1\right) + qs(s, p-1) \leq QS^U + \sum_{j \in J_s} V_j^U \left(1 - x(s,p)\right) \; \forall \; j \in J, \; p \in P, \; s \in S \tag{13}$$

According to Constraint (14), the starting time of a task that consumes state s at time point p must be equal to the finishing time of a task that produces state s at time point p-1, if both consuming and producing tasks are active at time point p and time point p-1 respectively and if there is no storage to store the amount of state s produced at time point p. In a case when storage is available to store a state s at time point p, then the starting time of the consuming task at time point p is not necessarily equal to the finishing time of a task producing state s at time point p-1. Constraints (13) and (14) reduce the number of time points and give a better objective value as can be seen in the case study.

$$tp\left(s_{in,j'}, p-1\right) \leq tp\left(s_{in,j}, p-1\right) + H\left(2 - y\left(s_{in,j'}, p\right) - y\left(s_{in,j}, p-1\right)\right) + H\left(x(s,p)\right)$$
$$\forall \; j \in J, \; p \in P, \; s_{in,j} \in S_{in,J}^{sp}, \; s_{in,j'} \in S_{in,J}^{sc} \tag{14}$$

Storage constraints

Constraint (15) indicates that the amount of state s stored at any time point must not exceed the maximum capacity of the storage. The state s that is produced at time point p-1 can be stored for a while in a unit that is producing it in the next time points until it is used, if the unit is not performing tasks.

$$q_s(s,p) \leq QS^U + \sum_{s_{in,j} \in S_{in,J}^{sp}} u\left(s_{in,j}, p\right) \quad \forall s \in S, \; p \in P, \; j \in J \tag{15}$$

Constraint (16) states that portion of the state that is produced at time point p can be stored in the unit for consecutive time points, if the unit is not active in those consecutive time points. Which is a similar concept to that of slot based formulations that allow task to continue in the next consecutive slots, indicating that the unit stores the states in those consecutive time points.

$$u\left(s_{in,j}, p\right) \leq \rho_{s_{in,j}}^{sp} mu\left(s_{in,j}, p-1\right) + u\left(s_{in,j}, p-1\right) \;\; \forall p \in P, \; j \in J \tag{16}$$

Constraint (17) ensures that if a state is stored at time point p in the unit then the unit should not be active to start any other task.

$$u\left(s_{in,j}, p\right) \leq V_j^U \left(1 - \sum_{s_{in,j} \in S_{in,j}^*} y\left(s_{in,j}, p\right)\right) \;\; \forall p \in P, \; j \in J \tag{17}$$

Time horizon constraints

$$t_u\left(s_{in,j}, p\right) \leq H, \; \forall \; s_{in,j} \in S_{in,J}, \; p \in P, \; j \in J \tag{18}$$

$$t_p\left(s_{in,j}, p\right) \leq H, \; \forall \; s_{in,j} \in S_{in,J}, \; p \in P, \; j \in J \tag{19}$$

Tightening constraints

$$\sum_{s_{in,j} \in S_{in,j}^*} \sum_{p} \left(\tau\left(s_{in,j}\right) y\left(s_{in,j}, p\right) + \beta\left(s_{in,j}\right) mu\left(s_{in,j}, p\right)\right) \leq H, \; \forall \;\; p \in P, \; j \in J \tag{20}$$

Objective function

The objective of scheduling problem is to maximize the product throughput if the target is not known *a priori* or to minimize the makespan, if the target is known beforehand, as given in Constraint (21).

$$\text{maximize} \sum_{s} price\left(s^P\right) qs\left(s^P, p\right), \; \forall \; p = P, \; s^P \in S^P$$

Or

$$\text{minimize } H \tag{21}$$

The model ML1 consists of constraints 1-8, 10, 13-15, 18-21. This model does not cater for a situation where the sequence constraint for different task in different units is to be relaxed in a case where the state is not used. For this model the sequences Constraints (10) and (15) can be written as:

$$tu(s_{in,j'}, p) \geq tp(s_{in,j}, p) - H(1 - y(s_{in,j}, p-1)), \; \forall j \in J, \; p \in P, \; s_{in,j} \in S^{sp}_{in,J}, \; s_{in,j'} \in S^{sc}_{in,J} \quad (10)$$

$$q_s(s,p) \leq QS^U \quad \forall s \in S, p \in P \quad (15)$$

The model ML2 consists of constraints 1-21.

Case study

At first Sundaramoorthy and Karimi (2005) studied this case study which is relatively more complex problem and later often used in the literature to check the efficiency of models in terms of optimal objective value and CPU time required to get to optimal objective value. The plant accommodates many common features of multipurpose batch plants such as unit performing multiple tasks, multiple units suitable for a task and dedicated units for specific tasks. This case study is solved for FIS policy using the presented model ML1 and ML2. The STN representation of this case study is depicted in Figure 1. Data required for this case study is depicted in Table 1.

The processing time of a task i on unit j is assumed to be linearly dependent, $\alpha_i + \beta_i B$ of its batch size B. Where α_i is constant term of processing time of task i and β_i is coefficient of variable term of processing time of task i. The batch size B of a task should be in the range of V_i^L and V_i^U representing the minimum and maximum batch size respectively. The initial stock levels for all intermediates are assumed to be zero except for state $s6$ and $s7$ with initial stock of 50 (mu). The stock is available as required for states $s1$, $s2$, $s3$ and $s11$. The price of product 1 and 2 is \$5/mu. This motivating example is solved for FIS operational policy. For the case of FIS the storage capacity for states $s3$, $s4$, $s5$, $s6$, $s7$, $s9$, $s10$ are 100mu, 100mu, 300mu, 150mu, 150mu, 150mu, 150mu respectively. The storage capacity for raw material and product states, $s1$, $s2$, $s8$, s11, $s12$ and $s13$ are assumed unlimited.

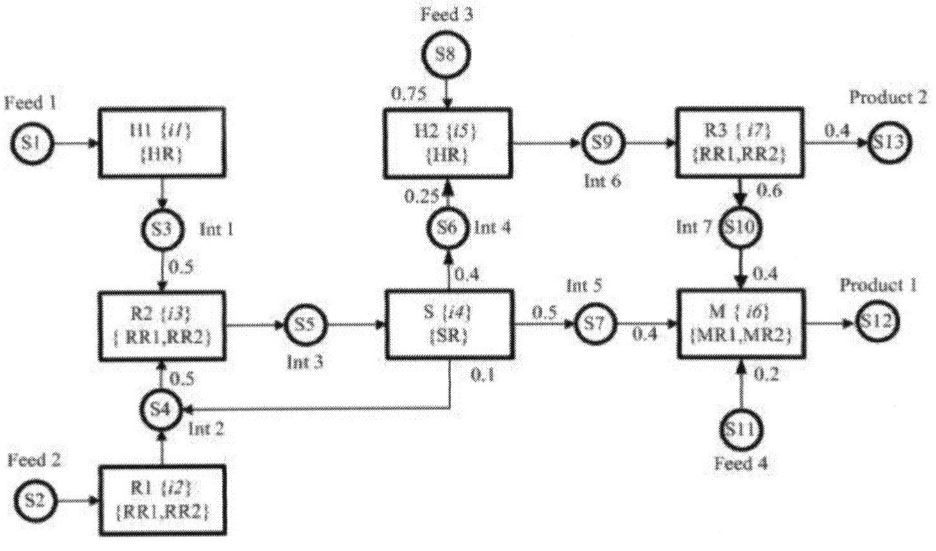

Figure 1 STN representation for the case study.

Table 1: Data required for the case study.

Task	Lable i	unit	Lable j	α_i	β_i	$\gamma_i^L - \gamma_i^U$
Heating-1	$i1$	Heater	HR	0.667	0.00667	0-100
Heating-2	$i5$	Heater	HR	1	0.01	0-100
Reaction-1	$i2$	Reactor 1	RR-1	1.333	0.01333	0-100
	$i2$	Reactor 2	RR-2	1.333	0.00889	0-150
Reaction-2	$i3$	Reactor 1	RR-1	0.667	0.00667	0-100
	$i3$	Reactor 2	RR-2	0.667	0.00445	0-150
Reaction-3	$i7$	Reactor 1	RR-1	1.333	0.0133	0-100
	$i7$	Reactor 2	RR-2	1.333	0.00889	0-150
Separation	$i4$	Separator	SR	2	0.00667	0-300
Mixing	$i6$	Mixer 1	MR-1	1.333	0.00667	20-200
	$i6$	Mixer 2	MR-2	1.333	0.00667	20-200

Results and Discussion

Recently this example is solved for FIS operational philosophy by Susarla et al. (2010) for the objective of maximization of profit for a time horizon of 10 h by using different literature models. These models are the unit-specific slot based models of Susarla et al. (2010) (SLK2&SLK1), Shaik and Floudas (2009) (S&F), Sundaramoorthy and Karimi (2005) (S&K) and Maravelias and Grossmann (2003) (M&G). Table 2 provides a comparison of different models statistics for 10 h time horizon.

Table 2.Computational results for the case study under maximization of profit with FIS

Model	P	CPU time (s)	Nodes	RMILP	MILP	B.V	C.V	Constraints	Non-zeros	Relative Gap (%)
					(H = 10)					
SLK2	9	10,000	872,204	3618.6	2337.36	136	853	1428	4970	5.07
SLK1	9	10,000	711,493	3618.6	2337.36	136	799	1380	4886	11.6
SF										
($\Delta n = 0$)	8	115	1160,988	3618.6	2292.5	88	382	884	2847	-
($\Delta n = 1$)	8	10,000	2256,858	3618.6	2337.36	165	459	1658	6187	1.68
SK	10	10,000	2666,604	3473.9	2260.9	153	874	1121	3380	10.7
MG	10	10,000	1415,516	3473.9	2137.1	220	1,151	2691	10,710	23.9
ML1	6	1.34	3,279	2871.9	2292.5	108	397	811	2283	-
ML2	6	8	18,260	2871.9	2345.3	141	538	1147	3422	-

The models SLK2, SLK1 and S&F ($\Delta n=1$) required 9 event points to get the optimal objective value of $2337.36. The models SK and MG needed 10 event points to give an inferior optimal objective value of $2260.9 & $2137.1 respectively. The better objective value by SLK2, SLK1 and S&F (2009) ($\Delta n=1$) is attributed to fact that the models are based on multi grid time representation and allow task to span over multiple event points.

The proposed model ML2 requires 5 slots ($p = 6$), which is 4 slots less when compared to single grid models of MG and SK and 3 slots less when compared to the multi grid models of SLK1&2 and SF. The model ML2 gives a better objective value $2345. The model ML2 gives the lowest RMILP value of $2871.9. A CPU time of 8 s is required by ML2 to solve the problem to a 0% relative gap which is much less when compared to the specified CPU time of 10,000 s for all models in literature. Figure 2 shows the Gantt chart for the time horizon of 10 h obtained by the proposed model ML2.

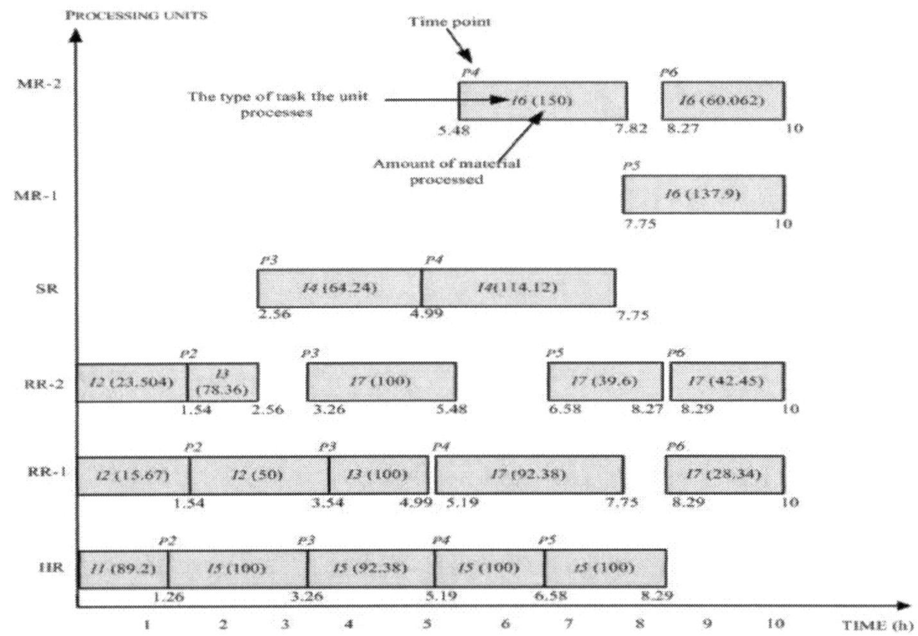

Figure 2 Gant chart for the time horizon of 10 h.

Conclusions

In this paper a novel scheduling technique for multipurpose batch plants has been presented. The technique involves two mathematical formulations that are based on SSN representation. Both presented models exhibit MILP structure for which global optimality can be proven. The proposed models allow nonsimultaneous transfer of material to get a better objective value. The models in this paper require less number of time points as a result giving a better CPU time as compared to existing models from the literature. Improved objective value is also obtained as compared to values obtained from models in literature.

Notation

Sets

S $\{S \mid S$ any state other than a product$\}$

I $\{i \mid i$ is a task$\}$

J $\{j \mid j$ is a unit$\}$

$S^{sc}_{in,J}$ $\{s^{sc}_{in,j} \mid s^{sc}_{in,j}$ task which consume state $s\}$

$S^{*}_{in,j}$ $\{s^{*}_{in,j} \mid s^{*}_{in,j}$ tasks performed in unit $j\}$

$S_{in,J}$ $\{s_{in,j} \mid s_{in,j}$ is an effective state representing a task$\}$

J_s $\{j_s \mid j_s$ is a unit producing state $s\}$

$S^{usc}_{in,J}$ $\{s^{usc}_{in,j} \mid s^{usc}_{in,j}$ task which consume unstable state $s\}$

P $\{p \mid p$ is a time point$\}$

$S^{usp}_{in,J}$ $\{s^{usp}_{in,j} \mid s^{usp}_{in,j}$ task which produce unstable state $s\}$

S^{p} $\{s^{p} \mid s^{p}$ a state which is a product$\}$

$S^{s^p p}_{in,J}$ $\{s^{s^p p}_{in,j} \mid s^{s^p p}_{in,j}$ task which produce state s which is a product$\}$

$S^{sp}_{in,J}$ $\{s^{sp}_{in,j} \mid s^{sp}_{in,j}$ task which produce state s other than a product$\}$

Variables

$t_p(s_{in,j}, p)$ time at which task ends at time point p, $s_{in,j} \in S_{in,j}$

$t_u(s_{in,j}, p)$ time at which task starts at time point p, $s_{in,j} \in S_{in,j}$

$mu(s_{in,j}, p)$	amount of material processed by a task at time point p
$q_s(s, p)$	amount of state s stored at time point p
$y(s_{in,j}, p)$	binary variable for assignment of task at time point p
$t(j, p)$	binary variable associated with usage of state produced by unit j at time point p
$t(j, s, p)$	binary variable associated with usage of state s produced by unit j at time point p if the unit produces more than one intermediate at time point p
$x(s, p)$	binary variable associated with availability of storage for state s at time point p
$u(s_{in,j}, p)$	amount of material stored in unit j at time point p

Parameters

V_j^U	maximum capacity of unit j
$V_{s_{in},j}^U$	maximum capacity of unit j to process a particular task
$V_{s_{in},j}^L$	minimum capacity of unit j to process a particular task
H	time horizon of interest
QS^o	initial amount of state s stored
QS^U	maximum capacity of storage to store a state s
$\tau(s_{in,j})$	coefficient of constant term of processing time of a task
$\beta(s_{in,j})$	coefficient of variable term of processing time of a task
$\rho(s_{in,j}^{sp})$	portion of state s produced by a task
$\rho(s_{in,j}^{sc})$	portion of state s consumed by a task

References

1 Ferrer-Nadal, S., Capón-Garćia, E., Méndez, C.A., Puigjaner, L., 2008. Material transfer operations in batch scheduling. A critical modeling issue. Industrial and Engineering Chemistry Research 47, 7721-7732.

2 Ierapetritou, M.G., Floudas, C.A., 1998. Effective continuous-time formulation for short-term scheduling: 1. Multipurpose batch processes. Industrial and Engineering Chemistry Research 37, 4341-4359.

3 Liu, Y., Karimi, I.A., 2008. Scheduling multistage batch plants with parallel units and no interstage storage. Computer and Chemical Engineering 32, 671- 693.

4 Liu, Y., Karimi, I.A., 2008. Scheduling multistage batch plants with parallel units and no interstage storage. Computer and Chemical Engineering 32, 671- 693.

5 Majozi, T., Zhu, X.X., 2001. A novel continuous-time MILP Formulation for multipurpose bach plants. Industrial and Engineering Chemistry Research 40, 5935-5949.

6 Maravelias, C.T., Grossmann, I. E., 2003. New general continuous-time state-task network formulation for short-term scheduling of multipurpose batch plants. Industrial and Engineering Chemistry Research 42, 3056-3074.

7 Méndez, C.A.;Cerdá, J., 2004. Short-term scheduling of multistage batch processes subject to lmited finite resources. Computer Aided Chemical Engineering 15B:984-989.

8 Shaik, M., Floudas, C., 2009. Novel unified modeling approach for short term scheduling. Industrial Engineering Chemistry Research 48, 2947-2964.

9 Shaik, M.A., Floudas, C.A., 2008. Unit-specific event-based continuous time approach for short-term scheduling of batch plants using RTN framework. Computer and Chemical Engineering32, 260-274.

10 Shaik, M.A., Janak, S.L., Floudas, C.A., 2006. Continuous-time models for short-term scheduling of multipurpose batch plants: A comparative study. Industrial and Engineering Chemistry Research 45, 6190-4209.

11 Sundaramoorthy, A., Karimi, I.A., 2005. A simpler better slot-based continuous-time formulation for short-term scheduling in multipurpose batch plants. Chemical Engineering Science 60, 2679-2702.

12 Susarla,N., Li, J., Karimi, I.A., 2010. A novel approach to scheduling of multipurpose batch plants using unit slots. AICHE Journal 56, 1859-1879.

(624h) Using Layer of Protection Analysis to Mitigate Supply Chain Risk

Russell A. Ogle[1], Dennis M. McCann[2] and Mark J. Viz[2], (1)Exponent, Lisle, IL, (2)Exponent, Wood Dale, IL

Over the last 25 years the chemical process industry has invested a great deal of effort to promote process safety and security in their production facilities. In that same time frame, the chemical supply chain has become increasingly more vulnerable to disruptions. Business factors such as reduced inventories ("lean manufacturing"), sole-source contracting, globalization, and aging transportation infrastructure have conspired to make supply chains extremely vulnerable to delays or interruptions in the flow of materials. The challenge of managing supply chain disruptions is complex since they can be caused by a broad array of events including accidents, natural hazards, and sabotage. Thus, while the threats to production capacity from hazards have been brought under a higher level of control, the potential frequency and severity of external hazards has actually increased. In this paper we illustrate how some of the tools developed for chemical process hazard analysis can be adapted for SCRM. Specifically, we demonstrate how Layer of Protection Analysis (LOPA) can be used to select safeguards for controlling supply chain risk.

SCRM consists of three stages: risk identification, risk assessment, and risk control. In our case study we demonstrate the use of checklists for risk identification. Risk assessment is facilitated with risk matrix that defines different levels of frequency and severity of consequence. Then LOPA is used to develop a system of safeguards commensurate with the desired level of risk reduction. The case study is based on a petrochemical facility that receives raw materials by truck, rail, barge, and pipeline. Products are shipped out by truck and rail. We consider representative supply chain disruptions due to accidents, natural hazards, and terrorist attack.

See more of this Session: Poster Session: Computers In Operations and Information Processing
See more of this Group/Topical: Computing and Systems Technology Division

(624i) Simultaneous Optimization of Hoist Scheduling and Production Line Arrangement

Chuanyu Zhao and Qiang Xu, Dan F. Smith Department of Chemical Engineering, Lamar University, Beaumont, TX

Simultaneous Optimization of Hoist Scheduling and Production Line Arrangement†

Chuanyu Zhao and Qiang Xu*

Dan F. Smith Department of Chemical Engineering

Lamar University, Beaumont, TX 77710, USA

Abstract

In multi-stage material handling processes, such as electroplating and polymeric coating, productivity maximization is always an objective. Hoist schedule development or hoist scheduling is the most relevant aspect to improve the production efficiency. Hoist is a controlled robot following a pre-set movement schedule to move various jobs in a production line for processing based on their recipes. It is reported that as high as 20% reduction in mean job waiting time and 50% improvement in standard deviation of cycle time can be achieved by hoist scheduling. In previous works, how to feasibly and timely organize cyclic or dynamic schedules has been broadly studied. However, the productivity of a multi-stage process not only depends on the hoist schedule, but also substantially relies on the production line arrangement, i.e., production units allocation spatially. This is because a hoist needs to travel among units for different manufacturing steps with or without a job and the travelling time depends on unit locations, which primarily affects the scheduling time or productivity. Thus, the design and operational scheduling should be simultaneously considered for the best performance of such multi-stage material production line.

In this paper, a systematic methodology coupling optimization of cyclic hoist scheduling and production line arrangement for maximum productivity has been developed. It first generates an MINLP model addressing all the major concerned scheduling issues, such as cyclic scheduling for large amounts of jobs with multiple recipes, various production unit capacities, and customized production portfolios. Then, the MINLP model is fully linearized as an MILP model to ensure the global optimal of the solution. The methodology can satisfy any customized demand on product ratios among different job types.

The efficacy of the proposed methodology is demonstrated by a case study of electroplating system.

Three types of jobs with different recipes are processed by a production line, where the overall production rate needs to be maximized. The discussion for hoist scheduling with and without consideration of production line arrangement will also be addressed.

Keywords: Cyclic Hoist scheduling, Production line arrangement, Optimization, MILP

See more of this Session: Poster Session: Computers In Operations and Information Processing
See more of this Group/Topical: Computing and Systems Technology Division

(624j) Real-Time Data Estimation Method for LNG Terminal Using Conventional Process Simulator

Sangho Lee, Chul-Jin Lee, Sungwoo Cho, Jeongnam Kim, **Chonghun Han** and Youngsub Lim, Chemical and Biological Engineering, Seoul National University, Seoul, South Korea

For analyzing the plant status exactly, it is necessary to get enough information from plant facilities by physical sensors such as flowmeters and thermocouples. But if every sensor were installed, the cost problem becomes serious therefore gathering sufficient data for analysis is unrealistic situation. Usuallly process simulation is an possible answer to solve this problem, thus many conventional simulators such as ASPEN PLUS, HYSYS and Pro/II offer various mathematical process model and with these models and sensored data, the unmeasured information can be calculated. Nevertheless the process simulation derives useful data, it is meaningless in field operation because the simulation couldn't be real-time.

LNG terminal which takes a part of storage and gasifying LNG have suffered from this issue. Especially on BOG(Boil-off Gas) estimation of storage tank inlet flow, real-time data about composition, flow rate, temperature and pressure are required to figure out exact amount of BOG flow rate though only temperature from a few thermocouple in the pipeline and pressure from the manometer inside the storage tank are available in real-time.

In this research, focusing on LNG terminal inlet pipeline, a simulation model that satisfies the steady state case's process data was built at first. After building a model, the model is validated in various cases - unloading, no-ship unloading and recirculation. Finally the model is solved during the sensored data was used as constraints of the dynamic simulation model and through this, BOG flow rate in the inlet pipeline which is the most important was estimated in real-time.

See more of this Session: Poster Session: Computers In Operations and Information Processing
See more of this Group/Topical: Computing and Systems Technology Division

(624k) Simulation Based Optimization of Ventilation Strategies for Subway Tunnel Fires

Kyungjun Park, Dongil Peter Shin and Ki Jun Lee, Dept. of Chemical Engineering, Myongji University, Yongin, Kyunggido, South Korea

As the population density in large cities of the world is getting higher and the traffic volume is increasing continuously, the roads cannot accommodate such traffic increase any more so that construction of new subways or extension work of existing subways are constantly going on. With this higher subway traffic, the subway stations are needed to be better prepared for fire. But the preparation for fire emergency is lacking in subway tunnels at the present. In Korea, the risk of subway fire has been widely recognized since the arson fire took place at the Daegu Jungangro subway station in 2003, resulting in 340 casualties. As a result of this accident, people recognized problems related to fire equipment, the emergency management agency and ventilation operation. Since most subway stations are easily exposed against the possibility of danger such as, it is necessary to set up substantial prevention measures against any possible fire or terror attack.

Many studies on the prediction of fire damage have been performed at subway platform. However, the research of the simulation on subway car fire have lacked in tunnels. Unlike other building fire on the ground, the combustion is unstable because fire on underground tunnel may be limited to get O_2 from the outside, therefore mass toxic gas would be released. Considering most death from suffocation by the fumes, passengers in the tunnel are more exposed to the hazard from toxic gas. The better adaptive ventilation strategies have been needed to help people survive in dangerous situations.

This study is focused on the heat and toxic gas characteristics in an underground subway tunnel where burning train is stopping in, by using Fire Dynamics Simulator (FDS). The tunnel, which is a part of Seoul Metro Line 1, has 6 sessions for ventilation. Due to structural limitation, only four fans were installed for forced ventilation and other two fans operate by natural ventilation. Usually when a fire occurs in a tunnel, they control forced ventilation facilities to set up the fire escape route for people: fans at the side of escape route are operated on the supply mode. Fans at the other side are operated on the exhaust mode for the removal of smoke. But this tunnel cannot use ordinary ventilation strategy because of one-sided ventilation facilities. Therefore, this study carries out a variety of simulations to find out the best ventilation control strategy according to the fire situations.

To simulate the combustion of real-size subway tunnel model in detail for the prediction of combustion gas behavior, we should calculate large amount of numerical analysis due to model of the complex structure and variety of oxidation, heat transfer as turbulent flow. For this reason, we used parallelized FDS, which based on Message Passing Interface (MPI), running on a high performance computer cluster. The objective of simulation is comparing each simulation result about air flow, temperature, visibility, toxic gas according to the selected method of fan operations. Then the best ventilation method for each fire situation is to be decided by the results. The result of this study will contribute to the improved saving of passengers in case of fire accidents in subway stations with optimized ventilation system design and operations.

See more of this Session: Poster Session: Computers In Operations and Information Processing
See more of this Group/Topical: Computing and Systems Technology Division

(624l) Application of Monitoring, Fault Detection and Diagnosis for Next-Generation Fire Detection

Dongil Peter Shin, **Kyungjun Park** and Ki Jun Lee, Dept. of Chemical Engineering, Myongji University, Yongin, Kyunggido, South Korea

Process fault monitoring, detection and diagnosis and fire detection and isolation technologies have been developed separately and have rich histories individually. However, there are many common characteristics between them and the amount of technological overlapping is being increased: Mitigation of sensor failures, context sensitivity, and minimization of Type I, II errors are important problems. Use of wireless sensors and merge of heterogeneous information are getting the focus en route to the next-generation technologies.

This paper discusses how the abnormal situation management technologies, studied in the process systems engineering perspective, would contribute in developing next-generation fire detection and management technologies and improve the safety of process plants and intelligent buildings as well. And the lessons from the fire detection research that could be applied for better monitoring, detection and diagnosis of processes are also discussed.

See more of this Session: Poster Session: Computers In Operations and Information Processing
See more of this Group/Topical: Computing and Systems Technology Division

State Estimation with Delayed Measurements In Polymer Processes Using Unscented Transform Filters

Ruben Galdeano, Mariano Asteasuain and Mabel C. Sánchez,
Planta Piloto de Ingeniería Química (UNS-CONICET), Bahía Blanca, Argentina

Introduction

The problem of state estimation in nonlinear processes has been covered extensively in the past. Typical examples of traditional chemical systems in which these problems arise are polymerization processes. Dynamic operation of polymerization reactors is an extremely complex task. Phenomena such as multiple steady states in continuous processes, parametric sensitivity, highly exothermic reactions, changes in viscosity and complex heat transfer dynamics show the difficulty and importance of a good knowledge of the state variables.
A widespread state estimation technique in process control is the Extended Kalman Filter (EKF). In spite of its popularity, this strategy may present problems in the case of highly nonlinear systems .A relatively new method, known as the Unscented Kalman Filter (UKF), has been developed for this type of processes. It is based on the unscented transform technique (UT), a mechanism for propagating the mean and covariance of a random variable through a nonlinear transformation.[1]
The process measurements that are available for estimation purposes are usually of the following types: continuous measurements which can be obtained online at high sampling frequencies, measurements provided with constant delays by different devices that are available at a fixed sampling rate, and manual laboratory measurements which are available at unequal intervals and with varying delays.[2] In polymer processes in particular, several critical quality variables belong to the last two of these types .[3]
In spite of its importance, few works related to state estimation in polymerization control systems refer to the use of delayed measurements and how they affect the quality of the estimate. Some methodologies have been presented for the optimal treatment of time delayed measurements in the framework of linear Kalman filters.[4] These works use a scheme of multi-rate state estimation, in which different measurements are available at different rates. For example, Ellis et al. [5]used a multirate EKF to estimate delayed measurements of average molecular weights in batch polymerization. Mutha et al. [6] evaluated another multirate EKF filter for a complex emulsion copolymerization process. Then Tatiraju et al. [7] compared this scheme with another multirate observer which does not require any linear approximation. Ogunnaike [8] used a modified EKF for on-line multirate estimation in an industrial terpolymerization reactor. Furthermore a multivariable multirate nonlinear model predictive control was proposed by Prasad et al [9]. The algorithm incorporated a multirate EKF to handle state variable and parameter estimation for the styrene polymerization in a CSTR, taking into account properties such as the number average molecular weight and polydispersity.
Also some authors have applied a procedure which works in two time scales. In this strategy, the delayed measurements are stored together with the output data, and the filter is re-run from the time when the measurement sample was drawn from the process. Then, the status is updated with these new values, which are the starting points for the next estimation.[10] In a previous work we successfully applied this approach to the UKF.[11] A comparative study between the two-time scale procedure using different UKF-family filters and the multirate EKF methodology showed that the use of UT-based filters has a good performance for online implementation. However, it was observed that the re-estimation step can be time-consuming under certain situations, which may cause that the estimated data are unavailable at the right time.

In this work, an approach for the treatment of delayed measurements within the framework of the UKF is presented. It is based on a fusion strategy in which the filter algorithm is modified in order to account for the delayed measurements. The gain matrix is changed in such a way that the update of the state vector and covariance matrix can be done in a single step.

Delayed-Measurements Filtering Strategy

A state-space representation is usually used for describing physical systems. The approach that follow will be concerned with the operation of a continuous polymerization process, which is described by a set of differential algebraic equations. The general representation of this continuous time nonlinear system can be written as

$$\frac{d\mathbf{x}}{dt} = \mathbf{f}(\mathbf{x},\mathbf{u},t) + \mathbf{w}(t), \quad \mathbf{w}(t) \sim N[0,\mathbf{Q}(t)] \tag{1}$$

$$\mathbf{y_k} = \mathbf{h}(\mathbf{x_k}) + \mathbf{v_k}, \quad \mathbf{v_k} \sim N[0,\mathbf{R_k}] \tag{2}$$

where $\mathbf{x}$ is the n-dimensional state vector, $\mathbf{f(x,u,}t\mathbf{)}$ is the n-dimensional vector-valued function, $\mathbf{h(x)}$ is the m-dimensional measurements function, $\mathbf{y}$ is the m-dimensional observation vector, $\mathbf{u}$ is a vector of known manipulated inputs, $\mathbf{w}(t)$ is the n-dimensional plant noise vector, and $\mathbf{v}$ is the m-dimensional observation noise vector. Furthermore $\mathbf{Q}(t)$ is an $(n\times n)$ constant diagonal matrix, representing the covariance of the process model errors, and $\mathbf{R}$ is the $(m\times m)$ covariance matrix of the measurement errors. The white noise vectors $\mathbf{v}$ and $\mathbf{w}$ are assumed to be independent of each other.
The UKF [13] is a recursive estimation technique, such as the EKF, but it is based on the UT. This transformation is devoted to calculate the statistics of a variable $\mathbf{z}$ that results from applying a nonlinear transformation to a random variable $\mathbf{x}$ of known statistics, i.e., $\mathbf{z} = \mathbf{g}(\mathbf{x})$. In the Appendix a brief review of the UT is presented. The application of the UT has the advantage that the state estimates and their error covariance matrix can be calculated using the exact nonlinear process and measurement models.
Assuming that delayed measurements are present, the estimated state values and covariance matrix are in general defined as:

$$\hat{\mathbf{x}}_{(k|k-\tau)} = E\left[\mathbf{x}_k \middle| \mathbf{y}_k \ldots \mathbf{y}_{k-\tau}\right] \tag{3}$$

$$\mathbf{P}_{(k|k-\tau)} = \mathbf{E}\left\{\left[\mathbf{x}_k - \mathbf{x}_{(k|k-\tau)}\right]\left[\mathbf{x}_k - \mathbf{x}_{(k|k-\tau)}\right]^{\mathrm{T}} \middle| \mathbf{y}_k \ldots \mathbf{y}_{k-\tau}\right\} \tag{4}$$

where $\hat{\mathbf{x}}_{(k|k-\tau)}$ is the estimate of the state vector at time k considering the series of measurement vectors obtained from the system during the time interval $[(k-\tau), k]$, that is $[\mathbf{y}_{(k-\tau)} \ldots \mathbf{y}_k]$, and $\mathbf{P}_{(k|k-\tau)}$ represents the covariance matrix of $\hat{\mathbf{x}}_{(k|k-\tau)}$.
To calculate both expected values, the UKF procedure uses the common prediction-update structure. In the prediction stage, the state vector $\mathbf{x}$ is approximated by a matrix of sigma points χ selected as shown in Eq. (5), and these sigma points are propagated as indicated in Eq. (6).

$$\chi_k = \left[\hat{\mathbf{x}}_k \quad \hat{\mathbf{x}}_k + \left(\sqrt{(n+\lambda)\mathbf{P}_k}\right)_i \quad \hat{\mathbf{x}}_k - \left(\sqrt{(n+\lambda)\mathbf{P}_k}\right)_i\right], \quad i = 1,\ldots,n \tag{5}$$

$$\chi_k^- = \mathbf{F}[\chi_k, \mathbf{u}_k] \tag{6}$$

The sigma points χ_i are associated with their corresponding weights W_i. The mean and covariance of $\mathbf{x}$ are calculated as:

$$\hat{\mathbf{x}}_k^- = \sum_{i=0}^{2n} W_i^m \chi_k^- \tag{7}$$

$$\mathbf{P}_k^- = \sum_{i=0}^{2n} W_i^c \left[\boldsymbol{\chi}_{i,k}^- - \hat{\mathbf{x}}_k^-\right]\left[\boldsymbol{\chi}_{i,k}^- - \hat{\mathbf{x}}_k^-\right]^{\mathrm{T}} \tag{8}$$

In this work a modification of the update stage of the conventional UKF is proposed, that consists in executing different procedures depending on the availability of delayed measurements. If only on-line measurements are available, the update stage follows the standard UKF algorithm:

$$\hat{\mathbf{x}}_k^+ = \hat{\mathbf{x}}_k^- + \mathbf{K}_k\left[\mathbf{y}_k - \hat{\mathbf{y}}_k\right] \tag{9}$$

$$\mathbf{P}_k^+ = \mathbf{P}_k^- - \mathbf{K}_k \mathbf{P}_{y_k,y_k} \mathbf{K}_k^T \tag{10}$$

where $\mathbf{K}_k$ is the gain matrix at t_k for on-line measurements defined as

$$\mathbf{K}_k = \mathbf{P}_{x_k,y_k} \mathbf{P}_{y_k,y_k}^{-1} \tag{11}$$

If both on-line and delayed measurements are available at time t_k, the actualization of states is performed as follows:

-first the estimate and covariance are updated with the on-line measurements as shown in Eqs. (16)-18), obtaining $\hat{\mathbf{x}}_k^+$ and $\mathbf{P}_k^+$.

-secondly, these values are corrected using the information of the delayed measurements according to Eqs. (19) and (20).

$$\hat{\mathbf{x}}_{k,d} = \hat{\mathbf{x}}_k^+ + \mathbf{K}_{k,d}\left(\mathbf{y}_{k-\tau} - \hat{\mathbf{y}}_{k-\tau}\right) \tag{12}$$

$$\mathbf{P}_{k,d} = \mathbf{P}_k^+ - \mathbf{P}_{k,d} \mathbf{P}_{k,d,y}^{-1} \mathbf{P}_{k,d}^T \tag{13}$$

where $\mathbf{K}_{k,d}$ is the gain matrix at t_k for delayed measurements. For convenience we let

$$\mathbf{K}_{k,d} = \mathbf{P}_{k,d} \mathbf{P}_{k,d,y}^{-1} \tag{14}$$

In order to apply Eqs. (12)-(14) the covariance of the retrodicted measurements using delayed measurements has to be determined. This is carried out considering the conditional statistic of $\mathbf{x}_k$ given $\mathbf{y}_{k-\tau}$. A suboptimal approximation to the solution of the state estimation problem is obtained, which yields the following equations for $\mathbf{P}_{k,d,y}$ and $\mathbf{P}_{k,d}$:

$$\mathbf{P}_{k,d,y} = \sum_{i=0}^{2n} W_i^c \left[\mathbf{h}_{k-\tau}\left(\boldsymbol{\chi}_{i,k-\tau}\right) - \mathbf{y}_{k-\tau}\right]\left[\mathbf{h}_{k-\tau}\left(\boldsymbol{\chi}_{i,k-\tau}\right) - \mathbf{y}_{k-\tau}\right]^T + \mathbf{R}_k \tag{15}$$

$$\mathbf{P}_{k,d} = \sum_{i=0}^{2n} W_i^c \left[\boldsymbol{\chi}_{i,k-\tau} - \hat{\mathbf{x}}_{k-\tau}\right]\left[\mathbf{h}_{k,d}\left(\boldsymbol{\chi}_{i,k-\tau}\right) - \hat{\mathbf{y}}_{k-\tau}\right]^T \tag{16}$$

where

$$\hat{\mathbf{y}}_{k-d} = \sum_{i=0}^{2n} W_i^c \mathbf{h}_{k-d}\left(\boldsymbol{\chi}_{i,k-\tau}\right) \tag{17}$$

$\mathbf{P}_{k,d,y}$ is the conditional covariance of observation statistics and $\mathbf{P}_{k,d}$ represents the conditional cross covariance of states with the delayed observations.[14]

Case Study

The process selected as case study is a solution polymerization in a CSTR with a recycle loop. This is a system that has been chosen in several works dealing with state estimation and process control applications.[15] **Figure 1**

shows a diagram of the reactor. The comonomers, MMA and vinyl acetate (VA), are continuously added, together with AIBN as initiator, benzene (B) as solvent and acetaldehyde as chain transfer agent (CTA). The monomer stream may also contain inhibitors such as m-dinitrobenzene. These are the components of the fresh feed stream (F_1). This is combined with the recycle stream (F_2) to form the reactor feed stream (F_3). Polymer, solvent, unreacted monomers, initiator, and CTA flow out of reactor to a separator (F_4), from which unreacted monomers and solvent continue to the purge point (F_7). After the purge (F_9) they are stored in a recycle hold tank.
This system is represented by a first principles model consisting of a set of differential algebraic equations (DAE), whose most important equations are:

Reactor Mass and Moment Balances:

$$\frac{dC_k}{dt} = \frac{C_{k,\mathrm{f}} - C_k}{\theta_\mathrm{r}} + R_k \quad C_k(0) = C_{k,0} \quad k = \mathrm{MMA, VA, AINBN, B, CTA, inh} \tag{18}$$

$$\frac{d\lambda_k}{dt} = \frac{\lambda_{k,\mathrm{f}} - \lambda_k}{\theta_\mathrm{r}} + S_k \quad \lambda_k(0) = \lambda_{k,0} \; k = \mathrm{MMA, VA} \tag{19}$$

$$\frac{d\psi_k}{dt} = \frac{\psi_{k,\mathrm{f}} - \psi_k}{\theta_\mathrm{r}} + V_k \quad \psi_k(0) = \psi_{k,0} \quad k = 0,1,2 \tag{20}$$

where subscript f stands for feed stream, C_k the molar concentration of species k and λ_k is the molar concentration of monomer k in the copolymer, Ψ_k the kth-order moment of the copolymer MWD, θ_r is the residence time of reactor, and R_k, S_k, and V_k are reaction rates.

Energy Balance:

$$\begin{aligned}\frac{dT_\mathrm{r}}{dt} &= \frac{T_{\mathrm{r,f}} - T_\mathrm{r}}{\theta_\mathrm{r}} + \frac{(-\Delta H_{\mathrm{paa}}) k_{\mathrm{paa}} C_{\mathrm{MMA}} C^*_{\mathrm{MMA}} + (-\Delta H_{\mathrm{pba}}) k_{\mathrm{pba}} C_{\mathrm{MMA}} C^*_{\mathrm{VA}}}{\rho_\mathrm{r} c_{\mathrm{p,r}}} \\ &+ \frac{(-\Delta H_{\mathrm{pab}}) k_{\mathrm{pab}} C_{\mathrm{VA}} C^*_{\mathrm{MMA}} + (-\Delta H_{\mathrm{pbb}}) k_{\mathrm{pbb}} C_{\mathrm{VA}} C^*_{\mathrm{VA}}}{\rho_\mathrm{r} c_{\mathrm{p,r}}} - \frac{U_\mathrm{r} S_\mathrm{r} (T_\mathrm{r} - T_\mathrm{j})}{V_\mathrm{r} \rho_\mathrm{r} c_{\mathrm{p,r}}} \qquad T_\mathrm{r}(0) = T_{\mathrm{r},0}\end{aligned} \tag{21}$$

where C^*_{MMA} and C^*_{VA} denote the global concentration of radicals with a MMA and a VA terminal unit, respectively, ρ_r and $c_{p,r}$ represent the density and heat-capacity of the reaction mixture, respectively, ΔH_{paa}, ΔH_{pba}, ΔH_{pab} and ΔH_{pbb} are reaction enthalpies, and k_{paa}, k_{pba}, k_{pab} and k_{pbb} are kinetic constants of the propagation reactions; U_r and S_r are the overall heat-transfer coefficient and transfer area, respectively, and T_j is the temperature of the reactor jacket.

Separator and hold tank balances:

$$\frac{dC_{k,\mathrm{s}}}{dt} = \frac{C_{k,\mathrm{s,f}} - \frac{F_n}{F_4} C_{k,\mathrm{s}}}{\theta_\mathrm{s}} \qquad C_{k,\mathrm{s}}(0) = C_{k,\mathrm{s},0} \quad k\text{=MMA,VA,AIBN,CTA,inh} \tag{22}$$

where θ_s is the residence time in the separator and n=7 for k=AIBN, CTA, inh, and n=6 for k=MMA, VA, B.

Hold Tank Balances:

$$\frac{dC_{k,\mathrm{h}}}{dt} = \frac{C_{k,\mathrm{h,f}} - C_{k,\mathrm{h}}}{\theta_\mathrm{h}} \qquad C_{k,\mathrm{h}}(0) = C_{k,\mathrm{h},0} \quad k\text{=MMA,VA,B} \tag{23}$$

where θ_h is the residence time in the hold tank.

Average Molecular Weights, Composition and Conversion:

$$\bar{M}_n = \frac{\psi_1}{\psi_0} \tag{24}$$

$$\bar{M}_w = \frac{\psi_2}{\psi_1} \tag{25}$$

$$Y_{MMA} = \frac{\lambda_{MMA}}{\lambda_{MMA} + \lambda_{VA}} \tag{26}$$

$$\text{Conv} = \frac{\psi_1}{\left(\psi_1 + M_{\text{MMA}} C_{\text{MMA}} + M_{\text{VA}} C_{\text{VA}}\right)} \tag{27}$$

where Y_{MMA} is the number-average composition of MMA in the copolymer and M_{MMA} and M_{VA} are the molecular weights of MMA and VA, respectively. This mathematical model is supported by experimental validation. More details on the model equations, such as algebraic equations for stream connections, reaction rates, global concentration of radicals, and parameters can be found in ref.[13; 15]

For this case study, it is considered that the reactor temperature T_r and input flow are measured online with a sampling interval of 5min, while the total conversion Conv, the copolymer composition Y_{MMA}, the number-average molecular weight $\bar{M}_n$ and the weight-average molecular weight $\bar{M}_w$ are delayed measurements with a certain delay time.

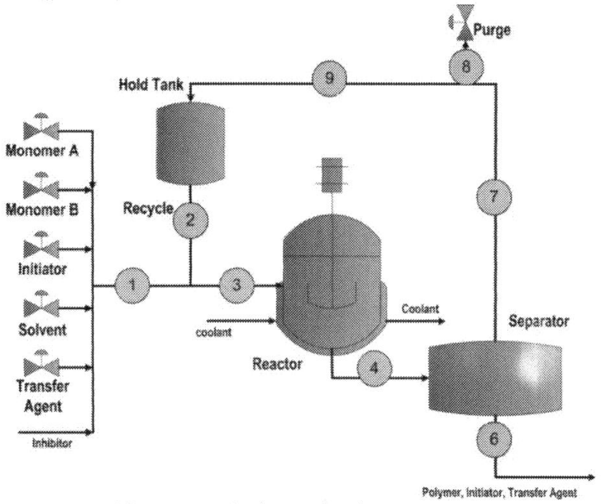

Figure 1. Polymerizations process

Results and Discussions

In this study it is considered that the on-line measurements are measured with a sample time of 5 min, while off-line measurements are sampled every 30 min, and that the measurement values are received with a delay of 30 min. Simulation runs for step changes at time 24h in the inlet feed of initiator, were performed. The process was sampled 100 times. The filter algorithm was implemented in Matlab with an embedded process model developed in gPROMS, in PC with a Intel Core i7 (3GHz) processor with 8 GB of RAM.

The performance of the proposed UKF for the estimation of $\bar{M}_w$ and Conv. are shown in **Figure 2** and **Figure 3** . It can be seen that the filter responds satisfactory in terms of accuracy and stability to a system change.

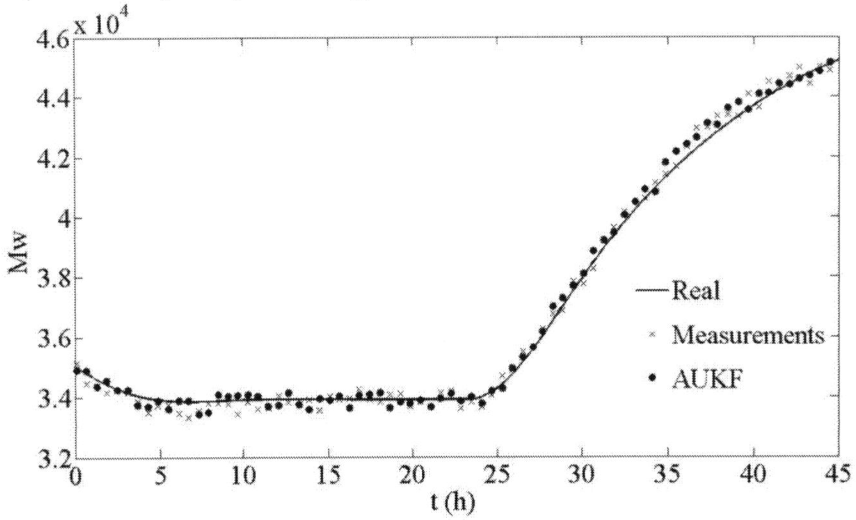

Figure 2. Estimation of the weight-average molecular weight with the proposed AUKF.

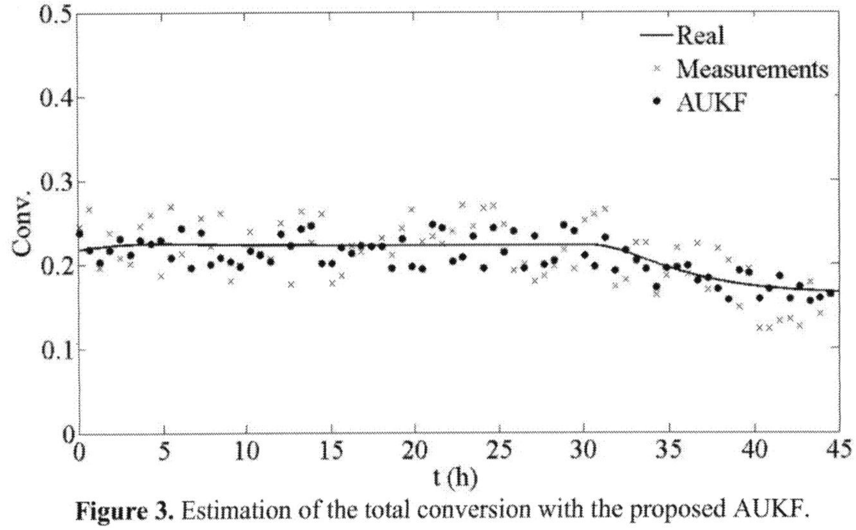

Figure 3. Estimation of the total conversion with the proposed AUKF.

The performance of proposed UKF (AUKF) was calculated by root mean square errors (RMS) in $\bar{M}_w$ and the computational time required for the state actualization step. The results are shown in Table1. It can be seen that the quality of the estimations are very satisfactory for continuous process and on-line application.

Table 1. ARSE and computational time.

	AUKF	ASUKF
ARSE ($\bar{M}_w$)	1236	1098
Propagation Time (s)	0.62	3.35
Update Time (s)	6.88	9.25

Conclusions

The proposed UKF algorithms were applied to the state estimation in a solution polymerization reactor. The computational load and the square root error were employed to evaluate the performance. The results show that these two techniques ensure quality and stability of the estimation. Also, the proposed methodologies reduce the computation time adequately, making it an attractive alternative for on-line application.

References

[1] Julier, S. J. and J. K. Uhlmann (1997). "New extension of the Kalman filter to nonlinear systems". *Proceedings of SPIE - The International Society for Optical Engineering*, Orlando, FL, USA, Society of Photo-Optical Instrumentation Engineers.

[2] Schei, T. S. (2008). "On-line estimation for process control and optimization applications." *Journal of Process Control* 18(9): 821-828.

[3] Fonseca, G. E., M. A. Dubé, et al. (2009). "A Critical Overview of Sensors for Monitoring Polymerizations." *Macromolecular Reaction Engineering* 3(7): 327-373.

[4] Bar-Shalom, Y. (2002). "Update with out-of-sequence measurements in tracking: exact solution." *IEEE Transactions on Aerospace and Electronic Systems* 38(3): 769-777.

[5] Ellis, M. F., T. W. Taylor, et al. (1994). "On-line molecular weight distribution estimation and control in batch polymerization." *AIChE Journal* 40(3): 445-462.

[6] Mutha, R. K., W. R. Cluett, et al. (1997). "On-Line Nonlinear Model-Based Estimation and Control of a Polymer Reactor." *AIChE Journal* 43(11): 3042-3058.

[7] Tatiraju, S., M. Soroush, et al. (1999). "Multirate nonlinear state estimation with application to a polymerization reactor." *AIChE Journal* 45(4): 769-780.

[8] Ogunnaike, B. A. (1994). "On-line modelling and predictive control of an industrial terpolymerization reactor." *International Journal of Control* 59(3): 711 - 729.

[9] Prasad, V., M. Schley, et al. (2002). "Product property and production rate control of styrene polymerization." *Journal of Process Control* 12(3): 353-372.

[10] Kim, K. J. and K. Y. Choi (1991). "On-line estimation and control of a continuous stirred tank polymerization reactor." *Journal of Process Control* 1(2): 96-110.

[11] Galdeano, R., M. Asteasuain, et al. (2011). "Unscented Transformation-Based Filters: Performance Comparison Analysis for the State Estimation in Polymerization Processes with Delayed Measurements." *Macromolecular Reaction Engineering*: n/a-n/a.

[12] van der Merwe, R., E. Wan, et al. (2004). "Sigma-Point Kalman Filters for Nonlinear Estimation and Sensor-Fusion: Applications to Integrated Navigation". *Proceedings of the AIAA Guidance, Navigation & Control Conference*.

[13] Adebekun, D. K. and F. J. Schork (1989). "Continuous solution polymerization reactor control. 2. Estimation and nonlinear reference control during methyl methacrylate polymerization." *Industrial & Engineering Chemistry Research* 28(12): 1846-1861.

[14] Galdeano, R., M. Asteasuain, et al. (2011). "Nonlinear State Estimation with Delayed Measurements. Application to Polymer Processes". Computer Aided Chemical Engineering. M. C. G. E.N. Pistikopoulos and A. C. Kokossis, Elsevier. Volume 29: 592-596.

[15] Congalidis, J. P., J. R. Richards, et al. (1989). "Feedforward and feedback control of a solution copolymerization reactor." *AIChE Journal* 35(6): 891-907.

Appendix

Let us consider the propagation of the random variable $\mathbf{x}$, of dimension n, through the set of nonlinear functions, and assume that $\mathbf{x}$ has mean $\bar{\mathbf{x}}$ and covariance $\mathbf{P}_x$. To calculate the statistics of $\mathbf{z}$, a matrix $\boldsymbol{\chi}$ composed of $(2n+1)$ sigma vectors $\boldsymbol{\chi}_i$ is defined as follows

$$\boldsymbol{\chi}_0 = \overline{\mathbf{x}} \quad \text{(A1)}$$

$$\boldsymbol{\chi}_i = \overline{\mathbf{x}} + \left(\sqrt{(n+\lambda)\mathbf{P}_\mathbf{x}}\right)_i, \quad i = 1, \ldots, n \quad \text{(A2)}$$

$$\boldsymbol{\chi}_i = \overline{\mathbf{x}} - \left(\sqrt{(n+\lambda)\mathbf{P}_\mathbf{x}}\right)_{i-n}, \quad i = n+1, \ldots, 2n \quad \text{(A3)}$$

$$\lambda = \alpha^2 (N + \kappa) - N \quad \text{(A4)}$$

where λ represents a scaling parameter, α is a constant which determines the spread of the sigma points around the mean state value $\overline{\mathbf{x}}$ and is usually set to a small positive value, κ provides an extra degree of freedom to fine tune the higher orders of the approximation, and $\left(\sqrt{(n+\lambda)\mathbf{P}_\mathbf{x}}\right)_i$ is the *i-th* row of the matrix square root of $(n+\lambda)\mathbf{P}_\mathbf{x}$, calculated using a stable numeric algorithm such as the Choleski decomposition.

Then the set of transformed sigma points $\boldsymbol{\gamma}_i$ is obtained by propagating vectors $\boldsymbol{\chi}_i$ through functions **g**

$$\boldsymbol{\gamma}_i = \mathbf{g}(\boldsymbol{\chi}_i), \qquad i = 1, \ldots, 2n \quad \text{(A5)}$$

The mean and convariance of **z** are calculated by the following weighted sums

$$\overline{\mathbf{z}} \approx \sum_{i=0}^{2n} W_i^m \boldsymbol{\gamma}_i, \quad \text{(A6)}$$

$$\mathbf{P}_\mathbf{z} \approx \sum_{i=0}^{2n} W_i^c [\boldsymbol{\gamma}_i - \overline{\mathbf{z}}][\boldsymbol{\gamma}_i - \overline{\mathbf{z}}]^\mathrm{T}, \quad \text{(A7)}$$

$$W_0^m = \frac{\lambda}{n+\lambda}, \quad \text{(A8)}$$

$$W_0^c = \frac{\lambda}{n+\lambda} + 1 - \alpha^2 + \beta, \quad \text{(A9)}$$

$$W_i^m = W_i^c = \frac{\lambda}{2(n+\lambda)}, \quad i = 1, \ldots, 2n \quad \text{(A10)}$$

where β is used to incorporated prior knowledge of the distribution of **x**, for Gaussian distributions, β=2 is optimal.

(624n) Automatic Generation of Optimal System Configurations and Maintenance Policies for Protective Systems

Chuei-Tin Chang and Edwin Wibisono, Chemical Engineering, National Cheng Kung University, Tainan, Taiwan

Generally speaking, a protective system is used to perform two basic functions, namely alarm and shutdown. The former is facilitated by independent sensors. Based on online measurements, a predetermined logic can be applied to make alarm decision. The latter function is usually performed with actuators, e.g., solenoid valves. Since sensor and valve failures are basically random events, the availability of a protective system is highly dependent upon its configuration and the corresponding maintenance policies.

Any hardware item can fail either safely (FS) or dangerously (FD). The FS malfunctions are recoverable because they are mostly caused by noisy signals, whereas the FD failures often require repair or replacement. To achieve a desired availability level, a common practice in the process industries is to introduce hardware redundancy in design. Specifically, each critical process condition may be monitored with more than one independent sensor, and the measurements signals are sent to a voting gate to generate alarm decision. These sensors are assumed to be maintained with spare-supported maintenance program in this study. On the other hand, it should be noted that the FD failure of a shutdown unit cannot usually be detected online. A preventive strategy must be applied to enhance its availability, that is, every unit is required to be inspected regularly at constant time intervals. The broken equipment must be replaced or repaired when detected, whereas the normal ones are kept online until the next inspection. Therefore, the length of the inspection interval should be considered as a design variable.

There have been a few attempts in the past to generate the optimal configurations and/or the corresponding maintenance policies with a mathematical programming model, e.g., Liang and Chang (2008), and Liao and Chang (2010), in which the failure rates of all compoenets were assumed to be constant. The objective of the present study is to develop a generalized mathematical programming model to produce the optimal design specifications and maintenance policies of a comprehensive multichannel and/or multilayered protective system by considering the time-variant nature of the failure mechnisms. By solving the corresponding mathematical program, the optimal configurations of sensors and shutdown units, the best corrective and preventive maintenance policies and alarm/shutdown logics can all be identified automatically.

References:

Liang K.H. and Chang C.T., 2008, A simultaneous optimization approach to generate design specifications and maintenance policies for the multilayer protective systems in chemical processes, *Ind. Eng. Chem. Res.* **47**, 5543.

Liao Y.C. and Chang C.T., 2010, Design and maintenance of multi-channel protective systems, *Ind. Eng. Chem. Res.* **49**, 11421.

See more of this Session: Poster Session: Computers In Operations and Information Processing
See more of this Group/Topical: Computing and Systems Technology Division

Multi-threading Parallelism on Multi-core Processor PC and its application on Interval Analysis

Gang Xu IOM/SimSci-Esscor
Mark Stadtherr U. Noter Dame

AIChE Annual, 2011, Minneapolis, MN

Objective

- A framework of workload-steal multi threading parallelism is provided
  - Introducing Mutex (Mutual exclusive event) as a general tool for thread safety
  - Emphasize on workload-steal for **severely unbalanced** initial distribution
  - Also highlights **dynamically generated workload** problems, as well as how to parallel such problem efficiently using workload-steal
- Tested with
  - interval analysis, a global solution solving technique, which is identified as a typical "**dynamically generated workload**" problem
  - Feigenbaum's equation solving, **severely unbalanced** binary tree

Agenda

- Introduce thread safety issue and Mutex
- A mechanism of efficient cross thread communication by hiding individual mutex under each thread is proposed
- Multi-threading Parallelism for **severe unbalanced** and **dynamically generated** workload problem is tested (**the worst case** for parallelism)

Introducing thread safety and mutex

- Accessing a shared resource by multiple threads is dangerous

- Mutex guarantees mutually exclusive access to a shared resource at any given time
  - Threads want to access “mutex protected” resource must wait until the current active thread is finished and unlock the mutex
  - Single global mutex
    - Would introduce **inefficiency** if multiple shared resources exist
  - Multiple mutexes
    - Dependency needs to be intelligently managed to avoid **dead-lock**

Multiple mutex deadlock issue

- Cyclic-dependency may cause <u>dead-lock</u>

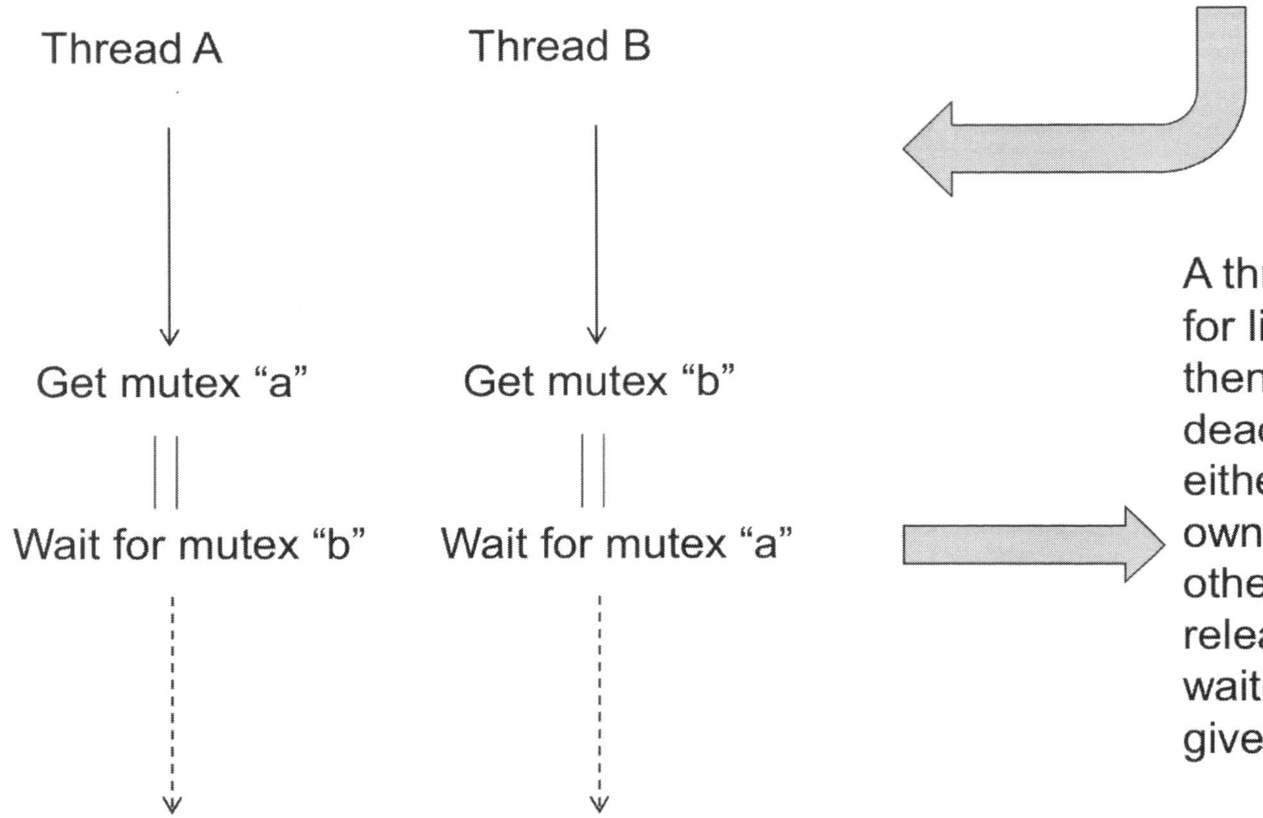

A thread must “wait” for limited time only, then break the dead-lock by either gain mutex ownership after the other thread releases the been waited mutex, or give up waiting

Introducing interval analysis and Feigenbaum's equation

- Interval analysis, a global solution algorithm which search through a dynamically generated binary tree

- Feigenbaum's equation
  - Free to scale up (higher dimension more roots)
  - All roots squashed within [0,1] region, causing the binary tree to be severely unbalanced

Introduction: Interval Arithmetic

- If X=[a,b] and Y=[c,d], then

$$X + Y = [a + c, b + d]$$
$$X - Y = [a - d, b - c]$$
$$X \times Y = [\min(ac, ad, bc, bd), \max(ac, ad, bc, bd)]$$
$$X / Y = X \times [1/d, 1/c], \qquad 0 \notin Y$$

- Interval operations may introduce overestimation

$$X - X = [a - b, b - a]$$

- To reduce overestimation, Berz et. al. introduced interval Taylor polynomial with tradeoff on the cost of additional real value operations

$$X - X = (x + (X - x)) - (x + (X - x)) = 0 + 0(X - x) = 0$$

Introduction: Interval Parametric function

- If $F[X, Y]$ represents interval function F in respect of interval variables X and Y, then F can be regarded as parametric function in respect of only one of X and Y, with the other variable be treated as parameter

- Such interval parameter function *also has Taylor model* (assume interval value, Y, is a parameter, and X is the variable)

$$F[X,Y] = \sum_{i=0}^{n-1} \frac{1}{i!}\left[(X-x)\cdot\nabla\right]^i F[x,Y] + \frac{1}{n!}\left[(X-x)\cdot\nabla\right]^n F\left[x+\theta(X-x),Y\right]$$

- Such interval parametric function provide additional means to reduce overestimation

Introduction: Interval Parametric function

- Example of tighter interval enclosure using Taylor model (2nd order) on real-parameter function
  - f[x]=x^2-4x+1 — x in [1,7]
  - F[X]=[1,49]-[4,28]+1= [-26,46] — natural interval extension
  - F[X]=f[x]+(X-x) F'[x+θ(X-x)]= [-29,31] — Taylor model 2nd order x=4
  - Taylor model provides tighter extension

- Example of tighter interval enclosure using Taylor model on interval-parameter function
  - f[x]=x^2-[3.9,4.1]x+1 — x in [1,7]
  - F[X]=[1,49]-[3.9,28.7]+1= [-26.7,46.1] — natural interval extension
  - F[X]=f[x]+(X-x) F'[x+θ(X-x)] = [-29.7,30.3] — Taylor model 2nd order x=4
  - Taylor model provides tighter extension

Introduction: Global solution algorithm by Interval Analysis

- Interval Newton Generalized Bisection (INGB) iteration
  - Interval Gauss-Seidel iteration: Precondition has to be done to "shrink" Interval Jacobian (Henson's book)

$$\left[\frac{\partial F(X_k)}{\partial X_k}\right][X_{k+1} - x_k] = -F(x_k)$$

$$[Y]\left[\frac{\partial F(X_k)}{\partial X_k}\right][X_{k+1} - x_k] = -[Y]F(x_k)$$

$$[Y] = \left[\frac{\partial F(x_k)}{\partial x_k}\right]^{(-1)}$$

INGB Method: Interval Newton Test

- There is no solution in **X** (upper right chart)
- There is unique solution in **N** (middle right chart)
- Solution may or may not exist (lower right chart)
- This is a Binary tree searching problem

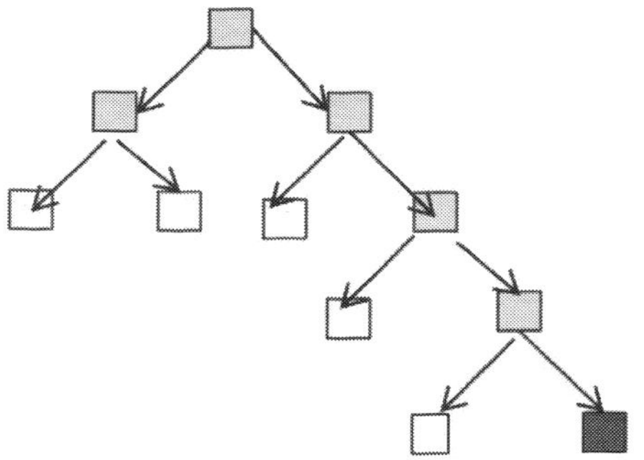

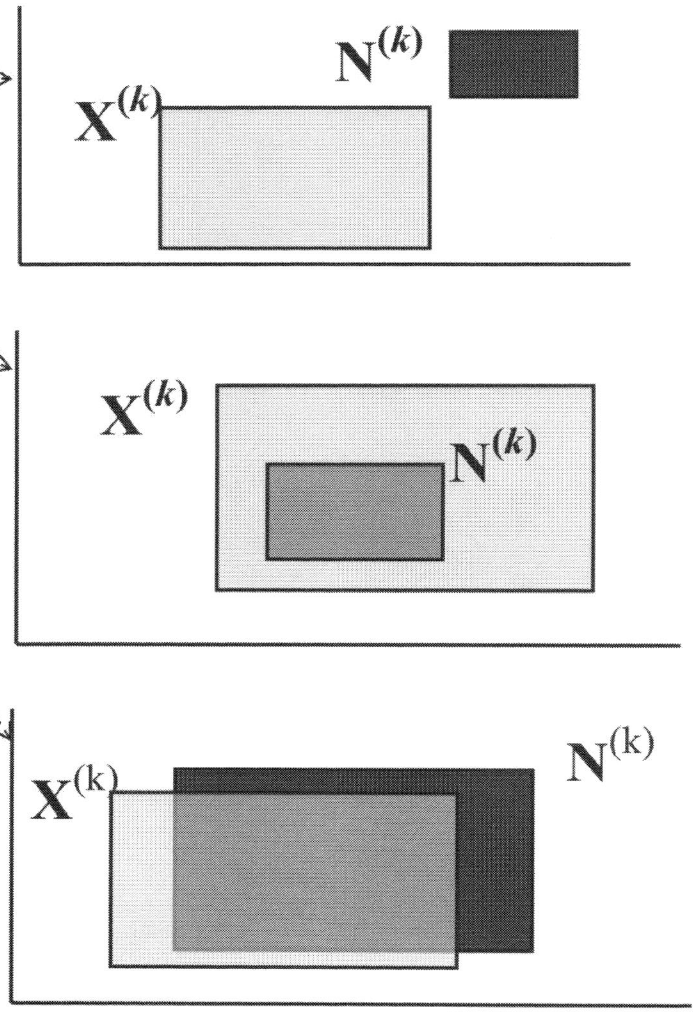

Single thread serialism

- Feigenbaum equation system:
  1. Free to scale up and down
  2. Binary tree
     - Severely unbalanced
     - Dynamically generated

$$-3.84x_k^2 + 3.84x_k - x_{k+1} = 0, \qquad k = 1, \dots, n-1$$
$$-3.84x_n^2 + 3.84x_n - x_1 = 0$$

Binary tree structure for Feigenbaum equation

- According to Herbort & Ratz (1997), and verified by our work on single thread computation, with starting box $[0,100]^n$, number of roots are

dimension	3	5	7	9	11	13
Roots	8	12	30	80	200	521
Tree-nodes	125	616	3533	20564	108249	549001

- Almost all those roots are packed within [0,1] region (unbalanced tree)
- Multithread parallelism is to execute interval analysis on

Workload-steal parallelism

- Virtual networks, such as 2D-ring, 2D-mesh, and 3D-mesh… can provide efficient cross-notes communication
  - Only “neighbors” has workload steal communication
  - Reduce overhead of cross communication

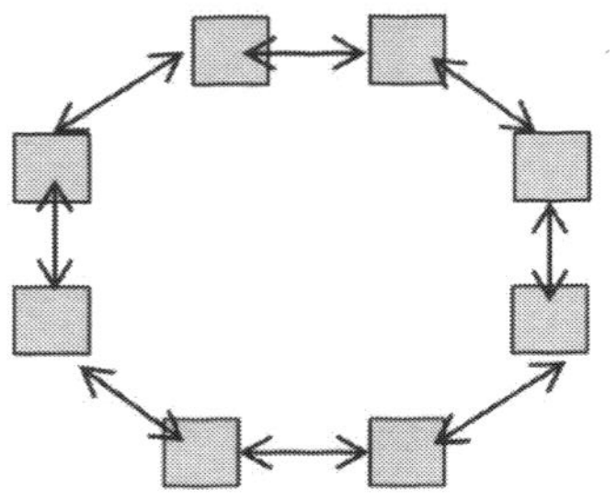
2D-ring

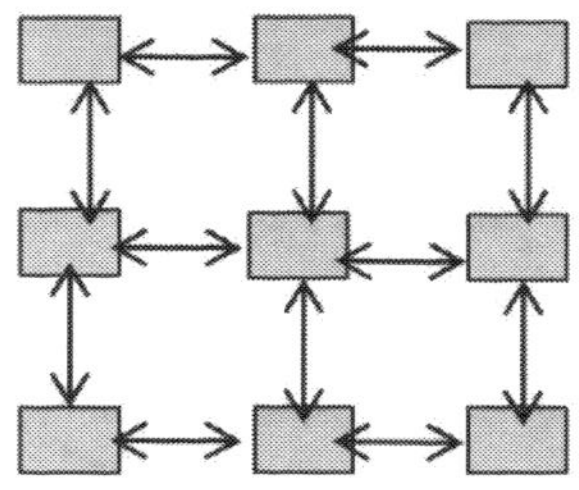
2D-mesh

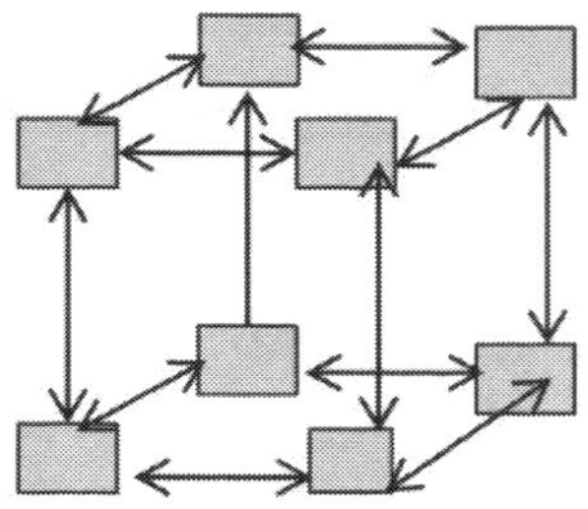
3D-mesh

Mutexes management

- 2D-mesh as example
- Each node is an individual thread
- Each node has four neighbors
- If more than one neighbors are stealing workload on current node,
  - Only one neighbor be granted with owner ship of mutex under current node
  - Others neighbors would wait for limited time, then
    - either give up and look for its other neighbors that have remaining work load to steal
    - or gain owner ship of mutex under current node after previous neighbor released the mutex ownership

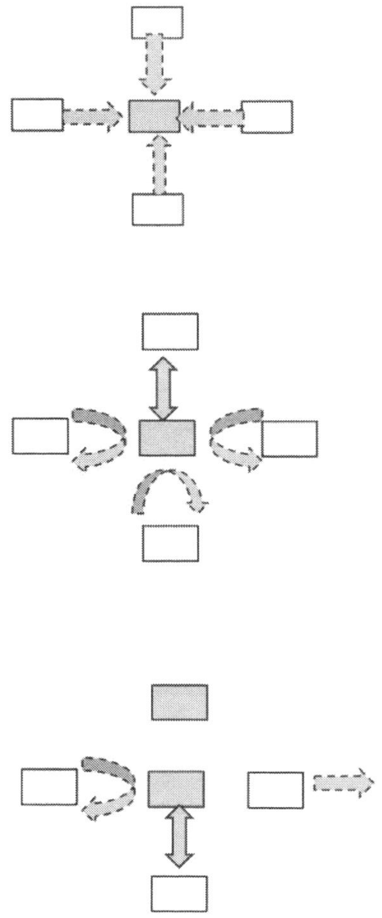

Network efficiency issue

- One to all and all to one communication is not efficient
- 2D-ring, 2D-mesh, and 3D mesh offers neighbor to neighbor communication
  - 2D-ring is simple, but some node along the ring may never be able to steal work load successfully
  - 2D-mesh and 3D-mesh allow a "starving" node to have more chance to steal successively across the network

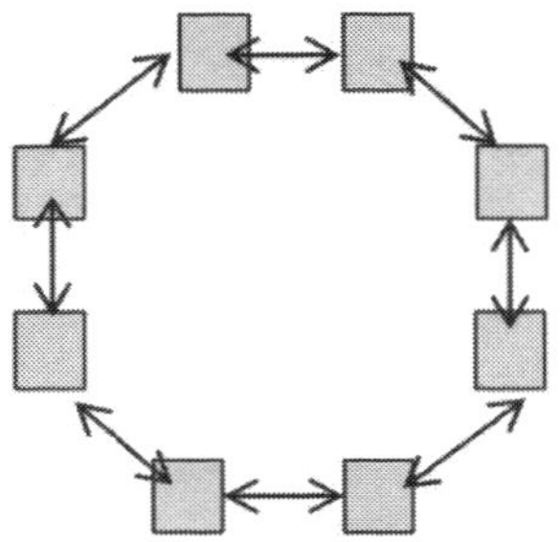

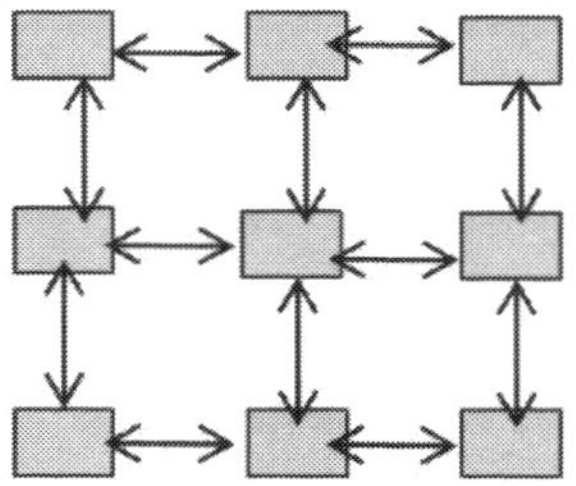

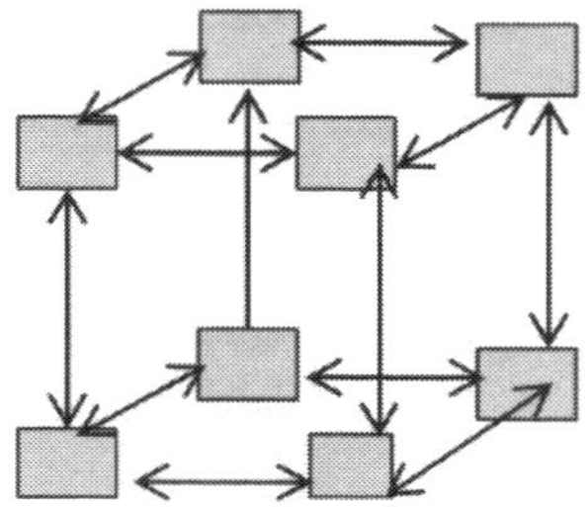

Multi-core processor machine

- Hp pavilion dv7
  - Intel® Core™ i7 CPU 1.60GHz
  - 4 cores, each runs two threads at a time
  - Recognized as 8 cores

Testing result

- Dimension 13 Feigenbaum (521 roots found, about 549001 tree-nodes executed)

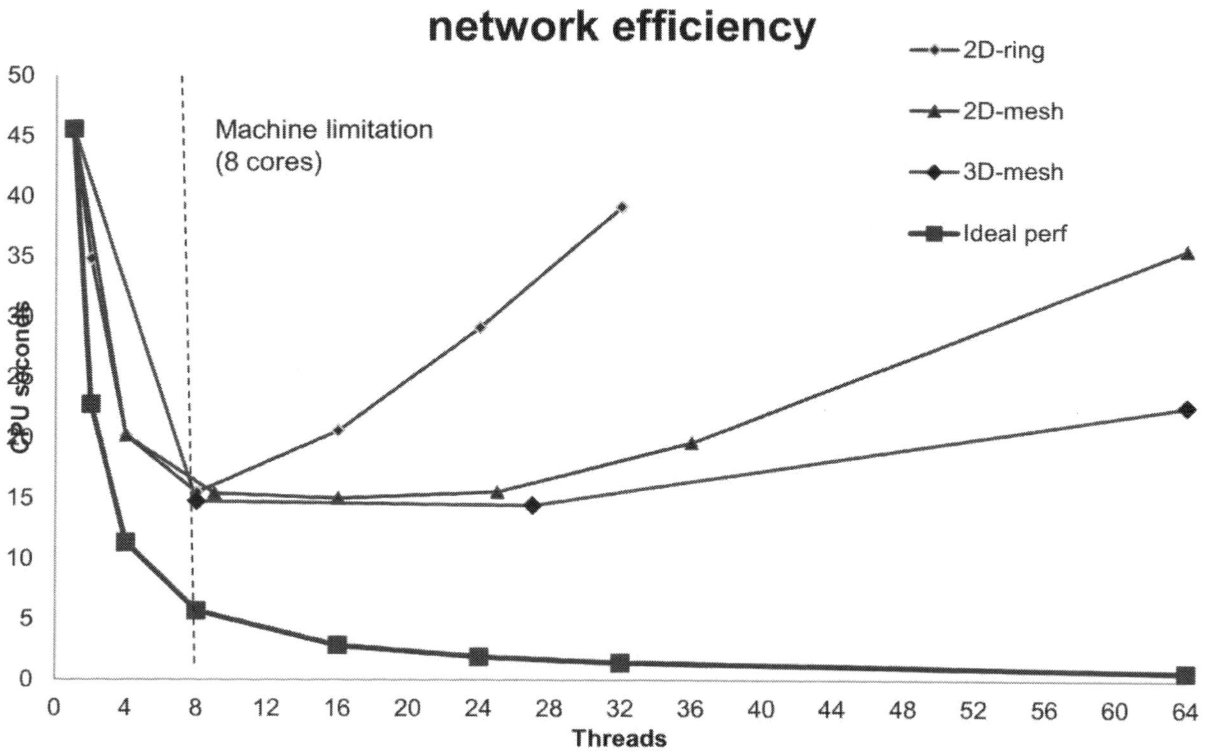

Conclusion:

- Parallelism on severely unbalanced and dynamically generated binary tree structure, evenly workload is impossible. Thus, one may never meet “ideal” performance (even with zero synchronizationoverhead).
- For threads number >8, performance can no longer to be improved due to machine limit
- However, performance on efficient virtual network can be extended on better multi-core machines (such as GPU)
- Individual mutex management can be applied to any area where parallel computation is promising

INVERSION OF PERTURBED MATRIX: THE KEY STEP TOWARDS GENERIC PARAMETRIC PROGRAMMING

Rajab Khalilpour, I. A. Karimi

School of Chemical and Biomolecular Engineering,
National University of Singapore, Singapore
E-mail: khalilpour@gmail.com

11AIChE
Annual Meeting, Minneapolis, MN

2011 AIChE Annual Meeting: Minneapolis, MN, October 16-21

Abstract

Parametric optimization with uncertainties on the objective function (OF) or on the so-called "right-hand-side" (RHS) of the constraints has been successfully addressed in recent years. However, very little work exists on the same with uncertainties on the left-hand-side (LHS) of the constraints or in the coefficients of the constraint matrix. This gap in the research exists even though there are numerous real industrial problems where the LHS parameters are subject to some uncertainties.

When the uncertainty is on the RHS or OF, we deal with a vector of uncertain parameters. However, if the uncertainty is on the LHS, the problem is a parametric matrix which is one of the complex mathematical problems. In this presentation, we will introduce an algorithm for a linear program with a multicomponent single parameter on the left hand side [1].

[1] KHALILPOUR, R., Parametric Optimization with Uncertainty on the Left Hand Side, July 2007, Research Report, Singapore: National University of Singapore.

(646a) Dynamics and Control of High Duty Counter-Current Heat Exchangers

Seongmin Heo, Sujit S. Jogwar and Prodromos Daoutidis, Chemical Engineering and Materials Science, University of Minnesota, Minneapolis, MN

Energy integration in chemical plants is crucial as it provides a reduction in external utility consumption which, in turn, results in cost savings. Heat exchangers play an important role in energy integration by transferring energy from hot streams which need to be cooled to cold streams which need to be heated. Heat exchanger networks are now ubiquitous in chemical plants, and a lot of research has been done on synthesizing and optimizing such networks, and evaluating their controllability and operability. Yet, control of heat exchanger networks has not been studied extensively.

As a typical design objective of energy integrated networks is to maximize energy recovery, counter-current exchangers are preferred over co-current exchangers. Also, in order to recover as much energy as possible, high duty exchangers, which are characterized by high Stanton numbers, are frequently used. Several papers have studied the (mostly) control of heat exchangers under different assumptions on the modeling of such systems. None however has dealt with high duty exchangers explicitly.

In this talk, we present a dynamic analysis and nonlinear controller design of such heat exchangers. The dynamics of counter-current heat exchangers are described by first order hyperbolic partial differential equations (PDEs). In the case of high duty exchangers, the governing equations become stiff, and the system shows a potential of multi-time scale dynamics. We generalize singular perturbation-based model reduction for the stiff PDE model describing such exchangers to derive a nonlinear non-stiff PDE model which captures the dynamics in the slow time scale. Several numerical issues arise in the numerical simulation using finite differences due to the counter-current nature of the system. Using order of magnitude analysis, we derive a proper boundary for the spatial discretization step to ensure consistency of the numerical simulation of the full and reduced models. Subsequently, distributed nonlinear controllers are designed to allow tracking and regulation of an outlet temperature by manipulating an inlet flowrate. The advantages of using the reduced PDE model as the basis of controller design, and the efficacy of the proposed controller design are illustrated with simulations.

See more of this Session: Dynamics, Reduction and Control of Distributed Parameter Systems
See more of this Group/Topical: Computing and Systems Technology Division

(646b) Control of the Detached Bridgman Crystal Growth Process

Andrew Yeckel, Prodromos Daoutidis and Jeffrey J. Derby, Chemical Engineering and Materials Science, University of Minnesota, Minneapolis, MN

Detached Bridgman growth represents a radical change over existing industrial crystallization technology. In this process, first observed in space-based growth, the melt dewets the wall, allowing the crystal to pull away from the ampoule. Detached growth eliminates deleterious interactions between the growing crystal and the ampoule, dramatically improving crystal quality. However, the promising early results of microgravity experiments have been difficult to advance in terrestrial growth systems due to a number of instabilities that can be manifest during growth. In particular, the growth of most semiconductor crystals is inherently unstable due to shape instabilities associated with thermal-capillary interactions set by growth and ampoule wetting angles.

In this presentation, we discuss the dynamics, operability limits, and tuning of several feedback controllers to stabilize detached vertical Bridgman crystal growth. The controlled variable is the pressure difference between upper and lower vapor spaces, and the output variable is gap width at the triple-phase line. Open and closed loop dynamics of step changes in these state variables are analyzed under both shape stable and shape-unstable growth conditions. Effects of step changes in static contact angle and growth angle are also studied. Proportional and proportional-integral control can stabilize unstable growth, but only within tight operability limits imposed by the narrow range of allowed meniscus shapes. These limits are used to establish safe operating ranges of controller gain. Strong nonlinearity of the capillary model restricts the range of perturbations that can be stabilized, and under some circumstances stabilizes a spurious operating state far from the set point. Stabilizing detachment at low growth angle proves difficult and becomes impossible at zero growth angle. Substantially better performance is shown to arise from a model-based control scheme implemented with a simple analytical expression derived from capillary statics.

See more of this Session: Dynamics, Reduction and Control of Distributed Parameter Systems
See more of this Group/Topical: Computing and Systems Technology Division

(646c) Output Feedback Control Design for Transport Reaction Processes with Partial Sensor Information

Sivakumar Pitchaiah, Department of Chemical Engineering, The Pennsylvania State University, University Park, PA and **Antonios Armaou**, Chemical Engineering, The Pennsylvania State University, University Park, PA

Many industrially relevant diffusion-convection-reaction processes are characterized by the presence of strong spatial variations due to coupling between diffusive and convective mechanisms with chemical reactions. Typical examples of such processes include fluidized-bed and packed-bed reactors, rapid thermal processing, plasma reactors, and chemical vapor deposition. The feedback control problem for these processes is nontrivial owing to the spatially distributed nature of their dynamics. This leads to infinite dimensional dynamic descriptions in appropriate spaces which necessitate the use of model reduction in order to design practically implementable controllers. The key step in this approach is the computation of basis functions that are subsequently utilized to obtain finite dimensional approximations to the original infinite dimensional system. A common approach for this task, when the spatial dynamics are described by nonlinear operators, is the proper orthogonal decomposition (POD) combined with method of snapshots. However, this approach requires the a priori availability of a sufficiently large ensemble of PDE solution data (snapshots) which excites all of the possible spatial modes in the solution of the PDE system, a requirement which may be difficult to satisfy. It is also important to note that the complete spatial profile of the state is needed. Recently, the adaptive POD methodology [1] was proposed as a solution to the above problem. In this methodology the basis functions required for the model reduction were computed recursively as new snapshots from the process becomes available. Initially basis functions were computed using a relatively small number of snapshots; we then kept track of the dominant eigenspace of the covariance matrix which was subsequently utilized to compute the empirical basis functions required for model reduction.

In this talk, we will be presenting our recent results on stabilization of distributed-parameter-systems with partial state measurements (i.e., there are spatial regions where information is unavailable) employing our adaptive proper-orthogonal-decomposition (APOD) methodology. Initially, the incomplete measurements are reconstructed using a gappy-reconstruction procedure and are then utilized to derive and periodically update a reduced-order-model (ROM). The use of APOD methodology along with the gappy procedure allows the development of accurate low-dimensional ROMs for controller synthesis thus resulting in a computationally-efficient alternative to using large-dimensional models with global validity. The proposed gappy-APOD methodology is applied to address the problem of stabilization of an open-loop unstable system modeled by Kuramoto-Sivashinksy equation. The proposed controller successfully stabilizes the system at a spatially uniform steady-state even though only partial information of the state profiles is known.

Reference:

1. Pitchaiah, and A. Armaou, "Feedback control of dissipative distributed parameter systems using adaptive model reduction," Ind. & Eng. Chem. Res., **55**, 906-918, 2010.

See more of this Session: Dynamics, Reduction and Control of Distributed Parameter Systems
See more of this Group/Topical: Computing and Systems Technology Division

Modeling Spatial Variations In Thin-Film Processing of Copper Oxide Films for Solar Energy Applications and Hydrogen Production

David Arana Chavez and Raymond A. Adomaitis
University of Maryland, College Park, MD

Introduction

Solar Hydrogen has become an alternative energy source of great promise in recent years. However, the expensive and toxic Platinum used for its generation makes the industrial scale production not feasible. For this reason, a variaty of new materials for the electrodes are being researched [8]- [10]. Among such materials copper oxide has shown potential in splitting water under solar irradiation [2].

Experiments show that, under certain operating conditions, copper oxide forms two different phases on the same substrate. One containing only cuprous oxide Cu_2O, and the other containing only cupric oxide CuO with a sharp transition between them [6]- [7]. The sharpness of this transition is what makes this system particularly interesting. Understanding the nature of such transition is vital for the design and operation of CVD copper electrode producing reactors.

The origin of the sharp transtion as well as the basic thermodynamics of the system were first studied by Hong and Komiyama [7]. They concluded that the oxide molecules formed in the gas to then coallesce and precipitate on the surface. A later study [6] suggested that the kinetics of the substrate's surface played a dominant role on the growth of the film and that the sharpness of the transition is due to the limited degrees of freedom of the oxides-oxygen system. Indeed this system has only one degree of freedom, which can translate into a critical oxygen partial pressure.

In this our aim is to identify the dominant dynamics of the system, i.e. whether it is the gas phase reactions or the surface kinetics what controls the deposition process. To do this we developed a transport phenomena simulation based on a two dimensional axysymmetric model of the tubular hot-wall CVD reactor were the experiments were carried out.

Methodology

To estimate of the oxygen concentration reaching the substrate at the wall of the reactor, we added radial diffusion to the 1D model making it into a 2D axisymetric model. The schematic of this model is shown in Figure .

From a material balance we obtain the governing equation for the oxygen concentration in the reactor

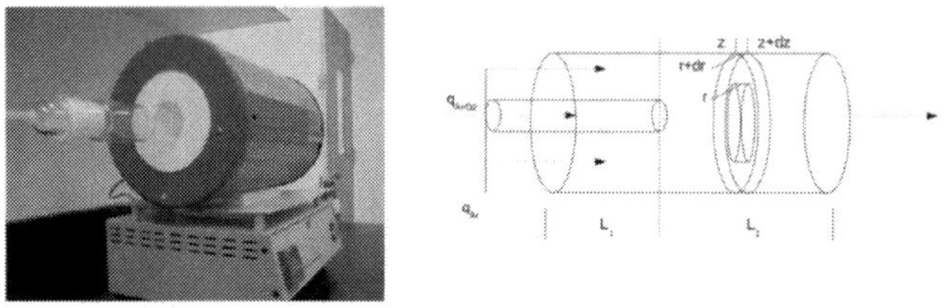

Figure 1: Schematic of the reactor in two dimensions. Angular symmetry is assumed and radial diffusion is included in this model. Nevertheless the boundary conditions for the axial coordinate do not change with respect to the 1D model.

$$\frac{1}{r}\frac{\partial}{\partial r}\left(r\frac{\partial C}{\partial r}\right)+\frac{\partial^2 C}{\partial z^2}-\frac{v}{D}\frac{\partial C}{\partial Z}=-\frac{f}{D}=-Q_{O_2}\frac{\delta\left(r,z;0,L_1\right)}{D}=Q_{O_2}\frac{\delta(r)\delta(z-L_1)}{D2\pi r} \tag{1}$$

Subject to the boundary conditions

$$\begin{aligned} J=-D\tfrac{\partial C}{\partial z}+vC=0 \quad & @ \quad z=0 \\ -D\tfrac{\partial C}{\partial z}=0 \quad & @ \quad z=L \\ -D\tfrac{1}{r}\tfrac{r\partial C}{\partial r}=0 \quad & @ \quad r=R_o \\ |C|<\infty \quad & @ \quad r=0 \end{aligned} \tag{2}$$

We assume that the solution to Equation 1 can be approximated by a series of global eigenfunctions as shown below

$$C=\sum_{i,j}^{M,N}\alpha_{i,j}\psi_i(r)\phi_j(z) \tag{3}$$

where the eigenfunctions, $\psi_i(r)$ and $\phi_j(z)$ are given by the following expressions

$$\frac{1}{r}\frac{d}{dr}\left(r\frac{d\psi}{dr}\right)=\lambda\psi \tag{4}$$

$$\frac{d^2\phi}{dz^2}=\omega\phi \tag{5}$$

which under the action of the boundary conditions at $r=0$ and $z=0$ yield

$$\psi_i(r)=J_0\left(\lambda_i r/R_o\right) \tag{6}$$

$$\phi_j(z)=\frac{v}{D\omega_j}\sin\left(\omega_j z\right)+\cos\left(\omega_j z\right) \tag{7}$$

The eigenvalues λ_i and k_j are found by using the boundary conditions at $r=R_o$ and $z=L$, respectively, to give

$$J_0'\left(\lambda_i r/R_o\right)|_{r=R_o} = -J_1\left(\lambda_i\right) = 0 \tag{8}$$

$$\frac{D}{vL}\lambda_i = \tan\left(\lambda_i L\right) \tag{9}$$

To find the coefficents $\alpha_{i,j}$, we project the approximated solution onto the eigenfunctions and make it obey the governing equation

$$\left\langle \sum_{i,j}^{M,N} \left[\alpha_{i,j}\psi_i''\phi_j + \alpha_{i,j}\frac{\psi_i'}{r}\phi_j + \alpha_{i,j}\psi_i\phi_j'' - \frac{v}{D}\alpha_{i,j}\psi_i\phi_j' + \frac{Q_{O_2}}{2\pi D}p_i q_j \psi_i\phi_j \right], \psi_u\phi_v \right\rangle = 0 \tag{10}$$

Using equations 4 and 5 to replace the difussive terms, the above equation simplifies to Equation 11. Where p_i and q_j are the coefficients of the projection of each eigenfunction with its corresponding Dirac delta function of the source term in directions r and z, respectively. Taking advantage of the orthogonality of the eigenfunctions Equation 10 becomes

$$\sum_{i,j}^{M,N} -\alpha_{i,j}\left(\left(\lambda_i + \omega_j\right)\delta_{j,n} + \frac{v}{D}\left\langle \phi_j', \phi_n \right\rangle\right)\delta_{i,m} = \sum_{i,j}^{M,N} \frac{Q_{O_2}}{2\pi D}p_i q_j \delta_{i,m}\delta_{j,n} \tag{11}$$

Then we have $M \cdot N$ equations like Equation 11 to be solved simultaneously. To simplify the notation we group some terms and use index notation

$$\beta_{u,v,i,j}\alpha_{i,j} = \gamma_{u,v} \tag{12}$$

Then we have a linear system of $N^2 \times M^2$ equations into the same number of variables. To solve the system we multiply (12) by the inverse of β and find the expansion coefficients of the solution

$$\alpha_{i,j} = \left(\beta_{u,v,i,j}\right)^{-1}\gamma_{u,v} \tag{13}$$

To compute the inverse of the four-index tensor β we collpased pairs of indices, to transform the matrices α and γ into vectors and β into a matrix. After solving for the coefficients, we recover the matrix solution from the vector solution by expanding the index based on the collapsing order.

Results and Discussions

The model described in the previous section was solved in MATLAB^{TM}. The resulting profile is shown in Figure . The tubular reactor is oriented from left to right, the carrier gas, Argon, enters through the left and the oxygen feed enters through an inner tube at a position $z = 0.18cm$ (indicated in the figure with a vertical white line). The reactor is heated with an external furnace to keep the temperature constant at 750°C, the Argon flow is also kept constant at a rotameter value of 60 (equivalent to 570 sccm), and the flow of oxygen is varied between experiments.

From Figure we can observe that the partial pressure of oxygen grows sharply near the feed and remains constant downstream the reactor. It can also be seen that some of the oxygen diffuses backwards causing slightly elevated concentrations of oxygen upstream the feed location. To obtain

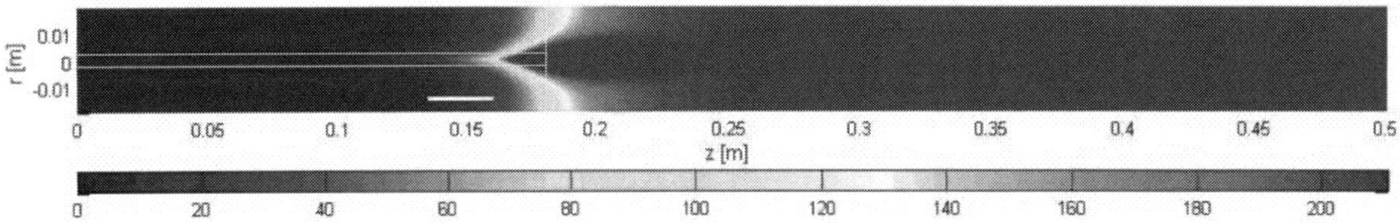

Figure 2: Oxygen profile inside the CVD reactor. Approximated by Equation 3 with the coefficients $\alpha_{i,j}$ obtained from Equation 13. The carrier gas, Ar, flows from left to right and the oxygen is fed at z=18cm through an inner tube shown in white lines. The substrate is shown as a white horizontal bar. It is placed on a position such that the critical Oxygen partial pressure can cross it. The color bar's scale is on Pa.

Figure 3: Oxygen profile projected on the substrate located inside the CVD reactor at the position shown in Figure . The color bar's scale is on Pa. The $P^c_{O_2} = 24$ is found inside the substrate, which suggests that such substrate will have both oxide phases deposited.

the estimate oxygen concentration profile on top of the substrate, the radial concentration was projected along equidistant axial points defined by the location of the substrate.

Figure shows the Oxygen partial pressure profile on the substrate placed on the location shown on Figure . In this trail computation the location of the substrate was determined by the position of the critical partial pressure of Oxygen so that the substrate will show a profile of concentration with potential to contain the two oxides. In this case the substrate is located before the oxygen feed which indicates that there is a consumption that is being ignored by the model yet. Such consumption comes obviously from the oxidation reactions and the deposition on the substrate and walls. Before addressing this problem we compared the results of our simulations to those yielded for the same system using the software COMSOL^{TM} to make sure our calculations were consistent.

Comparison with COMSOL

A three dimensional COMSOL model of the CVD reactor was created and solved. To compare this COMSOL model to the 2D axysymmetric we created before we compare the oxygen partial pressure profile at the wall, since it is were the substrate is and were the deposition takes place. Both profiles are shown in Figure . The figure shows that both profiles are pratically the same, and thus it is clear that the excess of oxygen from the solution in Figure is due to the lack of a sink in the model.

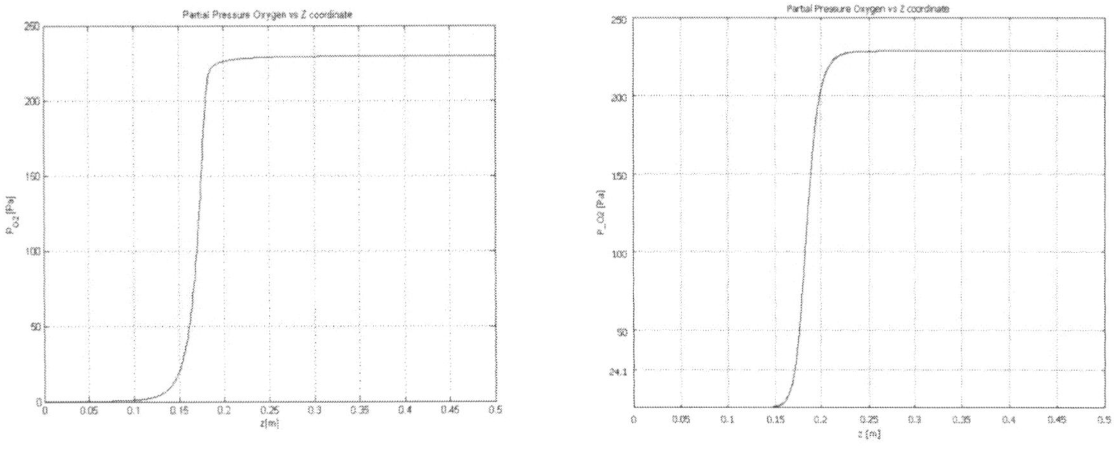

Figure 4: Comparison of the oxygen profiles at the wall yielded by our 2D axysymmetric model (A) and by simulations in COMSOL^{TM} (B).

As a first approach to solve this issue we considered two limiting cases: the first one when there are no surface reactions and all the oxides are formed on the gas phase and then deposited. The second case is when there are no reactions on the gas phase and thus the deposition is controlled by the events ocurring at the substrate as well as the reactor walls. These two modifications to the model are discussed in the following subsection.

Gas reactions vs Surface kinetics

To account for consumption in the case of gas phase reactions, we added a global one order oxygen consumption reaction to the mass balance

$$\frac{1}{r}\frac{\partial}{\partial r}\left(r\frac{\partial C_{O_2}}{\partial r}\right)+\frac{\partial^2 C_{O_2}}{\partial z^2}-\frac{v}{D}\frac{\partial C_{O_2}}{\partial Z}-k\cdot C_{O_2}=-\frac{f}{D} \tag{14}$$

with this addition the eigenvalues for each basis of eigenfunctions are still the same and only some terms are added to the matrix β. On the other hand, in the case were we want to account for surface kinetics a boundary condition is modified. Particularly the boundary condition at the wall, which defines the eigenvalues of the radial eigenfunctions and thus the solution must be obtained on the radial direction for each of the axial discretization. The new boundary condition on $r = Ro$ is

$$V\cdot k_{O_2,ads}\cdot P_{O_2}=-A\cdot J_{O_2}=A\cdot D_{O_2}\frac{\partial C_{O_2}}{\partial r} \tag{15}$$

Figure shows representative profiles for relatively high values of k and k_{ads}, 4000 and 0.05 respectively. In the figure for the gas phase case it is seen that the global consumption term efectively decreases the oxygen concentration in the reactor, as expected, whereas the surface consumption case creates a high gradient among the walls and the nearby gas that keeps the oxygen concentration still very high in the vicinity of the walls were the substrate is placed during experiments.

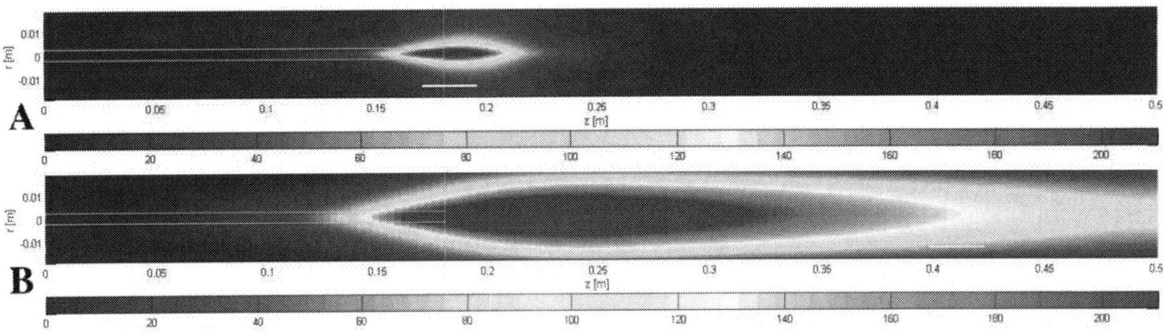

Figure 5: Comparison of the oxygen profiles generated by models considering only consumption reaction on the gas phase (A) or only consumption by surface reactions (B).

Figure provides a more general view of the behaviour of the system for various values of k and $k_{O_2,ads}$. In the plots the position at the walls were the critial oxygen partial pressure $P^c_{O_2} = 24\ Pa$ is located when the simulation is run at different values of the constants of reaction and adsorption.

In this figure we can see that the gas phase reaction case may eventually reach a rate of consumption at which the critial concentration of oxygen is punctual near the feed, which may originate, under certain conditions, substrates with alternate oxide phases. Evenmore, the consumption rate can be increased to the point that no oxygen reaches the walls.

In the case where the surface effects dominate, the critical pressure's locations are attracted in principle. However after a critical adsoprtion constant's value the critical pressure locations begin to grow apart, which can be expected since for high values of the constant the concentration

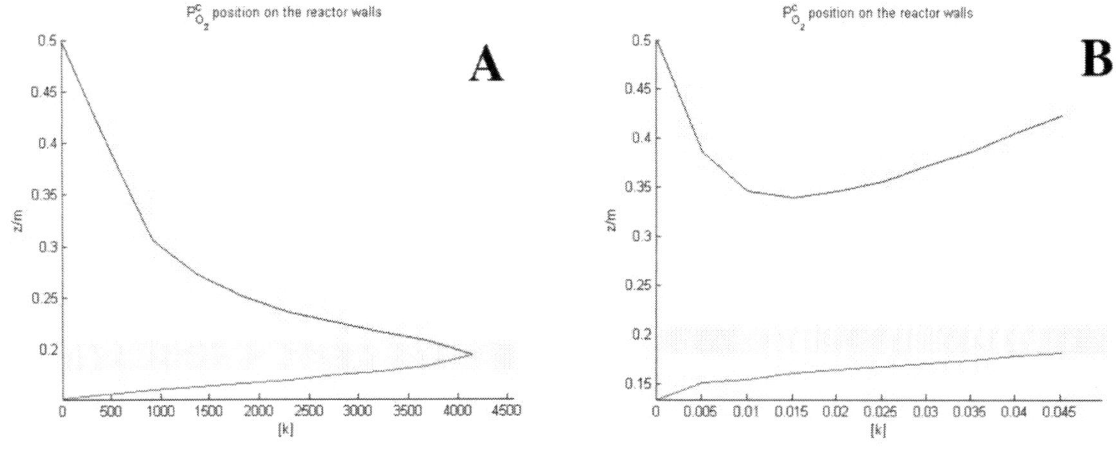

Figure 6: Comparison of the behavior fo the models considering only consumption reaction on the gas phase (A) or only consumption by surface reactions (B) for different reaction constant and adsorption constan values. The colored zone represents the position of the substrate in the experiments.

gradient generated will be such that the radial diffusion become comparable with the axial diffusion and part of convection generating high concentration areas around the oxygen feed until the vicinity of the walls. We can also notice that the location of the critical Oxygen partial pressure would require very high values of the global adsorption constant, therefore a single substrate with two oxide phases seems less likely under this limiting case.

Conclusions

A 2D axysymmetric transport model has been created to study the deposition of copper oxide on a hot-wall chemical vapor deposition reactor. The model was tested against a COMSOL simulation to establish the validity of the global collocation method used to solve the model. Both the collocation and the COMSOL simulations produced the same oxygen partial pressure profile on the reactor walls.

Based on previous work a critical oxygen partial pressure was identified as the separator of the two oxides usually observed on single substrates in experiments. The axysymmetric model was tested to locate the position of this critical pressure on the walls and thus the profile that the substrate would have. However, it was found that kinetic information is vital for the model. And as an initial approach two limiting cases were considered: only gas phase reactions and only surface reactions. The analysis of both cases is not enough to discern which are the dominant dynamics. However, they lay the bases for comparison with experiments, which is ongoing, to compare the reaction and adsorption constants directly with experiments and evaluate the proximity of the real system to each case.

References

[1] A. Gasparotto, D. Barreca, P. Fornasiero, V. Gombac, O. I. Lebedev, C. Maccato, T. Montini, E. Tondello, G. Van Tendeloo, E. Comini, G. Sberveglieri (2009), Multi-Functional Copper Oxide Nanosystems for H2 Sustainable Production and Sensing , *ECS Transactions.* **25**, pp 1169.

[2] D. Barreca, P. Fornasiero, A. Gasparotto, C. Maccato, V. Gombac, E. Tondello (2009), The Potential of Cu2O and CuO Nanosystems on the Photocatalytic H2 Production, *ChemSusChem.* **2**, pp 230.

[3] W. Siripala, A. Ivanovskaya, T. F. Jaramillo, S.-H. Baeck, E. W. McFarland (2003), A Cu2O/TiO2 heterojunction thin film cathode for photoelectrocatalysis, *Sol. Energy Mater. Sol. Cells.* **77**, pp 229.

[4] P. E. de Jongh, D. Vanmaekelbergh, J. J. Kelly (2000), Photoelectrochemistry of Electrodeposited Cu2O,*J. Electrochem. Soc.* **147**, pp 486.

[5] M. Hara, T. Kondo, M. Komoda, S. Ikeda, K. Shinohara, A. Tanaka, J. Kondo, K. Domen (1998), Cu2O as a photocatalyst for overall water splitting under visible light irradiation, *Chem. Commun.* **357**.

[6] Arana-Chavez D., Toumayan E., Lora F., McCaslin C. (2010), Adomaitis R.A. Modeling the Transport and Reaction Mechanisms of Copper Oxide *CVD. Journal of Chemical Vapor Deposition.* **16**, pp 336-345.

[7] L. S. Hong, H. Komiyama (1991), Chemical Vapor Deposition of CuOx Films by CuI and O2: Role of Cluster Formation on Film Morphology , *J. Am. Ceram. Soc.* **74**, pp 1597.

[8] Pradnya Pillewan, Shrabanti Mukherjee, Tarit Roychowdhury, Sera Das, Amit Bansiwal, Sadhana Rayalu (2011), Removal of As(III) and As(V) from water by copper oxide incorporated mesoporous alumina, *Journal of Hazardous Materials.* **186**, pp 367-375.

[9] Abdu, Y. and Musa, A.O (2009), Copper (I) oxide (Cu2O) based solar cells - a review , *Bayero Journal of Pure and Applied Sciences.* **2**(2), pp 8-12.

[10] Yatendra S. Chaudhary, Anshul Agrawal, Rohit Shrivastav, Vibha R. Satsangi, Sahab Dass (2004), A study on the photoelectrochemical properties of copper oxide thin films, *Internation Journal of Hydrogen Energy.* **29** pp 131-134.

(646e) Constrained Sensor Fault-Tolerant Control of Distributed Process Systems

Zhiyuan Yao and Nael H. El-Farra, Department of Chemical Engineering and Materials Science, University of California, Davis, Davis, CA

The majority of research studies on fault-tolerant process control have focused on spatially homogeneous processes modeled by systems of ordinary differential equations; yet, many chemical processes, such as transport-reaction processes and fluid flows, are characterized by inherent spatial variations and are more suitably modeled by systems of partial differential equations. The need to develop fault-tolerant control methods for distributed parameter systems has motivated a number of recent works in this area (see, for example, [1-3]). The focus of these studies, however, has been on the diagnosis and handling of control actuator faults. Sensor faults, on the other hand, are equally critical in terms of their influence on the overall process stability and performance, and require attention in the fault-tolerant control system design, especially as dense deployments of sensor networks are increasingly used in industrial systems. The measurement errors introduced by sensor malfunctions can deteriorate the overall control quality and may result in more catastrophic effects and economic losses if not explicitly handled. Systematic methods for handling sensor faults in distributed control systems remain lacking at this stage.

In this paper, we focus on the control of distributed process systems modeled by highly dissipative PDEs subject to sensor faults, control constraints and external disturbances. The key objective is to design a fault-tolerant control system that guarantees graceful degradation in process performance under sensor faults. To this end, we first obtain, via model reduction techniques, a suitable finite-dimensional model that captures the dominant dynamics of the infinite-dimensional system, and use it to investigate the closed-loop stability properties under different actuator and sensor configurations. The stability properties are influenced by the control constraints which limit the set of initial conditions starting from where stability can be achieved (the stability region), as well as the disturbances and sensor faults which constrain the set of terminal states that the system can be steered to in finite time (the terminal region). Using Lyapunov analysis techniques, an explicit characterization of the stability and terminal regions for each actuator and sensor configuration is obtained in terms of the constraints, and the available bounds on the sizes of the disturbances and sensor faults. This characterization is then used to devise switching strategies that ensure minimal deterioration in the size of the terminal region under a given sensor fault. The idea is to identify the location of the closed-loop state relative to the stability and terminal regions following fault detection, and have the supervisor either switch to an actuator configuration that recovers the fault-free terminal region (thus maintaining operation under the faulty sensor) if feasible, or switch to a new sensor configuration with a larger, but still acceptable, terminal region. Precise conditions for the implementation of the proposed switching strategies on the infinite-dimensional system are derived using singular perturbation techniques. Finally, the theoretical results are illustrated using a diffusion-reaction process example.

References:

[1] El-Farra, N. H. and S. Ghantasala, "Actuator fault isolation and reconfiguration in transport-reaction processes", *AIChE J.*, **53**, 1518-1537, 2007.

[2] Armaou, A. and M. Demetriou, "Robust detection and accommodation of incipient component faults in nonlinear distributed processes", *AIChE J.*, **54**, 2651-2662, 2008.

[3] Ghantasala, S. and N. H. El-Farra, Robust actuator fault isolation and management in constrained uncertain parabolic PDE systems, *Automatica*, **45**, 2368-2373, 2009.

See more of this Session: Dynamics, Reduction and Control of Distributed Parameter Systems
See more of this Group/Topical: Computing and Systems Technology Division

(646f) Stability Condition and Discretization Scheme for the Population Balance

Juan Du and B. Erik Ydstie, Department of Chemical Engineering, Carnegie Mellon University, Pittsburgh, PA

The main contribution of this work is two-fold. First, a global stability analysis is developed to derive the stability condition for population balance in the infinite dimensional space. The general form of population balance is a partial differential equation in the particle phase space which consists of external and internal coordinates. The external coordinates show the spatial location of each particle whereas the internal coordinates provide the unique information of particle such as size, shape, change of distribution. An entropy based energy function is developed to analyze the stability property of population balance. In order to formulate the entropy function, extensive variables and corresponding intensive variables are defined based on the characteristics of population balance. It is proven that the resulting energy function is a valid energy function and stability is demonstrated using classical energy methods. The resulting stability conditions provide guidelines for inventory control system design.

Second a novel discretization scheme is presented. The discretization scheme approximates population balance by establishing mass balance and number balance in each discrete interval. The continuous infinite dimension of particle phase space is reduced to finite dimension such that the partial differential equation is replaced with a set of ordinary different equations. Matrix analysis is employed to derive the stability property of the discretized population balance and the result is compared with that obtained for the entropy based energy analysis of the continuous population balance. Numerical simulations show that the matrix condition converges to analytical condition for continuous population balance if the number of discrete size intervals is large than a threshold value which can be obtained via simulations. Simulations results support the analytical stability condition derived from both approaches. The results of the paper are applied to crystallization and compared with Doherty's results.

See more of this Session: Dynamics, Reduction and Control of Distributed Parameter Systems
See more of this Group/Topical: Computing and Systems Technology Division

(647a) Process Optimization of Bioethanol Production Via Hydrolysis of Switchgrass

Mariano Martn, Chemical Engineering, University of Salamanca, Salamanca, Spain and Ignacio E. Grossmann, Department of Chemical Engineering, Carnegie Mellon University, Pittsburgh, PA

Processoptimization of bioethanol production via Hydrolysis of Switchgrass

Mariano Martn[a,b],Ignacio E. Grossmann[b][1]

mariano.m3@usal.es; grossmann@cmu.edu

[a] Departamentode Ingeniera qumica y textil. Universidad de Salamanca. Plz. Caidos1-5 37008, Salamanca (Spain)

[b]Department of ChemicalEngineering. Carnegie Mellon University 5000 Forbes Avd. Pittsburgh PA 15213

TheEnergy act in 2007 established the increase in the production of bioethanolusing lignocellulosic raw materials to help reduce the dependence on crude oil.This was based on the fact that among the different possibilities only biomassprovides an alternative fuel that can be implemented in the short-term for the transportationsector due to its compatibility with currentautomobile engines (Cole, 2007) and to the fact that it can take advantage ofthe existing supply chain of liquid fuels that is already well established. Currentlythe production of ethanol from lignocellulosic raw materials still faces technical,economic and commercial barriers (Huang, 2008).

Twotypes of process technologies can be used to transform lignocellulosic rawmaterials into ethanol. The first one is based onthe gasification of the raw material into syngas, which is used to obtainethanol either via Fischer-Tropsch based catalytic reaction or via fermentationof the syngas (Phillips et al, 2007; Huhnke, 2008;Piccolo and Bezzo, 2009; Zhuet al, 2009). The second oneis based on the hydrolysis of the raw material to break down the physical andchemical structure of the crops to expose the sugars that are fermented toethanol. Due to its similarity with the current production of ethanol and theexpected lower capital cost, this technology has received the attention of manyresearchers, e.g. Hamelinck et al. (2005); Cardona& Snchez (2006); Zhang et al. (2009); Keshwani& Cheng (2009).

In thispaper we develop a conceptual design for the production of ethanol fromlignocellulosic raw materials with hydrolytic pretreatment of the biomass usingmathematical programming techniques

(Daichendt & Grossmann, 1997, Grossmannet al., 1999). The process consist of three steps, raw material pretreatment toexpose the sugars, sugar fermentation to ethanol and ethanol dehydration tofuel grade. Wepropose a superstructure optimization approach where we first construct aflowsheet embedding the various process units involved in ethanol productionconsidering a number of alternatives for switcgrass pretreatment such as ammoniafiber explosion or dilute acid pretreatment (Sung and Cheng, 2002; Keshwani & Cheng,2009) and for differenttechnologies that can operated in parallel and/ or in sequence for dehydratingthe ethanol like rectification, adsorption in corn grits, molecular sieves andpervaporation. These units are interconnected to each other through networkflows and other utility streams. The goal is to optimize the structure by minimizingthe energy input in the ethanol production process. The optimization of thesystem is formulated as a mixed-integer nonlinear programming (MINLP) problem,where the model involves a set of constraints representing mass and energybalances for all the units in the system. We then substitute the distillation columnsby multieffect column to reduce the consumption on cooling water and steam and wedesign the optimal heat exchanger network using SYNHEAT. The heat recoverynetwork, together with a modified distillation column design, further reducesthe energy consumption in the plant and in turn decreases the unit productioncost of ethanol. Finally, we perform an economic evaluation. For the optimalflowsheet we design the optimal water network based on the work by Ahmetovic,et al (2010). First, we identify the sources of water within the process(distillation columns, utilities units), the sinks (fermentor, pretreatment)and the process units (unitities units), and optimize the correspondingsuperstructure for reuse and recycle of water and treatment units so as tominimize the fresh water consumption.

Theresults of the optimized conceptual design indicates that under the currentyields reported for the pretreatment processes (Sung and Cheng, 2002) the optimalprocess for the production of ethanol from switchgrass involves the use ofdilute acid pretreatment, followed by enzymatic hydrolysis and sugarfermentation and the use of molecular sieves for the final dehydration ofethanol to fuel grade. The predicted production cost of this process is $0.8/galwith an investment cost of $161MM, while the consumption of water is1.6gal/gal. These results are compared with the themal based processes thatrequires a higher investment of $335MM, but has a reduced production cost of$0.41/gal and water consumption of 1 gal/gal (Martin & Grossmann, 2011,Martin et al ,2010)

References

Ahmetovic, E.; Martn, M.; Grossmann,I.E. (2010) "Optimization of Water Consumption in Process industry: Corn -based ethanol case study" Ind. Eng. Chem Res. 49 (17) 7972- 7982

Cardona, C.A., Snchez. O.J. (2006)Energy consumption analysis of integrated flowsheets for production of fuelethanol from lignocellulosic biomass. Enrgy , 31, 2447- 2459

Cole, D. E. (2007) Issues facing theAuto Industry: Alternative Fuels, Technologies, and Policies ACP Meeting EagleCrest Conference Center June 20, 2007

Daichendt, M. M., and I. E. Grossmann,(1997) Integration of Hierarchical Decomposition and Mathematical Programmingfor the Synthesis of Process Flowsheets, Comp. Chem. Eng., 22, 147.

Grossmann, I. E.; Caballero, J. A.;Yeomans, H. (1999) Mathematical Programming Approaches to the Synthesis ofChemical Process Systems, Korean J. Chem. Eng., 16, 407-426.

Keshwani, D. R., Cheng, J.J. (2009)Switchgrass for bioethanol and other value-added applications: A reviewBioresource Technology 100, 1515-1523

Hamelinck, C. N., Geertje vanHooijdonk, G., Faaij, A.P.C. (2005) Ethanol from lignocellulosic biomass: techno-economic performance in short-, middle- and long-termBiomass andBioenergy 28, 384-410

Huang, J. Qiu,H. and Scott Rozelle, S., (2008), More pain ahead for China's food prices, FarEastern Economic Review, 171, 5, 8-13.

Huhnke, R. L. (2008) Cellulosicethanol using gasification-fermentation. Resource:Engineering & Technology for a Sustainable World

http://www.articlearchives.com/energy-utilities/renewable-energy-biomass/896186-1.html

Martn, M.; Ahmetovic, E.; Grossmann,I.E. (2010) "Optimization of Water Consumption in Second Generation bio-EthanolPlants " I&ECR doi: 10.1021/ie101175p

Martn, M., Grossmann, I.E. (2011)"Energy optimization of lignocellulosic bioethanol production via gasification"accepted AIChE J. | DOI: 10.1002/aic.12544

Phillips, S., Aden, A., Jechura, J.and Dayton, D., Eggeman, T (2007) Thermochemical Ethanol via IndirectGasification and Mixed Alcohol Synthesis of Lignocellulosic Biomass TechnicalReport, NREL/ TP-510-41168, April 2007

Piccolo, C., Bezzo, F., (2009) Atechno-economic comparison between two technologies for bioethanol productionfrom lignocelluloses. Biomass and Bioenergy 33 (2009) 478 - 491

Sun, Y., Cheng, J., (2002) Hydrolysisof lignocellulosic materials for ethanol production: a review. Bioresour.Technol., 83, 1-11

Zhu, Y., Gerber, M.A., Jones, S.B.,Stevens, D.J (2009) Analysis of the effects of compositional andconfigurational assumptions on product costs for the thermochemical cornversionof lignocellulosic biomass to mixed alcohols FY 2007 Progress report. DOEPNNL.17949 Revision 1.

[1]Corresponding author. Tel.:+1-412-268-3642; Fax: +1-412-268-7139.

Email address:grossmann@cmu.edu (I.E. Grossmann)

See more of this Session: Energy Systems Design and Alternative Energy Sources I
See more of this Group/Topical: Computing and Systems Technology Division

(647b) Optimization Framework for the Process Synthesis and Simultaneous Heat and Power Integration of a Thermochemical Coal, Biomass, and Natural Gas to Liquids Facility

Richard Baliban, Josephine A. Elia and Christodoulos A. Floudas, Chemical and Biological Engineering, Princeton University, Princeton, NJ

Thermochemical conversion of biomass or coal proceeds via either direct liquefaction which produces a synthetic crude oil or indirect liquefaction which produces synthesis gas (syngas) that can be converted to liquid hydrocarbons directly via the Fischer-Tropsch (FT) process [1]. Thermochemical conversion of natural gas generally occurs through indirect liquefaction using an autothermal reactor. A coal, biomass, and natural gas to liquids (CBGTL) process was recently developed [2,3] that utilizes input hydrogen to have an almost-100% conversion of the carbon in the feedstock to liquid fuels. While the use of hydrogen in the CBGTL process can produce near-zero CO_2 quantities without the need for carbon sequestration, it remains unclear if that is the most cost-effective option to produce liquid fuels. To facilitate the transition from petroleum-based transportation fuels to new carbon-based systems, it is essential to determine the process topology that can deliver the fuels at the lowest levelized cost while using the limited supply of coal, biomass, and natural gas available throughout the country.

To address the above problem, a process synthesis model is proposed to find the CBGTL refinery system topology that provides the lowest levelized cost of transportation fuel production [4]. The model utilizes a combination of continuous variables to model stream flow rates, process operating conditions, and thermodynamic properties and binary variables to model the logical use of a process unit or a process stream. Several constraints dictating the conservation of mass and energy, equilibrium of reactor species, split fractions, and extents of reaction are imposed to ensure that a feasible solution is found. The model is formulated as a mixed-integer non-linear non-convex optimization problem (MINLP). The MINLP model includes simultaneous heat and power integration utilizing heat engines to recover electricity from the process waste heat. Four case studies are presented to investigate the effect of CO_2 sequestration (CCS) and GHG reduction targets on the process topology along with detailed parametric analysis on the role of biomass and electricity prices. Optimal process topologies, the complete heat and power integration, and the overall costs of each case study are compared.

[1] National Academy of Sciences, National Academy of Engineering, and National Research Council. Liquid Transportation Fuels from Coal and Biomass: Technological Status, Costs, and Environmental Issues. Prepublication. Washington, D. C., EPA, 2009.

[2] R. C. Baliban, J. A. Elia, and C. A. Floudas. Toward Novel Biomass, Coal, and Natural Gas Processes for Satisfying Current Transportation Fuel Demands, 1: Process Alternatives, Gasification Modeling, Process Simulation, and Economic Analysis. Ind. Eng. Chem. Res., 49:7343-7370, 2010.

[3] J. A. Elia, R. C. Baliban, and C. A. Floudas. Toward Novel Biomass, Coal, and Natural Gas Processes for Satisfying Current Transportation Fuel Demands, 2: Simultaneous Heat and Power Integration. Ind. Eng. Chem. Res., 49:7371-7388, 2010.

[4] R. C. Baliban, J. A. Elia, and C. A. Floudas. Optimization Framework for the Simultaneous Process Synthesis, Heat and Power Integration of a Thermochemical Hybrid Biomass, Coal, and Natural Gas Facility. Comp. Chem. Eng., doi:10.1016/j.compchemeng.2011.01.041,2011.

See more of this Session: Energy Systems Design and Alternative Energy Sources I
See more of this Group/Topical: Computing and Systems Technology Division

(647c) Process Synthesis of Biodiesel Production Plant Using Artificial Neural Networks As the Surrogate Models

Ismail Fahmi, Chemical Engineering, The University of Tulsa, Tulsa, OK and Selen Cremaschi, Department of Chemical Engineering, The University of Tulsa, Tulsa, OK

Biofuel from biomass is very appealing due to its abundance and renewability to replace the usage of the depleting fossil-based fuels. Among many types of biofuels, biodiesel is probably one of the most attractive one because it can be used directly with the current traditional petro diesel engine, either as a substitute or as a blending aggregate [1-3]. Biodiesel is commonly obtained through transesterification of bio oil, either in a form of triglycerides or hydrolyzed fatty acid. There are several alternatives that can be considered for biodiesel production: transesterification with homogeneous catalysts [2, 4] or solid base catalysts [5-7], or by using supercritical methanol without the presence of catalyst [8]. In this work, superstructure optimization is performed to synthesize the optimum biodiesel production plant, i.e., the one that gives the minimum net present sink. Given the high computational resource required to solve the resulting Mixed Integer Nonlinear Programming (MINLP) formulation, using surrogate models is suggested [9, 10]. Among many kinds of surrogate models, Artificial Neural Network (ANN) has been shown to be a powerful tool both for complexity reduction and accurate mapping-function modeling [11, 12]. In this presentation, the feasibility of synthesizing the optimum biodiesel production plant using ANN as the surrogate models for the required unit operations is investigated. The optimum solution of using the alkali-catalyzed reactor with the total cost of USD 41 million was obtained in about 5 CPU seconds. When the solution was modeled in a process simulator, the resulting total cost of the simulation only differs about 1% attesting to the accuracy of the ANNs predictions. However, this comes with a cost of having to provide a large amount of training data for the ANN.

Keywords: Biodiesel Production, Artificial Neural Network, Optimization, Process Synthesis, Surrogate Models

References:

[1] D. Tapasvi, *et al.*, "Process model for biodiesel production from various feedstocks," *Transactions of the American Society of Agricultural Engineers,* vol. 48, pp. 2215-2221, 2005.

[2] Y. Zhang, *et al.*, "Biodiesel production from waste cooking oil: 1. Process design and technological assessment," *Bioresource Technology,* vol. 89, pp. 1-16, 2003.

[3] J. M. Lujan, *et al.*, "Comparative analysis of a DI diesel engine fuelled with biodiesel blends during the European MVEG-A cycle: Preliminary study (I)," *Biomass and Bioenergy,* vol. 33, pp. 941-

947, 2009.

[4] L. F. Bautista, *et al.*, "Optimisation of FAME production from waste cooking oil for biodiesel use," *Biomass and Bioenergy,* vol. 33, pp. 862-872, 2009.

[5] S. Yan, *et al.*, "Oil transesterification over calcium oxides modified with lanthanum," *Applied Catalysis A: General,* vol. 360, pp. 163-170, 2009.

[6] P. D. Patil and S. Deng, "Transesterification of Camelina Sativa Oil Using Heterogeneous Metal Oxide Catalysts," *Energy & Fuels,* vol. 23, pp. 4619-4624, 2009.

[7] D. E. Lopez, *et al.*, "Esterification and transesterification using modified-zirconia catalysts," *Applied Catalysis A: General,* vol. 339, pp. 76-83, 2008.

[8] K. T. Tan, *et al.*, "Production of FAME by palm oil transesterification via supercritical methanol technology," *Biomass and Bioenergy,* vol. 33, pp. 1096-1099, 2009.

[9] C. A. Henao and C. T. Maravelias, "Surrogate-Based Process Synthesis," in *Computer Aided Chemical Engineering*. vol. Volume 28, S. Pierucci and G. B. Ferraris, Eds., ed: Elsevier, 2010, pp. 1129-1134.

[10] N. V. Queipo, *et al.*, "Surrogate-based analysis and optimization," *Progress in Aerospace Sciences,* vol. 41, pp. 1-28, 2005.

[11] H. Seung-Soo and G. S. May, "Using neural network process models to perform PECVD silicon dioxide recipe synthesis via genetic algorithms," *Semiconductor Manufacturing, IEEE Transactions on,* vol. 10, pp. 279-287, 1997.

[12] I. Sabuncuoglu and S. Touhami, "Simulation metamodelling with neural networks: an experimental investigation," *International Journal of Production Research,* vol. 40, pp. 2483-2505, 2002.

See more of this Session: Energy Systems Design and Alternative Energy Sources I
See more of this Group/Topical: Computing and Systems Technology Division

(647d) An Integrated Biofuels Strategy: Catalytic Conversion of Lignocellulosic Biomass to Liquid Hydrocarbon Fuels

Sercan Murat Sen[1], Carlos A. Henao[2], Elif I. Gurbuz[1], David Martin Alonso[1], James A. Dumesic[1] and Christos Maravelias[3], (1)Chemical and Biological Engineering, University of Wisconsin-Madison, Madison, WI, (2)Chemical and Biological Engineering, University of Wisconsin - Madison, Madison, WI, (3)University of Wisconsin, Madison, WI

An integrated biofuels strategy: Catalytic conversionof lignocellulosic biomass to liquid hydrocarbon fuels

Replacement of fossil fuels with new sustainableresources becomes crucial due to depletion of petroleum reserves, increasingglobal energy demand and arising environmental concerns. Lignocellulosicbiomass provides sustainable and environmentally friendly ways of producingchemicals and fuels as an alternative for fossil fuels. One critical step isthe conversion of lignocellulosic biomass to versatile intermediates such aslevulinic acid (LA), which can be transformed into liquid fuels, fuel additivesand even other specialty chemicals. In this respect, we studied a LA-basedcatalytic process to convert lignocellulosic biomass into liquid hydrocarbonfuels for use in the transportation sector. Using experimental results for allassociated reactions, we synthesized an integrated biomass-to-fuels strategythat has a number of advantages over existing strategies.

The first step of this process is the removal ofhemicellulose fraction of the biomass by using dilute acid pretreatment. After thehemicellulose fraction (that mainly contains xylose) is removed, the celluloseportion of the biomass is converted to levulinic acid (LA) and formic acid in acellulose deconstruction reactor using sulfuric acid. The insoluble lignin and C_5-sugarpolymers are sent to a boiler/ turbogenerator to produce heat and electricity,covering the utility requirements of our process. Excess electricity is sold tothe grid. LA is converted to γ-valerolactone (GVL) over a Ru/C or RuRe/Ccatalyst in the presence of sulfuric acid. GVL is extracted from the sulfuricacid and GVL aqueous solution using butyl acetate solvent and sulfuric acid isrecycled back to the cellulose deconstruction reactor. Finally, GVL isseparated from butyl acetate by using a distillation column and purified GVL isconverted to butene and to butene oligomers.

To determine the economic potential of thisstrategy, we carried out detailed process simulation (based on experimentalresults) and capital/operational cost calculations. A comparison with alignocellulosic ethanol production facility reveals the potential feasibilityof the LA-based catalytic approach. Various feedstock alternatives (cornstover, sugarcane, wheat straw, hybrid poplar, switchgrass, loblolly pine andaspen wood) were analyzed and compared for cost-effective processing. Loblollypine was identified to be the most cost-effective feedstock due to its high C_6sugar content. Using loblolly pine as the feedstock, the minimum sellingprice (MSP) of butene oligomers was calculated as $4.31 per gallon of gasolineequivalent. We also performed sensitivity analysis studies for several processparameters (e.g., feedstock cost, equipment cost etc.) as well as economicparameters (e.g. equipment lifespan, income tax rate, return on investmentdiscount rate) to determine the bottleneck of the base case design. Feedstockprice appears to be the major cost driver: a 20%change results in slightly more than 8% change in the MSP of oligomers . Itwas also shown that the MSP of alkenes are sensitive to variations in the costof the turbogenerator and economic parameters (equipment life span and the ROIdiscount rate). Finally,we present the results of an energy efficiency analysis for the proposedprocess. The energy efficiency of biofuel production from C_6 sugarcontents was equal to 44.5%.

The presence of sulfuric acid in the GVLproduction step causes a dramatic decrease in the catalytic activity of the Ru/Ccatalyst which is preferred because it is cheaper than the RuRe/C catalyst. Toaddress this shortcoming, we considered the production of hydrophobic esters (Sec-Butyl Levulinate and Sec-ButylFormate) from the reaction of levulinic acid and formic acid with butene. Thesource of reacting butene is mainly provided from the conversion of GVL tobutene and CO_2 by catalytic decarboxylation over an acid catalyst ina later step. Hydrophobic esters can be separated from sulfuric acid withoutneed of an energy intensive distillation and they are converted to GVL over adual-catalyst-bed system. This sulfuric acid management strategy provides downstream catalytic processing in theabsence of sulfuric acid and therefore with higher yields. We investigate theeconomic impact of integration of the proposed reactive extraction strategywith the original process.

See more of this Session: Energy Systems Design and Alternative Energy Sources I
See more of this Group/Topical: Computing and Systems Technology Division

(647e) A Life Cycle-Based Analysis and Optimization Framework for Designing Sustainable, Multi-Product Biomass-to-Bioproducts Value Chains

Paritosh K. Sharma, Chemical Engineering, Louisiana State University, Baton Rouge, LA and Jose A. Romagnoli, Chemical Eng, Louisiana State University, Baton Rouge, LA

A life cycle-based analysis and optimization framework for designing sustainable, multi-product biomass-to-bioproducts value chains

In recent years there has been a marked surge in the search for alternative sources of energy that wean the world off of dependence on fossil fuels and reduce our carbon foot-print on this world. After a boom in U.S. corn-based ethanol in the early part of the 21st century, the interest has gradually shifted towards more viable sources of biofuels and biochemicals. Cellulosic ethanol, biodiesel, and syngas are examples of such fuels that are extremely attractive owing to the fact that the raw materials can be composed completely of "left-over" wastes of food crops and forest harvests that don't interfere with the human food chain and the natural ecosystem.

Even with increased research emphasis on biofuel and biochemical production technologies, large scale production is still hampered by many factors. Some of these factors that have received considerable interest in recent times include, cultivating more robust feedstock, genetically engineering enzymes and microbes that are more efficient and resistant to poisoning by by-products and improving process technologies that can render large scale production of biofuels a more profitable option for private entities. One area that has not received enough research interest in recent times is enterprise wide modeling and optimization in order to make the entire enterprise more competitive in the energy sector.

Modeling an enterprise that produces renewable-based fuels and chemicals can provide valuable insight into the inter-play of the bioproduct supply-chain. This idea had been motivated by a recent surge in the research areas concerning supply-chain optimization and how cost-cutting measures throughout a supply-chain can render an otherwise sluggish enterprise, profitable.

In the current formulation, we apply a distributed, systematic approach to model, optimize, and analyze three aspects of sustainability for a biomass-to-bioproducts value chain; business, environmental, and social sustainability. Life cycle analyses are conducted to generate forecasts for environmental parameters, infrastructural and employment data, input costs and supplies, product yields, and output market conditions, which are integrated with integer-based stochastic optimization models that yield strategic decisions such as feedstock and supplier selection, transportation logistics design, technology and product selection, waste mitigation design, and capacity design under uncertainty. A two-stage optimization strategy is used to increase computational efficiency while extracting the biggest value drivers for the successful construction and operation of a biorefinery.

The first stage in optimization is termed *portfolio screening*; a deterministic integer programming model is formulated to screen out economically and environmentally unfavorable choices of feedstock, conversion technologies, and products. Stakeholder value of the entire value chain, including suppliers and customers, is maximized in the model with a rolling-time horizon of 12 years. The model equations and constraints included are material, energy, capacity, cash, and debt balances, transformation equations, supply and demand constraints, and capacity and resource constraints. Carbon taxes and mitigation credits are used to represent environmental implications of biorefinery activities. The outputs of the model include integer decisions regarding feedstock, suppliers, technologies, and product selection. Furthermore, sensitivity analysis is carried out on the final supply chain and process configuration to determine process, supply chain, and market parameters that are the greatest economic, environmental, and social value drivers for the biorefinery.

The second stage of optimization is formulated as a *Real-options* stochastic optimization model. The purpose of the model is to design a timing strategy for feedstock production and pricing, incremental technology acquisition, and capacity design given uncertainty in input costs and supplies, technology costs, and product demands and prices. The model is similar in form and structure to the deterministic MIP from the first stage with stochastic forecasts as inputs. Two types of options are modeled; Growth options to model capacity expansion, and switching options to represent flexible biorefining systems with the ability to switch product volumes between different products. A binomial decision tree is used to discretize continuous stochastic forecasts for inputs. The final objective function is a probability weighted stakeholder value whose form is derived from the deterministic model. The screened portfolio choices that are obtained from the first-stage optimization are used as inputs. Integer decisions for the model include capacity establishment and expansion, investment in new technologies, and product switching by re-allocation of feedstock to different product groups.

The resultant framework is applied to a prospective biomass refinery in Southeastern United States. The feedstock choices include energy crops, namely, switchgrass, miscanthus, and energy cane. The feedstock decisions are affected by capital investment for land acquisition, fixed and operating costs for crop growth and harvest, transportation expenses, seasonal availability, environmental life cycle impact, and biomass and sugar yields of the crop. The initial product portfolio includes ethanol and butanol as fuels, and succinic and lactic acid and value-added chemicals. The conversion and separation technologies are demarcated into three groups; pretreatment, hydrolysis (dilute acid (DA), ammonia-fiber explosion (AFEX)), and fermentation (simultaneous saccharification and fermentation (SSF) and separate hydrolysis and fermentation (SHF)) which are common for all products, and separation and product recovery (distillation, molecular sieves, pervaporation, membrane separation) which are specific to each product. The product-technology set choices are affected by capital costs, operating costs, process yields, environmental life cycle impact, and market potential.

The screening model results yield miscanthus as the feedstock choice owing to high biomass yields and ease of processing, and ethanol and succinic acid as the products owing to great market potential and favorable profit margins. Dilute acid pretreatment, enzymatic hydrolysis, and SSF are selected as the conversion technologies owing to favorable capital cost structure and process economics including the energy and environmental load. Butanol and lactic acid not chosen primarily due to the high cost of product recovery while distillation and molecular sieves (ethanol) and membrane separation (succinic acid) are chosen for the products that are selected.

The results are then fed to the second stage of optimization with uncertainty being represented in the marginal returns that are generated by capacity and production decisions. The uncertainty in the return implies uncertain revenue and cost streams. The individual components of the uncertain revenue and cost streams include uncertain product prices, uncertain input costs and supplies, and uncertain process yields. The uncertain returns are assumed to capture all uncertainty during optimization while the distribution of returns is generated separately using the stochastic forecasts for the individual components. The results indicate an incremental capacity design to mitigate and manage supply, capital, operational, and market risks with the overall expected stakeholder value at $MM 3.4 indicating that positive value is being created. The incremental capacity design is compared to economies of scale based capacity design where all capacity is established during the first planning year. The stakeholder value for economies of scale based design is $MM -30.3 indicating a $MM 33.7 advantage to using incremental capacity design. Product switching capabilities are recognized in the final two time periods (4 years) where a price shock to the upside in succinic acid prices is simulated. Capacity is re-allocated to succinic acid production to exploit the favorable market conditions with a 2% increase in the stakeholder value thus depicting the advantages of designing flexible production systems for biomass refineries.

See more of this Session: Energy Systems Design and Alternative Energy Sources I
See more of this Group/Topical: Computing and Systems Technology Division

(647f) Decomposition Strategy for the Global Optimization of Energy Polygeneration Systems

Yang Chen[1], Xiang Li[1], Thomas A. Adams II[2] and Paul Barton[1], (1)Chemical Engineering, Massachusetts Institute of Technology, Cambridge, MA, (2)Chemical Engineering, McMaster University, Hamilton, ON, Canada

Coal- and biomass-based polygeneration with multiple energy products is a promising technology to supplement the current oil-based energy processes. A certain degree of operational flexibility can be introduced into a polygeneration process by over-sizing equipment so that one can alter the production rates of individual products in response to changing market conditions. The resulting flexible polygeneration plant achieves potentially higher profit and lower economic risk than the static plant [1].

In the design of flexible polygeneration systems, the optimal trade-off between operational flexibility and equipment capacity (or capital cost) needs to be determined. This problem can be formulated as a multi-period optimization problem, which is a potentially large-scale nonconvex mixed-integer nonlinear program (MINLP) and cannot be solved to global optimality by state-of-the-art global optimization solvers, such as BARON [2], within a reasonable time. A duality-based decomposition method, recently developed for the design and operation of natural gas production networks [3,4], can exploit the special structure of this mathematical programming problem and enable faster solution. In this method, the nonconvex MINLP is relaxed into a convex lower-bounding MINLP which can be further reformulated into a relaxed master problem according to the principles of projection, dualization and relaxation. The relaxed master problem yields an increasing sequence of lower bounds for the original problem, but these lower bounds may converge to a global optimum slowly due to the high nonconvexity of the problem.

In this work, an enhanced decomposition method with an improved relaxed master problem is developed by introducing additional dual information into the relaxed master problem, which requires solving additional nonconvex NLPs. The solutions of these nonconvex NLPs lead to stronger dual cuts that can significantly improve the convergence rate of the algorithm. The enhanced decomposition algorithm guarantees to find an ε-optimal solution in a finite number of iterations.

The computational advantages of the enhanced decomposition method are demonstrated via several case studies on the optimal design and operation of a flexible energy polygeneration system using coal and biomass as feedstocks and co-producing electricity, liquid fuels (naphtha and diesel) and chemicals (methanol). The computational results show that the enhanced decomposition algorithm achieves much faster convergence than both BARON and the original decomposition algorithm, and it solved the large-scale nonconvex MINLPs to ε-optimality in practical times (e.g., a problem with 70 binary variables and 14712 continuous variables was solved within 5 hours).

Reference

[1] Chen Y, Adams TA, Barton PI. Optimal design and operation of flexible energy polygeneration systems. *Industrial & Engineering Chemistry Research.* 2011;50(8):4553-4566.

[2] Tawarmalani M, Sahinidis NV. Global optimization of mixed-integer nonlinear programs: A theoretical and computational study. *Mathematical Programming.* 2004;99(3):563-591.

[3] Li X, Armagan E, Tomasgard A, Barton PI. Stochastic pooling problem for natural gas production network design and operation under uncertainty. *AIChE Journal.* DOI: 10.1002/aic.12419.

[4] Li X, Tomasgard A, Barton PI. Decomposition strategy for natural gas production network design under uncertainty. *Proceedings of the 49th IEEE Conference on Decision and Control*, 188-193 (2010).

See more of this Session: Energy Systems Design and Alternative Energy Sources I
See more of this Group/Topical: Computing and Systems Technology Division

(652a) Tighter Integration Between Planning and Engineering Models At BP

Thomas L. Bowman[1], Don L. Glass[2], Kheng Lau[2] and **Ajay Lakshmanan**[3], (1)BP, Houston, TX, (2) Process Modeling, Optimization & Strategic Support, BP, Naperville, IL, (3)Math Solvers and Frameworks, Aspen Technology, Inc., Burlington, MA

Accurate modeling can offer valuable insight into understanding the intricacies and trade-offs involved in a process plant. Using consistent models in design, operation, optimization and production planning can result in improved decision making and enterprise operations management. This paper details the use of consistent reduced order nonlinear models derived from rigorous first principles based engineering models within refinery planning applications at BP.

The motivation behind the use of reduced order models derived from engineering within the refinery planning application at BP includes the ability to separate the planning exercise from detailed modelling tasks, lower model development costs, and improve knowledge management through a standardized work process where models are created and maintained by a group of modeling experts and distributed to model users across the company.

This paper describes the challenges addressed during the design and implementation of the CatSim model for fluidized catalytic cracking and the shortcut distillation model for crude fractionation at BP's Texas City refinery. The paper includes a description of the overall refinery model that is used for long term planning, capital studies, crude selection, short-term plant optimization, and for reviewing past performance. The implementation is based on aspenONE including Aspen PIMS for refinery planning, Aspen Custom Modeler for model authoring, and the framework [1] that enables model sharing and reuse. The paper also details the benefits achieved at BP through tighter integration of planning and engineering models.

[1] Lakshmanan, A., Paules, G., and Mahalec, V., *"A Modeling Framework That Enables Process Synthesis, Design, Analysis, Optimization, and Planning,"* ***Annual American Institute of Chemical Engineers Meeting*** *(60c), San Francisco, CA, November 2006*

See more of this Session: Industrial Applications In Design and Operations
See more of this Group/Topical: Computing and Systems Technology Division

(652b) Scheduling for Performance-Decaying Cracking Furnace Operation with Consideration of Inherent Process Upset Reduction

Chuanyu Zhao and Qiang Xu, Dan F. Smith Department of Chemical Engineering, Lamar University, Beaumont, TX

Scheduling for Performance-Decaying Cracking Furnace Operation with Consideration of Inherent Process Upset Reduction†

Chuanyu Zhao and Qiang Xu*

Dan F. Smith Department of Chemical Engineering

Lamar University, Beaumont, TX 77710, USA

Abstract

Thermal cracking furnace system is the critical starting sector of an ethylene plant. Its multiple furnaces periodically experience hot and cold shutdowns for maintenance due to performance-decaying operational characteristic, which inherently upsets the downstream flowrate of almost the entire plant. In previous work, most studies are focused on the pyrolysis process simulation and control, or the cracking furnace system scheduling of profit enhancement. The operational scheduling with consideration of inherent upset reduction is totally ignored. In this paper, a mixed-integer linear programming (MILP) model is developed to obtain the optimal schedule for any customized scheduling period with two objective functions: 1) maximum average profit of the cracking furnace system per time: 2) minimum inherent process upset from the furnace system. The idea of inherent process upset reduction is to group shutdown operations of different furnaces together during the manufacturing so that the downstream will have less frequency to suffer upsets from the furnace system.

The MILP model addresses all the major scheduling issues of a cracking furnace system, such as multiple feeds, multiple furnaces, various product prices and manufacturing costs, hot shutdowns for decoking, cold shutdowns for mechanical maintenance, semi-continuous operation, cracking performance decaying, non-simultaneous shutdowns, and secondary ethane cracking. With the given customized scheduling period, optimal scheduling solutions for the developed MILP model can be obtained to provide quantitative answers to the following questions: i) what is the maximum average profit of the cracking furnace system within given time period? ii) how to allocate different feeds into different furnace for cracking? iii) how to setup each batch operation and determine its run length? iv) how to determine the best decoking sequence and maintenance location among multiple furnaces? and

v) how to manipulate shutdowns from all the furnaces to reduce the inherent upsets to downstream process.

A case study shows that the significant flowrate upsets from a cracking furnace system can be reduced to two during 150 days scheduling period with the developed methodology, compared to six upsets without the consideration of inherent upset reduction at the cost of affordable profit loss. It demonstrates the efficacy of the developed methodology in terms of its significant economic potential and operational benefits. Meanwhile, this work can be applied to dynamic scheduling as well.

Keywords: Scheduling, MILP, Cracking furnace, Shutdown, Optimization, Upset reduction

See more of this Session: Industrial Applications In Design and Operations
See more of this Group/Topical: Computing and Systems Technology Division

(652c) A New MINLP Model for Optimal Planning of Offshore Oil and Gas Field Infrastructure with Production Sharing Agreements

Vijay Gupta and Ignacio E. Grossmann, Department of Chemical Engineering, Carnegie Mellon University, Pittsburgh, PA

Optimal investment and operations planning of offshore oil and gas field facilities is very critical problem given that it involves huge investments and high potential profits [1-2]. Therefore, an efficient model that on one hand captures the realistic reservoir profiles, interaction among various fields and facilities, wells drilling limitations and other practical trade-offs involve in the offshore development planning, on the other hand it should be able to extend to include other complexities especially that arises due to the fiscal rules [3] and uncertainties [4-5].

In this paper, we present an efficient strategic/tactical planning model for offshore oilfield development problem that includes Production Sharing Agreements and fairly generic to be extended to other complexities. The proposed multiperiod non-convex MINLP model for multi-field site includes three components (oil, water and gas) explicitly in the formulation using 3rd and higher order polynomials avoiding bilineairties and other nonlinear terms. With the objective of maximizing total NPV for long-term planning horizon, the model involves decisions related to FPSO (floating production, storage and offloading) installation and expansion schedule and respective oil, liquid and gas capacities, connection between the fields and FPSO's, well drilling schedule and production rates of these three components in each time period. The resulting model can be solved with DICOPT in an efficient way for realistic instances and gives good quality solutions. Furthermore, it can be reformulated into an MILP model after piecewise linearization and other linearization techniques that can be solved globally in an efficient way.

Including fiscal considerations, especially progressive Production Sharing Agreements (PSA's), as part of the investment and operation decisions for this problem can significantly impact the optimal solution and required computational time. In particular, these specific rules determine the profit that the operating company can keep as well as the royalties, taxes that are paid to the government through complex calculations. Therefore, an effective extension of the proposed model to include the complex rules in Production Sharing Agreements (PSA's) for multiple field case will also be presented. Solutions of realistic instances involving 10 fields, 3 FPSO's and 20 years planning horizon will be reported, as well as comparisons between the computational performance of the proposed MINLP and MILP formulations. The difference in solutions and computational impact of adding the PSA constraints versus using a simple NPV optimization is also reported.

References

[1] Iyer, R. R.; Grossmann, I. E.; Vasantharajan, S.; Cullick, A. S. Optimal planning and scheduling offshore oilfield infrastructure investment and operations. *Ind. Eng. Chem. Res.* 1998, *37*, 1380–1397.

[2] van den Heever, S. A.; Grossmann, I. E. An iterative aggregation/disaggregation approach for the solution of a mixed integer nonlinear oilfield infrastructure planning model. *Ind. Eng. Chem. Res.* 2000, *39*, 1955–1971.

[3] van den Heever, S. A.; Grossmann, I. E. A Lagrangean Decomposition Heuristic for the Design and Planning of Offshore Hydrocarbon Field Infrastructures with Complex Economic Objectives. *Ind. Eng. Chem. Res.* 2001, *40*, 2857–2875.

[4] Goel, V.; Grossmann, I. E. A stochastic programming approach to planning of offshore gas field developments under uncertainty in reserves. *Comput. Chem. Eng.* 2004, *28* (8), 1409–1429.

[5] Tarhan, B., Grossmann, I.E.,& Goel, V. Stochastic programming approach for the planning of offshore oil or gas field infrastructure under decision- dependent uncertainty. *Ind. Eng. Chem. Res.* 2009, 48(6), 3078–3097.

See more of this Session: Industrial Applications In Design and Operations
See more of this Group/Topical: Computing and Systems Technology Division

(652d) Towards Continuous Operations In Pharmaceutical Manufacturing

Arun Giridhar, Girish Joglekar, Venkat Venkatasubramanian and Gintaras V. Reklaitis, Chemical Engineering, Purdue University, West Lafayette, IN

The manufacture of pharmaceutical products has in recent times started shifting from a batch processing paradigm to a continuous manufacturing one. Motivations for this shift include lower cost and time of manufacturing for drug products, the availability of new online sensing and instrumentation technologies, and competition from generic drug product manufacturers. Continuous manufacturing of pharmaceutical products also entails certain challenges in real-time process management. One set of challenges arises from the use of bulk powders, gels, suspensions, and other media that show non-ideal behaviors. Processing such media requires novel approaches in process control techniques. A second set of challenges arise in integrating and managing a very large amount of knowledge about process conditions and product formulations, which is needed to flexibly and reliably scale processes and transfer knowhow. The nature and structure of such knowledge makes it ill-suited for DBMS systems, and required novel techniques in informatics.

In this work, we present results and case studies from the Engineering Research Center for Structure Organic Particulate Systems (ERC-SOPS), a collaborative effort between academia and industry in improving pharmaceutical manufacturing. We show our advances in real-time process management, including model predictive control, fault detection & diagnosis, real-time optimization, as well as the knowledge management techniques that enable the rapid deployment and scaleup of process management strategies. We present case studies in the manufacture of various dosage forms, focusing especially on tablets and drop-on-demand products. Finally we summarize the challenges still ahead and offer general conclusions for the operations and systems communities in enabling continuous manufacture for pharmaceutical products.

See more of this Session: Industrial Applications In Design and Operations
See more of this Group/Topical: Computing and Systems Technology Division

Optimal measurement selection for controlled variables for Kaibel Distillation Column

Ramprasad Yelchuru*
Sigurd Skogestad
Deeptanshu Dwivedi

*Presenting author

MIQP - Mixed Integer Quadratic Programming

NTNU
Norwegian University of Science and Technology

Outline

1. Problem formulation, $c = Hy$
2. Convex formulation (full H)
3. CVs with structural constraints(Structured H)
4. MIQP formulations
5. Kaibel Column Case study
6. Conclusions

NTNU
Norwegian University of Science and Technology

CV – Controlled Variables
MIQP - Mixed Integer Quadratic Programming

1. Problem Formulation

Assumptions:

(1) Active constraints are controlled
(2) Quadratic nature of economic cost function around optimum
(3) Active constraints remain same throughout the analysis

Optimal steady-state operation

$$\min_u J(u,d)$$

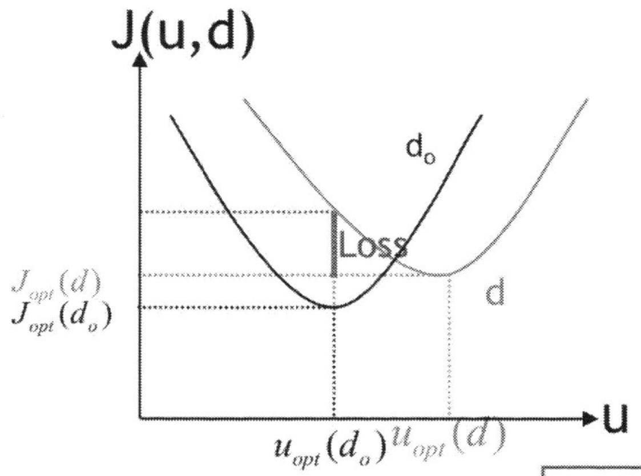

$$L = J(u,d) - \boxed{J_{opt}(u_{opt}(d),d)}$$

Real time optimization

$$J(u,d) = J(u_{opt},d) + J_u(u-u_{opt}) + \frac{1}{2}(u-u_{opt})^T J_{uu}(u-u_{opt}) + \zeta^3$$

NTNU
Norwegian University of
Science and Technology

1. Problem Formulation

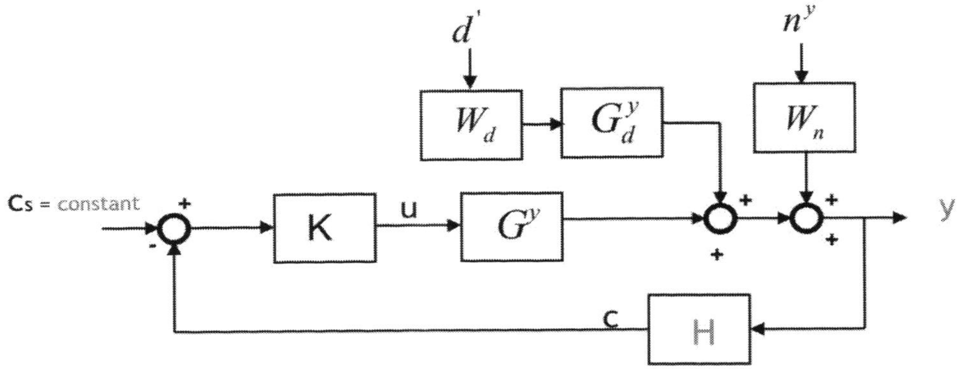

Loss $L = f(H, d', n^{y'})$

$$for \left\| \begin{bmatrix} d' \\ n^{y'} \end{bmatrix} \right\|_2 \leq 1$$

d′, ny′ as random variables

Controlled variables, $c = Hy$

$$L_{avg} = \left\| J_{uu}^{1/2} (HG^y)^{-1} HY \right\|_F^2$$

$$Y = [FW_d \quad W_n]$$

$$F = \frac{\partial y^{opt}}{\partial d} = G^y J_{uu}^{-1} J_{ud} - G_d^y$$

Ref: Halvorsen et al. I&ECR, 2003
Kariwala et al. I&ECR, 2008

NTNU
Norwegian University of Science and Technology

www.ntnu.no Ramprasad Yelchuru, Optimal controlled variable selection with structural constraints, 4

2. Convex formulation (full H)

$$\min_{H} \left\| J_{uu}^{1/2} (HG^{y})^{-1} HY \right\|_F$$

Seemingly
Non-convex
optimization problem

D : any non-singular matrix

$$H_1 = DH$$

$$(H_1 G_y)^{-1} H_1 = (DHG_y)^{-1} DH = (HG_y)^{-1} D^{-1} DH = (HG_y)^{-1} H$$

Objective function unaffected by D.
So can choose HG^y freely.

We made H unique by adding a constraint as $HG^y = J_{uu}^{1/2}$

$$\min_{H} \| HY \|_F$$

subject to $HG^y = J_{uu}^{1/2}$

Problem is convex in decision matrix H

Full H
Convex
optimization problem

Global solution

NTNU
Norwegian University of
Science and Technology

Vectorization

$$\min_{H} \|HY\|_F$$

subject to $HG^y = J_{uu}^{1/2}$

$$H = \begin{bmatrix} x_1 & x_2 & \cdots & x_{ny} \\ x_{ny+1} & x_{ny+2} & \cdots & x_{2*ny} \\ \vdots & \vdots & \ddots & \vdots \\ x_{(nc-1)*ny+1} & x_{(nc-1)*ny+2} & \cdots & x_{nc*ny} \end{bmatrix}_{nu \times ny}$$

$$X \square H^T$$

$$X = \begin{bmatrix} x_1 & x_{ny+1} & \cdots & x_{(nc-1)*ny+1} \\ x_2 & x_{ny+2} & \cdots & x_{(nc-1)*ny+2} \\ \vdots & \vdots & \ddots & \vdots \\ x_{ny} & x_{2*ny} & \cdots & x_{nc*ny} \end{bmatrix}_{ny \times nu}$$

$$X_\delta = \begin{bmatrix} x_1 \\ x_2 \\ \vdots \\ x_{nu*ny} \end{bmatrix}_{(nu*ny)\times 1}$$

$$\min_{X_\delta} \quad X_\delta^T Y_\delta Y_\delta^T X_\delta$$

$$st. \quad G_\delta^T X_\delta = J_\delta$$

Problem is convex QP in decision vector X_δ

NTNU
Norwegian University of
Science and Technology

Controlled variable selection

Optimization problem :

$$\min_H \left\| J_{uu}^{1/2}(HG^y)^{-1}HY \right\|_F \Longleftrightarrow \begin{array}{l} \min\limits_{H} \|HY\|_F \\ \text{st. } HG^y = J_{uu}^{1/2} \end{array} \Longleftrightarrow \begin{array}{ll} \min\limits_{X_\delta} & X_\delta^T Y_\delta Y_\delta^T X_\delta \\ st. & G_\delta^T X_\delta = J_\delta \end{array}$$

Minimize the average loss by selecting H and CVs as

(i) best individual measurements

(ii) best combinations of all measurements

(iii) best combinations with few measurements

Full H

Measurements from different sections

(i) best individual measurements

(ii) best combinations of all measurements

(iii) best combinations with few measurements

NTNU
Norwegian University of Science and Technology

Kaibel Column : Full H

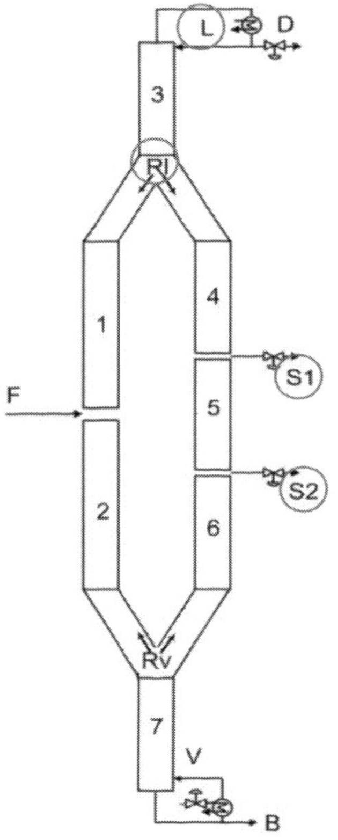

Kaibel column

c = Hy

$$c = \begin{bmatrix} c_1 \\ c_2 \\ c_3 \\ c_4 \end{bmatrix} \qquad y = \begin{bmatrix} T_1 \\ T_2 \\ \vdots \\ T_{71} \end{bmatrix}$$

$$c_1 = h_{1,1}T_1 + h_{1,2}T_2 + \cdots + h_{1,71}T_{71}$$
$$c_2 = h_{2,1}T_1 + h_{2,2}T_2 + \cdots + h_{2,71}T_{71}$$
$$c_3 = h_{3,1}T_1 + h_{3,2}T_2 + \cdots + h_{3,71}T_{71}$$
$$c_4 = h_{4,1}T_1 + h_{4,2}T_2 + \cdots + h_{4,71}T_{71}$$

$$H = \begin{bmatrix} h_{1,1} & h_{1,2} & \cdots & h_{1,20} & \cdots & h_{1,30} & \cdots & h_{1,71} \\ h_{2,1} & h_{2,2} & \cdots & h_{2,20} & \cdots & h_{2,30} & \cdots & h_{2,71} \\ h_{3,1} & h_{3,2} & \cdots & h_{3,20} & \cdots & h_{3,30} & \cdots & h_{3,71} \\ h_{4,1} & h_{4,2} & \cdots & h_{4,20} & \cdots & h_{4,30} & \cdots & h_{4,71} \end{bmatrix}$$

Find H that minimizes

$$L_{avg} = \left\| J_{uu}^{1/2} (HG^y)^{-1} HY \right\|_F$$

NTNU
Norwegian University of
Science and Technology

4. MIQP Formulation (full H)

$$H = \begin{array}{c} \begin{matrix} \sigma_1 & \sigma_2 & \cdots & \sigma_{ny} \end{matrix} \\ \begin{bmatrix} x_1 & x_2 & \cdots & x_{ny} \\ x_{ny+1} & x_{ny+2} & \cdots & x_{2*ny} \\ \vdots & \vdots & \ddots & \vdots \\ x_{(nc-1)*ny+1} & x_{(nc-1)*ny+2} & \cdots & x_{nc*ny} \end{bmatrix}_{nu \times ny} \end{array}$$

$$\sigma_i \in \{0,1\} \quad i = 1, 2, \cdots, ny$$

$$x_{aug} = \begin{bmatrix} X_\delta \\ \sigma_1 \\ \sigma_2 \\ \vdots \\ \sigma_{ny} \end{bmatrix}_{(nu*ny+ny)\times 1}$$

$$\min_{x_{aug}} \quad \left(x_{aug}^T F x_{aug} \right)$$

$$st. \quad G_{new}^{y^T} x_{aug} = J_\delta$$

$$p x_{aug} = n$$

Big M approach

$$\left\{ \begin{array}{c} \begin{bmatrix} -M & 0 & 0 & \cdots \\ 0 & -M & 0 & \cdots \\ \vdots & \vdots & \ddots & \ddots \\ 0 & 0 & \cdots & -M \end{bmatrix} \sigma_i \le \begin{bmatrix} x_i \\ x_{ny+i} \\ \vdots \\ x_{(nu-1)*ny+i} \end{bmatrix} \le \begin{bmatrix} M & 0 & 0 & \cdots \\ 0 & M & 0 & \cdots \\ \vdots & \vdots & \ddots & \ddots \\ 0 & 0 & \cdots & M \end{bmatrix} \sigma_i \\ \sigma_i \in \{0,1\} \end{array} \right\} for\, i = 1, 2, \cdots, ny$$

$$F = [Y_\delta Y_\delta^T \;\; zeros(ny, ny)]$$

$$G_{new}^{y^T} = [G_\delta^T \;\; zeros(nu*ny, ny)]$$

$$p = [zeros(1, nu*ny) \;\; ones(1, ny)]$$

$$upper\,bound\;\; for\, M = \max(J_\delta) / \min(G^y)$$

We solve this MIQP for n = n_u to n_y

NTNU
Norwegian University of Science and Technology

5. Case Study : Kaibel Column

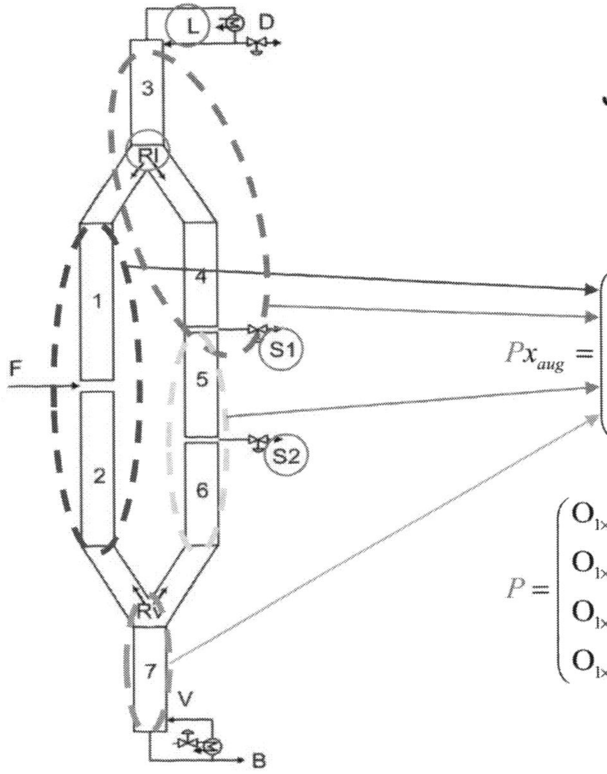

Kaibel column

J = impurities in product streams

$$Px_{aug} = \begin{pmatrix} 2 \\ 2 \\ 2 \\ 2 \end{pmatrix};$$

$$P = \begin{pmatrix} O_{1\times(nu*ny)} & ones(20,1) & 0 & 0 & 0 \\ O_{1\times(nu*ny)} & 0 & ones(20,1) & 0 & 0 \\ O_{1\times(nu*ny)} & 0 & 0 & ones(20,1) & 0 \\ O_{1\times(nu*ny)} & 0 & 0 & 0 & ones(11,1) \end{pmatrix}$$

NTNU
Norwegian University of
Science and Technology

5. Case Study : Kaibel Column

$$L_{avg} = \frac{1}{2}\left\|(J_{uu}^{1/2}(HG^{y})^{-1}HY)\right\|_F^{2}$$

$$Y = [FW_d \quad W_n]$$
$$F = G^{y}J_{uu}^{-1}J_{ud} - G_d^{y}$$

Data

$$G^{y} \in \mathbb{R}^{71\times 4}; G_d^{y} \in \mathbb{R}^{71\times 7}$$

$$J_{uu} \in S_{+}^{4}; J_{ud} \in \mathbb{R}^{4\times 7}$$

$$W_d \in \mathbb{R}^{7\times 7}; W_n \in \mathbb{R}^{71\times 71}$$

NTNU
Norwegian University of
Science and Technology

5. Kaibel Column : Results

No. Meas	Optimal Meas	Loss	CPU time (min)
4			
5	13 41 51 65 71	0.4607	180
6	12 39 41 51 54 65	0.0728	180

NTNU
Norwegian University of
Science and Technology

6. Conclusions

Using steady state economics of the total plant, the optimal controlled variables are obtained as

- optimal individual measurements
- combinations of optimal fewer measurements
- combinations of measurements from different sections

using MIQP based formulations.

The computational time required for CVs as combinations of measurements from different sections is less as the alternatives are lesser.

NTNU
Norwegian University of
Science and Technology

Thank You

NTNU
Norwegian University of
Science and Technology

(652f) A Comparative Study of Nonlinear CDU Models for Refinery Planning Optimization

Abdulrahman M. Alattas, Chemical Engineering, Carnegie Mellon University, Pittsburgh, PA, Ignacio E. Grossmann, Chemical Engineering Department, Carnegie Mellon University, Pittsburgh, PA and Ignasi Palou-Rivera, Argonne National Laboratory, Argonne, IL

Historically, the petroleum industry has used linear programming (LP) to address its planning and optimization needs (Favennec, 2001; Li et al 2005). The CDU yield prediction is modeled using linear functions of the crude feed. The linear function can be the simple fixed-yield equation or the enhanced swing cut equation (Zhang, 2001, Trierwiler, & Tan, 2001). The latter approach is currently used in many refineries. The simplicity, robustness and convenience of the linear approach are tradeoffs for the true optimal and accurate solution to the planning model. Attempts to add nonlinearity to CDU yield prediction in the planning models focused on fitting parameters of different crude oil feedstocks in addition to parameters for the operating variables (Alhajri et al, 2008, Lopez et al, 2009, Gueddar et al, 2009).

The goal of our work is to incorporate and evaluate nonlinear modeling approaches for CDU yield prediction. Our proposed CDU models range from the simple fractionation index (FI) model, to the moderate aggregate distillation model to the more rigorous tray-by-tray distillation model. These models are designed so they are fit for the optimization model of refinery planning by incorporating the appropriate simplifying assumptions and initialization procedure. The models are lastly evaluated to assess their impact and benefit in terms of accuracy, robustness and simplicity and computation effort.

The FI model is our first and simplest CDU model. The model employs the fractionation index defined by Geddes (Jakob, 1971). The CDU is represented by a series of separation units where the bottom products are the CDU product streams and the top products are fed to the next stage. Each stage will have two FI parameters representing the stripping and the rectifying sections of the separation unit (Wagner, 1978; Gilbert, 1966). The model is an equivalent short-cut approach to calculate the component distribution in the top and bottom product streams of each unit. The model lacks more details on the operating conditions, but its benefits include calculating the cut point temperatures while using column-characteristic parameters.

The next complex model, the aggregate model, decomposes the CDU into a set of simple and thermodynamically equivalent cascade of conventional distillation columns and steam distillation columns. The conventional distillation column uses reboilers as the energy-separating agent. On the other hand, steam replaces the reboiler in the steam distillation columns as a mass-separating agent. Steam distillation uses different mechanism for generating the vapor phase, and therefore, it displays unique characteristics from the more common conventional distillation (Suphanit, 1999). The conventional distillation column can be modeled using an aggregate model approach based on the work of Caballero & Grossmann (1999) for the synthesis of distillation columns. The principle of their approach is to treat the column sections above and below the feed tray as two integrated heat and mass exchangers. On the other hand, the unique characteristics of steam distillation require modifying the aggregate model of the conventional distillation. The modified aggregate model for steam distillation is significantly more difficult and requires more steps in the initialization phase for the model to converge. The results of the model successfully predict the inverse temperature profile in the stripping section of steam distillation columns.

The last CDU model in our comparison is the rigorous tray-by-tray calculations. The CDU decomposition used for the aggregate model is extended to the tray-by-tray model to simplify the calculations. Careful initialization and simplification assumptions are used to ensure convergence. The results of the three proposed models are presented and compared to CDU simulation model.

The proposed nonlinear CDU models, FI model, aggregate model and tray-by-tray model, are separately integrated into a refinery production planning model. The results of the new NLP production planning model are compared to each others and with the current LP models. The benefits of introducing the NLP models and the level of NLP complexity are assessed in terms of accuracy, robustness and simplicity and computation effort. As will be shown, the main advantage of the proposed nonlinear models is their potential of predicting higher profit in refinery planning operation.

References

(1) Alhajri, I., Elkamel, A., Albahri, T., & Douglas, P. L. (2008). A nonlinear programming model for refinery planning and optimisation with rigorous process models and product quality specifications. *Int. J. Oil, Gas and Coal Technology, 1*(3), 283-307.

(2) Caballero, J. A., & Grossmann, I. E. (1999). Aggregated Models for Integrated Distillation Systems. *Ind. Eng. Chem. Res., 38*(6), 2330-2344.

(3) Favennec, J.-P. (2001). *Refinery Operation and Management.* Paris: Editions TECHNIP.

(4) Gilbert, R. J. H., Leather, J., & Ellis, J. F. G. (1966). The application of the Geddes fractionation index to crude distillation units. *AIChE Journal, 12*(3), 432-437.

(5) Gueddar, T., & Dua, V. (2009). *Artificial Neural Networks Approach for Refinery-Wide Optimisation.* Proceedings from Centre for Process Systems Engineering - 20th Industrial Consortium Meeting.

(6) Jakob, R.R. (1971). Estimate number of crude trays. *Hydrocarbon Processing, 50(5), 149-152.*

(7) Li, W., Hui, C.W., & Li, A.X. (2005). Integrated CDU, FCC and product blending models into refinery planning. *Computers and Chemical* Engineering, *29(9),* 2010-2028.

(8) López, D.-C., Mahecha, C.-A., Hoyos, L.-J., Acevedo, L., & Villamizar, J.-F. (2009). Optimization Model Of A System Of Crude Oil Distillation Units With Heat Integration And Metamodeling. *C.T.F Cienc. Tecnol. Futuro, 3*(5), 159-173.

(9) Suphanit, B. (1999). *The Design of Complex Distillation Systems. UMIST*. Manchester.

(10) Trierwiler, D.; Tan, R.L. (2001) Advances in crude oil LP modeling. *Hydrocarbon Asia, 8, 52-58*

(11) Wagner, H. (1978) Thermodynamic design of multi-component rectifications. *International Chemical Engineering, 18 (3), 391-394.*

(12) Zhang, J.; Zhu, X.X.; Towler, G.P. (2001). A simultaneous optimization strategy for overall integration in refinery planning. *Industrial & Engineering Chemistry Research, 40, 2640-2653.*

See more of this Session: Industrial Applications In Design and Operations
See more of this Group/Topical: Computing and Systems Technology Division

(652g) Optimal Chlorine Booster Station Placement for Contamination Response In Water Distribution Systems Using Stochastic Programming

Gabriel Hackebeil[1], Angelica V. Wong[1], Katherine A. Klise[2], William Hart[2] and Carl D. Laird[1], (1) Chemical Engineering, Texas A&M University, College Station, TX, (2)Sandia National Laboratories, Albuquerque, NM

Accidental or intentional contamination of water distribution systems by chemical and biological agents can pose significant health risks to consumers, as well as have social and economic impacts. Several researchers have proposed the development of early-warning detection systems based on an installed sensor network within the distribution system. However, detecting the presence of a contaminant is only a part of the overall contamination problem. Adequate emergency response mechanisms must also be developed in conjunction with these early-warning systems, which include detailed cleanup and control strategies to return the water distribution system to an operational state. We have been working actively with researchers at Sandia National Laboratories and engineers at PUB Singapore to develop an integrated optimal response management package for protection of water distribution systems based on a series of mixed-integer programming formulations.

Here, we present large-scale mixed-integer linear program (MILP) formulations for performing optimal emergency response techniques. Various techniques are studied such as performing source inversion, obtaining optimal locations for additional manual sampling, choosing optimal placement of booster stations for rapid distribution of useful chemical agents, and evaluating network flushing techniques for removing unwanted agents.

In previous work, we focused on the development of effective source inversion techniques, including mixed-integer formulations for optimal manual sampling. In this present research, we have developed a mixed integer linear programming formulation for optimal placement of chlorine booster stations considering uncertainty. This produces a large-scale stochastic programming formulation with discrete decisions for booster station placement, and hundreds of thousands of scenarios for different potential realizations of contamination locations and times. This formulation presents two significant challenges. First, calculating all of the contamination scenarios can be time consuming, and we have developed a novel water quality model that scales exceptionally well to large networks, allowing for efficient solution of large sets of scenarios. The accuracy and computational complexity of this model is compared against existing water quality simulation tools. Second, the solution of such a large-scale mixed-integer linear programming problem can be very difficult. In this particular problem formulation, and based on certain scenario parameters, we show that a large number of the scenarios can be collapsed into several blocks, yielding an exact mathematical formulation with a tractable number of explicit scenarios. These two techniques allow for efficient formulation and solution of this challenging problem.

These tools have been integrated within a software framework for optimal response management of contamination events. The effectiveness of this package is demonstrated on two real-world water distribution networks, including a large-scale network model with approximately 13,000 nodes.

Sandia National Laboratories is a multi-program laboratory operated by Sandia Corporation, a wholly owned subsidiary of Lockheed Martin Corporation, for the U.S. Department of Energy's National Nuclear Security Administration under contract DE-AC04-94AL85000.

See more of this Session: Industrial Applications In Design and Operations
See more of this Group/Topical: Computing and Systems Technology Division

(669a) Cause and Effect Dynamic Modeling of Real Processes Under Freely Existing Data Collection

Derrick K. Rollins Sr.[1], **Stephanie Loveland**[2], Peggy Lee[2] and Yiying Khor[1], (1)Chemical and Biological Engineering, Iowa State University, Ame, IA, (2)Iowa State University, Ames, IA

Due to continual advancements of sensor technology, computertechnology and electronic data storage, the number of variables and theirfrequency of sampling and storage are growing at tremendous rates. While theplant data bases utilizing this growing technology have proven invaluable fordiagnostic and health monitoring, there are two critical reasons that they havenot been widely exploited in the development of *cause and effect* modeling which would allow direct implementationinto manual or automatic model-based control strategies. The first reason isthe low signal-to-noise ratio since the data arecollected under operating conditions where the objective is to minimize processvariability. The second reason is the highly correlated natural behavior of thevariables that is called multicollinearity thatimpedes the ability to associate a specific change in one variable with aspecific amount of change in the response of interest. To overcome thesecritical challenges, modelers rely on the use of experimental design where anintelligent sequence of changes are made to the inputs to maximize informationcontent both by the size of the changes and type of changes. However, there aretwo major drawbacks of this approach. The first one is the cost to run theexperiments due to production interruption; which often leads to unusableproducts since the changes have to be outside of normal operating regions. Thesecond one is the limited duration of the plant test because cost increaseswith duration. Limited duration impacts the ability to fully include thecyclical nature of unmeasured disturbances in model development. Thislimitation results in poor model performance as these disturbances move awayfrom the conditions when the model was developed.

To overcome these challenges, thecontinuous-time Wiener block-oriented modeling approach developed by Bḥandariand Rollins (2003) is extended to discrete-time modeling and input correlationis minimized by passing each input through a separate dynamic block and using astatic linear function on the outputs from these blocks to maintain accuracyfor mild extrapolation. By modeling over a sufficiently long period which ispossible from archived data, information content is strengthened and the impactof unmeasured disturbances on measurement bias is minimized.

This talkwill present results of this approach for a model of top tray temperature of apilot distillation column on at least 10 independent runs over a three yearperiod. The model was built from nine inputs, thatwere highly correlated in some cases, under open loop conditions. The fittedcorrelation coefficient (r_{fit})for training and validation were 0.96 and 0.97, respectively. For the test datasets, which were run under closed-loop control, r_{fit} ranged from 0.61to 0.93, with an average greater than 0.8, supporting the ability of thisapproach to develop accurate models with long term stability from data under adifferent correlation structure. In addition, a method based on principalcomponent analysis (PCA) is presented for elimination of cases representingextreme extrapolation.

Bhandari, N. and D. K. Rollins, A Continuous-Time MIMO Wiener ModelingMethod,Industrial and Engineering ChemistryResearch, **42**(22), pp. 5583-5595(2003).

See more of this Session: Process Modeling and Identification
See more of this Group/Topical: Computing and Systems Technology Division

(669b) Adaptive Data-Based Model Predictive Control of Batch Systems

Siam Aumi and Prashant Mhaskar, Chemical Engineering, McMaster University, Hamilton, ON, Canada

Batch (and fed-batch) processes have found widespread applications in a variety of sectors in the chemical industry, specifically those with an emphasis on quality control. The primary control objective in batch processes is to reach a target product quality, which may not be an equilibrium point, by batch termination. However, key characteristics of batch processes, such as the absence of measurements related to the final end-use quality coupled with strong nonlinear and time-varying dynamics over a wide range of operating conditions, complicate the control problem and limit the control performance achievable by implementing controllers designed for continuous processes (characterized by control at a steady-state and small variations in operating conditions). The industry response to these issues has been to operate batch processes with an entirely open-loop control policy (with incorporation of process knowledge and/or history) or indirectly pursue the primary control objective through process variable trajectory tracking approaches.

In trajectory tracking methods, trajectories for a set of measurable process variables related to the end-use quality are generated off-line or re-calculated at specific time points during the batch by solving a dynamic optimization problem. These trajectories are subsequently tracked using local model-based controllers with the most common control strategy being model predictive control (MPC). MPC performance is contingent on the accuracy of the underlying model's predictions. When available, a deterministic state-space model with well identified parameters can be used; however, there are situations, particularly in industry, when a good first-principles model is unavailable. This has motivated the use of data-based or empirical models, particularly those identified via latent variable tools (e.g. see [1,2]) for formulating computationally tractable MPC designs. For a batch system, the available 'training' data for building an empirical model is essentially limited to historical databases (possibly appended with one or two identification experiments) of previous batches. Further complicating the model identification task is the fact that batch process dynamics are highly nonlinear and time-varying, rendering conventional system identification approaches, where a single input-output linear model is identified around one operating condition, ill-suited for identifying an accurate model. One general strategy to describe nonlinear behavior while retaining the simplicity of linear models is to partition/cluster the training data into a number of different regions, identify local linear models for each region, and combine them with appropriate weights to describe the global nonlinear behavior. This idea was formalized in our previous work [3,4] that unified the concepts of auto-regressive exogenous (ARX) modeling, latent variable regression techniques, fuzzy c-means clustering, and multiple local linear models in an integrated framework capable of capturing the nonlinearities and multivariate nature of batch data. The key delineating aspects of the work in [3, 4] from existing multi-model approaches (e.g. see [2], [5], and [6]) were the clustering algorithm used to partition the training data, the use of latent variable tools to estimate the model parameters, and the derivation of a generalized continuous weighting function that is entirely data dependent and does not require precise process knowledge.

In this work, we generalize our previously developed modeling framework to account for time-varying dynamics by incorporating online learning ability into the model, making it adaptive. First, the standard recursive least squares (RLS) algorithm with a forgetting factor (to account for time-varying dynamics) is applied to update the model parameters. However, applying the standard RLS algorithm leads to a simultaneous or global update of all the models, which may be unnecessary depending on the operating conditions of the process. We address this issue by developing a probabilistic RLS (PRLS) estimator (also with a forgetting factor) for each model that takes the probability of the model being representative of the current plant dynamics into account in the update. The main advantage of adopting this local update approach is adaptation tuning flexibility. Specifically, the forgetting factor, which controls the speed and smoothness of adaptation, for each local PRLS estimator can be specified independently. More importantly, the model adaptations can be made more aggressive while maintaining better parameter precision (i.e., low variances) compared to the the standard RLS algorithm. The benefits from incorporating both RLS algorithms are demonstrated via simulations of a nylon-6,6 batch polymerization reactor. Closed-loop simulation results illustrate how using an adaptive model in a trajectory tracking predictive controller can substantially improve tracking performance (over the non-adaptive model based design). The adaptation of the model is also shown to be crucial for achieving acceptable control performance when encountering large disturbances in the initial conditions.

Reference

[1] Flores-Cerrillo J, MacGregor JF. Latent variable MPC for trajectory tracking in batch processes. J Process Control. 2005;15(6):651–663.

[2] Fletcher N, Morris A, Montague G, Martin E. Local Dynamic Partial Least Squares Approaches for the Modelling of Batch Processes. Can J of Chem Eng. 2008;86:960–970.

[3] Aumi S, Mhaskar P. Integrating Data-Based Modeling and Nonlinear Control Tools for Batch Process Control. AIChE J. 2011;(submitted).

[4] Aumi S, Mhaskar P. Integrating Data-Based Modeling and Nonlinear Control Tools for Batch Process Control, (accepted). In: Proc. of the American Control Conference (ACC). 2011; .

[5] Ferrari-trecate G, Muselli M, Liberati D, Morari M. A Clustering Technique for the Identification of Piecewise Affine Systems. Automatica. 2003;39:205–217.

[6] Takagi T, Sugeno M. Fuzzy Identification of Systems and Its Applications to Modeling and Control. IEEE Trans Syst Man Cybern. 1985;15(1):116–132.

See more of this Session: Process Modeling and Identification

Data Mining of Historic Data for Process Identification ★

Daniel Peretzki [a], Alf J. Isaksson [a,b], André Carvalho Bittencourt [a], Krister Forsman [c]

[a] *Linköping University, Department of Electrical Engineering, SE-581 83 Linköping, Sweden*

[b] *ABB AB, Corporate Research, SE-721 78 Västerås, Sweden*

[c] *Perstorp AB, SE-284 80 Perstorp, Sweden*

Abstract

Performing experiments for system identification is often a time-consuming task which may also interfere with the process operation. With memory prices going down, it is more and more common that years of process data are stored (without compression) in a history database. The rationale for this work is that in such stored data there must already be intervals informative enough for system identification. Therefore, the goal of this project was to find an algorithm that searches and marks intervals suitable for process identification (rather than performing completely automatic system identification). For each loop, 4 stored variables are required; setpoint, manipulated variable, process output and mode of the controller.

The proposed method requires a minimum of knowledge of the process and is implemented in a simple and efficient recursive algorithm. The essential features of the method are the search for excitation of the input and output, followed by the estimation of a Laguerre model combined with a chi-square test to check that at least one estimated parameter is statistically significant. The use of Laguerre models is crucial to handle processes with deadtime without explicit delay estimation. The method was tested on three years of data from more than 200 control loops. It was able to find all intervals in which known identification experiments were performed as well as many other useful intervals in closed/open loop operation.

Key words: Data Mining, Data Segmentation, System Identification, Excitation, Condition numbers, Laguerre filters

1 INTRODUCTION

In the process industry, models are relevant for different purposes, such as optimizing production, improving control performance and supervision. In many occasions the task of building a process model is complemented with an identification procedure, where the model parameters are identified from measured data. Performing dedicated experiments for system identification is often a time-consuming task which may also interfere with the process operation. Nowadays, it is common to store measurement data from the plant operation (without compression) in a history database. Such data is a very useful source of information about the plant, and might contain suitable data to perform process identification. Due to the size of such databases, searching for data intervals suitable for identification is a challenging task. Preferably, this task should be supported by a data scanning algorithm that automatically searches and marks data intervals of interest.

Relatively little can be found in the literature that directly addresses this problem. In [1], a data removal criterion is presented that uses the singular value decomposition (SVD) technique for discarding data which is only noise dependent and leads to a bigger mean square error (MSE) of the estimated model parameters. Horch introduced in [2] a method for finding transient parts of data after a setpoint change, specifically targetting the identification of time delay [3]. In [4], the authors discuss persistence of excitation for on-line identification of linear models but do not deal with finding intervals of data that are persistently exciting. Data mining techniques have been proposed to give a fully automated modeling and identification, based solely on data. In [5], the authors proposed a method to discover the topology of a chemical reaction network, whereas in [6] a method is proposed to find the dynamical model that generated the

★ Patent pending.

Email addresses: d.peretzki@gmx.net (Daniel Peretzki), alf.isaksson@liu.se (Alf J. Isaksson), andrecb@isy.liu.se (André Carvalho Bittencourt), krister.forsman@perstorp.com (Krister Forsman).

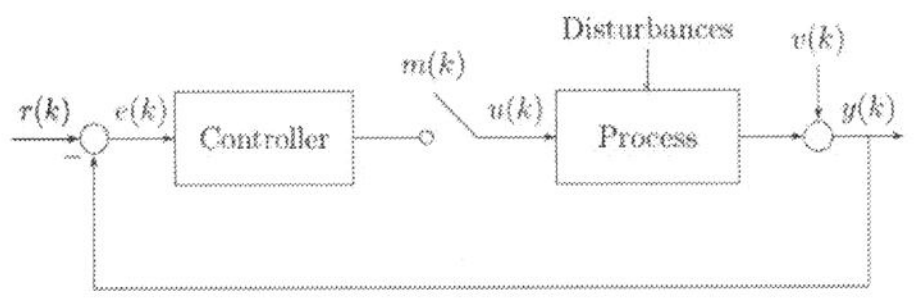

Fig. 1. Control loop.

data using symbolic regression. In [7], the authors consider the use of historical data to achieve process models for inferential control, some guidelines are suggested on how to select intervals of data to build models, but no algorithm is proposed with this objective.

Process plants have specific characteristics that make a fully automated modeling and identification a challenging task, see e.g. [8]. This work focuses on models that are suitable for the design/tuning of low order controllers like PI and PID. For this purpose, it is clear from the above considerations that a fully automated identification is challenging. Instead, **the objective** of this work is to develop a data mining algorithm that retrieves intervals of data from a historic database that are suitable for process identification. The method outputs the intervals together with a quality indicator. The user can then decide on the model and identification method to be used.

1.1 Problem Formulation

Consider the control loop in Fig. 1, at time k the operation mode $m(k)$ is either manual or automatic. In manual mode, the input to the process $u(k)$ is decided by the user. In automatic mode, $u(k)$ is given by the controller. The controller is driven by the control error $e(k)$, formed from the setpoint $r(k)$ subtracted by the measured process output $y(k)$, which is corrupted by noise $v(k)$. It is considered that the system can be described by an unknown model $\mathcal{M}(\theta)$, which is a function of the parameters θ.

A collection of data $Z^N = [Z(1)^T, \cdots, Z(N)^T]^T$ is available, where $Z(k) = [m(k),\ r(k),\ u(k),\ y(k)]$. The objective is to find time intervals $\Delta = [k_{\text{init}}, k_{\text{end}}]$ where the data in Z^N is suitable to perform identification of the process parameters.

For its practical use, the following characteristics are sought:

I **Minimal knowledge about the plant is required.** That is, none (or little) input is expected from the user.

II **The resulting algorithm should process the data quickly.** For example, a database containing data from a month of a large scale plant operation should not take longer than a few minutes to be processed.

III **For each interval found, a numeric measure of its quality should be given.** This can be used by the user in order to select which intervals to use for identification.

In order to achieve both I and II, some simplifying assumptions are taken:

Assumption 1.1 (SISO) *The complex interconnections present in a plant are disregarded and it is assumed that only SISO control loops are to be estimated.*

Assumption 1.2 (Linear models) *It is assumed that the process can be well described by a linear model $\mathcal{M}(\theta)$.*

For a process in operation there are mainly two scenarios to hope for that may result in data informative enough for system identification (see [9,2]):

- The process is operating in manual mode and the input signal $u(k)$ is varied enough to be exciting the process.
- The controller is in automatic and there are enough changes in the setpoint $r(k)$ to make identification possible.

As a consequence, the method developed below is treating these two cases separately.

2 SYSTEM IDENTIFICATION PRELIMINARIES

Consider first the case of open loop operation. Even under Assumptions 1.1 and 1.2, there are many model structures and numerical identification methods possible. Since we have a clear requirement on low computational complexity it is natural to focus on model structures based on linear regression:

$$y(k) = \varphi^T(\bar{Z}^{k-1})\,\theta + v(k), \qquad (1)$$

where the data $\bar{Z}^{k-1}$ contains past inputs $u(k-1), \ldots, u(1)$ and outputs $y(k-1), \ldots, y(1)$. The vector φ is a regressor and its choice defines the model structure $\mathcal{M}$ of the process. The n dimensional vector θ contains the unknown parameters and $v(k)$ is white noise with variance γ.

A common choice of identification approach is the prediction error method, where the prediction error $\varepsilon(k, \theta) = y(k) - \varphi^T(\bar{Z}^{k-1})\theta$ is minimized according to some criterion. Using a least squares criterion, the estimate is given by

$$\hat{\theta}_N = \arg\min_{\theta} \frac{1}{N} \sum_{k=1}^{N} \left[y(k) - \varphi^T\left(\bar{Z}^{k-1}\right)\theta \right]^2, \qquad (2)$$

which has the solution, [9],

$$\hat{\theta}_N = \hat{R}_N^{-1} \frac{1}{N} \sum_{k=1}^{N} \varphi(\bar{Z}^{k-1}) y(k), \tag{3}$$

$$\hat{R}_N \triangleq \frac{1}{N} \sum_{k=1}^{N} \varphi(\bar{Z}^{k-1}) \varphi(\bar{Z}^{k-1}), \tag{4}$$

The feasibility of this solution depends on whether the *information matrix* $\hat{R}_N$ is invertible.

Assume that the true system is described by a linear regression with true parameters θ_0, and with a true noise variance γ_0. Then, as $N \to \infty$, the estimate $\hat{\theta}_N$ will be asymptotically normally distributed, [9]. More precisely

$$\sqrt{N}(\hat{\theta}_N - \theta_0) \in As\mathcal{N}(0, P), \quad P \triangleq \gamma_0 \left[\lim_{N \to \infty} \hat{R}_N \right]^{-1}. \tag{5}$$

This means that for finite number of data, N,

$$\hat{\theta}_N \sim \mathcal{N}(\theta_0, P_N), \tag{6}$$

where an estimate of the covariance matrix $\hat{P}_N$ is given by

$$\hat{P}_N = \frac{1}{N} \hat{\gamma}_N \, [\hat{R}_N]^{-1}, \quad \hat{\gamma}_N = \frac{1}{N} \sum_{k=1}^{N} \varepsilon^2(k, \hat{\theta}_N). \tag{7}$$

The matrix $\hat{R}_N$ therefore determines the quality of the estimate $\hat{\theta}_N$. Notice that $\hat{R}_N$ is a function of the data Z^N as well as the model structure $\mathcal{M}$. In order to make the covariance small, $\hat{R}_N$ should be made large in some sense. This idea is explored, for instance, in experiment design (choosing u) for system identification. A data set Z^N is therefore suitable for identification of the model structure $\mathcal{M}$ if it is such that the matrix $\hat{R}_N$ is large. Another important piece of information in (5) is the size of $\hat{\gamma}_N$, i.e. how small the optimal prediction errors are. To test how informative an interval of data is in Sec. 3 we define quantities that relate to these measures of identification quality.

However, before defining test quantities, two other relevant issues should be addressed. The first is related to the solution of $\hat{\theta}_N$ and the related quantities. A solution based on explicitly forming the inverse as in (3) is not numerically well conditioned. In the proposed method this is overcome with a numerical solution based on QR factorization, which is presented in Sec. 2.1. The second issue is related to the fact that any condition based on $\hat{R}_N$ demands knowledge of the model structure $\mathcal{M}$, which is assumed to be unknown by the Requirement I. An idea to circumvent this is to use a model structure flexible enough to explain the input-output relation of a large variety of processes. Such issues are addressed in Sec. 2.2.

2.1 *Solution to the Least Squares Problem based on QR factorization*

An equivalent formulation of (2) is

$$\hat{\theta}_N = \arg\min_{\theta} \|Y - \Phi\theta\|_2^2, \tag{8}$$

where

$$Y^T = \left[y^T(1) \cdots y^T(N) \right], \Phi^T = \left[\varphi(1) \cdots \varphi(N) \right]. \tag{9}$$

Because the norm in the minimization is not affected by an orthonormal transformation, apply the QR factorization as $[\Phi \; Y] = QR$, where Q is orthonormal, $QQ^T = I$. R is a matrix of the form

$$R = \begin{bmatrix} R_0 \\ \cdots \\ 0 \end{bmatrix}, \quad R_0 = \begin{bmatrix} R_1 & R_2 \\ 0 & R_3 \end{bmatrix}, \tag{10}$$

where the matrix R_0 is square upper triangular of dimension $n+1$ and R_3 is a scalar. Applying the orthonormal transformation Q^T from the left in (8), then

$$\|Y - \Phi\theta\|_2^2 = \left\|Q^T (Y - \Phi\theta)\right\|_2^2 = \tag{11}$$

$$\left\| \begin{bmatrix} R_2 \\ R_3 \end{bmatrix} - \begin{bmatrix} R_1\theta \\ 0 \end{bmatrix} \right\|_2^2 = \|R_2 - R_1\theta\|_2^2 + |R_3|^2, \tag{12}$$

Hence, the $\hat{\theta}_N$ which minimizes (12) is given by the solution to

$$R_1 \hat{\theta}_N = R_2, \tag{13}$$

Since R_1 is upper triangular, solving for $\hat{\theta}_N$ in (13) is easier and better numerically conditioned than using (3). Similarly,

$$\hat{\gamma}_N = \frac{1}{N} \|Y - \Phi\hat{\theta}_N\|_2^2 = \frac{1}{N} |R_3|^2. \tag{14}$$

Furthermore

$$\hat{R}_N = \frac{1}{N} \Phi^T \Phi = \frac{1}{N} R_1^T R_1. \tag{15}$$

2.2 *A Flexible Model Structure*

Even restricting to linear regression, there are many potential model structures to choose from. A common choice is the so-called ARX model structure leading to regressors φ containing sequences of past inputs $u(k-1-d), \ldots, u(k-n_u-d)$ and outputs $y(k-1), \ldots, y(k-n_y)$, where d corresponds to the process delay and n_u, n_y are the model orders. For processes with a nonzero delay (commonly found in the process industry), the delay d needs to be known, in order to form the regressor, which is not the case. Since deadtime estimation is in itself a complicated matter (see e.g. [10]), alternative model structures are considered.

Another appealing model structure is Finite Impulse Response (FIR) models, resulting in regressors only containing lagged values of the input u. In the presence of deadtime an FIR model will lead to close to zero estimates of the leading parameters. The drawback with FIR is that a process with slow dynamics requires very many parameters leading to a very large, and impractical, size of $\hat{R}_N$.

Luckily, there are model structures available that combine the advantages of ARX and FIR models, without suffering from their drawbacks. One such model structure is the Laguerre model:

$$y(k) = \sum_{i=1}^{n} \theta_i \, L_i(q, \alpha) u(k) \tag{16}$$

where $L_i(q, \alpha)$ is the Laguerre filter

$$L_i(q, \alpha) = \frac{\sqrt{1-\alpha^2}}{q-\alpha} \left(\frac{1-\alpha q}{q-\alpha} \right)^{i-1}. \tag{17}$$

Hence, the Laguerre filter of order i consists of one low-pass filter cascaded with $(i-1)$ first-order all pass filters, which acts effectively as a delay approximation of order $i-1$. The substitution of the delay operator with a Laguerre filter has important characteristics that makes it a more suitable choice than an FIR model [11].

The parameter α determines the transient response of the low pass filter. It controls the settling time of the first Laguerre output $L_1(q, \alpha)$ and should be set as equal to the largest time constant in the system [12]. The maximum delay $\bar{d}$ a Laguerre model can explain can be found by comparing a Padé approximation of a delay with the all pass part of the Laguerre filters [10,13] and is given by

$$\bar{d} = -2(n-1)T_s / \log \alpha \tag{18}$$

where T_s is the sampling interval. If the real pole α and order n are selected properly, then the Laguerre model can efficiently approximate a large class of linear systems [2]. In general, the performance of the identification is relatively insensitive to the choice of α [14].

Plants with an integrator

Because a Laguerre model has a finite gain at low frequencies, it is not a good approximation for plants with an integrator. This is in fact an important limitation of Laguerre models since many processes in process industry (most notably level control) have an integrating behavior. To overcome this, it is *assumed known* whether a plant has an integrator or not. For integrating processes, a Laguerre model between the integrated input $\bar{u}(k)$ and output sequences is considered instead:

$$y(k) = \sum_{i=1}^{n} \theta_i L_i(q, \alpha) \bar{u}(k), \quad \bar{u}(k) = \frac{u(k)}{1-q^{-1}}. \tag{19}$$

3 DATA FEATURES FOR PROCESS IDENTIFICATION

Based on the previous section, the testing of three data features with increasing computational complexity and theoretical justification are presented below.

3.1 Variability in the data

The first test of potential excitation is to check that there is an actvitity in the input and output signals at all. An empirical and simple solution is therefore to monitor the signals' variability over time k. Changes in the signal variances can be used with this purpose.

3.2 Numerical Conditioning of $\hat{R}_N$

A more theoretically based approach is to monitor $\hat{R}_N$. From (3), the least squares problem solution is well posed only if $\hat{R}_N$ is invertible. A near singular matrix $\hat{R}_N$ might occur when the input data is not exciting enough to fit a model of order n. The condition number, $\kappa(M)$ of a matrix M is defined as the ratio of the largest and smallest singular values of M. The accuracy with which the model parameters can be estimated are related to the condition number [1]. An information matrix with condition number close to 1 means that the least squares problem is numerically well-conditioned. If a QR algorithm is used, then the relationship $\hat{R}_N = \Phi^T \Phi = R_1^T R_1$ gives

$$\kappa(\hat{R}_N) = \kappa(R_1)^2, \tag{20}$$

This is easily seen by taking the SVD of R_1 and comparing the singular values of R_1 and the ones of $\hat{R}_N = R_1^T R_1$.

3.3 Statistical Significance of $\hat{\theta}_N$

The first two tests proposed above do not consider the actual correlation between the input and output sequences, and whether it appears as it is possible to fit a linear model between them. A more conclusive test is to in fact compute $\hat{\theta}_N$ and check whether any of the estimated parameters are significantly non-zero.

According to the null hypothesis ($\theta_0 = 0$) and (5), the estimate $\hat{\theta}_N$ is asymptotically normal $\mathcal{N}(0, P)$, i.e.

$$\hat{\mathcal{X}}_N \triangleq \hat{\theta}_N^T \hat{P}_N^{-1} \hat{\theta}_N \in \mathcal{X}_n^2 \tag{21}$$

where the degrees of freedom n for the chi-square distribution is the dimension of the parameter vector θ. Hence checking whether $\hat{\theta}_N$ is non-zero, rejecting the null hypothesis, corresponds to comparing the quantity $\hat{\mathcal{X}}_N$ with the ones in a table of the chi-square distribution.

If the QR solution is used, then

$$\hat{\mathcal{X}}_N = \hat{\theta}_N^T \hat{P}_N^{-1} \hat{\theta}_N = [R_1^{-1} R_2]^T \frac{N}{\hat{\gamma}_N} \hat{R}_N [R_1^{-1} R_2]$$
$$= \frac{1}{\hat{\gamma}_N} R_2^T R_2 = \frac{N}{|R_3|^2} R_2^T R_2 = \left\| \frac{\sqrt{N}}{|R_3|} R_2 \right\|_2^2 .$$

Notice that the quantity $\hat{\mathcal{X}}_N$ behaves directly proportional to the statistical significance of the the parameters. In fact, this quantity can be used to compare the quality of different data sets (larger meaning better).

4 METHOD OUTLINE

At this point, it is possible to define a method to search for suitable data to perform process identification. Laguerre models are used between the input-output data and the idea is to monitor the data features described in the previous section to consider whether the data is relevant. The signals are first scaled and the operating points are removed since linear models are being considered.

The features discussed in Sec. 3 have very different computational complexity. In order to avoid excessive computations, the features are computed conditionally in a cascade of events as

```
Compute variances of u(k) and y(k), v_u(k) and v_y(k).
if v_u(k) and v_y(k) are large then
    Compute κ(R̂_N).
    if κ(R̂_N) is large then
        Compute X̂_N.
        if X̂_N is large then
            Mark data interval as useful.
        end if
    end if
end if
```

The ordering also considers that the variances check is more easily satisfied than the condition number, and that the statistical significance is the most demanding test.

Since we are interested in finding relevant changes in the data, it is natural that the features are computed recursively, over a window of data or in a forgetting filter scheme. An efficient implementation can be achieved with an exponential moving average (EMA). For instance, the variance of the output signal, $\mathsf{v}_y(k)$, can be estimated as

$$\hat{\mathsf{v}}_y(k) = \frac{2 - \lambda_{\mathsf{m},y}}{2} \Big[\lambda_{\mathsf{v},y} \left[y(k) - \mathsf{m}_y(k) \right]^2 + (1 - \lambda_{\mathsf{v},y})\, \mathsf{v}_y(k-1) \Big] \tag{22a}$$
$$\hat{\mathsf{m}}_y(k) = \lambda_{\mathsf{m},y}\, y(k) + (1 - \lambda_{\mathsf{m},y})\, \mathsf{m}_y(k-1) \tag{22b}$$

where m_y is the estimate of the mean and $0 < \lambda_{\mathsf{v},y}, \lambda_{\mathsf{m},y} < 1$ are tuning parameters which control the effective size of the window. A moving average is also used to update the information matrix recursively.

The algorithm flowchart for open-loop data is shown in Fig. 2, where $\mathbf{L}_k = [L_1(q,\alpha)u(k), \cdots, L_n(q,\alpha)u(k)]$, $U = [u(1), \cdots, u(N)]$ and η are thresholds. After loading/scaling the data and removing the operating points, the sample where the input is first changed, k_0 is searched to avoid unnecessary computations. If the plant is integrating, the input signal is integrated. Several quantities used are computed from k_0 to N; the Laguerre outputs $\mathbf{L}$, the regression matrix Φ and the variance estimates of the input and output (computed with an EMA).

The algorithm enters in a loop, searching for excitation in the data k_0 to N. Before any criteria are checked, it is required that $\mathsf{v}_{L_1}(k)$ exceeds a minimum threshold, indicating that excitation might have started. This sample is marked as a candidate for the start of an interval k_{init}. The first check for data excitation is then performed using the estimated variances. Only if passed, the information matrix $\hat{R}_k$ is update using an EMA, followed by checking if the condition number of $\hat{R}_k$ is small enough.

Finally, if all tests so far were successful, the statistical significance of the estimates are compute using QR factorization and the interval is marked up to the current sample, $\Delta = [k_{\text{init}}, k]$. If any test fails, the algorithm moves to the next sample and continues until the data is over. In case any useful data was found, the algorithm outputs the interval Δ and the value of $\hat{\mathcal{X}}_{k_{\text{init}}:k}$ as an estimate of the data quality.

There are a total of 11 design parameters: The order of the Laguerre model, n, and its pole α, the moving average filters coefficients $[\lambda_{L_1}, \lambda_{\mathsf{v},y}, \lambda_{\mathsf{m},y}, \lambda_R]$ and the

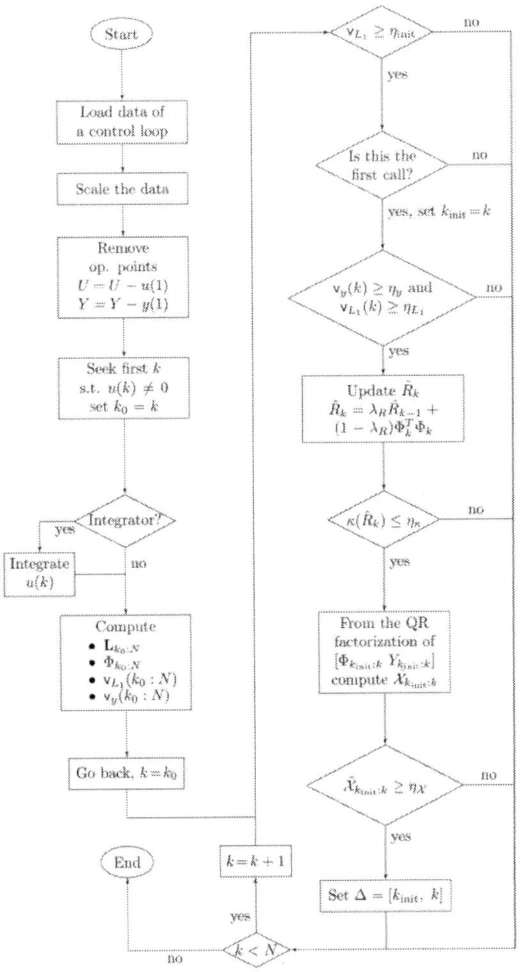

Fig. 2. Algorithm flowchart.

thresholds $[\eta_{\text{init}}, \eta_{L_1}, \eta_y, \eta_\kappa, \eta_{\mathcal{X}}]$. Notice that the same values of these design parameters are used, for any type of control loop or operational mode.

4.1 Closed Loop Data

For data collected in automatic mode, we first search for excitation in the setpoint, i.e. change u to r in the first 2 tests. Then, for the final statistical test, $\hat{\mathcal{X}}_\Delta$ is computed for a tentative model between u to y, because at the end of the day it is an input-out model we aim to identify.

5 TEST DATA EVALUATION

To test the developed method, historic data from a chemical plant was used. It contains data from 211 control

Loops where any Δ was found by mode	
closed loop	143 (67.7%)
open loop	185 (87.7%)
both	190 (90.1%)
Average length of Δ's found (samples) by mode	
closed loop	102.8
open loop	125.3
both	114.1
Average Δ's found by loop type	
Density	239
Flow	660
Concentration	84
Level	130
Conductivity	0
Temperature	35.3
Pressure	100

Table 1
Some quantities characterizing the performance of the method when applied to the test data.

loops of density, flow, concentration, level, conductivity, temperature and pressure types. The loops have considerably different dynamics but most of them can be modeled as a first order model and a delay, and it is not hard to tell beforehand which ones that will have an integrator. The delays can vary up to 10 min. The data is mainly from closed loop operation, but there is also open loop data. The database contains data of nearly 37 months of operation, sampled every 15s, in almost 1.1G samples and requires 6.7GB of memory to be stored.

The proposed method is applied to these data, taking approximately 1.5h to process it all. Table 1 summarizes the results. In total, about 1.5% of data was found to be useful for process identification.

The 3 year database contains stretches of data where we know that the control group at Perstorp conducted identification experiments, performed in manual mode with a sequence of steps in $u(k)$. All of these intervals and many others were found with the proposed algorithm. The models built using the data intervals selected by the new method were found very similar to those obtained during the identification experiments. The related $\hat{\mathcal{X}}_\Delta$ were also consistently large. Fig. 3 presents two examples of intervals found, in open/closed loop operation. It also illustrates the quantities used to infer the data quality to process identification.

6 CONCLUSIONS

From the quite extensive testing it seems that the method developed in this paper can successfully find intervals of data relevant for system identification. It requires minimal knowledge of the control loops, namely, whether the process is integrating or not. It is implemented efficiently in a recursive manner. A day's worth of data, from all loops in the test plant, takes less than

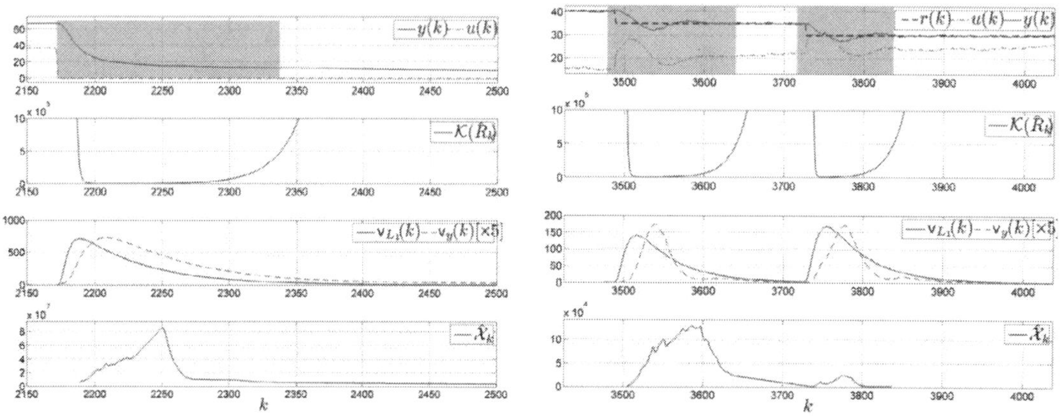

(a) Temperature, open loop.

(b) Density, closed loop

Fig. 3. The shaded regions are the found identification data intervals.

5s to be processed despite that we have not yet fully optimized the algorithm for computational speed.

The developed method is based on classical results from identification theory of linear systems. As an initial screening, it checks that input and output signals are varying at all. Then it forms an information matrix and checks its condition number. Finally, it estimates a provisional model using a Laguerre model structure. If the parameter estimates of this provisional model are found to be statistically non-zero the data interval is marked as potentially useful for system identification and a quality measure is also provided.

7 ACKNOWLEDGEMENTS

The authors gratefully acknowledge the financial support from the Swedish Foundation for Strategic Research (SSF) as part of the Process Industry Centre Linköping (PIC-LI). The authors would also like to thank Professor Lennart Ljung for encouraging and inspirational meetings early in the project.

References

[1] Carrette P, Bastin G, Genin Y, Gevers M. Discarding data may help in system identification. *IEEE Transactions on Signal Processing.* 1996;44(9):2300–2310.

[2] Horch A. Condition Monitoring of Control Loops. Ph.D. thesis, KTH, Signals, Sensors and Systems. 2000.

[3] Isaksson A, Horch A, Dumont G. Event-triggered deadtime estimation from closed-loop data. *American Control Conference, Arlington VA, June 25-27.* 2001;pp. 3280–3285.

[4] Green M, Moore J. Persistence of excitation in linear systems. *Systems & Control Letters.* 1986;7(5):351–360.

[5] Cho YJ, Ramakrishnan N, Cao Y. Reconstructing chemical reaction networks: data mining meets system identification. In: *Proceeding of the 14th ACM SIGKDD international conference on Knowledge discovery and data mining*, KDD '08. New York, NY, USA: ACM. 2008; pp. 142–150.

[6] Schmidt MD, Lipson H. Data-Mining Dynamical Systems: Automated Symbolic System Identification for Exploratory Analysis. *ASME Conference Proceedings.* 2008; 2008(48364):643–649.

[7] Amirthalingam R, Sung SW, Lee JH. Two-step procedure for data-based modeling for inferential control applications. *AIChE Journal.* 2000;46(10):1974–1988. URL `http://dx.doi.org/10.1002/aic.690461010`

[8] Ng YS, Srinivasan R. Data Mining for the Chemical Process Industry. In: *Encyclopedia of Data Warehousing and Mining*, edited by Wang J, pp. 458–464. IGI Global. 2009;.

[9] Ljung L. *System identification: Theory for the user.* Prentice-Hall Englewood Cliffs, NJ, 2nd ed. 1998.

[10] Björklund S, Ljung L. A review of time-delay estimation techniques. In: *Decision and Control, 2003. Proceedings. 42nd IEEE Conference on*, vol. 3. 2003; pp. 2502 – 2507 Vol.3.

[11] Wahlberg B. System Identification using Laguerre Models. *IEEE Trans on Automatic Control.* 1991;.

[12] Wang L, Cluett W. Building transfer function models from noisy step response data using the Laguerre network. *Chemical Engineering Science.* 1995;50(1):149–161.

[13] Isaksson M. A Comparison of Some Approaches to Time-Delay Estimation. *Master's Thesis ISRN LUTFD2/TFRT--5580--SE*, Department of Automatic Control, Lund University, Sweden. 1997.

[14] Park H, Sung S, Lee I, Lee J. On-line process identification using the Laguerre series for automatic tuning of the proportional-integral-derivative controller. *Ind Eng Chem Res.* 1997;36(1):101–111.

Calorimetric estimation employing Unscented Kalman Filtering for a batch emulsion polymerization reactor

Franklin D. Rincon[1], Marcelo Esposito[2], Galo A. C. Le Roux[1], Cláudia Sayer[2], Pedro H.H. de Araújo[2]

[1]Universidade de São Paulo, São Paulo, Brasil

[2]Universidade Federal de Santa Catarina, Santa Catarina, SC, Brasil

E-mail: pedro@enq.ufsc.br

Reaction calorimetry is a very useful tool to monitor exothermic polymerization reactions as it is based on the measurement of heat generated by the reaction, which can be then associated to the polymerization rate. Compared to other techniques, reaction calorimetry has the advantages of being a noninvasive technique, robust, fast (when compared to typical off-line characterization techniques) and low cost.[1-5]

Heat flow calorimetry is frequently used to monitor and control batch and semi-batch polymerization reactions.[6-16] Nonetheless, this technique requires the determination of the overall heat transfer coefficient between the reaction medium and the jacket (UA), which is a function not only of the reactor setup (reactor geometry and materials of the wall) and operation conditions (stirring rate and cooling fluid flow rate), but also of the properties of the reaction medium that in polymerizations changes throughout the reaction as the viscosity of the reaction medium can increase drastically and fouling at the reactor wall can occur.

Nonlinear state estimation can be employed to calculate the variation of UA during batch and semi-batch polymerization reaction. The well-known Kalman Filter is only suitable for linear systems and the Extended Kalman Filter (EKF) has become a standard formulation for nonlinear state estimation being employed to calculate UA for several polymerization reactions [2, 17, 18, 19].

However, EKF may cause significant error for highly nonlinear systems because of the propagation of uncertainty through the nonlinear system.

The Unscented Kalman Filter (UKF) is an enhanced filtering technique that effectively addresses the aforementioned issues [20, 21] .The idea is to produce several sampling points (Sigma points) around the current state estimate based on its covariance. Then, propagating these points through the nonlinear map to get more accurate estimation of the mean and covariance of the mapping results. In this way, in fact, it computes the Jacobian but without making use of a linear approximation, hence incurs only the similar computation load as the EKF.

Despite the apparent advantages of the UKF over the EKF, the former has not been yet used widely in polymer reaction engineering. Moreover, the analysis of the filter (UKF) has been done with simulated data[22, 23] that add a known noisy to the system, in addition, the same model that produced the output was provided to the algorithm that estimate the state, this is inconvenient because in the real world, the estimation is provided with model mismatch or unknown-model phenomenon. Finally, the covariance plays an important role in the stochastic filters. Usually, these matrices (Q and R) are defined by default in literature, but with real data measurement from the process tuning the covariance results in a difficult task. Recently, an off-line estimation of the two matrix has been proposed for nonlinear system[24], the method linear time-varying autocovariance least-squares (LTV-ALS) use a linear version of the model to obtain the full version of the matrices, also the technique can be limited to obtain the diagonal elements of the covariance matrices Q and R.

The objective of this work is to implement UKF to estimate the variation of UA and of the total heat loss (Q_{loss}) in order to calculate the heat generated by the reaction during a batch emulsion polymerization reaction by heat flow calorimetry. To evaluate the performance of this procedure, in-line monitoring of batch vinyl acetate emulsion polymerization reactions were carried out in a jacketed stainless steel tank reactor with an internal volume of 5 liters instrumented to operate as a calorimetric reactor at different reaction conditions and variable cooling fluid flow rate (4.0 - 24 L/min).

Results show that the use of UKF is a practical alternative to the online calorimetric estimation without the use of intrinsic equation or privileged information (*e.g.* gravimetric conversion experimental data obtained during the reaction and used to correct the estimation parameters). It was observed that different operation mode (low or high cooling fluid flow rate through the jacket) that affects the overall heat transfer coefficient and influences the term that represented the diagonal covariance for UA, which make it necessary to tune the covariance for the different operation conditions. As the cooling fluid flow rate is a known variable, an strategy is proposed in which the covariance is changed according to this variable making the estimation of UA using UKF possible, in order to calculate the heat of polymerization.

[1] S. BenAmor, D. Colombié, T. McKenna, *Ind. Eng. Chem. Res.* **2002**, *41*, 4233;

[2] R. Gesthuisen, S. Krämer, G. Niggemann, J.R. Leiza, J.M. Asua, *Comp. & Chem. Eng.* **2005**, *29*, 349;

[3] F. Lavanchy, S. Fortini, T. Meyer, *Org. Proc. Res. & Develop.* **2004**, *8*, 504;

[4] S. Krämer, R. Gesthuisen, *Chem. Eng. Sci.* **2005**, *60*, 4233;

[5] G. E. Fonseca, M. A. Dube, A. Penlidis, *Macromol. React. Eng.* **2009**, *3*, 327;

[6] L.M. Gugliotta, J.R. Leiza, M. Arotzarena, P.D. Armitage, J.M. Asua, *Polymer* **1995**, *36*, 2019;

[7] I. Sáenz de Buruaga, P.D. Armitage, J.R. Leiza, J.M. Asua, *Ind. Eng. Chem. Res.* **1997**, 36, 4243;

[8] C. Sayer, E.L. Lima, J.C. Pinto, G. Arzamendi, J.M. Asua, *J. Polym. Sci. A* **2000**, *38*, 1118;

[9] M. Vicente, S. Benamor, L.M. Gugliotta, J.R. Leiza, J.M. Asua, *Ind. Eng. Chem. Res.* **2001**, *40*, 218;

[10] K. Tauer, H. Müller, C. Schellenberg, L. Rosengarten, *Colloids Surfac. A* **1999**, 153, 143;

[11] T.F. McKenna, S. Othman, G. Fevotte, A.M. Santos, H. Hammouri, *Polym. React. Eng.* **2000**, *8*, 1;

[12] C.A. Mantelis, R. Barbey, S. Fortini, T. Meyer, *Macromol. React. Eng.* **2007**, *1*, 78;

[13] M. Esposito, C. Sayer, R.A.F. Machado, P.H.H. Araújo, *Macromol. Symp.* **2008**, *271*, 38;

[14] J.G.R. Poço, M. Danese, R. Giudici, *Macromol. React. Eng.* **2010**, *4*, 145;

[15] I.S. Altarawneh, V.G. Gomes, M.H. Srour, *Polymer International*, **2009**, *58*, 1427;

[16] M. Esposito, C. Sayer, P.H.H. Araújo, *Macromol. React. Eng.,* **2010**, 4, 682;

[17] B. Guo, A. Jiang, X. Hua, A. Jutan, *Chem. Eng. Sci.* **2001**, *56*, 6791;

[18] S. Krämer, R. Gesthuisen, *Chem. Eng. Sci.* **2005**, *60*, 4233;

[19] S. Fan, E. Alpay, *Che, Eng. Sci.* **2004**, *59*, 2811.

[20] S. J. Julier, J. Uhlmann, H. F. Durrant-Whyte, *IEEE Trans. Autom. Control.* **2000**, *45*, 477.

[21] M. Nørgaard, N. K. Poulsen, O. Ravn, *Automatica.* **2000**, *36*, 1627.

[22] C. Qu; J. Hahn, *Journal of Loss Prevention in the Process Industries.* **2009**, *22*, 703.

[23] C. Qu; J. Hahn, *J. Process Control.* **2009**, *19*, 358.

[24] F. V. Lima; J. B. Rawlings, *AIChE Journal.* **2011**, *57*, 996.

(669e) Selection of Similarity Measure for Locally Weighted Partial Least Squares Regression

Ryota Okajima, Sanghong Kim, Manabu Kano and Shinji Hasebe, Dept. of Chemical Engineering, Kyoto University, Kyoto, Japan

Soft-sensors are widely used in various industrial processes to realize high product quality and productivity. However, their estimation accuracy deteriorates due to changes in processes characteristics and operating condition. In fact, the recent questionnaire survey confirms that model maintenance is the most important issue concerning soft-sensors (Kano and Ogawa, 2010). Since model maintenance or reconstruction is demanding for operators and engineers, practical, adaptive modeling techniques that can maintain high estimation accuracy need to be developed.

Locally weighted partial least squares (LW-PLS) is an adaptive modeling technique that builds a local linear regression model based on the similarity between a query and sample data stored in the database when output estimation is required. LW-PLS combines the capabilities of locally weighted regression and PLS; thus it can cope with the nonlinearity, the time-varying nature, and the colinearity. In the recent years, LW-PLS has been applied to various industrial processes including chemicals, semiconductor, and pharmaceuticals.

The estimation accuracy of LW-PLS depends on the definition of similarity between a query and samples. Thus, the present research focuses on the selection of similarity measure. The similarity is generally defined on the basis of the distance such as Euclidean distance and Mahalanobis distance; it is determined by using exponential function of the distance in conventional methods. In this work, new weighted distances are investigated; each input variable is weighted by its corresponding regression coefficient derived by LW-PLS or conventional PLS. In addition, various functions such as Sigmoid function are investigated to improve the prediction performance.

Various LW-PLS methods were applied to actual operation data of the ethylene fractionator at Showa Denko K.K. in Japan. The objective of LW-PLS is to estimate ethane concentration in the ethylene product. As a result, the best estimation accuracy was achieved by the LW-PLS model based on the new similarity determined by Sigmoid function of the weighted distance in which each input variable is weighted by its corresponding regression coefficient derived by conventional PLS; the estimation accuracy was improved by 20 % in Root Mean Square Error of Prediction (RMSEP) and by 6.9 % in multiple correlation coefficient compared with the LW-PLS model based on the conventional similarity determined by exponential function of Euclidean distance. The results clearly show the proposed method is superior to the conventional method.

References:

M. Kano and M. Ogawa: The State of the Art in Chemical Process Control in Japan: Good Practice and Questionnaire Survey. J. Proc. Cont., 20, 969-982 (2010)

See more of this Session: Process Modeling and Identification
See more of this Group/Topical: Computing and Systems Technology Division

(669f) Loss Method: A Static Estimator Applied for Product Composition Estimation From Distillation Column Temperature Profile

Maryam Ghadrdan[1], Chriss Grimholt[1], Sigurd Skogestad[1] and Ivar J. Halvorsen[2], (1)Department of Chemical Engineering, Norwegian Univeristy of Science and Technology, Trondheim, Norway, (2)ICT, SINTEF, Trondheim, Norway

Loss method: A staticestimator applied for product composition estimation from distillation columntemperature profile

Maryam Ghadrdan[1], Chriss Grimholt[1] ,Sigurd Skogestad[1], Ivar J. Halvorsen[2]

[1] Department ofChemical Engineering, Norwegian University of Science andTechnology, N-7491 Trondheim, Norway, Email: ghadrdan@nt.ntnu.no, grimholt@stud.ntnu.no,skoge@nt.ntnu.no

[2] SINTEF ICT, Applied Cybernetics, N-7465 Trondheim, Norway, Email: ivar.j.halvorsen@sintef.no

Assuringthe quality of products is a necessity for many processes in order to optimize plant economy andreduce unwanted emissions. Unfortunately, because of delay in compositionanalysers or their offline operations, the product qualities cannot bemonitored inreal time. To overcome the lack of instrumentation for direct measurement ofsuch parameters,techniques capable of estimating them based on the information from othermeasurable variables are developed and are commonly referred as soft sensoramong othernomenclatures.

In thiswork, we have a presented a static estimator to estimate such immeasurableparameters from other measurements in the plant. Our estimation method (we callit Loss method) is a reformulation of the well-developed Self-optimizing methodproposed by Skogestad (2000) to make it adequate for thepurpose of estimation. The idea behind self-optimising control isto find a variable which characterise operation at the optimum, and the valueof this variable at the optimum should be less sensitive to variations indisturbances than the optimal value of the remaining degrees of freedom. Thusif we close a feedback loop with this candidate variable controlled to asetpoint, we should expect that the operation will be kept closer to optimumwhen a disturbance occur, and minimising control loss will be equivalent tominimising estimation error.

Figure 1 shows the block-diagram of the Lossmethod. The objective is to find the optimal H matrix which minimizes theprediction error which is defined as the difference between the estimated andreal values of the primary variables.

We haveformulated the Loss method for open-loop and closed-loop estimator. With the term "open-loop" estimator, it isimplied that the predicted variables are not used for control purposes. Itshould be noted that this is not the same as implying that variables in a givensystem are uncontrolled. We could use an "open-loop" estimator to predict aprimary variable that are in fact controlled by some other means than theprediction.

From this, we can think of three main types of"open-loop" estimators

1. Predicting primaryvariables from a open-loop system (no control ofvariables).

2. Predicting primaryvariables from a closed-loop system where the primary variables are controlled bymanipulating suitable input variables.

3. Predicting primaryvariables from a closed-loop system where the secondary variables arecontrolled (not by the predicted variable).

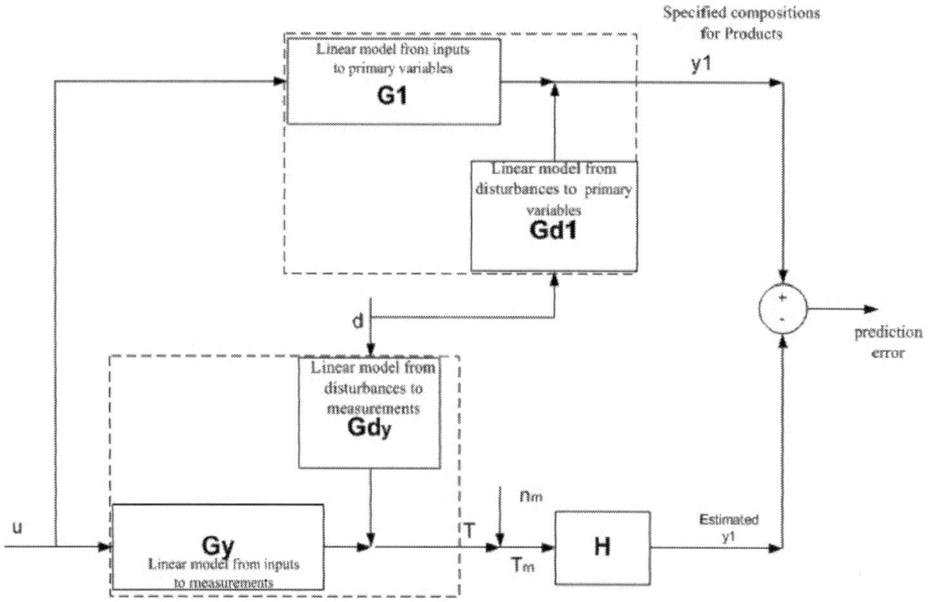

Figure 1.Block diagram for the Loss Method.

The Loss Method is also compared to PartialLeast Squares or Projection to Latent Structures (PLS) which is a family ofmultivariate analysis techniques used to extract useful information fromcorrelated data. This method is used to compress the predictor data into a set of latent variable. There are severaldifferent algorithms generating bases but which all give the same predictor. Wehave chosen the interpretation of Di Ruscio (2000) as the PLS solution wecompared with our method, because their interpretation of the method is theclosest and most comparable to our method. Instead of introducing scores andloadings, they present a non-iterative solution based on some weights which arethe only degrees of freedom in their method. We have applied this method andcompared it with partial least square on a distillation column case. In thiscase, the predictors are the temperatures sensed by thermocouples and theoutput variables are the composition of the products. Despite the fact thatthese two methods are developed in two different contexts, it was interestingto find out that the results from both are very similar.

References

Di Ruscio, D. (2000). "Aweighted view on the partial least-squares algorithm." Automatica **36**: 831-850

Skogestad, S. (2000). "Plantwidecontrol: the search for the selfoptimizing controlstructure." J. Proc. Control **10**: 487-507

See more of this Session: Process Modeling and Identification
See more of this Group/Topical: Computing and Systems Technology Division

(669g) Incremental Model Identification of Reaction and Mass Transfer Kinetics In a Liquid-Liquid Reaction System - An Experimental Study

Nimet Kerimoglu[1], Marcel Picard[2], Adel Mhamdi[1], Lasse Grenier[2], Walter Leitner[2] and Wolfgang Marquardt[1], (1)Process Systems Engineering, RWTH Aachen, Aachen, Germany, (2)Institute for Technical and Macromolecular Chemistry, RWTH Aachen, Aachen, Germany

Due to the efficient product separation and catalyst immobilization, multi-phase catalytic systems are favoured in industrial applications [1]. Nevertheless, it is usually difficult to decouple reaction and mass transfer kinetics such that the experimental determination of reaction kinetics is masked by mass transfer effects. A novel incremental identification methodology tailored to homogeneous reaction systems [2] and extended to multi-phase systems in previous works [3] allows for the decoupling of reaction kinetics and mass transfer and thus avoids the uncertainty in reaction kinetics identification by construction. While the method has been demonstrated in a simulation case study before [3] it is applied in a real experimental study of a multiphase system for the first time in this work.

The chemical system studied comprises a Friedel Crafts acylation of anisole [4]. It follows a complex catalytic reaction mechanism with two reactants and two products. Several reaction rate models, both elementary and complex, were analyzed. The quality of the candidate models has been assessed by the residual sum of squares (RSOS) serving as an objective function and the Akaike Information Criterion (AIC). Optimal experiments were designed to improve model quality using a novel AWDC criterion [5]. It was found out that a reaction rate model comprising only two rate constants for the forward and backward reactions respectively fits the best with a small confidence interval in contrast to the suggested mechanism in literature [6].

Since, mass transfer and chemical reaction can be systematically decoupled in the identification procedure, the best fitting mass transfer rate model of the 4 species involved can also be determined from the same experimental data set. Several mass transfer models of increasing complexity [7] were tested. Our preliminary results show that a simple model which neglects the diffusion cross effect fits the experimental data best. An optimal design of experiments will be conducted next to improve the reliability of the mass transfer model.

References:

[1] Cornils B, Hermann WA, Vogt D, Horvath I, Olivier-Bourbignion H, Leitner W, Mecking S. Multiphase Homogeneous Catalysis. Wiley-VCH, 2005.

[2] Brendel M, Bonvin D, Marquardt W: Incremental identification of kinetic models for homogeneous reaction systems. Chemical Engineering Science, 2006, 61, 5404-5420

[3] Michalik C, Brendel M, Marquardt W: Incremental identification of fluid multi-phase reaction systems. AIChE Journal, 2009, 55(4), 1009-1022

[4] Zayed F, Greiner L, Schulz P S, Lapkin A, Leitner W: Continuous catalytic Friedel–Crafts acylation in the biphasic medium of an ionic liquid and supercritical carbon dioxide. Chemical Communication, 2007, 79-81.

[5] Michalik C, Stuckert M., Marquardt W: Optimal experimental design for discriminating numerous model candidates - The AWDC Criterion. Industrial & Engineering Chemistry Research, 2009, 49(2), 913-919

[6] Dzudza A, Marks J: Lanthanide triflate-catalyzed arene acylation . Relation to classical Friedel-Crafts acylation. The Journal of Organic Chemistry, 2008, 73(11), 4004-4016

[7] Taylor R, Krishna R. Multicomponent Mass Transfer. New York: Wiley, 1993

See more of this Session: Process Modeling and Identification
See more of this Group/Topical: Computing and Systems Technology Division

(679a) A Computational Methodology for Learning Low-Complexity Surrogate Models of Processes From Experiments or Simulations

Alison Cozad[1], Nick Sahinidis[1] and David C. Miller[2], (1)Chemical Engineering, Carnegie Mellon University, Pittsburgh, PA, (2)U.S. Department Of Energy, National Energy Technology Laboratory, Morgantown, WV

Costly and/or insufficiently robust simulations or experiments can often pose difficulties when their use extends well beyond a single evaluation. This is case with the numerous evaluations of uncertainty quantification, when an algebraic model is needed for optimization, as well as numerous other areas. To overcome these difficulties, we generate an accurate set of algebraic surrogate models of disaggregated process blocks of the experiment or simulation. We developed a method that uses derivative-based and derivative-free optimization alongside machine learning and statistical techniques to generate the set of surrogate models using data sampled from experiments or detailed simulations.

Our method begins by building a low-complexity surrogate model for each block from an initial sample set. The model is built using a best subset technique that leverages a mixed-integer linear problem formulation to allow for very large initial basis sets. The models are then tested, exploited, and improved through the use of derivative-free solvers to adaptively sample new simulation or experimental points. The sets of surrogate models from each disaggregated process block are then combined with heat and mass balances around each disaggregated block to generate a full algebraic model of the process. The full model can be used for cheap and accurate evaluations of the original simulation or experiment or combined with design specifications and an objective for nonlinear optimization.

See more of this Session: Advances In Computational Methods and Numerical Analysis
See more of this Group/Topical: Computing and Systems Technology Division

(679b) Manifold Learning Techniques and Model Reduction for Dissipative Dynamics

Benjamin Sonday[1], Karthikeyan Rajendran[2], William Gear[2], Amit Singer[3] and **Ioannis G. Kevrekidis**[2], (1)Program in Applied and Computational Mathematics, Princeton University, Princeton, NJ, (2) Chemical and Biological Engineering, Princeton University, Princeton, NJ, (3)Mathematics & PACM, Princeton University, Princeton, NJ

Many chemical systems -for instance, large chemical reaction networks- are characterized by the existence of widely different time scales. Reduction of large systems of ODEs with such *separation of time scales* is one of the commonly studied topics in dynamical systems theory. Such systems often contain a slow, attracting, invariant manifold to which trajectories quickly decay. If this inertial manifold is low-dimensional, one can hope to obtain an efficient reduction of the system dynamics. This can be achieved by first "learning the manifold" empirically using sampled data and projecting the dynamics onto the empirically discovered slow manifold. The POD-Galerkin method [1,2], for instance, uses Principal Component Analysis (PCA [3]) to find a linear hyperspace containing the slow manifold and uses a Galerkin projection of the dynamics on the leading principal components of the sampled data. In highly non-linear systems, however, the intrinsic manifold dimension can be much smaller compared to the lowest dimensional linear subspace containing the slow manifold.

In this work, we consider a *non-linear extension* to the POD-Galerkin method. We propose using a non-linear machine learning technique -diffusion maps (DMAPs [4]), in particular- to empirically obtain the non-linear slow manifold of the dynamics. The ability to obtain a reduced model now rests on the successful definition of maps from the physical coordinates to the reduced (DMAP) coordinates and vice versa. The former is accomplished using the well-known *Nystrom extension*. The implementation of the latter, i.e., finding a consistent set of physical coordinates on the slow manifold given the reduced (DMAP) coordinates is one of the crucial points of discussion in this work. We illustrate our approach by considering two prototypical examples: a textbook singularly perturbed reacting system, and the truncated spectral discretization of a dissipative reaction-diffusion PDE. We thus link nonlinear manifold learning techniques for data analysis with model reduction techniques for evolution equations with separation of time scales.

References

[1] G. Berkooz, P. Holmes, and J. L. Lumley, *The proper orthogonal decomposition in the analysis of turbulent flows*, Annual Review of Fluid Mechanics, **25** (1993), pp. 539–575.

[2] K. Kunisch and S. Volkwein, *Galerkin proper orthogonal decomposition methods for a general equation in fluid dynamics*, SIAM Journal on Numerical Analysis, **40** (2003), pp. 492–515.

[3] I. T. Jolliffe, *Principal component analysis*, Springer-Verlag, 2002.

[4] R. Coifman, S. Lafon, A. Lee, M. Maggioni, B. Nadler, F. Warner, and S. Zucker, *Geometric diffusions as a tool for harmonic analysis and structure definition of data: Diffusion maps*, PNAS, **102** (2005), p. 7426.

See more of this Session: Advances In Computational Methods and Numerical Analysis
See more of this Group/Topical: Computing and Systems Technology Division

(679c) Application of Singular Value Decomposition to Enhance Proper Generalized Decomposition In Mildly Non-Linear 2-D Partial Differential Equation Problems

Stephen Edie and Antony N. Beris, Chemical Engineering, University of Delaware, Newark, DE

We consider the recently-developed Proper Generalized Decomposition (PGD) method [1] for the solution of certain classes of partial differential equations (PDEs). The PGD method approximates the PDE solution by a summation in which each term is a product of single-variable functions in each dimension. By applying a weighted residuals approach similar to the Galerkin method used in traditional Finite Element Methods, successive terms may be computed iteratively by repeatedly solving 1-D problems in each dimension. Use of the PGD method has therefore the potential to drastically reduce the computational requirements in high dimensional problems. Example applications include the Fokker-Planck equations in polymer solution models [1] and the chemical master equation [2]. A potential limitation, however, is that its application requires solving non-linear systems which admit multiple solutions, even when the original system of PDEs is linear. A recent theoretical analysis [3] considers this issue in detail and highlights the inherent difficulties that arise when trying to assess the convergence properties of PGD. It is readily shown that the term computed within a particular iteration of the PGD is not necessarily optimal in the sense that it leads to the best possible convergence of the sum to the true solution. Thus, we are interested in assessing the performance of PGD in real-world examples and exploring enhancements to improve its convergence characteristics, especially in association to high accuracy spectral approximations. By decomposing the original multidimensional problem to a series of n-by-m dimensional problems, all the way to 2-D problems at the end, we can explore a powerful tool, the singular value decomposition (SVD). The SVD when applied to a matrix representing the solution in an n-by-m dimensional space yields an optimum expression of the solution in terms of a sum of products of solutions in n and m dimensions, resepectively, in the sense that this approximation has optimum convergence with number of terms to the actual solution. As such the SVD can provide a benchmark against which the solutions obtained with various variants of PGD can be compared. Additionally, the SVD suggests a strategy for increasing the efficiency of PGD by periodically reducing the obtained sums through its application. We studied the PGD method applied to the 2-D pseudo-conformal map of a sinusoidally varying undulating channel on a straight channel. The problem is mildly nonlinear, with the non-linearity appearing only in one of the boundary conditions. We used pseudospectral discretization methods to solve the 1-D problems and perform the required integrations. We obtained solutions approaching machine accuracy on a 64x64 point pseudospectral grid having sums with an order of 100 terms. This is compared against a much more compact sum obtained using SVD on the full solution of only 15 terms. Various convergence acceleration options are discussed.

References

[1] A. Ammar, B. Mkdad, F. Chinesta and R. Keunings, J. Non-Newtonian Fluid Mech., 139:153-176 (2006).

[2] F. Chinesta, A. Ammar, E. Cueto, European J. Computational Mechanics, 19:53-64 (2010).

[3] C. Le Bris, T. Lelievre, Y. Maday, Constr. Approx. 30:621-651 (2009).

See more of this Session: Advances In Computational Methods and Numerical Analysis
See more of this Group/Topical: Computing and Systems Technology Division

(679d) ODE Model Parameter Identification Via Global Optimization Techniques

Jeremy A. Conner, Chemical & Biomolecular Engineering, University of California, Los Angeles, Los Angeles, CA and Vasilios Manousiouthakis, Chemical Engineering Department, University of California at Los Angeles, Los Angeles, CA

In this work we present our progress on the development of a globally optimal method of determining parameters from experimental data for systems described by ordinary differential equations. For cases where the ODE models cannot be analytically integrated, non-linear regression techniques are typically employed in the solution. These methods, however, can only guarantee local optimality of the solution, and also fail to ascertain whether or not the proposed problem parameterization is appropriate for the available data. To address these shortcomings, we propose a novel method for parameter identification that is guaranteed to identify the global optimum of the non-linear regression problem and is also able to deliver ranges for the model parameters for which the proposed model can describe the available data within a predetermined level of accuracy. The method is illustrated in a case study involving the Trambouze reaction scheme.

See more of this Session: Advances In Computational Methods and Numerical Analysis
See more of this Group/Topical: Computing and Systems Technology Division

(679e) A Novel Computational Framework for Reactive Flow and Multiphysics Simulations

James C. Sutherland, Chemical Engineering, The University of Utah, Salt Lake City, UT and **Tony Saad**, Chemical Engineering, Institute for Clean and Secure Energy, University of Utah, Salt Lake City, UT

Reactive flow simulations represent a challenging endeavour in computational science. The underlying coupling between fluid transport and chemical kinetics necessitates careful attention to algorithm design thus reducing code modularity and increasing software complexity. We present a framework that facilitates the implementation of transport equations using modern programming techniques. This framework consists of two C++ libraries that allow the abstraction of a differential equation using discrete operators and generic expressions. Then, by utilizing graph theory, the expressions are mapped into a graph and a solution algorithm is automatically generated. This functionality grants developers the capability to rapidly prototype new ideas and generate reliable simulation software without attention to algorithmic details. Furthermore, our approach exposes additional levels of parallelism such as algorithm decomposition and vectorized parallelism on field and stencil operations. The result is a transparent and easy to use tool that allows end users to develop scalable and highly efficient simulation code without specialized hardware knowledge.

See more of this Session: Advances In Computational Methods and Numerical Analysis
See more of this Group/Topical: Computing and Systems Technology Division

(679f) Computing Interval Bounds On the Solutions of Nonlinear Index One DAEs

Joseph K. Scott, Chemical Engineering, Massachusetts Institute of Technolgy, Cambridge, MA and Paul I. Barton, Process Systems Engineering Laboratory, Dept. of Chemical Engineering, Massachusetts Institute of Technology, Cambridge, MA

A method is presented for computing tight interval bounds onthe solutions of nonlinear semi-explicit, index one differential-algebraicequations (DAEs) subject to given intervals of permissible initial conditionsand parameters. Differential-algebraic equations are used to model a wide varietyof important dynamic chemical processes through the combination of material andenergy balances, thermodynamic relationships, reaction kinetics and empiricalcorrelations. Moreover, such models often contain parameters which are notknown precisely, such as physical constants, factors in empirical correlations, process disturbances and model uncertainties. Accordingly, one often requiresinformation about an entire family of possible solutions, parameterized bythese uncertainties, rather than a single nominal solution.

Assuming that the uncertain or unknown parameters ofinterest lie within known intervals, the proposed method computes time-varyingupper and lower bounds on the solutions of the given DAEs attainable withparameters in these intervals. Similar methods for bounding the solutions ofordinary differential equations (ODEs) have been studied extensively, withapplications in uncertainty analysis, state and parameter estimation, safetyverification, fault detection, global optimization, validated numericalintegration, and controller synthesis. On the other hand, producing guaranteedbounds on the solutions of DAEs has received much less success and remains anopen challenge.

The proposed approach for bounding DAE solutions combinesconcepts from differential inequalities and interval Newton-type methods. Thefirst key result is an interval inclusion test which verifies the existence anduniqueness of DAE solutions over a given time step, and provides a crudeinterval enclosure. This test combines a well-known interval inclusion test forsolutions of ODEs (used in standard Taylor methods) with an interval inclusiontest for solutions of systems of nonlinear algebraic equations from theliterature on interval Newton methods. The second key result is a set ofsufficient conditions, in terms of differential inequalities, for twotime-varying trajectories to bound the differential state variables; i.e.,those state variables whose time derivatives are given explicitly by the DAEequations. Using this result, refined, time varying bounds on the statevariables over the given time step are computed using a technique whichsimultaneously applies differential inequalities to the differential variablesand an interval Newton-type method to the algebraic variables.

Based on these key results, an efficient numericalimplementation using interval computations and a standard numerical integrationcode is presented. Strengths and shortcomings of the proposed algorithm arediscussed in the context of a DAE model for a simple distillation process.

(679g) Discrete Particle Swarm Optimization for the Selection of Input Variables of Feed Forward Neural Networks

Alex Kalos, Process Optimization & Modeling, Engineering Sciences, The Dow Chemical Company, Freeport, TX

The selection of input variables to an artificial neural network is typically a manual, trial-and-error endeavor. Given that the optimization of the weights of any single neural network is generally very time-consuming, the process of variable selection can be quite tedious. This work describes an approach for the automatic selection of relevant input variables. It uses discrete particle swarm optimization to select variables as well as other neural network structural elements (number of hidden layers and nodes and input lag structure) in order to optimize the topology of non-linear, multivariate time series networks. The application of the technique will be illustrated with a case study using data available in the public domain.

See more of this Session: Advances In Computational Methods and Numerical Analysis
See more of this Group/Topical: Computing and Systems Technology Division

(689a) Real-TimeWater and Energy Management In Power Plants and Implications In Electricity Markets

Victor M. Zavala[1], Juan M. Salazar[2], Emil Constantinescu[1] and Urmila Diwekar[2], (1)Mathematics and Computer Science, Argonne National Laboratory, Argonne, IL, (2)Vishwamitra Research Institute, Center for Uncertain Systems: Tools for Optimization and Management, Clarendon Hills, IL

We propose a real-time optimization framework to integrate water and energy management in power plants. The objective of the framework is to construct optimal bidding curves (power output vs. price) using a first-principles model of the power plant and of the cooling towers under uncertain weather conditions. The weather uncertainty is quantified in the form of ensembles using the state-of-the-art numerical weather prediction model WRF running at Argonne National Laboratory. The ensembles are used for real-time stochastic optimization performed by the BONUS algorithm and coupled to a steady-state power plant model implemented as a CAPEOPEN compliant capability. We analyze the effects of cooling capacity constraints and weather forecasts on the market participation of power plants. We use a pulverized coal power plant case study to demonstrate the potential of the framework.

See more of this Session: Energy and Sustainability In Operations
See more of this Group/Topical: Computing and Systems Technology Division

(689b) Optimal Production Planning Under Time-Sensitive Electricity Prices for Continuous Power-Intensive Processes

Sumit Mitra[1], Ignacio Grossmann[1], Jose M. Pinto[2] and Nikhil Arora[3], (1)Chemical Engineering, Carnegie Mellon University, Pittsburgh, PA, (2)Praxair Inc., Danbury, CT, (3)Praxair, Inc., Tonawanda, NY

Power-intensive processes can lower operating expenses when adjusting production planning according to time-dependent electricity pricing schemes, which exist due to fluctuations in power demand. Different researchers dealt with the operational problem of the so-called demand response for power-intensive processes.

The first line of research proposes a control approach. For the economic optimization of air separation plants, Zhu et al. (2010) developed a model based on heat and mass balances, including nonlinear thermodynamics. Huang (2010) proposed the application of economically-oriented nonlinear model predictive control. While these approaches have the advantage of providing an accurate process representation, they cannot incorporate discrete decisions such as equipment shutdowns and startups and corresponding operational constraints such as enforcing minimum up- or downtimes.

The second line of research comes from the area of planning and scheduling. An initial approach by Daryanian et al. (1989) shows the economic benefit of demand response for air separation plants but does not consider discrete operating decisions such as equipment shutdowns and startups. Ierapetritou et al. (2002) as well as Karwan and Keblis (2007) extend the methodology to incorporate discrete operating decisions. However, their models do not capture the transient behavior between - operating modes. Moreover, some of their logic constraints show room for improvement with respect to the tightness of the linear relaxation. For cement plants, Castro and co-workers (2009, 2011) extend the Resource Task Network (RTN) formulation to incorporate hourly changing electricity prices.

In this work, we describe a deterministic MILP model that allows optimal production planning for continuous power-intensive processes using a discrete time formulation. First, we generalize approaches taken by Ierapetritou et al. (2002) - and Karwan and Keblis (2007). We discuss the concept of transitional modes to allow for a more detailed modeling of transitional behavior that is due to plant dynamics. Further, we enhance the computational efficiency of their formulation by improving the logic constraints. Properties on the tightness of several logic constraints are proved. We successfully apply the model to two different real-world air separation plants that supply to the liquid merchant market, as well as cement plants.

For the time horizon of one week and hourly changing electricity prices, we solve an industrial case

study on air separation plants, where transitional modes help us capturing the ramping behavior. The operational optimization model created savings of more than 10% when compared to a simple heuristic. Furthermore, we learned that operational flexibility, in terms of production and storage capacity, is the key to lower operating expenses.
We also solve problem instances on cement plants that are reported in Castro et al. (2011). We observe that the introduction of transitional modes results in a large-scale model. Nevertheless, the model is superior compared to a smaller model that only had single product modes and thus lacked control of occurring transitions as well as could not be solved to optimality. Despite the large size of the MILP model with transitional modes, the required solution times to obtain the optimal solution were small for all test cases. Furthermore, the obtained schedules were practical to implement because we were able to control the occurring transitions.

References:

Castro, P.M.; Harjunkoski I.; Grossmann I.E. (2009) New Continuous-Time Scheduling Formulation for Continuous Plants under Variable Electricity Cost. Ind. Eng. Chem. Res., 48:6701–6714

Castro, P.M.; Harjunkoski I.; Grossmann I.E. (2011) Optimal scheduling of continuous plants with energy constraints. Comp. Chem. Eng., 35:372– 387

Daryanian, B.; Boln, R.E.; Tabors, R.D. (1989) Optimal demand-side response to electricity spot prices for storage-type customers. IEEE Transactions on Power Systems, 4:897-903

Huang, R. Nonlinear Model Predictive Control and Dynamic Real Time Optimization for Large-scale Processes. PhD thesis, Carnegie Mellon University, 2010.

Ierapetritou, M.G.; Wu, D.; Vin, J.; Sweeny P.; Chigirinskiy M. (2002) Cost Minimization in an Energy-Intensive Plant Using Mathematical Programming Approaches. Ing. Eng. Chem. Res., 41:5262–5277

Karwan, K.; Keblis M. (2007) Operations planning with real time pricing of a primary input. Computers & Operations Research, 34:848–867

Zhu, Y.; Legg S.; Laird, C.A Multiperiod Nonlinear Programming Approach for Operation of Air Separation Plants with Variable Power Pricing.AIChE Journal, accepted, 2010.

See more of this Session: Energy and Sustainability In Operations
See more of this Group/Topical: Computing and Systems Technology Division

(689c) Design of Networks for the Large-Scale Deployment of CO2 Capture, Transport and Storage Using Multi-Period Optimization Models: The Case for the Netherlands

N.V.S.N. Murthy Konda[1], Nilay Shah[1] and Nigel P. Brandon[2], (1)Centre for Process Systems Engineering, Department of Chemical Engineering, Imperial College London, London, United Kingdom, (2)Department of Earth Sciences, Imperial College London, London, United Kingdom

Carbon capture and sequestration (CCS) is widely accepted as one of the key technologies in the portfolio of actions that can help mitigate CO_2 emissions. CCS can play significant role especially in the short/medium term as it offers a seamless and cheaper transition route towards a more sustainable energy economy (e.g., renewables) in the long-term while allowing the continued use of fossil fuels environmentally benignly. However, one of the main stumbling blocks to realize large-scale deployment of carbon capture and storage (CCS) is the huge upfront costs involved. Hence, in this contribution we have proposed a comprehensive optimization framework, that is spatially and temporally explicit, to design the least-cost CCS networks and their optimal evolution with time over the next four decades (i. e., until 2050). This framework establishes optimal transport links amongst most potential sources and sinks (while leaving out the not-so-important ones from a global perspective), as and when they are necessary, to achieve a pre-set CO_2 reduction target that is to be met by CCS in every decade within this time horizon. We have then demonstrated the applicability and usefulness of our approach with a real case study by applying it to design CCS networks for the Netherlands.

As demonstrated by Konda et. al., 2011, CCS must be an integral part of the Dutch CO_2 mitigation portfolio to comply with the local and regional (i.e., EU level) CO_2 mitigation targets. Furthermore, the availability of a number of large-scale CO_2 point-sources and large storage capacity makes CCS an attractive CO_2 mitigation option for the Netherlands. Potential CO_2 sources considered within our framework include the existing power plants (including coal/gas-fired plants and Integrated Gasification Combined Cycle plants) and industrial sources (including refineries, cement, steel and chemical plants such as ammonia and hydrogen production plants). In addition to the existing plants, we have also considered the prospective plants within the scope of this study – this is an one of the salient features of our models, and as we will demonstrate, this is an important aspect as establishing a CO_2 capture facility at an existing/old power plant can be more expensive compared to establishing it at a new/prospective facility. This is because new plants can be built in accordance with the requirements so as to make its integration with CO_2 capture facility easier and cheaper (for instance, they can be capture-ready plants) and to minimize the additional capital burden and efficiency losses/penalties. Potential sinks considered include depleted gas reservoirs (including the Groningen fields) onshore and offshore, enhanced coal-bed methane recovery (ECBM) sites and saline aquifers. Within the Dutch context, depleted oil reservoirs' estimated storage capacity is rather limited and hence they are not considered. Extensive data on all the point sources (i.e., CO_2 emissions/yr, location, type etc) and sinks (i.e., sink capacity, location, type etc.) are obtained from latest data-sources (including official sources whenever possible) as it is important to ensure the quality of the data used in such a large-scale real case-study. CO_2 is transported by pipelines (both onshore and offshore) and pipeline economies of scale is explicitly accounted for (by piecewise linearization) in our models. We have already obtained interesting results that are practically useful/significant demonstrating the need and usefulness of such a holistic systems based multi-period optimization approach to design optimal and realistic large-scale CCS networks. The results, model and the salient features will be discussed in detail during the presentation. (Disclaimer: This work is done while the first author was at Imperial College London and the content/opinions/results discussed do not necessarily represent those of the associated institutions).

References: Konda, N. V. S. N. M.; Shah, N.; Brandon, Nigel P., Optimal transition towards a large-scale hydrogen infrastructure for the thansport sector: The case for the Netherlands. International Journal of Hydrogen Energy (36) 4619 -4635

See more of this Session: Energy and Sustainability In Operations
See more of this Group/Topical: Computing and Systems Technology Division

(689d) Global Optimization for Refinery Operations Planning Using Complex Unit Models

Miguel Bagajewicz[1], **Mesude Ozturk**[1] and DuyQuang Nguyen[2], (1)Chemical Engineering, University of Oklahoma, Norman, OK, (2)University of Oklahoma, Norman, OK

Refinery operations require careful planning, which is handled these days by different MINLP formulations. Existing Commercial software does not guarantee global optimality. Incentives to search for global optima are reflected in the fact that just 1% improvement in the gross refinery margin represents sizable profit. In addition, we recognize that several units in the refinery are reactors, which, if modeled in detail, are represented by differential equations. Therefore, any improvement in the direction of more rigorous modeling requires the incorporation of these differential equations in the planning model. Instead of attempting the explicit incorporation of these differential equations, we resort to creating input-output models, which are used in a global optimization method based on domain and image partitioning. Domain partitioning takes place the usual way. Image partitioning simply creates images of the discrete partitioning points. Once this is performed, a lower bound model and a recently developed bound contraction model is used to obtain the global solution.

See more of this Session: Energy and Sustainability In Operations
See more of this Group/Topical: Computing and Systems Technology Division

(689e) Economic and Environmental Optimization of the Biomass-to-Biofuel Supply Chain In the Midwest

W. Alex Marvin[1], Lanny D. Schmidt[1], Saif Benjaafar[2] and Prodromos Daoutidis[1], (1)Chemical Engineering and Materials Science, University of Minnesota, Minneapolis, MN, (2)Industrial and Systems Engineering, University of Minnesota, Minneapolis, MN

Biofuels are currently the only demonstrated replacement for liquid transportation fuel, owing largely to the expansion of the U.S. and Brazilian bioethanol industries. Biofuels have the potential to reduce petroleum imports, support domestic agriculture and forestry, and reduce greenhouse gas emissions relative to gasoline. In the U.S., the Renewable Fuels Standard (RFS) Program phases in the use of biofuels by setting annual blending mandates for renewable, advanced and cellulosic biofuels. The mandates for renewable fuel have so far been met largely by corn-based bioethanol, but cellulosic biofuel is lagging behind. For 2011, the original cellulosic biofuel mandate of 250 million ethanol-equivalent gallons was reduced by the EPA to 6 million gallons based on expected industry production [1]. Tools are required to analyze how the industry may develop to produce the 16 billion gallons of cellulosic biofuels required by RFS in 2022 [2].

This study aims to address this issue by using computational optimization to generate robust supply chains using the real-world distribution of biomass resources, existing options for biomass collection and transportation, and proven biomass-to-biofuels technologies. The Midwest is chosen as the study region due to the abundance of biomass there and its expected importance in fulfilling the RFS mandates. Forestry and agricultural residues are included in the optimization as available biomass resources with county-level resolution. Two cellulosic biofuel technologies are allowed to compete in the study region, including co-fermentation (biochemical) and gasification Fischer-Tropsch (thermochemical).

The problem is formulated as a multi-objective mixed integer linear program (MoMILP). The competing objectives are to maximize the net present value (NPV) and minimize the total greenhouse gas emissions of the whole supply chain by taking into account cash flows and emissions for biomass production, collection and transportation, biorefinery construction and operation, and biofuel sales. The effect of uncertainty in biomass availability and in economic parameters is examined by Monte Carlo sampling.

References:

[1] U.S. Environmental Protection Agency, Regulation of Fuels and Fuel Additives: 2011 Renewable Fuel Standards, 2010.

[2] U.S. Environmental Protection Agency, EPA Finalizes Regulations for the National Renewable Fuel Standard Program for 2010 and Beyond, 2010.

See more of this Session: Energy and Sustainability In Operations
See more of this Group/Topical: Computing and Systems Technology Division

(689f) Dimensionality Reduction Techniques Coupled with Multi-Objective Optimization Applied to the Design of Bioethanol Supply Chains In Argentina

Gonzalo Guilln-Goslbez, Departament d'Enginyeria Qumica, Universitat Rovira i Virgili, Tarragona, Spain

Energy security and environmental concerns have triggered the interest on bio-based sources of energy. Bioethanol, in particular, is the world's leading transportation biofuel with a worldwide production in 2009 of almost 74 billion liters [1]. This scenario has created a clear need for systematic tools to assist in the design and operation of biofuels facilities.

The overwhelming majority of the optimization studies for designing bioethanol infrastructures have concentrated on improving the economic performance. Few studies have accounted for the environmental impact of these systems. Zamboni *et al.* [4] formulated a multi-objective optimization model to reduce GHG emissions associated with the future corn-based Italian bioethanol network. Later, Giarola *et al.* [5] extended this model by considering second generation bioethanol production technologies. As pointed out in the literature[6,7],approaches focusing on minimizing GHG emissions as unique criterion are somehow limited, as they can lead to solution where this impact is mitigated at the expense of increasing others. Hence, the environmental assessment of bioethanol infrastructures requires the application of holistic methods capable of accounting for several impacts occurring throughout the entire production chain.

The aim of this work is to provide a decision-support tool for the strategic planning of integrated bioethanol-sugar supply chains (SCs) that accounts for the simultaneous optimization of several environmental impacts. The design task aims to determine the number, location and capacities of the SC facilities to be set up in each sub-region of a given country, their expansion policy for a given forecast of prices and demands over the planning horizon, the transportation links and number of trucks that need to be established in the network, and the production rates and flows of the involved feedstocks, wastes and final products.

The problem is mathematically formulated as a mixed-integer linear program (MILP). To encompass all possible conversion pathways, the proposed model includes production facilities of two types: sugar mills and distilleries. Depending on the utilized technology, the sugar mills can give two main by-products: molasses or honey, both of which can be fermented to obtain bioethanol. The liquid and solid materials require different storage conditions, so the model considers two options of warehouses depending on the physical state of the stored material. The region of interest (i.e., Argentina) is divided into a number of sub-regions, where the SC facilities can be installed. Different trucks are also considered for transporting materials between sub-regions of the country.

The environmental impact is quantified *via* a set of life cycle assessment (LCA) metrics[8] that measure the environmental performance of the integrated sugar-bioethanol SC in eleven damage categories. To tackle this high dimensional multi-objective problem, we follow a combined approach that relies on the use of a MILP-based dimensionality reduction technique recently developed by the authors. This strategy allows identification of redundant environmental objectives that can be left out of the analysis, shedding light on the interactions between the environmental damages caused by the bioethanol infrastructure.

Numerical results show that several environmental effects of the sugar-bioethanol network are highly correlated, which makes it possible to focus our attention on a reduced set of damages. Our tool allows uncovering relationships between environmental impacts, providing valuable insight into the design problem and suggesting process alternatives for holistic environmental improvements.

References

1. ENERS Energy Concept Ltd., (2010). Production of biofuels in the world
2. Dias Leite A., (2009). Energy in Brazil: towards a renewable energy dominated system. Earthscan: London
3. Renewable Fuels Association, (2009). Ethanol industry overview
4. Zamboni A., Bezzo F., Shah N., (2009). Spatially explicit static model for the strategic design of future bioethanol production systems. 2. Multi-objective environmental optimization. *Energy & Fuels* 23, 5134–5143
5. Giarola S., Zamboni A. , Bezzo F., (2011). Spatially explicit multi-objective optimisation for design and planning of hybrid first and second generation biorefineries. *Computers & Chemical Engineering*, DOI: 10.1016/j.compchemeng.2011.01.020
6. Niven R., (2005). Ethanol in gasoline: environmental impacts and sustainability review article. *Renewable & Sustainable Energy Reviews* 9 (6), 535–555.
7. Scharlemann J., Laurance W., (2008). How green are biofuels? Science 319 (5859), 43–44
8. PRe´-Consultants, The Eco-indicator 99, A damage oriented method for life cycle impact assessment.Methodology Report and Manual for Designers. Technical Report, PRe´ Consultants, Amersfoort, The Netherlands, 2000.

See more of this Session: Energy and Sustainability In Operations
See more of this Group/Topical: Computing and Systems Technology Division

(689g) Modeling and Optimization of Next Generation Feedstock Development for Chemical Process Industry

Selen Cremaschi, Department of Chemical Engineering, The University of Tulsa, Tulsa, OK

Biomass, as a renewable and locally available resource, has great potential for weaning our chemical process industry (CPI) from fossil based feedstocks. There are many different routes to transform biomass into feedstock for the bulk chemicals production. The switch from our current fossil fuel based CPI to a future CPI that utilizes biomass feedstock will require substantial amounts of research & development (R&D) and capital investments for technology development both by the CPI and the government. As such there is great opportunity for investigating how these investments will impact the evolution of the biomass feedstock system. This problem has two unique characteristics: uncertain decision-dependent endogenous technology evolutions and game theoretic issues, which limit the applicability of the existing representations and methods. This talk will explain the problem characteristics in detail, and give an overview of our research activities on developing novel frameworks and large-scale solution capabilities for this problem.

See more of this Session: Energy and Sustainability In Operations
See more of this Group/Topical: Computing and Systems Technology Division

(700a) Computer Aided Flowsheet Synthesis and Design Under Uncertainty In Vegetable Oil Production

Alberto Quaglia, CAPEC, Department of Chemical and Biochemical Engineering, Technical University of Denmark, Kgs. Lyngby, Denmark, Bent Sarup, Alfa Laval Copenhagen A/S, Soeborg, Denmark, Grkan Sin, CAPEC, Department of Chemical and Biochemical Engineering, Technical University of Denmark, Kgs. Lyngby, DK-2800, Denmark and Rafiqul Gani, CAPEC, Department of Chemical and Biochemical Engineering, Denmark Technical University (DTU), Kgs. Lyngby, Denmark

Although being a mature and well established industry segment, over the last few decades the vegetable oil industry has been facing many important new challenges due to emerging new products (such as biodiesel and nutraceuticals compounds), as well as new trends and regulations with regards to sustainability, environment and health.

Moreover, since the agricultural commodities market is highly volatile due to seasonal variations in supply and demand as well as intensive trading, the decision making process is complicated by uncertainties in price. These uncertainties in product and feedstock prices, in turn, are propagated to the economical and financial indicators, which are normally used to support and take decisions.

Most of the time the industry responds to these challenges by taking decisions based on heuristic approaches, which arguably are not fit to manage neither the complexity of the entire value chain nor price uncertainties. All these considerations underline the need for Computer-Aided Flowsheet Synthesis and Design tool in vegetable oil industry. Such a tool should be robust, cover all the value chain, ideally be able to handle new product and process development problems as well as retrofitting problem of existing operations and deal with price uncertainties. Moreover, it should be complemented with all data, models and algorithm databases necessary for its use.

In this paper, a systematic framework for Computer-Aided Flowsheet Synthesis and Design(CAFD) and resources allocation under price uncertainties for the vegetable oil sector is presented.

In the framework, the flowsheet synthesis and design problem under uncertainty is cast as a Stochastic Mixed Integer Non Linear Programming (SMINLP) problem, which is then solved using appropriate SMINLP solvers.

The SMINLP problem is formulated and solved in a systematic manner by following a 4 steps procedure: (i) problem definition (ii) data collection and superstructure definition (iii) models selection, development and validation (iv) model solution & analysis.

In the first step, the synthesis/ design problem is defined by stating the purpose (new process, retrofit) and defining the optimality criteria.

The second step deals with the collection and systematization of the existing knowledge (industrial know-how, engineering insights, commercial knowledge...) and data (both scalar value and distribution functions of uncertain parameters) relevant to the problem.

Particular emphasis is given to reconciliation and systematization of this multidisciplinary knowledge, as well as to the development of an appropriate infrastructure for efficient data management, composed by a superstructure representing different flowsheet alternatives, structural constraints list, compounds and unit operations inventory and databases of reconciled data, structured in such a way to be accessible from all project levels [1].

The third step aims at collecting all the submodels needed for the multiscale model formulation. These include physical properties, unit operations, operational and investment cost and sustainability models; a systematic model generation framework is used to ensure consistency among the different models and scales.

The fourth step deals with the CAFD problem solution, which requires the solution of the SMINLP formulated in the above described steps. Because of the size of the problem and of the combinatorial nature introduced by the binary variables, this task is often not trivial and computational intensive. Therefore particular emphasis is given to the selection of the optimal solution method among the different algorithms [2] [3] [4].

Depending on the nature and size of the problem and on the type and number of uncertain parameters, some of the existing algorithm may not be suitable for the solution of the problem, because (i) the size of the CAFD problem is too large, or (ii) correlation structure of the uncertain parameters cannot be handled or (iii) they don't exploit shortcut calculation to facilitate the convergence to the solution. A rule based method for the selection of the appropriate solver as a function of some key problem characteristic is provided.

The SMINLP solution is the optimal flowsheet and the optimal material flow to each process block, as well as the expected value of the objective function and of all the economical and sustainability metrics calculated at optimality. In addition to the expected value, other conditional value such as value at risk can be also calculated in order to provide additional information for decision making.

In order to test and highlight the features of the methods and tools presented here, a soybean processing case study has been formulated and solved under price uncertainty assumption to determine the optimal processing network for vegetable oil extraction and refining (including various options for byproducts valorization), as well as the optimal material flows to each processing step under the condition of raw material, product and utility price uncertainty.

1. R. Singh, K. Gernaey, R. Gani, Computers and Chemical Engineering 33 (2009) 22–42.
2. G. E. Paules, C. A. Floudas, Computers and Chemical Engineering 16 (1992) 189–210.
3. Z. N. Pintaic, Z. Kravanja, Computers and Chemical Engineering 24 (2000) 195-201.
4. V. Duaa, K.P. Papalexandri, E.N. Pistikopoulos, Computers and Chemical Engineering 23 (1999) 19-22.
5. G. Sin, R. Gani, Chemical Engineering transactions. In press.

See more of this Session: Model-Based Engineering for Design
See more of this Group/Topical: Computing and Systems Technology Division

(700b) Why Using Ontologies In Process Modelling

Heinz A. Preisig, Chemical Engineering, Norwegian University of Science and Technology, Trondheim, Norway

Designing and operating a plant makes extensive use of process models. Currently the computational tools used for these activities are segmented into approximately five vertical domains: design, control design, identification, monitoring, and optimization. In each domain, methods are reasonably mature and good tools are available, but the domains are disconnected and since there are many overlaps in the form of models, methods and algorithms, individual tasks in the respective domains require similar efforts multiple times. More concrete, models are built, identified, validated and improved on a case-to-case base and for each operation in design, operation and control separately, which makes the overall engineering process extremely expensive and time consuming.

Ontology-based modelling makes it possible to generate model component libraries that are internally consistant satisfying the concept defined in the ontology. Controlling the combination of model components using the same ontology guarantees also composite models to satisfy the same concepts. This mechanism yields process models that satisfy the basic principles such as the conservation principles, transport phenomena and physical-chemical principles. Adding the ability to also handle order-of-magnitude assumptions in the same framework provides the means to generate multi-facetted model libraries for the same process covering different domains of application, thus integrating the now separated computational domains and providing across-the-board consistancy in the model-based computation engineering activities associated with all aspects of design and operation of a plant.

See more of this Session: Model-Based Engineering for Design
See more of this Group/Topical: Computing and Systems Technology Division

(700c) Model Generation: Massive Parallel Computations for the Next Generation of Biomedical Engineering Applications

Andreas Linninger, Laboratory for Product and Process Design, University of Illinois at Chicago, Chicago, IL and Madhawa Hettiarachchi, University of Illinois at Chicago, Chicago, IL

Distributed systems models occur in life science applications such as drug delivery, intracranial dynamics or whole-body pharmacokinetics. These models lead to very large computational meshes due to the geometrical complexity of the problem domain typically obtained from medical images as well as the need for dense computational domains to proof mesh-independence in the spatio-temporal discretization scheme which involves multiple time and length scales.

Due to the huge problem sizes, a massive parallel approach for the solution of the patial differential equation systems is often required. In this presentation, we emphasize model generation as an approach to seamlessly port distributed application between various hardware solution architectures. Instead of insisting on solving partial differential equations in a single monolithic software tool, model generation synthesizes a set of algebraic equations in desired target languages or syntax. Especially large scale parallel processing require low level user-supplied residual functions and derivatives; thus the novel model generation paradigm serves as an important bridge enabling consistent specification, and solution of large scale application required in biomedical engineering. The presentation will demonstrate the state-of-the-art model generation tools for the solutions of mass and momentum transfer problems in drug delivery and simulation of intracranial dynamics.

See more of this Session: Model-Based Engineering for Design
See more of this Group/Topical: Computing and Systems Technology Division

Automatic reaction network generation using chemo-informatics

Nick M. Vandewiele, Kevin M. Van Geem, Marie-Françoise Reyniers, Guy B. Marin

Laboratory for Chemical Technology, Krijgslaan 281 (S5), 9000 Gent, Belgium

Large-scale detailed kinetic models find increasing use in the modeling of combustion processes, atmospheric chemistry, soot formation, and other areas of industrial or environmental interest. The enormous amount of reaction possibilities in these chemical processes lead to complex reaction networks consisting of hundreds of unique species and thousands of unique reactions. The manual construction of these networks is therefore in most cases impossible, and in the few cases it is possible it remains tedious and error-prone. Hence, automatic generation of a chemical reaction network is particularly useful for these complex reactive processes such as pyrolysis, gasification, oxidation, combustion, etc.

One of the most important aspect of computer tools that enable the automated execution of chemical transformations is the representation of chemical species. Graphs, consisting of vertices and edges, are an excellent mathematical abstraction of chemical species and were the starting basis for the majority of the tools. Moreover, the rise of chemo-informatics, i.e. the use of informatics methods to solve chemical problems, at the intersection between chemistry and computer science, responded to the needs to chemical community to store and analyze the ever-growing amount of chemical data available. Chemo-informatics unlocked a broad number of algorithms emanating from graph theory to chemists and chemical engineers, leading among others to unique molecule identifiers (SMILES [1], InChI [2]) and powerful substructure matching algorithms. However, these advanced representations and algorithms are only rarely picked up by researchers in the field of reaction engineering. Therefore we have developed a new automatic reaction network generation tool entitled 'Genesys' that takes full advantage of the recent advances made in the field of chemo-informatics. In Genesys graph theory algorithms are applied originating from open-source chemo-informatics libraries such as the *Chemistry Development Kit [3]*. Genesys generates a reaction network consisting of elementary reactions based on a number of user-defined reaction families. Each reaction family consists of a recipe containing the elementary actions to convert reactants into product species, a description of the required sub-molecular pattern inside candidate molecules and a set of constraints, structural features of species that prevent candidate molecules from undergoing a specific reaction family.

Sub-molecular patterns are unambiguously defined using the SMARTS (SMILES Arbitrary Target Specification) [4] language. The identification of these sub-molecular patterns allows to implement Benson's group additivity scheme [5] for the estimation of thermodynamic properties of chemical species in a straight forward way via a combination of group-additive values, ring-strain and non-nearest neighbor corrections.

The application of the reaction network generation program and its general applicability to both catalytic and non-catalytic processes will be discussed. For illustrative purposes different reaction networks have been automatically generated for the pyrolysis of ethane/toluene. The chemical knowledge of the user is reflected by the user-defined set of reaction families and allows to control reaction network size and generation time. Post-processing options such as species and reactions visualization, compatibility with reactor modeling tools such as Chemkin, and tools to analyze the generated reaction mechanism will be discussed.

References

[1]. Weininger, D., SMILES, A CHEMICAL LANGUAGE AND INFORMATION-SYSTEM .1. INTRODUCTION TO METHODOLOGY AND ENCODING RULES. *Journal of Chemical Information and Computer Sciences* **1988,** 28, (1), 31-36.
[2]. Heller, S. R.; Stein, S. E.; Tchekhovskoi, D. V., InChI: Open access/open source and the IUPAC international chemical identifier. *Abstracts of Papers of the American Chemical Society* **2005,** 230, 60-CINF.
[3]. Steinbeck, C.; Han, Y.; Kuhn, S.; Horlacher, O.; Luttmann, E.; Willighagen, E., The Chemistry Development Kit (CDK): an open-source Java library for Chemo- and Bioinformatics. *J Chem Inf Comput Sci* **2003,** 43, (2), 493 - 500.
[4]. Daylight Chemical Information Systems, I. Daylight Theory Manual. http://www.daylight.com/dayhtml/doc/theory/theory.smarts.html
[5]. Cohen, N.; Benson, S. W.; Patai, S.; Rappoport, Z., The thermochemistry of alkanes and cycloalkanes. In *The chemistry of alkanes and cycloalkanes*, Wiley: Chichester, 1992; pp 215-288.

(700e) Model Based Experimental Design and Parameter Estimation for An Industrial Hydrophobic Interaction Chromatography Step From a Quality by Design (QbD) Perspective

Edward J. Close, Department of Biochemical Engineering, University College London, London, United Kingdom, Jeff Salm, Biotherapeutic Pharmaceutical Sciences, Pfizer, Inc, Andover, MA, Jennafer Lyons, Biotherapeutic Pharmaceutical Sciences, Pfizer Inc, Andover, MA, Daniel Bracewell, Biochemical Engineering, University College London, London, United Kingdom and Eva Srensen, Chemical Engineering, University College London, London, United Kingdom

Integrated modelling and experimentation is an exciting approach to the design of chromatographic processes for the manufacture of therapeutic proteins. The approach has significant advantages over traditional process development methods, especially considering the desire to move to a Quality by Design (QbD) processing paradigm. The development of a mechanistic model derived from first principles is, however, often experimentally expensive as well as time consuming which is a major reason for their limited use in a biopharmaceutical industry dominated by time to market pressure.

In this work we begin to address these issues when considering a hydrophobic interaction step from an industrial purification process during the manufacture of a commercial therapeutic protein. Our work considers model based experimental design to optimally direct high throughput experimentation and column studies for the parameter estimation of a detailed first principles chromatography model.

The proposed methodology determines the experimental conditions, parameters to measure, and when to measure those parameters, in order to maximise the information gathered from the measurements in the context of utilisation for estimating model parameters. The methodology enables a model of the desired level of accuracy to be developed quicker than otherwise possible, and at minimum experimental cost. The results can be used to complete a range of highly desirable design tasks, such as resin screening, design space exploration and process optimisation, all key to the successful implementation of a Quality by Design (QbD) process development strategy.

See more of this Session: Model-Based Engineering for Design
See more of this Group/Topical: Computing and Systems Technology Division

(700g) Linking Simulation Tools Using the Pharmahub Work-Flow Management Functionality

Jos Miguel Lanez[1], Michael J. McLennan[2], Linas Mockus[1] and Gintaras V. Reklaitis[1], (1)School of Chemical Engineering, Purdue University, West Lafayette, IN, (2)ITaP, Purdue University, West Lafayette, IN

The pharmaHUB (www.pharmahub.org) is a cyber-infrastructure created for research collaboration and knowledge sharing in the pharmaceutical domain. One of the objectives of pharmaHUB is to contribute and facilitate innovations in the product development cycle for pharmaceuticals. The pharmaHUB is implemented on the HUBzero2 platform which is an open source package that builds a web-based infrastructure to create a collaboration environment where researchers, educators, and students can access and share tools and information. One of the main features of this platform is the capability of performing interactive simulation and modeling tools, shared by the user community, via a web browser. Currently, the pharmaHUB encompasses a range of contributed models and simulations, including particle-surface adhesion, DEM modeling of rotating drums, hopper discharge and high shear mixing, compartment modeling of particle blending, roller compaction, cake filtration, tablet dissolution, visualization of molecular crystals and multipurpose plant production planning among others. These tools address different portions of the product development cycle and are deployed independently. However, it is usually required along the product development cycle to have the outputs of some of these tools as partial inputs for another one. Here, we describe one of the latest functionalities of the pharmaHUB which allows connecting different independent simulation tools using a workflow management model. An example related to the process of individualizing drug dosage regimens is presented to demonstrate the capabilities of this functionality.

See more of this Session: Model-Based Engineering for Design
See more of this Group/Topical: Computing and Systems Technology Division

Computer-Aided Design for a Demonstration-Scale Packed-Bed Tubular Reactor Producing Epichlorohydrin

*Woohyun Kim[1], Choamun Yun[2], Ki Taek Jung[3], Sunwon Park[*1] and Sae Heon Kim[3]*

1. Department of Chemical & Biomolecular Engineering, Korea Advanced Institute of Science and Technology, 335 Gwahak-ro, Yuseong-gu, Daejeon, 305-701, South Korea

2. Doosan Heavy Industries & Construction, 463-1 Jeonmin-dong, Yuseong-gu, Daejeon, 305-811, South Korea

3. Hanwha Chemical R&D Center, 6 Sinseong-dong, Yuseong-gu, Daejeon, 305-804, South Korea

Introduction

Temperature control is crucial when designing a catalytic tubular reactor for an exothermic reaction because hot spots in a packed-bed tubular reactor affect conversion, selectivity and lifespan of solid catalysts (Fogler, 1999; Bartholomew, 2001; Singh et al., 2008). In order to resolve the hot spot problems in scale-up of a packed-bed tubular reactor, a computer-aided design method combining a process modeling software package, heat exchanger design software and computational fluid dynamics (CFD) analysis is proposed. The proposed design method is composed of three steps as follows: Firstly, the length and the number of packed-bed tubes are determined to achieve a target production rate by the simulation results of the pilot-scale reactor model. Secondly, the detailed geometry of the scaled-up reactor comprising multiple tubes is determined using commercial heat exchanger design software. Finally, optimal operating conditions are designated to control the hot spot in the reactor tubes based on the CFD studies. The design method is applied to the scale-up of the previously developed pilot-scale reactor, where epichlorohydrin (ECH) is produced and the production rate is approximately 0.54 kg/h, so that a demonstration-scale multi-tube reactor producing 100 kg ECH/h is designed and the optimal operating conditions of the reactor are also determined.

Methodology

The design methodology that we propose in this work combines three mutually complementary computer-aided strategies: modeling in equation-oriented process modeling and simulation software, e.g. Aspen Custom Modeler and gPROMS, determining macroscopic design with process equipment design software and optimizing operating conditions based on CFD studies. The concept of the methodology and its application to solving the hot spot problem of a packed-bed tubular reactor are explained according to the following steps:

Step 1) Determine the length and the number of tubes based on the simulation of a reactor model: Using a commercial equation-oriented process modeling software package, a dynamic model for a single-tube reactor can be developed in the process modeling and simulation software. The reactor model includes kinetic

---

[*] To whom correspondence should be addressed.

E-mail address: sunwon@kaist.ac.kr; Tel: 82-42-350-3920; Fax: 82-42-862-5961 models and mass and energy balance equations. The optimal length of the tube to maintain the desired conversion level is decided by the simulations of the developed model and the number of the tubes in a multi-tube reactor can also be determined to achieve a target production rate.

Step 2) Determine the geometry of the reactor: Commercial software to design heat exchangers, such as Aspen Shell & Tube Exchanger, can be used to determine the detailed geometry of the multi-tube reactor, e.g. tube layout, geometry of baffles, etc., since the structure and the construction cost of the multi-tube reactor without catalysts are considered to be similar to the ones of shell-and-tube heat exchangers (Fogler, 1999).

Step 3) Optimize operating conditions: Because the heat exchanger design software cannot consider the effects of reaction heat, a high-fidelity systematic method to assess the temperature controllability should be required. CFD analysis is used as a feasible approach for analyzing the performance of process equipment (Dixon & Nijemeisland, 2001; Zeiser et al., 2001; Joshi & Ranade, 2003; Guardo et al., 2004; Marín et al., 2010; Ozden & Tari, 2010). The CFD model of the multi-tube reactor which includes the kinetic models and the heat transfer models in the reactor should be developed so that the reactor design and the appropriate operating conditions affecting the temperature profiles of the tubes are simultaneously analyzed using the model.

Since the proposed method combines a process simulation software package and a process equipment design program, the effects of changes in both design parameters and operating conditions can be analyzed. Therefore, the optimum design and optimal operating conditions are simultaneously determined by the proposed computer-aided design method.

Application

The pilot-scale reactor has a single tube filled with titanium silicate-1 (TS-1) catalyst pellets, where TS-1 catalyzed epoxidation reaction of allyl chloride proceeds with hydrogen peroxide in methanol/water solvent (Gao et al., 1996; Pandey & Kumar, 2007; Kim et al., 2011). The configuration of the reactor is shown in the previously published paper (Kim et al., 2011). The proposed method is applied to scaling up the pilot-scale reactor to a demonstration-scale multi-tube reactor that produces 100 kg ECH/h. When designing the demonstration-scale reactor comprising multiple tubes, the main concern is to make sure that the temperatures of the hot spots caused by the exothermic reaction remain below 313 K. Since temperature higher than 313 K can stimulate the decomposition of hydrogen peroxide, the reactant of the main reaction to form ECH, the productivity of the reactor inevitably decreases (Chen et al., 1998). Hence, the design purpose is to determine an appropriate geometry of a cooling system and optimal operating conditions to control the hot spots affecting the performance of the reactor.

A. Defining the length and the number of the tubes

In the previous work, the mathematical model of the pilot-scale reactor was developed, where the relevant characteristics of the TS-1 catalyst pellets, i.e. catalyst deactivation and effectiveness factor, and the kinetic models were also taken into account in the model. Thus, the pilot-scale reactor model was used to show the complicated dynamic behaviors of the reactor so that the minimum length of the tube to maintain both conversion and selectivity at or above 96 % during 90 days was determined as 8 m long (Kim et al. 2011).

Based on the previous works, the number of the packed-bed tubes and the basic geometry of the multi-tube reactor can be determined to produce 100 kg ECH/h. Since the production rate of the 8-m-long reactor having a single tube is around 0.54 kg ECH/h, approximately 190 - 200 packed-bed tubes are required in order to achieve the target production rate of the demonstration-scale reactor.

To design an efficient cooling system for the hot spot temperature control of the multi-tube reactor, the locations of the hot spots are identified. According to the pilot-scale experiments and the simulation results, the hottest spot is always observed within 2 m from the inlet of the tube. The intensity of other hot spots observed in the rest part of the tube is always weaker than the first one because of the following reasons:

- Lower concentrations of the reactants: The concentrations of the reactants, hydrogen peroxide and allyl chloride, decrease as they flow through the catalyst bed so the rates of the reaction is reduced by the low concentrations of the reactants. Therefore, heat generated from the reactions inevitably decreases.

- Deactivation of the catalyst activity in the long-term operation: The hottest region moves toward the end of the tube in the long-term operation. However, the hot spot found in the rest part of the tube is not as hot as one found near the inlet in the early operation time because the catalytic activity continuously decreases after start-up.

Thus, for efficient cooling of the hottest spot, the cooling system of the demonstration-scale reactor is designed to have three separated cooling jackets: two consecutive 2-m-long jackets followed by one 4-m-long jacket as depicted in **Fig. 1**. Since the hottest spot will be found in the first 2-m-long part, the main purpose of the following work is to determine the optimal design and operating conditions for the first 2-m-long part.

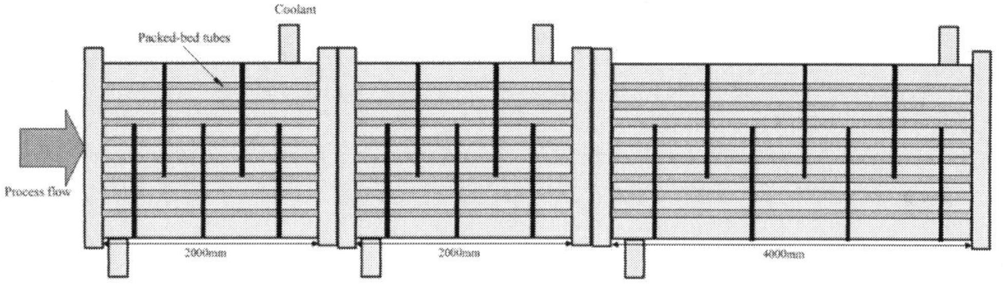

Fig. 1. Conceptual design of a demonstration-scale reactor

B. Defining the geometry of the reactor

The geometry and the construction cost of a multi-tube reactor are similar to those of shell-and-tube heat exchangers. Thus, Aspen Shell & Tube Exchanger, which is well-established commercial software for optimal design of shell-and-tube heat exchangers, is used to determine the geometry of the first 2-m-long reactor (web page for Aspen Exchanger Design & Rating (EDR)). The detailed design parameters of the multi-tube reactor are determined based on the previously determined length of the tubes and the amount of heat exchanged between the process flow and the cooling water of the pilot-scale reactor. The exchanged heat is calculated about 525 W within 2 m from the inlet of the pilot-scale reactor. Since the demonstration-scale reactor has at least 190 tubes, the multi-tube heat exchanger, which is 2 m long, is designed to exchange about 100 kW between the tubes and the cooling water. Based on this information, the geometry of the 2-m-long reactor is determined by Aspen Shell & Tube Exchanger. The relevant design parameters are summarized in **Table 1**.

Table 1. Summary of the geometry of the shell-and-tube reactor

Heat exchanged		100 kW
Shell side	Inner diameter	500 mm
	Outer diameter	510 mm
	Number of baffles	7
Tube side	Tube pattern	30-triangle
	Tube pitch	31.25 mm
	Number of tubes	200

C. CFD modeling

In this section, the CFD model for the 2-m-long reactor having 200 tubes is developed and the simulations of the model are carried out to analyze the effects of different operating conditions, i.e. the direction and the flow rate of the cooling water flowing in the shell-side of the reactor. The developed CFD model includes the catalytic reaction kinetics, the mass and energy balance models in the tubes and the heat transfer model between the tubes and the cooling water. The modeling and the simulation only focus on temperature profiles of the tubes in the early operation time without consideration of the effect of catalytic deactivation because the hottest spot observed within 30 min after start-up is the most difficult to control due to high activity of the fresh catalyst. Thus, the cooling system can control the hot spot temperature during the whole operating period if the hottest temperature, observed within 30 min after start-up, below 313 K. It should be noted that all the cases are assumed to be steady because the temperature profile of the tube seems to be almost steady in the pilot-scale experiments without changes in operating conditions before catalytic deactivation occurs and affects the performance of the reactor. Based on the simulation results, the hot spot temperature control problem of the designed reactor is resolved and the optimal operating conditions are also determined. For the CFD modeling and simulation, the CFD software ANSYS FLUENT 12.0 is used and all the simulations are carried out in an Intel Xeon E5520 processor machine with 16.0 GB RAM.

C. 1. Grid generation and boundary conditions: According to the geometry of the multi-tube reactor designed in Aspen Exchanger Design & Rating, the grid of the reactor is created using DesignModeler 12.0 of the ANSYS 12.0 package (see Figure 4. 5). It has 2,711,040 cells, 8,521,288 faces and 2,601,694 nodes. The maximum cell volume is $1.0717X10^{-6}$ m^3 and the minimum one is $6.4999X10^{-9}$ m^3 (see **Fig. 2**).

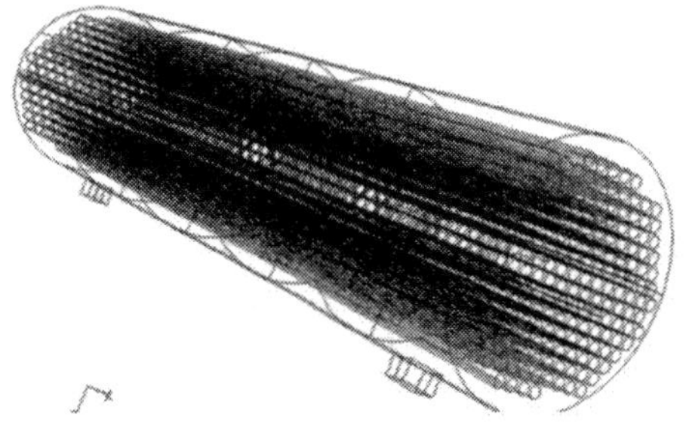

Fig. 2. Grid of the 2-m-long reactor

The boundary condition of the cooling water inlet is specified as a velocity inlet and its velocity magnitude varies at each case. On the other hand, the boundary condition of the cooling water outlet is specified as a pressure outlet and the pressure is set to be 3.0 bar. All the baffles and the shell are specified as a wall without heat transfer and all the shear conditions at the wall are set to be a no-slip condition. For the boundary condition of the tube surface, a heat transfer model is developed to describe energy transport phenomena between the tubes and the cooling water. The model is based on the kinetic models of the reactions and the model describing heat transfer phenomenon between the cooling water and the tubes. They are programmed as a user-defined function (UDF) for the FLUENT solver. Therefore, the CFD model can be used to calculate the amount of the heat generated by the reactions and the heat flux from the tubes to the cooling water with respect to different operating conditions.

C. 2. Modeling details for shell-side: The cooling water flowing in the shell-side is considered as fluid in the FLUENT solver so the governing equations of the cooling water flow comprise both the continuity equation and the momentum conservation equation. To describe energy transport phenomena in the flow, energy equation is also used. These governing equations are shown below without time dependent terms because all the cases are assumed to be steady.

Continuity equation: $\nabla \cdot \left(\rho \vec{v}\right) = 0$ (1)

Momentum conservation equation: $\nabla \cdot \left(\rho \vec{v}\vec{v}\right) = -\nabla p + \nabla \cdot \vec{\tau} + \rho \vec{g}$ (2)

Energy equation: $\nabla \cdot \left(\rho E \vec{v}\right) = -p\nabla \cdot \vec{v} + \nabla \cdot \left(k \nabla T\right) + q + \Phi$ where $E = H - p/\rho + \left|\vec{v}\right|^2 / 2$ (3)

The viscous model for the cooling water flow is specified as a laminar model since the Reynolds numbers of the flow are mostly less than 3,000 in the shell-side. The viscous heating option is also enabled so that the viscous dissipation term in Eq. (3) is calculated.

C. 3. Modeling details for tube-side: The model for the tube-side flow is developed as a one-dimensional model considering only axial direction. The one-dimensional mass balance model includes the reaction kinetic models, the axial diffusion equation and the convective term. All the reaction kinetic models that we used in this model are from our previous work (Kim et al., 2011). The steady-state mass balance model including the kinetic models is shown as Eq. (4).

Mass balance: $D_e \dfrac{d^2 C_i}{dx^2} + R_i = U \dfrac{dC_i}{dx}$ (4)

To describe heat transport phenomena, an energy balance model is also developed. In this energy balance model, it is assumed that all the heat transfer in the tubes is described as conduction so the effective conductivity, k_e, is used and it is experimentally estimated (Bahrami et al., 2006; Kim et al., 2011). The energy balance model for each tube is defined as follows.

Energy balance: $k_e a \dfrac{dT_{center}}{dx} + \sum \Delta H_{Rx,i} R_i - \left[\sum\left(-D_e \dfrac{dC_i}{dx} + UC_i\right)\right] C_p \dfrac{dT_{center}}{dx} = Q_w$ (5)

where $a = \dfrac{\pi d_i L}{\pi d_i^2 L / 4} = \dfrac{4}{d_i}$ and $\dfrac{Q_w}{a} = k_e \dfrac{\left(T_{center} - T_{in,tube}\right)}{d_i / 2} = k_w \dfrac{\left(T - T_{out,tube}\right)}{t_{wall}} = h_o\left(T_{out,tube} - T_c\right)$

In the energy balance model, Q_w is heat flux between the tubes and the cooling water. All the balance equations and the reaction models for the tube-side are programmed as a UDF so the effect of the reaction heat is effectively reflected in the simulation results.

D. Optimization of operating conditions

To find optimal operating condition for the cooling system of the multi-tube reactor, the case studies of the CFD simulations have been carried out. Thus, the optimal operating conditions for managing the hot spot below 313 K are determined. In the case studies, the effects of the flow rate and the direction of the cooling water flowing in the shell-side are investigated. The present section comprises two parts as follows:

Case 1: Counter-current cooling

Case 2: Co-current cooling

In each case, the minimum flow rate to control the hot spot is defined.

D. 1. Case 1: Counter-current cooling: In this case, the performance of the counter-current cooling system is investigated. To determine the optimum flow rate of the cooling water, the flow rate is increased from 5.06 kg/s and the temperature of the cooling water flowing in is set to be 295 K. As the simulation results of this cooling system are shown in **Fig. 3**, the hot spot is observed near the inlet of the process flow. The temperature of the hot spot eventually decreases as the flow rate of the cooling water increases. The temperature of the hot spot remains below 313 K when the flow rate of the cooling water reaches around 11 kg/s.

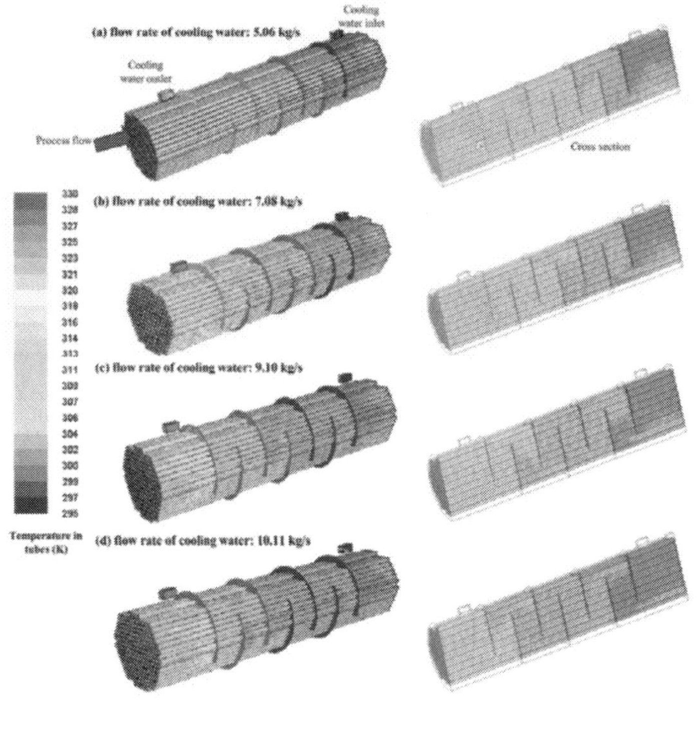

Fig. 3. Temperature contours in the tubes for different flow rate of cooling water (counter-current cooling)

D. 2. Case 2: Co-current cooling: Since the hottest spot is located near the inlet of the tubes in the reactor employing a counter-current cooling system, a co-current cooling system could be more effective to control the hot spot. Thus, the minimum flow rate of the cooling water for this system is also defined by carrying out the simulations of the CFD model and the simulation results of this case are compared with Case 1. The CFD simulation results of Case 2 represented in **Fig. 4** show that the hot spots are found near the outlet of the 2-m-long reactor unlike the previous case. The minimum flow rate of the cooling water to maintain the hottest temperature below 313 K is calculated to be about 11 kg/s, which is the same as the result in Case 1.

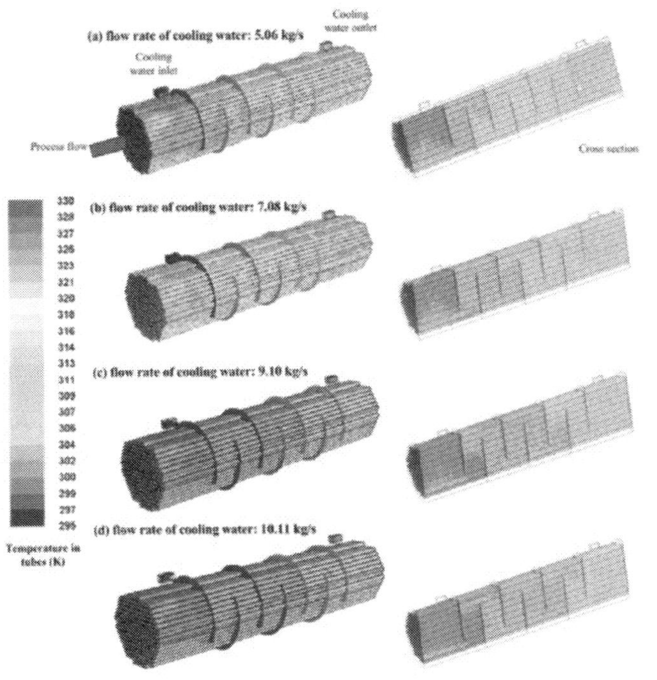

Fig. 4. Temperature contours in the tubes for different flow rate of cooling water (co-current cooling)

Conclusions

To resolve the design and control problems related to hot spots in a catalytic tubular reactor, the integrated computer-aided design method was proposed. This proposed design method was practically applied to scaling up the previously developed pilot-scale reactor to demonstration-scale reactor producing 100 kg ECH/h. In conclusion, the demonstration-scale reactor having 200 packed-bed tubes was successfully designed and the effects of the changes in the operating conditions, such as the direction and the flow rate of the cooling water, were also investigated to determine the optimal operating conditions. Therefore, the proposed computer-aided design method can be used for scaling up of catalytic tubular reactors in the chemical industry.

Nomenclature

C_i: concentration of component i

C_p: heat capacity of mixture

D_e: effective diffusivity

d_i: tube inner diameter

h_o: convective heat transfer coefficient

k: thermal conductivity

k_e: effective conductivity

k_w: conductivity of tube

L: length of a tube

R_i: rate of the formation of the consumption of component i

T: temperature of process flow

T_c: cooling water temperature

T_{center}: temperature at the center of the tube

$T_{in,tube}$: temperature at the inner surface of the tube

$T_{out,tube}$: temperature at the outer surface of the tube

U: velocity of fluid in axial direction

Q_w: heat flux from the outer surface of the tube to the cooling water

Greek letters

$\Delta H_{Rx,i}$: Heat of reaction forming component i

η : effectiveness factor of TS-1 catalyst pellets

τ : shear stress

Φ : dissipation function

References

Aspen Technology, Inc., Aspen Exchanger Design & Rating: http://www.aspentech.com/core/aspen-edr.aspx

Bahrami, M., Yovanovich, M. M., & Culham, J. R. (2006). Effective thermal conductivity of rough spherical packed beds. Int. J. Heat Mass Tran., 49 (19-20), 3691-3701.

Dixon, A. G. & Nijemeisland, M. (2001). CFD as a design tool for fixed-bed reactors. Ind. Eng. Chem. Res., 40 (23), 5246-5254.

Fogler, H. S. (1999). Elements of Chemical Reaction Engineering (3rd ed.). Upper Saddle River, New Jersey: Prentice-Hall, Inc.

Gao, H., Lu, G., Suo, J., Li, S. (1996). Epoxidation of allyl chloride with hydrogen peroxide catalyzed by titanium silicalite 1. Appl. Catal. A-Gen., 138 (1), 27-38.

Guardo, A., Coussirat, M., Larrayoz, M. A., Recasens, F., & Egusquiza, E. (2004). CFD flow and heat transfer in nonregular packings for fixed bed equipment design. Ind. Eng. Chem. Res., 43 (22), 7049-7056.

Joshi, J. B., & Ranade, V. V. (2003). Computational fluid dynamics for designing process equipment: Expectations, current Status, and path forward. Ind. Eng. Chem. Res., 42 (6), 1115-1128.

Kim, W., Yun, C., Kim, Y., Park, J., Park, S., Jung, K. T., Lee, Y. H., & Kim, S. H. (2011). Modeling of a tubular reactor producing epichlorohydrin with consideration of reaction kinetics and deactivation of titanium silicate-1 catalyst. Ind. Eng. Chem. Res., 50 (3), 1187-1195.

Marín, P., Ordóñez, S., & Díez, V. F. (2010). Rational design of heating elements using CFD: Application to a bench-scale adiabatic reactor. Comput. Chem. Eng., in press.

Ozden, E., & Tari, I. (2010). Shell side CFD analysis of a small shell-and-tube heat exchanger. Energ. Convers. Manage., 51 (5), 1004-1014.

Pandey, R. K., & Kumar, R. (2007). Eco-friendly synthesis of epichlorohydrin catalyzed by titanium silicate (TS-1) Molecular Sieve and Hydrogen Peroxide. Catal. Commun., 8 (3), 379-382.

Singh, S., Lal, S., & Kaistha, N. (2008). Case study on tubular reactor hot-spot temperature control for throughput maximization. Ind. Eng. Chem. Res., 47 (19), 7257-7263.

Strasser, W. (2010). CFD study of an evaporative trickle bed reactor: Mal-distribution and thermal runaway induced by feed disturbances. Chem. Eng. J., 161 (1-2), 257-268.

Zeiser, T., Lammers, P., Klemm, E., Li, Y. W., Bernsdorf, J., & Brenner, G. (2001). CFD-calculation of flow, dispersion and reaction in a catalyst filled tube by the lattice Boltzmann method. Chem. Eng. Sci., 56 (4), 1697-1704.

(710a) Online Outlier Detection with a Bayesian Supervisory Approach for Recursive Soft Sensor Update

Hector Galicia[1], Qinghua He[2] and Jin Wang[1], (1)Department of Chemical Engineering, Auburn University, Auburn, AL, (2)Department of Chemical Engineering, Tuskegee University, Tuskegee, AL

Partial least squares (PLS) based soft sensors that predict the primary variables of a process by using the secondary measurements have drawn increased research interests recently [1]. Such data-driven soft sensors are easy to develop and only require a good historical data set. In our previous work [2], a reduced-order dynamic PLS (RO-DPLS) soft sensor was developed to address some limitations of the traditional DPLS soft sensor when applied to processes with large transport delays. By taking the process characteristics into account, RO-DPLS soft sensor can significantly reduce the number of regressor variables and improve prediction performance.

As industrial processes often experience time-varying changes, it is desirable to update the soft sensor model with the new process data once the soft sensor is implemented online [3]. Recently, we extended the RO-DPLS soft sensor to its online adaptation version in order to track process changes [4]. Since our focus in [4] was to investigate the properties of different recursive updating schemes and data scaling methods, we preprocessed the industrial datasets to remove all outliers before subjecting them to different experiments.

However, it has been recognized that the PLS algorithms are sensitive to outliers in the dataset [5]. Therefore, outlier detection and handling plays a critical role in the development of the PLS based soft sensors, and there exist extensive studies on outlier detection for off-line model building. Despite many published results on outlier detection, it is still an unsolved problem, and it is often recommended to accompany any outlier detection method with a graphical inspection of the residual space and model parameters [6] to eliminate any possible outlier masking effect (i.e., outliers are classified as consistent samples) and outlier swamping effect (i.e., consistent samples are classified as outliers).

In this work, we propose multivariate approaches for both off-line outlier detection (for initial soft sensor model building) and online outlier detection (for soft sensor model recursive update). Specifically, for off-line outlier detection we combine leverage and y-studentized residuals; while for online outlier detection, we use squared prediction error indices for **X** and **Y** to monitor the independent variable and dependent variable space, respectively. For online outlier detection, to differentiate the outliers caused by erroneous reading from those caused by process changes, we propose a Bayesian supervisory approach to further analyze and classify the identified outliers. Both simulated and industrial case studies of a Kamyr digester show that the Bayesian supervisory approach is very effective in differentiating the outliers caused by erroneous readings from those caused by a process change, which enables the superior performance of the soft sensor with Bayesian supervisory approach.

[1] Kadlec, P.; Gabrys, B. & Strandt, S. Data-driven Soft Sensors in the process industry. *Computers & Chemical Engineering*, 2009, 33, 795-814.

[2] Galicia, H. J.; He, Q. P. & Wang, J. A reduced order soft sensor approach and its application to a continuous digester. *Journal of Process Control*, 2011, 21, 489-500.

[3] Kadlec, P.; Grbic, R. & Gabrys, B. Review of adaptation mechanisms for data-driven soft sensors. *Computers & Chemical Engineering*, 2011, 35, 1-24.

[4] Galicia, H. J.; He, Q. P. & Wang, J. Comparison of the performance of a reduced-order dynamic PLS soft sensor with different updating schemes for digester control. Submitted to *Control Engineering Practice*, 2011.

[5] Hubert, M. and Branden, K. V. Robust methods for partial least squares regression. *Journal of Chemometrics*, 2003, 17, 537-549.

[6] Martens, H. and Naes, T. Multivariate Calibration. *John Wiley and Sons Ltd.*, 2002.

See more of this Session: Process Monitoring and Fault Detection
See more of this Group/Topical: Computing and Systems Technology Division

(710b) MPC Performance Monitoring and Evaluation Principles

Luo Ji and James B. Rawlings, Chemical and Biological Engineering, University of Wisconsin-Madison, Madison, WI

Abstract

Over the past 30 years, Model Predictive Control (MPC) has become a dominant control technology and utilization of MPC loops has increased rapidly in the process and manufacturing industries [5]. The widespread utilization of MPC naturally calls for the need of performance assessment, monitoring and analysis [1]. A control performance problem usually consists of the goal-achievement evaluation of a single controller, performance rank ordering of multiple controllers, and corresponding performance maintenance or trouble shooting. However, some surveys indicate that only about one third of controllers actually provide an acceptable level of performance [2], and there is no well established, systematic control theory or technology that enables MPC practitioners to routinely evaluate the operations of their MPC systems. Therefore, industrial practice has a large and growing need for MPC controller performance monitoring and analysis.

For general control technologies other than MPC, there are some methods of performance evaluation. The most famous one is the Harris Index, or the minimum variance (MV) benchmark [3], which is based on the assumption that the smaller the variance of the output around its desired setpoint, the better the performance. Theoretically, the variance of an ideal controlled output has only the effect of the measurement noise. Comparison of the actual plant output variance with this theoretical minimum is a measure of control performance. There are also some other types of benchmarks, such as the historical benchmark, the LQG (Linear Quadratic Gaussian) benchmark, and the user-defined benchmark [7, 6, 8]. However, these benchmarks are not suitable for MPC controllers because of the complexity of MPC system models, the process noise in the system dynamics, limitations in current algorithms, and the use of suboptimal MPC optimization. Even with the current best MPC algorithm, those complexities would prevent the practical MPC output variances to be close to the theoretical minimum value. Therefore, these benchmarks usually provide an unrealistic and unreachable performance standard from the practical viewpoint.

This presentation introduces a new strategy of MPC performance evaluation, which takes advantage of the system model information as well as the autocovariance least squares (ALS) technology [4]. ALS is able to provide estimates of the covariances of both the process and measurement noise from provided plant input/output data. An MPC simulation can be performed using these estimated noises to mimic the plant operating condition. This simulated performance is referred to as the average achievable performance of a perfectly tuned MPC controller using current algorithms and in the presence of the plant noises. To numerically evaluate the performance, a key performance index (KPI) is defined as a quadratic function of the output and input tracking error. Comparison between the plant KPI and the simulated KPI indicates the current plant performance condition. Different calculation methods of KPI are discussed for linear models applied to unconstrained and constrained cases. Results verify that the simulated KPI converges to the plant KPI with perfect modeling. The simulated KPI fails to converge to the plant KPI when there is some process model mismatch or noise model mismatch, indicating the plant performance is worse than the expectation. Furthermore, for the trouble shooting of poor-performing systems, a performance evaluation scenario is proposed: four levels of simulated KPI [1] as well as the plant KPI will be calculated; the value difference of adjacent KPI levels implies what prevents our current system to have a better performance and how to improve it.

Two industrial case studies are investigated. The first one is a linear and unconstrained system, but with large nonzero-mean deterministic disturbance detected in the output data. The expected performance shown by the simulated KPI is much better than the actual plant performance, because the effect of the deterministic disturbance is not considered in the nominal simulated KPI. To address this problem, a deterministic-disturbance-affected KPI is defined and calculated using the sampled disturbance data, providing a more realistic performance standard.

The second case study is a nonlinear and constrained system, controlled with PI controllers. With the entire system model provided, we show that both a simulated PI KPI and a simulated MPC KPI could be calculated simultaneously. The comparison between simulated PI KPI and plant KPI suggests the plant controller is not well-tuned and what plant performance could be expected with improved PI control; while the comparison between simulated PI KPI and simulated MPC KPI illustrates how much plant performance improvement could be expected if an MPC layer is installed to replace the PI layer. This information will help in deciding if an MPC controller implementation is worthwhile.

References

[1] M. Bauer and I.K. Craig. Economic assessment of advanced process control–A survey and framework. J. Proc. Cont., 18(1):2–18, 2008.

[2] D.B. Ender. Process control performance: Not as good as you think. J. Proc. Cont., 40(10):180–190, 1993.

[3] Thomas J. Harris. Assessment of control loop performance. Can. J. Chem. Eng., 67:856–861, 1989.

[4] Brian J. Odelson, Murali R. Rajamani, and James B. Rawlings. A new autocovariance least-squares method for estimating noise covariances. Automatica, 42(2):303–308, February 2006.

[5] S. Joe Qin and Thomas A. Badgwell. A survey of industrial model predictive control technology. Control Eng. Prac., 11(7):733–764, 2003.

[6] S.J. Qin and J. Yu. Recent developments in multivariable controller performance monitoring. J. Proc. Cont., 17(3):221–227, 2007.

[7] J. Schafer and A. Cinar. Multivariable MPC system performance assessment, monitoring, and diagnosis. J. Proc. Cont., 14(2):113–129, 2004.

[8] S.L. Shah, R. Patwardhan, and B. Huang. Multivariable controller performance analysis: methods, applications and challenges. Chem. Proc. Cont., 98:190–207, 2002.

1
Four simulated KPIs are: linear and unconstrained KPI, linear and constrained KPI, linear and constrained with identified deterministic disturbance KPI, nonlinear and constrained KPI.

See more of this Session: Process Monitoring and Fault Detection
See more of this Group/Topical: Computing and Systems Technology Division

(710c) Nonlinear State Estimation Using Parent and Nested Particles In Sequential Monte Carlo

Sridhar Ungarala, Chemical and Biomedical Engineering, Cleveland State University, Cleveland, OH

State estimation is a fundamental element of engineering operations because it provides the knowledge for implementing process monitoring, control, optimization and safety related tasks. Many engineering systems tend to be nonlinear in nature and the uncertainties may be characterized by non-Gaussian pdfs. The recursive Bayesian solution to estimation is available for half a century, but faithful implementation is restricted to linear Gaussian systems.

Recently there has been a surge of interest in solving the nonlinear non-Gaussian conditional mean estimation, following the introduction of sequential Monte Carlo approach or particle filters. It is possible to design fast and accurate particle filters for nonlinear state estimation problems in engineering, tracking and guidance, meteorology and oceanography. There has been a flurry of recent research activity in chemical engineering regarding nonlinear and constrained estimation using the Monte Carlo approach with various simplifications such as Gaussianity and statistical linearization. Optimistic projections for the spread of particle filters in industrial practice are based on the availability of cheap computing resources. While computational cost is a crucial factor, the Monte Carlo approach can also gain in performance by focusing on a multiscale view of the state space. The effects of computationally expedient approximations can be limited to localized state space at a smaller scale while significantly alleviating the effects on the global state space of the filter.

The particle filter is a generic sequential Monte Carlo implementation of the optimal Bayesian filter. It approximately retrieves the properties of the a posteriori density by using a set of samples (particles) from the density. However, it is generally not feasible to compute the a posteriori density in closed-form, such that samples may be drawn from it. The particle filter is a combination of importance sampling and resampling techniques in lieu of sampling the conditional density. The importance sampling technique provides a means to sample a set of relevant support points (particles) from a chosen importance density that is amenable to sampling. The support points are assigned appropriate weights that reflect their relative importance such that the particles and associated weights provide an approximate discrete representation of the unknown a posteriori density.

The importance density is clearly a crucial filter design choice. Any probability measure with support including the support of the a posteriori density may be chosen as the importance density. The optimal importance density minimizes the variance of the weights conditioned on states and measurements. However, the optimal importance density is generally not available, which leads to suboptimal choices. The transition density is the most frequently used suboptimal importance density because it is readily available. Although easy to use, the transition density can be a poor choice in cases where there is little overlap between the importance density and the likelihood function. This can be particularly problematic when highly accurate measurements yield narrow and peaked likelihood functions. Furthermore, the transition density has no knowledge of the latest measurement information, as a result of which many of the the support points may not be naturally located in the high likelihood region suggested by the latest measurement. In practice, one needs to resort to simplifications such as linearization of nonlinear transformation and/or fixed shape Gaussian approximation to estimate the support of practically useful importance densities. Both approximations can be detrimental in revealing the importance support when applied for approximation on a global scale, i.e. over all of the appropriate state space. These effects may be mitigated by searching for multiple importance densities supported on state space that is localized in scale instead of a single importance density on a global scale.

It is proposed here that nonlinear transformation of the conditional density and Bayesian update with measurement is achieved by localized importance sampling, performed over multiple subsets of state space (local scale) that results in sampling the support of the desired density (global scale). The proposed approach uses nested particle sets related to the parent particles of the filter. The purpose of the nested child particles is to generate local (nested) importance densities with a knowledge of the measurement information as well as the local behavior of nonlinearity in state space. These local scale importance densities in turn generate the parent particles as support points for the determination of the global importance weights. The forward propagations of the nests in time are representative of the local behavior of the system nonlinearity in state space and the corresponding local transition density. It is conceivable that the propagated child particles may be corrected with the measurement information using various forms of approximate nonlinear filters, including the particle filter itself. The nesting approach broadly subsumes previously known local linearization extended Kalman particle filter and the the unscented particle filter as well as lay out a framework for the inclusion of other nonlinear filters in the particle filter approach.

This paper investigates the formulation of local importance densities as Gaussian approximations derived from the nested particles. More generally, it is also proposed here that kernel density estimation may be used to formulate more accurate and appropriate importance densities by avoiding the Gaussian approximation. Unlike the existing approaches of local linearized particle filters, the proposed nesting approach allows for two-way interaction between the nested particles (localized in state space) and the parent particles (over global state space of the filter). The gains in performance are demonstrated with a benchmark nonlinear estimation problem from the literature. Preliminary results indicate that the estimation error can be reduced by orders of magnitude with small increase in computational cost due to additional particles in the nests.

See more of this Session: Process Monitoring and Fault Detection
See more of this Group/Topical: Computing and Systems Technology Division

(710d) A Novel Framework for Multi-Mode Process Modeling and Monitoring

Zhibo Zhu[1], Zhihuan Song[1] and A. Palazoglu[2], (1)Control Science and Engineering, Zhejiang University, Hangzhou, China, (2)Chemical Engineering & Materials Science, UC Davis, Davis, CA

With vast amounts of process data collected by the distributed control systems, multivariate statistical process control (MSPC) has been gaining significant momentum as a viable response to process monitoring and fault diagnosis issues in process industries. In most of these applications, the process is often assumed to operate at a single design point subject to disturbances and faults. In practice, however, chemical processes frequently operate at multiple regimes associated with product changeovers, capacity modifications and others. In the presence of such dynamic transitions that are part of the normal process behavior, detection and diagnosis of faults often become confounded.

In this paper, a novel framework for process pattern construction and multi-mode monitoring is proposed. To identify process patterns, the framework utilizes a clustering method that consists of an ensemble moving window strategy along with an ensemble clustering solution strategy. Three graphical visualization tools are proposed to validate the ensemble clustering solution. Next, a new *k*-principal component analysis-independent component analysis (*k*-ICA-PCA) modeling is developed to capture the relevant process patterns in corresponding clusters and facilitates the validation of ensemble solutions. Following pattern construction, the proposed framework offers an adjoined multi-ICA-PCA model for detection of faults under multiple operating modes.

The Tennessee Eastman (TE) benchmark process is used to demonstrate the salient features of the method. Specifically, the proposed method will be shown to have superior performance compared to the previously reported *k*-PCA models clustering method.

See more of this Session: Process Monitoring and Fault Detection
See more of this Group/Topical: Computing and Systems Technology Division

(710e) Data-Based Monitoring and Retuning of Low-Level PID Control Loops

Akradej Leosirikul[1], David Chilin[2], Jinfeng Liu[3], Jim Davis[4] and Panagiotis D. Christofides[3], (1) UCLA, Los Angeles, CA, (2)Chemical and Biomolecular Engineering, Univ. of California, Los Angeles, Los Angeles, CA, (3)Department of Chemical and Biomolecular Engineering, University of California, Los Angeles, Los Angeles, CA, (4)Chemical and Biomolecular Engineering, UCLA, Los Angeles, CA

Model predictive control (MPC) is widely used in industry because of its ability to handle input/state constraints and to incorporate optimization considerations in a single formulation. In general, in the calculation of the optimal input trajectories for the manipulated inputs via MPC, the dynamics of the corresponding control actuators that will implement the control actions computed by the MPC are neglected and the MPC-computed control actions are assumed to be directly implemented by the control actuators. However, in practice, these control actuators have their own specific dynamics. As a result of this, there are always discrepancies (i.e., time lags, magnitude differences, etc.) between the actual control actions applied to the process by the control actuators and the actions requested by the MPC. To mitigate the influence of these discrepancies in closed-loop performance, PID controllers (typically called "low-level" PID controllers) are usually implemented on the control actuators to regulate the outputs of the actuators at the values requested by the MPC. In this case, the tuning of the PID controllers is critical for the overall control actuator and closed-loop system performance. An actuator with a well-tuned PID controller can effectively implement the actions requested by the MPC; whereas, an actuator with a poorly-tuned PID controller may reduce the performance of the closed-loop system dramatically or may even cause instability of the closed-loop system.

In the present work, we design a monitoring system to detect poorly-tuned PID controller used to regulate control actuators. The monitoring system employs a model-based dynamic filter used to compute suitable residuals that are influenced by PID controller behavior and threshold-levels that are computed from historical closed-loop process data under normal operation to detect and isolate poorly-tuned low-level PID controllers. The monitoring scheme captures the difference between the response of the outputs of the closed-loop process as obtained by the measurement sensors and the expected behavior under well-tuned PID controllers in order to detect poor PID tunings in the actuators. Some criteria are also developed to differentiate a poor PID tuning from an actuator fault based on the evolution of the magnitude and rate of change of the with respect to time. An automatic strategy for retuning the poorly-tuned PID controllers is also discussed. An example of a nonlinear reactor-separator process network under an MPC controller with its actuators under low-level PID controllers is used to demonstrate the approach.

See more of this Session: Process Monitoring and Fault Detection
See more of this Group/Topical: Computing and Systems Technology Division

(710f) Control Loop Performance Assessment Using Detrended Fluctuation Analysis

Babji Srinivasan[1], **Tim Spinner**[2] and Raghunathan Rengasamy[2], (1)Department Chemical Engineering, Texas Tech University, Lubbock, TX, (2)Department of Chemical Engineering, Texas Tech University, Lubbock, TX

Across the process industries there exist millions of automatic control loops, numbering several hundred for each process control engineer . Up to 60% of these controllers have less than acceptable levels of performance, with a corresponding deleterious effect on profits that could range into the hundreds of millions of dollars annually . The first step to improving control loop performance is to determine which loops are performing poorly . Process control engineers must divide their time between implementing new control assets and maintaining existing ones, so any detection method for poor performance should be fully automated. Integral of squared error and output variance are commonly monitored measures of loop performance, but these statistics are essentially meaningless unless compared against a benchmark . Devries & Wu and Harris have proposed benchmarking the mean squared error of the output against the theoretical minimum variance. Signicantly, the method of Harris uses knowledge of the process time delay along with routine operating data to find the controller invariant part of the process disturbance in order to calculate the minimum variance. This benchmark is based on the theoretical framework of minimum variance control developed by Astrom. However, time delay determination (for computation of minimum variance) in most cases will require the control engineers to schedule a identication test with operating personnel. This is because routine plant operating data will not in general contain sufficient information for time delay determination, unless there have been abrupt control signal changes or external excitations during data collection.

In this work, a new index is developed for assessing the performance of a single-input single-output (SISO) linear feedback control loop. The proposed metric is a specic scaling of the generalized Hurst exponent, computed through the method of detrended fluctuation analysis (DFA). We refer to this scaled exponent as the Hurst index. The new method compares favorably with the widely used minimum variance index (MVI), with both indices showing similar trends under changes in controller tunings during closed-loop simulations. The main advantage of the Hurst index over MVI and other existing performance measures is that its determination does not require a priori knowledge of any loop parameters. Instead, computation of the index relies solely upon process output data collected during routine plant operation. Therefore, this new technique could potentially allow engineers to more efficiently identify problematic control loops.

See more of this Session: Process Monitoring and Fault Detection
See more of this Group/Topical: Computing and Systems Technology Division

(710g) Fault Detection and Fault-Tolerant Control of Particulate Processes with Sampled and Delayed Measurements

Trina Napasindayao and **Nael H. El-Farra**, Department of Chemical Engineering and Materials Science, University of California, Davis, Davis, CA

The development of systematic methods for fault detection and fault-tolerant control of particulate processes is a fundamental problem whose practical significance encompasses a wide range of important processing industries including agricultural, chemical, food, minerals, and pharmaceuticals. A key motivation for studying this problem is the fact that the loss of control authority due to control system failures can impact the particle size distribution, and ultimately harm the end-product quality if faults are not properly diagnosed and handled. Major bottlenecks in the design of model-based fault-tolerant control systems for particulate processes include the infinite-dimensional nature of the process model as well as the complex nonlinear and uncertain dynamics of particulate processes. An effort to address these problems was initiated in [1] where a methodology for the detection and handling of actuator faults in particulate processes was developed on the basis of appropriate low-order models that capture the dominant process dynamics. These results were subsequently generalized in [2] to address the problems of fault isolation and model uncertainty.

Beyond the problems of nonlinearities and uncertainty, there is a host of practical implementation issues that need to be accounted for in the design of the fault-tolerant control system, including the issues of measurement sampling and delays. In particulate processes, measurements of the principal moments of the particle size distribution (obtained, for example, using light scattering techniques) and measurements of the continuous phase variables (e.g., solute concentration and temperature in a crystallizer) are typically available from the sensors at discrete time instances and are delayed. The frequency and times at which the measurements are received by the control system are constrained by the inherent limitations on the data collection, processing and transmission capabilities of the measurement sensors. These limitations impose restrictions not only on the implementation of the controller but also on the ability to accurately monitor the process evolution, which in turn can erode the diagnostic and control system reconfiguration capabilities of the fault-tolerant control system if not explicitly accounted for in the monitoring and control system design.

In this work, we develop a model-based framework for actuator fault detection and reconfiguration in particulate processes modeled by population balance equations with discretely-sampled and delayed measurements. Initially, model reduction techniques are used to derive an approximate finite-dimensional system that captures the dominant dynamics of the particulate process. An observer-based output feedback controller is then designed on the basis of this system to stabilize the process in the absence of faults. To compensate for the lack of continuous measurements, an inter-sample model predictor is included within the control system to provide the observer with an estimate of the process output when measurements are not available from the sensors. The model state is then updated when measurements are received at discrete times. To compensate for the measurement delay, the control system includes a propagation unit that uses the low-order model together with the past values of the control input to calculate an estimate of the current output from the received delayed measurements. This estimate is then used to update the inter-sample model predictor which, together with the controller, generates the control signal for the process. By formulating the closed-loop system as a combined discrete-continuous system, an explicit characterization of the minimum allowable sampling rate that guarantees stability in the absence of faults is obtained in terms of the plant-model mismatch, the controller and observer design parameters, the size of the measurement delay, and the choice of the control configuration. The characteristic fault-free closed-loop behavior obtained from this analysis is used as the basis for deriving appropriate rules for fault detection and control system reconfiguration. The idea is to use the state observer as a fault detection filter and compare its output with the estimate of the current plant output generated by the propagation unit at the sampling times. The discrepancy is used as a residual and compared against a time-varying alarm threshold obtained from the stability analysis to determine the fault or health status of the control actuators. Once a fault is detected in the operating control configuration, the control system is prompted to switch to one of the feasible fall-back control configurations under the given measurement sampling rate and delay to preserve stability and minimize performance deterioration. Finally, the proposed fault-tolerant control framework is illustrated using a simulated model of a continuous crystallizer with a fines trap.

References:

[1] El-Farra, N. H. and A. Giridhar, ``Detection and Management of Actuator Faults in Controlled Particulate Processes Using Population Balance Models," *Chem. Eng. Sci.*, 63:1185-1204, 2008.

[2] Giridhar, A. and N. H. El-Farra, ``A Unified Framework for Robust Fault Detection, Isolation and Compensation in Uncertain Particulate Processes," *Chem. Eng. Sci.*, 64:2963-2977, 2009.

See more of this Session: Process Monitoring and Fault Detection
See more of this Group/Topical: Computing and Systems Technology Division

(723a) Robust Hybrid Control of Nonlinear Process Systems Subject to Control and Communication Constraints

Ye Hu and Nael H. El-Farra, Department of Chemical Engineering and Materials Science, University of California, Davis, Davis, CA

Many chemical processes are characterized by strong nonlinearities, uncertain dynamics and input constraints. Not surprisingly, these problems have been the subject of significant research work within process control over the past two decades, and led to the development of systematic methods for robust nonlinear and constrained control (see, for example, [1] for some results and references in this area). A close examination of the available results, however, shows that the control problem is typically formulated and addressed within the traditional feedback control setting where the sensors are assumed to transmit the collected measurements continuously and directly to the controller. With the increasing complexity of the process/controller interface in modern control systems, there is a fundamental need to re-examine the classical feedback control paradigm in ways that take into account the additional information-processing devices involved. Emerging issues such as resource constraints, processing and communication delays, data sampling and losses, measurement quantization and real-time scheduling constraints, challenge many of the assumptions in traditional process control methods and need to be integrated explicitly in the control system design.

In this contribution, we consider the problem of robust stabilization of nonlinear process systems subject to control and communication constraints, where the control constraints are due to the physical limitations on the capacity of the control actuators, and the communication constraints are the result of inherent limitations on the measurement sensor sampling rate. To compensate for the unavailability of measurements between sampling instants, an inter-sample model predictor of the closed-loop system is used to provide estimates of the process state that the controller can use to calculate the control action. However, the co-presence of control and sampling rate constraints creates a conflict in the control design objectives in that the set of initial conditions starting from which stability can be achieved (i.e., the stability region) is limited by the control constraints, while the invariant region that the process state can eventually be steered to (i.e., the terminal set) is largely determined by the model uncertainty (the discrepancy between the dynamics of the actual process and that of the inter-sample model predictor), and it is usually the case that the control configuration that has the largest stability region does not have the smallest terminal set. As a result, steering the process state to a small neighborhood of the nominal steady-state of the system starting from a large initial condition cannot, in general, be achieved using a single control configuration. To address this problem, we propose a hybrid control strategy that switches between different control configurations. Specifically, a Lyapunov-based nonlinear controller that enforces practical closed-loop stability is initially synthesized for each control configuration, and the corresponding stability region and terminal set are then characterized in terms of the control constraints, the sampling period and the model uncertainty. To resolve the conflict mentioned earlier, a switching strategy that orchestrates a finite number of transitions between the different control configurations is devised based on the location of the closed-loop state with respect to the stability regions and terminal sets at any given time. In addition, the minimum number of transitions needed for the state to evolve from a given initial condition to a given terminal set is also characterized. Finally, the implementation of the developed hybrid control method is demonstrated using a chemical process example.

References

[1] P. D. Christofides and N. H. El-Farra, *Control of Nonlinear and Hybrid Process Systems: Designs for Uncertainty, Constraints and Time-Delays*. Berlin/Heidelberg: Springer-Verlag, 2005.

See more of this Session: Advances In Process Control
See more of this Group/Topical: Computing and Systems Technology Division

(723b) Lyapunov-Based Model Predictive Control of Stochastic Nonlinear Processes

Maaz Mahmood and Prashant Mhaskar, Chemical Engineering, McMaster University, Hamilton, ON, Canada

Incorporation of the complexity of system dynamics in the control design is central to achieving improved performance and closed–loop stability. Such characteristics can include highly nonlinear behavior, uncertainty (typically in the form of additive disturbances and/or uncertain model parameters), and input constraints. Neglecting these characteristics within the control design can lead to performance degradation or even closed-loop instability. Owing to the constraint handling ability of nonlinear model predictive control (NMPC), along with the ability to incorporate an explicit system model, the NMPC framework has been widely utilized to design robust, constrained optimization based controllers. In most NMPC approaches, the manipulated input trajectory is computed at each sampling time via dynamic optimization, where a cost function is minimized subject to a nonlinear dynamic system and input/state constraints. Several research studies dealing with NMPC have focused on issues such as feasibility, stability, constraint satisfaction, and uncertainty [1, 2] including Lyapunov-based NMPC (LMPC) designs [3] that provide *a priori* (i,e, before controller implementation or testing for feasibility), an explicit characterization of initial conditions from where stability and feasibility of the closed–loop system is guaranteed in the presence of constraints and bounded uncertainty.

All the aforementioned work on NMPC is however, dominated by use of deterministic system models with very few results using the stochastic nature of the process in the control design. In the existing results, stochastic noise is handled via use of inherently 'worst-case' robust NMPC schemes (e.g. [4]) where the uncertainty term is assumed to be bounded, however, such formulations are typically very numerically expensive, and as a result can impede on-line implementation. Moreover, the assumption of bounded disturbance, without the use of statistical information about the disturbances can lead to conservative control laws and thus degrade system performance. A natural alternative is to use stochastic unbounded system noise in the controller design.

This alternative approach has recently been addressed under the framework of stochastic MPC where the disturbances are modeled as random variables and the expected value of a cost function is minimized [5, 6, 7]. Although stochastic MPC circumvents the challenge of determining an a priori bound on the noise and also the conservatism originating from the use of a worst-case framework, it gives rise to several other challenging issues. Namely, the optimization problem is generally a stochastic program which induces significant computational burden. For example, the cost function requires the explicit calculation of a conditional expectation and/or probability associated with multi-dimensional random variables. For general nonlinear systems, this is a non-trivial task, and often requires resorting to probability density approximation techniques. Another conceptual challenge in the extension of MPC to stochastic systems is the concept of stability. Developments in probabilistic robust control, have shown that instead of stability guarantees under worst-case realizations of the uncertainty, control system performance with stochastic uncertainty can be improved by introducing a well-defined risk of instability. That is, the closed–loop trajectory will be only be able to reach a desired target region with an associated probability. Yet, such developments have not been applied to the MPC design. Moreover, Lyapunov techniques for stability analysis and control design for stochastic nonlinear systems do exist, albeit not as finely polished as their deterministic counterparts. The key hurdle in the use of Lyapunov techniques for stochastic systems is the presence of an additional Hessian term within the stochastic derivative. Nevertheless, there do exist stabilizing (in a suitable stochastic sense) control laws using stochastic Lyapunov techniques that provide explicitly-defined regions of attraction (in a probabilistic sense) for the closed–loop system [8, 9]. In fact, such results shadow the deterministic counterpart, and have recently been used to derive regions of attraction with well defined risk measures [10].

Deterministic Lyapunov-based control designs have been recently united with predictive control schemes to provide an explicit characterization of the states from where closed–loop stability is guaranteed in the presence of constraints and bounded uncertainty [11]. The analogous unification has yet to be addressed within the context of stochastic unbounded uncertainty. Hence, the MPC approach does stand to gain from the theoretical development in the area of stochastic Lyapunov-based bounded controllers. The incorporation of explicit optimality considerations in the control design in the MPC framework, together with explicit characterizations of states with well defined risk of instability measures which are derived using stochastic Lyapunov techniques, therefore becomes a meaningful goal.

Motivated by these considerations, in this work we propose a stochastic Lyapunov-based model predictive controller (SLMPC) for nonlinear systems with unbounded disturbances (with Ito noise and subject to input constraints). In particular, we utilize the information on the distribution of the uncertain variables (instead of traditionally used worst-case bounds) to develop model predictive controllers that yield less conservative (albeit probabilistic, with well characterized probabilities) stability region estimates while handling uncertainty. The key idea is to use stochastic Lyapunov-based feedback controllers, with well characterized stabilization in probability regions to design constraints in the SLPMC. The SLMPC scheme inherits all the stability and robustness properties of the Lyapunov-based feedback controller, while also incorporating optimality consideration. The SLMPC design provides an *a priori* characterization of a lower bound on the probability of achieving stability for initial conditions within an explicitly characterized set of initial conditions.

First, a general class of Lyapunov-based feedback controllers is introduced, and key properties pertaining to the sample and hold implementation of the Lyapunov-based feedback controller are derived. The SLMPC is then formulated and shown to inherit all the the stability properties from the Lyapunov-based feedback controller and simultaneously incorporates optimality considerations which improve closed-loop performance. Finally, the theoretical results are demonstrated on a chemical reactor example.

References

[1] H. Michalska and D.Q. Mayne. Robust receding horizon control of constrained nonlinear systems. *Automatic Control, IEEE Transactions on*, 38(11):1623 –1633, nov 1993.

[2] D. Q. Mayne. Nonlinear model predictive control: An assessment. In *Proceedings of 5th International Conference on Chemical Process Control*, pages 217–231, Tahoe City, CA, 1997.

[3] M. Mahmood and P. Mhaskar. Enhanced stability regions for model predictive control of nonlinear process systems. *AIChE J.*, 54:1487–1498, 2008.

[4] D. Q. Mayne, J. B. Rawlings, C. V. Rao, and P. O. M. Scokaert. Constrained model predictive control: Stability and optimality. *Automatica*, 36:789–814, 2000.

[5] P. Hokayem, E. Cinquemani, D. Chatterjee, J. Lygeros, and F. Ramponi. Stochastic Receding Horizon Control with Output Feedback and Bounded Control Inputs. In *IEEE Conference on Decision and Control*, pages 6095–6100, 2010.

[6] M. Cannon, B. Kouvaritakis, and Xingjian Wu. Probabilistic Constrained MPC for Multiplicative and Additive Stochastic Uncertainty. *IEEE Transactions on Automatic Control*, 54(7):1626–1632, 2009.

[7] Dennis Van Hessem and Okko Bosgra. Stochastic closed-loop model predictive control of continuous nonlinear chemical processes. *Journal of Process Control*, 16(3):225–241, 2006.

[8] H Deng, M Krstic, and R.J Williams. Stabilization of stochastic nonlinear systems driven by noise of unknown covariance. *IEEE Transactions on Automatic Control*, 46(8):1237–1253, 2001.

[9] R. Chabour and M. Oumoun. On a universal formula for the stabilization of control stochastic nonlinear systems. *Stochastic Analysis and Applications*, 17:359–368, 1999.

[10] Battilotti Stefano and Alberto De Santis. Stabilization in probability of nonlinear stochastic systems with guaranteed region of attraction and target set. *IEEE Transactions on Automatic Control*, 48:1585–1599, 2003.

[11] Prashant Mhaskar, Nael El-Farra, and Panagiotis Christofides. Techniques for uniting Lyapunov-based and model predictive control. In Rolf Findeisen, Frank Allgц wer, and Lorenz Biegler, editors, *Assessment and Future Directions of Nonlinear Model Predictive Control*, volume 358 of *Lecture Notes in Control and Information Sciences*, pages 77–91. Springer Berlin / Heidelberg, 2007.

See more of this Session: Advances In Process Control
See more of this Group/Topical: Computing and Systems Technology Division

(723c) Exploiting Market Fluctuations and Price Volatility Through Feedback Control

Christopher V. Rao, Chemical and Biomolecular Engineering, University of Illinois, Urbana Champaign, Urbana, IL and Donald J. Chmielewski, Chemical and Biological Engineering, Illinois Institute of Technology, Chicago, IL

Many products of the chemical and petroleum industries are fungible goods. They are highly liquid and, in many case, actively traded on centralized exchanges. Examples include not only petroleum distillates, such as reformulated gasoline blend-stock for oxygen blending (RBOB) gasoline, but also common plastics such as polyethylene and polypropylene. While this liquidity reduces market friction, it also subjects producers to rapid changes and potential shocks in market prices. Even in cases where the products themselves are not explicit commodities, the feedstocks are -- petroleum and natural gas being prime examples -- and further subject the producer to the whims of the market and potential dislocations in prices and spreads. While financial instruments such as future contracts and swaps can be used to hedge against these potentially disruptive market forces, they may not always exist. In addition, they may require exceptionally large premiums due to limited volume and the absence of market makers.

In this work, we wished to explore how feedback control can be used to make the chemical producer responsive to market forces through dynamic operating policies as opposed to through the use of financial instruments To explore this problem, we developed a toy model of a marginal chemical producer with variable capacity operating in a dynamic market environment. While the model is admittedly simple, we nonetheless demonstrate that it can be used to develop and analyze a number of different operating strategies for not only dealing with market risk but also exploiting it through arbitrage. In the process, we further demonstrate how core ideas from finance can be injected into control.

See more of this Session: Advances In Process Control
See more of this Group/Topical: Computing and Systems Technology Division

(723d) Multiple Model Predictive Control of Air Separation Unit As Part of IGCC Power Plant During Rapid Load Changes

Priyadarshi Mahapatra, National Energy Technology Laboratory, Morgantown, WV, B. Wayne Bequette, Chemical and Biological Engineering, Rensselaer Polytechnic Institute, Troy, NY and Stephen E. Zitney, Collaboratory for Process and Dynamic Systems Research, National Energy Technology Laboratory, U.S. Department of Energy, Morgantown, WV

A cryogenic air separation unit (ASU) plays a vital role in future oxygen-blown combustion/gasification-based power plant. It is subject to periods of significant changes in production demand due to which an effective control structure is needed to minimize the impact of transient operation on product purity. In a typical ASU, the column pressure, temperature and composition dynamics float in tune with the feed-air flowrate which, in turn, is feedforwarded/ratioed with gasifier oxygen demand. Since the amount of column reflux/reboil is tied-up and is a function of this flowrate, the system inherently entails significant non-linearity, even for operation close to nominal operating point. This paper extends previous studies[1,2] using single linear model for predictive control over entire operating range to using multiple linear models, each corresponding to different part-load conditions. Each linear model is chosen to represent a discretesubspace of the overall nonlinear operating space. When all n models are combined, the resulting bank of linear models spans the entire nonlinear operating space. Hence, at any operating point of the ASU, the multiple model predictive control (MMPC) algorithm naturally and continuously adapts by promoting the dominant model(s), through model-weight calculation, based on plant measurements only, thus providing a better 'pseudo-nonlinear' model prediction.

This study has been done using a fully pressure-driven Aspen Plus Dynamics model of the ASU, acting as a surrogate 'plant-model', whereas multiple 'control-models' have been devised using a simultaneous PRBS multiple input excitation to the Aspen 'plant-model' at different operating load points. The MMPC approach has been shown to provide better composition control at high ramp-rates compared to MPC approach. This in-turn allows for higher ramp-rates of ASU, without violating the oxygen-purity constraints. Since the ASU poses a bottleneck in the entire IGCC plant operation by limiting the ramp-rate during load-following operation, the proposed control strategy provides a better response to the plantwide IGCC load-following problem compared to conventional PID and MPC approaches, preventing the need for larger liquid oxygen/air storage requirements. In addition, a linear optimization problem being solved at each time step requires much less computational resources compared to a nonlinear MPC approach.

References:

1. P. Mahapatra and B. Wayne Bequette. Oxygen Purity Control in the Air Separation Unit of an IGCC

Power Generation System during rapid production fluctuation. In *Annual International Pittsburgh Coal Conference 2009 CDROM Proceedings*, 20-23 Sept, 2009.
2. P. Mahapatra and B. Wayne Bequette. Process design and control studies of Elevated-Pressure Air Separation Unit for IGCC power plants. In *2010 American Control Council Annual Meeting, Baltimore, DC*, June, 2010.

See more of this Session: Advances In Process Control
See more of this Group/Topical: Computing and Systems Technology Division

(723e) Modeling and Control of An Industrial Hydrocracker Using the Discrete and Continuous Lumping Methods and Model Predictive Control

Yaman Arkun[1], Hasan Sıldır[1], Ummuhan Canan[1], Berna Cakal[2], Dila Gokce[2] and Emre Kuzu[2], (1)Chemical and Biological Engineering, Koc University, Istanbul, Turkey, (2)TUPRAS Refineries, Kocaeli, Turkey

Modeling and Control of an Industrial HydrocrackerUsing the Dicrete and Continuous Lumping Methods and Model Predictive Control

Hasan Sildir, Ummuhan Canan,Yaman Arkun.Department of Chemical and Biological Engineering, Koc University, IstanbulTurkey.

Berna Cakal, Dila Gokce, Emre Kuzu,TUPRASRefineries, Korfez-Kocaeli,Turkey

Hydrocracking is a catalytic chemical processwhich converts high-boiling heavy petroleum fractions such as vacuum gas oilinto lighter and more valuable products like naphta, diesel, kerosene,gasoline, and LPG. Hydrocracking takes place in the presence of rich hydrogenat elevated temperatures and pressures. The products are free of sulphur andnitrogen compunds which are hydrogenated into hydrogen sulfide and ammonia andwhich are subsequently removed. The aim of this work is to develop a model foran industrial hydrocracking reactor for optimization and control purposes. Ourresearch is centered around the hydrocraking unit (HCU) of TUPRAS refinerieswhich consists of four catalytic beds with interstage cooling by hydrogenquench. The overall reaction is exothermic and tight control of bedtemperatures is crucial for achieving the optimal product distribution.

For modeling purposes the methods of discreteand continuous lumping [1,2] are used. In continuous lumping the reactionmixture is treated as a continuum in which the reaction rate constant *k*is a continuous function of the true boiling point of the mixture. A yielddistribution function *p(k,K)* is introduced to formulate the amount ofspecies with reactivity *k* formed from cracking the species withreactivity *K*. The existing HCU models of this type are steady statemodels and they do not explicitly include the heat effects. Our continuouslumping model is a pseudohomogeneous non-steady-state plug flow reactor modelwhich includes both the material and energy balances. As such the model isoriginal. In the case of discrete lumping the reaction mixture is characterizedin terms of pseudocomponents that are defined for the lumped species boiling ina particular temperature range (i.e.cut) [1]. The two sets of lumping methodsare fundamentally different approaches, and their joint development providesadditional insight into uderstanding the behavior of HCU and arriving at afinal reactor model for optimization and control. Model parameters wereestimated using parameter estimation methods. With optimally tuned parametervalues, the predicted reactor bed temperatures, hydrogen consumption,conversion and product distributions match TUPRAS` actual plant data veryclosely. This is demonstrated for different feedstosks in our training andvalidation data sets.

The developed HCU model is used for twopurposes: 1) To compute the optimal product distribution and the optimalreactor inlet temperature setpoint values. This economic optimization isperformed under steady state conditions. 2) The dynamic model is used by aModel Predictive Controller (MPC) to control the product amounts at theiroptimal set points. When the product amounts (light naphta, heavy naphta,diesel, kerosene and bottoms) deviate from their optimal values due tofeedstock changes, catalyst deactivation and other disturbances, MPC makes thenecessary adjustments in the reactor beds inlet temperature setpoints.Temperature increase in each bed and the weighted average bed temperatures(WABT) are the addional variables which are kept within limits by MPC tomaintain a desired level of conversion and uniform catalyst deactivation acrossthe beds. Reactor inlet temperatures are changed to their new setpoints by theregulatory PID loops which adjust the hydrogen quenches between beds. Thiscascade arrangement of MPC and PIDs provides both optimizing and regulatoryactions as shown in the figure below.

In the presentation, modeling, optimization andcontrol simulations will be given and compared with the present status in theplant.

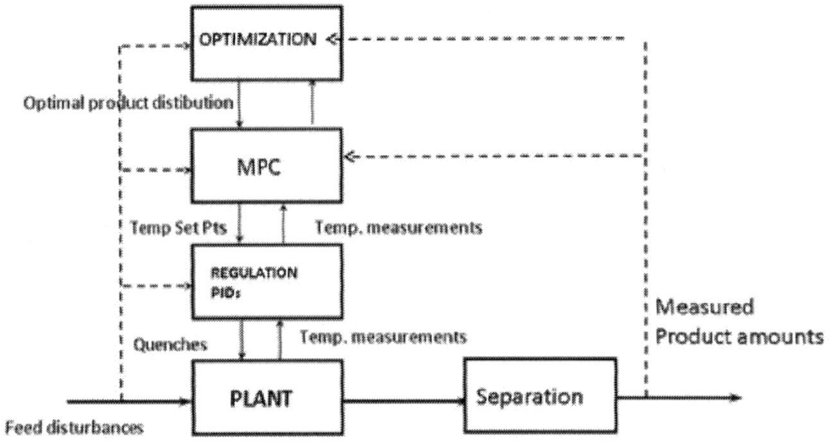

1.Mohanty, S.; Saraf, D. N.; Kunzru, D. Modeling of a Hydrocracking

Reactor.*Fuel Process. Technol*. 29, 1, 1991.

2.Chou M. Y.; Ho, T. C. Continuum Theory for Lumping Nonlinear

Reactions. *AIChE J*. 34, 1519, 1988.

See more of this Session: Advances In Process Control
See more of this Group/Topical: Computing and Systems Technology Division

(723f) Heat-Exchanger Bypass Control

William L. Luyben, Chemical Engineering, Lehigh University, Bethlehem, PA

Many methods are used for controlling temperatures in heat exchanger systems. Direct manipulation of the flowrate of either the hot or the cold stream is most often used when that stream is a utility (cooling water, steam, hot oil or refrigerant). When the flowrates of both streams are set by process requirements, heat-exchanger bypassing is widely used. A portion of one of the streams (either hot or cold) is sent through the heat exchanger and the remainder is bypassed around the exchanger. The temperature of the mixed steam is controlled by valves in each path. This system provides very tight temperature control since the dynamics of blending a hot stream and a cold streams are very fast.

This paper explores the design and control issues when heat-exchanger bypassing is used. The design optimization variables included the fraction of bypassing, the area of the heat exchanger and the design pressure drops over the control valves. Dynamic rangeability requires heat-transfer rates to be adjustable over a wide range. As expected, results demonstrate that larger area and more bypassing improve the ratio of maximum-to-design heat transfer rates, which is important for dynamic controllability. An unexpected, counter-intuitive result is that control valve design pressure drops have little effect on rangeability in heat-exchanger bypass systems in which variable-speed pumps are used to maintain total flowrates.

See more of this Session: Advances In Process Control
See more of this Group/Topical: Computing and Systems Technology Division

(723g) An Alternative Approach to Online Model Identification and PID Controller Tuning

Nataliya Baran, Gnter Wozny and Harvey Arellano-Garcia, Chair of Process Dynamics and Operation, Berlin Institute of Technology, Berlin, Germany

Currently most controllers used in the chemical industry are based on conventional PID algorithms. Priorly recognized for their simplicity and robust structure it is difficult to find good controller settings even though PID controllers only have a few adjustable parameters [1]. In this work, we present a new approach which enables online tuning of conventional controllers. The advantage it has over conventional methods lies in the online model identification by closed-loop experiments. These enable us to avoid undesirable state variable changes. As a result, we are able to identify the model parameters more quickly, which in turn reduces the number of necessary experiments and the amount of measured data. Moreover, this leads to a faster calculation of controller settings and reduced commissioning time. Thus, in this contribution, we describe in detail how the proposed approach can also be automated by eliminating the need to manually tune and retune PID controllers, when they are deployed or subject to system behavior changes.

In most cases a good process model is required for controller tuning. Controller settings are first determined by identifying the process behavior, which are usually calculated by time-consuming and expensive experiments. Therefore, it is preferable to design experiments in such a way that the resulting data yields maximum information with respect to model parameters. For this purpose, we use the approach of optimal experimental design [5]. By calculating optimal conditions, it maximizes information yield produced by experiments in order to increase the accuracy of estimated model parameters. The challenge of our approach is the need for online model identification by closed-loop experiments. For this purpose, we use a technique proposed by Körkel and Arellano-Garcia [6]. According to this concept, parameter estimation is applied in order to match the model to a real process. Applied to our domain, the outcome is a sequential approach to calculate model parameters in three steps. In the first step, the experiment is planned. Here, the optimal value of the control variable throughout the experiment is determined. Next, the planned experiment is executed and the measured data is analyzed. In the last step, the quality of the model is validated and new model parameters are calculated by applying the least square approach [5]. After the process model is identified, we calculate the controller settings and implement them in the online mode.

This work introduces the proposed solution strategy for the online controller tuning and closed-loop model parameter identification algorithm. In order to validate the approach and supply a comparison to common tuning methods, experimental results will also be presented. Furthermore, we will discuss the automated method's performance and influence to commissioning time.

References:

[1] Shamsuzzoha M. and Skogestad S., The setpoint overshoot method: A simple and fast closed-loop approach for PID tuning, J. Process Control, 2010, 20, 1220-1234.

[2] Rivera D.E., Morari M., Skogestad S., Internal model control. 4. PID controller design, Ind. Eng. Chem. Res. 1986, 25(1), 252-265.

[3] Skogestad S., Simple analytic rules for model reduction and PID controller tuning, J. Process Control, 2003, 13, 291-309.

[4] Ziegler J.G., Nichols N.B., Optimum settings for automatic controllers, Trans. ASME, 1942, 64, 759-768.

[5] Barz T., Arellano-Garcia H., Wozny G., Handling uncertainty in model-based optimal experimental design, Ind. Eng. Chem. Res., 2010, 49, 5702-5713

[6] Körkel S., Arellano-Garcia H., Online Experimental Design for Model Validation, 10th International Symposium on Process Systems Engineering – PSE2009, 2009

See more of this Session: Advances In Process Control
See more of this Group/Topical: Computing and Systems Technology Division

(735a) A Multi-Objective Mixed-Integer Dynamic Optimization Approach to Oil Spill Response Planning

Fengqi You, Department of Chemical and Biological Engineering, Northwestern University, Evanston, IL and Zhixia Zhong, Northwestern University, Evanston, IL

Catastrophic oil spills [1], such as the recent *Deepwater Horizon/BP* oil spill in the Gulf of Mexico [2], have demonstrated the importance of developing responsive and effective oil spill response planning strategies for the oil industry and the government. Although a few models have been developed for oil-spill response planning, response operations and the oil weathering process are usually considered separately [3-6]. Yet significant interactions between them exist throughout the response. Oil-spill cleanup activities change the volume and area of the oil slick and in turn affect the oil transport and weathering process, which also affects coastal protection activities and cleanup operations (e.g., performance degradation and operational window of cleanup facilities) [6- 8]. Therefore, it is of great importance to integrate the response planning model with the oil transport and weathering model.

In this work, we propose a bi-criterion optimization approach that seamlessly integrates the oil-spill response planning decisions with the oil transport and weathering process under the economic and responsiveness criteria. The economic criterion is measured by the total response cost, and responsiveness is measured by the time span of the entire response operations. A mixed-integer dynamic optimization (MIDO) model is proposed that simultaneously predicts the time trajectories of the oil volume and slick area and the optimal response cleanup schedule and coastal protection plan, by taking into account the time-dependent oil physiochemical properties, spilled amount, hydrodynamics, weather conditions, facility availability, performance degradation, variational operational window, and regulatory constraints. To solve the MIDO problem effectively, we reformulated it as a mixed-integer nonlinear programming (MINLP) problem using orthogonal collocation on finite elements. We also developed an approximate mixed-integer linear programming (MILP) model for the initialization for solving the nonconvex MINLP problem. The results of the multi-objective optimization problem yield a Pareto-optimal curve, which reveals how the optimal total cost, oil spill cleanup operations, and coastal protection plans change under different specifications of the response time span. The application of the proposed optimization approach is illustrated through an case study based on the *Deepwater Horizon/BP* oil spill.

References:

[1] NOAA Oil spill case histories 1967-1991: summaries of significant U.S. and international spills; Seattle, Washington, 1992.

[2] http://www.restorethegulf.gov/

[3] Reed M, Johansen I, Brandvik PJ, Daling P, Alun Lewisà, Fioccoà R, Mackay D, Prentki R. Oil spill modeling towards the close of the 20th century: overview of the state of the art. Spill Science & Technology Bulletin. 1999; 5:3-16.

[4] Wilhelm WE, Srinivasa AV. Prescribing tactical response for oil spill cleanup operations. Management Science. 1997; 43:386-402.

[5] Psaraftis HN, Ziogas B. A tactical decision algorithm for the optimal dispatching of oil spill cleanup equipment. Management Science. 1985; 31:1475-1491.

[6] Brebbia CA. Oil spill modeling and processes. WIT Press: UK, 2001.

[7] Fingas M. The basics of oil spill cleanup. Lewis: New York, 2001.

[8] Ornitz B, Champ M. Oil spills first principles: prevention and best response. Elsevier: Netherlands, 2003.

See more of this Session: Dynamic Simulation and Optimization
See more of this Group/Topical: Computing and Systems Technology Division

(735b) Simultaneous Design and Control of Pressure Swing Adsorption Processes

Harish Khajuria, Chemical Engineering, Imperial College, London, United Kingdom and Stratos Pistikopoulos, Chemical Engineering, Imperial College London, London, United Kingdom

Simultaneous Design and Control of PressureSwing Adsorption Processes

In the lastfew decades, pressure swing adsorption (PSA) has evidenced substantial growthin terms of size, versatility and complexity. In addition to handlingmulti-component separation and purification, it offers tremendous flexibilityat the design stage, requiring careful selection of key decision variablesincluding number of beds, time duration of process steps, bed length anddiameter. Further challenges are posed by the fact that the PSA operation isperiodic in nature and never attains a true steady state. Traditional approachfor the design of such process systems usually employs a two step,first-design-and-then-control, sequential treatment. In contrast, integrateddesign and control (IDC) approach takes into account the fact that it is the designof the system which determines its dynamic controllability. Its potential forsignificant design improvements in terms of superior real time operability hasbeen realized from quite some time now. However, early contributions to thisfield mostly performed analysis on the linear dynamic representation of theoriginal system. The application of IDC approach to inherently dynamic andhighly nonlinear system like PSA remains a challenging task and has never beenattempted before.

In this work,a systematic and rigorous methodology for the simultaneous design and controlof PSA systems, incorporating a comprehensive treatment of its design andoperational constraints is presented. Towards this goal, a detailed firstprinciple based model is first developed for a double bed, 6 step PSA system(Khajuria and Pistikopoulos, 2010). In the next step, the full scale integrateddesign and control dynamic optimization problem is formulated and solvedincorporating all operational constraints (Bansal, et.al., 2000; Mohideen,et.al., 1996; Sakizlis, et.al., 2004). The design objective for the PSA systemseparating 70 % H_2 - 30 % CH_4 mixture into high purityhydrogen, is to maximize the H_2 recovery, while the controllerobjective is to minimize the integral square error (ISE) in purity from a setpoint value of 99.99 %, for time varying disturbances in feed temperature andfeed rate. Furthermore, detailed studies are also conducted to compare theoptimal design obtained from integrated design and control case in terms of hydrogenrecovery and disturbance rejection, as compared to sequential design andcontrol case.

References:

Bansal,V.; Perkins, J.; Pistikopoulos, E.; Ross, R. & van Schijndel, J. (2000),'Simultaneous design and control optimisation under uncertainty', *Computers& Chemical Engineering* **24**(2-7), 261 - 266.

Khajuria,H. & Pistikopoulos, E. N. (2011), 'Dynamic modeling andexplicit/multi-parametric MPC control of pressure swing adsorption systems', *Journalof Process Control* **21**(1), 151 - 163.

Mohideen,M. J.; Perkins, J. D. & Pistikopoulos, E. N. (1996), 'Optimal Design ofDynamic Systems under Uncertainty', *AlChE Journal* **42**(8),2251-2272.

Sakizlis,V.; Perkins, J. D. & Pistikopoulos, E. N. (2004), 'Recent advances inoptimization-based simultaneous process and control design', *Computers &Chemical Engineering* **28**(10), 2069 - 2086.

See more of this Session: Dynamic Simulation and Optimization
See more of this Group/Topical: Computing and Systems Technology Division

(735c) An Economic Nonlinear Model Predictive Control Formulation for Gas Pipeline Optimization

Ajit Gopalakrishnan and Lorenz T. Biegler, Chemical Engineering, Carnegie Mellon University, Pittsburgh, PA

The most common objective of model predictive control is to track a set-point by minimizing a least-squares objective function subject to future predictions from a dynamic model of the process. The set-point is provided by a real-time optimization (RTO) layer that operates on top of the MPC, and minimizes a desired objective using a steady state process model. However, for certain energy intensive processes like gas pipeline distribution networks, the optimal operating point needs to be changed frequently based on the time-varying demands of the consumers. In such cases, it might be both challenging and sub-optimal to apply the traditional two-layer MPC approach i.e. taking the pipeline operation to an RTO set-point might be sub-optimal when we consider information about the economics of the process. As a result, an economically oriented nonlinear model predictive control (NMPC) formulation which directly considers an economic objective function along with the nonlinear dynamic model of the process is a preferred approach.

In this work, we present an economic NMPC formulation for the optimal operation of gas pipeline networks subject to transient demands of the consumers. The compression of gas accounts for the most significant portion of the energy costs, and is the objective function minimized. A detailed fully open nonlinear model of the pipeline network [1] is considered, and a novel feature of the model is the smoothing operations to handle flow reversals (due to possible compressor outages) and flow transitions (between laminar and turbulent flow regimes), that are otherwise discontinuous and/or non-differentiable. The gas demands of the consumers are assumed to be sinusoidal in nature, which is an approximation to a real-demand curve. Due to the diurnal nature of the demands, the system goes towards not a fixed, but a cyclic-steady state. This requires an NMPC formulation with an appropriately defined terminal region to ensure nominal stability, and we use the approach described in [2].

Several case studies are presented, including the incorporation of time-varying electricity prices into the NMPC objective. We show how the pipeline inventory (linepack) can be manipulated to achieve significant compression cost savings. We also show how, in the presence of disturbances in the network, the NMPC scheme recalculates the optimal compression profiles while satisfying consumer flow demands and contract pressures. The results of this work are generalizable to any energy intensive process or a process with a variable inventory.

References

[1] B.T. Baumrucker and L.T. Biegler, MPEC strategies for cost optimization of pipeline operations, Computers & Chemical Engineering, 34(6), 900-913.

[2] R. Huang, E. Harinath and L. T. Biegler, Lyapunov stability of economically oriented NMPC for cyclic processes, Journal of Process Control, 21(4), 501-509.

See more of this Session: Dynamic Simulation and Optimization
See more of this Group/Topical: Computing and Systems Technology Division

AGE DEPENDENT DYNAMIC OPERATIONAL RISK ASSESSMENT

Shubharthi Barua, Xiaodan Gao and M.Sam Mannan
Mary Kay O'Connor Process Safety Center, Artie McFerrin Department of Chemical Engineering, Texas A&M University, College Station, TX

ABSTRACT

Conventional risk assessment methodologies are not able to quantify dynamic changes in process. To bridge this gap, dynamic operational risk assessment methodology provides a mean to measure operational risk due to dynamic behavior of process. In dynamic risk assessment, ageing of equipment/component is an important time-dependent criterion that can increase equipment/component's failure rate, which hasn't been addressed in previous researches. In this paper, ageing factors are integrated in the framework of dynamic operational risk assessment methodology to analyze the effect of system degraded behavior on dynamic risk. A case study on commonly used oil/gas separators on offshore plants illustrates this model.

INTRODUCTION

Oil/gas and petrochemical plants are complicated and dynamic in nature. Among dynamic characteristics, ageing of equipment/components, season changes, stochastic processes, operator response times, inspections and testing time intervals, and timing of safety system operations are time dependent criteria that can influence the dynamic processes. But conventional quantitative risk assessment techniques have limited ability to address these kinds of time dependent effects. These methods cannot describe the variation of operational risks as time-dependent deviations or changes happening in the process. For example, the fault tree/event tree methodology provides a static logical relationship between a component output event and its effect on another component output event. But it is not able to reflect the time-dependent effects. Therefore, risk assessment methodology that integrates time factors for dynamic processes is more beneficial to concern those dynamic factors.

Dynamic Operational Risk Assessment (DORA) is a risk assessment technique proposed for complicated and dynamic processes which treats the testing/inspection interval and repair time as a critical parameter for evaluating the operational failure [1]. The probabilistic modeling part of DORA consists of stochastic modeling and process dynamics modeling. DORA considers abnormal events to evaluate the scenarios and is not dependent on conventional quantitative risk assessment methodology. It is assumed that a component can have three states in its lifetime: normal operating state, abnormal event undetected, and abnormal event detected and repaired. The time a component stays in a state is defined as the sojourn time and for sojourn time distribution, it is assumed that the component failure rate will remain constant for its lifetime. The Monte Carlo simulation technique for system-state trajectory provides ideas for system state transitions and the effects of testing/inspected interval on the probability and frequency of the component's abnormal events over a prolong time period have been studied.

A study by UK Health and Safety Executive [2] reported ageing as a significant factor in process industry accidents. Ageing refers to the condition of the equipment/plant and its changing behavior with respect of time. Corrosion, erosion, build up of dirt and debris, fatigue, vibration, wear, friction, ineffective inspection and maintenance schedule etc., can affect the ageing of a component. During ageing of equipment/component, the equipment performance degrades and signs of damages appear. Studies have identified that the component failure probabilities significantly increases with an increase in ageing of the components, if effective tests and maintenance is absent [3]. Therefore, to understand operational risk due to the increasing rate of degradation of equipment condition and to able to plan effective maintenance and inspection schedule, the effects of ageing should be considered in risk assessment method.

METHODS & EXPECTED RESULT

In the Dynamic Operational Risk Assessment (DORA) study, the component failure rate is considered constant and remains the same value for all component ages. But as it is described, it is important to consider the effects of ageing on failure rates. Therefore, the objective of this paper is to present how the ageing of equipment and components can be integrated to assess operational risks within the framework of DORA and to analyze the effect of ageing factors. As, in DORA the failure rate is constant, hence, probability density function for sojourn time distribution of normal operating condition is an exponential distribution. However, exponential distribution can describe equipment condition only when its failure rates remain constant. As with ageing the failure rate of equipment increases, exponential distribution will not provide real picture. Therefore, for age-dependent DORA study, Weibull distribution is considered to describe sojourn time distribution of normal operating condition. The sojourn time distribution between normal condition and the abnormal event undetected condition is also changed. The sojourn time distribution for abnormal undetected and abnormal event detected is obtained by fitting industrial data to different distributions. The system state trajectory simulation is performed considering the component's failure rate as a function of component's increasing age. The simulation is done using matlab computer programming. The effects of the testing/inspection interval are also evaluated.

The age dependent DORA study provides system degraded behavior due to the effects of ageing on abnormal events of a component. A case study on commonly used oil/gas separators on offshore plants illustrates this model.

REFERENCES

1. Xiaole Yang and M. Sam Mannan; The development and application of dynamic operational risk assessment in oil/gas and chemical process industry, Reliability Engineering and System Safety 95 (2010) page- 806-815
2. UK Health and Safety Executive; Plant ageing study-phase 1 report, 2010
3. W. E. Vesely, Approaches for Age-Dependent Probabilistic Safety Assessments with Emphasis on Prioritization and Sensitivity Studies, NUREG/CR-5587 SAIC-92/1137 RG, R1, R9, August 1992
4. TWI Ltd, ABB Engineering Services, SCS (INTL) Ltd, Allianz Cornhill Engineering for UK Health and Safety Executive; Plant ageing: Management of equipment containing hazardous fluids or pressure

(735e) Simulation Based Optimization of Expensive Flowsheet Models for Continuous Pharmaceutical Manufacturing

Fani Boukouvala, Chemical and Biochemical Engineering, Rutgers University, Piscataway, NJ and Marianthi G. Ierapetritou, Department of Chemical and Biochemical Engineering, Rutgers, The State University of New Jersey, Piscataway, NJ

The main purpose of this work is to develop an efficient strategy for simulation-based design and optimization for a pharmaceutical tablet manufacturing process. Simulation-based optimization is a research area that is currently attracting a lot of attention in many industrial applications, where expensive simulators are used to characterize, design and optimize real systems. Common optimization techniques cannot be applied due to the lack of knowledge of analytic derivatives, possible discontinuities caused by if-then operations, high computational cost which prohibits the realization of multiple function evaluations, and finally due to the presence of various types of noise [1]. Synthesis and design of integrated continuous manufacturing of pharmaceutical products is a field that is currently emerging through the development of large-scale flowsheet models [2]. The development of such flowsheet simulations for powder-based products require the integration of various types of models, ranging from inexpensive black-box models to costly first-principle based models, resulting in hybrid simulation models. Moreover, pharmaceutical products are typical examples of high cost products which involve expensive processes and raw materials, while at the same time must satisfy strict quality regulatory specifications, leading to the formulation of challenging and expensive optimization problems.

Tablet manufacturing, is a complex process line comprised of several integrated units which aim to efficiently blend different ingredients and compress them into consistent and high-quality tablets with desired dissolution properties. A typical continuous manufacturing line for oral solid dosage products may consist of feeding, mixing, roller compaction, granulation, milling and tablet compaction processes. A common practice in the simulation-based optimization literature is surrogate based optimization [3], in order to overcome problems of computational cost and lack of knowledge of input-output first principles. Using this approach, the expensive simulation is run for a number of sample points based on which a surrogate model is built as a fast approximation of the underlying process. Surrogate-based optimization, on the other hand, has raised another significant research topic- design and analysis of computer experiments- which aims not only to efficiently plan computer experiments for global optimization of surrogate model approximations, but also to minimize sampling cost [4]. An integrated system of the aforementioned candidate processes has a large number of variables which consist of design parameters, integer decision variables, and operating conditions. The first required step for optimization of an expensive simulation using metamodeling approximations, involves the selection of the important variables and parameters which will form the design space of the underlying system and should be included as inputs to the surrogate model used for optimization. Screening techniques [5] or sensitivity analysis methods [6] are suitable for the identification of the critical inputs and can reduce the dimensionality of the problem by elimination of insignificant variables that do not contribute considerably to the overall variance of the outputs or the objective function.

Once the dimensionality of the problem is identified, the feasible region of the system should also be explored in order to limit, when possible, the optimization problem inside the feasible region. Techniques in forms of penalty terms, or objective function deformation methods [7] are explored in order to drive the optimization procedure around the feasible space. In cases where the process constraints are not known in closed-form, black-box optimization techniques can be used [8]. Next, the initial experimental design and the surrogate technique used to approximate the expensive simulation must be chosen. Sampling the flowsheet model at a small number of design points based on a space-filling design (Latin Hypercube sampling), combined with a Kriging-based metamodel generation is a well established combination, proven to have high-quality results in surrogate-based optimization [9]. Kriging is a data-driven interpolating method which has the ability to provide an average error associated with each prediction [10]. Subsequently a metric must be chosen based on which new promising sampling locations are identified. Due to the inherently noisy nature of powder handling processes, the Kriging model as well as the figure of merit used should be stochastic, thus the Sequential Kriging Optimization (SKO) approach [11] which uses the stochastic Expected Improvement function (EI) as the figure of merit, is found to be most suitable for this process. The EI must concurrently favor regions where the kriging approximated objective function value is optimal (local search) but also regions where the kriging uncertainty and/or the noise is high (global search). Through an iterative procedure, the sampling set is enhanced and the optimum is updated, by locating promising new points which have a maximum EI function value. Finally, one of the critical aspects of the proposed methodology is the use of a robust optimization technique for reliable global optimization. A commonly used method for expensive optimization is the Nedler-Mead algorithm, however, its performance in the current problem which is noisy and highly non-linear is not found to be very reliable. The performance of Sequential Quadratic Programming (SQP) method is found to be more efficient for this case study.

Clearly, many opportunities lie in the research for simulation-based optimization techniques in (a) designing an appropriate computer experiment, (b) identifying an efficient surrogate approximation method which requires the minimum number of function evaluations and can also handle uncertainty caused by various sources of noise, and (c) distinguishing the suitable gradient-free optimization technique which can successfully identify the global optimum with a limited number of function evaluations in the presence of noise and possible feasibility constraints. All of the above aspects must be linked to the nature of the actual underlying process which, in this work, refers to continuous tablet manufacturing. The application of the proposed steps is shown through the optimization of a direct compression tablet manufacturing process, where the objective function is formed such that the variability of the final product properties is minimized while the constraints assure that their values are always within predefined ranges set by the Food and Drug Administration (FDA). This work aims to be one of the first attempts to formulate a multi-scale integrated pharmaceutical process as an optimization problem. The advantages of process optimization techniques which are described in this work are so far foreign to the pharmaceutical industry; however, this study aims to prove that simulation-based optimization techniques can have significant applications towards designing processes which will produce high-quality and consistent pharmaceutical products.

References:

1. Bertsimas, D., O. Nohadani, and K.M. Teo, *Robust Optimization for Unconstrained Simulation-Based Problems.* Oper. Res., 2010. **58**(1): p. 161-178.

2. Boukouvala, F., et al., *Computer aided design and analysis of continuous pharmaceutical manufacturing processes.* European Symposium of Computer Aided Process Engineering 21, Chalkidiki, Greece, 2011.

3. Jones, D.R., *A Taxonomy of Global Optimization Methods Based on Response Surfaces.* Journal of Global Optimization, 2001. **21**(4): p. 345-383.

4. Papalambros, P., P. Goovaerts, and M.J. Sasena, *Exploration of Metamodeling Sampling Criteria for Constrained Global Optimization.* Engineering Optimization, 2002: p. 263-278.

5. Schonlau, M. and W. Welch, *Screening the Input Variables to a Computer Model Via Analysis of Variance and Visualization*, in *Screening*, A. Dean and S. Lewis, Editors. 2006, Springer New York. p. 308-327.

6. Queipo, N.V., et al., *Surrogate-based analysis and optimization.* Progress in Aerospace Sciences, 2005. **41**(1): p. 1-28.

7. Lucia, A., et al., *A Barrier-Terrain Methodology for Global Optimization.* Industrial & Engineering Chemistry Research, 2008. **47**(8): p. 2666-2680.

8. Banerjee, I. and M.G. Ierapetritou, *Design Optimization under Parameter Uncertainty for General Black-Box Models.* Industrial & Engineering Chemistry Research, 2002. **41**(26): p. 6687-6697.

9. Jones, D.R., M. Schonlau, and W.J. Welch, *Efficient Global Optimization of Expensive Black-Box Functions.* Journal of Global Optimization, 1998. **13**(4): p. 455-492.

10. Cressie, N., *Statistics for Spatial Data (Wiley Series in Probability and Statistics).* 1993: Wiley-Interscience.

11. Huang, D., et al., *Global Optimization of Stochastic Black-Box Systems via Sequential Kriging Meta-Models.* Journal of Global Optimization, 2006. **34**(3): p. 441-466.

(735f) Global Optimization of Nonsmooth Dynamic Systems

Kamil A. Khan[1], Mehmet Yunt[2] and Paul I. Barton[1], (1)Process Systems Engineering Laboratory, Dept. of Chemical Engineering, Massachusetts Institute of Technology, Cambridge, MA, (2)Process Systems Engineering Laboratory, Dept. of Mechanical Engineering, Massachusetts Institute of Technology, Cambridge, MA

When dynamic optimization problems exhibit nonsmoothness in their embedded differential equations, this nonsmoothness can hinder the application of conventional optimization methods relying on differentiability assumptions. Nonsmoothness of the differential equations is generally a consequence of modeling autonomous changes in the physical regime of a system, such as flow transitions, phase changes, irregularities in the geometry of vessels, the action of nonreturn valves, etc. In this presentation we examine differential equations with continuous but nondifferentiable right-hand sides.

Extended stationarity conditions are presented for a class of dynamic optimization problems with nonsmooth right-hand sides. Satisfaction of these stationarity conditions is necessary for local optima. These conditions make use of the linear Newton approximation [1]: a generalization of the gradient that satisfies several calculus properties sharply. A nonsmooth single shooting method is developed to locate such points, and consists of two main elements: the Objective and Constraint Evaluator, and the Proximal Modified Bundle Method. During each iteration of the method, the Objective and Constraint Evaluator solves the embedded ODE system at a candidate parameter value, in order to compute an element of the linear Newton approximation of the objective function at this parameter value. The Proximal Modified Bundle Method then uses a subset of the linear Newton approximation elements obtained so far to determine if the candidate parameter value satisfies the extended stationarity conditions. If so, the procedure terminates; if not, the method furnishes a new candidate parameter value for the next iteration.

Next, a technique is presented for constructing concave and convex relaxations of the objective function of dynamic optimization problems with nonsmooth right-hand sides over interval subsets of parameter space. This technique is an extension of the relaxation theory developed in [3, 4]. These techniques are embedded in a branch-and-bound framework to yield an algorithm that solves such optimization problems to global optimality. Within this branch-and-bound framework, each lower-bounding subproblem is a convex nonsmooth NLP obtained as a convex relaxation of the original problem, which can be solved to global optimality using the nonsmooth single shooting method described above. Each upper-bounding subproblem may be either simply an objective function evaluation, or any point satisfying the necessary conditions for local minima, obtained using the nonsmooth single shooting method. Convergence of the method is discussed.

The presented techniques are applied to several examples for illustration: including a nonsmooth dynamic optimization problem based on an electric circuit system [2] which has several sub-optimal local maxima.

[1] F. Facchinei, J.-S. Pang, *Finite-dimensional variational inequalities and complementarity problems: Volume II*, Springer-Vergag New York, Inc., New York, 2003.

[2] E. M. Izhhikevich, *Dynamical systems in neuroscience: The geometry of excitability and bursting*, The MIT Press, Cambridge, 2007.

[3] J. K. Scott, M. D. Stuber, P. I. Barton, *Generalized McCormick relaxations*, J. Glob. Optim., In press (2011).

[4] J. K. Scott, B. Chachuat, P. I. Barton, *Nonlinear Convex and Concave Relaxations for the Solutions of Parametric ODEs*, submitted (2009).

See more of this Session: Dynamic Simulation and Optimization
See more of this Group/Topical: Computing and Systems Technology Division

(735g) Global Optimization of Detailed Non-Linear Kinetic Models of Metabolic Networks Through Recasting Into Canonical Power-Law Representation

Carlos Pozo[1], **Gonzalo Guilln-Goslbez**[1], Laureano Jimnez[1] and Albert Sorribas[2], (1)Departament d'Enginyeria Qumica, Universitat Rovira i Virgili, Tarragona, Spain, (2)Basic Medical Sciences Department, Unversity of Lleida, Lleida, Spain

Identifying appropriate modulation of gene expressions through optimization is a rational approach to design engineered organisms for biotechnological purposes. This strategy would greatly benefit from the development of detailed kinetic models of the target metabolism. Such models should incorporate regulatory influences, for example through a set of ordinary differential equations accounting for the system's dynamics. Models based on systematic approximated kinetic representations, such as power-laws, Saturating and Cooperative or convenience kinetics, provide uniform forms that are guaranteed to be accurate over a wide range of conditions.

Unfortunately, non-convexities included in such models may lead to the existence of multiple local optima in which standard optimization algorithms may get trapped during the search. Several stochastic and deterministic global optimization methods have been proposed to overcome this limitation, but only the second group can rigorously guarantee global optimality. Deterministic global optimization algorithms rely on the use of convex underestimators to formulate lower-bounding convex problems that yield lower bounds on the global optimum. Most of these methods are general purpose and originally devised to exploit special structures in the continuous terms of the formulation. Because of this, they can encounter numerical difficulties in some instances, particularly when a large number of non-convexities of different nature are found, as is the case in metabolic engineering problems.

In previous works [1-3], the authors developed highly efficient optimization algorithms and a set of related strategies for understanding the evolution of adaptive responses in cellular metabolisms described by the GMA canonical representation. GMA models provide a more realistic representation than stoichiometric linear models. The Saturable and Cooperative (SC) formalisms [4] are more accurate representations valid for a wider range of values. Unfortunately, the strategies proposed for GMA models are not directly applicable to the SC formalisms. In order to side-step this shortcoming and extend the global optimization methods developed for power-law models, we explore herein the use of systematic techniques for recasting SC models into GMA ones. Recasting permits the exact transformation of a continuous non-linear model of arbitrary form into a canonical GMA model [5-6]. This transformation is performed at the expense of increasing the number of variables of the original model. Through this technique, GMA models become a canonical form that can be used for simulation and optimization purposes, which opens the door for effectively extending the optimization and feasibility analysis originally devised for GMA models to other detailed kinetic models.

In this work, we shall show the practical utility of recasting SC models into GMA models for optimization purposes. This technique is similar to the symbolic reformulation algorithm proposed by Smith and Pantelides [7]. In contradistinction with this method, we focus on obtaining a power-law representation rather than bringing the model to a standard form containing linear constraints and a set of nonlinearities corresponding to bilinear product, linear fractional, simple exponentiation and univariate function terms. Our results show that recasting non-linear kinetic models into GMA models is indeed an appropriate strategy that helps overcoming some of the numerical difficulties that arise during the global optimization task.

References

1. Guilln-Gosálbez G, Sorribas A: Identifying quantitative operation principles in metabolic pathways: A systematic method for searching feasible enzyme activity patterns leading to cellular adaptive responses. BMC Bioinformatics 2009, 10.
2. Guillen-Gosálbez G, Pozo C, Jimnez L, Sorribas A: A global optimization strategy to identify quantitative design principles for gene expression in yeast adaptation to heat shock. Computer Aided Chemical Engineering 2009, 26:1045-1050.
3. Pozo C, Guilln-Gosálbez G, Sorribas A, Jimnez L: A Spatial Branch-and-Bound Framework for the Global Optimization of Kinetic Models of Metabolic Networks. Industrial and Engineering Chemistry Research 2010.
4. Sorribas A, Hernandez-Bermejo B, Vilaprinyo E, Alves R: Cooperativity and saturation in biochemical networks: a saturable formalism using Taylor series approximations. Biotechnology and bioengineering 2007, 97(5):1259-1277.
5. Savageau MA, Voit EO: Recasting nonlinear differential equations as S-systems: a canonical nonlinear form. Mathematical Biosciences 1987, 87:83-115.
6. Voit EO: Recasting nonlinear models as S-systems. Mathematical and Computer Modelling 1988, 11(C):140-145.
7. Smith EMB, Pantelides CC: A symbolic reformulation/spatial branch-and-bound algorithm for the global optimisation of nonconvex MINLPs. Computers and Chemical Engineering 1999, 23(4-5):457-478.

See more of this Session: Dynamic Simulation and Optimization
See more of this Group/Topical: Computing and Systems Technology Division

(738a) Optimization of IGCC with CFD Based ROM

Yi-dong Lang, Department of Chemical Engineering, Carnegie Mellon University, Pittsburgh, PA, Stephen E. Zitney, U.S. Department of Energy, National Energy Technology Laboratory, Morgantown, WV and Lorenz T. Biegler, Chemical Engineering, Carnegie Mellon University, Pittsburgh, PA

Integrated gasification combined cycle (IGCC) plants have significant advantages for efficient power generation with carbon capture. Moreover, with the development of accurate CFD models for gasification and combined cycle combustion, key units of these processes can now be modeled more accurately. However, the integration of CFD models within steady-state process simulators, and subsequent optimization of the integrated system, still presents significant challenges. This study describes the development and demonstration of a reduced order modeling (ROM) framework for these tasks. The approach builds on the concepts of co-simulation and ROM development for process units described in earlier studies. Here we show how the ROMs derived from both gasification and combustion units can be integrated within an equation-oriented simulation environment for the overall optimization of an IGCC process. In addition to a systematic approach to ROM development, the approach includes validation tasks for the CFD model as well as closed-loop tests for the integrated flowsheet. This approach allows the application of equation-based nonlinear programming algorithms and leads to fast optimization of CFD-based process flowsheets. The approach is illustrated on two flowsheets based on IGCC technology.

See more of this Session: Energy Systems Design and Alternative Energy Sources II
See more of this Group/Topical: Computing and Systems Technology Division

(738b) Optimal Design of Oxy-Fuel Combustion for Carbon Capture and Sequestration

Hussam Zebian, Nicholas D Mancini and **Alexander Mitsos**, Mechanical Engineering, Massachusetts Institute of Technology, Cambridge, MA

Oxy-fuel combustion is a thermodynamically attractive concept that seeks to mitigate the penalties associated with CO2 capture from power plants. The main disadvantage is the need to separate oxygen from atmospheric air. This can be achieved using conventional cryogenic air separation or novel membranes. In either case, careful integration of the air separation, power generation and CO2 treatment is required to achieve high conversion efficiencies. Moreover, in order to obtain meaningful prediction of efficiencies detailed modeling of all processes involved is mandated, but this is often ignored in the literature. The talk describes modeling, design and optimization of various oxycombustion concepts.

In the first part of the talk, a semi-commercial pressurized oxy-coal combustion is considered. Multi-variable optimization is performed involving high-fidelity modeling of the power plant units. In particular, the model accounts for heat losses, flow leaks, pressure drops, irreversibilities, and several technological and economical considerations. Continuous and discrete variables are varied simultaneously to maximize thermal efficiency while satisfying techno-economic constraints. The optimization leads to significant efficiency improvement and more moderate operating conditions (lower pressure) compared to prior studies based on single-variable sensitivity analysis.

In the second part, integrated oxygen ion transport membranes (ITMs) are considered for oxygen separation for the oxycombustion of natural gas. Oxygen separation in an ITM system consists of many distinct physical processes, ranging from complex electrochemical and thermochemical reactions, to conventional heat and mass transfer. The dependence of ITM performance on power cycle operating conditions and system integration schemes must be captured in order to conduct meaningful process flow and optimization studies. An axially distributed, quasi two-dimensional model is presented based on fundamental conservation equations, semi-empirical oxygen transport equations obtained from the literature, and simplified fuel oxidation kinetic mechanisms. Several power cycles from the literature are considered, along with new proposals, including separation in a reactive environment. The detailed modeling demonstrates that several literature claims are overly optimistic since they neglect irreversibilities and operational constraints. Moreover, the model is used to examine the feasibility of novel concepts to overcome some of the challenges associated with the ITM. Finally, the merit of partial CO2 emission concepts is discussed.

See more of this Session: Energy Systems Design and Alternative Energy Sources II
See more of this Group/Topical: Computing and Systems Technology Division

(738c) Modeling and Optimization of Cathode Layer Microstructures In Polymer Electrolyte Fuel Cells

Sesha Hari Vemuri[1], Parag Jain[2], Myung S. Jhon[1] and Lorenz T. Biegler[1], (1)Chemical Engineering, Carnegie Mellon University, Pittsburgh, PA, (2)UOP LLC - A Honeywell Company, Des Plaines, IL

The optimal design of catalyst layer (CL) sub-component is critical for achieving commercialization of polymer electrolyte fuel cells (PEFCs). Typically, a CL is designed as a heterogeneous porous composite of Platinum catalyst over carbon support, void space, and polymer electrolyte membrane (PEM) materials, through which heat & mass transport processes occur with electrochemical reaction. To overcome performance limitations within the CL, a detailed understanding of these transport and reaction processes is required. Especially, an accurate modeling & optimization tool that establishes relationship between CL micro-structures to overall PEFC performance can significantly expedite experimental efforts towards developing tailored CL structures. The non-homogeneous media theories [1-3] estimate the effective molecular transport coefficients through averaged phase hold-up and/or tortuosity parameters which do not account for the microscopic structural information. On the other hand, detailed sub-continuum theories may account for detailed CL structural geometry, they however, may not be suitable for systems level analysis.

In our earlier study [4], we successfully constructed a compact modeling and optimization framework comprising a macro-/meso-scale CL agglomerate model capturing essential features transport & reaction processes and linked to our in house state-of-the-art interior point optimization algorithm, IPOPT [5]. This framework was further exploited to examine PEFC model and optimization sensitivity for various CL agglomerate shapes (sphere, cylinder, and plate) and sizes [6].

Here, we analyze a generalized class of spherical shell agglomerate models originating from the approach by Aris [7], with a goal of providing a basis for realistic CL structures, and amenable to systems analysis.

First, we introduce non-homogeneity in distribution of triple phase boundaries (required for electrochemical reaction) by solving multi-zone spherical agglomerate reaction-diffusion problem, possessing radial variation in reaction rates. We examine sensitivity to PEFC performance in both simulation and optimization results for different zone thicknesses. *Second*, instead of the heuristic 'shape-factor' approach in Aris [7], we derive effectiveness of a spherical shell from first principles by introducing a Robin boundary condition, expressed in terms of Biot number (*Bi*), which plays an important role in estimating the effectiveness of CL layer. This generalized model gives a good agreement with Aris [7] for various values of shell thickness as *Bi* values approach to zero. Also, the case $Bi = 0$ for a solid sphere reduces to our previous model [4]. The optimization based on maximum current density is obtained in detail for different shell thicknesses and various values of *Bi*. We examine how *Bi* influences the critical system design parameters including the I-V characteristics.

1. J. C. Maxwell, A Treatise on Electricity and Magnetism (Clarendon, Oxford, 1881), Vol. 1, p. 435.
2. D.A.G. Brugemann, *Ann. Phys.*, 24, 637 (1935).
3. J. Wang, J. K. Carson, M. F. North, D. J. Cleland, *Int. J. Heat Mass Tran.*, 49, 3075 (2006).
4. P. Jain, L.T. Biegler, and M.S. Jhon, *Electrochem. Solid St.*, 11, B193 (2008).
5. A. Wächter, L.T. Biegler, *Math. Program.*, 106, 25 (2006).
6. P. Jain, L.T. Biegler, and M.S. Jhon, *J. Electrochem. Soc.*, 157, B1222 (2010).
7. R. Aris, *Chem. Eng. Sci.*, 6, 262 (1957).

See more of this Session: Energy Systems Design and Alternative Energy Sources II
See more of this Group/Topical: Computing and Systems Technology Division

(738d) Exergy-Based Optimization for Mixed Refrigerant Systems

Jian Zhang, Meiqian Wang, Qiang Xu and Kuyen Li, Dan F. Smith Department of Chemical Engineering, Lamar University, Beaumont, TX

A refrigeration system generally works by indirectly transferring thermal energy from low-temperature sources to high-temperature sinks at the expense of electricity or mechanical work to lower the source temperature to a certain value. In nature gas liquefaction process, mixed refrigerant (MR) system is widely used to chill and condense nature gas from gas to liquid phase for easy shipment. In comparison with pure refrigerant system, an MR system provides refrigerant composite curve that better fits with natural gas cooling curve in the enthalpy-temperature diagram. Thus, MR system needs less shaft work of compressors. To quantify the "fitness" of composite curves, the concept of exergy is employed. Exergy is the maximum useful work possible during a process that brings the system into equilibrium with environment. The exergy of a stream equals to the enthalpy minus the product of entropy and ambient temperature.

In this work, a novel exergy-based optimization model is introduced. To disclose the thermodynamic instincts of a refrigeration system, an exergy-temperature diagram (B-T diagram) is employed, where the refrigerant exergy is divided into phase-exergy, pressure-exergy, and temperature-exergy. Phase-exergy means the exergy change due to solely phase change from liquid to gas. Pressure-exergy means the exergy change due to solely pressure change from current pressure to the ambient pressure P_0. Temperature-exergy means the exergy change due to solely temperature change from current temperature to the ambient temperature T_0. The transition among different types of exergy for a normal refrigeration cycle is perfectly explained. Based on the understanding, a mixed-integer nonlinear programming (MINLP) model is built. The model includes all equipment models used in the MR system and exergy related equations. The objective function is to maximize exergy efficiency, which is the ratio between useful exergy and the total consumed exergy. An industrial case study of liquefied natural gas (LNG) plant is presented. Results show that the optimized result could increase the exergy efficiency by about 10%.

See more of this Session: Energy Systems Design and Alternative Energy Sources II
See more of this Group/Topical: Computing and Systems Technology Division

Design of the smart community through the use of EV in a university

Yuu Notoji, Kazuya Shimohara, Tsuguhiko Nakagawa,
Graduate school of Okayama Prefectural University, Soja, Japan

Abstract

The CO_2 emissions which is said to be one of the cause of global warming have been researched and developed to reduce all over the world. The smart community that is combined the fossil fuel and the renewable energy to manage the energy supply and demand by the use of IT technology is one of the important system. In this study, the mathematical simulation model which is calculated the energy demand and supply of heat and electric power in the university has been developed on the basis of the more than a year investigation on the hourly demand and changes in the weather. The replacement of the air conditioning apparatuses and the energy demand changes have been calculated when photovoltaic power generation system and a lot of EV are introduced. As a result, if heat pomp type air conditioner is replaced the absorption type, it is possible to reduce 70% of the CO_2 emission. As the feature of Okayama Prefectural university, 80 % of the commutation means is a private car and the average distance is within 30km/day. So the battery capacity of the EV can be changed to a small size. In addition, the photovoltaic power generation system can be generated 200kWh/(m^2・year) of electricity around the university. So it is possible to construct the high cost performance system that is combined the EV with photovoltaic power generation. It has been proved that it is possible to reduce 36% or more CO_2 emission using the proposed integration system.

Introduction

Recently, cutting the CO_2 emission has been the worldwide problem. The Japanese government is promoting various means that are aimed at cutting the country's CO_2 emissions by 60% to 80% on 2050 as compared with 1990. In order to achieve the target, it is important to make the energy demand and supply visibly and the set up to optimize the energy management. As a concrete method, "Smart Community" has been researching all over the world.

In this paper, it is a target to build the Smart Community in the Okayama Prefectural University (hereafter referred to as "OPU"), we have been studied to develop a combination energy system of the energy conservation and the natural energy based on the demand and supply with OPU.

Analysis of the current energy system

Annual energy consumption of OPU

The energy consumption in 2010 is shown in **Table 1.** As can be seen in **Table 1**, 86% of the energy is electric power and 14% is the heavy oil. The OPU has consumed the energy which is 54.1 TJ per year and discharged 2,310 ton CO_2 in 2010.

Table 1 The University's annual energy consumption (2010)

	Amount of energy consumption	**Calorie (TJ/年)**	**CO_2 mission (t-CO_2/年)**
Electric power	4,710MWh/年	46.3	※1 1,760
Heavy oil	202 kL/年	7.8	※2 550
total		54.1	2,310

※1 CO_2 emission coefficient of electric power in Japan 0.3743t-CO_2/MWh

※2 CO_2 emission coefficient of heavy oil 2.71t-CO_2/kL

Innovation of energy system

An image of Smart Community that is a combination energy system of the energy conservation and the natural energy is shown **Fig. 1**. The solid line in the figure shows a current energy network. The short dashed line shows the energy network of the studying object. As shown in **Fig. 1**, we have considered that the future energy system will be using solar cell, fuel cell, biomass fuel and wind power.

In this paper, we discuss the most basic system that is the combination of ①replacing the air-conditioner from the absorption type to a heat pump type, ②introducing the photovoltaic power generation and ③changing the car that is used for commute from the internal-combustion engine type to an electric vehicle (EV).

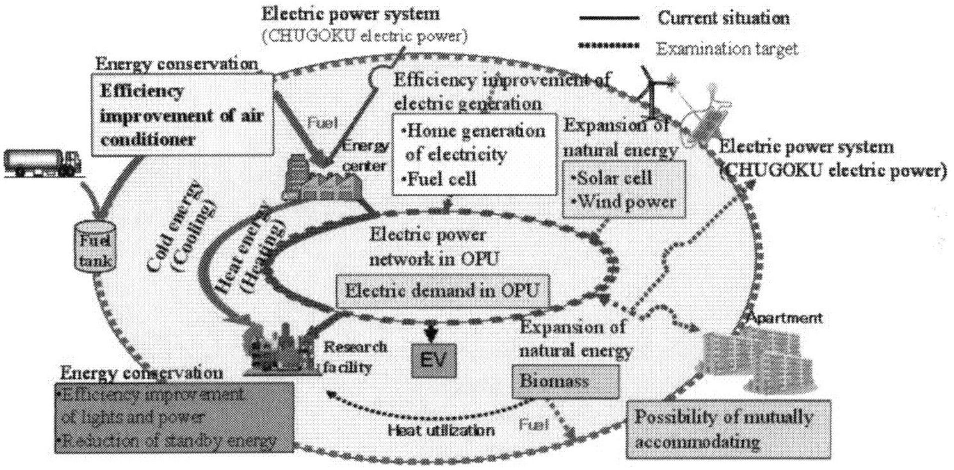

Fig. 1 Basic Smart Community image of OPU

Application of solar power

The capacity of the secondary battery that will need with the substitution rate of photovoltaic power has been presumed. For example, the calculated result that the substitution rate of photovoltaic power will be 100% is shown in **Fig. 2**. As can be seen in **Fig. 2**, the amount of the electricity accumulation becomes the maximum at the end of June. In this case, the battery capacity is calculated at 780MWh, however it is not realistic too huge.

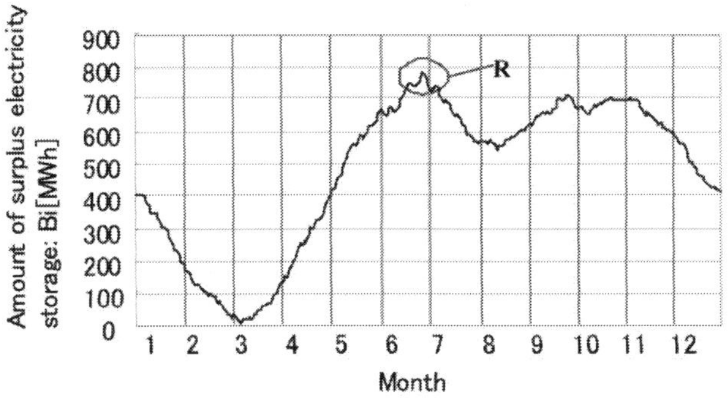

Fig. 2 Necessary capacity of the secondary battery with 100% of the photovoltaic power substitution rate

The relationship between the necessary of the secondary battery capacity with the substitution rate of photovoltaic power is shown in **Fig. 3**.

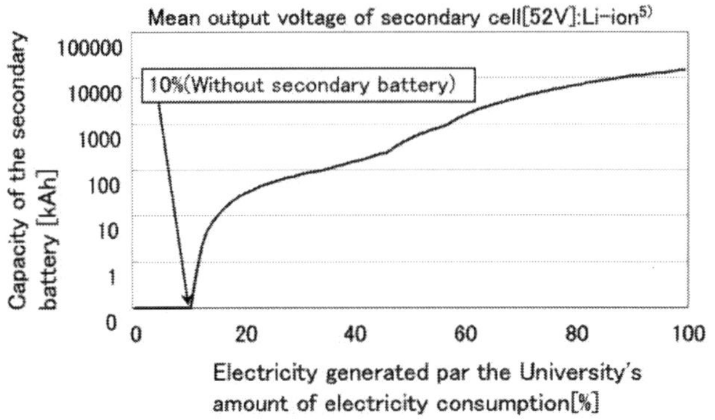

Fig. 3 Relationship between the necessary of secondary battery capacity with the substitution rate of photovoltaic power

As can be seen in **Fig. 3**, it is possible that the substitution rate of photovoltaic power will be 10.5% without using the secondary battery. If the substitution rate of photovoltaic power will increase, the secondary battery capacity will increase in the logarithmic.When the substitution rate of photovoltaic power will be 10.5%, the amount of the CO_2 will be reduced ▲537 [t-CO_2/year] and it means ▲26% reduction compared with 2010. In this case, the photovoltaic power will be generated 0.44MW and the payback period will become 20 years or more. So, we should consider a new system that could cope with both the CO_2 reduction and the payback period.

Combination system use of electric vehicles (EV)

We have thought in a broad concept to reduce the CO_2 emission. That is, we have considered to be related the amount of CO_2 that is emitted by commuting to the university of OPU. In the OPU, the most common commuting method has been driving the car that consumes gasoline and light oil as fuels. These cars will be changed into the electric vehicles (hereafter referred to as EV) and charged the electricity of photovoltaic power generation as the power source.

We have present the method that the generated electricity from photovoltaic power generation has given highest priority to charge EV in weekdays of daytime. In the holidays, the generated electricity has substituted for the basic electricity of OPU. We have studied with the same conditions of **Fig. 3**. In this case, the substitution rate of photovoltaic power is 10.5% and the amount of annual photovoltaic electricity generation is 537MWh/year. The distribution of the photovoltaic power generation is the photovoltaic electricity will be shared of 70% that will charge to EV and 30% that will supply for the demand of OPU.

The final result of all studies is shown in **Table 2.** As can be seen **Table 2**, In the case of replacing the air-conditioner from the absorption type to a heat pump type, the CO_2 emission will be reduced ▲14 %. In addition of introducing the photovoltaic power generation, the CO_2 emission will be reduced ▲20 %. In addition of changing the car that is used for commute from the internal-combustion engine type to an EV, the CO_2 emission will be reduced ▲36 %.

Conclusions

We have studied to develop a combination energy system of the energy conservation and the natural energy based on the demand and supply with OPU.

As a result, it is possible to cut off the 36% or more CO_2 emission with the combination of ①replacing the air-conditioner from the absorption type to a heat pump type , ②introducing photovoltaic power generation and ③changing the car that is used for commute from the internal-combustion engine type to an electric vehicle (EV).

Table 2 Amount of CO_2 emission and reduction ratio each equipment combination

		Current	①	①+②	①+②+③
Electric power	t-CO_2/year	1760	1920	1720	1850
	MWh/year	4710	5120	4590	4950
Heavy oil	t-CO_2/year	550	0	0	0
	kL	202	0	0	0
CO_2 reduction	t-CO_2/year	2310	1920	1720	1850
	%	—	17.0	26.0	20.0
Changing the car (361)	t-CO_2/year	580※1	580※1	580※1	0
	kL	248	248	248	0
Amount of CO_2 reduction	t-CO_2/year	**2890**	2500	2300	**1850**
	%	—	14.0	20.0	**36.0**

※1 CO_2 emission coefficient of gasoline 2.32t-CO_2/kL

①Heat pomp type air-conditioning
②Photovoltaic power generation
③EV

References

1 Kazuya Shimohara, Yuu Notoji, Seiji Shiba, Tuguhiko Nakagawa (2010), "A Formation of Distributed Energy Network and Energy Conservation in the University, The 15th National Symposium on Power and Enelgy System," SPES 2010, pp.267-268

2 SANYO Electric Company Ltd (2010), "Progress of Technology of Making to Highly Effective of Solar Battery that Uses Nanotechnology, " pp34-35

3 Yukinori Kuwano (2009), "Gaphic Eplanation All of the Latest Solar Cell, " pp100-105,

(738f) Design of a Self-Supporting Natural Gas Liquefaction Process for Accessing Stranded off-Shore Resources

Achim Wechsung[1], Audun Aspelund[2], Truls Gundersen[2] and Paul I. Barton[3], (1)Department of Chemical Engineering, Massachusetts Institute of Technology, Cambridge, MA, (2)Energy and Process Engineering, Norwegian University of Science and Technology, Trondheim, Norway, (3)Chemical Engineering, Massachusetts Institute of Technology, Cambridge, MA

Natural gas accounts for 24% of world energy consumption today, consumption is projected to grow by 44% until 2035 and natural gas is expected to become increasingly important since it combusts more cleanly and produces power more efficiently than other fossil fuels. On the one hand, close to 60% of the world's natural gas reserves are found in three countries: Russia, Iran, and Qatar. On the other hand, most of the future production growth is predicted to take place in non-OECD countries which may lead to increasingly strained and unreliable supply chains to consumers in developed nations. Stranded reserves account for the majority of the resource base and LNG will play a key role in exploiting these. Lastly, it is anticipated that public policy requiring capture and sequestration of carbon dioxide will become more widespread.

Stranded off-shore gas yields will require new technologies to make production economically viable. Traditional natural gas infrastructure for offshore production necessitates large capital expenditures to connect yields with onshore processing facilities. Ifnatural gas could be liquefied onboard dedicated platforms and then transported with LNG tankers to customers, most of the inflexible infrastructure becomes superfluous. Additionally, since LNG can be traded much more easily, long-term contracts couldbe replaced by marketing the gas on a global gas market.

Currently, liquefying natural gas uses large amounts of energy to provide refrigeration. At the receiving terminal, LNG is re-gasified using seawater or air as a heat source. However, the "cold" of the LNG could be recycled thus improving exergy efficiency of the supply chain. For example, nitrogen and captured carbon dioxide can be liquefied providing the required heat to gasify LNG. The liquefied N2 and CO2 can be loaded aboard the empty LNG tanker which returns to the off-shore site. There, the liquefied gases will provide the required cooling to liquefy natural gas and are vented to the atmosphere (N2) or sequestered in geological formation such as depleted gas and oil fields (CO2).

While N2 and CO2 vaporize at constant temperature, natural gas condenses over a temperature interval. It is possible to modify the composite curves by adding compressors and expanders to track the condensation curve of natural gas effectively. The design goal is to find a flowsheet that is self-contained in the sense that no additional power or cooling is necessary for its operation and uses the least nitrogen.

A MULTI-SCALE MODEL-BASED APPROACH TO THE SYNTHESIS AND HIGH-THROUGHPUT ANALYSIS OF BIOREFINERIES

Marinella L. Tsakalova, National Technical University of Athens, Greece
Aidong Yang, University of Surrey, United Kindom
Antonis C. Kokossis, National Technical University of Athens, Greece

Abstract

The paper highlights the pivotal role of systems technology to foster innovation, preview options, and support high-throughput computational experimentation, arguing that systems tools are largely under-deployed. Systems-enabled platforms could instead function as powerful environments to generate ideas for integrated designs and offer tremendous services to the complex and large problems produced by the numerous portfolios of feedstocks, unknown portfolios of products, multiple chemistries, and multiple processing paths. A decision support platform is developed to evaluate and screen options for producing chemicals from biomass feedstock in a multi-scale approach. The platform is based on the modelling of the so-called compartmental processing blocks (CPBs) and their connections. It includes a simulation tool and a synthesis/optimisation tool, both based on a central database which holds process and economic information about individual CPBs. The synthesis/optimisation tool performs screening over multiple production routes by means of modelling and optimisation of superstructures that represent possible feedstock, routes, and products in a given case of assessment. The current implementation uses Excel for the database and the simulation tool, while the synthesis/ optimisation tool is based on GAMS. Cases studies of green ethylene, ammonia and propylene glycol are reported to illustrate the application of the platform.

(764a) Stability Conditions for Adaptive Inventory Control Systems

Juan Du and B. Erik Ydstie, Department of Chemical Engineering, Carnegie Mellon University, Pittsburgh, PA

The main contribution of this work is three-fold. First we integrate an on-line adaptation with the inventory control framework using ideas from passivity based control. Second we derive the stability condition for adaptive inventory control system by connecting thermodynamic stability theory and passivity theory. Third we develop a case study to highlight the application of the theory. The bioreactor case study is characterized by input multiplicity. It is shown that the choice of input output pairing provides the key to achieve input output stability of the control system, convergence of estimated parameters and stability of the zero-dynamics.

The main advantage of the proposed approach is that the adaptive mechanism is incorporated into the passivity based inventory control scheme so that the unknown parameters can be estimated. Passive input-output pairs are derived from mass balance of each chemical component in the bioreactor and unknown model parameters are adapted online, so that the effect of unknown process dynamics can be compensated for. Simulation and stability results show that the zero-dynamics are stable and that the online estimator guarantees stability of the closed-loop system. Input-output pair selection has a significant impact on the inventory control system. If the input-output is chosen then the stability of the zero-dynamics is guaranteed.

The global stability analysis is achieved using the composite energy function consisting of controlled dynamics, zero-dynamics and estimated dynamics. A composite Lyapunov function based on nonlinear control theory is used to represent the energy of the controlled dynamics and estimated dynamics. An entropy based function, so called the availability in classic thermodynamics, describes the energy of the zero-dynamics. The global stability analysis provides a theoretical proof to the different behaviors of bioreactor under adaptive-inventory controller due to various input-output pairs.

See more of this Session: Control In Medicine and Biology
See more of this Group/Topical: Computing and Systems Technology Division

(764b) Data-Driven Network Reconstruction of Biological Systems: Comparison of Statistical and Optimization-Based Methods

Behrang Asadi, Department of Bioengineering, University of California, San Diego, La Jolla, CA, Mano R. Maurya, Department of Bioengineering, University of California San Diego, La Jolla, CA, Daniel M. Tartakovsky, Department of Mechanical and Aerospace Engineering, University of California, San Diego, La Jolla, CA and Shankar Subramaniam, Department of Bioengineering, San Diego Supercomputer Center, Department of Chemistry & Biochemistry, University of California San Diego, La Jolla, CA

Data-drivennetwork reconstruction of biological systems is an essential step towards extractinginformation from large volumes of biological data. There are several methodsdeveloped recently to reconstruct biological networks. However, to the best of ourknowledge, no systematic and comprehensive studies have been carried out to comparedifferent methods based on different properties of datasets in terms of their abilityto handle noisy data, different types of noise, level of correlation/ collinearity,size of the data set and incomplete data sets. In this study, we have comparedthree popular methods-principal component regression (PCR)[1], linear matrixinequalities (LMI) [2],and Least Absolute Shrinkage and Selection Operator (LASSO)[3] on both real/experimental and synthetic data sets. Each ofthese methods is a representative of a category of popular methods that can befound in the literature. Method of PCR is based on dimensionality reduction. InLASSO, the aim is to minimize the L-2 norm of the residual vector while satisfyinga parsimony constraint on the parameters. In LMI, the goal is to minimize an L-infinity-normof the residual vector, with the ability to simultaneously incorporate a prioriknowledge into the optimization problem. We have used three different metrics tocompare the performance of the methods: root-mean-squared-error (RMSE) inprediction, average fractional error in the value of estimated coefficients,and semi-binary evaluation metrics: accuracy, sensitivity, specificity, and thegeometric mean of sensitivity and specificity. This comparison enables us toestablish criteria for selection of an appropriate approach for networkreconstruction based on *a priori* properties of experimental data. Forexample, while PCR is the fastest method, LASSO and LMI perform better in termsof accuracy, sensitivity and specificity. Both PCR and LASSO are better thanLMI in terms of fractional error in the values of the computed parameters. Thesetrade-offs suggest that more than one aspect of each method needs to be taken intoaccount in selecting a methodology for network reconstruction.

References

1. Pradervand,S., M.R. Maurya, and S. Subramaniam, *Identification of signaling*

*componentsrequired for the prediction of cytokine release in RAW 264.7 macrophages.*Genome Biology, 2006. **7**(2): p. R11.

2. Cosentino,C., et al., *Linear matrix inequalities approach to reconstruction ofbiological networks.* IET Systems Biology, 2007. **1**(3): p. 164-173.

3. Tibshirani,R., *Regression shrinkage and selection via the Lasso.* Journal of theRoyal Statistical Society Series B-Methodological, 1996. **58**(1): p.267-288.

See more of this Session: Control In Medicine and Biology
See more of this Group/Topical: Computing and Systems Technology Division

(764c) A System Identification and Control Engineering Approach for Designing An Optimized Treatment Plan for Fibromyalgia

Sunil Deshpande, School of Electrical, Computer, and Energy Engineering, Arizona State University, Tempe, AZ, Naresh N. Nandola, ABB Global Industries and Services Limited, 560048, Bangalore, Karnataka, India, **Daniel E. Rivera**, School for Engineering of Matter, Transport, and Energy, Arizona State University, Tempe, AZ and Jarred W. Younger, Systems Neuroscience and Pain Laboratory, Stanford University School of Medicine, Palo Alto, CA

There is increasing interest in the medical community towards developing operationalized strategies for treating chronic diseases that take advantage of individual characteristics of the patient for optimizing treatment [1, 2]. In the field of behavioral health, the term "adaptive interventions" has been used to describe individually-tailored strategies for the prevention and treatment of chronic, relapsing disorders [3]. Control systems engineering principles applied to adaptive interventions have been proposed as enablers for more efficacious treatments that minimize waste and enhance intervention potency [4, 5, 6]. This paper will discuss how control engineering principles can be used to design an optimized plan for the treatment of fibromyalgia (FM), a disorder characterized primarily by chronic widespread pain.

The approach is based on a secondary analysis of data collected during a clinical trial using naltrexone, an opioid receptor antagonist, for the treatment of FM [7]. To obtain a dynamical model we apply system identification techniques to extensive daily diary reports completed by intervention participants. These diary reports include self-assessments of outcomes of interest (e.g., general pain symptoms, sleep quality) and additional external variables that affect these outcomes (such as stress, anxiety, and mood). We estimate a multi-input ARX (AutoRegressive with eXogenous inputs) model to best explain the output variance; the procedure is to begin with drug and placebo as inputs, and then systematically include other input variables which contribute to an improvement of the model fit [8]. The ARX model is then simplified to a continuous-time second-order model with zero for each input-output transfer function. Having obtained a set of dynamical models that describe both manipulated and measured disturbance variables, model predictive control is used as the decision algorithm for automated dosage selection of naltrexone over time. The categorical nature of the dosage assignment, i.e. that fact that the drug dosage levels can only assume discrete values over its domain, creates a need for hybrid model predictive control (HMPC) schemes. In lieu of the conventional tuning approach using objective function weights, a multiple degree-of-freedom formulation is evaluated that enables the user to adjust the speed of setpoint tracking, measured disturbance rejection and unmeasured disturbance rejection independently in the closed loop system [6, 9]. Simulation results for a representative participant in the presence of significant plant-model mismatch and disturbances (both measured and unmeasured) demonstrate the performance and broad-based applicability of a predictive control approach to this class of behavioral intervention problems.

In the final section of the paper, the problem of informative experimental designs that go beyond the traditional approach of applying fixed dosages in a clinical trial is discussed. Flexible dosage designs in literature have been typically related to drug toxicity concerns [10, 11] and do not explicitly target informative data for estimating a dynamical system. Since the proposed procedure relies on system identification for dynamical modeling, the information content of the input signal and its independence from other signals in the problem are crucial to the success of the overall treatment methodology. Input signals exceeding two levels are required to address nonlinearity, while the experimental protocol must agree with clinical constraints; for example, drug dosage levels must stay within designated settings and should not change abruptly from lowest to highest settings for a given intervention participant. A new class of optimal input design problems for system identification motivated by this application space that extend the ideas presented in [12] will be presented.

References

[1] Francis S. Collins. The future of personalized medicine. *NIH Medline Plus*, 5:2–3, 2010.

[2] Peter Wellstead, Eric Bullinger, Dimitrios Kalamatianos, Oliver Mason, and Mark Verwoerd. The role of control and system theory in systems biology. *Annual Reviews in Control*, 32(1):33– 47, 2008.

[3] L. M. Collins, S. A. Murphy, and K. L. Bierman. A conceptual framework for adaptive preventive interventions. *Prevention Science*, 5:185–196, 2004.

[4] D. E. Rivera, M. D. Pew, and L. M. Collins. Using engineering control principles to inform the design of adaptive interventions a conceptual introduction. *Drug and Alcohol Dependence*, 88(2):S31–S40, 2007.

[5] A. Zafra-Cabeza, D.E. Rivera, L.M. Collins, M.A. Ridao, and E. F. Camacho. A risk-based Model Predictive Control approach to adaptive behavioral interventions in behavioral health, *IEEE Transactions in Control Systems Technology*, in press, 2011.

[6] Naresh N. Nandola and Daniel E. Rivera. A novel model predictive control formulation for hybrid systems with application to adaptive behavioral interventions. In *Proceedings of the 2010 American Control Conference*, pages 6286 –6292, June 2010.

[7] Jarred Younger and Sean Mackey. Fibromyalgia symptoms are reduced by low-dose naltrexone: A pilot study. *Pain Medicine*, 10(4):663–672, 2009.

[8] Sunil Deshpande. A control engineering approach for designing an optimized treatment plan for fibromyalgia. Master's thesis, Arizona State University, 2011.

[9] W. Wang and D.E. Rivera. Model predictive control for tactical decision-making in semiconductor

manufacturing supply chain management. *Control Systems Technology, IEEE Transactions on*, 16 (5):841 –855, Sep. 2008.

[10] Stephen D. Durham, Nancy Flournoy, and William F. Rosenberger. A random walk rule for phase i clinical trials. *Biometrics*, 53(2):pp. 745–760, 1997.

[11] Lesley M. Arnold, Evelyn V. Hess, James I. Hudson, Jeffrey A. Welge, Sarah E. Berno, and Paul E. Keck. A randomized, placebo-controlled, double-blind, flexible-dose study of fluoxetine in the treatment of women with fibromyalgia. *The American Journal of Medicine*, 112(3):191 – 197, 2002.

[12] Daniel E. Rivera, Hyunjin Lee, Hans D. Mittelmann, and Martin W. Braun. Constrained multisine input signals for plant-friendly identification of chemical process systems. *Journal of Process Control*, 19(4):623 – 635, 2009.

See more of this Session: Control In Medicine and Biology
See more of this Group/Topical: Computing and Systems Technology Division

Experimental Design of Systems Involving Multiple Fluorescent Reporters

Loveleena Bansal, Eric Yang, Randall Nelson, Arul Jayaraman and Juergen Hahn

Department of Chemical Engineering, Texas A&M University, College Station, TX

Fluorescent proteins are now regularly used to monitor a variety of aspects of biological systems. Some uses of fluorescent proteins involve, but are not limited to, monitoring the expression of genes and tracking cells during the course of an experiment [1]. The main advantage that the use of fluorescent proteins has over other approaches is that the fluorescence can be monitored in real-time without destroying the sample, thereby allowing to monitor the location where fluorescence can be observed, taking several measurements to increase measurement accuracy, or to monitor aspects of an experiment over time. There are now a number of fluorescent proteins that can be commercially purchased (e.g., reference [2] lists 25 different fluorescent proteins). While each of these fluorescent proteins can be used separately, the main advantage of using fluorescent proteins, with different emissions spectra, together is that it is possible to simultaneously monitor different aspects of the same experiment. However, the emissions spectra of the different fluorescent proteins overlap and any measurement involves contributions from all the fluorescent proteins used [3, 4]. This overlap in the emissions spectra is a trivial issue if only two different fluorescent proteins are used, i.e., one chooses two fluorescent proteins with minimal overlap in their spectra [5]. However, if two or more fluorescent proteins are used in an experiment, extracting the contribution of individual types of fluorescent proteins from measurements becomes very challenging due to extensive overlap of spectrums. Thus, this work develops a technique for designing experiments involving different types of fluorescent proteins such that it becomes easier to extract the contribution of each protein from the observed fluorescence.

For efficient experimental design involving multiple reporters, there are two aspects that are addressed in this study. First, an optimization problem is formulated that determines the contribution of individual fluorescent proteins to the total fluorescence intensity of a mixture containing 2 or more proteins. In the second step, this optimization problem is extended to perform experimental design, such that the accuracy of the predicted levels of the different fluorescent proteins is maximized. This involves the decision of which different fluorescent proteins should be used in an experiment amongst the available proteins. These techniques have been validated using experimental data created from mixtures of different types of *E. coli* where each type of *E. coli* expressed a different fluorescent protein. The measurements were taken using a fluorescent plate reader at different emission wavelengths using a fixed excitation wavelength. Mixtures containing up to four fluorescent proteins have been analyzed. The intensity measurements for individual proteins and a mixture of red, yellow and green fluorescing E. Coli are shown in Figure **1**. It is evident that the overlap of spectrums makes it challenging to infer the composition of each protein from the intensity measurement of the mixture.

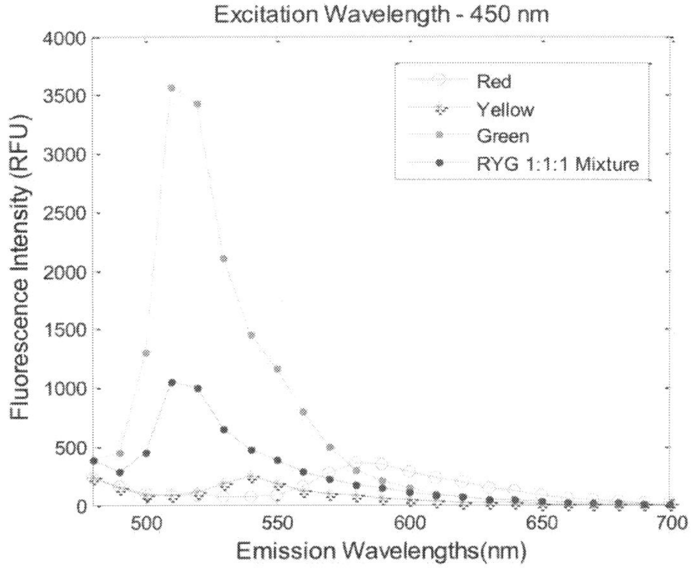

Figure 1 Fluorescence Intensity Measurements of Individual Proteins and their mixture using a Fluorescent Plate Reader

Using the formulated technique, it was possible to extract the contribution of each fluorescent protein in a mixture with a maximum error of 15-20 %. The results for the selection of proteins using the experiment design criterion are in good agreement with experimental observations as the errors observed on using the recommended proteins are significantly less than using a random selection of proteins from those available.

References

[1] Lippincott-Schwartz, J. and Patterson, G. H. (2003). Development and Use of Fluorescent Protein Markers in Living Cells. *Science*, Vol. 300, No. 5616, pp. 87-91

[2] http://www.clontech.com/products/detail.asp?product_id=10419&tabno=2

[3] Dickinson, M., Bearman, G., Tille, S., Lansford, R. and Fraser, S. (2001). Multi-spectral imaging and linear unmixing add a whole new dimension to laser scanning fluorescence microscopy. *Biotechniques,* Vol. 31, No. 6, pp. 1272-1279

[4] Lansford, R., Bearman, G. and Fraser, S. E. (2001). Resolution of multiple green fluorescent protein color variants and dyes using two-photon microscopy and imaging spectroscopy. *Journal of Biomedical Optics,* Vol. 6, No. 3, pp. 311-318

[5] Shaner, N. C., Steinbach, P. A. and Tsien, R. Y. (2005). A guide to choosing fluorescent proteins *Nature Methods,* Vol. 2, No. 12, pp. 905-909

(764e) Subject-Specific Multiple Input Block-Oriented Glucose Modeling of Several Type 1 Diabetic Subjects

Derrick K. Rollins Sr.[1], **Kaylee Kotz**[1], A. Cinar[2], Elizabeth Littlejohn[3] and Lauretta Quinn[4], (1) Chemical and Biological Engineering, Iowa State University, Ame, IA, (2)Chemical and Biological Engineering, Illinois Institute of Technology, Chicago, IL, (3)Section of Adult and Pediatric Endocrinology, Diabetes, & Metabolism, Institute for Endocrine Discovery and Clinical Care, Chicago, IL, (4)College of Nursing, University of Illinois at Chicago, Chicago, IL

Type 1 diabetics often experience extreme variations inglucose concentrations which can have adverse long- and short-term effects suchas severe hypoglycemia, hyperglycemia and organ degeneration. Studies haveestablished that there is a need to maintain the glucose levels within a normalrange to avoid complications caused by diabetes. However, initial attempts toregulate blood glucose levels using insulin infusion, multiple injections or acombination of the two have had limited success as they lack the ability todecide the appropriate rate and/or amount of insulin infusion based on thecurrent metabolic state of the body. An artificial pancreas consisting of acontinuous glucose monitor, an insulin infusion pump, and a control algorithmhas the potential to improve glucose regulation by intelligently deciding theproper amount of insulin delivery at the proper time. However, a critical keyin the success of the electro-mechanical pancreas is the ability to effectivelymodel and use this model to improve closed-loop control. Thus, the objective ofthis talk is to propose a modeling method that takes into account thesimultaneous and multiple effects of food, activity, stress and theirinteractions in developing subject-specific models for several type 1 diabeticsubjects.

This talk will present the results on 10 to 15 type 1diabetic subjects where food (3 variables), activity (7 variables), insulininfusion (2 variables), and time of day (TOD) (a total of 13 variables) arecollected for two weeks and modeled using the Wiener block-oriented method ofRollins et al. (2010). The data sets are split two ways. One way is intotraining and validation sets of 1 week each. The second way is into training (1week), validation (4 days) and testing (3 days). The method of Rollins et al.is modified to use an approach that determines dynamic parameters separatelyfrom the static parameters such that consistency in the fitted models aremaintained over all the data sets to guard against over-fitting the data. Threetypes of models are compared: input only (Model 1), input/output (Model 2), andoutput only (Model 3). Results are given for forecast predictions K-steps ahead (KSA) from 5 minutes to 3 hours. Model 1 uses only the13 input variables for prediction and technically it is not a KSA forecastmodel. Model 2 is a pre-whitened model for forecasting one step ahead and usesmeasured glucose and the 13 input variables. Model 3 is an auto-regressive-integratedmoving average (ARIMA) model that forecasts KSA using only glucosemeasurements. The results show that Model 2 approaches Model 1 as k increases and that this approach canvary considerably from subject to subject. They also show that Model 3consistently performs worse that Model 2 but its decay in performance as k increases can also vary considerablyfrom subject to subject. For each data set, results are given for models withonly food variables, only activity variables and only insulin variables.Results showing the importance of modeling interactions between input variablesare also given.

Rollins, D. K., N. Bhandari,J. Kleinedler, K. Kotz, A. Strohbehn, L. Boland, M. Murphy, D. Andre, N. Vyas, G.Welk and W. Franke, "Free-living inferential modelingof blood glucose level using only noninvasive inputs," Journal of ProcessControl **20** 95-107 (2010).

See more of this Session: Control In Medicine and Biology
See more of this Group/Topical: Computing and Systems Technology Division

(764f) Equilibrium Conditions In Tumors Undergoing Virotherapy

Ricardo Dunia, Chemical Engineering Department, The University of Texas at Austin, Austin, TX and Thomas F. Edgar, Chemical Engineering, The University of Texas at Austin, Austin, TX

Oncolytic viruses selectively infect and replicate inside tumor cells, which demonstrate great potential in cancer treatment [1][2][3]. However, there are several factors that diminish the effectiveness of virotherapy. Among these factors is the development of undesirable immune responses that attacks the virus inside the tumor [4]. Therefore, immune suppression drugs, like cyclophosphamide (CPA), have been used in conjunction with oncolytic viruses in order to maintain the effectiveness of virotherapy during treatment [5]. Nevertheless, the use of such drugs weakens the immune system, which promotes the development of infections in treated patients [6].

This work determines the minimum amount of CPA necessary to halt tumor growth under virotherapy. Such a limited amount provides the highest immune system levels that could be made available while cancer progress is hampered. Such a circumstance is referred as equilibrium or static condition. Two spatial virotherapy dynamic models are considered in this work for the calculation of such equilibrium conditions. The first model developed by Friedman et. al [4] and analyzed by Wang and Tian [7] considers a convective radial migration of immune system cells along the solid tumor. This model is based on the assumption that immune cells are large enough to be considered part of the tumor mass. A second model developed by Tao and Guo [8] considers the immune cells to penetrate and migrate by diffusion inside the tumor mass.

In these two models the tumor growth is based on the tumor cell velocity evaluated at the tumor surface, U(R), where R represents the tumor radius. A positive velocity at the boundary indicates tumor enlargement. In general, the tumor radius and the velocity are related by $dR/dt=U(R)$. This relation is used to define the equilibrium or static line (SL) for tumor growth, in which $dR/dt=0$. This SL is a function of the CPA drug delivery and can determine the minimum dose to prevent the tumor enlargement after few hours of virotherapy. Higher doses than the ones specified by SL will make the cancer shrink, while doses below the ones suggested by SL will make the tumor to grow. Because fast tumor shrinking might be achieved at the expense of weakening the immune system, only CPA doses a slightly above the one suggested by the SL are considered desirable for treatment.

References

[1] S. J. Ries, C. H. Brandts, Oncolytic viruses for the treatment of cancer: current strategies and clinical trials, Drug Discovery Today 9.

[2] M. Vaha-Koskela, J. E. Heikkila, A. E. Hinkkanen, Oncolytic viruses in cancer therapy, Cancer

Letters 254 (2007).

[3] L. M. Wein, J. T. Wu, D. H. Kirn, Validation and analysis of a mathematical model of a replication-competent oncolytic virus for cancer treatment: implications for virus design and delivery, Cancer Research 63 (2003).

[4] A. Friedman, J. P. Tian, G. Fulci, E. A. Chiocca, J. Wang, Glioma virotherapy: Effects of innate immune suppression and increase viral replication capacity, Cancer Research 66 (4) (2006).

[5] H. Wakimoto, G. Fulci, E. Tyminski, E. A. Chiocca, Altered expression of antiviral cytokine mrnas associated with cyclophosphamide's enhancement of viral oncolysis, Gene Therapy 11 (2004).

[6] H. Vladar, J. Gonzalez, Dynamic response of cancer under the influence of immunological activity and therapy, Journal of Theoretical Biology 227 (2004).

[7] J. Wang, J. Tian, Numerical study for a model of tumor virotherapy, Applied Mathematics and Computation 196 (2008).

[8] Y. Tao, Q. Guo, The competitive dynamics between tumor cells, a replication-competent virus and an immune response, Journal of Mathematical Biology 51 (2005).

See more of this Session: Control In Medicine and Biology
See more of this Group/Topical: Computing and Systems Technology Division

DESIGN AND SIMULATION OF A CONTROLLER SYSTEM FOR METABOLIC SHIFT REGULATION IN MAMMALIAN CELLS

Damián Baeza[1,3], Ziomara P. Gerdtzen[2,3,4], J. Cristian Salgado[1,2,3]

[1] Laboratory for Process Modeling and Distributed Computing, Department of Chemical Engineering and Biotechnology, Faculty of Physical and Mathematical Sciences, University of Chile, Santiago, Chile
[2] Millennium Institute for Cell Dynamics and Biotechnology, Department of Chemical Engineering and Biotechnology, Faculty of Physical and Mathematical Sciences University of Chile, Santiago, Chile
[3] Department of Chemical Engineering and Biotechnology, Faculty of Physical and Mathematical Sciences, University of Chile, Santiago, Chile
[4] Centre for Biochemical Engineering and Biotechnology, Department of Chemical Engineering and Biotechnology, Faculty of Physical and Mathematical Sciences University of Chile, Santiago, Chile

Background

Experimental data supports the possibility of multiple steady states in mammalian cell cultures, each with a distinct cellular metabolism [1]. Different metabolic states have different lactate to glucose stoichiometric ratios ($\Delta L/\Delta G$). Table 1 shows that the existence of multiple steady states involves the interaction between metabolic and gene [2].

Table 1. Gene level fold changes corresponding to low over high $\Delta L/\Delta G$ states

Enzyme	Real Time PCR	Microarray cDNA
Lactate Dehydrogenase	↓ 2.4	↓ 1.9
Pyruvate Kinase	↓ 2.9	↓ 2.0
Phosphofructokinase	↓ 2.0	↓ 1.3

Metabolic state variations in cell cultures have been found to be related to the glucose concentration in a reactor [3]. Experimental results further indicate that cell cultures in a low metabolic state present an immediate response to pulse additions of glucose.

Providing an optimized strategy for glucose feeding that allows control over a specific metabolic state is yet to be studied. A model based strategy for designing a control system for metabolic state regulation and maintenance, which considers the biological complexity of the regulation of the cellular system, is herein proposed.

Materials and methods

A detailed metabolic model for mammalian cell metabolism was formulated (complete model), which consists of a system of ordinary differential equations for the main metabolic variables of the following form:

$$\frac{d\vec{C}}{dt} = S \cdot \vec{r} - \vec{\tau}(\vec{C})$$

where $\vec{C}$ is a vector containing metabolite concentration, **S** is the stoichiometric matrix, $\vec{r}$ is a vector comprising reaction rates, and $\vec{\tau}(\vec{C})$ is a vector containing transport/consumption rates. The reaction rates can be described by:

$$r_i = r_i^{\max} \cdot f_i(\vec{C})$$

wherein $r_i^{\max}$ is the reaction rate constant for the ith reaction and $f_i(\vec{C})$, is a non-linear function that describes the reaction-reactant dependency. Furthermore, the cell growth rate was formulated by a monod function of the type:

$$\mu = \mu_{\max} \cdot \frac{C^e_{glc}}{K^X_{glc} + C^e_{glc}} \cdot \frac{I^X_{lac}}{I^X_{lac} + C^e_{lac}}$$

Model parameters that were not retrieved from literature were obtained through a fitting process to experimental data; said fitting process was by applying the Nelder-Mead simplex method to minimize the least squares [4].

Once the fitting process was completed for the detailed model; a simplified model was constructed and fitted, that includes the main metabolic reactions in CHO cells. Stability analysis was carried out on both the detailed and the simplified model in order to establish the number of possible steady states for both mathematical models.

In order to recreate the metabolic shift phenomenon observed experimentally, enzyme kinetics for lactate dehydrogenase were weighted with an enzyme factor *F*, which valued between 0 and 1:

$$r'_{LDH} = F \cdot r_{LDH}$$

The estimated *F* values were later used to adjust a Hill activation function that depends on the residual glucose concentration; said fitting was to obtain experimental curves of the culture presenting metabolic shift.

Results

Figures 1a and 1b show the results of the curve fitting process for the detailed and simplified models, respectively. From Figure 1d it can be concluded that the same model parameters cannot be used to simulate a cell culture that presents metabolic shift. Figure 1d shows the results of the use of the same parameters to simulate a MAK cell culture induced to an altered metabolic state.

Moreover, stability analysis confirmed the existence of only one steady state for both the simplified and the detailed model. The existence of a sole stable attractor supports the fact that metabolic regulation alone cannot regulate a metabolic shift in a cell culture. In addition, gene regulation is an element that should be considered in a model that correctly describes this phenomenon.

Figure 1. Simulation results for detailed and simplified model

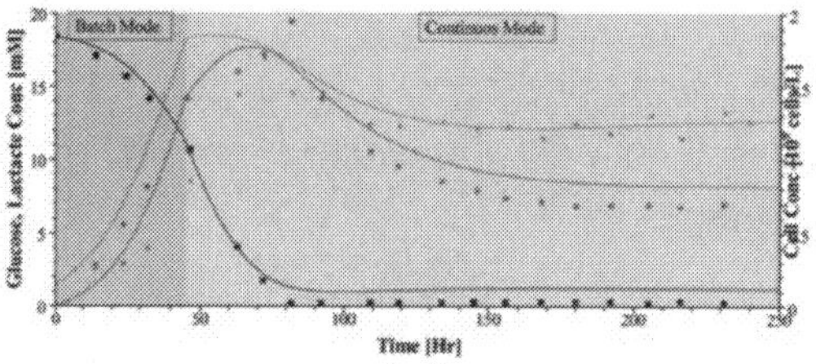

(a)

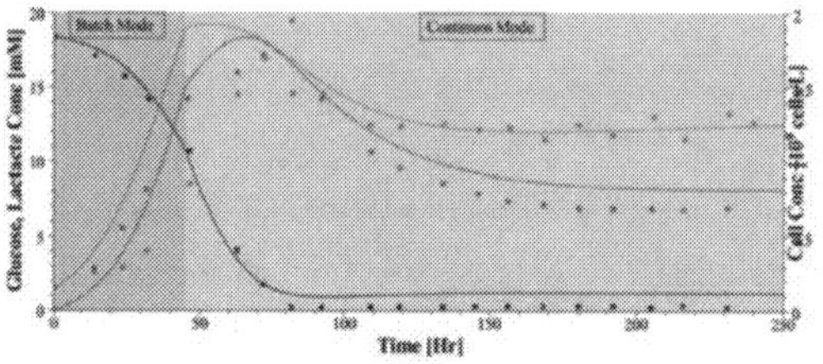

(b)

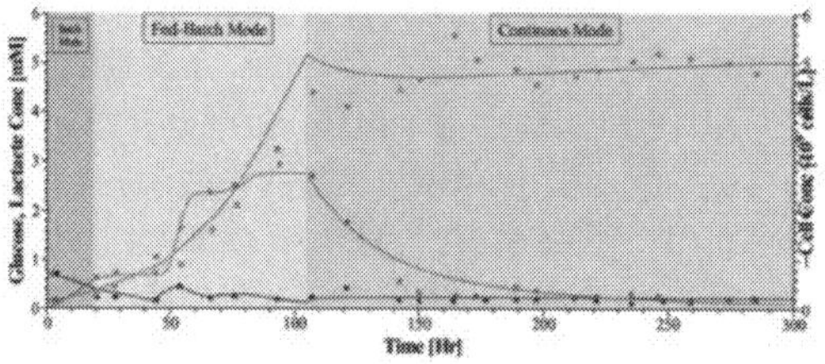

(c)

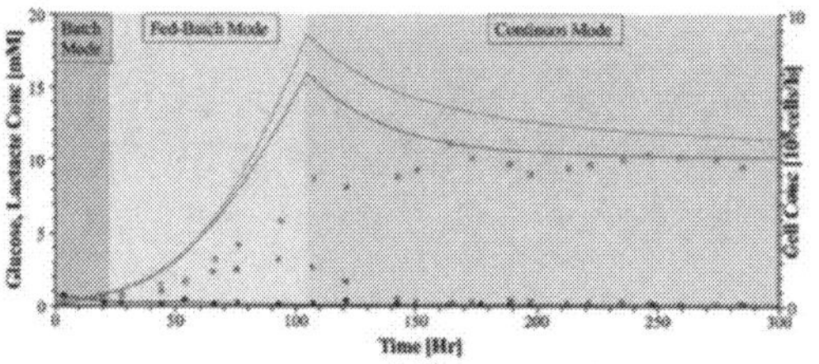

(d)

1

(a) Simulation results for complete model without metabolic shift, (b) Simulation results for unregulated simplified model without metabolic shift, (c) Simulation results for regulated simplified model with metabolic shift, (d) Simulation results for unregulated simplified model with metabolic shift. - : Simulated Lactate, - : Simulated Glucose, - : Simulated Cell Conc., ■ : Lactate, ■ : Glucose, ● : Cell Conc.

As previously indicated, the additional regulation model used consists of a Hill activation function. The implemented regulation function depends on the residual glucose concentration and modifies the existing enzyme kinetic for lactate dehydrogenase. The resulting Hill function is of the form:

$$r'_{LDH} = \beta \cdot \frac{\left(C^{e}_{glc}\right)^{n}}{\left(K^{X}_{glc}\right)^{n} + \left(C^{e}_{glc}\right)^{n}} \cdot r_{LDH}$$

wherein K_{glc} = 1.24 [*mM*] is the activation coefficient, $\beta = 1$ is the maximal expression level, C^{e}_{glc} is the residual glucose concentration, and n = 14.64 is the Hill coefficient. The results of the implementation of the above function are presented in Figure 1c. Both the maximal expression level and the activation coefficient were set, based on literature [3] and on initial function restrictions, respectively. The Hill coefficient was adjusted through a curve fitting process using Nelder-Mead simplex method. Additionally, the sum of square residuals for the regulated model's glucose, lactate and cell concentration curves was 15.74 when compared with another source of experimental data, which further supports the proposed metabolic model.

Conclusions

Experimental evidence and simulation results support that a metabolic shift in cell cultures is due to regulation both at gene expression and metabolic levels. Lower $\Delta L/\Delta G$ ratios are associated to a previous variation in the expression levels of certain glycolisis genes, which substantiate the implementation of an additional regulation function that takes values ranged between 0 and 1. The stability analysis carried out concluded that the current metabolic models are capable of reproducing only one steady state, which made explicit the need to implement a regulation function. The use of a continuous regulation function was due to the immediate response capability to glucose feed pulses observed in experimental data. A Hill activation function was used for supplying the gene regulation because of its structure; glucose concentration has been noted experimentally to be the main factor that produces a metabolic shift in mammalian cells and lactate production rates in BHK cells show a Michaelis-Menten dependency to the residual glucose concentration [3]. The weighing of initial enzyme kinetics leads to a lower $\Delta L/\Delta G$ ratio, thus to a different steady state.

Undergoing work

With the previously described metabolic model a controller will be designed in order to deliver the optimal feeding strategy capable of maintaining a cell culture in an altered metabolic state ($\Delta L/\Delta G$ ratio under 0.5), under continuous operation in spite of high load disturbances. A correct approach for the same is the manipulation of the glucose feed concentration in order to vary the $\Delta L/\Delta G$ ratio.

Open loop analysis of the bioreactor system response to a 10% variation in the dilution rate was concluded to be the most significant disturbance and will be used to design both modern and classic controllers. PID control has been ruled out because the controller's objective is to maintain the culture in a general altered state, which undermines the necessity of designing a controller that can adopt the optimal feeding strategy to keep a culture at a determined ΔL/ΔG ratio. Amongst the classic controllers to be designed are P and PD controllers. Wherein, controller parameters will be obtained through ITAE minimization of the response to a 10% variation in the dilution rate, while keeping the dilution rate control loop open. ITAE minimization will be used in order to obtain a quick response, which is an important design criterion.

Acknowledgments

Conicyt, This work has been supported by FONDECYT Initiation Grants 11080016 and 11090268

References

1. Europa, Anna F., Gambhir, Anshu, Fu Pend-Cheng, Hu Wei-Shou (2000), "Large scale gene expression profiling of metabolic shift of mammalian cells in culture," *Biotechnol bioeng*, 67: 25-34
2. Korke, Rashmi, Gatti, Marcela de Leon, Lau Ally L. Y., Lim, Justin W. E., Seow, Teck K., Chung, Maxey C. M., Hu Wei-Shou (2004), "Multiple steady states with distinct cellular metabolism in continuous culture of mammalian cells," *J Biotech*, 107: 1-17
3. Cruz, Helder J., Moreira, Jose L., Carrondo, Manuel J. T. (1999), "Metabolic shifts by nutrient manipulation in continuous cultures of BHK cells," *Biotechnol bioeng*, 66: 104-113
4. Lagarias, Jeffrey C., Reeds, James A., Wright, Margaret H., Wright, Paul E. (1998), "Convergence properties of the Nelder-Mead simplex method in low dimensions," *J Optim*, 9: 112-147

(765a) The Cascade Algorithm: a New Approach for Distributed Markov-Based Optimization

Antonis Kokossis, School of Chemical Engineering, National Technical University of Athens, Athens, Greece, Patrick Linke, Chemical Engineering, Texas A&M University at Qatar, Doha, Qatar and Siyu Yang, Process & Information Systems Engineering, University of Surrey, Guildford

This paper introduces a new stochastic optimisation approach in the form of a Cascade Optimisation Algorithm. The algorithm incorporates concepts from Markov processes whilst eliminating the inherent sequential nature that is a major obstacle preventing the exploitation of advances in distributed computing infrastructures. This method introduces partitions and pools to store intermediate solution and corresponding objectives. A Markov process increases the population of partitions and pools. The population is distributed periodically following an external certain. With the use of partitions and pools, multiple Markov processes can be launched simultaneously for different partitions and pools. The cascade optimisation algorithm holds a potential in two different fronts. One aims at its deployment in parallel and distributed computing environments. Through storage of solutions in the pools, the algorithm further offers cost-effective means to analyze intermediate solutions, to visualize progress and to integrate optimization with data and/or knowledge management techniques without additional burden to the process.

See more of this Session: Data Analysis: Design, Algorithms & Applications
See more of this Group/Topical: Computing and Systems Technology Division

(765b) High Performance Prediction of Molten Steel Temperature Through Gray-Box Model

Toshinori Okura[1], Manabu Kano[1], Shinji Hasebe[1], Hiroshi Kitada[2] and Noboru Murata[3], (1)Dept. of Chemical Engineering, Kyoto University, Kyoto, Japan, (2)Sumitomo Metal Industries, Ltd., Japan, (3) Waseda University, Tokyo, Japan

The steel industry faces a stiff competition in the global market; each steel company has to produce high quality products satisfying various customers demand. A key problem is variation in product quality. For example, surface cracks may occur and product quality may deteriorate if molten steel temperature is not controlled precisely in a continuous casting process. In the steel making process, molten steel temperature in the tundish is one of the key factors causing variation of product quality. It is crucial to develop a prediction and control system of molten steel temperature to achieve high quality steel production. However, the molten steel temperature cannot be directly controlled in the continuous casting process because there is no effective manipulated variable. In order to realize precise temperature control, it is necessary to adjust the molten steel temperature at the end of secondary refining. Therefore, a model relating the molten steel temperature in the continuous casting process and in the secondary refining process needs to be constructed. The model development is difficult because various factors, e.g. degradation of ladles and treating time at each station, affect molten steel temperature in the ladle and the tundish.

In the present research, a novel gray-box model is proposed to predict the molten steel temperature by combining a first-principles model and a statistical model. First, the first-principles model was developed for a converter, a secondary refining process, and a continuous casting process. This model takes account of heat balance among molten steel, slag, air, ladle, and tundish; the results of computational fluid dynamics (CFD) analysis were used to simplify the model and to achieve high prediction performance. Then, the statistical model was developed to compensate the prediction error of the first-principles model, because the prediction performance of the first-principles model was not sufficient for its industrial application. Finally, the temperature prediction is made by adding the output of the first-principles model and that of the statistical model.

To validate the developed prediction system, various types of models were compared: the first-principles model, a partial least squares (PLS) model as a linear model, locally weighted PLS (LW-PLS) model as a just-in-time model, a random forest (RF) model as a nonlinear model, and the gray-box models combining them. The prediction performance was evaluated by using real operation data provided by Sumitomo Metal Industries, Ltd. in Japan. As a result, it was confirmed that the gray-box model combining the first-principles model and RF could achieve the best prediction accuracy. Its root mean square error (RMSE) was improved by 35% and 37% in comparison with RF and the first-principles model, respectively. PLS and LW-PLS cannot realize the better prediction than RF. These results have clearly shown the advantage of the developed gray-box model over the other models. The gray-box model is now used to investigate the advanced control system of molten steel temperature.

See more of this Session: Data Analysis: Design, Algorithms & Applications
See more of this Group/Topical: Computing and Systems Technology Division

Performance Monitoring of Industrial Plant Alarm Systems Using Event Correlation Analysis

Tsutomu Takai and Masaru Noda

Graduate School of Information Science, Nara Institute of Science and Technology
8916-5 Takayama-cho, Ikoma, 630-0192 Japan

Abstract

Alarm System is essential to ensure the plant safety and effective operation, and the management efforts aiming at maintaining and improving alarm systems have recently intensified in process industries. Alarm management for existing plants is basically CAPDo approach that begins by evaluating the alarm system performance, and the result is quite important in order to find issues to be maintained and improved effectively. Conventional methods which evaluate alarm rates, alarm and event distributions, standing alarm times, etc. in quantity, remain a long way from an effective evaluation of alarm systems performance, because they do not evaluate each alarm as a signal requiring operator attention. Engineering Equipment & Materials Users' Association (EEMUA, 2007) says that every alarm presented to the operator should be useful and relevant to the operator. Thus, the relationship between alarm and the operator response is thought of as a new key performance indicator (KPI) of alarm system performance.

Takai *et al.* (2010) proposed an evaluation method of the relationship between them using operator questionnaire. Questionnaire is reasonable for working-level at plant operation but questionnaire-based evaluation is often subjective, and may contain bias arising from the format of the questionnaires or from individual respondents. In the paper, we propose a new KPI of evaluation alarm system performance and the calculation method by using an event correlation analysis (Nishiguchi and Takai, 2010) which is a data mining method to quantify the degree of similarity and time lag between two events by cross correlation function, from event log data composed of discrete alarms and operator actions with the time they occur.

In the event correlation analysis, events pairs separated by consistent time intervals are considered to be related, since the length of the time lags is determined by factors such as process dynamics and operator reaction time. A similarity measure between all event pairs is calculated from the event log data along with the probability distribution of correlation regarding independent event pairs. For similarities and intervals in all the combinations between event pairs are calculated, and groups with highly related events are identified using pair-wise similarities by the hierarchical clustering method. Then, each alarm is assayed if the based alarm has the corrective actions after the alarm occurs in the group. The relevant alarm rate is defined as the ratio of number of relevant alarms in all alarms. The effectiveness of the proposed method was validated with actual plant event data and simulation results.

References

(1) Engineering and Equipment Materials Users' Association (EEMUA), Alarm Systems – A Guide to Design, Management and Procurement Publication No. 191 2nd Edition, EEMUA, London (2007)

(2) Tsutomu Takai, Higuchi Fumitaka, Akirou Shimameguri, Masaru Noda and Taketoshi Kurooka,

A Comprehensive Evaluation Method of Alarm System from The Standpoint of 8 Characteristics, *Proceedings of PSE Asia 2010*, July 25-28, Singapore (2010)

(3) Junya Nishiguchi and Tsutomu Takai, IPL2&3 Performance Improvement Method for Process Safety Using the Event Correlation Analysis, *Computers & Chemical Engineering*, 34, pp. 2007-2013 (2010)

AIChE Annual Meeting 2011
Computing and Systems Technology Division
10C03 Data Analysis: Design, Algorithms & Applications

Event Correlation Analysis of Plant Operation Data with Large Variance in Time Lag for Alarm System Rationalization

Masaru Noda[1,2], Kohjiro Kurata[2], Yasunori Kikuchi[2], Masahiko Hirao[2]

[1]Graduate School of Information Science, Nara Institute of Science and Technology
8916-5 Takayama-cho, Ikoma, 630-0192 Japan
[2]Department of Chemical System Engineering, University of Tokyo,
7-3-1 Hongo, Bunkyo-ku, Tokyo 113-8656, Japan

Abstract

The advance of distributed control systems (DCSs) in the chemical industry has made it possible to install many alarms cheaply and easily. While most alarms help operators detect an abnormality and identify its cause, some are unnecessary. A poor alarm system might cause alarm floods and nuisance alarms, which reduces the ability of operators to cope with plant abnormalities because critical alarms are buried under a lot of unnecessary ones (EEMUA, 2007).

Nishiguchi and Takai (2010) developed an event correlation analysis method for reducing unnecessary alarms. Their analysis is a knowledge extraction method that detects statistical similarities among discrete events of alarms and operations. The method uses operation data from the plant to quantify the degree of similarity with its time lag between two events by evaluating the cross correlation function. By grouping correlated alarms and operations in accordance with the degree of similarities, the following types of nuisance alarms and operations can be found.

(1) Sequential alarms: When a group contains multiple alarms that occur in sequence, changing the alarm settings might effectively reduce sequential alarms.
(2) Routine operations: When many operations are included in a group and operations in the same group appear frequently in the event log data, they might be routine operations. These alarms and operations can be reduced by automating routine operations using a programmable logic controller.
(3) Unnecessary alarms: Alarms might be unnecessary when operations are not carried out in response to them.

The event correlation analysis method was applied to the operation data of an ethylene plant operated by Idemitsu Kosan in Japan. Unnecessary alarms and operations were accurately identified within a large amount of event log data by using the method (Higuchi *et al.*, 2010). However, the method sometimes failed to detect similarities between two physically related events when the variance in time lag between them was too big. Therefore, in this paper, we propose a new event correlation analysis method that is able to correctly identify similarities between two physically related events.

The proposed method first converts the plant event log data, which consists of occurrence times and tag names of alarms and operations, into sequential binary data using a certain size of time widow. Then, similarities between all combinations of any two events in the plant event log data are evaluated (Nishiguchi and Takai, 2010). If high similarity between two events is not detected, the time window size is doubled and the log data of two events are reconverted into sequential binary data using the new time window size. The expansion of the time window size and recalculation of the similarity continues until either high similarity is detected or the time window size becomes larger than the maximum pre-determined size.

We applied the proposed method to simulation data of an azeotropic distillation column. Results showed it was able to correctly identify similarities between two physically related events, even when the conventional method using a constant time window size failed due to the large variance in time lag. Using the proposed method, we can effectively identify unnecessary alarms and operations within a large amount of event data and reduce the number of unnecessary alarms and operations in chemical plants.

References

(1) Engineering and Equipment Materials Users' Association (EEMUA), Alarm Systems – A Guide to Design, Management and Procurement Publication No. 191 2nd Edition, EEMUA, London (2007)

(2) Junya Nishiguchi and Tsutomu Takai, IPL2&3 Performance Improvement Method for Process Safety Using the Event Correlation Analysis, *Computers & Chemical Engineering*, 34, pp. 2007-2013 (2010)

(3) Fumitaka Higuchi, Ichizo Yamamoto, Tsutomu Takai, Masaru Noda, Hirokazu Nishitani, Rationalization of Plant Alarm System using Event Correlation Analysis, *Proceedings of PSE Asia 2010*, pp. 1395-1401, Singapore (2010)

(765e) Operational Level Integration System Based On An Ontological Framework by Means of Master/Control Recipe Semantic

Edrisi Muoz, Elisabet Capon, Antonio Espua and Luis Puigjaner, Chemical Engineering, Universitat Politecnica de Catalunya, Barcelona, Spain

Process industries are highly involved systems which entail complex decision-making. They require some advanced methods and/or tools in order to integrate and optimize information from different decision levels. In this sense, semantic based decision support systems seem to offer an efficient tool for facilitating interoperability across multiple, heterogeneous systems. In addition, they allow to share information and experience as well as to learn from collective knowledge.

Regarding the integration of different decision levels, an ontology (semantic structure) allows to coordinate the information exchange among the different modeling paradigms/conventions currently used for the different decisions to be taken. Its key role consists of capturing the relevant distinctions of the enterprise structure (specific domain) at the highest level of abstraction, embodying the results of academic research, and offers an operational method to put theory to practice [1].

This work focuses on the integration within the operational decision levels (control and scheduling). A description of the informatics system of batch process ontology[2] is provided. An additional level of detail regarding to master/control recipe element, is represented inside the ontology for mining quality data which will be translated into useful information by the decision maker. It is important to remark that the generality should be maintained since it is a basic principle in any ontological framework.

The recipes contain a variety of information about available raw materials, processing requirements, the manufacturing of a single batch of a specific product, etc., according to the World Batch Forum schemas [3] which are based on the ANSI/ISA-88 standards [4 - 8].

The presented framework is applied to a benchmark scheduling problem formulated by Westerberg-Kallrath [9].

[1] Gruber, T., Springer-Verlag (ed.), Ontology, Encyclopedia of Database Systems, Ling Liu and M. Tamer özsu, 2008

[2] Munoz, E., Espuna, A., Puigjaner, L. Towards an ontological infrastructure for chemical batch process management. "Computers & chemical engineering", 2010, vol. 34, núm. 5, p. 668-682.

[3] Brandl, D., Emerson, D., September 2003. Batch markup language batchml.

[4] International Society for Measurement and Control. (1995). Batch control. Part 1. Models and terminology. International Society for Measurement and Control.

[5] International Society for Measurement and Control. (2001). Data structures and guidelines for languages. International Society for Measurement and Control.

[6] International Society for Measurement and Control. (2003). Batch control. Part 3. General and site recipe models and representation. International Society for Measurement and Control.

[7] International Society for Measurement and Control. (2006). batch Control part 4. Lot production registers. International Society for Measurement and Control.

[8] International Society for Measurement and Control. (2007). Batch control part 5 automated equipment control models & terminology. International Society for Measurement and Control.

[9] Kallrath, J. Planning and scheduling in the process industry. "OR Spectrum", 2002, vol. 24, 219-250

See more of this Session: Data Analysis: Design, Algorithms & Applications
See more of this Group/Topical: Computing and Systems Technology Division

(765f) Descriptor-Based Models From High Throughput Research Workflows for Architectural Coatings

Jonathan D. Moore, Core R&D - Materials Science, The Dow Chemical Company, Midland, MI, Guido Smits, Core R&D - Materials Science, The Dow Chemical Company, Terneuzen, Netherlands and Alan Chaput Jr., Dow Coating Materials, The Dow Chemical Company, Midland, MI

The coatings application area is one of critical strategic importance for high throughput research (HTR) at The Dow Chemical Company (Dow). HTR is being used to accelerate the design and development of new and differentiated products by facilitating the exploration and understanding of many more experimental variables than could be achieved with traditional approaches. It is enabled by an integrated set of advanced capabilities in both hardware and software. This presentation will focus on the advanced computational intelligence modeling and analysis techniques that are being used at Dow to provide insight into the critical raw-material physical properties, chemical structures, and formulation interactions that drive performance in architectural coatings.

See more of this Session: Data Analysis: Design, Algorithms & Applications
See more of this Group/Topical: Computing and Systems Technology Division

(765g) Design of Experiments Based On Computational Singular Perturbation

Siddhartha Kumar, Chemical and Materials Engineering, University of Alberta, Edmonton, AB, Canada and Vinay Prasad, Department of Chemical and Materials Engineering, University of Alberta, Edmonton, AB, Canada

Parameter estimation for large-scale systems has always been a difficult task, because of identifiability constraints on such systems in the vast majority of cases. Hence, preliminary sensitivity analysis and model reduction is inevitable in such systems. In order to derive reduced models, the conventional method used for kinetic systems is a criterion based on quasi-steady state (QSS) analysis, by using net rates of production/consumption of species, and identifying and removing the QSS species. In earlier work [1], we have investigated alternate methods of model reduction for kinetic systems such as singular value decomposition (SVD) and parameter clustering-based approaches; these were applied to steady-state data and models. In this contribution, we develop a method based on transient analysis of models of large-scale kinetic and catalytic systems that combines model reduction and the design of experiments.

Using computational singular perturbation (CSP), we separate the model of our system into dynamic modes and contributing reactions in a rigorous automated manner, and use this to build a reduced-order model at every instant. This reduced model is obtained by performing the modal decomposition, and separating the dynamic modes into active, exhausted and dormant modes. The exhausted and dormant modes do not contribute to the current dynamics of the process in a significant manner, and their contribution can be neglected; this provides us with a reduced-order model. This reduced model provides us with a subset of significant parameters that need to be estimated, i.e., the parameters of the reactions that contribute significantly to the active modes. Optimal input sequences are then designed (based on D optimal analysis) that provide the maximum sensitivity of the measured system outputs (outlet species concentrations and partial pressures) to these significant parameters, subject to identifiability constraints for the reduced order system. A sequential design protocol is devised to apply the model reduction and experiment design steps iteratively, until an acceptable model has been identified. Robustness of the modal decomposition and the experiment design to parametric uncertainty is achieved by including a stochastic optimization approach to obtaining the D optimal experimental design. The method is demonstrated for two catalytic systems, the decomposition of ammonia on ruthenium catalyst to produce hydrogen, and the preferential oxidation of carbon monoxide on platinum catalyst.

References:

[1] Kumar, S. Narasimhan, K. Patwardhan, S.C. Prasad, V. , 'Experiment design, identification and control in large-scale chemical processes', Proceedings of the 2010 International Conference on Modelling, Identification and Control (ICMIC)

(767a) Exploring Protein-Excipient Interactions for Optimal Design of Protein Stabilizers

Brock C. Roughton[1], **Anthony I. Pokphanh**[2], E. M. Topp[3] and K. V. Camarda[2], (1)Bioengineering Graduate Program, University of Kansas, Lawrence, KS, (2)Department of Chemical and Petroleum Engineering, University of Kansas, Lawrence, KS, (3)Department of Industrial and Physical Pharmacy, Purdue University, West Lafayette, IN

The number of proteins being considered as therapeutic candidates is rapidly increasing. Many obstacles are faced in the formulation of drug products containing proteins and a system approach to designing formulations is needed. A common and significant problem in protein therapeutic formulation is aggregation. Aggregation occurs through protein-protein interactions that lead to the formation of dimers, trimers, or larger aggregates. Protein aggregation can be caused by and also result in unfolding and loss of native conformation. Protein aggregates are often no longer effective as drug molecules and can elicit an immune response *in vivo*. Lyophilization, or freeze-drying, is a common method employed to minimize aggregation by limiting protein mobility. However, even in lyophilized formulations, aggregation can still occur in significant measures. Excipients, or additives, can be included in a formulation to stabilize the protein. Several excipients have been shown to minimize aggregation in lyophilized formulations, but the mechanisms are not well understood. Two basic hypotheses exist: protein aggregation is limited when protein mobility is limited by formation of a glass or amorphous solid and protein aggregation is limited by direct stabilizing interactions between an excipient and a protein. The current work explores direct excipient interactions with the surface of folded proteins and incorporates the information in the computational molecular design (CMD) of optimal excipients for aggregation minimization.

Protein-excipient interactions were studied using computer docking simulations to provide information for design of excipients using the CMD methodology. Several proteins were considered (calmodulin, lysozyme, myoglobin, and b-lactoglobin) along with carbohydrate (mannitol, sucrose, trehalose, and raffinose) excipients. The excipients chosen are commonly used as stabilizers in protein formulations. Blind docking simulations were performed to determine the most likely regions of direct interaction between excipients and proteins. In the blind docking experiments, the excipient was allowed to be flexible while the amino acid resides were held rigid. The amino acids that interacted with an excipient's docked conformation were recorded for fifty total docked conformations for each protein-excipient pair. The regions were compared to computationally predicted aggregation prone regions and experimental hydrogen/deuterium exchange (HDX) data. Regions where excipients exhibited protective effects in the HDX experiments were chosen as sites for direct docking simulations with the excipients. In the directed docking simulations, both the ligand and amino acid residues were allowed to be flexible.

Aggregation prone regions, or "hotspots", are areas on a protein which are likely to participate in aggregation reactions. The hotspots were calculated based on the amino acid sequence of the protein using two previously published algorithms available online: Aggrescan (http://bioinf.uab.es/aggrescan/) and PASTA (http://protein.cribi.unipd.it/pasta/). The calculated hotspot regions were compared to the regions of frequent protein-excipient interaction determined by the blind docking simulations. Hot spot regions calculated by the algorithms did not compare well with the docking results for most protein-excipient pairs. The amino acid sequence alone may not be sufficient to accurately predict regions prone to aggregation.

In HDX experiments, direct interaction between an excipient and protein limits the exchange of hydrogen on the protein backbone for deuterium. The use of peptic digests in HDX provides region specific information on exchange. Regions of reduced exchange in HDX data for protein-excipient pairs matched well with regions of frequent protein-excipient interaction in the blind docking simulations, indicating that the docking simulations gave reasonable estimates of regions of protein-excipient interaction that can lead to stabilization or protection of the protein. Directed docking simulations were performed for the regions of limited exchange for each protein-excipient pair. The docked free energies of the excipients were compared to the percent deuterium incorporated by the regions. The docked free energy provided an indication of how favorable the protein-excipient interaction was. An optimal stabilizer will bind preferentially to vulnerable regions on the protein's surface, minimizing contact between the protein and other protein molecules in the formulation. The docked free energy correlated well with the percent deuterium incorporated, indicating that the docking simulations could provide quantitative information about protein-excipient interaction. Regions with low deuterium uptake bound excipients with lower docked free energies. A quantitative structure-property relationship (QSPR) was developed to relate the docked free energy to the molecular topology of the excipients. The QSPR was then utilized in a computational molecular design framework to generate candidates that would bind protein regions with a minimum free energy. Prediction intervals at 95% were used to compare the candidates and estimate the uncertainty in binding energy prediction. By designing an excipient candidate that will optimally bind to a region on a protein, the region will be protected from interaction with other proteins and aggregation will be minimized. For the proteins considered, several such candidates have been designed using the computational molecular design methodology.

See more of this Session: Methodologies for Product Design
See more of this Group/Topical: Computing and Systems Technology Division

(767b) Comprehensive Computer-Aided Molecular Design Framework for Pure Component Design

Apurva Samudra and Nick Sahinidis, Chemical Engineering, Carnegie Mellon University, Pittsburgh, PA

Computer-aided molecular design (CAMD) has become a powerful tool to design novel molecules to meet product target property criteria. Over the years, the field has matured from neighborhood-based searches to fathoming the chemical space using optimization and enumeration. The techniques used in this field have been rooted in the basic group contribution (GC) property methods and thus more generic design problems have either been decomposed or projected onto GC properties to make them compatible with CAMD methods. These techniques have also been plagued by the exponential search-space of molecular structures which inhibit a generic enumeration based design approach.

We have developed an optimization-based framework for basic molecular design aimed to automate, augment and accelerate the steps in CAMD. This framework employs integer optimization efficiently and effectively to cut down on the computational cost of CAMD. Using property decomposition and model transformation, this framework achieves the flexibility required for extensive molecular design along with accuracy for basic group contribution properties. The framework is based on GC+ [1] property models for thermodynamic properties like (Tc,Pc,Tm,Tb,Hv etc.). The framework is split into three stages. In the first stage, mixed-integer linear programming (MILP) is used to generate molecular candidates that fit a relaxed range of defined property targets with approximate GC+ models. This approach tackles the exponentially increasing search-space very effectively by exploiting the linearity in property models. The second stage of the framework screens the solutions further by using structure-dependent corrections to the approximate property values. Detailed chemical structures are developed using a graph theory approach. Different descriptors in different GC+ models are used by estimating their contribution via connectivity indices. The last stage is used to validate the model using molecule-specific property techniques and to integrate the CAMD problem with many involved property prediction techniques that extend its applicability to diverse problems.

We present various extensions to the framework to include other property models for various applications. These extensions range from simple groups contributions for solvent design, e.g. Hansen solubility parameter, UNIFAC, to complex property models and simulations for other diverse applications. We demonstrate the simplicity of use of the framework. Simple methodology to extend the framework for desired application for the use is presented. We conclude by presenting many case studies illustrating efficiency, wide application domain and versatility of proposed framework.

[1] R. Gani, P. M. Harper, and M. Hostrup. Automatic Creation of Missing Groups through Connectivity

Index for Pure-Component Property Prediction. Industrial & Engineering Chemistry Research, 44:7262–7269, 2005.

See more of this Session: Methodologies for Product Design
See more of this Group/Topical: Computing and Systems Technology Division

(767c) Systematic Molecular Design of Biofuel Additives Using Hybrid Characterization and Group Contribution Based Techniques

Subin Hada, Charles C. Solvason and Mario Richard Eden, Department of Chemical Engineering, Auburn University, Auburn, AL

Any type of feedstock that contains free fatty acids and/or triglycerides such as vegetable oil, waste oil, animal fat, and waste grease can be converted into biodiesel. Because of its biological origin, biodiesel has many advantages compared to its petroleum derived counterparts. However, it does exhibit several limitations like oxidative stability and low temperature operability that have impaired its wide use and commercialization. Simultaneous solution of these problems has proven difficult as improvements in one area tend to impair another. As a result, design of an optimal fuel for all environments can be a quite difficult task. In addition, feedstock origin will impact the final quality of the biodiesel product. Therefore, to produce fuel grade biodiesel, the characteristics of the feedstock are very important during the initial research and production stage.

Several approaches are possible for improving the fuel properties of biodiesel. Modification of the fatty acid composition through physical processes or uses of additives are the most prevalent. Fuel additives, such as antioxidants, cetane enhancers, or cold-flow improvers have become indispensable tools not only to alleviate the drawbacks described above, but also to assure that any fuel blend will meet international and regional standards regardless of origin. However, an additive solution to one problem often aggravates another problem. Furthermore, the questions of additive compatibility, required additive levels, the effect on other properties, and whether these additives function as designed for biodiesel fuels with differing fatty acid profiles still remain challenges and that need further investigation.

Therefore, it is desired to molecularly design biodiesel additives that simultaneously account for the unintended effect on other fuel properties in the neat and the blend fuel in order to achieve the performance properties of the petroleum based fuel. In this way, biofuels can be formulated that are adaptable to a range or blend of feedstocks and the desirable fuel characteristics like oxidative stability and wide operating temperature range.

In this work, we aim to identify all possible compounds which, when added to off spec biodiesel, results in a fuel that satisfies performance standards such as ASTM D6751 in the US and EN14214 in Europe. To meet this end, multivariate characterization data obtained from IR spectroscopy of common additives were combined with decomposition and property clustering techniques in a reverse problem formulation. In this approach, the fuel additive property targets are identified in the first reverse problem followed by molecular design to match the targets. The characterization data consists of multitude of properties of interest (such as cetane number, melting point, and kinematic viscosity) to ensure adequate performance. To facilitate an efficient design we consolidated these various properties into a latent property domain using principle component analysis (PCA) techniques. Finally, characterization based molecular design using group contribution parameters are then used to build novel additives that match the fuel specifications in the latent property space.

See more of this Session: Methodologies for Product Design
See more of this Group/Topical: Computing and Systems Technology Division

(767d) Multiscale Approach to Colloidal Systems (emulsions) Design

Andrea Nieto, Diego Pradilla, Natalia Murcia and Oscar Alvarez, Universidad de Los Andes, Bogota, Colombia

Multiscale Approach to Colloidal Systems (emulsions) Design

A. Nieto, D.Pradilla, N. Murcia, O. Alvarez

Group of Process and Product Design (GDPP)

Department ofChemical Engineering

Universidad de losAndes, Bogota-Colombia

Preparation of emulsions (generally mixing ofimmiscible liquids, water and oil) involves several variables (emulsificationconditions, system formulation and composition), being the final properties aconsequence of the established level for each of these variables. To analyzeseparately the effect of a given process condition, formulation or compositionon emulsion's properties, facilitates the establishment of a relation betweenthe study variable and the changes in measured properties. However, thestatement above limits the comprehension of the existent interactions betweenformulation, composition and process.

In this paper, molecular interactions of thesystem water/mineral oil/non ionic surfactant used in the preparation of inverseand direct emulsions were studied, evaluating the effect on the microscopic (particle size) and macroscopic (apparent viscosity, elastic and loss moduli)properties of the system. To achieve this, different agitation geometries andemulsification processes were used in order to establish the influence of theamount of energy added during emulsion preparation. The molecular scalebehavior is evaluated using near-infrared spectroscopy NIR (FOSS 5000Smartprobe Analyzer), size distribution and particle size are settled using dynamiclight scattering technique (Zetasizer nano ZS), and finally the rheologicalbehavior is measured from steady state and oscillatory tests using a controlledstress rheometer (TA Instruments AR-G2).

The study of the phenomena that occur duringthe preparation of these emulsions at three different scales: macroscopic,microscopic and molecular, simultaneously with the effect of the amount ofenergy added, results in the understanding of the relations between process(emulsification conditions), product (formulation and composition) andproperties (rheology), allowing us to design emulsions with specificcharacteristics. For instance, it has been established for concentrated inverseemulsions (volumetric fraction of dispersed phase > 0.9) that elasticmodulus is independent of impeller type used during emulsification.

See more of this Session: Methodologies for Product Design
See more of this Group/Topical: Computing and Systems Technology Division

(767e) Teaching Chemical Product Design

Ed Cussler, Department of Chemical Engineering and Materials Science, University of Minnesota, Minneapolis, MN and **Geoff D. Moggridge**, Chemical Engineering and Biotechnology, University of Cambridge, Cambridge, CB2 3RA, United Kingdom

Teaching Chemical Product Design

E L Cussler and G D Moggridge

Universities of Minnesota and Cambridge

Since its inception around a century ago, the chemical industry has focussed on the manufacture of commodities. A commodity chemical, produced at over 1000 tons per year, is sold into a world market where the products are differentiated by price. Benzene, polypropylene and titanium dioxide are examples. This industry had its Golden Age from 1940 to 1980, with growth equivalent to that of the modern software industry.

More recently, as market growth slowed, chemical companies have moved towards higher value added products. These higher value added products, often produced at less than 10 tons per year, are often made of ingredients which cost a tenth or less of their selling price. They gain their value from a molecular or microstructure which gives them better performance. However, many chemical professionals are trained largely, if not exclusively, to serve the commodity chemical industry. These professionals now focus largely on the traditional question, "How should we make this commodity product?" These professionals must also consider what is increasingly the more relevant question, "What high value added products are we going to make?"

This paper describes teaching chemical product design, and so attempts to fill part of the educational gap between the commodity chemical industry and the new, more fragmented, high value added chemical product industry. Filling this gap has two parts. First, we present a four-step template for chemical product design of customer needs, product ideas, selection of the best idea, and product manufacturing. Second, we explore which step in this template is most important for which kind of product. As expected, the manufacturing step is the most important for commodities. Less obviously, the selection step is the most important for many non-commodity products.

We teach product design using two different strategies. Sometimes, we teach a separate course of lectures and smaller recitations, where the specific problems are discussed. The students develop ideas as teams, but write six to ten individual reports. Alternatively, we teach twelve classes which supplement to chemical process design. In this stategy, each student writes three individual essays. Both of these strategies work better for us than imitating the process design course.

We are interested that different experiences lead to trouble with different parts of product design. Inexperienced students are wonderful at generating new product ideas, but they have trouble making estimates which let them quickly select among possible alternative products. Experienced chemists and engineers have no trouble making quick estimates and sensible selections, but they are less effective at suggesting new ideas. Both groups have benefited from and enjoyed their efforts to get better at chemical product design. We challenge you, either as a professor or as a student: while this material is hard to teach and hard to learn, it can be the most satisfying part of your education. Have fun with your designs.

See more of this Session: Methodologies for Product Design
See more of this Group/Topical: Computing and Systems Technology Division

(772a) Automated Real-Time Diagnostics of Model Predictive Control Actions

Aditya Kumar, Chemical engineering, University of Houston, Houston, TX and Michael Nikolaou, Chemical Engineering, University of Houston, Houston, TX

Automated Real-Time Diagnostics of Model Predictive Control Actions

Aditya Kumar and Michael Nikolaou, University of Houston, Houston, TX

Model predictive control (MPC) is widely used in process industries. The design and closed-loop behavior of MPC systems can now be supported by well developed theoretical tools. However, it is well documented that many MPC installations may not function in the automatic mode for a substantial fraction of time. In many instances, this is caused by human operators switching MPC to the manual mode, because of poor understanding of MPC operation. In such case, all benefits from MPC are obviously lost. Questions that operators frequently raise about MPC include the following.

- Why is MPC taking such action?
- What will happen if I allow it to take such action?
- Do I need to intervene?
- How far can MPC move before I have to intervene?

While experienced operators or engineers may be able to answer some of the above questions, particularly if they spend enough time analyzing data, there is a need for an automated tool that would provide insight into MPC actions, by answering in real time questions such as raised above. The purpose of this presentation is to provide a diagnostic framework and specific tools for answering some of these questions and related variants.

The approach taken is algorithmic in nature, namely diagnostic information is generated by applying specific computational tools. The results of these computations can be easily presented in natural language that can be understood by operators.

The proposed approach is applied on data taken from a real MPC system with 24 controlled and 12 manipulated variables. The results demonstrate that a number of questions raised by operators can be answered satisfactorily. Some of the answers generated agree with the results produced by extensive off-line analysis of the same data by experienced engineers.

In future work, the proposed framework will be extended to include additional elements, and integration with natural language processing tools that will make it user-friendly will be undertaken.

See more of this Session: Smart Plants Operation and Control
See more of this Group/Topical: Computing and Systems Technology Division

(772b) Economic Cost Function Design and Model Predictive Control of a Chemical Process Network

Xianzhong Chen[1], Mohsen Heidarinejad[2], Jinfeng Liu[3] and Panagiotis D. Christofides[3], (1)Chemical and Biomolecular Engineering, Univ. of California, Los Angeles, Los Angeles, CA, (2)Electric Engineering, University of California, Los Angeles, Los Angeles, CA, (3)Department of Chemical and Biomolecular Engineering, University of California, Los Angeles, Los Angeles, CA

Maximizing profit has been and will always be the primary purpose of optimal process operation. Within process control, the economic optimization considerations of a plant are usually addressed via a real-time optimization (RTO) system. In general, an RTO system includes two different layers: the upper layer that optimizes process operation set-points taking into account economic considerations using steady-state system models, and the lower layer (i.e., process control layer) whose primary objective is to employ feedback control systems to force the process to track the set-points. In the past years, there have been several attempts to integrate MPC and economic optimization of processes. In particular, in our recent work [1], we developed Lyapunov-based economic MPC (LEMPC) designs which are capable of optimizing closed-loop performance with respect to general economic considerations for nonlinear systems. The design of the LEMPC is based on uniting receding horizon control with explicit Lyapunov-based nonlinear controller design techniques and allows for an explicit characterization of the stability region of the closed-loop system.

In the present work, we focus on the application of the theoretical results developed in [1] to a catalytic alkylation of benzene process network, which consists of four continuous stirred tank reactors and a flash separator. In the design of the economic cost function of the LEMPC, we take into three aspects. First, we try to maximize the production of the product ethylbenzene and minimize the production of the by-product 1,3-diethylbenzene. To account for this aspect, the reaction rates of the reactions in the reactors are involved in the cost function in an appropriate form. Second, we want to improve the quality of the separation process which is directly related to the quality of the product. The last aspect we consider is to reduce the energy consumption of the distillation column. This consideration is based on the fact that the reactions take place in the reactors are exothermic and the distillation column consumes most of the energy of the process. Based on this economic cost function, we design the LEMPC for the alkylation process network. Extensive simulations are carried out to compare the LEMPC with a traditional model predictive control design using a quadratic cost function, from stability and performance points of view.

[1] M. Heidarinejad, J. Liu and P. D. Christofides, "Economic Model Predictive Control of Nonlinear Process Systems Using Lyapunov Techniques," AIChE Journal, in press.

(772c) Risk Analysis In the Process Industries: A Perspective

Ankur Pariyani, Chemical and Biomolecular Engineering, University of Pennsylvania, Philadelphia, PA, Warren D. Seider, Chemical and Biomolecular Engineering, The University of Pennsylvania, Philadelphia, PA, Ulku Oktem, Risk Management and Decision Center, Wharton School,University of Pennsylvania, Philadelphia, PA and Masoud Soroush, Department of Chemical and Biological Engineering, Drexel University, Philadelphia, PA

Fortunately, accidents in the chemical process industries (CPIs) have very low frequencies of occurrence. Typically, the probability of an event is estimated using its observed frequency, but for rare events like accidents, data are too sparse. In such a case, risks of occurrence of these events are computed based on: (a) hazards identification, (b) their frequency estimation and (c) consequence modeling. Since the late 1970s and early 1980s, these concepts (leading to techniques like HAZOP, PHA, LOPA) have received much attention in the process and nuclear industries. A system (or plant) is decomposed into individual components – for example, power generators, pumps, and heat exchangers. Possible hazards are identified, their probabilities of failure under demand and likely consequences are estimated using reliability (which could be generic or prior) data and expert knowledge.

More recently, techniques have been developed to access the risk levels *dynamically* using alarm data recorded in plant databases. Because the accident data is limited, the *near-misses* that occur in day-to-day operations are tracked, and the failure probabilities of regulating and protection systems, and the probabilities of incidents are computed.

This presentation is intended to describe the advantages and disadvantages of many methods for analyzing alarm databases, including the estimation and online tracking of key indicators, and the use of leading indicators to project the occurrence of incidents sufficiently far in advance to permit operators to circumvent them. Also, methods are being explored to identify the special causes that lead to the abnormal events that cause alarm flooding – which often significantly distracts operators as risk levels build. In this presentation, a review of state-of-the-art techniques and practices for risk analysis is presented and promising new developments in future research are projected.

See more of this Session: Smart Plants Operation and Control
See more of this Group/Topical: Computing and Systems Technology Division

(772d) A Systematic Methodology for the Assessment and Troubleshooting of Control Strategies and Operational Problems In Distillation Systems

Miguel Mauricio-Iglesias[1], Kristoffer Johansen[2], Sten Bay Jrgensen[1] and Grkan Sin[1], (1)CAPEC, Department of Chemical and Biochemical Engineering, Technical University of Denmark, 2800 Kgs. Lyngby, Denmark, (2)NNE Pharmaplan, Sborg, Denmark

A systematic methodology for the assessment and troubleshooting of control strategies and operational problems in distillation systems

Miguel Mauricio-Iglesias[1], Kristoffer Johansen[2], Sten Bay Jørgensen[1], Gürkan Sin[1].

[1]CAPEC – Department of Chemical and Biochemical Engineering, Technical University of Denmark, 2800 Lyngby, Denmark;[2]NNE Pharmaplan, Vandtårnsvej 108-110, 2860 Søborg

In order to increase efficiency in resources and energy allocation and reduce capital costs, chemical processes tend to use more heat and mass integration, which increase interactions, and fewer surge vessels that could dampen disturbances. As a consequence, interactions between process units become more important, which has to be considered and managed properly in the design of the control layer for the process operation.

One particular area where process operation optimization and control challenges frequently are observed is separation systems. In particular for distillation processes, strategies for heat recovery have been extensively implemented in industrial processes during the last decades. Operation (including start-up and shut-down) and control of, heat-integrated systems become formidable challenges if not problematic since higher-order dynamics occur where different time-scale phenomena interact. Furthermore, if the dynamic state of the process does not lie close to the designed state, the amount of energy wasted increases, questioning the economics of the implementation of heat integration.

On the other hand, literature deals extensively with classic distillation columns whereas much less has been investigated on dynamics of heat integrated distillation systems and the significance of developing a proper plantwide control strategy to ensure optimal and stable operation. Classic control strategies may be unsuitable for such complex systems and, more importantly, the design and operating conditions of the process may lead to a hardly controllable process. The goal of this work is to present a systematic and process engineering oriented methodology for the assessment, troubleshooting and improvement of existing operation and control strategies for distillation systems. The operation and control strategy is assessed stepwise following a top-down and bottom-up approach (Skogestad 2003) according to the following steps:

(i) reviewing control objectives & strategy;

(ii) the model development & control degrees of freedom analysis strategy;

(iii) reviewing and generating pairing of control and manipulated variables and obtaining the subsequent fine-tuning parameters;

(iv) finally, evaluation of the new proposed strategy.

Particular attention was given to the analysis of existing operating conditions of the process and how their modification can lead to a more easily controllable system. To this end, the driving force concept -a powerful and relatively simple graphical method (Bek-Pedersen et al. 2010), is used that provides optimal operating set-points insensitive to disturbances and require minimum energy for the separation (Hamid et al. 2010).

The application of the methodology was highlighted using a real industrial case problem, for which a dynamic model for heat-integrated distillation systems is developed and used to analyze and solve operational and control problems for solvent recovery in a pharmaceutical industry.

The system studied in the industrial case is composed of two distillation columns in series (operated at 3.5 and 1.5 bar respectively) for recovery of ethanol (light key) from a liquid mixture, modeled as pseudo-binary (water is the heavy key). Heat recovery is carried out both backward (from the two bottoms flow to the feed) and forward (from the first condenser to the second reboiler). Finally, based on a numerical model of the system and using the methodology described above three solutions were generated for improving its control: i) a fine tuning of the existing control strategy, therefore suitable for immediate implementation; ii) a new control strategy considering improved pairings of variables in the existing plant; and iii) a new control structure, including sensors and actuators for long term modification of the system. Benefits of each alternative remedies have been evaluated and benchmarked against each other.

Bek-Pedersen E., Gani R., & Levaux O. (2000). *Computers and Chemical Engineering*, 24(2-7), 253-259.

Hamid M.K.A., Sin G., & Gani R. (2010). *Computers and Chemical Engineering*, 34(5), 683-699.

Skogestad S. (2002). *Computer Aided Chemical Engineering*, 10(C), 57-69.

See more of this Session: Smart Plants Operation and Control

(772e) Automatic Generation of Emergency Response Procedures for Batch Processes

Chuei-Tin Chang and Ming-Li Yeh, Chemical Engineering, National Cheng Kung University, Tainan, Taiwan

Quick response to mitigate the damaging effects of accidents has always been an important operational issue for chemical plants. In the present study, a systematic automata-based procedure is developed to automatically generate fail-safe operation steps in a given batch process. Specifically, three distinct sets of automata are first built off-line to respectively characterize the plant behaviors, the control specifications and also auxiliary constraints. By applying the parallel composition operation on the first two types of automata, an admissible supervisor for a given failure-induced system state can be generated online to steer the system away from hazardous conditions while still maintaining sustained production. In order to identify an efficient procedure, the auxiliary automata can be augmented with this supervisor to set the operation target(s) and to limit the total number of actuator actions. Two examples will be presented to demonstrate the effectiveness and feasibility of the proposed method.

See more of this Session: Smart Plants Operation and Control
See more of this Group/Topical: Computing and Systems Technology Division

(772f) Plantwide Control for Economically Optimal Operation of An Ethyl Benzene Process

Ashok S. Pathak and **Nitin Kaistha**, Chemical Engineering, IIT Kanpur, Kanpur, India

Plantwide Control for Economically Optimal Operation of an Ethyl Benzene Process

Ashok S Pathak and Nitin Kaistha

Department of Chemical Engineering

Indian Institute of Technology Kanpur 208016

In this work, plantwide control system design for economically optimal operation of an ethyl benzene process is studied for Luyben's recently developed base-case design. The process is depicted in Figure 1 and consists of two CSTRs followed by two simple distillation columns with two liquid recycle streams. Ethyl benzene (EB) is produced by the alkylation of fresh benzene with ethylene (Main Reaction: Bz + C_2 --> EB). The EB can further alkylate with ethylene to form diethyl benzene (DEB) (Side Reaction: EB + C_2 --> DEB). The reactors are therefore operated in excess benzene environment to suppress DEB formation. The effluent from the reaction section is separated to recover the unreacted benzene from the top of the first column, which is recycled back to the first CSTR. The bottoms from the first column is further distilled to recover nearly pure EB product as the distillate and DEB with some EB as the bottoms. The DEB stream is recycled to the second CSTR where DEB transalkylates with benzene to form EB (Transalkylation: DEB + Bz --> 2EB). The DEB is allowed to build in this second recycle stream to a level where the DEB formation by the side reaction and consumption by transalkylation exactly balance each other. The DEB is thus recycled to extinction.

The nine steady state operating degrees of freedom for the process are optimized for the given base-case throughput to maximize the yearly operating revenue. The constrained optimization is equivalent to minimizing energy consumption per kmol product for a fixed product quality. The three active constraint variables, one throughput specification and two product quality specifications leave 3 unconstrained dofs. The ethylene feed ratio to the reactor, the EB impurity in the benzene recycle stream and in the DEB recycle stream are chosen to specify these unconstrained dofs. For a ±20% throughput change around the base-case, the difference in the rigorously optimized yearly revenue and the revenue holding these three variables constant (at their base-case optimum values) is found to be very small (<0.5%). Process operation holding these three variables constant is thus deemed as economically optimal for a large change in throughput around the base-case.

Constrained optimization is also performed with the throughput itself as an optimization variable to maximize the product rate (maximum throughput) corresponding to the maximum achievable yearly process revenue. At the optimum solution, the six active constraints and two product purity specifications leave one unconstrained degree of freedom. The throughput shows a maximum with respect to this unconstrained degree of freedom, namely, the EB mol fraction in column 2 bottoms. The EB mol fraction versus throughput curve is however quite flat so that process operation at constant column 2 bottoms EB mol fraction is deemed acceptable for process operation close to maximum throughput.

A plantwide control scheme is devised to mitigate snowballing in the DEB recycle stream and transients in the active constraints at the maximum throughput steady state solution. Rigorous dynamic simulations are then performed to quantify the back-off in the active constraints and consequent economic loss for a worst-case feed composition disturbance. A quantitative comparison of the economic loss due to back-off using the synthesized control system with a conventional one which fixes the throughput manipulator at the ethylene fresh feed shows that the former results in significant economic benefit to the tune of 5%. The energy savings for process operation at the base-case throughput is also found to be ~ 3% higher. The study clearly demonstrates a "top-down bottom-up" plantwide control system design approach as being the most appropriate for economically optimal process operation.

References

1. Luyben, W. L. Design and control of the ethyl benzene process. *AIChE J.* **2011**, 57: 655-670.

2. Luyben, W. L. Snowball effects in reactor/separator processes with recycle. *Ind. Eng. Chem. Res.* **1994**, *33*, 299-305.

See more of this Session: Smart Plants Operation and Control
See more of this Group/Topical: Computing and Systems Technology Division

AUTHOR INDEX

AUTHOR INDEX

AUTHOR INDEX

AUTHOR INDEX

AUTHOR INDEX

AUTHOR INDEX

AUTHOR INDEX

AUTHOR INDEX

AUTHOR INDEX